COLD SPRING HARBOR SYMPOSIA
ON QUANTITATIVE BIOLOGY

VOLUME XLII—PART 2

COLD SPRING HARBOR SYMPOSIA ON QUANTITATIVE BIOLOGY

VOLUME XLII

Chromatin

COLD SPRING HARBOR LABORATORY

1978

COLD SPRING HARBOR SYMPOSIA ON QUANTITATIVE BIOLOGY
VOLUME XLII

© 1978 by The Cold Spring Harbor Laboratory
International Standard Book Number 0–87969–041–0
Library of Congress Catalog Card Number 34–8174

Printed in the United States of America

COLD SPRING HARBOR SYMPOSIA ON QUANTITATIVE BIOLOGY

Founded in 1933 by
REGINALD G. HARRIS
Director of the Biological Laboratory 1924 to 1936

Previous Symposia Volumes

I (1933) Surface Phenomena
II (1934) Aspects of Growth
III (1935) Photochemical Reactions
IV (1936) Excitation Phenomena
V (1937) Internal Secretions
VI (1938) Protein Chemistry
VII (1939) Biological Oxidations
VIII (1940) Permeability and the Nature of Cell Membranes
IX (1941) Genes and Chromosomes: Structure and Organization
X (1942) The Relation of Hormones to Development
XI (1946) Heredity and Variation in Microorganisms
XII (1947) Nucleic Acids and Nucleoproteins
XIII (1948) Biological Applications of Tracer Elements
XIV (1949) Amino Acids and Proteins
XV (1950) Origin and Evolution of Man
XVI (1951) Genes and Mutations
XVII (1952) The Neuron
XVIII (1953) Viruses
XIX (1954) The Mammalian Fetus: Physiological Aspects of Development
XX (1955) Population Genetics: The Nature and Causes of Genetic Variability in Population
XXI (1956) Genetic Mechanisms: Structure and Function

XXII (1957) Population Studies: Animal Ecology and Demography
XXIII (1958) Exchange of Genetic Material: Mechanism and Consequences
XXIV (1959) Genetics and Twentieth Century Darwinism
XXV (1960) Biological Clocks
XXVI (1961) Cellular Regulatory Mechanisms
XXVII (1962) Basic Mechanisms in Animal Virus Biology
XXVIII (1963) Synthesis and Structure of Macromolecules
XXIX (1964) Human Genetics
XXX (1965) Sensory Receptors
XXXI (1966) The Genetic Code
XXXII (1967) Antibodies
XXXIII (1968) Replication of DNA in Microorganisms
XXXIV (1969) The Mechanism of Protein Synthesis
XXXV (1970) Transcription of Genetic Material
XXXVI (1971) Structure and Function of Proteins at the Three-dimensional Level
XXXVII (1972) The Mechanism of Muscle Contraction
XXXVIII (1973) Chromosome Structure and Function
XXXIX (1974) Tumor Viruses
XL (1975) The Synapse
XLI (1976) Origins of Lymphocyte Diversity

The Symposium Volumes are published by the Cold Spring Harbor Laboratory, Cold Spring Harbor, New York 11724, and may be purchased directly from the Laboratory or through booksellers.

First row: S. Tonegawa/R. G. Roeder/G. P. Georgiev
Second row: N. Federoff/A. Efstratiadis, P. Wensink
Third row: J. Paul/S. Elgin, D. Carroll, T. Maniatis, M. Botchan

First row: M. Bellard/G. M. Rubin/I. Dawid
Second row: J. E. Darnell, Jr./P. Leder/M. L. Pardue
Third row: T. Martin/R. Reeder/R. Kornberg

Contents

Part 1

Nucleosome Structure IV: H1 Interactions

Chromosome Coiling and Assembly

Part 2

Transcriptionally Active Chromatin

Repeated DNA Sequences

Summary

Name Index

Subject Index

COLD SPRING HARBOR SYMPOSIA ON QUANTITATIVE BIOLOGY

VOLUME XLII—PART 2

Regulation of Gene Expression in Chick Oviduct

B. W. O'Malley, M.-J. Tsai, S. Y. Tsai, and H. C. Towle[*]

Department of Cell Biology, Baylor College of Medicine, Houston, Texas 77030

Estrogen-mediated growth and differentiation of the chick oviduct and the concomitant induction of ovalbumin has proven to be an excellent system for studying the mechanism of steroid hormone action (O'Malley and Means 1974; Jensen et al. 1974; O'Malley and Schrader 1976). Several lines of evidence indicate that steroid hormones act primarily to regulate oviduct gene expression (O'Malley and McGuire 1968; Cox et al. 1973; O'Malley et al. 1972; McKnight et al. 1975; Schwartz et al. 1975; S. Y. Tsai et al. 1975). During primary stimulation of immature chicks, estrogen has a dramatic effect on the level of endogenous RNA polymerase activity (O'Malley et al. 1969; Cox et al. 1973), nuclear RNA synthesis (O'Malley and McGuire 1968; Means et al. 1972; Harris et al. 1975), chromatin template activity (O'Malley et al. 1969; Cox et al. 1973; Spelsberg et al. 1973), and the number of initiation sites on chromatin available for in vitro RNA synthesis (Schwartz et al. 1975). In addition, the appearance of messenger RNA for a specific induced protein, ovalbumin, precedes the accumulation of that protein during estrogen-mediated growth (Means et al. 1972; Comstock et al. 1972; Rhoads et al. 1973). Withdrawal of estrogen leads to a decrease in the number of initiation sites, which correlates temporally with the decline in the level of nuclear-bound estrogen receptors (S. Y. Tsai et al. 1975). Readministration of estrogen to these withdrawn chicks results first in an increase of the level of nuclear-bound receptors, followed by a twofold increase in the level of chromatin initiation sites (S. Y. Tsai et al. 1975), and, finally, by the appearance of specific ovalbumin mRNA sequences ($mRNA_{ov}$) (Chan et al. 1973; Cox et al. 1974; Palmiter 1973; Harris et al. 1975). These combined results indicate that regulation of RNA synthesis by estrogen occurs at the level of transcription.

To date, considerable evidence has implicated non-histone proteins as important factors for regulating differential gene expression in eukaryotic cells (Stein and Kleinsmith 1975). However, the mechanism of action by which nonhistone proteins regulate the transcription of specific genes is largely unknown. To elucidate the requirements for gene restriction and expression in eukaryotes, it is necessary to develop a sensitive assay system which can assess the biological function of the nonhistone proteins. In this regard, several groups have used reconstituted chromatin coupled with an in vitro chromatin transcription system to demonstrate the biological significance of the nonhistone proteins. For instance, Paul and Gilmour (1968) and Spelsberg et al. (1971) have demonstrated that a substantial amount of the RNA transcribed in vitro from reiterated DNA sequences in reconstituted chromatin is tissue-specific. The specificity of the in vitro RNA synthesis was determined by the origin of the nonhistone proteins. Subsequently, Gilmour and Paul (1973), Barrett et al. (1974), and Chiu et al. (1975) reported that globin mRNA sequences were detected in the in vitro transcripts from reconstituted chromatin containing nonhistones from erythroid chromatin but not in those from chromatin containing nonhistones of nonerythroid tissues. Similarly, Stein et al. (1975) reported that histone mRNA sequences were transcribed from S-phase, but not G_1-phase, HeLa cells, and Park et al. (1976) demonstrated further that nonhistone proteins from S-phase cells were capable of rendering histone genes available for transcription. These results indicate that the nonhistone proteins of chromatin play a central role in the control of differential gene expression.

One of the problems inherent in studying in vitro chromatin transcription is that the level of endogenous mRNA sequences present in most chromatin preparations is usually quantitatively significant when compared to the level of newly synthesized RNA sequences of interest. To help minimize this problem, optimal reaction conditions for synthesis of RNA are normally maintained. Optimal conditions for overall RNA synthesis, however, may not be optimal for biologic fidelity of transcription. To investigate different reaction conditions, we have utilized mercurated UTP as one of the substrates for RNA polymerase (Dale et al. 1975; Dale and Ward 1975). The newly synthesized RNA containing Hg-UMP residues can be separated efficiently from endogenous RNA by means of sulfhydryl-Sepharose chromatography under specific conditions. Utilizing this technique, we have examined the in vitro transcription of the ovalbumin gene from oviduct chromatin.

MATERIALS AND METHODS

Deoxynucleoside triphosphates were purchased from Calbiochem and radiolabeled nucleotides from

---

[*] Present address: Department of Medicine, University of Minnesota, Minneapolis, Minnesota 55455.

Schwarz/Mann. Hg-dUTP was obtained from P-L Biochemicals. S$_1$ nuclease and RNase-free DNase were obtained from Miles Laboratories and proteinase K from EM Laboratories. Hg-UTP was synthesized as described by Dale et al. (1975). Sulfhydryl-Sepharose 4B was prepared by the method of Cuatrecasas (1970). *E. coli* RNA polymerase was isolated from ¾-log-phase *E. coli* K12 by a modified procedure of Burgess and Jendrisak (1975) as described by Towle et al. (1977).

Preparation of DNA and chromatin. Oviducts were obtained from chicks which had received daily subcutaneous injections of diethylstilbestrol (2.5 mg) for 14 days. Chromatin was prepared according to a modified procedure of Spelsberg and Hnilica (1971) as described by M. J. Tsai et al. (1975). Reticulocytes were obtained from chicks made anemic by subcutaneous injection of 2.5 mg of neutralized phenylhydrazine dissolved in 0.14 M NaCl with 10% ethanol, pH 7.0, for 4 consecutive days. On the fifth day, animals were decapitated and bled into bottles prerinsed with heparin. Reticulocytes were collected by centrifugation at 2800g for 10 minutes and washed twice in 0.15 M sodium citrate, 0.15 M NaCl containing 0.01% heparin, pH 7.0. Nuclei were isolated from washed reticulocytes by the method of Seligy and Neelin (1970), and chromatin was extracted as for oviducts. All chromatin preparations were analyzed for DNA, RNA, histone, and nonhistone protein content as described elsewhere (Spelsberg et al. 1971) and stored at approximately 1 mg/ml of DNA in 0.15 mM sodium citrate, 1.5 mM NaCl, pH 7.0, at −20°C. DNA was purified from chick tissues as reported previously (Rosen et al. 1973).

Purification of mRNA for ovalbumin and preparation of cDNA probes. Ovalbumin mRNA was prepared from whole hen oviduct and determined to be pure as described previously (Woo et al. 1975). Globin mRNA was extracted from reticulocytes as described by Towle et al. (1977).

In vitro synthesis and extraction of RNA. Conditions for the synthesis of RNA from chromatin were as follows: 50 mM Tris-HCl, pH 7.9, 5 mM MgCl$_2$, 10 mM 2-mercaptoethanol, 100 mM (NH$_4$)$_2$SO$_4$, 1 mM each of ATP, CTP, and GTP, 0.5 mM UTP, 0.5 mM Hg-UTP, and RNA polymerase and chromatin as indicated. Incubation was carried out for 2 hours at 37°C. Chromatin was removed by centrifugation at 12,000g for 30 minutes. The extract was treated with 40 μg/ml proteinase K for 15 minutes at 37°C, and RNA was extracted with an equal volume of 1:1:0.04 phenol:chloroform:isoamyl alcohol three times and then precipitated in ethanol. The precipitate was collected by centrifugation, dried, and redissolved in 1.5 ml H$_2$O. This sample was chromatographed on Sephadex G-50 (0.8 × 50 cm) in 10 mM Tris-HCl, pH 7.5, 1 mM EDTA. Void-volume fractions were pooled and denatured in a boiling water bath for 2 minutes before being applied to a sulfhydryl-Sepharose column (0.7 × 11 cm) previously equilibrated in 50 mM Tris-HCl, pH 7.5, 100 mM NaCl at a flow rate of less than 0.25 ml/min. After loading, the column was washed in the same buffer until the A$_{260}$ of the flow-through reached background. RNA was then eluted in the same buffer containing 0.1 M 2-mercaptoethanol. Peak fractions were pooled, adjusted to a pH of about 5.5 by addition of 2 M sodium acetate, pH 4.5, and precipitated with 2 volumes of ethanol. The precipitate was dissolved in H$_2$O and the amount of RNA determined by optical absorption at 260 nm (assuming 1 mg/ml RNA = 25 A$_{260}$/ml). The RNA was then lyophilized to dryness.

RNA-cDNA hybridization analysis. RNA-cDNA hybridization reactions were carried out as described previously (Harris et al. 1976). All reactions contained 1.5 ng of cDNA and varying amounts of RNA and were incubated at 68°C for 40–48 hours in 25 μl of hybridization mixture. These conditions yield a DNA C$_0$t value of 0.12 mole·sec/liter, which is more than enough to ensure completion of the hybridization reaction (Gilmour and Paul 1973; Young et al. 1974). Following hybridization, samples were treated with S$_1$ nuclease as reported previously.

cDNA$_{ov}$ was synthesized from purified ovalbumin mRNA by the procedure of Monahan et al. (1976b). Double-stranded ovalbumin DNA was synthesized from a complete copy of ovalbumin cDNA by the procedure of Monahan et al. (1976a), except that instead of dTTP, Hg-dUTP was incorporated into the second (anticoding) strand. Hg-dUTP can be used as a substrate for reverse transcriptase with a reduced rate of incorporation as compared with dTTP. By analysis, using alkaline sucrose gradient sedimentation, the mercurated cDNA has a size distribution similar to nonmercurated cDNA. Therefore, the reduced rate of cDNA synthesis when Hg-dUTP was substituted for dTTP was not a result of synthesis of shorter cDNA chains.

RESULTS AND DISCUSSION

Use of Hg-UTP as a Substrate for Chromatin Transcription

To separate RNA sequences transcribed from chromatin during the course of incubation from those endogenously bound to the chromatin, RNA was synthesized using an equimolar mixture of Hg-UTP and UTP in addition to the other three ribonucleotides. The presence of Hg-UTP inhibited the total synthesis of RNA by about 10–15% throughout a 2-hour incubation (data not shown), which is in accord with published reports (Smith and Huang 1976; Biessmann et al. 1976).

To obtain an accurate assessment of the effectiveness of the SH-Sepharose column for removing RNA contaminants from the Hg-RNA newly synthesized from chromatin, the experiments in Table 1 were done. Known amounts of ovalbumin mRNA were added to a standard incubation mixture containing oviduct chromatin prepared from hormone-withdrawn chicks, and Hg-RNA was synthesized utilizing an equimolar mixture of Hg-UTP and UTP. The RNA samples were then extracted and divided into three equal aliquots. Two of the three samples were subjected to SH-Sepharose chromatography to remove the ovalbumin mRNA from the Hg-RNA transcripts under various conditions. Hybridization was performed using $[^3H]cDNA_{ov}$ to quantitate the amount of $mRNA_{ov}$ present. By these means it is possible to determine the level of nonspecific trapping of $mRNA_{ov}$ which does not contain mercurated substitutions. As shown in Table 1 (column A), direct passage of the RNA sample through the SH-Sepharose column at either high-salt or low-salt conditions only removed about 20–25% of the ovalbumin mRNA sequences. However, denaturation of the RNA samples in 0.1 M NaCl, 50 mM Tris-HCl (TN) prior to loading on the SH-Sepharose column removed 70% of the ovalbumin mRNA sequences (Table 1). When the RNA sample was heated at 100°C for 2 minutes in a low-salt buffer (10 mM Tris-HCl, 1 mM EDTA [TE]) before loading onto the SH column, greater than 90% of the ovalbumin mRNA sequences was consistently removed. Under the same conditions, 95% of the Hg-RNA was retained by the SH column. Thus, the bulk of contaminating endogenous mRNA sequences in the Hg-RNA samples could be effectively removed under appropriate conditions. This additional denaturation step (heat + low salt) before application to the SH-Sepharose column was used in all of the following experiments.

Several possible reactions might contribute to the binding of endogenous contaminating $mRNA_{ov}$ to the SH-Sepharose column. First, anti-mRNA sequences might be synthesized by RNA polymerase via an RNA-dependent process. The resultant Hg-containing anti-mRNA sequences can form hybrids with the contaminating coding-strand RNAs. Thus, the endogenous contaminating RNA will be retained

by the SH column. Second, binding may be due to the addition of Hg-UMP to the 3′ end of preexisting contaminating RNA chains. Third, synthesis of poly-(U), which can form hybrids with the poly(A) sequences of the contaminating mRNA, may occur in vitro. Fourth, retention may simply be due to the aggregation of Hg-RNA with contaminating RNA.

To study the importance of RNA aggregation and RNA-dependent RNA synthesis in these experiments, we transcribed *E. coli* DNA and oviduct chromatin from withdrawn tissue under conditions as described in Methods, extracted Hg-RNA, and added 10 ng $[^{125}I]mRNA_{ov}$ to the extract. The mixture was kept at 4°C overnight in TN buffer and then applied to an SH column. As shown in Table 2, more than 50% of the $[^{125}I]mRNA$ was bound to the column. Since *E. coli* transcripts could not contain sequences complementary to the $mRNA_{ov}$, we conclude that this trapping to the SH column occurs by nonspecific aggregation.

We studied the extent of RNA-dependent RNA synthesis as shown in Table 3. Oviduct chromatin from stimulated chicks was transcribed in a reaction mixture containing $[^{125}I]mRNA_{ov}$ and Hg-UTP. RNA-dependent synthesis of Hg-RNA from the $[^{125}I]mRNA_{ov}$ templates would result in double-stranded, RNase-resistant molecules. After extraction from the chromatin, the RNA was tested for sensitivity to pancreatic RNase, followed by precipitation of undigested RNA by TCA. Table 3 shows that about half of the $[^{125}I]RNA$ was present as RNase-resistant duplexes. By boiling the samples

Table 1. Removal of $mRNA_{ov}$ from In Vitro RNA by SH-Sepharose Column Chromatography

Conditions	Percentage bound to SH column	
	A	B
SH	73	83
Heated, 2 min, 100°C → SH	31	7

$mRNA_{ov}$ (37.5 ng) was added to the reaction mixture containing 200 μg estrogen-withdrawn chromatin, 400 μg *E. coli* RNA polymerase, and Hg-UTP, as described in Methods. The reaction was carried out at 37°C for 2 hr. RNA was then extracted, chromatographed on Sephadex G-50 columns, and loaded onto SH columns in either 50 mM Tris-HCl, pH 7.5, 0.1 N NaCl (TN) as in A or 10 mM Tris-HCl, pH 7.5, 1 mM EDTA (TE) as in B. After loading, the SH columns were washed with TN and TN + 0.1 M 2-mercaptoethanol, as described by Towle et al. (1977).

Table 2. Interaction of $mRNA_{ov}$ with In Vitro Hg-RNA

Hg-RNA + $[^{125}I]mRNA$	$[^{125}I]mRNA_{ov}$ bound to SH column (%)
Chromatin$_w$ transcripts + $[^{125}I]mRNA_{ov}$ → 4°C overnight	47
E. coli DNA transcripts + $[^{125}I]mRNA$ → 4°C overnight	55
$[^{125}I]mRNA$ alone	< 1

Isolated Hg-RNA transcripts (30 μg) transcribed from withdrawn chromatin or *E. coli* DNA were incubated with 10 ng of $[^{125}I]mRNA$ (10^7 cpm/μg) at 4°C overnight in TN buffer before being applied to an SH-Sepharose column without the heating step as described in Methods.

Table 3. RNase Sensitivity of RNA after Incubation in Standard RNA Synthesizing Conditions

Conditions	RNase-resistant $[^{125}I]mRNA_{ov}$ (%)
No RNase	100
RNase (20 μg/ml, 30 min)	48
Heated at 100°C, 5 min, then RNase (20 μg/ml, 30 min)	4.6

$[^{125}I]mRNA_{ov}$ (10 ng) was added in chromatin transcriptional mixture. The RNA synthesized (Hg-RNA and $[^{125}I]mRNA$ mixture) was then treated with RNase as shown in the table.

before RNase treatment, all but about 5% of the ^{125}I became digestible. These results show that RNA-dependent RNA synthesis is an important side reaction and possible source of artifact. However, as shown in Table 1, heating of the extracts before the SH-column step effectively destroys these RNA-RNA duplexes and largely prevents adsorption of coding-strand contaminants. The SH column will, of course, adsorb Hg-anticoding strands which result from the RNA-directed synthesis, but these strands cannot hybridize with the $cDNA_{ov}$ probe and so are not detected in the $mRNA_{ov}$ assay. In conclusion, our data suggest that both aggregation of Hg-RNA with non–Hg-RNA and RNA-dependent RNA synthesis contribute to the binding of contaminating RNA to the column.

Next we studied the in vitro synthesis of ovalbumin messenger RNA from chick oviduct chromatin. RNA was synthesized from oviduct chromatin and isolated through an SH-Sepharose column as described above. RNA eluted from the column by 2-mercaptoethanol was then assayed for $mRNA_{ov}$ sequences by hybridization. As shown in Table 4, 0.006% of the RNA was homologous to ovalbumin mRNA sequences.

To prove that the RNA detected by hybridization does in fact represent newly synthesized RNA, the following criteria have to be satisfied. First, it is necessary to demonstrate that the transcription of these hybridizable mRNA sequences is sensitive to actinomycin D. Since proper transcription of chromatin should be DNA-dependent, the synthesis of new $mRNA_{ov}$ sequences should be sensitive to actinomycin D inhibition, whereas neither adsorption of contaminating RNA_{ov} sequences to the SH column nor RNA-dependent RNA synthesis should be affected. The amount of $mRNA_{ov}$ sequences detected is inhibited by 85% at an actinomycin D concentration of 8 $\mu g/ml$ (Table 4). When the actinomycin D concentration was increased to 150 $\mu g/ml$, 95% inhibition of $mRNA_{ov}$ sequences occurred. On the other hand, if $[^{125}I]mRNA_{ov}$ was added to the chro-

matin, the amount of contaminating $[^{125}I]mRNA_{ov}$ in the preparation was not reduced by the actinomycin D. In fact, the level of contaminating $[^{125}I]RNA$ increased slightly. Therefore, the chromatin $mRNA_{ov}$ transcripts arise largely from DNA-dependent synthesis.

Second, it is necessary to demonstrate that RNAs hybridizing to the $cDNA_{ov}$ probe are in fact mercurated. Newly synthesized $mRNA_{ov}$ should contain Hg, whereas contaminating $mRNA_{ov}$ sequences present in vivo would not contain this element. To test this, Hg-RNA was synthesized off chromatin and recovered from an SH column. It was then reacted with $[^{3}H]cDNA_{ov}$. At the end of hybridization the hybridizable $cDNA_{ov}$ sequences were then applied to a second SH column to test for the presence of Hg-UMP in the hybrid. As shown in Table 5, more than 85% of the S_1-resistant $cDNA_{ov}$ sequences were bound to the SH column. This shows that the RNA hybridizing to $cDNA_{ov}$ contained mercurated nucleotides and was not merely non-Hg endogenous RNA contaminants. These two experiments firmly demonstrate that the hybridizable sequences are indeed RNA product newly transcribed from oviduct chromatin by $E.\ coli$ RNA polymerase and not due merely to the artifact of endogenous RNA contamination. However, it must be emphasized that this experiment does not address itself to the question of RNA initiation in the chromatin. Short RNA primers, for example, could be elongated in a DNA-dependent reaction and/or RNA polymerase could initiate randomly on the ovalbumin DNA sequences. At the present time, all these possibilities still exist in the in vitro transcriptional system.

Transcription of Ovalbumin and Globin Genes from Oviduct Chromatin

To study the fidelity of transcription and the effects of varying the ratio of RNA polymerase to chromatin in vitro, we have chosen to examine the transcription of two genes, ovalbumin and globin. Both

Table 4. Actinomycin-D Inhibition of $mRNA_{ov}$ and Total RNA Synthesis

Actinomycin D ($\mu g/ml$)	Total RNA synthesis[a]		mRNA synthesized[b]		Percentage of added $[^{125}I]mRNA_{ov}$ binding to SH column[c]
	cpm	inhibition (%)	$mRNA_{ov}$ ($\times 10^3$) (%)	inhibition (%)	
0	21,424	—	6.7	—	9.2
4	9560	55	1.9	74	—
8	6928	68	1.2	86	14.8
50	1890	91.2	—	—	17.0
150	705	96.7	0.3	95	—

Details of in vitro RNA synthesis reaction are given in Methods.

[a] ^{3}H-labeled Hg-UTP (sp. act. 4 mCi/mmole) was added. After incubation at 37°C for 120 min, RNA was precipitated with cold 5% TCA, washed and collected on glass-fiber filters, and counted for ^{3}H.

[b] Companion incubations to those used to measure total RNA; nonradioactive Hg-UTP was used as precursor as described in Methods. $mRNA_{ov}$ was detected by $[^3H]cDNA$ assay as outlined in Methods.

[c] Companion incubation reactions were carried out as in b except 20 ng of $[^{125}I]mRNA$ was added as a contaminant during RNA synthesis. The amount of $[^{125}I]mRNA_{ov}$ bound to SH column was then measured.

Table 5. Rebinding of In Vitro Hg-RNA and cDNA$_{ov}$ Hybrid to SH-Sepharose Column

Input (cpm)	Bound (cpm)	Re-bound (%)
2000	1737	87

Table 7. In Vitro Transcription of Ovalbumin mRNA Sequences from Chick Oviduct Chromatin

Enzyme/chromatin ($\mu g/\mu g$)	RNA synthesized/DNA ($\mu g/\mu g$)	mRNA$_{ov}$ ($\times 10^2$) (%)
5	1.76	0.46
4	1.44	0.51
1.25	0.60	1.15
0.5	0.24	1.47

of these mRNA species are highly tissue-specific. Thus, globin mRNA would not be expected to be synthesized to a significant degree in chick oviduct tissue. To ensure that the in vitro chromatin transcription also maintained the same tissue specificity, Hg-RNA was synthesized from estrogen-stimulated oviduct chromatin. This RNA sample was then analyzed for the amounts of both globin and ovalbumin mRNA sequences. As can be seen in Table 6, ovalbumin mRNA sequences are transcribed from oviduct chromatin to a much greater extent than are globin mRNA sequences. The concentrations of ovalbumin and globin mRNA transcribed from oviduct chromatin were 0.0051% and 0.0005%, respectively. Thus, qualitatively, the chromatin preparations maintained the expected in vivo tissue specificity.

Next, we examined the effects of varying the ratio of RNA polymerase to chromatin on the transcription of specific genes. Chromatin from hormone-stimulated chick oviduct was transcribed at four different enzyme-to-DNA ratios—5:1, 4:1, 1.25:1, and 0.5:1. In these experiments, the concentration of chromatin was held constant by varying the final volume of the reaction mixture to eliminate possible effects of varying chromatin solubility (see Table 7). As the enzyme-to-DNA ratio was decreased, the level of mRNA$_{ov}$ sequences in the transcripts increased. Over the tenfold range of enzyme-to-DNA ratios tested, the percentage of mRNA$_{ov}$ sequences transcribed from oviduct chromatin increased more than threefold (Table 7). Thus, the efficiency with which bacterial RNA polymerase transcribed the ovalbumin gene increased relative to total RNA synthesis when the ratio of enzyme molecules to chromatin DNA was lowered.

The increase in the percentage of mRNA$_{ov}$ sequences could represent either an increase in the transcription of the ovalbumin gene or a decrease in the synthesis of other RNA species. To distinguish between these possibilities, parallel experiments utilizing radioactively labeled UTP were carried out to monitor accurately the effects of changing enzyme-to-chromatin DNA ratios on total RNA synthesis. (Yields of RNA recovered do not provide precise estimates of total RNA synthesis due to differences in recoveries of RNA during extraction procedures of samples of varying volumes.) As the enzyme-to-chromatin ratio decreased tenfold, the total amount of RNA synthesized decreased in a linear relationship. Thus, from an equivalent amount of RNA polymerase, the total amount of RNA synthesized was roughly equivalent (Table 7) and resulted in a threefold increase in the absolute amount of mRNA$_{ov}$ synthesized. Thus, the number of ovalbumin mRNA molecules synthesized per molecule of RNA polymerase must increase at the lower enzyme-to-DNA ratios.

Strand Selection during Transcription of the Ovalbumin Gene

Before we can study the strand selection of chromatin transcription, it is necessary to obtain a cDNA probe specific for sequences of RNA which are complementary to ovalbumin RNA, that is, sequences transcribed from the anticoding strand of the ovalbumin gene (Fig. 1). Basically, the procedure involved using the full-length, single-stranded cDNA$_{ov}$ as a template-primer for avian myeloblastosis reverse transcriptase. By using Hg-dUTP as a substrate, a double-stranded cDNA$_{ov}$ can be produced which contains mercury substituents only in the anticoding strand. The double-stranded cDNA can then

Table 6. In Vitro Transcription of Ovalbumin and Globin mRNA Sequences from Chick Oviduct Chromatin

mRNA sequences	Percentage in total RNA ($\times 10^3$)
Ovalbumin	5.1
Globin	0.5

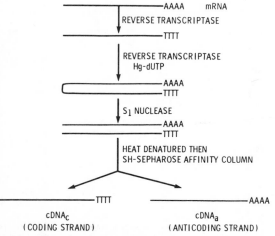

Figure 1. Flow diagram outlining steps for the synthesis and isolation of cDNA probes for transcripts of the coding and anticoding strands of the ovalbumin gene.

be clipped open with single-strand-specific S_1 nuclease and the anticoding strand isolated by sulfhydryl-Sepharose chromatography. The anticoding cDNA obtained after two passes through the SH columns had a size distribution similar to nonmercurated ovalbumin cDNA, by analysis using alkaline sucrose density gradient sedimentation (data not shown).

Before the coding and anticoding cDNA probes could be used to analyze in vitro chromatin transcripts, it was necessary to demonstrate the specificity of these coding and anticoding ovalbumin cDNAs. The coding cDNA should contain only the sequences complementary to ovalbumin mRNA and the anticoding cDNA should contain only those sequences hybridizable to coding cDNA. The experiments shown in Figure 2 were designed to show this specificity. Coding and anticoding cDNAs were hybridized with increasing amounts of ovalbumin mRNA or cDNA to ovalbumin mRNA (coding-strand cDNA). The presumptive coding cDNA hybridized only with ovalbumin mRNA and not with cDNA to ovalbumin mRNA (Fig. 2). In contrast, the presumptive anticoding ovalbumin cDNA hybridized only with coding cDNA and not with ovalbumin mRNA. These results demonstrated that the cDNA probes indeed represented uncontaminated preparations of coding or anticoding cDNAs. Since the Hg-UMP–containing anti-cDNA$_{ov}$ also had the same hybridization properties as the nonmercurated cDNA$_{ov}$ (Towle et al. 1977), these probes could be used to quantitate the ovalbumin mRNA and anti-ovalbumin mRNA present in chromatin transcripts.

The asymmetry of transcription has often been used as a measure of the fidelity of in vitro transcription. By comparing the amount of this anti-mRNA$_{ov}$ with the amount of mRNA$_{ov}$ sequences, the symmetry of transcription of the ovalbumin gene can be monitored. RNA was synthesized from chick oviduct chromatin at three enzyme-to-DNA ratios. At all three ratios tested, the percentage of anti-mRNA$_{ov}$ was approximately the same—0.0010% (Table 8). Since the percentage of mRNA$_{ov}$ sequences increased with decreasing enzyme-to-DNA ratio, the asymmetry of transcription must have increased. At the highest enzyme-to-chromatin ratio tested, the percentage of RNA from the noncoding strand was 16%. Thus, although there was a large amount of RNA transcribed from both strands of DNA, there was a predominance of RNA transcribed from the coding strand. As the enzyme-to-chromatin ratio decreased, the percentage of symmetry decreased from 16% to around 6%. Thus, at the lowest enzyme-to-DNA ratio we tested, the in vitro transcription of the ovalbumin gene by E. coli RNA polymerase was almost totally asymmetric (94%).

When in vitro RNA transcripts are hybridized to either coding or anticoding cDNA probes, there are two competing reactions which take place. In the case of the coding cDNA probe, the ovalbumin mRNA sequences can anneal to either cDNA$_{ov}$ or to the anti-mRNA sequences. With the anticoding cDNA probe, the anticoding-strand sequences can hybridize to the anti-cDNA$_{ov}$ or to ovalbumin mRNA sequences in the transcript. The degree of interference will be dependent on the ratio of mRNA to anti-mRNA sequences and the relative rates of hybridization of cDNA probe to RNA transcripts as opposed to the RNA-RNA hybridization. Since the level of ovalbumin mRNA sequences in the transcripts is higher than the anticoding-strand sequences, the effect of anti-ovalbumin mRNA sequences on the measurement of mRNA$_{ov}$ sequences will be minimal. To quantitate the effect of ovalbumin mRNA sequences on the measurement of anti-ovalbumin mRNA sequences, the following control experiment was performed. A mixture of ovalbumin mRNA and anti-ovalbumin RNA sequences was prepared by transcribing double-stranded cDNA$_{ov}$. These transcripts were then annealed to anti-

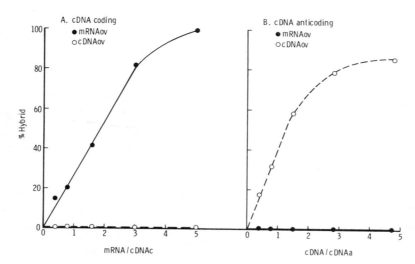

Figure 2. Hybridization of ovalbumin mRNA and cDNA to the coding and anticoding strands of the ovalbumin gene: 1.5 ng of coding ovalbumin cDNA *(A)* or anticoding strand ovalbumin cDNA *(B)* was hybridized to various amounts of ovalbumin mRNA (●———●) or ovalbumin cDNA (○--○) according to the procedure described in Methods.

Table 8. In Vitro Transcription of mRNA$_{ov}$ and Anti-mRNA$_{ov}$ Sequences from Chick Oviduct Chromatin

Enzyme/chromatin ($\mu g/\mu g$)	Anti-mRNA$_{ov}$ ($\times 10^2$) (%)	mRNA$_{ov}$ ($\times 10^2$) (%)	Asymmetry (%)
4.0	0.1	0.51	84
1.25	0.1	1.15	92
0.5	0.1	1.47	94

cDNA$_{ov}$ probe in a standard hybridization assay. Isolated mRNA$_{ov}$ sequences were then added to parallel hybridization reactions in ratios of 3:1 and 10:1 with the anti-mRNA$_{ov}$. The results demonstrated that there was some decrease in the estimated amount of anti-mRNA$_{ov}$ due to the presence of mRNA$_{ov}$. At a ratio of 3:1, the actual amount of anti-mRNA$_{ov}$ sequences was underestimated by about 20%, and at 10:1 the underestimate could be as high as 100%. Thus, the percentage of anti-mRNA$_{ov}$ that we have estimated in the oviduct transcript could be slightly lower than the real values. For instance, at the highest enzyme-to-DNA ratio we tested, the symmetrical transcription could increase from 16% to about 20%, whereas at the lowest enzyme-to-DNA ratio, the increase could be roughly from 6% to 10%. The interference of the mRNA, therefore, will only slightly increase the absolute value of symmetrical transcription. However, this does not affect the qualitative conclusion that the asymmetry of transcription increases as lower enzyme-to-chromatin ratios are used for RNA synthesis. It should also be emphasized that an enzyme:chromatin ratio of 0.5:1 still represents a 40-fold excess of enzyme molecules over that which normally exists in the intact nucleus.

The simplest explanation to account for changes in the concentration of a specific RNA in a total RNA population during transcription would involve initiation sites which differ in the efficiency with which they bind RNA polymerase and initiate RNA synthesis. At the highest enzyme-to-DNA ratios, the strongest initiation sites (those which bind and initiate most rapidly) would be completely saturated with enzyme. In addition, the large excess of enzyme molecules would also be able to utilize weaker secondary initiation sites. As the enzyme-to-DNA ratio was lowered, more competition would occur for initiation sites. Under these conditions, the stronger initiation sites, due to their higher efficiency of binding and initiation, would be responsible for synthesizing a large proportion of the RNA. This explanation would imply that the initiation site(s) which the *E. coli* RNA polymerase uses for synthesis of mRNA$_{ov}$ is a strong site and the one resulting in transcription of anti-mRNA$_{ov}$ sequences is a weak site. It must be emphasized, however, that at this time there is no direct evidence to indicate that the initiation sites utilized in vitro correspond to actual in vivo promoters.

SUMMARY

We have developed an in vitro cell-free transcriptional system capable of transcribing the ovalbumin gene in oviduct chromatin. By lowering the enzyme-to-DNA ratio, more than 90% of the ovalbumin gene transcripts were from the coding or correct strand.

Transcription of the ovalbumin gene was nonrandom since the proportion of ovalbumin mRNA synthesized was greater than other RNAs synthesized from chromatin in vitro. At limiting concentrations of RNA polymerase, preferential transcription of the ovalbumin gene increased further, thus supporting the concept of strong and weak initiation sites.

To study the specific interactions between these regulatory proteins and the ovalbumin gene, we attempted to synthesize, isolate, and amplify this gene. Using reverse transcriptase and the complete ovalbumin complementary DNA (cDNA$_{ov}$) as a template, a double-stranded DNA (the synthetic ovalbumin gene) was synthesized in vitro (Monahan et al. 1976a). This double-stranded synthetic gene has been successfully incorporated into bacterial plasmids and amplified (McReynolds et al. 1977b). By an alternative method designed to obtain DNA sequences adjacent to the coding portion of the ovalbumin gene, we have recently also purified the natural ovalbumin gene from total chick DNA by affinity chromatography. Using separate columns in which purified ovalbumin mRNA (mRNA$_{ov}$) and cDNA$_{ov}$ were covalently linked to phosphocellulose, the respective complementary ovalbumin DNA strands were purified from total chick DNA. Following reannealing, this "natural gene" can also be subjected to cloning and amplification in plasmids. It should be ideal for studies of regulation of gene expression since it is likely to contain both structural and regulatory sequences for the ovalbumin gene. It is hoped that our future efforts may allow us to reconstruct a "minichromosome" of the purified ovalbumin gene.

Note Added in Proof

Recently, we have further modified the procedure for running the SH-Sepharose column by adding 0.2% of SDS to the TN solution to wash the SH-Sepharose column before and after Hg-RNA samples were loaded onto the column. With this modified procedure we are able to reduce further the added [^{125}I]mRNA$_{ov}$ contaminant from 9–10% down to 1–2%. This method is especially applicable in the case of chromatin preparations that contain high levels of contaminant RNA or if in vitro synthesizing conditions are less than optimal.

Finally, we have developed a filter hybridization technique to determine directly the concentration of [^{3}H]mRNA$_{ov}$ sequences by hybridizing ^{3}H-labeled RNA synthesized in vitro from chromatin to pOV$_{230}$ DNA (pMB9 plasmid DNA containing ovalbumin gene insert is fixed to nitrocellulose filters) (McReynolds et al. 1977a). With estrogen-stimulated chromatin transcripts, we can detect $0.0044 \pm 0.001\%$ of $mRNA_{ov}$ sequences, which can be competed out completely with excess cold $mRNA_{ov}$ sequences. Since potential in vivo contaminating RNA sequences would not contain a radioactive label, these results again support the notion that isolated oviduct chromatin does support the synthesis of $mRNA_{ov}$ sequences in vitro.

Acknowledgments

The authors wish to thank Jim Cook and Valerie S. McMullian for excellent technical assistance. We also wish to thank Drs. Christina Chang, Savio Woo, and John Monahan for helpful discussion. This work was supported by grants HD-8188 and HD-7857 from the National Institutes of Health and by a grant from the Kelsey-Leary Foundation (958).

REFERENCES

BARRETT, T., D. MARYANKA, P. H. HAMLYN, and H. T. GOULD. 1974. Nonhistone proteins control gene expression in reconstituted chromatin. *Proc. Natl. Acad. Sci.* **71**: 5057.

BIESSMANN, H., R. A. GJERSET, B. W. LEVEY, and B. J. MCCARTHY. 1976. Fidelity of chromatin transcription *in vitro. Biochemistry* **15**: 4356.

BURGESS, R. R. and J. J. JENDRISAK. 1975. A procedure for the rapid large scale purification of *E. coli* DNA-dependent RNA polymerase involving polymin P precipitation and DNA-cellulose chromatography. *Biochemistry* **14**: 4634.

CHAN, L., A. R. MEANS, and B. W. O'MALLEY. 1973. Induction rates of specific translatable mRNAs for ovalbumin and avidin by steroid hormones. *Proc. Natl. Acad. Sci.* **70**: 1870.

CHIU, J. F., Y. H. TSAI, D. SAKUMA, and L. S. HNILICA. 1975. Regulation of *in vitro* mRNA transcription by a fraction of chromosomal proteins. *J. Biol. Chem.* **250**: 9431.

COMSTOCK, J. P., G. C. ROSENFELD, B. W. O'MALLEY, and A. R. MEANS. 1972. Estrogen-induced changes in translation and specific messenger RNA levels during oviduct differentiation. *Proc. Natl. Acad. Sci.* **69**: 2377.

COX, R., M. HAINES, and N. CAREY. 1973. Modification of the template capacity of chick oviduct chromatin for form B polymerase by estradiol. *Eur. J. Biochem.* **32**: 513.

COX, R. F., M. E. HAINES, and J. S. EMTAGE. 1974. Quantitation of ovalbumin mRNA in hen and chick oviduct by hybridization to complementary DNA. *Eur. J. Biochem.* **49**: 225.

CUATRECASAS, P. 1970. Protein purification by affinity chromatography: Derivations of agarose and polyacrylamide beads. *J. Biol. Chem.* **245**: 3059.

DALE, R. M. K. and D. C. WARD. 1975. Mercurated polynucleotides: New probes for hybridization and selective polymer fractionation. *Biochemistry* **14**: 2458.

DALE, R. M. K., E. MARTIN, D. C. LIVINGSTON, and D. C. WARD. 1975. Direct covalent mercuration of nucleotides and polynucleotides. *Biochemistry* **14**: 2447.

GILMOUR, R. S. and J. PAUL. 1973. Tissue-specific transcription of the globin gene in isolated chromatin. *Proc. Natl. Acad. Sci.* **70**: 3440.

HARRIS, S. E., J. M. ROSEN, A. R. MEANS, and B. W. O'MALLEY. 1975. Use of a specific probe for ovalbumin messenger RNA to quantitate estrogen-induced gene transcripts. *Biochemistry* **14**: 2072.

HARRIS, S. E., R. J. SCHWARTZ, M. J. TSAI, A. K. ROY, and B. W. O'MALLEY. 1976. Effect of estrogen on gene expression in the chick oviduct: *In vitro* transcription of the ovalbumin gene in chromatin. *J. Biol. Chem.* **251**: 524.

JENSEN, E. V., S. MOHLA, T. A. GORELL, and E. R. DE SOMBRE. 1974. The role of Estrophilin in estrogen action. *Vitam. Horm.* **32**: 89.

MCKNIGHT, G. S., P. PENNEQUIN, and R. T. SCHIMKE. 1975. Induction of ovalbumin mRNA sequences by estrogen and progesterone in chick oviduct as measured by hybridization to complementary DNA. *J. Biol. Chem.* **250**: 8105.

MCREYNOLDS, L. A., J. F. CATTERALL, and B. W. O'MALLEY. 1977a. The ovalbumin gene: Cloning of a complete double-stranded cDNA in bacterial plasmid. *Gene* (in press).

MCREYNOLDS, L. A., J. J. MONAHAN, D. W. BENDURE, S. L. C. WOO, G. V. PADDOCK, W. SALSER, J. DORSON, R. E. MOSES, and B. W. O'MALLEY. 1977b. The ovalbumin gene: Insertion of ovalbumin gene sequences in chimeric bacterial plasmids. *J. Biol. Chem.* **252**: 1840.

MEANS, A. R., J. P. COMSTOCK, G. C. ROSENFELD, and B. W. O'MALLEY. 1972. Ovalbumin messenger RNA of chick oviducts: Partial characterization, estrogen dependence and translation *in vitro. Proc. Natl. Acad. Sci.* **69**: 1146.

MONAHAN, J. J., L. A. MCREYNOLDS, and B. W. O'MALLEY. 1976a. The ovalbumin gene: *In vitro* enzymatic synthesis and characterization of the ovalbumin gene. *J. Biol. Chem.* **251**: 1960.

MONAHAN, J. J., S. E. HARRIS, S. L. C. WOO, D. L. ROBBERSON, and B. W. O'MALLEY. 1976b. The synthesis and properties of the complete complementary DNA transcript of ovalbumin mRNA. *Biochemistry* **15**: 223.

O'MALLEY, B. W. and W. L. MCGUIRE. 1968. Studies on the mechanism of estrogen-mediated tissue differentiation: Regulation of nuclear transcription and induction of new RNA species. *Proc. Natl. Acad. Sci.* **60**: 1527.

O'MALLEY, B. W. and A. R. MEANS. 1974. Female steroid hormones and target cell nuclei. *Science* **183**: 610.

O'MALLEY, B. W. and W. T. SCHRADER. 1976. The receptors of steroid hormones. *Sci. Am.* **234**: 32.

O'MALLEY, B. W., W. L. MCGUIRE, P. O. KOHLER, and S. G. KORENMAN. 1969. Studies on the mechanism of steroid hormone regulation of synthesis of specific proteins. *Recent Prog. Horm. Res.* **25**: 105.

O'MALLEY, B. W., G. C. ROSENFELD, J. P. COMSTOCK, and A. R. MEANS. 1972. Steroid hormone induction of a specific translatable messenger RNA. *Nat. New Biol.* **240**: 45.

PALMITER, R. D. 1973. Rate of ovalbumin messenger RNA synthesis in the oviduct of estrogen-primed chicks. *J. Biol. Chem.* **248**: 8260.

PARK, W. D., J. L. STEIN, and G. S. STEIN. 1976. Activation of *in vitro* histone gene transcription from HeLa S₃ chromatin by S-phase non-histone chromosomal proteins. *Biochemistry* **15**: 3296.

PAUL, J. and R. S. GILMOUR. 1968. Organ-specific restriction of transcription in mammalian chromatin. *J. Mol. Biol.* **34**: 305.

RHOADS, R. E., G. S. MCKNIGHT, and R. T. SCHIMKE. 1973. Quantitative measurement of ovalbumin messenger RNA activity. *J. Biol. Chem.* **248**: 2031.

ROSEN, J. M., C. D. LIARAKOS, and B. W. O'MALLEY. 1973. Effect of estrogen on gene expression in the chick oviduct. I. DNA-DNA renaturation studies. *Biochemistry* **12**: 2803.

SCHWARTZ, R. J., M. J. TSAI, S. Y. TSAI, and B. W. O'MALLEY. 1975. Effect of estrogen and gene expression in chick oviduct: Changes in the number of RNA polymerase binding and initiation sites in chromatin. *J. Biol. Chem.* **250**: 517.

SELIGY, V. L. and J. M. NEELIN. 1970. Transcription properties of stepwise acid-extracted chicken erythrocyte chromatin. *Biochim. Biophys. Acta* **213**: 380.

SMITH, M. M. and R. C. C. HUANG. 1976. Transcription *in vitro* of immunoglobulin kappa light chain genes in isolated mouse myeloma nuclei and chromatin. *Proc. Natl. Acad. Sci.* **73**: 775.

SPELSBERG, T. C. and L. S. HNILICA. 1971. Proteins of chromatin in template restriction. I. RNA synthesis *in vitro.* *Biochim. Biophys. Acta* **228**: 202.

SPELSBERG, T. C., A. W. STEGGLES, and B. W. O'MALLEY. 1971. Changes in chromatin composition and hormone binding during chick oviduct development. *Biochim. Biophys. Acta* **254**: 129.

SPELSBERG, T. C., W. M. MITCHELL, F. CHYTIL, E. M. WILSON, and B. W. O'MALLEY. 1973. Chromatin of the developing chick oviduct: Changes in the acidic proteins. *Biochim. Biophys. Acta* **312**: 765.

STEIN, G. S. and L. J. KLEINSMITH. 1975. *Chromosomal proteins and their roles in the regulation of gene expression.* Academic Press, New York.

STEIN, G., W. PARK, C. THRALL, R. MEANS, and J. STEIN. 1975. Regulation of cell cycle stage-specific transcription of histone genes from chromatin by nonhistone chromosomal proteins. *Nature* **257**: 764.

TOWLE, H. C., M. J. TSAI, S. Y. TSAI, and B. W. O'MALLEY. 1977. Effect of estrogen on gene expression in chick oviduct: Preferential initiation and asymmetrical transcription of specific chromatin gene. *J. Biol. Chem.* **252**: 2396.

TSAI, M. J., R. J. SCHWARTZ, S. Y. TSAI, and B. W. O'MALLEY. 1975. Effect of estrogen on gene expression in the chick oviduct: Initiation of RNA synthesis on DNA and chromatin. *J. Biol. Chem.* **250**: 5165.

TSAI, S. Y., M. J. TSAI, R. J. SCHWARTZ, M. KALIMI, J. CLARK, and B. W. O'MALLEY. 1975. Effect of estrogen on gene expression in chick oviduct: Nuclear receptor levels and initiation of transcription. *Proc. Natl. Acad. Sci.* **72**: 4228.

WOO, S. L. C., J. M. ROSEN, C. D. LIARAKOS, Y. C. CHOI, H. BUSCH, A. R. MEANS, B. W. O'MALLEY, and D. L. ROBBERSON. 1975. Physical and chemical characterization of purified ovalbumin mRNA. *J. Biol. Chem.* **250**: 7027.

YOUNG, B. D., P. R. HANISON, R. C. GILMOUR, G. D. BIRNIE, A. HELL, S. HUMPHRIES, and J. PAUL. 1974. Kinetic studies of gene frequency II. Complexity of globin complementary DNA and its hybridization characteristics. *J. Mol. Biol.* **84**: 555.

QUESTIONS/COMMENTS

Question by:

B. ALBERTS
University of California, San Francisco

The recent work by Schutz et al. (this volume) and Zasloff and Felsenfeld (*Biochem. Biophys. Res. Commun.* **75**: 598 [1977]) has revealed several artifacts which appear to be associated with the use of mercury-labeled nucleoside triphosphates. Also, others at this meeting have reported that traces of endogenous RNA can prime synthesis by *E. coli* RNA polymerase in such in vitro systems. It seems to me that in view of these results and your own, the fact that you report a partial inhibition of your detected RNA synthesis by actinomycin D is insufficient proof of true de novo RNA chain initiation by *E. coli* RNA polymerase in the manner claimed. Could the same problems have affected any of the transcription results which your laboratory has previously reported in the chicken oviduct system? For example, do the extensive modifications which you have now introduced into your RNA preparation procedure have any effects on your previous results and conclusions?

Response by:

B. W. O'MALLEY
Baylor College of Medicine

I just spent the first third of my own talk discussing this exact point, i.e., that the mercury labeling technique contains a "potential artifact" which, however, can be obviated by using an appropriate isolation procedure for the RNA. To clarify this issue let me again summarize the point in question.

Problem:

During transcription of chromatin in vitro in the presence of Hg-nucleotides, nascent Hg-RNA can form aggregates and hybrids to endogenous contaminating mRNA and thereby carry it as a complex through an SH-Sepharose column. In this way, one might unknowingly score this contaminating RNA as RNA which is synthesized in vitro.

Solution:

1. Utilize reaction conditions that will give a sufficiently large quantity of in vitro synthesized RNA to render contaminating sequences insignificant in amount. For example: we synthesize

~1.2–1.5 mg RNA/mg chromatin DNA, whereas Schutz synthesized 40 μg RNA/mg chromatin and Zasloff synthesized only 20 μg RNA/mg chromatin. These amounts are so small that contaminating sequences become a problem.

2. Use Mg^{++} instead of Mn^{++} in the in vitro reaction; this minimizes the potential of RNA-dependent RNA synthesis and greatly favors template DNA-dependent RNA synthesis (see also McCarthy et al., this volume).

3. Prior to application of the Hg-RNA to the SH-Sepharose column, one must dilute the sample into *low*-salt buffer and heat denature it. This will disrupt RNA-RNA hybrids and aggregates and eliminate artifactual carry-through of contaminating sequences.

Controls:

1. We have shown that $[I^{125}]mRNA_{ov}$ tracer added to the in vitro synthetic reaction can be eliminated (> 91%) by the procedure described above.

2. We have shown that large quantities (37.5 ng) of $mRNA_{ov}$ added to the reaction can also be eliminated (> 95%) by our method.

3. RNA hybridized to $cDNA_{ov}$ can be recovered after hybridization, and > 90% of this RNA can be shown to contain Hg by re-binding it to SH-Sepharose.

4. Most important, synthesis of the nascent $Hg\text{-}mRNA_{ov}$ can be shown to be inhibited by actinomycin D. The statement you made is not entirely correct. Rather than "a partial inhibition," we have shown that ~90% of the $mRNA_{ov}$ scored in the reaction is sensitive to actinomycin D at 8 μg/ml. Our opinion is that this is the most reliable test in that RNA-dependent RNA synthesis, aggregation, and polynucleotide phosphorylase activity are all insensitive to actinomycin D. This result further confirms our contention that the in vitro synthesis of $mRNA_{ov}$ is template (DNA)-dependent.

Relationship to previous results:

We have, in fact, published only one previous paper using the Hg-technique (*J. Biol. Chem.* **252**: 2396 [1977]). As I stated early in my talk, all slides (with one exception) contained data obtained using the new extract modification, which eliminates potential artifacts. As you can compare, the new technique does not result in significant changes in our previous conclusions as our background of contamination was very low. The reasons for this have been discussed above.

Question by:
M. ZASLOFF
National Institutes of Health

1. Can you demonstrate directly that the RNA sequences that remain in hybrid with the ovalbumin DNA probe after S_1 nuclease digestion are in fact mercury-substituted? Such a demonstration is critical for the interpretation of the data presented as it is unequivocal proof that the ovalbumin sequences detected are in fact newly synthesized.

2. As *E. coli* RNA polymerase will transcribe single-stranded RNA molecules under the reaction conditions you have used for in vitro chromatin transcription (Zasloff and Felsenfeld, *Biochem. Biophys. Res. Commun.* **75**: 598 [1977]), do you have evidence that the antisense ovalbumin sequences detected in the chromatin transcript have been copied off of a DNA template?

Response by:
B. W. O'MALLEY
Baylor College of Medicine

1. Yes. Despite the fact that control reactions show that some loss of Hg occurs in the RNA during incubation and S_1 digestion, we have recovered the RNA-$cDNA_{ov}$ hybrid following the S_1 nuclease digestion and can show that the majority of the RNA is mercury-substituted ($> 90\%$).

2. This is a good point. One can distinguish antisense ovalbumin mRNA sequences synthesized from DNA vs RNA template by their relative sensitivity to actinomycin D. Our conclusions are that some, but not all, of the antisense mRNA may be the result of RNA-dependent RNA synthesis rather than transcription of the antisense strand of the gene. Of course, this would result in an *under*estimate of our fidelity relative to transcription of the chromatin DNA template.

Question by:
K. YAMAMOTO
University of California, San Francisco

Experiments from your laboratory (O'Malley and McGuire, *J. Clin. Invest.* **47**: 654 [1968]) and some recent work of Cox (*Cell* **7**: 455 [1976]) show that acute in vivo administration of progesterone actually causes a slight *decrease* in total RNA synthesis in oviduct. How do you reconcile these results with your reports that exogenously added RNA polymerases reveal up to a threefold increase in polymerase start sites upon receptor addition? Also, do you have any data which demonstrates that the reported receptor-stimulated synthesis is selective for ovalbumin sequences?

Response by:
B. W. O'MALLEY
Baylor College of Medicine

I think there is some confusion here. Our report of changes in polymerase start sites upon in vivo administration of progesterone was under conditions of hormone injection to *withdrawn* chicks (*J. Biol Chem.* **251**: 5166 [1976]). In this state, the tissue responds dramatically to progesterone (or estrogen) by virtue of large increases in RNA synthesis. In the case of our report years ago (*Biochim. Biophys. Acta* **157**: 187 [1968]), we showed a "transitory" decrease in nuclear RNA synthesis followed by a rise. This study, however, was carried out in chicks which were already maximally stimulated by large doses of estrogen (5 mg/day, DES). It is possible that such a large dose of steroid hormone (10 mg: 5 mg E + 5 mg P) causes a transitory toxic depression of transcription. In any event, the two states cannot be compared.

With respect to the selective increase in ovalbumin synthesis, my slide showed very little change in the *total* RNA synthesis ($< 25\%$), whereas the increase in ovalbumin mRNA synthesis was greater than tenfold, from 1.9 pg ($\times 10^{-3}$) to 20.9 pg ($\times 10^{-3}$) per 400 μg of chromatin DNA.

Hormonal Control of Egg White Protein Messenger RNA Synthesis in the Chicken Oviduct

G. Schütz, M. C. Nguyen-Huu, K. Giesecke, N. E. Hynes,
B. Groner, T. Wurtz, and A. E. Sippel

Max-Planck-Institut für Molekulare Genetik, Berlin-Dahlem, West Germany

The chick oviduct is an attractive system for studying the control of specific protein synthesis by steroid hormones (Palmiter 1975; Rosen and O'Malley 1975; Schimke et al. 1975). Readministration of estrogens to estradiol-stimulated and subsequently withdrawn chickens leads to the synthesis of the major egg white proteins ovalbumin, ovomucoid, conalbumin, and lysozyme in the tubular gland cells. These four proteins account for approximately 75% of total protein synthesis in the fully induced oviduct. They are, however, synthesized at different rates: ovalbumin comprises about 50–60% of the cell protein synthesized, conalbumin accounts for about 10%, ovomucoid for about 8%, and lysozyme for 2–3% (Palmiter 1972). The common control, yet the different extent of their accumulation, makes the coordinated synthesis of these proteins attractive for the study of differential gene expression.

The steroid-controlled rate of specific protein synthesis was shown to be closely correlated to the accumulation of specific mRNAs by quantitation of mRNA activity in cell-free protein synthesis systems (Chan et al. 1973; Rhoads et al. 1973; Schütz et al. 1973). In the case of the ovalbumin mRNA, the increase in biological activity of the mRNA upon hormone administration results from an increased cellular concentration (Cox et al. 1974; Harris et al. 1975; McKnight et al. 1975; Palmiter et al. 1976). Likewise we have shown that the induction of ovomucoid and lysozyme is due to an increase in the respective mRNA concentrations (Hynes et al. 1977). The accumulation of the egg white protein mRNAs for ovalbumin, ovomucoid, and lysozyme might result from increased synthesis or reduced degradation. Thus, the rate of accumulation of these mRNAs might be proportionate to the relative rates of transcription of their genes. Alternatively, the mRNA sequences might be synthesized with similar rates prior to hormone treatment, with differential accumulation of the gene products resulting from selection mechanisms at some level after transcription, such as processing, transport, or stabilization. Determination of the rates of synthesis of the mRNA during hormonal induction would allow a distinction between these alternatives. Since it is impossible to measure the rate of specific mRNA synthesis in the animal, we chose isolated oviduct nuclei as the cell-free system which might best preserve the transcriptional state of the intact cell. Newly synthesized mRNA sequences were distinguished from preexisting molecules by two different procedures: (a) by the use of a mercurated nucleotide as a substrate for RNA synthesis, subsequent purification of the RNA by affinity chromatography and hybridization to labeled cDNA; (b) by labeling of the RNA to high specific activity and hybridization to unlabeled complementary DNA. Using either method we could show that ovalbumin mRNA sequences are synthesized in isolated oviduct nuclei in a highly selective manner. The accumulation of mRNAs during hormonal stimulation was found to be related to changes in the rate of transcription of the ovalbumin gene.

EXPERIMENTAL PROCEDURES

Animals. HNL chicks and laying hens were used in all experiments. Oviducts from laying hens and chronically hormone-withdrawn chickens were obtained as described previously (Hynes et al. 1977). Alternatively, diethylstilbestrol was applied subcutaneously in silicon tubings and acutely withdrawn by removal of the tubings. The animals were sacrificed 3 days later (acute withdrawal) or after secondary stimulation as indicated.

Isolation of total cellular RNA, polysomal poly(A)-containing RNA, and specific mRNAs. A 0.5-g sample of oviduct tissue from three chickens was homogenized in 10 ml of 10 mm Tris, pH 7.5, 5 mm $MgCl_2$, 25 mm NaCl, 5% sucrose, 1 mg/ml heparin. Desoxycholate and Triton X-100 were added to 1% followed by 10 ml of 5 mm EDTA, 1% SDS. The homogenate was extracted with phenol and chloroform. After ethanol precipitation, the nucleic acids were resuspended in 100 mm NaCl, 10 mm Tris, pH 7.5, 5 mm magnesium acetate and digested with 40 μg/ml pancreatic DNase, which had been pretreated with iodoacetate. The nucleic acids were extracted with phenol and chloroform and passed over a G-50 Sephadex column.

The ratio of RNA/DNA in the homogenate was used for the estimation of the number of specific mRNA molecules per tubular gland cell.

Poly(A)-containing mRNA and the mRNAs for ovalbumin, ovomucoid, and lysozyme were isolated as described previously (Hynes et al. 1977; Groner et al. 1977). The preparation of cDNA and the cDNA excess hybridization reactions were performed as described previously (Hynes et al. 1977).

Preparation of nuclei and chromatin and synthesis of RNA. Oviduct nuclei were prepared by the method of Ernest et al. (1976), with the buffers used by Marshall and Burgoyne (1976). Briefly, frozen oviducts were homogenized in 0.3 M sucrose in Buffer A containing 2 mM EDTA, 0.5 mM EGTA, 14 mM β-mercaptoethanol (Buffer A: 60 mM KCl, 15 mM NaCl, 0.15 mM spermin, 0.5 mM spermidin, 15 mM HEPES, pH 7.5). Crude nuclei were resuspended in the homogenization solution containing 0.1% Triton X-100 in a Teflon-glass homogenizer by five strokes at 500 rpm. The nuclei were then centrifuged through 2 M sucrose in Buffer A containing 0.1 mM EDTA and 0.1 mM EGTA. The nuclei were resuspended in 25% glycerol containing Buffer A and frozen in liquid nitrogen. Hg-RNA was synthesized in the incubation mixture described by Ernest et al. (1976), with the following changes: CTP or Hg-CTP was used at 0.5 mM, 14 mM β-mercaptoethanol replaced 2.5 mM dithiothreitol, and MgCl$_2$ replaced magnesium acetate. Nuclei were incubated at a concentration of 1.5 mg/ml DNA and [^{32}P]UTP (2 mCi/mmole) was used to label the RNA to low specific activity. The Hg-RNA was isolated and purified according to a modification (Nguyen-Huu et al. 1978) of the method of Dale and Ward (1975).

RNA labeled to high specific activity was synthesized using 20 μM [^{3}H]UTP (46 Ci/mmole). The RNA was extracted by the hot-phenol method, purified by G-50 Sephadex chromatography, and hybridized to ovalbumin Hg-cDNA which had been synthesized with 12.5% substitution of dTTP by Hg-dUTP. The hybridization was carried out in 50% formamide, 0.5 M NaCl, 25 mM HEPES, pH 6.8, 0.5 mM EDTA at 40°C. After hybridization for 2 hours, the nonhybridized RNA was digested with RNase A (40 μg/ml) 30 minutes at 37°C in 10 mM Tris, pH 7.5, 0.3 M NaCl, 1 mM EDTA, and 2 μg/ml yeast RNA and the hybrids purified by SH-agarose chromatography (Nguyen-Huu et al. 1977).

Chromatin was prepared by lysis of nuclei in low salt. RNA was synthesized in 10 ml containing 500 μg DNA and 200 μg *E. coli* RNA polymerase (a gift of Dr. H. Sternbach) with 30% Hg-CTP as substrate (Giesecke et al. 1977). The RNA was extracted, purified, and hybridized to labeled cDNA as described above.

RESULTS

Accumulation of Egg White Protein mRNA Sequences

The effect of steroid hormones may be reflected in the qualitative and quantitative composition of the mRNA population. We therefore compared the mRNA populations isolated from the oviducts of the laying hen and of the estradiol-stimulated and hormone-withdrawn chick (Hynes et al. 1977). By analysis of the hybridization kinetics of the total mRNA with its cDNA, we found that approximately 13,000 different mRNA species are present in the oviducts from laying hens and hormone-withdrawn chickens. Cross-hybridization reactions revealed that most, if not all, of the mRNA species present in the oviducts of laying hens are also present in the oviducts of hormone-withdrawn chickens. However, the concentrations of a few mRNAs increase dramatically in the oviducts of laying hens. The availability of the mRNAs for ovalbumin, ovomucoid and lysozyme, highly purified by immunoadsorption of specific polysomes (Groner et al. 1977; Schütz et al. 1977), allowed a precise determination of the concentration of these mRNAs in total, as well as polysomal, poly(A)-containing RNA in the oviducts of laying hens and hormone-withdrawn chickens.

Table 1 summarizes the results obtained when an excess of each specific cDNA was hybridized to total or polysomal poly(A)-containing RNA. Although the mRNAs for ovalbumin, ovomucoid, and lysozyme are present in the oviducts of both laying hens and hormone-withdrawn chickens, their concentrations are 3000-fold higher in the laying hen. These results, together with the complexity analysis (Hynes et al. 1977), reveal that the number and type of structural genes expressed in the laying hen oviduct are very similar to those expressed in the hormone-withdrawn chick oviduct and suggest that the inducing

Table 1. Ovalbumin, Ovomucoid, and Lysozyme mRNA Content in Oviduct RNA

Type and source of RNA	Percentage of mRNA		
	ovalbumin	ovomucoid	lysozyme
Polysomal poly(A)$^+$ RNA			
hormone-withdrawn chicken	0.011	0.0039	0.0046
hen	50.0	6.6	3.4
Total RNA			
hormone-withdrawn chicken	0.00012–0.00026	0.00007–0.00014	0.00004–0.00008
restimulated chicken, 12 hr	0.096	0.0064	0.0053
hen	0.83–1.03	0.15	0.11

steroids cause a quantitative, rather than a qualitative, change in the mRNA population.

Time Course of Accumulation of Egg White Protein mRNAs during Secondary Stimulation

The induction pattern of the ovalbumin, ovomucoid, and lysozyme mRNAs following estradiol administration to hormone-withdrawn chickens has been studied in the experiment depicted in Figure 1. All three mRNAs accumulate in a similar fashion. As previously observed for the induction of the ovalbumin mRNA (Cox et al. 1974; McKnight et al. 1975; Palmiter et al. 1976), the concentration of these mRNAs changes only slightly in the first 3 hours. After this lag period, the mRNAs accumulate rapidly at a constant rate. From the RNA/DNA ratio, it can be calculated that 5–10 molecules of ovalbumin mRNA are present in the oviduct cell prior to hormone treatment. Twelve hours after diethylstilbestrol administration, 3300 molecules are found per cell, an increase in concentration of approximately 500-fold. In this phase of rapid accumulation, ovalbumin, ovomucoid, and lysozyme mRNAs appear at rates of approximately 6.9, 0.9, and 1.1 molecules per minute per oviduct cell, respectively. Conalbumin mRNA induction follows a strikingly different pattern: initially present at a significantly higher concentration than ovalbumin mRNA, it increases without a significant lag period upon hormone administration (Palmiter et al. 1976).

The lag period in the induction of the ovalbumin, ovomucoid, and lysozyme mRNAs is not due to delayed entrance of the hormone into the cell, since the receptor sites in the nucleus are fully saturated within 30 minutes (Kalimi et al. 1976; Palmiter et al. 1976). It seems possible, therefore, that the estrogen-mediated accumulation of egg white protein mRNAs is not an immediate response to the translocation of the hormone receptor complex into the nucleus.

Synthesis of Ovalbumin mRNA Sequences in Isolated Nuclei

The cellular accumulation of the egg white protein mRNAs upon hormonal induction may be due to transcriptional activation of their respective genes or to posttranscriptional stabilization of their respective RNAs. To distinguish between these possibilities, the synthesis of ovalbumin mRNA sequences was studied in isolated oviduct nuclei. The analysis of specific mRNA synthesis in isolated nuclei or chromatin is difficult for two reasons: (1) the specific mRNA is expected to represent only a small fraction of the in-vitro-synthesized RNA; (2) the large amounts of preexisting mRNA molecules in nuclei and chromatin complicate the measurement of newly synthesized mRNA sequences. We have used two different approaches to distinguish newly synthesized from preexisting ovalbumin mRNA molecules (Fig. 2): (1) Mercurated nucleoside triphosphates were used for the synthesis of RNA (Dale et al. 1973; Dale and Ward 1975; Crouse et al. 1976; Smith and Huang 1976; Towle et al. 1977). The newly synthesized RNA was subsequently purified by affinity chromatography on SH-agarose and titrated for ovalbumin mRNA sequences with radioactive cDNA. (2) Highly labeled RNA was synthesized in nuclei. The RNA was hybridized to nonradioactive mercurated cDNA. RNase-A digestion and subsequent purification of the hybrids allowed the determination of the ovalbumin RNA content in the in vitro transcripts.

Mercury labeling of the newly synthesized RNA. Figure 3 shows that RNA polymerase A and B of isolated nuclei accept mercurated CTP as a substrate. For either enzyme, substitution of Hg-CTP for CTP reduces the rate of synthesis by 40% (Fig. 3a). RNA synthesis is inhibited by α-amanitin, actinomycin D, DNase, or omission of Hg-CTP (Fig. 3b). Newly synthesized RNA was separated from RNA endogenous to the isolated nuclei by affinity chroma-

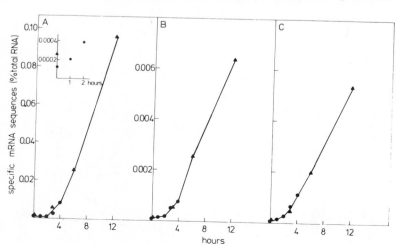

Figure 1. Accumulation of ovalbumin, ovomucoid, and lysozyme mRNA sequences during secondary hormone stimulation. Total cellular RNA was prepared from oviduct isolated from chicks stimulated for various times with diethylstilbestrol. The RNA was hybridized to ovalbumin *(A)*, ovomucoid *(B)*, and lysozyme cDNA *(C)* for a length of time sufficient to ensure completion of each hybridization reaction. The percentage of ovalbumin, ovomucoid, and lysozyme mRNA sequences in total RNA was estimated by comparison with a standard hybridization curve derived from hybridization of each purified mRNA with its respective cDNA. (●, ▲) Two sets of animals.

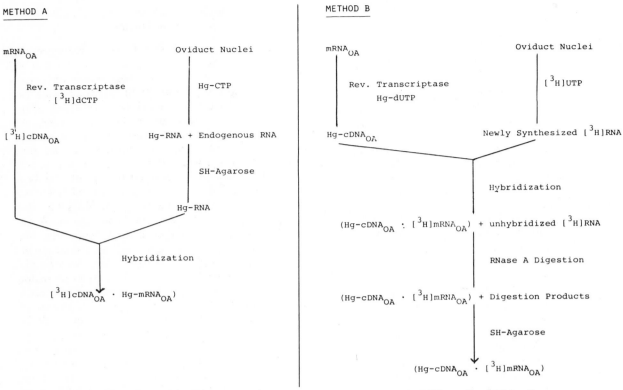

Figure 2. Schematic representation of the methods used for analysis of specific RNA synthesized in vitro.

tography on SH-agarose. To prevent copurification of preexisting with in-vitro-synthesized RNA, well-defined conditions for in vitro synthesis, extraction, and purification of Hg-RNA were developed. We could show that no binding of endogenous RNA occurred through unspecific adsorption, through aggregation to mercurated RNA, or through chemical mercuration (data not shown).

Figure 4 shows the hybridization analysis of isolated RNA synthesized with the same number of

nuclei under different incubation conditions. Ovalbumin mRNA sequences were synthesized during the incubation of the nuclei, whereas no ovalbumin mRNA was detectable in the SH-agarose eluate of the RNA sample prepared from unincubated nuclei. Actinomycin D and α-amanitin inhibited specific synthesis by 85–90%. The concentration of ovalbumin mRNA sequences within newly synthesized RNA was determined by comparison with a standard hybridization of ovalbumin mRNA with its

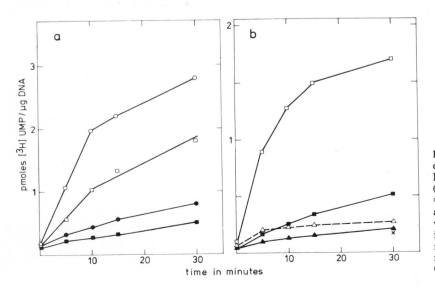

Figure 3. RNA synthesis in isolated oviduct nuclei. (a) Using [³H]UTP as label, RNA was synthesized with CTP (○) or Hg-CTP (□) in the absence (□, ○) or presence (■, ●) of 2.5 μg/ml α-amanitin. (b) RNA was synthesized with Hg-CTP in the absence of inhibitor (□) or in the presence of 5 μg/ml α-amanitin (■), 80 μg/ml actinomycin D (▲), or 80 μg/ml DNase (△). Omission of Hg-CTP (x).

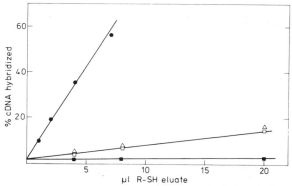

Figure 4. Hybridization of in-vitro-synthesized Hg-RNA to ovalbumin [^{3}H]cDNA. Hg-RNA was synthesized in isolated nuclei and purified by chromatography on SH-agarose. Aliquots of the fraction eluted with β-mercaptoethanol (R-SH eluate) were hybridized to ovalbumin cDNA. Nuclei were either not incubated (■) or incubated for 30 min without inhibitor (●), with 5 μg/ml α-amanitin (△), or with 80 μg/ml actinomycin D (□).

cDNA. Ovalbumin mRNA sequences represent 0.15% of the newly synthesized RNA. This value is probably an overestimate. Only elongation of in-vivo-initiated RNA chains occurs in isolated nuclei (Ernest et al. 1976). Thus, the Hg-labeled RNA molecules should contain in-vivo-initiated sequences contiguous to in-vitro-elongated sequences.

Radioactive labeling of the newly synthesized RNA. In an alternative approach, we determined the rate of synthesis of ovalbumin RNA in isolated hen oviduct nuclei through incorporation of [^{3}H]UMP of high specific activity into RNA. The labeled RNA was hybridized with excess mercurated cDNA and unhybridized RNA was digested with RNase A. The hybrids were subsequently chromatographed on SH-agarose to remove any nonhybridized RNase-resistant sequences and to reduce the background. Table 2 shows that 0.03% of the input ^{3}H-labeled RNA could be eluted from the SH-agarose column. No radioactive RNA was bound to the SH-agarose when cDNA was omitted from the hybridization reaction.

The specificity of the hybrid formation was tested by hybridization after addition of competing ovalbumin mRNA. At a 7:1 ratio of ovalbumin mRNA to cDNA, approximately 90% of the labeled RNA was displaced from the hybrid. The hybridized RNA was completely sensitive to digestion by RNase H (data not shown). Actinomycin D inhibits the synthesis of the ovalbumin-mRNA-specific sequences to the same extent as total RNA synthesis. α-Amanitin abolishes the specific RNA synthesis completely and reduces total RNA synthesis to only 40%. These results demonstrate that ovalbumin mRNA sequences have been synthesized in isolated hen oviduct nuclei with high selectivity. Approximately 0.08–0.1% of total or 0.12–0.15% of α-amanitin-sensitive RNA synthesis is ovalbumin mRNA synthesis. This percentage is lower than that determined by labeling the RNA with mercury-substituted CTP since this method measures only the in-vitro-synthesized sequences.

Approximately 10,000 RNA polymerase B molecules are engaged in transcription in the hen oviduct nucleus (Cox 1976). We thus can calculate that 6–8 polymerase molecules are transcribing the ovalbumin gene, if we assume that the elongation rate is identical on all genes, that both ovalbumin genes are being transcribed, and that all ovalbumin mRNA transcripts are conserved. This represents a packing density of polymerases of about one-third that found on rRNA cistrons.

Specific RNA Synthesis in Isolated Chromatin

It was of obvious interest to determine whether the selectivity of transcription is retained in isolated oviduct chromatin as has been found (Harris et al. 1976; Tsai et al. 1976; Towle et al. 1977). We therefore studied the synthesis of egg white protein mRNAs in isolated chromatin using *E. coli* polymerase with mercurated nucleotides. We found, however, that the analysis of specific RNA synthesis is complicated by an RNA-dependent RNA polymerase activity. Similar observations have been reported

Table 2. Hybridization of [^{3}H]RNA Synthesized in Isolated Nuclei to Ovalbumin Hg-cDNA

Transcription reaction	Hg-cDNA (ng)	Input RNA		mRNA$_{oa}$ added (ng)	RNase-A-resistant (cpm)	[^{3}H]RNA bound to SH-agarose		
		cpm	mRNA$_{oa}$ (ng)			cpm	%	% corrected
Complete	0	640,000	5.5	0	1130	0	0	0
Complete	72	640,000	5.5	0	1432	201	0.0314	0.094
Complete	72	640,000	5.5	500	1030	18	0.0028	0.0084
Complete	84	798,040	15	0	1512	204	0.026	0.078
+ Actinomycin D (40 μg/ml)	84	110,720	15	0	266	28	0.025	0.075
+ α-Amanitin (1 μg/ml)	84	299,760	15	0	596	0	0	0

[^{3}H]RNA synthesized in isolated oviduct nuclei for 30 min, using the conditions indicated, was hybridized to mercurated ovalbumin cDNA. In one hybridization reaction, 500 ng of ovalbumin mRNA was included as competitor; in another, Hg-cDNA was omitted from the reaction. After digestion with 40 μg/ml RNase A, the hybrids were purified by chromatography on SH-agarose. The corrected percentage of [^{3}H]RNA in the SH-agarose-bound hybrids was calculated from the efficiency of purification of ^{125}I-labeled ovalbumin mRNA hybridized to Hg-cDNA and chromatographed under the same conditions.

Table 3. Content of Ovalbumin- and Globin-specific RNA in Chromatin Transcripts

Transcription reaction	Total RNA synthesized (µg)	Ovalbumin-specific RNA		Globin-specific RNA	
		nonbound (ng)	bound (ng)	nonbound (ng)	bound (ng)
Chromatin + RNAP	24.0	220	6.0	0.04	<0.006
Chromatin + RNAP − Hg-CTP + CTP	28.5	213	<0.002	—	—
Chromatin + RNAP + rifampicin	0.57	229	1.1	—	—
Chromatin + RNAP + actinomycin D	3.2	180	6.2	—	—
Chromatin, no addition	0.28	205	0.065	—	—
Chromatin + RNAP + mRNA$_{oa}$	27.1	1250	18.4	—	—
Chromatin + RNAP + mRNA$_{gl}$	25.4	180	5.6	154	5.8

RNA was synthesized from oviduct chromatin (50.0 µg DNA/ml) using 20 µg/ml of *E. coli* RNA polymerase in the presence of 30% Hg-CTP. Transcription reactions were carried out with CTP, rifampicin (6 µg/ml), actinomycin D (50 µg/ml), without RNA polymerase or after addition of 1.2 µg ovalbumin mRNA and 0.2 µg globin mRNA. The RNA was isolated, chromatographed on SH-agarose, and the content of globin- and ovalbumin-specific sequences was determined in the nonbound and bound fraction by hybridization to [³H]cDNA.

by Zasloff and Felsenfeld (1977). The products of this side reaction lead to a copurification of the endogenous mRNA with newly synthesized RNA when the mercury-labeled RNA is purified on SH-agarose. We became aware of this side reaction of the *E. coli* RNA polymerase because the apparent production of the ovalbumin mRNA sequences could be inhibited by rifampicin, but not by actinomycin D (Table 3). When ovalbumin mRNA was added to the transcription reaction prior to incubation, an increase in the ovalbumin-specific RNA was detected. Furthermore, when globin mRNA was included into the transcription reaction, about 3% of the added RNA copurified with the newly synthesized RNA, although no globin-specific sequences could be detected in the transcripts of oviduct chromatin (Table 3). These results suggested to us that the RNA polymerase could use RNA as template for the synthesis of complementary RNA. When we used globin mRNA as template in the transcription reaction, more than 90% of the incorporated radioactivity was RNase-resistant but became sensitive to digestion with RNase after denaturation (data not shown), suggesting that the mRNA is used as template for the synthesis of complementary RNA. Table 4 shows that the complementary RNA leads to a retention of approximately 2% of the globin mRNA used as template. This retention does not occur when the RNA-dependent synthesis is inhibited by rifampicin.

It is not inhibited by actinomycin D, however. We believe that this mechanism causes contamination of the chromatin transcripts with endogenous RNA. We have attempted to prevent this copurification of endogenous with newly synthesized RNA by denaturation of the transcripts prior to SH-agarose chromatography. Denaturation of the RNA before affinity chromatography leads to a tenfold reduction of the ovalbumin mRNA sequences in the purified transcripts. These may represent products of in vitro transcription or may arise from contamination of the newly made RNA with endogenous RNA by a reaction additional to the RNA-dependent RNA polymerase activity. RNA isolated from transcription reactions containing actinomycin D and chromatographed on SH-agarose after heat denaturation did contain the same quantity of ovalbumin-specific sequences even though total RNA synthesis was inhibited by 90% (data not shown). Therefore, we conclude that most, if not all, of the selectivity of nuclear RNA transcription is lost in chromatin isolated and transcribed in these conditions.

Transcriptional Regulation of Ovalbumin mRNA Synthesis

Because transcription is likely to be the rate-limiting step in the steroid-induced accumulation of egg white protein mRNAs, we studied the synthesis of

Table 4. Retention of Globin mRNA Sequences on SH-Agarose after Incubation with RNA Polymerase and Hg-CTP

Condition of RNA synthesis	[³H]UMP incorporated (pmoles)	Globin RNA bound to SH-agarose (ng)
Complete	46.6	36
+ Rifampicin (6 µg/ml)	1.3	0.3
+ Actinomycin D (50 µg/ml)	47.0	30
− RNAP	0	<0.01
− Hg-CTP + CTP	50.3	<0.01

RNA was synthesized using 20 µg/ml *E. coli* RNA polymerase with 2 µg globin mRNA as template and Hg-CTP as substrate. The reaction mixture was phenol-extracted and then chromatographed on G-50 Sephadex and subsequently on SH-agarose. The amount of globin mRNA bound to the column was determined by hybridization to globin cDNA.

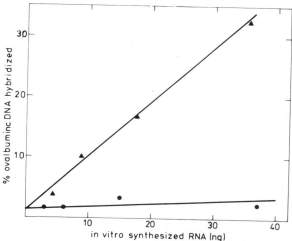

Figure 5. Hybridization of Hg-RNA synthesized in oviduct nuclei from laying hens (▲) and hormone-withdrawn chickens (●). Hg-RNA was purified on SH-agarose, and increasing amounts were hybridized with 0.15 ng of ovalbumin [³H]cDNA.

ovalbumin-mRNA-specific sequences in oviduct nuclei isolated from hormone-withdrawn chickens using both types of analysis for the specificity of transcription. In the experiment shown in Figure 5 specific RNA synthesis in isolated nuclei from hens and hormone-withdrawn chickens was compared. Ovalbumin mRNA sequences could not be detected in the RNA synthesized in nuclei from hormone-withdrawn chickens. Considering the level of sensitivity of this assay, this indicates at least a 50-fold decrease in the transcriptional activity of the ovalbumin gene in the absence of inducing steroids. We also determined the relative rates of ovalbumin mRNA synthesis early in secondary induction (Table 5). Highly labeled RNA was synthesized in oviduct nuclei isolated from chickens that had received either no or maximally inducing doses of diethylstilbestrol for 6 and 12 hours, and the specific sequence content was analyzed as outlined in Figure 2. Six hours after steroid administration, the relative rate of ovalbumin mRNA synthesis in vitro is similar to that found in laying hens. No ovalbumin mRNA synthesis could be detected in isolated nuclei from hormone-withdrawn chickens since no significant radioactivity was found above the rat-liver [³H]RNA background. The sensitivity of the assay would have allowed us to detect 1/20 as much labeled ovalbumin mRNA as in transcripts from restimulated oviduct nuclei. We conclude that the transcriptional efficiency of the ovalbumin gene is at least 20 times lower prior to hormonal induction.

SUMMARY

The number and type of structural genes expressed in the hen oviduct are very similar to those expressed in the hormone-withdrawn oviduct; however, a few specific mRNAs are present with highly altered concentrations. We have shown that the mRNAs coding for the egg white proteins, ovalbumin, ovomucoid, and lysozyme, increase in concentration several thousandfold during hormonal induction. This suggests that the steroid hormones effect a quantitative rather than a qualitative change in gene expression in the chick oviduct.

During the first 2 hours of hormone induction of the withdrawn oviduct, the concentration of the mRNAs specific for ovalbumin, ovomucoid, and lysozyme changes only slightly. Following this lag period, a constant rate of accumulation is attained.

Synthesis of ovalbumin mRNA sequences was studied in isolated nuclei and chromatin using methods which allowed a distinction of newly synthesized from endogenous RNA. The ovalbumin gene is transcribed in isolated nuclei with high selectivity by RNA polymerase B. Approximately 0.1% of newly synthesized RNA is ovalbumin RNA. The apparent specificity observed when chromatin is transcribed by *E. coli* RNA polymerase using mercurated nucleotides is mainly due to an RNA-dependent RNA polymerase activity.

A marked increase in synthesis of ovalbumin RNA in oviduct nuclei isolated from steroid-treated chickens as compared to hormone-withdrawn animals was observed. This suggests that the cellular accu-

Table 5. Synthesis of Ovalbumin mRNA Sequences in Oviduct Nuclei Isolated after Secondary Stimulation

Source of nuclei	Input RNA (cpm)	Hg-cDNA (ng)	RNase-A-resistant RNA (cpm)	[³H]RNA bound to SH-agrose		
				cpm	%	% corrected
Rat liver	1,810,000	50	13,200	21	0.0012	
Withdrawn oviduct	1,420,000	160	17,200	20	0.0014	0.0006
Restimulated oviduct, 6 hr	1,037,000	160	14,460	308	0.03	0.09
Restimulated oviduct, 12 hr	423,000	160	6720	105	0.026	0.08

RNA was synthesized for 10 min in oviduct nuclei isolated from acutely hormone-withdrawn and restimulated chicks. The ovalbumin-mRNA-specific sequences were determined as described in the note to Table 2. The background value (obtained by hybridization of [³H]RNA synthesized in isolated rat-liver nuclei to Hg-cDNA) was substracted for determination of the percentage of ovalbumin-specific RNA.

mulation of ovalbumin mRNA during hormonal stimulation is due primarily to transcriptional activation of the ovalbumin gene.

Acknowledgments

We would like to thank S. Jeep and M. Dzedoch for expert technical assistance and Dr. R. Sweet for his help in preparing the manuscript.

REFERENCES

CHAN, L., A. R. MEANS, and B. W. O'MALLEY. 1973. Rates of induction of specific translatable messenger RNAs for ovalbumin and avidin by steroid hormones. *Proc. Natl. Acad. Sci.* **70**: 1870.

COX, R. F. 1976. Quantitation of elongating form A and B RNA polymerases in chick oviduct nuclei and effects of estradiol. *Cell* **7**: 455.

COX, R. F., M. E. HAINES, and J. S. EMTAGE. 1974. Quantitation of ovalbumin mRNA in hen and chick oviduct by hybridization to complementary DNA. *Eur. J. Biochem.* **49**: 225.

CROUSE, G. F., E. J. B. FODOR, and P. DOTY. 1976. *In vitro* transcription of chromatin in the presence of a mercurated nucleotide. *Proc. Natl. Acad. Sci.* **73**: 1564.

DALE, R. M. K. and D. WARD. 1975. Mercurated polynucleotides: New probes for hybridization and selective polymer fractionation. *Biochemistry* **15**: 2458.

DALE, R. M. K., D. C. LIVINGSTON, and D. C. WARD. 1973. The synthesis and enzymatic polymerisation of nucleotides containing mercury: Potential tools for nucleic acid sequencing and structural analysis. *Proc. Natl. Acad. Sci.* **70**: 2238.

ERNEST, M. E., G. SCHÜTZ, and P. FEIGELSON. 1976. RNA synthesis in isolated hen oviduct nuclei. *Biochemistry* **15**: 824.

GIESECKE, K., A. E. SIPPEL, M. C. NGUYEN-HUU, B. GRONER, N. E. HYNES, T. WURTZ, and G. SCHÜTZ. 1977. An RNA-dependent RNA polymerase activity: Implications for chromatin transcription experiments. *Nucleic Acids Res.* (in press).

GRONER, B., N. E. HYNES, A. E. SIPPEL, S. JEEP, M. C. NGUYEN-HUU, and G. SCHÜTZ. 1977. Immunoadsorption of specific chicken oviduct polysomes: Isolation of ovalbumin, ovomucoid and lysozyme mRNA. *J. Biol. Chem.* **252**: 6666.

HARRIS, S. E., J. M. ROSEN, A. R. MEANS, and B. W. O'MALLEY, 1975. Use of a specific probe for ovalbumin mRNA to quantitate estrogen-induced gene transcripts. *Biochemistry* **14**: 2072.

HARRIS, S. E., R. J. SCHWARTZ, M. J. TSAI, B. W. O'MALLEY, and A. K. ROY. 1976. *In vitro* transcription of the ovalbumin gene in chromatin. *J. Biol. Chem.* **251**: 424.

HYNES, N.E., B. GRONER, A. E. SIPPEL, M. C. NGUYEN-HUU, and G. SCHÜTZ. 1977. A comparison of total mRNA complexity and ovalbumin, ovomucoid and lysozyme mRNA content in the oviduct of laying hens and estradiol-stimulated and withdrawn chicks. *Cell* **11**: 923.

KALIMI, M., S. Y. TSAI, M. J. TSAI, J. H. CLARK, and B. W. O'MALLEY. 1976. Correlation between nuclear bound estrogen receptor and chromatin initiation sites for transcription. *J. Biol. Chem.* **251**: 516.

MARSHALL, A. J. and L. A. BURGOYNE. 1976. Interpretation of the properties of chromatin extracts from mammalian nuclei. *Nucleic Acids Res.* **4**: 1101.

MCKNIGHT, G. S., P. PENNEQUIN, and R. T. SCHIMKE. 1975. Induction of ovalbumin mRNA sequences by estrogen and progesterone in chick oviduct as measured by hybridization to complementary DNA. *J. Biol. Chem.* **250**: 8105.

NGUYEN-HUU, M. C., A. E. SIPPEL, N. E. HYNES, B. GRONER, and G. SCHÜTZ. 1978. Preferential transcription of the ovalbumin gene in isolated hen oviduct nuclei by RNA polymerase B. *Proc. Natl. Acad. Sci.* (in press).

PALMITER, R. 1972. Regulation of protein synthesis in chick oviduct. *J. Biol. Chem.* **247**: 6450.

————. 1975. Quantitation of parameters that determine the rate of ovalbumin synthesis. *Cell* **4**: 188.

PALMITER, R. D., P. M. MOORE, E. R. MULVIHILL, and S. EMTAGE. 1976. A significant lag in the induction of ovalbumin messenger RNA by steroid hormones: A receptor translocation hypothesis. *Cell* **8**: 557.

RHOADS, R. E., G. S. MCKNIGHT, and R. T. SCHIMKE. 1973. Quantitative measurement of ovalbumin messenger RNA activity. *J. Biol. Chem.* **248**: 2031.

ROSEN, J. and B. W. O'MALLEY. 1975. Hormonal regulation of specific gene expression in the chick oviduct. In *Biochemical action of hormones* (ed. G. Litwack), vol. 3, p. 271. Academic Press, New York.

SCHIMKE, R. T., G. S. MCKNIGHT, and D. J. SHAPIRO. 1975. Nucleic acid probes and analysis of hormone action in oviduct. In *Biochemical action of hormones* (ed. G. Litwack), vol. 3, p. 245. Academic Press, New York.

SCHÜTZ, G., M. BEATO, and P. FEIGELSON. 1973. Messenger RNA for hepatic tryptophan oxygenase: Its partial purification, its translation in a heterologous cell-free system and its control by glucocorticoid hormones. *Proc. Natl. Acad. Sci.* **70**: 1218.

SCHÜTZ, G., S. KIEVAL, B. GRONER, A. E. SIPPEL, D. T. KURTZ, and P. FEIGELSON. 1977. Isolation of specific messenger RNA by adsorption of polysomes to matrix-bound antibody. *Nucleic Acids Res.* **4**: 71.

SMITH, H. M. and R. C. C. HUANG. 1976. Transcription *in vitro* of immunoglobulin kappa light chain genes in isolated mouse myeloma nuclei and chromatin. *Proc. Natl. Acad. Sci.* **73**: 775.

TOWLE, H. C., M. J. TSAI, S. Y. TSAI, and B. W. O'MALLEY. 1977. The effect of estrogen on gene expression in the chick oviduct. *J. Biol. Chem.* **252**: 2396.

TSAI, M. J., H. C. TOWLE, S. E. HARRIS, and B. W. O'MALLEY. 1976. Comparative aspects of RNA chain initiation on chromatin using homologous versus *E. coli* RNA polymerase. *J. Biol. Chem.* **251**: 1960.

ZASLOFF, M. and G. FELSENFELD. 1977. Use of mercury-substituted ribonucleoside triphosphates can lead to artefacts in the analysis of *in vitro* chromatin transcripts. *Biochem. Biophys. Res. Commun.* **75**: 598.

Mammary Tumor Virus DNA: A Glucocorticoid-responsive Transposable Element

K. R. YAMAMOTO, M. R. STALLCUP, J. RING AND G. M. RINGOLD

Department of Biochemistry and Biophysics, University of California, San Francisco, California 94143

Steroid hormones exert profound and specific effects on patterns of gene expression in virtually all metazoan organisms. In mammalian and avian tissues, where the biochemistry of the action of these hormones has been most extensively studied, the "interpretation" of the chemical structure of the steroid seems to be mediated by soluble receptor molecules; these proteins bind the appropriate hormone specifically and with high affinity and thereby initiate a process that results in the appearance of a specific set of new gene products. In effect, the steroid acts as an allosteric ligand; that is, the binding event somehow alters the properties of the receptor so as to increase its affinity for binding sites in the cell nucleus, resulting in a net translocation of the steroid-receptor complex from the soluble to the nuclear bound state. The interaction with components of the chromatin is then presumed to bring about the biological response (for references and detailed review, see Gorski and Gannon 1976; Yamamoto and Alberts 1976). According to this view, steroid receptors can be considered to be nonhistone chromosomal proteins of a known regulatory function.

It is assumed that all steroids act according to this general scheme, but a detailed understanding of the mechanisms involved is lacking; indeed, virtually every step in the pathway of receptor action is currently a topic of controversy at some level. In particular, the nature of the chromosomal binding sites for receptors is essentially unknown, despite extensive biochemical experimentation (e.g., see Spelsberg 1974; Yamamoto and Alberts 1974; Puca et al. 1975); there is some genetic evidence implicating DNA as comprising at least a part of the binding site (Gehring and Tomkins 1974; Yamamoto et al. 1974, 1976), but sequence-specific binding has not been demonstrated. Several plausible models have been put forth to account for the reaction which might actually trigger the hormonal response (Schwartz et al. 1975; Palmiter et al. 1976; Yamamoto and Alberts 1976), but there are no unequivocal data relating biochemical observations to the biological situation. It is likely that a part of the difficulty stems from the genetic and experimental complexity of the systems under examination. In our laboratory, we are therefore seeking to characterize in detail the glucocorticoid response of a well-defined and isolatable set of viral genes in mammalian cells.

Normal mouse cells contain multiple copies of DNA related to murine mammary tumor virus (MTV) covalently integrated into their genomes (Varmus et al. 1972); these copies presumably represent endogenous viral genes, transmitted through the germ line as part of the normal mouse gene complement. The virus particle itself contains an RNA genome. When MTV infects cells, the incoming viral RNA is thought to replicate via a double-stranded DNA intermediate synthesized by the virion-associated RNA-directed DNA polymerase (reverse transcriptase) (Temin and Baltimore 1972). Upon infection, viral DNA becomes covalently integrated into the host genome (Ringold et al. 1977a; Yamamoto and Ringold 1977). In most mammary tumors, the tumor cells contain somewhat elevated levels of integrated viral DNA, probably acquired via infection (Morris et al. 1977). Thus, uninfected nonmurine cells contain no detectable MTV genes, whereas normal mouse tissues carry endogenous viral DNA, and mouse mammary tumor cells may contain additional exogenously derived copies.

The expression of integrated MTV genes in cultured cell lines derived from murine mammary tumors (Parks et al. 1974, 1975; Ringold et al. 1975a,b), rat hepatoma (Ringold et al. 1977a), mink lung (Vaidya et al. 1976), or cat kidney (Vaidya et al. 1976) is strongly regulated by glucocorticoids. Using the viral reverse transcriptase to synthesize DNA complementary to the MTV genome (MTV cDNA), molecular hybridization reagents for the specific and direct quantitation of the hormone-inducible transcript (Parks et al. 1974, 1975; Ringold et al. 1975a,b) are readily prepared. Thus, this system has the technical advantage of bypassing the often difficult step of mRNA isolation and purification yet still allows direct cDNA hybridization measurements.

In this report we describe (1) the effects of dexamethasone, a synthetic glucocorticoid, on the expression of chromosomal sequences coding for MTV, focusing in particular on whole-cell and cell-free studies examining the effect of dexamethasone on viral RNA synthesis; and (2) evidence consistent with the idea that the MTV DNA sequence is a glucocorticoid-responsive transposable genetic element. Specifically, we shall speculate that MTV genes might exert direct effects on host expression in a manner dependent on their site of integration into the host genome; it is conceivable that functionally analogous transposable genetic elements could determine the nature and specificity of hormone responsiveness during normal development.

Glucocorticoids Stimulate Rate of MTV Gene Transcription

It has been clearly established in several steroid-responsive systems that the steady-state level of specific mRNA species is increased in hormone-treated cells or tissues (Harris et al. 1975; McKnight et al. 1975; Feigelson et al. 1975; Tata 1976). For example, in the murine mammary carcinoma cell line GR, the stimulation of viral RNA accumulation by dexamethasone is rapid, becoming detectable within 30 minutes and half-maximal by 2.5 hours (Ringold et al. 1975b). Based on assays measuring the kinetics of molecular hybridization of cellular RNA to ^{3}H-labeled MTV cDNA, MTV RNA comprises about 0.02% of the steady-state concentration of cell RNA in the absence of dexamethasone and about 0.5% in its presence (Ringold et al. 1975a,b). It is tempting to assume that the receptor-genome interaction selectively alters the rate at which steroid-regulated genes are transcribed, perhaps in a manner analogous to prokaryotic gene regulatory proteins (e.g., see Nakanishi et al. 1975). However, the apparent complexity of mRNA metabolism, especially in eukaryotes and their viruses (Lewin 1974; Weber et al. 1977), suggests many other points at which receptor action might affect the steady-state levels of a transcript.

In the case of the MTV genes it has been possible to approach this question directly. Pulse-labeling of RNA in whole cells indicates that the dexamethasone-stimulated accumulation of MTV RNA can be accounted for by a selective increase in the rate of viral gene transcription. For these studies we employed a novel hybridization assay that vastly reduces hybridization backgrounds, thereby enabling specific detection of low levels of labeled RNA (E. Stavnezer and J. M. Bishop, in prep.). Using this method to measure the fraction of pulse-labeled cell RNA which is virus-specific, we found that dexamethasone causes a 10–20-fold increase in the rate of MTV RNA synthesis, and that the change in rate is complete within the shortest period examined, 0–15 minutes (Ringold et al. 1977b). The overall rate of cellular RNA synthesis is not appreciably affected by the steroid. H. Young et al. (1977), using a different cell line and assay procedure, have reached similar conclusions.

The fact that the stimulated rate becomes maximal with no detectable lag is consistent with the idea that the receptor protein is acting directly at the genetic locus coding for MTV RNA. Moreover, earlier work has shown that dexamethasone induces MTV RNA accumulation normally in the presence of inhibitors of protein synthesis (H. Young et al. 1975; Ringold et al. 1975b); thus, it is unlikely that the receptor first induces production of a protein intermediate which in turn stimulates viral RNA synthesis. Finally, the magnitude of the change in transcriptional rate is similar to the magnitude of the change in the steady-state viral RNA concentration; thus, this effect alone may be sufficient to account for the hormonal response.

Having established the nature of an early hormonal event in whole cells, we would like to understand the actual biochemical reactions that trigger the specific alteration in transcriptional rate. As a first step toward this long-term goal, we have begun to characterize cell-free synthesis of RNA in isolated GR cell nuclei. If we are able to find conditions that mimic aspects of the whole-cell response, we might be able to use such a system for detailed biochemical studies.

In these initial experiments we have sought to establish conditions under which nuclei from hormone-treated and untreated cells synthesize viral RNA at rates characteristic of hormone-treated and untreated whole cells, respectively. Our studies have been greatly facilitated by the work of other investigators, particularly P. Feigelson and R. C. C. Huang and their coworkers (Marzluff et al. 1973; Ernest et al. 1976), who have examined cell-free RNA synthesis in other systems.

When GR cell nuclei are isolated and incubated at 25°C in the presence of ^{3}H-labeled CTP under the conditions described in the legend to Table 1, radioactivity accumulates into trichloroacetic acid-(TCA)-insoluble material for about an hour. Several lines of evidence support the idea that the incorporation observed represents DNA-directed synthesis of RNA mediated by the endogenous cellular RNA polymerases. First, incorporation is dependent on the presence of ribonucleoside triphosphates, and the TCA insolubility of the product is abolished by ribonuclease, but not by deoxyribonuclease or by Pronase (data not shown). Second, incorporation is inhibited ~90% by actinomcyin D (data not shown); this is an important control in view of recent findings that artifactual reactions occur under some conditions, resulting in DNA-independent (and actinomycin-D-insensitive) RNA synthesis (see below). Finally, the pattern of inhibition of the incorporation by α-amanitin (Schwartz et al. 1974) implies that RNA polymerases I and II, and perhaps RNA polymerase III (Roeder and Rutter 1970), are active under our conditions (Fig. 1).

Obviously, it is crucial to demonstrate that actual RNA synthesis is occurring in the cell-free system, and that the synthesis observed is occurring by reactions similar to those that occur in vivo. Direct tests of these questions are complex, especially since the in-vitro-synthesized RNA is usually only a small fraction of the total (in-vivo-synthesized) RNA con-

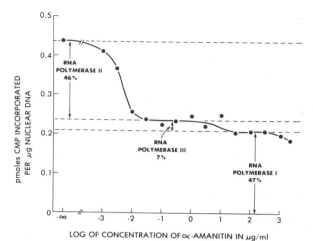

Figure 1. Inhibition of cell-free RNA synthesis by α-amanitin. RNA synthesis was carried out as described in the notes to Table 1, except that ^{3}H-labeled CTP was present at 1 Ci/mmole and total reaction mixtures were 50 μl. TCA-insoluble material was collected on glass-fiber filters and counted in toluene-based scintillation fluid containing 4 g/liter Omnifluor (New England Nuclear), 4 ml/liter H$_2$O and 25 ml/liter NCS tissue solubilizer; data shown have been corrected for an unincubated blank value of 0.04 pmole CMP/μg DNA. Varying concentrations of α-amanitin were included in each reaction mix. Schwartz et al. (1974) have studied in detail the inhibition of mammalian RNA polymerases by α-amanitin and find the general pattern of sensitivity to be II > III > I.

taminating the system. Use of mercurated pyrimidine nucleoside triphosphates (Dale and Ward 1975) in the cell-free reaction and subsequent isolation of the mercurated product on sulfhydryl-Sepharose columns has been widely employed to overcome these difficulties. Recently, use of this technique has helped to reveal an apparently artifactual side reaction, RNA-dependent RNA synthesis (Zasloff and Felsenfeld 1977; G. Schütz, pers. comm.); template-independent terminal addition may also occur (see Shih et al. 1977). These findings may necessitate reevaluation of the conclusions reached in many types of cell-free transcription studies.

To avoid these recently discovered difficulties, we label our newly synthesized RNA with radioactive nucleoside triphosphates and assay only that material which contains radioactivity (M. R. Stallcup et al., in prep.). This obviates the need to separate endogenous from newly synthesized RNA, provided steps are taken to avoid the artifacts noted above.

To estimate what fraction of the RNA synthesized in this cell-free system is MTV-specific, we again used a modification of the hybridization assay of E. Stavnezer and J. M. Bishop (in prep.) as described in detail by Ringold et al. (1977b). Briefly, the procedure utilizes unlabeled molecules of viral DNA which contain double-stranded regions and single-stranded "tails" complementary to the viral RNA; this "tailed duplex" DNA is the product of the endogenous viral reverse transcriptase (Ringold et al.

1976). When labeled viral RNA is annealed to the single-stranded regions of tailed duplex DNA, it can be selectively retained on hydroxylapatite by using conditions which allow only double-stranded DNA to bind to the column matrix (E. Stavnezer and J. M. Bishop, in prep.); the nonviral radioactive RNA passes through the column. Efficiency of hybridization is monitored internally by including in each reaction a tracer amount of ^{32}P-labeled MTV RNA; this corrects for contamination by endogenous RNA. The hybridization products are treated with RNase prior to hydroxylapatite chromatography to eliminate terminal-addition and RNA-templated reactants (M. R. Stallcup et al., in prep.).

Table 1 summarizes the results of experiments in which nuclei were prepared from untreated GR cells and from cells treated for 30 minutes with 10^{-6} M dexamethasone; the nuclei were incubated for 60 minutes in a reaction mixture containing ^{3}H-labeled CTP. Total RNA was extracted and purified, and labeled viral RNA was measured as described above. It can be seen that nuclei from cells exposed to dexamethasone for 30 minutes synthesize about 20-fold more MTV RNA than nuclei from untreated controls; total incorporation of ^{3}H-labeled CMP is unaffected by hormone treatment (data not shown). As an additional control, we added $1–50 \times 10^{-6}$ M estradiol, a potent estrogen which has no detectable affinity for the glucocorticoid receptor and no glucocorticoid activity. Estradiol alone caused little or no increase in the rate of MTV RNA synthesis and displayed no inhibitory effect on the dexamethasone-mediated induction (Table 1). Finally, of the total RNA synthesized in the cell-free system by nuclei from hormone-treated and untreated cultures, the fraction that is virus-specific is similar to the levels seen in comparably treated whole cells; the results of typical [^{3}H]uridine pulse-labeling experiments with whole cells (Ringold et al. 1977b) are given in Table 1 for comparison.

There is strong biochemical and genetic evidence that cellular mRNA is transcribed by RNA polymerase II, the enzyme that displays the greatest sensitivity to inhibition by α-amanitin (Lindell et al. 1970; Somers et al. 1975). Furthermore, synthesis of avian C-type viral RNA is also mediated by this enzyme (Rymo et al. 1974; Jacquet et al. 1974). Thus, it was of interest to examine the α-amanitin sensitivity of the synthesis of MTV RNA in isolated nuclei. For these experiments, cell-free RNA synthesis was performed in the absence of α-amanitin or in the presence of 0.3 or 300 μg/ml of the toxin. These concentrations are sufficient to inhibit, respectively, RNA polymerase II alone or RNA polymerases II and III in these nuclei (Fig. 1). The results reveal that virtually all (~97%) of the MTV RNA synthesis in isolated nuclei is blocked by the lower concentration of α-amanitin (Fig. 2), suggesting that RNA polymerase II is responsible for its synthesis.

It remains to be seen whether a system employing

Table 1. Relative Rates of Synthesis of MTV RNA

	$\dfrac{\text{MTV-specific [}^3\text{H]RNA}}{\text{total [}^3\text{H]RNA}} \times 100$	Relative MTV RNA synthesis
Isolated nuclei		
—	0.019 (0.006–0.030)	1
+ dex	0.406 (0.300–0.465)	21
+ estradiol	0.031	1.6
+ dex + estradiol	0.232	12
Whole cells		
—	0.036 (0.027–0.050)	1
+ dex	0.444 (0.240–0.530)	12

Nuclei were prepared using methods similar to those described by Marzluff et al. (1973) and Ernest et al. (1976). The cells were swollen in a hypotonic buffer, disrupted with a Dounce homogenizer in the presence of 0.1% NP-40, and the nuclei pelleted and resuspended in 1 M sucrose, 15 mM Tris-Cl, pH 8.0, 7.5 mM $MgCl_2$, 0.5 mM dithiothreitol. The nuclei were centrifuged through a 2 M sucrose pad containing the same buffer and resuspended in 25% glycerol, 50 mM HEPES, 5 mM $Mg(OAc)_2$, 0.5 mM dithiothreitol, and 1% crystalline bovine serum albumin. Conditions for cell-free RNA synthesis were essentially those described by Ernest et al. (1976) except that nuclei were used at a concentration of 1 mg nuclear DNA/ml and RNA was labeled with 60 μM ^{3}H-labeled CTP (10–21 Ci/mmole). Unlabeled ATP, UTP, and GTP were present at a concentration of 1 mM each in a buffer containing 50 mM HEPES, pH 8.0, 5 mM Mg $(OAc)_2$, 0.5 mM $MnCl_2$, 2.5 mM dithiothreitol, 150 mM NH_4Cl, 10% glycerol, 1% bovine serum albumin. The reactions were carried out at 25°C for 60 min in 0.5 ml, and total RNA was phenol-extracted and purified. The tritium-labeled MTV-specific RNA was measured using the hybridization procedure of E. Stavnezer and J. M. Bishop (in prep.) exactly as described by Ringold et al. (1977b). Briefly, [^{3}H]RNA was hybridized with MTV tailed duplex DNA and then treated with pancreatic ribonuclease and chromatographed on a column of hydroxylapatite under conditions in which only double-stranded DNA binds. MTV [^{3}H]RNA is thus indirectly bound to the column. Hybridization efficiency was internally determined and normalized by including tracer levels of MTV 70S [^{32}P]RNA; hybrid material not competed by excess unlabeled MTV 70S RNA was subtracted from each final value (0.01–0.04% of total incorporation). Values in parentheses represent the range of at least four determinations. Dexamethasone was used at 10^{-6} M and estradiol at 5×10^{-5} M.

Typical whole-cell labeling data are included for comparison and represent incorporation of [^{3}H]uridine into total and viral RNA in a 15-min pulse at 37°C; the whole-cell data are taken from Ringold et al. (1977b).

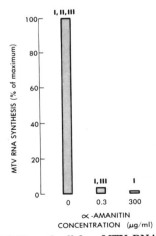

Figure 2. Inhibition of cell-free MTV RNA synthesis by α-amanitin. Cell-free RNA synthesis was carried out for 60 min as described in the notes to Table 1 but in the presence of 0, 0.3, or 300 μg/ml α-amanitin, conditions permissive for the activity of RNA polymerase I + II + III, I + III, or I, respectively (Fig. 1). Total RNA was then isolated and the labeled viral RNA measured by molecular hybridization as described in the notes to Table 1.

intact nuclei can be used to assay the biological activities of exogenously added regulatory components; RNA-polymerase-II-mediated initiation and processing of primary transcripts may be even more complex than earlier data had led us to suspect (e.g., see Berget et al.; Broker et al.; both this volume). It is likely that, at the minimum, the cell-free response will require faithful synthesis beginning at the correct initiation sites. To date, systems utilizing endogenous RNA polymerase II appear to function predominantly by completion of chains initiated in vivo (Cox 1973; Ferencz and Seifart 1975); we have not yet determined whether initiation is occurring in our system. Nevertheless, it appears that our conditions for nuclear isolation and subsequent RNA synthesis are able to maintain, at least for a short time, the patterns of transcription established in the whole cell. Recent technical advances (Reeve et al. 1977; Reeder et al. 1977) may allow initiation events to be measured with greater sensitivity and accuracy and should aid attempts to define conditions that support faithful initiation in this system.

Host and Viral Gene Expression in Infected Cells

It is likely that all normal mouse tissues contain multiple copies (~5–10/cell) of integrated MTV genes; in a given strain, the number of copies is invariant from tissue to tissue (Morris et al. 1977). However, expression of these genes in normal cells, measured by accumulation of intracellular viral RNA, has been detected only in lactating mammary epithelium (Varmus et al. 1973); direct assessment of the role of glucocorticoids in this case has not been obtained. Mammary tumors that produce MTV, and the continuous cell lines derived from these tumors, appear in every case to contain viral genes whose expression is stimulated by glucocorticoids (Dickson et al. 1974; Fine et al. 1974; Parks et al. 1974; L. Young et al. 1975). It is important to note that mammary tumor cells generally carry more integrated viral genomes than normal tissues (Morris et al. 1977); these extra copies are presumably acquired via infection resulting from vertical (germ-line) transmission of "virulent genes" or horizontal transmission of mature virions in the milk (Bentvelzen 1974).

Several general mechanisms could account for the tissue specificity of expression and hormone responsiveness of MTV genes. One possibility is that factors which block expression of viral genes might exist in all tissues except mammary epithelium. This possibility seems unlikely to be of widespread importance in gene regulation since, in the extreme, it would require a vast number of tissue-specific gene regulatory components. Alternatively, certain positive controlling factors might occur only in mammary epithelium. This possibility is particularly interesting in terms of the nature of hormone responsiveness; for example, virtually all somatic tissues contain glucocorticoid receptors and respond to these steroids, but the specific genes affected in any one cell type are completely different from those affected in other tissues (Baxter and Forsham 1972). Examination of the physical and hydrodynamic properties of the receptor proteins suggests that they may be identical rather than tissue-specific (Ringold et al. 1975b); studies with receptors for other steroids are also consistent with this idea, e.g., a mutation affecting androgen receptor activity in the mouse results in defective androgen receptors in all androgen-sensitive tissues (Attardi et al. 1976). Thus, the tissue-specific positive regulatory component may not be the receptor itself, but rather some related entity such as a chromosomal "acceptor" or a soluble factor with which the receptor interacts.

Still another way to account for tissue-specific hormonal effects on MTV expression is to assume that the proviral genes are integrated at different chromosomal sites in different cell types; thus, only in mammary epithelium would MTV be integrated in a potentially active and hormone-responsive con-figuration. In general, transposition of discrete DNA sequences to various chromosomal sites appears to be widespread and might be an important determinant in development (see below). The property of transposition is clearly crucial to viruses, which can be viewed as specialized sequences that have acquired the added capability of becoming packaged for export to other cells. Ketner and Kelly (1976) and Botchan et al. (1976) have presented direct evidence suggesting that SV40 DNA integrates into the mouse cell genome at many sites. Moreover, Cooper and Temin (1976) have proposed that expression of the endogenous sequences related to avian sarcoma viruses in chicken cells is specifically inhibited, and that only proviruses which are exogenously introduced (and presumably are integrated at different chromosomal loci) can be transcribed. Clearly, the finding that tumors expressing MTV RNA contain more proviral copies than normal tissues is consistent with this notion.

Studies with MTV-producing mammary tumor cells do not distinguish the mechanisms responsible for hormone responsiveness. In contrast, certain inferences can be drawn from experiments in which cultured cells derived from various tissues of a number of species are infected with MTV. In MTV-infected feline kidney, mink lung (Vaidya et al. 1976), rat hepatoma (Ringold et al. 1977a), and rat mammary (J. Ring et al., unpubl.) cells, dexamethasone stimulates increased production of MTV RNA. Moreover, a BALB/c mouse lymphoma cell line that does not normally produce any detectable MTV RNA expresses MTV genes in a hormone-responsive fashion after infection (M. R. Stallcup et al., unpubl.). In the simplest interpretation, these results fail to support the notion that tissue-specific regulatory factors, acting either positively or negatively, are necessary to control the expression of these genes. On the contrary, MTV genes in each infected cell population, including those in a murine line not expressing MTV genes prior to infection, become hormonally regulated.

To gain a more detailed view of the uniformity or variability of the expression and dexamethasone responsiveness of MTV genes in a given cell type, it is necessary to examine independent subclones of an infected cell population. For this purpose, HTC cells were exposed to MTV particles (10^3–10^5/cell) (Ringold et al. 1977a), grown for several days or weeks under virus-free conditions, and cloned in soft agar (Yamamoto and Ringold 1977). Isolated colonies containing ~10^3 cells were randomly selected after 10–12 days and then grown and handled independently. Determinations of viral genome equivalents per haploid genome were carried out on nuclear DNA by means of published procedures (Yamamoto and Ringold 1977; Yamamoto et al. 1977). Analyses of these DNAs after restriction enzyme cleavage, fractionation by gel electrophoresis, and hybridization with ^{32}P-labeled cDNA (Southern 1975) have been consistent with the copy numbers obtained by these procedures (G. M. Ringold et al., unpubl.). Intracellular viral RNA concentration was determined by measuring the kinetics of hybridization between cellular RNA and ^{3}H-labeled MTV cDNA, as described for the uncloned infected cells (Ringold et al. 1977a).

Table 2A summarizes the results of viral DNA and RNA assays in a series of infected clones containing between one and ten integrated viral genomes per haploid cell. It appears that clones containing many (3–10) copies of viral genes produce more viral RNA in the presence of dexamethasone than clones containing few (1–2) MTV DNA copies. However, it is clear from these data, as well as from other studies comparing levels of MTV RNA in clones that contain similar numbers of viral genes (Yamamoto and Ringold 1977; J. Ring and K. R. Yamamoto, unpubl.), that this correlation is not a strict one. Moreover, Table 2A also shows that the extent of MTV RNA induction by dexamethasone varies greatly from clone to clone (ranging from 50- to 1000-fold) in a manner which has no apparent relationship to the number of integrated viral genes. All clones that produce basal levels of MTV RNA are stimulated to produce increased quantities by dexamethasone. However, the fact that basal expression and glucocorticoid responsiveness of the viral genes are not related in a simple manner to viral gene dosage is consistent with the idea that not all of the integrated MTV genomes are being transcribed to an equivalent extent in the absence of hormone and that not all of them are stimulated to an equivalent extent in its presence. Finally, Table 2A shows that not all clones containing integrated MTV DNA express viral RNA. Recent restriction enzyme analyses of DNA from these clones suggest that they, like the other infected clones, contain an intact integrated viral genome (G. M. Ringold and P. Shank, unpubl.); thus, their failure to express viral RNA is probably not due to integration of only small fragments of the genome. These results show that integration of viral DNA into the host genome is not by itself sufficient to allow expression of these genes. Taken together with the previous data, one interpretation is that the site of integration of the viral DNA is a critical determinant affecting its activity.

When a new DNA sequence integrates into a region of the genome that encodes a regulatory locus or structural gene, that genetic region would cease normal function; that is, MTV would act as a mutagen at the site of integration. Prokaryotic transposable DNA sequences have been shown to exhibit strong effects on gene expression in extensive regions adjacent to the integration site (see below). If integration occurs completely at random, selective

Table 2. Induction of Viral RNA and Host Enzymes in Clones of MTV-infected HTC Cells

clone	A. MTV DNA and RNA in infected HTC cells			B. TAT and GS in infected HTC cells relative enzyme activity			
	MTV DNA (copies/haploid genome)	MTV RNA	(copies/cell)	TAT		GS	
		−	+ dex	−	+ dex	−	+ dex
HTC4.1	0	0	0	1	15 (11–19)	1	8 (5–11)
J2.15	1	0	0	1	15	1.5	6
J2.31	1	0	0	0.8	13.5	1	8
J2.17	1	<1	700	0.9	12	1	7
J0.1	1–2	<1	1000				
J1.3	2	<0.05	5				
M1.54	3	160	18,000	0.3	1.5	1	2
M1.7	4–7	3	3000	1	1.5	1	1.7
M1.20	5–8	65	8000	1	12	1	4.6
M1.60	10	130	6500	1	5	1	2.3

Cells were cloned by immobilization in soft agar. A 5-ml underlay of Dulbecco's modified Eagle's medium (supplemented with 10% horse serum, antibiotics, and fungizone) containing 0.45% molten Difco Noble Agar was dispensed into 60-mm petri dishes. After the agar had hardened, 2 ml of medium, containing 0.35% agar and 100–200 cells, was added as an overlay. After incubation at 37°C for 10–12 days, individual colonies, containing ~10[3] cells, were removed from the agar using sterile pasteur pipettes with drawn tips. Cloning efficiency was 75–100%. Fewer than six clones were selected from any one plate.

(A) MTV sequences integrated into cellular DNA were measured by C_0t analysis (Yamamoto et al. 1977) or by calibration curve analysis (Yamamoto and Ringold 1977). Intracellular MTV RNA was measured by molecular hybridization as described by Ringold et al. (1977a); the number of viral RNA copies per cell is based on a $C_r t_{1/2} = 2 \times 10^{-2}$ mole·sec/liter for pure viral RNA (Varmus et al. 1973).

(B) Cells (1–3 × 10[7]) were removed from culture dishes in 10 mM Tris, pH 8.0, 1 mM EDTA, 0.25 M sucrose and were lysed by freezing and thawing. Cell debris was removed by low-speed centrifugation, and enzymatic activities were determined on aliquots of the extract, using essentially the methods of Diamondstone (1966) and Kulka et al. (1972) for TAT and GS, respectively. Results are presented relative to basal activities in uninfected control (HTC4.1) cells. Range given for dexamethasone-induced activities in uninfected cells represents the extent of clonal variability seen in subclones of HTC4 (Yamamoto and Ringold 1977). Dexamethasone was used at 10[−6] M in all experiments.

cis-acting effects on specific host genes would not be detected. On the other hand, if the MTV genes integrate preferentially into certain sites or regions of the host genome, or if their expression leads to a strong *trans* effect on the expression of certain host genes, specific changes might be detectable.

This question was examined in two ways. First, we monitored the activities of two enzymes, tyrosine aminotransferase (TAT) and glutamine synthetase (GS), which are normally induced by dexamethasone in uninfected HTC cells (Thompson et al. 1966; Kulka et al. 1972). Second, we surveyed the relative rates of synthesis of ~10^3 major cellular proteins after pulse-labeling with [^{35}S]methionine and fractionation by high-resolution two-dimensional gel electrophoresis (O'Farrell 1975).

Table 2B summarizes the relative enzyme activities, in the presence and absence of dexamethasone, of TAT and GS in selected clones of infected HTC cells. The activities shown are normalized relative to the basal levels detected in uninfected subclone HTC4.1; it should be noted that the dexamethasone-induced activities of these enzymes are somewhat variable even among different subclones of HTC4, fluctuating over a 60–140% range for GS and a 75–125% range for TAT (Yamamoto and Ringold 1977; Aviv and Thompson 1972). Nevertheless, the average magnitude of the dexamethasone effect in uninfected cells (~15-fold for TAT; ~8-fold for GS) is sufficient for comparison with the infected clones.

Several of the infected clones described in Table 2B display altered host-gene inducibility by dexamethasone. It is interesting that in cases where inducibility of one enzyme is abnormal the other also appears to be affected. For example, dexamethasone elicits an increase in TAT activity that is less than twofold in clone M1.7, yielding a fully induced enzyme level only 10% of that seen in uninfected cells; analysis on two-dimensional gels shows that the reduction in inducible enzyme activity is due to a defect in TAT synthesis (Yamamoto et al. 1977). Induction of GS is also reduced in M1.7, reaching < 25% of the activity in HTC4.1. In some cases it appears that both the basal and induced level of expression is affected; M1.54, for example, expresses TAT at a level threefold below that in HTC4.1 in the absence of dexamethasone and tenfold below that in HTC4.1 in its presence. The observed changes in the induction pattern of these infected clones appear not to be due to alterations in the receptor protein since dose-response analyses suggest that receptor activity in these clones is indistinguishable from that in uninfected cells (Fig. 3).

We have examined ~20 clones containing 3 or more MTV genome equivalents per haploid amount of cell DNA; over half of these clones seem to differ from HTC4.1 in TAT and GS inducibility, although most of the differences are not as striking as those seen with M1.7 or M1.54 (e.g., see M1.60). Enzyme induction in clones containing just 1–2 copies of

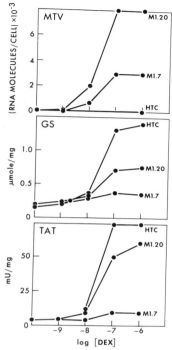

Figure 3. Dose-response of MTV RNA and TAT and GS enzymatic activities in uninfected and infected HTC subclones. Cells were grown in the presence of various concentrations of dexamethasone for 48–60 hr and then assayed for MTV RNA and activities of TAT and GS as described in the notes to Table 2.

MTV DNA per haploid cell has been measured in only a few cases; we have not yet detected any significant differences in TAT and GS activities in these clones relative to HTC4.1 (Table 2B).

Aside from these changes, a display of the pattern of expression of cellular proteins on two-dimensional gels indicates that there are very few differences between MTV-infected and uninfected clones. Moreover, the alterations seen appear to be specific: three proteins normally synthesized constitutively in HTC4.1 cells were consistently missing in the autoradiograms of randomly selected infected clones; another protein is found only in infected cells in the presence but not in the absence of dexamethasone (Yamamoto et al. 1977).

These data, while intriguing, are too preliminary to infer conclusions about the mechanism of this effect. Indeed, we emphasize that we have not rigorously shown that the presence of viral genes is required and responsible for the modulation of host-gene inducibility. One test of this hypothesis might be to demonstrate reversion to normal inducibility upon removal of the viral genome from the cell, a manipulation that has not yet been carried out successfully in cells infected with RNA tumor viruses.

Unintegrated MTV DNA in Infected Cells

We have assumed that the hormone responsiveness of MTV genes and the highly selective changes in host-cell gene expression after MTV infection are elicited directly or indirectly from viral genes integrated into the HTC cell genome. However, zonal alkaline sucrose gradient sedimentation and subsequent physical characterization of DNA from infected cell populations suggested that a significant fraction of the viral sequences in these cell populations is in the unintegrated state (Ringold et al. 1977c); in general, these findings are consistent with previous observations showing that unintegrated viral DNA is detected in cells infected by avian and mammalian C-type RNA viruses (Varmus et al. 1975; Gianni et al. 1975).

Our studies with uncloned populations of infected cells suggested that the unintegrated MTV sequences consist of double-stranded DNA in either the covalently closed circular (form I) or linear (form III) configuration; the estimated molecular weight of the form-I viral DNA was $\sim 6 \times 10^6$ daltons, consistent with the idea that it represents a complete subunit of the RNA genome (Ringold et al. 1977c). These molecules were detectable in HTC cells more than a year after infection; clearly, some mechanism exists in these cells for the continued production of full-length, unintegrated MTV DNA. It seemed conceivable that dexamethasone might affect the synthesis of the unintegrated sequences, or that their presence might somehow be responsible for apparent alterations in cellular metabolism. Moreover, isolated unintegrated viral DNA could be useful in analyses of proviral integration sites (see below). Thus, using clonal isolates from infected cell populations, we carried out a series of experiments to examine the synthesis, regulation, and biological consequences of unintegrated MTV DNA.

To determine whether dexamethasone affects the production of unintegrated viral DNA, M1.20 cells were grown in the presence or absence of dexamethasone (10^{-6} M) for 48 hours, crude cytoplasmic and nuclear fractions were prepared, and nuclear DNA was fractionated according to the procedure of Hirt (1967) in order to separate high- from low-molecular-weight DNA. Viral DNA was measured by hybridization with MTV 70S [^{32}P]RNA. Figure 4a shows that the cytoplasm from the hormone-treated cells contains > 20 times as much MTV DNA as that from untreated cells. The small amount of annealing in the control sample represents less than one molecule of viral DNA per cell and may simply reflect contamination of the cytoplasm with nuclear DNA (i.e., integrated viral DNA). Similar analyses of the nuclear DNA show that dexamethasone also stimulates the production of unintegrated viral DNA in the nucleus (Fig. 4b), whereas the amount of integrated viral DNA is not detectably altered by hormone treatment (Fig. 4c). When progesterone (5×10^{-5} M) and dexamethasone (10^{-7} M) were added simultaneously to M1.20 cells, induction of viral DNA and RNA were both inhibited (Table 3); progesterone competitively inhibits the dexamethasone-receptor interaction, thereby preventing dexamethasone-receptor-controlled events such as induction of tyrosine aminotransferase in HTC cells (Samuels and Tomkins 1970) and MTV RNA in GR cells (Ringold et al. 1975b). Thus, it appears that induction of viral DNA is mediated by the glucocorticoid receptor protein.

In contrast to these results, we were unable to detect unintegrated MTV DNA in the cytoplasm of two mouse mammary tumor lines (GR and CFZ), either in the absence or presence of dexamethasone; similarly, S49 mouse lymphoma cells, which contain high levels of MTV RNA (Ringold et al. 1975b; 1977b), produce no detectable unintegrated MTV

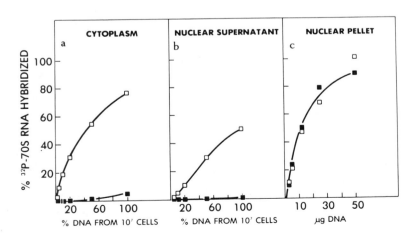

Figure 4. Dexamethasone-mediated induction of unintegrated MTV DNA. Approximately 3×10^7 M1.20 cells were grown in the absence (■) or presence (□) (10^{-6} M, 60 hr) of dexamethasone. Nuclear and cytoplasmic fractions were prepared, and DNA from the nuclei was fractionated by the SDS-NaCl procedure of Hirt (1967). DNAs from the cytoplasm, nuclear pellet, and nuclear supernatant were purified and hybridized with MTV 70S [^{32}P]RNA (~ 1000 cpm) for 48 hr in 0.6 M NaCl at 68°C. The amounts of unintegrated DNA from the cytoplasm and nuclear supernatant are plotted as a function of the original number of cells from which the DNA was derived. Relative concentrations of unintegrated DNA are estimated from the relative initial slopes of each set of curves (Varmus et al. 1974). The amount of hybridization was determined by the extent to which the [^{32}P]RNA could be digested by pancreatic ribonuclease in 2X SSC at 37°C.

Table 3. Effects of Inhibitors on Dexamethasone-mediated Induction of MTV RNA and DNA in M1.20 Cells

	Relative incorporation			Relative induction	
	[^{3}H]TdR (%)	[^{3}H]U (%)	[^{3}H]Leu (%)	intracellular MTV RNA (%)	cytoplasmic MTV DNA (%)
Dexamethasone	100	100	100	100	100
Dexamethasone + araC	2	55	100	> 80	< 5
Dexamethasone + actinomycin D	32	3	80	0	0
Dexamethasone + progesterone	N.T.	N.T.	N.T.	< 10	< 10

Inhibitors were added 2 hr prior to isotopic labeling or 30 min prior to dexamethasone treatment. Dexamethasone was used at 5×10^{-7} M, progesterone at 5×10^{-5} M, cytosine arabinoside at 10^{-3} M, and actinomycin D at 4 μg/ml. Isotope incorporation was determined as TCA-insoluble material. Intracellular MTV RNA concentration was measured by $C_r t$ analysis (Ringold et al. 1975b) and cytoplasmic MTV DNA by calibration curve analysis (Varmus et al. 1974). N.T. = not tested.

DNA (data not shown). The factors that restrict viral DNA synthesis in these mouse cells, or permit it in infected rat cells, are not known.

Further studies on the synthesis of unintegrated MTV DNA in clones of infected HTC cells have revealed the following (data not shown) (G.M. Ringold et al., in prep.): (1) Both strands of viral DNA are induced by dexamethasone, following a time course which closely parallels the induction of viral RNA. In this experiment, unintegrated DNA was isolated at various times after hormone treatment, and "plus" (the DNA strand with the same polarity as viral RNA) and "minus" (the DNA strand complementary to viral RNA) strands were measured using MTV [^{3}H]cDNA and MTV [^{32}P]RNA, respectively. (2) Both form-I and form-III MTV DNAs are induced by dexamethasone, and both disappear when the hormone is withdrawn. In this case, forms I and III were separated by agarose gel electrophoresis and visualized by filter hybridization (Southern 1975) with MTV [^{125}I]RNA. Interestingly, form III appears prior to form I during induction and disappears prior to form I after hormone is removed. These data are consistent with the idea that the linear molecules are precursors to the covalently closed ciruclar DNA; in the case of avian-sarcoma-virus-infected cells, there is some direct evidence that such a precursor-product relationship exists (Shank and Varmus 1977). (3) When DNA is density-labeled with bromodeoxyuridine, a major fraction of the unintegrated viral DNA sediments in cesium chloride gradients with heavy-heavy density (both strands substituted) after a labeling period in which no cellular DNA is substituted in either strand. One interpretation is that the unintegrated sequences might arise by some mechanism other than semiconservative replication.

Taken together, these results are consistent with the idea that unintegrated MTV DNA production occurs via reverse transcription of MTV RNA by RNA-dependent DNA polymerase. In this view, the primary action of dexamethasone is induction of viral RNA, whereas synthesis of MTV DNA increases as a secondary consequence of the increased template and enzyme concentration. Such synthesis would result in production of both strands of the linear molecule, which might then be followed by ligation and circularization to produce form-I molecules. Since dexamethasone is required for maintenance of MTV RNA synthesis (G. M. Ringold et al., in prep.), as well as its induction, removal of the hormone would lead to disappearance of template and a concomitant reduction in DNA synthesis.

Also consistent with this general interpretation are two experiments employing metabolic inhibitors (Table 3). Cytosine arabinoside, an inhibitor of DNA synthesis, blocks the appearance of viral DNA without affecting the induction of viral RNA. In contrast, an inhibitor of RNA synthesis, actinomycin D, prevents the induction of both viral RNA and DNA. Thus, it appears that the production of unintegrated viral DNA is dependent on RNA synthesis, whereas the induction of viral RNA is largely independent of unintegrated viral DNA. Clearly, a prediction of the "reverse transcription" hypothesis is that clones that produce viral RNA should also produce viral DNA (assuming the machinery for reverse transcription is active), whereas clones that fail to make viral RNA (e.g., J2.31, see Table 2A) should make no viral DNA. The clones that we have examined to date are consistent with such a pattern (J. Ring and M. R. Stallcup, unpubl.).

Finally, the experiments with cytosine arabinoside imply that unintegrated DNA is not a major template source for subsequent MTV RNA production. It seemed conceivable that a "gene amplification" process might be responsible in part for the large magnitude of the MTV RNA induction observed in the infected HTC cells (see Table 2A). Indeed, Schimke et al. (this volume) have shown that selective gene amplification in a different cultured mammalian cell system appears to account for increased RNA production for a specific gene product. In our case, although we cannot rule out the possibility that a small amount of transcription from unintegrated MTV DNA might contribute to viral RNA production, we have no evidence suggesting that this process is of major importance in our system.

Aside from the general mechanism by which viral DNA is produced, the unintegrated sequences are of interest for eventually examining the nature of the chromosomal loci into which the viral DNA sequences are inserted in different clones. It is likely that the unintegrated viral DNA, and specifically the covalently closed circular form, serves as the molecular intermediate to the integrated provirus; this appears to be the case with C-type retroviruses (Guntaka et al. 1975). Using the filter hybridization procedure of Southern (1975), the cleavage sites of 11 restriction endonucleases on unintegrated MTV DNA have been mapped (P. Shank et al., in prep.). An analysis of integrated sequences, using techniques similar to those described by Botchan et al. (1976), is now underway.

MTV and Transposable Genetic Elements

The variability from clone to clone of MTV proviral gene expression and hormone inducibility, together with the specific alterations detected in host-gene expression in MTV-infected HTC cells, can be explained by assuming that the chromosomal locus at which an MTV genome is located determines its phenotypic effects. According to this view, we might expect that (a) endogenous MTV sequences are integrated at loci that normally are potentially expressed only in lactating mammary epithelia; (b) expression of endogenous MTV sequences, or introduction of MTV sequences into new chromosomal loci, can occur concomitant with neoplastic growth of certain nonmammary tissues; (c) exogenously introduced MTV sequences can be nonrandomly integrated into the genomes of a wide variety of tissue types and species; in each case some of these sites are subject to glucocorticoid regulation; (d) integration of viral DNA can selectively alter the expression of certain host genes.

In effect, we propose that the MTV genome is a transposable genetic element whose biological effects depend on its position within the host genome; the site of integration might affect both *determinative* responsiveness (i.e., whether a given sequence will be transcribed in a given differentiated tissue) and *modulatory* responsiveness (i.e., ligand-mediated induction or repression of the level of transcription of that sequence in differentiated cells). The characteristics of known elements of this type have been described extensively recently (Bukhari et al. 1977; Kleckner 1977; Nevers and Saedler 1977), but it is worth mentioning briefly some of their features.

In bacteria, it has been clearly established that certain specific segments of DNA can move from place to place on the chromosome, and that these elements alter the pattern of expression of adjacent genes. Transposable elements were first directly detected in *E. coli* as short segments of DNA that insert into a coordinately controlled set of genes and thereby exert a strong polar effect, i.e., reduce expression of distal genes in the operon (Hirsch et al. 1972; Saedler et al. 1972). A number of distinct "insertion sequences" of this type (termed IS1, IS2, etc.) have been characterized; each probably exists in multiple copies in the *E. coli* chromosome. For example, Saedler and Heiss (1973) showed that about eight copies of IS1 and five copies of IS2 are found in each of four separate strains examined. Thus, although insertion sequences move to new loci, their total number remains roughly constant. Other segments of DNA sequences exhibiting similar biological characteristics have been characterized in various plasmid and viral DNAs (Heffron et al. 1975a; Berg et al. 1975). Thus, mechanisms exist for transmission of transposable elements to new host cells; Heffron et al. (1975b) have shown that such sequences can be passed efficiently between many different species.

Certain genetic phenomena in eukaryotes appear analogous to the behavior of prokaryotic transposable elements. Indeed, one of the first indications of the existence of such elements originated with McClintock's elegant genetic studies of mutable loci in *Zea mays*, in which she suggested that "controlling elements" regulate gene expression during plant development (see McClintock 1967 for review). She concluded that when a controlling element is present and active, it affects the activity of adjacent genes in a defined manner dependent on the state of the element. In addition, an unlinked second component can interact with the controlling element, thereby affecting its activity. "Presetting" the activities of the controlling element and modulating the activity of the second component could then produce a complex, regulated sequence of events governed by only two genetic regions. A straightforward analogy can be made between the actions of bacterial episomes and maize controlling elements; Fincham and Sastry (1974) and Nevers and Saedler (1977) have developed in detail the notion that controlling elements are specific transposable sequences.

Mutable loci have now been characterized in many systems (McClintock 1967), and it is not unreasonable to speculate that they might reflect the action of transposable elements. In *Drosophila*, Green (1969a,b; 1976) has examined a number of these loci, especially those operating on the *white* locus. It has been established that the mutability property maps adjacent to the affected genes (Green 1969a), and that transposition of this element to other chromosomes is accompanied by acquisition of the mutability property at the new integration site (Green 1969b). Green concludes that these changes are most easily explained by the integration of a "viruslike agent" into the *Drosophila* genome. Evidence for other transposable controlling elements in *Drosophila* has also been documented (Judd 1969; Green 1973; Ising and Ramel 1976).

If the effects of MTV infection of HTC cells are mechanistically related to the actions of known transposable elements in prokaryotes and those inferred in eukaryotes, several predictions can be made which suggest the direction of future experiments. First, effects on host-gene expression should require integration of viral DNA into host sequences and should act in *cis*. Second, the integration sites should be contiguous to genes whose activities are altered, assuming that the activity monitored is a primary effect of integration. Third, infection of these cells by viruses that are not responsive to dexamethasone (or infection by the appropriate MTV variants, if they can be selected) are likely to display a different set of biological effects. Fourth, specific integration loci, or a hierarchy of preferred integration loci, should exist in the host genome.

Currently, little is known about the integration sites of MTV. Battula and Temin (1977) have made the intriguing observation that the integration site for infectious DNA specifying spleen necrosis virus, an avian retrovirus, appears to be highly nonrandom and perhaps specific. It is important to note, however, that detailed analyses of the integration sites of certain prokaryotic transposable elements suggest that the actual insertion locus may be restricted to preferred *regions* of the genome and not necessarily to a specific base sequence within those regions (Heffron et al. 1975a; Barth and Grinter 1977; Kleckner 1977; Chadwell and Starlinger 1978). Insertion events of this type may occur by a mechanism analogous to that described for class-I restriction endonucleases (Horiuchi and Zinder 1972), in which a specific recognition sequence is required but enzyme action actually occurs at contiguous, nonspecific loci (Kleckner 1977; Chadwell and Starlinger 1978). Thus, the location of the recognition sequence would define the "integration region." Such regions, if sufficiently large, could severely complicate the interpretation of integration-site mapping with restriction endonucleases, since different integration sites within a single region might appear unrelated.

CONCLUSIONS

The extent of expression of integrated MTV genes is commonly regulated by glucocorticoids and their specific receptor proteins; that is, when viral DNA is present in the genomes of a variety of mammalian cell types and is transcribed at some basal level, dexamethasone appears to stimulate selectively viral RNA production. In this report, we have reviewed the evidence that the stimulation of viral RNA production is the result of an immediate shift to a new rate of transcription of the MTV genes and have presented preliminary results indicating

that RNA synthesis under cell-free conditions appears to reflect faithfully these different efficiencies of specific gene transcription.

In addition, we have presented a series of experiments with MTV-infected HTC cells which show that the extent of viral gene expression and hormone responsiveness varies from clone to clone, and that infected clones are altered in the expression of a few specific genes. Unintegrated viral DNA is produced in these cells, perhaps via reverse transcription, but is probably not extensively transcribed into RNA. It is perhaps overly optimistic to assume that a comparative analysis of proviral integration sites using restriction endonucleases will reveal some correlation between physical location and the various biological activities observed. Instead, it may prove important to localize the viral genes relative to some other specific gene or region of the genome; the technology for accomplishing such mapping with any degree of precision in mammalian cells is not yet available.

It is interesting to speculate about the evolutionary process that resulted in regulation of MTV by glucocorticoids; the fact that this regulation is observed even in infected cells from diverse species that are not normally exposed to MTV suggests that some aspects of the control phenomenon have been conserved even in the absence of direct selection. Perhaps the viral functions evolved originally as host genes important for some hormone-dependent phenomenon. An alternative explanation is that viruses per se may be important as transmissible genetic elements at specific stages in differentiation and development (Huebner et al. 1970); in this view, only under aberrant conditions will virus particles effect cell transformation. Finally, the suggestion that this genetic element might alter specific gene responsiveness to a hormonal signal could be a hint that hormone responsiveness is determined during normal tissue differentiation by insertion of functionally analogous sequences, perhaps otherwise unrelated to viruses, into regions near potentially responsive genes.

Acknowledgments

We acknowledge important contributions of valuable discussion and experimental effort by Harold Varmus and Peter Shank, with whom some of the viral DNA studies were carried out. We also thank Bonnie Maler for her meticulous cell-culture work, and Ed Stavnezer and J. Michael Bishop for introducing us to their hybridization assay prior to publication.

This work was supported by a grant (CA 20535) and a Research Career Development Award to K.R.Y. (CA 00347) from the National Cancer Institute of the National Institutes of Health, by a Damon Runyon-Walter Winchell Cancer Fund Fellowship to G.M.R., and by a postdoctoral fellowship from the National Institutes of Health to M.R.S.

REFERENCES

ATTARDI, B., L. N. GELLER, and S. OHNO. 1976. Androgen and estrogen receptors in brain cytosol from male, female, and testicular feminized (tfm/y ♀) mice. *Endocrinology* 98: 864.

AVIV, D. and E. B. THOMPSON. 1972. Variation in tyrosine aminotransferase induction in HTC cell clones. *Science* 177: 1201.

BARTH, P. T. and N. J. GRINTER. 1977. Map of plasmid RP4 derived by insertion of transposon C. *J. Mol. Biol.* 113: 455.

BATTULA, N. and H. M. TEMIN. 1977. Infectious DNA of spleen necrosis virus is integrated at a single site in the DNA of chronically infected chicken fibroblasts. *Proc. Natl. Acad. Sci.* 74: 281.

BAXTER, J. D. and P. H. FORSHAM. 1972. Tissue effects of glucocorticoids. *Am. J. Med.* 53: 573.

BENTVELZEN, P. 1974. Host-virus interactions in murine mammary carcinogenesis. *Biochim. Biophys. Acta* 355: 236.

BERG, D. E., J. DAVIES, B. ALLET, and J.-D. ROCHAIX. 1975. Transposition of R factor genes to bacteriophage λ. *Proc. Natl. Acad. Sci.* 72: 3628.

BOTCHAN, M., W. TOPP, and J. SAMBROOK. 1976. The arrangement of simian virus 40 sequences in the DNA of transformed cells. *Cell* 9: 269.

BUKHARI, A. I., J. A. SHAPIRO, and S. L. ADHYA, eds. 1977. *DNA insertion elements, plasmids, and episomes.* Cold Spring Harbor Laboratory, Cold Spring Harbor, New York.

CHADWELL, H. A. and P. STARLINGER. 1978. The specificity of integration of IS-elements. In *Microbiology—1978* (ed. D. Schlessinger). American Society for Microbiology, Washington, D. C. (In press.)

COOPER, G. M. and H. M. TEMIN. 1976. Lack of infectivity of the endogenous avian leukosis virus-related genes in the DNA of uninfected chicken cells. *J. Virol.* 17: 422.

COX, R. F. 1973. Transcription of high-molecular-weight RNA from hen-oviduct chromatin by bacterial and endogenous form-B RNA polymerases. *Eur. J. Biochem.* 39: 49.

DALE, R. M. K. and D. C. WARD. 1975. Mercurated polynucleotides: New probes for hybridization and selective polymer fractionation. *Biochemistry* 14: 2458.

DIAMONDSTONE, T. I. 1966. Assay of tyrosine aminotransferase activity by conversion of p-hydroxyphenylpyruvate and p-hydroxybenzaldehyde. *Anal. Biochem.* 16: 395.

DICKSON, C., S. HASLAM, and S. NANDI. 1974. Conditions for optimal MTV synthesis *in vitro* and the effect of steroid hormones on virus production. *Virology* 62: 242.

ERNEST. M. J., G. SCHUTZ, and P. FEIGELSON. 1976. RNA synthesis in isolated hen oviduct nuclei. *Biochemistry* 15: 824.

FEIGELSON, P., M. BEATO, P. COLMAN, M. KALIMI, L. A. KILLEWICH, and G. SCHUTZ. 1975. Studies on the hepatic glucocorticoid receptor and on the hormonal modulation of specific mRNA levels during enzyme induction. *Recent Prog. Horm. Res.* 31: 213.

FERENCZ, A. and K. H. SEIFART. 1975. Comparative effect of heparin on RNA synthesis of isolated rat-liver nucleoli and purified RNA polymerase A. *Eur. J. Biochem.* 53: 605.

FINCHAM, J. R. S. and G. R. K. SASTRY. 1974. Controlling elements in maize. *Annu. Rev. Genet.* 8: 15.

FINE, D. L., J. K. PLOWMAN, S. P. KELLEY, L. O. ARTHUR, and E. A. HILLMAN. 1974. Enhanced production of mouse mammary tumor virus in dexamethasone-treated, 5-iododeoxyuridine-stimulated mammary tumor cell cultures. *J. Natl. Cancer Inst.* 52: 1881.

GEHRING, U. and G. M. TOMKINS. 1974. A new mechanism for steroid unresponsiveness: Loss of nuclear binding activity of a steroid hormone receptor. *Cell* 3: 301.

GIANNI, A. M., D. SMOTKIN, and R. A. WEINBERG. 1975. Murine leukemia virus: Detection of unintegrated double-stranded DNA forms of the provirus. *Proc. Natl. Acad. Sci.* 72: 447.

GORSKI, J. and F. GANNON. 1976. Current models of steroid hormone action: A critique. *Annu. Rev. Physiol.* 38: 425.

GREEN, M. M. 1969a. Mapping a *Drosophila melanogaster* "controlling element" by interallelic crossing over. *Genetics* 61: 423.

_____. 1969b. Controlling element mediated transpositions of the *white* gene in *Drosophila melanogaster*. *Genetics* 61: 429.

_____. 1973. Some observations and comments on mutable and mutator genes in *Drosophila*. *Genetics* (Suppl.) 73: 187.

_____. 1976. Mutable and mutator loci. In *The genetics and biology of* Drosophila (ed. M. Ashburner and E. Novitski), vol. 1b, p. 929. Academic Press, New York.

GUNTAKA, R. V., B. W. J. MAHY, J. M. BISHOP, and H. E. VARMUS. 1975. Ethidium bromide inhibits appearance of closed circular viral DNA and integration of virus-specific DNA in duck cells infected by avian sarcoma virus. *Nature* 253: 507.

HARRIS, S. E., J. M. ROSEN, A. R. MEANS, and B. W. O'MALLEY. 1975. Use of a specific probe for ovalbumin messenger RNA to quantitate estrogen-induced gene transcripts. *Biochemistry* 14: 2072.

HEFFRON, F., C. RUBENS, and S. FALKOW. 1975a. Translocation of a plasmid DNA sequence which mediates ampicillin resistance: Molecular nature and specificity of insertion. *Proc. Natl. Acad. Sci.* 72: 3623.

HEFFRON, F., R. SUBLETT, R. W. HEDGES, A. JACOB, and S. FALKOW. 1975b. Origin of the TEM beta-lactamase gene found on plasmids. *J. Bacteriol.* 122: 250.

HIRSCH, H. J., H. SAEDLER, and P. STARLINGER. 1972. Insertion mutations in the control region of the galactose operon of *E. coli*. *Mol. Gen. Genet.* 115: 266.

HIRT, B. 1967. Selective extraction of polyoma DNA from infected mouse cell cultures. *J. Mol. Biol.* 26: 365.

HORIUCHI, K. and N. D. ZINDER. 1972. Cleavage of bacteriophage f1 DNA by the restriction enzyme of *Escherichia coli* B. *Proc. Natl. Acad. Sci.* 69: 3220.

HUEBNER, R. J., G. J. KELLOFF, P. S. SARMA, W. T. LANE, H. C. TURNER, R. V. GILDEN, S. OROSZLAN, H. MEIER, D. D. MYERS, and R. L. PETERS. 1970. Group-specific antigen expression during embryogenesis of the genome of the C-type RNA tumor virus: Implications for ontogenesis and oncogenesis. *Proc. Natl. Acad. Sci.* 67: 366.

ISING, G. and C. RAMEL. 1976. The behavior of a transposing element in *Drosophila melanogaster*. In *The genetics and biology of* Drosophila (ed. M. Ashburner and E. Novitski), vol. 1b, p. 947. Academic Press, New York.

JACQUET, M., Y. GRONER, G. MONROY, and J. HURWITZ. 1974. The *in vitro* synthesis of avian myeloblastosis viral RNA sequences. *Proc. Natl. Acad. Sci.* 71: 3045.

JUDD, B. H. 1969. Evidence for a transposable element which causes reversible gene inactivation in *Drosophila melanogaster*. *Genetics* 62: s29.

KETNER, G. and T. J. KELLY, JR. 1976. Integrated simian virus 40 sequences in transformed cell DNA: Analysis using restriction endonucleases. *Proc. Natl. Acad. Sci.* 73: 1102.

KLECKNER, N. 1977. Translocatable elements in procaryotes. *Cell* 11: 11.

KULKA, R. G., G. M. TOMKINS, and R. B. CROOK. 1972. Clonal differences in glutamine synthetase activity of hepatoma cells: Effects of glutamine and dexamethasone *J. Cell Biol.* **54**: 175.

LEWIN, B. 1974. *Gene expression-1.* Wiley, London.

LINDELL, T. J., F. WEINBERG, P. W. MORRIS, R. G. ROEDER, and W. J. RUTTER. 1970. Specific inhibition of nuclear RNA polymerase II by α-amanitin. *Science* **170**: 447.

McCLINTOCK, B. 1967. Genetic systems regulating gene expression during development. *Dev. Biol.* (Suppl.) **1**: 84.

McKNIGHT. G. S., P. PENNEQUIN, and R. T. SCHIMKE. 1975. Induction of ovalbumin mRNA sequences by estrogen and progesterone in chick oviduct as measured by hybridization to complementary DNA. *J. Biol. Chem.* **250**: 8105.

MARZLUFF, W. F., JR., E. C. MURPHY, JR., and R. C. C. HUANG. 1973. Transcription of ribonucleic acid in isolated mouse myeloma nuclei. *Biochemistry* **12**: 3440.

MORRIS, V., E. MEDEIROS, G. M. RINGOLD, J. M. BISHOP, and H. E. VARMUS. 1977. Comparison of mouse mammary tumor virus-specific DNA in inbred, wild and Asian mice and in tumors and normal organs from inbred mice. *J. Mol. Biol.* **114**: 73.

NAKANISHI, S., S. ADHYA, M. GOTTESMAN, and I. PASTAN. 1975. Selective effects of MgCl₂ and temperature on the initiation of transcription at *lac, gal* and λ promotors. *J. Biol. Chem.* **250**: 8202.

NEVERS, P. and H. SAEDLER. 1977. Transposable genetic elements as agents of gene instability and chromosomal rearrangements. *Nature* **268**: 109.

O'FARRELL, P. H. 1975. High resolution two-dimensional electrophoresis of proteins. *J. Biol. Chem.* **250**: 4007.

PALMITER, R. D., P. B. MOORE, E. R. MULVIHILL, and S. EMTAGE. 1976. A significant lag in the induction of ovalbumin messenger RNA by steroid hormones: A receptor translocation hypothesis. *Cell* **8**: 557.

PARKS, W. P., E. M. SCOLNICK, and E. H. KOZIKOWSKI. 1974. Dexamethasone stimulation of murine mammary tumor virus expression: A tissue culture source of virus. *Science* **184**: 158.

PARKS, W. P., J. C. RANSOM, H. A. YOUNG, and E. M. SCOLNICK. 1975. Mammary tumor virus induction by glucocorticoids. *J. Biol. Chem.* **250**: 3330.

PUCA, G. A., E. NOLA, U. HIBNER, G. CICALA, and V. SICA. 1975. Interaction of the estradiol receptor from calf uterus with its nuclear acceptor sites. *J. Biol. Chem.* **250**: 6452.

REEDER, R. H., B. SOLLNER-WEBB, and H. L. WAHN. 1977. Sites of transcription initiation *in vivo* on *Xenopus laevis* ribosomal DNA. *Proc. Natl. Acad. Sci.* (in press).

REEVE, A. E., M. M. SMITH, V. PIGIET, and R. C. C. HUANG. 1977. Incorporation of purine nucleoside 5'-[γ-S]triphosphates as affinity probes for initiation of RNA synthesis *in vitro. Biochemistry* **16**: 4464.

RINGOLD, G. M., P. B. BLAIR, J. M. BISHOP, and H. E. VARMUS. 1976. Nucleotide sequence homologies among mouse mammary tumor viruses. *Virology* **70**: 550.

RINGOLD, G. M., R. D. CARDIFF, H. E. VARMUS, and K. R. YAMAMOTO. 1977a. Infection of cultured rat hepatoma cells by mouse mammary tumor virus. *Cell* **10**: 11.

RINGOLD, G., E. Y. LASFARGUES, J. M. BISHOP, and H. E. VARMUS. 1975a. Production of mouse mammary tumor virus by cultured cells in the absence and presence of hormones: Assay by molecular hybridization. *Virology* **65**: 135.

RINGOLD, G. M., K. R. YAMAMOTO, J. M. BISHOP, and H. E. VARMUS. 1977b. Glucocorticoid stimulated accumulation of mouse mammary tumor virus RNA: Increased rate of synthesis of viral RNA. *Proc. Natl. Acad. Sci.* **74**: 2879.

RINGOLD, G. M., K. R. YAMAMOTO, P. R. SHANK, and H. E. VARMUS. 1977c. Mouse mammary tumor virus

DNA in infected rat cells: Characterization of unintegrated forms. *Cell* **10**: 19.

RINGOLD, G. M., K. R. YAMAMOTO, G. M. TOMKINS, J. M. BISHOP, and H. E. VARMUS. 1975b. Dexamethasone-mediated induction of mouse mammary tumor virus RNA: A system for studying glucocorticoid action. *Cell* **6**: 299.

ROEDER, R. G. and W. J. RUTTER. 1970. Specific nucleolar and nucleoplasmic RNA polymerases. *Proc. Natl. Acad. Sci.* **65**: 675.

RYMO, L., J. T. PARSONS, J. M. COFFIN, and C. WEISSMANN. 1974. *In vitro* synthesis of Rous sarcoma virus-specific RNA is catalyzed by a DNA-dependent RNA polymerase. *Proc. Natl. Acad. Sci.* **71**: 2782.

SAEDLER, H. and B. HEISS. 1973. Multiple copies of the insertion-DNA sequences of IS1 and IS2 in the chromosome of *E. coli* K-12. *Mol. Gen. Genet.* **122**: 267.

SAEDLER, H., J. BESEMER, B. KEMPER, B. ROSENWIRTH, and P. STARLINGER. 1972. Insertion mutations in the control region of the *gal* operon of *E. coli.* I. Biological characterization of the mutations. *Mol. Gen. Genet.* **115**: 258.

SAMUELS, H. H. and G. M. TOMKINS. 1970. The relation of steroid structure to enzyme induction in hepatoma tissue culture cells. *J. Mol. Biol.* **52**: 57.

SCHWARTZ, L. B., V. E. F. SKLAR, J. A. JAEHNING, R. WEINMANN, and R. G. ROEDER. 1974. Isolation and partial characterization of the multiple forms of deoxyribonucleic acid-dependent ribonucleic acid polymerase in mouse myeloma, MOPE 315. *J. Biol. Chem.* **249**: 5889.

SCHWARTZ, R. J., M.-J. TSAI, S. Y. TSAI, and B. W. O'MALLEY. 1975. Effect of estrogen on gene expression in the chick oviduct. V. Changes in the number of RNA polymerase binding and initiation sites in chromatin. *J. Biol. Chem.* **250**: 5175.

SHANK, P. R. and H. E. VARMUS. 1977. Virus-specific DNA in the cytoplasm of avian sarcoma virus infected cells is a precursor to covalently closed circular viral DNA in the nucleus *J. Virol.* (in press).

SHIH, T. Y., H. A. YOUNG, W. P. PARKS, and E. M. SCOLNICK. 1977. *In vitro* transcription of Moloney leukemia virus genes in infected cell nuclei and chromatin: Elongation of chromatin associated ribonucleic acid by *Escherichia coli* ribonucleic acid polymerase. *Biochemistry* **16**: 1795.

SOMERS, D. G., M. L. PEARSON, and C. J. INGLES. 1975. Isolation and characterization of an α-amanitin-resistant rat myoblast mutant cell line possessing α-amanitin-resistant RNA polymerase II. *J. Biol. Chem.* **250**: 4825.

SOUTHERN, E. M. 1975. Detection of specific sequences among DNA fragments separated by gel electrophoresis. *J. Mol. Biol.* **98**: 503.

SPELSBERG, T. C. 1974. The role of nuclear acidic proteins in binding steroid hormones. In *Acidic proteins of the nucleus* (ed. I. L. Cameron and J. R. Jeter, Jr.), p. 247. Academic Press, New York.

TATA, J. R. 1976. The expression of the vitellogenin gene. *Cell* **9**: 1.

TEMIN, H. M. and D. BALTIMORE. 1972. RNA-directed DNA synthesis and RNA tumor viruses. *Adv. Virus Res.* **17**: 129.

THOMPSON, E. B., G. M. TOMKINS, and J. F. CURRAN. 1966. Induction of tyrosine α-ketoglutarate transaminase by steroid hormones in a newly established tissue culture cell line. *Proc. Natl. Acad. Sci.* **56**: 296.

VAIDYA, A. B., E. Y. LASFARGUES, G. HEUBEL, J. C. LASFARGUES, and D. H. MOORE. 1976. Murine mammary tumor virus: Characterization of infection of nonmurine cells. *J. Virol.* **18**: 911.

VARMUS, H. E., S. HEASLEY, and J. M. BISHOP. 1974. Use of DNA-DNA annealing to detect new virus-specific DNA sequences in chicken embryo fibroblasts after infection by avian sarcoma virus. *J. Virol.* **14**: 895.

VARMUS, H. E., J. M. BISHOP, R. C. NOWINSKI, and N. H.

SARKAR. 1972. Mammary tumor virus specific nucleotide sequences in mouse DNA. *Nat. New Biol.* **238**: 189.

VARMUS, H. E., R. V. GUNTAKA, C. T. DENG, and J. M. BISHOP. 1975. Synthesis, structure and function of avian sarcoma virus-specific DNA in permissive and non-permissive cells. *Cold Spring Harbor Symp. Quant. Biol.* **39**: 987.

VARMUS, H. E., N. QUINTRELL, E. MEDEIROS, J. M. BISHOP, R. NOWINSKI, and N. SARKAR. 1973. Transcription of mouse mammary tumor virus genes in tissues from high and low tumor incidence mouse strains. *J. Mol. Biol.* **79**: 663.

WEBER, J., W. JELINEK, and J. E. DARNELL, JR. 1977. The definition of a large viral transcription unit late in Ad2 infection of HeLa cells: Mapping of nascent RNA molecules labeled in isolated nuclei. *Cell* **10**: 611.

YAMAMOTO, K. R. and B. M. ALBERTS. 1974. On the specificity of the binding of the estradiol receptor protein to deoxyribonucleic acid. *J. Biol. Chem.* **249**: 7076.

———. 1976. Steroid receptors: Elements of modulation of eukaryotic transcription. *Annu. Rev. Biochem.* **45**: 721.

YAMAMOTO, K. R. and G. M. RINGOLD. 1977. Glucocorticoid regulation of mammary tumor virus gene expression. In *Hormone receptors: Steroid hormones* (ed. B. W. O'Malley and L. Birnbaumer), vol. 1. Academic Press, New York. (In press.)

YAMAMOTO, K. R., M. R. STAMPFER, and G. M. TOMKINS. 1974. Receptors from glucocorticoid-sensitive lymphoma cells and two classes of insensitive clones: Physical and DNA-binding properties. *Proc. Natl. Acad. Sci.* **71**: 3901.

YAMAMOTO, K. R., U. GEHRING, M. R. STAMPFER, and C. H. SIBLEY. 1976. Genetic approaches to steroid hormone action. *Recent Prog. Horm. Res.* **32**: 3.

YAMAMOTO, K. R., R. D. IVARIE, J. RING, G. M. RINGOLD, and M. R. STALLCUP. 1977. Integrated mammary tumor virus genes: Transcriptional regulation by glucocorticoids and specific effects on host gene expression. In *Biochemical actions of hormones* (ed. G. Litwack), vol. 5. Academic Press, New York. (In press.)

YOUNG, H. A., E. M. SCOLNICK, and W. P. PARKS. 1975. Glucocorticoid-receptor interaction and induction of murine mammary tumor virus. *J. Biol. Chem.* **250**: 3337.

YOUNG, H. A., T. Y. SHIH, E. M. SCOLNICK, and W. P. PARKS. 1977. Steroid induction of mouse mammary tumor virus: Effect upon synthesis and degradation of viral RNA. *J. Virol.* **21**: 139.

YOUNG, L. J. T., R. D. CARDIFF, and R. L. ASHLEY, 1975. Long-term primary culture of mouse mammary tumor cells: Production of virus. *J. Natl. Cancer Inst.* **54**: 1215.

ZASLOFF, M. and G. FELSENFELD. 1977. Use of mercury-substituted ribonucleoside triphosphates can lead to artefacts in the analysis of *in vitro* chromatin transcripts. *Biochem. Biophys. Res. Commun.* **75**: 598.

Regulation of Gene Expression in the Chick Oviduct by Steroid Hormones

R. D. PALMITER, E. R. MULVIHILL, G. S. McKNIGHT, AND A. W. SENEAR

Department of Biochemistry, University of Washington, Seattle, Washington 98195

One of the central problems in our understanding of steroid hormone action is the absence of a molecular connection between the nuclear localization of hormone receptors in target cells and the subsequent accumulation of specific mRNAs in those cells. This paper deals with two aspects of this problem. First, we will describe experiments that demonstrate a pronounced lag between the time steroid receptors reach the nucleus and the induction of specific mRNAs. The existence of a lag has important implications because it suggests that there are as yet undefined, rate-limiting nuclear events interspersed between receptor binding in the nucleus and the dramatic acceleration of mRNA production that occurs a few hours later. This lag phenomenon is prevalent in steroid-responsive systems (see Palmiter et al. 1976 for references). However, it is not universal. For example, within the tubular gland cells of the chick oviduct, conalbumin mRNA (mRNA$_{con}$) is induced by estrogen with virtually no lag, whereas ovalbumin mRNA (mRNA$_{ov}$) production commences at an accelerated rate at about 3 hours. We will describe the kinetics of induction of these mRNAs with estrogen, progesterone, and their combination to illustrate the apparent complexity of the regulatory mechanisms involved.

The second aspect of steroid hormone action that we will touch on is why mRNA$_{ov}$ is produced in abundance in the oviduct but not in the liver, even though both tissues are target organs for estrogen and have nuclear estrogen receptors. The recent observations of Weintraub and Groudine (1976) that genes from diverse tissues show differential sensitivity to nucleases provide a handle for exploring the environment of specific genes. We will show that the ovalbumin gene, as compared to globin genes, is preferentially digested by DNase I in oviduct nuclei of laying hens and estrogen-treated chicks, in agreement with the results of Garel and Axel (1976). However, this preferential sensitivity appears not to be associated with transcription per se since the ovalbumin gene remains sensitive to DNase I even after estrogen action is blocked with the anti-estrogen, tamoxifen.

RESULTS

Kinetics of Ovalbumin and Conalbumin Induction with Estrogen

Figure 1 shows the induction of both ovalbumin and conalbumin synthesis during the first 10 hours of secondary stimulation of chicks with an optimal dose of estrogen. For each time point in this experiment, the magnum segment of the oviduct was divided into two portions. Part was incubated for 30 minutes in Hanks' salts with [^{3}H]leucine, and then the percentage of total labeled protein that was specifically immunoprecipitable by anti-ovalbumin or anti-conalbumin was determined. The remaining tissue was used to isolate RNA, which was then assayed for functional mRNA$_{ov}$ and mRNA$_{con}$ by measuring the incorporation of [^{3}H]isoleucine into the respective proteins in a cell-free, protein-synthesizing system derived from rabbit reticulocytes. Figure 1 shows that the increase in the synthesis of both proteins measured in intact cells is coincident with the accumulation of their respective mRNAs. This experiment demonstrates that the relative rate of ovalbumin and conalbumin synthesis is determined by the concentration of their respective mRNAs and rules out the possibility that the increase in the synthesis of these two proteins is due to a differential increase in the translational efficiency of their mRNAs. The early kinetics of ovalbumin and conalbumin induction have been studied on more than 15 occasions over the past 5 years. Similar results were obtained in three different laboratories with birds from three different farms. The salient features of the induction kinetics with an optimal dose of estrogen are outlined below.

(a) There is no discernible lag before conalbumin synthesis increases. The initial slope of the conalbumin induction curve averages 1.2% per hour (range 1.0–1.7% per hour), reaches a value of 5–8% of total protein synthesis within 8 hours and plateaus at 8–10% of total protein synthesis after several days of estrogen stimulation. In the absence of estrogen, conalbumin synthesis is maintained at about 1.3%

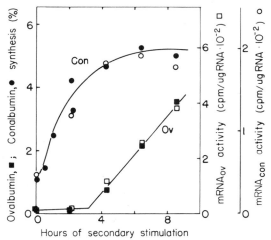

Figure 1. Induction of ovalbumin and conalbumin synthesis and mRNA by estrogen. Estradiol benzoate (2 mg) was administered to withdrawn chicks and at the times indicated chicks were killed. Part of the tissue was incubated with [³H]leucine, and the percentage of total protein synthesized that was specifically immunoprecipitated was determined. RNA was extracted from the remaining tissue and translated in a reticulocyte lysate; the incorporation of [³H]isoleucine into ovalbumin and conalbumin was determined immunologically. (Reprinted, with permission, from Palmiter et al. 1976.)

of total protein synthesis (range 0.7–1.7%), presumably by endogenous steroids.

(b) Ovalbumin synthesis, in contrast, is barely detectable (<0.1% of total protein synthesis) in the absence of estrogen. After estrogen is administered, there is a pronounced lag phase, ranging from 2.7–4.1 hours, during which there is a small increase in $mRNA_{ov}$ sequences. This is followed by a sharp transition to a rate of $mRNA_{ov}$ accumulation at least an order of magnitude greater than during the lag phase. The slope of the ovalbumin induction curve during the accumulation phase averages 0.7% per hour and ranges from 0.4–1.2% per hour. The relative rate of ovalbumin synthesis continues to rise for several days, becoming finally 40–50% of total protein synthesis (Palmiter 1972).

(c) A similar lag is observed if the accumulation of $mRNA_{ov}$ sequences is measured by hybridization of total RNA with complementary DNA (Cox et al. 1974; McKnight et al. 1975; Palmiter et al. 1976; see also Fig. 10). These experiments indicate that the lag is not due to the acquisition of functional mRNA. From these hybridization studies it is also possible to quantitate accurately the number of $mRNA_{ov}$ molecules. There are about 4 molecules per cell in the withdrawn oviduct, and this is equivalent to 30–40 molecules per tubular gland cell. After 8 hours of secondary stimulation, there are approximately 4000–7000 molecules per tubular gland cell, and the rate of accumulation is 15–25 molecules

per minute per tubular gland cell. The variability in these numbers stems from the uncertainty regarding the number of tubular gland cells. Nevertheless, the rate of $mRNA_{ov}$ production increases about 3 orders of magnitude upon estrogen stimulation. Assuming that the half-life of $mRNA_{ov}$ is 24 hours (Palmiter 1973) and does not change during early secondary stimulation and that the transit time of RNA polymerase along the gene is 45 seconds (1800 nucleotides at 40 nucleotides/sec), then there may be as many as nine polymerases transcribing each ovalbumin gene after the lag phase. In the absence of estrogen, each gene would be transcribed only nine times every 1000 minutes, or less than once per hour. This calculation will become important when considering the meaning of the DNase-I digestion studies described below.

Nuclear Estrogen Receptor Levels

Figure 2 shows the kinetics of accumulation of nuclear estrogen receptors after administration of an optimal dose of estrogen. Within 15–20 minutes the concentration of nuclear receptors increases about sevenfold and reaches a plateau of about 8000 molecules per tubular gland cell which is maintained for the next 6 hours. These results indicate that the lag in ovalbumin induction is not due to a slow or gradual accumulation of estrogen receptors. With suboptimal doses of estrogen, lower levels of nuclear receptors are achieved (Fig. 3A), and this is associated with lower rates of ovalbumin and conalbumin synthesis (Fig. 3B). Note that the lag phase for ovalbumin is virtually constant with dif-

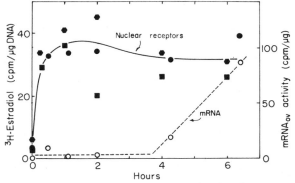

Figure 2. Kinetics of estrogen receptor localization in magnum nuclei and $mRNA_{ov}$ induction. Withdrawn chicks were given 2 mg 17β-estradiol-benzoate in vivo as usual. At the indicated times, chicks were killed, nuclei were isolated from magnum tissue, and the number of specific receptors for estradiol was estimated by an exchange assay with excess ³H-labeled 17β-estradiol. Different solid symbols represent different experiments; in one experiment (●), the exchange assay was performed on crude chromatin rather than nuclei. The $mRNA_{ov}$ activity (○) was measured with RNA isolated from the same tissue as one of the nuclear preparations (●). (Reprinted, with permission, from Palmiter et al. 1976.)

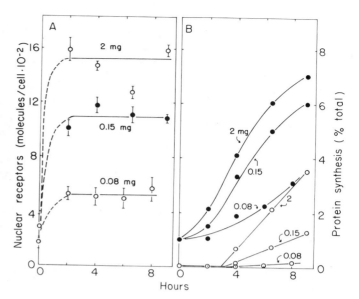

Figure 3. Effect of various doses of 17β-estradiol-benzoate on the kinetics of estrogen receptor appearance in magnum nuclei and ovalbumin and conalbumin induction. Various doses (80 μg, 150 μg, 2000 μg/bird) of estradiol-benzoate were administered to withdrawn chicks and at the times indicated chicks were killed. (A) Part of the magnum tissue was used to isolate nuclei and determine the level of specific estrogen receptors. (B) Part was incubated with [^{3}H]leucine, and the percentage of total protein synthesized that was specifically immunoprecipitated by anti-ovalbumin (○——○) and anti-conalbumin (●——●) was determined. (Reprinted, with permission, from Mulvihill and Palmiter 1977.)

ferent doses and that a lag becomes apparent for conalbumin at suboptimal doses of estrogen. The constancy of the lag phase for mRNA$_{ov}$ induction with different doses of estrogen is clearly depicted in Figure 4.

Conalbumin synthesis is induced by lower levels of estrogen than is ovalbumin. Figure 5 shows the correlation of nuclear estrogen receptor levels and the rate of ovalbumin and conalbumin synthesis

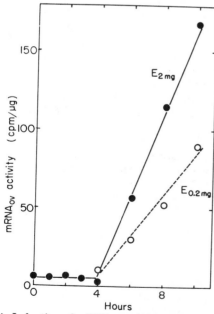

Figure 4. Induction of mRNA$_{ov}$ activity with optimal and suboptimal dosages of estrogen. Withdrawn chicks were given either 2 mg estradiol-benzoate (●) or 0.2 mg estradiol-benzoate (○); at the indicated times magnum RNA was isolated and mRNA$_{ov}$ activity determined by translation. (Reprinted, with permission, from Palmiter et al. 1976.)

achieved at 8 hours with varying doses of 17β-estradiol. In this experiment and others with different estrogens (Mulvihill and Palmiter 1977), it is apparent that the induction of conalbumin synthesis (mRNA) is directly proportional to the nuclear receptor level, whereas the induction of ovalbumin synthesis (mRNA) is related to nuclear receptor levels in a more complex manner. Half-maximal induction of ovalbumin occurs when the nuclear receptor levels are 80% of maximum.

Kinetics of Ovalbumin and Conalbumin Induction with Progesterone

Figures 6 and 8 show that, with an optimal dose of progesterone, mRNA$_{ov}$ accumulates at the same rate as with estrogen, but the lag is shortened to about 2 hours. Conalbumin synthesis is also induced by progesterone, but there is a lag of about 2 hours, and the initial rate of accumulation of synthetic activity is markedly reduced compared to that observed with estrogen. These experiments show that the rate of mRNA production and the lag phase are controlled independently. They also suggest that estrogen and progesterone act by slightly different mechanisms.

Figure 7 shows a typical dose-response relationship for induction of ovalbumin and conalbumin synthesis by progesterone. The curves are similar to those observed with estrogen (Fig. 5) except that maximum extent of conalbumin synthesis achieved at 8 hours is about half that obtained with estrogen. Again, conalbumin induction appears to correlate directly with nuclear progesterone receptor levels and ovalbumin induction is displaced to higher nuclear receptor levels.

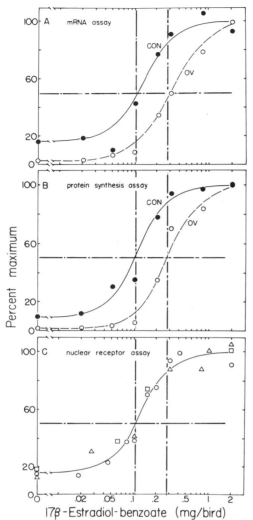

Figure 5. Dose-response relationships for 17β-estradiol-benzoate. Withdrawn chicks were injected with indicated doses of 17β-estradiol-benzoate and killed 8 hr later. *(A)* Accumulation of ovalbumin and conalbumin mRNA. RNA was isolated from the magnum and translated in the rabbit reticulocyte lysate. Incorporation of [³H]isoleucine into ovalbumin (○——○) and conalbumin (●——●). *(B)* Induction of ovalbumin and conalbumin synthesis. Magnum explants from the same tissue were incubated with [³H]leucine and the relative rate of ovalbumin (○——○) and conalbumin (●——●) synthesis (percentage of total protein synthesis) determined. *(C)* Nuclear receptor concentration. Nuclei were isolated from a third portion of the tissue, and the number of specific receptors for estrogen was determined by an exchange assay utilizing ³H-labeled 17β-estradiol. Different symbols represent different experiments. (Reprinted, with permission, from Mulvihill and Palmiter 1977.)

Steroid Hormone Interactions

The effect of a combination of optimal doses of estrogen and progesterone is shown in Figure 8. For ovalbumin, this combination results in a lag of 2 hours and a rate of accumulation greater than with either hormone alone. For conalbumin, the combina-

tion of hormones gives a response similar to progesterone alone; progesterone appears to dominate and inhibit the rapid accumulation of conalbumin synthetic activity observed with estrogen.

The dominance of progesterone over estrogen on conalbumin synthesis is also manifested when progesterone is given after estrogen. Figure 9 shows that progesterone temporarily interrupts estrogen-mediated accumulation of conalbumin synthesis activity; panel A shows the results when progesterone is given 2 hours after estrogen, and panel B shows an experiment in which progesterone was given 4 hours after estrogen. In both cases, the rate of conalbumin synthesis falls slightly and then, after about 4 hours, begins to rise again, so that by 8 hours it has reached the level attained with estrogen alone. This temporary inhibitory effect on conalbumin synthesis is not observed with ovalbumin. In contrast, there is no effect of progesterone for about 2 hours, and then the rate of $mRNA_{ov}$ accumulation increases about 1.5-fold compared to with estrogen alone (Fig. 9, insets). We have also observed that after progesterone administration there is a transient decrease in nuclear estrogen receptor levels that might be associated with this phenomenon.

Effects of Tamoxifen

Tamoxifen is a potent anti-estrogen that binds estrogen receptors competitively, allows nuclear localization of estrogen receptors, but confers no estrogenic activity on these nuclear receptors (Sutherland et al. 1977). Preliminary experiments established the dose of tamoxifen, administered simultaneously with estrogen, necessary to prevent estrogen action. These experiments showed that more tamoxifen was required to inhibit conalbumin induction than ovalbumin induction, as predicted by the results shown in Figure 5. To ascertain how quickly tamoxifen can act, an optimal dose of estrogen was administered to chicks, and 6 hours later 10 mg of tamoxifen was given. Figure 10A shows the accumulation of $mRNA_{ov}$ as measured by hybridization with cDNA. After tamoxifen was administered, $mRNA_{ov}$ continued to increase for about 2 hours and then began to fall. A more sensitive experiment than the one shown in Figure 10 involved giving tamoxifen at various times during the lag phase; the result is that $mRNA_{ov}$ was induced normally for 3 hours after the drug was given and then stopped abruptly. Conalbumin synthesis is inhibited by tamoxifen within an hour (data not shown). These results indicate that the time required for tamoxifen to repress $mRNA_{ov}$ and $mRNA_{con}$ accumulation is nearly the same as the time required by estrogen to induce them. If tamoxifen is administered first, then a 20-fold excess of estrogen can induce ovalbumin (Fig. 10B) and conalbumin (data not shown) with essentially the same kinetics as estrogen alone.

Tamoxifen has no effect on progesterone-mediated

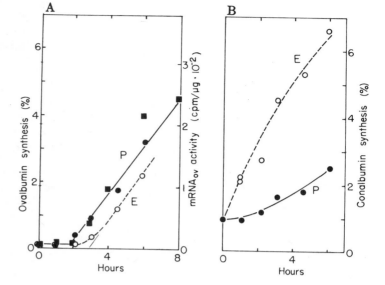

Figure 6. Induction of ovalbumin and conalbumin synthesis by estrogen or progesterone. Estrogen (E) (○) (2 mg) or progesterone (P) (●) (2 mg) was administered to chicks withdrawn for 11 days; at the times indicated, the chicks were killed, magnum explants were incubated with [³H]leucine, and the relative rates of (A) ovalbumin and (B) conalbumin synthesis (percentage of total protein synthesis) determined. Ovalbumin mRNA activity (■) was also measured with RNA from progesterone-treated chicks.

induction of mRNA$_{ov}$ and mRNA$_{con}$. These results not only show that tamoxifen is selective for estrogen receptors, but also demonstrate that its action on mRNA accumulation is not due to a general inhibitory effect on RNA synthesis.

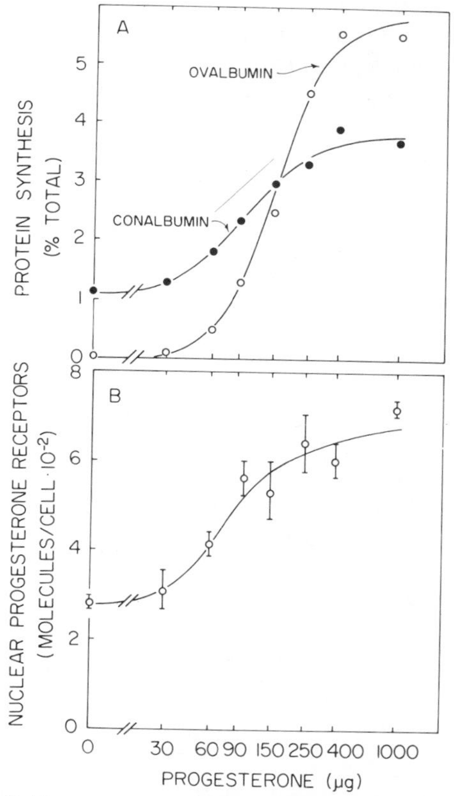

Sensitivity of the Ovalbumin Gene to DNase I

The ovalbumin gene in nuclei from laying hens is preferentially degraded by DNase I (see Garel and Axel; Bellard et al.; both this volume). To ascertain

Figure 7. Dose-response relationships for progesterone. Withdrawn chicks were injected with the indicated dosages of progesterone and 8 hr later oviduct magnums were isolated. (A) Relative rates of ovalbumin and conalbumin synthesis. (B) Quantitation of nuclear progesterone receptors by an exchange assay similar to that used for estrogen (Fig. 5).

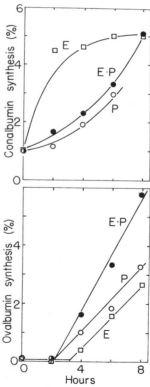

Figure 8. Effect of combination of estrogen and progesterone on conalbumin and ovalbumin synthesis. Chicks withdrawn for 16 days were given either 2 mg estrogen (□), 2 mg progesterone (○), or 2 mg estrogen and 2 mg progesterone (●). The relative rates of (A) conalbumin and (B) ovalbumin synthesis were determined as described in the legend to Fig. 1.

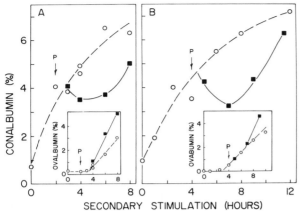

Figure 9. Effect of progesterone on conalbumin and oval-bumin synthesis when given after estrogen. Withdrawn chicks were given estradiol-benzoate (1 mg) and, at the times indicated, the rates of conalbumin and ovalbumin synthesis were measured (○——○). Progesterone (1 mg) was administered to some chicks at *(A)* 2 hr or *(B)* 4 hr after estrogen (■——■).

whether this is due to the high rate of transcription of the ovalbumin gene in this tissue, we investigated the sensitivity of this gene to DNase I before and after tamoxifen. These studies were performed with chicks that were given secondary stimulation with

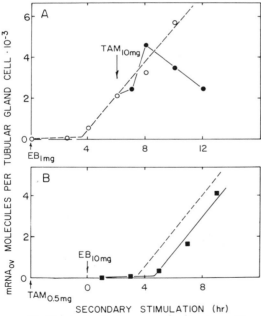

Figure 10. Reversal of estrogen action by tamoxifen. *(A)* Estradiol-benzoate (EB; 1 mg) was administered to chicks, and 6 hr later tamoxifen (TAM; 10 mg) was given. At the times indicated, oviducts were removed and used to measure the relative rate of ovalbumin synthesis (Fig. 9B inset) and RNA was isolated to measure $mRNA_{ov}$ sequences by hybridization with cDNA (McKnight and Schimke 1974). The number of molecules per tubular gland cell was calculated assuming 15% tubular gland cells. (○——○) Estradiol-benzoate (alone); (●——●) estradiol-benzoate followed by tamoxifen. *(B)* Tamoxifen was administered first and then overcome with excess estradiol-benzoate (■——■); the dashed line is transposed from upper panel.

estrogen (hexestrol pellets) for 6 days. Under these conditions, tubular gland cells comprise about 80% of the cell population, a sufficiently homogeneous population to allow a meaningful analysis by digestion experiments.

Figure 11 shows that a dose of 2.5 mg of tamoxifen daily completely counteracted the hexestrol pellet and by 3 days reduced ovalbumin and conalbumin synthesis to the same levels observed when the pellet was removed. The wet weight of the oviduct magnum also fell from 510 to 160 mg with 2.5 or 10 mg/day of tamoxifen. On the basis of these results, and the kinetics shown in Figure 10, a 10-mg dose of tamoxifen should inhibit $mRNA_{ov}$ production within 3 hours in these fully stimulated chicks.

Figure 12 depicts the results of DNase-I digestion of nuclei from oviduct and liver after various hormonal treatments. Nuclei were incubated with DNase I until 10–15% of total DNA was acid-soluble. Then total DNA was isolated and hydrolyzed with base to remove any contaminating $mRNA_{ov}$. Aliquots were hybridized with excess ³H-labeled cDNA made against either chicken globin mRNA ($cDNA_{gb}$) or $mRNA_{ov}$ ($cDNA_{ov}$). Figure 12A shows a control hybridization of both cDNAs with undigested chicken erythrocyte DNA; both cDNAs hybridize to the same extent. When oviduct nuclei from laying hens were digested with DNase I, hybridization of the $cDNA_{ov}$ was selectively reduced (Fig. 12B); typically, 2–3 times more DNA was required to achieve

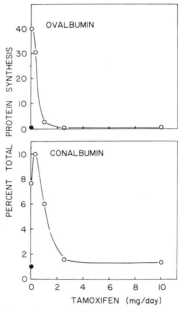

Figure 11. Dose-response for tamoxifen reversal of hexestrol action. Hexestrol pellets were implanted subcutaneously for 9 days. On the sixth, seventh, and eighth days, the indicated dosages of tamoxifen were injected subcutaneously; 64 hr after the first injection, the relative rates of ovalbumin and conalbumin synthesis were measured (○). The hexestrol pellets were removed from control birds 64 hr before sacrifice and measurement of protein synthesis (●).

Figure 12. Sensitivity of the ovalbumin gene to digestion by DNase I in liver and oviduct. Nuclei were isolated from liver or oviduct as described by Mulvihill and Palmiter (1977), resuspended in 10 mM Tris-Cl, pH 7.4, 10 mM NaCl, 3 mM MgCl$_2$ at 0.4–0.8 mg DNA/ml, and digested with 10 μg/ml DNase I for 2.5 min at 37°C to achieve 10–15% acid solubilization of the DNA. The DNA was isolated by SDS/proteinase-K digestion followed by phenol/chloroform extraction, base hydrolysis, and ethanol precipitation. Aliquots of DNA were hybridized with 30 pg of cDNA (500 cpm) in 20 μl of 0.6 M NaCl, 20 mM Tris-Cl, pH 7.5, 4 mM EDTA, 0.15% SDS for 24 hr at 68°C; the extent of hybridization was assayed using S$_1$ nuclease treatment (McKnight and Schimke 1974). In the absence of added DNA, less than 1% of the cDNA was resistant to S$_1$ nuclease; addition of excess mRNA allowed greater than 90% hybridization. (○——○) Hybridization with chick globin cDNA; (●——●) hybridization with ovalbumin cDNA. (A) Control chick erythrocyte DNA (Calbiochem)

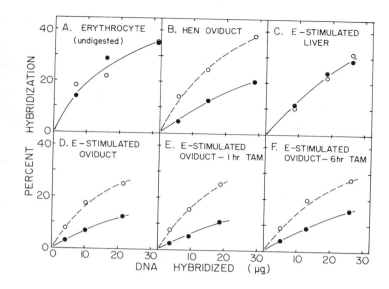

without DNase digestion. (B) Oviduct nuclei from laying hens. (C) Liver nuclei from the same chicks as in D. (D) Oviduct nuclei from chicks stimulated with hexestrol for 6 days. (E) Oviduct nuclei from chicks stimulated with hexestrol for 6 days, followed by 10 mg/chick tamoxifen for 1 hr. (F) Same as E, except tamoxifen was for 6 hr.

the same percentage hybridization of cDNA$_{ov}$ as cDNA$_{gb}$. However, when liver nuclei from estrogen-treated chicks were digested to the same extent as oviduct nuclei, there was no selective decrease in cDNA$_{ov}$ hybridization (Fig. 12C).

The digestion of oviduct nuclei from estrogen-treated chicks gave essentially the same results as did hen oviduct nuclei (compare Fig. 12 B and D). Administration of tamoxifen (10 mg) to estrogen-treated chicks for either 1 or 6 hours had no effect on the sensitivity of the ovalbumin gene to DNase I (compare Fig. 12 D, E, and F); in each case, DNase-I treatment reduced the hybridization of cDNA$_{ov}$ about 2.5-fold as compared to cDNA$_{gb}$. Since we established conditions such that mRNA$_{ov}$ production would have ceased within 3 hours, these results suggest that the increased sensitivity of the ovalbumin gene to DNase I in oviduct tissue is not due to the high rate of mRNA$_{ov}$ production per se, but may be related to an underlying feature of chromatin structure in chick oviduct.

DISCUSSION

In the chick oviduct, the regulation of specific protein synthesis is primarily at the level of mRNA production. For example, the relative rate of ovalbumin synthesis, i.e., percentage of total protein synthesis, is directly proportional to ovalbumin mRNA levels measured either by translation in a cell-free system or by hybridization. Furthermore, all the translatable mRNA$_{ov}$ is localized in polysomes (Palmiter 1973). The regulation of conalbumin synthesis is also directly related to the cellular levels of mRNA$_{con}$ as measured by translation (Fig. 1) or hybridization with cDNA$_{con}$ (D. C. Lee et al., in prep.). Superimposed upon this regulation of mRNA levels

is a substantial change in translational efficiency of the tissue, but this regulation appears not to be mRNA-specific (Palmiter 1975).

We have noted several dramatic differences in the regulation of mRNA$_{ov}$ and mRNA$_{con}$ induction by steroid hormones. Many of these observations are summarized in Table 1. In addition, the following points should be made. (a) Experiments with hydroxyurea indicate that DNA synthesis is not required for induction of either mRNA. Hydroxyurea also has the curious property of shortening the lag phase with either estrogen or progesterone while not affecting the rate of mRNA production (Palmiter et al. 1976). (b) Protein synthesis can be inhibited substantially by emetine in vivo without preventing mRNA induction (Palmiter et al. 1976); however, these experiments are complicated by the fact that emetine, as well as most other inhibitors of protein synthesis, also inhibits RNA synthesis in vivo. (c) Partial induction of mRNA$_{ov}$ production by a low dose of estrogen does not change the lag phase for full activation by an optimal dose of estrogen (Palmiter et al. 1976). (d) Progesterone activity is dominant over estrogen activity; this is especially obvious with conalbumin synthesis (Figs. 8 and 9). (e) Conalbumin mRNA levels plateau much sooner than mRNA$_{ov}$ levels after estrogen administration (Figs. 1, 7, and 8); this probably reflects a shorter half-life for mRNA$_{con}$ (2–4 hr) as compared to mRNA$_{ov}$ (~24 hr).

The challenge is to explain these observations by some rational scheme. In the absence of feasible genetic manipulation, studies of kinetics, hormonal interactions, and action of inhibitors provide some of the best clues. Considering the pronounced lag, one conclusion that comes from these observations is that there must be some intermediate step(s) be-

Table 1. Summary of Parameters Related to $mRNA_{ov}$ and $mRNA_{con}$ Induction
in Chick Oviduct

	Hormone	$mRNA_{ov}$	$mRNA_{con}$
Lag before pronounced increase in rate of mRNA production	E[a]	3.1 (2.7–4.0)[b]	<0.5
	P[a]	2.0 (1.8–2.5)	~2
	E + P	2.0	~1.5
		molecules/min/tubular gland cell[c]	
Initial rate of mRNA accumulation	E	20	20
	P	20	8
	E + P	32	12
		μg/chick (~160 g body weight)	
Dose required for half-maximal stimulation at 8 hr	E	290	105
	P	105 (90–165)	60 (45–85)

[a] E, 17β-estradiol-benzoate; P, progesterone; optimal doses except in bottom panel.
[b] Range of values from several experiments.
[c] Approximate because of uncertainty in percentage tubular gland cells and nonlinearity of initial rate with $mRNA_{con}$.

tween nuclear localization of steroid receptors and activation of $mRNA_{ov}$ production. This lag phenomenon is common to many specific proteins (mRNAs) that are induced by steroid hormones (see Palmiter et al. 1976 for a dozen examples). The phenomenon is observed both in vivo and in culture. Recently, we have been able to induce $mRNA_{ov}$ and $mRNA_{con}$ in oviduct explants in culture with either estrogen or progesterone; the lag persists, indicating that it is a property of the oviduct tissue and not due to the action of a second, serum-mediated signal from some other tissue (G. S. McKnight, in prep.).

There are basically two types of models that might be invoked to explain this regulatory system. One assumes that steroid receptors induce the synthesis or activation of some intermediate(s) that directly modulates gene expression. The lag is then a function of the time necessary to produce the intermediate(s). An alternative model is that the lag represents the time required for the receptor to reach the site regulating mRNA production. In either model, one must explain why the lag is variable for different gene products and with different hormones.

There are many possible variations of models invoking essential intermediates in steroid hormone action. One of the simplest models would predict that some short-lived intermediate accumulates with typical hyperbolic kinetics in response to nuclear steroid receptors. Then, if the rate of $mRNA_{con}$ production is directly proportional to the concentration of this intermediate, conalbumin synthesis would be induced immediately. The lag in $mRNA_{ov}$ induction and the displaced dose-response relationship could be explained if the rate of $mRNA_{ov}$ production depended on a cooperative interaction of the intermediate. Although this type of model satisfactorily explains the kinetics of both $mRNA_{ov}$ and $mRNA_{con}$ induction with maximal nuclear receptor concentrations, it predicts that the lag phase should

increase with decreasing nuclear receptor concentration, a result that is inconsistent with our observations of $mRNA_{ov}$ induction (Figs. 3 and 4). One would also have to postulate that progesterone acts via a different intermediate, one that induces $mRNA_{ov}$ faster and $mRNA_{con}$ slower than the estrogen intermediate.

The alternative model is exemplified by considering the possibility that there may be some unidirectional translocation of receptors (or some receptor-activated molecule) along chromatin from an initial specific binding site to another site that directly mediates mRNA production. In this scheme, the lag phase is a function of the distance and complexity of the translocation. An intriguing alternative to the receptor moving is a receptor-mediated propagation of histone modifications that is passed from nucleosome to nucleosome and allows transcription when it reaches a certain site (Yamamoto and Alberts 1976). This general model has been discussed in some detail (Palmiter et al. 1976) and continues to serve as a viable alternative to models invoking diffusable intermediates. Although the essential concepts remain the same, our more recent observations of hormonal interactions and dose-response relationships allow a refinement of the original suggestion. In particular, the observation that progesterone interferes with estrogen-mediated $mRNA_{con}$ induction (Fig. 9) could be accommodated by postulating that progesterone receptors interfere with estrogen receptor translocation along the stretch of chromatin regulating $mRNA_{con}$ production but act in concert with estrogen receptors on the stretch of chromatin involved in $mRNA_{ov}$ regulation. Also, the observation that higher concentrations of estrogen or progesterone nuclear receptors are required for $mRNA_{ov}$ induction as compared to $mRNA_{con}$ induction (Figs. 5 and 7) could reflect either different numbers of sites or different affinities of receptors for sites regulating the production of these two mRNAs.

Our experiments with tamoxifen are consistent with a receptor translocation mechanism. We envision that receptors carrying tamoxifen cannot bind specifically, and that receptors that are in the translocation mode either no longer carry their steroid or cannot exchange their steroid for tamoxifen and therefore are immune from the inhibitory action and continue to promote mRNA production until the last one reaches the productive site. Thus, it takes as long to shut off $mRNA_{ov}$ production as it does to turn it on.

At the present time we do not have definitive evidence for one model or another; rather, we feel that it is important to have alternatives to help generate meaningful experiments. The receptor translocation hypothesis serves such a function for us and has some interesting features which, if documented, would add a new dimension to chromatin protein dynamics. Perhaps the most important point to make at this time is that the regulation of mRNA production in the chick oviduct is not easily explained by a simple transcriptional control model in which receptors bind directly at promotor sites where they act as positive effectors by increasing specific mRNA transcription.

The other aspect of this work that requires some discussion is the DNase-I sensitivity of the ovalbumin gene. We embarked on these studies with the idea that they might tell us something about receptor-mediated chromatin changes. We approached the problem by investigating the shut-off of $mRNA_{ov}$ production with tamoxifen. The results show that the sensitivity of the ovalbumin gene to DNase I remains unaltered 1 and 6 hours after tamoxifen addition, times clearly shorter and longer than the time needed to halt $mRNA_{ov}$ production (Fig. 10). The level of transcription after tamoxifen is estimated to be less than one transcript per hour. The same result is obtained by simply removing the source of estrogen and then waiting until the nuclear receptor levels have fallen to uninduced levels (J. Shepherd et al., in prep.); thus, it is unlikely that the lack of effect is due to some peculiarity of tamoxifen-induced withdrawal. These results suggest that the sensitivity of the ovalbumin gene to DNase I is a function of some relatively stable aspect of chromatin structure rather than of hormone-mediated transcription per se. There are probably other changes in chromatin structure that are induced by transcription and that may be detectable under different conditions or with different enzymes. Our observations are in accord with the finding of Weintraub and Groudine (1976) that the globin gene remains sensitive to DNase I in the mature, nontranscribing chick erythrocyte. There may be permanent chromatin modifications set up during embryogenesis that correspond to the developmental phenomenon called cell determination. This chromatin modification may explain why the ovalbumin gene is not readily transcribed in the chick liver even though this tissue is a target organ for estrogen and has estrogen receptors. An important, but unanswered, question is whether this nuclease sensitivity is due to covalently modified nucleosomes or to other molecules, such as nonhistone proteins, which may be associated with specific genes in such a way as to expose the DNA to DNase I.

Acknowledgments

This work was supported by a grant (HD-9172) from the National Institutes of Health. R.D.P. is an investigator of the Howard Hughes Medical Institute. We thank Dr. W. C. Lesky of ICI United States, Inc., for a generous gift of tamoxifen.

REFERENCES

Cox, R. F., M. E. Haines, and S. Emtage. 1974. Quantitation of ovalbumin mRNA in hen and chick oviduct by hybridization to complementary DNA. Accumulation of specific mRNA in response to estradiol. *Eur. J. Biochem.* **49**: 225.

Garel, A. and R. Axel. 1976. Selective digestion of transcriptionally active ovalbumin genes from oviduct nuclei. *Proc. Natl. Acad. Sci.* **73**: 3966.

McKnight, G. S. and R. T. Schimke. 1974. Ovalbumin messenger RNA: Evidence that the initial product of transcription is the same size as polysomal ovalbumin messenger. *Proc. Natl. Acad. Sci.* **71**: 4327.

McKnight, G. S., P. Pennequin, and R. T. Schimke. 1975. Induction of ovalbumin mRNA sequences by estrogen and progesterone in chick oviduct as measured by hybridization to complementary DNA. *J. Biol. Chem.* **250**: 8105.

Mulvihill, E. R. and R. D. Palmiter. 1977. Relationship of nuclear estrogen receptor levels to induction of ovalbumin and conalbumin mRNA in chick oviduct. *J. Biol. Chem.* **252**: 2060.

Palmiter, R. D. 1972. Regulation of protein synthesis in chick oviduct. I. Independent regulation of ovalbumin, conalbumin, ovomucoid, and lysozyme induction. *J. Biol. Chem.* **247**: 6450.

———. 1973. Rate of ovalbumin messenger ribonucleic acid synthesis in the oviduct of estrogen-primed chicks. *J. Biol. Chem.* **248**: 8260.

———. 1975. Quantitation of parameters that determine the rate of ovalbumin synthesis. *Cell* **4**: 189.

Palmiter, R. D., P. B. Moore, E. R. Mulvihill, and S. Emtage. 1976. A significant lag in the induction of ovalbumin mRNA by steroid hormones: A receptor translocation hypothesis. *Cell* **8**: 557.

Sutherland, R., J. Mešter, and E.-E. Baulieu. 1977. Tamoxifen is a potent "pure" anti-oestrogen in chick oviduct. *Nature* **267**: 434.

Weintraub, H. and M. Groudine. 1976. Chromosomal subunits in active genes have an altered conformation. *Science* **193**: 848.

Yamamoto, K. R. and B. M. Alberts. 1976. Steroid receptors: Elements for modulation of eukaryotic transcription. *Annu. Rev. Biochem.* **45**: 721.

Amplification of Dihydrofolate Reductase Genes in Methotrexate-resistant Cultured Mouse Cells

R. T. SCHIMKE, F. W. ALT, R. E. KELLEMS, R. J. KAUFMAN, AND J. R. BERTINO*

Department of Biological Sciences, Stanford University, Stanford, California 94305

Our laboratory has been investigating the mechanism whereby cultured cells acquire resistance to the four amino analogs of folic acid, a phenomenon often associated both in cultured cells and in human neoplasms (Bertino et al. 1963) with elevated levels of dihydrofolate reductase activity. Resistant cell variants containing up to a 300-fold increase in dihydrofolate reductase activity can be obtained by growth of cells in progressively increasing concentrations of methotrexate (Hakala et al. 1961; Fisher 1961; Friedkin et al. 1962). Previous reports from this laboratory, using a resistant cell line (AT-3000) and several clones (R-1 to R-4) obtained from the AT-3000 line, all of which have been derived from a mouse sarcoma S-180 cell line, have demonstrated that the increase in dihydrofolate reductase activity results solely from an increased rate of enzyme synthesis (Alt et al. 1976), and that there is a proportionality between the rate of enzyme synthesis and the content of specific mRNA as assayed in a rabbit reticulocyte translation system (Kellems et al. 1976). Littlefield and his colleagues, employing a cell line derived from baby hamster kidney cells, have come to similar conclusions (Hanggi and Littlefield 1976; Chang and Littlefield 1976). One interesting difference between the cell lines we have studied and the cell line studied by Littlefield concerns the stability of the high enzyme level when cells are grown in methotrexate-free medium. Our results indicate that the property of high enzyme levels (and mRNA content) is unstable, whereas the cell line studied by Littlefield's group maintains high enzyme levels in the absence of methotrexate.

We summarize here studies in which cDNA complementary to the dihydrofolate reductase mRNA was employed to quantitate the number of mRNA sequences and the number of gene copies. These experiments were done with the AT-3000 cell line (high resistance), which reverts to sensitivity upon growth in methotrexate-free medium, as well as with a mouse lymphoma-derived resistant cell line, L1210, in which the properties of high resistance and high folate reductase levels are retained in the absence of methotrexate. In both the stable and unstable cell lines, we find a direct correlation between the content of enzyme and the relative amount of folate reductase mRNA. More importantly, we also find that this correlation extends to the number of gene copies. Thus, we conclude that the level of dihydrofolate reductase activity is directly proportional to the number of gene copies, i.e., amplification, and that in some cell lines the genes are subject to rapid loss, whereas in other cell lines they are fixed in the genome.

MATERIALS AND METHODS

Descriptions of the cells and culture methods have been presented elsewhere (Alt et al. 1976; Kellems et al. 1976). The figure legends provide certain details of experimental procedures, and extensive descriptions and corroborating data will be published elsewhere (F. W. Alt et al., in prep.).

Total poly(A) RNA was prepared by oligo(dT)-cellulose chromatography of total cytoplasmic RNA (F. W. Alt et al., in prep.).

cDNA:RNA hybridizations. Hybridizations were in 20 mM Tris-Cl, pH 7.7, 2 mM EDTA, and 0.2% sodium dodecyl sulfate. Reactions were overlaid with mineral oil in plastic tubes and incubated at 68°C. Approximately 50 pg of ^{3}H-labeled cDNA (500 cpm) was used in each reaction. Final $R_0 t$ values were corrected to standard conditions. At the end of the incubation, reactions were diluted to 1 ml with buffer containing 30 mM Na acetate, pH 4.5, 3 mM $ZnSO_4$, 300 mM NaCl, and 10 μg denatured calf thymus DNA. These samples were divided in two, and one was digested for 30 minutes at 45°C with 1000 units/ml of S_1 nuclease. After digestion, 100 μg of carrier calf thymus DNA was added to both tubes and nucleic acid precipitated with 10% trichloroacetic acid, 1% Na pyrophosphate at 4°C for 15 minutes. Precipitates were collected on Millipore filters, washed three times with 5% trichloroacetic acid, dried, and counted. Hybrid formation was scored as the amount of precipitable radioactivity remaining after S_1 digestion and expressed as a percentage of the undigested control. The cDNA was 1.5% S_1-resistant. In calculating $R_0 t$ values, we assumed an average value of 344 g of RNA nucleotides per mole.

DNA preparation. The 27,000g pellets (containing nuclei) resulting from standard RNA preparations

* Permanent address: Departments of Pharmacology and Medicine, Yale University, New Haven, Connecticut 06510.

(Kellems et al. 1976) were stored at −20°C. Approximately 5 ml of frozen nuclear pellet was thawed and gently homogenized in 50 ml of 0.15 M NaCl, 0.1 M EDTA, pH 8.0, 0.6 M sodium perchlorate, and 1% SDS by five strokes in a Dounce homogenizer (loose pestile). The homogenate was stirred slowly at 25°C for 30 minutes, extracted with two volumes of chloroform, and the DNA was spooled from the aqueous phase after the addition of two volumes of ice-cold ethanol.

Spooled DNA was dissolved in 10 mM Tris-Cl, pH 7.4, and then treated with 60 µg/ml pancreatic ribonuclease (boiled for 10 minutes in 20 mM NaCl prior to use) for 2 hours at 37°C. SDS and proteinase K were then added to a final concentration of 0.2% and 60 µg/ml, respectively, and the incubation continued for another 5 hours at 37°C. At this time, the solution was extracted with two volumes of phenol and the aqueous phase precipitated overnight at −20°C with two volumes of ethanol. Precipitated DNA was pelleted by centrifugation for 5 minutes at 2000g, lyophilized, and dissolved in 100 mM Na acetate (pH 7.8). DNA was then sheared to a uniform size by passing through a needle valve of a French pressure cell at a pressure of 20,000 psi. Divalent cations were removed by passing the sheared DNA preparations over a small (10 ml) column of Chelex (equilibrated with 100 mM Na acetate, pH 7.8), and the DNA was subsequently ethanol-precipitated as described above and redissolved in 20 mM Tris-Cl, pH 7.4, and 1 mM EDTA. Then 1 M NaOH was added to a final concentration of 0.3 M and the solution incubated for 22 hours at 37°C, at which time the base was neutralized by the addition of an equivalent amount of 1 N HCl.

These preparations were then stored at 4°C until subsequent use as described below. All of the DNA samples prepared in this fashion sedimented as symmetrical peaks on isokinetic alkaline sucrose gradients with a calculated size of approximately 450 base pairs.

DNA/DNA hybridization. DNA/DNA associations were done in reaction mixtures containing 25 mM Tris-Cl, pH 7.4, 1 mM EDTA, 300 mM NaCl, 50 pg ^{3}H-labeled cDNA (500 cpm), and 500 µg of cellular DNA (prepared as described above) in a final volume of from 0.05 ml to 1.1 ml. Reaction mixtures were overlaid with mineral oil in plastic tubes, heated to 102°C for 10 minutes in an H$_2$O/ethylene glycol bath, cooled, and incubated at 68°C for various times in order to achieve the desired C_0t values.

To generate C_0t curves of total cell DNA, single- and double-stranded DNA was fractionated by chromatography at 60°C on hydroxylapatite. Reaction mixtures were diluted into 5 ml of 0.12 M NaPO$_4$ (pH 6.8) and passed over a column containing 1 g of hydroxylapatite (boiled for 5 minutes in 5 ml of 0.12 M NaPO$_4$ prior to use and equilibrated in the same buffer). Single-stranded DNA was eluted with

0.12 M NaPO$_4$ (pH 6.8) and double-stranded material subsequently eluted with 0.5 M NaPO$_4$ (pH 6.8). The single- and double-stranded fractions were monitored for A$_{260}$, and the DNA was then precipitated by the addition of carrier calf thymus DNA to 25 µg/ml and 0.1 volume of 100% trichloroacetic acid (TCA). TCA-precipitated material was collected and counted as described above. To calculate DNA concentration, an A$_{260}$ reading of one was assumed to correspond to DNA concentrations of 43 µg/ml and 50 µg/ml, respectively, for single- and double-stranded DNA fractions. In calculating C_0t values, we assumed an average value of 332 g of DNA nucleotides per mole.

RESULTS

Figure 1 compares the pattern of pulse-labeled soluble protein in S-180 (sensitive) and AT-3000 (resistant) cells in which there is approximately a 200-fold difference in dihydrofolate reductase enzyme activity (Alt et al. 1976). As displayed by electrophoresis on SDS-acrylamide gels, there is no difference

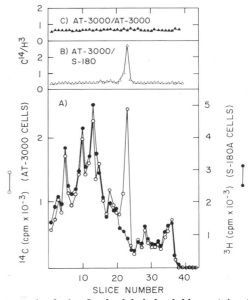

Figure 1. Analysis of pulse-labeled soluble proteins from sensitive and resistant cells. *(A)* Logarithmically growing cultures of S-180 and AT-3000 cells were pulse-labeled with [^{3}H]leucine (6 µCi/ml) and [^{14}C]leucine (2.5 µCi/ml), respectively. Cells were mixed and processed, and 100-µl aliquots of the cellular extract were precipitated with 10% trichloroacetic acid. The precipitates were dissolved and electrophoresed on SDS-polyacrylamide gels. Radioactivity in gels was determined separately for ^{3}H (●) or ^{14}C (○). *(B)* Each *point* gives the ratio of ^{14}C to ^{3}H for the corresponding point on the gel given in *A*. *(C)* Duplicate cultures of AT-3000 cells were pulse-labeled with either [^{3}H]leucine (6 µCi/ml) or [^{14}C]leucine (2.5 µCi/ml). The cells were mixed and processed, and aliquots of the extract were electrophoresed as described above. Each point gives the ratio of ^{14}C to ^{3}H in that gel slice. Details are given in Alt et al. (1976).

in the patterns of labeled proteins except for the prominent peak of ^{14}C-label (resistant cells), which we have shown in other experiments comigrates with dihydrofolate reductase. We conclude that the only detectable difference in the pattern of synthesis of proteins between the two cell lines can be ascribed to increased synthesis of the enzyme.

That the property of high enzyme synthesis is not stable in the AT-3000 cell line is shown in Figure 2, where both the parental line (AT-3000) and four separate clones (R-1, R-2, R-3, R-4) were grown in the absence of methotrexate. The high rate of enzyme synthesis was maintained for approximately ten cell doublings and then declined rapidly through succeeding cell doublings. This phenomenon is distinctly different from that observed with the mouse lymphoma L1210 line, as well as from that with the baby hamster cell line studied by Nakamura and Littlefield (1972), where resistance is stable in the absence of the drug.

We have observed a variety of parameters which alter dihydrofolate reductase levels, including the different clones of the AT-3000 line used, the duration of cell growth in the absence of methotrexate (Fig. 2), and the growth phase of the culture (Alt et al. 1976). When cytoplasmic RNA is extracted from such cells and quantitated for ability to code for dihydrofolate reductase synthesis in a rabbit reticulocyte lysate, we find a good correlation between the proportional rate of enzyme synthesis (percent of radioactivity incorporated into immunoprecipitable dihydrofolate reductase relative to incorporation into total soluble protein) and the amount of translatable mRNA (Fig. 3). Thus, by this relatively crude assay, we conclude that cells synthesize more enzyme because they contain more specific mRNA activity.

To obtain more accurate quantitation of the relative number of specific mRNA sequences, as well as to determine the number of gene copies, we have purified DNA complementary to dihydrofolate reductase mRNA. The mRNA for dihydrofolate reductase contains poly(A) sequences, but its size distribution relative to total poly(A) RNA is not sufficiently distinct to allow for meaningful purifications by size fractionation. Therefore we have purified dihydrofolate reductase cDNA by a combination of the two techniques outlined in Figure 4. We undertook an initial enrichment of the mRNA by means of immunoprecipitation of the specific polysomes (Shapiro et al. 1974). We have estimated that dihydrofolate reductase constitutes approximately 3–4% of total protein synthesis in these cells, and that immunoprecipitation has enriched for specific mRNA between five- and tenfold as assayed in a translation-immunoprecipitation assay (Kellems et al. 1976). Therefore, our conservative estimate is that the dihydrofolate reductase mRNA constitutes approximately 20–30% of the mRNA population in the immunoprecipitated poly(A) RNA. Clearly, this mRNA is not sufficiently purified to obtain specific cDNA employing this method only.

We have made use of the facts that there is more dihydrofolate reductase mRNA in resistant cells (Fig. 3) and that it would appear that only dihydrofolate reductase synthesis is vastly increased in resistant cells (Fig. 1) to remove the contaminating cDNA species by hybridization of the cDNA generated from

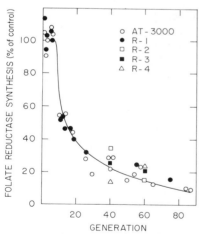

Figure 2. Effect of methotrexate on folate reductase synthesis. Resistant cells (AT-3000, R-1, R-2, R-3, and R-4) were split in duplicate into flasks containing medium with and without 50 μM methotrexate. At various cell doublings thereafter, mid-log-phase cells were pulse-labeled with [3H]leucine and the relative rate of isotope incorporation into immunoprecipitable folate reductase was measured. Each *point* represents the relative rate of folate reductase synthesis in cells grown in the absence of methotrexate, expressed as a percent of the corresponding value for control cells grown in the presence of methotrexate. Details are given in Alt et al. (1976).

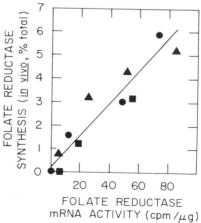

Figure 3. The relative rate of folate reductase synthesis in cells as a function of folate reductase mRNA activity. The data from several independent experiments are gathered in this figure to illustrate the relationship between folate reductase mRNA activity and in vivo folate reductase synthesis. Details concerning the measurement of folate reductase synthesis and mRNA activity are given in Kellems et al. (1976). (●) AT-3000 cells ± methotrexate; (■) R-1 cells ± methotrexate; (▲) AT-3000 cells, growth phase.

Step A

1. Poly(A)-containing RNA from resistant cells (enriched for dihydrofolate reductase sequences)

$$\downarrow$$

[³H]cDNA

$$\downarrow$$

2. Hybridized to 30-fold mass excess of poly(A)-containing RNA from sensitive cells (final R_0t = 1600 moles · sec/liter)

$$\downarrow$$

3. Nonhybridized [³H]cDNA recovered by hydroxylapatite chromatography

Step B

1. [³H]cDNA selected by step A hybridized to large excess of resistant cell poly(A)-containing RNA (final R_0t = 0.8 moles · sec/liter)

$$\downarrow$$

2. Hybridized [³H]cDNA recovered by hydroxylapatite chromatography

Figure 4. Purification of cDNA sequences complementary to dihydrofolate reductase mRNA.

the immunoprecipitated poly(A) RNA to a 40-fold (mass) excess of poly(A) RNA from sensitive cells. Basically, by hybridizing the cDNA to a limited excess of mRNA from sensitive cells and subsequently selecting, by chromatography on hydroxylapatite, the cDNA that did not hybridize, we remove from the cDNA preparation those cDNA sequences which are of similar abundance in mRNA from both sensitive and resistant cells. Second, we have hybridized the resulting cDNA preparation to poly(A) RNA from the resistant cells and selected those cDNA sequences whose rate of hybridization is very rapid, i.e., those mRNA sequences present in high abundance in the resistant cells. Details of the purification will be provided elsewhere (F. W. Alt et al., in prep.).

Figure 5a shows the hybridization kinetics of the final cDNA product (see Fig. 4) with poly(A) RNA from sensitive, resistant (200-fold increase in enzyme relative to sensitive cells), and resistant cells grown in the absence of methotrexate for 400 cell doublings, designated as revertant 400 (tenfold increase in enzyme over sensitive cells) (see Fig. 7 for methotrexate resistance). Three points are of note: (1) The hybridization approaches 100%. (2) The kinetics of the reactions are essentially defined by a single pseudo-first-order reaction characteristic of the interaction of a single reacting species of cDNA with a single reacting species of RNA (or a number of species of cDNA reacting with RNA present in equal concentrations). (3) The different RNA preparations contain different abundances of the RNA species reacting with the cDNA, and the $R_0t_{1/2}$ values of the reactions are consistent with the differences in dihydrofolate reductase levels (and the rate of enzyme synthesis) in the different cells. We therefore conclude that there is a direct correlation between the rate of enzyme synthesis and the abundance of mRNA complementary to the cDNA.

In studies to be published elsewhere (F. W. Alt et al., in prep.), we find that the cDNA probe hybridizes to 90%, with kinetics characteristic of unique sequence DNA, with DNA obtained from mouse liver. The melting of cDNA:DNA hybrids is sharp and identical with DNA obtained from both mouse liver and resistant cells. In addition, hybridization of the cDNA with poly(A) RNA from resistant cells that has been subjected to sucrose gradient centrifugation shows hybridization to RNA with a peak S-value of 14–15S, and this peak coincides with the RNA coding for dihydrofolate reductase synthesis in a translational assay.

Taking into account the method used to obtain the cDNA, the hybridization kinetics of Figures 5 and 6, and the experiments described in the above paragraph, we believe that the cDNA is highly specific for dihydrofolate reductase of murine origin. Further evidence supporting this conclusion will be

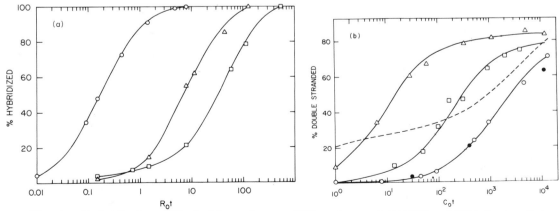

Figure 5. Kinetics of hybridization of dihydrofolate reductase cDNA probe to poly(A) RNA and DNA from methotrexate-sensitive and resistant mouse sarcoma 180 cells. The three cell lines are described in the text and their resistances to methotrexate are shown in Fig. 7. *(a)* cDNA:poly(A) RNA hybridizations; (□) sensitive, (△) revertant 400, (○) resistant. *(b)* cDNA:DNA hybridization; (○) sensitive, (□) revertant 400, (△) resistant, (●) sensitive with added poly(A) RNA from resistant cells (see text). (- - - -) C_0t curve of total DNA.

presented elsewhere (F. W. Alt et al., in prep.). The alternative, i.e., that there are a number of different mRNAs all coding for a similar molecular weight protein, that the mRNAs are present at equal frequencies in cells, and that RNA and DNA sequences complementary to the cDNA are all responsive to the methotrexate sensitivity-resistance phenomenon in the same manner in two independently derived and maintained mouse cell lines, seems highly improbable.

Figure 5b shows hybridization of the cDNA to DNA from the sensitive, resistant, and revertant cell lines of S-180, as well as the C_0t curve for total cell DNA (dotted line). The hybridizations approximate 80–90% completion with kinetics characteristic of a unique second-order reaction. Most important are the findings that there is an acceleration in the rate of hybridization when DNA from resistant and revertant cells is used and that the degree of acceleration is consistent with the amount of dihydrofolate enzyme and mRNA content in these cell lines. We conclude, therefore, that there are different numbers of dihydrofolate reductase genes, and, furthermore, that the number is directly proportional to the rate of enzyme synthesis. An important control is shown by the closed circles in Figure 5b, in which an amount of poly(A) RNA from resistant cells equivalent to that derived from the amount of cells used to obtain the DNA for each C_0t point was added to the DNA from sensitive cells prior to base hydrolysis (to remove RNA). This result shows that all of the RNA was destroyed. Hence the acceleration of the cDNA:DNA hybridization cannot be ascribed to contaminating RNA in the DNA preparations from the resistant cells.

Figure 6 a and b show comparable hybridization analyses with the mouse lymphoma-derived sensitive and resistant L1210 cell lines, in which the resistance and high enzyme levels are stable in the absence of methotrexate in the growth medium. The relative contents of mRNA and genes between sensitive and resistant cell lines are consistent with the approximately 50–100-fold difference in dihydrofolate reductase activity in these lines. Thus we find a direct relationship between the number of genes and the rates of specific protein synthesis in the L1210 lines as well as in the S-180 lines.

In the highly resistant lines of the S-180 cells, where high rates of dihydrofolate reductase synthesis decline on growth of cells in the absence of methotrexate, the question arises as to whether the loss occurs as an all-or-none phenomenon in individual cells, or whether all of the cells gradually lose high enzyme levels and resistance. We have approached this question in two ways.

Figure 7 shows a representative experiment in which we have determined the methotrexate sensitivity of four different groups of cells: the sensitive S-180 line, R-1 cells (a clone derived from AT-3000, see Fig. 2) grown in the presence of 50 μM methotrexate, the revertant 400 derivative of R-1 obtained by growth of R-1 in the absence of methotrexate for 400 cell doublings (5–10% of enzyme level of R-1), and R-1 cells grown for sufficient doublings to reduce enzyme levels to 50% of R-1 (revertant 80) and which synthesize dihydrofolate reductase at 50% of the rate of resistant cells. The sensitive, revertant 400, and revertant 80 cells have vastly different degrees of resistance to methotrexate, consistent with the differences in enzyme activity. The revertant 80 cells have a sensitivity to methotrexate that is significantly greater than that of resistant cells and which shows a killing curve of the same shape as those obtained with the other cell lines. The shape of this killing curve is to be contrasted with that obtained by the combination of equal numbers of revertant 400 and resistant cells, the aggregate of which also synthesizes enzyme at 50% the rate of the resistant cells. We conclude from this type of analysis that loss of high enzyme activity

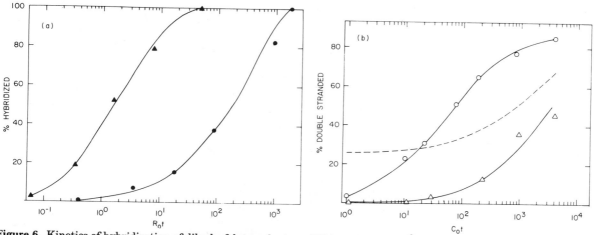

Figure 6. Kinetics of hybridization of dihydrofolate reductase cDNA probe to poly(A) RNA and DNA from methotrexate-sensitive and resistant mouse lymphoma L1210 cells. (a) cDNA:poly(A) RNA hybridizations; (▲) resistant cells, (●) sensitive cells. (b) cDNA:DNA hybridizations; (○) resistant cells, (△) sensitive cells. (- - - -) C_0t curve of total DNA.

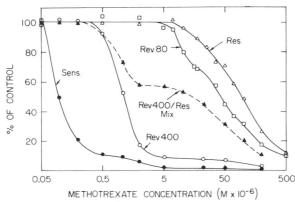

Figure 7. Methotrexate sensitivity of sensitive, resistant, and revertant cell lines. Sensitivity to methotrexate was determined by a growth assay in which 10^5 cells from log-phase cultures were inoculated into T-25 flasks containing different concentrations of methotrexate in Eagle's minimal essential media supplemented with 10% fetal calf serum and buffered with Hanks' salts. In parallel, a number of flasks were set aside in order to determine growth rates in the absence of methotrexate. All the flasks were fed with fresh media 3 days later. Under these conditions, the cells grew exponentially to a density of 3×10^6 cells/flask. When the number of cells in the cultures for growth-rate determination reached 2.2×10^6, the cells were washed 3 times with Hanks' balanced salts solution, scraped into Hanks' balanced salts, and the cell number determined using a Coulter counter. The number of cells at each concentration of methotrexate was calculated from duplicate flasks, all of which varied by less than 10%. The results were then expressed as a percentage of growth in the absence of methotrexate. The sensitive cell line had a doubling time of 16 hours, whereas all other lines had doubling times of 23 hours. The cell lines are described in the text.

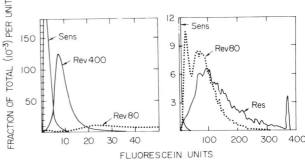

Figure 8. Fluorescence distributions of sensitive, resistant, and revertant cell lines after fluorescein-methotrexate uptake, Methotrexate was derivatized with fluorescein according to the procedure of Gapski et al. (1975) and then used to label various cell lines. Cells to be used for fluorescence analysis on the basis of fluorescein-methotrexate uptake were split into T-25 flasks and fed 4 times over 6 days with methotrexate-free medium. The late log phase cultures were then washed with 5 ml of medium, fed with 1 ml of medium containing 150 μM fluorescein-methotrexate, and incubated on a slow rocker panel at 37°C for 3.5 hr. This time and concentration were found to be optimal for saturating the uptake of the fluorescent derivative in the R-1 line. After incubation, the medium was removed, and the cells were rinsed with 5 ml of medium and incubated in 10 ml of medium for 15 min at 37°C to assure adequate washing. The cells were then gently trypsinized, suspended in 1.5 ml of medium, overlaid on 5 ml of cold isotonic sucrose, and centrifuged at 1000 rpm for 5 min in a desk-top Sorvall. After discarding the supernatant, the pellet was suspended in 2 ml of cold Hanks' balanced salts solution. The populations were analyzed using a fluorescence-activated cell sorter. The cell lines are described in the text. Note the expanded abscissa and contracted ordinate in the left panel relative to the right panel.

and resistance is a gradual process as opposed to an all-or-none phenomenon. The increase in methotrexate sensitivity in the revertant 400 cells relative to resistant cells on growth in the absence of methotrexate shows that the original cells (resistant) were not contaminated with cells resistant to methotrexate because of a stable alteration in drug transport.

We have also asked the same question by attempting to quantitate the amount of enzyme per cell. We have exploited the ability of a methotrexate-fluorescein conjugate to bind with a specificity and high affinity to dihydrofolate reductase similar to that observed with methotrexate itself (Gapski et al. 1975). Cells can be analyzed for fluorescence by the fluorescence-activated cell sorting apparatus (Herzenberg et al. 1976). Representative results are shown in Figure 8. The sensitive cells form a single, unique class of cells with essentially undetectable fluorescence (Fig. 8, left panel). The revertant 400 cells also form a discrete class of cells that is clearly different from sensitive cells. Significantly, there is no overlap between these two populations, and, in addition, the revertant 80 cells have essentially no overlap with either the sensitive or the revertant 400 cells. Thus, these represent uniquely different cell populations. The resistant cells (right panel)

would appear to be far more disperse in the amount of enzyme per cell than either the sensitive or revertant 400 populations. The revertant 80 cells display a complicated pattern: first, the average amount of enzyme per cell is less than with the resistant cells, and the shifting pattern is not one of the appearance of a unique peak of cells at the enzyme content characteristic of the revertant 400 cells (see left panel). Thus, we again conclude that the process of loss of enzyme activity is not an all-or-none process. Nevertheless, it is clear from the pattern that there is a large modality of cells with relatively low enzyme per cell. Our tentative conclusion is that the loss of enzyme activity may occur in various types of steps. We plan to resolve this question at a molecular level by using the separator mode of the analyzer to isolate discrete cell populations with differing enzyme contents.

DISCUSSION

The observation that under appropriate selection conditions the rate of synthesis of a specific protein, i.e., dihydrofolate reductase, can be altered by changes in the number of specific genes in cultured mammalian cells has, to the best of our knowledge, not been clearly documented previously. We know very little about the mechanism(s) of the process whereby this occurs, and our initial observations open up a number of experimental approaches to the problem. Modulation of the number of specific gene copies has been documented in various organisms, including the classical study of Sturtevant in 1925 involving the Bar locus in *Drosophila,* as well as more recently in various prokaryotic systems (Emmons et al. 1975; Folk and Berg 1971; Anderson et al. 1976). Amplification of the ribosomal genes in *Xenopus* during oocyte maturation is a classical example in higher organisms (Brown and Dawid 1968). That an increase in the number of specific genes can occur in such a wide spectrum of biological material suggests that it can occur in all organisms.

It is perhaps instructive to compare the alteration in gene number in other cases with our studies in cultured mammalian cells. Classically, in *E. coli,* the duplication of genes has been observed when cells have been placed under selection such that those cells with an increased number of gene copies (and gene expression) will have a growth advantage, e.g., *E. coli* with a defective aminoacyl tRNA synthetase selected for more rapid growth in limiting amino acid concentration (Folk and Berg 1971). Anderson et al. (1976) have suggested recently that in *S. typhimurium* there is a random duplication of genes, and hence the selection process will obtain those individual cells that are duplicated for the specific gene at any given time. The development of resistance to chloramphenicol in *P. mirabilis* is analogous to methotrexate resistance in cultured cells and results from an increased capacity for synthesis of chloramphenicol acetyltransferase and an increase in the number of genes specifying chloramphenicol resistance as well as other antibiotic resistances (Perlman et al. 1975).

Similarly, the presence of methotrexate in the growth medium subjects mammalian cells to selection for growth. Inhibition of dihydrofolate reductase inhibits de novo purine, thymidine, and glycine synthesis and hence selects for cells capable of increased enzyme synthesis. Selection of resistant cell lines requires a slow incremental increase in methotrexate concentration in the medium. Subjecting cells to a high drug concentration has uniformly failed to result in obtaining resistant cells. We would suggest that the progressive amplification of genes is a relatively rare event. Our preliminary estimates, as well as those of Hakala and Ishihara (1962), are that approximately one in 10^6 or 10^7 cells survive each incremental increase in methotrexate concentration, and cells must undergo a number of amplifications to attain high enzyme levels and high resistance. That the phenomenon of increased gene copies may not be unique to dihydrofolate reductase is suggested by recent studies of Stark and his colleagues (Kempe et al. 1976), who have observed the emergence of resistance to PALA, a specific inhibitor

of aspartate transcarbamylase, in cultured hamster cells. They have found that there is more enzyme activity in resistant cells. In both cases, the inhibitors are highly specific for enzymes required for cell growth. It will be interesting to determine if the same phenomenon (increased gene copies) can be observed with a variety of specific enzyme inhibitors, or whether the phenomenon is limited to certain genes. It should be noted that the inhibitor must be highly specific, since, if it inhibits several enzymes, the chances of two genes being amplified in the same cell at any specific time would be highly unlikely.

In the cases of increased gene numbers studied in *Drosophila* and prokaryotes, the phenomenon appears to be one of tandem duplication, and, in *Drosophila,* Sturtevant originally proposed that it occurred by unequal crossing over during DNA replication (Sturtevant 1925). Although we do not know if the genes in our cultured cells are tandemly duplicated, the karyology studies of Biedler and her colleagues are most interesting. They have reported both in hamster cells (Biedler and Spengler 1976) and in the L1210 (Biedler et al. 1965) cell line we have studied that there is a well-defined alteration in a specific chromosome in methotrexate-resistant cells, but not in sensitive cells. In situ hybridizations of the dihydrofolate reductase cDNA to mitotic chromosomes are currently in progress in our laboratory with both the S-180 and the L1210 cells to determine whether the chromosomal abnormality observed by Biedler's group corresponds to the dihydrofolate reductase genes, or whether the genes are dispersed at different sites within the chromosomes. If the genes appear to be localized to a specific chromosomal segment, it will be most interesting to determine by restriction fragment analysis and sequence studies of recombinant DNA clones whether the genes and their adjacent sequences are similar or not. If the former is observed, it would suggest that the amplification process occurs at specific DNA sequences. Such sequences would then have analogy with insertion sequences as observed in prokaryote cells (Cohen 1976) and would suggest a possible role in genome structure (and flexibility) in higher eukaryotes.

We have employed the term "amplification" to describe an increased number of gene copies in a context different from that employed to describe the process where there is amplification of ribosomal genes in *Xenopus* (Brown and Dawid 1968). In our case, there appears to be a selection pressure for obtaining those cells randomly in a state of specific gene duplication. We differentiate this process from that which is specifically regulated as part of a developmental sequence. It is of interest to note that the rDNA amplification in *Xenopus* occurs by amplification of a subset of rDNA sequences in a given oocyte (Bird, this volume) by a so-called rolling-circle process which results in the generation of extra-chromosomal rDNA segments (Hourcade et al. 1974). Gene amplification during a developmental sequence has not been observed in cases where large amounts of differentiated cell products are synthesized, e.g., globin (Packman et al. 1972), ovalbumin (Sullivan et al. 1973), silk fibroin (Suzuki et al. 1972), implying that regulation of these genes occurs by modulation of expression of a limited number of gene copies. However, it would appear that the rate of dihydrofolate reductase synthesis in the cultured cells we have studied is proportional to the number of specific genes, implying that little modulation of the activity of each gene has occurred.

The finding that certain cell lines "revert" to methotrexate sensitivity, whereas other lines are stable, is both interesting and puzzling. Reversion (gene loss) is a well-known phenomenon in instances of tandem duplication in prokaryotes, and a reversal of such a duplication process can be readily envisaged as a mechanism for gene loss in our cells. One can consider two general processes that would lead to "fixation" of genes in the genome. One is a secondary alteration in cells resulting in lack of an enzymatic process that "excises" genes. Perhaps more likely is an alteration in the sequence structure of the multiplied genes, either by translocation or loss of specific flanking sequences (insertion or recombinational sequences [Cohen 1976]), that prevents subsequent excision. The analogy between the generation of increased gene copies and their fixation in the genome in our cultured cells and the process of gene duplication, fixation, and random mutation selection in evolution (Ohno 1970) is striking and bears further study.

Our results add further evidence to support the emerging concept that the genome in higher organisms is not constant; it can undergo a variety of changes, including duplications, instances of translocation of the variable and constant regions of the immunoglobulin light chains (Hozumi and Tonegawa 1976), variable genomic structure in *Drosophila* (Ilyan et al.; Potter and Thomas; both this volume), fragmentation of macronuclear genomes in protozoa (Prescott et al. 1973) and ascaris (Tobler et al. 1972), and the loss of entire chromosomes in insect somatic cells (Geyer-Duszynska 1959). It is equally interesting to note that in the case of insects and protozoa where the genome disruption occurs, as well as in the vast majority of higher organisms, the germ cell line is sequestered (differentiated) early in the process of cell division and differentiation. Hence, those cells carrying the full complement of the genome and involved in sexual conjugation need not be subject to potential possible deletions, inversions, translocations, etc. While the striking alteration in the genomic structure of a limited number of species need not be characteristic of all differentiating organisms, less obvious alterations may underlie a number of controls of differentiation. Hence it would appear that a reanalysis of the con-

cept of the constancy of the genome during development (Gurdon 1968) is warranted.

Acknowledgments

These studies were supported by grants from the American Cancer Society (NP 148) and the National Cancer Institute, National Institutes of Health (CA 16318).

REFERENCES

ALT, F. W., R. E. KELLEMS, and R. T. SCHIMKE. 1976. Synthesis and degradation of folate reductase in sensitive and methotrexate-resistant lines of S-180 cells. *J. Biol. Chem.* **251**: 3063.

ANDERSON, R. P., C. G. MILLER, and J. R. ROTH. 1976. Tandem duplications of the histidine operon observed following generalized transduction in *Salmonella. J. Mol. Biol.* **105**: 201.

BERTINO, J. R., D. M. DONOHUE, B. SIMMONS, B. W. GABRIO, R. SILBER, and F. M. HUENNEKENS. 1963. The "induction" of dihydrofolate reductase activity in leukocytes and erythrocytes in patients treated with amethopterin. *J. Clin. Invest.* **42**: 466.

BIEDLER, J. L. and B. A. SPENGLER. 1976. Metaphase chromosome anomaly: Association with drug resistance and cell-specific products. *Science* **191**: 185.

BIEDLER, J. L., A. M. ALBRECHT, and D. J. HUTCHISON. 1965. Cytogenetics of mouse leukemia L1210. 1. Association of a specific chromosome with dihydrofolate reductase activity in amethopterin-treated sublines. *Cancer Res.* **25**: 246.

BROWN, D. D. and I. B. DAWID. 1968. Specific gene amplification in oocytes. *Science* **160**: 272.

CHANG, S. and J. W. LITTLEFIELD. 1976. Elevated dihydrofolate reductase mRNA levels in methotrexate-resistant BHK cells. *Cell* **7**: 391.

COHEN, S. N. 1976. Transposable genetic elements and plasmid evolution. *Nature* **263**: 731.

EMMONS, S. W., V. MACCOSHAM, and R. L. BALDWIN. 1975. Tandem genetic duplications in phage lambda. III. The frequency of duplication mutants in two derivatives of phage lambda is independent of known recombination systems. *J. Mol. Biol.* **91**: 133.

FISHER, G. A. 1961. Increased levels of folic acid reductase as a mechanism of resistance to amethopterin in leukemic cells. *Biochem. Pharmacol.* **7**: 75.

FOLK, W. R. and P. BERG. 1971. Duplication of the structural gene for glycyltransfer RNA synthetase in *Escherichia coli. J. Mol. Biol.* **58**: 595.

FRIEDKIN, K., E. S. CRAWFORD, S. R. HUMPHREYS, and A. GOLDIN. 1962. The association of increased dihydrofolate reductase with amethopterin resistance in mouse leukemia. *Cancer Res.* **22**: 600.

GAPSKI, G. R., J. M. WHITELEY, J. I. RADER, P. L. CRAMER, G. B. HENDERSON, V. NEEF, and F. M. HUENNEKENS. 1975. Synthesis of a fluorescent derivative of amethopterin. *J. Med. Chem.* **18**: 526.

GEYER-DUSZYNSKA, I. 1959. Experimental research on chromosome elimination in *Cecidomyidae* (Diptera). *J. Exp. Zool.* **141**: 391.

GURDON, J. B. 1968. Transplanted nuclei and cell differentiation. *Sci. Am.* **219**: 24.

HAKALA, M. T. and T. ISHIHARA. 1962. Chromosomal constitution and amethopterin resistance in cultured mouse cells. *Cancer Res.* **22**: 987.

HAKALA, M. T., S. F. ZAKRGEWSKI, and C. NICHOL. 1961. Relation of folic acid reductase to amethopterin resistance in cultured mammalian cells. *J. Biol. Chem.* **236**: 952.

HANGGI, U. J. and J. W. LITTLEFIELD. 1976. Altered regulation of the rate of synthesis of dihydrofolate reductase in methotrexate-resistant hamster cells. *J. Biol. Chem.* **251**: 3075.

HERZENBERG, L. A., R. G. SWEET, and L. A. HERZENBERG. 1976. Fluorescence-activated cell sorting. *Sci. Am.* **234**: 108.

HOURCADE, D., D. DRESSLER, and J. WOLFSON. 1974. The nucleolus and the rolling circle. *Cold Spring Harbor Symp. Quant. Biol.* **38**: 537.

HOZUMI, N. and S. TONEGAWA. 1976. Evidence for somatic rearrangement of immunoglobulin genes coding for variable and constant regions. *Proc. Natl. Acad. Sci.* **73**: 3628.

KELLEMS, R. E., F. W. ALT, and R. T. SCHIMKE. 1976. Regulation of folate reductase synthesis in sensitive and methotrexate-resistant sarcoma-180 cells. *J. Biol. Chem.* **251**: 6987.

KEMPE, T. D., E. A. SWYRYD, M. BRUIST, and G. R. STARK. 1976. Stable mutants of mammalian cells that overproduce the first three enzymes of pyrimidine nucleotide biosynthesis. *Cell* **9**: 541.

NAKAMURA, H. and J. W. LITTLEFIELD. 1972. Purification, properties, and synthesis of dihydrofolate reductase from wild type and methotrexate-resistant hamster cells. *J. Biol. Chem.* **247**: 179.

OHNO, S. 1970. *Evolution by gene duplication.* Springer-Verlag, Berlin.

PACKMAN, S., H. AVIV, J. ROSS, and P. LEDER. 1972. A comparison of globin genes in duck reticulocytes and liver cells. *Biochem. Biophys. Res. Commun.* **49**: 813.

PERLMAN, D., T. M. TWOSE, M. J. HOLLAND, and R. H. ROWND. 1975. Denaturation mapping of R-factor deoxyribonucleic acid. *J. Bacteriol.* **123**: 1035.

PRESCOTT, D. M., K. G. MURTI, and C. J. BOSTOCK. 1973. Genetic apparatus of *Stylonychia* sp. *Nature* **242**: 576.

SHAPIRO, D. J., J. M. TAYLOR, G. S. MCKNIGHT, R. PALACIOS, C. GONZALEZ, M. L. KIELY, and R. T. SCHIMKE. 1974. Isolation of hen oviduct and rat liver albumin polysomes by indirect immunoprecipitation. *J. Biol. Chem.* **249**: 3665.

STURTEVANT, A. H. 1925. The effects of unequal crossing over at the Bar locus in *Drosophila. Genetics* **10**: 117.

SULLIVAN, D., R. PALACIOS, J. STAVNEZER, J. M. TAYLOR, A. J. FARAS, M. L. KIELY, N. M. SUMMERS, J. M. BISHOP, and R. T. SCHIMKE. 1973. Synthesis of a deoxyribonucleic acid sequence complementary to ovalbumin messenger ribonucleic acid and quantification of ovalbumin genes. *J. Biol. Chem.* **248**: 7530.

SUZUKI, Y., L. P. GAGE, and D. D. BROWN. 1972. The genes for silk fibroin in *Bombyx mori. J. Mol. Biol.* **70**: 637.

TOBLER, H., K. D. SMITH, and H. URSPRUNG. 1972. Molecular aspects of chromatin elimination in *Ascaris lumbricoides. Dev. Biol.* **27**: 190.

Hormonal Modulation of α 2u Globulin mRNA: Sequence Measurements Using a Specific cDNA Probe

P. Feigelson and D. T. Kurtz

The Institute of Cancer Research and the Department of Biochemistry, College of Physicians and Surgeons, Columbia University, New York, New York 10032

α 2u Globulin is a protein of m.w. 20,000, which is synthesized in the liver of mature male rats, secreted into the serum, and excreted in the urine. This protein represents approximately 1% of hepatic protein synthesis in an adult male rat, and no detectable α 2u globulin synthesis occurs in females (Sippel et al. 1976). Our interest in this protein is based on the fact that its rate of hepatic biosynthesis is under multihormonal control: androgens, glucocorticoids, thyroid hormones, and pituitary growth hormone are necessary to maintain normal levels of α 2u globulin synthesis in male rats, and estrogens repress the synthesis of this protein (Sippel et al. 1975; Kurtz et al. 1976a,b). α 2u Globulin can be induced in female rats by ovariectomy and androgen administration (Sippel et al. 1975). Using a wheat-germ cell-free translation system, we have quantitated the level of translationally functional α 2u globulin mRNA contained in hepatic mRNA from rats in various endocrine states. For all of the endocrine states studied, it was found that the level of functional α 2u globulin mRNA paralleled the rate of synthesis of the protein (Sippel et al. 1975; Kurtz et al. 1976a,b) (see, e.g., Fig. 1). However, these studies provide no information as to whether the hormonal modulation of functional α 2u globulin mRNA is due to transcriptional control or to the conversion of nonfunctional gene transcripts to active messengers by modulation of RNA processing and transport mechanisms. Distinguishing between these alternatives requires the preparation of a labeled complementary DNA (cDNA) specific for α 2u globulin sequences. This cDNA probe can be used to measure genomic, nuclear, and cytoplasmic α 2u globulin sequences.

Preparation of α 2u Globulin cDNA

We have developed the following procedure for the preparation of a pure α 2u globulin cDNA.

1. Partial purification of α 2u globulin-synthesizing polysomes by incubation of male liver polysomes with rabbit anti-α 2u globulin, which binds to the nascent α 2u globulin chains. These polysome-antibody complexes are then adsorbed to goat anti-rabbit IgG which has been immobilized on *p*-aminobenzyl cellulose (Schutz et al. 1977).

2. Isolation of the poly(A)-containing mRNA from these α 2u globulin-enriched polysomes. This mRNA is now found to contain 30–35% α 2u globulin mRNA (Fig. 2). An α 2u globulin-enriched cDNA is then synthesized to this mRNA, using reverse transcriptase.

3. Removal of non-α 2u globulin cDNA species by preparative hybridization of the enriched cDNA to a R_0t of 1000 vs female liver mRNA, which, though it shares the vast majority of mRNA sequences in common with male liver, contains no α 2u globulin mRNA. The cDNA remaining unhybridized to female mRNA is collected using hydroxylapatite.

4. Hybridization of this cDNA to male mRNA to

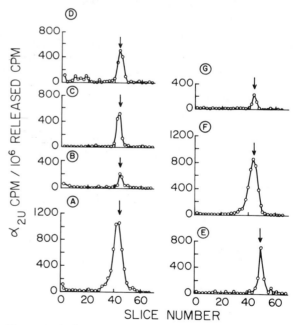

Figure 1. Androgenic control of functional α 2u globulin mRNA. Hepatic mRNA was translated in a wheat-germ cell-free system and the amount of α 2u globulin synthesized in vitro was quantitated by immunoprecipitation and SDS-polyacrylamide gel electrophoresis of the solubilized immunoprecipitate. α 2u Globulin synthesized by hepatic mRNA from *(A)* control males, *(B)* castrated males, and castrated males treated with dihydrotestosterone for *(C)* 1 day, *(D)* 2 days, *(E)* 4 days, *(F)* 8 days, and *(G)* 8 days mock injected. Arrows mark position of authentic α 2u globulin.

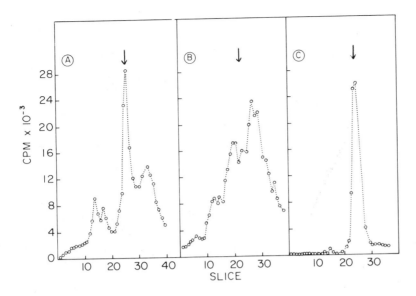

Figure 2. Cell-free translation of α 2u globulin-enriched mRNA. *(A)* Total released proteins synthesized in the wheat-germ system in response to 5 μg of α 2u globulin-enriched mRNA. *(B)* Total released proteins synthesized in response to 5 μg of unfractionated male liver mRNA. *(C)* Immunoprecipitated α 2u globulin synthesized by 5 μg of α 2u globulin-enriched mRNA. Arrows mark position of authentic α 2u globulin.

a $R_0 t$ of 10 and isolation of the hybridized cDNA to remove the short, nonhybridizable cDNA strands.

We have characterized the cDNA obtained using this procedure to show that it is indeed specific for α 2u globulin sequences.

This cDNA hybridizes with a single transition to male liver mRNA with a $R_0 t_{1/2}$ of 0.2, which indicates that the driver mRNA species represents approximately 1% of total male liver mRNA. This cDNA does not hybridize, by a $R_0 t$ of 400, to mRNA from female liver, or from male kidney or spleen, tissues

which synthesize no α 2u globulin (Fig. 3). The cDNA hybridizes to the α 2u globulin-enriched mRNA with a $R_0 t_{1/2}$ of 0.006, indicating that the hybridizing mRNA species is 33-fold purified in this enriched mRNA preparation, relative to male liver mRNA. As outlined above, translationally functional α 2u globulin mRNA can be induced in female rats by ovariectomy and androgen administration. The α 2u globulin cDNA hybridizes to hepatic mRNA from an androgen-treated castrated female with a $R_0 t_{1/2}$ of 1.5 (Fig. 3), indicating that the concentration of the hybridizing mRNA species in this mRNA sample is approximately 13% of the level found in normal

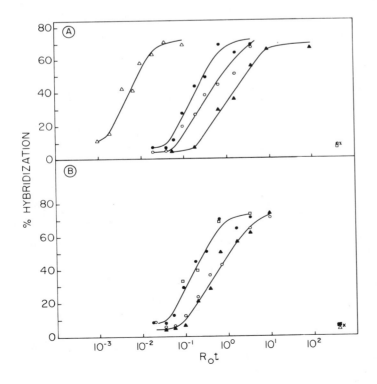

Figure 3. Hybridization of hepatic mRNAS to α 2u globulin cDNA. The α 2u globulin cDNA was hybridized in 0.4 M NaCl at 69°C under paraffin oil (Axel et al. 1976). *(A)* Hybridization to (●) male liver mRNA; (◐) female liver mRNA; (□) kidney mRNA; (x) spleen mRNA; (○) castrated male liver mRNA; (▲) liver mRNA from an ovariectomized female treated with androgens; (△) α 2u globulin-enriched mRNA. *(B)* Hybridization to (●) male liver mRNA, (■) thyroidectomized male liver mRNA (□) liver mRNA from a thyroidectomized male treated with thyroxine for 10 days; (△) liver mRNA from a prepubescent 20-day-old male; (▲) liver mRNA from a 45-day-old male; (○) liver mRNA from an adult male treated with estradiol-17β for 4 days; (x) hepatoma mRNA.

Table 1. α 2u Globulin Translational Activity vs Kinetics of Hybridization to α 2u Globulin cDNA

mRNA	Activity[a]	$R_0 t_{1/2}$[b]
Adult male liver	100	0.2
Female liver	0	>400
Ovariectomized female + 14 days androgens	12	1.5
Male kidney, spleen	0	>400
Thyroidectomized male	0	>400
Thyroidectomized male + 10 days thyroxine	110	0.2
Castrated male	30	0.4
Adult male + 4 days estradiol-17β	45	0.6
20-Day-old male	0	>400
45-Day-old male	35	0.6
Hepatoma 252	0	>400
α 2u Globulin-enriched mRNA	3000	0.006

[a] α 2u Globulin synthesis in vitro in the wheat-germ system was determined as described in the text. Values given are percent of adult male.

[b] $R_0 t_{1/2}$ determined from the hybridization curves shown in Fig. 3.

male hepatic mRNA. This is in good agreement with the level found translationally (Table 1). Our cDNA is thus specific for an mRNA sequence that is not present in normal female livers but which can be induced in females by castration and androgen treatment, characteristics expected for α 2u globulin cDNA.

If a fixed amount of the α 2u globulin cDNA is titrated with increasing amounts of male liver mRNA in a cDNA-excess hybridization, the initial slope, extrapolated to 100% hybridization, indicates that 0.1 ng of the cDNA would be protected from nuclease digestion by approximately 11.5 ng of male liver mRNA (Fig. 4), again indicating that the hybridizing mRNA species represents approximately 1% of the total mRNA population.

To demonstrate the specificity of our cDNA probe further, total male liver mRNA was fractionated on a sucrose gradient. The gradient RNA fractions were assayed for α 2u globulin translational activity in the wheat-germ system and were then used in cDNA-excess hybridizations vs the α 2u globulin cDNA. Protection of the cDNA from nuclease digestion was found only with those gradient fractions which contained α 2u globulin translational activity (Fig. 4). This correlation between hybridization and translational activity, and the absence of hybridization elsewhere in the gradient, indicates that our cDNA probe is specific for α 2u globulin sequences.

Hormonal Modulation of α 2u Globulin mRNA Sequences

The α 2u globulin cDNA was then used in RNA-driven hybridizations vs hepatic mRNA samples

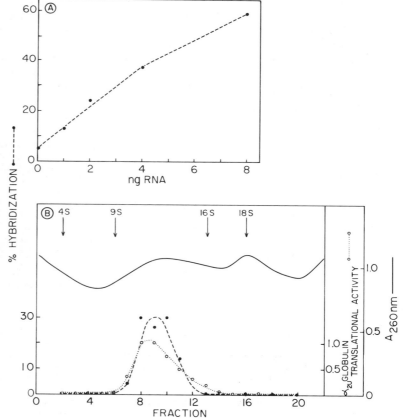

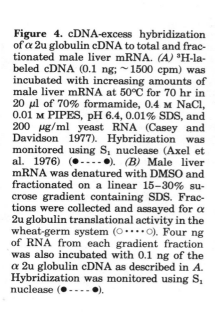

Figure 4. cDNA-excess hybridization of α 2u globulin cDNA to total and fractionated male liver mRNA. (A) ³H-labeled cDNA (0.1 ng; ~1500 cpm) was incubated with increasing amounts of male liver mRNA at 50°C for 70 hr in 20 μl of 70% formamide, 0.4 M NaCl, 0.01 M PIPES, pH 6.4, 0.01% SDS, and 200 μg/ml yeast RNA (Casey and Davidson 1977). Hybridization was monitored using S₁ nuclease (Axel et al. 1976) (●----●). (B) Male liver mRNA was denatured with DMSO and fractionated on a linear 15–30% sucrose gradient containing SDS. Fractions were collected and assayed for α 2u globulin translational activity in the wheat-germ system (○····○). Four ng of RNA from each gradient fraction was also incubated with 0.1 ng of the α 2u globulin cDNA as described in A. Hybridization was monitored using S₁ nuclease (●----●).

from rats in various endocrine and developmental states. In each case, it was found that the level of α 2u globulin mRNA sequences, as measured by the $R_0 t_{1/2}$ for hybridization to the α 2u globulin cDNA, paralleled the level of translatable α 2u globulin mRNA, as assayed in the wheat-germ system, which in turn paralleled the rate of synthesis of the protein (Table 1, Fig. 3). This correlation between the rate of synthesis of α 2u globulin and the level of its corresponding mRNA indicates that translational control is not a major factor in the hormonal modulation of the synthesis of this protein. The endocrine control is thus pretranslational and may be transcriptional.

α 2u Globulin Synthesis in Hepatomas

Several Morris minimal deviation hepatomas which we have tested synthesize no detectable α 2u globulin, whereas host liver synthesizes this protein at a normal rate (Sippel et al. 1976). We have shown that these hepatomas contain no translationally functional α 2u globulin mRNA, as assayed in Krebs II ascites or wheat-germ cell-free translational systems. Host liver contains a normal level of functional α 2u globulin mRNA. The α 2u globulin cDNA was used in RNA-excess hybridization vs hepatoma poly(A)-containing RNA. No hybridization was found by $R_0 t$ of 400, indicating that there is less than one copy of α 2u globulin mRNA per cell in this hepatoma.

One mechanism which would lead to the absence of α 2u globulin mRNA in these hepatomas is the deletion of the α 2u globulin gene during or subsequent to the transformation process. To test this possibility, total male liver and hepatoma DNA were sonicated to an average length of 400 nucleotides, denatured, and reannealed with the addition of α

2u globulin cDNA tracer. The cDNA tracer hybridized with identical kinetics to the liver and hepatoma DNA (Fig. 5) with a $C_0 t_{1/2}$ of ~ 800, indicating that the α 2u globulin gene is a single-copy sequence in both DNA samples. Thus, the α 2u globulin gene has not been deleted in this minimal deviation hepatoma but has become transcriptionally silent as a result of the transformation process.

DISCUSSION

Studies on hormonal modulation of the biosynthesis of specific proteins have been greatly facilitated by the availability of cDNA probes specific for the mRNAs coding for these proteins (McKnight et al. 1975; Ringold et al. 1977; Ryffel et al. 1977). These studies have dealt with viral mRNA species or mRNAs which represent a large percentage of the total mRNA in the tissue from which they originate. We have developed a procedure for the preparation of a cDNA specific for the mRNA coding for α 2u globulin, a male rat liver protein under multihormonal control, which represents only 1% of hepatic protein synthesis. This cDNA probe was used to quantitate the level of α 2u globulin mRNA in response to various endocrine manipulations. It was found that the level of α 2u globulin mRNA sequences paralleled the rate of synthesis of the protein in vivo. Our previous findings, which indicated that the hormonal modulation of α 2u globulin biosynthesis occurs via modulation of the level of translationally functional mRNA, are now extended to demonstrate that this modulation is not the result of activation of nontranslatable α 2u globulin mRNA sequences in response to hormones. These findings are consistent with hormonal control of specific gene transcription.

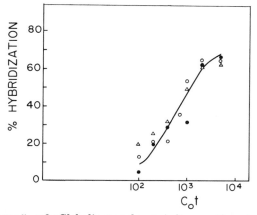

Figure 5. α 2u Globulin gene dosage in liver and hepatoma. Male liver ($\circ$) or hepatoma ($\bullet$) DNA was sonicated to an average length of 400 nucleotides, denatured, and reannealed with α 2u globulin cDNA tracer. Hybridization was monitored using S_1 nuclease. Annealing of total rat liver [^{3}H]DNA is shown for comparison ($\triangle$).

REFERENCES

AXEL, R., P. FEIGELSON, and G. SCHUTZ. 1976. Analysis of the complexity and diversity of mRNA from chicken liver and oviduct. *Cell* **7**: 247.

CASEY, J. and N. DAVIDSON. 1977. Rates of formation and thermal stabilities of RNA:DNA and DNA:DNA duplexes at high concentrations of formamide. *Nucleic Acids Res.* **4**: 1539.

KURTZ, D. T., A. E. SIPPEL, and P. FEIGELSON. 1976a. Effect of thyroid hormones on the level of the hepatic mRNA for α 2u globulin. *Biochemistry* **15**: 1031.

KURTZ, D. T., E. E. SIPPEL, R. ANSAH-YIADOM, and P. FEIGELSON. 1976b. Effect of sex hormones on the level of the messenger RNA for rat hepatic protein α 2u globulin. *J. Biol. Chem.* **252**: 3594.

McKNIGHT, G. M., P. PENNEQUIN, and R. T. SCHIMKE. 1975. Induction of ovalbumin mRNA sequences by estrogen and progesterone in chick oviduct as measured by hybridization to complementary DNA. *J. Biol. Chem.* **250**: 8105.

RINGOLD, G., R. D. CARDIFF, H. C. VARMUS, and K. YAMAMOTO. 1977. Infection of cultured rat hepatoma cells by mouse mammary tumor virus. *Cell* **10**: 11.

RYFFEL, G. U., W. WAHLI, and R. WEBER. 1977. Quantitation of vitellogenin messenger RNA in the liver of male

Xenopus toads during primary and secondary stimulation by estrogen. *Cell* 11: 213.

SCHUTZ, G., S. KIEVAL, B. GRONER, A. E. SIPPEL, D. T. KURTZ, and P. FEIGELSON. 1977. Isolation of specific messenger RNA by adsorption of polysomes to matrix-bound antibody. *Nucleic Acids Res.* 4: 71.

SIPPEL, A. E., P. FEIGELSON, and A. K. ROY. 1975. Hormonal regulation of the hepatic messenger RNA levels for α 2u globulin. *Biochemistry* 14: 825.

SIPPEL, A. E., D. T. KURTZ, H. P. MORRIS, and P. FEIGELSON. 1976. Comparison of *in vivo* translation rates and messenger RNA levels of α 2u globulin in rat liver and Morris hepatoma 5123D. *Cancer Res.* 36: 3588.

The Nucleolus, a Model for Analysis of Chromatin Controls

H. Busch, N. R. Ballal, R. K. Busch, Y. C. Choi, F. Davis, I. L. Goldknopf, S.-I. Matsui, M. S. Rao, and L. I. Rothblum

Department of Pharmacology, Baylor College of Medicine, Houston, Texas 77030

The nucleolus offers a unique opportunity for the study of chromatin function in a highly specialized system. The nucleolus is the sole source of preribosomal particles in the cell and constitutes the site of synthesis and processing of preribosomal RNA and also assembly of proteins into the preribosomal particles (Busch and Smetana 1970).

The nucleolus contains much DNA but it is the sole site of localization of the rDNA, the template for synthesis of rRNA. The simplicity of induction of biological variations in nucleolar function with drugs, hormones, and other factors that change growth rates provides a simple approach to physiological and pharmacological alteration of nucleolar activity.

Great progress has been made in the evolution of information on nucleolar components, including enzymes, preribosomal particle proteins, and nonhistone chromatin proteins. In addition, recent studies have shown that the nucleolus in situ and nucleolar chromatin both retain the capacity to produce essentially native preribosomal RNA. Thus the stage is set for much more refined studies on factors involved in control of nucleolar function.

The protein-synthesizing machinery of the cell in ¡ch the ribosomes play a key role is obviously nportance to its rate of growth (Busch 1976). availability of ribosomes is dependent upon the ivity of the nucleolus (Fig. 1), which is the sole site of synthesis of preribosomal RNA and of assembly of preribosomal particles. The high rate of protein synthesis is one reason why the attention of cancer biologists has been directed to analysis of nucleolar structure and function (Busch and Smetana 1970). An interesting feature of the nucleolus of cancer cells is that its size and shape vary markedly in relationship to the degree of malignancy, and, in the most actively growing tumors, the nucleoli are almost completely aneuploid (Caspersson 1950; Busch and Smetana 1970).

Our understanding of the many components of actively synthesizing ribosomes has increased enormously in the last decade. In addition to four RNA species, ribosomes contain almost 100 proteins that have been shown to be parts of the protein-synthesizing systems which include GTP, initiation factors, elongation factors, and termination factors.

Behind the mystery of the continued and uncontrolled growth of cancer cells are basic mechanisms that lack the sensitivity of feedback controls of nontumor cells. Whether these mechanisms constitute factors that are positive gene derepressors or factors that interfere with gene repression is still the subject of intense investigation. Both possibilities have been suggested. Possible repressors such as protein A24 and protein A11 (possibly a phosphorylated H1 histone) are markedly reduced in nucleoli of cells undergoing rapid growth and division. On the other hand, possible derepressors include nuclear antigens of tumor cells that are not present in nontumor cells. Interestingly, other nucleolar antigens have been found in nontumor cells that are not present in tumors.

There are multiple elements in the various feedback arms that may control the activity of specific chromatin elements. The task of sorting out these factors continues to be a complex one that is of importance in comprehension of the cancer problem as well as of normal growth and cell division.

Gene Controls in the Nucleolus

Much information has been developed on the structure of the nucleolus (Busch and Smetana 1970) and its function in the synthesis of ribosomes. However, little information is available on the controls of the many genes involved in the synthesis of the proteins and RNA of nucleolar rRNP particles. The nucleolus responds rapidly to cellular demands for growth products, hormonal stimuli, and toxic substances. However, the mechanisms are still unclear.

An enormous excess of DNA is associated with the nucleolus; its rDNA content is less than 0.5% of the total. The role of the other 99.5% of nucleolar DNA is unclear; since it is not read during RNA synthesis, it is "repressed" or "restricted."

Several possibilities exist for restriction of nucleolar gene readouts to rDNA. One is supercoiling (Table 1), which prevents physical interaction with RNA polymerase or factors for nucleolar readouts. Other evidence shows that "restriction" proteins specifically inhibit the readouts of DNA other than the rDNA or specifically activate rDNA readouts (Table 1). Both possibilities could also be true for rDNA, which varies in its rate of transcription in varying states of nucleolar activity.

CARCINOGENESIS   A

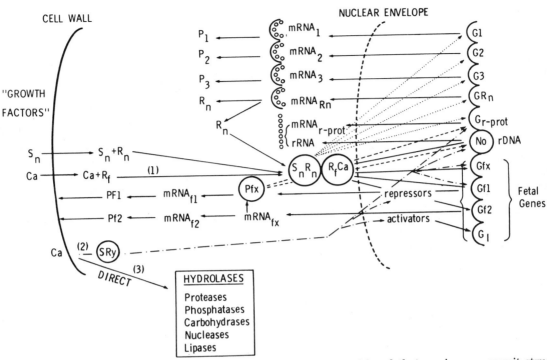

Figure 1. *(A)* Effect of carcinogenic agents on cellular responses. It is envisioned that carcinogens permit structural genes to function in the production of normal products but that, through several mechanisms, fetal genes are activated to produce a variety of fetal products, including Pf1 and Pf2, which are important to invasiveness and metastasis. The carcinogen may act with a fetal receptor to interact directly with the genome or may cause a new stimulus within the cell to interact with a receptor that will interact with the genome. Alternatively, the carcinogen may interfere with degradative reactions that are involved in normal growth controls. Ca: carcinogen; R_1–R_n: cytoplasmic receptor proteins; G1–GR_1: structural genes, including genes for receptor proteins (GR_1); Gf fetal genes with functions indicated; G_1: inhibitor genes; No: rDNA genes; and $G_{r\text{-}prot}$: genes for ribosomal proteins.

Restriction of transcription of the nucleolus-associated DNA is due to special proteins. When the nucleolus is deproteinized, the oligonucleotide fingerprints of its transcriptional products are markedly changed and the specific oligonucleotides of the fingerprint of the nucleolar readout are totally lost and replaced by random readouts. In addition, uncharacteristic fingerprints of other RNA products appear. Thus, one role of the nucleolar proteins is to specifically restrict transcription of nucleolar DNA other than rDNA.

What other mechanisms operate on expression of nucleolar function? Hildebrandt and Sauer (1977) reported a polyphosphate that inhibits nucleolar initiation of rRNA in *Physarum*. This inhibitor is selective and reversibly binds to RNA polymerase I. It is released during differentiation but not during growth. During starvation, this inhibitor increases in amount, and, after refeeding, it decreases in amount. Accordingly, control mechanisms in *Physarum* might involve transient changes in concentration of this inhibitor.

Positive controls of nucleolar function have been suggested by electron microscopic observations and labeling analyses of 45S RNA synthesis in lectin-

activated lymphocytes and other growing cells. The triggering events that result in the 15–50-fold increases in nucleolar rDNA readouts may include (1) polyamines and ornithine decarboxylase (ODC) which have been reported to enhance RNA polymerase I activity (Manen and Russell 1977), and (2) ribosomal proteins which produced a marked shift in the activity of nuclei to the synthesis of preribosomal products (Bolla et al. 1977) and tissue-specific proteins (Orrick et al. 1973).

A number of cautions must be exercised in evaluation of these results. Matsui et al. (1977) and Ballal et al. (1977) showed a huge increase in "nucleolar synthesis" when whole nucleolar DNA or *Escherichia coli* RNA polymerase was added to nucleoli. Unfortunately, a corresponding increase in the characteristic T_1 RNase oligonucleotides was not demonstrated by homochromatography. Accordingly, it is important to demonstrate fidelity of the product as well as increased labeling.

Nucleolar gene expression. Although in eukaryotes less than 10% of the total genome is expressed at a given time in any type of cell (Busch et al. 1975), the large variety of types of mRNA and other RNA

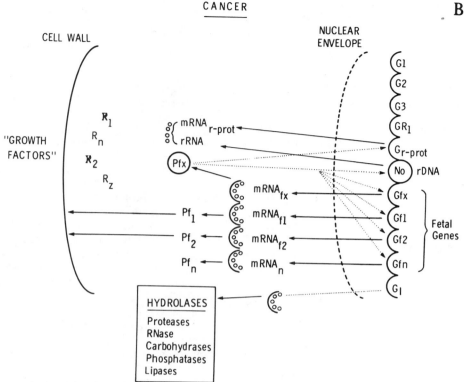

Figure 1 *(continued). (B)* Expression of cancer as a continuous production of gene products involved in growth, invasiveness, and metastasis. Such cells no longer produce R1, R2 (blocked R_1, R_2) or others that may have phenotypic specificity. It is envisioned that these gene products and their derepressors are produced or maintained in high concentration through mitosis and keep these genes activated during new cell formation. Moreover, the lack of fetal extracellular regulatory mechanisms does not permit these genes to be inactivated as they would be during fetal growth and development (Busch 1976).

species transcribed complicates the evaluation of expression of specific genes. For studies on specific gene expression, nucleoli and nucleolar chromatin have the advantage that rDNA is virtually the only active gene set in nucleoli and thus rRNA is the major product in vivo and in vitro (Busch and Smetana 1970; Blatti et al. 1971; Zylber and Penman 1971; Grummt and Lindigkeit 1973; Beebee and Butterworth 1975; Grummt 1975). An advantage of this system is that rRNA is distinguishable from many other RNA species by its high GC content. RNA polymerase I, which synthesizes rRNA, is readily isolated in high purity (Roeder and Rutter 1969,

1970; Blatti et al. 1971; Yu and Feigelson 1972; Liau et al. 1973; Chambon et al. 1974; Ferencz and Seifart 1975; Ballal and Rogachevsky 1976; Choi et al. 1976).

Nucleolar preribosomal 45S RNA (pre-rRNA) is the initial rDNA transcript (Busch and Smetana 1970), and improvements in homochromatography fingerprinting have simplified its characterization (Choi et al. 1976). This technique permits evaluation of the fidelity of transcription in vitro and analysis of the specific sequences of rRNA and nonconserved spacer regions. Also, the unique oligonucleotides resulting from complete or partial T_1 RNase digestion have been identified (Inagaki and Busch 1972; Eladari and Galibert 1975; Matsui et al. 1975; Fuke and Busch 1975, 1977; Choi et al. 1976; Ballal et al. 1977) and are absent from other RNA species (Woo et al. 1975). Isolated nucleoli synthesize RNA in vitro that resembles pre-rRNA in base composition, marker oligonucleotides, and hybridization (Yu and Feigelson 1972; Liau et al. 1973; Grummt 1975; Ferencz and Seifart 1975; Ballal and Rogachevsky 1976; Choi et al. 1976; Matsui et al. 1977). Recently, Grummt (1975) demonstrated by DNA-RNA hybridization experiments that nucleolar transcripts can be competed by 45S pre-rRNA; these experiments provide evidence that the major product of isolated nucleoli is rRNA.

Table 1. Nucleolar Control Factors

I. Chromatin structure
    A. Supercoiling
    B. Restriction proteins
II. rDNA activators
    A. Proteins
        1. preribosomal
        2. S-R complexes
        3. RNA polymerase I activation (rapid, slow)
    B. Soluble factors (polyamines, ODC)
III. Inhibitors
    A. Gene repressors (macromolecular)
    B. Polyphosphates (PPXPP analogs)

Fractionation of nucleolar proteins. Over 200 different species of nucleolar proteins could be found by two-dimensional gel electrophoresis (Orrick et al. 1973). Even in the highly purified nucleolar RNP particles, almost 100 proteins were present, including a large number of high-molecular-weight proteins which were clearly not ribosomal proteins. To fractionate the nucleolar proteins further into groups that were chemically functional, a study was undertaken of products extractable in various salt and EDTA buffers. These studies demonstrated that the usual NaCl-EDTA wash and the Tris wash for preparation of chromatin extracted more than 50% of all the nucleolar proteins including the vast bulk of RNA polymerase I and nucleolar protein kinase. In addition, the Tris wash extracted more of the nucleolar RNP particles. Interestingly, the residue fraction (nucleolar chromatin) retained its transcriptional specificity (Matsui et al. 1977; Ballal et al. 1977). Thus, the restriction proteins for rDNA remain in this fraction. The locus of the proteins which affect rates of readouts of RNA in nucleoli is not yet defined.

Antibodies to nucleolar proteins. Antibodies to nucleoli in this laboratory were prepared by immunization of rabbits with whole nucleoli of Novikoff hepatoma and normal rat liver (Busch and Smetana 1970; Busch et al. 1974). Immunofluorescence analysis studies showed specificity of localization of the antibodies to the nucleolus (Busch et al. 1974).

Immunological studies on chromatin proteins of tumors and other tissues demonstrated the presence of an antigen, NAg-1 (Yeoman et al. 1976), in tumors and fetal liver. These results are consistent with the presence in tumor nuclei of fetal proteins that were not detected in normal growing or nongrowing tissues (Chytil and Spelsberg 1971; Wakabayashi and Hnilica 1972; Zardi et al. 1973; Busch 1976; Yeoman et al. 1976).

With the techniques developed in studies on NAg-1 (Yeoman et al. 1976), a reinvestigation was made of the antigens in nucleoli and nucleolar chromatin. In addition to NAg-1, tumor nucleoli contain another antigen, noAg-1, which is different from NAg-1 and appears to be more limited in localization. No noAg-1 was found in normal liver nucleoli. Three antigens were found in normal liver that were not present in tumor nucleoli or nuclei.

For an analysis of the function of the various nucleolar proteins and nucleolar antigens, it seemed essential to establish a working nucleolar system. The first goal was to establish the fidelity of transcription in the nucleolus.

Fidelity of the Nucleolar Transcriptional System

RNA synthesis by isolated nucleoli and nucleolar chromatin. The chemical compositions of isolated nucleoli and nucleolar chromatin (Table 2) indicated

Table 2. Compositions of Nucleoli and Nucleolar Chromatin

	DNA	RNA	Protein acid-soluble[a]	Protein acid-insoluble[b]
Nucleoli	1.00	1.51	5.60	2.60
Nucleolar chromatin	1.00	0.30	2.35	0.79

Values are relative to DNA and averages of three different preparations. DNA was determined by a modified diphenylamine reaction and RNA by alkaline hydrolysis.
[a] Soluble in 0.25 N HCl.
[b] Insoluble in 0.25 N HCl.

that the preliminary extractions for chromatin preparation removed substantial amounts of RNA and protein from the nucleoli. Also, nucleolar chromatin is richer in RNA and acid-soluble proteins than is whole nuclear chromatin (Busch et al. 1975).

The kinetics of RNA synthesis by isolated nucleoli, nucleolar chromatin, and nucleolar DNA showed that after extensive washing with 0.075 M NaCl–0.025 M EDTA and 10 mM Tris-HCl, nucleolar chromatin retained sufficient RNA polymerase I activity to synthesize RNA at approximately one-fifth to one-tenth the rate of whole nucleoli (Table 3). When RNA polymerase I was added, the rate of transcription of chromatin varied from 20% to almost that of whole nucleoli (Table 3). Transcription was linear up to 10 minutes, the standard incubation time. The sedimentation peaks of incorporation were at 18S, 20S, and 25S for chromatin, DNA, and nucleolar transcripts, respectively.

Fidelity of in vitro rRNA synthesis by isolated nucleoli and nucleolar chromatin. Nucleotide compositions determined by incorporation of each labeled nucleotide showed that RNA synthesized by the nucleoli and nucleolar chromatin was rich in GMP and CMP; its composition was very similar to that of rRNA or preribosomal RNA (Table 4). When RNA polymerase I was added to the nucleolar chromatin, the rate of synthesis was increased; the purine content of the RNA product was essentially the same with whole nucleoli or chromatin as template. The pyrimidine content differed; less CMP and more UMP were incorporated. With DNA as template (Table 4), much more AMP and UMP were incorporated than with whole nucleoli or nucleolar chromatin.

RNA treated with DNase I to eliminate trace amounts of contaminating DNA was hybridized to nucleolar DNA fractionated on CsCl gradients. The nucleolar transcripts hybridized mainly to DNA of density 1.715 g/cm³; little hybridized to the main-band DNA (1.692 g/cm³). The base composition and hybridization of the RNA products (Table 4) showed that they contained ribosomal products and products of main-band DNA.

To determine the fidelity of transcription of rRNA more specifically, homochromatography fingerprint-

Table 3. Transcriptional Activity of Nucleoli and Nucleolar Chromatin

Components	pMoles GMP incorporated/μg DNA for 10 min	Percentage
Nucleoli	5.74	100
Nucleoli + α-amanitin (0.5 μg/ml)	6.79	118
Nucleoli + rifampicin AF/013 (60 μg/ml)	7.10	124
Nucleolar chromatin	0.46	8
Nucleolar chromatin + rifampicin AF/013	0.69	12
Nucleolar chromatin + RNA polymerase I (28 units)	0.72	12.5
Nucleolar chromatin + RNA polymerase I (28 units) + rifampicin AF/013	0.79	13.8
Nucleolar chromatin + E. coli RNA polymerase (25 units)	3.22	56.0

Incorporation of [^{3}H]GTP by isolated nucleoli or nucleolar chromatin was determined. Nucleolar RNA polymerase was purified through a DEAE-Sephadex column.

ing of T_1 RNase digestion products was used. Figure 2A shows the pattern obtained from Novikoff hepatoma nucleolar 45S pre-rRNA labeled in vivo with [^{32}P]orthophosphate (Choi et al. 1976). Structural studies have established the identities of many fragments (Fig. 2B) with fragments of the 45S pre-rRNA (Fuke and Busch 1975; Woo et al. 1975; Choi et al. 1976; Fuke and Busch 1977; Matsui et al. 1977).

A similar pattern was found for RNA transcribed in vitro (28–35S) with isolated nucleoli using α-^{32}P-labeled nucleotides. Four spots that contain five oligonucleotides of the in vivo product were not found (see Fig. 3B). Oligonucleotides Y_1-3 ($C_6A_6U_4,\psi$, GmA,UmC,G-18S; C_3,A_7,U_5,AmGmCmA,G-28S) and Y_1-4 (C_5,A_5,U_6,GmU,G-28S) each contain a Gm residue. Of the total, three spots were found in the in vitro transcripts and not in the 45S pre-rRNA. It is possible that the missing fragments were undermethylated (Nazar et al. 1975) and cleaved by T_1 RNase. Specificity of the expression of rDNA genes was retained in the nucleoli and nucleolar chromatin (Fig. 3).

Since the system containing nucleolar RNA polymerase I was not as efficient in transcription of nucleolar chromatin as whole nucleoli, the efficiency and fidelity of transcription were tested with E. coli RNA polymerase. Even though the rate of transcription of nucleolar chromatin was four to five times greater with E. coli polymerase than with RNA polymerase I (Table 3), it was still less than that of whole nucleoli. When the labeled RNA was subjected to homochromatography fingerprinting, the spot pattern (Fig. 4) indicated that all of the larger oligonucleotide markers of preribosomal RNA and oligonucleotides C_5G through C_7G were absent from these transcripts. Even in the lower-chain-length region of the map, instead of discrete spots in the tetra- and pentanucleotide region, only continuous stripes were visible, indicating that many sequences were transcribed other than those of pre-rRNA.

When the homochromatography patterns of RNA transcribed using nucleolar DNA and nucleolar RNA polymerase I were prepared, as with transcripts from nucleolar chromatin and bacterial RNA polymerase, these spot patterns lacked any large marker oligonucleotide spots of preribosomal RNA (Fig. 4). Heterogeneity and randomness were found in the regions containing smaller oligonucleotides.

Table 4. Nucleotide Composition Analysis by Relative Nucleotide Incorporation

Condition	No. of determinations	Relative nucleotide incorporation[a]				
		AMP	UMP	GMP	CMP	$\dfrac{A+U}{G+C}$
Nucleoli	3	13.0	2.10	35.5	30.5	0.52
Nucleolar 45S RNA[b]		14.6	20.5	35.1	29.7	0.54
Chromatin	2[c]	13.4	20.6	37.3	29.7	0.52
Chromatin + polymerase I	4[c]	14.5	24.4	37.9	23.2	0.66
DNA + polymerase I	2	24.9	32.3	21.9	20.9	1.34

Data from Matsui et al. (1977).
[a] Values based on 10-min transcription carried out in the presence of α-amanitin (10 μg/ml).
[b] Nucleotide composition of nucleolar 45S RNA determined by ultraviolet spectra analyses.
[c] The endogenous incorporation by chromatin was 0.46 pmoles of [^{3}H]GMP per μg of DNA per 10 min. With exogenous enzymes, the incorporation was 3.52 pmoles of [^{3}H]GMP per μg of DNA per 10 min.

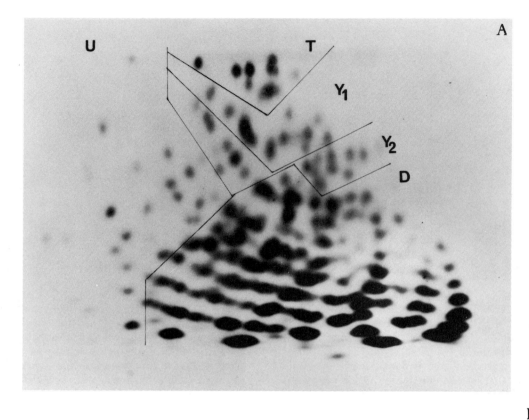

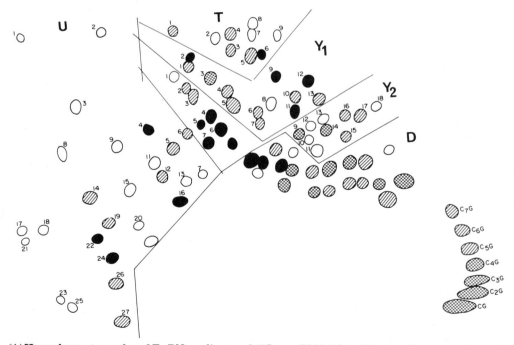

Figure 2. *(A)* Homochromatography of T_1 RNase digests of 45S pre-rRNA labeled in vivo. For convenience, the pattern was divided into regions: U (uridylate-rich); T (top triangle); Y_1 (Y-shaped); Y_2 (below Y_1); and D (approximately decanucleotides). *(B)* Diagrammatic map of the oligonucleotide homochromatogram in *A*. Closed circles represent oligonucleotide spots found in 18S rRNA; hatched circles represent fragments found in 28S rRNA; crosshatched circles represent fragments present in both 18S and 28S rRNA; open circles represent oligonucleotide fragments present in the spacer segments of pre-rRNA. Spot T-9 was a variable spot which occasionally was missing from the digests (Ballal et al. 1977).

670

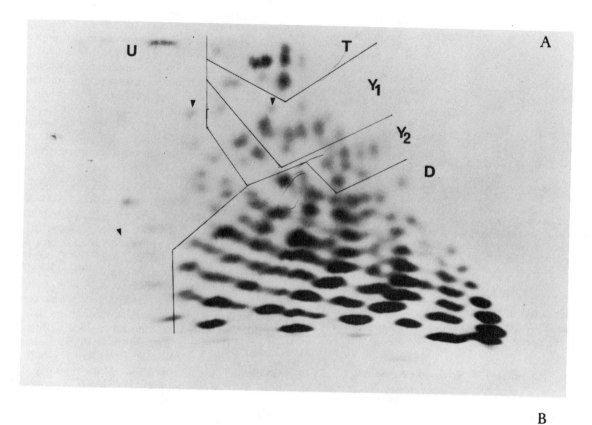

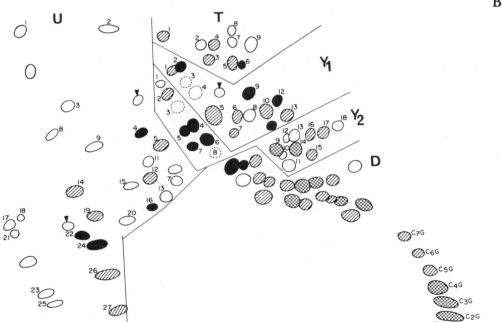

Figure 3. *(A)* Homochromatography of the T_1 RNase digestion products of RNA products synthesized in vitro in nucleolar chromatin with RNA polymerase I. The sedimentation coefficient of this RNA was approximately 18–28S. Arrowheads represent spots that are absent in 45S pre-rRNA. *(B)* Diagrammatic map of the oligonucleotide homochromatogram in *A*. See legend to Fig. 2. A similar pattern was obtained with *E. coli* polymerase transcription of whole nucleoli (Ballal et al. 1977).

Figure 4. Homochromatography of the T_1 RNase digestion products of RNA transcribed in vitro with nucleolar chromatin and *E. coli* RNA polymerase. A similar spot pattern was produced when naked nucleolar DNA was transcribed with nucleolar RNA polymerase I (Ballal et al. 1977).

These results demonstrate the requirements of the proper DNA, RNA polymerase, and restriction proteins for specific rDNA transcription.

Chemistry of Nucleolar Proteins

Numbers and types of nucleolar proteins. In the nucleolus, the ratio of protein to RNA or DNA is approximately 8. The proteins in the nucleolus are of many types and include the precursors of ribosomal proteins as well as enzymes, histones, and nonhistone chromatin proteins. The histones are present in an equal amount to DNA and account for one-eighth of the total nucleolar proteins (Busch et al. 1975).

Until the first two-dimensional gel system of Orrick et al. (1973), it was generally assumed that there were relatively few species of nucleolar acid-soluble proteins; however, the current evidence shows that there are approximately 200. Included in this group of proteins are approximately 40 phosphorylated proteins, of which proteins C23–C25 and C26 and C27 have been the most studied (Busch et al. 1975; Olson et al. 1975). They are typical nonhistone proteins with large quantities of glutamic and aspartic acid and, in addition, they are the first proteins in which clusters of acidic residues have been found associated with the phosphorylated sites (Mamrack et al. 1977).

During the nucleolar hypertrophy produced by thioacetamide treatment, a series of rapid changes occur in nucleolar proteins, some of which are markedly decreased; others, associated with synthesis of preribosomal ribonucleoprotein particles, are markedly increased in amount (Ballal et al. 1974, 1975). These changes were studied by one- and two-dimensional polyacrylamide gel electrophoresis of the 0.4

N H_2SO_4-soluble nucleolar proteins, which showed an overall increase in the ratio of nonhistone proteins to histones following thioacetamide treatment. Most spots that increased in size and density had electrophoretic mobilities of proteins of the preribosomal ribonucleoprotein particles. One spot (A25) remained constant in size and density during the course of the treatment. Interestingly, marked decreases were found very early in two protein spots (A11 and A24), while two other protein spots (C13 and C14) decreased slowly with time. These results indicate that the nucleolus rapidly exhibits multifaceted changes during alterations in cell function.

Protein A24. Isolation and initial chemical characterization indicated that protein A24 was a nonhistone chromosomal protein with approximately equal amounts of acidic and basic amino acids. It constituted approximately 1.9% of the sum of histones 2A, 2B, 3, and 4 (Goldknopf and Busch 1975). Protein A24 was found to contain the tryptic and chymotryptic peptides of histone 2A as well as additional peptides, and, accordingly, it was suggested that protein A24 contained a nonhistonelike polypeptide linked to a histone 2A molecule (Sugano et al. 1972; Goldknopf and Busch 1975; Schlesinger et al. 1975; Olson et al. 1976; Hunt and Dayhoff 1977; Goldknopf and Busch 1977). More recently (Olson et al. 1976; Goldknopf and Busch 1977), protein A24 was found to contain as one amino-terminal sequence:

```
1                5                      10
Met-Gln-Ile-Phe-Val-Lys-Thr-Leu-Thr-Gly-
11              15                     20
Lys-Thr-Ile-Thr-Leu-Glu-Val-Glu-Pro-Ser-
21              25                     30
Asp-Thr-Ile-Glu-Asn-Val-Lys-Ala-Lye-Ile-
31              35
Gln-Asp-Lys-Glu-Gly-Ile-Pro.
```

The identity of this structure to that of ubiquitin (Schlesinger et al. 1975) was very quickly noted by Hunt and Dayhoff (1977):

```
                                       10
NH2-Met-Gln-Ile-Phe-Val-Lys-Thr-Leu-Thr-Gly-
                                       20
    -Lys-Thr-Ile-Thr-Leu-Glu-Val-Glu-Pro-Ser-
                                       30
    -Asp-Thr-Ile-Glu-Asn-Val-Lys-Ala-Lys-Ile-
                                       40
    -Gln-Asp-Lys-Glu-Gly-Ile-Pro-Pro-Asp-Gln-
                                       50
    -Gln-Arg-Leu-Ile-Phe-Ala-Gly-Lys-Gln-Leu-
                                       60
    -Glu-Asp-Gly-Arg-Thr-Leu-Ser-Asp-Tyr-Asn-
                                       70
                                       80
    -Ile-Gln-Lys-Glu-Ser-Thr-Leu-His-Leu-Val-
    -Leu-Arg-Leu-Arg-COOH.
```

This sequence is not homologous to histone sequences.

The second amino terminus of protein A24.
Comparison of tryptic peptide maps of histone 2A and protein A24 revealed remarkable similarities (Sugano et al. 1972). A peptide designated 16 was not previously detected in protein A24 because it did not stain with the ninhydrin-cadmium or the fluorescamine procedure, both of which require the presence of a primary amino group. This peptide contained the blocked amino terminus of histone 2A (Sugano et al. 1972; Yeoman et al. 1972):

N-acetylserine-Gly-Arg.

The single carboxyl terminus of protein A24.
Carboxypeptidase A and B digestion indicated that protein A24 contains the carboxyl-terminal sequence (Olson et al. 1976; Goldknopf and Busch 1977) identical to that of histone 2A (Yeoman et al. 1972):

-His-His-Lys-Ala-Lys-Gly-Lys
123 124 125 126 127 128 129
(histone-2A residue number).

Quantitative hydrazinolysis released molar yields of carboxyl-terminal lysine of 1.01 and 0.88 for protein A24 and histone 2A, respectively; no other carboxyl-terminal amino acids were detected in protein A24.

The branched tryptic peptide of protein A24. The detection of two amino termini and one carboxyl terminus suggested that the protein A24 molecule was branched and that the nonhistone polypeptide was linked to histone 2A in a manner that prevented detection of its carboxyl terminus. The amino acid composition and carboxyl terminus of peptide 17 of histone 2A (Table 5) were the same as that reported for a peptide containing amino acid residue 119–125 of the histone 2A sequence (Yeoman et al. 1972). Edman degradation (Table 6) confirmed the identity and amino acid sequence of this peptide. The amino

Table 5. Amino Acid Ratios and Carboxyl Termini of Peptides 17 of Histone 2A and 17′ of Protein A24

	17	17′
Gly	—	1.64
Lys	1.67	1.68
His	1.71	1.61
Thr	0.91	0.93
Ser (= 1.00)	1.00	1.00
Glu	0.95	0.94
COOH terminus[a]	Lys (0.57)[b]	Lys (0.64)[b]
Molar yield[c]	0.67	0.74

Data from Goldknopf and Busch 1977.
[a] Data obtained by hydrazinolysis; only lysine was found.
[b] Data in parentheses are molar yield of carboxyl-terminal lysine.
[c] nMoles of peptide recovered from tryptic peptide maps per nmole of protein digested.

Table 6. Sequential Edman Degradation of Peptides 17 and 17′

Amino acid	Edman cycle[a]			
	1	2	3	4
Peptide 17 of histone 2A				
Gly				
Lys	0.5			
Thr		1.0		
Glu			0.9	
Ser				0.3
Peptide 17′ of protein A24				
Gly	1.6	0.5		
Lys	0.4	0.1		
Thr		1.1		
Glu			0.8	
Ser				0.3

Data from Goldknopf and Busch (1977).
[a] nMoles of amino acids per nmole of peptide released from thiazolinones after hydrolysis with H1

acid sequence analysis of tryptic peptide 17′ of protein A24,

$$
\begin{array}{c}
\overset{\displaystyle O}{\overset{\displaystyle \|}{} \;\; \overset{\displaystyle H}{\overset{\displaystyle |}{}} \qquad \overset{\displaystyle O}{\overset{\displaystyle \|}{}} \\
H_2N\text{-}CH_2\text{-}C\text{-}N\text{-}CH_2\text{-}C\text{-}NH \\
\text{(Gly)} \qquad \text{(Gly)} \quad | \\
(CH_2)_4 \\
| \qquad O \\
| \quad // \\
H_2N\text{-}CH\text{-}C\text{-}Thr\text{-}Glu\text{-}Ser\text{-}His\text{-}His\text{-}Lys \\
\text{(Lys 119)},
\end{array}
$$

showed that it contains tryptic peptide 17 of histone 2A, Lys-Thr-Glu-Ser-His-His-Lys. Lysine 119, the amino terminus of this peptide, which is derived from the histone 2A portion of protein A24, is linked by an isopeptide bond to the carboxyl group of a glycine residue. Accordingly, the branched structure of protein A24 proposed is:

$$
\begin{array}{c}
\overset{\displaystyle H}{\overset{\displaystyle |}{}} \qquad \overset{\displaystyle O}{\overset{\displaystyle \|}{}}\overset{\displaystyle H}{\overset{\displaystyle |}{}} \qquad \overset{\displaystyle O}{\overset{\displaystyle \|}{}} \\
Met\text{- - -}N\text{-}CH_2\text{-}C\text{-}N\text{-}CH_2\text{-}C\text{-}NH \\
\text{(Gly)} \qquad \text{(Gly)} \quad | \\
\overset{\displaystyle H}{\overset{\displaystyle |}{}} \quad (CH_2)_4 \\
\qquad O \\
N\text{-}acetylserine \text{- - - - - -} N\text{-}CH\text{-}C \text{- - -}Lys \\
\text{Histone 2A: 1} \qquad \text{Lys 119} \qquad 129.
\end{array}
$$

Possible role of protein A24 in chromatin supercoiling. The presence of protein A24 in chromatin in a ratio of approximately 1:10 to histone 2A suggested the possibility that this protein may be in one-fifth or one-sixth of the octameric nucleosomes. Therefore, it may be related to the "five- or sixfold nu-body symmetry" of coiled chromatin. Protein A24 may serve as a recognition site for a "zipper" enzyme that supercoils chromatin. If the "ubiquitin" moiety were cleaved off the A24 molecule, the chromatin might then uncoil to the "euchromatin" state, which could be transcribed readily by RNA polymerase I.

Fractionation of Nucleolar Proteins

Distribution of nucleolar proteins in salt fractions. When nucleoli were extracted with 0.075 M NaCl–0.024 M EDTA, 19% of the nucleolar proteins were removed (Table 7). Subsequent washes with 10 mM Tris extracted another 35% of the nucleolar proteins (L. I. Rothblum et al., in prep.). These two buffers extracted 54% of the nucleolar proteins (Fig. 5A,B; Table 7).

The remainder of the proteins were removed by extraction with salt solutions and may be more tightly bound to the DNA or chromatin; 0.15 M NaCl extracted 5% of the nucleolar proteins and 0.35 M NaCl, 0.6 M NaCl, and 3 M NaCl–7 M urea each extracted approximately 14% of the nucleolar proteins (Table 7).

Distribution of RNA polymerase and protein kinase. Both RNA polymerase I and protein kinase (Kang et al. 1974) activities were assayed in whole nucleoli and the various fractions (L. I. Rothblum et al., in prep.). The 0.075 M NaCl–0.025 M EDTA wash and the 10 mM Tris wash contained 40% and 60% of the extracted nucleolar protein-kinase activity, respectively. Only 13% of the extracted RNA polymerase I was in the NaCl–EDTA extracts; 87% of the RNA polymerase I was in the 10 mM Tris extracts (Table 8).

The nucleolar chromatin residue (after the NaCl–EDTA and 10 mM Tris extractions) contained only 22% of the nucleolar RNA polymerase I activity and 25% of the nucleolar protein kinase (Table 9); 0.15 M NaCl extracted 17% of polymerase I and 15% of the protein kinase from the chromatin residue. After two 0.35 M NaCl extractions, less than 3% of the polymerase activity remained associated with the chromatin.

Analysis of nucleolar proteins. The first two nucleolar extractions, i.e., NaCl–EDTA and 10 mM Tris, extracted more than 50% of the nucleolar proteins. The proteins in the NaCl–EDTA extract of nucleoli were heterogeneous; it contained a minimum of 25 polypeptides (Fig. 5A). In the C region, the most prevalent proteins were C23–C25, C6, C14, CI, and CG. In contrast, the 10 mM Tris extract consisted predominantly of proteins B18, B23–B25, and C23–C25 (Fig. 5B).

The initial RNA:DNA ratio of nucleoli was 1:1; following the NaCl–EDTA and 10 mM Tris extracts, the RNA:DNA ratio of the residue was 0.2:1. Accordingly, a significant amount of the nucleolar ribonucleoproteins had been extracted. The 10 mM Tris extracts were sedimented at 105,000g for 18 hours over a 1 M sucrose cushion. Ten percent of the extracted proteins were in the pelleted particles; they were extracted from the pellet with 3 M NaCl–7 M urea, 20 mM Tris-HCl, pH 8.0. The two-dimensional electrophoretogram (Fig. 6) of these proteins was essentially identical to those of the nucleolar ribonucleoprotein particles (Daskal et al. 1974; Prestayko et al. 1974).

The 0.15 M NaCl extract of chromatin consisted almost entirely of proteins with molecular weights greater than 30,000. The sample was of limited heterogeneity; proteins B17, B18, and C24 were the densest spots present. The 0.35 M NaCl extract of chromatin contained many proteins not extracted by 0.15 M NaCl. For example, proteins B24, B25, C14, C17, and C27, which were dense spots in the 0.35 M NaCl extract, were not found in the 0.15 M NaCl extract. On the other hand, proteins B35 and B17 were not present in the 0.35 M NaCl extract; they were in the group of proteins extracted by 0.15 M NaCl.

The H1 histone spots were dense and large in the 0.6 M NaCl extract (Fig. 7); proteins C17 and C24 also were dense spots. When the post-0.6 M NaCl chromatin residue was extracted with 3 M NaCl–7 M urea, histones other than the H1 histones were the major polypeptides present. In addition, several low-molecular-weight proteins, including A7, A15, A24, and A25, were also found in the 3 M NaCl–7 M urea extract. Additional dense spots in this fraction were B7, B9, B13, BJ, CA, C6, CM, CM', CI, CP, and CQ, which were previously found by Yeoman et al. (1975) in chromatin fraction II and by Olson et al. (1975) in nucleolar chromatin fraction II.

Table 7. Distribution of Nucleolar Proteins

Fraction	Percentage total nucleolar protein	Percentage total dissociated nucleolar chromatin protein
NaCl–EDTA	19 ± 3	
10 mM Tris	35 ± 5	
Chromatin	46 ± 5	100
0.15 M NaCl	5 ± 1	14 ± 1.4
0.35 M NaCl	13 ± 4	27 ± 4.0
0.6 M NaCl	14 ± 4	30 ± 4.8
3 M NaCl–7 M urea	14 ± 6	28 ± 7.6

Nucleoli were extracted as indicated above, and the amounts of protein extracted were determined and are represented here as the percent of the total amount extracted. Data presented are the mean ± the standard error of the mean.

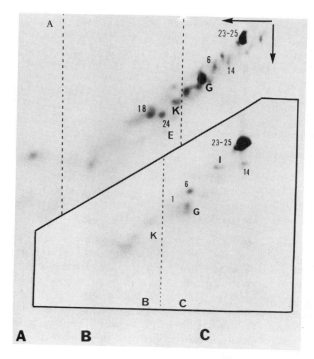

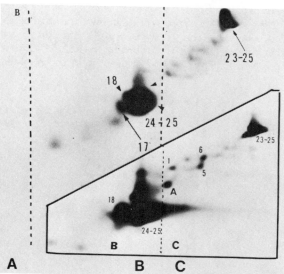

Figure 5. *(A)* Two-dimensional polyacrylamide gel electrophoresis of 200 μg Novikoff hepatoma nucleolar proteins extracted with NaCl-EDTA. Samples were run in the first dimension on disk gels of 10% acrylamide, 6 M urea, 0.9 N acetic acid. For the second-dimension electrophoresis, a 12% acrylamide, 0.1% SDS slab gel was run (referred to as 10/12 two-dimensional polyacrylamide gel electrophoresis). *(Inset)* Two-dimensional polyacrylamide gel electrophoresis of Novikoff hepatoma nucleolar proteins extracted with NaCl-EDTA. Samples were run in the first dimension on the disk gels of 6% acrylamide, 6 M urea, 0.9 N acetic acid. For the second dimension, an 8% acrylamide, 0.1% SDS slab gel was run (referred to as 6/8 two-dimensional gel electrophoresis) (L. I. Rothblum et al., in prep.). *(B)* Two-dimensional gel electrophoresis (10/12) of 110 μg of Novikoff hepatoma nucleolar proteins extracted with 10 mM Tris-HCl, pH 8.0. *(Inset)* Two-dimensional gel electrophoresis (6/8) of Novikoff hepatoma nucleolar proteins extracted with 10 mM Tris-HCl, pH 8.0 (Rothblum et al. 1977).

Table 8. Distribution of RNA Polymerase I and Protein Kinase in the NaCl–EDTA and Tris Extracts of Nucleoli

	Protein kinase[a] (cpm × 10⁻⁵)	RNA polymerase[b] (units × 10⁻³)
NaCl–EDTA	11.2 (40)	10.4 (12.5)
10 mM Tris	18.4 (60)	67.9 (87.5)

Enzyme activities are expressed as cpm of ^{32}P incorporated into casein for protein kinase and as units for RNA polymerase. Numbers in parentheses indicate the percent of the total extracted.
[a] Protein kinase represents 77% of the total nucleolar activity.
[b] RNA polymerase represents 73% of the total nucleolar activity.

Translational factors in nucleoli. Recent studies in our laboratory have established the presence of elongation factors in nucleoli (M. S. Rao et al., in prep.). The data in Table 10 show that the specific activity of EF-1α in nucleoli is higher than that of the whole homogenate. Inasmuch as the rDNA is a very small percentage of active nucleolar DNA, the concentration of the translational factors at rDNA loci may be much higher than indicated by the overall specific activity in terms of units/mg DNA. Preliminary studies indicate that both EF-2 and initiation factors may be concentrated in nucleoli. Whether they have a role in the assembly of preribosomal particles or as control elements in nucleolar function remains to be established by future studies.

Attempts to define small quantities of nucleolar proteins. In earlier studies one approach to separation and identification of small amounts of nucleolar proteins was undertaken by studies on phosphorylated nucleolar proteins (Kang et al. 1974; Busch et al. 1975; Olson et al. 1975; Mamrack et al. 1977). Another approach was the development of immunological probes (Busch et al. 1974). Initially, rabbits were immunized against whole nucleoli in an attempt to find which of the nucleolar proteins constituted effective antigens and to make use of immunofluorescence analysis. Subsequently, the following studies were carried out to characterize nucleolar immunogens further.

Immunology of Nucleolar Proteins

noAg-1. In previous studies (Yeoman et al. 1976) a chromatin antigen (NAg-1) was found in Novikoff hepatoma nuclei and fetal liver that was not found in normal liver cells. This antigen was purified and found to be a glycoprotein (Yeoman et al. 1976). Initial evidence (R. K. Busch and H. Busch, in prep.) suggested that tumor nucleolar chromatin contained a different antigen (noAg-1). Figure 8A shows the immunoprecipitin bands formed with tumor chromatin antigens (TCAg) and antibodies to tumor nuclei (TNAb), tumor chromatin (TcAb), tumor nu-

Table 9. Distribution of RNA Polymerase I and Protein Kinase in
Subnucleolar Fractions

| | RNA polymerase I | | | Protein kinase |
	1	2	3	
Nucleoli	20.8	106.6	170.6	38.4
Chromatin	4.8 (23)	22.5 (21)	38.6 (23)	8.8 (23)

RNA polymerase activity determined using: (1) endogenous template; (2) endogenous
template and added nucleolar DNA; and (3) extracted nucleoli or chromatin. Protein
kinase activity is expressed as cpm of ^{32}P incorporated into casein. Numbers in parentheses
indicate the percentage of the total nucleolar activity.

cleoli (TnAb), and liver nucleoli (LnAb). The lack
of antigenic identity of the bands (TCAg-TnAb) with
the tumor nucleolar antibodies and tumor chromatin antibodies (TCAg-TcAb) is apparent. Antibodies
to liver nucleoli did not form the same bands with
antigens from tumor chromatin as did antibodies
to tumor nucleoli (Fig. 8A,B).

Is noAg-1 distinct from NAg-1? To compare noAg-
1 and NAg-1, tumor nucleolar chromatin antigens
(TnCAg) were tested for immunoprecipitation with
tumor nucleolar antibodies (TnAb) and tumor chromatin antibodies (TcAb). As shown in Figure 8C,
three bands formed with TnAb and two with TcAb.
The inner band is common to both types of antibodies. The dense band between TcAb and TnCAg shows
that tumor nucleolar chromatin contains NAg-1 as

well as noAg-1, which is the middle of the three
bands between TnCAg and TnAb.

*Does NAg-1 form immunoprecipitin complexes with nu-
cleolar antibodies?* Figure 8D provides an analysis
of the interaction of a purified preparation of NAg-
1 in a concentrated preparation of tumor cytoplasmic proteins (TCyAg) with a variety of antibody
preparations. TCyAg was shown earlier to contain
NAg-1 (Yeoman et al. 1976). Figure 8D shows that
this preparation formed a dense band with TcAb.
However, with TNAb and TnAb only faint bands
were present. Since TnAb formed dense bands with
noAg-1 (Fig. 8C), these results show that antibodies
that reacted strongly with noAg-1 did not react
strongly with NAg-1. Accordingly, noAg-1 differs
from NAg-1. Thus, NAg-1 is present in the nucleus

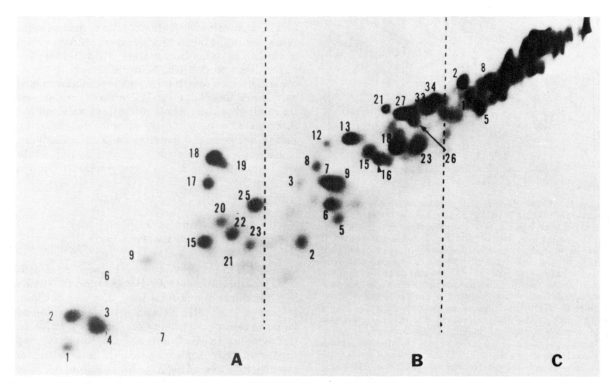

Figure 6. Two-dimensional gel electrophoresis (10/12) of 300 μg of Novikoff hepatoma nucleolar preribosomal particle
proteins extracted with 10 mM Tris-HCl, pH 8.0, (Rothblum et al. 1977).

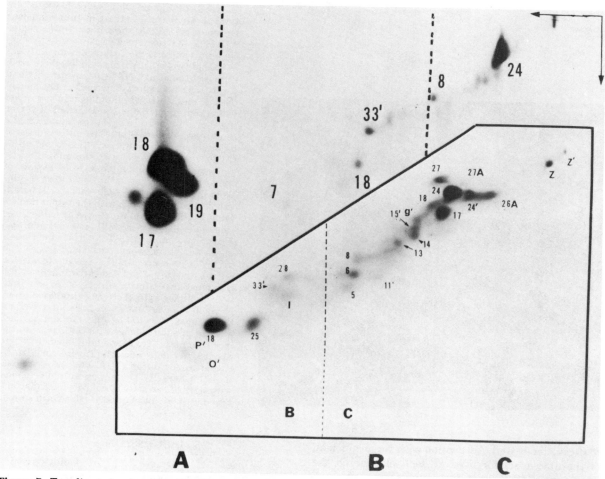

Figure 7. Two-dimensional polyacrylamide gel electrophoresis (10/12) of 575 μg of the proteins extracted with 0.6 M NaCl from nucleolar chromatin. *(Inset)* Two-dimensional electrophoresis (6/8) run (Rothblum et al. 1977).

Table 10. Elongation Factor 1 in Cellular Fractions

Fractions	Percentage distribution of EF-1α	Specific activity of EF-1α (units[a]/ mg protein × 10^{-2})	Specific activity of EF-1α (units/ mg DNA × 10^{-3})
Whole homogenate	100	2.2	23
Nuclei	14	6.2	3.1
Nucleoplasm	9.5	5.7	3.0
Nucleoli	0.9	6.4	3.5

Elongation factor 1α-assay incubation mixture included the following in a total volume of 100 μl: Tris-HCl, pH 7.5, 40 mM; MgAc, 7 mM; NH₄Cl, 50 mM; KCl, 50 mM; GTP, 50 μM; DTT, 2 mM; creatine perosphate, 1.25 mM; creatine phosphokinase, 2.5 μg, poly(U), 40 μg; salt-washed ribosomes, 0.8 A₂₆₀ units; and ³H-labeled phenylalanine-tRNA, 30 pmoles. The reaction is carried out at 37°C for 5 min with all the constituents except ³H-labeled phenylalanine-tRNA and enzyme. After preincubation, enzyme and tRNA were added and incubation was continued for 10 min. Eighty microliters of the assay mixture was filtered through Millipore filters and washed with 3 ml of ice-cold buffer containing 0.05 M Tris-HCl, pH 7.5, 0.05 M KCl, and 0.001 M DTT.

[a]One unit is the equivalent of 1 pmole of phenylalanyl-tRNA bound to ribosomes in 10 min at 37°C. Each value is an average of three independent experiments.

and cytoplasm, but noAg-1 is localized to the nucleolus and nucleolar chromatin.

Evidence that noAg-1 is tightly bound to tumor chromatin. noAg-1 was not extracted from tumor nucleoli with 0.075 M NaCl–0.025 M EDTA, pH 8.0, since tumor nucleolar antibodies did not react with proteins in this extract. Tumor nuclear antibodies (TNAb), however, formed a dense band with antigens of tumor nucleolar chromatin (TnCAg), but formed only a faint band with these soluble antigens. The band for NAg-1 was readily observed when tumor chromatin antibodies were reacted with either TnCAg or NaEAg. Thus, the saline-EDTA extraction (NaEAg) solubilized NAg-1 but not noAg-1.

Is antigen noAg-1 in normal liver? Evidence provided earlier indicated that NAg-1 was not present in a variety of normal adult growing and nongrowing tissues. Although antibodies to liver nucleoli and liver chromatin antigens formed bands with the tumor nucleolar antigens, they were diffuse and close

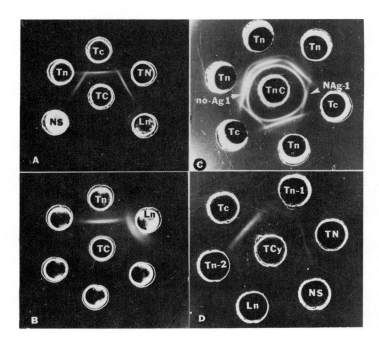

Figure 8. *(A)* Immunodiffusion plate which contains a 0.6 M NaCl extract of tumor nuclear chromatin (TCAg) as the antigen in the center well (300–400 μg). The antigen formed precipitin bands with 33 μl of the following antisera in the side wells: TnAb (tumor nucleolar); TcAb (tumor chromatin); TNAb (tumor nuclear); and LnAb (liver nucleolar). A normal serum control (NS) is negative. *(B)* Immunodiffusion plate which contains a 0.6 M NaCl extract of tumor nuclear chromatin (TCAg) as the antigen in the center well (300–400 μg). The antigen formed different precipitin bands with TnAb and LnAb (33 μl) in the side wells. *(C)* Immunodiffusion plate which contains a 0.6 M NaCl extract of tumor nucleolar chromatin (TnCAg) as the antigen in the center well (280–300 μg). The antigen formed three bands with the tumor nucleolar antibodies (TnAb) which was added to three side wells (33 μl per well). The antigen also formed two bands with TcAb, the tumor chromatin antiserum which was added to two side wells (33 μl per well). The specific bands for noAg-1 and NAg-1 are shown with arrows. *(D)* Immunodiffusion plate which contains a fraction of tumor cytosol TCyAg (prepared from Novikoff Nonidet nuclei) in the center well as the antigen (300–400 μg). The antigen formed a dense precipitin band with antibodies to tumor chromatin (TcAb) and faint bands with antibodies to tumor nucleoli (TnAb-1 and TnAb-2) and tumor nuclei (TNAb) (33 μl contained in each side well). There were no visible bands with normal serum (NS) or liver nucleolar antiserum (R. K. Busch and H. Busch, in prep.).

to the antibody well. Adsorption with tumor nucleoli of the antiserum for liver nucleoli eliminated the formation of these immune complexes.

Liver nucleolar antibodies formed several bands with the liver chromatin antigens. Evidence that these immunoprecipitin bands are not the same as those found in the corresponding tumor samples is shown in Figure 9. The single dense bands formed between the nucleolar (TnAb), chromatin (TcAb), and nuclear (TNAb) antibodies and the chromatin antigens of the tumor (TCAg) did not show identity with the bands formed between LnAb and LCAg.

These results indicate that the antigen recognized by the antinucleolar antibodies to the tumor was not demonstrable in normal liver nucleoli. Conversely, several liver nucleolar chromatin antigens were not found in the Novikoff hepatoma nucleoli.

Immunoelectrophoretic analysis in which the samples shown in Figure 10 were placed in the wells and a new antitumor nucleolar antisera was placed in the trough showed multiple arcs formed with tumor and liver nucleolar proteins extracted with 0.075 M NaCl–0.025 M EDTA. Of these arcs, the one marked by the arrow was uniquely found in the tumor nucleolar extract. This result supports the results shown above which indicate the presence of unique antigens in tumor nucleoli. It is noteworthy that these were not demonstrated in the fetal liver nuclei thus far.

DISCUSSION

This brief review of analytical and isolation studies in our laboratory reflects the overall progress in our knowledge of the various elements of nucleolar activity and nucleolar gene control. Nevertheless, there are many important questions fundamental to future understanding of nucleolar function, gene control, and tumor growth.

Restriction proteins: structure and function. Current studies in our and other laboratories (Beebee and Butterworth 1975; Ballal et al. 1977; Matsui et al. 1977) show that specificity of readout of rDNA genes persists in nucleolar residues after extraction of nucleoli with the Zubay-Doty buffer (0.075 M NaCl–0.025 M EDTA, pH 7.4) and with 0.01 M Tris, pH 8.0. Under these circumstances, more than 50% of the nucleolar proteins were extracted, including most of the RNA polymerase I, protein kinase, and most of the RNP particles (Rothblum et al. 1977). The less-soluble proteins that remain can be fractionated into two groups: (1) those that are referred to as the tight-binding proteins (TBP), (Busch et al. 1975) and (2) those soluble in 3 M NaCl–7 M urea. The latter contain the restriction proteins inasmuch as the DNA residue containing the TBP is read for genes other than rDNA. With this preparation, the rDNA readout was no longer specific. Further studies on the nature and specificity of the restriction

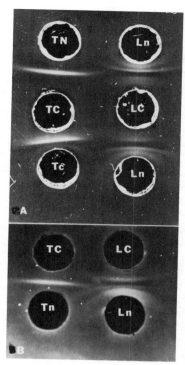

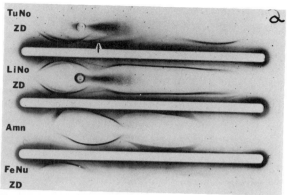

Figure 10. Immunoelectrophoretic profile of nucleolar antigens. Zubay-Doty extracts (0.075 M NaCl, 0.025 M EDTA, pH 8.0, containing 1 mM PMSF) were made from nucleoli isolated by the calcium-sucrose procedure from Novikoff hepatoma nucleoli (TuNoZD), from rat liver nucleoli (LiNoZD), and from nuclei prepared by the NP-40 sucrose procedure from livers of 19-day-old fetal rats (FeNuZD). Then 20 μg of each of these extracts and of the amniotic fluid of the fetal rats was analyzed by electrophoresis for 35 min at 75 V in 1% agarose (Immunoagaroslides, Millipore Biomedica, Acton, Mass.). Next 50 μl of an immunoglobulin fraction at 80 mg/ml, which was prepared from the serum of a rabbit immunized against whole Novikoff nucleoli, was placed in each trough, and precipitin arcs were allowed to develop for 24 hr. Precipitin arcs were stained with Coomassie blue. Note the presence of a precipitin arc (arrow) in the extract from Novikoff nucleoli which is not detected in the other fractions.

Figure 9. *(A)* Immunodiffusion plate which contains a 0.6 M NaCl extract of tumor nuclear chromatin (TCAg) in the left center well and liver chromatin (LCAg) in the right center well (300–400 μg) as antigens. The TCAg formed precipitin bands with tumor nuclear (TNAb) and tumor chromatin (TcAb) antibodies. LCAg formed at least three precipitin bands with the liver nucleolar (LnAb) antibodies. The antibody wells contain 33 μl. *(B)* Immunodiffusion plate which contains 0.6 M NaCl extracts of tumor nuclear chromatin (TCAg) in the top left well and liver nuclear chromatin (LNAg) in the top right well (300–400 μg). The tumor chromatin antigen formed a precipitin band with the tumor nucleolar antiserum (TnAb). LCAg formed at least three precipitin bands with the liver nucleolar antibodies (LnAb). The antibody wells contain 33 μl (R. K. Busch and H. Busch, in prep.).

proteins are in progress. These proteins must be isolated and reconstituted with the chromatin to determine which ones specifically restrict and how they function. Clearly, both the restriction proteins and the RNA polymerase affect the type of readout obtained (Reeder 1973; Honjo and Reeder 1974; Ballal et al. 1977).

Chromatin supercoiling. At present, methods are improving for separation of active and inactive chromatin. Proteins such as protein A24 may be important in the control of chromatin supercoiling. Alternatively, transpeptidases with the capacity to form isopeptide linkages in such proteins may be rate-determining. In this connection, it is particularly pertinent to recall that as nucleolar function increased, the concentrations of protein A24 and A11 rapidly diminished, and, accordingly, their possible role in gene control was anticipated. Protein A24 does not seem to be a specific repressor but rather

seems to be inversely related in amount to the total activity of the liver nucleolus. The structure and function of protein A11 are much less adequately characterized. In any event, the mechanisms of nucleolar chromatin supercoiling and uncoiling are of great interest and necessitate careful analysis of both active and inactive chromatin to determine what specific roles of individual proteins may be involved. Recently, I. L. Goldknopf et al. (in prep.) have found the relative concentrations of protein A24 in nucleosomes and chromatin to be the same.

Derepression. Nucleolar synthesis is subject to rapid change (Busch and Smetana 1970; Mauck 1977), e.g., in specific in vitro systems significant increases were found in 1 hour in RNA polymerase I. Such changes may reflect the activity of RNA polymerase I alone or interactions of repressors or stimulators with rDNA and its promoters. In bacterial systems, ppGpp inhibits rRNA synthesis (Travers 1973, 1976). Within the limits of the methods available, neither ppGpp, cAMP, cGMP, nor their analogs exert effects on eukaryotic nucleolar activities. The polyphosphate of Hildebrandt and Sauer (1977) reduced rates of rRNA synthesis; in growing or dividing cells, the rates may be increased by removal of such inhibitory cellular elements.

An event of relative rapidity is the phosphorylation or dephosphorylation of nucleolar proteins, such as RNA polymerase. Much effort has been devoted to the study of phosphorylated proteins C23–C25 and C26 and C27 (Mamrack et al. 1977). However, variation in their phosphorylation parallels rather than precedes changes in nucleolar activity. Phosphorylation of RNA polymerase I has been reported, but, like changes in nucleolar protein kinases, it is an associated rather than a primary event in nucleolar function.

Moment-to-moment controls. What then are the driving factors that influence the rate rather than the specificity of nucleolar functions? Both the supercoiling of nucleolar chromatin and its specificity of readout are probably long-acting rather than moment-to-moment control mechanisms.

Systems for analysis of controls of nucleolar function. A key problem in studies on mechanisms has been the lack of satisfactory systems for rDNA activation and inhibition. Recently, M. Andersen and coworkers (in prep.) have been analyzing rates of activation of nucleoli in thioacetamide-treated livers. A tenfold increase in nucleolar RNA synthesis occurred in the 7-hour period following administration of the drug. At 48–72 hours there was a return to the normal level. This system offers a simple tool for analysis of nucleolar function with the opportunity to determine correlates between reduction of ribosome levels in the cytoplasm (Busch and Smetana 1970), activation of cytoplasmic RNases, and "feedback circuitry" of the nucleolus.

Other systems that may be useful include the "refeeding" system recently described by Mauck (1977), in which rapid stimulation of RNA polymerase has been found. In this system, a rapid increase in rRNA synthesis occurs in cells that were stationary and grow after refeeding.

Feedback systems. It has been postulated that ribosomes or ribosomal products could be important feedback factors in nucleolar function (Busch and Smetana 1970). Studies in our laboratory (Wu et al. 1977) and those of Warner (1974) directed attention to the close correlation of the relationship of ribosomal protein synthesis and that of rRNA. In addition, both Higashi's group (1968) and Muramatsu's group (1970) have shown that inhibition of protein synthesis by cycloheximide produced cessation of nucleolar function.

Recently, Bolla et al. (1977) have done the interesting experiment of adding ribosomal proteins to nuclei of liver cells. They reported that these proteins caused a shift in synthesis from mRNA to pre-rRNA, suggesting that r proteins redirect nuclear to nucleolar synthetic activity.

At the same time, our group, which demonstrated that the nucleoli contain many ribosomal proteins (Prestayko et al. 1974), has been evaluating elongation and initiation factors in nucleoli. Recently, M. S. Rao and coworkers (in prep.) have found both initiation and elongation factors (both EF-1 and EF-2) in rat liver nucleoli in concentrations higher (activity/mg protein) than those of the nucleoplasm or cytoplasm. Since the nucleolus contains high concentrations of the preribosomal proteins and other preribosomal elements, the higher concentration of the initiation and elongation factors may be important to the overall biosynthetic and/or assembly reactions of the nucleolus.

In our earlier studies (Prestayko et al. 1974) the nucleolar RNP particles were shown to contain a number of high-molecular-weight proteins in addition to ribosomal proteins. These particles are easily extractable from the nucleoli (Busch and Smetana 1970); they are identical in size and ultrastructure to the large ribosomal subunit and are precursors of the ribosomes. These early findings have now been confirmed (Rothblum et al. 1977). Two-dimensional electrophoresis also showed that the RNP particles contain higher-molecular-weight proteins corresponding in migration to the initiation and elongation factors (M. S. Rao et al., in prep.) purified by Merrick's group (Kemper et al. 1976) and others (Weissbach and Ochoa 1976). Studies are now in progress to determine whether the 0.5 M KCl extract of nucleoli contains all the initiation and elongation factors of the cytoplasmic ribosomes (M. S. Rao et al., in prep.).

Reconstitution systems. Reconstitution techniques may provide the types of answers required for analysis of the rate-limiting factors involved in nucleolar function (Paul and Gilmour 1968; Bekhor et al. 1969; Chytil and Spelsberg 1971; Spelsberg et al. 1971; Stein and Farber 1972; Axel et al. 1975; Blüthmann et al. 1975; Matsui et al. 1975; Stein et al. 1975; Daubert et al. 1977). Further analysis of this technique is in progress in this and other laboratories.

Tumor nucleolar antigens (noAg-1). There have been many reports of unusual antigens in tumor cells, including carcinoembryonic antigens (Gold and Freedman 1965), α-fetoproteins (Abelev et al. 1959), preneoplastic antigens (Okita et al. 1975), as well as viral antigens including T antigen (Todaro et al. 1965) and the Epstein-Barr nuclear antigen (Suzuki and Himuna 1974). An extractable nuclear antigen has also been described (Morris et al. 1975). Chromatin extracted from tumor nucleoli contains an antigen (noAg-1) which appears to be limited in location to nucleoli (R. K. Busch and H. Busch, in prep.). This antigen was not found in normal liver nuclear or nucleolar chromatin, in cytoplasmic fractions of tumor, or in the 0.075 M NaCl–0.025 M EDTA and 0.01 M Tris-HCl extracts of nucleoli.

Although immunoprecipitin analyses show the presence of noAg-1 in tumor nucleoli, the fact that it was not detected in liver may reflect (1) its absence from liver, (2) a very small amount of the antigen in the liver, (3) masking of the antigen, or (4) association with other macromolecules altering its extractability.

Evidence that antigen noAg-1 differs from NAg-1 (R. K. Busch and H. Busch, in prep.) was obtained in studies showing that antibodies to tumor chromatin proteins soluble in 0.15 M NaCl–0.35 M urea did not form an immunoprecipitin band with this antigen. Although antibodies to whole nucleoli formed a dense immunoprecipitin band with this antigen, the two immunoprecipitin lines crossed over each other. The antinucleolar antibodies did not react with a more purified preparation of NAg-1 obtained from tumor cytosol, whereas the antichromatin antibodies produced a dense band with this preparation.

Interestingly, the antibodies raised against liver nucleoli reacted with different proteins than those of the Novikoff hepatoma nucleoli or nuclei. The immunoprecipitin bands formed did not exhibit identity with those formed with antigens of the Novikoff hepatoma nuclei, chromatin, nucleoli or nucleolar chromatin. This result is interesting because the normal liver nucleolus operates in a highly repressed state, i.e., the rDNA is transcribed at only one-tenth to one-fifteenth the rate of transcription in Novikoff hepatoma (Busch and Smetana 1970), and possibly these repressors are in the liver antigens. Also, some nucleolar enzymes, including RNA methylases (Liau et al. 1976) and protein kinases (Thomson et al. 1975), have been reported to differ in tumors and other tissues and may be among the antigens detected in these studies.

The possibility also exists that noAg-1 may serve as a derepressor protein which may increase readouts of rDNA (Busch 1976). Feedback stimulation of nucleolar function was suggested by studies on estrogen-stimulated systems (Arnaud et al. 1971; Cohen and Hamilton 1975; Dierks-Ventling and Bieri-Bonniot 1977). It may be that protein A11 is a phosphoprotein (such as phosphorylated lysine-rich histone F10) that serves a similar function as ppGpp (Travers 1973, 1976) or the polyphospate of Hildebrandt and Sauer (1977). If noAg-1 were a phosphatase that reduced its content, nucleolar activity might well reach the high levels that characterize these tumors.

Tests of the effects of various proteins which are currently being isolated in sufficient quantities for further studies are now being improved rapidly. The key development has been the evidence that the nucleolus in vivo and in vitro synthesizes unique gene products which can be simply assayed for by homochromatography fingerprinting of T_1 RNase digests. This development permits demonstration of fidelity and offers the opportunity to correct the lack of fidelity. The methods for reconstitution of nucleolar activity require further development and standardization. With this system and the vastly improved techniques for separating and isolating proteins, the next decade should bring rapid advances in our understanding of nucleolar gene control.

Acknowledgments

These studies were supported by Cancer Research Center Grant CA-10893 awarded by the National Cancer Institute, DHEW; the Davidson Fund; the Wolff Memorial Foundation; and a generous gift from Mrs. Jack Hutchins.

REFERENCES

ABELEV, G. I., Z. A. AVENIROVA, N. V. ENGEL-GARDT, Z. L. BAIDAKOVA, and G. I. STEPHANCHENOK-RUDNIK. 1959. An organ specific liver antigen absent in hepatoma. *Proc. Acad. Sci. USSR* (Engl. Transl.) **124**: 51.

ARNAUD, M., Y. BEZIAT, J. L. BORGNA, J. C. GUILLEUX, and M. MOUSSERON-CANET. 1971. Lerecet teur de l'oestradio, l'amp cyclique et al RNA polymeras nucleolaire dans l'uterus de genisse. Stimulation de la biosynthese de RNA in vitro. *Biochim. Biophys. Acta* **254**: 241.

AXEL, R., H. CEDAR, and G. FELSENFELD. 1975. The structure of the globin genes in chromatin. *Biochemistry* **14**: 2489.

BALLAL, N. R. and L. ROGACHEVSKY. 1976. Fidelity of preribosomal RNA synthesis *in vitro*. *Proc. Am. Assoc. Cancer Res.* **17**: 162.

BALLAL, N. R., Y. C. CHOI, R. MOUCHE, and H. BUSCH. 1977. Fidelity of synthesis of preribosomal RNA in isolated nucleoli and nucleolar chromatin. *Proc. Natl. Acad. Sci.* **74**: 2446.

BALLAL, N. R., I. L. GOLDKNOPF, D. A. GOLDBERG, and H. BUSCH. 1974. The dynamic state of liver nucleolar proteins as reflected by their changes during administration of thioacetamide. *Life Sci.* **14**: 1835.

BALLAL, N. R., Y. J. KANG, M. O. J. OLSON, and H. BUSCH. 1975. Changes in nucleolar proteins and their phosphorylation patterns during liver regeneration. *J. Biol. Chem.* **250**: 5921.

BEEBEE, T. J. C. and P. H. W. BUTTERWORTH. 1975. Transcription of isolated nuclei and nucleoli by exogenous RNA polymerases A and B. *Eur. J. Biochem.* **51**: 537.

BEKHOR, I., G. M. KUNG, and J. BONNER. 1969. Sequence-specific interaction of DNA and chromosomal protein. *J. Mol. Biol.* **39**: 351.

BLATTI, S. P., C. J. INGLES, T. J. LINDELL, P. W. MORRIS, R. F. WEAVER, F. WEINBERG, and W. J. RUTTER. 1971. Structure and regulatory properties of eukaryotic RNA polymerase. *Cold Spring Harbor Symp. Quant. Biol.* **35**: 649.

BLÜTHMANN, H., Z. MROZEK, and A. GIERER. 1975. Nonhistone chromosomal proteins. Their isolation and role in determining specificity of transcription *in vitro*. *Eur. J. Biochem.* **58**: 315.

BOLLA, R., H. E. ROTH, H. WEISSBACH, and N. BROT. 1977. Effect of ribosomal proteins on synthesis and assembly of preribosomal particles in isolated rat liver nuclei. *J. Biol. Chem.* **252**: 721.

BUSCH, H. 1976. A general concept for molecular biology of cancer. *Cancer Res.* **36**: 4291.

BUSCH, H. and K. SMETANA. 1970. *The nucleolus.* Academic Press, New York.

BUSCH, H., N. R. BALLAL, M. O. J. OLSON, and L. C. YEOMAN. 1975. Chromatin and its nonhistone proteins. In *Methods in cancer research* (ed. H. Busch), vol. XI, p. 43. Academic Press, New York.

BUSCH, R. K., I. DASKAL, W. H. SPOHN, M. KELLERMAYER, and H. BUSCH. 1974. Rabbit antibodies to nucleoli of Novikoff hepatoma and normal liver of the rat. *Cancer Res.* **34**: 2362.

CASPERSSON, T. O. 1950. *Cell growth and cell function. A cytochemical study.* Norton, New York.

CHAMBON, P., F. GISSINGER, C. KEDINGER, J. L. MANDEL, and M. MEILHAC. 1974. Animal nuclear DNA-dependent

RNA polymerases. In *The cell nucleus* (ed. H. Busch), vol. III, p. 269. Academic Press, New York.

CHOI, Y. C., N. R. BALLAL, R. K. BUSCH, and H. BUSCH. 1976. Homochromatographic and immunological analysis of controls of nucleolar gene function. *Cancer Res.* **36**: 4301.

CHYTIL, F. and T. C. SPELSBERG. 1971. Tissue-differences in antigenic properties of non-histone protein-DNA complexes. *Nat. New Biol.* **233**: 215.

COHEN, M. E. and T. H. HAMILTON. 1975. Effect of estradiol-17β on the synthesis of specific uterine nonhistone chromosomal proteins. *Proc. Natl. Acad. Sci.* **72**: 4346.

DASKAL, Y., A. W. PRESTAYKO, and H. BUSCH. 1974. Ultrastructural and biochemical studies of the isolated fibrillar component of nucleoli from Novikoff hepatoma ascites cells. *Exp. Cell Res.* **88**: 1.

DAUBERT, S., D. PETERS, and M. E. DAHMUS. 1977. Selective transcription of ribosomal sequences *in vitro* by RNA polymerase I. *Arch. Biochem. Biophys.* **178**: 381.

DIERKS-VENTLING, C. and F. BIERI-BONNIOT. 1977. Stimulation of RNA polymerase I and II activities by 17β-estradiol receptor on chick liver chromatin. *Nucleic Acids Res.* **4**: 381.

ELADARI, M.-E. and F. GALIBERT. 1975. Sequence determination of 5'-terminal and 3'-terminal T1 oligonucleotides of 18S ribosomal RNA of a mouse cell line (L5178Y). *Eur. J. Biochem.* **55**: 247.

FERENCZ, A. and K. H. SEIFART. 1975. Comparative effect of heparin on RNA synthesis of isolated rat-liver nucleoli and purified RNA polymerase A. *Eur. J. Biochem.* **53**: 605.

FUKE, M. and H. BUSCH. 1975. A T1 ribonuclease fragment present in 18S ribosomal RNA of Novikoff rat ascites hepatoma cells and absent from 18S ribosomal RNA of HeLa cells. *J. Mol. Biol.* **99**: 277.

———. 1977. Sequence analysis of T1 ribonuclease fragments of 18S ribosomal RNA by 5'-terminal labeling, partial digestion, and homochromatography fingerprinting. *Nucleic Acids Res.* **4**: 339.

GOLD, P. and S. O. FREEDMAN. 1965. Specific carcinoembryonic antigens of the human digestive system. *J. Exp. Med.* **122**: 467.

GOLDKNOPF, I. L. and H. BUSCH. 1975. Remarkable similarities of peptide fingerprints of histone 2A and nonhistone chromosomal protein A24. *Biochem. Biophys. Res. Commun.* **65**: 951.

———. 1977. Isopeptide linkage between nonhistone and histone 2A polypeptides of chromosomal conjugate-protein A24. *Proc. Natl. Acad. Sci.* **74**: 864.

GRUMMT, I. 1975. Synthesis of RNA molecules larger than 45S by isolated rat-liver nucleoli. *Eur. J. Biochem.* **57**: 159.

GRUMMT, I. and R. LINDIGKEIT. 1973. Pre-ribosomal RNA synthesis in isolated rat-liver nucleoli. *Eur. J. Biochem.* **36**: 244.

HIGASHI, K., T. MATSUSHITA, A. KITAO, and Y. SAKAMOTO. 1968. Selective suppression of nucleolar RNA metabolism in the absence of protein synthesis. *Biochim. Biophys. Acta* **166**: 388.

HILDEBRANDT, A. and H. W. SAUER. 1977. Transcription of ribosomal RNA in the life cycle of Physarum may be regulated by a specific nucleolar initiation inhibitor. *Biochem. Biophys. Res. Commun.* **74**: 466.

HONJO, T. and R. H. REEDER. 1974. Transcription of Xenopus chromatin by homologous ribonucleic acid polymerase. Aberrant synthesis of ribosomal 5S ribonucleic acid. *Biochemistry* **13**: 1896.

HUNT, L. T. and M. O. DAYHOFF. 1977. Amino-terminal sequence identity of ubiquitin and the nonhistone component of nuclear protein A24. *Biochem. Biophys. Res. Commun.* **74**: 650.

INAGAKI, A. and H. BUSCH. 1972. Structural analysis of nucleolar precursors of ribosomal ribonucleic acids. Sequence analysis of long oligonucleotides produced by

T1 RNase digestion of nucleolar and ribosomal 28S ribonucleic acid of Novikoff hepatoma ascites cells. *J. Biol. Chem.* **247**: 3327.

KANG, Y.-J., M. O. J. OLSON, and H. BUSCH. 1974. Phosphorylation of acid-soluble proteins in isolated nucleoli of Novikoff hepatoma ascites cells: Effects of divalent cations. *J. Biol. Chem.* **249**: 5580.

KEMPER, W. M., K. W. BERRY, and W. C. MERRICK. 1976. Purification and properties of rabbit reticulocyte protein synthesis initiation factors M2Bα and M2Bβ. *J. Biol. Chem.* **251**: 5551.

LIAU, M. C., M. E. HUNT, and R. B. HURLBERT. 1976. Role of ribosomal RNA methylases in the regulation of ribosome production in mammalian cells. *Biochemistry* **15**: 3158.

LIAU, M. C., J. B. HUNT, D. W. SMITH, and R. B. HURLBERT. 1973. Inhibition of transfer and ribosomal DNA methylases by polyinosinate. *Cancer Res.* **33**: 323.

MAMRACK, M. D., M. O. J. OLSON, and H. BUSCH. 1977. Negatively charged phosphopeptides of nucleolar nonhistone proteins from Novikoff hepatoma ascites cells. *Biochem. Biophys. Res. Commun.* **76**: 150.

MANEN, C.-A. and D. H. RUSSELL. 1977. Ornithine decarboxylase may function as an initiation factor for RNA polymerase I. *Science* **195**: 505.

MATSUI, S., M. FUKE, and H. BUSCH. 1975. Reconstitution of nucleolar chromatin-fidelity of rRNA readouts. *J. Cell Biol.* **67**: 266a.

———. 1977. Fidelity of rRNA synthesis by nucleoli and nucleolar chromatin. *Biochemistry* **16**: 39.

MAUCK, J. C. 1977. Solubilized DNA-dependent RNA polymerase activities in resting and growing fibroblast. *Biochemistry* **16**: 793

MORRIS, A. D., C. LITTLETON, L. C. CORMAN, J. ESTERLY, and C. G. SHARP. 1975. Extractable nuclear antigen effect on the DNA anti-DNA reaction and NZB/NZW mouse nephritis. *J. Clin. Invest.* **55**: 903.

MURAMATSU, M., N. SHIMADE, and T. HIGASHINAKAGAWA. 1970. Effect of cycloheximide on the nucleolar RNA synthesis in rat liver. *J. Mol. Biol.* **53**: 91.

NAZAR, R. N., T. O. SITZ, and H. BUSCH. 1975. Tissue specific differences in the 2'-O-methylation of eukaryotic 5.8S ribosomal RNA. *FEBS Lett.* **59**: 83.

OKITA, K., L. H. KLIGMAN, and E. FARBER. 1975. A new common marker for premalignant and malignant hepatocytes induced in the rat by chemical carcinogens. *J. Natl. Cancer Inst.* **54**: 199.

OLSON, M. O. J., E. G. EZRAILSON, K. GUETZOW, and H. BUSCH. 1975. Intranuclear compartmentalization of chromatin proteins: Localization and phosphorylation of nuclear, nucleolar and extranucleolar nonhistone proteins of Novikoff hepatoma ascites cells. *J. Mol. Biol.* **97**: 611.

OLSON, M. O. J., I. L. GOLDKNOPF, K. A. GUETZOW, G. T. JAMES, T. C. HAWKINS, C. J. MAYS-ROTHBERG, and H. BUSCH. 1976. The amino- and carboxyl-terminal amino acid sequence of nuclear protein A24. *J. Biol. Chem.* **251**: 5901.

ORRICK, L. I., M. O. J. OLSON, and H. BUSCH. 1973. Comparison of nucleolar proteins of normal rat liver and Novikoff hepatoma ascites cells by 2-dimensional polyacrylamide gel electrophoresis. *Proc. Natl. Acad. Sci.* **70**: 1316.

PAUL, J. and R. S. GILMOUR. 1968. Organ-specific restriction of transcription in mammalian chromatin. *J. Mol. Biol.* **34**: 305.

PRESTAYKO, A. W., G. R. KLOMP, D. J. SCHMOLL, and H. BUSCH. 1974. Comparison of proteins of ribosomal subunits and nucleolar preribosomal particles from Novikoff hepatoma ascites cells by two-dimensional polyacrylamide gel electrophoresis. *Biochemistry* **13**: 1945.

REEDER, R. H. 1973. Transcription of chromatin by bacterial RNA polymerase. *J. Mol. Biol.* **80**: 229.

ROEDER, R. G. and W. J. RUTTER. 1969. Multiple forms of

DNA-dependent RNA polymerase in eukaryotic organisms. *Nature* **224**: 234.

———. 1970. Specific nucleolar and nucleoplasmic RNA polymerases. *Proc. Natl. Acad. Sci.* **65**: 675.

ROTHBLUM, L. R., P. M. MAMRACK, H. M. KUNKLE, ØM. OJ. OLSON, and H. BUSCH. 1977. Fractionation of nucleoli. Enzymatic and two-dimensional polyacrylamide gel electrophoretic analysis. *Biochemistry* **16**: 4716.

SCHLESINGER, D. H., G. GOLDSTEIN, and H. D. NIALL. 1975. The complete amino acid sequence of ubiquitin, an adenylate cyclase stimulating polypeptide probably universal in living cells. *Biochemistry* **14**: 2214.

SPELSBERG, T. C., L. S. HNILICA, and A. T. ANSEVIN. 1971. Proteins of chromatin-template restriction. III. The macromolecules in specific restriction of the chromatin DNA. *Biochim. Biophys. Acta* **228**: 550.

STEIN, G. S. and J. FARBER. 1972. Role of nonhistone chromosomal proteins in the restriction of mitotic chromatin template activity. *Proc. Natl. Acad. Sci.* **69**: 2918.

STEIN, G. S., R. T. MANS, E. J. GABBAY, J. L. STEIN, J. DAVIS, and P. D. ADAWAKAR. 1975. Evidence for fidelity of chromatin reconstitution. *Biochemistry* **14**: 1859.

SUGANO, N., M. O. J. OLSON, L. C. YEOMAN, B. R. JOHNSON, C. W. TAYLOR, W. C. STARBUCK, and H. BUSCH. 1972. Amino acid sequence of the C-terminal portion of the AL-histone of calf thymus. *J. Biol. Chem.* **247**: 3589.

SUZUKI, M. and Y. HIMUNA. 1974. Evaluation of Epstein-Barr virus-associated nuclear antigen with various human cell lines. *Int. J. Cancer* **14**: 753.

THOMSON, J. A., J.-F. CHIU, and L. S. HNILICA. 1975. Nuclear phosphorprotein kinase activities in normal and neoplastic tissues. *Biochim. Biophys. Acta* **407**: 114.

TODARO, G. J., K. HABEL, and H. GREEN. 1965. Antigenic and cultural properties of cells doubly transformed by polyoma virus and SV40. *Virology* **27**: 179.

TRAVERS, A. 1973. Control of ribosomal RNA synthesis *in vitro*. *Nature* **244**: 15.

———. 1976. RNA polymerase specificity and the control of growth. *Nature* **263**: 641.

WAKABAYASHI, K. and L. S. HNILICA. 1972. Immunochemical and transcriptional specificity of chromatin. *J. Cell Biol.* **55**: 271a.

WARNER, J. R. 1974. The assembly of ribosomes in eukaryotes. In *Ribosomes* (ed. M. Nomura et al.), p. 461. Cold Spring Harbor Laboratory, Cold Spring Harbor, New York.

WEISSBACH, H. and S. OCHOA. 1976. Soluble factors required for eukaryotic protein synthesis. *Annu. Rev. Biochem.* **45**: 191.

WOO, S. L. C., J. M. ROSEN, C. D. LIARAKOS, Y. C. CHOI, H. BUSCH, A. R. MEANS, and B. W. O'MALLEY. 1975. Physical and chemical characterization of purified ovalbumin messenger RNA. *J. Biol. Chem.* **250**: 7027.

WU, B. C., M. S. RAO, K. K. GUPTA, L. I. ROTHBLUM, P. C. MAMRACK, and H. BUSCH. 1977. Evidence for coupled synthesis of mRNA for ribosomal proteins and rRNA. *Cell Biol. Int. Rep.* **1**: 31.

YEOMAN, L. C., C. W. TAYLOR, J. J. JORDAN, and H. BUSCH. 1975. Differences in chromatin proteins of growing and nongrowing tissues. *Exp. Cell Res.* **91**: 207.

YEOMAN, L. C., J. J. JORDAN, R. K. BUSCH, C. W. TAYLOR, H. E. SAVAGE, and H. BUSCH. 1976. A fetal protein in chromatin of Walker hepatoma and Walker 256 carcinosarcoma tumors that is absent from normal and regenerating rat liver. *Proc. Natl. Acad. Sci.* **73**: 3258.

YEOMAN, L. C., M. O. J. OLSON, N. SUGANO, J. J. JORDAN, C. W. TAYLOR, W. C. STARBUCK, and H. BUSCH. 1972. Amino acid sequence of the center portion of the arginine-lysine-rich histone from calf thymus and the total sequence. *J. Biol. Chem.* **247**: 6018.

YU, F.-L. and P. FEIGELSON. 1972. The rapid turnover of RNA polymerase of rat liver nucleolus, and of its messenger RNA. *Proc. Natl. Acad. Sci.* **69**: 2833.

ZARDI, L., J. LIN, and R. BASERGA. 1973. Immunospecificity to non-histone chromosomal proteins of antichromatin antibodies. *Nat. New Biol.* **245**: 211.

ZYLBER, E. A. and S. PENMAN. 1971. Products of RNA polymerases in HeLa cell nuclei. *Proc. Natl. Acad. Sci.* **68**: 2861.

Stimulation of Ribosomal RNA Synthesis in Isolated Nuclei and Nucleoli by Partially Purified Preparations of SV40 T Antigen

R. BASERGA, T. IDE, AND S. WHELLY

Fels Research Institute and Department of Pathology, Temple University School of Medicine, Philadelphia, Pennsylvania 19140

Under conditions restrictive for growth, such as high cell density or serum deprivation, human diploid cells (Wiebel and Baserga 1969; Westermark 1971) and a number of established rodent cell lines (Todaro et al. 1965; Nilhausen and Green 1965; Holley and Kiernan 1968; Bürk 1970) enter a stationary phase in which most of the cells have a DNA content equivalent to that of cells in the G_1 phase of the cell cycle (Wiebel and Baserga 1969; Bartholomew et al. 1976). When these resting cells, called G_0 cells by some investigators (Epifanova and Terskikh 1969; Sander and Pardee 1972; Baserga et al. 1973; Smets 1973), are stimulated to proliferate, they enter DNA synthesis after a lag period that varies from 10 hours to more than 40 hours. The entrance into S is then followed by a wave of mitosis. Under similar conditions restrictive for growth, virally transformed mammalian cells do not enter G_0 but instead continue to proliferate, albeit slowly, until they eventually die (Gierthy and Studzinski 1973; Pardee 1974; Burstin and Basilico 1975; Pardee and James 1975; Bartholomew et al. 1976; Schiaffonati and Baserga 1977).

During the lag period preceding the entrance of cells into the S phase (often called the prereplicative phase), the cells stimulated to proliferate undergo a series of biochemical events which include membrane changes, cytoplasmic events, and nuclear changes. Among the nuclear changes that occur in the early prereplicative phase of cells stimulated to proliferate, the most prominent have been an increase in transcriptional activity and changes in nuclear proteins.

Increased Transcriptional Activity in Cells Stimulated to Proliferate

The notion that increased transcriptional activity occurs in G_0 cells stimulated to proliferate goes back to the experiments of Lieberman and coworkers in 1963 (Lieberman et al. 1963), and since then a large body of evidence has accumulated in support of Lieberman's findings, evidence that has been analyzed in detail in reviews and even in books (Baserga 1976; Baserga et al. 1976). This increased transcriptional activity in nuclei (or chromatin) of cells stimulated to proliferate has been confirmed repeatedly in several systems and by many laboratories. Its extent, however, is somewhat surprising. For instance, in isolated nuclei, RNA synthesis increases 60–80% within a few hours after G_0 cells are stimulated to proliferate. It is hardly believable that the number of genes active in transcription may double in growing cells, and indeed, in the few instances in which the number and diversity of cytoplasmic mRNAs have been investigated, the differences between growing and resting cells have always been found to be small (Williams and Penman 1975; Grady and Campbell 1975; Getz et al. 1976). Recently it has been shown that most of the quantitative changes in transcriptional activity described in cells stimulated to proliferate can be attributed to the nucleolus. The observation of an increased synthesis of rRNA in growing cells again goes back several years, and to the laboratory of Irving Lieberman (Tsukada and Lieberman 1964), but it is only with the measurement of RNA synthesis (endogenous RNA polymerase activity) in isolated nucleoli that it has become possible to determine the magnitude of the increase accurately. A number of reports have now appeared indicating that the doubling in RNA synthesis in isolated nuclei can largely be attributed to a fourfold increase (or even more) of RNA synthesis in nucleoli (Schmid and Sekeris 1975; Baserga et al. 1977; Rossini et al. 1977). Since the synthesis of preribosomal RNA ordinarily accounts for about 35–40% of total nuclear RNA synthesis (Reeder and Roeder 1972), a fourfold increase in pre-rRNA synthesis can explain, without disturbing the laws of nature, most of the quantitative changes in transcriptional activity described in cells stimulated to proliferate. The increased activity of rRNA genes does not necessarily involve the activation of new genes, since the synthesis of rRNA, for instance in *Neurospora*, can be increased by simply increasing the number of RNA polymerase I molecules per rRNA gene (Alberghina et al. 1975). Alternatively, because of the large number of copies of rRNA genes, rRNA synthesis could be markedly increased by increasing the number of active rRNA genes, as seems to happen in *Drosophila* embryogenesis (McKnight and Miller 1976).

Nuclear Proteins

The second prominent group of changes described in nuclei of cells stimulated to proliferate has to do with nuclear proteins, both histones and nonhistone chromosomal proteins. These changes have also been the object of extensive reviews (Stein et al. 1974; Baserga 1976). When investigators first focused their attention upon nonhistone chromosomal proteins, it was hoped that a specific, nonhistone protein, phosphorylated or not, could be identified as being responsible for the initial triggering of G_0 cells into a cycling state, presumably by increasing transcriptional activity in nuclei (or nucleoli). Attempts to identify and isolate such a protein have failed thus far. Recently, an acidic protein that specifically stimulates rRNA in nucleoli of resting cells has been identified and almost completely purified. This protein is almost certainly the T antigen from SV40-transformed cells, and in the balance of this paper we will discuss the evidence accumulated in the past year on the effect of SV40 T antigen on RNA synthesis in isolated nuclei and nucleoli of quiescent cells.

The A-gene Product

It has been known for a number of years that infection by certain DNA oncogenic viruses can induce cellular DNA synthesis in the host cell (Dulbecco et al. 1965; Sauer and Defendi 1966; Zimmerman and Raska 1972). In some respects these viruses, which Weil has called mitogenic viruses (Weil et al. 1975), behave in the same way as serum in density-inhibited cultures. The stimulation of cellular DNA synthesis by DNA oncogenic viruses also occurs after a lag period, as in resting cells stimulated by serum, and is preceded by the appearance of a nuclear antigen which, in the case of SV40 and polyoma, has been called the T antigen (Oxman and Black 1966; Henry et al. 1966; Weil and Kara 1970; Mauel and Defendi 1971; Weil et al. 1975). There is considerable evidence that the A-gene product of the SV40 genome is required for (1) the initiation of SV40 DNA replication (Tegtmeyer 1972), (2) the transcription of late SV40 genes (Cowan et al. 1973), (3) the establishment and maintenance of transformation (Abrahams et al. 1975; Brugge and Butel 1975; Martin et al. 1975), and (4) the stimulation of cellular DNA synthesis that occurs after SV40 infection (Chou and Martin 1975; and see review by Levine 1976). There is also substantial evidence that the main, if not the only, product of the A gene is an 80,000- to 100,000-m.w. protein identifiable with the T antigen (Lewis et al. 1973; Tegtmeyer et al. 1975; Ahmad-Zadeh et al. 1976; Carroll and Smith 1976). Recently, Graessmann and Graessmann (1976) synthesized in vitro RNA from the early strand of SV40 DNA, which is believed to code for the A-gene product (see review by Acheson 1976).

This in vitro synthesized RNA (cRNA) was microinjected into primary mouse kidney cells and shown to cause the appearance of T antigen and the stimulation of cellular DNA synthesis in host cells. Graessmann and Graessmann (1976) concluded that SV40-specific T antigen provides the necessary information for the stimulation of cellular DNA synthesis. On the basis of these findings and our knowledge of the biochemical events occurring in the prereplicative phase of cells stimulated to proliferate, we decided to test whether T-antigen preparations could directly stimulate RNA synthesis in isolated nuclei from resting cells.

Stimulation of RNA Synthesis in Isolated Nuclei

The original experiment (Ide et al. 1977) is shown in Figure 1. Rat liver nuclei were preincubated with SV40 T-antigen preparations, and RNA synthesis was assayed by the method of Marzluff et al. (1973). Addition of T-antigen preparations causes a marked increase in the incorporation of [³H]UTP into acid-precipitable material. When a similar amount of protein extracted from the nuclei of nonvirally transformed cells (mock T) was used for the preincubation, there was no appreciable stimulation of RNA synthesis. Since that original experiment, the two questions we have been trying to answer are: (1) What species of RNA is synthesized in response to SV40 T-antigen preparations? (2) Is the stimulatory factor in the T-antigen preparations really the T antigen itself or some other contaminating protein?

The first question is easier to answer, as shown in Tables 1 and 2. The increase in RNA synthesis in isolated nuclei is 100% α-amanitin-resistant. Actinomycin D completely inhibits RNA synthesis,

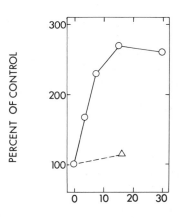

Figure 1. Stimulation of RNA synthesis in isolated rat liver nuclei by added T-antigen preparations from SV40-transformed cells. Conditions are described in Table 1. The abscissa gives micrograms of protein added and the ordinate represents RNA synthesis expressed in percent of control values (no T added). (○———○) SV40 T; (△----△) proteins from a similar preparation of nonvirally transformed cells.

Table 1. Effect of T-antigen Preparations from SV40-transformed Cells on RNA Synthesis in Isolated Nuclei and Nucleoli

	Cpm in RNA/μg DNA	
	−T	+T
Rat liver nuclei	241	501
+ α-amanitin (5 μg/ml)	150	451
+ actinomycin D (10 μg/assay)	0	0
+ RNase (5 μg/ml)	16	13
+ PMSF (1 mM)	241	511
Rat liver nucleoli	2822	6006
+ α-amanitin	3497	6116
+ RNase (5 μg/ml)	74	122

Nuclei were prepared and assayed for RNA synthesis by the method of Marzluff et al. (1973). Nucleoli were prepared and assayed by the procedures of Muramatsu et al. (1974) and Grummt (1975). T antigen was extracted and partially purified from SV80 cells using the method of Jessel et al. (1975). Nuclei and nucleoli were preincubated at 25°C with T antigen for 30 or 15 min, respectively. The assay was carried out at 37°C for 15 min, and the results express the incorporation of [³H]UTP into acid-insoluble material.

whereas the product synthesized in vitro by either nuclei or nucleoli is RNase-digestible. Proteolytic inhibitors, such as phenylmethylsulfonyl fluoride (PMSF), have no effect on either controls or stimulated nuclei. Stimulation can also be achieved in isolated nucleoli, and the nucleolar product in RNA-DNA hybridization tests hybridizes much more efficiently to nucleolar DNA than to nuclear DNA; the hybridization is effectively competed out by cold cytoplasmic rRNA. The only demonstrable product of nucleolar RNA synthesis is rRNA precursor (Grummt 1975) but, of course, absence of evidence is not evidence of absence and it may be possible that other RNA species are synthesized by the isolated nucleolus. The same thing can be said of the finding that the increased synthesis of RNA in isolated nuclei is 100% α-amanitin-resistant. The activation of a few extranucleolar genes can be missed completely in this kind of assay. However, although we cannot rule out an effect on some extranucleolar genes, we can say that SV40 T-antigen preparations definitely stimulate rRNA synthesis in both nuclei and nucleoli, and indeed that most of the stimulation must be due to the increased synthesis of rRNA precursors. Isolated nuclei have to be preincubated with T-antigen preparations for at least 20 minutes to obtain a good stimulation. This is not true of nucleoli, where a 5-minute preincubation is sufficient to produce the maximum stimulatory effect.

Evidence That T Antigen Stimulates rRNA Synthesis

In most of the experiments we reported (Ide et al. 1977), the source of our T-antigen preparations were SV80 cells that were kindly given to us by Dr. David Livingston, Sidney Farber Cancer Center, Boston, Massachusetts. The T-antigen preparations were partially purified by the method of Jessel et al. (1975). The presence of T was monitored by complement-fixation tests for T antigen by the microtiter technique of Sever (1961). These partially purified preparations of T contain other proteins. Two activities that ought to be ruled out quickly are RNA polymerase activity and DNase activity. Table 3 shows that our T-antigen preparations contain both RNA polymerase activity and DNase activity. Both of these activities are essentially eliminated by heating the T-antigen preparations at 50°C for 30 minutes. Under the condition of heating used in our experiments, wild-type T antigen is not inactivated by 30-minute heating at 50°C, both in terms of its complement-fixing activity and its ability to stimulate RNA synthesis in isolated nuclei or nucleoli. We should add that in other experiments we were able to separate, by column chromatography, the RNA polymerase activity of the preparations from their complement-fixing activity. The T-antigen-containing fractions, devoid of any RNA polymerase activity, are still capable of stimulating RNA synthesis in isolated nuclei and nucleoli (T. Ide et al., unpubl.).

Stimulation of RNA synthesis in isolated nuclei or nucleoli is inhibited by anti-T hamster antiserum, but not by normal hamster serum (Ide et al. 1977). More important are our experiments with tsA mutants that were kindly given to us by Dr. Robert G. Martin, National Institutes of Health. These are CHL cells that have been transformed by tsA mutants of SV40. Two such cell lines, tsA239 and tsA241, grew equally well at 34°C and 39.5°C, if in

Table 2. Hybridization of the In Vitro Product of the Nucleolus with Rat Liver DNA

	Cpm hybridized
Whole-cell DNA	not detectable
Nucleolar DNA	141
+ tRNA (30 μg)	151
+ rRNA (30 μg)	73

The nucleoli were incubated as described in Table 1, and the radioactive product (labeled with [³H]UTP) was hybridized to DNA by the procedure of Gillespie (1968). The input cpm was 1400 cpm with a background of 66 cpm, which was not subtracted from the above figures. The amount of DNA bound to the filter was 25 μg.

Table 3. Elimination of RNA Polymerase and DNase Activities from T-antigen Preparations

	Cpm in RNA/μg DNA[a]		
	−T	+T	+ heated T
Nuclei	203	461	576
Nucleoli	555	1089	1148
25 μg Poly[d(AT)]	0	2598	0
DNase activity[b]	6.1	23.0	4.6

[a] Nuclei and nucleoli were assayed as described in Table 1. Heated T refers to T-antigen preparations heated for 30 min at 50°C.

[b] DNase activity is expressed as percent of form-III DNA formed from circular SV40 DNA and determined by the method of Greene et al. (1974).

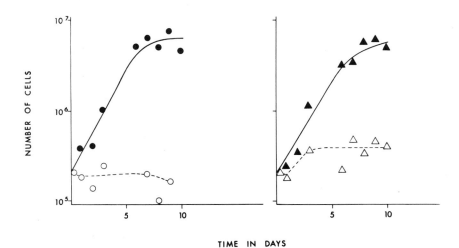

Figure 2. Growth of cells transformed by SV40 tsA mutants in 1% serum. (▲) tsA241 at 34°C; (△) tsA241 at 39°C; (●) tsA239 at 34°C; (○) tsA239 at 39°C. These cell lines were the kind gift of Dr. R. G. Martin, NIH.

10% serum. However, in 1% serum, the cells grow well at 34°C but very little, if at all, at 39.5°C (Fig. 2). The cell line we used for our studies is the cell line tsA239, which has a thermosensitive T antigen (Anderson et al. 1977). T-antigen preparations were partially purified from these cells, as well as from CHL wt15 cells, which are CHL cells transformed by wild-type SV40. Incubation of these preparations at 25°C, up to 2 hours, has no effect on either the wild type or the tsA T, both in terms of complement-fixing activity or ability to stimulate nucleolar RNA synthesis. When the same T-antigen preparations were heated at 50°C for a period of 2 hours, the complement-fixing activity of the tsA T decreased more profoundly and much more rapidly than the complement-fixing activity of the wild-type T (Ide et al. 1977). Table 4 shows that heated wild-type T preparations are still active in stimulating nucleolar RNA synthesis. When the tsA T preparations are used, preheating at 50°C for 2 hours rapidly abolishes their ability to stimulate nucleolar RNA synthesis.

Incidentally, nucleolar RNA synthesis can also be stimulated by cytoplasmic extracts prepared from J19L cells (kindly given to us by Dr. Janet Butel, Baylor University), a cell line which has a cytoplasmic T antigen.

Together, these results already lend strong support to the possibility that it is the T antigen in the partially purified preparations that stimulates

Table 4. Effect of T-antigen Preparations from tsA Mutants on RNA Synthesis in Isolated Nucleoli from Rat Liver

	Cpm in RNA/μg DNA		
Source of T	−T	+T	+heated T
CHL wt15	613	1089	1609
tsA239	613	2054	777

Nucleoli were prepared and assayed as described in Table 1. Heated T means T-antigen preparations heated at 50°C for 2 hr. CHL wt15 and tsA239 cells were the kind gift of Dr. Robert G. Martin, NIH. The former are CHL cells transformed by wild-type SV40; the latter are CHL cells transformed by a tsA mutant of SV40.

RNA synthesis in either isolated nuclei (Ide et al. 1977) or isolated nucleoli. Further experiments were carried out with more purified preparations that were kindly given to us by Dr. David Livingston and Dr. Robert G. Martin. The results are shown in Table 5. The preparation from the laboratory of Dr. Robert G. Martin was said to be 50% pure. The preparation from the laboratory of Dr. David Livingston was believed to be more than 90% pure. Both preparations stimulated RNA synthesis in isolated nucleoli from rat liver. Quite clearly, a final demonstration that SV40 T antigen does indeed stimulate rRNA synthesis must await the complete purification of T. In the meantime though, it is apparent that the evidence favors T. In the Discussion we will assume that T antigen is indeed the component

Table 5. Stimulation of RNA Synthesis in Isolated Rat Liver Nucleoli by T-antigen Preparations from Different Sources

Source of T	Protein (μg)	Percentage of control
No T added	—	100
SV80 T from our laboratory	13	211
T from Dr. Martin's laboratory	10	124
T from Dr. Livingston's laboratory	0.9	154

Nucleoli were prepared and assayed as described in Table 1.

of the T-antigen preparations that is capable of stimulating rRNA synthesis in isolated nuclei and nucleoli.

DISCUSSION

If it is indeed the T antigen from SV40 that stimulates rRNA synthesis in isolated nuclei and nucleoli, the results could throw some light on our understanding of the mechanism of cell proliferation, both in normal cells and in virally transformed cells. In the case of cellular DNA synthesis induced by infection with SV40, a reasonable explanation is that the A-gene product of SV40, the T antigen, may stimulate rRNA synthesis in much the same way as serum stimulates rRNA synthesis in density-inhibited cells. From there on, the other steps of the prereplicative phase could be essentially autonomous so that the resting cells, whether stimulated by serum or by SV40 infection, eventually would enter Weil et al. (1975) have previously reported that in cells infected with SV40 or polyoma virus, there is an increase in the amount of cellular RNA, and an increase in cellular RNA synthesis has also been reported in BHK cells infected by adenovirus 12 (Raska et al. 1971). Our results would fit in with the data available in the literature and essentially make the T antigen from SV40 the virally directed counterpart of the hypothetical protein that is induced by serum stimulation and that activates rRNA synthesis in density-inhibited normal cells stimulated to proliferate.

These findings could also explain, at least in part, the mechanism of viral transformation. In SV40-transformed cells, the T antigen is expressed, and one could hypothesize that its presence continually stimulates the nucleolus of the transformed cells. This continuous interaction of the T antigen with the nucleolus of transformed cells would keep the virally transformed cells from entering the G_0 phase that is known to occur in normal cells or, at least, in cells that are capable of density inhibition. The cells then would continue to grow until the nutritional conditions become such that they cannot further sustain growth and the virally transformed cells would then die. As mentioned earlier, this is indeed what has been reported in the literature. Extrapolation to cells transformed by chemical carcinogens or to neoplastic cells in the living animal is not justified at this point, but if one is allowed a speculation in order to find some relevance to human disease, the following thoughts are worthwhile considering. There are four neoplastic disorders in man in which nonrandom chromosomal aberrations have been described. These are chronic myeloid leukemia, in which there is a translocation from chromosome 22 to chromosome 9 (Rowley 1973); retinoblastoma, in which there is a partial deletion of chromosome 13 (Knudson et al. 1976); some forms of meningioma, in which there is a deletion of one G_{22} chromosome

(Mark 1974); and Burkitt's lymphoma, in which there is a translocation to chromosome 14 (Kaplan 1976). The rDNA genes in human chromosomes are located on chromosome 13, 14, 15, 21, and 22 (Evans et al. 1974); therefore, in all cases of neoplastic disorders in man in which there are nonrandom chromosomal aberrations, the chromosomal abnormalities involve a chromosome that has rDNA genes.

Although this could be a pure coincidence, the finding that T antigen may stimulate rRNA synthesis in nuclei and nucleoli of resting cells would certainly be of help in our understanding of the mechanisms by which quiescent cells are triggered into the growth cycle. To our laboratory it is especially gratifying that, after several years of futile attempts, the pieces of the puzzle seem to fall together. The increased transcriptional activity, both in the case of serum-stimulated cells and in the case of SV40-infected cells, can largely be attributed to the nucleolus, a phenomenon which is much easier to explain than the activation of a large number of unique copy genes. It also turns out that the stimulation of rRNA synthesis can be brought about by an acidic protein. We do not know, however, what is the mechanism by which T antigen stimulates rRNA synthesis, nor do we know whether a normal counterpart of T antigen exists in density-inhibited cells stimulated to proliferate.

Acknowledgment

This work was supported by U.S. Public Health Service Research Grants CA-08474 and CA-12923 from the National Cancer Institute.

REFERENCES

Abrahams, P. J., C. Mulder, A. Van De Voorde, S. O. Warnaar, and A. J. Van Der Eb. 1975. Transformation of primary rat kidney cells by fragments of simian virus 40 DNA. *J. Virol.* **16**: 818.

Acheson, N. H. 1976. Transcription during productive infection with polyoma virus and simian virus 40. *Cell* **8**: 1.

Ahmed-Zadeh, C., B. Allet, J. Greenblatt, and R. Weil. 1976. Two forms of simian-virus 40 specific T-antigen in abortive and lytic infection. *Proc. Natl. Acad. Sci.* **73**: 1097.

Alberghina, F. A. M., E. Sturani, and J. R. Gohlke. 1975. Levels and rates of synthesis of ribosomal ribonucleic acid, transfer ribonucleic acid, and protein in *Neurospora crassa* in different steady states of growth. *J. Biol. Chem.* **250**: 4381.

Anderson, J. L., C. Chang, P. T. Mora, and R. J. Martin. 1977. Expression and thermal stability of simian virus 40 tumor-specific transplantation antigen and tumor antigen in wild type- and ts mutant-transformed cells. *J. Virol.* **21**: 459.

Bartholomew, J. C., H. Yokota, and P. Ross. 1976. Effect of serum on the growth of Balb 3T3 mouse fibroblasts and an SV40 transformed derivative. *J. Cell. Physiol.* **88**: 277.

Baserga, R. 1976. *Multiplication and division in mammalian cells.* Marcel Dekker, New York.

Baserga, R., M. Costlow, and G. Rovera. 1973. Changes

in membrane function and chromatin template activity in diploid and transformed cells in culture. Pathology of transcription and translation. *Fed. Proc.* **32**: 2115.

BASERGA, R., C. H. HUANG, M. ROSSINI, H. CHANG, and P. M. L. MING. 1976. The role of nuclei and nucleoli in the control of cell proliferation. *Cancer Res.* **36**: 4297.

BASERGA, R., P. M. L. MING, Y. TSUTSUI, S. WHELLY, H. CHANG, M. ROSSINI, and C. H. HUANG. 1977. Nuclear control of cell proliferation. In *International cell biology 1976–1977* (ed. B. R. Brinkley and K. R. Porter). Rockefeller University Press, New York. (In press.)

BRUGGE, J. S. and J. S. BUTEL. 1975. Role of simian virus 40 gene A function in maintenance of transformation. *J. Virol.* **5**: 619.

BÜRK, R. R. 1970. One-step growth cycle for BHK21/13 hamster fibroblasts. *Exp. Cell Res.* **63**: 309.

BURSTIN, S. J. and C. BASILICO. 1975. Transformation by polyoma virus alters expression of a cell mutation affecting cycle traverse. *Proc. Natl. Acad. Sci.* **72**: 2540.

CARROLL, R. B. and A. E. SMITH. 1976. Monomer molecular weight of T-antigen from simian virus 40-infected and transformed cells. *Proc. Natl. Acad. Sci.* **73**: 2254.

CHOU, J. Y. and R. G. MARTIN. 1975. DNA infectivity and the induction of host DNA synthesis with temperature sensitive mutants of simian virus 40. *J. Virol.* **15**: 145.

COWAN, K., P. TEGTMEYER, and D. D. ANTHONY. 1973. Relationship of replication and transcription of simian virus 40 DNA. *Proc. Natl. Acad. Sci.* **70**: 1927.

DULBECCO, R., L. H. HARTWELL, and M. VOGT. 1965. Induction of cellular DNA synthesis by polyoma virus. *Proc. Natl. Acad. Sci.* **53**: 403.

EPIFANOVA, O. I. and V. V. TERSKIKH. 1969. Review of the resting period in the cell life cycle. *Cell Tissue Kinet.* **2**: 75.

EVANS, H. J., R. A. BUCKLAND, and M. L. PARDUE. 1974. Location of the genes coding for 18S and 28S ribosomal RNA in the human genome. *Chromosoma* **48**: 405.

GETZ, M. J., P. K. ELDER, E. W. BENZ, JR., R. E. STEPHENS, and H. L. MOSES. 1976. Effect of cell proliferation on levels and diversity of poly(A)-containing mRNA. *Cell* **7**: 255.

GIERTHY, J. F. and G. P. STUDZINSKI. 1973. Absence of aminonucleoside-sensitive steps in the cell cycle of SV40-transformed human fibroblasts. *Cancer Res.* **33**: 2673.

GILLESPIE, D. 1968. The formation and detection of DNA-RNA hybrids. *Methods Enzymol.* **12B**: 641.

GRADY, L. J., and W. P. CAMPBELL. 1975. Non-repetitive DNA transcripts in nuclei and polysomes of polyoma-transformed and non-transformed mouse cells. *Nature* **254**: 356.

GRAESSMANN, M. and A. GRAESSMANN. 1976. "Early" simian-virus-40-specific RNA contains information for tumor antigen formation formation and chromatin replication. *Proc. Natl. Acad. Sci.* **73**: 366.

GREENE, P. J., M. C. BETLACH, H. W. BOYER, and H. M. GOODMAN. 1974. The *Eco*RI restriction endonuclease. In *DNA replication* (ed. R. B. Wickner), p. 87. Marcel Dekker, New York.

GRUMMT, I. 1975. Synthesis of RNA molecules larger than 45S by isolated rat-liver nucleoli. *Eur. J. Biochem.* **57**: 159.

HENRY, P., P. H. BLACK, M. N. OXMAN, and S. M. WISEMAN. 1966. Stimulation of DNA synthesis in mouse cell line 3T3 by simian virus 40. *Proc. Natl. Acad. Sci.* **56**: 1170.

HOLLEY, R. W. and J. A. KIERNAN. 1968. "Contact inhibition" of cell division in 3T3 cells. *Proc. Natl. Acad. Sci.* **60**: 300.

IDE, T., S. WHELLY, and R. BASERGA. 1977. Stimulation of RNA synthesis in isolated nuclei by partially purified preparations of SV40 T-antigen. *Proc. Natl. Acad. Sci.* **74**: 3189.

JESSEL, D., J. HUDSON, O. T. LANDAU, D. TENEN, and D.

M. LIVINGSTON. 1975. Interaction of partially purified simian virus 40 T antigen with circular viral DNA molecules. *Proc. Natl. Acad. Sci.* **72**: 1960.

KAPLAN, H. S. 1976. Hodgkin's disease and other human malignant lymphomas: Advances and prospects. *Cancer Res.* **36**: 3863.

KNUDSON, A. G., A. T. MEADOWS, W. W. NICHOLS, and R. HILL. 1976. Chromosomal deletion and retinoblastoma. *N. Engl. J. Med.* **295**: 1120.

LEVINE, A. J. 1976. SV40 and adenovirus early functions involved in DNA replication and transformation. *Biochim. Biophys. Acta* **458**: 213.

LEWIS, A. M., JR., A. J. LEVINE, A. S. CRUMPACKER, M. J. LEVINE, R. J. SAMAHA, and P. H. HENRY. 1973. Studies of nondefective adenovirus 2-simian virus 40 hybrid viruses. *J. Virol.* **11**: 655.

LIEBERMAN, I., R. ABRAMS, and P. OVE. 1963. Changes in the metabolism of ribonucleic acid preceding the synthesis of deoxyribonucleic acid in mammalian cells cultured from the animal. *J. Biol. Chem.* **238**: 2141.

MARK, A. 1974. The human menigioma: A benign tumor with specific chromosome characteristics. In *Chromosomes and cancer* (ed. J. German), p. 497. Wiley, New York.

MARTIN, R. G., J. Y. CHOU, J. AVILA, and R. SARAL. 1975. The semiautonomous replicon: A molecular model for the oncogenicity of SV40. *Cold Spring Harbor Symp. Quant. Biol.* **39**: 17.

MARZLUFF, W. F., E. C. MURPHY, JR., and R. C. C. HUANG. 1973. Transcription of ribonucleic acid in isolated mouse myeloma nuclei. *Biochemistry* **12**: 3440.

MAUEL, J. and V. DEFENDI. 1971. Infection and transformation of mouse peritoneal macrophages by simian virus 40. *J. Exp. Med.* **134**: 335.

McKNIGHT, S. L. and O. L. MILLER, JR. 1976. Ultrastructural patterns of RNA synthesis during early embryogenesis of *Drosophila melanogaster*. *Cell* **8**: 305.

MURAMATSU, M., Y. HAYASHI, T. ONISHI, M. SAKAI, K. TAKAI, and T. KASHIYAMA. 1974. Rapid isolation of nucleoli from detergent purified nuclei of various tumor and tissue culture cells. *Exp. Cell Res.* **88**: 345.

NILHAUSEN, K. and H. GREEN. 1965. Reversible arrest of growth in G_1 of an established fibroblast line (3T3). *Exp. Cell Res.* **40**: 166.

OXMAN, M. N. and P. H. BLACK. 1966. Inhibition of SV40 T-antigen formation by interferon. *Proc. Natl. Acad. Sci.* **55**: 1133.

PARDEE, A. B. 1974. A restriction point for control of normal animal cell proliferation. *Proc. Natl. Acad. Sci.* **71**: 1286.

PARDEE, A. B., and L. J. JAMES. 1975. Selective killing of transformed baby hamster kidney (BHK) cells. *Proc. Natl. Acad. Sci.* **72**: 4994.

RASKA, K., W. A. STROHL, J. HOLOWCZAK, and J. ZIMMERMAN. 1971. The response of BHK21 cells to infection with type 12 adenovirus. VI. Stimulation of cellular RNA synthesis and evidence for transcription of the viral genome. *Virology* **44**: 296.

REEDER, R. H. and R. G. ROEDER. 1972. Ribosomal RNA synthesis in isolated nuclei. *J. Mol. Biol.* **67**: 433.

ROSSINI, M., A. KANE, and R. BASERGA. 1977. Nuclear control of cell proliferation. In *29th Annual Symposium on Fundamental Cancer Research*, p. 180. Williams and Wilkins, Baltimore, Maryland.

ROWLEY, J. D. 1973. A new consistent chromosomal abnormality in chronic myelogenous leukemia identified by quinacrine fluorescence and Giemsa staining. *Nature* **243**: 290.

SANDER, G. and A. B. PARDEE. 1972. Transport changes in synchronously growing CHO and L cells. *J. Cell. Physiol.* **80**: 267.

SAUER, G. and V. DEFENDI. 1966. Stimulation of DNA synthesis and complementfixing antigen production by SV40 in human diploid cell cultures: Evidence for "abortive" infection. *Proc. Natl. Acad. Sci.* **56**: 452.

SCHIAFFONATI, L. and R. BASERGA. 1977. Different survival of normal and transformed cells exposed to nutritional conditions nonpermissive for growth. *Cancer Res.* **37**: 541.

SCHMID, W. and W. SEKERIS. 1975. Nucleolar RNA synthesis in the liver of partially hepatectomized and cortisoltreated rats. *Biochim. Biophys. Acta* **402**: 244.

SEVER, J. L. 1961. Application of microtechnique to viral serological investigations. *J. Immunol.* **88**: 320.

SMETS, L. A. 1973. Activation of nuclear chromatin and the release from contact-inhibition of 3T3 cells. *Exp. Cell Res.* **79**: 239.

STEIN, G. S., T. C. SPELSBERG, and L. J. KLEINSMITH. 1974. Nonhistone chromosomal proteins and gene regulation. *Science* **183**: 817.

TEGTMEYER, P. 1972. Simian virus 40 deoxyribonucleic acid synthesis: The viral replication. *J. Virol.* **10**: 591.

TEGTMEYER, P., M. SCHWARTZ, J. K. COLLINS, and K. RANDELL. 1975. Regulation of tumor antigen synthesis by simian virus 40 gene A. *J. Virol.* **16**: 168.

TODARO, G. J., G. K. LAZAR, and H. GREEN. 1965. The initiation of cell division in a contact-inhibited mammalian cell line. *J. Cell. Comp. Physiol.* **66**: 325.

TSUKADA, K. and I. LIEBERMAN. 1964. Metabolism of nucleolar ribonucleic acid after partial hepatectomy. *J. Biol. Chem.* **239**: 1564.

WEIL, R. and J. KARA. 1970. Polyoma "tumor antigen." An activator of chromosome replication? *Proc. Natl. Acad. Sci.* **67**: 1011.

WEIL, R., C. SALMON, E. MAY, and P. MAY. 1975. A simplifying concept in tumor virology: Virus-specific "pleitropic effectors." *Cold Spring Harbor Symp. Quant. Biol.* **39**: 381.

WESTERMARK, B. 1971. Proliferation control of cultivated human glia-like cells under "steady state" conditions. *Exp. Cell Res.* **69**: 259.

WIEBEL, F. and R. BASERGA. 1969. Early alterations in amino-acid pools and protein synthesis of diploid fibroblasts stimulated to synthesize DNA by addition. *J. Cell. Physiol.* **74**: 191.

WILLIAMS, J. G. and S. PENMAN. 1975. The messenger RNA sequences in growing and resting mouse fibroblasts. *Cell* **6**: 197.

ZIMMERMAN, J. E., JR. and K. RASKA. 1972. Inhibition of adenovirus type 12 induced DNA synthesis in G_1-arrested BHK21 cells by dibutyryl adenosine cyclic 3':5'-monophosphate. *Nat. New Biol.* **239**: 145.

Purification of a Protein from Unfertilized Eggs of *Drosophila* with Specific Affinity for a Defined DNA Sequence and the Cloning of This DNA Sequence in Bacterial Plasmids

H. Weideli, P. Schedl, S. Artavanis-Tsakonas, R. Steward, R. Yuan,* and W. J. Gehring

*Department of Cell Biology and *Department of Microbiology, Biozentrum, University of Basel, CH-4056 Basel, Switzerland*

In prokaryotes, gene activity is regulated at the transcriptional level by proteins which bind to specific DNA sequences and thereby control the expression of an adjacent gene or a group of genes (Jacob and Monod 1961; Gilbert and Mueller-Hill 1966, 1967; Ptashne 1967a,b; Engelsberg et al. 1969; Eron et al. 1971; Guha et al. 1971). In eukaryotes, there is little information about the mechanism of gene regulation, but it seems likely that higher organisms use a similar regulatory mechanism to which, in the course of evolution, other control mechanisms were added. On the basis of this working hypothesis, we have attempted to isolate from *Drosophila melanogaster* proteins which bind to specific DNA sequences.

The unfertilized egg was chosen as a source of material for several reasons: (1) Large quantities of eggs uncontaminated with other cell types can be obtained for biochemical studies. (2) During embryogenesis, the egg goes through rapid cycles of nuclear divisions until there are about 6000 nuclei, which subsequently become incorporated into cells. Since there is little protein synthesis during this time, it seems that the egg requires a presynthesized pool of DNA-binding proteins needed for early development, as has been shown for DNA polymerase (Margulies and Chargaff 1973). (3) Since there is only one nucleus and a large amount of cytoplasm, such DNA-binding proteins are probably located in the cytoplasm rather than being bound to chromatin, from which they would be difficult to dissociate. (4) It has been shown that there is some determination as early as the cellular blastoderm stage (Chan and Gehring 1971; Wieschaus and Gehring 1976; Steiner 1976; Lawrence and Morata 1977) immediately following the series of rapid nuclear divisions. It is therefore reasonable to assume that DNA-binding proteins that may be needed during early development are also stored in the unfertilized egg.

Using a filter assay (Riggs et al. 1970), it is possible to detect binding of a protein to moderately repetitive sequences which have been implicated in gene regulation (Britten and Davidson 1969, 1971). Furthermore, it is possible to demonstrate the binding of a protein to a unique DNA sequence if this sequence can be isolated by cloning. In the present study, we report the isolation of a protein (DB-1) that binds specifically to a cloned DNA sequence. In situ hybridization indicates that this sequence is located in the nucleolus.

EXPERIMENTAL PROCEDURES

DNA isolation and labeling. *Drosophila* DNA was isolated from frozen pupae and from Kco tissue-culture cells according to the procedure of Robertson et al. (1969) with minor modifications. The Kco line is derived from Kc cells (Echalier and Ohanessian 1969, 1970) and grows in medium D 20, diluted by 10% double-distilled water, in the absence of serum. For labeling, the cells were grown in large (75 cm²) Falcon flasks to a cell density of 6×10^6 cells per flask. The medium was then replaced by medium containing dialyzed yeast extract and 150 μCi/ml [³H]thymidine (sp. act. 80–100 Ci/mmole) and the cells labeled until they reached confluency (3–4 days). They were then harvested, washed, and the DNA extracted as described above. A specific activity of approximately $2–3 \times 10^5$ cpm/μg of DNA was obtained.

Plasmid DNA was isolated by the procedure of Clewell and Helinski (1969) from *E. coli* HB101 carrying pSF2124 (So et al. 1975), pCR1 (Covey et al. 1976), or hybrid plasmids derived from these vectors. The plasmid DNA was labeled in vitro with deoxynucleotide [³²P]triphosphates by nick translation (Schachat and Hogness 1974), yielding a specific activity of $3–6 \times 10^6$ cpm/μg.

Purification of DNA-binding proteins from unfertilized **Drosophila** *eggs.* Unfertilized eggs were collected from a *l(1)E12^ts/ClB* stock of *Drosophila melanogaster,* carrying the sex-linked, temperature-sensitive, lethal *E12* mutation (Tarasoff and Suzuki 1970). At the restrictive temperature, this stock produces pure populations of virgin females laying unfertilized eggs, which were collected daily from large population cages and stored at −70°C. The frozen eggs were homogenized in a Vortex-Omnimix with 1.5 volumes of Buffer A (10 mM Tris-HCl, pH 7.4,

25 mM NaCl, 5 mM Mg acetate, 5 mM β-mercapto-
ethanol, 0.1 mM EDTA, and 10% glycerol) at 0°C.
The homogenate was centrifuged at 16,000g and the
pellet reextracted with 15 volumes of Buffer A. After
centrifugation at 16,000g, the combined supernatant
fractions were dialyzed overnight against Buffer A
and recentrifuged at 96,000g for 16 hours before
DNA-cellulose chromatography.

DNA-cellulose was prepared according to the procedure of Litman (1968) by coupling native Drosophila DNA extracted from pupae to purified cellulose (Munktell's No. 410, see Alberts et al. 1969)
by UV irradiation. DNA-cellulose was packed into
a column (10 × 25 cm) and the soluble egg proteins
passed through the column at 4°C in Buffer A, using
the procedures of Alberts et al. (1969) and Litman
(1968). After extensive washing with the same
buffer, the bound proteins were eluted in one step
with 0.1% sodium dodecyl sulfate (SDS) in Buffer
A at room temperature and concentrated by vacuum
dialysis.

In subsequent purification steps, the DNA-binding
proteins were first fractionated according to molecular weight on Bio-Gel and Sephadex columns in the
presence of SDS and by preparative SDS-polyacrylamide gel electrophoresis and then fractionated according to charge by isoelectric focusing in 6 M urea
(Allington and Aron 1971). Additional details of the
purification procedures will be published elsewhere
(H. Weideli and W. J. Gehring, in prep.). The proteins were renatured from SDS by the method of
Weber and Kufer (1971).

Filter binding assay. At several stages of the purification procedure, the renatured protein fractions
were tested for their binding specificity using the
filter binding assay developed by Riggs et al. (1970).
The binding was measured by incubating tritiated
DNA and protein fractions at varying concentrations for 20 minutes at room temperature in Buffer
A and filtering in duplicate at unit gravity through
nitrocellulose filters (Gelman, Metricel GN-6, which
were pretreated by boiling in distilled water for 15
min). For purified DB-1 protein, a modified Buffer
B (10 mM Tris-HCl, pH 7.4, 10 mM KCl, 10 mM Mg
acetate, 5 mM β-mercaptoethanol, and 10% glycerol)
was used. The filters were washed in the same buffer,
dried, and then counted in a liquid scintillation
counter. For competition assays, Drosophila and
salmon sperm DNA (Serva) were sheared by sonication to a size of 200–600 nucleotides.

Construction of hybrid plasmids. DNA fragments
obtained by treatment with restriction endonuclease
RI or by random shearing were inserted into either
the ampicillin-resistant plasmid pSF2124 (So et al.
1975) or the kanamycin-resistant plasmid pCR1
(Covey et al. 1976) and cloned in E. coli strain HB101
in the appropriate selective media. Two procedures
were used for inserting the Drosophila DNA into
the plasmids, the RI-ligase method (Mertz and Davis

1972; Cohen et al. 1973; Glover et al. 1975) and the
poly(dA-dT) connector method (Jackson et al. 1972;
Lobban and Kaiser 1973; Wensink et al. 1974). All
hybrid plasmids were propagated in EK1 host-vector
systems in a P2 laboratory according to the National
Institutes of Health Guidelines issued in June 1976.

In situ hybridization. Chromosome squash preparations were made from salivary glands of giant (gt/
gt^x11) larvae by standard methods. In situ hybridization was performed according to the method described by Pardue and Gall (1975) as modified by
Bonner and Pardue (1976). Nick-translated, heat-denatured DNA was used as a probe at 2×10^5 cpm
in 20 μl of hybridization buffer per slide (sp. act.
$3-6 \times 10^6$ cpm/μg). The exposure time was 5–7 days.

RESULTS

Purification of DNA-binding Proteins

From 1 kg of unfertilized eggs, 40 g of soluble
proteins were extracted (as described under Experimental Procedures) and passed over a DNA-cellulose
column. Approximately 2.5% of the total soluble
protein was retained on the column and eluted with
0.1% SDS buffer. To avoid aggregation, the subsequent purification steps were performed in the presence of SDS. By means of Bio-Gel columns and preparative polyacrylamide gel electrophoresis, a large
number of fractions were purified to apparent homogeneity, as judged by SDS-polyacrylamide gel electrophoresis (Laemmli 1970). At this stage, these protein fractions were renatured and their binding to
DNA measured by the filter binding assay (Riggs
et al. 1970). Proteins which apparently showed some
sequence specificity of binding were purified further
by preparative isoelectric focusing in a supporting
sucrose gradient.

By means of these procedures, we have been able
to purify several different DNA-binding proteins.
One of these is DNA-binding protein 1 (DB-1). Protein DB-1 gives a single band on SDS-polyacrylamide
gels and has an apparent molecular weight of approximately 30,000 daltons relative to standard molecular-weight markers. Isoelectric focusing yields
a major component, containing at least 80% of the
total protein, with an isoelectric point of 6.1–6.3.
In addition, at least four minor components with
isoelectric points between 6.5–8.2 can be resolved.
These minor components show DNA binding specificity similar to that of the main band.

Binding to Total Drosophila DNA

The filter binding assay (Riggs et al. 1970) is based
on the finding that DNA-protein complexes generally bind to nitrocellulose filters, whereas native
DNA passes through the filter. The binding can be
quantitated by using radioactively labeled DNA and
determining the amount of DNA bound to the filter.

The sequence specificity of the binding can be determined by saturation binding studies or by competition assays. The saturation level of binding is determined by adding increasing amounts of protein to a fixed amount of DNA. Proteins with general affinity for DNA, e.g., histones, show a high saturation level of binding to *Drosophila* DNA (R. Aten and W. J. Gehring, unpubl.) since they bind to all the DNA fragments generated by shearing forces during the isolation procedure. Under our conditions, histones bind 80–90% of the input DNA. Proteins binding to specific DNA sequences are expected to show a low saturation level of binding to *Drosophila*, or any other complex DNA, depending on the number and distribution of the binding sequences. Since our isolation procedure yields DNA with an average molecular weight of approximately 30×10^6 daltons (Fig. 1), it can be calculated that a protein binding to a unique DNA sequence in the *Drosophila* genome (1.1×10^{11} daltons) would bind less than 0.03% of the total DNA. This value is below the detection level of the filter binding assay. However, binding to moderately repetitive DNA sequences can be detected. It was also necessary to take into consideration that the saturation level of binding depends on interactions between the protein and the nitrocellulose as well as the stability of the DNA-protein complex.

The sequence specificity of the binding can also be tested in competition assays, in which increasing amounts of unlabeled competitor DNA are added to the incubation mixture containing the labeled DNA probe and the protein at subsaturating levels. Both salmon and *Drosophila* DNA were used as competitors. It was found that the use of *Drosophila* DNA results in complete competition, whereas the

extent of competition with salmon DNA depends on the binding specificity of the protein.

Using these criteria, the various protein fractions were tested for their binding properties, and fractions which apparently showed sequence specificity of binding were purified further. In the case of the highly purified DB-1 fraction, the saturation level was approximately 0.5% of the input DNA. This value is close to the detection level of the assay. Since the saturation level is so low, competition experiments could not be done and therefore it could not be shown conclusively that DB-1 binds to a specific sequence.

Binding to DNA Restriction Fragments

In view of the difficulty in demonstrating specific binding with total DNA by the filter assay, we attempted to enrich for specific binding sequences. This would increase the proportion of the DNA bound and enable us to perform competition experiments. For this purpose, the *Drosophila* DNA was cleaved with restriction endonuclease *Eco*RI and subsequently fractionated on agarose gels. This approach has the additional advantage that binding of a protein to a specific fraction of restriction fragments provides an alternative demonstration of recognition of specific DNA sequences.

As shown in Figures 1b and 2, ³H-labeled *Drosophila* DNA was cleaved with *Eco*RI (Greene et al. 1974), fractionated by agarose gel electrophoresis and cut into 30–50 fractions, and the DNA eluted and purified by DEAE-cellulose chromatography. Binding of protein DB-1 to the *Eco*RI fragments was measured by incubating equal amounts of DNA from each fraction with a subsaturating amount of pro-

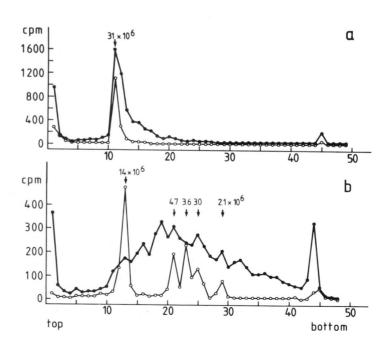

Figure 1. *(a)* Tritiated *Drosophila* DNA from the Kco cell line was fractionated on a 0.6% agarose gel, sliced, and counted (●). ³²P-labeled λ DNA was included as an internal molecular-weight marker (○). *(b)* Tritiated *Drosophila* DNA from the Kco cell line was completely digested with *Eco*RI, separated on a 0.8% agarose gel, sliced, and counted (●). ³²P-labeled λ DNA was included as an internal standard (○).
Ordinate: cpm per fraction; *abscissa:* fraction number. The molecular weights (in daltons) of the standards are indicated above the respective peaks.

tein and subsequent filtration. DB-1 binds preferentially to fractions that have a molecular weight of approximately $6–8 \times 10^6$ daltons (Fig. 2). In contrast, proteins that do not recognize specific DNA sequences (for example, histones) show a uniform level of binding to all DNA fractions (data not shown).

In competition experiments, at least 50% of the original binding to restriction fragments (fraction 17 in Fig. 2) was observed in the presence of as much as a 50-fold excess of salmon DNA. At saturating concentrations of DB-1, 3% of the input DNA is bound. Several other proteins tested in this way were found to bind to different fractions of restriction fragments, and in all cases the saturation level was five to ten times higher than with total unfractionated DNA. Since protein DB-1 appears to bind to specific restriction fragments, it would be valuable to isolate and study this sequence in detail.

Isolation of Hybrid Plasmids Containing the Binding Sequences

For the isolation of the DNA sequences containing the binding site for DB-1, RI restriction fragments of the peak binding fractions (Fig. 2) were inserted into pCR1 and pSF2124 plasmids and cloned in *E. coli* HB101 (see Experimental Procedures). Since these fractions contain several hundred different types of restriction fragments, between 100 and 400 hybrid clones of each kind were isolated.

Individual hybrid colonies grown on agar plates were pooled in groups of 100–200 and incubated for 60–90 minutes in trypton broth before chloramphenicol amplification and purification of the plasmid DNA. To select those plasmids which contain

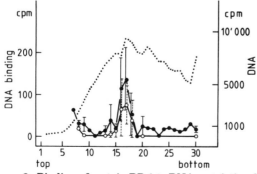

Figure 2. Binding of protein DB-1 to DNA restriction fragments. ³H-labeled *Drosophila* DNA was digested to completion with restriction endonuclease *Eco*RI and fractionated by electrophoresis on a 0.8% agarose gel (dotted line; right ordinate). The gel was cut into slices, from which the DNA was eluted and purified. Equal amounts of DNA from each fraction (10,000 cpm) were incubated with a constant amount of protein DB-1, and binding was assayed by the filter binding technique. *Ordinate* (left): cpm bound to the filters; *abscissa:* fraction number. DNA bound in the absence (●) and in the presence (○) of a 50-fold excess of unlabeled salmon DNA. All points have background values subtracted.

the binding site for DB-1, we used a modification of the Riggs' filter binding assay. Supercoiled hybrid plasmid DNA (0.1 μg) was incubated in Buffer A with limiting amounts of DB-1 protein in the presence of a 50-fold excess of salmon DNA for 20 minutes at room temperature. After filtration, the DNA bound to the nitrocellulose filter was eluted with 100 mM Tris-HCl (pH 9) containing 0.1% SDS. A control filter was treated in the same way except that no protein was added to the incubation mixture. The eluted DNA was then used for transformation of HB101 cells. In five out of eight groups of plasmids, 2–15 times more transformants were recovered with DNA from the protein-bound eluate than with the control DNA. This selection procedure was repeated three to five times successively for each batch of plasmids in order to enrich for those hybrid plasmids containing the binding sequences for DB-1. Finally, plasmid DNA from single colonies was isolated and nick-translated to test for binding specificity.

Using the same selection procedure, a collection of hybrid plasmids, obtained by inserting nonenriched, randomly sheared DNA into pSF2124 by the poly(dA-dT) connector method, was also screened in groups of 100 to 200 for plasmids binding to DB-1. From 2 out of 14 groups, an increased number of transformants (as compared to the control) was obtained in the first selection cycle. These two groups were subjected to three additional cycles of selection as described above.

Binding of Protein DB-1 to Cloned DNA Sequences

By repeated selection cycles with the modified filter binding assay, three plasmids which bind specifically to DB-1 have so far been isolated: plasmid K23 was obtained by inserting RI fragments from peak binding fractions (16 and 17 in Fig. 2) into pCR1, and plasmids A24 and A25 were both generated by insertion of randomly sheared DNA fragments into pSF2124. Protein DB-1 binds up to 40% of A24 DNA at saturation (Fig. 3). This is roughly comparable to the binding of *lac* repressor to operator DNA (Riggs et al. 1970). In contrast, only 3% of the input DNA is bound by DB-1 when the A3 plasmid containing a different *Drosophila* DNA segment is used. This suggests that DB-1 binds with high specificity to a DNA sequence contained in plasmid A24.

These results were confirmed by competition experiments under conditions of DNA excess. As shown in Figure 4, a 100-fold excess of unlabeled homologous A24 DNA reduced the amount of binding of ³H-labeled A24 DNA to 4%, whereas the same amount of heterologous pSF2124 DNA reduced the binding only to 32%. Therefore, the protein has a much higher affinity for A24 DNA than for pSF2124. The competition obtained with a large excess of heterologous DNA probably reflects a nonspecific interaction of DB-1 with DNA. This is also observed in the case of the *lac* repressor.

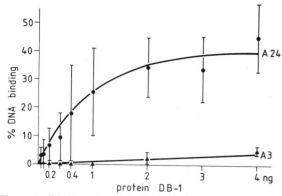

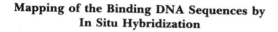

Figure 3. Binding of protein DB-1 to tritiated A24 plasmid DNA. *Ordinate:* DNA binding to nitrocellulose filters in percent of input DNA (20,000 cpm = 5 ng of DNA; sp. act. 4×10^6 cpm/μg). *Abscissa:* protein DB-1 concentration. (●) Binding to plasmid A24 DNA; (▲) binding to plasmid A3 DNA (control). Standard deviations calculated from eight experiments. Background values were subtracted.

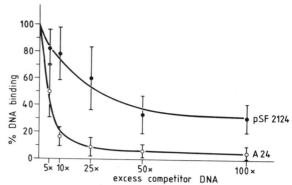

Figure 4. Competition between plasmid A24 and pSF2124 DNA for binding to DB-1 protein. Tritiated A24 plasmid DNA (5 ng; 20,000 cpm) was added to increasing amounts of cold pSF2124 DNA (●) or cold A24 plasmid DNA (○) and 2 ng of DB-1. Binding of tritiated A24 plasmid DNA is plotted as the percentage of the amount bound with no competitor.

Mapping of the Binding DNA Sequences by In Situ Hybridization

DNA of plasmids A24, A25, and K23 was ^{3}H-labeled by nick translation to high specific activity (see Experimental Procedures) and hybridized in situ to salivary gland chromosomes. All three plas-

mids hybridized specifically to the nucleolus (Fig. 5). No radioactivity was detected over other regions of the chromosomes. Another plasmid binding to a different protein (not shown here) did not hybridize to the nucleolus but to a single chromosome band. Therefore, the in situ hybridization experiments indicate that these three hybrid plasmids contain ribo-

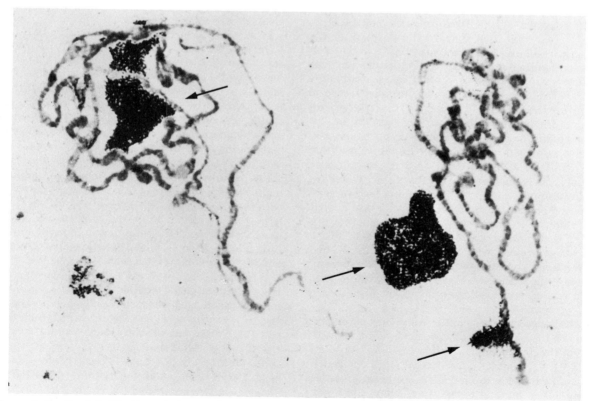

Figure 5. In situ hybridization of plasmid A25 DNA to salivary gland chromosomes. Two sets of chromosomes are shown. The radioactivity is localized over the nucleoli (arrows). Conditions were as described in Experimental Procedures. Magnification, 480X.

somal DNA sequences or sequences associated with the ribosomal genes. This result was confirmed by a filter hybridization experiment, in which it was shown that ^{32}P-labeled ribosomal 18S + 28S RNA hybridizes to A25 plasmid DNA.

DISCUSSION

Proteins involved in gene regulation, e.g., the *lac* repressor, bind with high affinity to a specific DNA sequence, but they also have a general affinity for DNA (Lin and Riggs 1975a,b). This latter property makes it possible to purify them by DNA-cellulose chromatography even though the DNA on the column may contain a very low concentration of the specific sequence to which the protein binds. We have used DNA-cellulose chromatography as the first step in the isolation of DNA-binding proteins from unfertilized *Drosophila* eggs. Since the subsequent steps of our isolation procedure require the presence of SDS in order to avoid aggregation, this purification procedure selects proteins that can be renatured by the procedure of Weber and Kufer (1971). Furthermore, proteins composed of different subunits may not be recovered in their native form. Thus, it is possible that protein DB-1 is part of a more complex protein in vivo. Although we regain DNA-binding activity using the renaturation procedure, we do not know whether a considerable fraction of the DB-1 molecules is inactive. Hence, it is not possible to determine the equilibrium dissociation constant of the DNA-protein complex accurately. An indication that a large fraction of the DB-1 protein may not be active comes from the saturation binding assay (Fig. 3). At saturation, the molar ratio of DNA to protein is approximately 1:400. The high concentration of protein needed to saturate the DNA possibly suggests that a large fraction of the protein is inactive. However, the saturation binding assays (Fig. 3) and the competition experiments indicate a considerable binding specificity, as would be expected for a protein recognizing a specific DNA sequence. By means of the isolated hybrid plasmids containing the binding sequences, it may be possible to isolate DB-1 in its native form by affinity chromatography. Like several other proteins that bind to specific DNA sequences, DB-1 is an acidic protein, with an isoelectric point 6.1–6.3. It remains to be seen whether the minor components copurifying with DB-1, which have different isoelectric points, are due to secondary modifications of the protein.

The large genome size in eukaryotic organisms makes the detection of specific binding difficult. From the haploid genome size of *Drosophila*, which is 1.1×10^{11} daltons (Rasch et al. 1971), it can be calculated that a unique DNA sequence gives less than 0.03% binding in the Riggs filter assay when randomly sheared DNA of 30×10^6 daltons molecular weight is used. This is below the detection level

of the assay. However, binding to moderately repetitive sequences is measurable with this assay. The filter assay proved to be useful in the initial screening of DNA-binding proteins for those with increased affinity for specific sequences, even though the values obtained were close to the detection level. However, the binding specificity can only be demonstrated conclusively in DNA preparations enriched for the binding sequence. This was first achieved by digestion of the DNA with restriction endonuclease RI and subsequent fractionation of the fragments by agarose gel electrophoresis. However, with complex DNA, the degree of purification is relatively small, so that DNA cloning techniques had to be used to demonstrate the specificity of the binding conclusively. For this purpose, the DNA fractions containing restriction fragments to which DB-1 binds preferentially were inserted into bacterial plasmids. The ability of DB-1 to bind specific DNA sequences was used to select hybrid plasmids containing such sequences. The specific DB-1 plasmid complexes are preferentially retained on nitrocellulose filters, and after elution and purification of the plasmid DNA, transformants containing hybrid plasmids with the binding sequences can be isolated. This selection procedure can also be applied to a random collection of hybrid plasmids, and two specifically binding plasmids (A24 and A25) were isolated in this way. The efficiency of our selection procedure will be tested in "reconstruction experiments" in which specifically binding and nonbinding plasmids are mixed in certain ratios.

In situ and filter hybridization indicate that DB-1 binds to ribosomal 18S + 28S DNA or to sequences associated with rDNA, because hybridization is observed over the nucleolus exclusively, and A25 plasmid DNA hybridizes to 18S + 28S ribosomal RNA. It remains to be seen whether DB-1 binds to the coding sequences, to the spacer, or to other sequences closely associated with rDNA. As expected, the binding sequences belong to the moderately repetitive class. The size of the RI restriction fragments containing the binding sequences, approximately $6–8 \times 10^6$ daltons (Figs. 1 and 2), coincides with the known major repeat length of rDNA, which is 7.1×10^6 daltons (Glover and Hogness 1977; Wellauer and Dawid 1977). From the data in Figure 2, it cannot be determined whether there is also binding to minor rDNA repeats.

It remains to be shown whether DB-1 is involved in gene regulation or whether it is a structural protein of the nucleolus. Histochemical experiments suggest that the nucleolar organizer region in various eukaryotic chromosomes may contain specific nonhistone proteins that can be visualized as specifically staining regions designated as N-bands (Funaki et al. 1975). However, in the giant chromosomes of *Drosophila melanogaster*, N-bands are located within the chromocenter, whereas the binding sequences of DB-1 are located specifically in the nucleolus. A functional assay will have to be used to test for the possible effect of DB-1 on gene expression. Since there is no detectable rRNA synthesis prior to blastoderm formation, when nucleoli are first formed (Zalokar 1976), the injection of DB-1 into fertilized eggs around the time of the onset of rRNA transcription may provide an assay for the possible regulatory role of this protein.

Acknowledgments

The contribution of Dr. Raymond F. Aten to the early phases of this work is gratefully acknowledged. P.S. is a postdoctoral fellow of the Helen Hay Whitney Foundation, and S.A-T. was supported by a postdoctoral fellowship of the European Molecular Biology Organization. The project was supported by grant 3.499.75 from the Swiss National Science Foundation and by the Kanton Basel-Stadt.

REFERENCES

ALBERTS, B. M., F. J. AMADIO, M. JENKINS, E. D. GUTMANN, and F. L. FERRIS. 1969. Studies with DNA-cellulose chromatography. I. DNA-binding proteins from *Escherichia coli. Cold Spring Harbor Symp. Quant. Biol.* **33**: 289.

ALLINGTON, W. B. and C. G. ARON. 1971. Isoelectric focussing with ISCO density gradient electrophoresis equipment. *ISCO Application Res. Bull.* 4. ISCO, Lincoln, Nebraska.

BONNER, J. J. and M. L. PARDUE. 1976. The effect of heat shock on RNA synthesis in *Drosophila* tissues. *Cell* **8**: 43.

BRITTEN, R. J. and E. H. DAVIDSON. 1969. Gene regulation for higher cells: A theory. *Science* **165**: 349.

———. 1971. Repetitive and nonrepetitive DNA-sequences and a speculation on the origins of evolutionary novelty. *Q. Rev. Biol.* **46**: 111.

CHAN, L.-N. and W. GEHRING. 1971. Determination of blastoderm cells in *Drosophila melanogaster. Proc. Natl. Acad. Sci.* **68**: 2217.

CLEWELL, D. B. and D. R. HELINSKI. 1969. Supercoiled circular DNA-protein complex in *Escherichia coli:* Purification and induced conversion to an open circular DNA form. *Proc. Natl. Acad. Sci.* **62**: 1159.

COHEN, S. N., A. C. Y. CHANG, H. W. BOYER, and R. B. HELLING. 1973. Construction of biologically functional bacterial plasmids *in vitro. Proc. Natl. Acad. Sci.* **70**: 3240.

COVEY, C., D. RICHARDSON, and J. CARBON. 1976. A method for the deletion of restriction sites in bacterial plasmid DNA. *Mol. Gen. Genet.* **145**: 155.

ECHALIER, G. and A. OHANESSIAN. 1969. Isolement, en cultures *in vitro,* de lignées cellulaires diploides de *Drosophila melanogaster. C. R. Acad. Sci. Ser. D.* (Paris) **268**: 1771.

———. 1970. *In vitro* culture of *Drosophila melanogaster* embryonic cells. *In Vitro* **6**: 162.

ENGELSBERG, E., C. SQUIRES, and F. MERONK, JR. 1969. The L-arabinose operon in *Escherichia coli B/R:* A genetic demonstration of two functional states of the product of a regulator gene. *Proc. Natl. Acad. Sci.* **62**: 1100.

ERON, L., R. ATTARDI, G. ZUBAY, S. CONNAWAY, and J. R. BECKWITH. 1971. An adenosine 3':5'-cyclic monophosphate-binding protein that acts on the transcription process. *Proc. Natl. Acad. Sci.* **68**: 215.

FUNAKI, K., S. MATSUI, and M. SASAKI. 1975. Location of nucleolar organizers in animal and plant chromosomes by means of an improved N-banding technique. *Chromosoma* **49**: 357.

GILBERT, W. and B. MUELLER-HILL. 1966. Isolation of the *lac*-repressor. *Proc. Natl. Acad. Sci.* **56**: 1891.

———. 1967. The *lac*-operator is DNA. *Proc. Natl. Acad. Sci.* **58**: 2415.

GLOVER, D. M. and D. S. HOGNESS. 1977. A novel arrangement of 18S and 28S sequences in a repeating unit of *Drosophila melanogaster* rDNA. *Cell* **10**: 167.

GLOVER, D. M., R. L. WHITE, D. J. FINNEGAN, and D. S. HOGNESS. 1975. Characterization of six cloned DNAs from *Drosophila melanogaster,* including one that contains the genes for r-RNA. *Cell* **5**: 149.

GREENE, P. J., M. C. BETLACH, H. M. GOODMAN, and H. W. BOYER. 1974. The Eco Rl restriction endonuclease. *Methods Mol. Biol.* **7**: 87.

GUHA, A., W. SZYBALSKI, W. SALSER, A. BOLLE, E. P. GEIDUSCHECK, and J. F. PULLITZER. 1971. Controls and polarity of transcription during bacteriophage T4 development. *J. Mol. Biol.* **59**: 329.

JACKSON, D., R. SYMONS, and P. BERG. 1972. A biochemical method for inserting new genetic information into SV40 DNA: Circular SV40 DNA molecules containing lambda phage genes and the galactose operon of *E. coli. Proc. Natl. Acad. Sci.* **69**: 2904.

JACOB, F. and J. MONOD. 1961. Genetic regulatory mechanisms in the synthesis of proteins. *J. Mol. Biol.* **3**: 318.

LAEMMLI, U. K. 1970. Cleavage of structural proteins during the assembly of the head of bacteriophage T4. *Nature* **227**: 680.

LAWRENCE, P. A. and G. MORATA. 1977. The early development of mesothoracic compartments in *Drosophila.* An analysis of cell lineage and fate mapping and an assessment of methods. *Dev. Biol.* **56**: 40.

LIN, S. and A. D. RIGGS. 1975a. A comparison of *lac*-repressor binding to operator and to non-operator DNA. *Biochem. Biophys. Res. Commun.* **14**: 3238.

———. 1975b. The general affinity of *lac* repressor for *E. coli* DNA: Implications for gene regulation in procaryotes and eucaryotes. *Cell* **4**: 107.

LITMAN, R. M. 1968. A deoxyribonucleic acid polymerase from *Micrococcus luteus* isolated on deoxyribonucleic acid-cellulose. *J. Biol. Chem.* **243**: 6222.

LOBBAN, P. and D. KAISER. 1973. Enzymatic end-to-end joining of DNA molecules. *J. Mol. Biol.* **78**: 453.

MARGULIES, L. and E. CHARGAFF. 1973. Survey of DNA polymerase activity during the early development of *Drosophila melanogaster. Proc. Natl. Acad. Sci.* **70**: 2946.

MERTZ, J. E. and R. W. DAVIS. 1972. Cleavage of DNA by RI restriction endonuclease generates cohesive ends. *Proc. Natl. Acad. Sci.* **69**: 3370.

PARDUE, M. L. and J. G. GALL. 1975. Nucleic acid hybridization to DNA. *Methods Cell Biol.* **10**: 1.

PTASHNE, M. 1967a. Isolation of the λ phage repressor. *Proc. Natl. Acad. Sci.* **57**: 306.

———. 1967b. Specific binding of the λ phage repressor to λ DNA. *Nature* **214**: 232.

RASCH, E. M., H. J. BARR, and R. W. RASCH. 1971. The DNA content of sperm of *Drosophila melanogaster. Chromosoma* **33**: 1.

RIGGS, A., H. SUZUKI, and S. BOURGEOIS. 1970. *lac*-Repressor operator interaction. I. Equilibrium studies. *J. Mol. Biol.* **48**: 67.

ROBERTSON, F. W., M. CHIPCHASE, and NGUYEN THI MÂN. 1969. The comparison of differences in reiterated sequences by RNA-DNA hybridization. *Genetics* **63**: 639.

SCHACHAT, F. H. and D. S. HOGNESS. 1974. Repetitive sequences in isolated Thomas circles from *Drosophila melanogaster. Cold Spring Harbor Symp. Quant. Biol.* **38**: 371.

SO, M., R. GILL, and S. FALKOW. 1975. The generation of

ColE1-Ap^r cloning vehicle which allows detection of inserted DNA. *Mol. Gen. Genet.* **142**: 239.

STEINER, E. 1976. Establishment of compartments in the developing leg imaginal discs of *Drosophila melanogaster. Wilhelm Roux's Arch. Dev. Biol.* **180**: 9.

TARASOFF, M. and D. T. SUZUKI. 1970. Temperature-sensitive mutations in *Drosophila melanogaster:* VI. Temperature effects on development of sex-linked recessive lethals. *Dev. Biol.* **23**: 492.

WEBER, K. and D. J. KUFER. 1971. Reversible denaturation of enzymes by sodium dodecyl sulfate. *J. Biol. Chem.* **246**: 4504.

WELLAUER, P. K. and I. B. DAWID. 1977. Structural organization of ribosomal DNA in *Drosophila melanogaster. Cell* **10**: 193.

WENSINK, P., D. J. FINNEGAN, J. E. DONELSON, and D. S. HOGNESS. 1974. A system for mapping DNA sequences in the chromosomes of *Drosophila melanogaster. Cell* **3**: 315.

WIESCHAUS, E. and W. J. GEHRING. 1976. Clonal analysis of primordial disc cells in the early embryo of *Drosophila melanogaster. Dev. Biol.* **50**: 249.

ZALOKAR, M. 1976. Autoradiographic study of protein and RNA formation during early development of *Drosophila* eggs. *Dev. Biol.* **49**: 425.

The Structure of the Transcriptionally Active Ovalbumin Genes in Chromatin

A. GAREL* AND R. AXEL

Institute of Cancer Research and Department of Pathology, College of Physicians and Surgeons, Columbia University, New York, New York 10032

Control of gene expression is likely to result, at least in part, from the specific interaction of chromatin proteins with DNA in such a way as to permit the transcription of a given set of genes in one tissue while restricting their expression in other tissues. It is therefore possible that structural changes occur within the chromatin complex which are responsible for the induction and maintenance of the transcription of specific genes. In our laboratory, we have been interested in the distribution of proteins about the ovalbumin gene in chromatin and the possible role of this nucleoprotein structure in regulating the level of expression of this gene. The magnitude and specificity of the response of the ovalbumin gene to hormonal stimuli results, at least in part, from control at the level of transcription and therefore permits us to examine the organization of this gene in chromatin in both highly active and inactive states.

The experimental approach we have adopted for the study of the structure of the ovalbumin gene in chromatin involves dissection of the chromosome with deoxyribonucleases. Examination of the products of digestion reveals characteristic nucleoprotein complexes that reflect an aspect of the structure of the total genome. After utilizing ovalbumin cDNA in molecular hybridization reactions with the DNA products of digestion, we then ask whether this gene is organized in a manner analogous to that of total chromatin. This approach permits us to ask: (1) Do proteins reside on the transcriptionally active ovalbumin genes? (2) Do these proteins organize the DNA into periodic repeating structures characteristic of the bulk of the genome? (3) Are the proteins organized randomly with respect to the ovalbumin gene sequence? (4) Can we discern a difference between the conformation of the ovalbumin gene in chromatin in its transcriptionally active and inactive states?

The Repeating Subunit about the Ovalbumin Genes

The histones of the chromosome organize the DNA into discrete repeating subunits consisting of about 200 base pairs of DNA and eight histone molecules

(numerous papers, this volume). This particulate or nucleosomal structure encompasses the bulk of the genomic DNA. Therefore, it was of interest to determine whether such a periodic array can be discerned about the ovalbumin gene in cells in which this gene is actively transcribed. The experimental approach we have chosen involves a titration of the ovalbumin gene content in the monomeric and multimeric nucleosomal particles generated following mild microccoccal nuclease digestion of oviduct nuclei (Axel and Garel 1976). To this end, oviduct nuclei were subjected to mild nuclease digestion solubilizing only 8% of the nuclear DNA. The resultant nucleoprotein fragments were then fractionated on sucrose velocity gradients. The absorbance profile shown in Figure 1 reveals the discrete peaks of nucleoprotein that correspond to the monomeric and multimeric forms of the 11S nucleosomal subunit. DNA was extracted from the monomeric, dimeric, and trimeric peaks, and the ovalbumin gene content was determined by annealing this DNA to ovalbumin cDNA. In these reactions, nucleosomal DNA is in vast excess over cDNA, such that the rate of annealing is dependent on the concentration of the ovalbumin genes in nucleosomal DNA. The kinetics of reassociation revealed in all cases a single, second-order kinetic transition. From the $C_0 t_{1/2}$ of these reactions we can determine the ovalbumin gene content in each of the three populations of nucleosomal DNA. We find that the ovalbumin gene is retained in these nucleoprotein particles in concentrations virtually identical to that observed in total, unfractionated DNA.

These studies indicate that nuclease cleavage sites exist at periodic intervals about the ovalbumin gene, liberating this gene as nucleoprotein particles whose distribution follows that of the nucleosomal subunits which comprise the bulk of the chromosome. Since it has been demonstrated that the histone molecules are responsible for the generation of the nucleosomal repeat, it is tempting to assume from these studies that histones are present about the ovalbumin gene in chromatin in a manner analogous to the distribution of these proteins about the transcriptionally inert segments of the genome. It is possible, however, that a unique set of proteins reside about the ovalbumin gene which are capable of generating a repeating structure analogous to the nucleosomal

---

* Visiting scientist from the Institut de Biologie Moléculaire et Cellulaire, 15 Rue Descartes, Strasbourg, France.

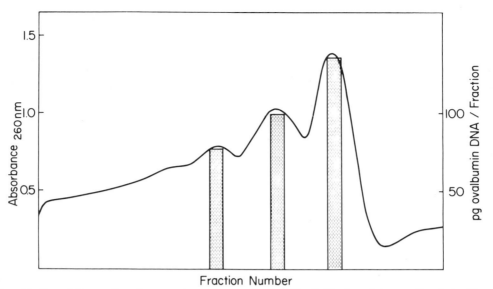

Figure 1. Quantitation of the ovalbumin genes in nucleosomal subunits. Oviduct nuclei were digested with micrococcal nuclease to 8% solubilization of DNA. The resultant nucleosomal subunits were fractionated on a sucrose gradient and the DNA was purified from the monomeric, dimeric, and trimeric nucleosomal fractions and annealed with ovalbumin cDNA. The solid curve represents the absorbance profile of fractionated nucleosomes. The bars represent the quantities of ovalbumin sequences in the three nucleoprotein fractions determined by annealing kinetics (Axel and Garel 1976).

subunit. Nevertheless, it is apparent from these studies that proteins reside about the ovalbumin gene and that the mere presence of protein along a specific gene is not sufficient to prevent its transcription.

Although these studies strongly suggest that histone molecules reside about the ovalbumin gene and are organized in such a way as to generate a nuclease site at 200-base-pair intervals about this gene, they do not prove that a discrete 11S particle exists about the ovalbumin gene. It is certainly possible that the particulate nucleosomal structure is extended in such a way as to retain a periodic, nuclease-sensitive site without the maintenance of an intact nucleosome. If these regions of the chromosome are physically restrained in an extended conformation, nuclease action may serve to relieve this restraint, thereby allowing the nucleosomes to refold over the liberated fragments. Such a model would still require the presence of histones over active genes; however, it would imply that these genes are not organized in a compact subunit, but only assume the more stable, condensed nucleosomal conformation following digestion. This interpretation is supported by the recent data indicating that extension of the chromatin fiber need not be associated with a loss of the regularly repeating, nuclease-sensitive sites (Jackson and Chalkley 1975; Oudet, Spadafora, and Chambon, this volume).

Core Particles Contain the Entire Ovalbumin Gene

Extensive digestion of the chromosome, or of isolated monomeric particles, results in the generation of a metastable intermediate consisting of 140 base pairs of DNA and two each of the core histones H2A, H2B, H3, and H4 (Axel 1975; Sollner-Webb and Felsenfeld 1975; Noll and Kornberg 1977; Shaw et al. 1976). This trimmed nucleosome, or nucleosomal core, is free of linker DNA, and although significant variation in the nucleosomal repeat length has been reported among different organisms, tissues, and even within a given cell type, the DNA content within the core particle appears to remain constant among all eukaryotes examined to date (Noll 1976; Compton et al. 1976; Lohr et al. 1977; Thomas and Thompson 1977). It was therefore of interest to determine whether this core particle could be generated upon cleavage of the transcriptionally active ovalbumin gene. These experiments could provide further evidence that histones reside upon this active gene and organize the DNA, at least at one basic level, in a manner analogous to the majority of inactive sequences in the genome. In addition, these experiments can provide preliminary information relevant to the distribution of chromatin proteins with respect to nucleotide sequence.

Three possible models of alignment of histone proteins about the ovalbumin gene are depicted in Figure 2. The ovalbumin gene is considered as a series of sequences A, A′, . . . , N, N′. Sequences A through N are assumed to be bound to chromatin proteins in such a way that they resist nuclease treatment and would therefore be present in the 140-base-pair nucleosomal core. Sequences A′ through N′ bridge adjacent resistant particles and will be digested by nuclease in the preparation of core particles. The average repeat length in the chicken oviduct is 198

A. Phased with a single initiation

B. Phased with multiple initiations

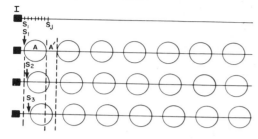

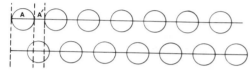

C. Random

Figure 2. Possible schemes of nucleosome organization along the ovalbumin gene. For explanation, see text.

base pairs, whereas the core particle contains 140 base pairs. Thus, the linker sequences comprise about 30% of the genomic DNA. If, as depicted in model A, the assembly of nucleosomes about the ovalbumin gene is phased with the initiation of nucleosome formation proceeding at a single site I with subsequent nucleosomes organized at fixed and regular intervals along the gene, then cleavage to core particles should result in the specific digestion of 30% of the ovalbumin sequences. At the other extreme, model C depicts a random distribution of nucleosomes about the ovalbumin gene such that no fixed initiation site for nucleosome assembly is present, and in different cells of the same cell type, nucleosomes exist in a fixed array without regard to nucleotide sequence. Such a model would predict that isolated DNA from core particles would contain all of the ovalbumin sequences in equal concentrations. A third, more complex alternative is depicted in model B. In this model an initiation site does exist for nucleosome assembly; however, the placement of the first nucleosome need not occur at the phasing signal I, but may begin at a series of sites at regular intervals from I (denoted as sites S_1 through S_j). This model of phasing with multiple initiation sites cannot be distinguished from a random model by a simple analysis of the ovalbumin gene content in isolated core particles.

These experiments require that we isolate a purified preparation of core DNA fragments 140 base pairs in length. To this end, oviduct nuclei were extensively digested with micrococcal nuclease to solubilize 20% of the nuclear DNA. The resultant monomeric particles were isolated by velocity sedi-

mentation and the subpopulation of trimmed particles was purified by differential salt extraction in KCl (Olins et al. 1976), followed by a second velocity gradient. Two parameters we have used to assess the purity of these particles include the length of the DNA contained within the purified cores and the presence or absence of histone H1. Figure 3 demonstrates the 198-base-pair nucleosomal repeat generated upon micrococcal nuclease cleavage of oviduct nuclei, along with the 140-base-pair fragment obtained from our purified core preparations. Analysis of the histones present in our core populations reveals the complete absence of histone H1.

The DNA from these isolated core particles was then annealed to ovalbumin cDNA, along with control annealings with total unfractionated avian DNA. Analysis of the saturation values with core and total DNA permits a determination of the extent of representation of ovalbumin sequences in the core particle. From the kinetics of hybridization, we can directly determine the relative ovalbumin gene content in the isolated core particle. The kinetics of these annealing reactions are shown in Figure 4. The annealing of cDNA to either unfractionated, sheared DNA or oviduct core DNA are superimposable, revealing a single, second-order transition with both reactions saturating at over 80% duplex formation. The concordance of these curves indicates that the ovalbumin gene content in isolated cores is virtually identical to its representation in the genome, and no specific sequences appear to be deleted following core purification. These experiments appear to rule out the simple phasing model depicted in Figure 2A.

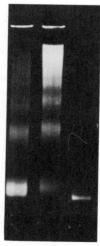

Figure 3. A 3% polyacrylamide gel electrophoresis of DNA derived from nucleosome cores. Nuclei were digested with micrococcal nuclease and core particles were purified by sucrose velocity centrifugation followed by differential extraction in 0.1 M KCl followed by a second velocity gradient. Samples include (left to right) DNA obtained from 12% and 4% digests of nuclei and DNA isolated from purified core particles.

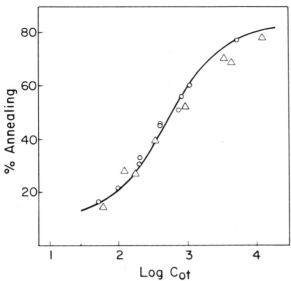

Figure 4. Kinetics of annealing of ovalbumin cDNA with nucleosomal core DNA and unfractionated total DNA. Core particles were obtained from micrococcal nuclease digest of oviduct nuclei. The DNA was purified of protein and annealed with ovalbumin cDNA. Annealing reactions contained 3 mg of either core DNA (○) or total sonicated DNA (△) and 0.12 ng of ³H-labeled ovalbumin cDNA and were performed at 69°C. Duplex formation was monitored with S₁ nuclease.

Selective Digestion of Transcriptionally Active Genes

The previous studies employing micrococcal nuclease indicate that the proteins reside about the transcriptionally active ovalbumin gene in the oviduct. These studies further indicate that the proteins about these genes organize the DNA into a periodic repeating structure which, upon extended digestion, results in the generation of a 140-base-pair core fragment of DNA. This suggests that at least at one level the organization of proteins about these genes is quite similar to the gross organization of histones about the bulk of the genome. We would predict, however, that structural changes accompany transcriptional activation, but that these changes are not detected in the sort of analyses described above. Perhaps the most convincing evidence for the presence of an altered structure about active genes derives from electron microscopic observations, which reveal sites of RNA synthesis within the chromosome in a more diffuse and extended configuration. A striking example of this phenomenon is reflected in the puffs observed in the polytene chromosomes of Diptera (Daneholt 1975). More recent biochemical studies have indicated that transcriptionally active genes are exceedingly sensitive to attack by pancreatic DNase, thus providing a consistent biochemical parameter distinguishing active from inactive chromatin (Weintraub and Groudine 1976; Garel and Axel 1976).

The ovalbumin gene in the oviduct following a hormonal induction is transcribed with high frequency, and in initial experiments we therefore examined the sensitivity of this gene to attack by DNase I. To this end, oviduct nuclei were digested with DNase I for increasing lengths of time to generate 3%, 11%, and 24% solubilization of nuclear DNA. The kinetics of annealing of ovalbumin cDNA to these DNase-treated DNAs, along with total unfractionated DNA from the oviduct, are shown in Figure 5. Ovalbumin cDNA anneals to a vast excess of total nuclear DNA, with a $C_0t_{1/2}$ of 550. Examination of the kinetics of annealing to DNase-I-treated DNA, however, indicates that as the digestion proceeds, the $C_0t_{1/2}$ increases to a limiting value of 2300 at 24% digestion of nuclear DNA. Further solubilization with DNase I results in no further increase in the $C_0t_{1/2}$ of annealing with ovalbumin cDNA. Proof of the selectivity of the digestion process for the transcriptionally active ovalbumin gene requires that we demonstrate that tissues inactive in ovalbumin RNA synthesis do not show the selective sensitivity of this gene. The kinetics of annealing of ovalbumin cDNA to DNase-I-treated liver DNA and oviduct DNA were therefore compared. The data in Figure 6 indicate that ovalbumin cDNA anneals to a 15% digest of liver DNA with a $C_0t_{1/2}$ of 600, whereas the $C_0t_{1/2}$ of an equivalent digest of oviduct DNA is 2000. These data indicate that the ovalbumin genes are specifically sensitive to DNase-I digestion only in those tissues actively engaged in ovalbumin mRNA synthesis.

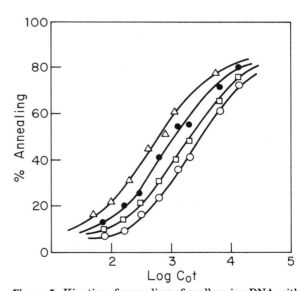

Figure 5. Kinetics of annealing of ovalbumin cDNA with DNase-I-treated oviduct nuclear DNA. Oviduct nuclei were digested for various times with DNase I, and the resistant DNA was purified of proteins and RNA. ³²P-labeled ovalbumin cDNA (0.12 ng) was incubated with 2.4 mg of sonicated total oviduct DNA (△), and DNA was extracted from DNase-I-treated nuclei after 3% (●), 11% (□), and 24% (○) solubilization.

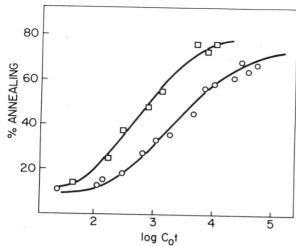

Figure 6. Kinetics of annealing of ovalbumin cDNA with DNase-I-treated liver of oviduct nuclear DNA. ^{3}H-labeled ovalbumin cDNA (0.18 ng) was annealed with 4 mg of DNA obtained following 15% digestion of either oviduct (○) or liver (□) nuclei.

The DNA used in the annealing reactions described above were examined on urea-polyacrylamide gels (Fig. 7). Digestion of oviduct nuclei with DNase I results in the generation of a set of DNA products which upon denaturation reveal discrete, single-stranded fragments at ten-nucleotide intervals, from 20 to 160 base pairs (Noll 1974). This array of fragments appears very early in the digestion process, and the generation of these fragments proceeds at the expense of the poorly resolved, high-molecular-weight DNA. Even at early times in the digestion process, when only a few percent of the nuclear DNA has been solubilized, significant cleavage of the intranucleosomal DNA has occurred. This finding indicates that DNA within the nucleosome may be as accessible to cleavage by DNase I as the DNA bridging the individual nucleosomal subunits. It is apparent that there is a striking reduction in the average molecular weight as the cleavage process proceeds. Since the kinetics of DNA reassociation are sensitive to the molecular weight of the reacting species, it is important to verify that the reduction in the rate of reassociation observed following DNase-I digestion does not result merely from a decrease in the size of the annealing species. We therefore examined the kinetics of DNase-I digestion of the inactive globin gene in the oviduct. Experiments were performed in which ^{32}P-labeled ovalbumin cDNA and ^{3}H-labeled globin cDNA were annealed in the presence of a vast excess of DNase-I-treated oviduct nuclear DNA. The kinetics of reassociation with the ovalbumin probe have already been presented in Figure 5. In Figure 8 we observe that DNase-I treatment results in virtually no selective degradation of the globin genes in the oviduct. Twenty-four percent solubilization of DNA results in a 20% reduction in the annealing rate with globin cDNA, whereas the annealing rate with ovalbumin cDNA is reduced 80%. These experiments further support the conclusion that transcriptionally active segments of the genome are organized by the chromatin proteins in a conformation which renders them exceedingly sensitive to nuclease attack.

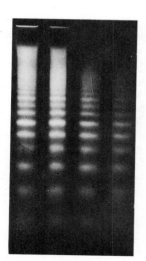

Figure 7. Polyacrylamide gel electrophoresis of DNA products following digestion of oviduct nuclei. Oviduct nuclei were digested with DNase I (20 μg/ml) for various times, and the DNA was purified, denatured, and applied to a 7 M urea, 10% polyacrylamide gel. Samples depict (left to right) 3%, 7%, 11%, and 24% solubilization. These DNA preparations were used in the annealing reactions described in the text.

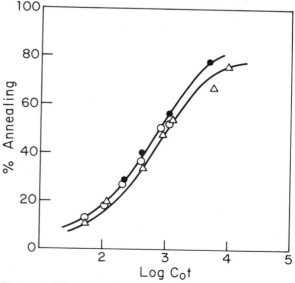

Figure 8. Kinetics of annealing of globin cDNA with DNase-I-treated oviduct nuclear DNA. ^{3}H-labeled globin cDNA (0.12 ng) was annealed with 2.4 mg of sonicated total DNA (●) and DNA extracted from DNase-I-treated oviduct nuclei after 7% (○) and 24% (△) solubilization of nuclear DNA.

An additional control was performed in which oviduct DNA was purified of chromatin proteins and reconstituted with a purified preparation of bull histones. This reconstitution permits the reassembly of nucleosomes along the DNA but will not result in the generation of an active conformation about the ovalbumin genes. This reconstituted complex was then subjected to digestion with DNase I, and the resulting DNA fragments were annealed with ovalbumin cDNA. A 30% DNase digest shows only a 20% decrease in the rate of annealing to ovalbumin cDNA when compared with total unfractionated oviduct DNA, further indicating that the observed decrease in the rates of annealing does not merely result from a nonspecific reduction in the size of the annealing fragments. We conclude from these studies that the ovalbumin gene is organized by chromatin proteins into an active conformation in the oviduct but not in liver, a tissue in which it is not expressed, and that this altered conformation is recognized by DNase I, resulting in the rapid and selective digestion of these genes. This selectivity is not observed for the transcriptionally inert globin genes in the oviduct.

Conformation of the Rare mRNA Genes in Chromatin

At present we have no information identifying those factors responsible for the induction and maintenance of an active conformation about transcribing segments of the chromosome. At one extreme we could postulate that some component of the transcriptional apparatus involving either polymerase or nascent ribonucleoprotein is responsible for the enhanced DNA sensitivity of active genes. This view predicts that selective DNase sensitivity should only be observed when a specific gene is actually engaged in the process of transcription. Alternatively, we could argue that a subset of genomic sequences within the chromosome is maintained in an active conformation independent of the presence of either RNA polymerase or nascent ribonucleoprotein. One approach we have chosen to assess the role of the transcriptional apparatus in the maintenance of the active conformation involves a comparative analysis of the specific sensitivity of sets of genes with widely differing transcriptional rates. In the previous experiments we have analyzed the kinetics of digestion of the ovalbumin gene, a gene which is transcribed in the oviduct with an exceedingly rapid rate of initiation (Palmiter 1974; Schütz et al.; Bellard et al.; both this volume). We therefore examined the DNase-I sensitivity of the gene transcribed in the oviduct with frequencies far below that of ovalbumin.

In previous studies from our laboratory we analyzed the complexity and diversity of mRNA present in the induced oviduct (Axel et al. 1976). These data

indicate that oviduct poly(A) mRNA distributes into three discrete frequency classes, the most abundant consisting of the ovalbumin mRNA species present approximately 100,000 times per cell. The least abundant class of mRNA consists of 14,000 different sequences, each present only five times per cell. Whatever mechanisms are invoked to explain the maintenance of these mRNA levels, it is likely that the maintenance of this enormous difference in the levels of these two classes of RNAs results, at least in part, from large differences in the rates of initiation of transcription. The distribution of polymerase on the ovalbumin gene, therefore, should be significantly greater than the distribution of polymerases on genes within the least abundant, highly complex class of mRNA. If the mere presence of polymerase were responsible for the observed sensitivity to DNase I, we would predict that these genes would be far less sensitive to DNase-I attack than the rapidly transcribing ovalbumin genes.

These experiments require that we isolate a purified cDNA preparation which is a copy of the complex mRNA and is free of ovalbumin sequences. This was accomplished by preparing a total cDNA copy of oviduct mRNA. This cDNA was then allowed to undergo several cycles of hybridization of R_0t values of 2, which will permit the reassociation of only the more abundant species of mRNA. The cDNA that did not anneal at these R_0t values was then purified by hydroxylapatite chromatography. This cDNA now anneals as a single first-order transition with an $R_0t_{1/2}$ of 35, consistent with the fact that it is a copy of the most complex, least abundant class of oviduct mRNA. Less than 5% of this cDNA anneals to purified preparations of ovalbumin mRNA. The kinetics of reassociation of this cDNA to DNase-I-treated oviduct DNA preparations used in the previous experiments is shown in Figure 9. We again observe that as the digestion proceeds, the relative rate of annealing of this rare-sequence cDNA diminishes by a factor of 3. A limiting C_0t value of 1600 is obtained at 20% solubilization of nuclear DNA; further DNase-I digestion results in no further decrease in the rate of reassociation. In Figure 10 we compare the rates of digestion of three sets of genes in the oviduct: the globin genes, which are never transcribed in this organ; the ovalbumin genes, which are transcribed with an initiation frequency of about 20 per minute; and a set of 14,000 genes represented only five times per cell, which are likely to be transcribed several thousand times less frequently than the ovalbumin gene. The data indicate that although no selective digestion of the globin gene is observed with the enzyme DNase I, the ovalbumin and rare-sequence genes are recognized and cleaved by DNase I at quite similar rates. Maintenance of the steady-state level of ovalbumin mRNA in the oviduct requires an exceedingly high initiation frequency, and it is therefore likely that this gene is continuously packed with a high density of

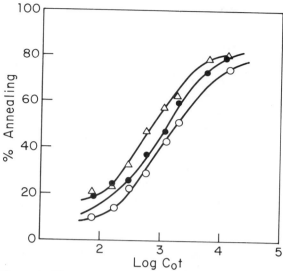

Figure 9. Kinetics of annealing of oviduct rare-sequence cDNA to DNase-I-treated nuclear DNA. cDNA complementary to the complex, rare class of oviduct polysomal RNA was purified from cDNA synthesized from total poly(A)-containing polysomal RNA by several cycles of hybridization as described in the text. ^{3}H-labeled rare-sequence cDNA (0.6 ng) was annealed with DNA extracted from DNase-I-treated oviduct nuclei after 3% ($\triangle$), 11% ($\bullet$), and 24% ($\bigcirc$) solubilization of nuclear DNA.

polymerases. In contrast, the probability that a given rare-sequence gene will be in the process of transcription at any given point in time is exceedingly low. These data would therefore indicate that the mere presence of polymerase about a given gene

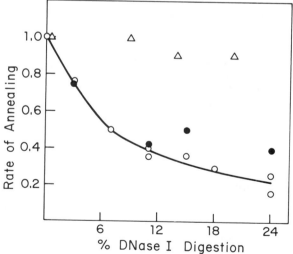

Figure 10. Comparison of the rates of digestion of ovalbumin, globin, and rare-sequence genes following DNase-I digestion of oviduct nuclei. Composite of annealing data in Figs. 6, 9, and 10. The relative rate of annealing is determined by comparison of the $C_0 t_{1/2}$ of DNase-I-treated samples with the $C_0 t_{1/2}$ of total sonicated DNA with globin ($\triangle$), ovalbumin ($\bigcirc$), and rare-sequence ($\bullet$) cDNAs. The percent DNase-I digestion reflects the level of acid solubilization of total nuclear DNA.

is not sufficient to render it selectively sensitive to DNase-I attack. This is in accord with observations in mature erythrocytes which demonstrate the specific sensitivity of the globin genes to DNase-I attack long after transcription of this gene has ceased (Weintraub and Groudine 1976). These results suggest further that the subpopulation of gene sets in an active conformation may represent the transcriptional potential of a given cell type. DNase-I digestion, therefore, would reflect a conformation which may be necessary, but not sufficient, to permit transcription.

SUMMARY

Digestion of the eukaryotic chromosome with deoxyribonucleases has revealed the presence of periodic cleavage sites which liberate the nucleosome or basic subunit of the chromosome. A second set of nuclease cleavage sites exist within the nucleosome which reflect an aspect of the internal structure of the monomeric subunit. In our studies we have probed the DNA generated upon nuclease treatment of chromatin with cDNA copies of specific mRNA sequences to study the structure and organization of transcriptionally active genes in chromatin. In initial studies we demonstrated that the transcriptionally active ovalbumin gene in the oviduct is associated with chromatin proteins to generate a structure with nuclease-sensitive sites at 200-base-pair intervals along the DNA. Further digestion liberates these active genes as trimmed nucleosomal cores containing 140 base pairs of DNA. Furthermore, we find that the entire ovalbumin gene can be detected in the nucleosomal core. These results strongly suggest that histone molecules are present about the ovalbumin gene during transcription. The mere presence of histone about a specific gene is therefore not sufficient to prevent its transcription. Our results do not permit us to comment on the degree of condensation of the DNA about the ovalbumin gene and therefore provide no information on the maintenance of higher-order structure about active genes. It is likely, however, that active genes are maintained in an extended conformation which still maintains the repeating nuclease-sensitive sites characteristic of the bulk of the genome.

In accord with this prediction, we find that the ovalbumin genes in the oviduct appear to be organized by chromatin proteins in such a way that they are rendered exceedingly sensitive to digestion by the enzyme DNase I. This sensitivity is not observed in the liver, a tissue in which these genes are transcriptionally inert. Furthermore, the transcriptionally inactive globin genes in the oviduct are not sensitive to nuclease attack and are digested five times slower than the ovalbumin genes in this tissue. These studies indicate that some component of the chromatin complex is responsible for maintaining transcriptionally active segments of the chromosome in a conformation which renders them accessible to attack by DNase I only in tissues in which these genes are transcribed. We do not know whether the digestion process results in the solubilization of the transcriptionally active genes or whether these genes are rapidly cleaved at 10-base-pair intervals to fragments which anneal at strikingly reduced rates.

In initial experiments we examined whether the accessibility of active genes reflects the distribution of polymerases about these genes. To this end, the accessibilities of active-gene sets with widely different transcription rates were compared. Estimates of the initiation frequency required to maintain a steady-state level of ovalbumin mRNA observed in the fully induced oviduct require from 10 to 20 initiations per minute along this gene. If we assume conservation of mRNA sequences, then maintenance of the steady-state level of the rare or complex class of mRNAs in the oviduct cytoplasm would require initiation frequencies several-thousandfold lower than that observed for the ovalbumin gene. Analysis of the rates of accessibility of these two sets of genes reveals that they are recognized and cleaved by DNase I at similar rates. The presence of polymerase, therefore, does not seem to be responsible for the maintenance of an active conformation about transcriptionally active genes. Furthermore, it is unlikely that a significant number of these rare-sequence genes would be in the act of transcription at the moment of sampling. This active conformation, therefore, appears to be maintained between transcriptional events. Our results suggest that the induction and maintenance of an active conformation about specific genes may be necessary, but not sufficient, to permit transcription and may reflect the transcriptional potential of a given cell type.

Acknowledgments

We thank Drs. Gary Felsenfeld and Pat Williamson for providing globin cDNA. This investigation was supported by grant CA-16346, awarded by the National Cancer Institute, DHEW.

REFERENCES

AXEL, R. 1975. Cleavage of DNA in nuclei and chromatin with staphylococcal nuclease. *Biochemistry* 14: 2921.

AXEL, R. and A. GAREL. 1976. The structure of the ovalbumin gene in chromatin. In *Biochemical actions of progesterone and progestins* (ed. E. Gurpide), vol. 286, p. 135. New York Academy of Science, New York.

AXEL, R., P. FEIGELSON, and G. SCHÜTZ. 1976. Analysis of the complexity and diversity of mRNA from chicken liver and oviduct. *Cell* 7: 247.

COMPTON, J. L., M. BELLARD, and P. CHAMBON. 1976. Biochemical evidence of variability in the DNA repeat length in the chromatin of higher eukaryotes. *Proc. Natl. Acad. Sci.* 73: 4382.

DANEHOLT, B. 1975. Transcription in polytene chromosomes. *Cell* 4: 1.

GAREL, A. and R. AXEL. 1976. Selective digestion of transcriptionally active ovalbumin genes from oviduct nuclei. *Proc. Natl. Acad. Sci.* 73: 3966.

JACKSON, V. and R. CHALKLEY. 1975. The effect of urea on staphylococcal nuclease digestion of chromatin. *Biochem. Biophys. Res. Commun.* 67: 1391.

LOHR, D., J. CORDEN, K. TATCHELL, R. T. KOVACIC, and K. E. VAN HOLDE. 1977. Comparative subunit structure of HeLa, yeast, and chicken erythrocyte chromatin. *Proc. Natl. Acad. Sci* 74: 79.

NOLL, M. 1974. Internal structure of chromatin subunit. *Nucleic Acids Res.* 1: 1573.

———. 1976. Differences and similarities in chromatin structure of *Neurospora crassa* and higher eucaryotes. *Cell* 8: 349.

NOLL, M. and R. D. KORNBERG. 1977. Action of micrococcal nuclease on chromatin and the location of histone H-1. *J. Mol. Biol.* 109: 393.

OLINS, A. L., R. D. CARLSON, E. B. WRIGHT, and D. E. OLINS. 1976. Chromatin ν bodies: Isolation, subfractionation and physical characterization. *Nucleic Acids Res.* 3: 3271.

PALMITER, R. D. 1974. Quantitation of parameters that determine the rate of ovalbumin synthesis. *Cell* 4: 189.

SHAW, B. R., T. M. HERMAN, R. T. KOVACIC, G. BEAUDREAU, and K. E. VAN HOLDE. 1976. Analysis of subunit organization in chicken erythrocyte chromatin. *Proc. Natl. Acad. Sci.* 73: 505.

SOLLNER-WEBB, B. and G. FELSENFELD. 1975. A comparison of the digestion of nuclei and chromatin by staphylococcal nuclease. *Biochemistry* 14: 2915.

THOMAS, J. O. and R. J. THOMPSON. 1977. Variation in chromatin structure in two cell types from the same tissue: A short DNA repeat length in cerebral cortex neurons. *Cell* 10: 633.

WEINTRAUB, H. and M. GROUDINE. 1976. Chromosomal subunits in active genes have an altered conformation. *Science* 193: 848.

QUESTIONS/COMMENTS

Comment by:
Y. SUZUKI
Carnegie Institution

I would like to add the silk fibroin gene in the posterior silk gland of *Bombyx mori* as another example of active genes that are more susceptible to DNase than the bulk of DNA. When a high-molecular-weight DNA (60×10^6 and 30×10^6 daltons for double- and single-strand molecular weight, respectively) was extracted by a nuclei isolation method, about 50% of fibroin gene was found to be smaller than 10×10^6 daltons, leaving the remaining 50% as large as the bulk DNA (see Fig. 1, Suzuki and Ohshima, this volume).

Structure of *Xenopus* Ribosomal Gene Chromatin during Changes in Genomic Transcription Rates

R. REEVES

Department of Zoology, University of British Columbia, Vancouver, B. C., Canada V6T 1W5

The relationship of chromatin structure to functional genomic activity is unclear at the present time. Nevertheless, the apparently ubiquitous presence of histone-containing chromatin subunits (Kornberg 1974; Elgin and Weintraub 1975) or nucleosomes (Oudet et al. 1975) within the nuclei of most higher eukaryotic cells has naturally led to the question of the role of these structures in nuclear metabolism.

Not surprisingly, most recent data have supported the view that transcriptionally quiescent DNA is generally associated with nucleosomes (Elgin and Weintraub 1975; Reeves and Jones 1976; Reeves 1976). These results are consistent with the now classical concept that histones play a role in the nonspecific repression of gene activity (Stedman and Stedman 1957; Huang and Bonner 1962). However, a considerable body of more recent evidence has accumulated, suggesting that the true in vivo role of histones may be far more complex and subtle than that of a simple generalized inhibitor of genomic function. For example, numerous biochemical studies have indicated that transcriptionally active genes may also be partially, or perhaps transiently, associated with histone aggregates in the form of nucleosomes (Reeder 1975; Reeves and Jones 1976; Reeves 1976; Mathis and Gorovsky 1976; Piper et al. 1976; Gottesfeld et al. 1976; Weintraub and Groudine 1976; Reeves 1977; Higashinakagawa et al. 1977; Matsui and Busch 1977). Furthermore, indirect evidence has also been advanced indicating that during replication at least half of the DNA present in the replication fork may be associated with nucleosomes (Seale 1976). Thus, the available biochemical evidence seems to indicate that a complex relationship exists between the structure of chromatin and its functional activity.

In any attempt to investigate the nature of the apparently far-from-simple relationship that exists between gene form and function, a well-defined system is required for experimental analysis. The ribosomal genes of the amphibian *Xenopus laevis* offer such a system.

The Ribosomal Genes of *Xenopus*

In terms of both biochemistry and ultrastructural morphology, the ribosomal genes of *Xenopus* have been among the most thoroughly studied of all higher eukaryotic genes (for recent reviews of this subject consult Gurdon 1974; Brown and Stern 1974; Reeder 1974; Davidson 1976; Reeder et al. 1976). In addition, these reiterated genes have, in the past, proven to be extremely useful experimentally because of the existence of various strains of animals that carry different deletion mutations involving the ribosomal gene DNA. By using these genetic deletion mutations in various biochemical and morphological studies, investigators have been able to investigate meaningfully problems that might otherwise have been refractory to analysis (Brown and Gurdon 1965; Gurdon 1974). I thought these same genes might provide useful material for an investigation of the structure of chromatin during active gene transcription.

There are about 450–500 adjacent copies of the ribosomal genes in the haploid complement of wild-type *X. laevis* DNA (Brown and Weber 1968a; Dawid et al. 1970; Birnstiel et al. 1971). This corresponds to about 0.2% of the nuclear DNA of diploid wild-type somatic cells (Brown and Weber 1968a). Each of these tandemly repeated ribosomal gene copies includes DNA which codes for a 40S RNA transcript that is a precursor to 28S and 18S ribosomal RNA (rRNA), as well as DNA for a nontranscribed "spacer" region of high deoxyguanylic and deoxycytidylic acid content (Brown and Weber 1968b). In these diploid wild-type animals, the 900–1000 or so genes are distributed equally between two nucleolar organizer regions located on homologous chromosomes (Wallace and Birnstiel 1966). Such animals usually have two nucleoli per nucleus during early embryonic larval stages and have the genetic constitution usually designated as +/+ nu (Gurdon 1974).

A number of deletion mutations also exist in certain strains of *Xenopus*. For example, the anucleolate mutation (Elsdale et al. 1958) apparently involves the total elimination of the repeated ribosomal genes from the nucleolar organizer region of one of the chromosomes of the normal haploid set. In the diploid homozygous recessive condition (O/O nu animals), this mutation is lethal because embryos carrying this deletion on both chromosomes lack any detectable ribosomal DNA and consequently never synthesize any rRNA (Brown and Gurdon 1965). On the other hand, diploid animals heterozygous for this mutation (O/+ nu individuals) are quite viable even though they possess only half of the normal complement of ribosomal genes of wild-type animals. These heterozygous O/+ nu individuals synthesize rRNA at about the same rate as wild-type animals but can be distinguished from them cytologically because they have only one nucleolus per nucleus in all of their cells. In addition to the anucleolate mutation, there are also partial deletion mutations in other *Xenopus* strains which reduce the number of ribosomal gene copies within a single chromosomal nucleolar region to below the normal haploid value but do not represent total deletions of the ribosomal gene DNA (Miller and Gurdon 1970; Miller and Knowland 1970, 1972; Knowland and Miller 1970). These genetic variants have allowed for critical analysis of the ribosomal genes of this animal (Gurdon 1974).

The Nucleosome Structure of Wild-type (+/+ nu) Ribosomal Genes

In recent years a number of reports from several different laboratories have suggested that the nuclear chromatin of cells actively engaged in ribosomal RNA synthesis can be cleaved by the enzyme micrococcal endonuclease (E.C. 3.1.4.7) into nucleosome-size fragments which contain pieces of ribosomal gene DNA (Reeder 1975; Reeves and Jones 1976; Reeves 1976, 1977; Mathis and Gorovsky 1976; Piper et al. 1976; Higashinakagawa et al. 1977). These results are consistent with, but do not prove, the suggestion that transcriptionally active ribosomal gene chromatin has a nuclease-resistant structure similar to that of the nuclear chromatin as a whole. The ribosomal genes of *Xenopus* somatic cells also seem to contain nucleosomes in both the active and inactive states.

Figure 1 demonstrates that the ribosomal chroma-

tin from both transcriptionally active and inactive cells of wild-type +/+ nu *Xenopus* contains nucleosome-protected regions of DNA that are not readily accessible to short-term micrococcal nuclease attack. In this experiment nuclei were isolated from either a permanent line of embryonic *Xenopus* tissue-culture cells (X58) growing exponentially in culture (and demonstrably very active in rRNA synthesis) or from adult red blood cells that are transcriptionally quiescent (Hilder and Maclean 1974). The nuclear isolations and all subsequent operations were conducted under conditions that minimize both histone rearrangements on the chromatin and histone degradation (Reeves and Jones 1976; Reeves 1976). The isolated nuclei were then digested with micrococcal nuclease for various times so that monomer-size nucleosomes (monosomes), as well as larger chromatin fragments, were released. The nuclei were then homogenized to release the soluble chromatin, and the nuclear membranes and other insoluble material were removed by centrifugation. The chromatin-containing supernatant was then loaded onto 10–30% sucrose gradients and, after appropriate lengths of centrifugation, the monosome peaks were isolated from the gradients as described previously (Reeves and Jones 1976). The monosome-peak DNA (containing about 200 base pairs of double-stranded DNA as determined by gel electrophoresis using appropriate markers; Reeves and Jones 1976) was then isolated by the phenol techniques of Brown and Weber (1968a). In other experiments to be described shortly, the monosome-size DNA fragments were isolated directly from agarose gels after electrophoretic separation rather than by velocity sedimentation. In either case, the monosome DNA fragments were then used for saturation nucleic acid hybridization studies with isotopically labeled rRNA following the procedures of Gillespie (1968) for nitrocellulose filter hybridizations (Reeves

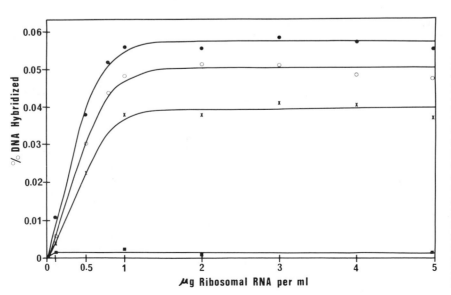

Figure 1. Saturation profiles for hybridizations obtained from one experiment in which all DNA-containing filters were annealed under identical conditions with increasing concentrations of the same batch of electrophoretically purified [^{32}P]RNA (18S + 28S), as described in the text. (●) Sheared +/+ nu DNA from *Xenopus* X58 tissue-culture cells; (○) nuclease-derived monomer nucleosome DNA from adult red blood cells; (X) nuclease-derived monomer nucleosome DNA from transcriptionally active X58 tissue-culture cells; (■) sheared O/O nu DNA from homozygous anucleolate embryos.

and Jones 1976). The conditions chosen for these filter hybridization experiments (0.3 M NaCl, 0.03 M trisodium citrate, pH 7.8, 50% formamide, 45°C) correspond to incubation at 25°C below the T_m of *Xenopus* rRNA-DNA hybrids and yet allow for both a high specificity of annealing and fast reaction rates (Miller and Knowland 1970). The fidelity of annealing and reproducibility using these techniques have been reported on previously (Reeves 1976; Reeves and Jones 1976).

The results of these experiments in which saturating amounts of ribosomal RNA (18S + 28S) were hybridized to monomer nucleosome DNA fragments derived either from transcriptionally active or inactive cells are shown in Figure 1. The fragments of monosome DNA derived from red blood cells contain about 86% of the amount of hybridizable DNA contained in sheared (non-nuclease-treated) wild-type +/+ nu DNA. Thus, most of the ribosomal genes present in transcriptionally quiet chromatin appear to be protected from short-term nuclease digestion by association with nucleosomes. On the other hand, the chromatin derived from the synthetically active tissue-culture cells has more of its gene sequences accessible to digestion by the nuclease since the monosome DNA fragments hybridize to only about 70% of the level found in sheared +/+ nu DNA.

At face value, these experiments might be interpreted to suggest that many of the ribosomal genes present in transcriptionally active cells probably have a major portion of their DNA associated with nucleosomes and are therefore protected from nuclease attack. Such a conclusion seems even more plausible when the additional evidence is presented that these results cannot be reasonably explained on the basis of histone rearrangements or histone proteolysis (Reeves and Jones 1976) or as a result of the presence of small basic proteins other than histones protecting the DNA from nuclease attack (Reeves 1976). Furthermore, the 200-base-pair fragments of DNA used for these hybridization studies are much too large to be the result of protection by association with RNA polymerase molecules (Reeves 1977). Finally, purified somatic ribosomal gene DNA can be reassociated with high-salt-extracted histones to form nuclease-resistant nucleosome fragments similar to those used in this study (Reeves 1977).

However, even given all of this indirect evidence supporting such a hypothesis, considerable caution must be exercised in assessing the validity of the assertion that transcriptionally active chromatin contains, at least partially, histones in the form of nucleosomes. There are several technical and theoretical reasons for this reservation, with perhaps the foremost being the inherent difficulty in interpreting hybridization data concerning the transcriptional activity of genes that are present in multiple copies within a genome. For instance, it is not known what fraction of the total number of reiterated genes

is functional at any given time within a random population of growing cells. Similarly, even though rRNA is known to be synthesized continuously through all of the cell cycle (except during mitosis; Mitchison 1971), it is not at all clear that all of the genes coding for ribosomal RNA need to be active in transcription at the same time. Thus, these results from wild-type cells could also be interpreted as indicating that the nucleosomes derived from the starting population of tissue-culture cells could have been released from transcriptionally inactive gene sequences present within this mixed population.

The Nucleosome Structure of Heterozygous (O/+ nu) Ribosomal Genes

In an attempt to circumvent the technical problem posed by the possibility of transcriptionally inactive genes being present as a significant contaminant in a starting population of cells, I have taken advantage of the genetic deletion mutations present in certain strains of *Xenopus* to artificially produce animals in which all of the ribosomal genes are probably active. Such a condition seems very likely to be the case in animals that are heterozygous for the anucleolate mutation (O/+ nu animals). As noted previously, these heterozygotes have only about half of the number of ribosomal genes of wild-type diploid cells and yet synthesize rRNA at the same rate as +/+ nu cells (Gurdon and Brown 1965; Knowland and Miller 1970). This finding has been interpreted to mean either that the O/+ nu animals have doubled their rate of rRNA transcription or that the same number of active genes are present in both wild-type and mutant animals (Knowland and Miller 1970). In any event, the amount of dosage compensation of which the ribosomal gene complement is capable in these heterozygous mutants is very limited in nature. For example, it has been demonstrated by using a series of partial-deletion (p^1/O nu) mutations in the nucleolar genes that a reduction in the number of genes below the heterozygote O/+ nu level in these p^1/O nu animals results in larval death before the feeding-tadpole stage of development (stage 46 of Nieuwkoop and Faber 1956) (Miller and Gurdon 1970). The most likely cause for this lethality is the inability of these embryos to synthesize rRNA at a rate fast enough to maintain viability after they have used up their store of maternally derived ribosomes (Miller and Knowland 1970, 1972; Knowland and Miller 1970; Reeder 1974). This conclusion is further supported by the finding that in p^1/O nu embryos the rate of synthesis of rRNA is directly proportional to their number of ribosomal genes, but they are not able to compensate this rate up to the level found in O/+ nu which is needed for continued viability (Miller and Knowland 1970). Thus it seems quite likely that in heterozygous O/+ nu animals there is a physiological need for all of the ribosomal genes to be transcriptionally active in order to maintain viability. It should be noted, however, that there is no formal proof of this assertion. Nevertheless, the heterozygous O/+ nu embryos offer an excellent opportunity for investigating the structure of ribosomal genes that must have rates of transcriptional activity near maximal level by the time these embryos reach advanced tadpole stages of development.

Another feature of *Xenopus* embryonic development that allows for experimental investigation of the structure-function relationship of chromatin is the fact that ribosomal gene transcription is very precisely controlled during early embryonic stages in wild-type +/+ nu embryos (Brown and Littna 1964a; Gurdon 1968; Woodland and Gurdon 1968; Knowland 1970; Gurdon 1974). A similar regulation is seen in heterozygous O/+ nu embryos as shown in Figure 2. In this experiment, tritiated uridine was microinjected into embryos of various ages, and the relative amounts of radioisotopically labeled ribosomal RNAs synthesized per embryonic cell per hour were then determined as shown in Table 1. For technical reasons discussed elsewhere (Reeves 1976), the early embryo stages were a mixture of three phenotypes, 1-Nu, O-Nu, and 2-Nu, which correspond to the genotypes O/+ nu, O/O nu, and +/+ nu, respectively. After the tail-bud stage, however, the em-

bryos were typed cytologically as being heterozygous O/+ nu (1-Nu). It is readily seen from Figure 2 and Table 1 that detectable ribosomal RNA synthesis first begins at around the gastrula stage of development and then increases rapidly until around stages 40–42 (Nieuwkoop and Faber 1956) when it appears to be approaching a maximum plateau. These results for O/+ nu embryos are consistent with the findings of others for wild-type embryos of the same ages.

With this information as background, experiments were designed to determine whether there is a detectable relationship between the changing rates of rRNA synthesis observed during early embryogenesis of heterozygous O/+ nu animals and the nucleosome structure of the ribosomal gene chromatin. Monomer nucleosome DNA was isolated as previously described from blastula embryos (stages 8 and 9) which have not yet started rRNA synthesis, from neurula embryos (stages 16–18) which are actively synthesizing rRNA but not at maximal rates, and from stage-46 tadpoles engaged in near-maximal rates of synthesis. Ribosomal RNA was annealed to these DNA fragments and the resulting saturation hybridization curves shown in Figure 3 were obtained.

A number of significant points are evident in Figure 3. Among these are: (1) In transcriptionally inactive blastula cells, most (about 84%) of the ribosomal DNA is protected by nucleosome packing—a level of protection similar to that seen before in the synthetically moribund adult red blood cells. (2) In stage-46 tadpoles that are synthesizing rRNA at a near-maximal rate, there is still a significant portion (about 60%) of the genes protected by nucleosomes from nuclease attack. (3) In neurula embryos, where the cells are engaged in rRNA synthesis at only about 60% of the maximal rate seen in older stages, an intermediate level of nucleosome protection between that of the blastula stage and that of tadpoles is seen.

These results suggest that there may be a reciprocal relationship between the relative rates of rRNA synthesis found at various developmental stages and the amount of protection from nuclease digestion afforded the ribosomal gene DNA by association with nucleosomes. Thus, the transcriptionally inactive blastula cells have most of their ribosomal genes protected by nucleosomes, and as the rate of ribosomal RNA synthesis increases, proportionally more of the rDNA becomes accessible to nuclease hydrolysis. However, there appears to be a definite limit to the amount of rDNA that can be digested even in the stage-46 tadpoles where most, if not all, of the genes are probably maximally active. This limited nuclease accessibility strongly suggests that at least some of the transcriptionally active ribosomal genes in these tadpole cells are partially, or perhaps transiently, associated with histones in the form of

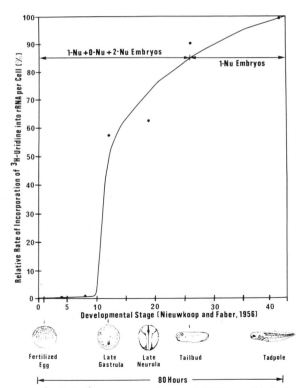

Figure 2. Relative rates of ribosomal RNA synthesis in heterozygous O/+ nu embryos calculated on a per cell basis during various stages of development.

Table 1. Relative Rates of Ribosomal RNA Synthesis in O/+ nu and Mixed Genotype Embryos

Stage labeled	Labeling time (hr)	Embryo stages[a]	Cells/embryo[b]	[3H] dpm incorporated into rRNA[c]	Acid-soluble dpm/embryo[d]	Average dpm/embryo[d] / acid-soluble dpm	Average corrected[d] dpm/cell/hr	Relative rate of synthesis
Fertilization to 8 cells	1–3	1–4	8	0	88,104	—	—	0
Middle to late blastula	4–6	6–8	10,000	0	90,075	—	—	0
Middle to late gastrula	12–14	11–13	38,000	328	88,815	3.69×10^{-3}	4.86×10^{-8}	58.6
Middle to late neurula	19–21	15–19	89,700	803	88,416	9.08×10^{-3}	0.51×10^{-7}	60.8
Early tail bud	30–32	26	170,000	2215	85,780	2.58×10^{-2}	0.76×10^{-7}	91.6
Tadpole	80–82	42	755,000	9872	78,551	0.125	0.83×10^{-7}	100

[a] Embryo stages classified according to Nieuwkoop and Faber (1956). Embryos were microinjected with 50 nl of [3H]uridine at 20 mCi/ml and incubated in pond water for 2 hr at 20°C before rRNA extraction and purification.
[b] Values taken from the data of Woodland and Gurdon (1968).
[c] 18S + 28S + 40S ribosomal RNA purified by electrophoresis as described in text.
[d] Each value is the average of five determinations.

nucleosomes. Furthermore, the data imply a dynamic association between chromatin structure and transcriptional activity.

$C_o t$ Reassociation Curves of Nucleosome DNA

If, as the data above suggest, nucleosomes can be associated with transcriptionally active DNA, then $C_o t$ DNA reassociation curves of monomer nucleosome DNA fragments should be very similar to $C_o t$ curves of sheared (non-nuclease-treated) DNA from the same source. This is indeed the case with DNA fragments derived from these heterozygous O/+ nu

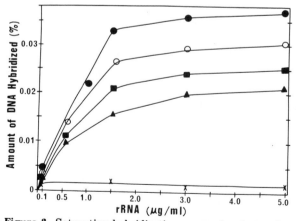

Figure 3. Saturation hybridization curves for electrophoretically purified [32P]rRNA (18S + 28S) annealed to DNA from various developmental stages of heterozygous O/+ nu embryos (see text for explanation of genotypes involved in these experiments). (●) O/+ nu sheared DNA (stage-42 tadpoles); (○) O/+ nu × O/+ nu monosome DNA from stage-8 and stage-9 blastula embryos; (■) O/+ nu × O/+ nu monosome DNA from stage-16 to stage-18 neurula embryos; (▲) O/+ nu monosome DNA from stage-46 tadpoles; (X) O/O nu sheared DNA (stage 40). (Reprinted, with permission, from Reeves 1976).

tadpoles (Reeves 1977). Thus, heterozygous O/+ nu monosome DNA contains all of the gene-sequence-frequency classes (unique, intermediate repetitive, fast, and zero-time binding sequences) in about the same relative proportions as are found in whole O/+ nu DNA. These results suggest that during nuclease digestion nucleosomes are released from both active and inactive gene sequences and that no individual gene-frequency class is more susceptible to hydrolysis than any other. The results shown in Figure 4 reinforce this conclusion.

In this experiment, intermediate repetitive and nonrepetitive (unique-sequence) DNA from monosomes of O/+ nu tadpoles were isolated individually by selective hydroxylapatite chromatography and then chemically labeled with 125I as described elsewhere (Reeves 1977). These isotopically labeled DNA fractions were then annealed individually in DNA-driven reactions with a vast excess of sheared wild-type (+/+ nu) DNA. The results of these experiments depicted in Figure 4 indicate that the $C_o t_{1/2}$ values for reassociation of both the unique-sequence O/+ nu nucleosome DNA ($C_o t_{1/2} = 1.47 \times 10^3$ moles · sec/liter) and the intermediate repetitive O/+ nu sequences ($C_o t_{1/2} = 0.61$ mole · sec/liter) are very similar to those of the same frequency classes in sheared +/+ nu DNA (Reeves 1977). These results are consistent with the hypothesis that a major portion of the ribosomal DNA of heterozygous O/+ nu chromatin exposed to nuclease digestion is recoverable as nucleosome-size fragments even from those stages that are most active in rRNA synthesis.

The Relative Rates of Ribosomal RNA Synthesis in Oocytes

In *Xenopus*, as in many other organisms, the ribosomal RNA genes are amplified during the pachytene stage of oogenesis soon after the female tad-

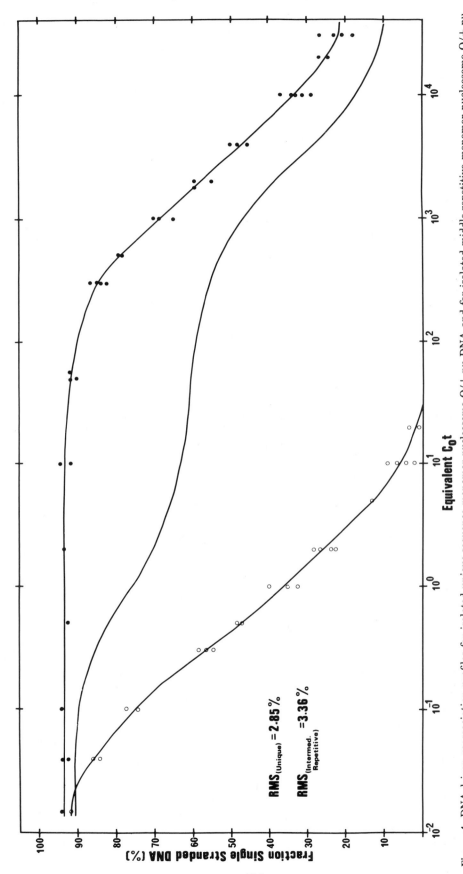

Figure 4. DNA-driven reassociation profiles for isolated unique-sequence monomer nucleosome O/+ nu DNA and for isolated middle-repetitive monomer nucleosome O/+ nu DNA in the presence of a vast excess of sheared +/+ nu DNA. (●) Unique-sequence monomer nucleosome O/+ nu DNA; (○) middle-repetitive-sequence monomer nucleosome O/+ nu DNA. The solid line is the DNA reassociation curve for mechanically sheared O/+ nu DNA. (Reproduced, with permission, from Reeves 1977.)

poles have metamorphosed to froglets. This phenomenon has been reviewed extensively elsewhere (see, e.g., Brown 1967; Brown and Dawid 1968; Gall 1968, 1969; Hourcade et al. 1974; Davidson 1976). The end result of rDNA replication is the production of about 1500 extrachromosomal nucleoli, each containing several sets of ribosomal genes often with detectably different lengths of nontranscribed spacer DNA separating the coding sequences (Wellauer et al. 1974). The total number of these gene sets has been estimated to be between 3×10^3 and 5×10^3 (Perkowska et al. 1968; Brown and Dawid 1968; Gall 1969). Since each haploid ribosomal gene set contains about 450–500 copies of the individual ribosomal genes (Brown and Weber 1968a), the total number of these genes in the oocyte is 1.5×10^6 to 2.5×10^6, and their combined mass is about 25–30 pg per nucleus (Brown and Dawid 1968; Perkowska et al. 1968). This amount of amplified ribosomal DNA is in considerable excess over the amount of pachytene 4C chromosomal DNA which has been estimated to be about 12 pg per nucleus (Dawid 1965).

During the extended growth period of amphibian oogenesis, the rates of transcription of rRNA from the amplified extrachromosomal ribosomal genes are closely regulated and growth-stage dependent (Brown and Littna 1964b; Gurdon 1968; Ford 1972; La Marca et al. 1973; Denis 1974; Reeder 1974; Scheer et al. 1976a; Davidson 1976). It was thus of some interest to investigate the nucleosome structure of these amplified ribosomal genes during various oocyte growth stages when the rates of synthesis are changing dramatically. Before such experiments could be conducted meaningfully, however, it was first necessary to establish unambiguously the relative rates of rRNA synthesis in the various isolated oocyte stages used for chromatin structure studies. The results of these experiments are reported elsewhere (R. Reeves, in prep.) and are summarized in Figures 5 and 6 and Table 2.

Figure 5 shows the electrophoretic profiles of ribosomal RNAs isolated from various *Xenopus* oocyte size classes (classified according to Dumont 1972) which were labeled in culture with a medium containing [³H]uridine. Relative rates of incorporation of radioactivity into rRNA in different oocyte stages were determined in time-course kinetic experiments in which the labeled rRNA for each time point was extracted from the oocytes and then separated by gel electrophoresis into 18S, 28S, and 40S ribosomal RNA species which were then isolated from appropriate regions of the gels (the cross-hatched areas shown in Fig. 5). The amount of radioactivity in these peaks was determined and rates of incorporation were calculated (R. Reeves, in prep.). In addition, in the same experiments, the specific activity of the RNA precursor UTP pools within each of the oocyte size classes was determined so that the relative rates of synthesis in the different oocytes could be compared directly. Figure 6 shows the calibration

curves and the appropriate data for the determination of the UTP pool sizes in each of the oocyte stages analyzed (R. Reeves, in prep.). Determination of both UTP and ATP pool sizes was carried out by using *Escherichia coli* RNA polymerase with poly(dA-dT) as template under polymerization conditions in which either the ATP or the UTP present within oocyte extracts was the limiting component in the enzyme-catalyzed production of a ribonucleotide polymer (Sasvari-Szekely et al. 1975). The uptake and phosphorylation of [³H]uridine into [³H]UTP by oocytes were determined by nucleotide extraction and PEI-cellulose thin-layer chromatographic analysis as described elsewhere (R. Reeves, in prep.). Table 2 summarizes the pertinent findings from these pool studies and also gives the relative rates of rRNA synthesis determined for the various oocyte classes investigated.

From the data in Table 2 it is evident that the rates of rRNA synthesis vary markedly in different classes of oocytes, and from arguments presented elsewhere (R. Reeves, in prep.) it seems very likely that this variation is due to changing rates of initiation of transcription rather than resulting from other factors such as rRNA turnover or changes in elongation rates (see also Denis 1974; Gurdon 1974; Scheer et al. 1976a; Davidson 1976). It is thus evident that stage-I previtellogenic oocytes synthesize rRNA at very low levels (about 0.3 ng/oocyte/hr) compared to stage-II and stage-III oocytes which have started to accumulate yolk and which contain prominent lampbrush chromosomes that are also synthetically active (Dumont 1972). These stage-II and stage-III oocytes were observed to be maximally active in rRNA synthesis, accumulating about 6.2 ng/oocyte/hr under the experimental conditions employed. In later-stage oocytes (stage IV and stages V and VI), the relative rates of synthesis are seen to drop slightly but synthesis never ceases completely. For a comparison of these values with those obtained by others, see Davidson (1976).

The Nucleosome Structure of Oocyte Amplified Ribosomal Genes

Does the structure of the extrachromosomal ribosomal genes in oocytes vary as a function of changes in relative rates of transcription? To answer this question, various size classes of oocytes were obtained in bulk (R. Reeves, in prep.) and their nuclei (germinal vesicles) isolated. The nuclei were exposed to micrococcal nuclease, and the monosome DNA fragments were recovered and used for nucleic acid hybridization experiments as described.

In connection with these experiments, it should be noted that the cytoplasm of developing oocytes contains anywhere from 30 to 100 times more mitochondrial DNA than the total DNA present within the nucleus as chromosomal and amplified ribosomal DNA (Dawid 1965; Webb and Smith 1977). Thus,

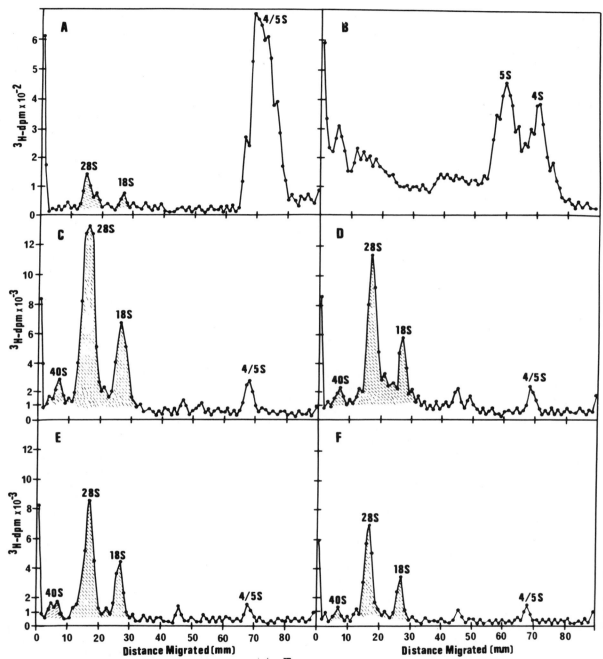

Figure 5. Gel electrophoretic analysis of radioactively labeled RNA extracted from different oogenic stages of *Xenopus*. Oocytes were cultured in vitro with [³H]uridine as described by R. Reeves (in prep.). After various lengths of labeling, the oocytes were isolated from their surrounding follicle cells and separated into five size classes (after Dumont 1972) and the RNA extracted from each class and electrophoresed on polyacrylamide gels. The RNA from the regions of the gels shown with hatching was then isolated for each class of oocytes and the amount of acid-insoluble radioactivity determined. From such measurements, the rates of incorporation of isotope into rRNA for each oocyte class were determined. *(A)* Stage-I (50–300-μm diameter) oocyte RNA, 2.7% polyacrylamide gels; *(B)* stage-I oocyte RNA electrophoretically separated on 8% polyacrylamide gels to distinguish 4S from 5S RNA; *(C)* stage-II and -III (300–600-μm diameter) oocyte RNA, 2.7% polyacrylamide gels; *(D)* stage-IV (600–1000-μm diameter) oocyte RNA, 2.7% polyacrylamide gels; *(E)* stage-V (1000–1200-μm diameter) oocyte RNA, 2.7% polyacrylamide gels; *(F)* stage-VI (1200–1300-μm diameter, white band) oocyte RNA, 2.7% polyacrylamide gels.

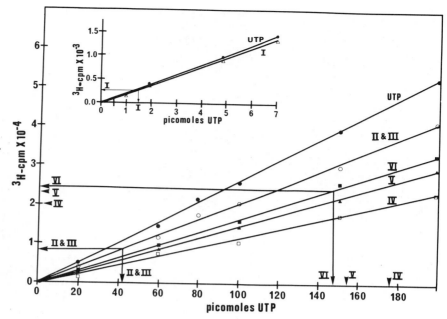

Figure 6. Standard RNA polymerase incorporation calibration curves for the determination of the endogenous UTP pool sizes of different classes of *Xenopus* oocytes (see text and R. Reeves [in prep.] for experimental details and calculations). By the use of these standard curves, the amount of UTP in the various oocyte-stage extracts (roman numerals) was calculated from the radioactivity incorporated in the presence of cell extract alone without exogenous UTP added (horizontal and vertical arrows).

trace contaminations of isolated nuclei by cytoplasm containing mitochondria could easily lead to marked variations in any ribosomal RNA saturation hybridization values obtained for oocyte DNA. For this reason, in each of the oocyte nuclear isolation experiments, a known quantity of mitochondria obtained from *Xenopus* tissue-culture cells and labeled in vitro with [14C]thymidine was added to the starting oocyte population to monitor the resulting amount of contamination of the isolated nuclear preparations (R. Reeves, in prep.). The amount of mitochondrial DNA contamination for each oocyte stage was then adjusted to the average amount of mitochondrial DNA contamination found in stage-II and stage-III oocyte DNA preparations (54.5%) which were routinely used as reference standards for all of the saturation hybridization experiments (R. Reeves, in prep.).

Figure 7 shows the profiles seen under ultraviolet light of the purified double-stranded monosome DNAs obtained from various oocyte growth stages

and separated by electrophoresis on agarose gels containing ethidium bromide. DNA molecular-weight markers obtained from either tissue-culture cells or red blood cells flank the oocyte DNA in this photograph. The results of saturation hybridization experiments annealing these various oocyte monosome DNAs to ribosomal RNA are shown in Figure 8 and are summarized in Table 3.

From these data it is quite clear that in stage-I previtellogenic oocytes (which have very low rates of rRNA synthesis) most of the extrachromosomal rDNA sequences (about 84%) are associated with nucleosomes—again this is about the same level of protection seen in inactive somatic cells (see above). On the other hand, in stage-II and stage-III oocytes, which are maximally active in transcription, the amount of nucleosome protection has dropped to around 37% of the amount of rDNA found in control stage-II and stage-III DNA not exposed to nuclease. In these experiments the hybridization values are considered to reflect mainly the condition of the am-

Table 2. Oocyte Nucleotide Pool Sizes and Relative Rates of Ribosomal RNA Synthesis

Stage[a]	Diameter (µm)	UTP pool[b] (pmoles)	ATP pool[b] (pmoles)	Range of rates of rRNA synthesis (ng/hr/oocyte)	Average rate[c] of rRNA synthesis (ng/hr/oocyte)	Relative rate of rRNA synthesis (%)	Amount of nucleosome protection (%)
I	50–300	1.05	8.4	0.23–0.4	0.315	5.1	84
II, III	300–600	37.0	114	3.62–8.7	6.16	100	37
IV	600–1000	174	504	3.45–7.92	5.69	92.3	48
V	1000–1200	317	634 ⎫	2.80–5.13	4.06	65.9	57
VI	1200–1300	306	650 ⎭				

[a] Dumont (1972).
[b] Each value is the average of four determinations.
[c] See R. Reeves (in prep.) for calculations.
[d] Rates are given as values relative to stage-II and -III oocytes.

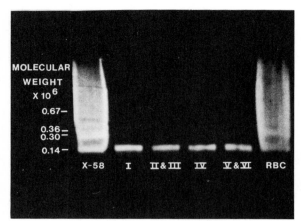

Figure 7. Patterns obtained when monomer nucleosome double-stranded DNA fragments from various oocyte classes were electrophoresed on a 1.4% agarose slab gel, stained with ethidium bromide, and photographed under ultraviolet light. Nuclease digestion times and conditions will be given elsewhere (R. Reeves, in prep.). The oocyte stages (Dumont 1972) from which the DNA was obtained are indicated by roman numerals. Marker DNAs (with apparent molecular weights calibrated as described previously [Reeves and Jones 1976]); from X58 tissue-culture cells and from adult red blood cells (RBC) were electrophoresed on the same gel for size comparisons. Oocyte DNAs from which these samples were obtained were used for hybridization studies.

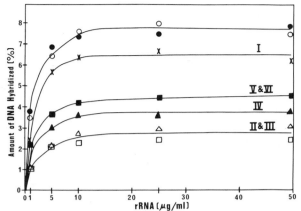

Figure 8. Saturation hybridization curves for electrophoretically purified [³H]RNA (18S + 28S) annealed to DNA from various oocyte stages. (●) Sheared nuclear DNA from stage-II and -III oocytes (Dumont 1972); (○) sheared nuclear DNA from stage-V and -VI oocytes; (X) monomer nucleosome DNA from stage-I oocyte nuclei; (■) monomer nucleosome DNA from stage-V and -VI oocyte nuclei; (▲) monomer nucleosome DNA from stage-IV oocyte nuclei; (△,□) monomer nucleosome DNA from stage-II and -III oocyte nuclei taken from two entirely independent experiments. Each point shown represents the average of three separate determinations.

plified rDNA since the chromosomal ribosomal genes represent only about 0.09% of all of the hybridizable DNA. It is therefore significant that in these maximally active cells a sizable portion of the amplified rDNA is still associated with histones in

the form of nucleosomes. While this work was in progress, Higashinakagawa et al. (1977) also reported having found micrococcal nuclease-resistant fragments of rDNA in nucleoli isolated from these same stages of *Xenopus* oocytes. Furthermore, these same workers have reported finding appreciable quantities of what appear to be all of the major

Table 3. Saturation Hybridization of Oocyte Germinal Vesicle DNA with [³H]RNA

DNA[a]	Percentage DNA hybridized by			rDNA as percentage of germinal vesicle rDNA	Nucleolar equivalents[c] of rDNA per oocyte
	18S + 28S[b]	18S[b]	28S[b]		
Stages II and III, undigested	7.76	2.48	4.89	100	—
Stages V and VI, undigested	7.68	2.46	5.07	100	—
Average oocyte, undigested	7.72	2.47	4.98	100	1616
Stage I, monosome	6.48	2.27	4.25	83.9	1357
Stages II and III, monosome	2.85	0.94	1.82	36.9	596
Stage IV, monosome	3.71	1.29	2.53	48.1	755
Stages V and VI, monosome	4.41	1.37	2.95	57.1	923

Data from R. Reeves (in prep.).

[a] Stages of Dumont (1972).

[b] Each value is the average of three separate experiments.

[c] Calculations based on the following assumptions: (1) Each germinal vesicle (nucleus) contains 25 pg of extrachromosomal rDNA and 12 pg of chromosomal DNA (Brown and Dawid 1968). (2) Each haploid somatic cell nucleus contains 3.1 pg of DNA (Dawid 1965). (3) Somatic cells contain 0.057% of their DNA complementary to rRNA (Brown and Weber 1968a). (4) 28S RNA has a molecular weight of 1.5×10^6 daltons and 18S RNA has a molecular weight of 0.75×10^6 daltons (Dawid et al. 1970). (5) The values given for the amount of DNA hybridized (%) to rRNA represent only half of the mass of total rDNA (not counting the nontranscribed spacer DNA) present in the preparations. (vi) A weight of nuclease-derived monomer nucleosome DNA equal to the weight of whole DNA in a normal stage-VI oocyte nucleus is a germinal-vesicle equivalent.

classes of histones associated with these isolated nucleoli. All of these results are consistent with the hypothesis (Reeves 1976) that nucleosome (and hence histone) association with DNA may not necessarily preclude its transcription. However, as noted here, such protein-DNA interaction appears to be a labile complex that can vary relative to changes in transcriptional activity of this DNA, as seen in Figure 8 and Table 3. For example, near the end of oogenesis, stage-V and stage-VI oocytes have considerably reduced their rates of rRNA synthesis compared to the earlier maximally active stages and, significantly, the average amount of nucleosome-protected rDNA has correspondingly risen from the earlier minimum value of 37% protection up to about 57% protection near the end of oocyte growth.

Thus, the results from experiments with amplified ribosomal genes seem to support the conclusions drawn earlier from the hybridization studies with heterozygous O/+ nu embryos. Both sets of data indicate that there appears to be a labile relationship between chromatin nucleosome structure and DNA template activity. The main observable difference between the ribosomal genes of somatic cells and the amplified extrachromosomal genes of oocytes is the markedly increased nuclease sensitivity of the latter. The reason for this difference is unknown, but it is known that the DNA of the former contains the modified nucleotide 5-methyl deoxycytidine whereas the latter does not. There is no evidence relating differential nuclease sensitivity to this modification, however.

DISCUSSION

Figure 9 summarizes the findings relating the relative rates of ribosomal RNA synthesis to the relative amounts of micrococcal nuclease-resistant nucleosomes found in different developmental stages of *Xenopus* oocytes or O/+ nu embryos. From these results it is quite evident that, in addition to the variations noted in the amount of nucleosome protection from stage to stage, there is no stage in which there is complete accessibility of the ribosomal DNA to micrococcal nuclease hydrolysis. These results strongly suggest the following: (1) Nucleosome association with DNA is part of a dynamic process that appears to be closely regulated and is somehow related to changing rates of gene transcription. (2) At no time during development are all of the many reiterated transcriptionally active ribosomal genes completely free of nucleosomes. (3) The association of histones with DNA may not necessarily preclude template activity. (4) Transcriptionally inactive DNA is almost completely protected by nucleosomes, although a small but rather constant fraction (about 15%) of the DNA is always unprotected. (5) As chromatin becomes increasingly active in transcription, the amount of histones (nucleosomes) tightly bound

to DNA decreases as monitored by increases in nuclease sensitivity. (6) This increased sensitivity to nuclease attack found in maximally active chromatin is probably variable in nature and may possibly reverse itself when the rates of transcription subsequently decline. (7) There appears to be a definite limit to the amount of transcriptionally active DNA that is available to short-term micrococcal nuclease digestion, and this limit varies with different cell types. (8) Even though chromatin structure and functional activity are closely correlated, the molecular mechanisms mediating this association are probably much more complex than previously anticipated.

All of the results presented here suggest that the nucleosome structure of chromatin, as monitored by micrococcal nuclease sensitivity, is in a dynamic state of flux and can possibly be altered in either biochemical or perhaps morphological ways as rates of gene transcription change. Two additional points about such a postulated labile chromatin structure need to be mentioned, however. First, even though completely "free" or unprotected ribosomal gene DNA has not been detected in any of the experiments reported here or elsewhere using micrococcal nuclease digestions, the dynamic structure of chromatin proposed here does not a priori rule out this physical possibility for a minority of the population of reiterated ribosomal genes. For example, consider the situation found in stage-II and stage-III oocytes where only about 37% of the rDNA is observed to be protected by nucleosomes. These data do not allow for unambiguous distinction between the situation in which 37% of each of the multiple copies is protected and one in which 37% of all of the genes are completely protected and 63% completely free or open to nuclease attack. However, considering the evidence presented above for the heterozygous O/+ nu mutants, the extreme situation where a large fraction of all of the reiterated genes are completely unprotected seems rather unlikely. Thus, a more reasonable interpretation of the data might be that there is probably a great deal of heterogeneity among the multiple copies of the ribosomal genes (see Scheer et al. 1976a,b) with respect to form, function, and nuclease sensitivity and that 37% protection from nuclease digestion represents an average value for all of these genes. This would then theoretically allow for a possible structural (and/or functional) continuum to exist between one extreme, where a few genes might be almost completely devoid of tightly bound nucleosome structures, and the other extreme, where transcriptionally inactive genes would be almost completely covered with firmly associated histones. Most of the genes in normal metabolic situations, where there is a certain fixed capacity for regulation of transcription rates in response to physiological needs, might be somewhere between these two extremes in their nuclease sensitivity.

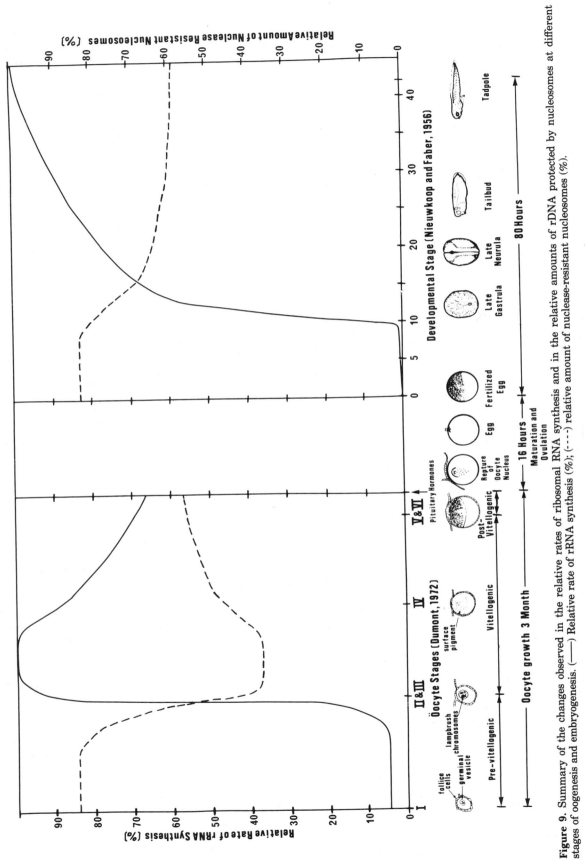

Figure 9. Summary of the changes observed in the relative rates of ribosomal RNA synthesis and in the relative amounts of rDNA protected by nucleosomes at different stages of oogenesis and embryogenesis. (——) Relative rate of rRNA synthesis (%); (---) relative amount of nuclease-resistant nucleosomes (%).

The second point to be made is that the decreasing degrees of "protection" by nucleosomes observed in the more transcriptionally active chromatins need not necessarily signify the physical absence of histones (or even nucleosomes) from the accessible DNA. Their are several reasons for making this assertion. For example, "accessibility" of DNA to micrococcal nuclease hydrolysis does not seem to be an all-or-none phenomenon and may vary with the digestion conditions and lengths of time the chromatin is exposed to nuclease (Bellard et. al., this volume). Furthermore, differences in nuclease accessibility may also indicate that some nucleoproteins are less firmly associated with DNA than are other neighboring proteins. Thus, the data are not incompatible with the idea that almost all of the DNA within a nucleus may be associated with histones, but it also suggests that regional differences might exist both in the biochemical and structural arrangements of the histones with each other and in the degree of their binding to the DNA itself. What the data do seem to rule out is the probability that there is a total absence of histones from all of the reiterated copies of the transcriptionally active ribosomal genes of *Xenopus*. This is a testable prediction.

In conclusion, the question remains open as to the meaning of the apparently close correlation between transcription rates and chromatin nucleosome structure reported here. Does the structure of chromatin "loosen up" or change form before initiation of gene transcription can begin, or does the process of transcription itself induce the changes in the chromatin observed here? Since the transcriptionally active nucleolar chromatin can now be isolated from *Xenopus* oocytes in relatively pure form (Higashinakagawa et al. 1977), it would be of considerable interest to compare the histones and nonhistone chromatin proteins associated with nucleosomes from this material with those of nucleosomes from transcriptionally inactive cells. Perhaps such a comparison might give some insight into the cause-effect relationship between structure and function during active gene transcription.

Acknowledgments

This work was supported in part by a grant from the National Research Council of Canada and in part by a grant from The President's Research Fund, University of British Columbia. I thank Dr. J. B. Gurdon for the generous gift of the O/+ nu animals.

REFERENCES

BIRNSTIEL, M., M. CHIPCHASE, and J. SPEIRS. 1971. The ribosomal RNA cistrons. *Prog. Nucleic Acid Res. Mol. Biol.* **11**: 351.

BROWN, D. D. 1967. The genes for ribosomal RNA and their transcription during amphibian development. *Curr. Top. Dev. Biol.* **2**: 47.

BROWN, D. D. and I. DAWID. 1968. Specific gene amplification in oocytes. *Science* **160**: 272.

BROWN, D. D. and J. B. GURDON. 1964. Absence of ribosomal RNA synthesis in the anucleolate mutant of *X. laevis. Proc. Natl. Acad. Sci.*, **51**: 139.

BROWN, D. D. and E. LITTNA. 1964a. RNA synthesis during the development of *Xenopus laevis,* the South African clawed toad. *J. Mol. Biol.* **8**: 669.

————. 1964b. Variations in the synthesis of stable RNA's during oogenesis and development of *Xenopus laevis. J. Mol. Biol.* **8**: 688.

BROWN, D. D. and R. STERN. 1974. Methods of gene isolation. *Annu. Rev. Biochem.* **43**: 667.

BROWN, D. D. and C. S. WEBER. 1968a. Gene linkage by RNA-DNA hybridization. I. Unique DNA sequences homologous to 4S RNA, 5S RNA and ribosomal RNA. *J. Mol. Biol.* **34**: 661.

————. 1968b. Gene linkage by RNA-DNA hybridization. II. Arrangement of the redundant gene sequences for 28S and 18S ribosomal RNA. *J. Mol. Biol.* **34**: 681.

DAVIDSON, E. H. 1976. *Gene activity in early development.* Academic Press, New York.

DAWID, I. B. 1965. Deoxyribonucleic acid in amphibian eggs. *J. Mol. Biol.* **12**: 581.

DAWID, I. B., D. D. BROWN, and R. H. REEDER. 1970. Composition and structure of chromosomal and amplified ribosomal DNAs of *Xenopus laevis. J. Mol. Biol.* **51**: 341.

DENIS, H. 1974. Nucleic acid synthesis during oogenesis and early embryonic development of the amphibians. In *MTP International review of science* (ed. J. Paul), vol. 9, p. 95. Butterworths, London.

DUMONT, J. N. 1972. Oogenesis in *Xenopus laevis* (Daudin). I. Stages of oocyte development in laboratory maintained animals. *J. Morphol.* **136**: 153.

ELGIN, S. and H. WEINTRAUB. 1975. Chromosomal proteins and chromatin structure. *Annu. Rev. Biochem.* **44**: 725.

ELSDALE, T., M. FISCHBERG, and S. SMITH. 1958. A mutation that reduces nucleolar number in *Xenopus laevis. Exp. Cell Res.* **14**: 642.

FORD, P. J. 1972. Ribonucleic acid synthesis during oogenesis in *Xenopus laevis.* In *Oogenesis* (ed. J. D. Biggers and A. W. Schuetz), p. 167. University Park Press, Baltimore, Maryland.

GALL, J. G. 1968. Differential synthesis of the genes for ribosomal RNA during amphibian oogenesis. *Proc. Natl. Acad. Sci.* **60**: 553.

————. 1969. The genes for ribosomal RNA during oogenesis. *Genetics* (Suppl.) **61**: 121.

GILLESPIE, D. 1968. The formation and detection of DNA-RNA hybrids. *Methods Enzymol.* **12B**: 641.

GOTTESFELD, J. M., G. BAGI, B. BERG, and J. BONNER. 1976. Sequence composition of the template-active fraction of rat liver chromatin. *Biochemistry* **15**: 2427.

GURDON, J. B. 1968. Nucleic acid synthesis in embryos and its bearing on cell differentiation. *Essays Biochem.* **4**: 25.

————. 1974. *The control of gene expression in animal development.* Oxford University Press, Oxford.

HIGASHINAKAGAWA, T., H. WAHN, and R. H. REEDER. 1977. Isolation of ribosomal gene chromatin. *Dev. Biol.* **55**: 375.

HILDER, V. A. and N. MACLEAN. 1974. Studies on the template activity of "isolated" *Xenopus* erythrocyte nuclei. I. The effects of ions. *J. Cell Sci.* **16**: 133.

HOURCADE, D., D. DRESSLER, and J. WOLFSON. 1974. The nucleolus and the rolling circle. *Cold Spring Harbor Symp. Quant. Biol.* **38**: 537.

HUANG, R. C. C. and J. BONNER. 1962. Histone, a suppressor of chromosomal RNA synthesis. *Proc. Natl. Acad. Sci.* **48**: 1216.

KNOWLAND, J. 1970. Polyacrylamide gel electrophoresis of nucleic acids synthesized during the early development of *Xenopus laevis* (Daudin). *Biochim. Biophys. Acta* **204**: 416.

KNOWLAND, J. and L. MILLER. 1970. Reduction of ribosomal

RNA synthesis and ribosomal RNA genes in a mutant of *Xenopus laevis* which organizes only a partial nucleolus. I. Ribosomal RNA synthesis in embryos of different nucleolar types. *J. Mol. Biol.* **53**: 321.

KORNBERG, R. D. 1974. Chromatin structure: A repeating unit of histones and DNA. *Science* **184**: 868.

LaMARCA, M. J., L. D. SMITH, and M. STROBEL. 1973. Quantitative and qualitative analysis of RNA synthesis in stage 6 and stage 4 oocytes of *Xenopus laevis*. *Dev. Biol.* **34**: 106.

MATHIS, D. J. and M. A. GOROVSKY. 1976. Subunit structure of rDNA-containing chromatin. *Biochemistry* **15**: 750.

MATSUI, S. and H. BUSCH. 1977. Isolation and characterization of rDNA-containing chromatin from nucleoli. *Exp. Cell Res.* **109**: 151.

MILLER, L. and J. B. GURDON. 1970. Mutations affecting the size of the nucleolus in *Xenopus laevis*. *Nature* **227**: 1108.

MILLER, L. and J. KNOWLAND. 1970. Reduction of ribosomal RNA synthesis and ribosomal RNA genes in a mutant of *Xenopus laevis* which organizes only a partial nucleolus. II. The number of ribosomal genes in animals of different nucleolar types. *J. Mol. Biol.* **53**: 329.

———. 1972. The number and activity of ribosomal RNA genes in *Xenopus laevis* embryos carrying partial deletions in both nucleolar organizers. *Biochem. Genet.* **6**: 65.

MITCHISON, J. M. 1971. *The biology of the cell cycle.* Cambridge University Press, Cambridge, England.

NIEUWKOOP, P. D. and J. FABER. 1956. *Normal table of Xenopus laevis (Daudin).* North-Holland, Amsterdam.

OUDET, P., M. GROSS-BELLARD, and P. CHAMBON. 1975. Electron microscopic and biochemical evidence that chromatin structure is a repeating unit. *Cell* **4**: 281.

PERKOWSKA, E., H. C. MACGREGOR, and M. L. BIRNSTIEL. 1968. Gene amplification in the oocyte nucleus of mutant and wild-type *Xenopus laevis*. *Nature* **217**: 649.

PIPER, P. W., J. CELIS, K. KALTOFT, J. C. LEER, O. F. NIELSEN, and O. WESTERGAARD. 1976. Tetrahymena ribosomal RNA gene chromatin is digested by micrococcal nuclease at sites which have the same regular spacing on the DNA as corresponding sites in the bulk nuclear chromatin. *Nucleic Acids Res.* **3**: 493.

REEDER, R. H. 1974. Ribosomes from eukaryotes: Genetics. In *Ribosomes* (ed. M. Nomura et al.), p. 489. Cold Spring Harbor Laboratory, Cold Spring Harbor, New York.

———. 1975. The structure of active ribosomal gene chromatin. *J. Cell Biol.* **62**: 357a.

REEDER, R. H., T. HIGASHINAKAGAWA, and O. MILLER, JR. 1976. The 5′ → 3′ polarity of the *Xenopus* ribosomal RNA precursor molecule. *Cell* **8**: 449.

REEVES, R. 1976. Ribosomal genes of *Xenopus laevis*: Evidence of nucleosomes in transcriptionally active chromatin. *Science* **194**: 529.

———. 1977. Analysis and reconstruction of *Xenopus* ribosomal chromatin nucleosomes. *Eur. J. Biochem.* **75**: 545.

REEVES, R. and A. JONES. 1976. Genomic transcriptional activity and the structure of chromatin. *Nature* **260**: 495.

SASVARI-SZEKELY, M., M. VITEZ, M. STAUB, and F. ANTONI. 1975. Determination of UTP and ATP pool sizes in human tonsillar lymphocytes by using *Escherichia coli* RNA polymerase. *Biochim. Biophys. Acta* **395**: 221.

SCHEER, U., M. F. TRENDELENBURG, and W. W. FRANKE. 1976a. Regulation of transcription of genes of ribosomal RNA during amphibian oogenesis. *J. Cell Biol.* **69**: 465.

———. 1976b. Regulation of transcription of ribosomal RNA-genes during amphibian oogenesis. In *Progress in differentiation research. Proceedings of the Second International Conference on Differentiation,* Copenhagen (ed. N. Muller-Berat et al.), p. 105. North-Holland, Amsterdam.

SEALE, R. L. 1976. Studies on the mode of segregation of histone nu bodies during replication in HeLa cells. *Cell* **9**: 423.

STEDMAN, E. and E. STEDMAN. 1957. The basic proteins of cell nuclei. *Philos. Trans. R. Soc. Lond. B* **235**: 565.

WALLACE, H. R. and M. L. BIRNSTIEL. 1966. Ribosomal cistrons and the nucleolus organizer. *Biochim. Biophys. Acta* **114**: 296.

WEINTRAUB, H. and M. GROUDINE. 1976. Chromosomal subunits in active genes have an altered conformation. *Science* **193**: 848.

WEBB, A. C. and L. D. SMITH. 1977. Accumulation of mitochondrial DNA during oogenesis in *Xenopus laevis*. *Dev. Biol.* **56**: 219.

WELLAUER, P., R. REEDER, D. CARROLL, D. BROWN, A. DEUTCH, T. HIGASHINAKAGAWA, and I. DAWID. 1974. Amplified ribosomal DNA from *Xenopus laevis* has heterogeneous spacer lengths. *Proc. Natl. Acad. Sci.* **71**: 2823.

WOODLAND, H. R. and J. B. GURDON. 1968. The relative rates of synthesis of DNA, sRNA and rRNA in the endodermal region and other parts of *Xenopus laevis* embryos. *J. Embryol. Exp. Morphol.* **19**: 363.

Modulation of Ribosomal RNA Synthesis in *Oncopeltus fasciatus:* An Electron Microscopic Study of the Relationship between Changes in Chromatin Structure and Transcriptional Activity

V. E. Foe*

Department of Zoology, University of Washington, Seattle, Washington 98195

I have used an electron microscopic approach to study transcriptional control of ribosomal RNA (rRNA) synthesis in the milkweed bug *Oncopeltus fasciatus* (Dallas). This insect was chosen for investigation because biochemical studies indicated that the rate of rRNA synthesis is extensively modulated during embryonic development. Harris and Forrest (1967) determined, for *O. fasciatus* embryos of various developmental stages, both the absolute amount of rRNA per embryo and the relative rates of incorporation of [³H]uridine into RNA molecules that sediment at the speed of rRNA. They detected no rRNA synthesis in embryos 20 hours old (syncytial blastoderm stage; Butt 1949). Incorporation of [³H]uridine into rRNA was found to occur at low levels in 44-hour-old embryos (mid-germ band stage) and then to increase dramatically, reaching a peak at 68 hours of development (during neurulation). Afterwards, rRNA synthesis declines and is barely detectable in embryos more than 116 hours old. The data of Harris and Forrest (1967) are replotted in Figure 1A, which shows the magnitude of these developmentally modulated changes in rRNA synthetic activity.

Subsequent biochemical studies have failed to resolve whether the modulation of rRNA synthesis in *O. fasciatus* results from changes in the number of genes in a transcriptionally competent state or from changes in the rate of transcription of a fixed number of genes. Harris and Forrest (1970) found that chromatin from 68-hour embryos (the developmental stage with the highest rate of rRNA synthesis) is about twice as effective as chromatin from 140-hour embryos as a template for total RNA synthesis using exogenous RNA polymerase from *Escherichia coli.* Although it was not demonstrated that the RNA synthesized was ribosomal, they concluded that production of rRNA precursor might be controlled via the transcriptional accessibility of ribosomal DNA (rDNA). The same workers (Harris and Forrest 1971) also determined that total endoge-

nous RNA polymerase activity is greater in nuclei from 68-hour embryos than in equivalent amounts of nuclei from younger or older developmental stages. From this result they concluded that rRNA synthesis in *O. fasciatus* embryos might also be regulated by changes in the amount of available RNA polymerase. Kamine and Forest (1975) extended this work and examined separately the activities of form-I (α-amanitin-resistant, nucleolar) and form-II (α-amanitin-sensitive, nuclear) RNA polymerases. They found that there is about three times more extractable form-I RNA polymerase activity per nucleus in 68-hour embryos than in 140-hour embryos. However, both sets of workers (Harris and Forrest 1971; Kamine and Forrest 1975) studied the level of RNA polymerase activity present in isolated nuclei rather than the level of total available RNA polymerase in whole embryos. Since unbound RNA polymerase molecules may leak out of nuclei, it is still uncertain how the rates of rRNA synthesis in *O. fasciatus* embryos correlate with the actual availability of RNA polymerase.

In an electron microscopic study of *O. fasciatus* embryos (Foe et al. 1976) three morphological criteria for identifying active ribosomal transcription units[1] were established: (1) *length* — the distance along the chromatin from the inferred transcriptional initiation site to the site at which the completed transcript is released measures 2.4 ± 0.4 μm; (2) *organization* — the active ribosomal transcription units usually occur as tandem repeats separated by lengths of nontranscribed chromatin; and (3) *chromatin morphology* — the chromatin underlying the arrays of nascent ribosomal transcripts is nonbeaded, whereas, in the same preparations, the remainder of the chromatin (including active nonribosomal transcription units) has a beaded morphology.

I report here on an electron microscopic study

---

* Present address: Department of Biochemistry and Biophysics, University of California, San Francisco, California 94143.

[1] The general term *transcription unit* is used here, in accordance with Foe et al. (1976) and Laird et al. (1976), to designate any segment of chromatin which is bounded by signals for transcriptional initiation and transcript release and which therefore would be traversed as a unit by RNA polymerase molecules engaged in synthesis of RNA.

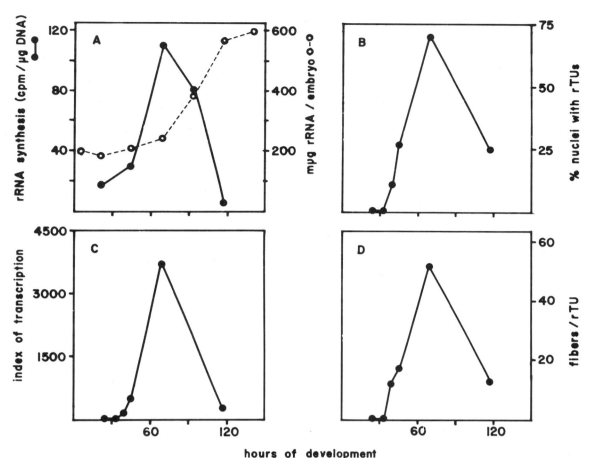

Figure 1. Assays of rRNA synthesis in *O. fasciatus* embryos. *(A)* The amounts of ribosomal RNA present in embryos (○) and the rate of incorporation of [³H]uridine into ribosomal RNA per unit of DNA (●). (Data from Harris and Forrest 1967.) *(B)* The percent of nuclei from each developmental stage in which at least one ribosomal transcription unit was observed. It is assumed (see Experimental Procedures and text for further discussion) that the probability of observing an active transcription unit is proportional to the number of transcription units active. Thus, this graph is interpreted as indicating stage-specific changes in the relative number of active ribosomal transcription units per embryo. *(C)* Indices of stage-specific transcriptional activity derived from electron microscopic data. The index of transcriptional activity was derived by multiplying the mean number of transcripts per ribosomal transcription unit (Fig. 1D) by the percent of nuclei in which at least one ribosomal transcription unit was observed (Fig. 1B; see text and notes to Table 1 for further discussion). *(D)* The mean number of fibers (nascent transcripts) per ribosomal transcription unit at each developmental stage. If transcriptional elongation proceeds at a constant rate in embryos of all developmental stages, then this graph depicts the stage-specific changes in mean rate of transcriptional initiation on active ribosomal genes.

of the changes that ribosomal transcription units undergo during embryogenesis in *O. fasciatus*. The data show that modulation of rRNA synthesis results from changes in the mean number of active ribosomal genes per nucleus as well as from changes in the level of transcriptional activity of each activated gene.

A striking observation made during this study was that only beaded chromatin is seen in 32-hour (late blastoderm stage) embryos, whereas 6 hours later (in early germ band stage, which is shortly before ribosomal RNA synthesis is first detected biochemically) nonbeaded stretches of ribosomal chromatin are seen, many of which completely lack transcripts. Thus, the data strongly suggest that transcriptional

activation of ribosomal genes in 32–38-hour *O. fasciatus* embryos is *preceded* by a major change in chromatin structure that encompasses the entire gene.

EXPERIMENTAL PROCEDURES

Rearing of milkweek bug embryos. *Oncopeltus fasciatus* (Dallas) were obtained from the stock colonies of H. S. Forrest at the University of Texas at Austin and maintained as described previously (Butt 1949). Eggs (which showed good developmental synchrony as assayed by time interval from oviposition to hatching) were collected at 2-hour intervals, incubated at 21°C until the appropriate developmental age was reached, and their nuclei prepared for electron microscopy (EM). Data described below as representing embryos N hours old are for embryos which range in age from $N-1$ to $N+1$ hours old; this distribution of stages is narrower than that used for the biochemical studies of Harris and Forrest (1967; $N \pm 5$ hr) with which the data in this paper are compared.

Preparation of chromatin for electron microscopy. Embryonic nuclei were prepared for observation as has been described previously (Foe et al. 1976), except that the chromatin dispersal was carried out at 27°C and the dispersal period was extended to 25 minutes. Also, to make chromatin morphology easier to see, some grids were rotary-shadowed with platinum. The concentration of tissue being prepared was adjusted for each embryonic stage so that the chromatin masses from individual nuclei did not overlap each other on the EM grid. The large size of the *O. fasciatus* genome ($N = 2.8 \times 10^{12}$ daltons; Lagowski et al. 1973) makes it technically difficult to observe the entire mass of chromatin in each nucleus (similar limitations would exist for most eukaryotic nuclei); consequently, the absolute number of transcription units per nucleus was not determined in this study. Under the conditions employed, there was a region at the periphery of each electron-dense nucleus where chromatin was adequately dispersed to allow observation of individual chromatin strands and detection of the ribonucleoprotein (RNP) fibers associated with some of these strands. However, despite constancy of preparative conditions, variation in chromatin dispersal was observed among grids. Therefore, grids were selected for quantitative analysis only if (1) there was a region, approximately 10 μm across, of well-dispersed chromatin around the nuclei and (2) the morphology of the dispersed chromatin strands was distinct (i.e., the chromatin was well stained and not aggregated). I estimate that the region of well-dispersed chromatin for nuclei thus selected corresponds to between 0.2% and 8% of the nuclear contents.

Chromatin analysis. On those grids selected for analysis according to the conditions described above, every nuclear mass that was completely visible (i.e., not partially obscured by a grid bar) was scored for the presence of ribosomal transcription units, whose distinctive morphological properties are as identified by Foe et al. (1976) and as described in the introduction to this paper. Positively scored regions were photographed, and the RNP fibers per ribosomal transcription unit were counted. For transcription units with >30 fibers, measurements were made of (1) the length of chromatin underlying the fiber arrays (measured from the first short fiber of each array to the last long one) and (2) the length of fiber-

free chromatin connecting adjacent arrays (measured from the last long fiber of one transcription unit to the first short fiber of the adjacent transcription unit). Alternatively, for transcription units with < 30 fibers, these same two parameters were measured as (1) the lengths of the segments of nonbeaded chromatin and (2) the lengths of beaded chromatin separating tandem regions of nonbeaded chromatin. Data are reported as mean values ± the standard deviation. The sample sizes (n) of transcription units, nuclei, or embryos analyzed are reported in the text or in Table 1.

RESULTS

Morphology of the Ribosomal Transcription Units as a Function of Transcriptional Activity

Chromatin during active rRNA synthesis (68-hr embryos). Tandemly repeated arrays of ribonucleoprotein (RNP) fibers were frequently seen associated with the chromatin of 68-hour embryos. Electron micrographs of such arrays are shown in Figure 2 A and B. Arrays are composed of 51 ± 22 fibers ($n = 84$) during this developmental stage, have a definite polarity with respect to RNP fiber length, and subtend chromatin lengths of 2.4 ± 0.4 μm ($n = 25$). The morphology of the chromatin underlying the fiber arrays generally cannot be seen clearly, being obscured by the abundance of RNP fibers. Previous analyses of *O. fasciatus* chromatin indicated that fiber arrays with the organization described above result from transcription of ribosomal DNA (Foe et al. 1976; Laird et al. 1976). The term *ribosomal transcription unit* (rTU) was therefore applied to the chromatin underlying such an array.

Tandem rTUs are connected by segments of fiber-free chromatin, hereinafter referred to as spacers. Figure 3 is a histogram displaying the lengths of those spacers analyzed. The most commonly observed spacer length is 0.6 ± 0.2 μm ($n = 45$). Spacers usually have a beaded morphology (e.g., see Fig. 2B), but a few resemble the nonbeaded chromatin of the active rTUs of 116-hour embryos, which are described below.

Chromatin that has undergone partial deactivation of rRNA synthesis (116-hr embryos). Chromatin preparations from 116-hour embryos revealed tandemly repeated fiber arrays which resemble in organization the arrays described above for 68-hour embryos and which therefore are also designated as ribosomal transcription units. However, there are fewer RNP fibers per array (13 ± 10; $n = 56$) for these older embryos (and, as discussed below, active rTUs occur less frequently). The lower fiber frequency makes it possible to see that the chromatin underlying these fiber arrays is different from the remainder

Table 1. Electron Microscopic Data on Levels of rDNA Transcription in *O. fasciatus* Embryos

Grid[a] no.	Age of embryos (hr)	Nuclei scored	Nuclei with active transcription units (rTUs)	rTUs per nucleus[b]	Percent[c] nuclei with rTUs	Fibers per rTU (mean value)	Index of transcriptional activity[d] = fibers per rTU × percentage of nuclei with rTUs
1	25	11					
2	25	6					
3	25	18	none found	none found	0	0	0
4	25	6					
5	25	3					
		total = 44					
6	32	6					
7	32	15	none found	none found	0	0	0
8	32	14					
9	32	6					
		total = 41					
10	38	8	1	2	13		
11	38	10	2	1, 5	20		
12	38	6	1	> 5	17	12 ± 11 (n = 20)	132
13	38	7	0	none found	0		
14	38	9	2	> 15, 8	22		
15	38	14	0	none found	0		
		total = 54	total = 6		mean = 11%		
16	44	18	5	5, 6, 3, > 15, 2	28		
17	44	11	4	> 15, 9, 9, 3	36		
18	44	16	6	> 15, > 5, 2, > 5, > 10	38	17 ± 12 (n = 18)	476
19	44	15	5	2, 3, 5, 2, 6	33		

Embryo[a]	Age (hr)	No. nuclei scored	No. nuclei with active rTUs	Min. no. active rTUs per nucleus[b]	% nuclei with active rTUs[c]	Mean no. active rTUs	Transcriptional index[d]
20	44	14	1	> 10	7		
total / mean		total = 74	total = 21		mean = 28%	51 ± 22 (n = 84)	3570
21	68	9	7	> 20, > 20, 6, 12, > 20, > 20, > 20, > 20	78		
22	68	6	6	> 10, > 10, > 30, > 20, > 20, > 15	100		
23	68	5	0	none found[e]	0		
24	68	6	4	> 15, > 20, > 30, > 15	67		
25	68	4	1	> 15	25		
26	68	8	7	> 20, > 5, 3, 8, 2, 10, > 20	88		
27	68	7	5	> 15, 4, > 20, > 20, 1	71		
28	68	15	7	> 20, > 20, > 15, > 10, > 10, > 4	47		
29	68	9	8	> 10, > 10, > 5, 3, > 30, 7, 7, 10, > 20, 6	89		
30	68	15	10	> 10, > 5, > 15, > 2, 5, > 20, > 40, > 15	67		
total / mean		total = 84	total = 55		mean = 70%		
31	116	7	2	4, 5	29		
32	116	47	14	4, 3, 7, 4, 6, > 15, 3, 5, 6, > 30, 1, 4, 4, 8	30		
33	116	8	1	> 10	13		
34	116	17	4	6, 9, 3, 2	24		
35	116	8	0	none found	0		
total / mean		total = 87	total = 21		mean = 24%	13 ± 10 (n = 56)	312

[a] Each grid contains nuclei from only one embryo.

[b] When the number of active rTUs observed was large or the rTUs were tangled, only their minimum number was recorded. The data clearly show that the number of active ribosomal genes detected per positively scored nucleus varies with development, and only in the 68-hr embryos does it frequently exceed 15. Note that the probability of observing a given number of active rTUs in one dispersed region of a nucleus is a function not only of the number of active rTUs present in the whole nucleus, but also of the degree to which active rTUs are linked. Consequently, the percentage of nuclei in which active rTUs were detected and not the number of rTUs observed per nucleus was used to estimate the relative number of active rTUs per embryo.

[c] As discussed in the text, the mean percentage of nuclei in which active rTUs were detected was taken as a relative measure of the number of active ribosomal genes per embryo. However, it should be noted that the frequency with which active rTUs are detected could vary among nuclei of the same embryo because of (1) cell-specific differences in developmental type that involve differences in level of rDNA transcription, (2) differences in ploidy or cell cycle phase of the nuclei, and (3) variations in the proportion of the nuclear chromatin that is sufficiently dispersed to permit detection of transcription units (see Experimental Procedures).

[d] Embryos with active rTUs detected in 100% of their nuclei shift the transcriptional index towards an underestimate of transcriptional activity.

[e] It was assumed that this embryo was nonviable, since neither ribosomal nor nonribosomal transcription units were observed. Data from this embryo are omitted from the mean.

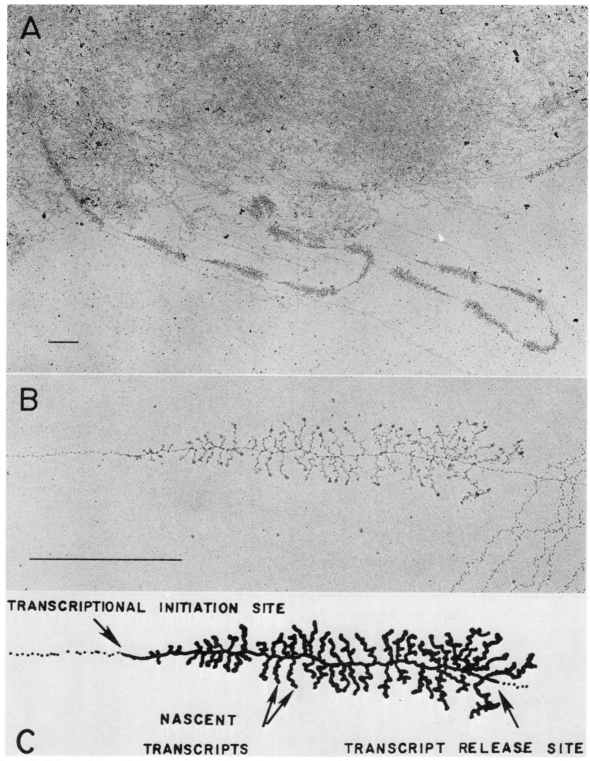

Figure 2. Interpretation of electron micrographs: ribosomal transcription units from 68-hr (neurula stage) *O. fasciatus* embryos. *(A)* Region of dispersed chromatin at a nuclear periphery. In this electron micrograph there are 16 or more ribosomal transcription units that are associated with fibers (and that therefore are interpreted as having been transcriptionally active at the time of preparation). The fiber-free spacers that flank ribosomal transcription units are heterogeneous in length. *(B)* Detail of a ribosomal transcription unit from a 68-hr embryo (not the same preparation as in *A*). The chromatin morphology of the transcription unit is not clearly visible because of the presence of fibers (nascent transcripts). The spacer chromatin is beaded. Preparations shown in *A* and *B* were stained with phosphotungstic acid (pH 2.5, unadjusted), but not platinum-shadowed. Scale bar indicates 1 μm. *(C)* Interpretative drawing of *B*.

728

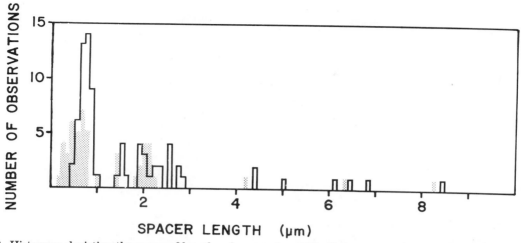

Figure 3. Histogram depicting the range of lengths of spacer chromatin that connect adjacent ribosomal transcription units in 68-hr embryos (open bars) and in 116-hr embryos (stippled bars). In the case of the 68-hr embryos, measurements were made from the last long fiber of one transcription unit to the first short fiber of the next one. For the 116-hr embryos, which have fewer fibers per transcription unit, the lengths of beaded chromatin that connect adjacent nonbeaded ribosomal transcription units were measured from one transition in chromatin morphology to the next. The distribution of spacer lengths is probably biased in favor of the shorter spacers, since the longer ones are more likely to be tangled and hence unmeasurable. In both 68- and 116-hr embryos, spacers longer than 10 μm were observed but not measured because of chromatin tangling.

of the dispersed chromatin in the same preparations. The distinctive appearance of the ribosomal chromatin is evident in Figure 4A, which is an electron micrograph of rTUs from a 116-hour embryo. Whereas the majority of the embryonic chromatin is clearly beaded, the rTUs have a morphology that I will refer to as nonbeaded.[2] The mean length of the segments of nonbeaded ribosomal chromatin which underlie the fiber arrays is 2.6 ± 0.5 μm ($n = 12$). As previously reported (Foe et al. 1976), the ribosomal chromatin stains with phosphotungstic acid (pH 2.5 unadjusted), which suggests that the rDNA is associated with basic proteins, and when thus stained has a width of 73 ± 17 Å (a width of 20 Å would be expected for naked DNA, which is not stained by phosphotungstic acid under my preparative conditions). These observations do not determine whether proteins are associated with the rDNA in vivo or become bound to it during the preparation for electron microscopy. The former conclusion is indicated by experiments in which exogenous T4 DNA was added to the embryonic lysates and coprepared with the embryonic chromatin. In prepa-

rations that were platinum-shadowed (to permit visualization of the free DNA), the DNA is conspicuously thinner than the ribosomal chromatin in the same preparation. The electron micrograph in Figure 4B shows both free DNA and ribosomal chromatin from one such preparation.

Tandem rTUs of 116-hour embryos are connected by intervals of fiber-free chromatin which are usually of beaded morphology; as illustrated in the histogram of Figure 3, these intervals of beaded chromatin show a range of lengths similar to that of the 68-hour ribosomal spacers.

Chromatin undergoing induction of rRNA synthesis (25-, 32-, 38-, and 44-hr embryos). In nuclear preparations from 44-hour embryos, tandem fiber arrays were observed which are analogous in length (2.5 ± 0.5 μm [$n = 14$]) to the rTUs seen in 68- and 116-hour embryos. These rTUs have 17 ± 12 ($n = 18$) fibers per array. The chromatin underlying these arrays has the same nonbeaded structure described above for ribosomal chromatin from 116-hour embryos. The nonbeaded morphology of rTUs distinguishes them from most ribosomal spacers and from *transcribed* and nontranscribed *nonribosomal* chromatin, all of which are beaded under my preparative conditions. In Figure 5, electron micrographs of chromatin from a 44-hour *O. fasciatus* embryo illustrate the different morphologies of ribosomal and nonribosomal transcription units. The beaded structure of the ribosomal spacers is also shown.

[2] In my preparations, ribosomal chromatin that has been stained with phosphotungstic acid generally looks smooth and nonbeaded (for other examples, see Figs. 5C and 8C). However, examination of shadowed ribosomal chromatin at higher magnification frequently reveals a particulate substructure (see, e.g., Fig. 8A,B). Detailed data on the morphology of these particles (which are smaller than nucleosomes) will be published elsewhere. Also, it should be noted that it is not uncommon to see an *occasional* bead of nucleosomal diameter (136 Å ± 26 Å; Laird et al. 1976) on the ribosomal transcription units.

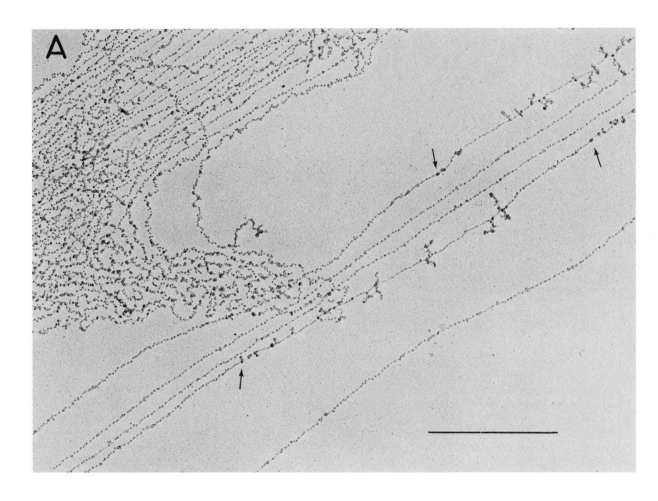

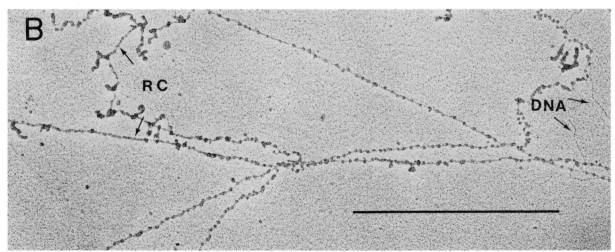

Figure 4. Interpretation of electron micrographs: ribosomal transcription units from 116-hr (organogenesis stage) embryos. *(A)* The arrows mark electron-dense particles thought to be RNA polymerase molecules near the transcriptional initiation sites of the ribosomal transcription units. The low number of nascent transcripts permits observation of the distinctive chromatin morphology of the ribosomal transcription units. In contrast, the spacers have a beaded morphology which is similar to that of nontranscribed chromatin. *(B)* The nonbeaded chromatin characteristic of active ribosomal genes (denoted RC) is much thicker than naked T4 DNA (denoted DNA) added to the same preparation. Preparations shown in both *A* and *B* were platinum-shadowed. Scale bars indicate 1 μm.

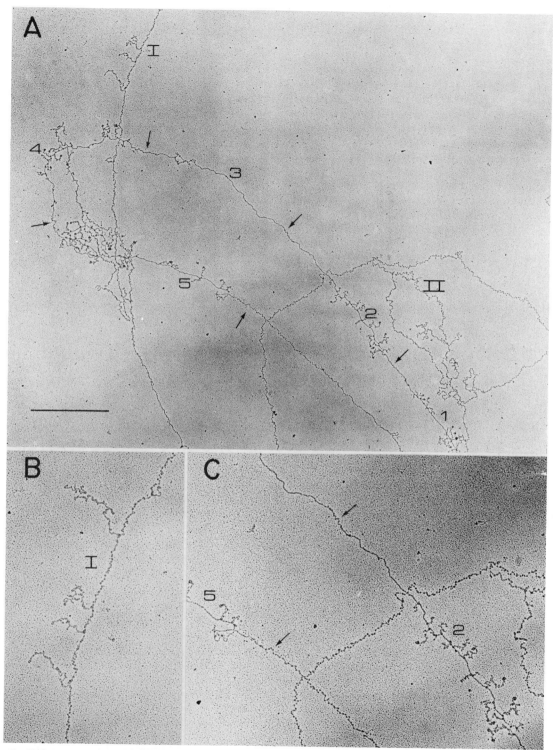

Figure 5. *(A)* Interpretation of electron micrographs: ribosomal and nonribosomal transcription units from a 44-hr (mid-germ band stage) embryo. The ribosomal transcription units (characterized by their tandem organization, the nonbeaded morphology of their chromatin, and their length [~2.5 μm]) are numbered 1–5. These ribosomal transcription units are associated with few nascent transcripts and are connected by short regions of beaded spacer chromatin (arrows). Ribosomal transcription unit 5 is flanked by a long segment of beaded fiber-free chromatin. Regions of chromatin that are interpreted as active nonribosomal transcription units are numbered I and II; the chromatin between the RNP fibers of these transcription units is beaded. Scale bar indicates 1 μm. *(B)* Enlargement of a region of *A* showing the beaded morphology of active nonribosomal chromatin. *(C)* Enlargement of a region of *A* showing the nonbeaded morphology of active ribosomal transcription units and the beaded morphology of the ribosomal spacers (arrows). This chromatin preparation was platinum-shadowed.

In preparations of nuclei from slightly younger (38-hr) embryos, rTUs were also observed; however, as discussed below, they occur less frequently than in the 44-hour embryos. At this developmental stage the mean number of RNP fibers per ribosomal transcript unit is low (12 ± 11; $n = 20$). As discussed below, rTUs totally lacking nascent transcripts are occasionally seen at all the developmental stages described above; however, fiber-free rTUs occur with greatest frequency in nuclei from 38- and 44-hour embryos.

In chromatin preparations from even younger (32-hr, cellular blastoderm stage and 25-hr, syncytial blastoderm stage) embryos, neither tandemly repeating fiber arrays nor segments of nonbeaded chromatin were observed. All dispersed chromatin from these developmental stages has a beaded morphology. All fiber arrays observed in these blastoderm-stage embryos were classifiable as nonribosomal (V. E. Foe et al., in prep.).

Distribution of RNP Fibers along the Ribosomal Transcription Units

The majority of the rTUs observed at all developmental stages have RNP fibers distributed fairly uniformly along the length of the transcription unit. In some nuclei, however, rTUs (identified by their length, chromatin morphology, and tandem organization) are observed completely free of associated RNP fibers, while others are associated with fibers that are nonuniformly distributed along the ribosomal chromatin. Ribosomal transcription units with uniform fiber arrays (diagramed in Fig. 6A), those with irregular distributions of fibers (the most frequently observed of these distributions are diagramed in Fig. 6B,C,D), and those totally lacking fibers (diagramed in Fig. 6E) sometimes occur together in the same nucleus. Indeed, I have observed in 38–55-hour embryos (when the de novo appearance of much of the nonbeaded ribosomal chromatin occurs) that in a small percentage of the nuclei (1–5% in the preliminary estimates) more than 50% of the rTUs have "abnormal" fiber distributions. In 68-hour embryos (the embryonic stage displaying peak levels of rRNA synthesis) no nuclei were observed in which so high a percentage of the rTUs were abnormal. However, in some nuclei from all developmental stages examined, including those from 68-hour embryos, occasional rTUs that were fiber-free or had abnormal fiber distributions were observed. Of the abnormal transcription units, those in the major class have no nascent transcripts, and the rest are distributed among the classes diagramed in Figure 6 B, C, and D. (A more detailed analysis of the frequency of occurence of these aberrant rTUs is in progress.)

Figure 7 A and B shows electron micrographs of

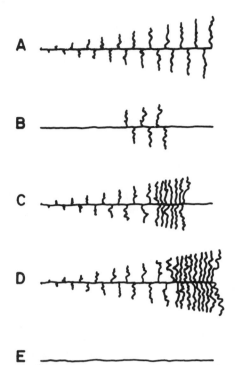

Figure 6. Diagramatic representations of the different organizations of fibers seen in ribosomal transcription units. *(A)* Fiber organization that would result from transcriptional activity if transcriptional initiation, transcript elongation, and transcript release were occurring simultaneously. *(B)* Fiber organization that could result if transcriptional initiation had been switched off, then on, then off again. *(C)* Fiber organization that could result if movement of the active RNA polymerase molecules along the chromatin were temporarily and locally impeded. *(D)* Fiber organization that could result if transcript release from the chromatin were temporarily blocked. *(E)* Fiber-free ribosomal chromatin that could result if transcriptional initiation were repressed. Although fiber distributions of the types diagramed in *B* and *E* could be attributed to removal of nascent transcripts during preparation, those diagramed in *C* and *D* (with greater than average fiber frequencies) cannot be thus explained. It is the observation of fiber distribution classes *C* and *D* in the same nuclei that contain classes *A, B,* and *E* (see, e.g., Figs. 7 and 8) that suggests that all these aberrant fiber distributions occur in vivo.

38- and 55-hour embryonic nuclei that contain clusters of rTUs with all the aberrations of fiber distribution diagramed in Figure 6. Electron micrographs of 55- and 68-hour rTUs with abnormal fiber distributions are displayed in Figure 8 A, B, and C at higher magnification, which permits a more detailed examination of their morphology. Interpretations, in terms of transcriptional activity, of the specific aberrant RNP fiber distributions diagramed in Figure 6 are presented in the legend to that figure. Possible causes for the existence of perturbations in rRNA synthetic activity, which could give rise to the abnormal RNP fiber distributions, are discussed below in relation to a discussion of the control of transcription of ribosomal DNA.

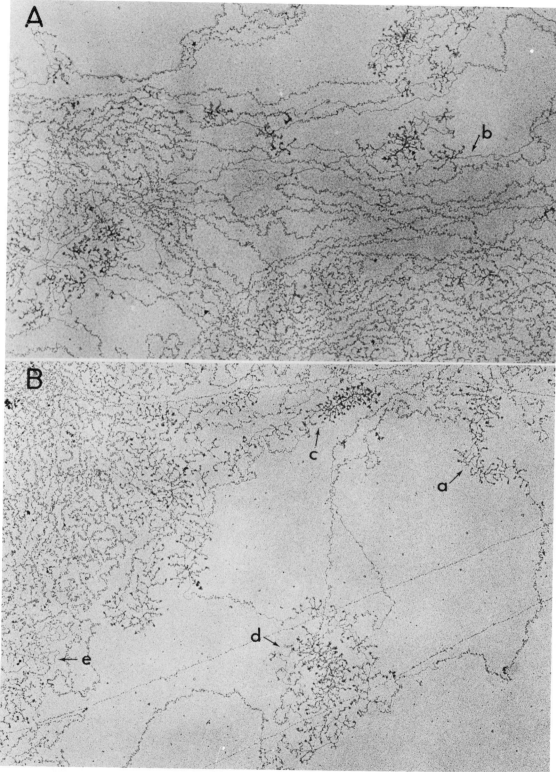

Figure 7. Interpretation of electron micrographs: ribosomal transcription units during transcriptional induction. The chromatin preparations from a 38-hr embryo *(A)* and a 55-hr embryo *(B)* show nonbeaded ribosomal chromatin with nonuniform distributions of the nascent transcripts. A letter (a, b, c, d, or e) indicates that the transcription unit thus marked has a fiber organization similar to the diagram designated by the same letter in Fig. 6 (only one transcription unit of each class has been marked, but many others are present). Interpretations in terms of transcriptional activity of the various fiber organizations are indicated in the legend to Fig. 6. The preparations in *A* and *B* were stained with phosphotungstic acid and were not shadowed with platinum. Magnification, 11,000X.

733

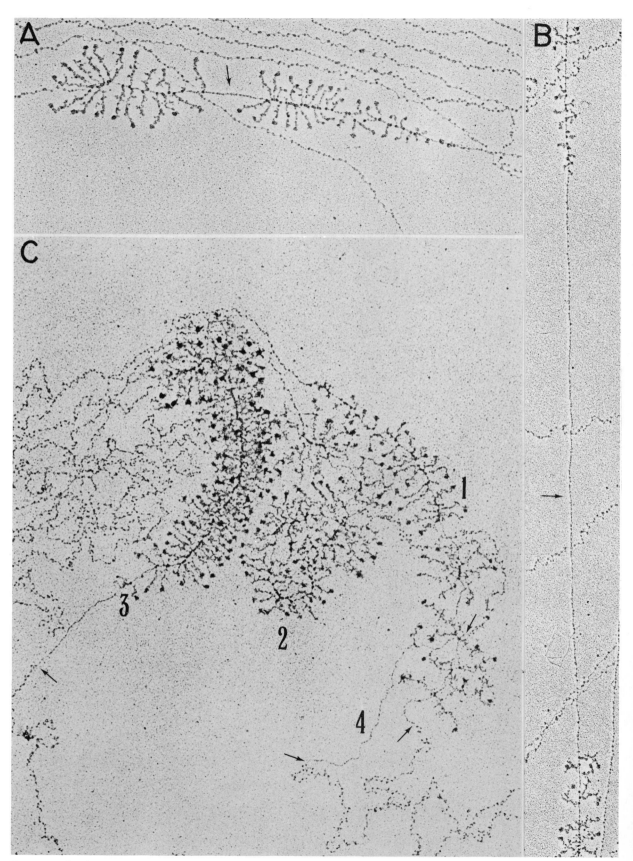

Figure 8. *(See facing page for legend.)*

Number of Active
Ribosomal Transcription Units
per Nucleus

In addition to studying the morphology of the rTUs of different developmental stages, I used electron microscopic data to estimate the relative levels of rRNA precursor synthesized at these stages. To determine relative transcriptional activity from electron microscopic data, I combined data on the relative number of active ribosomal genes and data on the relative level of activity on those genes for each developmental stage under study.

As noted in Experimental Procedures, the data presented in this paper were derived from examining the region of dispersed chromatin around each nucleus (a region almost certainly corresponding to less than 8% of the nucleus) and not from studies of the entire nuclear mass. If one assumes that the same specific region of the genome, *or* that similar amounts but of a randomly selected region of the genome, is dispersed for all embryonic stages, then the probability of observing at least one rTU per nucleus is a function of the number of rTUs present in a nucleus. As can be seen from the data recorded in Table 1 and summarized in Figure 1B, the frequency with which one (or more) active rTU was observed per nucleus varied with developmental stage. It is thus inferred that the number of active rTUs changes during development. This conclusion is also supported by the changes in the number of clustered rTUs observed in nuclei of different developmental stages. These data are also recorded in Table 1.

To quantify the transcriptional activity of those rTUs detected, I make the assumption that the number of RNP fibers on a transcription unit in an electron microscopic preparation is proportional to the rate of transcriptional initiation before isolation. Also, I assume that polymerization occurs at a constant rate in embryos of different developmental stages. Then it follows that (except for transcription units in which transcript release is limiting, which apparently is a small fraction of the rTUs in *O. fasciatus;* see preceding section) the number of RNP fibers per rTU is a measure of the number of rRNA precursor molecules generated per transcription unit per unit time. Data on the mean number of RNP fibers per transcription unit for the various developmental stages analyzed (see above) are summarized in Table 1 and shown graphically in Figure 1D.

To obtain a relative value for the total transcriptional activity of each embryonic stage, the mean number of fibers per rTU (Table 1 and Fig. 1D) was multiplied by the mean percentage of the nuclei in which active rTUs were observed (Table 1 and Fig. 1B). The quantitative estimates of relative stage-specific transcriptional activity thus derived from the electron microscopic data (Table 1 and Fig. 1C) are strikingly consistent with the biochemical data of Harris and Forrest (1967) on transcription rate, which were described earlier (see Fig. 1A).

DISCUSSION

Definitions

The majority of the dispersed chromatin in my preparations has a beaded morphology which is interpreted to result from the organization of the DNA into nucleosomes (Van Holde et al. 1974; Finch et al. 1975). In a previous study (Foe et al. 1976), a DNA packing ratio (defined as the length of B-structure DNA [3.4 Å/base pair] per unit length of chromatin) of 1.6 to 2.3 was calculated for the *O. fasciatus* chromatin of beaded morphology in electron microscopic preparations; the term *nu chromatin* was proposed to denote this chromatin. It was also determined that the nonbeaded ribosomal chromatin, denoted as *rho chromatin,* had a lower DNA packing ratio of 1 to 1.2. The terms nu chromatin and rho chromatin will be used to facilitate the discussion below.

Figure 8. Ribosomal transcription units whose transcript distributions indicate that transcriptional initiation and transcript release are independently controlled on individual genes. *(A)* A ribosomal transcription unit (from a 68-hr embryo) with an internal transcript-free region (arrow). The discontinuity in transcript distribution could have resulted from a transient repression of the transcriptional initiation signal. This preparation was platinum-shadowed. Magnification, 43,000X. *(B)* A segment of nonbeaded chromatin completely free of nascent transcripts (~2.4 μm long) in tandem with two ribosomal transcription units with transcripts (from a 68-hr embryo). The fiber-free ribosomal transcription unit (arrow) is connected to the adjacent transcription units by beaded spacer chromatin. A fiber-free ribosomal transcription unit (arrow) could result if transcriptional initiation on that transcription unit (and not on the adjacent two) were repressed. This preparation was platinum-shadowed. Magnification, 38,000X. *(C)* Four ribosomal transcription units from the same nucleus (from a 55-hr embryo) which show different transcript frequencies. Points of transition in chromatin morphology are marked by arrows. Two ribosomal transcription units (1 and 2; 2 is partially obscured by tangling) have uniform transcript distributions. One transcription unit (3) is transcript-free for the third of the transcription unit proximal to the initiation site and is densely packed with transcripts over its distal two-thirds; this transcript distribution could result if both the transcriptional initiation site and the transcript release site on this gene were blocked. Segment 4 of nonbeaded chromatin (partially obscured) is only about half of the length expected for a ribosomal transcription unit; this may represent a ribosomal transcription unit that is part in a beaded and part in a nonbeaded chromatin structure, perhaps because it was changing chromatin structure at the time of preparation. There is only one visible transcript on 4, indicating that transcriptional initiation on this interval of chromatin has been derepressed only transiently. This preparation was stained with phosphotungstic acid and was not platinum-shadowed.

Modulation of Transcriptional Activity during Embryogenesis

My observations concerning the pattern of rDNA transcription during *O. fasciatus* embryogenesis can be summarized as follows: (1) During blastoderm stage (25- and 32-hr embryos), there is no evidence of ribosomal transcription and no evidence of non-beaded rho chromatin. (2) Later, during germ band stage (38- and 44-hr embryos), segments of non-beaded chromatin, which are inferred to correspond to rTUs, appear. It is noteworthy that, during these early embryonic stages when they are first detected, some of the segments of rho chromatin are associated with no nascent transcripts. (3) As development proceeds, both the mean number of transcripts per rTU and the mean number of active rTUs per embryo increase, reaching a peak during neurulation (68 hr of development). (4) By organogenesis (116-hr embryos), the mean transcript frequency has declined, as has the mean number of active transcription units per nucleus. It thus appears that both the number of ribosomal genes that are active and the level at which an average active ribosomal gene is transcribed are modulated during development.

The Distinctive Structure of Active Ribosomal Chromatin; Electron Microscopic and Biochemical Studies

The present and previous electron microscopic studies have indicated that active ribosomal chromatin has a structure different from that observed for most chromatin. (For discussion of *O. fasciatus* ribosomal chromatin see Foe 1975 and Foe et al. 1976; for *Triturus alpestris, T. cristatus, Acetabalaria mediterranea,* and *A. cliftonii,* see Franke et al. 1976 and this volume; for *Notophthalmus viridescens,* see Woodcock et al. 1976). For organisms in which the abundance of nascent transcripts obscures the morphology of the underlying chromatin, the low DNA packing ratio of ribosomal chromatin in electron microscopic preparations (1 to 1.2) is used to infer that these active rTUs exist in an extended (nonnucleosomal) configuration.

The present study indicates that, although all dispersed chromatin from blastoderm stage *O. fasciatus* embryos is nu chromatin, 6 hours later some rTUs occur in the rho-chromatin configuration. During subsequent developmental stages, as rRNA synthesis increases, the absolute number of ribosomal genes that exist in this rho configuration increases, while inactive ribosomal genes presumably remain as nu chromatin. Indeed, it is likely (though unproven) that some of the long spacers of nu chromatin that are frequently observed connecting active ribosomal genes (see Fig. 3) contain inactive ribosomal genes. In summary, the electron microscopic data presented here strongly indicate that transcriptionally active and inactive ribosomal genes have different chromatin structures, with the number of genes in the active structure changing during development.

Recent biochemical investigations on ribosomal chromatin provide data whose interpretation is more ambiguous but which I believe to be consistent with the above conclusions. It was first demonstrated that at least some ribosomal chromatin is protected from staphylococcal nuclease digestion in a manner similar to that for bulk chromatin; rDNA is found in the monomeric chromatin subunits (which contain about 200 base pairs of DNA) generated by short-term staphlococcal nuclease digestions (for *Xenopus laevis,* see Reeder 1975; for *Tetrahymena pyriformis,* see Mathis and Gorovsky 1976). Subsequently it was shown that the amount of rDNA thus recoverable in the 200-base-pair subunits decreases as the rRNA synthetic activity of nuclei increases (Reeves and Jones 1976; Reeves, this volume). However, even in tissues where all the ribosomal genes are thought to be active, a significant percentage (40%) of the rDNA is recovered in the monomeric subunit (Reeves 1976). Finally, Mathis and Gorovsky (this volume) have demonstrated that both DNase I and prolonged staphylococcal nuclease digestions generate limit DNA fragment patterns that are similar, but not identical, for bulk chromatin and for ribosomal chromatin (that is assumed, though not proven, to be transcriptionally active). Taken together, these results suggest that (1) the DNA of active ribosomal genes is organized into nuclease-resistant subunits which contain about 200 base pairs of DNA, but that (2) the accessibility of ribosomal genes to nuclease increases with increasing transcriptional activity, and that (3) the internal structure of the 200-base-pair ribosomal chromatin subunit is in some way distinctive.

Results presented in this volume (Oudet et al.; Woodcock and Frado) indicate that, in vitro, DNA in association with the four core histones can be made to assume either the commonly observed beads-on-a-string morphology or an extended linear morphology (assayed electron microscopically). Chromatin which has been in either conformation can be reduced to subunits that contain 200 base pairs of DNA by brief staphylococcal nuclease digestion. The nonbeaded ribosomal chromatin in my preparations is not naked DNA but contains associated basic proteins (Foe et al. 1976). These proteins could confer a subunit structure (with respect to nuclease attack) on the ribosomal DNA. In fact, it is not unreasonable to anticipate that rho chromatin may contain an altered histone complement, which causes the chromatin to adopt an extended conformation under my conditions of sample preparation for electron microscopy.

Transcriptional Control of Individual Genes

In principle, the rate of transcriptional initiation on ribosomal genes could be controlled only by the concentration of available form-I RNA polymerase or similar soluble factors needed to initiate transcription. Alternatively, there could also be constraints on transcriptional initiation that reside in the structure of the ribosomal transcription units per se. If only the former type of factor is modulated, then all ribosomal genes in a given nucleus might be expected to be equally affected and therefore to sustain similar rates of transcriptional initiation. Only if transcriptional initiation signals on the rDNA templates are modulated (or different for different ribosomal genes) should different ribosomal genes in the same nucleus sustain transcriptional initiation at greatly different rates.

My observations reveal that in the majority of active rTUs the RNP fibers are distributed at random along the ribosomal chromatin. Moreover, the number of transcripts per transcription unit is fairly constant among active rTUs from nuclei of the same developmental stage. Yet analysis of chromatin from different developmental stages shows that the mean number of ribosomal transcripts per transcription unit changes significantly during embryogenesis. These data (see Table 1 and Fig. 1D) suggest that transcriptional initiation is developmentally modulated in some global way — perhaps by changes, during development, in the levels of form-I RNA polymerase (or of some other diffusible factor required for transcriptional initiation). Other interpretations, though less likely, are also possible.[3]

My data also reveal that the number of ribosomal genes that are active is developmentally modulated; this alone indicates that transcriptional initiation is also controlled at the level of the individual ribosomal gene. Moreover, in some nuclei I find clusters of rTUs which are conspicuous for having extreme heterogeneity both in the number of transcripts per gene and in the distribution of those transcripts along the gene. The particular observed fiber distribution patterns sketched in Figure 6 B and E can be interpreted as resulting from transcriptional initiation signals with a binary (on/off) mode of function, which had switched shortly before chromatin preparation. Other fiber distribution patterns (see Fig. 6C,D) indicate that the ability of RNA polymerase to traverse chromatin and the transcript release signals are also regulatable in a binary fashion. Since rTUs with different aberrations in fiber distribution

tend to occur together in the same nucleus (see, e.g., Figs. 7A,B and 8C), it is tempting to postulate that all of the aberrant fiber distributions have a common cause. I suggest below that fluctuations in the chromatin structure of each transcription unit could be responsible for these distribution patterns.

Chromatin Structure and Transcriptional Competence

My study has indicated that at the stage when rTUs first appear in rho configuration (at 38 hr of development) they are associated with few or no nascent transcripts (see Table 1). These data eliminate the possibility that it is the presence of numerous RNA polymerase molecules which changes the structure of ribosomal genes. But these data cannot exclude the possibility that the activity of a single RNA polymerase molecule is adequate to effect a stable change in ribosomal chromatin structure. My own conclusion from my observations is that the change in chromatin structure may well *precede* the initiation of transcription on ribosomal genes. This interpretation, which is detailed below, has the advantage of providing a common explanation for fiber-free ribosomal chromatin and for genes with other aberrations in fiber distribution.

The model diagrammed in Figure 9 represents my tentative interpretation of the observations described in this paper. DNA organized by histones into nu chromatin (which has a beaded morphology in my electron microscopic preparations) is transcriptionally inaccessible to form-I RNA polymerase. Ribosomal gene activation involves a change to an altered structure, rho chromatin (detected electron microscopically by a nonbeaded morphology),

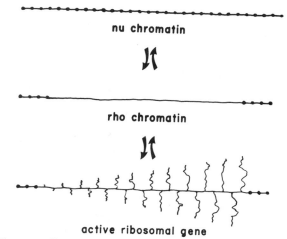

nu chromatin

rho chromatin

active ribosomal gene

Figure 9. Schematic representation of the proposed sequence of activation of ribosomal chromatin. Nu chromatin (beaded) is represented by beaded lines, and rho chromatin (nonbeaded) by solid lines. The conversion from nu to rho chromatin is postulated to precede transcriptional activation of the rDNA. In addition, the interconversion of nu and rho chromatin is postulated to be reversible.

---

[3] An alternative explanation of the stage-specific changes in transcript number is also tenable. RNA polymerase molecules might be present at constant concentration at all developmental stages, but the level of some factor needed for transcript elongation (e.g., ribonucleoside triphosphates) might change with development. If, during periods of elevated synthetic activity, the elongation factor became limiting, elongation might occur slowly, causing the transcription units to become densely packed with slow-moving RNA polymerase molecules. Likewise, when the elongation factor is not limiting, say during early development, elongation might be more rapid, causing genes to have fewer nascent transcripts. This explanation, though it cannot be eliminated by any current data, seems unlikely as it implies inefficient rRNA synthesis during periods of maximum demand.

along the entire length of a ribosomal transcription unit. The number of genes that are active is controlled by a reversible regulation of the number of genes in transcriptionally inert (nu) and transcriptionally competent (rho) chromatin configurations.

Ribosomal transcription units with nonuniform transcript distributions can be understood in terms of the model as resulting from instabilities generated during the conversions in chromatin structure. Perturbations in chromatin structure at the transcriptional initiation site might prevent initiation, which would be visualized as gaps in the arrays of nascent transcripts along the transcription unit if the perturbation were of short duration (Fig. 6B), or as fiber-free transcription units if the impediment to initiation were more long lasting (Fig. 6E). Perturbations in chromatin structure internal to the transcription unit or at the site of transcript release might hinder movement of the RNA polymerase molecules; transcription units that had been thus affected might display accumulations of nascent transcripts either along the gene (Fig. 6C) or against its distal end (Fig. 6D). The nuclei observed with a high percentage of aberrant ribosomal genes can be interpreted as undergoing a large-scale induction of rRNA synthesis. The fact that such nuclei occur in embryos that range in age from 38 to 55 hours can be explained by postulating that the exact time at which ribosomal genes are activated in different cells in a embryo differs, depending on the cell type (primary cell determination in *O. fasciatus* embryos apparently takes place at about 30 hr and before the induction of rRNA synthesis in any cells; V. E. Foe et al., in prep.). In addition, the fact that a few rTUs at all developmental stages show aberrant fiber distributions suggests that the number of ribosomal genes that are active may be under continuous modulation.

The sequence of events for *O. fasciatus* embryos described in this paper differs from those observed in *Drosophila melanogaster,* where the induction of rRNA synthesis has also been studied electron microscopically. McKnight and Miller (1976) observed that the ribosomal genes in *D. melanogaster* sustain an elevated level of transcriptional initiation from the moment that rRNA synthesis first begins. No change in chromatin structure was observed to precede that activation. The abundance of transcripts (> 130 fibers per rTU) obscures the morphology of the active genes. However, the fully active ribosomal genes in *D. melanogaster* have a low DNA packing ratio (1 to 1.2) like that of *O. fasciatus* rho chromatin. Thus it is unclear whether the pattern of transcriptional induction in *D. melanogaster* is substantially different or is merely a more accelerated version of what is reported above for *O. fasciatus.*

Many interesting questions are raised by the observations described above. Is the protein associated with rho chromatin unmodified histone, modified histone, or nonhistone protein? How are changes

from nu- to rho-chromatin structure propagated, and how are these changes restricted to rDNA sequences? Also, and most fundamentally, how are transcriptional activity and changes in chromatin structure functionally related?

Acknowledgments

This study began in the laboratory of C. D. Laird and was supported there by grant GM-19179 from the National Institutes of Health. The work was continued in the laboratory of B. M. Alberts where it was supported by grant GM-24020 and by an Institutional Training Grant (T32-CA-09270) from the National Institutes of Health to the Department of Biochemistry and Biophysics of the University of California, San Francisco.

I am very grateful to Bruce Alberts, Charles Laird, and Hugh Forrest for the extensive time, effort, and care they have given towards my development as a scientist. I also acknowledge the helpful suggestions of S. Tobin and S. L. McKnight.

REFERENCES

BUTT, F. H. 1949. *Embryology of the milkweed bug* Oncopeltus fasciatus *(Hemiptera).* Memoir 283. Cornell University, Agricultural Experimental Station, Geneva, New York.

FINCH, J. T., M. NOLL, and R. D. KORNBERG. 1975. Electron microscopy of defined lengths of chromatin. *Proc. Natl. Acad. Sci.* **72**: 3320.

FOE, V. E. 1975. "Activation of transcriptional units during the embryogenesis of *Oncopeltus fasciatus.*" Ph.D. thesis, University of Texas, Austin, Texas.

FOE, V. E., L. E. WILKINSON, and C. D. LAIRD. 1976. Comparative organization of active transcription units in *Oncopeltus fasciatus. Cell* **9**: 131.

FRANKE, W. W., U. SCHEER, M. F. TRENDELENBURG, H. SPRING, and H. ZENTGRAF. 1976. Absence of nucleosomes in transcriptionally active chromatin. *Cytobiologie* **13**: 401.

HARRIS, S. E. and H. S. FORREST. 1967. RNA and DNA synthesis in developing eggs of the milkweed bug, *Oncopeltus fasciatus* (Dallas). *Science* **156**: 1613.

―――. 1970. Template properties of embryo chromatin from eggs of the milkweed bug, *Oncopeltus fasciatus. Dev. Biol.* **23**: 324.

―――. 1971. RNA polymerase activity in isolated nuclei from eggs of *Oncopeltus fasciatus. J. Insect Physiol.* **17**: 303.

KAMINE, J. and H. S. FORREST. 1975. DNA-dependent RNA polymerase activities during embryogenesis in the bug, *Oncopeltus fasciatus. J. Insect Physiol.* **21**: 355.

LAGOWSKI, J. M., M.-Y. W. YU, H. S. FORREST, and C. D. LAIRD. 1973. Dispersity of repeat DNA sequences in *Oncopeltus fasciatus*, an organism with diffuse centromeres. *Chromosoma* **43**: 349.

LAIRD, C. D., L. E. WILKINSON, V. E. FOE, and W. Y. CHOOI. 1976. Analysis of chromatin-associated fiber arrays. *Chromosoma* **58**: 169.

MATHIS, D. J. and M. A. GOROVSKY. 1976. Subunit structure of rDNA-containing chromatin. *Biochemistry* **15**: 750.

McKNIGHT, S. L. and O. L. MILLER, JR. 1976. Ultrastructural patterns of RNA synthesis during early embryogenesis of *Drosophila melanogaster. Cell* **8**: 305.

REEDER, R. H. 1975. The structure of active ribosomal gene chromatin. *J. Cell Biol.* **67**: 357a.

REEVES, R. 1976. Ribosomal genes of *Xenopus laevis*:
Evidence of nucleosomes in transcriptionally active
chromatin. *Science* 194: 529.

REEVES, R. and A. JONES. 1976. Genomic transcriptional
activity and the structure of chromatin. *Nature* 260:
495.

VAN HOLDE, K. E., C. G. SAHASRABUDDHE, B. R. SHAW,

E. F. J. VAN BRUGGEN, and A. C. ARNBERG. 1974.
Electron microscopy of chromatin subunit particles.
Biochem. Biophys. Res. Commun. 60: 1365.

WOOKCOCK, C. L. F., L.-L. Y. FRADO, C. L. HATCH, and L.
RICCIARDIELLO. 1976. Fine structure of active ribosomal
genes. *Chromosoma* 58: 33.

QUESTIONS/COMMENTS

Question by:
E. N. MOUDRIANAKIS
Johns Hopkins University

First, I want to compliment you on the very elegant piece of work you have just presented. Second, I want you to know that we have reached similar conclusions while studying very different systems. We have recently published (Moudrianakis et al. 1977)* some of our results obtained several years ago from experiments designed to fractionate the chromatin fiber into "active" and "inactive" segments. The experiments were done with total chromatin from six different tissues of the chicken embryo and with calf thymus. Very similar results were obtained from all these systems. Through enzymatic, viscosimetric, and electron microscopic studies, we concluded that nontranscribable chromatin is in the form of a string of contiguous, ~100-Å diameter beads, whereas transcribable chromatin is in the form of relatively smooth fiber ca. 35 Å thick. Both types of fibers have the same histone complement. We proposed then that this differential condensation is etiologically underlying differential gene expression. This view appeared unacceptable four years ago because the ~100-Å bead was considered an omnipresent subunit of all chromatin fibers, but I sense that it may be in harmony with your views and those of P. Chambon as presented at this meeting.

Let me ask then: Can we now conclude that the nucleosome, operationally defined as that repeat length of chromatin that is protected from the action of micrococcal nuclease, *need not* always assume the morphology of the V-body, i.e., that of an ~100-Å bead? This, of course, will mean that the DNA compaction within these nonbeaded chromatin segments will be something less than six- to sevenfold, and we estimate it to be something like twofold.

Response by:
V. E. FOE
University of California, San Francisco

I completely agree with your early conclusion that the structure of chromatin, at least that of ribosomal chromatin is dynamic. Also, it seems likely, although formally unproven by the papers thus far presented, that this dynamism resides in the "nucleosome." Although not prepared to offer an alternate definition, I hesitate just a little at your definition of "nucleosome," as it is not yet clear that the same digestion patterns by micrococcal nuclease are characteristic of all the native chromatin states. For example, the data presented by Reeves (this volume) appear to indicate a differential susceptibility of active ribosomal chromatin to micrococcal nuclease.

Concerning chromatin compaction, let me say the following: In my electron microscopic preparations, for which chromatin is spread under conditions that remove H1 and where the majority of the chromatin shows the extended beads-on-a-string morphology with clearly visible linker DNA between beads, we calculate an average DNA/chromatin packing ratio of 2.3 for strands of nontranscribed chromatin (with a beaded morphol-

* Moudrianakis, E. N., P. L. Anderson, T. H. Eickbush, D. E. Longfellow, P. Pantazis, and R. L. Rubin. 1977. In *The molecular biology of the mammalian genetic apparatus* (ed. P. Ts'O), p. 301. Elsevier/North-Holland, Amsterdam.

ogy), a ratio of 1.6–1.9 for the transcriptionally active nonribosomal chromatin (which in the electron microscopic preparations also has a beaded nucleosomal morphology) and a ratio of 1.0–1.2 for the the transcribed nonbeaded ribosomal chromatin (Foe et al. 1976; Laird et al. 1976).† We have no data on the packing ratio of native chromatin, but clearly a closer association of nucleosomes (perhaps produced in native chromatin by the presence of H1) will result in a higher packing ratio.

---

† See Foe, this volume, for complete references.

Electron Microscopic Analysis of Chromosome Metabolism in the *Drosophila melanogaster* Embryo

S. L. McKnight,*† M. Bustin,‡ and O. L. Miller, Jr.*

*Department of Biology, Gilmer Hall, University of Virginia, Charlottesville, Virginia 22901; ‡Developmental Biochemistry Section, National Institute of Arthritis, Metabolism, and Digestive Diseases, National Institutes of Health, Bethesda, Maryland 20014

Information regarding the patterns of chromatin transcription and replication is important in understanding the mechanisms regulating genome metabolism. Electron microscopic procedures provide a useful tool for the study of chromosome function. We have chosen to apply the technique of chromatin spreading to *Drosophila melanogaster* as it is a well-studied biological system that is amenable to cytogenetic manipulation.

Here we present a quantitative description of the patterns of genome transcription at two stages of *D. melanogaster* embryogenesis and a morphological description of chromatin in transcribed and replicated segments of the *Drosophila* chromosomes. Also, chromatin has been probed by immunoelectron microscopy using anti-histone immunoglobulins. It is shown that histone proteins H2B and H3 remain associated with metabolically active chromatin.

EXPERIMENTAL PROCEDURES

Collection and staging of embryos. *Drosophila melanogaster* (Ore-R) embryos were collected on agar spoons coated with honey-yeast and incubated for up to 2.0 hours in moist chambers at 24°C. Embryos were placed on Scotch tape and dechorionated with #5 jewelers' forceps. Individual embryos were placed in a well slide containing an insect ringer solution and staged by phase-contrast microscopy. Figure 1 shows embryos photographed at the syncytial blastoderm stage (Fig. 1a) and at progressive intervals of the first cell cycle of the cellular blastoderm stage (Fig. 1b–d).

Specimen preparation and electron microscopy. Single, prestaged embryos were disrupted with #5 jewelers' forceps in 50 μl of 5 μM sodium borate buffer (pH 8.5). After 5 minutes at room temperature, the homogenate was transferred to a lucite microcentrifuge chamber containing a 0.1 M sucrose–10% formalin cushion above a carbon-coated electron microscope grid made hydrophilic by rinsing for 1–2 minutes in 95% ethanol. Each

preparation was centrifuged at 2500g for 5 minutes in a Sorvall GLC-1 tabletop centrifuge. After centrifugation, the grids were removed from the lucite chamber, rinsed in glass-distilled H_2O, dipped in a surface-tension reducing agent (0.5% Kodak Photoflo), and air-dried. Grids were routinely stained for 30 seconds with 1% phosphotungstic acid in 70% ethanol. In some cases, preparations were also rotary-shadowed with platinum metal in a Kinney vacuum evaporator at 0.1 μm Hg.

Original micrographs were taken at 60 kV using a JEM-100C transmission electron microscope. Contour-length measurements were obtained from photographic negatives projected with either a Durst enlarger or a Nikon profile projector, using a Numonics graphic calculator.

Immunological procedures. To obtain antihistone immunoglobulins, purified histone fractions obtained from calf thymus were complexed with RNA and injected into rabbits (Stollar and Ward 1970; Bustin 1973). The specificity of the antisera tested by the microcomplement-fixation technique has been described previously (Goldblatt and Bustin 1975). The antisera reacted strongly with histone fractions obtained from *Drosophila melanogaster* embryos (Bustin et al. 1977). The IgG fraction from anti-histone sera was obtained by chromatography on DEAE-cellulose eluted with a linear gradient from 0.015 to 0.3 M sodium phosphate buffer (pH 6.8). Antibodies were purified by affinity chromatography on Sepharose to which histones were covalently attached (Bustin and Kupfer 1976; Simpson and Bustin 1976).

To enhance detection of the anti-histone probe, ferritin-coupled anti-rabbit IgG was used (Bustin et al. 1976). Monomerically conjugated ferritin–anti-rabbit IgG was purified by molecular-weight sieving using Bio-Gel A-5m (Olsen and Prockop 1974). Conjugates were eluted with 0.1 M Tris-HCl buffer (pH 7.5) and then dialyzed against 0.05 M NaCl. Fractions were assayed by electron microscopy to locate monomeric conjugates and then tested for anti-rabbit specificity by Ouchterlony immunodiffusion.

Immunoelectron microscopic procedures. In each experiment, a single, pre-staged cellular blastoderm

---

† Present address: Department of Embryology, Carnegie Institution of Washington, 115 West University Parkway, Baltimore, Maryland 21210.

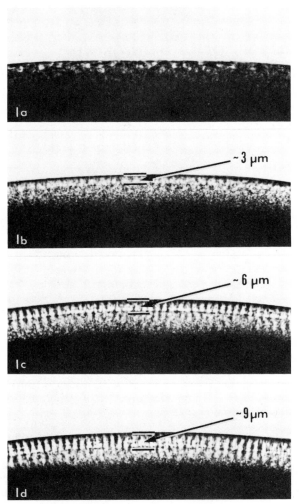

Figure 1. Phase-contrast micrographs of dechorionated *D. melanogaster* embryos. *(a)* An embryo photographed at the syncytial blastoderm stage approximately 15 min prior to the onset of cellular blastulation. *(b–d)* Embryos photographed at progressive intervals of a single cell cycle at the cellular blastoderm stage. Distances between lines (arrows) depict the amount of inward cell-membrane growth.

embryo was disrupted in 40 μl of glass-distilled H₂O and then reacted with 10 μl of either affinity-purified antihistone antibodies or nonimmune IgG. The final antibody concentration ranged between 0.01 and 0.1 mg/ml in 0.025 M sodium phosphate buffer (pH 6.8). After incubation for 10–30 minutes at room temperature, the antibody-reacted chromatin solution was mixed with 10–25 μl of 0.02–0.4 mg/ml (O.D. 410 nm) ferritin-conjugated goat anti-rabbit antibody, yielding a final salt concentration between 0.03 and 0.04 M. After an additional 5 minutes at room temperature, the solution was placed in a lucite microcentrifuge chamber above a 0.25 M sucrose–10% formalin cushion and centrifuged for 15 minutes at 5000g in a Sorvall RC2B refrigerated centrifuge. After centrifugation, the supernatant

was removed from the lucite chamber by means of a braking pipet and replaced with glass-distilled H₂O. This procedure was repeated four times to remove the nonreacted, ferritin-coupled antibody, which binds strongly to carbon support film. Grids were then removed, rinsed in glass-distilled H₂O, dipped in 0.5% Kodak Photo-flo, and air-dried. Samples were either positively stained for 15 seconds with 1% phosphotungstic acid in 70% ethanol or negatively stained with 1% uranyl acetate in 70% ethanol.

RESULTS AND DISCUSSION

Analysis of Transcriptional Configurations

Using electron microscopic procedures, we have recently initiated analytical studies of nuclear transcription in *Drosophila melanogaster* embryos. Since preliminary details of these investigations have been communicated elsewhere (McKnight and Miller 1976), our discussion here will be limited to salient points.

In preblastoderm embryos, a very high percentage of the *Drosophila* genome is transcriptionally inert. However, a distinct population of nonribosomal transcriptional gradients is detectable (see Fig. 2a–f). Two features characterize these gradients: (1) they are relatively short, ranging in contour length from 0.5 to 3.0 μm; and (2) they are tightly packed with RNA polymerase molecules. Figure 3a shows the contour-length and polymerase-frequency distributions of 52 precellular blastoderm transcription units. Active loci exhibiting only one to a very few attached ribonucleoprotein (RNP) fibers are also present at this stage, but active ribosomal genes are not observed.

Significant metabolic changes occur as the embryo enters the cellular blastoderm stage. The most marked is a large increase in the duration of the cell cycle (from ~10 min in preblastoderm embryos to well over 1 hr at the cellular blastoderm stage). The onset of this developmental period is evidenced by the formation of cellular membrane around each peripheral nucleus. Membrane synthesis progresses inward for approximately 45 minutes, at which time lateral synthesis effects the compartmentalization of each peripheral nucleus. Inward membrane synthesis can be viewed by phase-contrast microscopy (see Fig. 1b–d), and the extent of the process can be measured by means of an ocular micrometer. Such measurements are used to obtain embryos at staged intervals within the first true cell cycle. This staging procedure, when used in conjunction with standard electron microscopic procedures, allows the analysis of transcriptional patterns throughout a single cell cycle.

In thin-sectioned electron microscopic preparations, prominent nucleoli are first detected at the cellular blastoderm stage (Mahowald 1963). As was suggested by that observation, ribosomal RNA

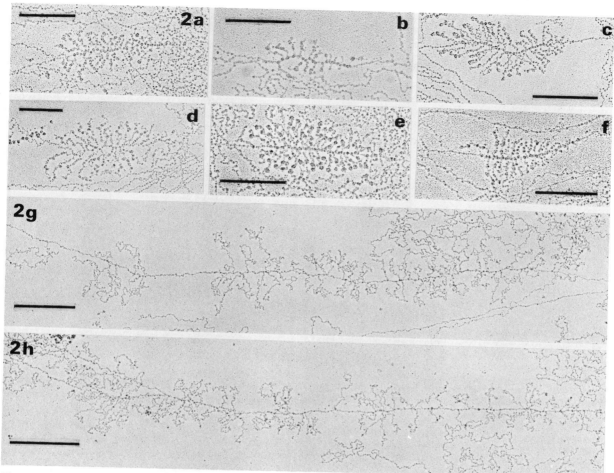

Figure 2. Electron micrographs of nonribosomal transcription units. *(a–f)* Ribonucleoprotein fiber gradients from precellular blastoderm embryos. Note the high density of nascent transcripts. Scale bars represent 0.5 μm. *(g,h)* Transcription units obtained from cellular blastoderm embryos. Note the intermediate frequency of nascent RNP fibers. Scale bars represent 1.0 μm. (Parts *a, c, e,* and *f* are reprinted, with permission, from McKnight and Miller 1976.)

(rRNA) is first synthesized at blastulation (McKnight and Miller 1976; Zalokar 1976). When nuclei of cellular blastoderm embryos are dispersed for ultrastructural analysis, tandemly linked rRNA transcription units are detected. Active rRNA genes prepared from embryos staged early in the cell cycle exhibit a characteristically dense population of nascent ribosomal ribonucleoprotein (rRNP) chains, but they are of highly variable contour length (see Fig. 4a–c). The simplest explanation for the variable lengths is that the shorter matrices are transcriptionally immature. That some genes have not reached a steady-state level, where the number of RNA polymerase initiation events equals the number of termination events, is supported by two observations. First, the short rRNP gradients are of variable length up to a 2.6 μm plateau, which is the mature gene length; and second, when the genome is inspected at later stages of the same cell cycle, the shorter rRNP matrices are rarely found. It should also be noted that early during this cell cycle

tandemly linked rRNP matrices may be of variable contour length (see Fig. 4c).

If one assumes that the rate of rRNA chain elongation is constant on adjacent ribosomal genes, these data imply that the activation of each ribosomal cistron is regulated independently. Also, the high frequency of nascent chains typical of immature rRNP gradients suggests that the rate-limiting step of rRNA production is not governed by the RNA polymerase-I pool size. The nature of rRNA gene activation in the *Drosophila* embryo may be somewhat more dynamic than the induction of rRNA synthesis in the amphibian oocyte (Scheer et al. 1976) and in the embryonic milkweed bug (Foe, this volume), where newly activated rRNA genes show lower frequencies of nascent rRNP fibers.

Extensive regions of RNP-free chromatin often bracket active rRNA genes during the cellular blastoderm stage (see Fig. 4b). Many of these segments probably harbor dormant ribosomal cistrons, as they are considerably larger than the predominant

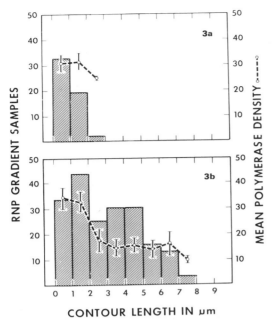

Figure 3. Contour-length and polymerase-density distributions of nonnucleolar transcription units. *(a)* Length and polymerase-density (frequency) distributions of 52 nonribosomal transcription units observed on the chromatin of precellular blastoderm embryos. *(b)* Distributions of 194 nonribosomal transcription units derived from cellular blastoderm embryos. RNA polymerase density values were estimated by measuring the frequency of nascent RNP fibers per unit length and assuming that a single RNA polymerase molecule is located at the intercept between each nascent transcript and the chromatin template. Individual transcriptional complexes are grouped on the abscissa in 1.0 μm intervals. Perpendicular bars on mean RNA polymerase density circles represent standard deviation measurements. (Data from McKnight and Miller 1976.)

4.5-kb nontranscribed "spacer" of *D. melanogaster* rDNA (Wellauer and Dawid 1977). Thus, although rRNA production is initiated very early in the cellular blastoderm stage, the rate of production appears to be regulated by the number of rRNA genes activated.

In addition to the induction of rRNA synthesis, a substantial increase in the appearance of newly synthesized poly(A)+ RNA is observed at cellular blastoderm formation in *Drosophila* (Lamb and Laird 1976). This increase presumably can be accounted for by the increased frequency at which nonnucleolar transcription units are observed at cellularization (McKnight and Miller 1976). A high proportion of nuclear transcription detected at cellular blastulation occurs on RNP fibril gradients which are longer than those observed at earlier stages, and which generally exhibit an intermediate frequency of nascent transcripts. Figure 2g,h shows examples of large nonribosomal transcription units.

Thus, as viewed ultrastructurally, nonribosomal transcriptional configurations of the cellular blastoderm genome can be arbitrarily subdivided into three classes: (1) relatively short (0.5–3.0 μm), polymerase-dense loci indistinguishable from, and

perhaps the same as, those present in earlier developmental stages; (2) longer gradients (3.0–8.0 μm) which exhibit intermediate polymerase frequencies; and (3) loci showing very low polymerase frequencies and no definitive RNP fiber gradients (see Fig. 6d; also see Laird and Chooi 1976). Since it is difficult to identify accurately the sites for RNA polymerase initiation and termination on the latter class of active loci, we have limited contour-length analyses of nonribosomal configurations to the first two categories of transcription units. Figure 3b tabulates the length and polymerase-frequency distributions of 194 transcription units visualized on genomes of cellular blastoderm embryos.

It is important that electron microscopic observations be related to biochemical information regarding genome organization. An initial step in this direction is the estimation of nonribosomal transcription-unit size in base pairs (bp) or kilobases (kb), rather than micrometers (μm). It has been established in a number of systems that the chromatin of transcription units saturated with RNA polymerases is extended in contour length to a degree approximately equivalent to B-form DNA (Miller and Beatty 1969; Laird et al. 1976; McKnight and Miller 1976; McKnight et al. 1976; Trendelenburg et al. 1976). We therefore estimate that the amount of DNA contained in polymerase-dense, nonribosomal transcription units of the *D. melanogaster* embryo is slightly more than 3.0 kb/μm of chromatin template.[1] Active rRNA genes and nonribosomal loci densely packed with RNA polymerases exhibit few, if any, nucleosomelike particles (Franke et al. 1976; Laird et al. 1976; Woodcock et al. 1976). On the other hand, the morphology of chromatin segments of nonribosomal loci exhibiting either intermediate or sparse RNA polymerase frequencies is distinctly beaded (McKnight and Miller 1975; Foe et al. 1976; Laird et al. 1976). If the beadlike structures characteristic of these chromatin segments are true nucleosomes, then the chromatin conformation cannot be extended to a degree equivalent to B-form DNA when analyzed by electron microscopy. However, if it can be established that the particulate

---

[1] The complete coding sequence of the *D. melanogaster* rRNA gene measures ~8.6 kb as extrapolated from the size of the rRNA precursor (Perry et al. 1970). Since the *D. melanogaster* rRNP matrix of an active rRNA gene measures ~2.6 μm (McKnight and Miller 1976), a micrometer of active rDNA chromatin contains ~3.3 kb/μm. Similar estimates of the rDNA packaging ratio in active rRNA chromatin have been obtained for other ribosomal gene types (Miller and Beatty 1969; Foe et al. 1976; Laird et al. 1976; Trendelenburg et al. 1976). It is possible that there are inherent differences in the structures of rDNA and nonribosomal chromatin (Foe et al. 1976; Foe, this volume). However, the chromatin of the silk fibroin gene is also extended in contour length during its active transcription (McKnight et al. 1976). The polymerase-saturated fibroin transcription unit measures ~5.4 μm in contour length. As the coding region of the fibroin gene is estimated to contain ~1.8 × 10⁴ bp (Lizardi 1976), the template chromatin of that gene also contains ~3.3 kb/μm during its active transcription. By definition, B-form DNA contains 2.9 kb/μm (Watson and Crick 1954). Since the chromatin of highly active transcription units contains ~3.3 kb/μm, it appears that the DNA of such chromatin segments remains slightly contracted (~1.1-fold relative to B-form DNA).

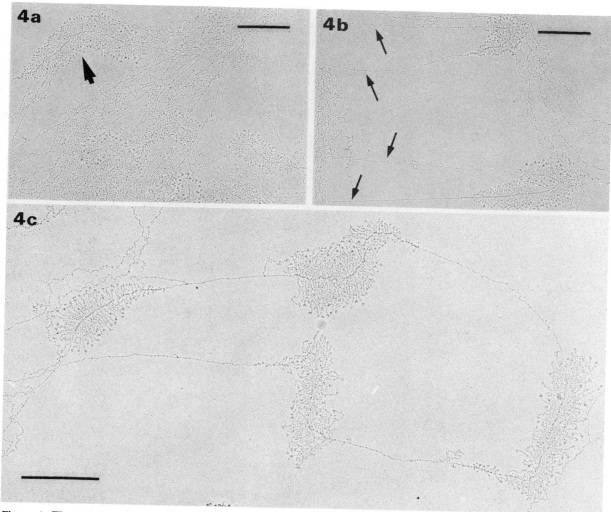

Figure 4. Electron micrographs of newly activated nucleolar genes. *(a)* A full-length rRNP matrix (large arrow) clustered with transcriptionally immature genes. *(b)* Extended regions of transcriptionally dormant chromatin (small arrows) adjacent to both a mature rRNP matrix and an immature matrix. *(c)* Direct tandem linkage of nucleolar matrices of varying contour length. All samples were prepared from cellular blastoderm embryos showing less than 4 μm of inward cell-membrane growth. Scale bars represent 1.0 μm. (Reprinted, with permission, from McKnight and Miller 1976.)

elements of nonribosomal chromatin are nucleosomes, it should be possible to estimate the DNA content of such regions (Laird et al. 1976). To approach the question of whether or not histones are present on active genetic loci, we have adapted the immunocytochemical procedures used to visualize the organization of histones in inactive chromatin (Bustin et al. 1976) to study the protein composition of transcribed genomic sequences.

When dispersed, inactive *Drosophila* chromatin is reacted with 0.1 mg/ml affinity-purified anti-H3 or anti-H2b immunoglobulins, then treated with ferritin-conjugated anti-rabbit IgG. The resulting electron microscopic preparations show extensively aggregated chromatin decorated with a high density of ferritin molecules (see Fig. 5b). Minimal aggregation occurs when samples are treated with 0.2 mg/ml anti-human IgG prior to anti-rabbit–ferritin incubation, indicating that the aggregation results from

specific cross-linking of histones by antihistone immunoglobulins. As Figure 5a shows, very few ferritin molecules are attached to chromatin incubated for 30 minutes with 0.2 mg/ml ferritin-coupled anti-rabbit IgG without prior addition of antihistone immunoglobulins.

The extensive aggregation which occurs at high antibody concentrations prohibits ultrastructural analysis. Therefore, a lower antibody concentration, which allows partial chromatin dispersion, was used. Figure 5c,d shows chromatin samples which were first incubated with 0.01 mg/ml anti-H2b and then treated with 0.05 mg/ml ferritin-conjugated anti-rabbit IgG. The antibody "tagging level," extrapolated from the frequency of ferritin-binding events, is highly variable even within the same electron microscope grid. At the lower reaction concentration, the "tagging level" usually ranges between two and six ferritin molecules/μm of chromatin contour

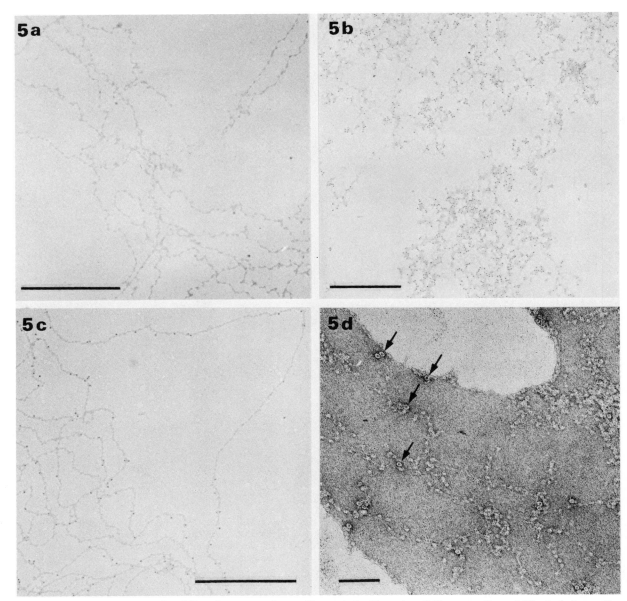

Figure 5. Electron micrographs of *D. melanogaster* chromatin reacted with various IgG preparations. *(a)* Control chromatin reacted for 30 min with 0.4 mg/ml ferritin-coupled anti-rabbit IgG. Note absence of chromatin aggregation and beadlike morphology of chromatin fibers. *(b)* Chromatin reacted for 30 min with 0.1 mg/ml anti-H2b and then treated for an additional 30 min with 0.2 mg/ml ferritin-coupled anti-rabbit IgG. *(c,d)* Chromatin reacted for 10 min with 0.01 mg/ml affinity-purified anti-H2b and for an additional 10 min with 0.02 mg/ml anti-rabbit–ferritin IgG. *(c)* Positive staining with phosphotungstic acid. Note that the nucleosome morphology typical of chromatin is altered. *(d)* Negative staining with 1% uranyl acetate in 70% ethanol. Arrows depict horse-spleen ferritin molecules. All samples were prepared from cellular blastoderm embryos. Scale bars represent 0.5 μm in *a–c* and 0.1 μm in *d*.

length. Since a micrometer of dispersed *Drosophila* chromatin consists of ~31 nucleosomes,[2] the anti-

---

[2] Chromatin prepared for electron microscopy by hypotonic shock is conformationally extended from its native form (Griffith 1975). When prepared by the methods reported here, inactive chromatin contains ~31 nucleosomes/μm. A single *D. melanogaster* nucleosome, including the nuclease-sensitive internucleosome fiber, is complexed with 185 bp (G. Wiesehahn and J. Hearst, pers. comm.). Therefore, we estimate that a micrometer of dispersed, transcriptionally inactive chromatin contains ~5.7 kb of DNA and is nearly twofold contracted from the 2.9-kb/μm conformation of B-form DNA (Watson and Crick 1954). For a more comprehensive analysis of this interpretation, see Laird et al. (1976).

histone "tagging level" ranges from 6% to 20% on a per nucleosome basis.

The incubation of unfixed transcriptional complexes with antibodies in solution poses a dilemma. While antibody-binding efficiency is optimal at physiological salt concentrations, chromatin spreading is best achieved in a minimum of sodium ions. The salt concentrations employed (0.03–0.04 M) result in a satisfactory antibody fixation level while still allowing reasonable chromatin spreading. However, some degree of precipitation of nascent RNP fibrils is unavoidable at salt concentrations above 0.01 M, a level which markedly affects the visualization of ribosomal transcription units and densely populated nonnucleolar genes. Thus, our immunological analysis of transcription has so far been limited to configurations exhibiting an intermediate or sparse population of nascent ribonucleoprotein fibrils.

Figure 6b–d shows antihistone-treated transcriptional units with ferritin molecules bound to the chromatin template. At the concentration of antibody used, not all active nonribosomal loci are tagged with ferritin. However, the "tagging level" of transcribed nonribosomal chromatin, when probed with either anti-H2b or anti-H3 immunoglobulins, is similar to that of inactive nucleosomal chromatin (see Fig. 6). From these observations, we conclude that histones H2b and H3 do remain associated with transcriptionally active, nonnucleolar

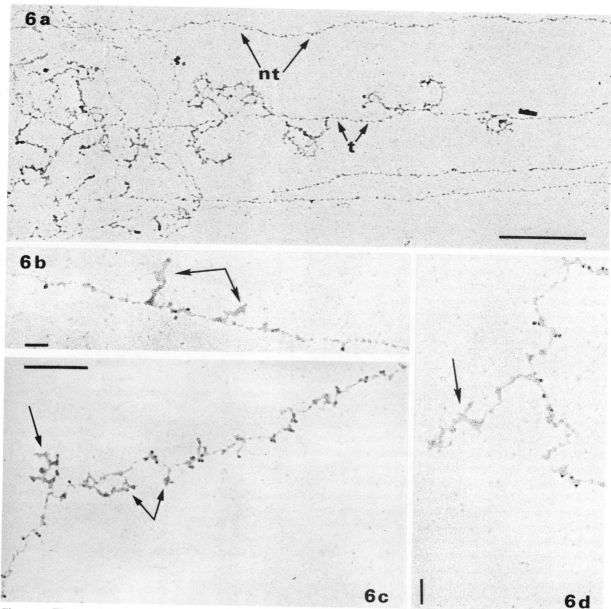

Figure 6. Electron micrographs of nonribosomal transcription units. (a) A nonribosomal transcription unit stained with phosphotungstic acid. Note that the chromatin morphology is similarly beaded in transcribed (t) and nontranscribed (nt) regions. (b–d) Nonribosomal transcription units reacted with anti-H3 (b and d) and anti-H2b (c) immunoglobulins and ferritin-coupled anti-rabbit IgG. Arrows depict nascent RNP fibers. The anti-histone "tagging level" observed on transcription units reacted with anti-H3 was 2.8 ± 1.42 ferritin molecules/μm ($N = 12$), whereas the level observed to react with transcription units incubated with anti-H2b immunoglobulins was 3.48 ± 1.72 ferritin molecules/μm ($N = 10$). Analysis of ferritin binding to inactive chromatin observed on the same preparations was 3.78 and 4.07 ferritin molecules/μm for anti-H3 and anti-H2b, respectively. All samples were prepared from cellular blastoderm *D. melanogaster* embryos. Scale bars represent 0.5 μm in *a* and *c*, and 0.1 μm in *b* and *d*.

genes. By means of a variety of biophysical approaches, D'Anna and Isenberg (1974) have documented that strong interactions occur between histones H3 and H4 and between histones H2b and H2a. On this ground, we propose that all four histone constituents of the nucleosome core are present in the spherical particles found on transcribed nonribosomal chromatin.

Three recent reports have shown that chromatin of transcriptionally active genome sequences is particularly sensitive to digestion by pancreatic DNase I (Weintraub and Groudine 1976; Garel and Axel 1976; Garel and Axel, this volume). These observations have led to the conclusion that transcribed chromatin segments exist in a structurally altered conformation relative to metabolically inert chromatin. Since we have not detected alterations in the histone composition of transcribed chromatin regions relative to inactive regions, we suggest that the preferential susceptibility of transcribed chromatin to pancreatic nuclease digestion is not due to an absence of any of the four histones of the nucleosome core. However, we have not yet determined whether histone H1 is associated with transcribed chromatin, nor are we able to determine whether any of the "core" histones of such regions are chemically modified (e.g., by phosphorylation, acetylation, or methylation).

How does an RNA polymerase molecule transcribe nucleosomal chromatin? We envision at least three possible mechanisms: (1) the nucleosome directly associated with an RNA polymerase may locally unfold without physical displacement of the histone proteins (i.e., a variation of a model proposed by Weintraub et al. 1976); (2) histones may be transiently displaced at the site of RNA polymerase function; or (3) the basic nucleosome structure may remain intact during the RNA polymerization process.

Either of the first two mechanisms would likely be accompanied by an extension of chromatin beyond the normally condensed conformation. Since the chromatin template of highly active transcription units of known size and function is extended to near-B-form, we favor one of the two former hypotheses. Reeder et al. (this volume) have documented in isolated nucleoli of the amphibian *Xenopus laevis* the presence of proteins which exhibit the electrophoretic mobilities of histones. Isolated nucleoli contain a pure population of rRNA genes which are transcriptionally active (Higashinakagawa et al. 1977). Also, a normal molar ratio (1:1) of histone to DNA has been found in the Mg++-soluble fraction of chromatin digested with DNase II, which is enriched for transcriptionally active DNA sequences (Gottesfeld and Bonner 1977). These observations provide additional evidence that histones remain associated with DNA sequences during transcription. Regardless of whether one assumes either localized nucleosome unwinding or very transient histone displacement during transcription, it is apparent that a dynamic equilibrium exists between the unwinding (or displacement) and re-formation events as nucleosomes are found closely bracketing each RNA polymerase molecule (see Fig. 6a).

If, as our observations suggest, transcribed chromatin segments remain in a nucleosome conformation and each RNA polymerase molecule effects a localized extension of chromatin, we should be able to calculate in base pairs the amount of DNA contained in nonribosomal transcription units. Laird et al. (1976) measured the bead frequency on chromatin fibers of nonribosomal transcription units of *Drosophila melanogaster* and *Oncopeltus fasciatus* and, by assuming the beaded element to represent nucleosomes, estimated such regions to contain ~4.5 kb/μm. Our observations are consistent with the data from which the 4.5 kb/μm value was obtained, and we fully agree with the procedures used to arrive at that estimate of DNA content. We extend the estimation procedures advanced by Laird and his colleagues by proposing that a linear continuum exists between the maximally extended conformation of chromatin saturated with RNA polymerases (3.3 kb/μm) and the condensed form of minimally transcribed nucleosomal chromatin (4.8 kb/μm).[3] The pregnant element of this proposal is that the degree of DNA condensation in a given transcription unit is assumed to be inversely proportional to the number of RNA polymerases associated with the chromatin template.

For example, take a chromatin fiber 1.0 μm in contour length visualized in the electron microscope. If very few RNA polymerases are associated with the segment (less than 2/μm), it should be assembled into 26 nucleosomes, and is therefore estimated to contain 4.8 kb. A 1.0-μm chromatin segment complexed with 15 RNA polymerases/μm, with nucleosomes present on the chromatin fiber between nascent transcripts, is estimated to contain 4.1 kb of DNA. A highly active nonribosomal transcription unit associated with 30 RNA polymerases/μm should be extended to near-B-form and is thus estimated to contain only 3.3 kb of DNA if measured at 1.0 μm. It is important to point out that this means of analysis has been devised to allow size estimation of transcription units in kilobases of DNA and is not meant to reflect the conformational dimensions of active chromatin in vivo.

Using this approach, we can estimate in kilobases

---

[3] Chromatin segments associated with nascent RNP fibers often appear "stretched" and exhibit a reduced nucleosome periodicity when compared to inactive chromatin segments (Laird et al. 1976). In our preparations, the mean nucleosome frequency observed on minimally transcribed chromatin fibers is 26 nucleosomes/μm. Therefore, such chromatin segments are estimated to contain a maximum of 4.8 kb/μm (26 nucleosomes/μm × 185 bp/nucleosome = 4.8 kb/μm). However, in some cases, transcribed chromatin shows a normal nucleosome frequency (~31/μm). Under the latter conditions, the maximum DNA/μm value should be elevated to 5.7 kb/μm in order to estimate the amount of DNA per micrometer of chromatin.

the size range of transcription units which exhibit variable RNA polymerase frequencies (see Figs. 2 and 3). The shortest nonribosomal transcription units, when complexed with a high frequency of nascent transcripts, are probably extended to near-B-form. Since transcription units of this type have been found as small as 0.5 μm, the DNA template segments of these shortest active loci are estimated to contain only 1.65 kb. The largest transcription units found in the *D. melanogaster* cellular blastoderm embryo ($\sim$8.0 μm), which are associated with as few as five RNA polymerases/μm and show nucleosomes on their chromatin templates, are estimated to contain more than 40 kb of DNA. The mean size of nonribosomal transcription units of the cellular blastoderm *D. melanogaster* embryo is $\sim$2.9 μm (McKnight and Miller 1976), and the mean polymerase frequency value for that transcription-unit population equals $\sim$20/μm. Using the analytical procedures described here, we estimate that, in the *Drosophila* embryo, an average nonribosomal transcription unit consists of approximately 13 kb of DNA. Although we submit that this is a legitimate procedure for estimating the DNA content of transcription units, substantiation of the accuracy of this method of analysis must await the identification of specific genes of known DNA content which show intermediate RNA polymerase frequencies.

Analysis of Chromosome Replication

The genome of *D. melanogaster* embryos is replicated by the simultaneous activation of highly multiple replication origins, from which DNA is synthesized bidirectionally (Blumenthal et al. 1974; Kriegstein and Hogness 1974). In electron microscopic preparations of purified DNA, replicative molecules were identified by the presence of "eye-form" structures (replicons) within which (1) single-stranded regions occur adjacent to most replication forks; (2) the contour length of each daughter chromatid is equivalent; and (3) daughter chromatids show similar denaturation patterns (Kriegstein and Hogness 1974).

Using the chromatin spreading technique, we have analyzed chromosome replication at the cellular blastoderm stage of *D. melanogaster* embryogenesis (McKnight and Miller 1977). S-phase genomes can be obtained reproducibly by selecting cellular blastoderm embryos exhibiting between 2 μm and 8 μm of inward membrane synthesis (see Fig. 1b–d). Replicative chromatin fibers are identified by the presence of replicon configurations (Fig. 7). Each replicon is delimited by two bifurcations along the chromatin fiber. Such structures are similar to the "eye-forms" reported by Kriegstein and Hogness (1974), although chromosomal proteins are retained in our procedures.

As shown in Figure 8a,b, the chromatin morphology observed on both daughter chromatid arms of replicon structures is distinctly beaded. Particulate structures also occur on chromatin regions directly in front of each replication fork site. Measurements of the bead diameter and periodicity on chromatin within replicons yield values equivalent to the same quantitative parameters of inactive nucleosomal chromatin (McKnight and Miller 1977).

To test whether these beadlike elements contain histones, we have utilized the immunological procedures described earlier in this report. Immunoelectron microscopic analysis of replicating chromatin is somewhat more straightforward than analysis of transcriptional complexes. Precipitation from salt exposure is not significant because nascent RNP fibers are not usually associated with replicon structures. Also, the spatial proximity of intrareplicon daughter chromatids allows antibody-mediated cross-linking, which can be used in addition to the ferritin probe as an assay for antibody binding.

Figure 8c–e shows chromatin replicons that had been reacted with antihistone immunoglobulins. Multiple intrareplicon cross-links by divalent antihistone IgG molecules reveal the presence of histone proteins H2b and H3 on both daughter chromatids. When S-phase chromatin is prepared in the absence of antihistone immunoglobulins, or in the presence of nonimmune IgG, cross-linking within replicons occurs only rarely. Also, as Figure 8e shows, following the double antibody reaction, ferritin molecules are found in regions directly adjacent to replication fork sites. These data show that at least two histone proteins of the nucleosome core are associated with newly synthesized DNA.

Two features of replicative chromatin are examined in this report: (1) the ultrastructural morphology and (2) the immunological reactivity to two antihistone immunoglobulin fractions. With respect to both of these properties, replicating chromatin is indistinguishable from inactive nucleosomal chromatin. We conclude, therefore, that, in the *D. melanogaster* embryo, histones need not be removed from chromatin segments prior to DNA replication and that, upon replication, DNA of both daughter chromatids is rapidly reassociated with histones, yielding a nucleosome configuration. The first conclusion is consistent with biochemical analyses of replicating chromatin (Weintraub 1976; Seale 1976). However, nuclease digestion studies have yielded conflicting results regarding the reconstitution of nucleosomes following DNA replication. On the one hand, it is reported that during chromosome replication in HeLa cells, newly replicated regions are preferentially susceptible to nuclease degradation (Seale 1975, 1976), whereas newly replicated chromatin of cultured erythroblasts is indistinguishable from inactive chromatin in resistance to nuclease digestion (Weintraub 1974, 1976). The inconsistency of the latter observations suggests the possibility that chromatin maturation may not occur at the same rate in different cell types.

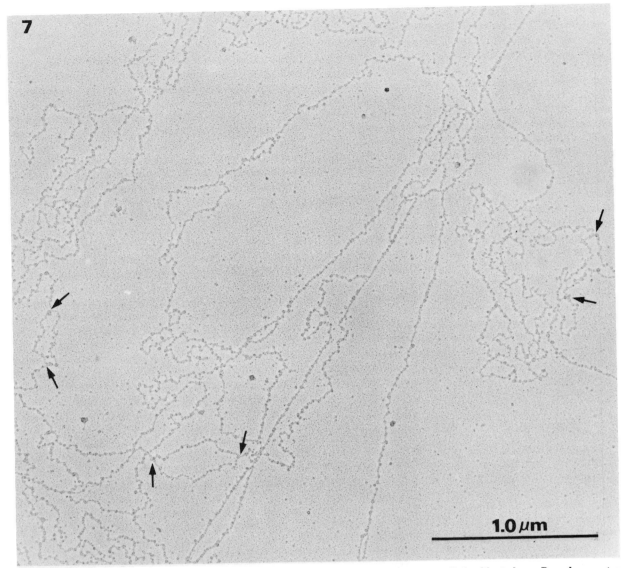

Figure 7. Electron micrograph of replicating chromatin. Sample was obtained from a cellular blastoderm *D. melanogaster* embryo which exhibited ~4 µm of inward membrane synthesis. Arrows depict six replication fork sites.

Interaction between Replication and Transcription

As we have shown in the two previous sections of this report, the *Drosophila* genome is functional in both replicative and transcriptional processes during cellular blastoderm formation. Electron microscopy permits analysis of temporal and structural interactions between these aspects of chromosome metabolism.

Figure 9a–c shows that transcription can be activated rapidly following replication of nonnucleolar chromatin segments. More than 50 configurations showing transcriptional gradients on newly replicated chromatin have been observed, and all have exhibited ribonucleoprotein fibrils on *both* daughter chromatids. These results indicate that daughter chromatids usually behave similarly with respect

to transcriptional activity (see Fig. 9b). The rapidity of transcriptional activation following replication raises the possibility that an "active" chromatin conformation can be inherited directly (Tsanev and Sendov 1971); however, we are unable to establish whether the chromatin underlying homologous nonnucleolar transcription units seen on daughter chromatids is "active" prior to replication.

It has been suggested to us (D. Hogness, pers. comm.) that the rapid activation of transcription on both newly replicated daughter chromatids might produce gene dosage effects. Informational loci proximal to replication origins could, for a certain amount of time, produce up to twice the normal gene product (RNA) relative to loci located more distal to replication origin sites. Such an effect would require that the doubling of a particular sequence

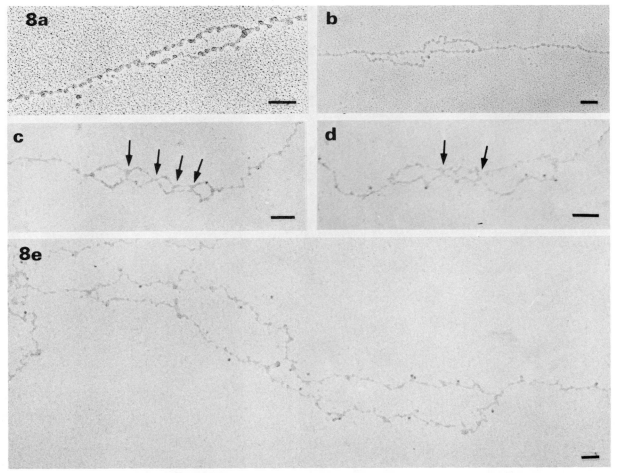

Figure 8. Electron micrographs of replicating chromatin. *(a,b)* Replicon configurations rotary-shadowed with platinum. Note the particulate morphology of chromatin regions adjacent to replication forks; *(c)* reacted with affinity-purified anti-H2b; *(d,e)* reacted with anti-H3 antibodies. Arrows point to intrareplicon cross-linking of daughter chromatids. All samples were prepared from cellular blastoderm embryos exhibiting less than 8 μm of inward cell-membrane growth. Scale bars represent 0.1 μm. (Parts *a* and *b* are reprinted, with permission, from McKnight and Miller 1977.)

not be counteracted by a reduction in the frequency of RNA polymerase initiation events on the newly replicated genes. If the pattern of genome replication occurs in an ordered manner which is constant from one embryo to the next, such a mechanism might play a transient, yet important, role in determination events which occur at cellular blastulation in *Drosophila* (see Capdevila and Garcia-Bellido 1974).

It is apparent that after replication RNA synthesis can be initiated rapidly on nonnucleolar genes. On the other hand, what occurs when a replication fork approaches an actively transcribed genomic sequence? The most suitable locus to investigate this question is the nucleolus organizer, since we can identify with relative confidence RNA polymerase initiation and termination sites on rRNA genes. Fourteen replicative configurations have been visualized adjacent to active nucleolar genes (see Fig. 10a–d). Of the samples in which both replication

forks are identifiable, all replicons are interpreted to have been activated in nontranscribed "spacers" (see Fig. 10a,b). At present, we do not have sufficient data to determine whether a specific site for DNA replication occurs within the "spacer" regions.

Following origin activation in the "spacer," what occurs when the two forks reach the initiation and termination sequences of the adjacent transcribed rDNA? We have visualized four configurations which suggest that when forks proceed in the same direction as RNA polymerization, they can replicate the transcribed portion of an active ribosomal gene (see Fig. 10b,c). When this interaction occurs, either the replicase complex physically dislodges RNA polymerases and nascent rRNP chains or passively follows the last RNA polymerase molecule through the locus. Six configurations have been visualized in which a replication fork has reached the termination region of rRNA synthesis in a convergent direction (see Fig. 10d). Contour-length analyses of these

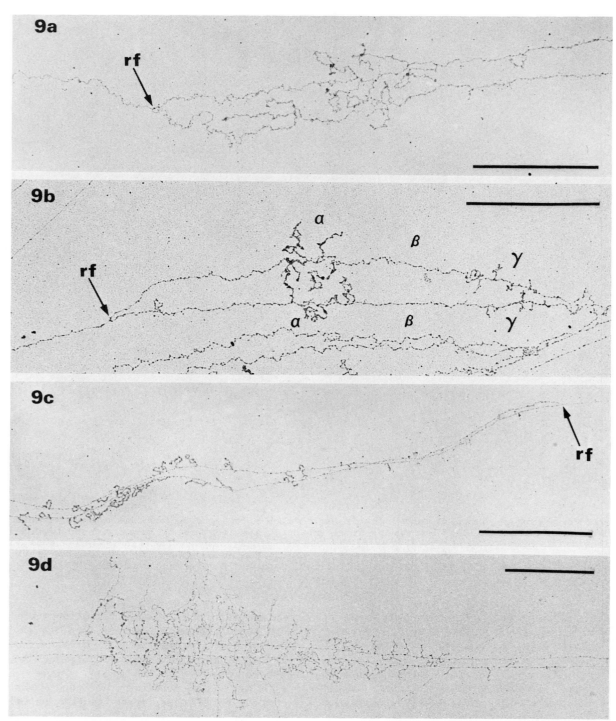

Figure 9. Electron micrographs of nascent RNP fibers on newly replicated chromatin. *(a–c)* Transcriptional configurations associated with chromatin strands within replicons. The positions of the nascent transcripts on both chromatin fibers of each micrograph are equidistant from the replication forks (rf). The striking pattern similarity indicates that each set of "homologous" RNP fiber arrays represents the transcription of identical template sequences on daughter chromatids. Such observations confirm the replicative identity of replicon configurations. No replicons exhibiting nascent transcripts on only one daughter chromatid were observed. *(b)* Three distinct regions are evident on both daughter chromatids: α, which exhibits a high frequency of long nascent RNP fibers; β, which is devoid of nascent transcripts; and γ, which is complexed with a high frequency of short RNP fibers. The configuration can be interpreted by assuming that transcription was initiated on both daughter chromatids shortly after replication, producing the long nascent fibers of the α regions. Shortly thereafter, RNA polymerase initiation was somehow suppressed, yielding the fiber-free β regions, and then reactivated, giving rise to the short nascent RNP fibers of the γ regions. The strict homology of the three regions on *both* daughter chromatids reveals coordinate effects on the transcription of both chromatid sequences. Chromatin of *a–c* was prepared from cellular blastoderm embryos which exhibited less than 6 μm of inward cell-membrane growth. *(d)* Homologous transcription units are visualized on daughter chromatid strands which do not show adjacent replication forks. Sample was prepared from a cellular blastoderm embryo exhibiting more than 10 μm of inward cell-membrane synthesis. Scale bars represent 1.0 μm.

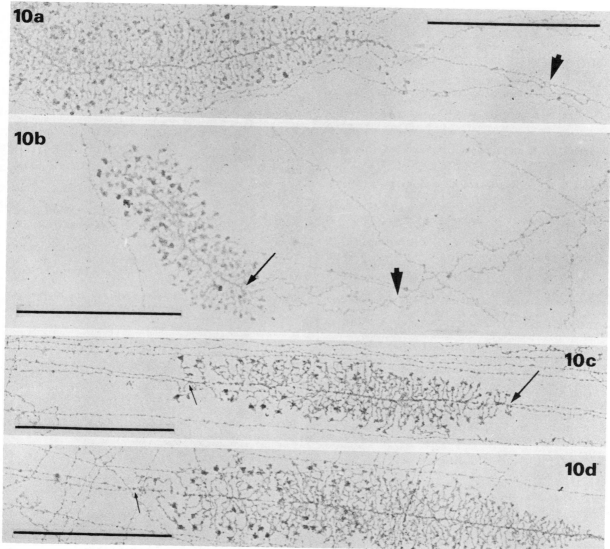

Figure 10. Electron micrographs of replicating ribosomal DNA chromatin. *(a,b)* Large arrows denote estimated origin of replication sites extrapolated by assuming bidirectional replication at a constant rate. *(b,c)* Forks (medium arrows) appear to be replicating the transcribed rDNA element in the same polarity of RNA chain propagation. *(c,d)* Forks (small arrows) occur at the estimated termination sites of rRNA transcription. All samples were prepared from cellular blastoderm embryos. Scale bars represent 1.0 μm.

rRNP matrices indicate that all measure approximately 2.6 μm, which is the mature gene length. Our failure to detect fork complexes which have entered the terminal portion of rRNP matrices suggests that a replication fork cannot enter an active ribosomal gene in a direction opposite to the polarity of RNA polymerase propagation.

Although these observations show clearly that no supraregulatory mechanisms suppress rRNA transcription prior to genome replication, clarification of the precise interaction between the two processes will require further investigation.

Acknowledgments

We are indebted to Drs. R. Reeder, R. Rodewald, and R. Simpson for helpful comments on experimental procedures; Dr. D. Hogness for comments on our observations regarding chromosome replication; Dr. J. Hearst for the communication of unpublished observations; and Ms. L. Blanks for excellent technical assistance. This work was supported by grants from the National Science Foundation and the National Institutes of Health awarded to O.L.M. and was carried out in part by S.L.M. as partial fulfillment of the requirements for the Ph.D. degree at the University of Virginia.

REFERENCES

BLUMENTHAL, A. B., H. J. KRIEGSTEIN, and D. S. HOGNESS. 1974. The units of DNA replication in *Drosophila melanogaster* chromosomes. *Cold Spring Harbor Symp. Quant. Biol.* **38**: 205.

BUSTIN, M. 1973. Arrangement of histones in chromatin. *Nat. New Biol.* **245**: 207.

BUSTIN, M. and H. KUPFER. 1976. Purification of anti-histone H1 antibodies and their use in measuring histone determinants in chromatin by radioimmunoabsorbance. *Biochem. Biophys. Res. Commun.* **68**: 718.

BUSTIN, M., D. GOLDBLATT, and R. SPERLING. 1976. Chromatin structure visualization by immuno-electron microscopy. *Cell* **7**: 297.

BUSTIN, M., R. H. REEDER, and S. L. McKNIGHT. 1977. Immunological cross-reaction between calf and *Drosophila* histones. *J. Biol. Chem.* **252**: 3099.

CAPDEVILA, M. P. and A. GARCIA-BELLIDO. 1974. Development and genetic analysis of *bithorax* phenocopies in *Drosophila*. *Nature* **250**: 500.

D'ANNA, J. A. and I. ISENBERG. 1974. A histone cross-complexing pattern. *Biochemistry* **13**: 4992.

FOE, V. E., L. WILKINSON, and C. D. LAIRD. 1976. Comparative organization of active transcription units in *Oncopeltus fasciatus*. *Cell* **9**: 131.

FRANKE, W. W., U. SCHEER, M. F. TRENDELENBURG, H. SPRING, and H. W. ZENTGRAF. 1976. Absence of nucleosomes in transcriptionally active chromatin. *Cytobiologie* **13**: 401.

GAREL, A. and R. AXEL. 1976. Selective digestion of transcriptionally active ovalbumin genes from oviduct nuclei. *Proc. Natl. Acad. Sci.* **73**: 3966.

GOLDBLATT, D. and M. BUSTIN. 1975. Exposure of histone antigenic determinants in chromatin. *Biochemistry* **14**: 1689.

GOTTESFELD, J. M. and J. BONNER. 1977. Isolation and properties of the expressed portion of the mammalian genome. In *The molecular biology of the mammalian apparatus* (ed. P.O.P. Ts'o), p. 381. Elsevier/North-Holland, Amsterdam.

GRIFFITH, J. D. 1975. Chromatin structure: Deduced from a minichromosome. *Science* **187**: 1202.

HIGASHINAKAGAWA, T., H. WAHN, and R. H. REEDER. 1977. Isolation of ribosomal gene chromatin. *Dev. Biol.* **55**: 374.

KRIEGSTEIN, H. J. and D. S. HOGNESS. 1974. Mechanism of DNA replication in *Drosophila* chromosomes: Structure of replication forks and evidence for bidirectionality. *Proc. Natl. Acad. Sci.* **71**: 135.

LAIRD, C. D. and W. Y. CHOOI. 1976. Morphology of transcription units in *Drosophila melanogaster*. *Chromosoma* **58**: 193.

LAIRD, C. D., L. E. WILKINSON, V. E. FOE, and W. Y. CHOOI. 1976. Analysis of chromatin-associated fiber arrays. *Chromosoma* **58**: 169.

LAMB, M. M. and C. D. LAIRD. 1976. Increase in nuclear poly(A)-containing RNA at syncytial blastoderm in *Drosophila melanogaster* embryos. *Dev. Biol.* **52**: 31.

LIZARDI, P. M. 1976. Biogenesis of silk fibroin mRNA: An example of very rapid processing? *Prog. Nucleic Acid Res. Mol. Biol.* **19**: 301.

MAHOWALD, A. P. 1963. Ultrastructural differentiations during formation of the blastoderm in the *Drosophila melanogaster* embryo. *Dev. Biol.* **8**: 186.

McKNIGHT, S. L. and O. L. MILLER, JR. 1975. Transcription and replication during early blastoderm formation in *Drosophila melanogaster*. *J. Cell Biol.* **67**: 276a.

———. 1976. Ultrastructural patterns of RNA synthesis during early embryogenesis of *Drosophila melanogaster*. *Cell* **8**: 305.

———. 1977. Electron microscopic analysis of chromatin replication in the cellular blastoderm *Drosophila melanogaster* embryo. *Cell* **12**: 795.

McKNIGHT, S. L., N. L. SULLIVAN, and O. L. MILLER, JR. 1976. Visualization of the silk fibroin transcription unit and nascent silk fibroin molecules on polyribosomes of *Bombyx mori*. *Prog. Nucleic Acid Res. Mol. Biol.* **19**: 313.

MILLER, O. L. and B. R. BEATTY. 1969. Visualization of nucleolar genes. *Science* **164**: 955.

OLSEN, B. R. and D. J. PROCKOP. 1974. Ferritin-conjugated antibodies used for labeling of organelles involved in the cellular synthesis and transport of procollagen. *Proc. Natl. Acad. Sci.* **71**: 2033.

PERRY, R. P., T. Y. CHENG, J. J. FREED, J. R. GREENBERG, D. E. KELLY, and K. D. TARTOF. 1970. Evolution of the transcription unit of ribosomal RNA. *Proc. Natl. Acad. Sci.* **65**: 609.

SCHEER, U., M. R. TRENDELENBURG, and W. W. FRANKE. 1976. Regulation of transcription of genes of ribosomal RNA during amphibian oogenesis. *J. Cell Biol.* **69**: 465.

SEALE, R. L. 1975. Assembly of DNA and protein during replication in HeLa cells. *Nature* **255**: 247.

———. 1976. Studies on the mode of segregation of histone nu-bodies during replication in HeLa cells. *Cell* **9**: 423.

SIMPSON, R. T. and M. BUSTIN. 1976. Histone content of chromatin subunits studied by immunosedimentation. *Biochemistry* **15**: 4305.

STOLLAR, B. D. and M. WARD. 1970. Rabbit antibodies to histone fractions as reagents for preparative and comparative studies. *J. Biol. Chem.* **245**: 1261.

TRENDELENBURG, M. F., U. SCHEER, H. ZENTGRAF, and W. W. FRANKE. 1976. Heterogeneity of spacer length in circles of amplified ribosomal DNA of two insect species, *Ditiscus marginalis* and *Acheta domesticus*. *J. Mol. Biol.* **108**: 453.

TSANEV, R. and B. SENDOV. 1971. Possible molecular mechanisms for cell differentiation in multicellular organisms. *J. Theor. Biol.* **30**: 337.

WATSON, J. D. and F. H. C. CRICK. 1954. The structure of DNA. *Cold Spring Harbor Symp. Quant. Biol.* **18**: 123.

WEINTRAUB, H. 1974. The assembly of newly replicated DNA into chromatin. *Cold Spring Harbor Symp. Quant. Biol.* **38**: 247.

———. 1976. Cooperative alignment of nu-bodies during chromosome replication in the presence of cycloheximide. *Cell* **9**: 423.

WEINTRAUB, H. and M. GROUDINE. 1976. Chromosomal subunits in active genes have an altered conformation. *Science* **193**: 848.

WEINTRAUB, H., A. WORCEL, and B. ALBERTS. 1976. A model for chromatin based upon two symmetrically paired half-nucleosomes. *Cell* **10**: 409.

WELLAUER, P. K. and I. B. DAWID. 1977. The structural organization of ribosomal DNA in *Drosophila melanogaster*. *Cell* **10**: 193.

WOODCOCK, C. L. F., L. L. FRADO, C. L. HATCH, and L. RICCIARDIELLO. 1976. Fine structure of active ribosomal genes. *Chromosoma* **58**: 33.

ZALOKAR, M. 1976. Autoradiographic study of protein and RNA formation during early development of *Drosophila* eggs. *Dev. Biol.* **49**: 425.

Morphology of Transcriptionally Active Chromatin

W. W. Franke, U. Scheer, M. Trendelenburg, H. Zentgraf,* and H. Spring

*Division of Membrane Biology and Biochemistry, Institute of Experimental Pathology, and *Institute of Virology, German Cancer Research Center, D-69 Heidelberg, Federal Republic of Germany*

Some decades ago it was noted by cytologists that within the interphase nucleus large portions of the transcriptionally ("genetically," in their terms) inactive chromosomal material are contained in aggregates of condensed chromatin, the "chromocenters," whereas transcriptionally active regions of chromosomes appear in a more dispersed form and are less intensely stained with DNA-directed staining procedures (Heitz 1929, 1932, 1956; Bauer 1933). The hypothesis that condensed chromatin is usually characterized by very low or no transcriptional activity, and that transcription occurs in loosely packed forms of chromatin (including, in most cells, the nucleolar chromatin) has received support from studies of ultrathin sections in the electron microscope and from the numerous attempts to separate transcriptionally active from inactive chromatin biochemically (for references, see Anderson et al. 1975; Berkowitz and Doty 1975; Krieg and Wells 1976; Rickwood and Birnie 1976; Gottesfeld 1977). Electron microscopic autoradiography has revealed that sites of RNA synthesis are enriched in dispersed chromatin regions located at the margins of condensed chromatin (Fakan and Bernhard 1971, 1973; Bouteille et al. 1974; Bachellerie et al. 1975) and are characterized by the occurrence of distinct granular and fibrillar ribonucleoprotein (RNP) structures, such as perichromatin granules and fibrils. The discovery that, in most eukaryotic nuclei, major parts of the chromatin are organized in the form of nucleosomes (Olins and Olins 1974; Kornberg 1974; Baldwin et al. 1975) has raised the question whether the same nucleosomal packing of DNA is also present in transcriptionally active chromatin strands. Recent detailed examination of the morphology of active and inactive chromatin involving a diversity of electron microscopic methods, particularly the spreading technique by Miller and coworkers (Miller and Beatty 1969; Miller and Bakken 1972), has indicated that the DNA of some actively transcribed regions is not packed into nucleosomal particles but is present in a rather extended form within a relatively thin (4–7 nm) chromatin fiber.

OBSERVATIONS

Structure of Transcriptionally Inactive Condensed Chromatin

Ultrathin sections through regions of in situ fixed, condensed chromatin, including the bulk of the chromatin of relatively inactive nuclei (such as avian and amphibian erythroblasts and erythrocytes, lymphocytes, and sperm cells and spermatids of sea urchins and fishes), reveal 18–26-nm granules, which are intensely stained with most of the conventional heavy-metal stains. Such predominant, large granules, particularly striking in their ordered arrays in the peripheral layers of chromatin subjacent to the nuclear envelope (Figs. 1 and 2a), have been interpreted as cordlike linear arrangements of tightly packed large granules (e.g., Zentgraf et al. 1969, 1975; Chentsov and Polyakov 1974; Franke and Scheer 1974; Franke 1977). This interpretation is somewhat at variance with that of Davies and coworkers, who suggested that this peripheral condensed chromatin is organized in cylindrical threads of uniform width (Davies and Small 1968; for references, see Davies and Haynes 1976). We interpret the so-called "thick" (20–30 nm) chromatin fibers, which frequently exhibit a knobby appearance, to represent strands of tightly packed large granules, perhaps with somewhat obscured contours due to the specific preparative methods (Gall 1966; Rae 1966; Wolfe and Grim 1967; Ris 1975; Zirkin 1975; Brasch 1976; Finch and Klug 1976). The organization of condensed chromatin in 18–26-nm large granules is also observed in sections through metaphase chromosomes (cf. Plate 7 in Franke 1977). Such large granular units of condensed chromatin can also be seen in spread preparations of disrupted nuclei (Fig. 2b–d), in particular in association with fragments of the nuclear envelope (Franke et al. 1976b). In an accompanying paper in this volume, Olins shows that such granular structures are also observed from freshly spread disrupted nuclei and are clearly shown to consist of close-packed nucleosomes. Depending upon the ionic strength of the

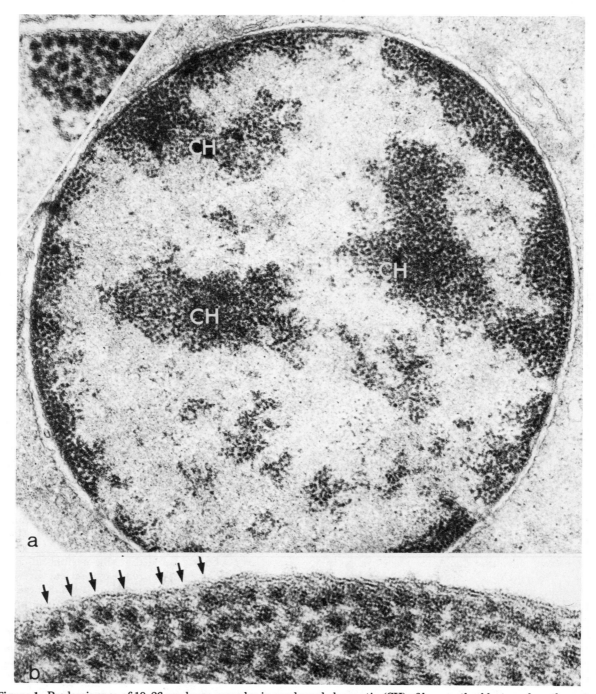

Figure 1. Predominance of 19–26-nm large granules in condensed chromatin (CH) of hen erythroblasts and erythrocytes as revealed in ultrathin sections through cells fixed in warm (40°C; *a*) or cold (5°C; *b*) solutions of glutaraldehyde (for details and references, see Zentgraf et al. 1975). Note the particularly regular and close packing of these granules in the peripheral layers of chromatin attached to the inner nuclear membrane (insert in *a*, arrows in *b*). *(a)* 50,000×; insert, 135,000×; *(b)* 200,000×.

Figure 2. Appearance of the large (20–26 nm) granules of peripheral condensed chromatin in cultured cells (*a*, murine 3T3 cells; *b–d*, murine sarcoma 180 cells) in ultrathin sections after fixation in monolayer *(a)* and in spread preparations after disruption of cells and/or isolated nuclei and dispersal of chromatin in low-salt concentrations *(b–d)*. These large granules are usually particularly conspicuous in the chromatin layer associated with the nuclear envelope (NE), whereas thinner and more loosely arranged chromatin fibrils are often recognized in the nuclear interior *(a)*. Cords of linear arrays of such large chromatin granules are identified not only in tangential sections (cf. Zentgraf et al. 1975), but also in moderately dispersed, positively stained and/or metal-shadowed spread preparations of chromatin (e.g., *b* and *c*; DCA, dense chromatin aggregate). Upon more extensive dispersion, the majority of the chromatin fibrils appear in typical nucleosomal arrays of 11–14-nm large "beads on a chain" *(d)*. Transitions between arrays of large and nucleosomal granules in the same chromatin fibril are occasionally noted (arrows in the insert in *c*). Magnifications: *(a)* 130,000×; *(b)* 45,000×; *(c)* 30,000×; insert, 65,000×; *(d)* 27,000×.

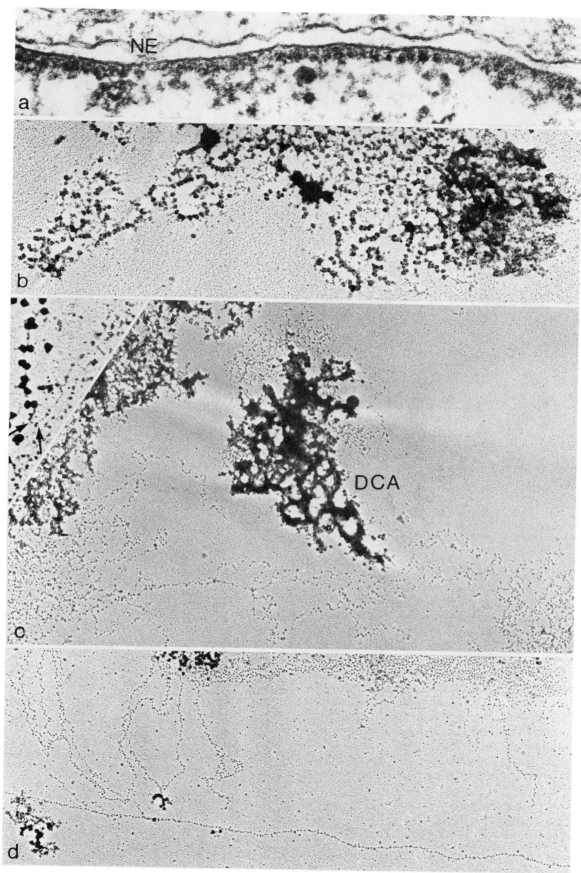

Figure 2 *(see facing page for legend)*

757

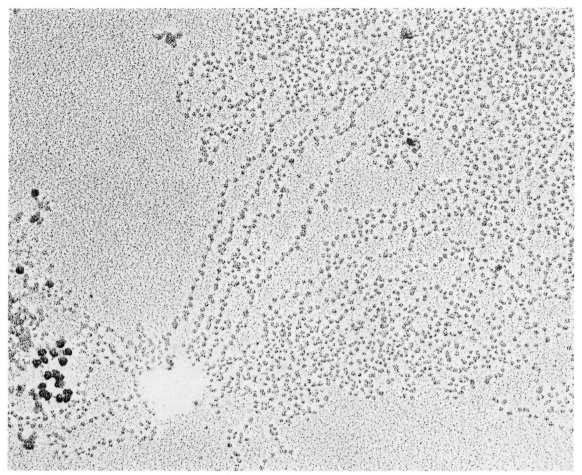

Figure 3. Large chromatin granules (e.g., in the lower left) are only rarely seen in extensively dispersed and swollen chromatin (from cultured murine sarcoma 180 cells, preparation as described in Fig. 2d) in which nearly all the chromatin appears in the nucleosomal form of fibers. Magnification, 75,000X.

specific medium used and the concentration of divalent cations, large granules of a supranucleosomal order of chromatin package can be isolated in a rather stable form (Kiryanov et al. 1976; Renz et al. 1977). In media of very low ionic strength, especially in the absence of divalent cations, large granular arrays of condensed chromatin are progressively unravelled into chromatin strands with the typical nucleosomal appearance, i.e., 9–13-nm "beads-on-a-string" (Fig. 2a–d). However, even after prolonged incubation in low-salt media in which the vast majority of the chromatin appears in nucleosomal configuration, occasional large granules may still be observed (Fig. 3). This characteristic form of chromatin organization seems to represent the higher order

of packing of nucleosomal chromatin (Franke et al. 1976b; Kiryanov et al. 1976; Renz et al. 1977). It is worth emphasizing that this aggregation into large granules does not seem to result in the appearance of corresponding preferential DNA-length subunits upon cleavage with micrococcal nuclease (H. Zentgraf et al., unpubl.), indicating that within these large granules the nucleosomal cleavage sites are accessible to this enzyme.

Structure of Transcriptionally Active Chromatin in Ultrathin Sections

The above-described 18–26-nm large granules, characteristic of inactive condensed chromatin, are

Figure 4. Ultrathin sections showing the predominance of nucleolonema organization in transcriptionally active nucleoli of a cultured rat kangaroo (PtK2) cell *(a)* and a vitellogenic oocyte of the newt *Triturus alpestris (b)*. The 0.1–0.2-μm thick cords most likely represent the transcriptionally active chromatin containing rDNA, and the dense ribonucleoprotein granules located in the periphery of these cords contain the precursors to rRNAs. Note that the central portions of these nucleolonema are characterized by finely filamentous material (e.g., indicated by arrows in *b*) and do not show the 20–26-nm large chromatin granules. Magnifications: *(a)* 100,000X; *(b)* 60,000X.

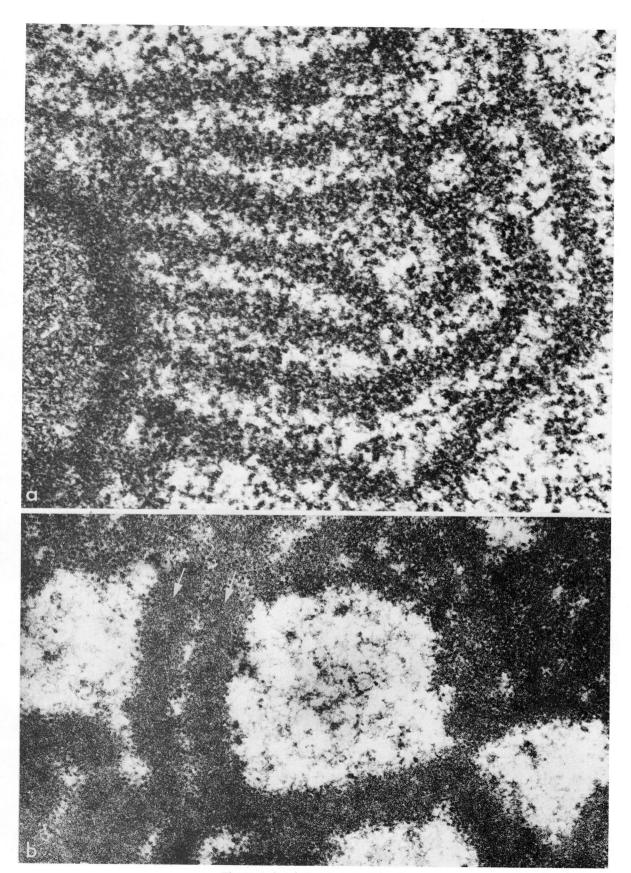

Figure 4 *(see facing page for legend)*

759

not observed in chromatin regions of known transcriptional activity. Granular structures of comparable size, which are often associated with dispersed chromatin, can be demonstrated by cytochemistry not to represent deoxyribonucleoproteins (DNP) but rather RNP material (Monneron and Bernhard 1969; Bouteille et al. 1974; Heumann 1974; Bachellerie et al. 1975). A particularly instructive example is nucleolar chromatin (Fig. 4) in which the nucleolonema — structures containing the transcribed rDNA-chromatin (i.e., the equivalent structures in situ to the matrix units[1] described below) — do not reveal such large, dense granules in their cores but are associated with RNP granules at their periphery. Similar organization has been described by conventional and scanning transmission electron microscopy in the loop axes of lampbrush chromosomes in amphibian oocytes and the green alga *Acetabularia* (Malcolm and Sommerville 1974; Mott and Callan 1975; Spring et al. 1975; H. Spring and W. Franke, unpubl.) and in the axes of Balbiani rings in *Chironomus* (Case and Daneholt 1977). Although thin-section studies do not allow one to make conclusions as to the presence of nucleosomes in heavily transcribed chromatin regions, they clearly rule out the presence of the supranucleosomal granules that are characteristic of inactive chromatin.

Structure of Transcribed Nucleolar Chromatin as Revealed by the Spreading Technique

The organization of inactive chromatin into nucleosomal "beads-on-a-string" was first discovered by a technique in which the chromatin was dispersed in low-salt concentrations (Olins and Olins 1974; for additional references, see Woodcock et al. 1976a). When inactive chromatin is dispersed under such conditions and centrifuged or adsorbed on thin films, with or without fixation in aldehyde or treatment with ethanol, the typical beaded chains are recognized by a variety of staining methods and in dark field illumination (Olins and Olins 1974; for references, see Franke et al. 1976b). In this connection, it should be kept in mind that the special conditions and rates of optimal dispersal are somewhat specific for the type of chromatin; for example, constitutive

heterochromatin and metaphase chromosomes are notoriously more resistant to low-salt dispersal (e.g., see Rattner et al. 1975).

Actively transcribed chromatin portions are especially suitable for dispersal and spreading, most likely a consequence of their relatively dispersed state in vivo (Miller and Beatty 1969; Miller and Bakken 1972; Scheer et al. 1975, 1976a,b). Probably the best characterized category of large eukaryotic genes are those coding for pre-rRNA molecules, which occur in relatively high numbers of copies, are clustered in distinct aggregates, and, at least in some cell systems, are present at high transcriptional activity. Especially favorable objects for studies of rDNA chromatin and its transcription are the masses of extrachromosomal amplified nucleolar units that are present in oocytes of some insects, amphibia, and fishes and which go through a natural cycle of transcriptional activity with a maximum at the time of vitellogenesis (for references, see Scheer et al. 1976a; Trendelenburg et al. 1973, 1977; Trendelenburg 1977). Figure 5 is representative of the appearance of maximally transcribed nucleolar chromatin from amphibian oocytes in spread preparations; the typical tandem arrangement of alternating sequences containing transcribed pre-rRNA genes (pre-rRNA matrix units) and apparently nontranscribed spacer units is present (Fig. 5b,c). Another nucleus with fully transcribed nucleolar chromatin suitable for analysis by this technique is the giant primary nucleus of certain green algae, the Dasycladaceae (Fig. 6a,b; Spring et al. 1974, 1976; Trendelenburg et al. 1974; Berger and Schweiger 1975).

Nucleolar chromatin with a very high transcriptional activity of the pre-rRNA genes is characterized by matrix units with a high density of lateral RNP fibrils that contain the nascent pre-rRNA (Fig. 5b,c; Figs. 6–8; cf. references cited above and Foe et al. 1976; Laird and Chooi 1976; McKnight and Miller 1976; as to mammalian nucleoli see also Miller and Bakken 1972; Puvion-Dutilleul et al. 1977). The bases of these lateral fibrils are mostly accentuated by a 12–14-nm large granule that contains the RNA polymerase (Fig. 6c,d). Usually all granules attached to the axis of such a densely fibril-covered matrix unit are also associated with a lateral fibril and are arranged in almost close packing. This demonstrates their nature as transcriptional complexes of the RNA polymerase A (polymerases A and B have diameters, in the dehydrated state, of about 12 nm; P. Chambon and P. Oudet, pers. comm.) and also shows that there are no additional nucleosomal particles in such spread matrix units. The granules attached to the matrix unit axis are not "decoration artifacts" resulting from stain deposition or precipitation with ethanol since they are also observed in preparations made without the use of any heavy-metal stain and ethanol (Fig. 6e).

The number of nucleotides of DNA per pre-rRNA

---

[1] The following morphological terms are used: *Matrix unit* — the intercept on the chromatin axis which is covered by a series of lateral fibrils increasing in length from one point (the starting point) or, at least, all of which are longer than the fibril at the starting point (the latter to allow for potential processing events or higher packing density of the nascent RNA, which may lead to subsequent fibril shortening). *Transcriptional unit* — the intercept that is transcribed by an RNA polymerase into one covalent ribopolynucleotide, i.e., an intercept is limited by a promotor and a terminator site. A transcriptional unit may be identical with a matrix unit. *Apparent spacer unit* — the morphologically identified axial intercepts not covered with lateral fibrils, which lie between matrix units. *Repeating unit* — in nucleolar chromatin, the unit consisting of transcriptional unit for pre-rRNA and the adjacent (subsequent or preceding) spacer. *Transcriptional complex* — the chromatin-associated particle containing the RNA polymerase and the attached nascent RNP fibril.

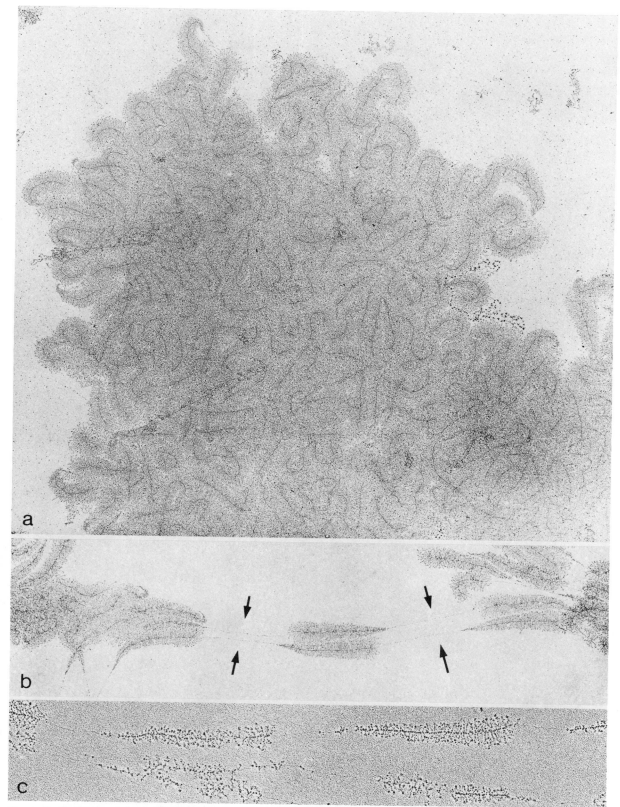

Figure 5. Positively stained *(a,b)* and metal-shadowed *(c)* preparations of moderately *(a)* and extensively *(b,c)* dispersed and spread transcriptionally active chromatin of amplified, extrachromosomal nucleoli from oocytes of the newts *Triturus cristatus (a,b)* and *Pleurodeles waltli (c)*. Note that apparently all pre-rRNA genes are transcribed at this stage of oogenesis (mid-lampbrush chromosome stage) and appear in the form of "matrix units" separated by intercepts free of lateral fibrils, the "spacer units" (arrows in *b*). The chromatin axis appears very thin (4–7 nm) with or without the use of detergents in the dispersion medium (the material shown in *c* has been prepared in the presence of 0.3% "Joy"; cf. Miller and Bakken 1972; Scheer et al. 1977). Magnifications: *(a)* 6500×; *(b)* 10,000×; *(c)* 20,000×.

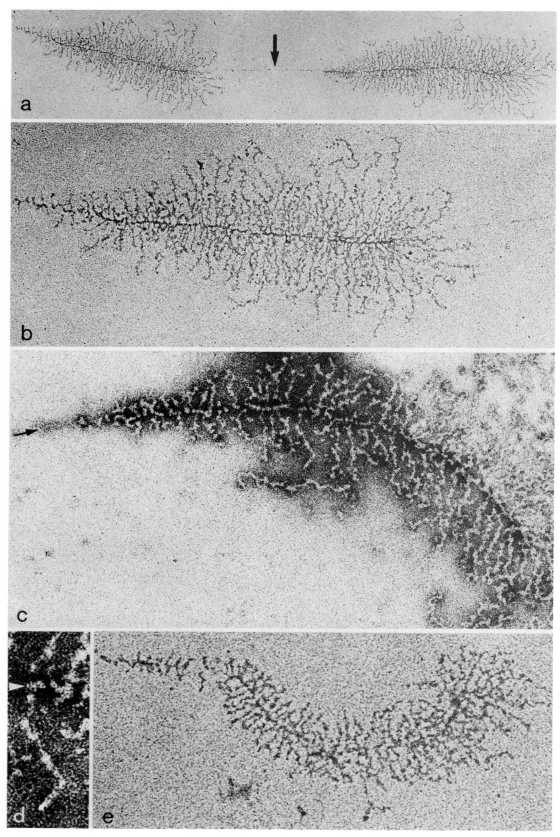

Figure 6 *(see facing page for legend)*

matrix unit should correspond to the molecular weight of the specific pre-rRNA species isolated from this organism, assuming that this is the primary transcriptional product or at least an only slightly processed transcript. This correlation has been demonstrated in some organisms (Scheer et al. 1973; Spring et al. 1974, 1976; for additional references, see Foe et al. 1976; Laird and Chooi 1976; McKnight and Miller 1976) but does not hold in all organisms (Trendelenburg et al. 1973, 1976) or in all matrix units of a nucleolus (for detailed discussion, see Franke et al. 1976a). The inclusion of special components in the preparation media, such as Sarkosyl NL-30 which removes most chromatin proteins, histones included (Gariglio 1976), or tRNA which removes histone H1, does not result in significant changes of matrix unit length (Scheer et al. 1977).

Pre-rRNA genes of reduced transcriptional activity can appear as matrix units with reduced lateral fibril density (Scheer et al. 1975, 1976a). Here the axial intercepts between the insertion sites of the lateral fibrils, i.e., between the transcriptional complexes, are relatively thin (4–8 nm). This shows that the absence of transcriptional complexes in an intercept of a transcribed gene does not result in the immediate formation of a stable nucleosome in this specific region, i.e., for a certain time between the transcriptional events (Franke et al. 1976b). The same conclusion can also be derived from our finding that matrix units of reduced lateral fibril densities are not foreshortened in an inverse proportion to the number of lateral fibrils present per unit (W. W. Franke and U. Scheer, unpubl.).

Structure of Nontranscribed Regions Adjacent to Transcribed Nucleolar Chromatin

Pre-rRNA matrix units are separated by intercepts which are usually free of lateral fibrils; these are the apparent spacer units which contain sequences of the so-called nontranscribed spacer and in some nucleolar strands exhibit considerable length heterogeneity (Scheer et al. 1973, 1977; Wellauer et al. 1974, 1976a,b; Wellauer and Reeder 1975; Foe et al. 1976; Spring et al. 1976; Trendelenburg et al. 1976). While these spacer regions appear as a rather thin (4–8 nm) and uniform DNP fibril (Fig. 7a; Foe et al. 1976; Franke et al. 1976b), granular particles of somewhat variable or

of uniform size, including particles with a diameter similar to that of nucleosomes, are sometimes observed in these spacer intercepts (Fig. 7b; Angelier and Lacroix 1975; Franke et al. 1976b; Woodcock et al. 1976c). It has been suggested by some authors that such spacer-unit-associated particles might represent nucleosomes (e.g., see Woodcock et al. 1976c). Our finding, however, that the number of particles per spacer unit is not inversely correlated to the length of the specific spacer unit and the demonstration that spacer-unit lengths are not significantly altered when the nucleolar material has been exposed to concentrations of Sarkosyl that are known to remove large portions of the histones (Franke et al. 1976b; Scheer et al. 1977) speak against this interpretation. On the contrary, there are observations suggesting that at least some of the spacer-unit-associated particles, in particular those located in regions preceding the beginnings of matrix units, represent RNA polymerase-containing complexes that are either inactive or are associated with very small nascent products. Some of the spacer-unit-attached particles show a resistance to treatment with Sarkosyl similar to the matrix unit RNA polymerase-containing particles (Fig. 7e; Franke et al. 1976b; Scheer et al. 1977). Occasionally one finds in apparent spacer regions of spread nucleoli small matrix units with their lateral fibril gradients discontinuous with those of the typical pre-rRNA matrix units (Fig. 7c,d; Scheer et al. 1973, 1977; Franke et al. 1976a). Such an occurrence indicates that transcriptional events can occur simultaneously with the synthesis of pre-rRNA, at least in some repeating units. The histograms of the lengths of spacer units with and without the small matrix units have been found to be almost superimposable, again indicating that the DNA in the chromatin of the apparent spacer regions is not in a form of nucleosomal packing but is extended as in transcribed regions.

The conclusion that apparent spacer regions of transcribed nucleolar chromatin are not foreshortened and packed into nucleosomes is also supported by the comparisons of total repeating units in purified rDNA with the lengths of the repeating units identified in spread preparations of transcribed nucleolar chromatin. For example, the distribution of repeating-unit-size classes of amplified rDNA of Xenopus laevis, as determined in the products obtained from cleavage with restriction endonuclease EcoRI (Wellauer et al. 1974, 1976a,b; Wellauer and

Figure 6. Dense packing of lateral ribonucleoprotein fibrils and RNA polymerase complexes in matrix units of transcriptionally active nucleolar chromatin from the giant primary nucleus of the green alga Acetabularia cliftonii (a,b; for details, see Franke et al. 1976b) and oocytes of Xenopus laevis (c,d) and Triturus helveticus (e) as revealed after positive staining (a,b), negative staining with 1% phosphotungstic acid adjusted to neutrality with NaOH (c,d), and without the use of ethanol or any staining reagent (e). Note also the absence of granules of nucleosomal size in "spacer intercepts" (arrows in a, c, and right part of b). (d) Detail of c showing at higher magnification the appearance of a lateral fibril containing the nascent pre-rRNA; arrowhead denotes the basal attachment to the axis and the polymerase complex, respectively. (e) Native electron density of matrix unit fibrils. Magnifications: (a) 23,000×; (b) 48,000×; (c) 75,000×; (d) 150,000×; (e) 50,000×.

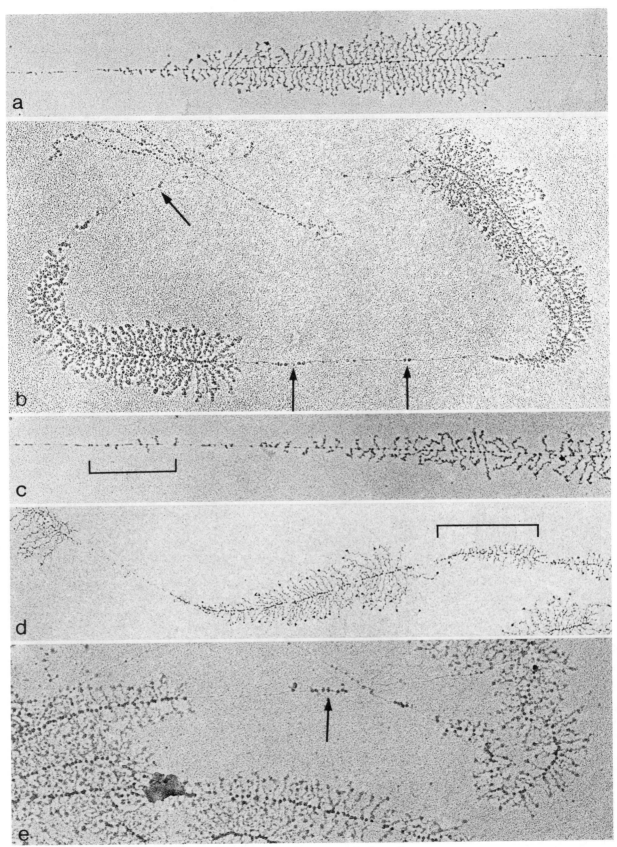

Figure 7 *(see facing page for legend)*

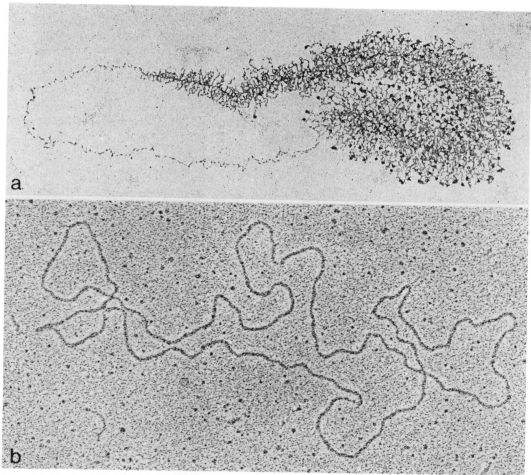

Figure 8. Comparison of the contour length of the smallest size class of extrachromosomal rings (oocytes of the house cricket *Acheta domesticus*) containing one gene for pre-rRNA in the form of transcriptionally active chromatin *(a)* and isolated rDNA *(b)*. For details, see Trendelenburg et al. (1976, 1977). Magnifications: *(a)* 27,000×; *(b)* 60,000×.

Reeder 1975), is similar to that found in spread nucleolar chromatin from vitellogenic oocytes (Franke et al. 1976b; Scheer et al. 1977). Likewise, the histograms of the ring sizes of purified rDNA from the oocytes of the water beetle *Dytiscus marginalis* and the house cricket *Acheta domesticus* are similar to those of the circumferences of rings identified in spread preparations of the transcribed extrachromosomal nucleolar chromatin (Trendelenburg et al. 1976). This correlation is especially clear in the smallest size class of rDNA rings, i.e., the rings containing one pre-rRNA gene and a spacer unit (Fig. 8). The DNA contained in these small "one-gene circles" is clearly extended in the spread transcribed nucleolar chromatin as one does not find the condensation of DNA expected if it were packed in nucleosomes (regarding packing ratios, cf., e.g., Carlson and Olins 1976; Germond et al. 1975; Griffith 1975; Oudet et al. 1975; Cremisi et al. 1976; Sperling and Tardieu 1976; Woodcock et al. 1976b; Varshavsky et al. 1976).

In states of reduced transcriptional activity of nu-

Figure 7. Details of the morphology of matrix and spacer units in spread preparations of dispersed chromatin of transcriptionally active nucleoli from amphibian oocytes *(a–c, Triturus alpestris; d, Xenopus laevis; e, Triturus cristatus)*. Occasionally, small granules of subnucleosomal *(a)* or nucleosomal *(b,* arrows) size are found to accentuate the axis of spacer intercepts. The association of some of these particles with short lateral fibrils (e.g., in the region indicated by the brackets in *c*) indicates that they contain RNA polymerases and small transcripts. The occurrence of transcriptionally active RNA polymerase molecules in some apparent spacer regions is also demonstrated by the "prelude complexes" (one is denoted by the brackets in *d;* cf. Scheer et al. 1973, 1977). Like active RNA polymerases, some of the apparent spacer-associated particles resist treatment *(e,* arrow) even with relatively high concentrations of specific detergents which remove most other proteins, histones included *(e* shows an example of a preparation made in the presence of 0.3% Sarkosyl NL-30 [cf. Franke et al. 1976b; and Scheer et al. 1977]). Magnifications: *(a)* 35,000×; *(b)* 30,000×; *(c)* 38,000×; *(d)* 23,000×; *(e)* 62,000×.

cleoli, it is often noted in spread preparations that individual pre-rRNA genes appear to be completely inactive, i.e., free of lateral fibrils, whereas adjacent pre-rRNA genes show the active form, i.e., are densely covered with the nascent RNP fibrils (Fig. 9a–c; Scheer et al. 1975, 1976a; Scheer and Franke 1976). Such regions corresponding to transcriptionally inactive genes adjacent to actively transcribed ones mostly also reveal a thin (4–8 nm) and nonbeaded chromatin axis (Franke et al. 1976b; cf. Foe et al. 1976). Comparisons of lengths of such fibril-free regions with corresponding intercepts in adjacent regions containing fully fibril-covered matrix units (i.e., units containing a spacer unit plus a matrix unit plus the subsequent spacer unit) have not shown any foreshortening in the fibril-free pre-rRNA gene intercepts (Scheer and Franke 1976). This strongly suggests that the extended state of nucleolar chromatin is not simply a result of the ongoing transcription, but that the change of chromatin structure from the nucleosomal to the extended state can be dissociated in time and space from the transcriptional process as such.

Structure of Transcriptionally Inactive Nucleolar Chromatin as Identified in Spread Preparations

Certain processes of cell differentiation, such as late spermiogenesis and erythropoesis, are characterized by the disappearance of nucleolar structure, concomitant with a cessation of synthesis of precursors to ribosomal RNAs. Electron microscopic studies of ultrathin sections suggest that in such nuclei the rDNA is contained in the large, supranucleosomal granules described above, indicative of nucleosomal packaging. Spread preparations of chromatin from nuclei of early embryonic stages, such as of *Drosophila melanogaster* (McKnight and Miller 1976) and *Oncopeltus fasciatus* (Foe et al. 1976), and from nuclei in different stages of oogenesis in amphibia (Scheer et al. 1976a) and certain insects (Trendelenburg et al. 1977) allow one to follow the structural changes of rDNA-containing chromatin during transcriptional activation. Moreover, progressive inactivation can be studied, for example, after inhibition with certain drugs (Scheer et al. 1975) or, in a natural form, during late stages of oocyte maturation in amphibia (Scheer et al. 1976a). In such studies, we have noted that large portions of the nucleolar chromatin, in which the pre-rRNA genes are inactive, as judged by the absence of lateral fibrils, appear in the form of regularly packed "beads-on-a-string" arrays, in contrast to other regions of the same nucleolus in which both transcriptional activity and a "smooth and thin" appearance of the DNP axis is noted (Fig. 10a,b) (Franke et al. 1976b). This occurrence of both forms of nucleolar chromatin in the same area of the electron microscopic specimen grid also contradicts the argument that the nonbeaded form described in the transcrip-

tionally active chromatin is the result of a general artifact of this preparation. On occasion, we have also noted situations in which a chromatin axis can be traced from a transcribed pre-rRNA gene region with a "thin and smooth" fiber appearance into an adjacent intercept with a beaded appearance, suggesting nucleosomal organization (Fig. 10c). Similar observations are presented in this volume by Foe from studies of embryonic stages of *Oncopeltus*.

Structure of Transcriptionally Active Non-nucleolar Chromatin

Matrix units of non-nucleolar chromatin that are densely covered with lateral fibrils, indicating high transcriptional activity, have been studied in detail in the loops of the lampbrush chromosomes of amphibian oocytes (Miller and Bakken 1972; Angelier and Lacroix 1975; Scheer et al. 1976b) and in the primary nucleus of the green alga *Acetabularia* (Spring et al. 1975; Scheer et al. 1976b). These matrix units are very heterogeneous in length, some of them far exceeding the length of pre-rRNA matrix units; this may indicate the formation of very large primary transcriptional products in these regions (for references pertaining to the occurrence of similar structures in some insect spermatocytes, see Scheer et al. 1976b). In these matrix units of lampbrush chromosome loops, the basal granules attached to the loop chromatin axis are usually associated with a lateral fibril and an almost close-packing arrangement, at least in some regions (Miller and Bakken 1972; Franke et al. 1976b; Scheer et al. 1976b). Intercepts within matrix units which are free of lateral fibrils appear relatively thin and do not show a nucleosomelike beaded appearance (Fig. 11a–c; cf. Fig. 7 of Angelier and Lacroix 1975). This seems to hold even for matrix units with a rather sparse coverage by lateral fibrils (Fig. 11c; for further demonstrations, see Franke et al. 1976b). Fibril-free intercepts corresponding to apparently nontranscribed spacer regions adjacent to matrix units have also been described in chromosome loops (Scheer et al. 1976b) and likewise do not show a regular "beads-on-a-string" appearance. Another densely fibril-covered type of matrix unit has been described by O. Miller (pers. comm.) in silk gland cells of *Bombyx mori* and has been tentatively identified as the transcriptional unit of the gene coding for silk fibroin.

A different form of transcribed non-nucleolar chromatin seems to be represented by chromatin axes with arrays of lateral fibrils of increasing lengths in which the fibril density is comparatively low. Such chromatin has been described in embryonic cells of *Drosophila* and *Oncopeltus*, in HeLa cells, in rat liver, and in different stages of mouse spermiogenesis (Miller and Bakken 1972; Kierszenbaum and Tres 1975; Foe et al. 1976; Laird and Chooi 1976; Laird et al. 1976; Puvion-Dutilleul et al. 1977).

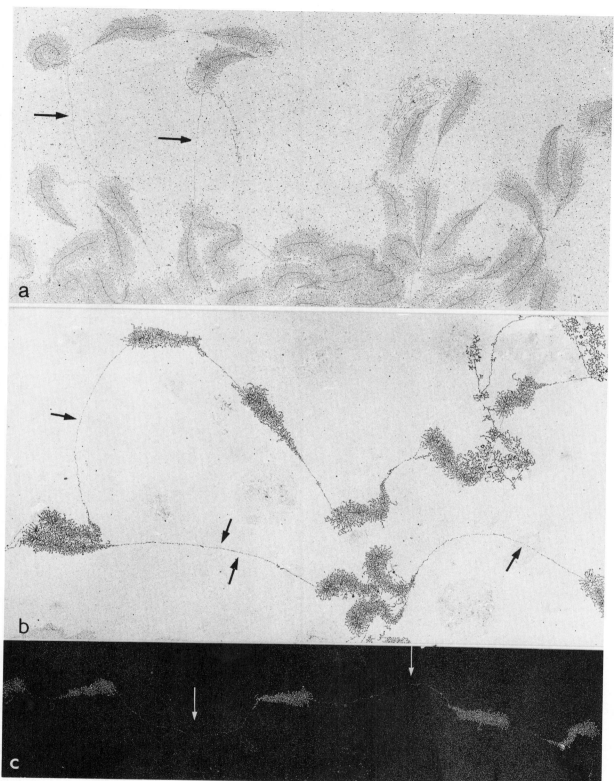

Figure 9. Occasionally, exceedingly long axial intercepts free of lateral fibrils are recognized in spread preparations of nucleolar chromatin from amphibian oocytes *(a, Triturus cristatus; b,c, T. alpestris)*, indicative of the inactive state of one individual gene adjacent to apparently fully transcribed genes (arrows). Such unusually long fibril-free intercepts are particularly frequent in states of reduced transcriptional activity (cf. Scheer et al. 1976a). Comparison of the lengths of such regions with the corresponding intercepts containing a transcribed pre-rRNA gene, i.e., one matrix unit plus two spacer units, shows a similar distribution, suggesting that the DNA in such fibril-free regions is not considerably foreshortened (cf. Scheer and Franke 1976). Magnifications: *(a)* 7000X; *(b)* 9000X; *(c)* 6000X.

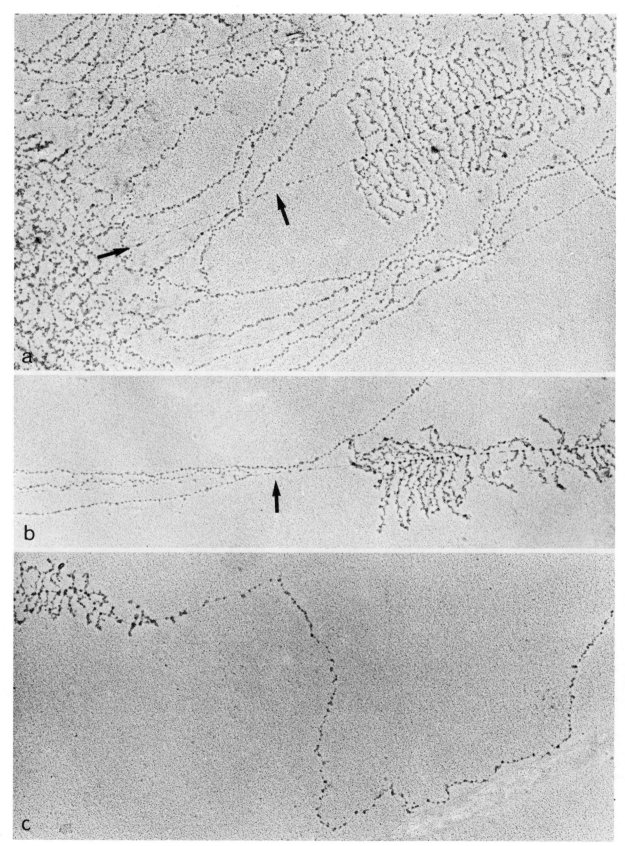

Figure 10 *(see facing page for legend)*

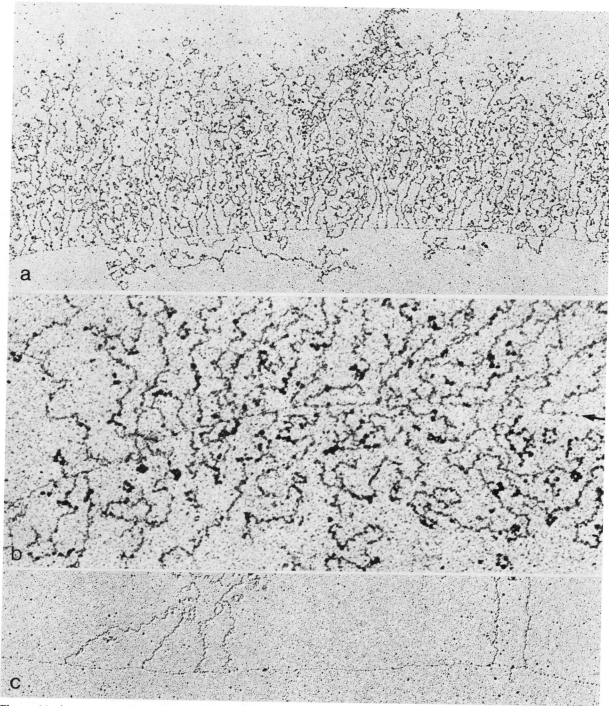

Figure 11. Appearance of transcriptionally active non-nucleolar chromatin in loops of lampbrush chromosomes from the urodelan amphibia *Pleurodeles waltli (a)* and *Triturus cristatus (b,c)* as revealed in positively stained spread preparations *(a,b)* (the preparation shown in *c* has in addition been metal-shadowed). Note the thin and smooth appearance of the loop axis *(a)*, in particular in between the transcriptional complexes *(b)* (the axis is denoted by the arrow in the right). The width of the axis is only slightly reduced after treatment with Sarkosyl (for details, see Scheer et al. 1976b) as illustrated in an example *(c)* of a loop with reduced density of lateral fibrils. Magnifications: *(a)* 23,000×; *(b)* 45,000×; *(c)* 30,000×.

Figure 10. Appearance of nucleolar chromatin in positively stained and metal-shadowed spread preparations made from nearly mature oocytes of the Alpine newt *Triturus alpestris,* i.e., in stages of greatly reduced transcriptional activity (Scheer et al. 1976a). In such preparations, a large proportion of the nucleolar chromatin fibers appears in the typical beaded form indicative of nucleosomal organization *(a,b)*, whereas the fibril-free axial intercepts within matrix units (e.g., *b*) as well as adjacent to matrix units (arrows in *a* and *b*) usually reveal distinctly different and much smoother contours. Only rarely are situations encountered in which a fibril-free, long axial intercept adjacent to a transcriptional unit shows a beaded structure with nucleosome-size granules *(c)*. Magnifications: *(a)* 53,000×; *(b)* 46,000×; *(c)* 48,000×.

In axial intercepts between the sparse lateral fibrils in such arrays, the chromatin has been said to be "beaded" by some of these authors. In our opinion, there is still some uncertainty as to the nature of these particles, although some authors have suggested they are nucleosomal (Kierszenbaum and Tres 1975; Laird et al. 1976; Foe et al. 1976; McKnight et al., this volume). This interpretation has been confirmed in "retracting" loops of lampbrush chromosomes of maturing amphibian oocytes by the disappearance of such granules after treatment with Sarkosyl (U. Scheer, unpubl.).

DISCUSSION

Our observations suggest that the DNA in transcriptionally active chromatin regions is in an extended state and is not packed in nucleosomal particles (Franke et al. 1976b). In addition, these findings indicate that the conformational change from the nucleosomal to the extended state takes place somewhat before the beginning of transcription, and the reverse change does not occur immediately after transcription, but rather during long-term inactivation (Franke et al. 1976b). Most of our observations have been made in chromatin material that has been dispersed briefly in media of very low ionic strength, i.e., conditions known to be favorable to the unfolding of nucleosomal structure (Griffith 1975; see also articles by Zama et al. and by Oudet et al., II, both this volume; for related changes in chromatin structure see also Tsanev and Petrov 1976). So we cannot rule out the possibility that the low-salt treatment has selectively altered the nucleosomal arrangement in transcribed chromatin, but not in nontranscribed chromatin. However, we consider this rather unlikely, particularly since we find similar lengths of chromosome loops and matrix and spacer units in preparations made at physiological salt concentrations, though with much less clarity of structural detail (unpubl. obs.). The dimension of the extended chromatin fiber as one sees it in regions of transcribed chromatin indicates that it represents deoxyribonucleoprotein, most likely including the nucleosomal histone complement (cf. McKnight et al., this volume). The presence of histones in transcriptionally active chromatin has also been reported from a variety of biochemical studies (for references, see Pospelov et al. 1975; Weintraub and Groudine 1976; Higashinakagawa et al. 1977), although it is still unclear whether histones occur in the immediate vicinity of the transcriptional complex as such.

The concept of a different conformation in transcribed regions of chromatin also concurs with the demonstrated selective susceptibility of transcribed chromatin to digestion by deoxyribonuclease I (see, e.g., Garel and Axel 1976; Weintraub and Groudine 1976; Gottesfeld 1977). The numerous reports claiming that chromatin of actively transcribed genes can be digested by micrococcal nuclease into nucleoso-

mal units (Brown et al. 1976; Garel and Axel 1976; Kuo et al. 1976; Leer et al. 1976; Mathis and Gorovsky 1976; Piper et al. 1976; Reeves 1976; Reeves and Jones 1976; Tata and Baker 1976; Weintraub and Groudine 1976; Gottesfeld 1977) are not in conflict with the idea of an extended nucleohistone fiber in actively transcribed regions, since this enzyme apparently induces a nucleosomelike cleavage pattern of DNA even in nucleosomes unfolded in very low salt concentrations or in urea; neither are they evidence for the presence of a beaded chromatin arrangement (for references on urea effects, see Carlson et al. 1975; cf. Oudet et al., II; Woodcock and Frado; Zama et al.; all this volume). Rather, the results obtained with micrococcal nuclease suggest that during the unfolding of the nucleosomal state to the extended state, the arrangement of the histones with the DNA is not markedly altered.

Acknowledgments

We thank Mrs. Erika Schmid for valuable technical assistance. This work has been supported by the Deutsche Forschungsgemeinschaft.

REFERENCES

ANDERSON, K. M., H. CHANCE, and N. KADOHAMA. 1975. Separation of transcriptionally active chromatin from less active rat ventral prostate chromatin. *Exp. Cell Res.* **94**: 176.

ANGELIER, N. and J. C. LACROIX. 1975. Complexes de transcription d'origines nucléolaire et chromosomique d'ovocytes de *Pleurodeles waltlii* et *P. poireti* (Amphibiens, Urodèles). *Chromosoma* **51**: 323.

BACHELLERIE, J. P., E. PUVION, and J. P. ZALTA. 1975. Ultrastructural organization and biochemical characterization of chromatin-RNA-protein complexes isolated from mammalian cell nuclei. *Eur. J. Biochem.* **58**: 327.

BALDWIN, J. P., P. G. BOSELEY, E. M. BRADBURY, and K. IBEL. 1975. The subunit structure of the eukaryotic chromosome. *Nature* **253**: 245.

BAUER, H. 1933. Die wachsenden Oocytenkerne einiger Insekten in ihrem Verhalten zur Nuklealfärbung. *Z. Zellforsch.* **18**: 254.

BERGER, S. and H. G. SCHWEIGER. 1975. 80 S ribosomes in *Acetabularia major*. Redundancy of rRNA cistrons. *Protoplasma* **83**: 41.

BERKOWITZ, E. M. and P. DOTY. 1975. Chemical and physical properties of fractionated chromatin. *Proc. Natl. Acad. Sci.* **72**: 3328.

BOUTEILLE, M., M. LAVAL, and A. M. DUPUY-COIN. 1974. Localization of nuclear functions as revealed by ultrastructural autoradiography and cytochemistry. In *The cell nucleus* (ed. H. Busch), vol. 1. p. 3. Academic Press, New York.

BRASCH, K. 1976. Studies on the role of histones H1 (f1) and H5 (f2c) in chromatin structure. *Exp. Cell Res.* **101**: 396.

BROWN, I. R., J. J. HEIKKILA, and N. A. STRAUS. 1976. Organization and transcriptional activity of DNA in brain chromatin subunits. *J. Cell Biol.* **70**: 121a.

CARLSON, R. D. and D. E. OLINS. 1976. Chromatin model calculations: Arrays of spherical ν-bodies. *Nucleic Acids Res.* **3**: 89.

CARLSON, R. D., A. L. OLINS, and D. E. OLINS. 1975. Urea

denaturation of chromatin periodic structure. *Biochemistry* **14**: 3122.

CASE, S. T. and B. DANEHOLT. 1977. Cellular and molecular aspects of genetic expression in *Chironomus* salivary glands. *Int. Rev. Biochem.* (in press).

CHENTSOV, JU. S. and W. JU. POLYAKOV. 1974. *Ultrastruktura kletochnovo jadra (Ultrastructure of the cellular nucleus).* NAUKA Publishers, Moscow.

CREMISI, C., P. F. PIGNATTI, O. CROISSANT, and M. YANIV. 1976. Chromatin-like structures in polyoma virus and simian virus 40 lytic cycle. *J. Virol.* **17**: 204.

DAVIES, H. G. and M. E. HAYNES. 1976. Electron-microscope observations on cell nuclei in various tissues of a teleost fish: The nucleolus-associated monolayer of chromatin structural units. *J. Cell Sci.* **21**: 315.

DAVIES, H. G. and J. V. SMALL. 1968. Structural units in chromatin and their orientation on membranes. *Nature* **217**: 1122.

FAKAN, S. and W. BERNHARD. 1971. Localisation of rapidly and slowly labelled nuclear RNA as visualized by high resolution autoradiography. *Exp. Cell Res.* **67**: 129.

———. 1973. Nuclear labelling after prolonged ³H-uridine incorporation as visualized by high resolution autoradiography. *Exp. Cell Res.* **79**: 431.

FINCH, J. T. and A. KLUG. 1976. Solenoidal model for superstructure in chromatin. *Proc. Natl. Acad. Sci.* **73**: 1897.

FOE, V. E., L. E. WILKINSON, and C. D. LAIRD. 1976. Comparative organization of active transcription units in *Oncopeltus fasciatus. Cell* **9**: 131.

FRANKE, W. W. 1977. Structure and function of nuclear membranes. *Biochem. Soc. Symp.* **42**: 125.

FRANKE, W. W. and U. SCHEER. 1974. Structures and functions of the nuclear envelope. In *The cell nucleus* (ed. H. Busch), vol. 1, p. 219. Academic Press, New York.

FRANKE, W. W., U. SCHEER, H. SPRING, M. F. TRENDELENBURG, and G. KROHNE. 1976a. Morphology of transcriptional units of rDNA. *Exp. Cell Res.* **100**: 233.

FRANKE, W. W., U. SCHEER, M. F. TRENDELENBURG, H. SPRING, and H. ZENTGRAF. 1976b. Absence of nucleosomes in transcriptionally active chromatin. *Cytobiologie* **13**: 401.

GALL, J. G. 1966. Chromosome fibers studied by a spreading technique. *Chromosoma* **20**: 221.

GAREL, A. and R. AXEL. 1976. Selective digestion of transcriptionally active ovalbumin genes from oviduct nuclei. *Proc. Natl. Acad. Sci.* **73**: 3966.

GARIGLIO, P. 1976. Effect of sarkosyl on chromatin and viral RNA synthesis. The isolation of SV 40 transcription complex. *Differentiation* **5**: 179.

GERMOND, J. E., B. HIRT, P. OUDET, M. GROSS-BELLARD, and P. CHAMBON. 1975. Folding of the DNA double helix in chromatin-like structures from simian virus 40. *Proc. Natl. Acad. Sci.* **72**: 1843.

GOTTESFELD, J. M. 1977. Structure of transcriptionally active chromatin. *Philos. Trans. R. Soc. Lond. B* (in press).

GRIFFITH, J. D. 1975. Chromatin structure: Deduced from a minichromosome. *Science* **187**: 1202.

HEITZ, E. 1929. Heterochromatin, Chromocentren, Chromomeren. *Ber. Dtsch. Bot. Ges.* **47**: 274.

———. 1932. Die Herkunft der Chromocentren. *Planta* **18**: 571.

———. 1956. Die Chromosomenstruktur im Kern während der Kernteilung und der Entwicklung des Organismus. In *Chromosomes: Lectures held at the Conference on Chromosomes, Wageningen*, p. 5. N.V. Uitgevers-Maatschappij W.E.Z. Tjeenk Willink-Zwolle, The Netherlands.

HEUMANN, H. G. 1974. Electron microscope observations of the organisation of chromatin fibers in isolated nuclei of rat liver. *Chromosoma* **47**: 133.

HIGASHINAKAGAWA, T., H. WAHN, and R. H. REEDER. 1977. Isolation of ribosomal gene chromatin. *Dev. Biol.* **55**: 375.

KIERSZENBAUM, A. L. and L. L. TRES. 1975. Structural and transcriptional features of the mouse spermatid genome. *J. Cell Biol.* **65**: 258.

KIRYANOV, G. I., T. A. MANAMSHJAN, V. JU. POLYAKOV, D. FAIS, and JU. S. CHENTSOV. 1976. Levels of granular organization of chromatin fibres. *FEBS Lett.* **67**: 323.

KORNBERG, R. D. 1974. Chromatin structure: A repeating unit of histones and DNA. *Science* **184**: 868.

KRIEG, P. and J. R. E. WELLS. 1976. The distribution of active genes (globin) and inactive genes (keratin) in fractionated chicken erythroid chromatin. *Biochemistry* **15**: 4549.

LAIRD, C. D. and W. Y. CHOOI. 1976. Morphology of transcription units in *Drosophila melanogaster. Chromosoma* **58**: 193.

LAIRD, C. D., L. E. WILKINSON, V. E. FOE, and W. Y. CHOOI. 1976. Analysis of chromatin-associated fiber arrays. *Chromosoma* **58**: 169.

LEER, J. C., O. F. NIELSEN, P. W. PIPER, and O. WESTERGAARD. 1976. Isolation of the ribosomal RNA gene from *Tetrahymena* in the state of transcriptionally active chromatin. *Biochem. Biophys. Res. Commun.* **72**: 720.

MALCOLM, D. B. and J. SOMMERVILLE. 1974. The structure of chromosome-derived ribonucleoprotein in oocytes of *Triturus cristatus carnifex* (Laurenti). *Chromosoma* **48**: 137.

MATHIS, D. J. and M. A. GOROVSKY. 1976. Subunit structure of rDNA-containing chromatin. *Biochemistry* **15**: 750.

McKNIGHT, S. L. and O. L. MILLER. 1976. Ultrastructural patterns of RNA synthesis during early embryogenesis of *Drosophila melanogaster. Cell* **8**: 305.

MILLER, O. L. and A. H. BAKKEN. 1972. Morphological studies of transcription. *Acta Endocrinol. Suppl.* **168**: 155.

MILLER, O. L. and B. R. BEATTY. 1969. Extrachromosomal nucleolar genes in amphibian oocytes. *Genetics* (Suppl.) **61**: 134.

MONNERON, A. and W. BERNHARD. 1969. Fine structural organization of the interphase nucleus in some mammalian cells. *J. Ultrastruct. Res.* **27**: 266.

MOTT, M. R. and H. G. CALLAN. 1975. An electron-microscope study of the lampbrush chromosomes of the newt *Triturus cristatus. J. Cell Sci.* **17**: 241.

OLINS, A. L. and D. A. OLINS. 1974. Spheroid chromatin units (ν-bodies). *Science* **183**: 330.

OUDET, P., M. GROSS-BELLARD, and P. CHAMBON. 1975. Electron microscopic and biochemical evidence that chromatin structure is a repeating unit. *Cell* **4**: 281.

PIPER, P. W., J. CELIS, K. KALTOFT, J. C. LEER, O. F. NIELSEN, and O. WESTERGAARD. 1976. *Tetrahymena* ribosomal RNA gene chromatin is digested by micrococcal nuclease at sites which have the same regular spacing on the DNA as corresponding sites in the bulk nuclear chromatin. *Nucleic Acids Res.* **3**: 493.

POSPELOV, V. A., A. A. SOKOLENKO, and G. L. DIANOV. 1975. Investigation of DNA bound to histones in chromatin. *Mol. Biol.* (USSR) **9**: 691.

PUVION-DUTILLEUL, F., A. BERNADAC, E. PUVION, and W. BERNHARD. 1977. Visualization of two different types of nuclear transcriptional complexes in rat liver cells. *J. Ultrastruct. Res.* **58**: 108.

RAE, P. M. M. 1966. Whole mount electron microscopy of *Drosophila* salivary chromosomes. *Nature* **212**: 139.

RATTNER, J. B., A. BRANCH, and B. A. HAMKALO. 1975. Electron microscopy of whole mount metaphase chromosomes. *Chromosoma* **52**: 329.

REEVES, R. 1976. Ribosomal genes of *Xenopus laevis:* Evidence of nucleosomes in transcriptionally active chromatin. *Science* **194**: 529.

REEVES, R. and A. JONES. 1976. Genomic transcriptional activity and the structure of chromatin. *Nature* **260**: 495.

RENZ, M., P. NEHLS, and J. HOZIER. 1977. Involvement of

histone H1 in the organization of the chromosome fiber. *Proc. Natl. Acad. Sci.* (in press).

RICKWOOD, D. and G. D. BIRNIE. 1976. Preparation, characterization and fractionation of chromatin. In *Subnuclear components* (ed. G. D. Birnie), p. 129. Butterworths, London.

RIS, H. 1975. Chromosomal structure as seen by electron microscopy. *Ciba Found. Symp.* **28** (new series): 8.

SCHEER, U. and W. W. FRANKE. 1976. Transcriptional complexes in nucleolar genes. In *Proceedings 6th European Congress on Electron Microscopy* (ed. Y. Ben-Shaul), vol. II, p. 26. TAL International, Israel.

SCHEER, U., M. F. TRENDELENBURG, and W. W. FRANKE. 1973. Transcription of ribosomoal RNA cistrons. *Exp. Cell Res.* **80**: 175.

———. 1975. Effects of actinomycin D on the association of newly formed ribonucleoproteins with the cistrons of ribosomal RNA in *Triturus* oocytes. *J. Cell Biol.* **65**: 163.

———. 1976a. Regulation of transcription of genes of ribosomal RNA during amphibian oogenesis. *J. Cell Biol.* **69**: 465.

SCHEER, U., W. W. FRANKE, M. F. TRENDELENBURG, and H. SPRING. 1976b. Classification of loops of lampbrush chromosomes according to the arrangement of transcriptional complexes. *J. Cell Sci.* **22**: 503.

SCHEER, U., M. F. TRENDELENBURG, G. KROHNE, and W. W. FRANKE. 1977. Lengths and patterns of transcriptional units in the amplified nucleoli of oocytes of *Xenopus laevis*. *Chromosoma* **60**: 147.

SPERLING, L. and A. TARDIEU. 1976. The mass per unit length of chromatin by low-angle X-ray scattering. *FEBS Lett.* **64**: 89.

SPRING, H., U. SCHEER, W. W. FRANKE, and M. F. TRENDELENBURG. 1975. Lampbrush-type chromosomes in the primary nucleus of the green alga *Acetabularia mediterranea*. *Chromosoma* **50**: 25.

SPRING, H., G. KROHNE, W. W. FRANKE, U. SCHEER, and M. F. TRENDELENBURG. 1976. Homogeneity and heterogeneity of sizes of transcriptional units and spacer regions in nuclear genes of *Acetabularia*. *J. Microsc.* (Paris) **25**: 107.

SPRING, H., M. F. TRENDELENBURG, U. SCHEER, W. W. FRANKE, and W. HERTH. 1974. Structural and biochemical studies of the primary nucleus of two green algal species, *Acetabularia mediterranea* and *Acetabularia major*. *Cytobiologie* **10**: 1.

TATA, J. R. and B. BAKER. 1976. Specific release of active nuclear transcriptional units as polynucleosomes by micrococcal nuclease. *J. Cell Biol.* **70**: 219a.

TIEN KUO, M., C. G. SAHASRABUDDHE, and G. F. SAUNDERS. 1976. Presence of messenger specifying sequences in the DNA of chromatin subunits. *Proc. Natl. Acad. Sci.* **73**: 1572.

TRENDELENBURG, M. F. 1977. "Elektronenmikroskopische Darstellung der rDNA Transkriptionseinheiten in nukleolärem Chromatin verschiedener Eukaryonten." Ph.D. thesis, University of Heidelberg, Germany.

TRENDELENBURG, M. F., W. W. FRANKE, and U. SCHEER. 1977. Frequencies of circular units of nucleolar DNA in oocytes of two insects, *Acheta domesticus* and *Dytiscus marginalis*, and changes of nucleolar morphology during oogenesis. *Differentiation* **7**: 133.

TRENDELENBURG, M. F., U. SCHEER, and W. W. FRANKE. 1973. Structural organization of the transcription of ribosomal DNA in oocytes of the house cricket. *Nat. New Biol.* **245**: 167.

TRENDELENBURG, M. F., U. SCHEER, H. ZENTGRAF, and W. W. FRANKE. 1976. Heterogeneity of spacer lengths in circles of amplified ribosomal DNA of two insect species, *Dytiscus marginalis* and *Acheta domesticus*. *J. Mol. Biol.* **108**: 453.

TRENDELENBURG, M. F., H. SPRING, U. SCHEER, and W. W. FRANKE. 1974. Morphology of nucleolar cistrons in a plant cell, *Acetabularia mediteranea*. *Proc. Natl. Acad. Sci.* **71**: 3626.

TSANEV, R. and P. PETROV. 1976. The substructure of chromatin and its variations as revealed by electron microscopy. *J. Microsc.* (Paris) **27**: 11.

VARSHAVSKY, A. J., V. V. BAKAYEV, P. M. CHUMAKOV, and G. P. GEORGIEV. 1976. Minichromosome of simian virus 40: Presence of histone H1. *Nucleic Acids Res.* **3**: 2101.

WEINTRAUB, H. and M. GROUDINE. 1976. Chromosomal subunits in active genes have an altered conformation. *Science* **193**: 848.

WELLAUER, P. K. and R. H. REEDER. 1975. A comparison of the structural organization of amplified ribosomal DNA from *Xenopus mulleri* and *Xenopus laevis*. *J. Mol. Biol.* **94**: 151.

WELLAUER, P. K., I. B. DAWID, D. D. BROWN, and R. H. REEDER. 1976a. The molecular basis for length heterogeneity in ribosomal DNA from *Xenopus laevis*. *J. Mol. Biol.* **105**: 461.

WELLAUER, P. K., R. H. REEDER, I. B. DAWID, and D. D. BROWN. 1976b. The arrangement of length heterogeneity in repeating units of amplified and chromosomal ribosomal DNA from *Xenopus laevis*. *J. Mol. Biol.* **105**: 487.

WELLAUER, P. K., R. H. REEDER, D. CARROLL, D. D. BROWN, A. DEUTCH, T. HIGASHINAKAGAWA, and I. B. DAWID. 1974. Amplified ribosomal DNA from *Xenopus laevis* has heterogeneous spacer lengths. *Proc. Natl. Acad. Sci.* **71**: 2823.

WOLFE, S. L. and J. N. GRIM. 1967. The relationship of isolated chromosome fibers to the fibers of the embedded nucleus. *J. Ultrastruct. Res.* **19**: 382.

WOODCOCK, C. L. F., J. P. SAFER, and J. E. STANCHFIELD. 1976a. Structural repeating units in chromatin. I. Evidence for their general occurrence. *Exp. Cell Res.* **97**: 101.

WOODCOCK, C. L. F., H. E. SWEETMAN, and L. L. FRADO. 1976b. Structural repeating units in chromatin. II. Their isolation and partial characterization. *Exp. Cell Res.* **97**: 111.

WOODCOCK, C. L. F., L. L. FRADO, C. L. HATCH, and L. RICCARDIELLO. 1976c. Fine structure of active ribosomal genes. *Chromosoma* **58**: 33.

ZENTGRAF, H., B. DEUMLING, and W. W. FRANKE. 1969. Isolation and characterization of nuclei from bird erythrocytes. *Exp. Cell Res.* **56**: 333.

ZENTGRAF, H., H. FALK, and W. W. FRANKE. 1975. Nuclear membranes and plasma membranes from hen erythrocytes. IV. Characterization of nuclear membrane attached DNA. *Cytobiologie* **11**: 10.

ZIRKIN, B. R. 1975. The ultrastructure of nuclear differentiation during spermiogenesis in the salmon. *J. Ultrastruct. Res.* **50**: 174.

Structure of rDNA-containing Chromatin of *Tetrahymena pyriformis* Analyzed by Nuclease Digestion

D. J. MATHIS AND M. A. GOROVSKY

Department of Biology, University of Rochester, Rochester, New York 14627

The gene coding for ribosomal RNA (rDNA) in *Tetrahymena* macronuclei is an amplified, extrachromosomal, palindromic dimer capable of replicating independently from the bulk of the macronuclear DNA (Table 1). Engberg et al. (1974b) have described a protocol for starving and refeeding *Tetrahymena pyriformis* (a micronucleate strain, GL), during which time the rDNA is preferentially synthesized. We, and others, have utilized this technique to label *Tetrahymena* rDNA-containing chromatin (rChr) preferentially and to probe its structure by digestion with nonspecific deoxyribonucleases (Mathis and Gorovsky 1976; Piper et al. 1976; Mathis 1977). The advantage of such an approach for studying the properties of chromatin containing a specific, presumably transcriptionally active (see below) gene is that the small DNA fragments produced by these digestions can be studied directly. The disadvantage of this system is that it is impossible to prove that the genes which have been labeled are, in fact, transcriptionally active. In this regard, it should be noted that with the exception of the experiment shown in Figure 1, all of the results described here have also been obtained with cells chased into logarithmic growth (see Methods). It is also possible that genes actively transcribing ribosomal RNAs (ribosomal transcription units) are not typical active genes; their properties could differ from those of nonribosomal transcription units (Foe et al. 1976). Moreover, though we consider it unlikely, the properties of *Tetrahymena* rChr may not be typical of ribosomal transcription units of other organisms.

METHODS

The basic experimental protocol we have used (Mathis and Gorovsky 1976; Mathis 1977) consists of growing cells for six to ten generations in a medium containing ^{14}C-labeled thymidine to label the bulk DNA. Cells are then starved for 20–24 hours in 0.01 M Tris (pH 7.5), refed, and labeled with [3H]thymidine for an appropriate period (usually 60–120 min) after refeeding. DNA extracted from nuclei isolated immediately after [3H]thymidine labeling contains 45–80% of the tritium label in rDNA (see Fig. 1A; Mathis and Gorovsky 1976; Mathis 1977). If cells are chased with a large excess of nonradioactive thymidine after labeling, they resume growth with only slight loss of specificity of the tritium label.

RESULTS

Digestion of Ribosomal Chromatin with Staphylococcal Nuclease

When nuclei labeled as described above are digested with staphylococcal nuclease, the rates and extents of digestion of ^{3}H and ^{14}C are indistinguishable (Mathis and Gorovsky 1976). If, after brief digestion with staphylococcal nuclease, DNA is isolated and analyzed on 1.5% agarose gels, rChr exhibits a periodic structure indistinguishable from that of bulk chromatin. This repeating structure is not found in free rDNA and does not result from protein exchange during nuclease digestion (Mathis and Gorovsky 1976). Nuclei isolated from growing cells give similar digestion patterns, suggesting that actively transcribing rChr exhibits a periodic struc-

Table 1. Properties of *Tetrahymena* rDNA

A. Macronucleus
  1. Amplified
    a. multiple copies/haploid genome[1,2]
    b. extrachromosomal[3,4]
    c. independent replication[5,6]
  2. Structure
    a. m.w. = $\sim 12.5 \times 10^6$ (strain GL) – $\sim 13.5 \times 10^6$ (syngen 1)[7]
    b. palindrome[7,8]
    c. 2 genes/molecule (dimer)[7,8]
    d. 17S coding region toward center of molecule[8]
  3. Transcription unit
    a. 65–70% transcribed; 30–35% untranscribed spacer[9]
    b. promoter towards center of molecule[10]
  4. Length homogeneity[11,12,13]
    a. within a strain of a single species
    b. between strains of a single species

B. Micronucleus
  1. Probably 1 copy/haploid genome[2,14]
  2. Integrated into chromosome[14]
  3. Not a palindrome[14]

[1] Engberg and Pearlman (1972); [2] Yao et al. (1974); [3] Gall (1974); [4] Engberg et al. (1974a); [5] Charret (1969); [6] Engberg et al. (1972); [7] Karrer and Gall (1976); [8] Engberg et al. (1976); [9] calculated assuming a primary transcript of 2.2×10^6 daltons (Eckert et al. 1975); [10] R. Grainger, pers. comm. [11] Yao and Gorovsky (1975); [12] Yao (1975); [13] Karrer (1976); [14] Yao and Gall (1977).

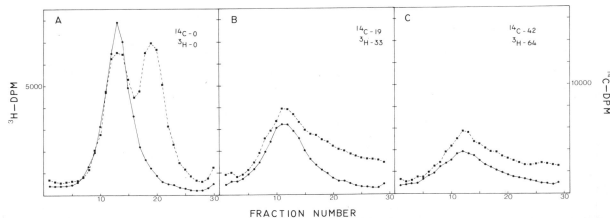

Figure 1. Equilibrium centrifugation of nuclear DNA from starved-refed cells on CsCl-Hoechst density gradients. CsCl (4.5 ml, starting density = 1.638) containing 1 μg Hoechst 33258 per μg DNA was centrifuged in a Beckman 50 rotor for 24 hr at 42,000 rpm followed by 48 hr at 33,000 rpm. Fractions were collected from the top of the gradient using a Buchler Auto-Densiflow. (●——●) [^{14}C]thymidine-labeled DNA; (■——■) [^{3}H]thymidine-labeled (starved-refed) DNA. *(A)* Undigested; *(B)* digested with DNase I for 10 units · min/ml; *(C)* digested with DNase I for 100 units · min/ml. Numbers in upper right hand corner of each figure refer to percentage TCA-soluble for each isotope.

ture. Comparable results have been obtained by others for ribosomal chromatin of *Tetrahymena* (Piper et al. 1976) and of other organisms (Higashinakagawa and Reeder 1975; Reeves 1976; Reeves and Jones 1976).

Prolonged digestion of macronuclei with staphylococcal nuclease results in a limit-digest pattern of DNA fragments indistinguishable from that of similarly digested calf thymus nuclei (Mathis 1977). Tritium-labeled rChr exhibits a different pattern, characterized by an enrichment for higher-molecular-weight fragments (Fig. 2). These differences cannot be attributed to the labeling procedure since they also can be demonstrated in growing cells. Dou-

bling the time of digestion also has little effect on the pattern.

Thus, there are sites in rDNA-containing chromatin that are readily accessible to staphylococcal nuclease and have the same periodicity as in bulk chromatin. However, there appear to be fewer sites available which are susceptible to prolonged digestion. Unfortunately, though the staphylococcal limit digest appears to be a reflection of chromatin organization (Camerini-Otero et al. 1976; Camerini-Otero and Felsenfeld 1977), its precise interpretation is unclear (Noll and Kornberg 1977). Therefore, although the differences between the staphylococcal limit digests of rChr and bulk chromatin suggest

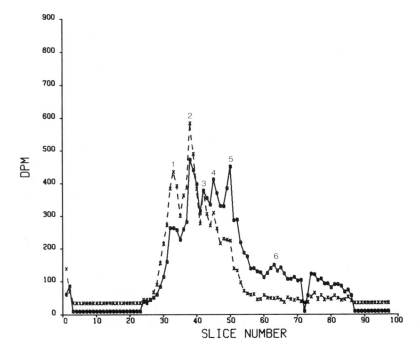

Figure 2. Electrophoretic analysis of staphylococcal nuclease limit digest from nuclei of starved-refed cells. Nuclei were digested for 160 min with 50 units/ml nuclease. DNA digests were electrophoresed on 10% acrylamide gels for 5½ hr. ^{3}H-counts were multiplied by 0.8 to normalize with respect to ^{14}C-counts. (●——●) [^{14}C]DNA; (x- -x) [^{3}H]DNA.

that the structure of (presumably active) rChr is altered, the precise nature of the structural change is uncertain.

Digestion of Ribosomal Chromatin with DNase I

Noll (1974) first demonstrated that pancreatic DNase I cleaves DNA in chromatin to give fragments which differ by ten bases when analyzed in acrylamide gels under denaturing conditions. A ten-base pattern is also obtained when isolated nucleosomal core particles are cleaved with DNase I (Simpson and Whitlock 1976). Therefore, DNase I is a sensitive probe of the internal structure of the nucleosome.

Weintraub and Groudine (1976) and Garel and Axel (1976) have shown that transcriptionally active genes are considerably more sensitive to digestion by DNase I than are inactive genes. This differential sensitivity of active genes to DNase-I digestion could reflect differences in the internal subunit structure; if so, DNA fragments produced by DNase-I digestion of active genes might differ from the ten-base series derived from bulk (mostly inactive) chromatin. Alternatively, increased sensitivity of active genes to DNase-I digestion could reflect some higher-order structural feature, which simply affects the susceptibility of active genes to digestion. In this latter case, DNase-I digestion fragments of active genes should show the ten-base series.

We have examined the kinetics of digestion of rChr by DNase I and have analyzed the digestion products on denaturing polyacrylamide gels. When nuclei from starved-refed (or starved-refed-chased) *Tetrahymena* are digested with DNase I, tritium-labeled DNA digests more rapidly than ¹⁴C-labeled bulk DNA (Fig. 3). If the digestion products are analyzed by equilibrium centrifugation on a CsCl-Hoechst gradient (Fig. 1), tritium label in the riboso-

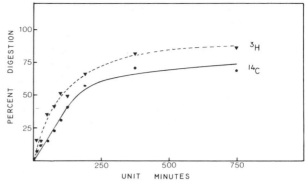

Figure 3. Kinetics of DNase-I digestion of DNA in nuclei of starved-refed cells. Nuclei were digested 0–750 units·min/ml with DNase I at 37°C. At selected points, the percentage of DNA that was TCA-soluble was determined. (●——●) [¹⁴C]DNA; (▼--▼) [³H]DNA.

mal gene appears to be more sensitive to digestion than bulk DNA. However, analysis of DNA at later times of digestion (Fig. 1C) suggests either that some ribosomal genes digest more slowly than others or (more likely) that some part of all the ribosomal genes (perhaps the untranscribed spacer) digests more slowly. We conclude from these studies that *Tetrahymena* rChr labeled by our procedures is preferentially digested by DNase I, suggesting that we are, in fact, studying active (or potentially active; see Foe, this volume) genes.

When rapidly digested ³H-labeled DNA fragments are analyzed on denaturing polyacrylamide gels, they do not show a typical ten-base spacing. Figure 4 shows gel patterns of ³H-labeled DNA (A) and ¹⁴C-labeled DNA (B) after comparable extents of digestion. A ten-base pattern is apparent in ¹⁴C-labeled bulk DNA, but not in ³H-labeled DNA which is highly enriched in ribosomal genes. DNA isolated from more extensively digested ³H-labeled chroma-

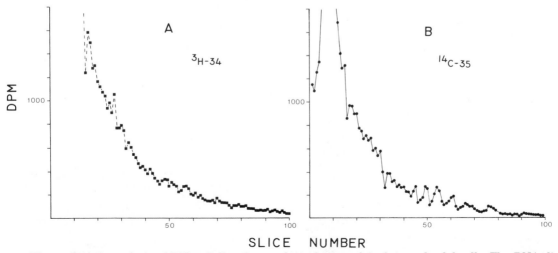

Figure 4. Electrophoretic analysis of DNase-I digestion products from nuclei of starved-refed cells. The DNA digests were denatured and analyzed on 7 M urea–10% acrylamide gels at 22°C, 150 V for 6 hr (Maniatis et al. 1975). *(A)* ³H-labeled (starved-refed) DNA digested to 34% TCA solubility; *(B)* ¹⁴C-labeled (bulk) DNA digested to 35% TCA solubility.

tin does reveal a ten-base pattern (Fig. 5A). This is to be expected since a significant portion of the tritium label is in bulk DNA (see Fig. 1A). Figure 5A also demonstrates that there is no artifact inherent in [³H]thymidine labeling which prevents demonstration of the ten-base pattern. Figure 5B shows that after extensive digestion, a ten-base pattern can be demonstrated in ³H-labeled material even at low count levels. Thus, failure to detect a ten-base pat-

tern in ³H-labeled DNA at earlier times of digestion is not due to difficulties in detecting low levels of radioactivity. We have been able to detect a characteristic ten-base pattern in bulk chromatin of macro- or micronuclei after only 5–10% of the DNA becomes acid-soluble (Gorovsky et al., this volume; M. A. Gorovsky and J. B. Keevert, unpubl.). We consistently fail to detect this pattern in ³H-labeled chromatin from starved-refed (or starved-refed-

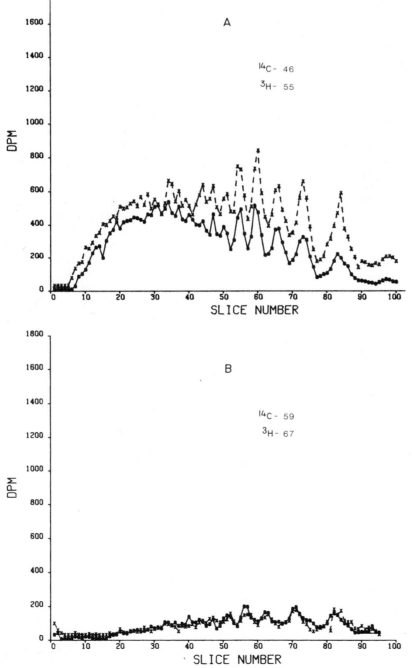

Figure 5. Electrophoretic analysis of DNase-I digestion products from nuclei of starved-refed cells. For experimental details, see Fig. 4. (●——●) [¹⁴C]DNA; (x- -x) [³H]DNA. *(A)* Digestion for 750 units · min/ml; *(B)* digestion for 10,000 units · min/ml. Numbers in upper right hand corner of each figure refer to percentage TCA-soluble for each isotope.

chased) cells until more than 30–40% of the DNA has been digested. Thus, DNA isolated from rapidly digested chromatin of cells labeled with [³H]-thymidine during starvation-refeeding, which consists mainly of ribosomal genes, does not show a ten-base series of fragments characteristic of bulk chromatin. We conclude from this that the structure of the (presumably active) rChr, and not simply its accessibility to DNase I, is distinctly different from that of the majority of (presumably inactive) genes.

DISCUSSION

Brief digestion of rChr with staphylococcal nuclease demonstrates a typical, 200-base-pair periodicity, suggesting that histones are associated with this gene and have a spacing similar to that of histones in bulk chromatin. On the other hand, our studies of staphylococcal nuclease limit and DNase-I digests indicate that rChr is organized differently from bulk chromatin.

It is likely that the absence of a ten-base pattern in DNA isolated from chromatin rapidly digested by DNase I is correlated with the absence of typical particulate (nucleosome) structure. Electron microscopic analyses of ribosomal transcription units in other organisms have demonstrated that they are smooth fibers and that their lengths are much greater than that expected for chromatin compacted into nucleosomes (Hamkalo et al. 1974; Foe et al. 1976; Woodcock et al. 1976; Foe, this volume). Thus, active (or potentially active, see Foe, this volume) ribosomal genes appear to have a periodic structure (assayed by brief staphylococcal nuclease digestion) but not a particulate structure (as viewed in the electron microscope or assayed by DNase-I digestion). Results presented in this volume (Woodcock et al.; Oudet et al., II) also demonstrate that, in vitro, the periodic structure of chromatin can be maintained in the absence of particulate (nucleosome) structure. A more precise interpretation of differences between the staphylococcal nuclease limit and DNase-I digests of Tetrahymena ribosomal chromatin and bulk chromatin requires a greater understanding of the mechanism of action of these enzymes on chromatin. Finally, it remains to be seen whether the digestion patterns observed with these enzymes on Tetrahymena rChr are a unique property of ribosomal transcription units in Tetrahymena, of ribosomal transcription units in general, or of all transcriptionally active genes.

Acknowledgments

This research was supported by U.S. Public Health Service-National Institutes of Health (USPHS-NIH) research grant GM-21793 and a Research Career Development Award to M. G.; D. M. was a USPHS-NIH predoctoral trainee.

REFERENCES

CAMERINI-OTERO, R. D. and G. FELSENFELD. 1977. Supercoiling energy and nucleosome formation: The role of the arginine-rich kernel. Nucleic Acids Res. 4: 1159.

CAMERINI-OTERO, R. D., B. SOLLNER-WEBB, and G. FELSENFELD. 1976. The organization of histones and DNA in chromatin: Evidence for an arginine-rich histone kernel. Cell 8: 333.

CHARRET, R. 1969. L'ADN nucléolaire chez Tetrahymena pyriformis: Chronologie de sa replication. Exp. Cell Res. 54: 353.

ECKERT, W. A., W. W. FRANKE, and U. SCHEER. 1975. Nucleocytoplasmic translocation of RNA in Tetrahymena pyriformis. Exp. Cell Res. 94: 31.

ENGBERG, J. and R. PEARLMAN. 1972. The amount of ribosomal RNA genes in Tetrahymena pyriformis in different physiological states. Eur. J. Biochem. 26: 393.

ENGBERG, J., G. CHRISTIANSEN, and V. LEICK. 1974a. Autonomous rDNA molecules containing single copies of the ribosomal RNA genes in the macronucleus of Tetrahymena pyriformis. Biochem. Biophys. Res. Commun. 59: 1356.

ENGBERG, J., D. MOWAT, and R. E. PEARLMAN. 1972. Preferential replication of the ribosomal RNA genes during a nutritional shift-up in Tetrahymena pyriformis. Biochim. Biophys. Acta 272: 312.

ENGBERG, J., P. ANDERSSON, V. LEICK, and J. COLLINS. 1976. Free ribosomal DNA molecules from Tetrahymena pyriformis GL are giant palindromes. J. Mol. Biol. 104: 455.

ENGBERG, J., J. R. NILSSON, R. E. PEARLMAN, and V. LEICK. 1974b. Induction of nucleolar and mitochondrial DNA replication in Tetrahymena pyriformis. Proc. Natl. Acad. Sci. 71: 894.

FOE, V. E., L. E. WILKINSON, and C. D. LAIRD. 1976. Comparative organization of active transcription units in Oncopeltus fasciatus. Cell 9: 131.

GALL, J. G. 1974. Free ribosomal RNA genes in the macronucleus of Tetrahymena. Proc. Natl. Acad. Sci. 71: 3078.

GAREL, A. and R. AXEL. 1976. Selective digestion of transcriptionally active ovalbumin genes from oviduct nuclei. Proc. Natl. Acad. Sci. 73: 3966.

HAMKALO, B. A., O. L. MILLER, JR., and A. H. BAKKEN. 1974. Ultrastructure of active eukaryotic genomes. Cold Spring Harbor Symp. Quant. Biol. 38: 915.

HIGASHINAKAGAWA, T. and R. H. REEDER. 1975. Purification and preliminary characterization of active ribosomal gene chromatin. Fed. Proc. 34: 2051 (Abstr.).

KARRER, K. M. 1976. "The palindromic structure of ribosomal DNA in Tetrahymena pyriformis." Ph.D. thesis, Yale University, New Haven, Connecticut.

KARRER, K. and J. G. GALL. 1976. The macronuclear ribosomal DNA of Tetrahymena pyriformis is a palindrome. J. Mol. Biol. 104: 421.

MANIATIS, T., A. JEFFREY, and H. VAN DESANDE. 1975. Chain length determination of small double- and single-stranded DNA molecules by polyacrylamide gel electrophoresis. Biochemistry 14: 3787.

MATHIS, D. J. 1977. "Nuclease digestion analyses of rDNA-containing chromatin in Tetrahymena pyriformis." Ph.D. thesis, University of Rochester, New York.

MATHIS, D. J. and M. A. GOROVSKY. 1976. Subunit structure of rDNA-containing chromatin. Biochemistry 15: 750.

NOLL, M. 1974. Internal structure of the chromatin subunit. Nucleic Acids Res. 1: 1573.

NOLL, M. and R. D. KORNBERG. 1977. Action of micrococcal nuclease on chromatin and the location of histone H1. J. Mol. Biol. 109: 393.

PIPER, P. W., J. CELIS, K. KALTOFT, J. C. LEER, O. F. NIELSEN, and O. WESTERGAARD. 1976. Tetrahymena ribosomal RNA gene chromatin is digested by micrococcal nuclease at sites which have the same regular spacing on the DNA as corresponding sites in the bulk nuclear chromatin. *Nucleic Acids Res.* **3**: 493.

REEVES, R. 1976. Ribosomal genes of *Xenopus laevis:* Evidence of nucleosomes in transcriptionally active chromatin. *Science* **194**: 529.

REEVES, R. and A. JONES. 1976. Genomic transcriptional activity and the structure of chromatin. *Nature* **260**: 495.

SIMPSON, R. T. and J. P. WHITLOCK, JR. 1976. Mapping DNase I susceptible sites in nucleosomes labeled at the 5' ends. *Cell* **9**: 347.

WEINTRAUB, H. and M. GROUDINE. 1976. Chromosomal subunits in active genes have an altered configuration. *Science* **193**: 848.

WOODCOCK, C. L. F., L.-L. Y. FRADO, C. L. HATCH, and L. RICCIARDIELLO. 1976. Fine structure of active ribosomal genes. *Chromosoma* **58**: 33.

YAO, M.-C. 1975. "Comparison of the sequence composition of macro- and micronuclear DNA of *Tetrahymena pyriformis.*" Ph.D. thesis, University of Rochester, New York.

YAO, M.-C. and J. G. GALL. 1977. A single integrated gene for ribosomal RNA in a eucaryote, *Tetrahymena pyriformis. Cell* **12**: 121.

YAO, M.-C. and M. A. GOROVSKY. 1975. Length homogeneity of rDNA of the macronuclei of *Tetrahymena pyriformis. J. Cell Biol.* **67**: 467a.

YAO, M.-C., A. R. KIMMEL, and M. A. GOROVSKY. 1974. A small number of cistrons for ribosomal RNA in the germinal nucleus of a eukaryote, *Tetrahymena pyriformis. Proc. Natl. Acad. Sci.* **71**: 3082.

Nucleosome Structure III: The Structure and Transcriptional Activity of the Chromatin Containing the Ovalbumin and Globin Genes in Chick Oviduct Nuclei

M. Bellard, F. Gannon, and P. Chambon

Laboratoire de Génétique Moléculaire des Eucaryotes du CNRS, U. 44 de l'INSERM - Institut de Chimie Biologique, Faculté de Médecine, 67085 Strasbourg Cédex, France

The demonstration that the bulk of eukaryotic chromatin is organized in a repeating subunit structure, the nucleosome (for references, see Kornberg 1977), raised the question as to whether all genes are packaged in a similar manner, since any model of chromatin structure should also account for the selective transcription of only a minor part of the genome. Studies based on micrococcal nuclease digestion of chromatin have suggested that actively transcribed ribosomal RNA genes (rRNA genes) are indeed in the bulk nucleosome structure (Reeves 1976; Reeves and Jones 1976; Mathis and Gorovsky 1976; Piper et al. 1976). However, in most of these studies the interpretation of the results is complicated by the fact that not all of the rRNA genes are necessarily active at a given time. The examination of specific structural gene sequences coding for mRNA provides a sharper focus on the question, although in this case, also, the possible heterogeneity of the cell population under study has to be taken into account when interpreting the results. It was concluded from such studies that micrococcal nuclease digestion does not reveal any distinct feature in transcribing chromatin and that nucleosomes are present on transcribing structural genes (Axel et al. 1975; Lacy and Axel 1975; Garel and Axel 1976; Weintraub and Groudine 1976; Tien Kuo et al. 1976). In contrast, digestion studies with DNase II (Gottesfeld et al. 1975) or DNase I (Berkowitz and Doty 1975; Weintraub and Groudine 1976; Garel and Axel 1976; Levy W. and Dixon 1977) have shown that, in chromatin, actively transcribing genes are in a different structure from the bulk of inactive genes. There is, therefore, an apparent discrepancy between the results obtained from micrococcal nuclease digestion on the one hand and DNase-I or DNase-II digestions on the other. However, all of the micrococcal nuclease digestion studies of specific structural gene sequences were carried out at late times of digestion, and the possibility that chromatin subunits containing active genes could be released faster than those of bulk DNA chromatin was not investigated. It should also be pointed out that the nucleoprotein particles, and not the DNA fragments, were fractionated in these micrococcal nuclease studies, and, therefore, that it was not demonstrated

unequivocally that actively transcribed genes participate in the basic repeat structure characteristic of bulk chromatin. This is obviously an important question in view of the recent finding of the variability of the chromatin DNA repeat length according to cell type (Compton et al. 1976; for other references, see Kornberg 1977).

In the present study we have attempted to answer the following questions: (1) Could micrococcal nuclease digestion reveal specific features of transcribing chromatin? (2) Is the chromatin DNA repeat length of an actively transcribed gene the same as that of the bulk chromatin in a given cell type? The ovalbumin and the globin genes of laying hen oviduct were chosen for this study, since indirect evidence was available showing that the ovalbumin gene is transcribed at a high rate in the tubular gland cells of laying hen oviduct (Palmiter 1975), and we thought that the globin genes should be examples of nontranscribed genes in this tissue. However, since our micrococcal nuclease digestion results cannot be interpreted validly without knowing the exact transcription status of the ovalbumin and globin genes, we have also determined the number of RNA polymerase molecules transcribing the ovalbumin and the globin genes in mature oviduct.

METHODS

Preparation of nuclei. Nuclei from the magnum portion of the oviducts of laying hens were prepared according to the method of Hewish and Burgoyne (1973), slightly modified (last sucrose cushion was 2 M instead of 2.2 M). The yield of hen oviduct nuclei was 50–60% as determined by the assay of Burton (1956). Nuclei of adult hen erythrocytes were isolated as described by Wilhelm and Champagne (1969).

Nuclease digestions. Nuclease digestions were performed on freshly prepared nuclei. Micrococcal nuclease digestion was as follows: nuclei (about 1 mg DNA/ml) were incubated at 20°C in Buffer A (Hewish and Burgoyne 1973) containing 1 mM CaCl$_2$ and 1200 units micrococcal nuclease/ml (Worthington). DNase-I digestion was carried out under the following conditions: nuclei prepared as described above were washed twice with 10 mM Tris-HCl buffer (pH 7.4) containing 10 mM NaCl and 3 mM MgCl$_2$ and suspended in the same buffer at about 1 mg DNA/ml. DNase I (8–10 μg; Sigma, DN-Cl) was added and the incubation run at 37°C.

After various times of incubation with micrococcal nuclease or DNase I, aliquots to which 5 mM EDTA (final conc.) was added were taken for DNA extraction and DNA acid-solubility determination. For acid-solubility determination, NaCl (2 M final conc.) and perchloric acid (1 N final conc.) were added at 0°C. After centrifugation, DNA was determined in the supernatant by UV absorption at 260 nm and in the pellet by the Burton (1956) assay. In general, there was good agreement between the amount of DNA degraded to acid-soluble material as determined directly in the supernatant or indirectly by determining the amount of DNA which remained acid-insoluble in the pellet.

DNA purification after nuclease digestion. DNA was extracted from the aliquots taken at various times of the nuclease digestion. In the case of DNase I, the undigested DNA was purified directly from the aliquots or from the nuclear pellet obtained from the aliquots by centrifuging for 10 minutes at 4000g. Essentially the same results were obtained by each method, no acid-insoluble DNA being released from the nuclei during the centrifugation.

DNA was purified as follows: Sodium dodecyl sulfate (SDS) (0.5% final conc.) and proteinase K (50 μg/ml) were added to the samples, which were incubated overnight at 37°C. After phenol-extraction (1 vol. phenol saturated with 500 mM Tris-HCl, pH 8) followed by chloroform-isoamyl alcohol (24:1) extraction, the DNA was precipitated in the presence of 2% sodium acetate at −20°C by the addition of 2 volumes of ethanol. After centrifugation, the pellet was dried and dissolved at about 2 mg DNA/ml in 10 mM Tris-HCl, pH 8, 1 mM EDTA, 10 mM NaCl. Pancreatic RNase (50 μg/ml) and RNase T$_1$ (2 μg/ml) (previously treated for 20 min at 80°C in 0.15 M NaCl to inactivate DNase) were added and the sample incubated for 1 hour at 37°C. The DNA was then extracted with phenol and chloroform as described above after a 3-hour incubation at 37°C with proteinase K in the presence of 0.5% SDS. After ethanol-precipitation, the DNA was dried by lyophilization.

Fractionation of micrococcal nuclease-digested DNA by agarose gel electrophoresis. About 0.5 mg of purified, digested DNA (in 0.40 ml of 10 mM Tris-HCl, pH 8, 1 mM EDTA, 5% sucrose) was loaded on a vertical agarose slab gel (L × H × W: 165 × 140 × 4 mm; 2% agarose in 40 mM Tris-HCl, pH 7.8, 20 mM sodium acetate, 1 mM EDTA). Electrophoresis was run at room temperature in the same buffer at 20 mA per gel (about 20 V) until the bromophenol blue tracking dye had migrated about 12 cm. DNA was

stained with ethidium bromide (1 μg/ml), the gels were photographed under UV transillumination (negatives were scanned to determine the amount of DNA present in the various bands), and the DNA bands were sliced as described in Results. The DNA was then eluted from the agarose slices (which could be stored at −20°C) essentially as described by Lewis et al. (1975). Typically, about 60–100 g of agarose gel was dissolved at 60°C in 2 volumes of 5 M sodium perchlorate and passed over a hydroxylapatite column (1.5 cm long and 1.5 cm in diameter) at the same temperature. After washing at 60°C with 40 ml of 5 M sodium perchlorate and 40 ml of 50 mM sodium phosphate (pH 6.8) to remove the agarose, the DNA was eluted at 60°C with 400 mM sodium phosphate (pH 6.8) (DNA recovery was usually 70–80%). Phosphate buffer was removed from the DNA peak, located by UV absorption at 260 nm (about 4 ml), by overnight dialysis against 1 liter of 100 mM NaCl, 2 mM EDTA (pH 7.5). After ethanol precipitation, the DNA was dissolved in 1 ml water and dialyzed again against 2 × 1 liter of the same solution until all of the phosphate was removed (assayed as described by Ames 1966). After ethanol precipitation, the DNA pellet was dissolved in H$_2$O at a concentration of about 500 μg/ml, sonicated if required (see below), and denatured by alkaline treatment (0.3 M NaOH, 2 hr at 37°C), which also hydrolyzes any RNA that could be left. After neutralization, the DNA was ethanol-precipitated, dissolved in 20 mM Tris-HCl, pH 7.5, 1 mM EDTA, its concentration was determined according to the method of Burton (1956), and appropriate dilutions for hybridization were prepared. The overall yield of DNA through all of these steps was usually about 50%, and we checked that no specific losses of ovalbumin or globin genes had occurred.

cDNA preparation. All glassware was siliconized and all solutions were sterilized by autoclaving to destroy any contaminating nucleases. Full-length ovalbumin [³H]cDNA with a specific activity of about 60,000 cpm/ng was prepared as follows: Ovalbumin mRNA in the form of immunoprecipitated polysomal RNA (2 mg/ml) (Humphries et al. 1977) was reverse-transcribed in the presence of 50 mM Tris-HCl, pH 7.5, 10 mM MgCl$_2$, 2 mM dithiothreitol, 400 μM dGTP, 400 μM dATP, 60 μM [³H]dCTP (23 Ci/mmole), 60 μM [³H]dTTP (86 Ci/mmole), 0.2 mg/ml actinomycin D, 45 μg/ml oligo(dT)$_{10-12}$, and 263 units/ml of AMV reverse transcriptase (a gift of Dr. Beard). Ovalbumin cDNA with a lower specific activity (about 15,000 cpm/ng) was also prepared with [³H]dCTP only as label. After a 60-minute incubation at 41°C, the reaction mixture was adjusted to 25 mM EDTA, 10 mM Tris-HCl, pH 7.5, 200 mM NaCl and ethanol-precipitated in the presence of tRNA (10 μg/ml). After centrifugation for 1 hour at 16,000g, the pellet was washed at 0°C with 65% ethanol, 60 mM NaCl, dried by lyophilization, and the

RNA was hydrolyzed with 300 mM NaOH (1 hr at 37°C). EDTA (5 mM final conc.) and NaCl (800 mM final conc.) were added, NaOH was adjusted to 200 mM, and the mixture was centrifuged through a 4-ml alkaline sucrose gradient (5–20% in the same solution as that of the cDNA sample, Spinco SW56 rotor, 40,000 rpm for 14 hr at 20°C, nitrocellulose tubes presoaked in 50 mM EDTA). Thirty fractions were collected from the bottom of the tube, aliquots were counted after acid precipitation, and only the fast-sedimenting fractions of the cDNA peak (about 20% of the total radioactivity) were pooled, neutralized with HCl, brought to 200 mM Tris-HCl (pH 7.5) and ethanol-precipitated in the presence of 10 µg/ml of E. coli tRNA (Boehringer). The pellet was dissolved in 10 mM Tris-HCl, pH 7.5, 1 mM EDTA and stored at −20°C. The length of the ovalbumin [³H]cDNA prepared in this way was determined as described previously (Humphries et al. 1977) and was found to be "full-length" (1900–2000 nucleotides).

"Full-length" globin [³H]cDNA with a specific activity of about 15,000 cpm/ng was prepared in a similar way with the following modifications: purified 9S globin mRNA (gifts of Drs. Hendrick and Scherrer) was at 17 µg/ml, dCTP only was tritium-labeled (23 Ci/mmole), and the alkaline sucrose gradient was 5–15%. Again, only the fast-sedimenting fractions (about 50% of total acid-soluble radioactivity) corresponding to "full-length" globin [³H]cDNA were collected, and these were shown to be about 600 nucleotides in length.

Hybridization conditions. Total undigested DNA purified from hen erythrocyte nuclei as described above for digested DNA, purified DNA from DNase I or micrococcal nuclease total digests, and DNA eluted from agarose gels after gel electrophoresis fractionation, were hybridized with ovalbumin or globin "full-length" [³H]cDNA under the following conditions: denatured DNA samples and [³H]cDNA were mixed in a siliconized tube in a final volume of 20 µl containing 20 mM Tris-HCl, pH 8, 5 mM EDTA, 600 mM NaCl, 0.1% SDS. When the amount of DNA in the reaction was lower than 10 µg, 10 µg of sonicated denatured calf thymus DNA was added. The mixture was overlaid with paraffin oil and incubated at 70°C for 96 hours. (We checked in all instances that the hybridization reaction was fully completed during this time). In every case, a control reaction was run without the hen DNA but in the presence of an equivalent amount of calf thymus DNA. At the end of the hybridization period, 700 µl of a solution containing 30 mM sodium acetate (pH 4.5), 300 mM NaCl, 3 mM ZnCl₂, 10 µg/ml of sonicated calf thymus DNA, and 400–600 units of nuclease S₁ (Miles) were added. After 1 hour at 50°C, the hybridized [³H]cDNA was acid-precipitated (10% trichloroacetic acid in the presence of 200 µg of carrier calf thymus DNA), collected on GF/C filters

(Whatman), and dissolved in 0.2 ml soluene (Packard) before counting. Blank values ([³H]cDNA hybridized with calf thymus DNA) were 1–2% of the input cDNA.

Total undigested DNA, total nuclease-digested DNA, and DNA fragments corresponding to the tetramer, pentamer, and larger (see Results) were reduced in size prior to hybridization by sonication (average length below 600 nucleotides). All hen DNA samples were pipetted in the hybridization mixture with identically calibrated micropipettes as used for DNA determination in order to ensure that all titration curves for the specific genes can be validly compared.

Determination of number of RNA polymerase molecules actively transcribing specific genes. The oviducts of chickens which had received a secondary stimulation (Palmiter 1975) by estrogen for 7 consecutive days were excised and used for the preparation of nuclei (Tata et al. 1972), which were then stored in liquid nitrogen until used. Incubations of nuclei (0.2–1 mg DNA/ml) were at 37°C for 40 minutes with 300 mM ammonium sulfate, 0.8–4.0 mg/ml heparin, 3 mM Mn⁺⁺, 1 mM β-mercaptoethanol, 80 mM Tris-HCl, pH 7.5, 13% glycerol, 1 mM ATP, 1 mM GTP, 0.125 mM α-[³²P]CTP (30 cpm/pmole), and 0.125 mM UTP with 30% of the UTP in the form of Hg-UTP prepared by the method of Dale et al. (1975). Under these conditions, approximately 5 ng RNA were synthesized per 1 µg of nuclear DNA. The incorporation of Hg into the in vitro synthesized RNA permits its separation from the endogenous RNA by affinity column (Dale and Ward 1975). To control for possible incomplete separation of the two populations of RNA, nuclei were incubated in parallel with no Hg-UTP in the reaction mixture but with [³²P]CMP-labeled Hg-RNA synthesized from calf thymus DNA using E. coli polymerase and 30% of the UTP as Hg-UTP added to monitor the elution of the endogeneous RNA from affinity columns. After the incubation, the nuclei (control and test) were treated with proteinase K (100 µg/ml) in 0.5% SDS, 5 mM EDTA at 37°C for 1 hour, extracted with phenol-chloroform, and precipitated with ethanol with yeast RNA (15 µg/ml) as carrier. The precipitate was dialyzed against 10 mM Tris-HCl (pH 8) for 2 hours before treating with RNase-free DNase (Brison and Chambon 1976), phenol-chloroform extraction, and ethanol precipitation. Unincorporated nucleoside triphosphates were removed by passage over a Sephadex G-75 column and, after ethanol precipitation, the RNA was dissolved in 1 mM Tris-HCl (pH 8), heated at 65°C for 12 minutes, cooled rapidly to 0°C, adjusted to 50 mM Tris-HCl (pH 8) and 50 mM NaCl, and loaded onto 15-ml columns of thiol-Sepharose 4B (Pharmacia). The sample was cycled through the column at 4°C for 1 hour before elution with 2 column volumes of 50 mM Tris-HCl (pH 8) containing 50 mM NaCl, 2 column volumes of 50 mM Tris-HCl (pH 8) containing 1 M NaCl, 1 column volume of 50 mM Tris-HCl (pH 8) containing 250 mM NaCl, and finally with 50 mM Tris-HCl (pH 8) containing 250 mM NaCl and 100 mM β-mercaptoethanol. The Hg-RNA eluted by the β-mercaptoethanol was ethanol-precipitated immediately and further purified by a second passage on a thiol-Sepharose 4B column under the same conditions. Recoveries from the purification up to the affinity column step were typically 75% (based on the recovery of acid-precipitable ^{32}P), and recovery of the RNA from each affinity column step was approximately 60%. The purified RNA was dissolved in 22 μl of 1 μM β-mercaptoethanol; 2 μl were used for the determination of RNA recovery and 2-, 5-, and 10-μl aliquots were hybridized with 500 cpm (approximately 30 pg) of the appropriate "full-length" [^{3}H]cDNA for 3 days at 70°C in 0.6 M NaCl, 4 mM EDTA, 20 mM Tris-HCl buffer (pH 7.5), 10 μM β-mercaptoethanol, 0.1% SDS (final vol. 20 μl). After hybridization, the samples were digested with S$_1$ nuclease (as above) and then heated at 80°C for 12 minutes in 0.3 M NaOH in the presence of 75 μg of calf thymus DNA. This alkaline digestion was performed to hydrolyze the α-^{32}P-labeled RNA that might complicate the counting of the [^{3}H]cDNA hybridized with the RNA. The hybridized [^{3}H]cDNA was counted as described above. In every experiment, a reference curve was established in parallel, using varying known quantities of the specific mRNA and the appropriate [^{3}H]cDNA. This curve was then used to determine the amount of that messenger present in the sample under study.

RESULTS AND DISCUSSION

Digestion of the Ovalbumin Gene in Oviduct Chromatin

Garel and Axel (1976) have shown previously that the ovalbumin gene is not selectively degraded in oviduct nuclei when 15% of the DNA has been rendered acid-soluble by micrococcal nuclease digestion. This observation is confirmed by the results presented in Figure 1, which shows that no selective degradation of the ovalbumin DNA to unhybridizable fragments could be detected, irrespective of the time of the micrococcal nuclease digestion at which the resistant DNA was titrated for its ovalbumin gene content. Since "full-length" ovalbumin cDNA was used as a probe, the identity of the titration curves for the DNA extracted at various times of the micrococcal nuclease digestion demonstrates that no selective degradation had occurred within the ovalbumin gene. We have also carried out digestions of oviduct nuclei with DNase I since, as an extension of the original observation of Weintraub and Groudine (1976) on globin genes, Garel and Axel (1976) have shown recently that the ovalbumin gene of laying hen oviduct chromatin is selectively cleaved by DNase I. In Figure 2 we have plot-

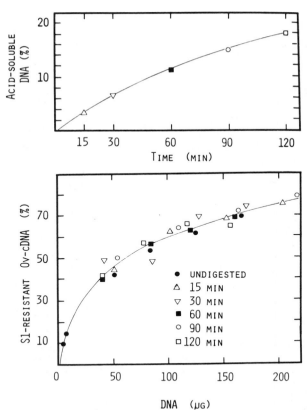

Figure 1. Micrococcal nuclease digestion of the ovalbumin gene in laying hen oviduct nuclei. *(Top)* Conversion of DNA to acid-soluble material (Methods) as a function of the time of incubation. *(Bottom)* Titration curves of the ovalbumin gene in the DNA extracted at various times of digestion with ovalbumin cDNA (Ov-cDNA). Tritiated Ov-cDNA (0.033 ng) was hybridized with increasing quantities of micrococcal nuclease-resistant DNA extracted at various times of the digestion, and duplex formation was monitored by S$_1$ nuclease digestion (Methods). Incubation times of oviduct nuclei with micrococcal nuclease are given in the lower panel.

ted the relative ovalbumin gene content (as estimated by titration with ^{3}H-labeled ovalbumin cDNA) of the DNase-I-resistant DNA as a function of time of DNase-I digestion. It is clear that about 70% of the ovalbumin genes are selectively and rapidly degraded to unhybridizable fragments, whereas 30% of the oviduct ovalbumin genes are degraded at the same rate as the bulk chromatin DNA. It is likely, as previously discussed by Garel and Axel (1976), that these more resistant ovalbumin DNA sequences correspond to the nontubular gland cells, in which the ovalbumin gene is most likely not transcribed (Palmiter 1975). For comparison, we have titrated the globin genes in the DNase-I-resistant DNA. The results (Fig. 2) indicate that the rate of DNase-I digestion of these inactive genes (see below for evidence that globin genes are not transcribed in mature oviduct) parallels the rate of digestion of the bulk chromatin DNA, confirming the original observation of Weintraub and Groudine (1976).

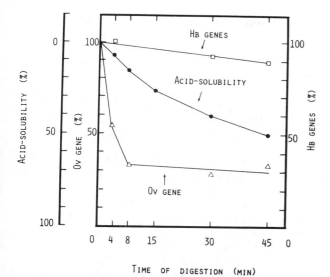

Figure 2. Kinetics of DNase-I digestion of the ovalbumin and globin genes in laying hen oviduct nuclei. The degradation of DNA to acid-soluble material (●) was monitored as described in Methods. The relative amounts of the ovalbumin gene (△) and the globin genes (□) which remained hybridizable at various times of the digestion were determined by titrating these genes in the DNA extracted at various times of the digestion (Methods). Titration (curves [not shown] similar to those in Fig. 6) was carried out in the presence of large cDNA excess. To establish the titration curves, 0.04 ng of [3H]-labeled ovalbumin cDNA or 0.04 ng of [3H]-labeled globin cDNA was annealed with increasing quantities (between 1 and 3 μg) of DNase-I-resistant DNA extracted at various times of digestion as indicated in the figure. 100% corresponds to the ovalbumin- or globin-gene content of the undigested chromatin DNA.

The Ovalbumin Gene Is Cleaved Faster by Microccocal Nuclease Than the Globin Gene in Laying Hen Oviduct Chromatin

Since the above results demonstrate that the ovalbumin gene of oviduct chromatin is not degraded to unhybridizable fragments by microccocal nuclease, we next fractionated the digested DNA in an attempt to answer two basic questions: (1) Is the chromatin DNA repeat length of the actively transcribed ovalbumin gene the same as that of the bulk ovalbumin chromatin? (2) Could microccocal nuclease digestion reveal any distinctive property of the actively transcribing ovalbumin gene chromatin? After various times of microccocal nuclease digestion, the DNA was extracted and the multimeric DNA fragments separated by preparative agarose gel electrophoresis (see Methods). Figure 3 shows the DNA band pattern after 15 minutes of digestion. DNA band 1 (the monomer), band 2 (the dimer), band 3 (the trimer), bands 4 + 5 (tetramer plus pentamer), and DNA larger than the pentamer (> 4 + 5) were cut out and the DNA eluted and purified as described in Methods. Figure 4 shows the titration curves of the ovalbumin gene (left panel) and of the globin genes (right panel) in the various DNA prepa-

rations corresponding to the bands of Figure 3. The titration curves of undigested hen erythrocyte DNA and of the DNA of the total microccocal nuclease digest before fractionation by agarose gel electrophoresis are given for comparison. In all cases, hybridizations were run in the presence of an excess of specific [3H]cDNA under conditions that allow completion of the annealing reaction (Methods). It is clear that the ovalbumin gene sequences are not equally distributed in the various multimeric bands, when compared to bulk chromatin DNA, the monomer being the richest (Fig. 4, left panel). The results (Fig. 4, right panel) indicate that a reverse situation applies to the globin genes, the monomeric and dimeric fragments having the lowest content in globin gene sequences.

There are two possible explanations for these results: (1) The repeat lengths of the microccocal nuclease DNA fragments of the two specific genes are different from that of the bulk of the multimeric DNA fragments; or (2) the DNA repeat length of the ovalbumin gene chromatin is very similar to that of bulk chromatin, but the ovalbumin gene chromatin is, in fact, cleaved faster than the bulk chromatin. The first alternative was excluded by the following results: First, DNA bands 1, 2, and 3 (Fig. 3) were sliced horizontally in their middle to yield upper and lower halves which were then titrated for their ovalbumin gene content (Fig. 5). Clearly, the ovalbumin gene content of each pair of lower

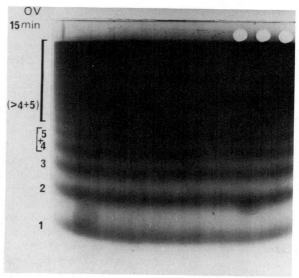

Figure 3. Separation by agarose gel electrophoresis (Methods) of the DNA fragments extracted from laying hen oviduct nuclei incubated for 15 min with microccocal nuclease (Methods). *1, 2, 3, (4 + 5)* and *(>4 + 5)* correspond to the monomer, dimer, trimer, tetramer plus pentamer DNA bands, and to the DNA present above the pentamer band, respectively. The percentage of DNA rendered acid-soluble by the microccocal nuclease digestion was 2.4%, and the percentages of the DNA present in the various bands were as follows: *1,* 5.7%; *2,* 9%; *3,* 11.2%; *(4 + 5),* 15.3%; *(>4 + 5),* 57%.

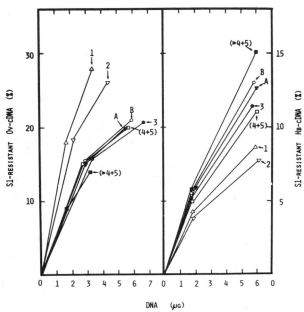

Figure 4. Titration of the ovalbumin *(left)* and the globin *(right)* genes in the various micrococcal nuclease multimeric DNA bands. ^{3}H-labeled ovalbumin cDNA (0.033 ng) or ^{3}H-labeled globin cDNA (0.04 ng) was annealed (Methods) with increasing quantities of DNA eluted from the various DNA bands resolved in agarose gels as shown in Fig. 3. cDNA duplex formation was monitored by S_1 nuclease digestion (Methods). *1* ($\triangle$), *2* ($\triangledown$), *3* ($\bullet$), *(4 + 5)* ($\square$), and *(>4 + 5)* ($\blacksquare$) correspond to the DNA bands of Fig. 3. *A* ($\blacktriangledown$) and *B* ($\bigcirc$) are the titration curves of undigested hen erythrocyte DNA and of the DNA of the micrococcal nuclease digest before fractionation, respectively.

(a) and upper (b) halves is the same, suggesting that, within the sensitivity of the technique, the DNA repeat length of the transcribed ovalbumin chromatin is identical to that of bulk chromatin DNA. This conclusion has been supported further by titration of the ovalbumin gene in the DNA eluted from the interband region between bands 1 and 2 (Fig. 3). Its relative ovalbumin gene content was similar to those of bands 1 and 2 (result not shown). Since there is very little DNA between the bands, this result indicates that the ovalbumin gene sequences occur in a much reduced amount between the bands. These quantitative results have recently been confirmed qualitatively by hybridizing a nick-translated, ^{32}P-labeled ovalbumin cDNA plasmid (Humphries et al. 1977) to the micrococcal nuclease multimeric DNA fragments after their transfer from the agarose gel to a nitrocellulose filter (M. Bellard and R. Breathnach, unpubl.): the labeled ovalbumin multimeric hybrid DNA bands (as revealed by autoradiography) were superimposable on the bulk of the multimeric DNA bands (as revealed by ethidium bromide staining).

To establish the enrichment of monomer and dimer DNA bands (Figs. 3 and 4) in ovalbumin gene sequences, we have carefully titrated the DNA prep-

arations corresponding to the various bands in the presence of a larger excess of ^{3}H-labeled ovalbumin cDNA under conditions where there is a linear relationship between the ovalbumin gene content of the DNA preparation and the percentage of protected cDNA. It can be estimated from the curves shown in Figure 6 that after 15 minutes of nuclease digestion, the monomer DNA band is about 3.5-fold enriched in ovalbumin DNA sequences, whereas the relative ovalbumin gene content of the material above the pentamer is only 70% of that of the DNA of the total unfractionated micrococcal nuclease digest.

From the above results, we conclude that the DNA repeat length of actively transcribed ovalbumin gene chromatin is very similar to that of bulk chromatin; however, the ovalbumin chromatin is more readily attacked by the nuclease than is the bulk chromatin and, more specifically, than is the inactive globin gene chromatin, accounting for the relative decrease in globin genes in the monomeric and dimeric DNA fragments (Fig. 4). In addition, the relative increase in globin gene sequences in the DNA fragments larger than the pentamer, which accounts for about 60% of the total DNA (Fig. 3),

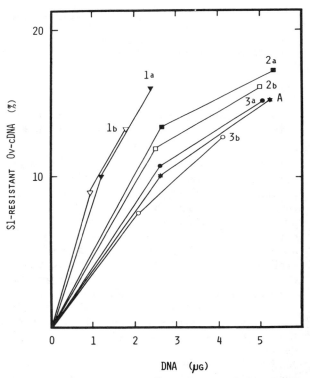

Figure 5. Titration of the ovalbumin gene in the horizontal halves (see text) of the monomer, dimer, and trimer bands shown in Fig. 3. The titration curves were obtained as described in legend to Fig. 4, but using 0.04 ng of [^{3}H]cDNA. *1a* ($\blacktriangledown$), *1b* ($\triangledown$), *2a* ($\blacksquare$), *2b* ($\square$), *3a* ($\bullet$), and *3b* ($\bigcirc$) correspond to the lower *(a)* and upper *(b)* halves of bands 1, 2, and 3. *A* ($\bigstar$) corresponds to the titration curve of undigested hen erythrocyte DNA.

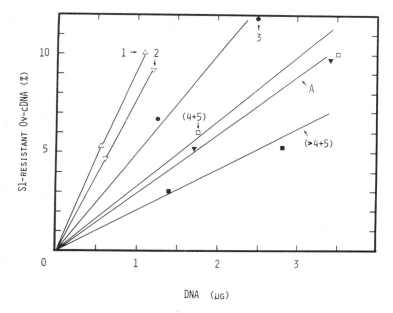

Figure 6. Titration of the ovalbumin gene in the multimeric DNA bands (Fig. 3) in the presence of larger excess of cDNA. The titration curves were obtained as described in the legend to Fig. 4, but at a higher cDNA/DNA ratio. ^{3}H-labeled ovalbumin cDNA (0.04 ng) was used in each annealing reaction. Each point corresponds to the average of three independent reactions which agreed within 15%. The symbols are the same as in Fig. 4.

suggests that other genes, presumably also actively transcribed, are more readily attacked by the nuclease.

To confirm the results obtained after 15 minutes of micrococcal nuclease digestion, similar studies were performed at later times of the reaction, when more of the chromatin DNA had been converted to the lower oligomer bands. Figure 7 shows a gel electrophoresis DNA band pattern after 55 minutes of digestion. At this time, two additional bands, corresponding to the 160- and 140-base-pair bands (Noll and Kornberg 1977), are visible in the lower part of the gel. No attempt has been made to elute them individually, and their DNA was extracted as a large band called "0" (Fig. 7). Figure 8 shows the titration curves of the ovalbumin (left panel) and the globin (right panel) genes in the various DNA bands extracted from the agarose gel represented in Figure 7. In the left panel it is clear that again the ovalbumin gene fragments are not distributed in the various bands in the same proportion as the bulk chromatin DNA fragments. At this point of the micrococcal nuclease digestion, the DNA present in the region called "0" is highly enriched in ovalbumin sequences, whereas the DNA present in the tetramer, pentamer, and higher multimeric bands has a lower ovalbumin gene content relative to the undigested hen DNA or to the DNA of the unfractionated micrococcal nuclease digest. Again, there is a marked difference in the distribution of the ovalbumin and globin genes (Fig. 8, right panel) in the different regions of the agarose gel, the DNA present in the higher multimeric bands larger than the pentamer ($>4+5$) being the richest in globin sequences. A careful titration of the ovalbumin gene DNA in the DNA of the various multimeric bands of Figure 7 is shown in Figure 9. From these curves, one can estimate that the DNA contained in region

"0," which presumably arises from the monomeric bands (Noll and Kornberg 1977), is about 3.5-fold enriched in ovalbumin gene sequences, whereas the DNA of the upper bands (tetramer, pentamer, and larger) have a relatively lower ovalbumin gene content (in the order of 60% of that of the DNA of the total unfractionated micrococcal nuclease di-

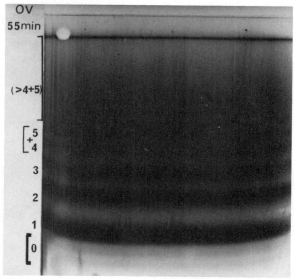

Figure 7. Separation by agarose gel electrophoresis (Methods) of the DNA fragments extracted from laying hen oviduct nuclei incubated for 55 min with micrococcal nuclease. *1, 2, 3, (4 + 5),* and *(>4 + 5)* are the same as in Fig. 3. *"0"* corresponds to the region of the gel below the monomer and contains the 140- and 160-base-pair fragments (see text). The percentage of DNA rendered acid-soluble by the micrococcal nuclease digestion was 9%, and the percentages of the DNA present in the various DNA bands were as follows: *0,* 2.5%; *1,* 20%; *2,* 15%; *3,* 13%; *(4 + 5),* 20%; *(>4 + 5),* 30%.

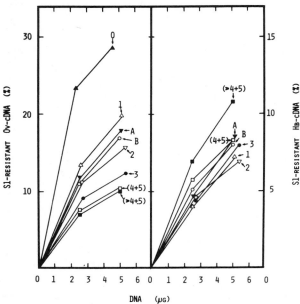

Figure 8. Titration of the ovalbumin *(left)* and globin *(right)* genes in the various micrococcal multimeric DNA bands (Fig. 7). The titration curves were obtained as described in the legend to Fig. 4. The symbols are the same as in Fig. 4 except for the additional *"0"* region (▲), which corresponds to the DNA present below the monomer DNA band (see Fig. 7).

mains to be seen whether the ovalbumin sequences are digested into both the 160- and 140-base-pair fragments as is the bulk monomer (Noll and Kornberg 1977).

The Ovalbumin and the Globin Genes of Hen Erythrocyte Chromatin Are Cleaved at the Same Rate by Micrococcal Nuclease

Since the adult hen erythrocyte genome is transcriptionally silent, one would expect that both ovalbumin and globin chromatin would be cleaved at the same rate as the bulk chromatin, if faster digestion by micrococcal nuclease is a property of actively transcribed genes. One would therefore predict that for both genes there would not be a difference in the gene content of the various multimeric DNA bands isolated by gel electrophoresis after micrococcal nuclease digestion. This is indeed the case, as is shown in Figure 10, where the DNA of the multimeric bands of an erythrocyte micrococcal nuclease digestion (similar to the 55-min oviduct digest shown in Fig. 7 in terms of degree of digestion) was titrated for the globin and the ovalbumin genes. These results demonstrate that the faster cleavage rate of ovalbumin chromatin in laying hen oviduct is a property of the actively transcribing ovalbumin chromatin and not of the ovalbumin gene DNA per se.

Transcriptional Activities of Ovalbumin and Globin Genes in Oviduct

The results presented thus far show a difference in the structural characteristics of chromatin of the ovalbumin gene and the globin genes in mature oviducts. In this tissue, one would anticipate that globin

gest). These results support our previous conclusion concerning the relatively higher micrococcal nuclease susceptibility of an actively transcribed chromatin when compared to the bulk of the chromatin or to the inactive globin gene chromatin. Furthermore, we conclude that this increased susceptibility remains even when the nuclease digestion proceeds further than the monomeric DNA fragment. It re-

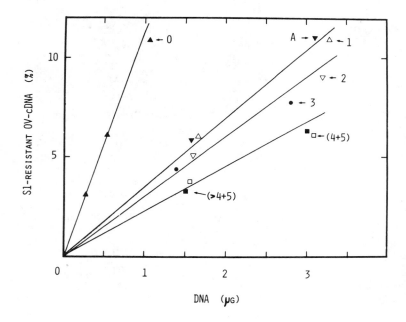

Figure 9. Titration of ovalbumin gene in the multimeric DNA bands of the 55-min micrococcal nuclease digest (Figs. 7 and 8) in the presence of a larger excess of cDNA. The titration curves were obtained with ³H-labeled ovalbumin cDNA as described in the legend to Fig. 6.

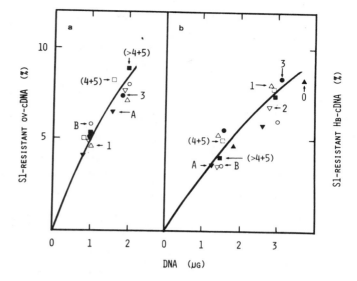

Figure 10. Titration of the ovalbumin *(a)* and the globin *(b)* genes in the various multimeric DNA bands isolated after micrococcal nuclease digestion (90 min) of adult hen erythrocyte nuclei. The percentage of DNA rendered acid-soluble was 6.5%, and the percentages of DNA in the various bands were as follows: *0*, 2.5%; *1*, 14.5%; *2*, 14.6%; *3*, 13%; *(4 + 5)*, 16%; *(>4 + 5)*, 40%. (The definitions of *0*, *1*, *2*, *3*, *(4 + 5)*, and *(>4 + 5)* are the same as in Fig. 7.) The titration curves were obtained as described in the legend to Fig. 4, but the cDNA/DNA ratios were higher. ^{3}H-labeled ovalbumin cDNA (0.04 ng) or ^{3}H-labeled globin cDNA (0.04 ng) was used in the annealing reactions. The symbols are the same as in Fig. 8.

is either synthesized in very small quantities or not at all, whereas many studies have shown that ovalbumin mRNA is present in large amounts (Palmiter 1975). However, finding or the failure to find, at a given time, a specific mRNA in a cell is not, in itself, proof for the transcription or the nontranscription of that gene at that time, as post-transcriptional processing could result either in a drastic increase in its stability in the cytoplasm or in its rapid destruction, possibly even before its release from the DNA template. The interpretation of our results therefore demands a direct demonstration of the transcriptional status of the ovalbumin and the globin genes in mature oviducts. To obtain this information, we have incubated oviduct nuclei in a medium suitable for in vitro RNA synthesis but containing both high salt and heparin. These drastic conditions permit elongation of the RNA chains by RNA polymerase molecules which were engaged in transcription in vivo but which inhibit initiation in vitro completely (Gissinger et al. 1974; Ferencz and Seifart 1975; Cox 1976). The transcription observed in vitro is therefore a direct consequence of the genomic location of RNA polymerase molecules in vivo, since it is unlikely that initiation of RNA chains could take place during the purification of the nuclei. In addition, these incubation conditions are so severe that it is unlikely that enzymes responsible for the post-transcriptional processing of the RNA could be active. The quantity of heparin used, for instance, is known to inhibit general nuclear RNase activity (Cox 1976).

In addition to having conditions where neither reinitiation nor processing of RNA chains occurs, another requirement for these transcription studies is that the only RNA chains which are measured are those which are elongated under these restrictive conditions, i.e., that RNA molecules elongated in vitro are separated from the RNA molecules that had been synthesized previously and released from

the template in vivo. To achieve this aim, we have incorporated Hg-UTP into the RNA synthesized in vitro and subsequently separated the mercurated RNA from that which was not mercurated at all (i.e., fully synthesized in vivo) by passage of the purified RNA over a thiol-Sepharose column (Methods). Although this approach permits a potentially complete separation of the two populations, it also can give rise to artifacts, particularly when RNA chain initiation could occur in vitro (Konkel and Ingram 1977; Zasloff and Felsenfeld 1977). The experiments described here were performed after a rigorous study of the validity of the technique. Important points to note are the use of two successive affinity columns and the heating of the RNA in very low salt before using each column (Methods). Even with these precautions, there remains a background of hybridizable material (in studies on ovalbumin gene transcription) in control nuclei (i.e., incubated without Hg-UTP), which was up to 10% of the test nuclei and for which the results in Table 1 were corrected. Another imperfection in this technique is that the recovery which we find for each affinity column step is approximately 60%. This raises the question of the randomness of the loss. However, since essentially identical results were obtained after two or three passages over the affinity column, we feel that no RNA species selection occurred.

As most published studies of in vivo transcription in the oviduct have been performed with immature chickens that had received a defined treatment with estrogen, we first examined the relative activities of the ovalbumin and globin genes in the nuclei prepared from oviducts of chickens that had received a secondary stimulation of seven successive daily injections of estrogen (experiments 1–7, Table 1). The result pertinent to the first part of this paper is that, in the nuclei prepared after this estradiol treatment, there are at least 15 times more RNA polymerase molecules transcribing the ovalbumin

Table 1. Number of RNA Polymerase Molecules Transcribing the Oviduct Ovalbumin and Globin Genes

| Exp. no. | Type of RNA analyzed | Specific RNA elongated in vitro | | Nuclei in the incubation $\times 10^{-7c}$ | Specific RNA chains elongated in vitro per nucleus[d] |
		pg[a]	molecules $\times 10^{-7b}$		
7-Day secondary stimulation					
1	ovalbumin	320	29.5	4.8	6.1
2[e,f]	ovalbumin	103	9.5	2.4	4.0
3[e,g]	ovalbumin	109	10.0	2.4	4.2
4[e]	ovalbumin	249	23.0	4.8	4.8
5	ovalbumin	505	46.3	10.8	4.3
6[h]	ovalbumin	≤50	≤4.6	7.2	≤0.6
7	globin	≤23	≤3.5	10.8	≤0.3
Laying hen					
8	ovalbumin	3900	360	39.2	9.2
9[i]	ovalbumin	1538	143	19.6	7.3
10[j]	ovalbumin	1538	143	19.6	7.3
11	ovalbumin	2350	217	21.3	10.0

[a] RNA elongated in vitro (Hg-RNA) was purified by its affinity for thiol-Sepharose columns (Methods). The quantity of the specific RNA present in the purified sample was determined by hybridization with the appropriate [³H]cDNA and comparison with a titration curve of defined quantities of this RNA with [³H]cDNA. The values were corrected for endogenous RNA which copurified (determined from a control Hg-UTP-free incubation) and for the recovery of the synthesized RNA (see text and Methods).

[b] Based on a molecular weight value for ovalbumin mRNA of 650,000 (Woo et al. 1975) or 400,000 for globin mRNAs, and Avogadro's Number (6×10^{23} molecules per mole).

[c] Assuming 2.5 pg DNA per oviduct cell (Cox 1976).

[d] See text for factors which might modify these numbers.

[e] The RNA was passed over three thiol-Sepharose columns.

[f] After purification, the RNA was first self-hybridized to a $C_{ro}t$ of 0.064, and then diluted and heated at 100°C for 2 min before hybridization with ovalbumin [³H]cDNA.

[g] As noted in footnote f but not heated before hybridization with ovalbumin [³H]cDNA.

[h] α-Amanitin (1 μg/ml) was present during the incubation.

[i] As noted in footnote f but self-hybridization to a $C_{ro}t$ of 0.014.

[j] As noted in footnote i but not heated before hybridization with ovalbumin [³H]cDNA.

gene than there are RNA polymerase molecules transcribing the globin genes. This figure must be taken as a minimum as there was in fact no significant hybridization observed over the background with the globin cDNA, and the figure quoted is based on the sensitivity of the assay and not on a detected value. As globin cDNA made from total globin mRNA corresponds to several genes (Knochel et al. 1976; Longacre and Rutter 1977) and as hybridization evidence suggests that the ovalbumin gene is a single-copy gene (Palmiter 1975), then there must be at least a 15-fold difference in the transcriptional activities of the ovalbumin and globin genes in the estrogen-stimulated oviduct.

More recently, we have initiated studies on transcription in nuclei prepared from the oviducts of laying hens; these results are presented in Table 1, experiments 8–11. Again, the ovalbumin gene is shown to be actively transcribed and the status of the globin gene is presently being examined (though from a physiological standpoint it is difficult to imagine that it is induced and transcribed in the laying hen when this had not occurred in nuclei from 7-day secondarily stimulated chicks).

In addition to demonstrating that the ovalbumin gene in the oviduct can be considered to be an actively transcribed gene in these studies (and globin an inactive gene), the analysis of transcription of the ovalbumin gene in the nuclei from estrogen-stimulated chicks and laying hen shows that: (1) ovalbumin mRNA synthesis is sensitive to α-amanitin at 1 μg/ml (experiment 6, Table 1), i.e., the in vivo transcription of the ovalbumin gene is most likely catalyzed by RNA polymerase B (Chambon 1975); and (2) the transcription of the ovalbumin gene is asymmetric in vivo. (Experiments 2–3 and 9–10, Table 1, give the same result, showing that self-annealing of the RNA had not occurred.)

Since, as mentioned above, the incubation conditions used for RNA synthesis prohibit chain initiation in vitro, the value for the number of specific RNA chains per nucleus reflects the number of RNA polymerase molecules actively transcribing the ovalbumin gene in vivo. Table 1 (experiments 1–6) shows that there are over four to six ovalbumin mRNA molecules elongated in vitro per nucleus in the 7-day estrogen-stimulated oviduct; therefore, assuming two ovalbumin genes per diploid genome and that both genes are expressed, that corresponds to two to three transcribing RNA polymerase molecules per gene. In the laying hen this figure is clearly higher, with possibly five RNA polymerase molecules actively transcribing each ovalbumin gene in vivo (experiments 8–11, Table 1). For comparison, a calculation based on the size of RNA polymerase (about 120–160 Å; P. Oudet, pers. comm.) and the length of the ovalbumin gene (Palmiter 1975) shows that, theoretically, about 40 enzyme molecules could be tightly packed on the gene. However, based on a variety of indirect quantitative data for ovalbumin mRNA production and degradation, Palmiter (1975) has calculated that a model in which there were 8–12 transcribing RNA polymerase molecules per gene would be in keeping with experimental observations (assuming an RNA chain elongation rate of about 50 nucleotides/sec). Our direct determination of the number of transcribing molecules yields a lower number. However, two to three RNA polymerase molecules per gene for 7-day estrogen-stimulated chicks or five molecules for laying hen is likely to be a minimal number for several reasons. First, it is known that in the oviduct, ovalbumin is synthesized in tubular gland cells, and that at least 20% of the cells are not tubular gland cells in the 7-day stimulated oviduct used in our transcription experiments. Second, we have observed that the use of Hg-UTP instead of UTP leads to a 30% inhibition of RNA synthesis (not shown). In addition, there are also several unknown factors which would tend to give a low estimate of the number of transcribing RNA polymerase molecules per gene: (1) the possible inactivation during the preparation of nuclei of RNA polymerase molecules transcribing in vivo; (2) the failure in vitro to synthesize "full-length" ovalbumin mRNA, since in the calculation of the number of ovalbumin mRNA molecules per nucleus, the molecular weight of full-length mRNA was used (see legend to Table 1) (If the average length of the ovalbumin mRNA synthesized was less than that, then the number of molecules calculated would be too low. In experiment 8 [Table 1], hybridization of the full-length cDNA with the in vitro elongated Hg-RNA reached saturation, and the curve was almost superimposable on that for a standard purified ovalbumin mRNA, suggesting that the RNA synthesized in vitro contained only full transcripts of the ovalbumin gene. However, a convincing proof that this is the case would require the use of hybridization probes specific for each end of the ovalbumin mRNA.); (3) the loss of weakly mercurated RNA molecules due to the fact that only 30% of the UTP present in the reaction mixture was Hg-UTP (to avoid inhibition of eukaryotic RNA polymerases) (Control experiments in which radioactively labeled [^{203}Hg] UTP was used have indicated that RNA with less than 10% of the UMP in the mercurated form was retained by the thiol-Sepharose columns with lower efficiency. Therefore, molecules which were synthesized for the most part in vivo would probably not be detected. Thus again, the calculated number of RNA polymerase molecules per gene is probably an underestimate.); (4) similarly, any traces of RNase activity which would separate the mercurated part of the RNA from the nonmercurated in vivo synthesized part of the RNA molecules would result in an underestimate.

Whereas our determination of the number of transcribing RNA polymerase molecules per ovalbumin gene could be a low estimate for the above reasons, there is one additional possibility to consider that would make it an overestimate: the ovalbumin gene could be transcribed in part by RNA polymerase molecules not initially located on the ovalbumin gene and which would transcribe it by a "read-through" mechanism. Contrary to this suggestion are the facts that transcription of the globin genes is not seen (Table 1, experiment 7) and that the transcription of the ovalbumin gene is asymmetric in these nuclei (Table 1, experiments 2–3, 9–10).

At this point in our analysis of transcription it seems premature to discuss the possible causes or consequences of the apparently greater number of transcribing RNA polymerase molecules per ovalbumin gene in the laying hen as compared to the artificially stimulated tissue. In any case, our data clearly show that the ovalbumin gene is transcribed in the oviduct, whereas the globin genes are probably not expressed in this tissue.

CONCLUSIONS

Recent electron microscopic studies have indicated that the expression of nonribosomal eukaryotic genes could be regulated at the level of transcription, since morphological transcription units delimited by initiation and termination sites have been visualized (Foe et al. 1976; Franke et al. 1976; McKnight and Miller 1976). However, the results of these studies do not exclude that, in fact, any given gene (the globin gene, for instance) could be transcribed in all cell types, but that its RNA transcripts could be very rapidly degraded and not accumulate unless posttranscriptional stabilization would occur. Our present results lead to the conclusion that this possibility is very unlikely, since we did not find elongating globin RNA molecules on the globin genes of mature hen oviduct under experimental conditions where posttranscriptional events are unlikely to occur. Supporting the conclusion that regulation of gene expression in eukaryotic cells occurs, at least in part, at the level of transcription is the recent finding (F. Gannon, unpubl.) that the number of elongating ovalbumin RNA molecules in the nuclei of unstimulated chicken oviduct is much less than that found on the ovalbumin genes of 7-day estradiol-stimulated chicken oviduct or laying hen oviduct. Taking into account the multiple factors that could affect the determination of the number of transcribing RNA polymerases and lower it (see Results and Discussion), our results are in very good agreement with the calculation of Palmiter (1975), who has deduced from the rates of synthesis and degradation of ovalbumin mRNA that 8–12 RNA polymerase molecules should at any time transcribe the ovalbumin gene, to account for ovalbumin mRNA accumulation. In addition to providing, for the first time, a direct estimate of the number of RNA polymerase molecules transcribing the ovalbumin gene, our results demonstrate that the ovalbumin gene is transcribed asymmetrically in vivo by α-amanitin-sensitive class-B RNA polymerase molecules (Chambon 1975).

Our results indicating that after micrococcal nuclease digestion the ovalbumin DNA sequences are found in the same multimeric DNA bands as the bulk chromatin DNA represent the first unequivocal demonstration that an actively transcribed nonribosomal gene is cleaved by micrococcal nuclease in repeating units containing the same DNA length as the repeating unit of bulk chromatin. Since the binding of histones to DNA is responsible for the specific micrococcal nuclease digestion pattern (for references, see Kornberg 1977), it follows that an actively transcribed gene is associated with histones in a manner very similar to that of nontranscribed genes. Does it mean that, in chromatin, the DNA of an actively transcribed gene is packaged in regular nucleosomes, defined as units in which about 200 base pairs of DNA are compacted about sixfold over their extended length (Oudet et al. 1975; see also Kornberg 1977)? Such a possibility seems rather unlikely, since it is difficult to envisage how all of the ovalbumin gene DNA could be compacted in nucleosomes while at least five RNA polymerase molecules (which occupy at least 15% of the gene) transcribe it. In fact, there are several indications from previously published studies that the repeating subunit of transcribed chromatin is different from bulk nucleosomes. Transcribed genes are indeed selectively sensitive to DNase-I digestion (our present results; see introduction for other references), and we have shown here that the chromatin subunits of an actively transcribed gene (the ovalbumin gene in mature oviduct) are released faster by micrococcal nuclease than are those of bulk chromatin, including specific nontranscribed genes (the globin genes). We propose that the modified subunit of transcribing chromatin corresponds to an extended (open) form of chromatin, since we have shown conclusively that a periodic subunit structure, as revealed by nuclease digestion, is retained in extended chromatin in the absence of any particulate nucleosomal structure (Oudet et al., this volume). The presence of actively transcribed genes in such an extended chromatin structure would explain their selective accessibility to nucleases and also the electron microscopic controversy concerning the presence (Foe et al. 1976) or the absence (Franke et al. 1976) of nucleosomes in nonribosomal transcription units, since the in vivo state may be a dynamic equilibrium which can be pushed in one way (the extended form) or the other (the compact nucleosomal form), depending on the conditions used for chromatin isolation and visualization.

Acknowledgments

We thank the Viral Cancer Program, National Cancer Institute (Dr. Beard) for gifts of AMV DNA polymerase, and Drs. Hendrick (Berlin) and Scherrer (Paris) for gifts of globin mRNA. The excellent technical assistance of Mrs. M. Acker, Miss C. Lambs, and Mr. G. Dretzen, and the help of Mrs. B. Chambon in the preparation of the manuscript, are gratefully acknowledged. This work was supported by grants of the CNRS (ATP différenciation, Contract No 2117; ATP internationale, Contract No 2054), of the INSERM (Contract CRL 76.5.099.1), and of the Fondation pour la Recherche Médicale Française. F. G. is supported by a long-term EMBO fellowship.

REFERENCES

AMES, B. N. 1966. Assay of inorganic phosphate, total phosphate and phosphatases. *Methods Enzymol.* **8**: 115.

AXEL, R., H. CEDAR, and G. FELSENFELD. 1975. The structure of the globin genes in chromatin. *Biochemistry* **14**: 2489.

BERKOWITZ, E. M. and P. DOTY. 1975. Chemical and physical properties of fractionated chromatin. *Proc. Natl. Acad. Sci.* **72**: 3328.

BRISON, O. and P. CHAMBON. 1976. A simple and efficient method to remove ribonuclease contamination from pancreatic deoxyribonuclease preparations. *Anal. Biochem.* **75**: 402.

BURTON, K. 1956. A study of the conditions and mechanism of the diphenylamine reaction for colorimetric estimation of deoxyribonucleic acid. *Biochem. J.* **62**: 315.

CHAMBON, P. 1975. Eukaryotic RNA polymerases. *Annu. Rev. Biochem.* **44**: 613.

COMPTON, J. L., M. BELLARD, and P. CHAMBON. 1976. Biochemical evidence of variability in the DNA repeat length in the chromatin of higher eukaryotes. *Proc. Natl. Acad. Sci.* **73**: 4382.

COX, R. F. 1976. Quantitation of elongating form A and B RNA polymerases in chick oviduct nuclei and effect of oestradiol. *Cell* **7**: 455.

DALE, R. M. K. and D. C. WARD. 1975. Mercurated polynucleotides: New probes for hybridization and selective polymer fractionation. *Biochemistry* **14**: 2458.

DALE, R. M. K., E. MARTIN, D. C. LIVINGSTONE, and D. C. WARD. 1975. Direct covalent mercuration of nucleotides and polynucleotides. *Biochemistry* **14**: 2447.

FERENCZ, A. and K. H. SEIFART. 1975. Comparative effect of heparin on RNA synthesis of isolated rat liver nucleoli and purified RNA polymerase A. *Eur. J. Biochem.* **53**: 605.

FOE, V. E., L. E. WILKINSON, and C. D. LAIRD. 1976. Comparative organization of active transcription units in *Oncopeltus fasciatus*. *Cell* **9**: 131.

FRANKE, W. W., U. SCHEER, M. F. TRENDELENBURG, H. SPRING, and H. ZENTGRAF. 1976. Absence of nucleosomes in transcriptionally active chromatin. *Cytobiologie* **13**: 401.

GAREL, A. and R. AXEL. 1976. Selective digestion of transcriptionally active ovalbumin genes from oviduct nuclei. *Proc. Natl. Acad. Sci.* **73**: 3966.

GISSINGER, F., C. KEDINGER, and P. CHAMBON. 1974. Animal DNA-dependent RNA polymerases. X. General enzymatic properties of purified calf thymus RNA polymerases AI and B. *Biochimie* **56**: 319.

GOTTESFELD, J. M., R. F. MURPHY, and J. BONNER. 1975. Structure of transcriptionally active chromatin. *Proc. Natl. Acad. Sci.* **72**: 4404.

HEWISH, D. R. and L. A. BURGOYNE. 1973. Chromatin substructure. The digestion of chromatin DNA at regularly spaced sites by a nuclear deoxyribonuclease. *Biochem. Biophys. Res. Commun.* **52**: 504.

HUMPHRIES, P., M. COCHET, A. KRUST, P. GERLINGER, P. KOURILSKY, and P. CHAMBON. 1977. Molecular cloning of extensive sequences of the *in vitro* synthesized chicken ovalbumin structural gene. *Nucleic Acids Res.* **2**: 2389.

KNOCHEL, W., D. LANGE, and D. HENDRICK. 1976. Molecular weights, poly(A)-content and partial separation of chick globin mRNAs. *Mol. Biol. Rep.* **3**: 143.

KONKEL, D. and V. M. INGRAM. 1977. RNA aggregation during sulfhydryl-agarose chromatography of mercurated RNA. *Nucleic Acids Res.* **4**: 1979.

KORNBERG, R. D. 1977. Structure of chromatin. *Annu. Rev. Biochem.* (in press).

LACY, E. and R. AXEL. 1975. Analysis of DNA of isolated chromatin subunits. *Proc. Natl. Acad. Sci.* **72**: 3978.

LEVY, W. B. and G. H. DIXON. 1977. Renaturation kinetics of cDNA complementary to cytoplasmic polyadenylated RNA from rainbow trout testis. Accessibility of transcribed genes to pancreatic DNase. *Nucleic Acids Res.* **4**: 883.

LEWIS, J. B., J. F. ATKINS, C. W. ANDERSON, P. R. BAUM, and R. F. GESTELAND. 1975. Mapping of late adenovirus genes by cell-free translation of RNA selected by hybridization to specific DNA fragments. *Proc. Natl. Acad. Sci.* **72**: 1344.

LONGACRE, S. S. and W. J. RUTTER. 1977. Isolation of chicken hemoglobin mRNA and synthesis of complementary DNA. *J. Biol. Chem.* **252**: 2742.

MATHIS, D. and M. A. GOROVSKY. 1976. Subunit structure of rDNA-containing chromatin. *Biochemistry* **15**: 750.

MCKNIGHT, S. L. and O. L. MILLER. 1976. Ultrastructural patterns of RNA synthesis during early embryogenesis of *Drosophila melanogaster*. *Cell* **8**: 305.

NOLL, M. and R. D. KORNBERG. 1977. Action of micrococcal nuclease on chromatin and the location of histone H1. *J. Mol. Biol.* **109**: 393.

OUDET, P., M. GROSS-BELLARD, and P. CHAMBON. 1975. Electron microscopic and biochemical evidence that chromatin structure is a repeating unit. *Cell* **4**: 281.

PALMITER, R. D. 1975. Quantitation of parameters that determine the rate of ovalbumin synthesis. *Cell* **4**: 189.

PIPER, P. W., J. CELIS, K. KALTOFT, J. C. LEER, O. F. NIELSEN, and O. WESTERGAARD. 1976. *Tetrahymena* ribosomal RNA gene chromatin is digested by micrococcal nuclease at sites which have the same regular spacing on the DNA as corresponding sites in the bulk nuclear chromatin. *Nucleic Acids Res.* **3**: 493.

REEVES, R. 1976. Ribosomal genes of *Xenopus laevis*: Evidence of nucleosomes in transcriptionally active chromatin. *Science* **194**: 529.

REEVES, R. and A. JONES. 1976. Genomic transcriptional activity and the structure of chromatin. *Nature* **260**: 495.

TATA, J. R., M. J. HAMILTON, and R. D. COLE. 1972. Membrane phospholipids associated with nuclei and chromatin: Melting profile, template activity and stability of chromatin. *J. Mol. Biol.* **67**: 231.

TIEN KUO, M., C. G. SAHASRABUDDHE, and G. F. SAUNDERS. 1976. Presence of messenger specifying sequences in the DNA of chromatin subunits. *Proc. Natl. Acad. Sci.* **73**: 1572.

WEINTRAUB, H. and M. GROUDINE. 1976. Chromosomal subunits in active genes have an altered conformation. *Science* **193**: 848.

WILHELM, F. X. and M. H. CHAMPAGNE. 1969. Dissociation de la nucléoprotéine d'érythrocytes de poulets par les sels. *Eur. J. Biochem.* **10**: 102.

WOO, S. L. C., J. M. ROSEN, C. D. LIARAKOS, Y. C. CHOI, H. BUSCH, A. R. MEANS, and B. W. O'MALLEY. 1975. Physical and chemical characterization of purified ovalbumin messenger RNA. *J. Biol. Chem.* **250**: 7027.

ZASLOFF, M. and G. FELSENFELD. 1977. Use of mercury-substituted ribonucleoside triphosphates can lead to artifacts in the analysis of *in vitro* chromatin transcripts. *Biochem. Biophys. Res. Commun.* **75**: 598.

Structure and Function of the Low-salt Extractable Chromosomal Proteins. Preferential Association of Trout Testis Proteins H6 and HMG-T with Chromatin Regions Selectively Sensitive to Nucleases

B. LEVY W., N. C. W. WONG, D. C. WATSON, E. H. PETERS, AND G. H. DIXON

Department of Medical Biochemistry, Faculty of Medicine, University of Calgary, Calgary, Alberta, T2N 1N4, Canada

Several lines of evidence suggest that the structures of regions of the genome active and inactive in transcription of RNA are different. The observation that *E. coli* RNA polymerase is capable of selective transcription from chromatin templates (Axel et al. 1973; Gilmour and Paul 1973; Biessmann et al. 1976) together with the enhanced susceptibility of transcribed genes to DNase I (Weintraub and Groudine 1976; Garel and Axel 1976; Levy W. and Dixon 1977c) and DNase II (Gottesfeld et al. 1975) make it likely that the heterologous enzymes can recognize some alteration in chromatin structure, correlated with transcriptional activity. Recently, Weintraub and Groudine (1976) have demonstrated that specific globin sequences from erythroid nuclei and also sequences complementary to nuclear RNA are preferentially digested by DNase I. In addition, the same authors have shown that the ovalbumin gene is resistant to digestion in cells in which the gene is inactive. Garel and Axel (1976) have extended these studies by showing that similar treatment of hen oviduct nuclei with DNase I selectively digests the ovalbumin genes that are actively expressed in that tissue.

These observations have led to the conclusion that the conformation of active genes in a given tissue is such that it renders the region of DNA in which they occur particularly accessible to DNase I. In an attempt to broaden these studies, we have asked whether the selective sensitivity to DNase I is a common feature of all DNA sequences which are transcribed in a given tissue. To this end, we have used as a general probe a cDNA representative of the population of cytoplasmic polyadenylated RNA from a terminally differentiating tissue, trout testis.

Our results agree with those previously reported in other systems and show that the DNA sequences involved in the production of polyadenylated mRNA are very susceptible to digestion by DNase I.

In an attempt to search for the distinctive structural features characteristic of active chromatin regions, we digested trout testis nuclei and chromatin with three different endonucleases — DNase I, DNase II, and micrococcal nuclease — and made a comparison of the acid-soluble proteins solubilized by these various nucleases. Each of these enzymes shows a characteristic pattern of digestion in chromatin. DNase I and DNase II exhibit specificity towards transcribed regions (Gottesfeld et al. 1975; Weintraub and Groudine 1976; Levy W. and Dixon 1977c), whereas micrococcal nuclease cleaves the DNA in a random manner in regard to base-sequence (Axel et al. 1975; Tienkuo et al. 1976), with the most accessible spacer regions between core nucleosomes being attacked first (Noll 1974).

In all cases, we observed that proteins of the low-salt-extractable group, or high-mobility group as it has been called by Johns' group, had been preferentially released. A protein which we described some years ago as a characteristic trout histone (histone T), or H6 as it has been called more recently, is the major component solubilized by DNase I, and HMG-T, a protein homologous in properties with calf thymus HMG-1 and -2, is the major protein released upon micrococcal nuclease digestion. Both of these proteins were also preferentially solubilized by DNase II action under conditions similar to those described by Gottesfeld et al. (1975) for solubilizing regions of chromatin enriched in active genes.

The generality of these findings was tested by performing similar experiments in another eukaryotic tissue calf thymus. The results were consistent with those of trout testis and argue for a fundamental role for the low-salt-extractable HMG proteins in the structural organization of chromatin.

METHODS

Trout testis. Testis were collected at a late stage of maturation (October 1974) from freshly killed trout (Dantrout, Brande, Denmark), immediately frozen on dry ice, and stored frozen at −70°C.

Preparation of cytoplasmic polyadenylated RNA. Cytoplasmic RNA was prepared as described previously (Levy W. and Dixon 1977a,b,c,). Polyadenylated RNA was prepared by two cycles of chromatography on oligo(dT)-cellulose, essentially as described by Aviv and Leder (1972). Polyadenylated RNA eluting with H_2O after the second oligo(dT) column was adjusted to 0.2 M ammonium acetate and precipitated overnight with 2 volumes of 95% ethanol at −40°C. The RNA was recovered by centrifugation at 10,000 rpm for 1 hour at −10°C in a Sorvall HB-4 rotor.

Synthesis of complementary DNA. cDNA complementary to polyadenylated cytoplasmic RNA was synthesized and purified as described before (Levy W. and Dixon 1977c).

DNA extraction from trout testis nuclei. DNA for hybridization experiments was prepared as described earlier (Levy W. and Dixon 1977c).

Isolation of DNA fragments resistant to DNase I. After incubation with DNase I, the nuclear suspension was pelleted by centrifugation for 5 minutes at 3000 rpm in a Sorvall SS-34 rotor. The pellet was used as a source of resistant DNA fragments. The DNA was extracted from the pellet as described above.

cDNA/RNA hybridization reactions and cDNA/DNA reassociation reactions were performed as described previously (Levy W. and Dixon 1977c).

The purification of trout testis nuclei and chromatin, the conditions for digestion with the various nucleases, the procedures for acid extraction of proteins and their analysis by gel electrophoresis have been described previously (Levy W. et al. 1977). The techniques employed for the purification and characterization of proteins H6 and HMG-T have been described in detail elsewhere (Watson et al. 1977; G. H. Dixon et al., in prep.).

RESULTS

DNase-I Susceptibility of Chromatin Regions Containing Transcribed Genes

Polyadenylated cytoplasmic RNA was prepared from trout testis at a late stage of spermatogenesis and used as a template for the synthesis of highly labeled complementary DNA by AMV reverse transcriptase. The complexity of this RNA population was determined by hybridization of the cDNA with a vast excess (~2000-fold) of template poly(A⁺) RNA. The results are shown in Figure 1. A detailed analysis of the hybridization kinetics revealed a complexity of about 3.6×10^9 daltons (Levy W. and Dixon 1977b) for this RNA population, corresponding to some 6000 different sequences of average length 600,000 daltons (Levy W. and Dixon 1977a). The cDNA representative of this population of cytoplasmic polyadenylated RNA was used as a probe to examine the sensitivity to DNase I of those genes which were transcribed.

Trout testis nuclei were incubated with DNase I for 10 minutes and the DNase-resistant DNA purified as described in Methods. The DNA obtained in this manner was used as a driver in renaturation experiments with the cDNA representative of the

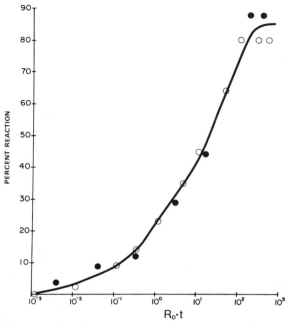

Figure 1. Complexity of the trout testis polyadenylated cytoplasmic RNA. Hybridization between cDNA and a vast excess of polyadenylated RNA, prepared either by chromatography on poly(U)-Sepharose (○) or oligo(dT)-cellulose (●), was performed as described in Methods. The points represent the mean value of six determinations. The mean size of the cDNA used in this hybridization was determined by alkaline sucrose gradient centrifugation to be approximately 7S (ca. 400 nucleotides).

polyadenylated RNA population. As a control, we examined the kinetics of the reaction of DNA extracted from nuclei incubated under identical conditions, but in the absence of DNase I. Figure 2 shows the results of these experiments. The kinetics of the reaction driven by the DNase-resistant DNA is slower than that of the control reaction, as evidenced by a displacement to higher $C_o t$ values. At a $C_o t$ value of 10^4 moles·sec, at which the control reaction has achieved completion of 77% saturation of the cDNA, the reaction of the resistant DNA has reached only 50%. Even though we were unable to achieve higher saturation values than 55%, even after carrying the reaction to $C_o t$ values higher than 20,000, we are not certain whether this represents a real plateau since we observed a decrease in the counts of our controls without S₁ nuclease when the reaction was carried out to $C_o t$ values higher than 2×10^4 moles·sec, indicating that the long period of incubation at 70°C (needed to reach such high $C_o t$ values due to the large analytical complexity of the trout genome) had caused some degradation of our cDNA probe.

Even if this reaction were likely to achieve the same level of saturation as the control reaction when carried out at much higher $C_o t$ values, our data still show that upon DNase I treatment in which only 10% of the total DNA is digested, there is a depletion

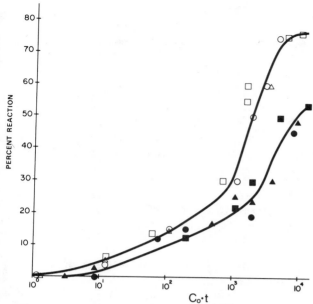

Figure 2. Reassociation kinetics of cDNA with DNase-resistant DNA. The reactions were performed as described in Methods. Each symbol represents a different experiment with a different batch of DNA. (○, △, □) Reassociation of the cDNA with DNA from untreated nuclei; (●, ▲, ■) reaction of the cDNA with DNase-resistant DNA. Both fractions of DNA used to drive the reactions were found to be 300–400 nucleotides in length by polyacrylamide gel analysis and comparison with DNA markers of known length.

teins, which were present in the chromatin, released into the supernatant and also those which remained bound to the DNA resistant to DNase I. Figure 3 shows a starch gel pattern of the acid-soluble proteins from the DNase I soluble and insoluble fractions. The supernatant fraction contained almost exclusively H6 and small amounts of HMG-T. The pellet fraction contained all of the other histones and was depleted in H6.

In every experiment a control was performed in which the reaction was stopped immediately after the addition of the DNase I. Control samples incubated in parallel at 37°C, but in the absence of DNase I, were also included. In both types of controls, we detected the solubilization of traces of protein, as judged by the absorbance of 230 nm, but no band of H6 could be detected when these fractions were separated on gels. In another experiment using polyacrylamide gel electrophoresis, which allowed us to obtain a quantitative densitometer scan, H6 constituted 90% of the additional protein released into the supernatant fraction (data not shown). The identification of individual protein bands from both starch and polyacrylamide gels was by reference to purified protein standards in parallel slots of the same slab gels. Several kinetic experiments were performed in which identical aliquots of nuclei and DNase I were incubated at 37°C for different periods of time, and a progressive release of H6 into the

of sequences coding for cytoplasmic polyadenylated RNA in the DNase-resistant DNA. Unfortunately, the lack of a suitable probe representative of an inactive portion of the trout genome in our system has not allowed us to perform experiments to study the DNase I sensitivity of unexpressed genes. Therefore, we cannot extend our conclusions to say that "only" those genes in an active conformation in trout testis nuclei are preferentially attacked by DNase I. Instead, we can say that, in trout testis, a large proportion of those genes that are actively transcribed, giving rise to cytoplasmic polyadenylated RNA, are very susceptible to DNase I attack.

These studies extend and confirm the results of Weintraub and Groudine (1976) for the globin genes and those of Garel and Axel (1976) for the ovalbumin genes, and reveal that the template-active segments are organized in chromatin in such a way that they are very sensitive to DNase I attack.

Analysis of the Acid-soluble Proteins Solubilized by DNase I

In order to examine the proteins solubilized upon degradation of this limited fraction of the DNA apparently enriched in active genes, we performed DNase I digestions of trout testis nuclei for various periods of time and analyzed the acid-soluble pro-

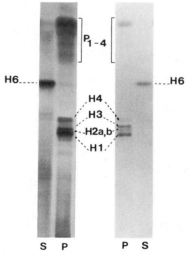

Figure 3. DNase I action on trout testis nuclei. Nuclei were incubated for 10 min with DNase I (to the extent of 10% digestion of the input DNA) and the acid-soluble proteins from the solubilized fraction (S) and insoluble fraction (P) were analyzed by starch gel electrophoresis as described in Methods. *(Left)* Top portion of the gel stained with the arginine-sensitive stain; *(right)* bottom portion of the gel stained with amido black. The identification of the protein bands was accomplished by comparison with the mobilities of purified protein standards run in the same slab gels. P_{1-4}, protamines; H6, component T, protein H6; H4, histone 4; H3, histone 3; H2a,b, histones 2A and 2B; H1, histone 1.

supernatant fraction was consistently observed. None of the nucleosomal histones or H1 were released. Identical results were also obtained when the same experiments were performed with purified chromatin rather than nuclei (Fig. 4).

Purification and properties of protein H6. Due to its high mobility on starch and polyacrylamide gels (it migrates faster than H4, about midway between it and protamine), H6 was at first thought to be a new histone characteristic of trout tissues (Wigle and Dixon 1971). Initially it was called histone T and later, after the revolution in histone nomenclature, H6. However, its composition was unusual (Table 1): the aromatic amino acids histidine, tyrosine, phenylalanine, and tryptophan were absent, as were the sulfur-containing ones, cysteine and methionine; and there was no isoleucine. The most frequent amino acids were Ala, Pro, and Lys, reminiscent of the composition of H1, but there was also a high content of aspartic and glutamic acids (or their amides), more characteristic of a nonhistone protein. The N-terminal sequence of 29 residues was determined by Huntley and Dixon (1972) and although quite basic, was clearly different from that of any other single histone, although short homologies with several other histones were present. We now know the rest of the sequence of H6; it shares many properties with a recently described calf thymus HMG protein, HMG-17 (Fig. 5a,b), sequenced by Walker

Table 1. Properties of Trout Testis H6 (Histone T)

1. Extracted with 0.35 M NaCl or 5% TCA
2. Unusual amino acid composition—no histidine, tyrosine, phenylalanine, tryptophan, methionine, cysteine (or cystine), or isoleucine.
3. Molecular weight 7231; 72 residues
4. $Asp_5Thr_1Ser_4Glu_4Pro_7Gly_5Ala_{19}Val_3Leu_1Lys_{19}Arg_4$

et al. (1976b). There are major regions of homology in the N-terminal regions of these two proteins (positions 1–4 in both proteins and positions 10–36 of H6 with 18–45 of HMG-17) and also in the C-terminal region (47–53 and 67–72 of H6 with 69–75 and 84–89, respectively, of HMG-17). It should also be noted that H6 is adenosine diphosphoribosylated probably at the glutamyl residue at position 48 (Fig. 5b).

As will be described below, HMG-17 is also released preferentially upon digestion of calf thymus nuclei with DNase I. This implies functional, as well as structural, similarities between these proteins.

DNase II Action on Trout Testis Chromatin

It has been established that, under defined experimental conditions, DNase II preferentially attacks a minor fraction of chromatin DNA; the amount of DNA solubilized is variable depending upon the source of the chromatin but corresponds closely to the template activity of the particular chromatin for RNA synthesis as measured with *E. coli* RNA polymerase (Gottesfeld et al. 1975). If the structural feature recognized in chromatin by both DNase I and DNase II were the same and also if the preferential association of H6 with DNase I-sensitive regions in chromatin was in any way related to transcriptional activity, we might also expect to find an enrichment in the content of H6 in the chromatin fraction solubilized by DNase II. Accordingly, we fractionated chromatin into DNase II soluble (S_1) and insoluble (P_1) fractions, using a slight modification of the original procedure of Gottesfeld et al. (1975). The acid-soluble proteins obtained from fractions S_1 and P_1 were analyzed by starch gel electrophoresis (Fig. 6). S_1 was clearly enriched ($\sim$2-fold) in H6 and $\sim$1.5-fold in HMG-T (Levy W. et al. 1977), but the result was not as clear cut as with DNase I since both S_1 and P_1 contained qualitatively the same protein bands, i.e., all of the histones plus protamines, H6, and HMG-T. Therefore, we conclude that DNase II can also solubilize selectively a fraction of trout testis chromatin enriched in H6 (and, to a lesser extent, HMG-T).

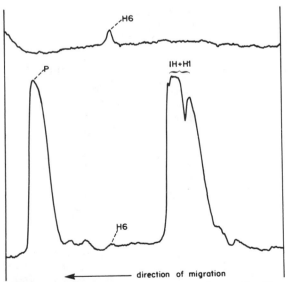

Figure 4. DNase I action on trout testis chromatin. Chromatin was incubated for 30 min with DNase I, exactly as indicated for nuclei in the Methods section. Equal amounts of A_{230} of acid-soluble proteins derived from the soluble fraction *(upper line)* and insoluble fraction *(lower line)* were analyzed on starch gels, stained with amido black. The gels were photographed and the negatives scanned with a Joyce Loebl microdensitometer. P, protamines; H6, protein H6; IH + H1, inner histones plus histone 1.

Micrococcal Nuclease Action on Trout Testis Nuclei

To investigate whether the enrichment in H6 of chromatin fractions selectively sensitive to DNases

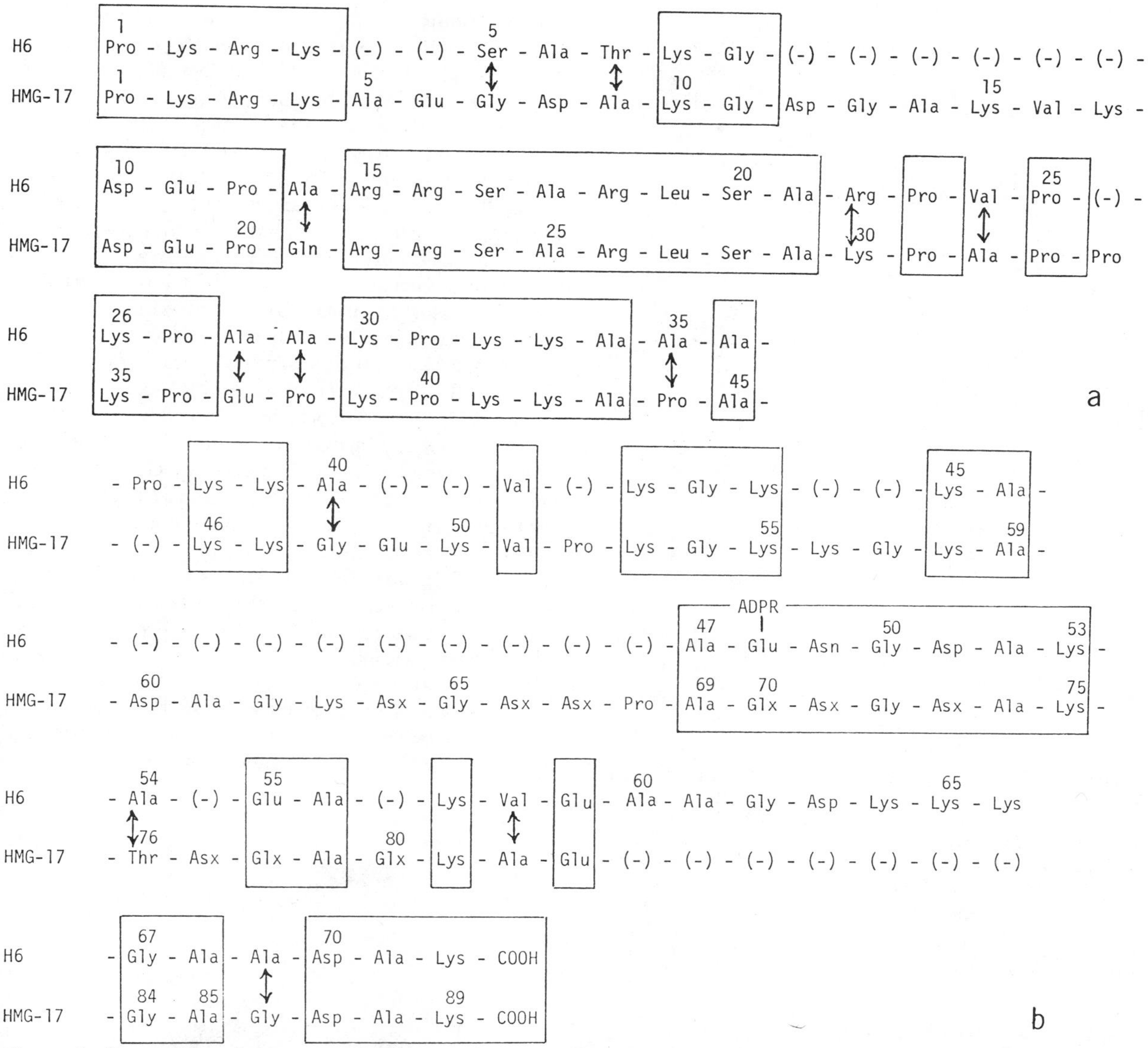

```
          1                    5
H6      Pro - Lys - Arg - Lys - (-) - (-) - Ser - Ala - Thr - Lys - Gly - (-) - (-) - (-) - (-) - (-) - (-) -
          1                5               ↕           ↕      10
HMG-17  Pro - Lys - Arg - Lys - Ala - Glu - Gly - Asp - Ala - Lys - Gly - Asp - Gly - Ala - Lys - Val - Lys -

          10                                    15                      20
H6      Asp - Glu - Pro - Ala - Arg - Arg - Ser - Ala - Arg - Leu - Ser - Ala - Arg - Pro - Val - Pro - (-) -
                         ↕                          25                  ↓30          ↕
HMG-17  Asp - Glu - Pro - Gln - Arg - Arg - Ser - Ala - Arg - Leu - Ser - Ala - Lys - Pro - Ala - Pro - Pro

          26                30                35
H6      Lys - Pro - Ala - Ala - Lys - Pro - Lys - Lys - Ala - Ala - Ala -
                    ↕     ↕                                ↕
          35                40                     45
HMG-17  Lys - Pro - Glu - Pro - Lys - Pro - Lys - Lys - Ala - Pro - Ala -
                                                                                                              a

                     40
H6      - Pro - Lys - Lys - Ala - (-) - (-) - Val - (-) - Lys - Gly - Lys - (-) - (-) - Lys - Ala -
                46        ↕           50                                55                      59
HMG-17  - (-) - Lys - Lys - Gly - Glu - Lys - Val - Pro - Lys - Gly - Lys - Lys - Gly - Lys - Ala -

                                                                    ┌─────── ADPR ───────┐
                                                                    47    |       50            53
H6      - (-) - (-) - (-) - (-) - (-) - (-) - (-) - (-) - (-) -   Ala - Glu - Asn - Gly - Asp - Ala - Lys -
          60                          65                            69    70                      75
HMG-17  - Asp - Ala - Gly - Lys - Asx - Gly - Asx - Asx - Pro -   Ala - Glx - Asx - Gly - Asx - Ala - Lys -

          54          55                                  60
H6      - Ala - (-) - Glu - Ala - (-) - Lys - Val - Glu - Ala - Ala - Gly - Asp - Lys - Lys - Lys
          ↓76                      80                ↕
HMG-17  - Thr - Asx - Glx - Ala - Glx - Lys - Ala - Glu - (-) - (-) - (-) - (-) - (-) - (-) - (-)

          67                70
H6      - Gly - Ala - Ala - Asp - Ala - Lys - COOH
                         ↕
          84    85                89
HMG-17  - Gly - Ala - Gly - Asp - Ala - Lys - COOH
                                                                                                              b
```

Figure 5. Comparison of the amino acid sequences of trout testis protein H6 (G. H. Dixon et al., unpubl.) and calf thymus HMG-17 (J. M. Walker et al. 1976 and unpubl.). *(a)* N-terminal regions; *(b)* C-terminal regions.

I and II could also be observed in those regions of chromatin most accessible to micrococcal nuclease, intact trout testis nuclei were digested with this enzyme. Micrococcal nuclease does not show selective specificity towards chromatin regions enriched in template activity but digests the chromatin in a non-sequence-specific manner (Axel et al. 1975), with the accessible spacer regions between nucleosomes being attacked first (Noll 1974). The first supernatant fraction after digestion under the present conditions is equivalent to that of the DNase I experiments. The major protein released into this soluble fraction was HMG-T (Fig. 7). This nonhistone protein contains two cysteine residues and previously has been observed to convert upon standing to two disulfide forms, an intramolecular disulfide, HMG-T′, which

migrates faster in starch gels, and a dimer which migrates more slowly (Watson et al. 1977). In the portion of the gel stained with the arginine-sensitive stain, we could also observe a small release of H6 at later time points, accompanied by the release of the nucleosomal histones (data not shown).

After the removal of the first supernatant, the nuclear pellet was disrupted by homogenization in 0.2 mM EDTA. This procedure has two effects: it breaks the nuclear membrane, and allows the release into the supernatant of those proteins which were also released by the enzyme but which remained insoluble, due probably to their nonspecific aggregation to the chromatin pellet promoted by the divalent cations present in the solution. After treatment of digested nuclei with EDTA, it was apparent

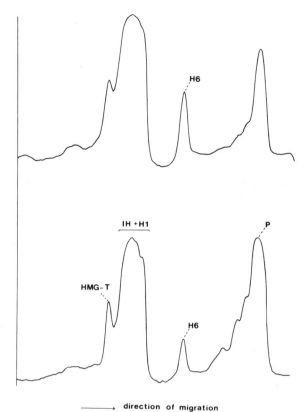

Figure 6. DNase II action on trout testis chromatin. Chromatin was incubated for 5 min at 24°C with DNase II, as described in Methods. Equal amounts of acid-soluble proteins obtained from the soluble fraction S₁ *(upper line)* and the insoluble fraction P₁ *(lower line)* were analyzed on starch gels, stained with amido black. The gels were photographed and the negatives scanned on a Joyce Loebl microdensitometer. P, protamines; H6, component T; HMG-T, component R; IH + H1, inner nucleosomal histones plus histone 1.

that several other chromatin proteins had been released from the chromatin but had remained bound to the nuclear pellet. This group included ~60% of the H6, most of the remainder of the HMG-T, together with substantial quantities of the nucleosomal histones and traces of protamine. The insoluble pellet after EDTA treatment, representing chromatin resistant to micrococcal nuclease, contained nucleosomal histones, the bulk of the protamines, and only traces of H6 and HMG-T. Thus, in contrast to DNases I and II, the action of micrococcal nuclease leads to the preferential solubilization of HMG-T. However, a substantial portion of the chromatin is also solubilized, as can be clearly observed when purified chromatin is employed as a substrate (Fig. 8).

Purification and Characterization of HMG-T

HMG-T was first observed by Marushige and Dixon (1971) in 0.35 M sodium chloride extracts of trout testis chromatin and termed protein R. That this nonhistone chromosomal protein is not a degra-

dation product of any of the five major histone fractions of trout testis is established by several lines of evidence. It possesses a molecular weight of approximately 28,700 based upon the SDS-polyacrylamide method (Weber and Osborn 1969; Laemmli 1970), thereby making it a substantially larger molecule than any of the known histone fractions. HMG-T also possesses an unusually high acidic amino acid content, whereas the known histone fractions are predominantly basic in nature.

The amino acid composition of HMG-T shows that, like the histones, it contains approximately 22% basic amino acids; but in complete contrast to the histones, it also contains 20.5% of acidic amino acids, for a total of 42.5% charged amino acids (Watson et al. 1977). This composition is similar to that reported by Goodwin and Johns (1973) for two proteins extracted with 0.35 M sodium chloride from calf thymus chromatin, HMG-1 and HMG-2. These proteins both show high contents of both acidic (27–29%) and basic (26–27%) amino acids; thus, the trout testis protein HMG-T shows a generally similar distribution of amino acids to the calf thymus proteins. There are fairly large differences in detail, however (Watson et al. 1977). Table 2 shows a direct comparison of the sequence of the amino-terminal region of trout testis HMG with HMG-1 and HMG-2 extracted from calf thymus chromatin as described by Goodwin and Johns (1973).

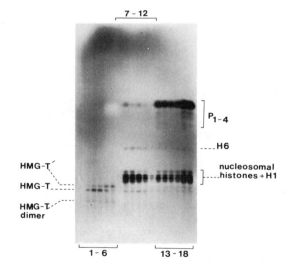

Figure 7. Micrococcal nuclease action on trout testis nuclei. Nuclei were incubated at 37°C with micrococcal nuclease for various periods of time, as indicated in Methods. Acid-soluble proteins from fractions S₁, S₂, and P₁ were analyzed by starch gel electrophoresis. Slots 1, 2, 3, 4, 5, 6 represent the acid-soluble proteins present in fraction S₁ after 0, 2, 5, 10, 30, and 65 min incubation with the nuclease, respectively. Slots 7–12 show the acid-soluble proteins present in fraction S₂ after incubation with the enzyme for the time periods indicated above. Slots 13–18 show the acid-soluble proteins that remain bound to the insoluble fraction P₁ after micrococcal nuclease treatment for the time periods indicated above. The identification of individual protein bands was performed as described in the legend to Fig. 3.

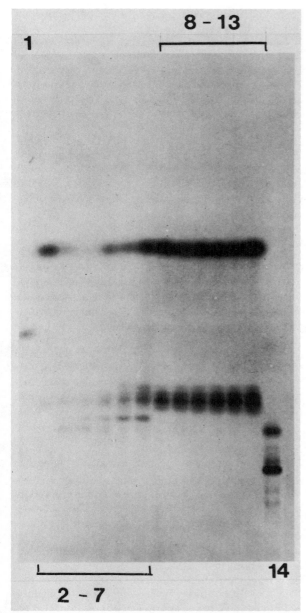

Figure 8. Micrococcal nuclease action on trout testis chromatin. Chromatin was incubated at 37°C with micrococcal nuclease for 0, 5, 10, 20, 45, and 90 min and the acid-soluble proteins analyzed as described in the legend to Fig. 6. *(Left to right)* Slot 1 is a standard sample of H6; slots 2–7 represent the acid-soluble proteins appearing in the supernatant after incubation with the enzymes for the time periods given above. Slots 8–13 represent the acid-soluble proteins remaining bound to the insoluble pellet after digestion. Slot 14 is an HMG-T standard.

Table 2. Comparison of NH₂-terminal Sequences of Trout Testis and Calf Thymus HMGs

Residue no.	Trout HMG	C.T. 1	C.T. 2
1.	Pro	Gly	Gly
2.	Gly	Lys	Lys
3.	Lys	Gly	Gly
4.	Asp	Asp	Asp
5.	Pro	Pro	Pro
6.	Asn	Lys	Asp
7.	Lys	Lys	Lys
8.	Pro	Pro	Pro
9.	Lys	Arg	Arg
10.	Gly	Gly	Gly
11.	Lys	Lys	Lys
12.	Thr	Lys	Lys
13.	Ser	Ala	Ala
14.	Ser	Ser	Ala
15.	Ser	Tyr	Tyr
16.	Ala	Ala	Ala
17.	Phe	Arg	Phe
18.	Phe	Phe	Phe
19.	Val	Val	Val
20.	Ala	Glu	Gly
21.	Val	Thr	Thr
22.	Arg	Arg	Arg
23.	Arg	Arg	Arg
24.	Glx	Glu	Ala
25.	Glx	—	Glu
26.	His	—	—

Brackets indicate sequence homologies.

by Walker et al. (1976a). This conclusion is based on the following similarities: the method of extraction, electrophoretic mobility, amino acid composition, and the homology of amino acid sequence in the N-terminal region of 25 residues. In a study in progress, we have determined 80–90% of the total amino acid sequence of HMG-T, so that when further sequence analyses of calf thymus HMG-1 and -2 become available, further extension of the N-terminal homology may become possible.

To investigate whether the preferential location of proteins of the HMG group in chromatin regions selectively sensitive to nucleases could also be observed in other eukaryotes, we performed experiments similar to those described above using calf thymus. Our results were similar to those in trout testis; HMG-17, the calf thymus protein analogous to trout H6, was the major protein solubilized by DNase I, but small amounts of HMG-1 and -2 and H1 were also solubilized (data not shown). On the other hand, the action of micrococcal nuclease on calf thymus nuclei led to a solubilization of HMG-1, HMG-2, and H1, implying that these proteins might be located in the linker region between core nucleosomes (B. Levy W. and G. H. Dixon, in prep.).

DISCUSSION

We have examined the distribution of acid-soluble proteins in localized regions of chromatin from rainbow trout testis by selective digestion with the endonucleases DNase I, DNase II, and micrococcal nu-

Homologies between the sequences of trout testis and calf thymus proteins occur in short sequences throughout the amino-terminal region and have been carefully analyzed in a previous report (Watson et al. 1977).

HMG-T clearly belongs to the HMG class of non-histone chromosomal proteins previously described by Goodwin et al. (1973) and partially characterized

clease. Each of these enzymes shows a characteristic pattern of digestion of chromatin. Micrococcal nuclease cleaves DNA at sites regularly spaced along the repeating chromatin subunit structure (Noll 1974) in a nonspecific manner with regard to base sequence (Axel et al. 1975). At early times of digestion, the cleavage is restricted to the DNA located in the spacer regions between subunits (Noll 1974). In contrast, with DNase I the DNA within the nucleosomes is as accessible as the DNA in the spacer regions (Burgoyne et al. 1976). Under our experimental conditions, DNase I preferentially digests the DNA of regions of the chromatin that are transcribed in vivo (Weintraub and Groudine 1976; Garel and Axel 1976; Levy W. and Dixon 1977c). DNase II also shows selectivity towards transcribed regions and has been employed to fractionate chromatin into active and inactive regions (Gottesfeld et al. 1975).

The experiments with DNases I and II, and also those using sheared chromatin, suggest strongly that H6 is preferentially located in the domain of active chromatin. It is interesting to note that H6 is the only major acid-soluble protein solubilized by DNase I action which can be detected in our gel electrophoresis system. DNase II, on the other hand, solubilizes a substantial portion of the nucleosomal histones together with H6, HMG-T, and traces of protamine. However, H6 is enriched by ~2-fold and HMG-T is enriched by ~1.5-fold in this fraction as compared with the insoluble fraction (Levy W. et al. 1977). These findings suggest that the sites of cleavage within the chromatin may be different for DNase I and DNase II and also that DNase I may cleave the DNA preferentially at regions in intimate contact with H6, thus accounting for its rapid release.

The selective solubilization of H6 by DNase I, but not by micrococcal nuclease, in turn suggests that it might be bound to the nucleosome core. If so, it is possible that this protein might be located on the outside of the nucleosome, as judged by its unusual amino acid composition, characterized by the lack of a hydrophobic core and by an overall high charge density. In addition, protein H6 is also highly modified by poly-ADP ribosylation, a modification which so far has only been described for outer nucleosomal proteins like H1 and protamine (Wong et al. 1977).

The selective and rapid solubilization of HMG-T by micrococcal nuclease suggests that it may be located in the linker region between nucleosomes. Partial amino acid sequence determinations of both HMG-T and H6 have indicated that these proteins are related closely to the group of low-salt-extractable proteins (HMG) or high-mobility-group proteins described by Johns et al. (1975) in calf thymus chromatin. The major difference between these groups of proteins in the two tissues is that in trout testis the population is less heterogeneous and its two major member proteins, H6 and HMG-T, are easily detected on gels in the presence of the nucleosomal histones.

Similar nuclease digestions were performed on calf thymus nuclei. The results of experiments using DNase I confirmed the association of HMG proteins with chromatin regions selectively sensitive to this nuclease (B. Levy W. and G. H. Dixon, in prep.).

These susceptible chromatin regions are likely to be enriched in DNA sequences that are transcribed (Weintraub and Groudine 1976; Garel and Axel 1976; Levy W. and Dixon 1977c). Therefore, the presence of HMG proteins in those chromatin regions might confer upon them different physicochemical properties, which could act as recognition signals not only for DNase I but also for other nonhistone proteins and enzymes, such as DNase II and RNA polymerases.

Since it is likely that "nucleosomes" exist in both transcriptionally active and inactive chromatin, it is possible that these chromatin regions enriched in their content of HMG proteins might be represented by a subpopulation of nucleosomes of slightly different structure, in which the HMG proteins would be a major component. Alternatively, HMG proteins could be located nonrandomly in linker regions between core nucleosomes. In this case, nucleosome cores from both active and inactive chromatin would be both qualitatively and quantitatively very similar in their protein composition, consisting mainly of the four inner histones: H2A, H2B, H3, and H4.

In any event, the question of whether the various HMG proteins are located in the core nucleosome or in the linker region between adjacent cores remains unanswered. It has recently been established that the length of the linker region between nucleosomes varies over a certain range in transcriptionally active chromatins (Lohr et al. 1977), whereas inactive chromatins exhibit a relatively regular spacing of nucleosomes. It is possible, therefore, that the presence of HMG-type proteins, perhaps in various combinations with H1 and other proteins in linker regions, might determine the position and length of these linker regions and contribute to the unique conformation of transcriptionally active regions of chromatin.

Acknowledgments

This research was supported by a grant from the Medical Research Council of Canada. B.L.W. is the recipient of a postdoctoral fellowship from the Medical Research Council of Canada.

REFERENCES

AVIV, H. and P. LEDER. 1972. Purification of biologically active globin messenger RNA by chromatography on oligothymidilic acid-cellulose. *Proc. Natl. Acad. Sci.* **69**: 1408.

AXEL, R., H. CEDAR, and G. FELSENFELD. 1973. Synthesis of globin ribonucleic acid from duck reticulocyte chromatin *in vitro. Proc. Natl. Acad. Sci.* **70**: 2029.

————. 1975. The structure of the globin genes in chromatin. *Biochemistry* **14**: 2489.

BIESSMANN, H., R. A. GJERSET, B. LEVY W., and B. J. MCCARTHY. 1976. Fidelity of chromatin transcription *in vitro*. *Biochemistry* **15**: 4356.

BURGOYNE, L. A., J. D. MODDS, and A. J. MARSHALL. 1976. Chromatin structure: A property of the higher structure of chromatin and the time course of its formation during chromatin replication. *Nucleic Acids Res.* **3**: 3293.

GAREL, A. and R. AXEL. 1976. Selective digestion of transcriptionally active ovalbumin genes from oviduct nuclei. *Proc. Natl. Acad. Sci.* **73**: 3966.

GILMOUR, R. S. and J. PAUL. 1973. Tissue specific transcription of the globin gene in isolated chromatin. *Proc. Natl. Acad. Sci.* **70**: 3440.

GOODWIN, G. H. and E. W. JOHNS. 1973. Isolation and characterization of two calf-thymus chromatin non-histone proteins with high contents of acidic and basic amino acids. *Eur. J. Biochem.* **40**: 215.

GOODWIN, G. H., C. SAUNDERS, and E. W. JOHNS. 1973. A new group of chromatin associated proteins with a high content of acidic and basic amino acids. *Eur. J. Biochem.* **38**: 14.

GOTTESFELD, J., R. F. MURPHY, and J. BONNER. 1975. Structure of transcriptionally active chromatin. *Proc. Natl. Acad. Sci.* **72**: 4404.

HUNTLEY, G. H. and G. H. DIXON. 1972. The primary structure of the NH_2-terminal region of histone T. *J. Biol. Chem.* **247**: 4916.

JOHNS, E. W., G. H. GOODWIN, J. M. WALKER, and C. SAUNDERS. 1975. Chromosomal proteins related to histones. *Ciba Found. Symp.* **28**: 95.

LAEMMLI, U. K. 1970. Cleavage of structural proteins during the assembly of the head and bacteriophage T4. *Nature* **227**: 680.

LEVY W., B. and G. H. DIXON. 1977a. Diversity of sequences of polyadenylated cytoplasmic RNA from rainbow trout testis and liver. *Biochemistry* **16**: 958.

————. 1977b. Changes in the sequence diversity of polyadenylated cytoplasmic RNA during testis differentiation in rainbow trout *(Salmo gairdnerii)*. *Eur. J. Biochem.* **74**: 61.

————. 1977c. Renaturation kinetics of cDNA complementary to cytoplasmic polyadenylated RNA from trout testis. Accessibility of transcribed genes to pancreatic DNase. *Nucleic Acids Res.* **4**: 883.

LEVY W., B., N. C. W. WONG, and G. H. DIXON. 1977. Selective association of the trout specific H6 protein with chromatin regions susceptible to DNase I and DNase II. Possible location of HMG-T in the spacer region between core nucleosomes. *Proc. Natl. Acad. Sci.* **74**: 2810.

LOHR, D., J. CORDON, K. TATCHELL, R. T. KOVACIC, and K. E. VAN HOLDE. 1977. Comparative subunit structure of HeLa, yeast and chicken erythrocyte chromatin. *Proc. Natl. Acad. Sci.* **74**: 79.

MARUSHIGE, K. and G. H. DIXON. 1971. Transformation of trout testis chromatin. *J. Biol. Chem.* **246**: 5799.

NOLL, M. 1974. Subunit structure of chromatin. *Nature* **251**: 249.

TIEN KUO, M., C. G. SAHASRABUDDHE, and G. F. SAUNDERS. 1976. Presence of messenger specific sequences in the DNA of chromatin subunits. *Proc. Natl. Acad. Sci.* **73**: 1572.

WALKER, J. M., G. H. GOODWIN, and E. W. JOHNS. 1976a. The similarity between the primary structure of two non-histone chromosomal proteins. *Eur. J. Biochem.* **62**: 461.

WALKER, J. M., J. R. B. HASTINGS, and E. W. JOHNS. 1976b. The partial amino acid sequence of a non-histone chromosomal protein. *Biochem. Biophys. Res. Commun.* **73**: 72.

WATSON, D. C., E. H. PETERS, and G. H. DIXON. 1977. The purification, characterization and partial sequence determination of a trout testis non-histone protein, HMG-T. *Eur. J. Biochem.* **74**: 53.

WEBER, K. and M. OSBORN. 1969. The reliability of molecular weight determinations by dodecyl sulfate polyacrylamide gel electrophoresis. *J. Biol. Chem.* **244**: 4406.

WEINTRAUB, H. and M. GROUDINE. 1976. Chromosomal subunits in active genes have an altered conformation. *Science* **193**: 848.

WIGLE, D. T. and G. H. DIXON. 1971. A new histone from trout testis. *J. Biol. Chem.* **246**: 5636.

WONG, N. C. W., G. G. POIRIER, and G. H. DIXON. 1977. Adenosine diphosphoribosylation of certain basic chromosomal proteins in isolated trout testis nuclei. *Eur. J. Biochem.* **77**: 11.

Transcription of Polytene Chromosomes and of the Mitochondrial Genome in *Drosophila melanogaster*

J. J. BONNER,* M. BERNINGER, AND M. L. PARDUE

Department of Biology, Massachusetts Institute of Technology, Cambridge, Massachusetts 02139

TRANSCRIPTION IN SALIVARY GLAND NUCLEI

Polytene chromosomes, such as the ones found in *Drosophila* salivary glands, are particularly suited for cytological studies on RNA and DNA synthesis. During polytenization, DNA strands do not separate following replication; the resulting giant chromosomes provide a level of cytological resolution that can be obtained in few systems. It is possible to identify specific genetic regions throughout the cell cycle because the chromosomal structure remains intact during and between periods of DNA replication. However, the local chromosomal structure may change during periods of RNA synthesis. Many chromosomal regions actively engaged in RNA transcription undergo a phenomenon known as puffing, in which the banding structure becomes diffuse and the chromosomal diameter increases.

Changes in puffing pattern occur in response to various stimuli, including nutritional (Beermann 1973), developmental (Beermann 1952; Becker 1962; Berendes 1965, 1966; Ashburner 1967), metabolic (Ritossa 1962; Berendes et al. 1965; Ashburner 1970), and hormonal (Clever and Karlson 1960; Berendes 1967; Ashburner 1972a,b). Puffs are sites of RNA accumulation; thus, analysis of puffing patterns provides one way of studying RNA synthesis. In polytene nuclei, it is also possible to use incorporation of [³H]uridine over individual sites as a measure of chromosomal activity (Pelling 1959, 1964). We will refer to the uridine incorporation assay by the term "transcription autoradiogram," although this type of preparation may not be an accurate measure of transcription. The transcription autoradiogram visualizes both nascent RNA chains and any completed RNA molecules that may be associated with the chromosome during the preparation of the autoradiogram.

In addition, it is possible to extract radioactive RNA from salivary glands incubated in medium containing [³H]uridine and to analyze the populations of RNAs by in situ hybridization to polytene chromosomes. In situ hybridization to polytene chromosomes can also be used to analyze [³H]RNA extracted from nonpolytene cells. Thus, this technique permits direct comparisons of genetic activity in diploid tissues with such activity in polytene tissues in which many genetic responses have been studied at the level of the polytene chromosome.

In studying RNA synthesis in polytene nuclei, we have compared in situ hybridization patterns with the pattern of chromosomal uridine incorporation and with the morphological expression of RNA synthesis, puffing. We have focused on the early response to ecdysone so that these studies could help to relate our work on the ecdysone response in imaginal discs (Bonner and Pardue 1976b; Bonner 1977) to studies of the ecdysone response of polytene chromosomes (Clever and Karlson 1960; Berendes 1967; Ashburner 1972a,b). The three assay techniques measure somewhat different aspects of RNA metabolism and therefore comparison of the results of the assays provides additional information about RNA in polytene cells.

MATERIALS AND METHODS

Fly stocks. Two stocks of *Drosophila melanogaster* were used throughout this study, gt wᵃ/gt wᵃ and yˢᶜgtˣ¹¹/FM6, referred to as g-1 and g-X11, respectively. The cross g-X11 × g-1 generates the giant larvae used for salivary gland chromosome cytological preparations; g-1 larvae were used for all RNA preparations. Flies were grown on cornmeal-agar medium at 22°C.

Salivary gland cultures. Pre-ecdysone larvae were selected as described previously (Bonner and Pardue 1976b). Salivary glands were cultured at 22°C in a minimal medium derived from Robb's medium, called "MRM" (Fristrom et al. 1973; Robb 1969), in Falcon plastic petri dishes as described previously (Bonner and Pardue 1976a). For RNA extractions, 30 larvae were dissected in MRM and their salivary glands stockpiled in MRM. At zero time, the glands were transferred to MRM made up to contain 400 μCi/ml 5,6-[³H]uridine, 400 μCi/ml 5-[³H]cytidine, 0.1% ethanol; β-ecdysone, if present, was at 2 μg/ml. RNA was extracted after the appropriate labeling time as described in the text.

---

* Present address: Department of Biochemistry and Biophysics, University of California Medical School, San Francisco, California 94143.

RNA preparation. Salivary glands and culture medium were taken up in 0.5 ml chilled lysis buffer (100 mM NaCl, 20 mM $CaCl_2$, 30 mM Tris, pH 7.8) and homogenized several strokes in the presence of 50 μg/ml *E. coli* ribosomal RNA (added as carrier) and 1% diethylpyrocarbonate in a Dounce-type homogenizer with a loose-fitting pestle. Triton X-100 was added to 0.5%, the material was swirled vigorously (Vortex Genie mixer), and the nuclei sedimented at 1400*g* for 4 minutes. The supernatant was recentrifuged; the first pellet was resuspended in 0.5 ml lysis buffer containing 50 μg/ml *E. coli* RNA. Both nuclear and cytoplasmic preparations were made 0.5% in sodium dodecyl sulfate and 15 mM in EDTA and extracted with phenol. The resulting aqueous phases were precipitated in 2 volumes of 95% ethanol overnight at $-20°C$.

The precipitates were collected by centrifugation at 20,000*g* for 30 minutes. The cytoplasmic RNA was resuspended and reprecipitated twice from 0.1 M Na acetate (pH 5.0). The nuclear RNA was rinsed twice in 80% ethanol, 0.1 M NaCl, and taken up in 0.5 ml MES buffer (0.05 M 2-[*N*-morpholine]-ethane-sulfonic acid, 0.002 M Mg acetate, pH 7.0). DNase I was added to 50 μg/ml. After 15 minutes on ice, the RNA was extracted with phenol, ethanol-precipitated, and rinsed twice in 0.1 M Na acetate (pH 5.0). The final RNA pellets were dissolved in 15 μl each of 2×TNS (TNS: 0.15 M NaCl, 0.01 M Tris, pH 6.8) for hybridization.

Hybridization was performed as described previously (Bonner and Pardue 1976b), except that the hybridization volume was 5 μl/slide, under 9-mm-square cover slips (prepared from Corning 18-mm-square cover slips).

Transcription autoradiograms. Salivary glands (from the giant cross) were cultured as described above. At zero time, β-ecdysone was added to 2 μg/ml. At 4 hours postecdysone, 5,6-[³H]uridine was added to 25 μCi/ml. Ten minutes later, glands were fixed in 45% acetic acid and squashed. Cover slips were removed on dry ice. The preparations were dehydrated in ethanol and air-dried. Unincorporated uridine was removed by rinsing the slides in cold 5% trichloroacetic acid for 15 minutes, followed by ethanol dehydration. Slides were then autoradiographed.

RESULTS AND DISCUSSION

Comparison of In Situ Hybridization with Transcription Autoradiograms

Figures 1 and 2 show the comparison of two chromosomal regions. Figures 1A and 2A are autoradiograms of nuclear RNA hybridized in situ to polytene

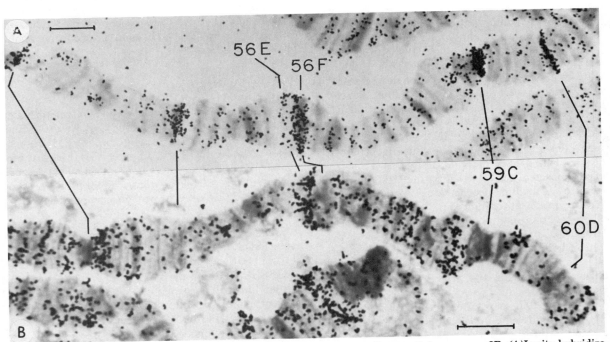

Figure 1. Comparison of in situ hybridization with a "transcription autoradiogram," chromosome 2R. *(A)* In situ hybridization of nuclear RNA isolated from glands labeled for 4 hr in the presence of ecdysone (27,000 cpm/slide, 12 weeks exposure). *(B)* Transcription autoradiogram from a gland labeled for 10 min after 4 hr of culture with ecdysone, then fixed and prepared for autoradiography. The puffs are the major sites of uridine incorporation but not the only sites. Most of the sites labeled by "transcription" are labeled by hybridization, though at different relative levels. The puffs are not major sites of hybridization. There are also additional sites (e.g., 60D and 59C) labeled by hybridization which do not appear to label in the transcription autoradiogram. Bar = 10 μ.

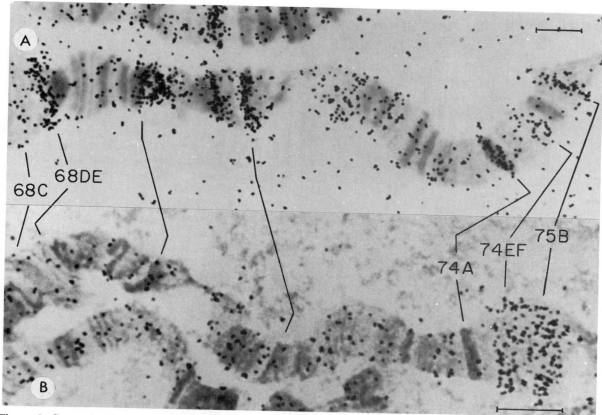

Figure 2. Comparison of in situ hybridization with a "transcription autoradiogram," chromosome 3L. Labeling was performed as described in the legend to Fig. 1. *(A)* In situ hybridization. *(B)* Transcription autoradiogram. Region 68C is the site of a puff that is repressed by ecdysone; 74EF and 75B are puffs that are induced by ecdysone. Bar = 10 μ.

chromosomes; 1B and 2B are transcription autoradiograms. From a purely qualitative standpoint, it appears that the chromosomal bands that label in the transcription autoradiogram also label in the hybridization autoradiogram. It is difficult to make an exact statement, since the minor bands are not sufficiently resolved to allow a detailed analysis: this is especially true of the puffs, where banding structure is lost over a wide region.

Surprisingly, there are several bands which label only by hybridization; 60D and 59C are good examples. We have not seen these bands label in a transcription autoradiogram, even after long labeling times or long autoradiographic exposure; these bands are probably not transcribed in salivary glands. However, the in situ hybridization pattern implies that these bands have some sequence homology with other bands which are transcribed.

We have extended the comparison of in situ hybridization with the transcription autoradiogram as shown in Figure 3. The histogram presents the data as percentage of chromosomes labeled for each chromosomal locus. Data were compiled from slides with a lower degree of labeling than those shown in Figures 1 and 2. Any chromosomal site exhibiting three

or more grains on or touching a single band was scored as labeled. This technique provides a satisfactory method for examining the entire chromosome set for gross differences in labeling characteristics (see also Bonner and Pardue 1976b; Bonner 1977).

Most chromosomal loci label to some extent in both preparations. There are many sites which, like 60D and 59C described above, label only by hybridization. There are a few sites (77E, 79E, 92A), however, which label only in the transcription autoradiogram. These latter sites may reflect the observation, described below, that hybridization to puff sites does not necessarily occur over the entire puff, but may occur at one end. Thus we can make the generalization that most, if not all, sites that label in the transcription autoradiogram can be detected by in situ hybridization of salivary gland nuclear RNA, and that the qualitative differences tend to be in the form of sites that hybridize but do not appear to transcribe.

In comparing the pattern of uridine incorporation measured by the transcription autoradiogram with the pattern of hybridization by RNA isolated from salivary glands, it is important to realize that the transcription autoradiogram represents essentially

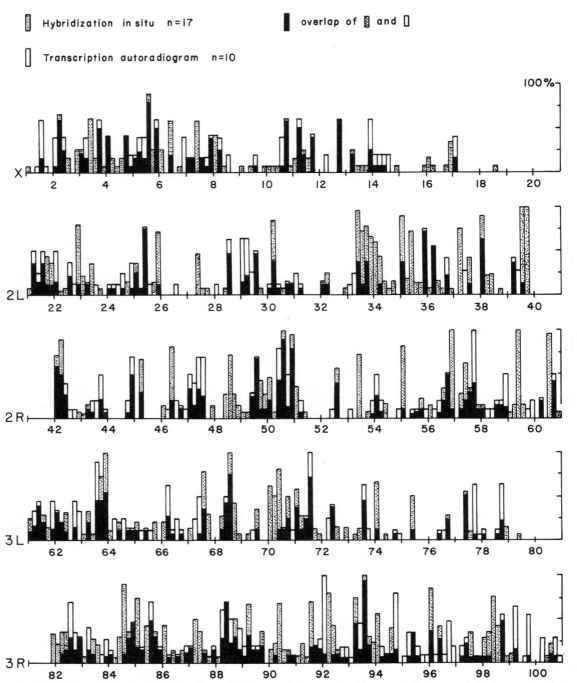

Figure 3. Comparison of in situ hybridization with "transcription autoradiograms." Data are plotted as percentage of chromosomes labeled against chromosomal locus. Data are derived from nuclear RNA labeled for 4 hr in the absence of ecdysone and hybridized in situ at 19,000 cpm/slide, 89 days exposure; transcription autoradiogram labeled for 10 min after 4 hr of culture without ecdysone, 1 week exposure.

a static picture of chromosomal labeling, whereas the hybridization assay measures a population of RNA molecules that have accumulated during a relatively long labeling period. Thus, the comparison is between the actual hybridization data and the hybridization data that would be predicted from the transcription autoradiogram. Only if all post-transcriptional events are similar for different RNA

species will the actual and predicted hybridization be similar.

Differences in the metabolism of individual RNA species would be expected to affect the fidelity with which the in situ hybridization pattern reflects the uridine incorporation assay. It is likely that there are differences among RNA species in their life-times, turnover rates, and abundancies; these variables would all affect the hybridization assay. As yet, we know little about these variables for any single RNA species. RNAs may also differ with respect to their degree of chromosomal retention. It is interesting to note that retention of RNA has been suggested as one of the functions of puffs (Beermann 1965; Berendes 1972). Storage of RNA at puff sites could result in relatively heavy labeling of puffs by uridine incorporation.

It seems likely that gene reiteration may also play a part in determining the pattern of in situ hybridization. Hybridization to a tandem array of repeated sequences would be likely to produce a major site of labeling, even though the RNA species were "average" in its other characteristics. In addition, sites that are not transcribed at all in salivary glands may show hybridization if they share DNA sequences with other regions that are transcribed in these cells.

The Puffs

It was shown initially by Pelling (1959, 1964) that puffs are sites of active uridine incorporation. Puff sites label more intensely by uridine incorporation, but many nonpuffed regions also appear to be active in RNA transcription (Berendes 1972; Zhimulev and Belyaeva 1975). Thus, morphological puffing appears to be restricted to a certain subset of active loci, though the basis for the restriction is unknown.

In general, hybridization to puffs occurs in a restricted region of the puff. The band(s) that hybridize may be in the center of the puff, or at one end (Fig. 4). If the puff and the hybridizing band are fortuitously not given the same letter on the chromosome map, then the labeling data will not coincide in a histogram like that shown in Figure 3. This situation is reminiscent of the heat-shock RNA, which is transcribed from a puff in the 87B region (Ashburner 1970), but which hybridizes to 87C1 (Spradling et al. 1975). Band 87C1 forms one end of the puff that also includes most of region 87B.

In terms of quantitative labeling, there are significant differences between in situ hybridization and the apparent transcription pattern. It may be seen from the comparisons presented in Figures 1 and 2 that the puffs are major sites of uridine incorporation; however, they are not major sites of hybridization. The observation that puff sites are not the most intense sites of hybridization can perhaps be explained by a combination of effects of gene reiteration and RNA metabolism, as discussed above. However, it is significant that for all of the puffs we have studied, high levels of hybridization can be detected only with RNA samples labeled while the puffs were active.

The Ecdysone-sensitive Loci

Experimental application of the steroid hormone ecdysone to third instar larvae or salivary glands

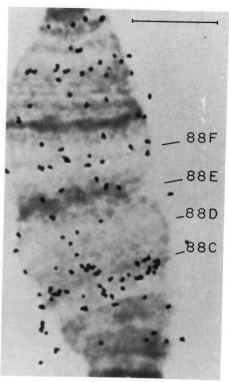

Figure 4. Hybridization to puff sites. Hybridization at puff 88D occurs at the end of the puff region (in locus 88C); hybridization at puff 88EF occurs in the center of the puff region. (Nuclear RNA, 43,000 cpm/slide, 11 weeks exposure.) Bar = 10 μ.

will elicit some of the same puffing changes that occur during normal development (Clever and Karlson 1960; Berendes 1967; Ashburner 1972a,b). In the initial response to ecdysone, two puffs regress and approximately six are induced (Ashburner 1972b). We will discuss below the puffs at 25AC and 68C, which regress, and 74EF and 75B, which are induced.

Figure 5 presents hybridization data from salivary gland RNA populations labeled in organ culture in the presence and in the absence of ecdysone. The data were compiled by counting the numbers of silver grains over the various loci, and, for each nucleus, normalizing the grain counts to the value obtained for the locus 50F. 50F was chosen because it exhibited similar numbers of grains for all samples. In Figure 5, we have plotted the means of the individually normalized values. Such normalization is necessary to eliminate the variability introduced by differences in the degree of polyteny of different nuclei (Bonner and Pardue 1976a).

Regions 50D, 50F, 60D, and 68DE are typical of most hybridization sites—they label similarly whether ecdysone is present or not. They differ relative to each other in their degree of detectability in nuclear and cytoplasmic RNA. Of these sites, only 50D is involved in puffing activity during the labeling period. A puff is induced at region 50CD in re-

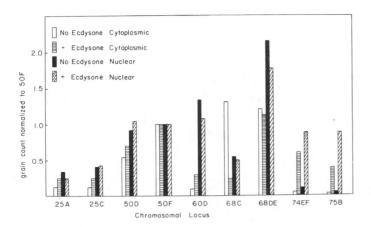

Figure 5. In situ hybridization of salivary gland RNA. Silver grains were counted over selected chromosomal loci. The grain counts for individual nuclei were normalized to the grain count of 50F; the normalized values were used to calculate the mean. Bars in each sample are, left to right: cytoplasmic RNA labeled 4 hr in the absence of ecdysone (12,000 cpm/slide, 12 weeks exposure); cytoplasmic RNA labeled 5 hr in the presence of ecdysone (25,000 cpm/slide, 11 weeks exposure); nuclear RNA labeled 4 hr in the absence of ecdysone (19,000 cpm/slide, 12 weeks exposure); nuclear RNA labeled 5 hr in the presence of ecdysone (43,000 cpm/slide, 11 weeks exposure).

sponse to in vitro incubation (Ashburner 1972a); the puff shows no response to ecdysone.

Region 25AC shows only low levels of hybridization by all RNA preparations. We cannot at this time explain why we find so little RNA complementary to this region in any of the preparations. For the other puff that regresses in response to ecdysone, 68C, the correlation is much better. Cytoplasmic RNA labeled in the presence of ecdysone shows a low level of hybridization, whereas cytoplasmic RNA labeled in the absence of ecdysone shows a high level of hybridization at 68C. There is less hybridization at 68C by nuclear RNA; this may reflect the slow regression of the puff which is seen during the labeling period in culture (Ashburner 1973). The nuclear RNA present at the end of the labeling period is probably recently synthesized, since much of the RNA made earlier would have been degraded or transported to the cytoplasm. Table 1 shows results of hybridization at 68C with RNA prepared from salivary glands after either 1 or 4 hours in culture without ecdysone. The change in the nuclear/cytoplasmic distribution is consistent with the regression of the puff in culture.

The ecdysone induction of puffs at 74EF and 75B is also reflected in the hybridization data; significant hybridization at these sites is detectable only by the RNA samples labeled in the presence of ecdysone.

Metabolism of Puff RNAs

Hybridization data will be described below for chromosomal sites 50D, 68C, 74EF, and 75B. It is tempting to suggest that the RNAs hybridizing to these sites are the "puff RNAs," transcribed from the puffs at these sites. The ecdysone-sensitivity of hybridization at these loci parallels that of the puffs.

The puffs at these four locations have been described by Ashburner (1972b). In our culture system they behave similarly. The puff at 50D is induced at the onset of in vitro culture and remains puffed during the labeling period. The puff at 68C is present prior to the beginning of the culture period; it regresses rapidly upon the addition of ecdysone and slowly when cultured in the absence of the hormone. The puffs at 74EF and 75B are never induced when cultured in the absence of ecdysone; in the continuous presence of ecdysone these puffs are immediately

Table 1. Labeling of 50D and 68C in the Absence of Ecdysone, Relative to 50F

RNA	Labeling time (hr)	Grains	Grains	Ratio	Nuclear/ cytoplasmic	4 hr/ 1 hr
		68C	*50F*	*68Cᵃ/50F*		
Nuclear	1	7.3 ± 2.6	9.3 ± 3.4	$0.78 \pm .28$	1.32	nuc.
Cytoplasmic	1	5.1 ± 1.0	8.6 ± 3.2	$0.59 \pm .21$		0.69
Nuclear	4	6.7 ± 3.5	12.4 ± 5.9	0.54 ± 30	0.52	cyto.
Cytoplasmic	4	16.3 ± 5.5	13.0 ± 2.6	$1.04 \pm .26$		1.76
		50D	*50F*	*50Dᵃ/50F*		
Nuclear	1	10.4 ± 5.3	9.3 ± 3.4	$1.08 \pm .32$	1.56	nuc.
Cytoplasmic	1	5.6 ± 2.3	8.6 ± 3.2	$0.69 \pm .37$		0.84
Nuclear	4	11.3 ± 3.8	12.4 ± 5.9	$0.91 \pm .31$	1.68	cyto.
Cytoplasmic	4	7.0 ± 2.7	13.0 ± 2.6	$0.54 \pm .20$		0.78

Silver grains were counted over 50D, 68C, and 50F. The grain counts for individual nuclei were normalized to the grain count of 50F; it is the mean of the normalized values that is shown in the fourth column. Data are from hybridization of nuclear RNA (36,000 cpm/slide) and cytoplasmic RNA (12,500 cpm/slide) fractionated from one set of glands labeled for 1 hr (autoradiographic exposure, 9 weeks) and from hybridization of nuclear RNA (19,000 cpm/slide) and cytoplasmic RNA (12,000 cpm/slide) fractionated from another set of glands labeled for 4 hr (autoradiographic exposure, 12 weeks).

ᵃ Normalized individually by nucleus.

induced, reach maximal activity at about 4 hours, and regress significantly by 5 hours postecdysone. The behavior of the RNAs hybridizing to these four sites is markedly different.

Due to the uncertainties in specific radioactivity, concentration, and hybridization efficiencies of the different RNAs, it is difficult to make absolute quantitative measurements. However, variations in the detectability of different RNA species relative to one another can provide some clues about the metabolic fates of those RNAs. As a measure of RNA metabolism, we can look at the relative concentrations of RNA species in the nuclear and cytoplasmic RNA populations.

Tables 1 and 2 present hybridization data for several labeling times. For any particular labeling time, the nuclear and cytoplasmic RNA samples are from the same batch of glands, and the data represent hybridization by RNA from equal numbers of cells. Hybridization at regions 50D and 68C was measured for RNA labeled in the absence of ecdysone, for 1 hour or 4 hours. To compare the relative hybridization levels, we chose to normalize the data first to hybridization at a third site; we used 50F, which showed similar hybridization for nuclear and cytoplasmic RNA samples. The nuclear/cytoplasmic distribution of 68C decreases with increased labeling time, while the distribution of 50D remains essentially constant.

The data presented in Table 1 suggest that the RNA species hybridizing to 68C may be relatively stable in salivary glands, since the RNA concentration appears to build up in the cytoplasm. The RNA hybridizing to 50D, however, behaves quite differently, even in the same cells. The data suggest that 50D RNA may have reached a steady-state concentration in both nucleus and cytoplasm within the first hour of labeling.

Table 2 presents data on hybridization to 74EF and 75B by RNA labeled in the presence of ecdysone for 4 or 5 hours. Very little RNA hybridizing to

either site is found in the cytoplasm after 4 hours of labeling. However, by 5 hours, the nuclear/cytoplasmic distribution has changed. The data suggest that the RNA hybridizing to these sites remains within the nucleus during the first 4 hours of ecdysone culture (the time the puffs at these sites are active), and that the RNA is first transported at about the time the puffs begin to regress. The significance of this apparent delay in exit from the nucleus is not clear at this time.

From these results it appears that puffs are not the sites of synthesis of a uniform class of RNAs. While the data presented above cannot address the question of absolute quantities of particular RNA species, they do demonstrate that RNAs hybridizing to different puffs do not have identical metabolic characteristics. Two puffs, 50D and 74EF, are both induced at the beginning of the labeling period (50D by in vitro culture, 74EF by ecdysone), yet RNA hybridizing to them can be detected in the cytoplasm at very different times — by 1 hour for 50D, not until after 4 hours for 74EF. RNA hybridizing to the puff at 68C seems relatively stable in the cytoplasm of salivary glands, while in the same cells RNA hybridizing to 50D appears to have reached a steady-state level by the first hour of labeling. Thus, the simple fact that a chromosomal region has formed a puff — or labels well in the transcription autoradiogram — tells us very little about the RNA transcribed at that locus. In spite of the differences in their patterns of RNA metabolism, the puffs that we have studied are similar in one respect: significant levels of RNA from these loci are detected by the hybridization assay only if the period of [³H]uridine labeling occurs while the locus is puffed.

TRANSCRIPTION OF MITOCHONDRIAL DNA

The *Drosophila melanogaster* mitochondrial genome is a circular molecule of approximately 18,000 base pairs, 4600 of which are contained in a region

Table 2. Labeling of 74EF and 75B in the Presence of Ecdysone, Relative to 50F

RNA	Labeling time (hr)	Grains	Grains	Ratio	Cytoplasmic/ nuclear
		74EF	*50F*	*74EF[a]/50F*	
Cytoplasmic	4	1.9 ± 1.9	13.4 ± 6.6	0.14 ± .26	
Nuclear	4	12.4 ± 4.5	11.2 ± 1.9	1.11 ± .26	0.13
Cytoplasmic	5	11.0 ± 3.7	18.2 ± 4.7	0.60 ± .27	
Nuclear	5	11.4 ± 3.3	14.0 ± 4.5	0.87 ± .33	0.69
		75B	*50F*	*75B[a]/50F*	
Cytoplasmic	4	1.1 ± 1.5	13.4 ± 6.6	0.08 ± .23	
Nuclear	4	10.4 ± 2.6	11.2 ± 1.9	0.93 ± .20	0.09
Cytoplasmic	5	6.5 ± 2.6	18.2 ± 4.7	0.37 ± .16	
Nuclear	5	11.8 ± 4.6	14.0 ± 4.5	0.87 ± .34	0.42

Data were compiled, as described in the notes to Table 1, from hybridization of nuclear RNA (27,000 cpm/slide) and cytoplasmic RNA (10,000 cpm/slide) fractionated from one set of glands labeled for 4 hr (autoradiographic exposure, 12 weeks) and from hybridization of nuclear RNA (43,000 cpm/slide) and cytoplasmic RNA (25,000 cpm/slide) fractionated from another set of glands labeled for 5 hr (autoradiographic exposure, 11 weeks).

[a] Normalized individually by nucleus.

extremely rich in adenine and thymine (A+T) (Polan et al. 1973; Peacock et al. 1974; Bultmann and Laird 1973; Klukas and Dawid 1976). The size of the A+T-rich region varies considerably among different species of *Drosophila*, whereas the rest of the mitochondrial genome remains fairly constant in size (Fauron and Wolstenholme 1976). If one assumes that the A+T-rich region does not contain sequences which are transcribed, approximately 13,400 base pairs are available in the *D. melanogaster* mitochondrial genome to code for its products.

In several eukaryotes, mitochondrial DNA has been shown to code for the RNAs of the two subunits of the mitochondrial ribosome and for a number of transfer RNAs (Nass and Buck 1970; Aloni and Attardi 1971; Borst 1972; Dawid 1972; Angerer et al. 1976). In addition to coding for structural RNAs of the protein-synthesizing machinery, mitochondrial DNA codes for a small number of poly(A)+ RNAs (Hirsch and Penman 1973; Ojala and Attardi 1974). These RNAs apparently are messenger RNAs since they sediment with polysomes and are released from this association by puromycin (Perlman et al. 1973). It is interesting that the number and size of these poly(A)+ RNAs show a strong evolutionary conservation. Animals as distantly related as human, hamster, mosquito, and *Drosophila* all show similar patterns of poly(A)+ RNA on polyacrylamide gels (Hirsch et al. 1974). It is not known what proteins these poly(A)+ RNAs code for. The most extensive genetic and biochemical studies have been carried out in yeast, where it appears that the mitochondrial genome codes for about a dozen hydrophobic polypeptides, several of which are components of the enzyme systems of the mitochondrial inner membrane (Schatz and Mason 1974).

The lack of mitochondrial mutations in higher eukaryotes requires that analysis of the structure of the genome be by physical methods. The first step in this analysis is the mapping of the transcription products to various regions of the mitochondrial DNA molecule.

It is possible to obtain radiochemically pure preparations of mitochondrial RNA transcripts from *D. melanogaster* cultured cells that have been labeled with [³H]uridine in the presence of 5 µg/ml actinomycin D. Under these conditions, actinomycin D almost completely inhibits transcription of nuclear DNA but apparently does not enter the mitochondria. When such RNA preparations are separated on low-percentage polyacrylamide gels, they yield a set of 12 radioactive RNA bands. Eleven of the RNA species bind to oligo(dT)-cellulose, suggesting that they contain poly(A) sequences. One species of RNA, the small mitochondrial ribosomal RNA, shows little binding to oligo(dT)-cellulose. The [³H]RNA species from cells labeled in the presence of actinomycin D comigrate on polyacrylamide gels with the RNAs obtained from mitochondria isolated from cells labeled in the absence of any drug (Sprad-

ling et al. 1977). We have used such radiochemically pure populations of mitochondrial RNAs in combination with restriction enzyme cleavage of mitochondrial DNA to map the sequences complementary to the mitochondrial transcripts.

MATERIALS AND METHODS

Mitochondrial RNA. Radioactive mitochondrial RNA was prepared from a *D. melanogaster* cultured cell line that had been adapted for growth in medium lacking yeast extract (Lengyel et al. 1975). RNA was labeled with tritiated uridine (25 Ci/mmole, 100 µCi/ml) in the presence of 5 µg/ml actinomycin D for 4 hours at 3×10^7 cells/ml (cells concentrated 10 times). RNA was extracted from cells, fractionated on oligo(dT)-cellulose (Collaborative Research) into poly(A)+ and poly(A)− fractions, and electrophoresed on 2.5–5% polyacrylamide gradient gels by the methods of Spradling et al. (1977). The gels were prepared for fluorography (Bonner and Laskey 1974), dried down, and fluorographed for analytical purposes (Fig. 7). Gels were fluorographed while wet if RNA was to be recovered from the gel for hybridization. Fluorographs of the undried gels were used to determine the location in the gel of a given RNA; that part of the gel was excised and the RNA was eluted from the piece of excised gel as described by Spradling et al. (1977).

Mitochondrial DNA. Mitochondrial DNA was prepared from 6–18-hour *D. melanogaster* embryos (the generous gift of Sarah Elgin) by the method of Bultmann and Laird (1973). Restriction enzymes were purchased from New England Biolabs. Digestions were carried out in the buffers suggested by the suppliers. The fragments resulting from these digestions were separated on 1% or 1.4% agarose (Sigma) gels as described by Southern (1975a). The gels were stained with 0.5 µg/ml ethidium bromide for 10 minutes and photographed under UV light through a Kodak 23A filter on Polaroid 57 or 55 P/N film. The *Hae*III and *Hind*III fragments of *D. melanogaster* mitochondrial DNA, which have been measured by electron microscopy (Klukas and Dawid 1976), were used as standards in determining the lengths of other fragments. In all cases, the sum of the sizes of the fragments of mitochondrial DNA equaled the size of the entire genome, and the fragments appeared to be in equimolar amounts.

Transfer of DNA to nitrocellulose filters and hybridization with mitochondrial RNA. DNA was transferred from the agarose gel to nitrocellulose filters (Millipore) and the filters were prepared for hybridization according to the procedures of Southern (1975b). Hybridizations were carried out at 37°C for 48 hours in a volume sufficient to soak the filter. The hybridization mix consisted of 5×10^3 to 3×10^4 cpm of gel-purified mitochondrial RNA (depending on the experiment) and *E. coli* tRNA at a concentration of 100 μg/ml dissolved in 0.45 M NaCl, 0.045 M Na citrate, 0.16 M Tris, 0.02 M EDTA, 30% formamide, 0.1% SDS, pH 8.0 (Klukas and Dawid 1976). Filters were washed in 100 ml of hybridization buffer at 37°C for 1 hour; they were then treated with pancreatic RNase (50 μg/ml) at room temperature in 0.3 M NaCl, 0.03 M Na citrate, pH 7.0, for 1 hour and washed in two changes of 100 ml of 0.3 M NaCl, 0.03 M Na citrate, 0.5% SDS, pH 7.0, for 1 hour at room temperature. The filters were dried, dipped in 20% PPO (2, 5-diphenyloxazole) in toluene (Southern 1975b), and fluorographed at −70°C on Kodak XR-1 X-ray film that had been prefogged to an optical density of 0.25. Exposures were from 1 week to 6 weeks.

RESULTS AND DISCUSSION

The Restriction Enzyme Map of the Mitochondrial DNA

Cleavage sites for the *Hae*III and *Hind*III restriction enzymes relative to the A+T-rich region and to each other have been determined by Klukas and Dawid (1976). We have confirmed their findings and mapped the sites of cleavage of several additional restriction enzymes: *Hind*II, *Bgl*II, *Pst*I, *Hpa*I, *Hpa*II, and *Eco*RI. The map of these sites on the *D. melanogaster* mitochondrial DNA (Fig. 6) was constructed by analysis of the sizes of fragments generated by restriction enzymes used singly and in combination. The sizes of fragments were determined by electrophoresis on agarose slab gels. In each digestion, the sum of the lengths of the restriction enzyme cleavage products was equal to the length of the mitochondrial genome; however, the measurements of the fragment sizes are not precise enough to distinguish between a single cleavage site by a given enzyme and two or more cleavage sites that are located very close to one another.

Mapping of the Mitochondrial RNAs

The mapping of various species of mitochondrial RNA was determined by hybridization of each of the radiochemically pure RNAs to preparations of mitochondrial DNA that had been cleaved with several restriction enzymes. Although the RNA used in these experiments was prepared from total cytoplasmic RNA, only mitochondrial transcripts had been labeled with tritium because the actinomycin D in the labeling medium prevented transcription of high-molecular-weight RNA from nuclear DNA. When such RNA, labeled in the presence of 5 μg/ml actinomycin D, is run on a 2.5–5% polyacrylamide gradient gel and detected by its radioactivity, twelve RNA bands can be identified (Fig. 7). The bands are sufficiently well resolved that most can be cut individually from the polyacrylamide gel and eluted free of contamination by neighboring bands as determined by electrophoresing the eluted RNA on a second polyacrylamide gel. It has not yet been possible to separate with confidence RNAs B-1 and B-2; also fractions of RNA B-3 are contaminated with material from band B-4.

Each of the purified RNAs was found to hybridize to a particular set of DNA fragments generated by restriction enzyme cleavage (Fig. 8). When compared to the map of restriction enzyme sites, the set of fragments to which a given RNA hybridized was always found to be consistent with a single location on the map for the sequences homologous to that RNA. Our results with RNAs B-4 and B-10, the large and small mitochondrial ribosomal RNAs, respectively, confirm the mapping of these RNAs estab-

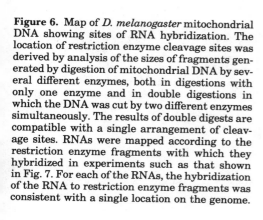

Figure 6. Map of *D. melanogaster* mitochondrial DNA showing sites of RNA hybridization. The location of restriction enzyme cleavage sites was derived by analysis of the sizes of fragments generated by digestion of mitochondrial DNA by several different enzymes, both in digestions with only one enzyme and in double digestions in which the DNA was cut by two different enzymes simultaneously. The results of double digests are compatible with a single arrangement of cleavage sites. RNAs were mapped according to the restriction enzyme fragments with which they hybridized in experiments such as that shown in Fig. 7. For each of the RNAs, the hybridization of the RNA to restriction enzyme fragments was consistent with a single location on the genome.

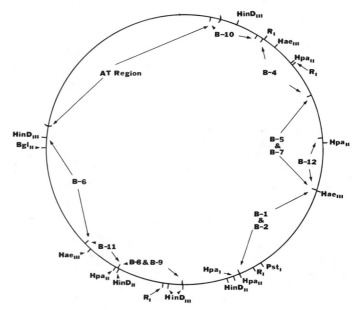

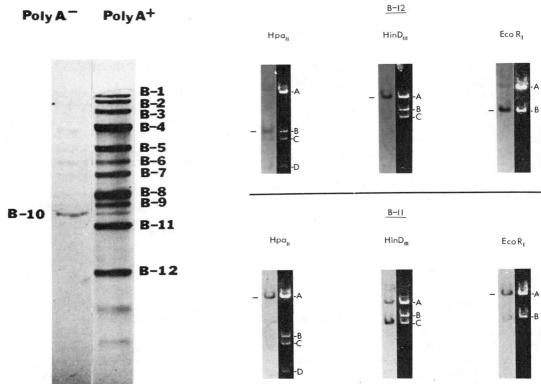

Figure 7. *D. melanogaster* mitochondrial RNA. RNA was prepared as described in Materials and Methods and electrophoresed on a 2.5–5% polyacrylamide gradient slab gel containing 7.0 M urea. Either poly(A)$^+$ or poly(A)$^-$ RNA (5 × 10^4 cpm) was loaded in each well. The poly(A)$^-$ material represents RNA prepared from 3.2 × 10^7 cells and the poly(A)$^+$ material represents RNA prepared from 1.6 × 10^8 cells. The gel was fluorographed and dried. Exposure was for 24 hr. Most of the radioactive RNA in the poly(A)$^-$ fraction is of low molecular weight and migrates off the gel during electrophoresis. Occasional fractionations on oligo(dT)-cellulose exhibit limited binding of the B-10 (small mitochondrial ribosomal) RNA as may be seen in the poly(A)$^+$ fraction in this experiment.

Figure 8. Hybridization of RNA to *D. melanogaster* mitochondrial DNA fragments generated by restriction enzyme cleavage. RNA eluted from bands B-11 and B-12, cut from 2.5–5% gradient polyacrylamide gels, was hybridized to filters prepared by the method of Southern (1975b) from 1.4% agarose gels containing the *Hpa*II, *Hind*III, and *Eco*RI restriction enzyme cleavage fragments of mitochondrial DNA and the *Eco*RI cleavage fragments of bacteriophage λ DNA. The B-11 RNA hybridization mix contained 1.1 × 10^4 cpm of [^{3}H]RNA and 100 μg/ml of *E. coli* tRNA. The B-12 RNA hybridization mix contained 1.3 × 10^4 cpm of [^{3}H]RNA and 100 μg/ml of *E. coli* tRNA. The pattern of restriction enzyme cleavage fragments of mitochondrial DNA separated on the 1.4% agarose gel from which the DNA was transferred to nitrocellulose filters is shown beside the pattern of exposure produced by that filter following hybridization and fluorography. The smallest *Hind*III fragment and the two smallest *Eco*RI fragments were not transferred to the filter in this experiment. Exposure of the filter was for 14 days. No hybridization was observed to the fragments of λ DNA (not shown).

lished by Klukas and Dawid (1976). The sites of hybridization by the other mitochondrial RNAs that we have mapped are widely dispersed around the mitochondrial genome, although none have bound to the A+T-rich region of the DNA. In each of our experiments, the heavy hybridization to one (or two) fragments of DNA that was characteristic of a given RNA fraction was accompanied by a low level of hybridization to the other fragments of mitochondrial DNA in the restriction enzyme digest. We believe that the low-level hybridization is specific because there was no hybridization to fragments of bacteriophage λ DNA run in the same experiments. However, we have not yet determined whether this hybridization could be due to a small amount of sequence homology among the RNAs or to low levels of cross-contamination between the RNAs.

In several cases, two or more RNA species map within the same region of the genome but have not

yet been ordered within that region. This is the case for RNAs B-1 and B-2, RNAs B-5, B-7, and B-12, and RNAs B-8 and B-9. Each of these groups of RNAs binds to DNA fragments, the lengths of which are equal to less than the sum of the lengths of the RNAs in the group, as estimated by gel mobilities. These results suggest the possibility that the RNAs within a group might share identical or complementary sequences.

Acknowledgments

This work was supported by grant 5-RO1-GM21874 from the National Institutes of Health (to

M. L. P.) and by National Institutes of Health Predoctoral Training Grants to the Department of Biology at the Massachusetts Institute of Technology (for J. J. B. and M. B.).

REFERENCES

ALONI, Y. and G. ATTARDI. 1971. Expression of the mitochondrial genome in HeLa cells. IV. Titration of mitochondrial genes for 16S, 12S and 4S RNA. *J. Mol. Biol.* **55**: 271.

ANGERER, L., N. DAVIDSON, W. MURPHY, D. LYNCH, and G. ATTARDI. 1976. An electron microscope study of the relative positions of the 4S and ribosomal RNA genes in HeLa cell mitochondrial DNA. *Cell* **9**: 81.

ASHBURNER, M. 1967. Patterns of puffing activity in the salivary gland chromosomes of *Drosophila*. I. Autosomal puffing patterns in a laboratory stock of *D. melanogaster*. *Chromosoma* **21**: 398.

———. 1970. Patterns of puffing activity in the salivary gland chromosomes of *Drosophila*. V. Responses to environmental treatments. *Chromosoma* **31**: 356.

———. 1972a. Patterns of puffing activity in the salivary gland chromosomes of *Drosophila*. VI. Induction by ecdysone in salivary glands of *D. melanogaster* cultured *in vitro*. *Chromosoma* **38**: 255.

———. 1972b. Puffing patterns in *Drosophila melanogaster* and related species. *Results Probl. Cell Differ.* **4**: 101.

———. 1973. Sequential gene activation by ecdysone in polytene chromosomes of *Drosophila melanogaster*. I. Dependence upon ecdysone concentration. *Dev. Biol.* **35**: 47.

BECKER, H. J. 1962. Die Puffs der Speicheldrüsenchromosomen von *Drosophila melanogaster*. II. Die Auslosung der Puffbildung, ihre Spezifität, und ihre Benziehung zur Funktion der Ringdrüse. *Chromosoma* **13**: 341.

BEERMANN, W. 1952. Chromosomenkonstanz und Spezifische Modifikationen der Chromosomenstruktur in der Entwicklung und Organdifferenzierung von *Chironomous tentans*. *Chromosoma* **5**: 139.

———. 1965. Operative Gliederung der Chromosomen. *Naturwissenschaften* **52**: 365.

———. 1973. Directed changes in the pattern of Balbiani ring puffing in *Chironomous*: Effects of a sugar treatment. *Chromosoma* **41**: 297.

BERENDES, H. D. 1965. Salivary gland function and chromosomal puffing patterns in *Drosophila hydei*. *Chromosoma* **17**: 35.

———. 1966. Gene activities in the Malpighian tubules of *Drosophila hydei* at different developmental stages. *J. Exp. Zool.* **162**: 209.

———. 1967. The hormone ecdysone as effector of specific changes in the pattern of gene activity of *Drosophila hydei*. *Chromosoma* **22**: 274.

———. 1972. The control of puffing in *Drosophila hydei*. *Results Probl. Cell Differ.* **4**: 181.

BERENDES, H. D., F. M. A. VAN BREUGAL, and T. K. H. HOLT. 1965. Experimental puffs in salivary gland chromosomes of *Drosophila hydei*. *Chromosoma* **16**: 35.

BONNER, J. J. 1977. "Ecdysone-stimulated RNA synthesis in tissues of *Drosophila melanogaster*. Assay by *in situ* hybridization." Ph.D. thesis, Massachusetts Institute of Technology, Cambridge, Massachusetts.

BONNER, J. J. and M. L. PARDUE. 1976a. The effect of heat shock on RNA synthesis in *Drosophila* tissues. *Cell* **8**: 43.

———. 1976b. Ecdysone-stimulated RNA synthesis in imaginal discs of *Drosophila melanogaster*. Assay by *in situ* hybridization. *Chromosoma* **58**: 87.

BONNER, W. M. and R. A. LASKEY. 1974. A film detection method for tritium-labelled proteins and nucleic acids in polyacrylamide gels. *Eur. J. Biochem.* **46**: 83.

BORST, P. 1972. Mitochondrial nucleic acids. *Annu. Rev. Biochem.* **41**: 333.

BULTMANN, H. and C. D. LAIRD. 1973. Mitochondrial DNA from *Drosophila melanogaster*. *Biochim. Biophys. Acta* **299**: 196.

CLEVER, U. and U. P. KARLSON. 1960. Induktion von Puffänderungen in den Speicheldrüsenchromosomen von *Chironomous tentans* durch Ecdyson. *Exp. Cell Res.* **20**: 623.

DAWID, I. B. 1972. Mitochondrial RNA in *Xenopus laevis*. I. The expression of the mitochondrial genome. *J. Mol. Biol.* **63**: 201.

FAURON, C. M.-R. and D. R. WOLSTENHOLME. 1976. Structural heterogeneity of mitochondrial DNA molecules within the genus *Drosophila*. *Proc. Natl. Acad. Sci.* **73**: 3623.

FRISTROM, J. W., W. R. LOGAN, and C. MURPHY. 1973. The synthetic and minimal culture requirements for evagination of imaginal discs of *Drosophila melanogaster in vitro*. *Dev. Biol.* **33**: 441.

HIRSCH, M. and S. PENMAN. 1973. Mitochondrial polyadenylic acid-containing RNA: Localization and characterization. *J. Mol. Biol.* **80**: 379.

HIRSCH, M., A. SPRADLING, and S. PENMAN. 1974. The messenger-like poly(A)-containing RNA species from the mitochondria of mammals and insects. *Cell* **1**: 31.

KLUKAS, C. K. and I. B. DAWID. 1976. Characterization and mapping of mitochondrial ribosomal RNA and mitochondrial DNA in *Drosophila melanogaster*. *Cell* **9**: 615.

LENGYEL, J., A. SPRADLING, and S. PENMAN. 1975. Methods with insect cells in suspension culture. II. *Drosophila melanogaster*. *Methods Cell Biol.* **10**: 195.

NASS, M. M. K. and C. A. BUCK. 1970. Studies on mitochondrial tRNA from animal cells. II. Hybridization of aminoacyl-tRNA from rat liver mitochondria with heavy and light complementary strands of mitochondrial DNA. *J. Mol. Biol.* **54**: 187

OJALA, D. and G. ATTARDI. 1974. Identification and partial characterization of multiple discrete polyadenylic acid-containing RNA components coded for by HeLa cell mitochondrial DNA. *J. Mol. Biol.* **88**: 205.

PEACOCK, W. J., D. BRUTLAG, E. GOLDRING, R. APPELS, C. W. HINTON, and D. L. LINDSLEY. 1974. The organization of highly repeated DNA sequences in *Drosophila melanogaster* chromosomes. *Cold Spring Harbor Symp. Quant. Biol.* **38**: 405.

PELLING, C. 1959. Chromosomal synthesis of ribonucleic acid as shown by incorporation of uridine labeled with tritium. *Nature* **184**: 655.

———. 1964. Autoradiographische Untersuchungen an *Chronomous tentans*. *Chromosoma* **15**: 71.

PERLMAN, S., H. T. ABLESON, and S. PENMAN. 1973. Mitochondrial protein synthesis: RNA with the properties of eukaryotic messenger RNA. *Proc. Natl. Acad. Sci.* **70**: 350.

POLAN, M. L., S. FRIEDMAN, J. G. GALL, and W. GEHRING. Isolation and characterization of mitochondrial DNA from *Drosophila melanogaster*. *J. Cell Biol.* **56**: 580.

RITOSSA, F. 1962. A new puffing pattern induced by heat shock and DNP in *Drosophila*. *Experientia* **18**: 571.

ROBB, J. 1969. Maintenance of imaginal discs of *Drosophila melanogaster* in a chemically defined media. *J. Cell Biol.* **41**: 876.

SCHATZ, G. and T. L. MASON. 1974. The biosynthesis of mitochondrial proteins. *Annu. Rev. Biochem.* **43**: 51.

SOUTHERN, E. M. 1975a. Long range periodicities in mouse satellite DNA. *J. Mol. Biol.* **94**: 51.

———. 1975b. Detection of specific sequences among DNA fragments separated by gel electrophoresis. *J. Mol. Biol.* **98**: 503.

SPRADLING, A., M. L. PARDUE, and S. PENMAN. 1977. Messenger RNA in heat-shocked *Drosophila* cells. *J. Mol. Biol.* **109**: 559.

SPRADLING, A., S. PENMAN, and M. L. PARDUE. 1975. Analysis of *Drosophila* mRNA by *in situ* hybridization: Sequences transcribed in normal and heat shocked cultured cells. *Cell* 4: 395.

ZHIMULEV, I. F. and E. S. BELYAEVA. 1975. ^{3}H-uridine labeling patterns in the *Drosophila melanogaster* salivary gland chromosomes X, 2R, and 3L. *Chromosoma* 49: 219.

QUESTIONS/COMMENTS

Question by:
  E. ZUCKERKANDL
  Linus Pauling Institute

Thin bands contain as little as 5000 base pairs of DNA. This is insufficient for establishing a structure above the level of the nucleosome "solenoid" (or whatever structures are found at that level). Thick bands might function in gene regulation in a qualitatively different way. They alone can, in principle, accommodate the next higher order structure. Perhaps puffing is predominantly the melting of this next higher order structure. In that case, notable puffs would occur only in the thicker bands, and the thin bands would be those that transcribe without much puffing. Since there are more thin bands than thick bands, this would imply that the fraction of bands which transcribe without notable puffing is important. You mentioned bands that are transcribed without puffing. Do you happen to know what their proportion is?

Response by:
  M. L. PARDUE
  Massachusetts Institute of Technology

We haven't done that type of analysis. However, the two bands for which we have actually identified cytoplasmic RNAs at times when there is no evident puffing, the 5S locus at 56F and the histone locus at 39DE, are both thin bands, although 39DE also includes some thin bands.

Modified Histones in HeLa and Friend Erythroleukemia Cells Treated with *n*-Butyrate

M. G. RIGGS, R. G. WHITTAKER,* J. R. NEUMANN, AND V. M. INGRAM

Massachusetts Institute of Technology, Department of Biology, Cambridge, Massachusetts 02139

n-Butyrate at low concentrations (1–2 mM) causes Friend erythroleukemia cells to begin hemoglobin synthesis (Leder and Leder 1975). Prasad and Sinha (1976) have summarized the effects of *n*-butyrate on neuroblastoma, HeLa, and other cell types. They and others (Ginsburg et al. 1973; Wright 1973; Griffin et al. 1974; Simmons et al. 1975; Ghosh et al. 1975; Fishman et al. 1976; Altenburg et al. 1976; Henneberry and Fishman 1976) have seen morphological modifications, reversible inhibition of proliferation, decrease of DNA content, and increases in the activity of specific enzymes, such as adenylate cyclase, alkaline phosphatase, and a sialyltransferase. We have also studied in vivo and in cell-free systems the reversible inhibition of DNA synthesis in chick embryo fibroblasts and in HeLa cells (Hagopian et al. 1977). We now wish to describe rapid, extensive, and reversible increases in histone acetylation caused by low concentrations of *n*-butyrate. Some of these results have been reported elsewhere (Riggs et al. 1977).

METHODS

HeLa cells were maintained at 37°C in suspension culture at 0.1–0.5 × 10⁶/ml in Joklik's modified MEM containing 7% horse serum. Friend erythroleukemia cells (clone 745-PC-4) (Gusella et al. 1976) were grown at 37°C in suspension in α medium (Stanners et al. 1971) lacking nucleosides and supplemented with 15% fetal calf serum. They were maintained at 0.05–0.5 × 10⁶/ml to permit log-phase growth.

Aliquots of a stock solution of 1 M Na *n*-butyrate in Ca⁺⁺- and Mg⁺⁺-free PBS (final pH 7.3) were added to cultures 24 hours before histone extraction. Preparations of nuclei and histones were carried out at 0–4°C. Nuclei were prepared from washed cells by Dounce homogenization and histones extracted in 0.4 N H₂SO₄ (Panyim et al. 1971). Histones were precipitated overnight at −40°C with 10 volumes of acetone, dried, and electrophoresed on acid-urea-15% acrylamide gels (Moss et al. 1973) at 20°C for 5 hours; the gels were prepared in 0.125% ammonium persulfate. Alternatively, histones, prepared

as before, were run on 15% gels containing 6 M urea and 0.38% Triton DF-16, according to the method of Alfageme et al. (1974). Staining was with 0.1% amido black in 40% methanol, 7% acetic acid. The gels were scanned at 615 nm.

RESULTS

Histone Modifications

Nuclei from *n*-butyrate-treated HeLa and Friend erythroleukemia cells contain greatly increased amounts of modified forms of histone H4 (Figs. 1, 2, and 3). Total histones from nuclei of cells after 24 hours of treatment show several strong, slower moving H4 bands when examined in acid-urea gels and in Triton DF-16–acid-urea gels. Their electrophoretic positions correspond to the mono-, di-, tri-, and tetraacetylated H4 histones which occur during normal histone biosynthesis (Ruiz-Carrillo et al. 1975). For this reason and because of the experiments described below, we believe that the modified H4 histones are acetylated. In control cells, only a small amount of monoacetyl-H4 is present, whereas this is the dominant form in treated cells. In HeLa cells, the extent of modification, although marked at 1 mM, increases at higher *n*-butyrate concentrations (Fig. 1). H4 acetylation and H3 modification in HeLa cells are already noticeable after only 1 hour in 5 mM *n*-butyrate (Fig. 2), the earliest effect of *n*-butyrate reported. Friend cells also show the acetylated H4 pattern after 24 hours in 1 mM *n*-butyrate (Fig. 3); it becomes more pronounced at 3 mM. In both HeLa and Friend erythroleukemia cells, the histone H4 pattern reverts completely to normal when cells treated with *n*-butyrate for 24 hours are shifted back to control medium for a further 24 hours (Fig. 3).[1]

In both cell lines there is also a clear alteration of the histone H3 band with an increase of slower moving material. This is a broadening of the H3 band rather than the appearance of new distinct bands. There is a striking increase in the relative amounts of H1 (Figs. 1, 2, and 3). These changes are also reversible within 24 hours after removal of *n*-butyrate.

---

* Present address: University of New South Wales, School of Biochemistry, P.O. Box 1, Kensington, New South Wales 2033, Australia.

[1] Note added in proof: More recent experiments have shown almost complete reversal of the histone modifications at 30 minutes after return of the cells to control medium.

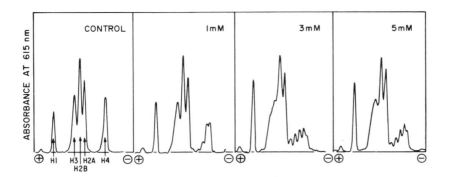

Figure 1. Effect of different concentrations of *n*-butyrate on histone modification in HeLa cells at 24 hr, examined on acid-urea gels. The scans of the gels were normalized to peak H2B.

To investigate the nature of the histone modification, total acid-extracted histones from HeLa cells treated for 24 hours with 5 mM *n*-butyrate were completely digested with trypsin and pronase (Gershey et al. 1968). After separation from the proteolytic enzymes by Sephadex G25 column gel filtration, the free amino acids were analyzed on a Durrum analyzer, Model D500. A peak corresponding to ϵ-*N*-acetyl-lysine was found just in front of glycine in the position of a standard prepared from acetylated bovine serum albumin (Wong and Marushige 1976). This peak was absent in the sample prepared from control cells. No evidence for ϵ-*N*-*n*-butyryl-lysine was found between alanine and valine in the position of an ϵ-*N*-*n*-butyryl-lysine standard. The amount of ϵ-*N*-acetyl-lysine was 1.7 moles/mole H4. It remains to be determined whether these acetyl groups are all on H4 or whether they are more generally distributed. Quantitative analysis of the scans of gels shown in Figure 2 indicated that the sum of the H4 components in *n*-butyrate-treated cells was equal to the sum of the H4 components in control cells and that no H4 had been lost. The proportion of acetylated H4 reaches approximately 80% of total H4 at 24 hours in 3 mM (Fig. 1) or 5 mM *n*-butyrate (Fig. 2). In addition, we believe that the modification is acetylation and not phosphorylation, since incubation of the modified histones with *E. coli* alkaline phosphatase (Thomas and Hempel 1976) leaves the modified H4 pattern intact.

It is possible that the extensive histone acetylation in the presence of *n*-butyrate is involved in the shutdown of DNA synthesis, although we do not as yet have direct proof of such a mechanism. However,

in HeLa cells, the time courses for extent of histone acetylation and for decrease of DNA synthesis (to 10% or less) are roughly parallel in the first 24 hours. In other cell-free experiments, we have obtained evidence that nuclei from *n*-butyrate-treated cells are inactive in DNA synthesis even in control cytosol, suggesting a modification of nuclear structure (Hagopian et al. 1977).

DISCUSSION

We are particularly interested in the effect of *n*-butyrate on the proliferative capacity of the malignant Friend erythroleukemia cell and of the HeLa cell, which is also derived from a malignant cell. In the Friend cell, differentiation is blocked. *n*-Butyrate induces the continuation of differentiation and the production of specific products, as well as the striking accumulation of acetylated histones. We do not know whether there is a causal relationship between histone acetylation and these observations. Darzynkiewicz et al. (1976) and Nudel et al. (1977) have reported subtle changes in the structure of Friend-cell chromatin using either DMSO or *n*-butyrate as the inducing agent.

It is not yet known whether *n*-butyrate affects tumorigenicity of the Friend cell. If it behaves like the original inducing agent, DMSO, we would predict a decrease in tumorigenicity. Moore and Chalkley (1977), using cultured hepatoma cells, suggest an interdependence of the level of histone H4 acetylation and the rate of cell division. We do not observe acetylation changes in the histone pattern of DMSO-treated Friend cells (J. R. Neumann, unpubl.); per-

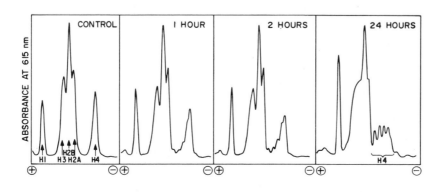

Figure 2. Time course of *n*-butyrate effect on HeLa histones, examined on acid-urea gels, as in Fig. 1. The scans of the gels were normalized to peak H2B.

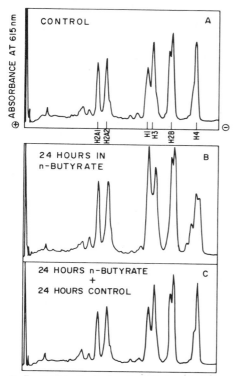

Figure 3. Reversal of n-butyrate effect on histones from Friend erythroleukemia cells examined on Triton–acid–urea gels. The scans of the gels were normalized to peak H2B. Untreated Friend cells *(A)*, cells exposed to 1 mM n-butyrate for 24 hr *(B)*, and 24-hr n-butyrate-treated cells, washed, and resuspended in fresh control medium for 24 hr *(C)*.

haps the two agents act at different points in these complex processes. Further experimental work is needed to look for a causal relationship between the accumulation of polyacetylated histones in the presence of n-butyrate on the one hand and the inhibition of DNA synthesis and the switching of malignant cells to nonmalignant differentiating cells on the other.

Acknowledgments

We thank Professor L. A. Steiner and Erlinda Capuno for their help with the amino acid analyses. We also wish to acknowledge the technical assistance of Barbara Blanchard. The Friend erythroleukemia cells were grown in Professor David Housman's laboratory, Center for Cancer Research, Massachusetts Institute of Technology; the expert help of Virginia Weeks is gratefully acknowledged. This work was supported by a grant from the National Institutes of Health (AM 13945).

REFERENCES

ALFAGEME, C. R., A. ZWEIDLER, A. MAHOWALD, and L. H. COHEN. 1974. Histones of *Drosophila* embryos. *J. Biol. Chem.* 249: 3729.

ALTENBURG, B. C., D. P. VIA, and S. H. STEINER. 1976. Modification of the phenotype of murine sarcoma virus-transformed cells by sodium butyrate. *Exp. Cell Res.* 102: 223.

DARZYNKIEWICZ, A., F. TRAGANOS, T. SHARPLESS, C. FRIEND, and M. R. MELAMED. 1976. Nuclear chromatin changes during erythroid differentiation of Friend virus induced leukemic cells. *Exp. Cell Res.* 99: 301.

FISHMAN, P. H., R. M. BRADLEY, and R. C. HENNEBERRY. 1976. Butyrate-induced glycolipid biosynthesis in HeLa cells: Properties of the induced sialyltransferase. *Arch. Biochem. Biophys.* 172: 618.

GERSHEY, E. L., G. VIDALI, and V. G. ALLFREY. 1968. Chemical studies of histone acetylation. *J. Biol. Chem.* 243: 5018.

GHOSH, N. K., S. I. DEUTSCH, M. J. GRIFFIN, and R. P. COX. 1975. Regulation of growth and morphological modulation of HeLa$_{65}$ cells in monolayer culture by dibutyryl cyclic AMP, butyrate and their analogs. *J. Cell. Physiol.* 86: 663.

GINSBURG, E., D. SALOMON, T. SREEVALSAN, and E. FREESE. 1973. Growth inhibition and morphological changes caused by lipophilic acids in mammalian cells. *Proc. Natl. Acad. Sci.* 70: 2457.

GRIFFIN, M. H., G. H. PRICE, K. L. BAZZELL, R. P. COX, and N. K. GHOSH. 1974. A study of adenosine 3':5'-cyclic monophosphate, sodium butyrate and cortisol as inducers of HeLa alkaline phosphatase. *Arch. Biochem. Biophys.* 164: 619.

GUSELLA, J., R. GELLER, B. CLARKE, V. WEEKS, and D. HOUSMAN. 1976. Commitment to erythroid differentiation by Friend erythroleukemia cells: A stochastic analysis. *Cell* 9: 221.

HAGOPIAN, H. K., M. G. RIGGS, L. A. SWARTZ, and V. M. INGRAM. 1977. Effect of n-butyrate on DNA synthesis in chick fibroblasts and HeLa cells. *Cell* 12: 855.

HENNEBERRY, R. C. and P. H. FISHMAN. 1976. Morphological and biochemical differentiation in HeLa cells. *Exp. Cell Res.* 103: 55.

LEDER, A. and P. LEDER. 1975. Butyric acid, a potent inducer of erythroid differentiation in cultured erythroleukemic cells. *Cell* 5: 319.

MOORE, M. R. and R. CHALKLEY. 1977. The effect of perturbations of the environment on histone acetylation and its turnover in hepatoma tissue culture cells. *Fed. Proc.* 36: 785 (Abstr.).

MOSS, B. A., W. G. JOYCE, and V. M. INGRAM. 1973. Histones in chick embryonic erythropoiesis. *J. Biol. Chem.* 248: 1025.

NUDEL, U., J. E. SALMON, M. TERADA, A. BANK, R. A. RIFKIND, and P. A. MARKS. 1977. Differential effects of chemical inducers on expression of β globin genes in murine erythroleukemia cells. *Proc. Natl. Acad. Sci.* 74: 1100.

PANYIM, S., D. BILEK, and R. CHALKLEY. 1971. An electrophoretic comparison of vertebrate histones. *J. Biol. Chem.* 246: 4206.

PRASAD, K. N. and P. K. SINHA. 1976. Effect of sodium butyrate on mammalian cells in culture: A review. *In Vitro* 12: 125.

RIGGS, M. G., R. G. WHITTAKER, J. R. NEUMANN, and V. M. INGRAM. 1977. n-Butyrate causes histone modification in HeLa and Friend erythroleukemia cells. *Nature* 268: 462.

RUIZ-CARRILLO, A., L. J. WANGH, and V. G. ALLFREY. 1975. Processing of newly synthesized histone molecules. *Science* 190: 117.

SIMMONS, J. L., P. H. FISHMAN, E. FREESE, and R. O. BRADY. 1975. Morphological alterations and ganglioside sialyltransferase activity induced by small fatty acids in HeLa cells. *J. Cell Biol.* 66: 414.

STANNERS, G. P., G. L. ELICEIRI, and H. GREEN. 1971. Two types of ribosomes in mouse-hamster hybrid cells. *Nat. New Biol.* 230: 52.

THOMAS, G. and K. HEMPEL. 1976. Correlation between histone phosphorylation and tumor ageing in Ehrlich ascites tumor cells. *Exp. Cell Res.* **100**: 309.

WONG, T. K. and K. MARUSHIGE. 1976. Modification of histone binding in calf thymus chromatin and in the chromatin-protamine complex by acetic anhydride. *Biochemistry* **15**: 2041.

WRIGHT, J. A. 1973. Morphology and growth rate changes in Chinese hamster cells cultured in presence of sodium butyrate. *Exp. Cell Res.* **78**: 456.

The Effect of Heat Shock on Gene Expression in
Drosophila melanogaster

M.-E. Mirault, M. Goldschmidt-Clermont, L. Moran, A. P. Arrigo, and A. Tissières

Département de Biologie Moléculaire, Université de Genève, 1211 Geneva, Switzerland

When *Drosophila melanogaster* is exposed to 37°C, a series of specific genes is activated, whereas most of the other genes, active at 25°C before this heat shock, appear to be repressed.

The induction of about eight or nine new puffs on the salivary gland chromosomes occurs very shortly after the temperature shift to 37°C, and at the same time most of the puffs active at 25°C before the heat shock rapidly regress at 37°C (Ritossa 1962; Ashburner 1970). The same new puffs are induced under a variety of other stress conditions unrelated to temperature (Berendes 1972; Lewis et al. 1975). The heat shock also induces the rapid synthesis of a small number of proteins, whereas the rate of synthesis of most cellular proteins, normally made at 25°C, is strongly reduced (Tissières et al. 1974; Lewis et al. 1975; McKenzie et al. 1975). This phenomenon is observed in different tissues (Tissières et al. 1974), in many different wild-type strains (L. Moran and A. P. Arrigo, unpubl.), and in different *Drosophila melanogaster* tissue-culture cell lines (McKenzie et al. 1975; Moran et al. 1977). These observations suggest that new messenger RNAs, coding for the heat-shock proteins, are synthesized at the high temperature, whereas most preexisting messenger RNAs, made at 25°C, are no longer translated. This is further suggested by the rapid disappearance of preexisting polyribosomes in cells shifted to high temperature, followed by the appearance of new and larger heat-shock polyribosomes (McKenzie et al. 1975). New species of RNA are synthesized at the high temperature and they were shown to hybridize in situ at heat-shock puff sites (McKenzie et al. 1975; Spradling et al. 1975, 1977).

Here we report the analysis and characterization of the heat-shock-induced polypeptides and messenger RNAs from tissue-culture cells. The heat-shock-induced messenger RNAs are shown to direct the synthesis of the heat-shock-induced polypeptides. Moreover, evidence is presented to show that messenger RNAs, normally synthesized and translated at 25°C, are still present in the cytoplasm after heat shock, although not efficiently translated in the cells.

RESULTS

Synthesis of Heat-shock-induced Proteins

Tissue-culture cells or salivary glands were pulse-labeled with [^{35}S]methionine, both at 25°C and after a heat shock at 37°C; the left panel in Figure 1 shows the electrophoretic pattern of the labeled proteins. A small number of new protein bands appear after heat shock, but the synthesis of most of the proteins synthesized at 25°C is strongly reduced, as observed earlier (Tissières et al. 1974; Lewis et al. 1975; McKenzie et al. 1975). The overall rate of incorporation of [^{35}S]methionine is roughly the same at the two temperatures. Salivary gland and tissue-culture cell labeling patterns after heat shock are qualitatively similar (see Fig. 1), and similar patterns are observed if [^{3}H]leucine is substituted for [^{35}S]methionine (data not shown).

The apparent molecular weights of the heat-shock polypeptides, as given in Figure 1, were determined from their migration in sodium dodecyl sulfate (SDS)-polyacrylamide gels of different concentrations, calibrated with protein standards of known molecular weight. The molecular weights of several heat-shock polypeptides were dependent on gel concentrations, which could be due to chemical modifications. Thus the molecular weights should be considered as nominal values.

Kinetics of synthesis of heat-shock proteins. A pulse-labeling with [^{35}S]methionine of about 10 minutes from the start of the heat shock was necessary to detect the heat-shock proteins in tissue-culture cells. Within this time, synthesis of all heat-shock proteins is initiated. We could not observe a clear-cut sequence of appearance in contrast to the observation of Lewis et al. (1975) with *Drosophila hydei*, although our experiments cannot rule out an asynchrony of a few minutes.

The maximum rate of synthesis of the heat-shock proteins in tissue-culture cells is reached within 90–120 minutes and subsequently declines to about 50% of its initial rate after 6–8 hours of prolonged incubation at 37°C. In contrast, when the tissue-culture cells are brought back to 25°C after a heat shock

819

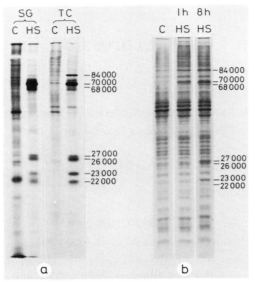

Figure 1. The effect of heat shock on protein synthesis.
(a) Salivary glands (SG) dissected from wild-type
Drosophila melanogaster (Kolmar) raised at 25°C were labeled with [³⁵S]methionine (40 μCi/ml, New England Nuclear, 300–500 Ci/mM) for 30 min at 25°C (control, C) and
for the same time at 25°C following a heat shock of 20
min (HS) in a medium consisting of one part 10% ethanol
and five parts of Grace's (1962) medium without methionine. A tissue-culture cell line KC161 from Dr. Echalier
(Echalier and Ohanessian 1969) was adapted to grow in
suspension in the D22 medium of Echalier and Ohanessian
(1970), supplemented with 2% fetal calf serum; the cells
were grown at 25°C to concentrations between 2 and 8 × 10⁶
cells per ml. Tissue-culture cells (TC) were labeled for 1
hr at 25°C (C) or for the same time at 37°C following a
2-hr preincubation at 37°C (HS) in medium without cold
methionine, yeast extract, and serum. The proteins were
electrophoresed in SDS-12.5% polyacrylamide gels according to the method of Laemmli (1970) and Studier (1973),
and the labeled proteins were detected by autoradiography.
(b) Coomassie blue staining patterns of proteins from
tissue-culture cells grown at 25°C (C) and from cells incubated for 1 or 8 hr at 37°C (HS) showing the accumulation
of heat-shock proteins with time.

of 1 hour and the rate of synthesis of heat-shock
proteins is examined by pulse-labeling at various
times thereafter, we find a gradual decrease of the
rate of synthesis of these proteins until, by 8 hours,
no synthesis of heat-shock protein is detected, and
normal protein synthesis has resumed; the cells continue to grow normally and can respond to a second
heat shock.

The stability of the heat-shock proteins was investigated in pulse-chase experiments with salivary
glands and tissue-culture cells. The proteins were
labeled with [³⁵S]methionine for 1 hour at 37°C and
chased for various times at 25°C with 200 μg/ml
unlabeled methionine. The intensity of each of the
labeled heat-shock protein bands seen on autoradiographs of SDS-polyacrylamide gels did not show any
significant variation for periods up to 20 hours, and
the total amount of radioactivity in acid-precipitable
material remained constant throughout the experi-

ment. We conclude that, under these conditions, the
heat-shock proteins are stable.

When tissue-culture cells are shifted to 37°C and
incubated at this temperature, the heat-shock proteins are synthesized at high rates, and after 1 to 2
hours they can already be detected on gels stained
with Coomassie blue. They accumulate at 37°C until,
by 8 hours, they represent about 10% of the total
protein of the cells, as estimated from the densitometer tracing of the stained gels (Fig. 1, right panel).

Fingerprint Analysis of Heat-shock-induced Polypeptides

How many distinct heat-shock-induced polypeptides are there? In an attempt to answer this question, [³⁵S]methionine-labeled protein bands from
heat-shocked salivary glands and tissue-culture cells
were eluted from the gels, oxidized with performic
acid, and digested with trypsin. After separation by
two-dimensional chromatography, the methionine-
labeled peptides were detected by autoradiography.
The results are shown in Figure 2. The 84,000-,
70,000-, 68,000-, 26,000-, 23,000-, and 22,000-dalton
bands (see Fig. 1) all give clearly different fingerprints. The fingerprint of the 27,000-dalton band displays all the spots found in the fingerprint of the
26,000-dalton band, plus additional spots. Due to
technical difficulties in separating these two protein
bands completely, we cannot say whether these proteins share some common sequences. The results indicate that each of the other heat-shock protein
bands is likely to represent a distinct polypeptide.
On the basis of this analysis, we conclude that the
synthesis of at least six distinct polypeptides is induced by heat shock.

*Two-dimensional gel analysis of the heat-shock
polypeptides.* Since each fingerprint of Figure 2
could result from a possible mixture of different
polypeptides of the same or similar molecular
weights, we carried out a two-dimensional gel analysis according to the method of O'Farrell (1975). This
method combines isoelectric focusing in 8 M urea
in the first dimension and SDS-polyacrylamide
gel electrophoresis in the second. The effect of
heat shock on the two-dimensional gel patterns of
[³⁵S]methionine-labeled polypeptides is striking, and
the data confirm the observations already made in
the one-dimensional gel analysis. The major new
finding, however, is the complexity of the 70,000-
dalton heat-shock band, which now displays five distinct spots on the two-dimensional gel (Fig. 3, HS).
Although the last 70,000-dalton spot on the right
appears in the 25°C two-dimensional gel pattern
(Fig. 3, C) at roughly equal intensity, the four other
70,000-dalton protein spots are strongly induced by
heat shock. The 27,000-dalton band also appears as
multiple spots. On the other hand, the single spots
corresponding to the 22,000-, 23,000-, 26,000-, 68,-

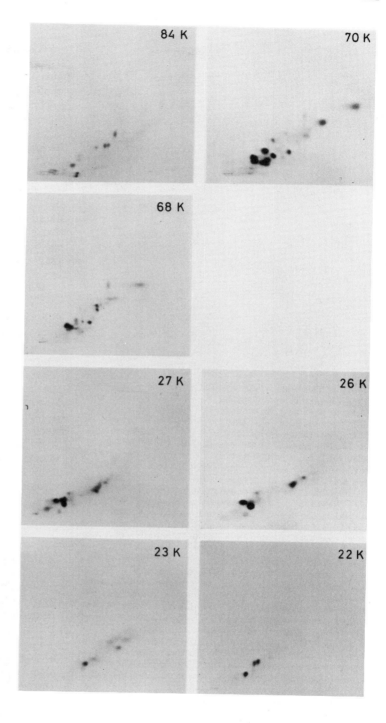

Figure 2. Tryptic fingerprints of the heat-shock-induced polypeptides. Tissue-culture cells were labeled with [^{35}S]methionine for 1 hr at 37°C, and the total proteins were fractionated by SDS-polyacrylamide gel electrophoresis as described in the legend to Fig. 1, except that the electrophoretic migration was increased in order to improve the separation of the 68,000- and 70,000-dalton polypeptides. The radioactive heat-shock-induced polypeptides were eluted from the gel by electrophoresis, oxidized with performic acid, and digested with trypsin as described by Allet et al. (1973). The tryptic peptides were separated by two-dimensional, thin-layer chromatography (first dimension, pyridine/isoamyl alcohol/water: 7/7/6; second dimension, butanol/pyridine/acetic acid/water: 5/4/1/4), and the labeled spots were detected by autoradiography.

000-, and 84,000-dalton protein bands are likely to originate from unique polypeptides, the synthesis of which is strongly induced by heat shock. Tryptic fingerprint analysis indicates that the two major spots in the 70,000 molecular weight range represent two polypeptides with very similar, if not identical, primary structures. These two spots account for over 90% of the protein synthesis at 70,000 daltons.

The differences in isoelectric points may result from posttranslational modification. Note that a di-

rect comparison of spot intensities in Figure 3 might lead to an overestimation of minor spots, since the major spots have been largely overexposed.

In several instances when the proteins were labeled at 25°C, we could detect weak spots in the two-dimensional gel pattern, corresponding precisely to heat-shock spot positions, e.g., (6.7; 1.85), (7.0; 1.85), (9.75; 1.35). They represent probably a low level of heat-shock protein synthesis at 25°C.

In conclusion, the evidence obtained so far sug-

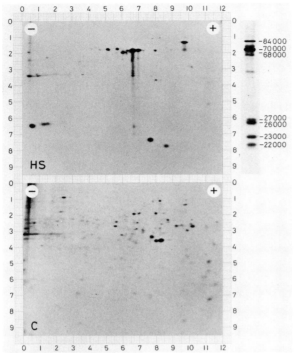

Figure 3. Two-dimensional gel analysis of [^{35}S]methionine-labeled heat-shock polypeptides. A cell culture was divided into two portions and each portion labeled with 50 μCi/ml [^{35}S]methionine. One portion was labeled for 2 hr at 37°C following a 1-hr incubation at 37°C in medium without methionine and yeast extract (HS); the other was labeled for 2 hr at 25°C following a 1-hr incubation at 25°C in the same medium (C). About 2.5 × 10^7 cells of each portion were washed in ice-cold isotonic saline and lysed in 0.1 ml of a solution containing 10 mM HEPES, pH 7.4, 0.5% NP-40, 50 μg/ml pancreatic ribonuclease A, 50 μg/ml pancreatic deoxyribonuclease I, and 20 mM β-mercaptoethanol. Each lysate was incubated under nitrogen for 10 min at 30°C followed by a further incubation of 5 min in the presence of 80 mM EDTA, pH 7.4. The samples were then treated according to the method of O'Farrell (1975), and an aliquot corresponding to about 2.5 × 10^5 cells was analyzed by isoelectric focusing in the first dimension followed by electrophoresis in 12.5% polyacrylamide-SDS gels in the second dimension (O'Farrell 1975). Fluorographic exposure was for 30 hr at −70°C with preflashed Kodak XR-5 films, as described by Laskey and Mills (1975). The heat-shock (HS) and control (C) patterns are shown in the figure. An autoradiogram of the heat-shock proteins analyzed by electrophoresis in the SDS dimension only is shown on the right as reference. The 70,000, 68,000, 27,000, and 26,000 heat-shock protein bands display spots with the following coordinates: (5.2; 1.8), (5.7; 1.8), (6.65; 1.85), (7.0; 1.85), (8.0; 1.85), (6.05; 2.0), (1.3; 6.35), (1.5; 6.35), (0.6; 6.5). Evidence suggests that the reproducible streaking-like pattern seen in the SDS dimension of the figure (HS) results from some partial proteolytic digestion of the major 70,000-dalton heat-shock protein, possibly occurring within the cells at 37°C.

gests that heat shock strongly induces the synthesis of at least six distinct polypeptides, and possibly as many as nine.

Heat-shock-induced Messenger RNAs

Polyribosomes from heat-shocked cells consist of a major fraction composed of large polyribosomes with 20 to 30 ribosomes and a minor fraction of smaller polyribosomes (McKenzie et al. 1975). The newly synthesized poly(A)-containing RNA found in these polyribosomes was analyzed by sedimentation in sucrose gradients and compared with the RNA from polyribosomes of the corresponding size from non-heat-shocked cells (Fig. 4). Polysomal RNA, labeled with [^{3}H]uridine at 37°C, sediments essentially as two peaks: a 20S RNA fraction predominant in the large polyribosomes and a 12S RNA fraction found mostly in the small polyribosomes. In contrast, the poly(A)-containing RNA labeled at 25°C sediments quite heterogeneously throughout the gradient, and very little of this labeled RNA is seen in the heat-shock RNA sedimentation profiles.

Several RNA fractions isolated from a preparative sucrose gradient and sedimenting at 20S, 18S, 15S, and 12S were assayed for messenger RNA activity in vitro, using the messenger-dependent reticulocyte lysate described by Pelham and Jackson (1976). The analysis of the products synthesized in vitro are shown in Figure 5. The 20S heat-shock RNA directed the synthesis of at least three polypeptides of 84,000, 70,000 and 68,000 daltons, plus traces of smaller products. The 12S heat-shock RNA directed the synthesis of at least three polypeptides comigrating with the small heat-shock polypeptides labeled in vivo. Electrophoretic migration, fingerprint analysis of each individual protein band labeled in vivo and in vitro, respectively, and two-dimensional gel analysis in several cases suggest that probably all of the heat-shock polypeptides have been faithfully synthesized in vitro. RNA fractions sedimenting at 18S and 15S directed the synthesis of small amounts of products comigrating with polypeptides synthesized in vivo at 25°C. The products synthesized in vitro by the 20S, 18S, 15S, and 12S RNA fractions from cells kept at 25°C (Fig. 5, left panel) correspond to those observed after in vivo labeling at 25°C.

Further purification of individual heat-shock messenger RNAs was attempted by polyacrylamide gel electrophoresis. Figure 6 shows the electrophoretic pattern of [^{3}H]uridine-labeled 20S heat-shock RNA in a 7 M urea-polyacrylamide gel. Three RNA peaks have been partially resolved. When assayed in vitro, fraction III directed the synthesis of a 68,000-dalton polypeptide; fraction II, the most abundant heat-shock mRNA, directed predominantly the synthesis of a 70,000-dalton polypeptide; and fraction I directed the synthesis of an 84,000- as well as a 70,000-dalton polypeptide. This last fraction was still contaminated by messengers present in fraction II.

Similar electrophoretic and translation patterns were obtained with 20S heat-shock RNA fraction-

Figure 4. Sucrose gradient sedimentation of polysomal poly(A)-containing RNA synthesized at 37°C *(a,b)* and at 25°C *(c,d)*. Tissue-culture cells were labeled with [³H]uridine (100 μCi/ml) for 60 min at 37°C following a 15-min incubation at this temperature, and for 60 min at 25°C in a parallel incubation. The cells were quickly chilled to 0°C and lysed according to the method of McKenzie et al. (1975). Fractions of large and small polyribosomes were pooled from sucrose gradients, as described previously, and precipitated with 1 volume of ethanol (Moran et al. 1977). The pellets (100–500 A₂₆₀ units) were first dissolved in 10 ml of a solution containing 0.02 M NaCl, 0.04 M EDTA, 0.02 M Tris-HCl, pH 7.5, 1% SDS, and 1 mg proteinase K (Merck, Germany), incubated 90 sec at 100°C, treated again with 1 mg proteinase K for 15 min at 25°C, and phenol-extracted according to the method of Aviv and Leder (1972). The RNA was fractionated on oligo(dT)-cellulose, and the poly(A)-containing RNA samples were analyzed by sedimentation in linear 5–20% sucrose gradients according to the method of Haines et al. (1974). Centrifugation was carried out for 150 min in the SW56 rotor of the Beckman centrifuge at 56,000 rpm and 20°C. *(a)* RNA from large, heat-shock

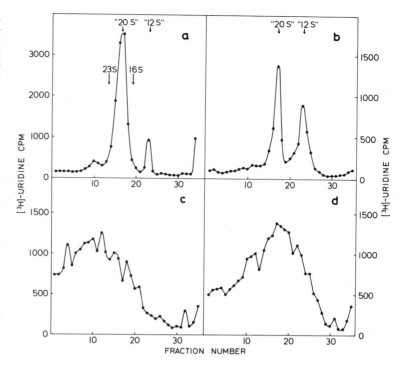

polyribosomes; *(b)* RNA from small, heat-shock polyribosomes; *(c)* RNA from large polyribosomes of cells labeled at 25°C; *(d)* RNA from small polyribosomes of cells labeled at 25°C.

Figure 5. Fluorogram of [³⁵S]methionine-labeled polypeptides synthesized in vitro by poly(A)-containing RNA fractions isolated from preparative sucrose gradients. The RNA was translated in vitro in the messenger RNA-dependent reticulocyte lysate (Pelham and Jackson 1976). Incubation was carried out for 60 min at 30°C at rate-limiting RNA concentrations, and the [³⁵S]methionine-labeled polypeptides were analyzed by SDS-polyacrylamide gel electrophoresis and detected by fluorography. *(Center panel)* The polypeptides labeled in vivo for 1 hr at 25°C (C) or during a 1-hr heat shock at 37°C (HS) are shown at two different exposures. *(Right and left panels)* The polypeptides synthesized in vitro by 20S, 18S, 15S, and 12S poly(A)-containing RNA fractions from heat-shock cells (HS) and from 25°C control cells (C), respectively. Fluorographic exposure was at −70°C for 5 hr.

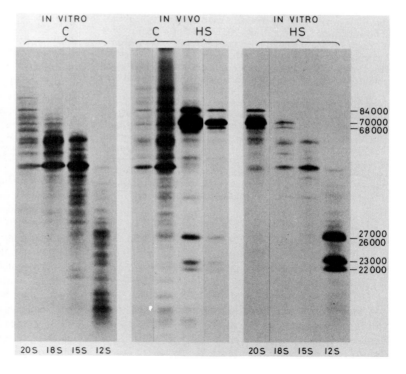

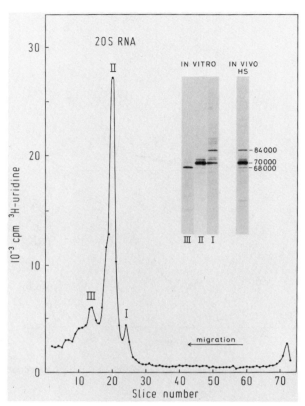

Figure 6. Electrophoretic pattern of 20S heat-shock poly(A)-containing RNA labeled with [³H]uridine. The 20S heat-shock poly(A)-containing RNA was isolated from sucrose gradients as in Fig. 4 and electrophoresed in 3.3% polyacrylamide gels in 7 M urea, according to the method of Reijnders et al. (1973) as modified by Spradling et al. (1977), at 14 V/cm for 22 hr. The cylindrical gels (14 cm long, 0.6 cm diameter) were cut frozen in 2-mm slices, and the RNA was eluted by shaking each slice in 0.5 ml 0.01 M Tris-HCl, pH 7.4 and 0.1% SDS at 4°C for 48 hr. This elution step was repeated once and the total TCA-precipitable radioactivity was determined. RNA fractions to be translated in vitro were repurified on oligo(dT)-cellulose prior to in vitro translation. More than 90% of the radioactivity was then recovered as poly(A)-containing RNA. After several ethanol precipitations to remove salt and SDS, suitable amounts of the RNA were dissolved directly into 15 µl messenger RNA-dependent reticulocyte lysate (Pelham and Jackson 1976) supplemented with [³⁵S]methionine. The lysates were incubated for 90 min at 30°C. Samples (2 µl) were analyzed by SDS-polyacrylamide gel electrophoresis. The fluorograms of the polypeptides synthesized in vitro by the 20S RNA peak fractions I, II, and III are shown next to the heat-shock polypeptides labeled in vivo.

ated in nondenaturing gels (Moran et al. 1977). The molecular weight of the major component in 20S RNA was estimated to be about 9×10^5 by electrophoresis in 98% formamide gels under fully denaturing conditions according to the method of Spohr et al. (1975). The fingerprint of the 70,000-dalton polypeptides synthesized in vitro by fraction II is shown in Figure 7, together with that of the 70,000-dalton heat-shock polypeptides synthesized in vivo. Both fingerprint patterns are identical. Thus, taken to-

gether, the results indicate that three heat-shock-induced messenger RNA species synthesized at 37°C each code for one of the three major heat-shock polypeptides of large molecular weight.

The electrophoretic pattern of the poly(A)-containing RNA from small heat-shock polyribosomes, labeled with [³H]uridine at 37°C, is shown in Figure 8. The 12S heat-shock RNA has been partially resolved into two peaks. The RNA from the minor peak directed predominantly the in vitro synthesis of a 23,000-dalton polypeptide. RNA from the major peak gave rise to three protein bands corresponding to the small heat-shock polypeptides labeled in vivo. Similar results were obtained with nondenaturing gels (Moran et al. 1977). The partial resolution of the 12S heat-shock RNA suggests the existence of at least three distinct messenger RNAs coding for the small heat-shock polypeptides.

Estimates of the quantity and specific activity of both the 20S and 12S RNAs clearly indicate that the major part of these heat-shock messenger RNAs is synthesized after transfer of the cells to the high temperature. Thus, the synthesis of these RNAs is induced by heat shock.

In summary, we have so far identified six heat-shock-induced messenger RNAs, each of which codes for a distinct heat-shock polypeptide.

Many messenger RNAs normally translated at 25°C are not efficiently translated after heat shock. The rapid shut-off in the synthesis of many proteins (Figs. 1, 3, and 9) and the concomitant disappearance of most 25°C polyribosomes after heat shock (McKenzie et al. 1975) suggest that most of the messenger RNA present at 25°C is either quickly degraded or not efficiently translated into proteins at high temperature.

When poly(A)-containing RNA prepared from the cytoplasm of heat-shocked tissue-culture cells is translated in vitro, the labeled products include not only the heat-shock polypeptides, but also those normally synthesized in vivo at 25°C (Fig. 9). Essentially the same results were obtained when in vitro translation was done under saturating, rather than rate-limiting, RNA concentrations, whether at 25°C or 37°C (R. J. Jackson, pers. comm.). In contrast, we have seen that purified poly(A)-containing RNA prepared from heat-shock polyribosomes directs mainly the synthesis of the heat-shock proteins.

We conclude that much of the messenger RNA made and normally translated at 25°C is still present in the cytoplasm of heat-shocked cells but not efficiently translated. The active polyribosomes contain mainly heat-shock-induced messenger RNAs. We have observed that at least some histones and several other proteins continue to be synthesized efficiently after heat shock, suggesting that some messenger RNAs are more resistant to the translation shut-off caused by heat shock.

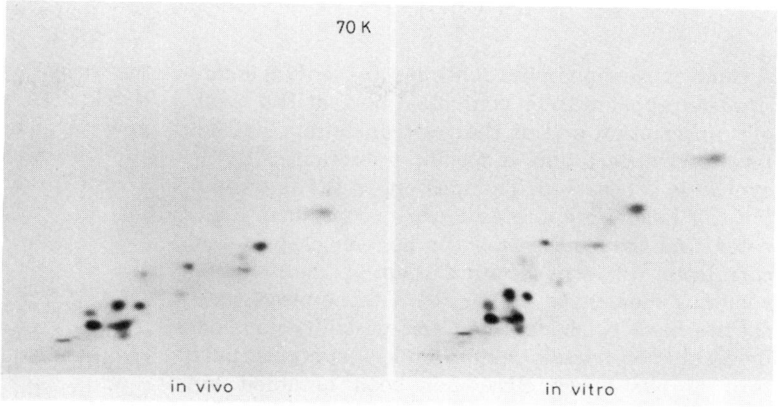

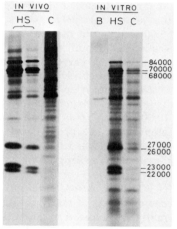

Figure 7. Tryptic fingerprints of the 70,000-dalton heat-shock polypeptides synthesized in vivo or in vitro as directed by the 20S fraction-II heat-shock mRNA (see Fig. 6). The [³⁵S] methionine-labeled polypeptides were eluted from the gels, digested with trypsin, and analyzed as described in the legend to Fig. 2.

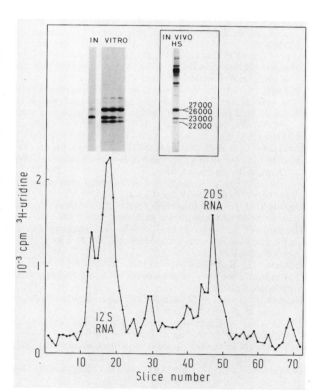

Figure 8. Electrophoretic pattern of poly(A)-containing RNA from small heat-shock polyribosomes. Electrophoresis was carried out in 3.5% polyacrylamide gels in 7 M urea and the RNA was recovered and translated in vitro as described in the legend to Fig. 6. The fluorograms of the polypeptides synthesized in vitro by discrete 12S RNA fractions and separated by electrophoresis are displayed on top of the corresponding gel slices. In vivo labeled heat-shock polypeptides analyzed in the same gel are shown as references.

Figure 9. In vitro translation products of cytoplasmic poly(A)-containing RNA from the cytoplasm of heat-shocked cells and of cells kept at 25°C. A culture of cells grown at 25°C was divided into two equal parts of about 2.5 × 10⁹ cells each. One part was incubated for 80 min at 37°C (HS) while the other part was kept at 25°C (C). The cells were lysed according to the method of McKenzie et al. (1975) and the postmitochondrial supernatants (2 ml) were phenol-extracted at room temperature (Aviv and Leder 1972) in 12 ml of a solution containing 0.02 M NaCl, 0.04 M EDTA, 0.02 M Tris-HCl, pH 7.5, 1% SDS, and 2 mg proteinase K (Merck, Germany). Phenol extraction and proteinase-K treatment were repeated until the interphase precipitates completely disappeared. After ethanol precipitation, the RNA was fractionated on oligo(dT)-cellulose and the poly(A)-containing RNA was reprecipitated twice with 3 volumes of ethanol to remove traces of SDS before final dissolution in sterile bidistilled water. The poly(A)-containing RNA was translated in vitro and the products displayed as in Fig. 5. *(Left)* Patterns of proteins synthesized in vivo: heat-shock proteins at two dilutions (HS) and proteins synthesized at 25°C (C). *(Right)* Patterns of proteins synthesized in vitro: B, blank with no RNA added; HS, with cytoplasmic poly(A)-containing RNA from heat-shocked cells; C, with cytoplasmic poly(A)-containing RNA from cells kept at 25°C.

DISCUSSION

Gene expression in heat-shocked *Drosophila melanogaster* appears to be controlled both at the level of transcription and at that of translation. At the level of transcription, a specific induction of RNA synthesis is observed. The messenger RNAs coding for the heat-shock proteins are produced at high rates, and the synthesis of the heat-shock proteins correlates with the accumulation of their corresponding messenger RNAs. Six of these messenger species have been identified and partially purified by gel electrophoresis. A similar electrophoretic pattern of heat-shock RNA has been obtained by Spradling et al. (1977), who have hybridized in situ specific RNA fractions to salivary gland chromosomes. Combining the results of in situ hybridization and in vitro translation reported here, a tentative correlation between puffs and polypeptides can be drawn in several cases. Assuming that the A1, A2, and A3 RNA species of Spradling et al. (1977) are the 20S-I, -II, and -III species described here, the corresponding heat-shock polypeptides of 84,000, 70,000, and 68,000 apparent molecular weight originate from sequences at puff sites 63C, 87A–87C1, and 95D, respectively. No correlation can be drawn as yet for the small heat-shock polypeptides as their messengers have not yet been adequately resolved.

Our results (Fig. 3 and fingerprint analysis of the two major 70,000-dalton spots shown in Fig. 3) are not inconsistent with the presence of two major 70,000-dalton polypeptides, with slight differences in primary structures, arising from two different genes.

In parallel with the induction of specific RNAs, shut off in the synthesis of many messenger RNA species normally synthesized at 25°C is observed. Normal developmental puffs regress after heat shock (Ashburner 1970), as does the synthesis of the corresponding messengers (see Fig. 4 together with the results of in situ hybridizations by McKenzie et al. [1975] and Spradling et al. [1977]).

Finally, the finding of messenger RNA not translated in heat-shocked cells but nevertheless present in the cytoplasm and translatable in vitro suggests an additional type of control at the level of translation. Although a more thorough analysis of this phenomenon remains to be done, we can conclude that an appreciable fraction of the messenger RNAs made at 25°C is still present but not translated efficiently in vivo at high temperature. The degree of translation shut-off varies from experiment to experiment, but we have observed a consistent decrease in the rate of synthesis of most cellular proteins in heat-shocked cells. Although the corresponding messenger RNAs may undergo some degradation, our in vitro translation results (Fig. 9) show that degradation alone cannot account for the translation shut-off observed in vivo at high temperature. The untranslated messengers could simply be outcompeted by the heat-shock-induced messenger RNAs. This has not been observed in vitro in the reticulocyte lysate system, even under heat-shock RNA saturating conditions. We do not know, however, whether such mRNA competition does occur within the cells, accounting for the shut off observed in vivo.

Acknowledgments

We wish to thank Mrs. Jacqueline Molliet, for excellent technical assistance, Dr. R. J. Jackson for helpful comments on the manuscript, and Dr. G. Echalier for a strain of *Drosophila melanogaster* tissue-culture cells. This work was supported by grant No. 3.491.075 from the Swiss National Foundation for Scientific Research.

REFERENCES

ALLET, B., K. J. KATAGIRI, and R. F. GESTELAND. 1973. Characterization of polypeptides made *in vitro* from bacteriophage lambda DNA. *J. Mol. Biol.* **78**: 589.

ASHBURNER, M. 1970. Patterns of puffing activity in the salivary gland chromosomes of *Drosophila*. V. Responses to environmental treatments. *Chromosoma* **31**: 356.

AVIV, H. and P. LEDER. 1972. Purification of biologically active globin messenger RNA by chromatography on oligothymidylic acid-cellulose. *Proc. Natl. Acad. Sci.* **69**: 1408.

BERENDES, H. D. 1972. The control of puffing in *Drosophila hydei*. In *Developmental studies on giant chromosomes* (ed. W. Beermann et al.), p. 186. Springer Verlag, New York.

ECHALIER, G. and A. OHANESSIAN. 1969. Isolement, en culture *in vitro*, de lignées cellulaires diploïdes de *Drosophila melanogaster*. *C. R. Acad. Sci. D* **268**: 1771.

———. 1970. *In vitro* culture of *Drosophila melanogaster* embryonic cells. *In Vitro* **6**: 162.

GRACE, T. D. C. 1962. Establishment of four strains of cells from insect tissues grown *in vitro*. *Nature* **195**: 788.

HAINES, M. E., N. H. CAREY, and R. D. PALMITER. 1974. Purification of ovalbumin messenger RNA. *Eur. J. Biochem.* **43**: 549.

LAEMMLI, U. K. 1970. Cleavage of structural proteins during the assembly of the head of bacteriophage T4. *Nature* **227**: 680.

LASKEY, R. A. and A. D. MILLS. 1975. Quantitative fluorography of acrylamide gels. *Eur. J. Biochem.* **56**: 335.

LEWIS, M., P. J. HELMSING, and M. ASHBURNER. 1975. Parallel changes in puffing activity and patterns of protein synthesis in salivary glands of *Drosophila*. *Proc. Natl. Acad. Sci.* **72**: 3604.

McKENZIE, S. L., S. HENIKOFF, and M. MESELSON. 1975. Localisation of heat-induced polyribosomal RNA. *Proc. Natl. Acad. Sci.* **72**: 1117.

MORAN, L., M.-E. MIRAULT, A. P. ARRIGO, M. GOLDSCHMIDT-CLERMONT, and A. TISSIÈRES. 1977. Heat shock of *Drosophila melanogaster* induces the synthesis of new messenger RNAs and proteins. *Proc. R. Soc. Lond. B* (in press).

O'FARRELL, P. H. 1975. High resolution two-dimensional electrophoresis of proteins. *J. Biol. Chem.* **250**: 4007.

PELHAM, H. R. B. and R. J. JACKSON. 1976. mRNA-dependent translation system from reticulocytes. *Eur. J. Biochem.* **67**: 247.

REIJNDERS, P., P. SLOOF, J. SIVAL, and P. BORST. 1973. Gel

electrophoresis of RNA under denaturing conditions. *Biochim. Biophys. Acta* **324**: 320.

RITOSSA, F. 1962. A new puffing pattern induced by temperature shock and DNP in *Drosophila. Experientia* **18**: 571.

SPOHR, G., M.-E. MIRAULT, T. IMAIZUMI, and K. SCHERRER. 1975. RNA electrophoresis on gels in formamide. *Eur. J. Biochem.* **62**: 313.

SPRADLING, A., M. L. PARDUE, and S. PENMAN. 1977. Messenger RNA in heat-shocked *Drosophila* cells. *J. Mol. Biol.* **109**: 559.

SPRADLING, A., S. PENMAN, and M. L. PARDUE. 1975. mRNA in *Drosophila*-cultured cells. *Cell* **4**: 395.

STUDIER, F. W. 1973. Analysis of bacteriophage T7 early RNAs and proteins on slab gels. *J. Mol. Biol.* **79**: 237.

TISSIÈRES, A., H. K. MITCHELL, and U. M. TRACY. 1974. Protein synthesis in salivary glands of *Drosophila melanogaster:* Relation to chromosome puffs. *J. Mol. Biol.* **84**: 389.

Correlation of Structural Changes in Chromatin with Transcription in the *Drosophila* Heat-shock Response

H. Biessmann, S. Wadsworth, B. Levy W.,* and B. J. McCarthy

Department of Biochemistry and Biophysics, University of California, San Francisco, California 94143

When the temperature is raised from 25°C to 37°C, specific puffs are induced in the polytene salivary gland chromosomes of *Drosophila melanogaster* larvae (Ritossa 1962; Ashburner 1970). The same response occurs in other organs of the larvae, in explanted glands in vitro, and even in cultured cells. In tissue-culture cells, drastic changes occur in the population of newly synthesized RNA following a heat shock. The newly synthesized RNA hybridizes to only those few regions of the chromosome at which puffs are induced in heat-shocked larvae (McKenzie et al. 1975; Spradling et al. 1975; Bonner and Pardue 1976). These changes in RNA synthesis lead to an altered pattern of protein synthesis in larvae and in cultured cells (Tissières et al. 1974; McKenzie et al. 1975; Lewis et al. 1975). It therefore appears that heat shock rapidly leads to the production of a new set of messengers which encode specific new proteins. This heat-induced change in transcription may provide a useful system for elucidation of the mechanism of changes in gene expression and any concomitant modulations of chromosome structure.

Considerable evidence exists that isolated chromatin retains some of its transcriptional specificity. Even when heterologous *Escherichia coli* RNA polymerase is employed, the ability to express a tissue-specific gene such as globin or ovalbumin is retained (Axel et al. 1973; Gilmour et al. 1974; Harris et al. 1976) despite the fact that in vitro transcription does not occur with absolute fidelity (Reeder 1973; Honjo and Reeder 1974; Wilson et al. 1975). In the present study we ask whether the heat-shock induction of new mRNA in *Drosophila* cells is accompanied by a structural change in chromatin which is stable during isolation and is recognizable by *E. coli* RNA polymerase. We also report some preliminary results using deoxyribonuclease I (DNase I) as a probe of altered chromatin structure (Weintraub and Groudine 1976).

RESULTS

Preparation of Heat-shock RNA and Complementary DNA

Cultured *Drosophila* cells were incubated for 1 hour at 37°C and polyadenylated RNA was prepared from polysomes as described by McKenzie et al. (1975). This poly(A) RNA was used as template for the synthesis of cDNA by reverse transcriptase (Levy W. and McCarthy 1975). The reaction was dependent upon the addition of oligo(dT) as a primer, as expected for transcription of sequences adjacent to poly(A).

The complexity of the heat-shock-induced poly(A)-containing RNA sequences associated with polysomes was estimated by analyzing the hybridization kinetics of this cDNA preparation to an excess of its own template (Fig. 1). The observed $R_0 t_{1/2}$ value of 7×10^{-2} moles·sec indicates the presence of some 60 RNA sequences of an average molecular weight of 6×10^5 according to the calibration developed by Bishop et al. (1974). This is only about 1–2% of the complexity of total polyadenylated mRNA in the same cell line grown at 25°C (Levy W. and McCarthy 1975). It is evident, however, that only about 50% of sequences represented in the cDNA probe are heat-shock-specific since about half as much hybridization occurred with poly(A)-containing mRNA from normal cells.

The average repetition frequency of the RNA sequences represented in the heat-shock cDNA was determined by annealing with an excess of unlabeled *Drosophila* DNA (Fig. 2). The $C_0 t_{1/2}$ value was approximately 30. Recalling that under our experimental conditions single-copy *Drosophila* DNA renatures with a $C_0 t_{1/2}$ of 150, we conclude that the genes responsible for heat-shock-induced RNA occur with an average frequency of some five copies per haploid genome.

Preparation of a cDNA Probe Specific for Heat-shock-induced RNA

The data of Figure 1 show that the cDNA probe was not completely specific for heat-shock-induced RNA. A more specific probe was prepared by annealing this cDNA to polyadenylated RNA from normal cells and removing the reacted material. After this reaction reached completion, the unreacted cDNA was purified by chromatography on hydroxylapatite. About 13% of the labeled cDNA failed to bind to the column in 0.02 M phosphate buffer and was discarded. Of the remainder, approximately half was recovered as single-stranded (ss) material by elution with 0.14 M phosphate, and half as double-stranded

* Present address: Division of Medical Biochemistry, University of Calgary, Alberta, Canada.

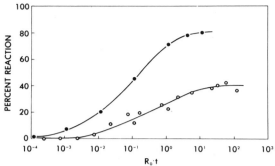

Figure 1. Kinetics of hybridization of cDNA prepared from poly(A) mRNA from heat-shock-induced polysomes to an excess of its own template (●) and to poly(A) mRNA from cells grown at 25°C (○). One hour after raising the temperature to 37°C, polysomes were layered on a 0.5–1.5 M sucrose gradient in 0.05 M Tris-HCl, pH 7.5, 0.25 M KCl, 0.025 M MgCl$_2$ and centrifuged for 75 min at 40,000 rpm in an SW41 rotor at 2°C. Fractions containing the heat-induced polysomes were pooled, and poly(A) mRNA was prepared by phenol/chloroform extractions and chromatography on poly(U)-Sepharose 4B. Hybridizations were done in 0.24 M phosphate buffer, 1 mM EDTA, 0.01% SDS at 70°C. The amount of cDNA in hybrid was determined by digestion with S$_1$ nuclease.

(ds) RNA/cDNA hybrid. The physical integrity of the fractionated cDNA was examined by reaction with excess *Drosophila* DNA. Both fractions reacted almost as well as the original total cDNA (Fig. 2), indicating that no appreciable degradation occurred during the fractionation.

To validate successful fractionation, the single-stranded and double-stranded materials were hybridized in a second cycle to poly(A)-containing mRNA from cells grown at 25°C (Fig. 3). The cDNA fraction which had formerly annealed to polyadenylated RNA from normal cells reacted to 75% when recycled with the same RNA. In contrast, the single-stranded material hybridized in the second cycle to

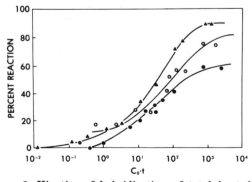

Figure 2. Kinetics of hybridization of total heat-shock cDNA to an excess of sheared *Drosophila* DNA (▲). After complete annealing to poly(A) mRNA from normal cells and separation of the hybrids from unreacted cDNA by chromatography on hydroxylapatite, the fractionated probe was hybridized again to an excess of unlabeled DNA in 0.3 M NaCl, 20 mM Tris-HCl, pH 7.5, 1 mM EDTA at 70°C. Single-stranded (ss) cDNA (○); double-stranded (ds) cDNA (●).

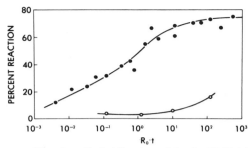

Figure 3. Kinetics of hybridization of the ds cDNA (●) and ss cDNA (○) to poly(A) mRNA from cells grown at 25°C.

mRNA from normal cells only at very high R_0t values, demonstrating removal of cross-reacting sequences from the cDNA probe. The failure to hybridize to mRNA of normal cells was not due to degradation of the single-stranded cDNA probe since it did react efficiently with total DNA (Fig. 2). This purified probe was employed to quantitate heat-shock-induced RNA transcribed in vitro.

In Vitro Transcription of Heat-shock-induced RNA

We next inquired whether chromatin isolated from heat-shocked cells would yield the specific heat-shock-induced RNA sequences when transcribed in vitro with *E. coli* RNA polymerase. The use of Hg-UTP as precursor facilitated the isolation of newly synthesized RNA by affinity chromatography on sulfhydryl-Sepharose 6B (Dale et al. 1975). Hybridization kinetics of RNA synthesized from normal and heat-shock chromatin with the total cDNA probe representing polyadenylated polysomal RNA of heat-shocked cells are illustrated in Figure 4. In both cases the R_0t values are shifted to about 30-fold higher values as compared with those obtained with polyadenylated polysomal RNA (see Fig. 1). This reflects the greater complexity of in-vitro-synthesized RNA previously reported (Biessmann et al. 1976). More important is the difference in plateau values attained. Transcripts from chromatin of heat-shocked cells gave a final reaction of 60%, whereas those from normal cells gave only a 35% reaction. This result demonstrates that sequences are tran-

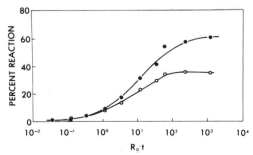

Figure 4. Kinetics of hybridization of in vitro transcript from chromatin which was isolated from cells after 1 hr of heat shock (●) and from cells grown at 25°C (○) to the total heat-shock cDNA.

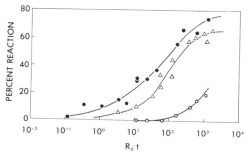

Figure 5. Kinetics of hybridization of heat-shock-specific ss cDNA with an excess of in vitro transcript from chromatin isolated from cells 1 hr after heat shock (●) and from cells grown at 25°C (○). Annealing of chromatin transcript from cells grown at 25°C to cDNA from poly(A) cytoplasmic RNA isolated from cells grown at 25°C (△).

scribed from chromatin of heat-shocked cells which are absent in the transcript from normal chromatin.

The validity of this conclusion was verified by repeating the same experiment with the cDNA probe enriched to represent only heat-shock-specific RNA (Fig. 5). RNA transcribed from chromatin of heat-shocked cells annealed with a $R_0t_{1/2}$ value of 30 to reach a plateau of almost 80% reaction. In contrast, transcript of control chromatin annealed with 100-fold slower kinetics.

Since complete replacement of UTP by Hg-UTP results in aberrant hybridization (Beebee and Butterworth 1976), the annealing of the same control chromatin transcript to cDNA from total poly(A) cytoplasmic RNA of cells grown at 25°C is also shown in Figure 5. The fact that the control transcript hybridizes to this cDNA with normal kinetics demonstrates that the observed kinetics with the heat-shock-specific probe are not an artifact of mercury content. Thus we conclude that heat-shock-specific sequences are transcribed from chromatin with 100-fold higher efficiency if cells are heat-shocked prior to chromatin isolation.

The use of mercurated UTP in the transcription reaction allows us to ask the important question of whether the RNA that hybridized to the labeled probe was in fact synthesized in vitro. If so, the resulting cDNA/RNA hybrids should contain Hg-UTP and should therefore bind to sulfhydryl-Sepharose. As a general approach to this problem, cDNA from cytoplasmic poly(A) RNA was hybridized to an excess of in vitro chromatin transcript in 50% formamide, 0.3 M NaCl, 20 mM Tris-HCl, pH 7.4, 1 mM EDTA at 35°C to avoid possible exchange of mercury (Dale and Ward 1975). After incubation to R_0t of 1100, the reaction mixture was treated with S_1 nuclease to remove unreacted cDNA. The S_1-resistant cDNA was then applied to a sulfhydryl-Sepharose 6B column, and 35% of the counts was retained and eluted by mercaptoethanol (Table 1). The fact that only 3% of the S_1-resistant cDNA was bound to the column when the hybrids were denatured shows that the cDNA was retained because it was hybridized to mercury-containing RNA. When the 35% was corrected for the efficiency of the sulfhydryl column, approximately 85%, it appeared that at least 42% of the S_1-resistant cDNA reacted with newly synthesized RNA containing mercurated nucleotides.

Digestion of Isolated Nuclei with DNase I

Since the chromatin transcription experiments revealed changes in chromatin structure, we also sought to detect such changes by means of sensitivity to DNase I (Weintraub and Groudine 1976). Ashburner (1970) showed that many puffs regress upon heat-shocking of salivary glands. One interpretation of this phenomenon is that the genes within the puffed regions are repressed by the heat shock. Our experiments attempted to detect a similar shut off of these genes in tissue-culture cells by testing the

Table 1. Retention of [³H]cDNA by Sulfhydryl-Sepharose 6B after Hybridization to Hg-containing In Vitro Chromatin Transcript

	After hybridization to in vitro chromatin transcript to R_0t of 1100 (a)	After hybridization to R_0t of 1100 and denaturation (b)
Percent of radioactivity recovered which was retained by the sulfhydryl column and eluted with mercaptoethanol	35	3

Chromatin from cells grown at 25°C was transcribed in the presence of Hg-UTP. This transcript was annealed in excess in 50% formamide, 0.3 M NaCl, 20 mM Tris, pH 7.4, 1 mM EDTA at 35°C with tritium-labeled cDNA from poly(A) cytoplasmic RNA from cells grown under the same conditions. The hybridization reaction was terminated at an R_0t of 1100 and treated with S_1 nuclease. After phenol extraction, the S_1-resistant material was ethanol-precipitated, redissolved in 10 mM Tris, pH 7.4, 100 mM NaCl, 1 mM EDTA, and split into two parts, one of which (a) was applied directly to a 10-ml sulfhydryl-Sepharose 6B column. The other aliquot (b) was denatured by heating for 7 min in a boiling water bath and then passed over the sulfhydryl-Sepharose column. Effluents were passed over the column twice, and unretained material was eluted with TNE. The bound material was eluted with TNE containing 0.1 M 2-mercaptoethanol. Fractions of 0.5 ml were collected and TCA-precipitable radioactivity was determined.

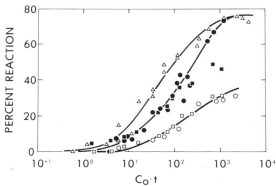

Figure 6. Kinetics of reassociation of ³H-labeled 25°C cDNA with DNA from nuclease-digested control or heat-shocked cells. Cells were heat-shocked by incubation in a controlled-temperature water bath such that the temperature of the culture reached 37°C within 10 min and was maintained at that temperature for 1 hr. Heat-shocked or control cultures were then processed identically, as described by Weintraub and Groudine (1976). Nuclei were resuspended routinely to an $OD_{630} = 32-33$, which was determined empirically to be a DNA concentration of 1 mg/ml. A standard curve of percent acid solubility vs time of incubation was constructed using cells labeled overnight with [³H]thymidine. The level of digestion of the samples used in the hybridization experiments was estimated from the time of incubation with DNase I. All DNA preparations were sheared in a French pressure cell at 16,000 psi. Undigested DNA (△); DNA from heat-shocked cells, 10% digested (■) and 20% digested (□); DNA from control cells, 10% digested (●) and 20% digested (○).

ability of DNase-I-digested DNA from control or heat-shocked cells to hybridize with labeled cDNA representing the genes active at 25°C. Nuclei were digested as described by Weintraub and Groudine (1976) to a level of 10% or 20% solubility; the DNA was extracted and used to drive labeled cDNA into hybrid (Fig. 6). Two points are to be made from the data. The level of DNase-I digestion required to remove significant amounts of active gene sequences is greater in *Drosophila* cells than in the chick erythrocyte cells used by Weintraub. This is probably due to the fact that a greater percentage of the *Drosophila* genome is devoted to actively transcribed gene sequences (Levy W. and McCarthy 1975). Compared to the hybridization kinetics of undigested DNA, removal of 10% of the sequences resulted in a 3-fold slower reaction, and of 20% in a 60-fold slower reaction, with the cDNA probe. This is in agreement with Weintraub's finding that active genes are preferentially attacked by DNase I. The more significant point is that we can detect no difference in the DNase-I sensitivity of the genes represented by the cDNA probe under the control or heat-shock condition. This result is not entirely unexpected since there is evidence for some residual transcription at the normally puffed loci even after 1 hour of heat shock (Bonner and Pardue 1976). We conclude that no detectable change occurs in the structure of chromosomal loci responsible for the bulk of the mRNA synthesized at 25°C as a result

of the heat shock. The question of whether the heat-shock loci become more sensitive to DNase I after induction is being explored.

DISCUSSION

Although the mechanism of heat-induced puff formation in dipteran polytene chromosomes is poorly understood (Berendes 1973), gene activation does seem to require cytoplasmic factors (see also Compton and Bonner, this volume). Isolated salivary gland nuclei do not respond to heat shock as do cells. However, once induced in the intact cell, the heat-shock loci remain in the puffed configuration after isolation of the chromosomes (Berendes and Boyd 1969). The simplest model which would account for this response involves activation of a cytoplasmic factor which then acts upon selected loci in the chromosomes, leading to a visible morphological change and to new transcription.

In addition to the visible puffing in giant chromosomes, it is evident that structural changes accompany active transcription even in interphase chromatin. For example, DNase II (Gottesfeld et al. 1974) and DNase I (Weintraub and Groudine 1976) preferentially attack those sections of interphase chromatin that are transcriptionally active. The fact that such sections of the chromatin are preferentially open and available to DNA-binding proteins probably also accounts for the ability of *E. coli* RNA polymerase to transcribe chromatin nonrandomly in a manner which reflects the pattern of transcription in the cells from which it was isolated. For example, it is well established that the globin genes are preferentially transcribed in erythroid cells (Axel et al. 1973; Gilmour and Paul 1973) just as they are preferentially digested by DNase I (Weintraub and Groudine 1976).

Diploid *Drosophila* cells also respond to heat shock with a change in RNA synthesis which parallels the pattern of puff formation in polytene chromosomes (McKenzie et al. 1975; Spradling et al. 1975, 1977). If the analogy between the visible puffs and altered structure of interphase chromosomes is valid, it seemed appropriate to ask whether the transcription of chromatin by *E. coli* RNA polymerase was altered by heat shock.

The question was approached through synthesis of a cDNA probe specific for heat-shock-induced RNA synthesis. The probe made using polysomal polyadenylated RNA as template proved not to be completely specific for heat-induced RNA either because of some residual synthesis of RNA also characteristic of normal cells or because some preexisting RNA molecules survived the 1-hour exposure to 37°C. In either event, the specificity of the probe was improved by challenging it with excess normal RNA and removing hybridized cDNA. This enriched probe proved to be essentially specific for heat-shock-induced RNA and was used to characterize the in vitro transcript. The kinetics of hybridization showed that chromatin from heat-shocked cells synthesizes RNA complementary to this probe 100 times more efficiently than does chromatin from normal cells. We cannot determine whether the 1% relative cross-reaction reflects the fact that heat-shock-specific sequences are transcribed at low levels in normal cells or whether the probe still contains some sequences that are not specific for the heat-shock response. Alternatively, it could be a manifestation of less-than-complete transcriptional fidelity.

Our data on the retention by the sulfhydryl-Sepharose column demonstrate that more than 40% of the RNA which had hybridized to cDNA from cytoplasmic poly(A) RNA contained mercury. Although the yield is higher than reported for a similar model experiment with ^{3}H-labeled SV40 DNA and its Hg-containing RNA transcript (Dale and Ward 1975), the incomplete binding efficiency of the hybrids remains to be explained. The unretained hybrids may have suffered removal of the mercury from the RNA during the hybridization. A second possibility arises from the observation that E. coli polymerase can use endogenous RNA in chromatin as a template (Zasloff and Felsenfeld 1977). Any resulting double-stranded RNA would bind to the sulfhydryl affinity column and contaminate the purified transcript of the chromatin template itself. However, the use of RNA as a template depends upon Mn^{++} ions (Fox et al. 1964) and is unlikely to account for a major portion of the RNA synthesized under our conditions of excess Mg^{++} over Mn^{++}, especially since relatively large amounts of RNA are produced.

In addition to E. coli RNA polymerase, we have begun to employ DNase I as a probe for changes in chromatin structure. Our present results show only that chromosomal loci responsible for mRNA synthesis at 25°C are not changed in terms of DNase sensitivity as a result of heat shock. This result is not unexpected and is reminiscent of the fact that the globin genes remain sensitive to DNase I even after RNA synthesis ceases in chicken erythroid cells (Weintraub and Groudine 1976). The possibility that the heat-shock genes experience a change in DNase sensitivity is a more interesting one and is being explored.

Although the conclusion that isolated chromatin retains structural features associated with transcriptional regulation is not without precedent, the Drosophila heat-shock system may offer some experimental advantages. The response is rapid and involves discrete parts of the genome which are mapped cytologically and genetically. These features, combined with the small Drosophila genome and the possibility of gene isolation, may make it feasible to define the structural features of the chromatin which are modified after heat shock and the putative cytoplasmic factors responsible for the induced configurational changes.

Acknowledgments

We wish to thank Mr. Ed Tischer for preparing the sulfhydryl-Sepharose 6B and Mrs. Brinilda Lum for help with the cell cultures. We are indebted to Drs. G. Felsenfeld and M. Zasloff for communication of unpublished results. This work was supported by research grants from the National Institutes of Health, U.S. Public Health Service Grant GM 20287, and National Science Foundation Grant PCM 72.

REFERENCES

ASHBURNER, M. 1970. Patterns of puffing activity in the salivary gland chromosomes of Drosophila. V. Responses to environmental treatments. Chromosoma 31: 356.

AXEL, R., H. CEDAR, and G. FELSENFELD. 1973. Synthesis of globin RNA from duck reticulocyte chromatin in vitro. Proc. Natl. Acad. Sci. 72: 1117.

BEEBEE, T. J. C. and P. H. BUTTERWORTH. 1976. The use of mercurated nucleoside triphosphate as a probe in transcription studies in vitro. Eur. J. Biochem. 66: 543.

BERENDES, H. D. 1973. Synthetic activity in polytene chromosomes. Int. Rev. Cytol. 32: 61.

BERENDES, H. D. and J. B. BOYD. 1969. Structural and functional properties of polytene nuclei isolated from salivary glands of Drosophila hydei. J. Cell Biol. 41: 591.

BIESSMANN, H., R. A. GJERSET, B. LEVY W., and B. J. MCCARTHY. 1976. Fidelity of chromatin transcription in vitro. Biochemistry 15: 4356.

BISHOP, J. O., J. G. MORTON, M. ROSBASH, and M. RICHARDSON. 1974. Three abundance classes in HeLa cell messenger RNA. Nature 250: 199.

BONNER, J. J. and M. L. PARDUE. 1976. The effect of heat shock on RNA synthesis in Drosophila tissues. Cell 8: 42.

DALE, R. K. M. and D. C. WARD. 1975. Mercurated polynucleotides: New probes for hybridization and selective polymer fractionation. Biochemistry 14: 2458.

DALE, R. K. M., E. MARTIN, D. C. LIVINGSTON, and D. C. WARD. 1975. Direct covalent mercuration of nucleotides and polynucleotides. Biochemistry 14: 2448.

FOX, C. F., W. S. ROBINSON, R. HASELKORN, and S. B. WEISS. 1964. Enzymatic synthesis of ribonucleic acid. III. The ribonucleic acid-primed synthesis of ribonucleic acid with Micrococcus lysodeikticus ribonucleic acid polymerase. J. Biol. Chem. 239: 186.

GILMOUR, R. S. and J. PAUL. 1973. Tissue-specific transcription of the globin gene in isolated chromatin. Proc. Natl. Acad. Sci. 70: 3440.

GILMOUR, R. S., P. R. HARRISON, J. D. WINDASS, N. A. AFFARA, and J. PAUL. 1974. Globin messenger RNA synthesis and processing during haemoglobin induction in Friend cells. I. Evidence for transcriptional control in clone M2. Cell Differ. 3: 9.

GOTTESFELD, J. M., W. T. GARRARD, G. BAGI, R. F. WILSON, and J. BONNER. 1974. Partial purification of the template active fraction of chromatin: A preliminary report. Proc. Natl. Acad. Sci. 71: 2193.

HARRIS, S. E., R. J. SCHWARTZ, M. J. TSAI, B. W. O'MALLEY, and A. K. ROY. 1976. Effect of estrogen on gene expression in the chick oviduct. J. Biol. Chem. 251: 524.

HONJO, T. and R. H. REEDER. 1974. Transcription of Xenopus chromatin by homologous ribonucleic acid polymerase: Aberrant synthesis of ribosomal and 5S ribonucleic acid. Biochemistry 13: 1896.

LEVY W., B. and B. J. MCCARTHY. 1975. Messenger RNA complexity in Drosophila melanogaster. Biochemistry 14: 2440.

834 BIESSMANN ET AL.

Lewis, M., P. J. Helmsing, and M. Ashburner. 1975. Parallel changes in puffing activity and patterns of protein synthesis in salivary glands of *Drosophila*. *Proc. Natl. Acad. Sci.* **72**: 3604.

McKenzie, S. L., S. Henikoff, and M. Meselson. 1975. Localization of RNA from heat-induced puff sites in *Drosophila melanogaster*. *Proc. Natl. Acad. Sci.* **72**: 1117.

Reeder, R. H. 1973. Transcription of chromatin by bacterial RNA polymerase. *J. Mol. Biol.* **80**: 229.

Ritossa, F. 1962. A new puffing pattern induced by heat shock and DNP in *Drosophila*. *Experientia* **18**: 571.

Spradling, A., M. L. Pardue, and S. Penman. 1977. Messenger RNA in heat shocked *Drosophila* cells. *J. Mol. Biol.* **109**: 559.

Spradling, A., S. Penman, and M. L. Pardue. 1975. Analysis of *Drosophila* mRNA by *in situ* hybridization: Sequences transcribed in normal and heat shocked cultured cells. *Cell* **4**: 395.

Tissières, A., H. K. Mitchell, and U. M. Tracy. 1974. Protein synthesis in salivary glands of *Drosophila melanogaster*: Relation to chromosome puffs. *J. Mol. Biol.* **84**: 389.

Weintraub, H. and M. Groudine. 1976. Chromosomal subunits in active genes have an altered conformation. *Science* **193**: 843.

Wilson, G. N., A. W. Steggles, and A. W. Nienhuis. 1975. Strand-selective transcription of globin genes in rabbit erythroid cells and chromatin. *Proc. Natl. Acad. Sci.* **72**: 4835.

Zasloff, M. and G. Felsenfeld. 1977. Use of mercury-substituted ribonucleoside triphosphates can lead to artefacts in the analysis of *in vitro* chromatin transcripts. *Biochem. Biophys. Res. Commun.* **75**: 598.

An In Vitro Assay for the Specific Induction and Regression of Puffs in Isolated Polytene Nuclei of *Drosophila melanogaster*

J. L. COMPTON AND J. J. BONNER

Department of Biochemistry and Biophysics, University of California, San Francisco, California 94143

Drosophila melanogaster has several characteristics favorable for studying the control of gene expression in eukaryotes. One of the greatest potential advantages is the reservoir of genetic information and the consequent sophistication of genetic techniques which may be employed. Genetic studies, in concert with other techniques, will provide the ultimate proof of control of expression. A second feature is the presence of highly polytene nuclei in some larval tissues. Autoradiographic techniques have exploited the cytology as a means of determining which regions of the genome are transcriptionally active and, coupled with hybridization, of determining from what regions a purified RNA was derived.

Recent work on gene expression in *Drosophila* has concentrated on two sets of genes which can be experimentally induced in excised salivary glands. The puffing pattern, originally described by Ritossa (1962) as a response to elevated temperature or exposure to dinitrophenol (DNP), can also be reproduced in isolated glands (Leenders and Berendes 1972; Ashburner 1972b). RNA transcribed in several tissues and in cultured cells during exposure to elevated temperatures hybridizes to a small number of loci which correspond to the sites of the temperature-induced puffs (McKenzie et al. 1975; Spradling et al. 1975; Bonner and Pardue 1976a). The patterns of protein synthesis under similar conditions change in a coordinate manner (Tissières et al. 1974; Lewis et al. 1975). Similarly, the normal developmental sequence of puffs induced by the molting hormone ecdysone (see Ashburner 1972b) can be reproduced in isolated glands under carefully controlled conditions (Ashburner et al. 1974; Richards 1976). Bonner and Pardue (1976b) and Bonner et al. (1977) examined the RNA produced by salivary glands and imaginal disks in response to incubation with ecdysone in vitro.

It is therefore possible to assay for differential gene activity in *Drosophila,* and the availability of cloned DNA fragments promises to facilitate this type of work (Wensink et al. 1974). However, because of the manipulations involved in the purification of RNA products and the subsequent hybridization reactions, these assays are not well suited to the screening of many preparations. With the ultimate objective of investigating the mechanisms of gene activation in the heat-shock and ecdysone systems, we set out to develop a cytological assay for gene expression based on the puffing of polytene chromosomes. We now report an in vitro assay capable of detecting the specific induction or regression of puffs in salivary gland polytene nuclei incubated in cytoplasm from *Drosophila* cultured cells exposed to different treatments.

The Assay

In order to use polytene nuclei as an assay system, it must be possible to isolate them undamaged from the salivary gland, to incubate them under conditions where they respond to functional signals, and then to squash them and score the results of the experiment. Kroeger (1960) was able to induce puffs in *Drosophila busckii* polytene nuclei by hand-dissecting and then transferring them under oil into diluted egg cytoplasm from *D. melanogaster;* but the tediousness of his procedure precludes its use as a routine assay. Berendes and Boyd (1969) reported the isolation of polytene nuclei from salivary glands of *Drosophila hydei* and showed that they would incorporate RNA precursors but would not puff in response to elevated temperatures or to ecdysone. After several unsuccessful attempts to find buffer conditions for the isolation of functional polytene nuclei from *D. melanogaster* salivary glands, we sought to devise procedures which maintained them in as natural an environment as possible. We are now able to prepare functional nuclei by disrupting salivary glands in a homogenate of *Drosophila* Kc cells.

Salivary glands were removed from *D. melanogaster* (Oregon R) third instar larvae in *Drosophila* PBS (Robb 1969) or Grace's insect medium (Pacific Biologicals) diluted with 0.2 volume H_2O (Ashburner 1972a). *Drosophila* Kc cells (Echalier and Ohanessian 1969; Dolfini 1971) were pelleted, washed once with 30 mM Tris, pH 8.0, 100 mM KCl, 1 mM $MgCl_2$, and repelleted in a Dounce homogenizer. After the buffer was removed, the cell pellet was disrupted

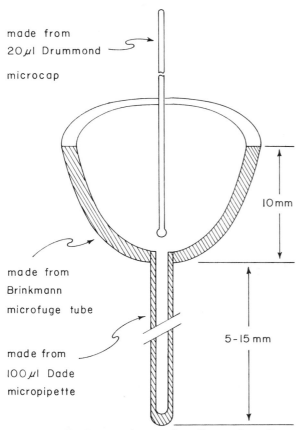

Figure 1. The homogenizer used to disrupt the salivary glands. The bowl was made from a 1.5-ml polypropylene tube designed for the Brinkmann microfuge, the barrel from a Dade 100-μl micropipette, and the pestle from a Drummond 20-μl microcap.

made from 20 μl Drummond microcap

made from Brinkmann microfuge tube

made from 100 μl Dade micropipette

10 mm

5–15 mm

with a tight pestle. The homogenate was either used intact or centrifuged for 2–3 minutes at ~ 8000g in a Brinkmann microfuge. The salivary glands were

pelleted in a small homogenizer (Fig. 1), washed once with the Kc homogenate or supernatant cytosol, resuspended, and disrupted with three to five gentle strokes. The volume of homogenate or cytosol used was approximately 20 times that of the gland pellet. After various times of incubation, aliquots were fixed for 10 minutes in 45% acetic acid on siliconized cover slips, stained for 5 minutes with the addition of 0.5 volume lactic-acetic-orein (22%, 39%, 2.2% in H$_2$O), and squashed by standard techniques.

The nuclei are extremely delicate, and excessive manipulations often result in the detachment of one or more of the chromosomes from the nuclear membrane. Damaged nuclei may degenerate during prolonged incubation and are nearly impossible to spread. In contrast, nuclei prepared and incubated by our procedure spread nearly as well as those isolated directly from intact glands, although the chromosomes seem slightly more resistant to stretching. We have incubated nuclei for up to 2 hours at 25°C in both total homogenates and supernatant cytosols, and they have retained their morphology. It is therefore possible to measure any induction or regression of puffs which has occurred during the incubation.

In Vitro Induction of Heat-shock Puffs

Salivary gland nuclei were prepared and incubated in homogenates of Kc cells which had been incubated at 35°C for 30 minutes immediately prior to homogenization. Figure 2 shows region 87 of chromosome IIIR of two nuclei after incubation. The sites of in vivo heat-shock puffs (87A and 87B) (Ashburner 1970) have been specifically induced to puff by the cytosol from heat-shocked Kc cells but remain uninduced in the control cytosol. Figure 3 shows a graph of the size distributions of 87B in isolated nuclei and in nuclei spread directly from glands. There appears to be a low level of induction by the control

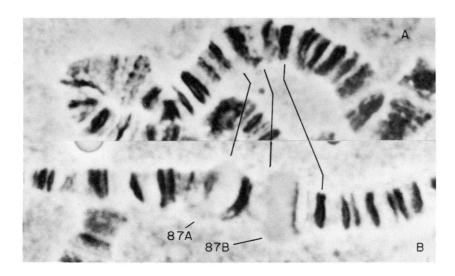

87A   87B

A

B

Figure 2. Region 87 of chromosome IIIR. Salivary glands were disrupted and incubated for 20 min at 25°C in (A) a cytosol prepared from Kc cells grown at 25°C and (B) a cytosol prepared from Kc cells incubated at 35°C for 30 min immediately before homogenization. Heat-shock puffs were induced at 87A and 87B by the cytosol prepared from the heat-treated cells.

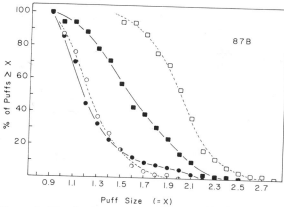

Figure 3. The size distribution of puffs at 87B. "Size" represents the ratio of chromosome width (measured with an ocular micrometer) of 87B and the unpuffed reference band 87Fl. The measurements plotted are the percentage of values greater than or equal to the size on the *x* axis. (○) Intact glands, 25°C; (□) intact glands incubated at 35°C for 15 min; (●) nuclei incubated for 20 min in cytosol prepared from Kc cells grown at 25°C; (■) nuclei incubated for 20 min in cytosol prepared from Kc cells incubated at 35°C for 30 min immediately before homogenization.

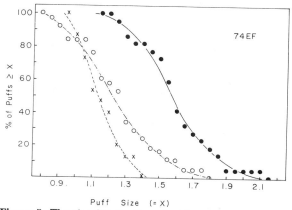

Figure 5. The size distribution of puffs at 74EF. Measurements were normalized with respect to 74Al and are plotted as the percentage of values greater than or equal to the size on the *x* axis. Nuclei from glands disrupted after 30-min incubation in 1 μg/ml ecdysone before (●) and after (○) 60-min incubation in Kc cytosol. (x) Intact glands after 30-min incubation with ecdysone, followed by 60-min incubation in the absence of ecdysone.

cytosol, and the induction by the heat-shock cytosol is incomplete. However, the isolated nuclei clearly respond differently to cytosols prepared from Kc cells treated in different manners. We have looked at the induction of puffing at 63BC, 67B, 87A, 87B, and 93D, loci which are induced to puff in vivo by a heat shock (Ashburner 1970), and found similar responses. This establishes that the nuclei retain at least a partial capacity to puff as they do in vivo.

Regression of Early Ecdysone Puffs In Vitro

Intact salivary glands were exposed to ecdysone for 30 minutes, washed, and then homogenized in

Kc cytosol. Figure 4 shows regions 74 and 75 of chromosome IIIL before and after incubation in vitro. Regions 74EF and 75B have partially regressed, as they do in vivo when ecdysone is removed prematurely (Ashburner et al. 1974). Figure 5 shows a graph of the size distributions of 74EF in isolated nuclei and in nuclei from intact glands. As in the case of puff induction, the response is incomplete; but it is clear that the early ecdysone puffs in isolated nuclei are at least partially competent to regress as they do in vivo.

DISCUSSION

The phenomenon of puffing in polytene nuclei includes both the specific induction and the specific

Figure 4. Regions 74 and 75 of chromosome IIIL. Salivary glands were incubated in 1 μg/ml ecdysone for 30 min and washed in Kc cytosol. *(A)* A nucleus fixed immediately after disruption; *(B)* a nucleus after 60-min incubation. The puffs induced at 74EF and 75B by the ecdysone treatment regressed to near normal sizes during incubation in the Kc cytosol.

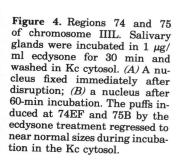
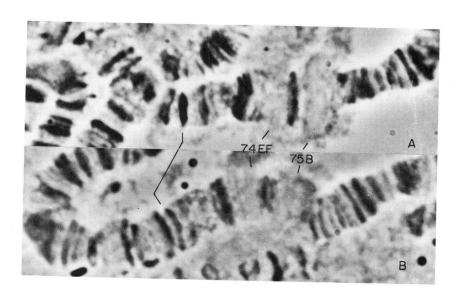

regression of puffs. Ideally, an in vitro system for measuring puffing activity should be competent for both. We have used the heat-shock system as a model for puff induction and the early ecdysone puffs as a model for puff regression, and we have demonstrated that polytene nuclei prepared from salivary glands under carefully controlled conditions retain at least partial ability to mimic their in vivo responses. We are currently using this system as an assay to investigate the mechanisms of the heat-shock- and ecdysone-induced changes in puffing patterns, and we believe that it will make the induction of gene activity in polytene chromosomes more amenable to study.

Acknowledgments

We are grateful to Drs. Brian McCarthy and Keith Yamamoto for the hospitality of their laboratories and for their interest in this work. J.L.C. and J.J.B. are supported by postdoctoral fellowships from the National Institutes of Health and the American Cancer Society, respectively. This research was supported by National Science Foundation Grant PCM 72 to B.J.M. and National Institutes of Health Grant CA-20535 to K.R.Y.

REFERENCES

ASHBURNER, M. 1970. Patterns of puffing activity in the salivary gland chromosomes of *Drosophila*. V. Responses to environmental treatments. *Chromosoma* 31: 356.

————. 1972a. Patterns of puffing activity in the salivary gland chromosomes of *Drosophila*. VI. Induction by ecdysone in salivary glands of *D. melanogaster* cultured *in vitro*. *Chromosoma* 38: 255.

————. 1972b. Puffing patterns in *Drosophila melanogaster* and related species. *Results Probl. Cell Differ.* 4: 101.

ASHBURNER, M., C. CHIHARA, P. MELTZER, and G. RICHARDS. 1974. Temporal control of puffing activity in polytene chromosomes. *Cold Spring Harbor Symp. Quant. Biol.* 38: 655.

BERENDES, H. D. and J. B. BOYD. 1969. Structural and functional properties of polytene nuclei isolated from salivary glands of *Drosophila hydei*. *J. Cell Biol.* 41: 591.

BONNER, J. J. and M. L. PARDUE. 1976a. The effect of heat shock on RNA synthesis in *Drosophila* tissues. *Cell* 8: 43.

————. 1976b. Ecdysone-stimulated RNA synthesis in imaginal discs of *Drosophila melanogaster*. Assay by *in situ* hybridization. *Chromosoma* 58: 87.

BONNER, J. J., M. BERNINGER, and M. L. PARDUE. 1977. Transcription of polytene chromosomes and the mitochondrial genome in *Drosophila melanogaster*. *Cold Spring Harbor Symp. Quant. Biol.* 42.

DOLFINI, S. 1971. Karyotype polymorphism in a cell population of *Drosophila melanogaster* cultured *in vitro*. *Chromosoma* 33: 196.

ECHALIER, G. and A. OHANESSIAN. 1969. Isolement, en cultures *in vitro*, de lignees cellulaires diploides de *Drosophila melanogaster*. *C. R. Acad. Sci.* (Paris) 268: 1771.

KROEGER, H. 1960. The induction of new puffing patterns by transplantation of salivary gland nuclei into egg cytoplasma of *Drosophila*. *Chromosoma* 11: 129.

LEENDERS, H. J. and H. D. BERENDES. 1972. The effect of changes in the respiratory metabolism upon genome activity in *Drosophila*. I. The induction of gene activity. *Chromosoma* 37: 433.

LEWIS, M., P. J. HELMSING, and M. ASHBURNER. 1975. Parallel changes in puffing activity and patterns of protein synthesis in salivary glands of *Drosophila*. *Proc. Natl. Acad. Sci.* 72: 3604.

McKENZIE, S. L., S. HENIKOFF, and M. MESELSON. 1975. Localization of RNA from heat-induced puff sites in *Drosophila melanogaster*. *Proc. Natl. Acad. Sci.* 72: 1117.

RICHARDS, G. 1976. Sequential gene activation by ecdysone in polytene chromosomes of *Drosophila melanogaster*. V. The late prepupal puffs. *Devel. Biol.* 54: 264.

RITOSSA, F. 1962. A new puffing pattern induced by heat-shock and DNP in *Drosophila*. *Experientia* 18: 571.

ROBB, J. 1969. Maintenance of imaginal discs of *Drosophila melanogaster* in chemically defined media. *J. Cell Biol.* 41: 876.

SPRADLING, A., S. PENMAN, and M. L. PARDUE. 1975. Analysis of *Drosophila* mRNA by *in situ* hybridization: Sequences transcribed in normal and heat shocked cultured cells. *Cell* 4: 395.

TISSIÈRES, A., H. K. MITCHELL, and U. M. TRACY. 1974. Protein synthesis in salivary glands of *Drosophila melanogaster:* Relation to chromosome puffs. *J. Mol. Biol.* 84: 389.

WENSINK, P. C., D. J. FINNEGAN, J. E. DONELSON, and D. S. HOGNESS. 1974. A system for mapping DNA sequences in the chromosomes of *Drosophila melanogaster*. *Cell* 3: 315.

Distribution Patterns of *Drosophila* Nonhistone Chromosomal Proteins

S. C. R. ELGIN,* L. A. SERUNIAN,* AND L. M. SILVER†

*Department of Biochemistry and Molecular Biology, and †Committee on Higher Degrees in Biophysics, Harvard University, Cambridge, Massachusetts 02138

The nonhistone chromosomal (NHC) proteins have been defined as those proteins, excluding the histones, which are isolated in association with DNA in purified chromatin or metaphase chromosomes. That most, if not all, of the NHC proteins so isolated are true constituents of the chromatin complex as it exists in vivo is now substantiated by several lines of evidence, including the experiments discussed below. (See Elgin et al. 1974 for a review of this topic.) Chromatin may be isolated from the embryos of *Drosophila melanogaster* by a modification of the technique of Marushige and Bonner (1966), involving lysis of the isolated nuclei in low-salt buffers and subsequent purification of the chromatin on a sucrose gradient. Nonhistone proteins are a major component of isolated chromatin; the mass ratios of the components of chromatin from *Drosophila* embryos are DNA:RNA:histone:NHCP, 1.0:0.06:0.8:1.2 (Elgin and Hood 1973). Careful analysis by two-dimensional gel electrophoresis has detected 450 different NHC proteins in HeLa cells, present in from 10^2 to 10^6 copies per haploid genome (Peterson and McConkey 1976). However, there are several major NHC proteins; probably a dozen or less make up about half of the mass of the fraction (Elgin and Bonner 1972; Peterson and McConkey 1976).

The limited evidence available (reviewed by Elgin and Weintraub 1975) indicates that the NHC protein fraction includes enzymes of chromosomal metabolism, specific activators of gene expression, and structural proteins involved in altering or organizing the histone-DNA fiber. The latter proteins are of particular interest because of their potential role in determining the different general structures characterizing the active and inactive states of chromatin. To learn more about the organization of chromosomes, we have developed an immunofluorescence staining technique to determine the distribution patterns of individual NHC proteins in polytene chromosomes. The use of polytene chromosomes of *Drosophila* allows one to attempt to correlate the patterns obtained with the known genetic map and with the observed patterns of gene activity. The technique is useful primarily in two ways. First, the distribution patterns obtained using antibodies against individual NHC proteins not otherwise studied may suggest the functional roles of these proteins. Second, the distribution patterns of chromosomal proteins already characterized in other terms (such as histones or RNA polymerases) may be utilized to test specific hypotheses concerning the functional organization of chromatin. We have used this approach recently to probe the structural differences between active and inactive loci of the polytene chromosomes.

EXPERIMENTAL PROCEDURES AND RESULTS

The Immunofluorescence Staining Technique

For initial studies to develop the immunofluorescence staining technique, antibodies were produced in rabbits against total NHC proteins from *Drosophila* embryo chromatin. Polytene chromosome squashes were prepared following fixation of the salivary glands of late-third-instar larvae in 2% formaldehyde, as described previously (Silver and Elgin 1976). The chromosomes were stained by incubation initially with the specific serum and subsequently with fluorescein-conjugated goat IgG directed against rabbit gammaglobulin. The stained chromosomes were photographed using both phase-contrast and UV darkfield optics (see Silver et al. [1977] for a detailed description of the procedure). In Figure 1, the chromosomes are prominently stained, with little or no staining of the cytoplasmic debris. This indicates that the proteins isolated with the DNA in the chromatin complex are indeed preferentially associated with the DNA in chromosomes in vivo. It is also apparent that the distribution of the NHC proteins as shown by this staining technique is nonuniform. This nonuniform pattern implies that individual NHC proteins will show specific distributions related to their functional roles. The resolution possible should allow one to detect staining of individual chromomeres, the potential units of gene organization (see Lefevre [1974] for review of the one band–one gene hypothesis). Several types of control experiments confirm that the staining is indeed a consequence of a specific antibody-antigen interaction involving the chromosomal proteins. No staining is observed if antiserum raised against bovine serum albumin, preimmune serum, or phosphate-buffered saline is used as the primary reagent. No

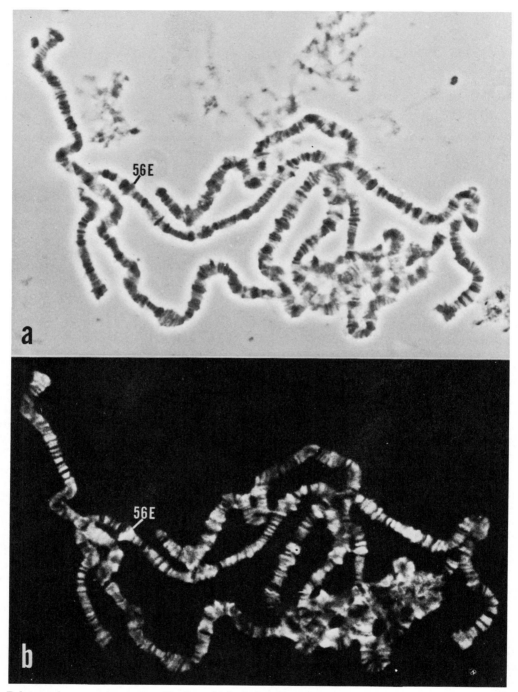

Figure 1. Polytene chromosomes prepared by formaldehyde fixation and stained using antiserum against total *Drosophila* embryo NHC proteins as described by Silver and Elgin (1976), and viewed by *(a)* phase-contrast and *(b)* UV dark-field optics. (Reprinted, with permission, from Elgin et al. 1977.)

staining is observed if the antiserum against NHC proteins is preabsorbed with chromatin. However, the staining pattern is not significantly altered if the antiserum is treated with an excess of *Drosophila* DNA or yeast RNA prior to use (data not shown).

It should be noted that the conditions of fixation

can have a significant effect on the results of such an antibody staining experiment. Polytene chromosome squashes are normally prepared using a squashing medium of 45% acetic acid. However, the acetic acid extracts significant amounts of histone and of some NHC proteins from the chromosomes. This can be prevented by prior fixation of the salivary glands with formaldehyde (Silver and Elgin 1976). Unfortunately, this fixation apparently alters or obscures some antigenic determinants. In particular, some very dense bands fail to stain with antiserum against histone 3 in formaldehyde-fixed chromosomes, probably because the bands are so cross-linked as to prevent access to the antibody probe (see Fig. 5). Consequently, it is both necessary and informative to examine the staining pattern obtained with a given antiserum using chromosomes prepared by a spectrum (at least two) of contrasting fixation techniques. Most NHC proteins are not susceptible to acetic acid extraction (Silver and Elgin 1976); consequently, the specific dominant staining patterns obtained using sera against NHC protein subfractions are generally maintained despite such variations in fixation conditions (L. M. Silver and S. C. R. Elgin, unpubl.).

Distribution Patterns of Nonhistone Chromosomal Proteins

We have prepared antisera against three different molecular weight subfractions of the NHC proteins and against three individual NHC proteins. The subfractions were prepared using SDS gel electrophoresis (Silver and Elgin 1977); the individual proteins were prepared using isoelectric focusing followed by SDS gel electrophoresis of the isolated bands (L. M. Silver and S. C. R. Elgin, in prep.). Antibodies were raised directly against the proteins in the acrylamide gel in rabbits or mice according to the method of Tjian et al. (1974). Results obtained using the sera against the molecular weight subfractions have been reported in detail (Silver and Elgin 1977) and are summarized here.

Fraction σ-φ. This antiserum, prepared against proteins of 140,000–280,000 molecular weight, caused extensive staining of the polytene chromosomes. The staining appeared to be the inverse of the phase-density pattern in many regions of the chromosomes. A few bands were very brightly stained.

Fraction π. This antiserum, prepared against proteins of 55,000–70,000 molecular weight, caused prominent staining of a limited set of chromomeres over a background of more general chromosome staining. Some variation in the staining pattern was noted, which was dependent on the developmental stage of the larva used. No correlation has been found between the set of bands stained and any known structural or functional parameter. For example, some puff loci are prominently stained (e.g., 63E) but others are not stained at all (e.g., 68C).

Fraction ρ. This antiserum, prepared against proteins of 80,000–110,000 molecular weight, produces a pattern of prominent staining of a limited set of bands of the polytene chromosomes (Fig. 2). In par-

ticular, all puffs are prominently stained. In a careful analysis of chromosome arms 3R and 3L, it was observed that 90% of the loci prominently and consistently stained using this antiserum are those known to form major puffs at some time during the third-instar-larval and prepupal developmental periods (scored by Ashburner 1972). A detailed study of several of these loci has shown that they are stained before, during, and after their puffing activity. A less detailed inspection of the rest of the chromosomes supports this observation.

Three antisera have been produced against individual NHC proteins. The staining pattern observed using the first antiserum was similar to that obtained using the π antiserum. The pattern was distinct but complex, with no obvious correlation with known structural or functional parameters of the polytene chromosomes. Use of the second antiserum produced a pattern of extensive staining similar to that obtained with the σ-φ antiserum. These two patterns have not yet been analyzed in detail. In the third instance, an antiserum prepared against a protein of isoelectric point 5.2 and molecular weight 21,000 daltons (NHCP 21/5.2) produced a pattern of extensive distribution (Fig. 3). The staining appears to correlate roughly with the phase-density pattern of the chromosomes; in general, dense chromomeres are prominently stained, whereas a very low level of staining is observed in the puffs. The chromocenter is prominently stained; the nucleolus is not. It should be noted that the protein 21/5.2 constitutes no more than a few percent of the total NHC protein (L. M. Silver and S. C. R. Elgin, in prep.).

Detailed comparisons of the staining patterns on the polytene chromosomes within a given squash (salivary glands of one larva) have shown that the staining patterns obtained are distinctive and reproducible. In comparisons of results obtained using chromosomes from different larvae, one observes that a small percentage of the pattern is dependent on the developmental stage of the test organism in all the above cases. Some variability is observed in the degree of low-level staining of interband regions (Silver and Elgin 1977).

Several general conclusions are suggested by these studies. It is apparent that individual NHC proteins have distinctive distribution patterns that provide indications of the role of these proteins in chromatin structure and function. The patterns of distribution range from widespread to very limited. For example, NHCP 21/5.2 is a protein having a widespread pattern of distribution (Fig. 3) and thus probably a very general role in chromatin structure. The D1 protein (localized at a few bands primarily on chromosome 4) is an example of a protein having a very limited pattern of distribution and thus a more specific role (Alfageme et al. 1976). NHC proteins are associated with chromosome structures such as the chromocenter, the nucleolus, and the telomeres. It is likely

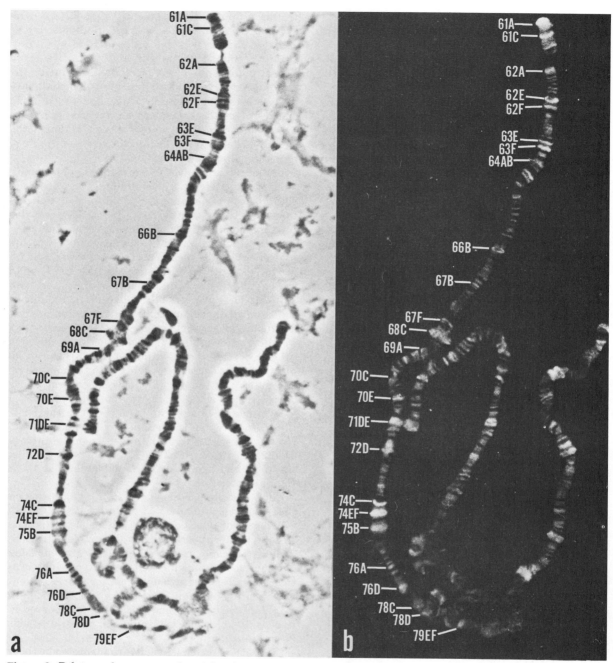

Figure 2. Polytene chromosomes (partial set) prepared by formaldehyde fixation and stained using antiserum against fraction ρ of the NHC proteins, and viewed by *(a)* phase-contrast and *(b)* UV dark-field optics. The prominently and consistently stained loci on chromosome arm 3L are labeled.

that proteins restricted to these sites will be identified and studied in the near future (e.g., see Busch et. al., this volume). Perhaps most interesting is the fact that it is clear that a subset of the NHC proteins is preferentially associated with a subset of potentially active chromomeres (the ρ case; see Fig. 2).

Puffs

It is of interest to inquire whether the ρ distribution pattern is indicative of a protein that serves

as a marker of loci active during the third-instar and prepupal developmental periods in *Drosophila,* or whether it is indicative of a general component of the chromatin structure of all active genes. To investigate this question, we have examined the staining characteristics of the most prominent heat-shock loci, 87A and 87B-C1. These loci, which are not normally active during third-instar-larval and prepupal development, can be induced to puff by subjecting the larvae to a "heat shock" of 37°C

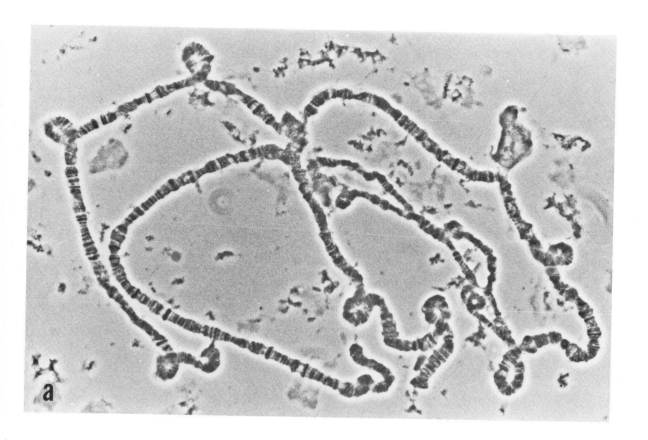

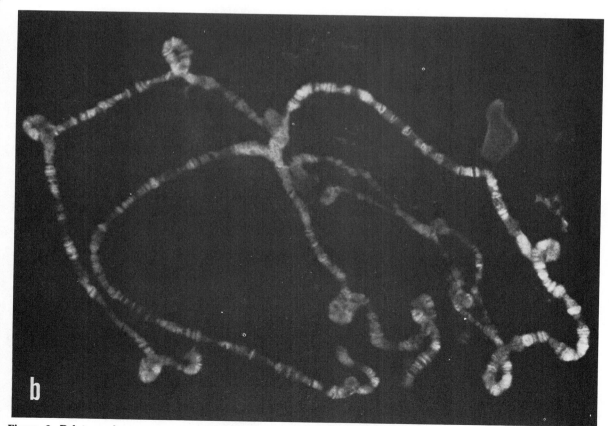

Figure 3. Polytene chromosomes prepared without formaldehyde fixation and stained using antiserum against NHC protein 21/5.2, and viewed by *(a)* phase-contrast and *(b)* UV dark-field optics.

(Ashburner 1972). That this puffing indeed indicates the induction of specific gene activity has been substantiated by studies correlating the puffing with the appearance of new mRNAs in the polysomes and the synthesis of new proteins. The new polysomal mRNA can be shown both to hybridize to the loci of the heat-shock puffs and to cause the synthesis of the induced proteins in an in vitro translation system (Tissières et al. 1974; McKenzie et al. 1975; Spradling et al. 1975; McKenzie 1976). In addition to the induction of the nine specific loci, heat shock results in a dramatic decrease in RNA synthesis at other loci, as shown by uridine incorporation studies (e.g., see Zhimulev and Belyaeva 1975).

The loci 87A and 87B-C1 are not significantly stained using the ρ antiserum on the chromosomes of non-heat-shocked larvae. If larvae are heat-shocked at 37°C for 20 minutes, the polytene

squashes obtained show the expected change in puffing pattern; in addition, all of the heat-shock loci (puffs) now stain brightly when the ρ antiserum is used. It should be noted that the developmental puffing sites, although now apparently much less active in RNA synthesis, continue to be stained when the ρ antiserum is used. The staining pattern and the uridine incorporation pattern for this region of a chromosome from a heat-shocked larva are shown in Figure 4. Thus, all puffs examined, not just those specified by the developmental program, are stained when the ρ antiserum is used. These results suggest that a particular chromatin configuration, as indicated by positive ρ staining, is a necessary but not sufficient characteristic of the active loci (Silver and Elgin 1977).

This different chromosomal structure might be the consequence of either a redistribution of a ρ

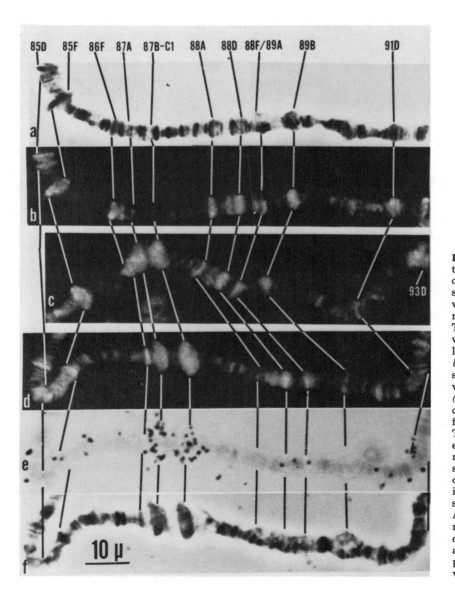

Figure 4. Staining and transcriptional activity in a middle section of chromosome 3R after heat-shock induction. All chromosomes were processed through the formaldehyde fixation techniques. The chromosomes in c and d–f were obtained from heat-shocked larvae; the chromosome in a and b was obtained from a non-heat-shocked larva. All chromosomes were stained using ρ antiserum. (a,f) Phase-contrast views of the chromosomes shown in UV darkfield views in b and d, respectively. The chromosome shown in d–f was exposed to tritiated uridine for 3 min, subsequent to a 20-min heat-shock treatment. After autoradiography and light Giemsa staining, this chromosome was observed with bright-field optics (e). Actual magnification of each chromosome is ± 15% of the value indicated by the bar. 87A, 87B-C1, and 93D are heat-shock loci. (Reprinted, with permission, from Silver and Elgin 1977.)

antigenic determinant, i.e., a new association of a specific protein(s) with the active sites, or, alternatively, a change in the chromatin configuration, making the ρ antigenic determinant(s) newly accessible to the antibody probe. The former interpretation is supported by the observation that bands 87A and 87B-C1 in the nonpuffed state are stained using antiserum against histone 3, suggesting that the chromatin fibers are accessible to antibody probes at these loci (Silver and Elgin 1977). However, experiments in which three different sera against histones are used indicate that the set of available antigenic determinants in soluble chromatin differs from that in the formaldehyde-fixed polytene chromosomes (L. M. Silver et al., in prep.). Consequently, the latter possibility cannot be dismissed on the basis of cytological evidence alone.

It is of interest to consider whether or not the antigenic determinants might be components of RNA polymerase. This appears unlikely for the following reasons: The ρ antiserum was prepared against a protein fraction of 80,000–110,000 molecular weight; there is no *Drosophila melanogaster* RNA polymerase II polypeptide within this molecular weight range (Greenleaf and Bautz 1975). The overall staining pattern obtained using an antiserum prepared against *Drosophila melanogaster* RNA polymerase II and the effects of different fixation procedures on the pattern are different from those observed using the ρ antiserum (Jamrich et al. 1977; L. M. Silver, unpubl.). For example, following heat shock, the staining observed using serum against RNA polymerase II at developmental puff sites such as 66B is significantly reduced, whereas that observed at those loci using anti-ρ serum is still prominent (see Fig. 7).

We have also used the fluorescent antibody staining technique to look for alterations in the histone-DNA association related to puffing. For these studies, antibodies were prepared against histone 1 of *Drosophila melanogaster* by the method of Tjian et al. (1974) (T. Ashley et al., unpubl.) and against histone 3 and histone 4 of calf thymus by the method of Stollar and Ward (1970). The use of a heterologous histone preparation is advantageous as it greatly reduces the possibility of obtaining a serum that cross-reacts with the *Drosophila* NHC proteins. The NHC proteins are generally much more immunogenic than the histones, so minor contaminants in the histone preparation can lead to serious problems. Fortunately, sera prepared against calf histones 3 and 4 react strongly with *Drosophila* histones 3 and 4 (Bustin et al. 1977), as might be expected from the conservation of amino acid sequences observed for these proteins. Using these sera against histones 1, 3, and 4, one obtains extensive staining of formaldehyde-fixed polytene chromosomes, as anticipated. Sites of both developmental and heat-shock puffs are stained, although at a reduced level relative to other regions of the chromo-

somes (Fig. 5). It should be noted that the nucleolus and many of the phase-dense chromomeres do not stain in these preparations. This is presumably because the formaldehyde fixation procedure denatures or obscures the antigenic determinants.

A different result is obtained with chromosomes prepared by squashing in 45% acetic acid. In this instance, no staining is observed using serum against histone 1; apparently all of this protein has been extracted, as can be shown to occur when soluble chromatin is so treated. Extensive staining is obtained using sera against histones 3 and 4, including staining of most bands and of the nucleolus. However, in contrast to the above observations, no staining is observed at the prominent heat-shock puffs (Fig. 6) (L. A. Serunian et al., in prep.). This striking difference was observed consistently at the heat-shock loci 63BC, 67B, 87A, 87B-C1, 93D, and 95D. The two smallest heat-shock puffs, loci 64F and 70A, continue to be stained. These results suggest that the histones associated with DNA at the major heat-shock puffs are preferentially susceptible to acid extraction. This phenomenon may be due in part to the less compact overall chromosome structure associated with puffing; however, the results suggest that the histones may be in an altered configuration in the rapidly induced heat-shock puffs. The question must be studied further by biochemical techniques.

DISCUSSION

The comparisons given in Figure 7 illustrate our observations concerning the altered chromatin configuration at active loci induced by heat shock. Following the stimulus, the heat-shock loci stain brilliantly when the ρ antiserum is used, suggesting a new association of an NHC protein with these loci. Those loci that are developmentally active, but show reduced RNA synthesis following heat shock, remain stained when the ρ antiserum is used. In contrast, the developmental puff sites are no longer brightly stained when serum against *Drosophila* RNA polymerase II is used. The fluorescent antibody distribution studies using formaldehyde-fixed polytene chromosomes indicate that the histones remain in association with the puffed loci. However, in the majority of the heat-shock puffs, at least histones 3 and 4 appear particularly susceptible to acid extraction. These results must now be substantiated by biochemical studies.

The results obtained using the ρ antiserum suggest that there exists a chromatin state defined by the set of loci that will be, are, or have been active in a given developmental period. This state includes, but is broader than, the set of loci that are active, i.e., those which are associated with RNA polymerase and are templates for RNA synthesis. It is of interest that similar conclusions are suggested by

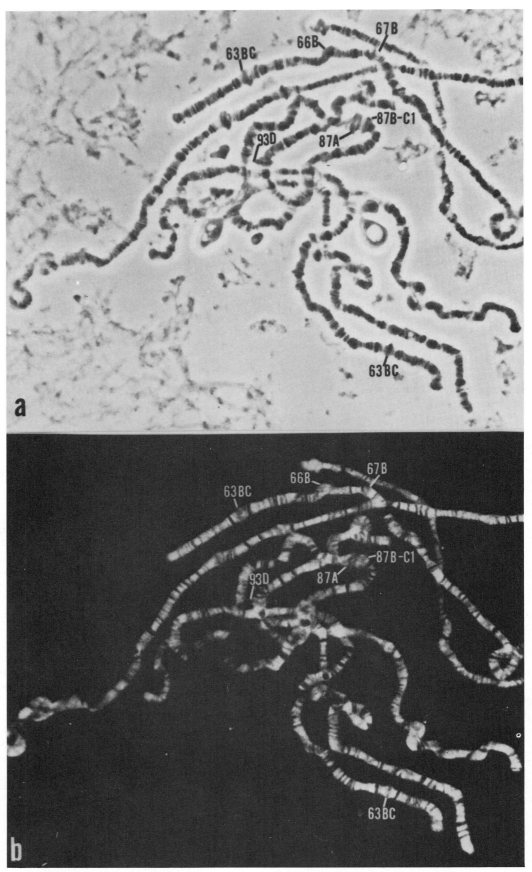

Figure 5. Polytene chromosomes (double set from a heat-shocked larva) prepared by formaldehyde fixation and stained using antiserum against histone 3, and viewed using *(a)* phase-contrast and *(b)* UV dark-field optics. Several of the heat-shock puffs and one developmental puff (66B) are labeled.

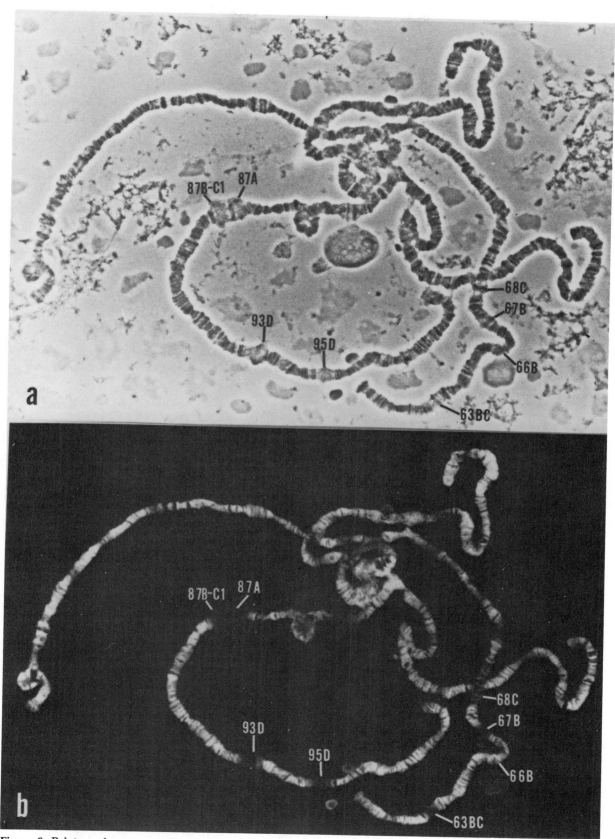

Figure 6. Polytene chromosomes from a heat-shocked larva prepared without formaldehyde fixation and stained using antiserum against histone 3, and viewed using (a) phase-contrast and (b) UV dark-field optics. Several of the heat-shock puffs and two developmental puffs (66B and 68C) are labeled.

847

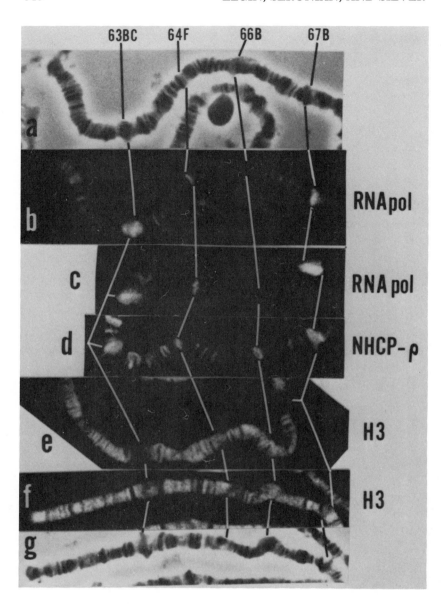

Figure 7. Staining in a section of chromosome 3L after heat-shock induction. *(a,g)* Phase-contrast views of the chromosomes shown in *b* and *f;* all others are UV dark-field views. The chromosomes were prepared with or without formaldehyde fixation and stained using antisera as follows: *(b)* formaldehyde fixation, RNA polymerase II antiserum; *(c)* no formaldehyde fixation, RNA polymerase II antiserum; *(d)* formaldehyde fixation, ρ antiserum; *(e)* no formaldehyde fixation, histone 3 antiserum; *(f)* formaldehyde fixation, histone 3 antiserum. Loci 63BC, 64F, and 67B are sites of heat-shock puffs; 66B is the site of a developmental puff.

the studies of the differential DNase-I sensitivity of chromatin. Weintraub and Groudine (1976) have observed that genes that are being transcribed are preferentially digested by pancreatic deoxyribonuclease I; this result has been confirmed by the studies of Garel and Axel (1976) and of Palmiter et al. (this volume). That this difference is not solely a consequence of the actual transcriptional events is indicated by two types of observations. First, DNase-I sensitivity, while associated with active genes being transcribed, is also found for those same loci at times after the actual transcriptional stage (Weintraub and Groudine 1976; Palmiter et al., this volume). Second, DNA sequences which are apparently transcribed with very different frequencies are equally susceptible to DNase-I digestion (Garel and Axel, this volume). Thus, the pattern of DNase-I sensitivity also suggests the existence of a chromatin state

which is necessary, but not sufficient, for active gene transcription.

It will be of interest to pursue the possibility of a correlation between DNase-I sensitivity and the pattern of distribution of the NHC proteins. The observation that a lysine-rich NHC protein is preferentially released from trout chromatin by DNase-I digestion (Levy W. et al. 1977) suggests experiments for which the *Drosophila* system should be ideal. Initial experiments with *Drosophila* embryos indicate that, following heat shock, the DNA sequences of loci 87A and 87B–C1 show increased sensitivity to digestion with nucleases (C. Wu and S. C. R. Elgin, in prep.).

Much work remains to be done. However, it is clear that the cytological techniques can be of great assistance, both in suggesting roles for the various NHC proteins and in testing hypotheses of function for known proteins. In conjunction with biochemical techniques, a knowledge of the distribution patterns of chromosomal proteins should aid us in establishing realistic models of the active and inactive states of the genome.

Acknowledgments

We thank Arno Greenleaf for providing us with antibodies against *Drosophila* RNA polymerase II and David Stollar for providing us with antibodies against calf thymus histones 3 and 4. This work was supported by grants GM 20779 from the National Institutes of Health and NP-184 from the American Cancer Society. L.M.S. is supported by an NIH predoctoral training grant to the Committee on Higher Degrees in Biophysics; S.C.R.E. is supported by an NIH Research Career Development Award.

REFERENCES

ALFAGEME, C. R., G. T. RUDKIN, and L. H. COHEN. 1976. Locations of chromosomal proteins in polytene chromosomes. *Proc. Natl. Acad. Sci.* **73**: 2038.

ASHBURNER, M. 1972. Puffing patterns in *Drosophila melanogaster* and related species. In *Developmental studies on giant chromosomes* (ed. W. Beermann et al.), p. 101. Springer Verlag, New York.

BUSTIN, M., R. H. REEDER, and S. L. McKNIGHT. 1977. Immunological cross-reaction between calf and *Drosophila* histones. *J. Biol. Chem.* **252**: 3099.

ELGIN, S. C. R. and J. BONNER. 1972. Partial fractionation and chemical characterization of the major nonhistone chromosomal proteins. *Biochemistry* **11**: 772.

ELGIN, S. C. R. and L. E. HOOD. 1973. Chromosomal proteins of *Drosophila* embryos. *Biochemistry* **12**: 4984.

ELGIN, S. C. R. and H. WEINTRAUB. 1975. Chromosomal proteins and chromatin structure. *Annu. Rev. Biochem.* **44**: 725.

ELGIN, S. C. R., L. M. SILVER, and L. SERUNIAN. 1977. Mapping with antibodies to nonhistone chromosomal proteins: A brief review. In *Molecular human cytogenics. UCLA-ICN Symposium on Molecular and Cellular Biology* (ed. R. Sparks et al.). Academic Press, New York. (In press.)

ELGIN, S. C. R., J. B. BOYD, L. E. HOOD, W. WRAY, and F. C. WU. 1974. A prologue to the study of the nonhistone chromosomal proteins. *Cold Spring Harbor Symp. Quant. Biol.* **38**: 821.

GAREL, A. and R. AXEL. 1976. Selective digestion of transcriptionally active ovalbumin genes from oviduct nuclei. *Proc. Natl. Acad. Sci.* **73**: 3966.

GREENLEAF, A. L. and E. K. F. BAUTZ. 1975. RNA polymerase B from *Drosophila melanogaster* larvae. Purification and partial characterization. *Eur. J. Biochem.* **60**: 169.

JAMRICH, M., A. L. GREENLEAF, and E. K. F. BAUTZ. 1977. Localization of RNA polymerase in polytene chromosomes of *Drosophila melanogaster*. *Proc. Natl. Acad. Sci.* **74**: 2079.

LEFEVRE, G., JR. 1974. The relationship between genes and polytene chromosome bands. *Annu. Rev. Genet.* **8**: 51.

LEVY W., B., N. C. W. WONG, and G. H. DIXON. 1977. Selective association of the trout-specific H6 protein with chromatin regions susceptible to DNase I and DNase II. Possible location of HMG-T in the spacer region between core nucleosomes. *Proc. Natl. Acad. Sci.* **74**: 2810.

MARUSHIGE, K. and J. BONNER. 1966. Template properties of liver chromatin. *J. Mol. Biol.* **15**: 160.

McKENZIE, S. L. 1976. "Protein and RNA synthesis induced by heat treatment in *Drosophila melanogaster* tissue culture cells." Ph.D. thesis, Harvard University, Cambridge, Massachusetts.

McKENZIE, S. L., S. HENIKOFF, and M. MESELSON. 1975. Localization of RNA from heat-induced polysomes at puff sites in *Drosophila melanogaster*. *Proc. Natl. Acad. Sci.* **72**: 1117.

PETERSON, J. L. and E. H. McCONKEY. 1976. Non-histone chromosomal proteins from HeLa cells. A survey by high resolution, two-dimensional electrophoresis. *J. Biol. Chem.* **251**: 548.

SILVER, L. M. and S. C. R. ELGIN. 1976. A method for determination of the in situ distribution of chromosomal proteins. *Proc. Natl. Acad. Sci.* **73**: 423.

———. 1977. Distribution patterns of three subfractions of Drosophila nonhistone chromosomal proteins: Possible correlations with gene activity. *Cell* **11**: 971.

SILVER, L. M., C. E. C. WU, and S. C. R. ELGIN. 1977. Immunofluorescent techniques in the analysis of chromosomal proteins. In *Methods in chromosomal protein research* (ed. G. Stein et al.). Academic Press, New York. (In press.)

SPRADLING, A., S. PENMAN, and M. L. PARDUE. 1975. Analysis of *Drosophila* mRNA by in situ hybridization: Sequences transcribed in normal and heat shocked cultured cells. *Cell* **4**: 395.

STOLLAR, B. D. and A. WARD. 1970. Rabbit antibodies to histone fractions as specific reagents for preparative and comparative studies. *J. Biol. Chem.* **245**: 1261.

TISSIÈRES, A., H. K. MITCHELL, and U. M. TRACY. 1974. Protein synthesis in salivary glands of *Drosophila melanogaster:* Relation to chromosome puffs. *J. Mol. Biol.* **84**: 389.

TJIAN, R., D. STINCHCOMB, and R. LOSICK. 1974. Antibody directed against *Bacillus subtilis* sigma factor purified by sodium dodecyl sulfate slab gel electrophoresis. *J. Biol. Chem.* **250**: 8824.

WEINTRAUB, H. and M. GROUDINE. 1976. Transcriptionally active and inactive conformations of chromosomal subunits. *Science* **193**: 848.

ZHIMULEV, I. F. and E. S. BELYAEVA. 1975. ³H-uridine labeling patterns in the *Drosophila melanogaster* salivary gland chromosomes 2R and 3L. *Chromosoma* **49**: 219.

QUESTIONS/COMMENTS

Question by:
E. ZUCKERKANDL
Linus Pauling Institute

Since ρ protein component(s) are found specifically at chromosomal sites that are going to puff, are puffing, or have puffed, it seems that this or these component(s) may be characteristic of what developmental biologists call determination, as distinct from differentiation. Determination is the singling out of a potential gene program that is subsequently carried out at another level of regulation. The ρ fraction would contain a gene-program-specific protein and be responsible, in part, for the appearance and preservation of a certain state of determination. This would be an important finding. Do you agree that this may be what you found?

Response by:
S. C. R. ELGIN
Harvard University

One must take into consideration the results of the heat-shock experiments.

The Expressed Portion of Eukaryotic Chromatin

J. Bonner, R. B. Wallace, T. D. Sargent, R. F. Murphy, and S. K. Dube*

Division of Biology, California Institute of Technology, Pasadena, California 91125

Approximately 10% of the single-copy sequences of rat-liver chromatin can be excised by DNase-II digestion followed by precipitation with 2 mM $MgCl_2$ (Marushige and Bonner 1971; Billing and Bonner 1972; Gottesfeld et al. 1974b, 1976). We have found by several criteria that the fraction not precipitated is the template-active portion of the genome. First, it contains the bulk of nascent RNA (Billing and Bonner 1972). Second, the DNA of this fraction hybridizes extensively to whole-cell RNA, whereas the remaining 90% does not (Gottesfeld et al. 1974b). Finally, the DNA of the soluble fraction constitutes a 10% subset of the single-copy sequences and a 15–20% subset of the families of repetitive sequences of whole-rat DNA.

That the expressed portion of the genome is more rapidly attacked by nucleases than the nonexpressed portion has also been reported by Weintraub and Groudine (1976) and by Garel and Axel (1976). When chromatin is digested with DNase I, which rapidly degrades DNA to acid-soluble material, the expressed portions of the genome are attacked much more rapidly than are the nonexpressed portions. This has been determined by the use of cDNA probes to specific expressed genes, such as hemoglobin and ovalbumin. Thus, it would appear that the expressed sequences of the genome are present in the nucleus in some conformation or arrangement different from that of the nonexpressed portions.

We have further studied our fractionation procedure with respect to a particular individual gene, the globin gene. Friend leukemia cells multiply exponentially in tissue culture until they are induced to undergo erythroid differentiation by substances such as dimethylsulfoxide. After one or two cell generations in the presence of inducer, the cells cease division, start producing messenger RNA for globin, and commence hemoglobin synthesis. We have asked the question: In chromatin from uninduced and induced Friend cells, how is the globin gene sequence distributed between the expressed and nonexpressed portions of the genome (Wallace et al. 1977a)? Friend cells (clone FSD-3) were induced with dimethylsulfoxide (1.5%), and after 24 hours the chromatin of the cells was harvested. At this time the globin gene is being actively transcribed (Ostertag et al. 1972). The chromatin was fractionated by the method previously described (Gottesfeld et al. 1974b), and the DNA of the DNase-II-susceptible fraction hybridized to a tritiated cDNA to mouse reticulocyte globin mRNA. Mass ratios of DNA to cDNA from 5×10^3 to 10^7 were used. The unlabeled driver DNA was maintained at a constant concentration of 5 mg/ml. The samples were denatured by heating and allowed to renature to an equivalent C_0t of 10,000 with respect to unlabeled DNA. As described in the legend to Figure 1, the fraction of cDNA in hybrid at a particular mass ratio is related to the frequency of the globin gene sequence in the driver DNA. Analysis of the data allows calculation of the point at which 50% of the maximal hybridization occurs. This is the point at which the concentration of globin sequences in the unlabeled driver is equal to the concentration of globin sequences in the globin cDNA. As shown in Figure 1, the DNA of the template-active fraction of induced Friend-cell chromatin is enriched sevenfold over whole Friend-cell DNA for the globin gene sequence.

The globin sequences are also contained in the putatively template-active fraction of uninduced FSD-3 cells. These results show that not only is a gene which is actually transcribed present in a DNase-II-susceptible conformation, but so also is the same gene in a cell which is inducible for the expression of that gene. As a control on the fractionation procedure, we have isolated template-active DNA from a Friend leukemia cell line which has lost the ability to be induced (the F4+ cell line). Saturation hybridization experiments show that the globin sequences are not contained in the DNA of this fraction. It is important to note that the globin gene is not deleted in F4+ cells.

Thus, it is possible to use our method of fractionation to separate two categories of DNA sequences, namely, those that are expressed or inducible and those that are nonexpressed and noninducible. Attempts to fractionate the globin gene from whole chromatin by other methods have been unsuccessful (Howk et al. 1975; Krieg and Wells 1976). These attempts have involved mechanical rather than enzymatic shearing. Mechanical shearing has been shown to alter many physical properties of chromatin and to result in histone rearrangement (Noll 1974; Doenecke and McCarthy 1976); it may also result in the isolation of fractions of chromatin which are random with respect to DNA sequence.

A vast amount of insight into the structure of

* Present address: Department of Biological Chemistry, Harvard School of Medicine, Boston, Massachusetts 02115.

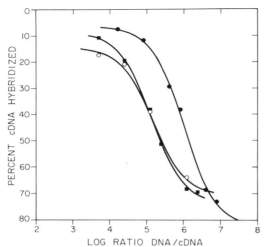

Figure 1. Titration of the globin gene sequences in DNA from template-active fractions of Friend leukemia cell chromatin and DNA from whole cells. Template-active fractions were prepared by the DNase-II technique. All DNA was sheared to 350 base pairs. DNA (5 mg/ml) was mixed with ³H-labeled globin cDNA (2×10^7 cpm/µg) in mass ratios varying from 5×10^3 to 1×10^7. DNA was denatured and allowed to renature for 40 hr at 70°C ($C_0 t 1 \times 10^4$). Hybridization of the ³H-labeled globin cDNA was detected by HAP chromatography.

The fraction of ³H-labeled globin cDNA in hybrid is related to the mass ratio of driver to cDNA by the relation

$$F(R) = A + \frac{1 - A}{1 + (R \cdot B)},$$

where $F(R)$ = fraction [³H]cDNA single-stranded at mass ratio R; A = fraction of cDNA unreactable (data not shown); B = frequency of globin gene sequences in the driver DNA expressed as fractional mass; R = ratio of driver to tracer. (●) Whole-cell DNA; (■) template-active DNA from induced FSD-3 cells; (○) template-active DNA from uninduced FSD-3 cells. (After Wallace et al. 1977a.)

chromatin has developed in the last several years. The bulk of interphase chromatin is organized into subunits 100–120 Å in diameter, consisting of DNA complexed with histones H2A, H2B, H3, and H4 (Olins and Olins 1974; Oudet et al. 1975). These four histones interact strongly with one another to form a tetramer, as first shown by the work of D'Anna and Isenberg (1974). It is possible that a second tetramer interacts with the first tetramer to form the histone core of the subunit. This complex may result

from the interaction of the hydrophobic portions of the histone molecules (DeLange et al. 1969), while the positively charged ends complex with DNA. We know from the work of many investigators that the amount of DNA so complexed is about 200 base pairs. Of these 200, approximately 40 are complexed with histone H1 (Noll and Kornberg 1977). These numbers are somewhat variable between different tissues in the same organism and between different organisms.

We have known for many years that the bulk of the DNA of interphase chromatin is not available for transcription by exogenous RNA polymerase. The fraction that is transcribable varies from about 2% to about 15–20%. We now ask what physical characteristics distinguish the chromatin available for transcription from the bulk of chromatin that is not. The rat-liver-chromatin template-active fraction contains all five histones and DNA in a ratio of about 0.8 to 1—slightly less than the ratio found for template-inactive chromatin (Table 1). Whether this is a real difference or the result of histone degradation or rearrangement, we do not know. The nonhistone chromosomal proteins are present at a much higher concentration in the template-active fraction than in the template-inactive fraction, as shown in Table 1.

Although expressed sequences appear to exist in a nucleosomelike conformation (Lacy and Axel 1975; Gottesfeld et al. 1975; Piper et al. 1976; Mathis and Gorovsky 1976; Reeves and Jones 1976; Kuo et al. 1976), many of the physical properties of isolated template-active chromatin approach those of free DNA. These include thermal denaturation behavior (Fig. 2), circular dichroism spectra, and quinacrine-binding fluorescence (Gottesfeld et al. 1974a; Bonner et al. 1974).

Marushige (1976) has reported that chemical acetylation of calf thymus chromatin increases its transcription by *Escherichia coli* RNA polymerase without any significant removal of histones from their DNA. These results suggest that acetylation of histones may cause activation of genes for transcription. That acetylation precedes such activation was suggested by Allfrey et al. as early as 1964. We have studied chromatin acetylation further and have found that chemical acetylation of rat-liver chromatin confers upon chromatin and upon nucleosomes

Table 1. Chemical Compositions of Rat-liver-chromatin Fractions

Sample	Composition relative to DNA (w/w)		
	histone	nonhistone	RNA
Whole chromatin	1.06	0.65	0.05
Template-inactive chromatin (P1)	1.15	0.58	0.05
Nucleosomes	1.03	<0.05	ca. 0
Template-active chromatin (S2)	0.82	2.3	0.33

After Gottesfeld et al. (1975).

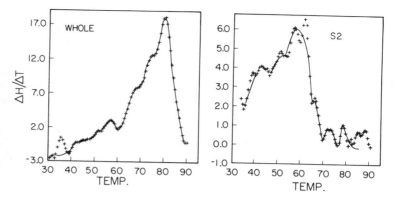

Figure 2. First-derivative melting profiles of whole rat-liver chromatin *(left)* and template-active (S2) fraction of chromatin *(right)*. Whole chromatin melts with a principal component of $T_m = 80°C$ and a minor component of $T_m = 57°C$. S2 chromatin melts with a major component of $T_m = 58°C$ and a minor component at 43°C. The solvent is 0.25 mM EDTA in which rat DNA has a T_m of 43°C.

physical properties similar to those of isolated template-active chromatin.

One of the most obvious physical changes that accompanies chemical acetylation of rat-liver chromatin is its solubilization. Under the conditions of the acetylation reaction (i.e., 0.15 M NaCl), chromatin is a condensed, insoluble precipitate. Chromatin is similarly condensed in 2 mM magnesium ion. Upon addition of acetic anhydride at 0.7 mM or 7 mM, the chromatin precipitate becomes noticeably more soluble. We have investigated this phenomenon by comparing the solubility of control and acetylated nucleosomes in various concentrations of magnesium (Fig. 3). The control nucleosomes are very insoluble, less than 10% remaining soluble in

10 mM magnesium. Acetylation increases the solubility of nucleosomes, the 0.7-mM- and 7-mM-acetylated material becoming almost completely soluble.

If acetylation is responsible for gene activation in vivo, in vitro acetylated chromatin might be expected to show greater nuclease sensitivity than unacetylated chromatin. Figure 4A and Table 2 show the kinetics of digestion of acetylated ^{14}C-labeled chromatin with DNase I. Increased acetylation of chromatin dramatically alters the sensitivity of the DNA to the nuclease. Chromatin acetylated with 0.14 mM acetic anhydride is digested four times faster than control chromatin, and 0.7-mM-acety-

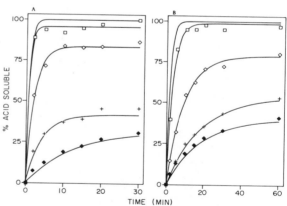

Figure 4. Nuclease sensitivity of acetylated chromatin. Chromatin was prepared from [^{14}C]thymidine-labeled cells and acetylated as indicated. The samples were dialyzed against 10 mM Tris-HCl, pH 7.4, the A_{260} adjusted to 0.4, and the sample adjusted to 10 mM NaCl, 3 mM Mg(OAc)$_2$, 1 mM CaCl$_2$. Then 0.5 μg/ml DNase I *(A)* or 0.14 μg/ml staphylococcal nuclease *(B)* was added and the digestions carried out at 24°C. Aliquots (0.5 ml) were removed at intervals and precipitated with 1 ml of cold 0.3 N perchloric acid. Insoluble material was removed by centrifugation. The supernatant was neutralized with NaOH, adjusted to 1% SDS, 50 mM Tris-HCl, pH 7.4, and counted in Aquasol 2. After 1 hr of digestion, the remaining chromatin was adjusted to 1% SDS, 50 mM Tris-HCl, pH 7.4, 0.2 M sodium perchlorate, and counted to determine the total radioactivity. (◆) Control; (+) 0.14 mM; (◇) 0.7 mM; (□) 7 mM. The lines represent nonlinear least-squares exponential fits to the data, with the upper line in *A* and *B* representing the digestion of deproteinized rat-liver DNA (see Table 1). (After Wallace et al. 1977b.)

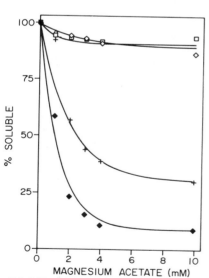

Figure 3. Solubility of acetylated chromatin in magnesium acetate. Nucleosomes were acetylated with the concentration of acetic anhydride indicated, dialyzed against 10 mM Tris-HCl, pH 7.4, 0.1 mM EGTA, and diluted to an A_{260} of 1.0. Aliquots were adjusted to the desired Mg(OAc)$_2$ concentration and centrifuged at 0°C for 10 min at 10,000 rpm in an SS-34 rotor (Sorvall RC 2-B). The A_{260} of the supernatant was measured and the values expressed as percent of total A_{260}. (◆) Control; (+) 0.14 mM; (◇) 0.7 mM; (□) 7 mM. The lines represent nonlinear least-squares exponential fits to the data. (After Wallace et al. 1977b.)

Table 2. Kinetic Parameters from Figure 4

	DNase I			Staphylococcal nuclease		
	K^a	M^b	R^c	K	M	R
Control	0.079	32.0	1.0	0.054	40.5	1.0
0.14 mM	0.237	41.6	3.9	0.054	53.9	1.3
0.7 mM	0.458	83.3	15.2	0.110	78.9	3.9
7 mM	1.300	95.6	49.4	0.322	98.8	14.4
DNA	1.290	100	51.3	0.653	100	29.6

After Wallace et al. (1977b).
[a] First-order rate constant (min^{-1}).
[b] Maximum percent digested.
[c] Relative initial rate of reaction. The initial rate for control chromatin was 0.01 A_{260} unit/ml/min with DNase I, and 0.0088 A_{260} unit/ml/min with staphylococcal nuclease.

Figure 5. Polyacrylamide gel electrophoresis of DNase-I-digested chromatin DNA. Control and acetylated chromatins were dialyzed against 10 mM Tris-HCl, pH 7.4, and the A_{260} adjusted to 10. Each was brought to 10 mM NaCl, 3 mM Mg(OAc)$_2$, 1 mM CaCl$_2$, and 10 μg/ml DNase I. Each sample was digested for various times at 24°C such that the percentages of acid-soluble A_{260} were 25, 30, 15, and 15 for control, 0.14 mM, 0.7 mM, and 7 mM, respectively. Reactions were stopped by bringing each to 1% SDS and the samples were phenol/chloroform-extracted. DNA was precipitated with ethanol and dissolved in gel buffer; 25 μg of each was brought to 50% formamide, boiled 3 min, and cooled on ice. Electrophoresis was performed on 12% acrylamide-7 M urea gels as described by Maniatis et al. (1975). From left to right are the DNAs from unacetylated chromatin and from 0.14-mM-, 0.7-mM-, and 7-mM-acetylated chromatin. (After Wallace et al. 1977b.)

lated chromatin 15 times faster. Maximally acetylated chromatin (7 mM) is as sensitive as deproteinized DNA to DNase I. Figure 4B shows the results of a similar experiment using staphylococcal nuclease. It can be seen that 0.14-mM-acetylated chromatin is slightly more sensitive than control chromatin to staphylococcal nuclease, and 0.7-mM-acetylated chromatin is digested four times as fast as control chromatin. Also, 7-mM-acetylated chromatin is digested approximately half as fast as deproteinized DNA. Thus, chemically acetylated chromatin is moderately sensitive to staphylococcal nuclease but extremely sensitive to DNase I.

Weintraub and Groudine (1976) report that during the digestion of active genes with DNase I, the digested DNA appears as multiples of 10 nucleotides. Figure 5 shows that both 0.14-mM- and 0.7-mM-acetylated chromatin are digested with DNase I to the same 10-nucleotide repeat pattern as control chromatin. This result suggests that the histone-DNA interaction which produces this periodic pattern is not disrupted by the acetylation. On the other hand, the 7-mM-acetylated chromatin does not have this

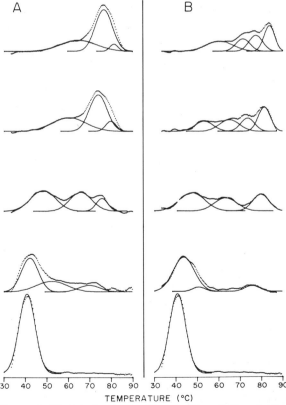

Figure 6. First-derivative melting profiles. Full scale for each profile is 3% change in hyperchromicity per °C. From top to bottom are unacetylated material, 0.14-mM-, 0.7-mM-, and 7-mM-acetylated material, and deproteinized rat-liver DNA. *(A)* Nucleosomes; *(B)* chromatin. (·····) dH/dT; (——) Gaussian components. (After Wallace et al. 1977b.)

same repeat pattern (although faint bands are apparent at 10-nucleotide intervals).

The thermal denaturation of control and acetylated chromatin and nucleosomes, as well as of deproteinized DNA, is shown in Figure 6. The data are presented as first-derivative melting profiles, fit to three or four Gaussian components to quantitate the transitions observed. The most obvious effect of chemical acetylation on thermal denaturation is the dramatic shift to lower melting temperatures of all the observed transitions. This is true for both chromatin and nucleosomes, although the melting profiles of control and acetylated chromatin are more complex.

Figure 7 shows the effect of acetylation on the sedimentation behavior of nucleosomes. The sedimentation of acetylated nucleosomes is dramatically

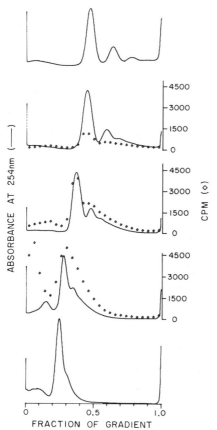

Figure 7. Sucrose gradient sedimentation of nucleosomes. Sedimentation is from left to right. From top to bottom are nucleosomes, 0.14-mM-acetylated nucleosomes, 0.7-mM-acetylated nucleosomes, 7-mM-acetylated nucleosomes, and SDS-treated nucleosomal DNA. (———) A_{260}; (◇) ^{3}H-acetate cpm. The sedimentation coefficients of the monomer peaks are 11.3, 10.6, 8.6, 6.2, and 5.4, respectively. The specific activities of the labeled samples were 19.1, 67.4, and 119 acetates per 200 base pairs of DNA for 0.14-mM-, 0.7-mM-, and 7-mM-acetylated nucleosomes. Under the conditions of acetylation, approximately half as much acetate is incorporated into nonhistones as into histones, and no detectable acetate is incorporated into DNA. (After Wallace et al. 1977b.)

retarded, compared to that of control nucleosomes, and approaches that of SDS-treated control nucleosomes. To determine whether all of these nucleosomes were acetylated equally and whether all of the acetate remained bound to the nucleohistone, the acetylation reaction was carried out with ^{3}H-labeled acetic anhydride. Figure 7 also shows the distribution of acetate counts across the gradients. It can be seen that from 0.14 to 0.7 mM acetic anhydride there is a fourfold increase in acetate incorporation per nucleosome, whereas from 0.7 to 7 mM acetic anhydride there is only a twofold increase. All nucleosomes, monomer through trimer, are acetylated. At 7 mM acetic anhydride, acetate counts are found at the top of the gradient, perhaps representing dissociated histone H1.

DISCUSSION

Those portions of chromatin which are transcribed in vivo exhibit many characteristics which distinguish them from untranscribed regions (Bonner et al. 1974). Of these, the most striking difference is the differential sensitivity to nucleases (Gottesfeld et al. 1974b; Weintraub and Groudine 1976; Garel and Axel 1976). In spite of this increased sensitivity to nuclease, it is clear that expressed sequences, when digested with staphylococcal nuclease or DNase II, are present in nucleosomes similar to those of total chromatin (Lacy and Axel 1975; Gottesfeld et al. 1975; Piper et al. 1976; Mathis and Gorovsky 1976; Reeves and Jones 1976; Kuo et al. 1976). However, when actively transcribed ribosomal RNA genes are visualized in the electron microscope, no nucleosomal subunits are apparent (Hamkalo et al. 1974). It is not clear whether the absence of nucleosomes is due to the failure of these structures to survive the spreading technique or to a true feature of these genes. In addition, the low thermal stability of expressed regions of chromatin (McCarthy et al. 1974) is uncharacteristic of nucleosomes.

As one approach to studying the properties of expressed regions of the genome, we have physically separated active and inactive chromatin. The template-active fraction is isolated by virtue of its greater sensitivity to nuclease and its solubility in MgCl$_2$, as discussed earlier. It has been shown to contain nucleosomelike particles (Gottesfeld et al. 1975; Gottesfeld and Bonner 1977) and to melt at lower temperatures (Fig. 2).

The suggestion that acetylation may play a role in gene regulation (Allfrey et al. 1964) and the observation that chemical acetylation in vitro increases template activity (Marushige 1976) prompted us to compare the physical properties of chemically acetylated chromatin to those attributed to the expressed portions of the genome. Acetylation dramatically increases the sensitivity of nucleohistone to nucleases, increases its solubility in salt and magnesium, and substantially decreases its thermal stability. All of these properties are similar to those of template-active chromatin. It is clear that chemical acetylation increases the accessibility of nucleosomal DNA to RNA polymerase and nucleases. The decreased sedimentation velocity of acetylated nucleosomes may reflect a more extended conformation which is responsible for this increased accessibility. Similar conformational changes may be responsible for the increased accessibility of sequences expressed in vivo. It should be pointed out that the sedimentation velocity of nucleosomes from active chromatin may be increased by their association with nascent RNA and nonhistone chromosomal proteins. We have previously reported that the nucleosomelike particles of template-active chromatin sediment more rapidly than those of unfractionated chromatin (Gottesfeld et al. 1975).

The process of gene regulation includes not only the facilitation or inhibition of transcription, but also mechanisms for the recognition of specific DNA sequences. The properties exhibited by chromatin following acetylation in vitro suggest that acetylation or analogous processes may be involved in the facilitation of transcription. However, the recognition processes which must precede this activation remain to be elucidated.

Acknowledgments

This work was supported in part by U.S. Public Health Service Grants GM 13762 and GM 20297. R. B. W. is a fellow of the Medical Research Council of Canada; T. D. S. is a predoctoral National Science Foundation fellow; and R. F. M. is a trainee supported by U.S. Public Health Service Training Grant GM 50086. S. K. D. has been supported as a Sherman Fairchild Distinguished Scholar.

REFERENCES

ALLFREY, V. G., R. FAULKNER, and A. E. MIRSKY 1964. Acetylation and methylation of histones and their possible role in the regulation of RNA synthesis. *Proc. Natl. Acad. Sci.* **51**: 786.

BILLING, R. J. and J. BONNER. 1972. The structure of chromatin as revealed by DNase digestion studies. *Biochim. Biophys. Acta* **281**: 453.

BONNER, J., W. T. GARRARD, J. GOTTESFELD, D. S. HOLMES, J. S. SEVALL, and M. WILKES. 1974. Functional organization of the mammalian genome. *Cold Spring Harbor Symp. Quant. Biol.* **38**: 303.

D'ANNA, J. A. and I. ISENBERG. 1974. A histone cross-complexing pattern. *Biochemistry* **13**: 4992.

DELANGE, R., D. M. FAMBROUGH, E. L. SMITH, and J. BONNER. 1969. Calf and pea histone IV. III. Complete amino acid sequence of pea seedling histone IV; comparison with the homologous calf thymus histone. *J. Biol. Chem.* **244**: 5669.

DOENECKE, D. and B. MCCARTHY. 1976. Movement of histones in chromatin induced by shearing. *Eur. J. Biochem.* **64**: 405.

GAREL, A. and R. AXEL. 1976. Selective digestion of transcriptionally active ovalbumin genes from oviduct nuclei. *Proc. Natl. Acad. Sci.* **73**: 3966.

GOTTESFELD, J. M. and J. BONNER. 1977. Isolation and properties of the expressed portion of the mammalian genome. In *The molecular biology of the mammalian ge-nome* (ed. P. O. P. Ts'o), p. 381. Elsevier/North-Holland, Amsterdam.

GOTTESFELD, J. M., R. F. MURPHY, and J. BONNER. 1975. Structure of transcriptionally active chromatin. *Proc. Natl. Acad. Sci.* **72**: 4404.

GOTTESFELD, J. M., G. BAGI, B. BERG, and J. BONNER. 1976. Sequence composition of the template-active fraction of rat liver chromatin. *Biochemistry* **15**: 2472.

GOTTESFELD, J. M., G. K. RADDA, and I. O. WALKER. 1974a. Biophysical studies on the mechanism of quinacrine staining of chromosomes. *Biochemistry* **13**: 2937.

GOTTESFELD, J. M., W. T. GARRARD, G. BAGI, R. F. WILSON, and J. BONNER. 1974b. Partial purification of the template-active fraction of chromatin. A preliminary report. *Proc. Natl. Acad. Sci.* **71**: 2193.

HAMKALO, B., O. MILLER, and A. H. BAKKEN. 1974. Ultrastructure of active eukaryotic genomes. *Cold Spring Harbor Symp. Quant. Biol.* **31**: 915.

HOWK, R. S., A. ANISOWIEZ, A. SILVERMAN, W. PARKS, and E. SCOLNICK. 1975. Distribution of murine type B and type C viral nucleic acid sequences in template active and template inactive chromatin. *Cell* **4**: 321.

KREIG, P. and J. R. E. WELLS. 1976. The distribution of active genes (globin) and inactive genes (keratin) in fractionated chicken erythroid chromatin. *Biochemistry* **15**: 4549.

LACY, E. and R. AXEL. 1975. Analysis of DNA of isolated chromatin subunits. *Proc. Natl. Acad. Sci.* **72**: 3978.

MANIATIS, T., A. JEFFREY, and H. VAN DESANDE. 1975. Chain length determination of small double- and single-stranded DNA molecules by polyacrylamide gel electrophoresis. *Biochemistry* **14**: 3787.

MARUSHIGE, K. 1976. Activation of chromatin by acetylation of histone site chains. *Proc. Natl. Acad. Sci.* **73**: 3937.

MARUSHIGE, K. and J. BONNER. 1971. Fractionation of liver chromatin. *Proc. Natl. Acad. Sci.* **68**: 2941.

MATHIS, D. J. and M. A. GOROVSKY. 1976. Subunit structure of rDNA-containing chromatin. *Biochemistry* **15**: 750.

MCCARTHY, B. J., J. T. NISHIURA, D. DOENECKE, D. S. NASSER, and C. B. JOHNSON 1974. Transcription and chromatin structure. *Cold Spring Harbor Symp. Quant. Biol.* **38**: 763.

NOLL, M. 1974. Subunit structure of chromatin. *Nature* **251**: 249.

NOLL, M. and R. D. KORNBERG. 1977. Action of micrococcal nuclease on chromatin and the location of histone H1. *J. Mol. Biol.* **109**: 393.

OLINS, A. and D. OLINS. 1974. Spheroid chromatin units (ν bodies). *Science* **181**: 330.

OSTERTAG, W., H. MELDERIS, G. STEINHEIDER, N. KLUGE, and S. DUBE. 1972. Synthesis of mouse hemoglobin and globin mRNA in leukemic cell cultures. *Nat. New Biol.* **239**: 231.

OUDET, P., M. GROSS-BELLARD, and P. CHAMBON. 1975. Electron microscopic and biochemical evidence that chromatin structure is a repeating unit. *Cell* **4**: 281.

PIPER, P. W., J. CELIS, K. KALTOFT, J. C. LEER, O. F. NIELSEN, and O. WESTERGAARD. 1976. Tetrahymena ribosomal RNA gene chromatin is digested by micrococcal nuclease at sites which have the same regular spacing on the DNA as corresponding sites in the bulk nuclear chromatin. *Nucleic Acids Res.* **3**: 493.

REEVES, R. and A. JONES. 1976. Genomic transcriptional activity and the structure of chromatin. *Nature* **260**: 495.

TIEN KUO, M., C. G. SAHASRABUDDHE, and G. F. SAUNDERS. 1976. Presence of messenger specifying sequences in the DNA of chromatin subunits. *Proc. Natl. Acad. Sci.* **73**: 1572.

WALLACE, R. B., S. K. DUBE, and J. BONNER. 1977a. Localization of the globin gene in the template active fraction of chromatin of Friend leukemia cells. *Science* (in press).

WALLACE, R. B., T. D. SARGENT, R. MURPHY, and J. BONNER. 1977b. Physical properties of chemically acetylated rat liver chromatin. *Proc. Natl. Acad. Sci.* **74:** 3244.

WEINTRAUB, H. and M. GROUDINE. 1976. Chromosomal subunits in active genes have an altered conformation. *Science* **193:** 848.

QUESTIONS/COMMENTS

Question by:
M. E. GOLDSMITH
National Institutes of Health

The argument that the S2 fraction contains an enrichment of actively transcribing genes hinges on the demonstration of the presence of nascent messenger RNA. The argument that the DNA of the S2 fraction contains the DNA sequences which code for a particular mRNA depends on the rigorous elimination of mRNA. What procedures have been used to eliminate messenger from the preparation?

Response by:
J. BONNER
California Institute of Technology

The S2 fraction of chromatin does bear the nascent RNA. The deproteinized DNA of the S2 fraction in our liver and ascites experiments was freed of RNA by extensive digestion with RNase followed by pronase to remove the RNase. In the case of the Friend-cell S2 fraction, the DNA was freed of endogenous RNA both by RNase digestion and by incubation in alkali. The final DNA contained no analytically detectable RNA.

"Native" Salivary Chromosomes of *Drosophila melanogaster*

R. J. HILL AND F. WATT

Genetics Research Laboratories, CSIRO, Division of Animal Production, Epping, N.S.W. 2121, Australia

Detailed morphology of *Drosophila melanogaster* salivary chromosomes survives exposure to aqueous acetic acid, the solvent used since the 1930s in cytological preparation of chromosomes in the classical squashing technique (Painter 1934). However, on return from aqueous acetic acid to simple aqueous solution, for example for treatment with antisera, much detail is often lost. Furthermore, there is direct evidence that acid fixatives can extract chromosomal proteins (Dick and Johns 1968; Brody 1974; Pothier et al. 1975) and destroy protein antigenicity (Stenman et al. 1975). These factors raise difficulties for biochemical studies of chromosome structure and function utilizing classical acetic acid squash preparations.

Although there may be a number of solutions to this problem, perhaps the most rigorous would be to isolate the polytene chromosomes by microdissection in saline and thereby avoid the use of aqueous acetic acid completely. Such a procedure, although requiring some skill in micromanipulation, is quite possible for *Drosophila melanogaster,* as evidenced by the following description.

EXPERIMENTAL PROCEDURES

Cytological preparation of salivary chromosomes. Larvae for dissection were grown in uncrowded, well-yeasted culture vials maintained at 18–19°C. Large late-third-instar larvae, staged by timed egg deposition, were selected for chromosome isolation.

Salivary glands were dissected out into a drop of 25 mM disodium glycerophosphate, 10 mM KH_2PO_4, 30 mM KCl, 10 mM $MgCl_2$, 3 mM $CaCl_2$, 162 mM sucrose, pH 6.8. Subsequent microdissection steps were carried out in salines based on that of Glancy (1946)—90 mM KCl, 60 mM NaCl, 1 mM $CaCl_2$, 5 mM sodium phosphate, pH 7.0, which shall be referred to as saline G. Glands were transferred to 1% Triton X-100, saline G for 20–30 seconds and then to 0.025% Triton X-100, saline G for 5 minutes. They were subsequently placed in a 3-μl drop of 0.01% Triton X-100, saline G on a microscope slide, and several cells were torn open with a tungsten needle to release nuclei. Clean nuclei were then transferred in 0.05% formaldehyde, saline G by micropipette or on a glass needle to a well in the slide.

A large hole was then torn in the nuclear membrane and the loosely adhering bundle of chromosomes removed with fine glass needles. The chromosome complement was unravelled, flattened, and allowed to attach to the glass, again using the glass needles. Once the chromosomes are attached, they generally remain so through several changes of buffer.

Preparation of embryo chromatin. This was prepared from dechorionated 2–20-hour embryos by a procedure based on that of Marushige and Bonner (1966) as described in detail by Hill and Watt (1977).

Immunofluorescence staining. Chromatin (ca. 4 A_{260} units) was emulsified with complete Freund's adjuvant and injected subcutaneously into mice at time zero, 4 weeks, 6 weeks, and 8 weeks. The animals were bled and sera collected 10 days after the final injection.

Chromosome preparations were immersed in saline G for 1 hour to remove traces of formaldehyde and then treated with antiserum diluted 1:30 in saline G. They were then washed for 1 hour to remove excess antibody and exposed for 30 minutes to 1:20 diluted rabbit anti-rat γ-globulin (Wellcome Reagents, Ltd.). They were finally washed for 1 hour and photographed through a Zeiss fluorescence microscope on Ilford HP4 film.

RESULTS

Salivary Chromosomes Isolated by Microdissection

Nuclei dissected out of third-instar salivary glands are illustrated in Figure 1. The most common chromosomal conformation for such nuclei is depicted in Figure 1a; in this general conformation, the chromosomes extend throughout the nuclear space, and there appear to be numerous attachments between the chromosomes and the nuclear membrane; these resist micromanipulative attempts to release intact chromosomes. Telomeres in particular are difficult to free from such nuclei. Holmquist and Steffenson (1973) briefly reported attachments between polytene chromosomes and the nuclear membrane and noted that there was a correspondence between late replicating bands and attachment sites. The present micromanipulative investigations have suggested

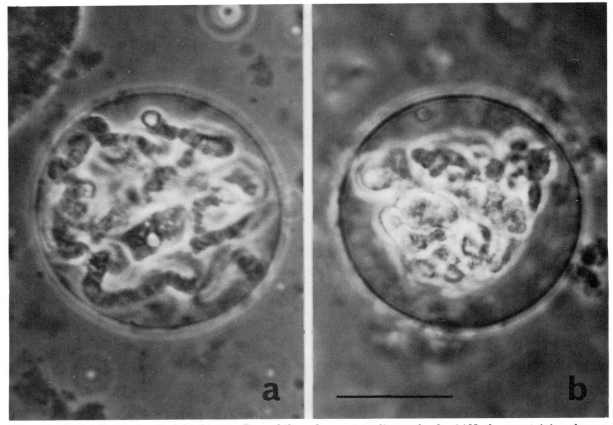

Figure 1. Nuclei dissected out of third-instar *Drosophila melanogaster* salivary glands. *(a)* Nucleus containing chromosomes in the most common configuration, namely, dispensed throughout the nuclear space; *(b)* nucleus in which the chromosome complement is well separated from the nuclear membrane. Bar = 20 μm.

that these physical attachments become fewer towards the end of third instar, in accord with the notion that they may be associated with DNA synthesis which is known to decrease at this stage (Rodman 1968; Rudkin 1972).

Although chromosomes and chromosomal segments can be dissected from nuclei whose chromosomes are in the conformation depicted in Figure 1a, the preparation of whole-genome spreads is greatly facilitated by the choice of nuclei in which the chromosome complement is well separated from the nuclear membrane (Fig. 1b). After the chromosome complement is removed through a large hole torn in the nuclear membrane and individual chromosome arms are unravelled, spreads such as that depicted in Figure 2 are obtained.

The degree of preservation of chromosome morphology possible without acid fixation is illustrated by the higher magnification micrographs in Figures 3 and 4. The segment shown in Figure 4 has been stretched slightly between glass needles to reveal detailed, fine band structure, which compares quite favorably with acid-fixed material taken from the standard photographic map of Lefevre (1976). Less

dense puff structures are also obtained in a good state of preservation; see, e.g., the large puff near the tip of the X chromosome (Fig. 3b,d).

Furthermore, the nucleolus is much better preserved than in classical preparations. During the micromanipulations it accompanies the chromosomes, remaining in the vicinity of the chromocenter. It is in an optical plane just above that of Figure 2 and is seen in focus in Figure 3a; it maintains an appearance virtually unchanged from that in the intact cell.

Micronucleolarlike structures are also sometimes apparent at other points (see Fig. 2). Whether these arise simply by fragmentation of the nucleolus itself or whether they have an independent existence has yet to be determined for these preparations. It may be of interest that one of these structures can be seen over region 56F (arrow in Fig. 2), the site of the 5S genes (Wimber and Steffenson 1970).

Reaction of Antiserum against Embryo Chromatin

Antisera raised against embryo chromatin were observed to contain antichromatin antibodies by spe-

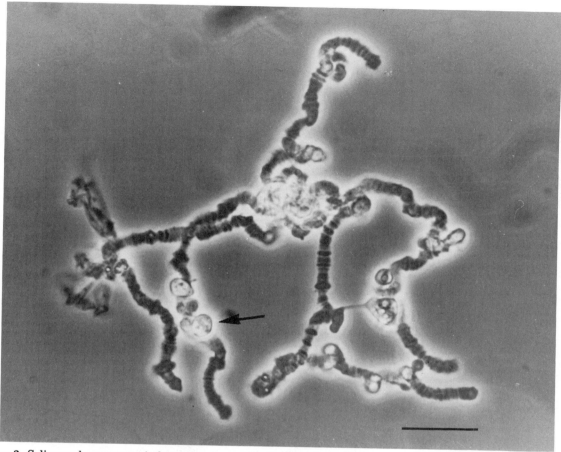

Figure 2. Salivary chromosomes isolated by micromanipulation. The nucleolus is in the vicinity of the chromocenter and is just above the optical plane of this micrograph. It is shown in focus in Fig. 3a. The arrow indicates a micronucleolus covering region 56F. Bar = 20 μm.

cific precipitin reactions on double-diffusion plates (Fig. 5). These precipitin reactions were not given by sera from preimmune animals, sham-injected animals, or animals injected with bovine serum albumin.

Figure 6 (a,b) records the result obtained on treating salivary chromosome preparations with antisera against embryo chromatin. The definite fluorescence over the chromosome contrasts strikingly with the barely detectable fluorescence over chromosomes treated with sera from control animals (Fig. 6c,d).

It is of considerable interest that the distribution of fluorescence along the chromosome is distinctly different from the distribution of total material; note, for example, the bands indicated by the arrow in Figure 6a. The difference between the phase-contrast and fluorescence patterns is a consistent characteristic of these reactions over different chromosome segments and using antisera raised in different animals. It is more obvious when chromosome segments are observed at higher magnification (Fig. 4a,b).

DISCUSSION

Since the pioneering studies of Glancy (1946), there have been a number of successful investigations of the chemical, structural, and functional properties of the polytene chromosomes of *Chironomus* isolated by microsurgical techniques which avoid acid fixation (Lezzi 1965; Robert 1971). Glancy also mentioned in passing "incidental observations" on *Drosophila* chromosomes obtained by micromanipulation. It has been demonstrated in this report that it is possible to prepare complete spreads of the polytene chromosome complement from *Drosophila melanogaster* salivary gland nuclei without the use of acid fixatives.

The micromanipulations involved in the release of *Drosophila* salivary chromosomes are greatly facilitated by prior brief exposure of the gland to 1% Triton X-100 (20–30 sec), followed by exposure to 0.025% Triton X-100 for 5 minutes. The mild detergent Triton X-100 was selected for this work because of the vast array of evidence that it does not cause

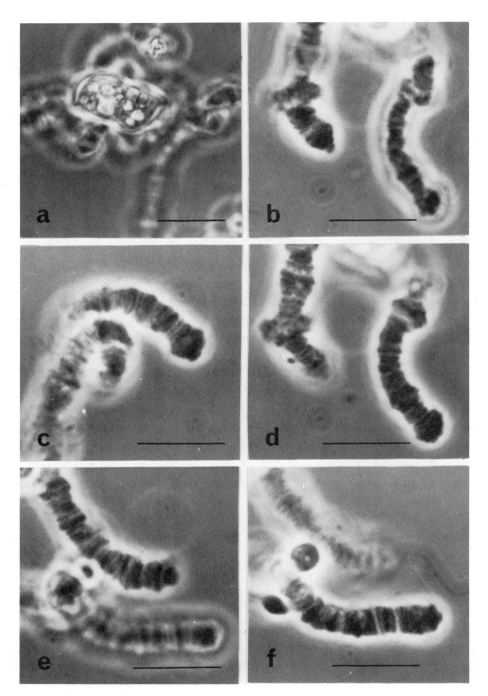

Figure 3. Higher power micrographs of some specific regions of Fig. 2. *(a)* Nucleolus; *(b)* distal segment of the X chromosome, in focus on the left; *(c)* distal segment of chromosome 2L; *(d)* the distal region of chromosome 2R *(right)* and the puff near the tip of the X chromosome *(left)*; *(e)* distal segment of chromosome 3L; *(f)* distal segment of chromosome 3R. Bar = 10 μm.

the loss of protein biological activity in the case of antigens, receptor proteins, and numerous enzymes (Helenius and Simons 1975). This is probably because the low critical micelle concentration for Triton X-100 limits the free concentration of the amphiphile to a level well below that required for the cooperative binding to protein necessary for denaturation (Makino et al. 1973).

To minimize any loss of protein from the *Drosophila* chromosomes, isolated nuclei were exposed

to a low concentration of formaldehyde (0.05%). Such low concentrations of formaldehyde have long been known to have only minimal, if any, effects on antigenicity (see, e.g., Stanley 1945). However, it should be mentioned that, with practice and favorable material, it is possible to isolate the chromosomes with no fixation at all. Moreover, it is of interest that the reactions of formaldehyde with proteins in general are largely reversible (Fox and Foster 1957); Jackson and Chalkley (1974) have shown that

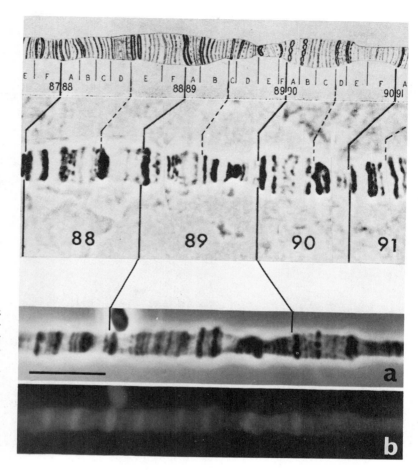

Figure 4. Higher-power phase-contrast micrograph *(a)* illustrating the preservation of fine band structure in a chromosome segment isolated without acid fixation. *(b)* The corresponding fluorescence micrograph illustrating the specificity of the reaction of chromatin antiserum with this chromosome segment. The upper portion of this figure, depicting the classical map and the corresponding photographic map for this segment of chromosome 3R, is taken from Lefevre (1976). Bar = 10 μm.

the cross-linking formaldehyde induces in nucleoproteins, in particular, is effectively completely reversible.

It has been reported above that antisera raised against *Drosophila* embryo chromatin react with larval polytene chromosomes, indicating the existence of antigenic determinants common to the chromosomes of embryos and larvae. These reactions are differential along the chromosome, reacting with some structures and not with others, consistent with a differential distribution of chromosomal protein antigens. Experiments currently in progress with purified chromosomal protein fractions further reinforce this conclusion. The absence of fluorescence

Figure 5. Double-diffusion plates showing specific antisera against chromatin. The plates were equilibrated with 0.1 M NaCl, 0.005 M sodium phosphate, pH 7.0. The central wells contain chromatin antiserum in *a* and preimmune serum in *b*. The peripheral wells contain, starting at 12 o'clock and proceeding clockwise, bovine serum albumin and then five different chromatin preparations.

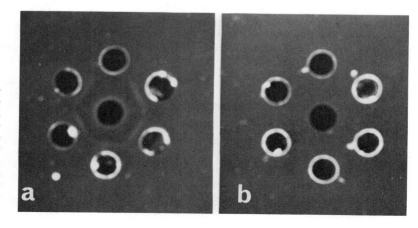

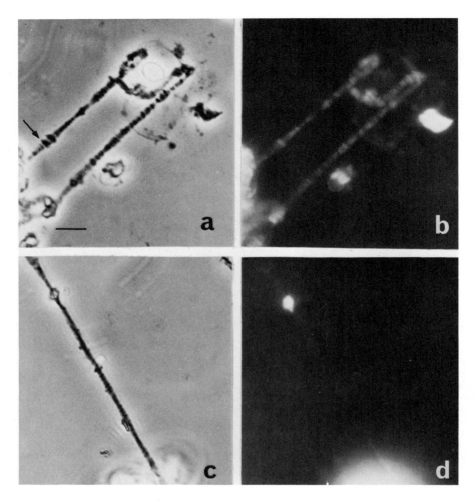

Figure 6. Phase-contrast (a,c) and fluorescence (b,d) photomicrographs depicting the reaction of antiserum against embryo chromatin with larval polytene chromosomes isolated by microdissection. (b) Reaction given with chromatin antiserum; (d) result obtained with control serum from a sham-injected animal. Bar = 10 μm.

from some chromosomal regions may result from the absence of specific larval nonhistones from the embryonic antigen population.

Drosophila salivary chromosomes isolated by microdissection should find a number of future applications, including the following: (1) investigation of chromosome physical structure; (2) localization of specific proteins by the immunofluorescence procedure; (3) asking ultimate questions about the native conformation of chromosomal proteins in situ; (4) studies on the binding of steroid hormones to chromosomes; and, perhaps, (5) analysis of the activation of specific chromosomal loci in vitro.

Acknowledgments

We wish to thank Dr. J. M. Rendel for encouragement and support. We would like to express our gratitude to Professor H. G. Callan, F.R.S., for introducing one of us (R.J.H.) to the basic art of microdissection within cell nuclei. The enthusiastic technical assistance of Mrs. D. MacPherson is gratefully acknowledged.

REFERENCES

BRODY, T. 1974. Histones in cytological preparations. *Exp. Cell Res.* **85**: 255.

DICK, C. and E. W. JOHNS. 1968. The effect of two acetic acid containing fixatives on the histone content of calf thymus deoxyribonucleoprotein and calf thymus tissue. *Exp. Cell Res.* **51**: 626.

FOX, S. W. and J. F. FOSTER. 1957. *Protein chemistry.* Wiley, New York.

GLANCY, E. A. 1946. Microsurgical studies on *Chironomus* salivary gland chromosomes. *Biol. Bull.* **90**: 71.

HELENIUS, A. and K. SIMONS. 1975. Solubilisation of membranes by detergents. *Biochim. Biophys. Acta* **415**: 29.

HILL, R. J. and F. WATT. 1977. Chromosomal proteins of *Drosophila melanogaster* and an approach for their localisation on polytene chromosomes. *Chromosoma* **63**: 57.

HOLMQUIST, G. and D. M. STEFFENSON. 1973. Evidence for a specific three dimensional arrangement of polytene chromosomes by nuclear membrane attachments. *J. Cell Biol.* **59**: 147a.

JACKSON, V. and R. CHALKLEY. 1974. Separation of newly synthesized nucleohistone by equilibrium centrifugation in cesium chloride. *Biochemistry* **13**: 3952.

LEFEVRE, G. 1976. A photographic representation and interpretation of the polytene chromosomes of *Drosophila melanogaster* salivary glands. In *The genetics and biology of* Drosophila (ed. M. Ashburner and E. Novitski), vol. 1a, p. 31. Academic Press, London.

LEZZI, M. 1965. Die wirkung von Dnase auf isolierte polytän-chromosomen. *Exp. Cell Res.* **39**: 289.

MAKINO, S., J. A. REYNOLDS, and C. TANFORD. 1973. The binding of deoxycholate and Triton X-100 to proteins. *J. Biol. Chem.* **248**: 4926.

MARUSHIGE, K. and J. BONNER. 1966. Template properties of liver chromatin. *J. Mol. Biol.* **15**: 160.

PAINTER, T. S. 1934. A new method for the study of chromosome aberrations and the plotting of chromosome maps in *Drosophila melanogaster*. *Genetics* **19**: 175.

POTHIER, L., J. F. GALLAGHER, L. E. WRIGHT, and P. R. LIBBY. 1975. Histones in fixed cytological preparations of Chinese hamster chromosomes demonstrated by immunofluorescence. *Nature* **225**: 350.

ROBERT, M. 1971. Einflub von ionenstärke und pH auf die differentielle Debondensation der Nucleoproeide isolierter speicheldrüsen—Zellberne und—chromosomen ion *Chironomus thummi*. *Chromosoma* **36**: 1.

RODMAN, T. C. 1968. Relation of developmental stage to initiation of replication in polytene nuclei. *Chromosoma* **23**: 271.

RUDKIN, G. T. 1972. Replication in polytene chromosomes. In *Developmental studies on giant chromosomes* (ed. W. Beermann et al.), p. 59. Springer Verlag, New York.

STANLEY, W. M. 1945. The preparation and properties of influenza virus vaccines concentrated and purified by differential centrifugation. *J. Exp. Med.* **81**: 193.

STENMAN, S., M. ROSENQVIST, and N. R. RINGERTS. 1975. Preparation and spread of unfixed metaphase chromosomes for immunofluorescence staining of nuclear antigens. *Exp. Cell Res.* **90**: 87.

WIMBER, D. and D. STEFFENSON. 1970. Localisation of 5S RNA genes on *Drosophila* chromosomes by RNA-DNA hybridization. *Science* **170**: 639.

The Size and Chromosomal Location of the 75S RNA Transcription Unit in Balbiani Ring 2

B. Daneholt, S. T. Case,* J. Derksen, M. M. Lamb, L. Nelson, and L. Wieslander

Department of Histology, Karolinska Institutet, S-10401 Stockholm 60, Sweden

During the past several years we have been interested primarily in the functional significance of Balbiani ring 2, a giant puff specific to the salivary glands of *Chironomus tentans*. It has been demonstrated that a 75S RNA transcript is produced within Balbiani ring 2 (BR2) and that this RNA molecule is transported via nuclear sap into cytoplasm (for review, see Daneholt 1975). Recently it has also been shown that 75S RNA molecules (Daneholt et al. 1977) and BR2 RNA sequences (Wieslander and Daneholt 1977) reside in large-size polysomes. Since, in cytoplasm, BR2 sequences are confined to the 75S RNA fraction (Lambert and Edström 1974), it was inferred that 75S RNA from BR2 is incorporated into polysomal structures (Daneholt et al. 1977). Furthermore, since 75S RNA constitutes by far the most abundant nonribosomal, high-molecular-weight RNA species in cytoplasm (Daneholt and Hosick 1973) and most of it is present in polysomes (Daneholt et al. 1977), it is suggested that 75S RNA corresponds to messenger RNA for the main protein product of these cells, the salivary polypeptides (e.g., Doyle and Laufer 1969; Grossbach 1974). These results are in good agreement with cytogenetic studies indicating that BRs in *Chironomus* salivary glands are coupled to the production of salivary polypeptides (Beermann 1961; Grossbach 1969, 1974). Thus, the available information strongly suggests that BR2 contains a structural gene heavily expressed in the salivary glands.

Due to the well-defined band-interband structure of polytene chromosomes, BR2 offers a suitable system for analyzing how a gene is related to the chromomere-interchromomere organization of the chromosomal material. It has been observed (for review, see Beermann 1962) that, upon gene activation, the tightly packed deoxyribonucleoprotein (DNP) filament of a chromomere is unfolded, a DNP loop is formed, and transcription takes place along the loop. After cessation of transcription, the DNP filament is folded back and constitutes a chromomere again. These cytological observations led Beermann (1964) to propose that, in general, a chromomere corresponds to a transcription unit. However, more information is needed before it can be decided whether this simple concept is correct or not. In our analysis of BR2, we are now approaching the question of how the 75S RNA transcription unit in BR2 is related to the chromomere-interchromomere structure of the BR2 region. In the present report, we present information on the size and chromosomal location of the 75S RNA transcription unit and consider what can be stated about the relationship between the 75S RNA transcription unit and the BR2 chromomere.

METHODS

Material. The dipteran *Chironomus tentans* was raised in the laboratory as outlined by Lambert and Daneholt (1975). Salivary glands of rapidly growing fourth-instar larvae were used and, in the cytological experiments, salivary glands and Malpighian tubules from old fourth-instar larvae or prepupae were used as well. Galactose treatment of larvae was carried out essentially according to the method of Beermann (1973).

Labeling conditions. Salivary gland RNA was labeled for 3 days in vivo as described previously (Daneholt and Hosick 1973), but only 100 μCi [³H]cytidine (>25 Ci/mmole) and 100 μCi [³H]uridine (>40 Ci/mmole) were added to the 20-ml incubation medium. When ¹⁴C-labeled precursors were used, the medium was supplied with 10 μCi [¹⁴C]cytidine (486 mCi/mmole) and 10 μCi [¹⁴C]uridine (527 mCi/mmole).

For in vitro labeling, six salivary glands were incubated for 45 minutes at 18°C in 50 μl modified Cannon's medium provided with tritiated cytidine and uridine (Lambert and Daneholt 1975).

Cytological methods. Salivary glands were fixed in 70% ethanol for 60 minutes at 4°C and transferred to ethanol-glycerol (1:1) for at least 60 minutes at 4°C. BR2s were isolated by microdissection (Edström 1964; Lambert and Daneholt 1975).

For morphological analysis, salivary glands and Malpighian tubules were isolated and subsequently fixed in ethanol-acetic acid (3:1), stained in aceto-orcein, squashed in 50% acetic acid, dehydrated, and mounted in Euparal as specified by Derksen (1977).

Extraction of salivary gland RNA and BR2 RNA. Salivary glands were solubilized in a sodium dodecyl

---

* Present address: Kline Biology Tower, Yale University, New Haven, Connecticut 06520.

sulfate (SDS)-pronase solution essentially according to the method of Case and Daneholt (1976), but the extraction medium contained 0.5% SDS, 1 mg/ml pronase, 0.01 M EDTA, and 0.02 M Tris-HCl (pH 7.4). When phenol extraction was included in the procedure, it was carried out twice using freshly distilled phenol saturated with the extraction medium (pronase excluded).

When BR2s were coextracted with fixed salivary glands, the same extraction method was applied but 0.002 M EDTA was used instead of 0.01 M EDTA to facilitate the release of RNA from the BRs.

Electrophoresis in agarose and agarose-formaldehyde gels. Electrophoresis in agarose was carried out in slabs as published previously (Daneholt et al. 1969). The electrophoretic buffer contained 0.5% SDS, 0.01 M EDTA, and 0.02 M Tris-HCl (pH 8.0), and the electrophoretic run was performed at room temperature. The troughs were made to accommodate 30 μl in analytical and 200 μl in preparative runs. The agarose concentrations used are given in the figure legends.

When denaturing conditions were required, the RNA sample was dissolved in 30 μl of a 50% formamide–2.2 M formaldehyde solution, treated at 60°C for 5 minutes, cooled, and fractionated in agarose gels containing 2.2 M formaldehyde, essentially according to the method of Lehrach et al. (1977). The slab-gel system (Daneholt et al. 1969) was used. The solubilization buffer, as well as the electrophoretic buffer, contained 0.01 M EDTA and 0.04 M triethanolamine-HCl (pH 8.0).

The distribution of radioactive RNA was determined in agarose gels as described previously (Daneholt and Hosick 1973).

Isolation of 75S RNA. Salivary gland RNA extracted from about 100 glands was fractionated in a preparative agarose gel (see above), and the 75S RNA region was cut out and electrophoretically eluted according to the method of Case and Daneholt (1976), with the modification that the elution buffer was the same as the buffer in the preparative gel of the present study. The eluted 75S RNA was then precipitated in ethanol.

Electron microscopic analysis of 75S RNA. The length of 75S RNA was determined with a urea-formamide spreading technique as specified in Kung et al. (1975). Purified 75S RNA was dissolved in a buffered urea-formamide solution, and cytochrome c was added just prior to the spreading. The final hyperphase contained 5.6 M urea, 70% formamide, 30 μg/ml cytochrome c, 1 mM Na₃-EDTA, 0.01 M Tris-HCl (pH 8.5), and the hypophase contained distilled water. The spread sample was collected on carbon-coated parlodion films, dehydrated, stained in alcoholic uranyl acetate, rinsed in ethanol, and dried. The grids were then rotary-shadowed with platinum-palladium. The RNA molecules were examined and photographed in a Phillips EM 300. The negatives were projected and the extended RNA molecules traced and their contour lengths measured. Purified 28S RNA from HeLa cells provided an internal, and 23S RNA from *E. coli* an external, length standard.

RESULTS

Estimation of the Size of the Transcription Unit in BR2

The size of a transcription unit can be determined from the size of the corresponding primary transcript. The BR2 system also offers another possibility. The primary transcript in BR2, 75S RNA, is transferred to cytoplasm without a measurable reduction in size (Daneholt and Hosick 1973; Lambert and Edström 1974). Moreover, the 75S RNA molecules accumulate in the cytoplasm and constitute an abundant RNA species. In the case of BR2, it is therefore possible to use a major cellular fraction, 75S RNA, in order to estimate the size of the corresponding transcription unit. Two different techniques have been used to determine the size of 75S RNA. First, we have analyzed 75S RNA in denaturing, agarose-formaldehyde gels. Second, we have purified 75S RNA and analyzed it by electron microscopic techniques.

Electrophoretic Determination of the Size of 75S RNA

Recently, Lehrach et al. (1977) have shown that agarose-formaldehyde gels are suitable for determining the molecular weight of large-size RNA molecules. However, the same authors have pointed out that if the gel concentration is too high and/or the voltage gradient is too steep, the migration of large RNA molecules is not strictly size-dependent. During such so-called exclusion conditions, the apparent molecular weight values will represent underestimates of the correct molecular weight. To decide whether or not 75S RNA migrated according to its size under our electrophoretic conditions, we determined the apparent molecular weight of 75S RNA in three different gel concentrations (0.3%, 0.5%, and 0.75% agarose) and at two voltage gradients (1.5 V/cm and 3 V/cm).

One example of an electrophoretic analysis of long-term-labeled salivary gland RNA in an agarose-formaldehyde gel is presented in Figure 1A (0.3% agarose). Prominent 75S RNA and 18S RNA fractions can be observed. A major 28S RNA species is lacking, most likely due to a conversion of 28S RNA to 18S RNA under the denaturing conditions used in the present study (cf. Rubinstein and Clever 1971). Three reference molecules, 45S RNA and 28S RNA from HeLa cells and 23S RNA from *E. coli*, were

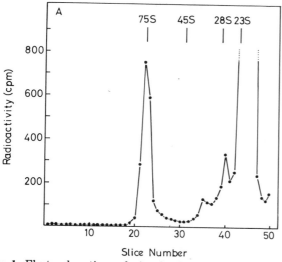

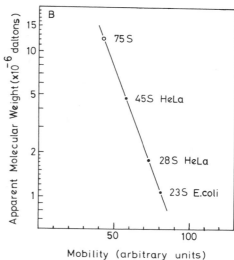

Figure 1. Electrophoretic analysis of salivary gland RNA in a 0.3% agarose gel containing 2.2 M formaldehyde (A) and a log molecular weight–mobility plot based on the electrophoretic data (B). 45S RNA and 28S RNA from HeLa cells and 23S RNA from *E. coli* were used as reference molecules. The apparent molecular weight of 75S RNA was estimated from the log molecular weight–mobility plot as shown in panel B.

analyzed in the same agarose gel and their positions are indicated in Figure 1A.

The mobilities of the reference RNA species, as well as of 75S RNA, were calculated and expressed in arbitrary units. The mobilities of 45S RNA, 28S RNA, and 23S RNA were plotted against the logarithms of their molecular weights (Fig. 1B). The molecular weight values used for 45S RNA (4.71 million daltons) and 28S RNA (1.76 million daltons) have been reported by Wellauer and Dawid (1974), and that for 23S RNA (1.07 million daltons) by Stanley and Bock (1965). For each electrophoretic condition, it was possible to draw a straight line through the three points (Fig. 1B), indicating that the three reference RNA molecules migrated according to their molecular sizes during the various experimental conditions used.

Knowing the mobility of 75S RNA for each condition, we could determine the apparent molecular weight for 75S RNA from plots similar to that presented in Figure 1B. The results are summarized in Figure 2. It should be noted that the apparent molecular weight of 75S RNA varied remarkably with the gel concentration, whereas the change in voltage gradient did not influence the apparent molecular weight value significantly. As expected for exclusion conditions, the apparent molecular weight increased with decreasing gel concentration: it amounted to 7.5–7.9 million daltons in 0.75% gels, 10.3–10.4 million daltons in 0.5% gels, and 11.9–12.1 million daltons in 0.3% gels. It is difficult to estimate the plateau value, but it seems as if the curve levels off at about 13 million daltons. We conclude, therefore, that the molecular weight for 75S RNA, as determined by electrophoresis in agarose-formaldehyde gels, is at least 12 million daltons and probably in the range of 12–14 million daltons.

The striking effect of gel concentration on the apparent molecular weight of 75S RNA in agarose-formaldehyde gels raised the question of whether our earlier experiments (Daneholt and Hosick 1973) using agarose gels (1% and no formaldehyde) were carried out under conditions of exclusion. If this were the case, the observed comigration of the primary transcript in BR2 and the major 75S RNA

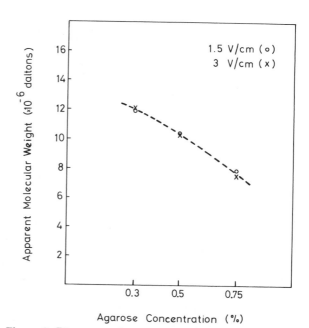

Figure 2. Diagrammatic presentation of the apparent molecular weight of 75S RNA as determined in agarose-formaldehyde gels at different gel concentrations and at two different voltage gradients. The apparent molecular weight of 75S RNA was estimated from log molecular weight–mobility plots as shown in panel B in Fig. 1.

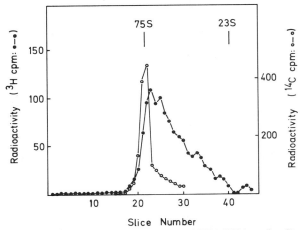

Figure 3. Electrophoretic analysis of BR2 RNA and salivary gland RNA in a 0.3% agarose-formaldehyde gel. The BR2 RNA was labeled with ³H for 45 min in vitro, and the salivary gland RNA was labeled with ¹⁴C for 3 days in vivo. A total of 140 isolated BR2s were coextracted with ten, fixed, ¹⁴C-labeled salivary glands.

fraction would not necessarily mean that they were of the same molecular size. It was therefore important to investigate whether the primary transcript in BR2 and the abundant cellular 75S RNA fraction do have the same mobility in gels that seem to give a size-dependent migration of molecules in the 75S RNA size range. ³H-labeled BR2 RNA and ¹⁴C-labeled salivary gland RNA were coextracted and analyzed in a 0.3% agarose-formaldehyde gel. The result is presented in Figure 3. The salivary gland 75S RNA shows a symmetric peak, whereas the BR2 RNA displays its characteristic asymmetric distribution (Daneholt 1972), which, at least to a large extent, can be attributed to a population of growing RNA molecules (for discussion, see Daneholt 1975). The peaks of the two distributions were almost at the same position in the gel, indicating a similarity in size of 75S RNA and the primary transcript in BR2.

Length Measurement of 75S RNA by Electron Microscopy

For electron microscopy, 75S RNA had first to be purified. Labeled RNA from about 100 glands was fractionated in a preparative 0.75% agarose gel. A longitudinal segment of the migration path was sliced and the radioactivity distribution determined (Fig. 4A). Subsequently, the 75S RNA band from the remaining part of the gel was cut out and the RNA electrophoretically eluted. This RNA was then tested for integrity in two different ways. First, the sample was rerun in a 0.75% agarose gel (Fig. 4B). The purified RNA moved as a distinct fraction with about the same mobility as the 75S RNA peak of total gland RNA (Fig. 4A). At least no major degradation of the RNA had occurred. In the second ex-

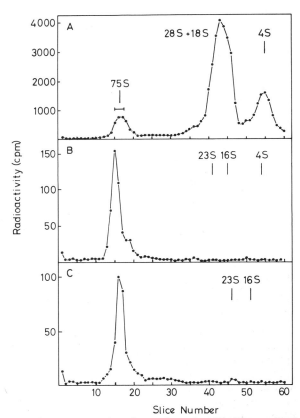

Figure 4. Preparative electrophoresis of salivary gland RNA in a 0.75% agarose gel *(A)*; analytical electrophoresis of isolated 75S RNA in a 0.75% agarose gel *(B)* and in a 0.5% agarose-formaldehyde gel after a brief treatment of the RNA sample at 60°C in a 50% formamide–2.2 M formaldehyde solution *(C)*. The segment of the preparative gel used for electrophoretic elution of the 75S RNA fraction is indicated in *A*. The positions of the reference molecules, 23S RNA, 16S RNA, and 4S RNA from *E. coli*, are also indicated.

periment, isolated 75S RNA was dissolved in a 50% formamide–2.2 M formaldehyde solution and kept at 60°C for 5 minutes. The sample was then run in a 0.5% agarose gel containing 2.2 M formaldehyde. Again a slowly moving RNA peak corresponding to 75S RNA was recorded (Fig. 4C). It could be concluded, therefore, that most of the 75S RNA molecules had retained their structural integrity during the isolation.

Purified 75S RNA was spread in a urea-formamide solution and examined in the electron microscope as described under Methods. Figure 5 shows an electron micrograph with three long 75S RNA molecules. More than 97% of all the analyzed 75S RNA molecules were essentially free of secondary structures. In some experiments, 28S RNA was used as an internal standard. Because the 28S RNA molecules display characteristic secondary structures (Wellauer and Dawid 1974), they can be distinguished easily from the 75S RNA molecules in the electron micrograph. Using the internal-standard

Figure 5. Electron micrograph of three 75S RNA molecules from *C. tentans,* spread in urea-formamide as described in Methods. Bar = 1 μm.

technique, we measured the length of 250 75S RNA molecules and 110 28S RNA molecules; the result is presented in Figure 6. The distribution of 75S RNA comprises mainly large molecules (6–10 μm), but a considerable number of smaller molecules were also recorded. These latter molecules probably represent 75S RNA degraded when the sample was spread for electron microscopy.

In a control experiment, essentially the same length distribution of 75S RNA molecules was obtained when the sample was heated at 80°C for 1 minute immediately prior to the spreading as described by Kung et al. (1975). The fact that the long molecules still remained after this treatment strongly suggests that they represent single, covalently linked molecules. In another control, it was shown that the molecules were sensitive to RNase. The sample was treated with RNase (2 μg/ml) at 0°C for 5 minutes, diluted 100-fold in the urea-formamide solution, and analyzed as above by electron microscopy. Only short molecules could then be detected.

The mean contour length of 75S RNA was calculated from the cross-hatched portion of the distribution in Figure 6 and was determined to be 7.75 μm. The 28S RNA length distribution (stippled area

in Fig. 6) was narrow, and the mean contour length was calculated to be 1.11 μm. Assuming a molecular weight of 1.76 million daltons for 28S RNA (Wel-

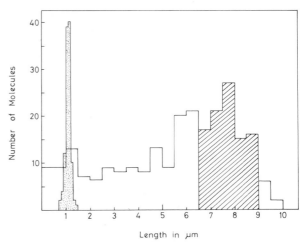

Figure 6. Length distribution of 75S RNA from *C. tentans* and 28S RNA from HeLa cells as determined by electron microscopy. The cross-hatched area designates the portion of the 75S RNA distribution used to calculate the mean contour length of 75S RNA. The dotted area represents the 28S RNA distribution.

lauer and Dawid 1974), we could estimate that 75S
RNA has a molecular weight of 12.3 million daltons.
This value corresponds to 37 kilobases.

The Chromosomal Location of the 75S RNA Transcription Unit

A cytological analysis was initiated by an investigation of the banding pattern in the BR2 region of
chromosome IV. The region studied is demarcated
in Figure 7 (upper part); the segmental designations
are the same as those used by Beermann (1952).
The result of the cytological analysis is presented
in detail in the lower part of Figure 7 and is essentially a confirmation of Beermann's 1952 map. We
have, however, included four additional minor bands
in the segment. To facilitate presentation of our
data, we have given the major bands particular symbols (bottom Fig. 7).

The chromosome map will not be extensively documented in this paper, but we will show the type of
material that has been used to construct the map.
Figure 8A shows the BR2 region in Malpighian tubules. In this tissue, the BR2 region is known to
lack giant-size puffs of the BR2 type (Beermann
1952). Note, however, that there is a BR to the left
of the BR2 region. This particular BR cannot be
found in the salivary glands; BRs are not present
exclusively in salivary glands. Figure 8B displays
the BR2 region in a salivary gland of a galactose-
treated larva. Galactose is known to cause regression of BR2 (Beermann 1973), and, after such a treatment, the structure of the BR2 region can be analyzed in detail.

When the banding pattern in the BR2 region of
the Malpighian tubules (Fig. 8A) is compared to that
in salivary glands (Fig. 8B), it is evident that the
pattern in the BR2 region is strikingly similar in
the two tissues. This is in accord with the general
conclusion reached by Beermann (1952) that the
chromosome banding pattern is the same in at least
the four tissues studied (salivary glands, Malpighian
tubules, small intestine, and rectum).

It is generally accepted (see, e.g., Beermann 1962)
that, when a puff is formed, the DNP in every chromomere of the corresponding band is uncoiled. The
activated band disappears as a distinct entity. The
chromomeres of the adjacent bands should not be
modified and should still constitute visible bands.
To locate the band involved in the puffing process
in BR2, we have used the detailed map of the BR2
region to find out which band is missing in a heavily
active BR2. The structure of a BR has been described
in detail by Beermann (for review, see Beermann
1962). The most important features can be seen in
Figure 9. When approaching the BR from both sides,
the solid chromosome breaks up into major branches
which, within the BR, split into minor ones and
eventually, in the periphery of the BR, pass over
into irregular, fuzzy structures. These latter structures have been characterized in detail in the electron microscope (Beermann and Bahr 1954; Stevens
and Swift 1966; Case and Daneholt 1977) and they
probably correspond to transcription complexes (for
discussion, see Case and Daneholt 1977). We have
studied the chromosome bands close to the transcription complexes and identified the bands in our map
of the BR2 region. The left, thick band (triangle)
and the three thinner, intermediate bands (open circles) of the BR2 region (Fig. 7) can be recognized
in Figure 9 on a major chromosome branch extending into the BR from the left. However, the broad
central band in segment 3B (open arrow in Fig. 7)
cannot be observed. It can also be noted that on
one arm coming into the BR from the right side,
the two thick bands in the right half of the map
(solid arrows) are present. Sometimes two additional thinner bands can be seen peripheral to these
two major bands, but again no broad central band
(open circle in Fig. 7) is seen. It can be inferred that
the only band always absent in a transcriptionally
active BR2 is the broad central band, which is therefore likely to be the main band involved in the puffing event and therefore likely to contain the 75S
RNA transcription unit(s). It also seems justified to
designate this broad band the BR2 band and the
corresponding chromomere the BR2 chromomere.

The Re-formation of the BR2 Band during Galactose Treatment

As shown in Figure 8B, the broad BR2 band can
be re-formed after galactose treatment. In the pres-

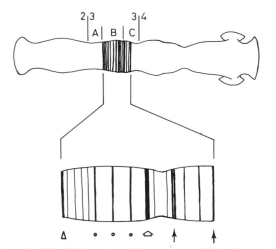

Balbiani Ring 2 Region

Figure 7. The BR2 region of chromosome IV in *C. tentans*
and an expansion of this segment displaying the banding
pattern in detail. The segmental designations are the same
as those adopted by Beermann (1952).

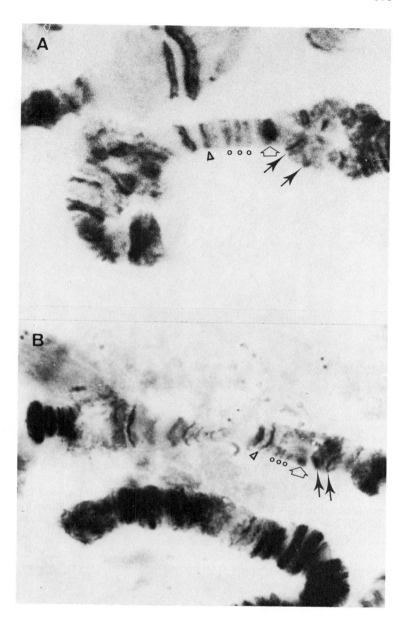

Figure 8. The BR2 region in Malpighian tubules *(A)* and in salivary glands from galactose-treated larvae *(B)*. The designations are the same as in Fig. 7.

ent context, we have been particularly interested in analyzing this process to see whether it can provide us with some information on the structure of the BR2 band. Figure 10 shows the structure of the BR2 region after galactose treatment before the broad band has appeared (cf. Fig. 8B). The bands adjacent to the BR2 band have been indicated in the figure (designations as in Fig. 7). While some transcription complexes are still present in the upper part of the BR (fuzzy areas), an essentially unpuffed chromosome arm is exposed in the central part of the BR. At the position of the broad BR2 band, a series of finer bands can be discerned. Therefore, it seems as if a number of thin bands fuse together and form the thick BR2 band. This observa-

tion indicates that, at the chromatid level, the BR2 chromomere is formed from a series of smaller packages of the DNP filament. It should be noted that in this paper we have defined a chromomere as the structure on the chromatid level corresponding to the chromosome band on the chromosome level.

DISCUSSION

In the present study, the size of the transcription unit in BR2 was estimated from determinations of the size of the major 75S RNA fraction in the salivary glands. This approach was suggested to us by the observation that the 75S RNA fraction had about the same electrophoretic mobility as the pri-

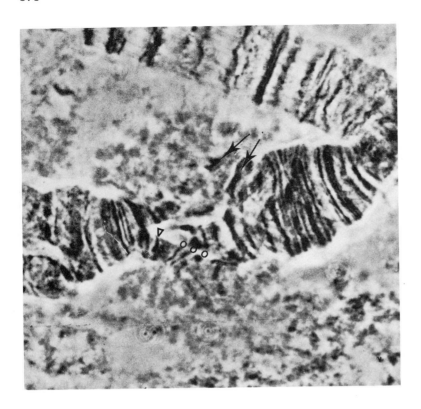

Figure 9. Balbiani ring 2. The band designations are the same as in Fig. 7. The broad central band denoted by an unfilled arrow in Fig. 7 is lacking, but the other major bands can be discerned.

mary transcript in BR2 (Daneholt and Hosick 1973) and hence should reflect the size of the corresponding transcription unit. This similarity as to mobility was observed initially in 1% agarose gels under nondenaturing conditions (Daneholt and Hosick 1973), but the result was verified in the present investigation, in which we used 0.3% agarose-formaldehyde gels, providing conditions for a size-dependent migration in the 75S RNA range. The approach to estimate the size of the transcription unit in BR2 from the size of the 75S RNA species in salivary glands therefore seems justified. From mobility measurements in agarose-formaldehyde gels, the molecular weight of 75S RNA was estimated to be at least 12 million daltons, and probably in the range of 12–14 million daltons. The molecular weight of

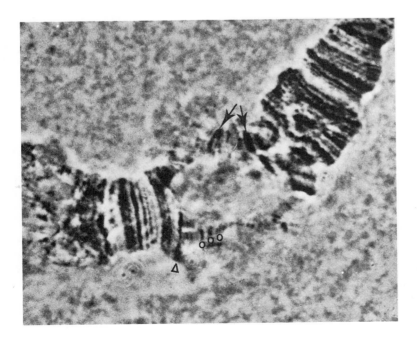

Figure 10. The BR2 region in a salivary gland of a galactose-treated larvae. Some transcriptive activity is indicated in the upper half of the Balbiani ring. A chromosome arm showing essentially no morphological signs of activity can be recognized in the lower part. Here the complete banding pattern of the BR2 region has almost been restored (cf. Fig. 7), but it should be noted that at the position of the broad central band in the completely repressed condition (cf. Fig. 8), a series of finer bands can be seen.

75S RNA, determined from contour length measurements in the electron microscope, was 12.3 million daltons (37 kilobases). Since this value agrees with the electrophoretic estimation, we accept this figure as a good value for the molecular weight of 75S RNA. This would then imply a size of the transcription unit of 37 kilobase pairs (kbp).

It was established from the cytological analysis of the BR2 region in puffed and unpuffed condition that the broad band in segment 3B (open arrow in Fig. 7) is the main band involved in the formation of BR2. A similar conclusion has been drawn by W. Beermann (unpubl.) in studies of the closely related species *Chironomus pallidivittatus* during galactose treatment. In the present study, the broad band is therefore designated the BR2 band and the corresponding chromomere, the BR2 chromomere. It should be pointed out, however, that the cytological method used reveals only the units active in transcription. If there are inactive transcription units in, for example, adjacent bands (cf. in situ hybridization results presented by Lambert [1975]), these have not been detected in the present study. Although we still have to leave open the question about the distribution of BR2 sequences, it can be concluded that the broad band in segment 3B is the main band engaged in the puffing process in BR2. Due to the limitations of the cytological technique, we cannot decide whether or not adjacent interbands are also involved.

Presently, the information on the 75S RNA transcription unit can be only roughly related to the BR2 chromomere because of the lack of data on the distribution of DNA in the BR2 region. On the basis of the thickness of the BR2 band, the BR2 chromomere has to be considerably larger than an average-size chromomere. For *C. tentans*, this implies that it has to contain considerably more DNA than 100 kbp (Daneholt and Edström 1967). Furthermore, the adjacent interchromomeres are not of exceptional length, and therefore their DNA content should not exceed the maximum value (5 kbp) calculated for average-size interchromomeres (Beermann 1972). When we relate the size of a 75S RNA transcription unit (37 kbp) to this information, we can draw some general conclusions. First, only a minor part of the unit can be accommodated in an adjacent interchromomere, which implies that even if there is only one 75S RNA transcription unit, most of it, if not the whole unit, is harbored in the BR2 chromomere. Moreover, it is evident that only a minor part of the BR2 chromomere has to be alloted to one 75S RNA transcription unit. We are then facing three major alternatives for the structure of the BR2 chromomere: (1) There is only one transcription unit, i.e., that for 75S RNA, in the BR2 chromomere; the functional significance of most of the BR2 chromomere DNA then still remains to be explained. (2) There are several different transcription units in the BR2 chromomere, with only the 75S RNA transcription unit being active in the salivary glands. (3) There are several 75S RNA transcription units in the BR2 chromomere.

There is no strong evidence favoring one alternative over the others. Two reports on RNA-DNA saturation hybridization experiments are available. Lambert (1972) found that 0.17% of the haploid genome was complementary to BR2 RNA, whereas Hollenberg (1976) obtained a value of less than 0.04%. The lower value is compatible with one 75S RNA transcription unit, but the higher figure would indicate as many as 15–20 units. We therefore need further information before we can make a more definite statement about the three major alternatives.

It is plausible that the large BR2 chromomere might turn out to be a complex structure. The observation that, during galactose treatment, the condensation of the unfolded DNP filament takes place at several points, is in agreement with such a possibility. Whether this behavior of the DNP filament is correlated to an organization of the BR2 chromomere into several transcription units can at present only be a matter of speculation. However, even if it is shown that the BR2 chromomere contains more than one transcription unit, it is still possible that the BR2 chromomere might be a functional unit of a higher order than a transcription unit of the conventional prokaryotic type.

SUMMARY

1. The size of the transcription unit in Balbiani ring 2 (BR2) was estimated to be 37 kilobase pairs.

2. A broad band in the 3B region of chromosome IV was found to be the main band involved in the formation of BR2. This band is designated the BR2 band, and the corresponding chromomere the BR2 chromomere.

3. Only a minor portion of a 75S RNA transcription unit can be accommodated in an interchromomere adjacent to the BR2 chromomere. Therefore, even if there is only one 75S RNA transcription unit in the BR2 region, most of it, if not the whole unit, is likely to be located within the chromomere itself.

4. On the other hand, only a minor part of the BR2 chromomere is required to harbor a single 75S RNA transcription unit. Various alternatives for the structure of the BR2 chromomere were considered briefly, including the possibility that the 75S RNA transcription unit is repeated within the BR2 chromomere.

Acknowledgments

We are indebted to Miss Eva Mårtenzon, Miss Jeanette Nilsson, and Mrs. Sigrid Sahlén for skillful technical assistance and to Miss Ann-Louise Grennstam for typing the manuscript. In Prof. R. Weber's laboratory (University of Bern, Switzerland), S.T.C. had the opportunity to learn the urea-formamide spreading technique under the expert guidance of Mr. T. Wyler. We are very much indebted to Prof. W. Beermann for his most valuable help in the cytological study and for his communication to us of unpublished results. Dr. Ulf Petterson (University of Uppsala, Sweden) kindly provided us with HeLa nuclear and cytoplasmic RNA. This research has been supported by grants from the Swedish Cancer Society, Magnus Bergvalls Stiftelse, and Karolinska Institutet (Reservationsanslaget). S.T.C. is a recipient of a National Institutes of Health National Research Service Award from the National Institute of General Medical Sciences, J.D of an EMBO fellowship, and M.M.L. of a postdoctoral fellowship from the American Cancer Society.

REFERENCES

BEERMANN, W. 1952. Chromomerenkonstanz und spezifische Modifikation der Chromosomenstruktur in der Entwicklung und Organdifferenzierung von Chironomus tentans. Chromosoma 5: 139.

———. 1961. Ein Balbiani-Ring als locus einer Speicheldrüsen-Mutation. Chromosoma 12: 1.

———. 1962. Riesenchromosomen. Protoplasmatologia 6D: 1.

———. 1964. Structure and function of interphase chromosomes. In Genetics today, p. 375. Pergamon Press, Oxford.

———. 1972. Chromomeres and genes. Results Probl. Cell Differ. 4: 1.

———. 1973. Directed changes in the pattern of Balbiani ring puffing in Chironomus: Effects of sugar treatment. Chromosoma 41: 297.

BEERMANN, W. and G. F. BAHR. 1954. The submicroscopic structure of the Balbiani ring. Exp. Cell Res. 6: 195.

CASE, S. T. and B. DANEHOLT. 1976. A simple method for the isolation of undegraded giant RNA molecules. Anal. Biochem. 74: 198.

———. 1977. Cellular and molecular aspects of genetic expression in Chironomus salivary glands. Int. Rev. Biochem. 15: 45.

DANEHOLT, B. 1972. Giant RNA transcript in a Balbiani ring. Nat. New Biol. 240: 229.

———. 1975. Transcription in polytene chromosomes. Cell 4: 1.

DANEHOLT, B. and J.-E. EDSTRÖM. 1967. The content of deoxyribonucleic acid in individual polytene chromosomes of Chironomus tentans. Cytogenetics 6: 350.

DANEHOLT, B. and H. HOSICK. 1973. Evidence for transport of 75S RNA from a discrete chromosome region via nuclear sap to cytoplasm in Chironomus tentans. Proc. Natl. Acad. Sci. 70: 442.

DANEHOLT, B., K. ANDERSSON, and M. FAGERLIND. 1977. Large-sized polysomes in Chironomus tentans salivary glands and their relation to Balbiani ring 75S RNA. J. Cell Biol. 73: 149.

DANEHOLT, B., J.-E. EDSTRÖM, E. EGYHÁZI, B. LAMBERT, and U. RINGBORG. 1969. Physico-chemical properties of chromosomal RNA in Chironomus tentans polytene chromosomes. Chromosoma 28: 379.

DERKSEN, J. 1977. Cytological analysis of Drosophila polytene chromosomes. Methods Cell Biol. 17: 133.

DOYLE, D. and H. LAUFER. 1969. Sources of larval salivary gland secretion in the dipteran Chironomus tentans. J. Cell Biol. 40: 61.

EDSTRÖM, J.-E. 1964. Microextraction and microelectrophoresis for determination and analysis of nucleic acids in isolated cellular units. Methods Cell Physiol. 1: 417.

GROSSBACH, U. 1969. Chromosomen-Ativität und biochemische Zelldifferenzierung in den Speicheldrüsen von Camptochironomus. Chromosoma 28: 136.

———. 1974. Chromosome puffs and gene expression in polytene cells. Cold Spring Harbor Symp. Quant. Biol. 38: 619.

HOLLENBERG, C. P. 1976. Proportionate representation of rDNA and Balbiani ring DNA in polytene chromosomes of Chironomus tentans. Chromosoma 57: 185.

KUNG, H.-J., J. M. BAILEY, N. DAVIDSON, M. O. NICOLSON, and R. M. McALLISTER. 1975. Structure, subunit composition, and molecular weight of RD-114 RNA. J. Virol. 16: 397.

LAMBERT, B. 1972. Repeated DNA sequences in a Balbiani ring. J. Mol. Biol. 72: 65.

———. 1975. The chromosomal distribution of Balbiani ring DNA in Chironomus tentans. Chromosoma 50: 193.

LAMBERT, B. and B. DANEHOLT. 1975. Microanalysis of RNA from defined cellular components. Methods Cell Biol. 10: 17.

LAMBERT, B. and J.-E. EDSTRÖM. 1974. Balbiani ring nucleotide sequences in cytoplasmic 75S RNA of Chironomus tentans salivary gland cells. Mol. Biol. Rep. 1: 457.

LEHRACH, H., D. DIAMOND, J. M. WOZNEY, and H. BOEDTKER. 1977. RNA molecular weight determinations by gel electrophoresis under denaturing conditions. A critical reexamination. Biochemistry 16: 4743.

RUBINSTEIN, L. and U. CLEVER. 1971. Non-conservative processing of ribosomal RNA in an insect, Chironomus tentans. Biochim. Biophys. Acta 246: 517.

STANLEY, W. M. and R. M. BOCK. 1965. Isolation and physical properties of the ribosomal ribonucleic acid of Escherichia coli. Biochemistry 4: 1302.

STEVENS, B. J. and H. SWIFT. 1966. RNA transport from nucleus to cytoplasm in Chironomus salivary glands. J. Cell Biol. 31: 55.

WELLAUER, P. K. and I. B. DAWID. 1974. Secondary structure maps of ribosomal RNA and its precursors as determined by electron microscopy. Cold Spring Harbor Symp. Quant. Biol. 38: 525.

WIESLANDER, L. and B. DANEHOLT. 1977. Demonstration of Balbiani ring RNA sequences in polysomes. J. Cell Biol. 73: 260.

Fate of Balbiani-ring RNA In Vivo

J.-E. Edström, E. Ericson, S. Lindgren, U. Lönn, and L. Rydlander

Department of Histology, Karolinska Institutet, S-10401 Stockholm, Sweden

Products of chromosome puffs can be traced to the cytoplasm (Daneholt and Hosick 1973; Lambert 1973; McKenzie et al. 1975; Spradling et al. 1975) but little is known about the extent of the transfer in vivo. Most of the nonribosomal, high-molecular-weight RNA (hnRNA) in mammalian cells seems to be degraded in the nucleus (reviewed by Darnell 1975) but it is not known to what extent this applies to sequences destined to become messenger RNA or to any defined puff product. Balbiani-ring (BR) RNA in *Chironomus tentans* offers a suitable material for the study of this question since it can be isolated and analyzed both at the site of production and in the cytoplasm (Edström and Beermann 1962; Daneholt and Hosick 1973) and since it has properties of a messenger, i.e., contains poly(A) (Edström and Tanguay 1974) and is localized in EDTA-sensitive structures (Daneholt et al. 1977).

In the work presented here, the production rate was estimated from the content of RNA in microdissected BR and from the transcription rate. The export was first determined relative to the export of ribosomal RNA (rRNA). The latter could then be assessed by measurements of growth rate and turnover rate. It was found that a considerable portion (50–100%) of the RNA produced in the BR is also exported in vivo.

METHODS

Animals. *Chironomus tentans* larvae, 7–9 weeks old and weighing around 20 mg (20 mm long), were used from a stock of animals obtained originally from Prof. W. Beermann, Tübingen (collection site Ploener See). They were kept at 18–19°C at a 16/8-hour day/night rhythm in aerated and deionized water containing 0.4 g/liter NaCl and cellulose tissue. Fermented nettle powder was given twice weekly.

Preparation of glands for microdissection. Glands were fixed for 10 seconds in ethanol:acetic acid, 3:1, followed by ethanol:0.2 M acetate buffer (pH 5.5), 24:1, for 1–2 minutes and 70% ethanol for 30 minutes. These solutions were kept on ice. The glands were then transferred to ethanol:glycerol, 1:1, and kept in this solution for at least 1 hour at room temperature before being transferred to an oil chamber for microdissection (Edström 1964).

Determinations of RNA content. Cells from the population consisting of the five to ten largest cells in each gland were used consistently. For determinations of the RNA content in nuclear components, the microdissected units were extracted with buffered ribonuclease solution after removal of free nucleotides. The ribonuclease extracts were transferred to quartz slides and the RNA content was determined by a microphotographic photometric procedure in ultraviolet light, as has been described previously (Edström 1964).

For determinations of the RNA content per cell, 25 cells were isolated from five glands for each determination (from the population consisting of the five to ten largest cells), and RNA was determined as described by Fleck and Begg (1965). This procedure is particularly suitable for cells with high concentrations of RNA, e.g., the salivary gland cells.

Determination of transcription time in vivo. Larvae were injected with 10 μCi [³H]uridine and 10 μCi [³H]cytidine and 4 hours later with 0.5 μg 5,6-dichloro-1-β-D-ribofuranosylbenzimidazole (DRB) (1 μl of a solution of 0.5 mg DRB per ml 12% ethanol), after which they were sacrificed at different time intervals. Glands were prepared according to the routine procedure for microdissection, and BR and nucleoli were isolated, transferred to 200 μl 0.02 M Tris-HCl buffer (pH 7.4), with 0.5% sodium dodecyl sulfate (SDS) and containing 20–30 μg *E. coli* carrier RNA. After 5 minutes at room temperature with intermittent shaking, the solutions were cooled on ice and an equal volume of 10% trichloroacetic acid (TCA) was added. After 15–20 minutes in the cold, the precipitates were spun down at 10,000 rpm during 10 minutes, the supernatant removed, and the precipitates dissolved in 0.1 ml Soluene (Packard), which, after 60 minutes at 60°C, was transferred to scintillation cocktail for counting.

RNA extraction and gel electrophoresis. RNA was extracted from microdissected components in 200 μl 0.02 M Tris-HCl buffer (pH 7.4), containing 0.5% SDS, 1 mg/ml pronase, and 20–30 μg *E. coli* carrier RNA, for 5 minutes at 25°C. The solution had previously been kept at 37°C for 30 minutes. RNA was then precipitated in the cold with 5 μl 4 M NaCl and 500 μl ethanol for at least 2 hours. It was then spun down at 10,000 rpm during 10 minutes and dissolved in 50 μl 0.02 M Tris-HCl buffer (pH 7.4) with 0.5% SDS. It was analyzed by electrophoresis in 1% agarose, essentially as described by Daneholt (1972), and the gels were washed in cold TCA and water before sectioning (Edström and Tanguay 1974).

Determinations of specific activities in BR and nucleolar RNA. Animals were given 10 μCi [³H]uridine and 10 μCi [³H]cytidine by injection and sacrificed after either 6 or 8.5 hours. Glands were prepared for microdissection and fourth chromosomes and nucleoli were isolated from three to six cells. RNA was extracted with ribonuclease and determined in the ultraviolet microscope (Edström 1964). Each extract was then transferred to a vial for scintillation counting.

Determination of ribosomal RNA turnover. Six-week-old animals were placed in normal culturing medium with 10 μCi [methyl-³H]methionine per ml with 2×10^{-5} M adenosine, guanosine, and thymidine (to suppress labeling in the purine rings and DNA) for 3 days in the absence of food. They were then washed and taken to medium with food, a 500-fold excess of unlabeled methionine and adenosine, guanosine, and thymidine in unchanged concentrations. After 11–18 days in this medium, one batch of seven to ten animals was taken for analysis of the specific activity in ribosomal RNA; after a further 6–10 days a second batch of animals was taken. The glands were fixed and transferred to ethanol:glycerol as described above and the secretion removed by dissection under a stereomicroscope. RNA was extracted and deproteinized as described previously (Edström and Rydlander 1976). The separations in agarose were recorded in a scanning densitometer at 254 nm and the ribosomal RNA contents determined. The separations were then sliced and the radioactivity under the ribosomal RNA peaks recorded. The specific activity was corrected for the decrease due to increase in glandular RNA content by growth. Controls were run with L-[³⁵S]methionine at comparable concentrations of radioactivity to ascertain that labeling due to protein was efficiently eliminated.

Radiochemicals and enzymes. [5-³H]cytidine, 25–29 Ci/mmole; [5,6-³H]uridine, 42–50 Ci/mmole; L-[methyl-³H]methionine, 10.3 Ci/mmole; and L-[³⁵S]-methionine, 600 Ci/mmole were obtained from the Radiochemical Centre, Amersham. Ribonuclease A, from bovine pancreas, type III-A, was from Sigma, St. Louis, Missouri, and pronase free of nucleases, lyophilized, B grade was obtained from Calbiochem, San Diego, California.

RESULTS

Production of BR RNA

Among hundreds of puffs in the salivary gland polytene chromosomes of *C. tentans,* three are noteworthy because of their size and designated Balbiani rings (BR1, BR2, and BR3). They are all localized on the small fourth chromosome. In terms of RNA content and synthesis, BR1 and BR2 are the dominating ones. Both produce a high-molecular-weight RNA, sedimenting at 75S (Daneholt 1972; Egyházi 1976). For BR1 and BR2, electron microscopy suggests that much of the RNA, which is present in 300–500-Å granules, is attached to DNA by stalks (Beermann and Bahr 1954). The electrophoretic distribution curve (Daneholt 1972) and inhibition studies with DRB, an inhibitor of transcription initiation (Egyházi 1974, 1975), also indicate that much or most of the BR DNA is in a nascent state, attached to the DNA template.

If a transcription site consists of an ideal population of continuously growing RNA chains and is uncontaminated by other RNA (Fig. 1a), it is possible to determine the amounts of RNA produced per hour *(s)* from the mass of RNA at this site *(m)* if the transcription time in minutes *(t)* is known, i.e., the time it takes for polymerases to travel from one end of the unit to the other end. The formula is $s = 2 \times m \times 60/t$. It applies whether the transcription site contains only one or several transcription

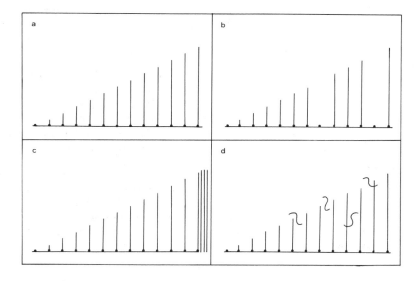

Figure 1. Diagram of different possible RNA compositions in a Balbiani-ring transcription unit. *(a)* A series of continuously growing RNA chains on a DNA template. The production rate is equal to twice the mass of RNA per transcription time. *(b)* Some of the chains are never finished; *(c)* finished chains are retained in the BR; *(d)* the BR contains RNA of non-BR origin. The latter three alternatives and combinations of alternatives lead to an overestimation of the rate of synthesis of BR RNA if the formula valid for *a* is applied.

units, as long as the transcription rate is similar in all the units. The two parameters m and t can be estimated for the BR1 + BR2.

There are three types of possible deviations from an ideal nascent distribution in BR isolated by micromanipulation (Fig. 1b–d). One possibility is that a fraction of the population of nascent chains never becomes finished (Fig. 1b). Such an error cannot at present be rigorously excluded and results in an overestimation if the formula is applied. A second possibility is that finished chains are retained for some time in the BR (Fig. 1c). This alternative would also lead to an overestimation but is not likely to exist for BR1 + BR2 because it would lead to a peaked electrophoretic distribution in the 75S range and this has so far not been observed. Finally, the rings, at least when isolated by microdissection, may contain non-BR RNA, e.g., ribosomal RNA (rRNA) and RNA from small adjoining puffs. The amounts of non-BR RNA can be determined from RNA content determinations in chromosome segments similar in volume to the isolated BR segments. This contamination is of the order of 25% for BR1 + BR2 and is assessed from determinations of the RNA content in chromosome 1, which is free of Balbiani rings. The isolated BR segments correspond to 20% of the volume of the fourth chromosome, or to 6.9% of the length (or DNA content) of the first chromosome (Daneholt and Edström 1967), which contained on the average 39 pg RNA per chromosome. The non-BR RNA content was therefore assumed to be 2.70 pg per Balbiani-ring segment, and 5.40 pg was therefore subtracted from the values for BR1 + BR2. If there is non-BR RNA specifically associated with the BR (cf. Ringborg and Rydlander 1971), even the corrected values will be an overestimate and the rate of production overestimated.

The RNA contents for BR1 and BR2 have been determined separately (Table 1). The sum of the values was 22.66 ± 2.51 pg (mean ± s.e.m.) before and 17.26 pg after correction as described above.

Estimations of transcription time by analysis of isotope uptake kinetics in explanted glands have been made for BR1 + BR2, indicating 20 minutes or less in a synthetic medium (Egyházi 1976) and closer to 30 minutes in hemolymph (Daneholt et al. 1969). Using the initiation inhibitor DRB, Egyházi (1975) followed the disappearance of RNA molecules of different size as a function of time after DRB administration. The elimination of labeled material starts with the small chains and ends with chains

of final size. The transcription time was estimated to be around 30 minutes in this report. However, since the elimination of label in the longest chains already starts during the preceding 10 minutes, it is likely that 30 minutes is an overestimate and that the DRB experiments indicate a time closer to 20 minutes (E. Egyházi, pers. comm.).

For the present work, it was of interest to learn whether the transcription time in vivo corresponds to the values for explanted glands. DRB in an alcoholic solution was injected into prelabeled animals as described under Methods. The two BRs were isolated from three cells, the nucleoli of which were also sampled. The microdissected components were dissolved and the TCA-insoluble activity determined. The ratio of radioactivity in BR RNA to that in nucleolar RNA was determined. Nucleolar RNA was chosen as a reference since it is unlikely to be much affected by DRB, at least during short exposure. DRB does not eliminate labeling of 38S RNA in the nucleolus (Egyházi et al. 1970), and the average residence time of RNA in the nucleolus lies far above the times used for drug exposure in the present experiments. The labeling of the BR reaches low levels after 25–32 minutes (Fig. 2). Control analyses of the total RNA content were made after 25- and 32-minute drug exposure to ascertain that the

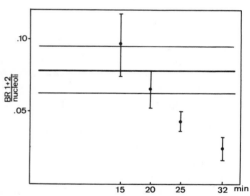

Figure 2. Ratio between labeled RNA in BR1 + BR2 and nucleoli at different times after injection of DRB. Larvae were injected with 10 μCi [³H]uridine and 10 μCi [³H]cytidine and 4 hr later with 0.5 μg DRB in 1 μl 12% ethanol, after which they were sacrificed at the times indicated. The animals were kept at 18–19°C throughout all operations. The dots represent mean values and the brackets s.e.m. Controls were injected with ethanol only. The horizontal heavy line indicates the level of the mean value for controls and the light lines the s.e.m. For each determination, three BR1 + BR2 and three pairs of nucleoli were isolated from one gland by micromanipulation and dissolved in 0.02 M Tris-HCl buffer (pH 7.4), with 0.5% SDS and containing 20–30 μg E. coli carrier RNA. After intermittent shaking for 5 min at room temperature, the solutions were cooled on ice and an equal volume of 10% TCA added. After 15–20 min in the cold, the precipitate was spun down at 10,000 rpm during 10 min, the supernatant removed, and the precipitates dissolved in 0.1 ml Soluene (Packard), which, after 60 min at 60°C, was transferred to scintillation cocktail for counting. Each value is based on at least four animals.

Table 1. RNA Content per Balbiani Ring (pg ± s.e.m.)

	Untreated animals	DRB-treated animals	
		25 min	32 min
BR1	7.8 ± 1.0	4.0	2.3
BR2	14.9 ± 2.0	4.1	2.5
BR1 + BR2	22.7 ± 2.5	8.1	4.8
Corrected for non-BR RNA	17.3	2.7	0

decrease in BR labeling corresponds to an elimination of the nascent RNA (Table 1). Comparison of the present data with results of Egyházi (1975) shows that the time required for elimination of nascent RNA is similar in vivo and in a synthetic medium.

An evaluation of available data for the rate of transcription in the BRs shows that an exact figure cannot as yet be given. We have presented four possible alternatives (Table 2), of which 20 minutes appears the most likely one at present, resulting in a production of about 2500 pg RNA per day. The 15-minute time probably can be considered as a lower limit and it would lead to an upper limit for the production of BR1 + BR2 RNA of 3300 pg per day.

Since the chain of BR2 RNA should correspond to about 37 kb (Daneholt et al., this volume) and BR1 RNA is similar in size, a transcription time of 15 minutes would lead to a maximal rate of 41 nucleotides/second. It appears unlikely that the rate could be higher because mammalian rates are around 100 nucleotides/second (Sehgal et al. 1976 for discussion) and since the *Chironomus* transcription takes place at an almost 20°C lower temperature. At least for microorganisms, the rate difference is about twofold per 10°C temperature difference.

Export of BR RNA

75S RNA is present in the cytoplasm of salivary gland cells (Daneholt and Hosick 1973) and shows in situ hybridization to BR1 and BR2 (Lambert 1973; Lambert and Edström 1974). With adequate techniques, it can be resolved into two components. The electrophoretically slow migrating peak hybridizes BR1, the fast migrating one the BR2. Material hybridizing BR1 or BR2 of a size other than 75S has so far not been detected in the cytoplasm (Edström and Rydlander 1976). Since RNA in the 75S range is specific for salivary glands like the BRs (Edström and Tanguay 1974), it is unlikely that non-BR RNA is present in the 75S RNA. It is also unlikely that BR3 RNA participates in this fraction since 75S RNA is absent in chromosome segments carrying the BR3 (Egyházi 1975).

Export of RNA from BR1 + BR2 occurs in response to growth and turnover, the latter factor being the dominating one. The turnover rate in vivo is not easily determined, however, because of label reutilization. Also the origin of the RNA from at least two sources could cause complications because of possible differences in half-life. The strategy was therefore applied to determine the export rate of BR RNA relative to that of rRNA after the export rate of rRNA had been assessed.

Export of rRNA also occurs both for growth and turnover. For the animals used (average weight 20 mg), the daily weight increase by growth was 0.30 ± 0.04 mg (mean $\pm$ s.e.m., $n = 36$). In this range, there is a more or less linear increase of RNA content in the glands with increasing animal weight (Lönn and Edström 1977), and 0.3 mg weight increase corresponds to 1.45% increase in RNA content. The total RNA content in cells of the size used for determination of production rates (from the population consisting of the five to ten largest cells of each gland) was $101,000 \pm 5000$ pg (mean $\pm$ s.e.m.). Electrophoretic analyses of the total RNA extracted from salivary glands (Edström et al. 1978) indicated that about 75% of the ultraviolet absorption is rRNA. From these data, the export of rRNA due to growth could be determined to be 1088 pg per day (Table 3).

The rRNA turnover was determined as described in Methods. The rate constant for the decay was found to be 0.0241 ± 0.0072 (mean $\pm$ s.e.m., $n = 4$). The rate constant expresses the decay rate at zero time per time unit, in this case days. With an rRNA content of about 75,000 pg per cell, the decay rate per day becomes 2.41% · 75,000 pg, or about 1810 pg of rRNA per day, which has to be restituted by export. (This rate of decay corresponds to a half-life of about 30 days.) Together with export due to growth, the total export of rRNA is thus estimated to be about 2900 pg per day (Table 3).

The ratio between the label in BR1 + BR2 RNA and rRNA (Fig. 3) was determined at different times after [³H]uridine injection and extrapolated to zero time. For the extrapolation, it was necessary to take into account that the pulse of tritiated precursors labels RNA over a period of several hours, and that it takes 2–3 hours for rRNA and BR RNA to enter the cytoplasm in vivo (Edström and Tanguay 1974). This means that the age in the cytoplasm of these RNA species in animals sacrificed, e.g., 18 hours after precursor injection, occupies a whole range less than 15–16 hours. From the accumulation curve published previously (Lönn and Edström 1977), the contribution hour by hour was determined and the BR RNA/rRNA ratio at each time point was read

Table 2. RNA Produced per Day in BR1 + BR2

Transcription time (min)	pg RNA per cell	Transcription rate (nucleotides/sec)
15	3314	41
20	2485	31
25	1988	25
30	1657	21

Table 3. Export per Day of Ribosomal RNA (Mean $\pm$ s.e.m.)

	Percentage of total RNA	pg RNA
Growth	$1.45 \pm .20$	1088
Turnover	$2.41 \pm .72$[a]	1810
Sum	3.86	2898

[a] Rate constant: 0.0241 ± 0.0072 (mean $\pm$ s.e.m., $n = 4$)

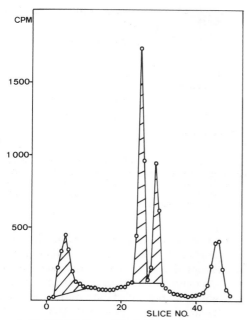

Figure 3. Radioactivity profile in 1% agarose gel of RNA from salivary glands 18 hr after injection of 25 μCi [³H]uridine. For determinations of the ratio between BR1 + BR2 RNA (75S) and rRNA, the area under the radioactivity peaks, indicated by hatching, was determined.

from an extrapolated curve. The ratio at an apparent zero time could then be determined from the sum of all contributions. This corresponds to a ratio of 49.6 radioactivity units of BR RNA for 100 units of rRNA, resulting in an apparent zero time 7.5 hours after precursor injection. Of this time, 2.5 hours represents the average nuclear processing time for the two types of nucleic acids. As can be seen in Figure 4, experiments carried out with two different doses of radioactivity do not give significantly different results.

The radioactivity ratio between BR1 + BR2 RNA and rRNA has to be converted to a mass ratio for

determination of rate of export. Therefore, it has to be known whether administered uridine is incorporated into uridine only or whether some conversion to cytidine occurs. Rubinstein and Clever (1972) did not find any measurable conversion in explanted glands during 60 minutes. To control any conversion in the present investigation, animals were injected with [³H]uridine and sacrificed after 24 hours; RNA was extracted, hydrolyzed in alkali, and analyzed by thin-layer chromatography (Randerath and Randerath 1967) with 1 N acetic acid as a solvent. The results showed a minor conversion to cytidylic acid, with radioactivities being recovered in cytidylic and uridylic acids in a ratio of 8:92.

To convert labeling ratios into mass ratios, one has to take into account the base compositions of BR1 + BR2 RNA and rRNA (Edström and Beermann 1962). The uridine and cytidine contents in BR RNA were calculated on the assumption that it contains one part of BR1 RNA and two parts of BR2 RNA (Table 1), and that in rRNA it is intermediate between nucleolar and cytoplasmic RNA. When compensation is performed for the difference in base composition, taking into account that 8% of the label is in cytidine and the remainder in uridine, the export ratio BR1 + BR2:rRNA becomes 69.5:100.

However, the labeling of RNA is also a function of the specific activities in the pools of free nucleotides from which BR1 + BR2 RNA and rRNA are labeled, and it is not known a priori whether these pools are identical. The present biological system offers a unique opportunity to determine directly the labeling from these pools by analyses of the specific activity of the RNA present in the BR and in the nucleoli. This makes it possible to avoid the uncertainty of whether a measured pool specific activity is relevant for the labeling of a certain kind of nucleic acid. However, it is necessary to observe the effect of the much longer time that preribosomal RNA spends in the nucleolus as compared to BR

Figure 4. Ratio between BR1 + BR2 RNA and rRNA at different times after injection of 25 μCi [³H]uridine (dots) or 5 μCi [³H]uridine (stars). Dots and stars indicate mean values and the brackets the s.e.m. From the interval between injection and sacrifice, 7.5 hr has been subtracted for the extrapolation to zero time (see text). Each value represents the mean of at least four animals. Glands from animals given 25 μCi were fixed according to Lönn and Edström (1977).

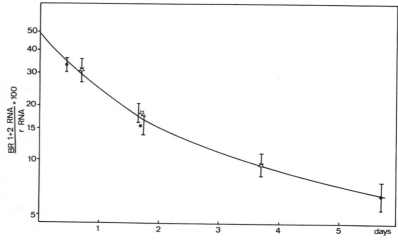

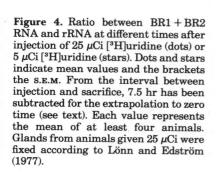

RNA in the BR. If there is a time-dependent change of specific activity levels in the labeling pools of precursors, this could give rise to differences in the overall specific activity in nucleolar and BR RNA which is due to this change, rather than to a labeling from different pools. The specific activity was therefore determined at two different time points so situated that there was reason to assume that the pool changes might be rather small over a period required to saturate both BR and nucleolar RNA with label. As a guideline, we used previously published determinations of TCA-soluble activities in glands after tritiated precursor injections (Lönn and Edström 1977), which show a maximum 6 hours after injection.

Table 4 shows that the specific activity is somewhat higher for chromosome 4 than for nucleolar RNA at the 6-hour time point if the effect of base composition is taken into account. This could, at least partially, be a result of rising precursor pool activities during the preceding time. At the 8.5-hour time point, the activities are closely similar. The results of this investigation thus do not give any grounds for assuming that BR RNA and nucleolar RNA are labeled from pools of different specific activity.

From the value for the export rate of rRNA, the export rate of BR1 + BR2 RNA, which is 69.5% as large, can be determined to be 2000 pg per day (Table 5). As previously discussed an upper limit for the production rate was determined to be 3300 pg per day (transcription time 15 min), but a more plausible level was estimated to be 2500 pg per day (transcription time 20 min). It is thus clear that a considerable fraction of the RNA produced in the BR, in the 50–100% range, is also exported in vivo.

DISCUSSION

The present data indicate that the BR1 and BR2 products, which are similar in size in the nucleus and the cytoplasm (Daneholt et al., this volume), are transferred to a large extent to the cytoplasm in vivo. If there is an intranuclear breakdown, it cannot be quantitatively dominating.

Table 5. Relation between Export of BR1 + BR2 RNA and rRNA

	BR1 + BR2 RNA	rRNA
Radioactive label at zero time	49.6	100
Conversion factor[a] to mass	1.40	1
Mass relations at zero time	69.5	100
pg RNA exported per day	2014	2898

[a] The conversion factor takes into account base compositions (uridylic acid: 18.3 and 26.4; cytidylic acid: 24.1 and 22.1, for BR1 + BR2 RNA and rRNA, respectively; molar proportions) and the relative incorporation of uridine into uridylic acid and cytidylic acid, 0.92 and 0.08, respectively.

We are well aware that, in particular, the calculation of the rate of production may suffer from systematic errors. These errors would, however, tend to lead to an overestimation of the rate of production and therefore cannot affect our main conclusions. Only if the transcription time is much shorter than 20 minutes, e.g., less than 10 minutes, would our conclusions be markedly influenced, but this appears unlikely since all determinations so far indicate a time of 20–30 minutes for explanted glands, and since it was found that about the same time is required in vivo as in explanted glands to eliminate RNA from the BR. Another reason why a transcription time of 10 minutes or less appears unlikely is that it would bring the rates close to rates for mammalian cells which are operating at 37°C, in contrast to the present cells which are kept at 18–19°C.

On the export side, one can distinguish between two sets of measurements, the quantitation of the relation between BR1 + BR2 RNA and rRNA export and the determination of rRNA export. It is true that the results of none of these can be accepted without critical examination. Thus, for the former set of measurements, animals are exposed to the trauma of injection and isotope administration. Unless bleeding occurs during the injection, the mortality of the animals is hardly affected by the procedures; furthermore, uridine was given in fivefold different amounts without any difference in results.

Table 4. Specific Activities in RNA Isolated from Chromosome 4 and Nucleoli after Injection of [³H]Uridine and [³H]Cytidine

Time after injection (hrs)	Chromosome 4 (cpm/pg)	Nucleoli (cpm/pg)	Ratio chromosome 4/nucleoli ± s.e.m.	
			uncorrected	corrected for base composition
6	0.58	0.50	1.15 ± 0.04	1.28 ± 0.04
8.5	1.02	1.26	0.83 ± 0.06	0.92 ± 0.07

Values for base composition for nucleoli were taken from Edström and Beermann (1962) and for chromosome 4 from data of J.-E. Edström and W. Beermann (unpubl.). The molar proportions of cytidylic acid and uridylic acid are 22.1 and 27.1 (nucleoli) and 20.9 and 23.7 (chromosome 4). Both [³H]cytidine and [³H]uridine were used for labeling the cells. The incorporation occurs in proportion to the specific activities at the doses used (unpubl. obs.) and these are related as 1:1.75 for cytidine:uridine. The correction factor for differences in base composition and different precursor specific activities is 1.11.

There would have to be either a specific increase in BR1 + BR2 RNA export or a specific decrease in rRNA export as a result of the injection procedure to affect the main conclusions. It is true that occasionally animals appear with a specifically decreased rRNA export. This is manifest as a distinctly lower than normal ratio between rRNA and 4S RNA; but no such animals were present among those used in the determinations of export ratios. This specifically decreased rRNA export is probably a normal functional variant rather than one induced by the experimental treatment since it can also be found when animals are given isotope by bathing. In the determinations of rRNA turnover, we cannot exclude some reutilization of the label as a possible source of error. If so, this would give too low a rate constant for the decay of rRNA and an underestimation of the rate of export, again an error that would not affect the main conclusions.

The half-life of BR1 + BR2 RNA in the cytoplasm is a function of the cytoplasmic content as well as of the export rate. Each of these cells contains, on average, 2500 pg of BR1 + BR2 RNA (Edström et al. 1978). An export rate of 2000 pg per day (out of which 1.5% is required for growth) results in an average half-life for BR1 and BR2 RNA close to 20 hours. It can be seen that the curve for the BR RNA to rRNA ratio approaches a half-life value of less than 24 hours as zero time is approached. This is a good agreement considering that this curve is likely to give an overestimation of the half-life of BR1 + BR2 RNA, since the labeling due to reutilization affects preferentially the more rapidly turning over RNA. Thus, even if rRNA export values were not available, we would be able to draw the same main conclusions regarding the extent of the export of BR1 + BR2 RNA.

It should be pointed out that the present type of analysis confers distinct advantages as compared to most other export studies based on kinetic analysis of the labeling of the RNA and the precursor pools. In the present case, the RNA is relatively homogeneous and it has a similar size in the nucleus and in the cytoplasm. Furthermore, the cells are fully differentiated and are functioning in the intact living animal. Thanks to the micromanipulatory approach, which permits direct isolation and analysis of specific activity in RNA at its source of origin, we can avoid the discussion about the representativeness of the precursor pools for a specific type of RNA synthesis.

Using explanted glands and considerably younger animals, Egyházi (1976) found much less export of BR1 + BR2 RNA to the cytoplasm than we have found. It is known from the work of Clever et al. (1969) that the stability of the messengers may vary with the age of the animals. We cannot at present decide whether the differences between the two investigations are due to techniques or biological material.

If our present results are generally applicable to messenger RNAs in differentiated cells, this would imply that the reported degradation of hnRNA applies to nuclear-specific portions of messenger precursors and/or an RNA population without a known cytoplasmic role.

Acknowledgments

This work was supported by grants from the Swedish Cancer Society and Karolinska Institutet. DRB was a gift from E. Egyházi.

REFERENCES

BEERMANN, W. and G. F. BAHR. 1954. The submicroscopic structure of the Balbiani ring. *Exp. Cell Res.* **6**: 195.

CLEVER, U., I. STORBECK, and C. G. ROMBALL. 1969. Chromosome activity and cell function in polytenic cells. Protein synthesis at various stages of larval development. *Exp. Cell Res.* **55**: 306.

DANEHOLT, B. 1972. Giant RNA transcript in a Balbiani ring. *Nat. New Biol.* **240**: 229.

DANEHOLT, B. and J.-E. EDSTRÖM. 1967. The content of deoxyribonucleic acid in individual polytene chromosomes of Chironomus tentans. *Cytogenetics* **6**: 350.

DANEHOLT, B. and H. HOSICK. 1973. Evidence for transport of 75 S RNA from a discrete chromosome region via nuclear sap to cytoplasm in Chironomus tentans. *Proc. Natl. Acad. Sci.* **70**: 442.

DANEHOLT, B., K. ANDERSSON, and M. FAGERLIND. 1977. Large-size polysomes in Chironomus tentans salivary glands and their relation to Balbiani ring 75 S RNA. *J. Cell Biol.* **73**: 149.

DANEHOLT, B., J.-E. EDSTRÖM, E. EGYHÁZI, B. LAMBERT, and U. RINGBORG. 1969. Chromosomal RNA synthesis in polytene chromosomes of Chironomus tentans. *Chromosoma* **28**: 399.

DARNELL, J. E. 1975. The origin of mRNA and the structure of the mammalian chromosome. *Harvey Lect.* **69**: 1.

EDSTRÖM, J.-E. 1964. Microextraction and microelectrophoresis for determination and analysis of nucleic acids in isolated cellular units. *Methods Cell Physiol.* **1**: 417.

EDSTRÖM, J.-E. and W. BEERMANN. 1962. The base composition of nucleic acids in chromosomes, puffs, nucleoli, and cytoplasm of Chironomus salivary gland cells. *J. Cell Biol.* **14**: 371.

EDSTRÖM, J.-E. and L. RYDLANDER. 1976. Identification of cytoplasmic RNA from individual Balbiani rings. *Biol. Zentralbl.* **95**: 521.

EDSTRÖM, J.-E. and R. TANGUAY. 1974. Cytoplasmic ribonucleic acids with messenger characteristics in salivary gland cells from Chironomus tentans. *J. Mol. Biol.* **84**: 569.

EDSTRÖM, J.-E., S. LINDGREN, U. LÖNN, and L. RYDLANDER. 1978. Balbiani ring RNA content and half-life in nucleus and cytoplasm of Chironomus tentans salivary gland cells. *Chromosoma* (in press).

EGYHÁZI, E. 1974. A tentative initiation inhibitor of chromosomal heterogeneous RNA synthesis. *J. Mol. Biol.* **84**: 173.

———. 1975. Inhibition of Balbiani ring RNA synthesis at the initiation level. *Proc. Natl. Acad. Sci.* **72**: 947.

———. 1976. Quantitation of turnover and export to the cytoplasm of hnRNA transcribed in the Balbiani rings. *Cell* **7**: 507.

EGYHÁZI, E., B. DANEHOLT, J.-E. EDSTRÖM, B. LAMBERT, and U. RINGBORG. 1970. Differential inhibitory effect of a substituted benzimidazole on RNA labeling in polytene chromosomes. *J. Cell Biol.* **47**: 516.

FLECK, A. and D. BEGG. 1965. The estimation of ribonucleic acid using ultraviolet absorption measurements. *Biochim. Biophys. Acta* **108**: 333.

LAMBERT, B. 1973. Tracing of RNA from a puff in the polytene chromosomes to the cytoplasm in *Chironomus tentans* salivary gland cells. *Nature* **242**: 51.

LAMBERT, B. and J.-E. EDSTRÖM. 1974. Balbiani ring nucleotide sequences in cytoplasmic 75 S RNA of *Chironomus tentans* salivary gland cells. *Mol. Biol. Rep.* **1**: 457.

LÖNN, U. and J.-E. EDSTRÖM. 1977. Movements and associations of ribosomal subunits in a secretory cell during growth inhibition by starvation. *J. Cell Biol.* **73**: 696.

McKENZIE, S. L., S. HENIKOFF, and M. MESELSON. 1975. Localization of RNA from heat-induced polysomes at puff sites in *Drosophila melanogaster*. *Proc. Natl. Acad. Sci.* **72**: 1117.

RANDERATH, K. and E. RANDERATH. 1967. Thin-layer separation methods for nucleic acid derivatives. *Methods Enzymol.* **12A**: 323.

RINGBORG, U. and L. RYDLANDER. 1971. Nucleolar-derived ribonucleic acid in chromosomes, nuclear sap and cytoplasm of *Chironomus tentans* salivary gland cells. *J. Cell Biol.* **51**: 355.

RUBINSTEIN, L. and U. CLEVER. 1972. Chromosome activity and cell function in polytene cells. V. Developmental changes in RNA synthesis and turnover. *Dev. Biol.* **27**: 519.

SEHGAL, P. B., E. DERMAN, G. R. MOLLOY, I. TAMM, and J. E. DARNELL. 1976. 5,6-Dichloro-1-β-D-ribofuranosylbenzimidazole inhibits initiation of nuclear heterogeneous RNA chains in HeLa cells. *Science* **194**: 431.

SPRADLING, A., S. PENMAN, and M. L. PARDUE. 1975. Analysis of Drosophila mRNA by in situ hybridization: Sequences transcribed in normal and heat-shocked cells. *Cell* **4**: 395.

The Packaging Proteins of Core hnRNP Particles and the Maintenance of Proliferative Cell States

W. M. LeStourgeon, A. L. Beyer, M. E. Christensen, B. W. Walker, S. M. Poupore, AND L. P. Daniels

Department of Molecular Biology, Vanderbilt University, Nashville, Tennessee 37235

In eukaryotes, the maintenance of correct information flow is dependent upon numerous events following the initiation of transcription. Mechanisms must be operative for transcript elongation, stabilization, and packaging of newly formed RNA, 5' "capping" and 3' adenylation of premessenger molecules, as well as for enzymatic cleavage and perhaps even splicing of the RNA (as described by Berget et al; Broker et al.; both this volume). Finally, mechanisms may be operative for intranuclear storage of pre-mRNA and eventually either complete degradation or efficient transport to the cytoplasm for translation.

Ultrastructural and biochemical studies have demonstrated that, in eukaryotes, proteins perhaps responsible in part for the above events, are deposited on nascent nonribosomal RNA during transcription (Miller and Hamkalo 1972; Augenlicht and Lipkin 1976). The ribonucleoprotein (RNP) complex that is visualized is of a repeating subunit structure of 200–300 Å spherical particles connected by ribonuclease-sensitive strands (Samarina et al. 1968; Kierszenbaum and Tres 1974; Mott and Callan 1975; Malcolm and Sommerville 1977). Monomer particles of this RNP complex from a variety of eukaryotes have been collected from 30–55S regions of sucrose density gradients and analyzed for protein and RNA composition. However, since the original studies of Georgiev and Samarina (1971) on "informofers" (more recently termed hnRNP particles), considerable confusion has existed in the literature concerning the protein components of these structures (see Beyer et al. 1977). Initially, it was reported that a single protein was responsible for the particle structure, but others have since reported various degrees of protein heterogeneity, up to 40 different polypeptides (Niessing and Sekeris 1971; Faiferman et al. 1971; Martin et al. 1974; Pederson 1974; Gallinaro-Matringe et al. 1975; Augenlicht and Lipkin 1976; Prestayko et al. 1976). Clearly before a complete understanding of the important events that follow transcription can be achieved, a more precise knowledge of nuclear RNP particle composition and structure must be available. Toward this objective we conducted the studies described in this and previous papers (Beyer et al. 1977; Christensen et al. 1977).

Considering the apparent structural analogy to the nucleosomes of chromatin, one might expect an analogous protein composition for the core matrix of nuclear RNP particles. If this were true, the bulk of particle protein mass would be comprised of a few cationic structural proteins of a conserved nature and present in a simple stoichiometric ratio. This general analogy, supported by the findings of Martin et al. (1974), is confirmed in our studies. Additional observations are the following:

1. hnRNP particles from rapidly dividing HeLa cells are composed of a simple stoichiometric complex of six proteins, which migrate in SDS gels as three groups of closely spaced doublets (groups A, B, and C). These proteins are generally conserved in several different mammalian cells. The rapidly labeled, nonribosomal RNA moiety associated with monomer particles averages near 600 nucleotides in length and is presumably a fragment from a larger in vivo molecule of hnRNA.

2. The group-A doublet (A_1 and A_2), with respective molecular weights of 32,000 and 34,000, constitutes 60% of the total HeLa particle protein mass. In HeLa, mouse embryo fibroblasts, rat liver, and hamster lung fibroblasts, these A-group proteins have identical respective molecular weights. In rapidly dividing cells, proteins A_1 and A_2 are present at a 1:1 ratio, but in slowly dividing or nondividing cells, such as normal adult liver or contact-inhibited cells, there is only about one-fifth as much A_1 as A_2. Proteins A_1 and A_2 have similar physical-chemical characteristics. Both proteins have basic isoelectric points of pH 9.2 and 8.4, respectively, and both contain high amounts of glycine (up to 28% total residue). These two proteins dissociate as a linked pair from the core particle structure at low ionic strength (0.25 M KCl); they are cationic and apparently bind RNA in an electrostatic manner.

3. The proteins of the group-B doublet in HeLa ($B_1 = 36,000$ daltons; $B_2 = 37,000$ daltons) are also basic but dissociate as a pair at slightly higher ionic strength than the group-A proteins (0.3 M). The group-B proteins in HeLa and rat liver comigrate in gels, but their percent composition in particles from these tissues varies somewhat. Approximately 20% of the arginine residues in protein B_1 are of an unusual methylated form, recently identified as N^G, N^G-dimethylarginine (Christensen et al. 1977).

This protein has an otherwise similar amino acid composition to the A-group proteins.

4. The group-C proteins in HeLa RNP particles (C_1 and C_2 of 42,000 and 44,000 daltons, respectively) have acidic isoelectric points and do not dissociate from RNA until an ionic strength of 0.60 M KCl is reached. The C-group doublet is of slightly higher molecular weight in mouse, rat, and hamster cells than in human cells.

5. The protein components of HeLa hnRNP particles may be dissociated from themselves and from the RNA moiety with 1.0 M NaCl, but all constituents reassociate on dialysis into low ionic strength solutions to form particles indistinguishable from nondissociated controls. The reconstitution is dependent upon an RNA "template" and argues for a simple, homogeneous "core" particle protein composition.

6. Proteins A_1, A_2, and B_1 appear to be the most highly conserved RNP particle proteins in all mammalian cell types studied, and, as stated above, these proteins are similar to each other. In the lower eukaryote *Physarum polycephalum*, a single 34,000-dalton protein possesses strikingly similar biochemical properties to this group of proteins, including N^G,N^G-dimethylarginine and high glycine, and is presumably functionally homologous. Like protein A_1, its presence is diagnostic of a rapidly dividing cell state, and it in fact disappears from quiescent nuclei of spores and cysts. Antibodies to the *Physarum* homolog cross-react with the proteins of mammalian hnRNP particles, and since the *Physarum* protein can be isolated in an undenatured state in large quantities (Christensen et al. 1977), it may prove to be an excellent model protein for the study of the physical properties of these RNA-binding proteins. CD spectral analysis suggests that the protein is composed of up to 30% β-structure, which is consistent with the high glycine content.

Although we have focused our initial efforts on the protein constituents of hnRNP particles, a sizeable body of evidence argues for the equivalence of hnRNA and the RNA associated with nuclear 40S hnRNP particles. This evidence includes labeling kinetics (Martin and McCarthy 1972), base composition (Georgiev and Samarina 1971), sequence complexity (Firtel and Pederson 1975), sensitivity to specific inhibitors such as actinomycin D (Pederson 1974; Beyer et al. 1977), and sensitivity to hormone stimulation such as increased incorporation of uridine in uterine nuclear RNP particles after estrogen stimulation (Knowler 1976). Kinniburgh and Martin (1976a) have shown that the majority, if not all, of the sequences complementary to total poly(A)-containing message of mouse acites cells are present in nuclear RNP particles. Similar results have been obtained by Sommerville and Malcolm (1976) for the RNA of nuclear RNP particles from amphibian oocytes.

RESULTS

The Proteins of "Core" hnRNP Particles

In an effort to eliminate heterogeneous chromatin and nuclear matrix fragments which result from nuclear sonication, we have obtained our crude nuclear hnRNP particles through the low-salt extraction technique of Samarina et al. (1968) using STM buffer (0.1 M NaCl, 1.0 mM $MgCl_2$, 10.0 mM Tris-HCl, pH 8.0) (for details, see Beyer et al. 1977). In our hands, this extraction method optimizes the recovery of monomer, as opposed to multimer, RNP particles. The sucrose gradient sedimentation pattern of the extracted material from HeLa and rat-liver nuclei is shown in Figure 1. All gradient fractions have been analyzed for protein distribution through high-resolution, thin slab gel electrophoresis. The proteins present in each sequential 1.0-ml gradient fraction appear below the respective OD_{260} value at that point. The RNA optical density peak is centered at approximately 40S, and our previous studies have shown that [³H]uridine counts coincide perfectly with this distribution following brief (15-min) labeling periods. Coincident with this RNA peak in the gradients is a group of specific major proteins, which have been enscribed in boxes in the figure. These are the protein components of core monomer hnRNP particles and they range in molecular weight between 32,000 and 44,000 daltons.

The six major proteins that interact with RNA in HeLa cells to form the 40S particles are arranged in three groups of closely spaced doublets, which form the basis for our nomenclature (groups A, B, and C in order of increasing molecular weight). There are seven major rat-liver RNP proteins in this molecular-weight range (see Fig. 1), including group-A and group-B proteins, that perfectly comigrate with the HeLa proteins but differ in stoichiometry. Clearly, in both cell types, these proteins are specifically enriched in the fractions that correspond to the distribution of RNA. Except for two or three minor high-molecular-weight proteins (to be discussed later), all other proteins peak with sedimentation coefficients different from that of the particles or in a completely dispersed manner throughout the gradient. This would not be apparent if only a 40S "cut" were examined for protein composition, and it forms the basis for our claim of a simple particle protein composition, which is confirmed by the salt dissociation-reconstitution experiments described in a later section. Labeling studies with [³H]thymidine indicate that low levels of DNA sediment throughout the gradient and many of the non-RNP proteins may be associated with randomly generated chromatin fragments arising during particle extraction. Sonication of nuclei leads to a much greater amount of this contaminant material in gradients.

Material collected from the 40S regions of the HeLa sucrose gradients appears in stained prepara-

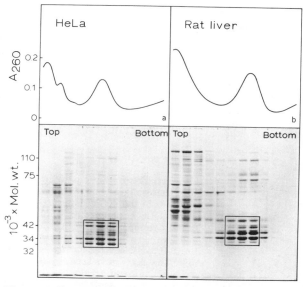

Figure 1. Sucrose density gradient sedimentation of *(a)* HeLa and *(b)* rat-liver nuclear hnRNP particles. The sucrose gradients (15–30%) were divided sequentially from bottom to top into 1.0-ml fractions and each fraction was analyzed for RNA and protein. In each panel the direction of migration is from left to right. The top panels show the distribution of RNA throughout the gradients (A_{260}), and the lower panels show the proteins present in corresponding fractions as resolved in consecutive wells of a single, thin, SDS-containing 8.75% polyacrylamide slab gel (for details, see Beyer et al. 1977; LeStourgeon and Beyer 1977). The peaks of RNA optical density roughly midway in the gradients are very near 40S and correspond to the RNA of monomer hnRNP particles. The protein bands enscribed in the boxes below are coincident in sedimentation with the RNA and represent the proteins of core hnRNP particles. Note that in the HeLa particles the proteins are arranged in three groups of closely spaced doublets. As demonstrated elsewhere, the same proteins are present in the rat-liver fraction but in differing stoichiometries.

tions as roughly spherical particles, with an average diameter of 210 Å when examined with the electron microscope (Fig. 2).

The RNA associated with each monomer particle is apparently a fragment of a larger transcript and presumably represents a length of RNA protected from the action of nucleases by its association with the packaging proteins. We have reported that the RNA associated with HeLa monomer particles sediments between 8–12S in sucrose gradients and migrates in SDS gels in a broad zone with a size corresponding to 400–900 nucleotides (Beyer et al. 1977). This size is similar to that reported for the RNA of nuclear particles from rat liver (Georgiev and Samarina 1971), human cancer cells (Augenlicht and Lipkin 1976), and *Dictyostelium* (Firtel and Pederson 1975). We have also found that low levels of actinomycin D (0.04 µg/ml), which inhibit 80% of the preribosomal RNA synthesis in HeLa cells, have no effect during 15-minute labeling periods on the incorporation of [³H]uridine into particle RNA.

Figure 3 shows the protein composition of HeLa, Chinese hamster lung fibroblast, and rat liver 40S hnRNP particles. As mentioned, the proteins appear to form three natural groupings in gels, and we have labeled them accordingly. The major 32,000- and 34,000-dalton components are the group-A proteins (A_1 and A_2, respectively). The group-B proteins (B_1 and B_2) have molecular weights of 36,000 and 37,000 daltons, respectively, and the group-C proteins are of 42,000 daltons (C_1) and 44,000 daltons (C_2). In HeLa RNP, the proteins in each doublet, except for the C group, are found in 1:1 ratios (i.e., $A_1 = A_2$, $B_1 = B_2$, $C_1 > C_2$), and our previous quantitative studies have shown that the group-A proteins alone in HeLa cells constitute 60% of the total particle protein mass (Beyer et al. 1977). In rat-liver and hamster fibroblasts, however, these relationships are not maintained. Reduction and carboxymethylation following standard procedures (Crestfield et al. 1963) do not change the electrophoretic patterns shown.

The A-group proteins and B_1 from three quite different cell types perfectly comigrate in SDS gels (Fig. 3). The proportion of protein A_1 in rat liver particles is significantly decreased, however. In logarithmically growing Chinese hamster cells in tissue culture, A_1 is present at a concentration equal to A_2, as it is in HeLa cells. Because of this, we suggest that the low amount of A_1 in rat liver does not reflect a fundamental difference between humans and lower mammals, but rather it reflects the low mitotic activity in liver tissues. Cell-state-dependent changes in intranuclear concentration of an homologous protein from the lower eukaryote *Physarum polycephalum* support this observation, as will be discussed in the final section.

In rat and hamster cells, B_1 is a more dominant protein. Although B_2 is not detectable in hamster particles, the single B protein may represent two different polypeptides, as in HeLa, except with indistinguishable molecular weights. As discussed later, A_1 may be a modified form of A_2 or B_1, since all three proteins have very similar properties. Proteins corresponding to the HeLa C group are of slightly higher molecular weight in rat and hamster RNP particles. The similarity in RNP particle proteins in these three eukaryotes argues for a conserved structural role for these proteins in RNA packaging. Certain analogies perhaps exist between these particles and nucleosomes.

In high-resolution gel electrophoresis studies (Beyer et al. 1977), we compared the proteins of HeLa chromatin, nucleoli, total nuclei, ribosomes, total histone, and 40S hnRNP particles to determine the "uniqueness" of the particle proteins. In these studies, it was determined that none of the group-A, B, or C proteins are constituents of ribosomes, nor do they migrate in gels with any of the histones. The A-group proteins can be easily seen in gels of

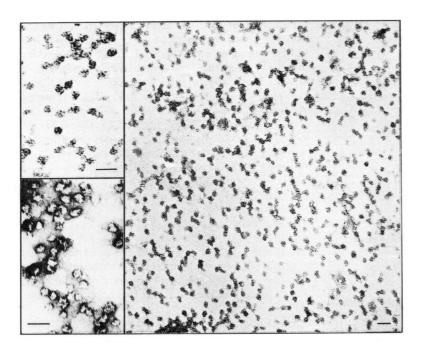

Figure 2. Electron micrographs of the 40S sucrose gradient fractions. The material in the 40S gradient fractions from HeLa is pelleted by high-speed centrifugation, fixed with formalin, and stained with uranyl acetate. The hnRNP particles exist mostly as monomer. The aggregates seen, in both linear and nonlinear arrays, are probably an artifactual result of fixation and staining, which were done after the particles were artificially aggregated by pelleting. Bars equal 500 Å.

total nuclear protein; but, due to the lower amount of the B and C proteins in particles, these proteins cannot be easily distinguished in gels of total nuclear protein. As would be expected, the major proteins of hnRNP particles can be easily distinguished in gels of total extranucleolar chromatin.

Dissociation of hnRNP Particles and Protein Affinities for RNA

Particle stability and the pattern of protein dissociation in solutions of increasing ionic strength may be indicative of the strength of specific ionic interactions between proteins and between proteins and RNA. The salt dissociation studies described below have provided preliminary information regarding particle assembly and a more precise identification of the true particle proteins. In these studies, hnRNP particles were divided into aliquots and sedimented in a series of sucrose gradients prepared with increasing concentrations of NaCl. The entire gradient in each case was analyzed for both RNA and protein distribution to detect the gradual dissociation of particle components and to identify more precisely the core particle constituents. An examination of Figure 4 reveals that as salt is increased from 0.1 M to 0.6 M, there is essentially a complete disassembly of the particle complex. It is evident that at 0.2 M NaCl, the group-A proteins are more than 50% dissociated. At 0.4 M salt, both A- and B-group proteins are dissociated and can be clearly seen solubilized on top of the gradients. An important observation (in this study and for our future understanding of particle structure and function) is the fact that the C-group proteins remain attached to the RNA fragments at 0.4 M salt and traces are

still attached at 0.6 M salt. As would be expected, the loss in protein groups A and B leads to a particle which no longer sediments at 40S (marked by the arrows in Fig. 4). Evidence that RNA is still associated with the group-C proteins in a salt-resistant structure at 0.3–0.4 M NaCl can be seen in Figure 5, which shows parallel fractionated gradients analyzed for [³H]uridine radioactivity. All of the radioactivity lost from the original 40S region is accounted for by acid-precipitable radioactivity in higher gradient fractions. In summary, as ionic strength is increased, there is a gradual dissociation first of the A- and B-group proteins, followed by the complete solubilization of the particle, the components of which then remain at the top of the gradients. In addition, there exists an intermediate complex of tightly associated RNA and group-C protein (0.4 M NaCl). In this complex, 70% of the RNA present initially in the hnRNP particles remains associated with the C-group proteins. The group-C protein-RNA complexes of a size greater than 500,000 daltons can be obtained in an essentially pure state simply by collecting the void volume of a BioGel A 0.5-m column loaded with a particle extract under conditions which dissociate the A- and B-group proteins (0.35 M NaCl) (Fig. 6). The elution of the C-group protein-RNA complex in the void volume confirms the greater affinity of these proteins for RNA and affords a rapid method of obtaining material for more detailed characterization. The affinity of the group-C protein for RNA will be discussed later in this paper.

Our results are consistent with the dissociation experiments of Stevenin and Jacob (1974), in which they report a progressive and linked dissociation of both RNA and protein from nuclear RNP, suggest-

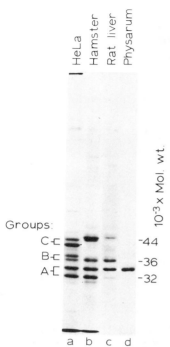

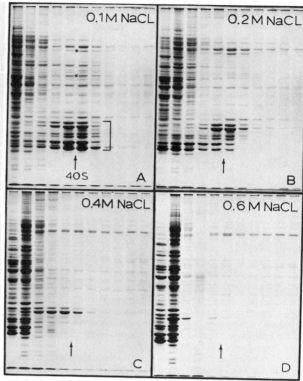

Figure 3. The proteins of various core hnRNP particles. hnRNP particles were collected from 40S regions of sucrose gradients containing the peak concentration of particle-associated OD_{260} material (see Fig. 1) and analyzed by SDS slab gel electrophoresis (LeStourgeon and Beyer 1977). The six major hnRNP particle proteins of HeLa migrate as three pairs of closely spaced doublets, which are labeled in order of increasing molecular weight as group A (A_1 and A_2 of 32,000 and 34,000 daltons), group B (B_1 and B_2 of 36,000 and 37,000 daltons), and group C (C_1 and C_2 of 42,000 and 44,000 daltons). Note that in Chinese hamster lung fibroblasts, the A-group proteins, like HeLa, are the most major proteins and are present in a 1:1 ratio. In adult rat liver, much less A_1 is present, and A_2 and B_1 are the two major proteins. In this figure, the liver sample is underloaded to show that A_2 and B_1 are not present in a 1:1 ratio. In Figure 1, more heavily loaded liver samples reveal three minor proteins between B_1 and the C-group proteins. In rodents, the C-group proteins are more tightly grouped and run with slightly higher molecular weights than the human proteins. Most of the minor high-molecular-weight proteins are not true components of core hnRNP particles. In gel *d* is shown the purified protein of *Physarum*, which appears to be a composite of the mammalian A- and B-group proteins (see text).

Figure 4. Salt dissociation of HeLa hnRNP particle protein. In these experiments, an initial particle extract (in 0.1 M NaCl) was divided into four equal aliquots and NaCl added to bring the salt concentration in three aliquots to 0.2, 0.4, and 0.6 M, respectively. Sucrose gradients (15–30%) were made with corresponding salt concentrations. After centrifugation, each gradient was fractionated into 1-ml fractions and analyzed in sequence using thin-slab SDS gel electrophoresis. Direction of particle sedimentation in the gradients is from left to right. The arrows in each gel mark the 40S region of the gradient. Panel A exists as a control sample in these experiments. The bracket in panel A marks the 32,000–44,000 molecular-weight range containing the major group-A, B, and C proteins. At 0.2 M NaCl, note the dissociation of the A-group and some of the B-group proteins. At 0.4 M salt the A- and B-group proteins are dissociated and remain at the top of the gradient. The C-group proteins, still attached to 70% of the particle RNA, move in the gradients with a lower sedimentation coefficient at 0.4 M NaCl. Note the heterodisperse sedimentation to the bottom of the gradient of most of the minor high-molecular-weight protein components even at 0.6 M salt, when all of the major core particle proteins and RNA (see Fig. 5) are completely dissociated. The asterisks in Frame A indicate the two proteins of approximately 75,000 and 110,000 molecular weight, whose sedimentation patterns indicate that they are associated with core 40S hnRNP particles. Similar minor high-molecular-weight proteins can also be seen in the liver particle preparation in Fig. 1.

ing an interspersed ribonucleoprotein structure. Our results are not supportive of the "informofer" model proposed by Lukanidin et al. (1972), which argues for a self-contained 30S protein core and external RNA. It is interesting to note that the three "pairs" of proteins, groups A, B, and C, seen in HeLa do in fact behave as pairs, not only in migration in polyacrylamide gels but also in their pattern of dissociation (Fig. 4).

Analysis of the individual protein components of 40S hnRNP particles in gradients of increasing ionic strength also provides an opportunity to monitor the behavior of the minor high-molecular-weight

proteins, which appear to be particle proteins if one examines only a single 40S "cut" from the gradients (Fig. 3). In Figure 4 it is clear that the majority of these proteins still sediment heterodispersely to the bottom of the gradient, even at 0.60 M NaCl, demonstrating that they are not true components of the

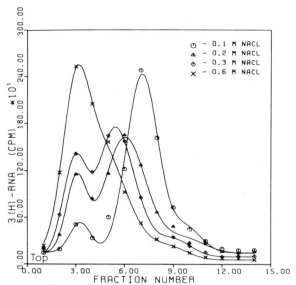

Figure 5. Salt dissociation of HeLa hnRNP particle RNA. In an experiment parallel to that described in Fig. 4, sucrose gradients of increasing ionic strength were fractionated and analyzed for [³H]uridine radioactivity. The HeLa cells had been labeled for 15 min prior to harvesting in the presence of 0.04 μg/ml actinomycin D. The curves are computer-derived and drawn, best Gaussian fits to the data points for the gradients of 0.1 M NaCl, 0.2 M NaCl, 0.3 M NaCl, and 0.6 M NaCl. Direction of migration is to the right. Note the progressive shift to the top of the gradient of the RNA peak as NaCl concentration is increased. See symbol code in the upper right corner of the figure. The major peak of RNA label is always associated with the C-group proteins seen in Fig. 4.

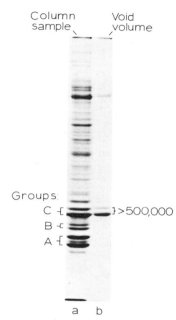

Figure 6. The elution of the group-C protein-RNA complex in a BioGel A 0.5-M column (molecular weight sieving chromatography). As shown in Figs. 4 and 5, the major A- and B-group proteins of HeLa hnRNP particles are dissociated at increased salt concentration from the more stable subparticle complex of group-C protein and particle RNA. When an initial hnRNP particle extract is brought to 0.35 M NaCl and loaded on a BioGel A 0.5-M molecular weight sieving column equilibrated at the same salt concentration, the group-C protein-RNA complex can be obtained in a much enriched state by collecting the void volume of the column. These findings confirm the affinity of group-C protein for particle RNA in 0.35 M NaCl and establish a minimum molecular weight of the complex at 500,000 daltons (exclusion limit of the column).

monomer 40S particles, which are completely dissociated at this ionic strength. Several other bands, for example, a 90,000-dalton protein, sediment originally at a lower S value than the monomer particles and are not concentrated in the 40S region. After carefully examining the gels, we feel that only three other proteins, in addition to the major A-, B-, and C-group proteins, meet the sedimentation criteria expected of monomer-particle-associated proteins. These are proteins of approximately 33,000, 75,000, and 110,000 daltons. The 33,000-dalton protein migrating between A$_1$ and A$_2$ may be a modified form of one of the major particle proteins as its presence is variable. Like the C-group proteins, the 110,000-molecular-weight protein seems more tightly associated with the hnRNP particle at 0.2 M NaCl than do the major A- and B-group proteins. One or both of these high-molecular-weight bands (75,000 and 110,000) may be equivalent to the poly(A)-binding proteins of similar molecular weight described by Schwartz and Darnell (1976) and Kish and Pederson (1975) in HeLa cells. Our previous quantitative studies indicate that neither of these minor proteins are represented in sufficient proportion in total hnRNP particle material to be present in one copy per monomer.

The above statements pertain to observations on HeLa 40S hnRNP particles; however, the same particles from rat liver are also sensitive to salt and are completely dissociated at 0.7 M NaCl.

hnRNP Particles — Fidelity of Reconstitution

Evidence supporting the existence of a core hnRNP particle structure composed of a simple stoichiometric complex of a few structural proteins has come from the observation that salt-dissociated particles naturally reassociate in low ionic environments. The reconstituted particles are indistinguishable in protein composition from nondissociated controls. Information concerning protein-RNA specificities and the nature of protein-RNA interaction in particles has also come from experiments where radioactive purified particle RNA, ribosomal RNA, and nonparticle protein were added to dissociated particle preparations before effecting reconstitution.

The reconstitution is shown in Figure 7, in which a nuclear hnRNP particle extract has been divided into three equal aliquots. One aliquot is used as a control (at 0.1 M NaCl), two aliquots are brought

Figure 7. Dissociation and reconstitution of hnRNP particles. In these experiments, an initial particle extract in buffered 0.1 M NaCl (pH 8.0) was divided into three equal portions and two portions brought to 1.0 M NaCl. One of the latter aliquots was dialyzed to equilibrium against low salt (0.1 M NaCl), and all three preparations were subjected to sucrose density sedimentation in the respective salt concentration. (*Upper panels*) Distribution of RNA in the three gradients as monitored by OD_{260} (——) and [³H]uridine counts (○——○) following a 15-min label in the presence of 0.04 μg/ml actinomycin D. (*Lower panels*) Proteins in the same gradients. Note that in the nondissociated control sample (panel *a*) the group-A, B, and C proteins sediment with the rapidly labeled RNA at about 40S. The efficiency of 1.0 M NaCl for complete particle dissociation is

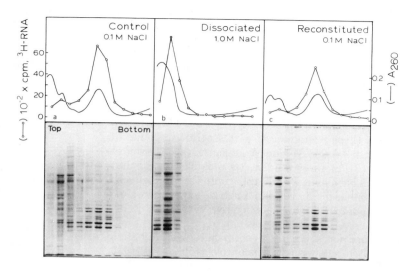

shown in panel *b* (RNA and protein on top of the gradients); panel *c*, when the dissociated preparation is brought back to low salt, 70% of the particles reassociate, with core protein composition and S value indistinguishable from control particles.

to 1.0 M NaCl, and one of the latter is dialyzed back to 0.1 M NaCl. As is evident in the figure, 1.0 M NaCl effectively dissociates the particles, as seen by the failure of either the particle RNA or the protein species to enter the gradient (Fig. 7b). Approximately 70% of the dissociated particles re-form in low ionic strength solution with the same protein stoichiometry and sedimentation coefficient as controls (Fig. 7c). If ribonuclease is used to digest the RNA while the particle is dissociated, reconstitution of RNP particles does not occur, nor do RNA-free protein cores form. As reported earlier (Beyer et al. 1977), if intact 40S particles are subjected to RNase digestion, particle dissociation results and the proteins again remain soluble on top of the gradients. Thus a 40S RNP particle is dependent on the presence of an 8–12S RNA molecule. The failure of particle proteins to enter gradients in these experiments suggests a lack of extensive protein-protein interaction in the absence of RNA

[³H]uridine-labeled particle RNA has been purified from the 40S RNP complex by standard phenol partitioning techniques. When this RNA is added to salt-dissociated hnRNP particles in amounts corresponding to 30% of the RNA already present, it is recovered in the reconstituted 40S particles. However, this same amount of purified particle RNA will also bind to nondissociated control particles, as similarly reported by Martin et al. (1974). In both of these cases, concomitant with the appearance of RNA label in the particle fraction there is a corresponding 30% increase in the OD_{260} peak at 40S. There are at this time several possible explanations for these events, including general protein affinity for RNA, RNA-RNA association, or accessible RNA binding sites on the particles due to nuclease activity

during extraction and resulting in RNA particles deficient in RNA relative to their in vivo state.

If purified HeLa ribosomal RNA is added to dissociated particles, the 40S RNP particle peak is diminished up to 60% relative to controls with no added RNA, and the hnRNP components are found in the gradient pellet with much of the rRNA. Whether this represents the formation of multimeric RNP particles on a longer RNA template or a nonspecific electrostatic aggregation of rRNA and hnRNP is not yet known. Other observations in our laboratory have indicated that extracted 40S RNP particles have a slight positive charge at pH 8.0 which may account for some of these RNA binding phenomena.

In preliminary studies where the purified native RNP protein homolog from *Physarum* is added to dissociated particles, it appears to prevent specifically the reassociation of the A-group proteins in the reconstituted mammalian RNP particles, perhaps through competition for the RNA. The same amounts of BSA or ovalbumin do not interfere with the fidelity of reconstitution. Numerous studies are now in progress to better characterize protein-protein and protein-RNA interactions during particle formation.

Characterization of the Basic Proteins of Mammalian hnRNP Particles

Our first efforts to purify and characterize each of the major core particle proteins have focused on the two group-A proteins of HeLa and protein A_2 and B_1 of rat liver. The unique physical-chemical properties of these proteins, their apparent dynamic involvement in cell-state transitions, and our recent circular dichroism studies on the purified *Physarum*

A-group homolog indicate that much information concerning particle function can come from a complete knowledge of the contributory roles of these proteins. A summary of our current knowledge concerning these proteins is presented below.

Proteins A_1 and A_2, are basic proteins with a general affinity for nucleic acids. In high-resolution, two-dimensional, O'Farrell-type gels (O'Farrell et al. 1975), these proteins run with isoelectric points of 9.2 and 8.4 for A_1 and A_2, respectively (Beyer et al. 1977). In these studies, we also discovered that B_1 was basic, but B_2 and the group-C proteins focus in the near acidic regions. In these experiments, no multiple-spot pattern associated with various degrees of carbamylation was apparent, nor was there evidence for a greater protein heterogeneity than indicated by single-dimension electrophoresis.

Further evidence for the cationic properties of A_1 and A_2 has come from cation exchange chromatography. Prior to BioRex-70 chromatography, nuclear RNP particle extracts were ribonuclease-digested to effect particle dissociation without protein denaturation. Thus, at pH 7.0 and low ionic strength, proteins that have cationic properties in a native state are expected to bind to the resin. Most nuclear proteins wash through at 0.05 M NaCl. As NaCl concentration is increased, protein A_2 elutes at 0.34 M NaCl, slightly before protein A_1. This is consistent with the isoelectric focusing data discussed above. Both A_1 and A_2 elute prior to H1 histone at 0.75 M NaCl.

A third finding consistent with the charged nature of these proteins has come from nucleic acid binding studies. HeLa group-A proteins bind to homologous DNA-cellulose with an affinity only slightly less than the two H1 histone species (B. W. Walker, unpubl.). This association is most likely electrostatic and nonspecific, as is that of the histones. In these binding studies, numerous acidic higher molecular weight proteins did not elute until salt concentrations greater than 0.8 M were reached. This suggests that these nonhistone proteins interact more specifically with DNA. An important finding from these experiments was that the group-C proteins have no affinity for homologous DNA. This is consistent with our finding that they are weakly acidic proteins and would not be expected to interact nonspecifically with the polyanion DNA as do the group-A proteins. The tight association of the C-group proteins with RNA in a salt-resistant complex and their lower intraparticle concentration suggest a rather specific binding mechanism in hnRNP particles. Augenlicht et al. (1976) described a 40,000-dalton RNP particle protein present in a nuclease-resistant complex that may be the same as our C-group proteins. With respect to the group-A proteins and the three types of studies so far performed, we can say that A_1 is slightly more basic than A_2 and both are less basic than the lysine-rich histones.

We have purified proteins A_1 and A_2 from HeLa to electrophoretic homogeneity (Fig. 8) by prepara-

Figure 8. Purified HeLa hnRNP proteins A_1 and A_2. Gel 1 shows the group-A, B, and C proteins of HeLa hnRNP particles collected from the peak 40S regions of a sucrose density gradient. In gels 2 and 3 are shown, respectively, the prep-gel-purified proteins A_2 and A_1 obtained as described elsewhere (LeStourgeon and Beyer 1977).

tive SDS gel electrophoresis (LeStourgeon and Beyer 1977) and, more recently, proteins A_2 and B_1 of rat liver have been purified through classical biochemical procedures. Table 1 shows the amino acid composition of total HeLa hnRNP particles, purified HeLa A-group proteins, A_2 and B_1 from rat liver, and the protein of *Physarum* homologous to the basic proteins of mammalian cells. The group-A proteins are quite similar in composition. The similarity of the total hnRNP particle composition to the group-A proteins reflects the fact that they constitute 60% of total particle protein. The analyses necessary to determine the amounts of glutamine and asparagine represented in the glutamic and aspartic acid figures have not been made; however, if an average relative proportion is assumed, based on the compiled analyses of 68 completely sequenced proteins (Jukes et al. 1975), the acidic to basic ratio is less than 1. This is consistent with the proteins' cationic properties. Glycine is present in significantly high amounts (up to 28% in A_2 of rat liver) as it is in the structural proteins collagen, chorion, and silk. We have previously identified the basic residue that elutes from Beckman PA-35 resin between ammonia and arginine as N^G,N^G-dimethylarginine (Christensen et al. 1977; Beyer et al. 1977). Sarasin (1969) commented on the appearance of this "unidentified" residue in rat-liver nuclear RNP and it has recently been identified from this source by Boffa et al. (1977). Other investigators have observed small amounts of an

Table 1. Amino Acid Composition of hnRNP Particle Proteins

Amino acid	Total HeLa particles	HeLa A₁	HeLa A₂	Rat liver A₂	Rat liver B₁	*Physarum* homolog
Lysine	6.9	7.7	6.0	4.1	6.1	6.0
Histidine	2.0	1.5	1.6	1.7	2.6	1.8
Arginine	5.9	3.3	5.3	6.1	4.6	4.7
Aspartic acid	11.4	7.6	10.3	11.8	9.5	6.6
Threonine	3.6	3.5	2.7	2.7	4.2	3.1
Serine	7.3	13.0	10.4	9.1	5.3	4.4
Glutamic acid	12.1	12.6	10.7	10.8	8.7	8.0
Proline	4.9	3.5	4.0	5.7	3.3	4.7
Glycine	17.1	23.2	24.9	27.9	18.6	22.6
Alanine	6.1	8.1	5.8	7.1	2.3	10.1
Cysteine	0.7	trace	trace	0.0	0.0	trace
Valine	4.1	3.2	2.5	2.9	5.0	6.9
Methionine	1.9	0.7	1.5	0.5	3.7	2.0
Isoleucine	2.6	3.3	2.7	1.3	3.0	3.6
Leucine	5.3	4.8	4.3	3.5	4.8	4.4
Tyrosine	3.7	1.3	2.6	0.0	9.9	2.4
Phenylalanine	3.9	2.6	4.3	2.9	7.6	4.6
N^G,N^G-Dimethyl-arginine[a]	0.5	0.1	0.4	0.0	1.3	4.1

Analyses expressed as moles/100 moles total amino acid. These single analyses are not corrected for hydrolytic loss.

[a] Content of N^G,N^G-dimethylarginine is based on the leucine color constant. The structure of this residue (Christensen et al. 1977) is:

"unidentified" residue in total RNP from duck liver (Martin et al. 1974) and rat liver (Krichevskaya and Georgiev 1969). The possible significance of this residue will be discussed later. The high content of glycine is consistent with considerable β-structure in proteins in general. Evidence for β-structure in these proteins comes from our CD data on the *Physarum* homolog to these proteins, as described in the next section.

The Physarum protein. In the lower eukaryote *Physarum polycephalum*, only one major nuclear protein with properties similar to the basic components of mammalian hnRNP particles exists in the molecular weight range of mammalian core particle proteins. We have purified this 34,000-dalton protein in a nondenatured state (Christensen et al. 1977) and have observed the following.

The *Physarum* hnRNP-like protein has an amino acid composition which seems to represent a composite of proteins A₁, A₂, and B₁ from HeLa and rat-liver particles (Table 1). More specifically, the *Physarum* protein, like A₁ and A₂, possesses high glycine (25 mole %) and, like B₁ from rat liver, a large amount of N^G, N^G-dimethylarginine (4.1 mole %). The *Physarum* protein is quite basic, with an isoelectric point near that of protein A₂ from HeLa. Proteins A₂ and B₁ from rat-liver particles are copurified when the protocol for purifying the

Physarum protein is followed. This is consistent with the biochemical similarities of these proteins. Like protein A₁ from mammalian cells, the *Physarum* protein is only present in major intranuclear concentrations in actively dividing cell states (LeStourgeon et al. 1974, 1975). As discussed in the next section, this observation first aroused our interest in the protein as a possible homolog. We have not been successful in obtaining 30–40S hnRNP particles from *Physarum,* perhaps due to endogeneous ribonuclease activity during nuclear isolation.

Clearly the most important experimental aspect of these comparative studies to date has been our recent observation that antibodies raised in rabbits to homogeneously purified *Physarum* protein cross-react with one or more of the basic proteins from mammalian particles (Fig. 9). This complements the biochemical data suggesting a homologous function for the *Physarum* protein in RNA binding and processing.

Because milligram quantities of the *Physarum* protein can be routinely purified in a nondenatured state, we have begun to characterize this protein more completely. With the thought that useful information concerning the function and manner of nucleic acid binding may come from a knowledge of secondary structure, we have performed a CD spectral analysis on the purified protein in solution. It has been estimated from the CD spectral data

Figure 9. Ouchterlony double-diffusion immunoprecipitation tests of antisera raised against the purified *Physarum* homolog to the basic proteins of hnRNP particles (see Fig. 3 and text). In both group *A* and *B*, antisera (50% ammonium sulfate cut) was placed in the center well. In group *A*, wells 2 and 4 contained the purified antigen, whereas wells 1, 3, and 5 contained other *Physarum* nuclear proteins as controls which partially copurify with the antigen. One of these proteins acts as an internal control since it remained as a 2% contaminant in the antigen samples injected in the rabbit. In group B, HeLa hnRNP particle proteins were tested for cross-reactivity with the antisera to the *Physarum* protein. The HeLa hnRNP particles in 0.1 M NaCl, 0.35 M NaCl, and 0.6 M NaCl were placed in wells 1, 3, and 5, respectively. The *Physarum* homolog was in wells 2 and 4. Note that an apparent cross-reaction has taken place. With increasing salt concentrations the complexity of the cross-reaction also appears to increase. This is consistent with the salt dissociation of the A- and B-group proteins of hnRNP particles. In controls where either normal rabbit serum or distilled water was placed in the center well, no reaction was observed with either the *Physarum* homolog or the HeLa hnRNP particle proteins.

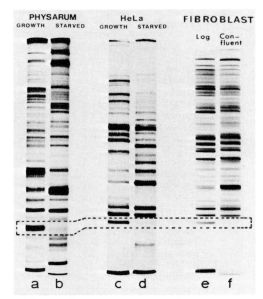

Figure 10. Electrophoretic resolution of total nonhistone protein extracted from three cell types during active growth and from nondividing cells. The protein enscribed in the dashed box is protein A$_1$ in HeLa and mouse embryo fibroblasts. In the *Physarum* samples, the enscribed protein is the protein apparently homologous to the mammalian protein. Note that in all cell types shown, a loss of this protein is associated with quiescent cells (see text for further explanation). Protein extraction and SDS gel electrophoresis is described elsewhere (LeStourgeon and Beyer 1977).

that the *Physarum* protein contains 10–11% α-helix and 23–31% β-structure. The large amount of β-structure is consistent with the high content of glycine in the protein.

Protein A$_1$ involvement in hnRNP particle function in proliferating cells. We have reported previously that when several types of cells cease dividing in response to various stimuli, specific changes occur in the complement of nuclear nonhistone proteins (LeStourgeon and Rusch 1971, 1973; LeStourgeon et al. 1973a,b, 1974, 1975; LeStourgeon 1977). When considerable morphological and biochemical changes are associated with the establishment of quiescent states, as in *Physarum* sporulation (Fig. 10a,b), or when senescence is involved in tissue-culture cells (Fig. 10c,d), numerous protein changes occur. In such cases, it is not possible to establish a functional correlation between specific nuclear proteins and the maintenance of cell proliferation. However, when mammalian cells cease dividing at confluency, only a few major quantitative changes occur in specific proteins (Figs. 10e,f, and 11), and it is more tempting to suggest a dynamic role for some of the proteins in the maintenance of genetic information flow. Protein A$_1$ of hnRNP particles is such a protein, since in all cell types examined it decreases or disappears entirely from nuclei of nondividing cells (see protein enscribed in dashed box in

Fig. 10). In mammalian cells, protein A$_1$ is one of the few nuclear proteins which undergo a significant quantitative change when cells cease dividing at confluency (Fig. 11).

In an independent study on the nonhistones of three mouse cell types (Loeb et al. 1976), a protein corresponding in molecular weight and intranuclear concentration to protein A$_1$ of hnRNP particles was found to vary as a function of cell proliferation. In rapidly dividing mouse teratocarcinoma cells and in mouse L cells, proteins corresponding to HeLa A$_1$ and A$_2$ were present in a 1:1 ratio. In mouse liver, however, the A$_1$ protein was essentially absent.

Earlier we called attention to the 1:1 ratio of A$_1$ and A$_2$ in rapidly dividing tissue-culture cells, whether of human or rodent origin. At the same time we pointed out that in adult mouse liver, a differentiated tissue with low mitotic index, little A$_1$ was present in hnRNP particles. Because of these findings and those described above, we call attention to the possibility that protein A$_1$ may be dynamically involved in hnRNP particle function and chromatin activity in general. Studies at in progress to determine if protein A$_1$ may be a cell-state-dependent modification of protein A$_2$ and whether it may function to maintain transcript elongation or to regulate other processing events.

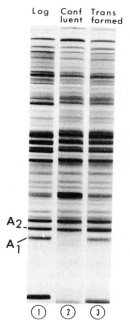

Figure 11. SDS-disc gel electrophoresis of total nonhistone protein extracted from logarithmically growing mouse embryo fibroblasts *(1)*, the same cells 48 hours after postconfluent inhibition of growth *(2)*, and in cells oncogenically transformed with nitrosoguanadine derivatives *(3)*. Note that the loss of protein A_1 is one of the few changes that occurs in confluent cells. Also note that the major hnRNP particle proteins A_1 and A_2 are major nuclear proteins and are easily identified in total nonhistone protein preparations. The major protein positioned below center in the gels which appears to increase in nondividing cells is actin (LeStourgeon et al. 1975).

DISCUSSION

In this paper we present evidence that, in the nucleus of mammalian cells, a few unique and conserved proteins interact with nonribosomal RNA in a stoichiometric fashion to form a homogeneous core hnRNP particle. Martin et al. (1974) have described a similar simple hnRNP particle structure, and, more recently, Patel and Holoubek (1977) have confirmed a limited core-protein complement for the purified monomer particles. The identification of these low-molecular-weight structural proteins should aid in distinguishing other proteins involved in the enzymatic modification of nascent transcripts associated with hnRNP particles.

Many of our findings are stated in a brief tabulated form in the introduction to this paper. In addition, we have previously discussed these observations with respect to the existing literature on nuclear hnRNP particles, hnRNA packaging and processing, and RNA-protein interactions (Beyer et al. 1977; Christensen et al. 1977). Stated below are some of our tentative views of particle formation and function, as well as areas deserving more in-depth study.

The salt-resistant group-C protein-RNA complex

in HeLa hnRNP particles is likely to be of structural significance. In addition to the high-salt-resistant group-C protein-RNA complex, we have observed a similar resistant RNP substructure containing the C-group proteins after limited RNase digestion or detergent dissociation (0.1% Na deoxycholate) of 40S particles (A. L. Beyer, unpubl.). Based on our observations of hnRNP particle dissociation, reconstitution, and specific protein affinities for RNA, it is possible that particle formation in vivo may be a cooperative and ordered event. In such a scheme, the tight-binding, nonbasic C-group proteins may bind RNA in a specific manner and then organize the adsorption of the major cationic A- and B-group proteins, which by themselves show little specificity for nucleic acid types. The ability to obtain a pure RNA–C-group complex, as described here, will aid in approaching these questions, with the characterization of the associated RNA molecule being of primary interest.

The proteins which constitute the majority of particle protein mass, and which are the most highly conserved, and similar particle proteins in the different cell types studied (A_1, A_2, and B_1) may be derived from a single ancestral sequence. In a lower eukaryote, the acellular slime mold *Physarum polycephalum*, a single protein has features that represent a composite of these mammalian particle proteins. Further secondary-structure and primary-sequence studies will address this possibility and will also provide information as to the nature of RNA-protein interaction. With the information at hand, however, it is tempting to speculate on several features of this interaction. Considerable efforts to build models for specific nucleic acid-protein interactions have resulted in only a few feasible associations, two of which are relevant to our findings. The first is that arginine is one of a few amino acids capable of specific and stable interactions with double-stranded nucleic acids (Seeman et al. 1976; Mansy et al. 1976). We have identified an unusual, posttranslational methylation of arginine in the major hnRNP particle proteins, which introduces a significant charge and steric change at the putative interaction site, i.e., the guanidinium group. Of course, this modification would necessarily be a dynamic one if it is operative in destablizing this interaction, but the fact that it is an unusual modification and has been found to a significant extent only in other RNP structures (Chang et al. 1976; Boffa et al. 1977) suggests strongly that it is indeed involved in dynamic protein-RNA interactions.

The second RNA-protein interaction which has been suggested, and which is relevant to our data, is the possibility of protein β-pleated sheet interaction with the minor groove of double-stranded RNA (Carter and Kraut 1974). The β-sheet is a repeating protein secondary structural feature, which assumes a right-handed twist at its lowest free-energy state (Chothia 1973) and matches that of the RNA helix.

Our circular dichroism studies indicate that the major RNP proteins may contain up to 30% β-structure, and the high glycine content of the proteins is consistent with this possibility. This electrostatic and non-sequence-specific interaction might also explain the affinity of these proteins for DNA. It is interesting to note with respect to these two possible models for RNA-protein interaction that some evidence exists for the presence of double-stranded RNA in hnRNP particles (Kinniburgh and Martin 1976b) and that hnRNA in general contains a considerable amount of double-stranded regions (Jelinek et al. 1974).

For a complete understanding of hnRNP particle function, it is important to determine what percentage, or perhaps class, of hnRNA is associated with these proteins in an RNP particle structure. In our nuclear extracts we consistently obtain 80% of HeLa nonribosomal nuclear RNA, which is labeled by a 10–15-minute incubation with [³H]uridine. The majority of this RNA (70%) is recovered in the 40S particle fraction, the rest being found on top of the gradients in an apparently degraded form. (A small amount of free particle protein is also observed on top of gradients, perhaps representing an intranuclear pool of nonassociated protein or perhaps the result of nuclease activity during particle extraction.) Our evidence, as well as that of others (Georgiev and Samarina 1971; Martin et al. 1974; Pederson 1974), indicates that the majority of hnRNA is recoverable in this RNP particle form and is presumably associated with these proteins in vivo. Ultrastructural electron microscope studies of transcription in amphibian oocytes have also indicated that nascent RNA is apparently nonspecifically packaged with protein into repeating 200-Å RNP particles (Malcolm and Sommerville 1974; Mott and Callan 1975). These data indicating a general packaging mechanism for hnRNA, together with the histonelike properties of the particle proteins, initially were suggestive of the analogy between this nucleoprotein complex and that of nucleosomes. In addition, in both cases, the nucleic acid is considerably condensed in a repeating particle structure and stabilized against general nucleolytic activity. It is probable that hnRNA is still available for specific enzymatic processing while condensed in this form, as DNA is in nucleosomes.

Results presented at this symposium have amply demonstrated the dynamic changes in the histone molecules and in nucleosome configuration characteristic of, and perhaps responsible for, active transcriptional states. Although hnRNP particle research is at a much earlier stage, our findings, and those of others (Albrecht and Van Zyl 1973; Patel and Holoubek 1976), suggest specific dynamic changes in hnRNP particle protein composition that are perhaps responsible for altered particle structure and thus altered hnRNA processing in different cell states. One of our most interesting findings is the quantitative change in protein A_1 concentration in the particle, characteristic of the quantitative genetic expression of the cell. The importance of post transcriptional nuclear events in the maintenance of correct information flow has been well demonstrated, and these proteins, which are intimately associated with the RNA, are implicated in these control processes. It has been shown that in the switch from a resting to a growing cell state, the important control event seems not to be a change in hnRNA production, sequence complexity, or nuclear polyadenylation, but rather an alteration in the efficiency of nuclear hnRNA processing and its subsequent release to the cytoplasm (Johnson et al. 1975, 1976; Williams and Penman 1975). In addition, a characteristic event in the transformation of a normal cell to a tumor cell is the loss of selective RNA release to the translational machinery of the cytoplasm (Shearer and Smuckler 1972). It is perhaps these gross changes in hnRNA transport and processing, rather than more specific enzymatic processing events, that are mediated through dynamic changes in hnRNP particle structure.

Acknowledgments

We thank Dr. P. W. Melera for providing isolated Chinese hamster cell nuclei (subline DC-3F/MQ-19). We also thank Dr. D. J. Puett for assistance with the circular dichroism studies and Mr. B. Luten for computer analyses. This research was supported by National Science Foundation grant BMS 75–03105 to W.M.L. and by U.S. Public Health Service traineeships to A.L.B. and S.M.P.

REFERENCES

ALBRECHT, C. and I. M. VAN ZYL. 1973. A comparative study of the protein components of ribonucleoprotein particles isolated from rat liver and hepatoma nuclei. *Exp. Cell Res.* **76**: 8.

AUGENLICHT, L. H. and M. LIPKIN. 1976. Appearance of rapidly labeled, high molecular weight RNA in nuclear ribonucleoprotein. *J. Biol. Chem.* **251**: 2592.

AUGENLICHT, L. H., M. McCORMICK, and M. LIPKIN. 1976. Digestion of RNA of chromatin and nuclear ribonucleoprotein by staphylococcal nuclease. *Biochemistry* **15**: 3818.

BEYER, A. L., M. E. CHRISTENSEN, B. W. WALKER, and W. M. LeSTOURGEON. 1977. Identification and characterization of the packaging proteins of core 40S hnRNP particles. *Cell* **11**: 127.

BOFFA, L. C., J. KARN, G. VIDALI, and V. G. Allfrey. 1977. Distribution of N^G,N^G-dimethylarginine in nuclear protein fraction. *Biochem. Biophys. Res. Commun.* **74**: 969.

Carter, C. W., Jr. and J. KRAUT. 1974. A proposed model for interaction of polypeptides with RNA. *Proc. Natl. Acad. Sci.* **71**: 283.

CHANG, F. N., I. J. NAVICKAS, C. N. CHANG, and B. M. DANCIS. 1976. Methylation of ribosomal proteins in HeLa cells. *Arch. Biochem. Biophys.* **172**: 627.

CHOTHIA, C. 1973. Conformation of twisted β-pleated sheets in proteins. *J. Mol. Biol.* **75**: 295.

CHRISTENSEN, M. E., A. L. BEYER, B. WALKER, and W. M. LeSTOURGEON. 1977. Identification of N^G,N^G-dimethylarginine in a nuclear protein from the lower eukaryote

Physarum polycephalum homologous to the major proteins of mammalian 40S ribonucleoprotein particles. *Biochem. Biophys. Res. Commun.* **74**: 621.

CRESTFIELD, A. M., S. MOORE, and W. H. STEIN. 1963. The preparation and enzymatic hydrolysis of reduced and S-carboxymethylated proteins. *J. Biol. Chem.* **238**: 622.

FAIFERMAN, I., M. G. HAMILTON, and A. O. POGO. 1971. Nucleoplasmic RNP particles of rat liver. II. Physical properties and action of dissociating agents. *Biochim. Biophys. Acta* **232**: 685.

FIRTEL, R. A. and T. PEDERSON. 1975. Ribonucleoprotein particles containing heterogeneous nuclear RNA in the cellular slime mould *Dictyostelium discoideum*. *Proc. Natl. Acad. Sci.* **72**: 301.

GALLINARO-MATRINGE, H., J. STEVENIN, and M. JACOB. 1975. Salt dissociation of nuclear particles containing DNA-like RNA. Distribution of phosphorylated and nonphosphorylated species. *Biochemistry* **14**: 2547.

GEORGIEV, G. P. and O. P. SAMARINA. 1971. D-RNA containing ribonucleoprotein particles. *Adv. Cell Biol.* **2**: 47.

JELINEK, W., G. MOLLOY, R. FERNANDEZ-MUNOZ, M. SALDIT, and J. E. DARNELL. 1974. Secondary structure in heterogeneous nuclear RNA: Involvement of regions from repeated DNA sites. *J. Mol. Biol.* **82**: 361.

JOHNSON, L. F., R. LEVIS, H. T. ABELSON, H. GREEN, and S. PENMAN. 1976. Changes in RNA in relation to growth of the fibroblast. IV. Alterations in the production and processing of mRNA and rRNA in resting and growing cells. *J. Cell Biol.* **71**: 933.

JOHNSON, L. F., J. G. WILLIAMS, H. T. ABELSON, H. GREEN, and S. PENMAN. 1975. Changes in RNA in relation to growth of the fibroblast. III. Post-transcriptional regulation of mRNA formation in resting and growing cells. *Cell* **4**: 69.

JUKES, T. H., R. HOLMQUIST, and H. MOISE. 1975. Amino acid composition of proteins: Selection against the genetic code. *Science* **189**: 50.

KIERSZENBAUM, A. L. and L. L. TRES. 1974. Transcription sites in spread meiotic prophase chromosomes from mouse spermatocytes. *J. Cell Biol.* **63**: 923.

KINNIBURGH, A. J. and T. E. MARTIN. 1976a. Detection of mRNA sequences in nuclear 30S ribonucleoprotein subcomplexes. *Proc. Natl. Acad. Sci.* **73**: 2725.

———. 1976b. Oligo(A) and oligo(A)-adjacent sequences present in nuclear ribonucleoprotein complexes and mRNA. *Biochem. Biophys. Res. Commun.* **73**: 718.

KISH, V. M. and T. PEDERSON. 1975. Ribonucleoprotein organization of polyadenylate sequences in HeLa cell heterogeneous nuclear RNA. *J. Mol. Biol.* **95**: 227.

KNOWLER, J. T. 1976. The incorporation of newly synthesized RNA into nuclear ribonucleoprotein particles after oestrogen administration to immature rats. *Eur. J. Biochem.* **64**: 161.

KRICHEVSKAYA, A. A. and G. P. GEORGIEV. 1969. Further studies on the protein moiety in nuclear DNA-like RNA containing complexes. *Biochim. Biophys. Acta* **164**: 619.

LESTOURGEON, W. M. 1977. The nonhistone proteins of chromatin during growth and differentiation in *Physarum polycephalum*. In *Eukaryotic microbes as model developmental systems* (ed. D. O'Day and P. Horgen), p. 34. Marcel Dekker, New York.

LESTOURGEON, W. M. and A. L. BEYER. 1977. The rapid isolation, high resolution electrophoretic characterization and purification of nuclear proteins. In *Methods in chromosomal protein research* (ed. G. Stein et al.), p. 387. Academic Press, New York.

LESTOURGEON, W. M. and H. P. RUSCH. 1971. Nuclear acidic protein changes during differentiation in *Physarum polycephalum*. *Science* **174**: 1233.

———. 1973. Localization of nucleolar and chromatin residual acidic protein changes during differentiation in *Physarum polycephalum*. *Arch. Biochem. Biophys.* **155**: 144.

LESTOURGEON, W. M., C. NATIONS, and H. P. RUSCH. 1973a. Temporal synthesis and intranuclear accumulation of the nuclear acidic proteins during periods of chromatin reactivation in *Physarum polycephalum*. *Arch. Biochem. Biophys.* **159**: 861.

LESTOURGEON, W. M., R. TOTTEN, and A. FORER. 1974. The nuclear acidic proteins in cell proliferation and differentiation. In *Acidic proteins of the nucleus: A volume of monographs on cell biology* (ed. I. L. Cameron and J. R. Jeter), p. 159. Academic Press, New York.

LESTOURGEON, W. M., W. WRAY, and H. P. RUSCH. 1973b. Functional homologies of acidic chromatin proteins in higher and lower eukaryotes. *Exp. Cell Res.* **79**: 487.

LESTOURGEON, W. M., A FORER, Y. YANG, J. S. BERTRAM, and H. P. RUSCH. 1975. Contractile proteins: Major components of nuclear and chromosome nonhistone proteins. *Biochim. Biophys. Acta* **379**: 529.

LOEB, J. E., E. RITZ, C. CREUZET, and J. JAMI. 1976. Comparison of chromosomal proteins of mouse primitive teratocarcinoma, liver and L cells. *Exp. Cell Res.* **103**: 450.

LUKANIDIN, E. M., E. ZALMANZON, L. KOMAROMI, O. SAMARINA, and G. P. GEORGIEV. 1972. Structure and function of informofers. *Nat. New Biol.* **238**: 193.

MALCOLM, D. B. and J. SOMMERVILLE. 1974. The structure of chromosome-derived ribonucleoprotein in oocytes of *Triturus cristatus carnifex*. *Chromosoma* **48**: 137.

———. 1977. The structure of nuclear ribonucleoprotein of amphibian oocytes. *J. Cell Sci.* **24**: 143.

MANSY, S., S. K. ENGSTROM, and W. L. PETICOLAS. 1976. Lasar raman identification of an interaction site on DNA for arginine containing histones in chromatin. *Biochem. Biophys. Res. Commun.* **68**: 1242.

MARTIN, T. E. and B. J. MCCARTHY. 1972. Synthesis and turnover of RNA in the 30S nuclear ribonucleoprotein complexes of mouse ascites cells. *Biochim. Biophys. Acta* **277**: 354.

MARTIN, T., P. BILLINGS, A. LEVEY, S. OZARSLAN, T. QUINLAN, H. SWIFT, and L. URBAS. 1974. Some properties of RNA protein complexes from the nucleus of eukaryotic cells. *Cold Spring Harbor Symp. Quant. Biol.* **38**: 921.

MILLER, O. L., JR. and B. A. HAMKALO. 1972. Visualization of RNA synthesis on chromosomes. *Int. Rev. Cytol.* **33**: 1.

MOTT, M. R. and H. G. CALLAN. 1975. An electron microscope study of the lamp brush chromosomes of the newt *Triturus cristatus*. *J. Cell Sci.* **17**: 241.

NIESSING, J. and C. E. SEKERIS. 1971. Further studies on nuclear ribonucleoprotein particles containing DNA-like RNA from rat liver. *Biochim. Biophys. Acta* **247**: 391.

O'FARRELL, P. H. 1975. High resolution two-dimensional electrophoresis of proteins. *J. Biol. Chem.* **250**: 4007.

PATEL, N. T. and V. HOLOUBEK. 1976. Protein composition of liver nuclear ribonucleoprotein particles of rats fed carcinogenic dyes. *Biochem. Biophys. Res. Commun.* **73**: 112.

———. 1977. Dependence of the composition of the protein moiety of nuclear ribonucleoprotein particles on the extent of particle purification as studied by electrophoresis including a 2-dimensional procedure. *Biochim. Biophys. Acta* **474**: 524.

PEDERSON, T. 1974. Proteins associated with heterogeneous nuclear RNA in eukaryotic cells. *J. Mol. Biol.* **83**: 163.

PRESTAYKO, A. W., P. M. CRANE, and H. BUSCH. 1976. Phosphorylation and DNA binding of nuclear rat liver proteins soluble at low ionic strength. *Biochemistry* **15**: 414.

SAMARINA, O. P., E. M. LUKANIDIN, J. MOLNAR, and G. P. GEORGIEV. 1968. Structural organization of nuclear complexes containing DNA-like RNA. *J. Mol. Biol.* **33**: 251.

SARASIN, A. 1969. Particules ribonucleoproteiques 40S des

noyaux de foie de rat. Proprietes des proteines de ces particules. *FEBS Lett.* **4**: 327.

SCHWARTZ, H. and J. E. DARNELL. 1976. The association of protein with the polyadenylic acid of HeLa cell messenger RNA; evidence for a "transport" role of a 75,000 molecular weight polypeptide. *J. Mol. Biol.* **104**: 833.

SEEMAN, N. E., J. M. ROSENBERG, and A. RICH. 1976. Sequence-specific recognition of double helical nucleic acids by proteins. *Proc. Natl. Acad. Sci.* **73**: 804.

SHEARER, R. W. and E. A. SMUCKLER. 1972. Altered regulation of the transport of RNA from nucleus to cytoplasm in rat hepatoma cells. *Cancer Res.* **32**: 339.

SOMMERVILLE, J. and D. B. MALCOLM. 1976. Transcription of genetic information in amphibian oocytes. *Chromosoma* **55**: 183.

STEVENIN, J. and M. JACOB. 1974. Effects of sodium chloride and pancreatic ribonuclease on the rat-brain nuclear particles: The fate of the protein moiety. *Eur. J. Biochem.* **47**: 129.

WILLIAMS, J. and S. PENMAN. 1975. A comparison of the messenger RNA from growing and resting mouse fibroblasts by molecular hybridization. *Cell* **6**: 197.

Substructure of Nuclear Ribonucleoprotein Complexes

T. MARTIN, P. BILLINGS, J. PULLMAN, B. STEVENS, AND A. KINNIBURGH*

Department of Biology, University of Chicago, Chicago, Illinois 60637

Nascent heterogeneous nuclear RNA (hnRNA) synthesized in regions of active chromatin rapidly becomes associated with protein; the presence of protein on the growing RNA chains has been visualized in spread chromatin by electron microscopy (EM) (see, e.g., Miller and Bakken 1972; Malcolm and Sommerville 1974; Laird et al. 1976; McKnight et al., this volume). It is reasonable to presume that this protein folds the lengthening RNA chains, facilitates their removal from the template, and later participates in the processing and maturation events which lead to the turnover of nucleus-restricted sequences while mature mRNA molecules are transported to the cytoplasm. There have been very extensive studies on the pathway of nuclear RNA molecules from synthesis to the appearance of mRNA on cytoplasmic polyribosomes. It is fair to say that, at present, the wealth of information produced promises rather than achieves a satisfying description of this complex process. These studies have centered on the characteristics of purified RNA molecules. Far fewer attempts have been made to examine the native forms of nuclear RNA molecules undergoing processing, i.e., as ribonucleoprotein (RNP) complexes, which are presumably the true substrates for the processing enzymes.

Questions of Structure and Function of hnRNP

We have been concerned for a number of years with extending and refining the observations of Georgiev and his colleagues that rapidly labeled hnRNA could be extracted from nuclei of mammalian cells in the form of RNP complexes, thus enabling a biochemical analysis of the structures (Samarina et al. 1966). The view that has emerged from these studies is that the bulk of newly synthesized RNA in the eukaryotic nucleus consists of large molecules folded by protein to form a chain of linked RNP substructures which, when cleaved by endogenous or exogenous nuclease, yield 30S particles (Martin and McCarthy 1972; Kinniburgh et al. 1976). Such a model was originally proposed by Samarina et al. (1968). The protein responsible for maintaining the 30S RNP substructure of hnRNA has a very simple polypeptide composition consisting

primarily of two to four distinct species, all having molecular weights in the range 35,000–40,000 (Martin et al. 1974; Billings and Martin 1977; see also LeStourgeon et al., this volume). These polypeptides appear to have been conserved in size and amino acid composition during the evolution of eukaryotes (Martin et al. 1974). Most of our studies, including all of the experiments described in this paper, have employed the Taper hepatoma of mouse in the ascites form as a matter of convenience.

The rapidity and extent of binding of newly synthesized RNA into a 30S subcomplex form suggest that the specific proteins associate with and fold the RNA chain while still in nascent form. The bulk of the bound RNA turns over with the half-life characteristic of hnRNA (Martin and McCarthy 1972), but the nuclear 30S RNP structures also contain the majority (and probably all) of sequences represented in cytoplasmic mRNA (Kinniburgh and Martin 1976a). We therefore infer that the simple set of polypeptides present in the 30S complex act as generalized "folding proteins" for most nascent RNA and aid in the formation of the native template for processing.

If the polypeptide components are to fold RNA effectively into distinctive unit structures, we may expect them to have definable protein-protein interactions. Furthermore, if the RNA-protein complex formed is to act as a substrate for specific processing events, then some hierarchy of binding affinity of specific types of nuclear RNA sequence for the protein complex would seem likely. Certain regions of the large mRNA molecules should be preferentially bound or exposed by the protein complex if it is to participate in regulating the specificity of processing events. The presence of nuclear poly(A) in the form of a 15S subcomplex, very different in size and in polypeptide composition to the 30S particle (Quinlan et al. 1974, 1977), yields an obvious example of the in vivo specificity of nuclear RNA-protein interactions. More relevant to hnRNA processing, however, would be the definition of interactions of nuclear proteins with specific internal regions of hnRNA such as oligo(U), oligo(A), and double-stranded (ds) RNA sequences. We have begun to approach this problem with reference to the 30S RNP (Kinniburgh et al. 1976), and the following experiments represent further attempts to understand the actual and potential protein-protein and RNA-protein interactions in this subunit of the hnRNP complex.

* Present address: McArdle Laboratory for Cancer Research, University of Wisconsin, Madison, Wisconsin 53706.

Morphology of Nuclear 30S and 15S RNP

Nascent RNP fibers observed by electron microscopy of spread chromatin have shown varying proportions of beaded and folded regions, presumably dependent upon the conditions employed (Miller and Bakken 1972; Malcolm and Sommerville 1974; Laird et al. 1976; McKnight et al., this volume). The micrographs imply a high degree of particulate character in the fibers, although admittedly it has not been as apparent in this case as in electron microscope studies of polyribosomes or, more recently, of the nucleosome form of chromatin. In regions not confused by secondary folding, the fibers have generally been assigned a diameter of approximately 200 Å.

We previously reported a diameter of 150–250 Å for negatively stained 30S RNP particles from mouse ascites cells (Martin et al. 1974), and recently we have begun a detailed electron microscope study of nuclear RNP complexes that can be sufficiently purified to justify such analysis. One fact emerges immediately from attempts to observe unfixed 30S particles and that is the far greater lability of these structures when compared with ribosomes or ribosomal subunits. In the absence of fixative, 30S particles appear as amorphous or "unfolded" structures (results not shown). When fixed with glutaraldehyde and negatively stained with uranyl acetate, the particulate character of 30S RNP preparations is readily apparent (Fig. 1A,B). Under these conditions the majority of 30S RNP appear as regular compact structures of varying morphology but relatively consistent size, 235×195 Å average dimensions. As expected, the morphology of the 15S poly(A)-containing RNP subcomplex is distinctly different. After glutaraldehyde fixation and negative staining, these particles appear as ellipsoids with dimensions of 165×120 Å (Fig. 1C). The 15S RNP are more difficult to purify than 30S complexes, and the preparation also shows the presence of elongated and crescent structures as well as 100-Å rings and "stacked disks" from the adjacent 17S peak (see Quinlan et al. 1974).

Shadowed 30S particles (prepared by freeze-drying) appear somewhat larger; our results for rotary shadowing with platinum give values of 250×195 Å, whereas those for unidirectional shadowing are $280 \times 250 \times 215$ Å (results not shown). The latter estimates indicate that the 30S RNP are slightly flattened structures, though this may arise from distortion of a relatively flexible particle in binding to the EM grid film. The generally larger size values obtained by shadowing compared with negative staining techniques are consistent with results from other systems, notably for ribosomes (Huxley and Zubay 1960; Vasiliev 1971).

We have begun to catalog the forms apparent in the negatively stained preparations of 30S RNP. Among the various shapes observed, the most clearly defined were round, ellipsoidal, rectangular, and triangular (Fig. 2a–e). The similarity among the members of each category, both in negatively stained and in shadowed preparations, suggests a limited range of potential morphological forms. The particles which do not fit into these categories appear as intermediate forms, e.g., rectangles with rounded corners. It is unlikely that all the profiles represent different projections of a single three-dimensional configuration; the shadows cast by particles of different morphologies cannot be rationalized on the basis of a single model, but rather suggest a limited diversity of shapes. If, in fact, these structures are subsections of the nascent RNA fibers, a variety of configurations would be consistent with the somewhat irregular folding apparent in electron micrographs of spread chromatin (see, e.g., Laird et al. 1976). Interactions between the polypeptides of the complexes and RNA sequences having distinctive properties could yield a set of conformational isomers. It is not surprising that RNP complexes of such a dynamic metabolic character are less rigid and consistent structures than ribosomes, for example.

In attempting to understand the internal structure of these nuclear particles, it may be of value to carry out a more detailed analysis of the unfolded forms of 30S RNP apparent in negatively stained particles (Fig. 2f). We have not yet completed such a study, which may be expected to contribute to a decision between the models of 30S RNP either as consisting of RNA coiled on a protein core or as the result of a regular folding of an RNP fibril. These latter alternatives arise from the investigation of protein-protein and protein-RNA interactions such as described in the following sections.

Protein-Protein Interactions and the Integrity of 30S RNP

Models have been constructed in which the native hnRNP structure has been interpreted as either particulate or as folded fibril. In the first model (Samarina et al. 1968; Lukanidin et al. 1972b), nascent hnRNA is thought to complex with surface binding sites on a number of stable globular protein complexes ("informofers"). Cleavage between particles would yield individual informofers, each carrying a segment of RNA, and would thus explain the nature and origin of 30S RNP (Fig. 3a). An alternative model implies that smaller protein complexes bind to the nascent RNA, forming an RNP fibril which is subsequently periodically folded into a more compact structure (Stevenin and Jacob 1974) and possibly maintained in this conformation by protein-protein interactions of varying degrees of stability (Fig. 3b). These various alternatives for the organization of nascent RNP fibers have been discussed by Malcolm and Sommerville (1974) in the context of amphibian lampbrush-loop fine structure.

In considering the interactions that maintain the integrity of the 30S RNP subcomplex, several observations tend to diminish the importance of the RNA

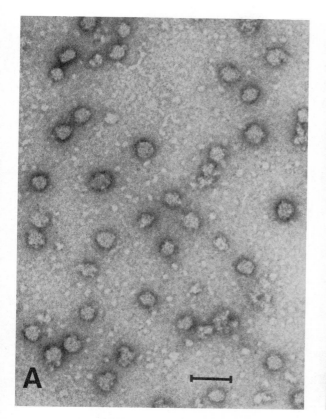

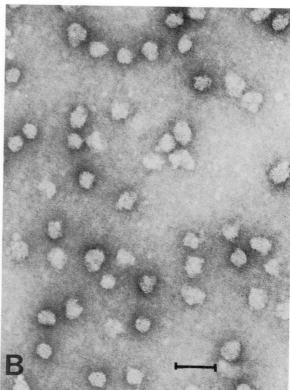

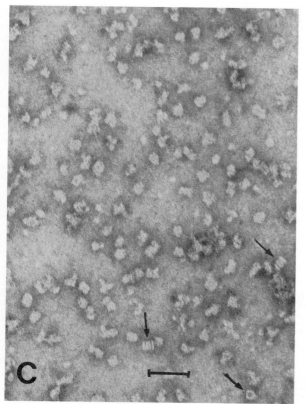

Figure 1. Electron micrographs of fields of 30S and 15S RNP particles from mouse ascites cells (Quinlan et al. 1974). Specimens were fixed with 0.5% glutaraldehyde and negatively stained with 1% uranyl acetate. *(A)* Crude 30S preparation; *(B)* purified 30S preparation; *(C)* 15S preparation. Lamellar structures and disks with central holes (arrows) are contaminants from the 17S peak. Scale bars represent 500 Å.

a   b   c   d   e   f

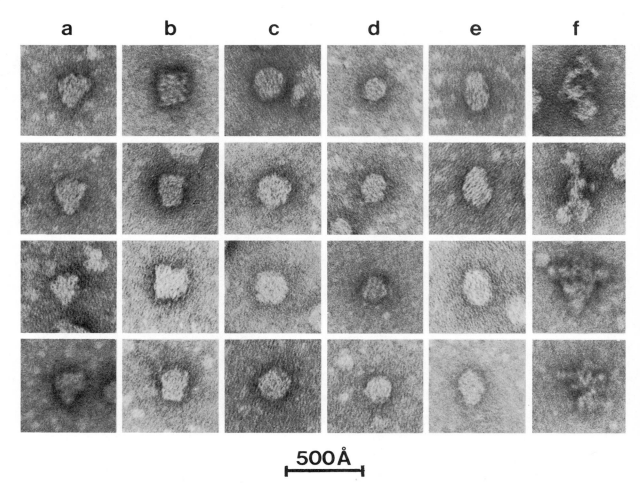

$\underset{\longmapsto\longmapsto}{500\text{Å}}$

Figure 2. Gallery showing the different projections of negatively stained 30S particles that were observed: *(a)* Triangular forms; *(b)* rectangular and square forms; *(c)* circular forms; *(d)* small circular forms (found mainly in material from the light shoulder of the 30S peak); *(e)* ellipsoidal forms; *(f)* disintegrating forms.

component. First, a large part of the RNA is relatively sensitive to RNase (see below). This suggests, at least operationally, a surface location of the nucleic acid. At moderate salt concentrations, 30S RNP proteins aggregate on treatment with nuclease, implying a considerable tendency for protein-protein interactions. Second, the 30S RNP contains approxi-

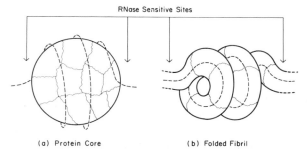

Figure 3. Two distinct models for the structure of the 30S RNP subcomplex of hnRNP: *(a)* RNA coiled on the surface of a protein core (the informofer of Samarina et al. [1968]); *(b)* condensed or folded region of a continuous RNP fibril. These alternative forms have also been discussed by Malcolm and Sommerville (1974).

mately 1000 nucleotides of RNA (Samarina et al. 1968; Martin et al. 1974); however, whether this is relatively intact or internally nicked to 50–100-nucleotide segments does not affect the integrity of 30S RNP at physiological salt concentrations (Kinniburgh and Martin 1976a). Thus, given the lack of a requirement for RNA continuity within the particle, our attention is focused on the protein component.

The concept of a reutilizable protein core complex (the informofer of Samarina et al. [1968]) is an attractive one and resembles in many ways the more recently evolved model for the nucleosome histone core in chromatin structure (see, e.g., Kornberg 1974; other papers, this volume). Major support for this model of the formation of hnRNP, however, comes from the observation that chemically radioiodinated 30S RNP protein cores, stripped of RNA and maintained in solution by high concentrations of salt, continue to sediment in approximately the same position in sucrose gradients as native 30S RNP (Lukanidin et al. 1972b). The stability of these cores suggests that the protein complexes could be reutilized intact in the nucleus.

In our first experiments on the salt stability of 30S RNP, we employed the lactoperoxidase-catalyzed radioiodination technique of Marchalonis (1969) to monitor the fate of the protein moiety. The 30S RNP were iodinated either directly after extraction from the nuclei (crude 30S RNP) or after resuspending pelleted RNP recovered from the 30S region of a first gradient (purified 30S RNP). RNP particles were then centrifuged on a sucrose gradient as usual to separate them from enzyme and free iodine. Portions of such 30S RNP were adjusted to a final salt concentration varying from 0.1 M to 2 M NaCl before analysis on sucrose gradients containing the same salt concentration. We found that iodinated 30S RNP-protein was recovered from such gradients in approximately the 30S region as reported earlier by Lukanidin et al. (1972b), indicating that under these conditions the protein moiety of 30S RNP remained intact even in salt concentrations as high as 2 M NaCl. (An iodinated 30S RNP gradient profile in which the sucrose gradient contained 0.5 M NaCl is shown in the top panel of Fig. 4.) Parallel experiments with uridine- or adenosine-labeled 30S RNP demonstrated that RNA was progressively lost from the 30S region as the salt concentration was raised above 0.2 M NaCl. The RNA displaced by high salt was recovered in a very slowly sedimenting peak at the top of the gradient (see top panel in Fig. 5).

Several observations, however, were somewhat inconsistent with this salt-resistant protein core model. First, the degree of stability as indicated by radioactivity in iodinated crude 30S RNP was invariably considerably less than that of iodinated purified particles. This was originally interpreted as reflecting a variable recovery of some minor, high-molecular-weight, cosedimenting or loosely bound proteins, perhaps more accessible to the enzyme, but which are largely removed in pelleting and repurifying 30S RNP (Martin et al. 1974). However, SDS-polyacrylamide gel analysis of the unstable protein displaced from the 30S region on high-salt sucrose gradients indicated that it was representative of the total protein composition of such 30S RNP and not enriched in high-molecular-weight species. Furthermore, the specific activity of the dissociated protein was not notably higher than that of the corresponding protein remaining with stable particles. Second, attempts to isolate RNA-free protein cores labeled in vivo with [³H]leucine clearly indicated that the bulk of the protein was easily dissociated in 0.5 M NaCl. These results suggested that iodination of the protein somehow rendered the protein complex stable to high salt.

We have now monitored the distribution of 30S RNP polypeptides across high-salt gradients by SDS-acrylamide gel electrophoresis for both iodinated and noniodinated preparations (Figs. 4 and 5). The results verify the conclusions of the [³H]leucine-labeling experiments, namely, that noniodinated RNPs do not contain a salt-stable core, but rather

that, in high salt, the 35,000–40,000-dalton proteins are released to the top of the gradient coordinately with [³H]uridine-labeled RNA (Fig. 5).

In contrast, 30S RNP iodinated by the lactoperoxidase procedure does contain a salt-stable protein complex. However, most of the iodinated protein fails to enter the separation gel; instead it is trapped

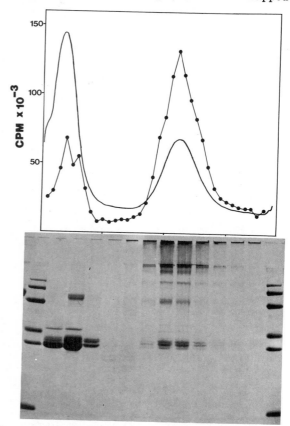

Figure 4. Iodinated 30S RNP centrifuged on sucrose gradients containing 0.5 M NaCl. Pelleted 30S RNPs were resuspended in 1 ml of STM, pH 8 (Martin and McCarthy 1972), and iodinated on ice with carrier-free [¹²⁵I]NaI by the lactoperoxidase-H_2O_2 procedure of Marchalonis (1969). The iodination reaction was terminated after 1 hr with 2-mercaptoethanol, and the suspension was adjusted to 0.5 M NaCl and recentrifuged on 15–30% sucrose gradients containing 0.5 M NaCl (SW 27 rotor, 25,000 rpm for 15 hr). The gradient was collected in 36 fractions. *(Top panel)* Acid-insoluble radioactivity of 20-μl aliquots of each fraction is shown (●——●) superimposed on the A_{254} absorption profile. Remaining fractions were pooled in 3-ml portions and TCA was added to 7%. The precipitates were collected by centrifugation, washed with ethanol, and solubilized by boiling for 10 min in buffered 2% SDS containing 8 M urea and 0.1 M dithiothreitol. Equal portions of each sample across the gradient were analyzed on 10% polyacrylamide SDS gels (shown in the lower panel beneath the equivalent pooled fractions). The 5% polyacrylamide stacker gel at the top is retained to demonstrate the cross-linked protein not entering the separation gel. Molecular-weight marker proteins in the first and last slots are *(top to bottom)* β-galactosidase (130,000); phosphorylase a (92,500); serum albumin (68,000); ovalbumin (43,000); glyceraldehyde-3-phosphate dehydrogenase (36,000); and globin (15,500). Lactoperoxidase appears in pooled fraction 2 as a polypeptide doublet of molecular weight ~78,000.

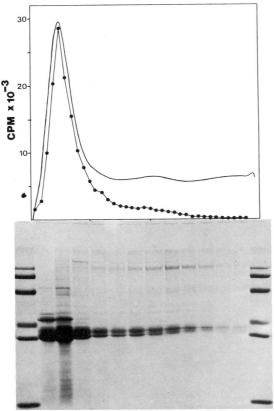

Figure 5. RNA-labeled 30S RNP centrifuged on sucrose gradients containing 0.5 M NaCl. Mouse ascites cells were pulse-labeled for 30 min in vitro with 15 μCi/ml [³H]uridine (Quinlan et al. 1974). Nuclear extracts were made and centrifuged on sucrose gradients in STM, pH 8 (Martin et al. 1974). The 30S RNPs recovered were pelleted overnight and resuspended in STM, pH 8. After the final salt concentration was adjusted to 0.5 M NaCl, the particle suspension was recentrifuged on sucrose gradients containing 0.5 M NaCl as in Fig. 4. (●——●) One-milliliter fractions were recovered and 0.1-ml aliquots were assayed for acid precipitable radioactivity; (——) UV absorption at 254 nm. The distribution of protein across the gradient is shown in the lower panel. Remaining fractions were pooled (1–3, 4–6, etc.), precipitated on ice with TCA, and dissolved in 2% SDS buffer. Equal volumes were assayed on SDS-polyacrylamide gels with molecular-weight markers in the first and last slots as described in Fig. 4.

in the stacking gel or at the interface of the stacking and separation gels (Fig. 4). Dimer and higher-order aggregates of the 35,000–40,000-dalton polypeptides are visible on the gel. By autoradiography or protein stain, it can be estimated that more than 90% of the protein, which is stable to high salt and is recovered in the 30S region, cannot be dissociated to complexes small enough to migrate into a 10% polyacrylamide gel even after prolonged boiling in 2% SDS, 0.1 M dithiothreitol, and 8 M urea. The relatively small amount of protein that is dissociated from iodinated 30S RNP on high-salt sucrose gradients fails to show this extensive aggregation or cross-linking (Fig. 4).

Although the exact nature of the protein-protein

cross-linking induced by the iodination conditions is still under investigation, its mediation by free radicals appears likely since they have been implicated in peroxidase-H_2O_2 reactions including iodination (Gross and Sizer 1959; Yamazaki et al. 1960; Yip and Hadley 1967). The demonstration that diphenyl dimers may be formed under similar reaction conditions (Gross and Sizer 1959) suggests that if the steric arrangement of tyrosine residues in the RNP-protein complexes allows sufficiently close approach, stable dityrosyl bridges may be formed. Dityrosine bridges may also be produced by UV-irradiation (Lehrer and Fasman 1967), and such cross-linking has been used to advantage in probing histone-histone binding sites in nucleosomes (Martinson and McCarthy 1976).

Finally, we have examined the fate of the pulse-labeled RNA of 30S RNP in which the proteins have been cross-linked by iodination to form a salt-stable complex. After determining that the iodination procedure did not affect the RNA remaining bound to cross-linked particles on gradients containing low salt, we monitored the distribution of labeled RNA on high-salt sucrose gradients. As found with un-cross-linked 30S RNP on identical gradients, most of the RNA label is dissociated from the 30S region where the cross-linked protein cores were recovered (data not shown). These data suggest that most if not all RNA binding sites are exposed on the surface of the protein core and may be easily disrupted by moderate changes in the ionic strength without the necessity of completely unfolding the particle for release. Whether a surface location of RNA binding sites reflects an in vivo folding of RNP fibrils after protein is bound, bringing protein-protein interaction sites into apposition, or whether preformed protein complexes bind RNA, perhaps with additional protein associated with the interparticle RNase-sensitive regions, remains to be determined. Although the salt stability of the iodinated protein core (informofer) appears to be an artifact of the in vitro labeling, the ability to zero-length cross-link a large proportion of the 20 or so component 35,000–40,000-dalton polypeptides of the 30S complex in this manner indicates a very close approach of these polypeptides in the native structure. The intimate protein-protein interactions responsible undoubtedly contribute substantially to the conformation and integrity of the RNP complex.

Specificity of the Interaction of Nucleic Acid Sequence Classes with RNP-Proteins

As mentioned earlier, there is evidence that a large part of the RNA of 30S RNP complexes is on the surface of the particles, at least in an operational sense. In addition, there appear to be unoccupied sites for RNA binding on the RNP as isolated from sucrose gradients. The binding of added purified RNA has previously been shown to exhibit a limited degree of specificity (Samarina et al. 1967a,b;

Martin et al. 1974). The capacity to bind RNA is increased greatly if the salt concentration is raised to a point where it appears that the complex dissociates (see preceding section), the RNA added, then the salt concentration lowered. In the case of direct addition and in the latter "reconstitution" experiments, the bound RNA sediments in 30S RNP form. Although the 30S RNPs have been shown to have low affinity for mature ribosomal RNA and tRNA, they bind pulse-labeled hnRNA, cytoplasmic poly(A)⁺ mRNA, and histone mRNA quite avidly (Martin et al. 1974). We have recently attempted to extend the analysis of RNA sequence affinities to 30S RNP proteins in order to understand more precisely the RNA-protein interaction and to assess which regions of the large hnRNA molecules would be expected to be tightly bound and which would be exposed from the surface or between the 30S subcomplexes.

Since we have shown that the poly(A) segment of nuclear RNA is bound in vivo in a 15S subcomplex form having a completely different polypeptide composition to the 30S RNP (Quinlan et al. 1974), we have sought in vitro evidence of this simple difference in binding specificity. Experiments in which the binding of poly(A) to 30S RNP is compared with that of other RNA species has revealed a much lower affinity than that of poly(U), hnRNA, or mRNA (results not shown). More dramatic evidence of the selectivity of binding is obtained when [³H]adenosine-labeled RNA is extracted and purified from 15S and 30S RNP and then each is incubated in tracer amounts with fresh nuclei under normal RNP extraction conditions. It is found that the poly(A) of 15S RNP is bound once more into a 15S form, while the 30S RNP-RNA sediments at 30S (Fig. 6). Thus a high degree of selectivity of this simple kind is apparent under these competitive conditions. It is not known at present whether the relatively low affinity of 30S RNP-proteins for homopolymer(A) will affect the orientation of hnRNA-oligo(A) sequences in the complexes. In this regard, however, it is worth noting that approximately 45% of the nuclear oligo(A) of the mouse ascites cells is found in 30S RNP form, a somewhat lower fraction than the value (70%) for extractable hnRNA sequences in general (Kinniburgh et al. 1976).

In considering other possible sequences in hnRNA which may be expected to have potential significance in nuclear functions, double-stranded (or at least complementary) RNA sequences emerge as strong candidates. Double-stranded RNA (dsRNA) regions have been shown to be sites of ribosomal and messenger RNA processing in prokaryotes (Dunn and Studier 1973; Robertson et al. 1977), and a similar role has been suggested for the cleavage of mRNA from eukaryotic hnRNA (Darnell 1976; Ryskov et al. 1976). We have therefore analyzed the distribution of dsRNA regions in sucrose gradients of the nuclear extracts containing 30S RNP complexes. In these experiments dsRNA is determined by resistance to

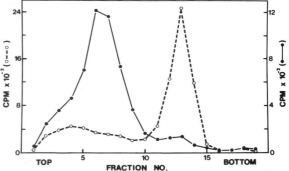

Figure 6. Specificity of association of 15S and 30S RNP-RNAs during nuclear extraction. Cells from 5 ml of mouse ascites fluid were labeled in vitro with 20 μCi/ml [³H]adenosine for 35 min at 37°C, and a nuclear extract was prepared and centrifuged as in Quinlan et al. (1974). The 15S and 30S regions of the sucrose gradient were extracted with chloroform-phenol-SDS and precipitated with yeast tRNA; the resulting purified RNP-RNAs were each dissolved in 0.5 ml of distilled water. Unlabeled nuclei were isolated from 10 ml of ascites fluid, divided into two equal fractions, and each fraction resuspended in 1.7 ml STM, pH 9. Each fraction was immediately mixed with 0.25 ml of either [³H]RNP-RNA (10S–20S) or [³H]RNP-RNA (30S) and extracted for 4 hr at 0°C. The nuclear extracts were centrifuged through identical 15–30% (w/v) sucrose gradients in STM (pH 8) in a SW 27 rotor for 17 hr at 26,000 rpm, 5°C, and each gradient was fractionated into 19 2-ml samples. Aliquots (1 ml) of each fraction were diluted with 20 mM KCl buffer and passed through HAWP Millipore filters. Both 15S RNP-RNA (●——●) and 30S RNP-RNA (○- - -○) were mixed with isolated nuclei during extraction (Quinlan 1976).

pancreatic, T₁, and T₂ RNases and subsequent electrophoresis on polyacrylamide gels. The relative distributions of total RNA pulse-labeled with [³H]adenosine and of dsRNA are shown in Figure 7. As expected, the total pulse-labeled RNA has the characteristic peak at 30S with some radioactivity in the region of the 15S poly(A) complex. The dsRNA, amounting to 0.85% of the total labeled RNA, has a distinctly different distribution. Although there is some dsRNA in the RNP region of the gradient, there is no peak at 30S, and although 1.5% and 1.25% of the labeled RNA in the 10S and 40S fractions of the gradient are scored as dsRNA, only 0.4% of the 30S fraction appears double-stranded. Essentially similar results have been obtained using [³²P]phosphate as label. We do not know at present whether the dsRNA found in our nuclear extracts is protein-associated, although the presence of significant amounts sedimenting at 40S suggests that at least part of these sequences must be in ribonucleoprotein form. However, the relative exclusion of pulse-labeled dsRNA from the 30S RNP subcomplex is obvious; the large proportion of dsRNA at the top of gradients of nuclear extracts might therefore arise in part from the cleavage of these sequences from the nuclease-sensitive interparticle regions of large hnRNP.

To obtain an assessment of the affinity of 30S RNP for double-stranded nucleic acids in cell-free sys-

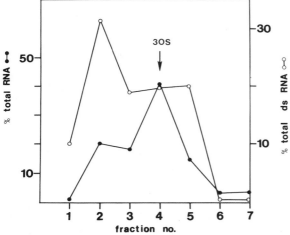

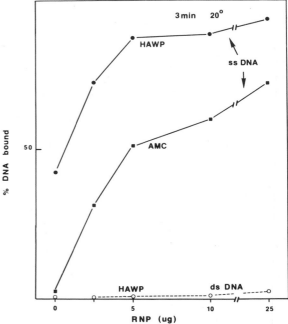

Figure 7. The distribution of total pulse-labeled and double-stranded (ds) RNA in a nuclear RNP extract. The nuclear RNP extract was prepared from cells labeled for 40 min with [³H]adenosine (Quinlan et al. 1974) and centrifuged on a 15–30% sucrose gradient. The gradient was fractionated and RNA prepared from each fraction by chloroform-phenol-SDS extraction and ethanol precipitation. Total radioactivity was analyzed by TCA precipitation of one-tenth of each gradient fraction. The remaining part of each fraction was treated with 2 μg/ml pancreatic, 10 units/ml T₁, and 2 units/ml T₂ RNases for 30 min at 37°C in 2 × SSC. The digested RNA was purified by chloroform-phenol-SDS extraction and ethanol precipitation. Each RNA sample was run on a 12% polyacrylamide gel, and the radioactivity in the 40–200 nucleotide region of the gel was summed to estimate the amount of dsRNA present.

tems, we have resorted to the simple expedient of using nucleosome-size double-stranded (ds) and single-stranded (ss) DNA. The reasons for this choice, apart from the simplicity of producing the nucleic acid probe, are that nucleosome DNA provides a perfectly paired double-stranded molecule of distinct size (ca. 180 nucleotide pairs in the case of our mouse DNA), having few, if any, single-strand nicks, and being completely insensitive to the low levels of RNase activity retained by 30S RNP even after extensive purification. The size homogeneity of this nucleic acid is particularly important since the length parameter is a significant factor in the binding of RNA to 30S RNP. The double-stranded length of the dsDNA is also within the range observed for dsRNA regions in hnRNA (Fedoroff et al. 1977).

We have examined the binding of the dsDNA and ssDNA to 30S RNP both by sucrose gradient analysis and by a filter-binding assay. The results of the latter, shown in Figure 8, indicate clearly that under conditions where 30S RNP can bind essentially all of the added ssDNA, there is barely detectable binding of dsDNA. Under conditions of maximal binding, the ssDNA cosediments with the 30S RNP complex (results not shown). Taken together with the previous observations of the failure of 30S RNP-proteins to bind tRNA, the results reported here would imply that double-stranded RNA regions, if they exist as such, in hnRNA would not form part of the compact 30S RNP regions of the hnRNP fiber and would be

Figure 8. Binding of double-stranded and single-stranded nucleosome-size DNA to 30S RNP complexes. Mouse ascites cell DNA was labeled with [³H]thymidine by a series of six injections of 100 μCi into a tumor-bearing animal over a 48-hr period prior to harvesting cells. Mononucleosomes were prepared by incubation of the cell nuclei with micrococcal nuclease and fractionation of the digested extract on sucrose gradients. Gel electrophoresis of the purified DNA (sp. act. 12,000 cpm/μg) from these nucleosomes indicated a single-strand size of 180 nucleotides. Increasing amounts of purified 30S RNPs were added to 1 μg of dsDNA or ssDNA in 1 ml of STM (pH 8), and after 3 min at 20°C the complexes were collected and washed on Millipore (HAWP) or Amicon (AMC) filters (Martin et al. 1974; Quinlan et al. 1974). The HAWP filters have high backgrounds (ca. 30–40%) for ssDNA in contrast to the AMC filters; both exhibit negligible binding of nucleosome-size dsDNA in STM, pH 8 (0.1 M NaCl). Sucrose gradient analysis showed that ssDNA was bound to 30S structures.

exposed to potential interaction with other proteins including processing enzymes.

Ribonuclease-resistant Regions in 30S RNP

There are at least three distinct levels of RNase sensitivity that can be defined in the hnRNP fiber of mammalian cells. The first, and most sensitive, includes the regions of the fiber not associated with the 30S RNP subcomplexes and presumably lying between them. Unless considerable precautions are taken (Samarina et al. 1968), endogenous nucleases cleave at these sites. However, the 30S complex itself appears to contain nuclease-sensitive and -insensitive regions. When pulse-labeled 30S RNPs are incubated at 24°C with increasing concentrations of pancreatic RNase, there is a rapid loss of labeled RNA even at low levels of enzyme (Fig. 9). At concentrations above 0.5 μg/ml, approximately 75% of the pulse label is lost, but this value does not increase

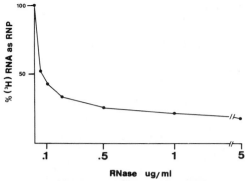

Figure 9. Ribonuclease resistance of 30S RNP complex RNA. The 30S RNPs were prepared from 40-min [³H]adenosine-labeled mouse ascites cells as previously described (Martin and McCarthy 1972). Aliquots of the isolated 30S RNPs were incubated with varying concentrations of pancreatic RNase for 10 min at 24°C. The samples were chilled on ice and filtered through HAWP Millipore filters to bind protein-associated RNA (Quinlan et al. 1974). Similar digestion patterns were obtained with [³H]uridine as label.

substantially with a tenfold increase in RNase (whether the label is in adenosine or uridine). Thus it would appear that about 20% of 30S RNP-RNA is reasonably well protected by its association with protein since deproteinized RNA from the complex does not share this nuclease resistance (Quinlan et al. 1974, 1977). It will be necessary to further define the basis of this protection of a limited region of RNP-RNA before attempting to assess its role in the structure of the 30S complex or in hnRNA processing. In particular, we need to know if there is sequence specificity to the protection and whether we can induce the protection of regions of exogenous RNA in dissociation-reconstitution experiments. Our results thus far cannot exclude the possibility that at least a short segment (ca. 100 nucleotides) of the hnRNA binds tightly to protein subunits and is necessary to the in vivo assembly of the 30S subcomplex, the RNA segment becoming "internalized" in the process. However, the observations with crosslinked protein complexes (described above) indicate that these at least can be obtained essentially free of any "structural" RNA.

SUMMARY

Update on a Working Model for hnRNP

It is certainly premature to propose a definitive model for nuclear ribonucleoprotein complexes containing hnRNA and pre-mRNA. Biochemical studies have necessarily focused on the more prominent nuclear species. In view of the fact that the 30S RNP form contains a large proportion of nucleus-restricted hnRNA sequences (Kinniburgh and Martin 1976a), as well as the failure to detect 30S RNP-proteins associated with cytoplasmic mRNA (Lukanidin et al. 1972a), it is apparent that the true transport complex for mRNA remains elusive. We may assume the complexes that we and most others have studied derive from early processing stages of hnRNA. If a definitive description of the structure and function of the hnRNP complexes cannot be given at the present time, it seems worthwhile to attempt to collate the available circumstantial evidence into a summarizing schematic model, which may at least serve as an Aunt Sally for future experiments. We have attempted such a rationalization for our own data (Fig. 10).

We consider that as the hnRNA strand is synthesized it becomes associated with the simple set of 30S RNP-polypeptides either in the form of an intact large protein complex or as small subunits which fold the RNA strand into compact 30S subcomplexes (see models in Fig. 3). The process of reutilization of the RNP-proteins upon turnover or transport of the RNA may involve modifications of the polypeptides, such as phosphorylation (Martin et al. 1974). As discussed above, the choice between the two major alternatives for the in vivo association between hnRNA and 30S RNP-proteins remains open. There is now no evidence for a salt-stable protein core resembling the histone complex of nucleosomes (Weintraub et al. 1975; other papers, this volume); however, as we have shown, intimate protein-protein contacts are involved in maintaining the integrity of the complex. Our experiments give no strong evidence for a structural role of RNA, although they do not completely exclude the possibility at this point.

Figure 10. Simple hypothetical model for the structure of hnRNP fibers, indicating the 30S RNP substructure and the presumed relationship of identifiable RNA sequences to the 30S particles. Oligo(A) and double-stranded regions are shown as is the poly(A) of hnRNA which is bound in a 15S subunit containing a completely distinct group of proteins from the simple set of 35,000–40,000-dalton polypeptides which, as a multimer, form the 30S complex. The model is an extension of previous proposals (Samarina et al. 1968; Malcolm and Sommerville 1974; Kinniburgh et al. 1976).

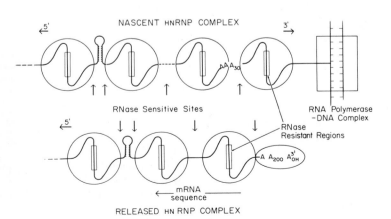

Instead, the data suggest that most of the RNA is on the surface of the 30S complex, being relatively exposed to exogenous RNase attack, which leads to aggregation of the particles either by removal of surface charge or by exposing the protein core to further possibilities for protein-protein interaction. The more RNase-resistant 20% of pulse-labeled RNA demands further attention with regard to sequence specificity and the nature of its distinctive interaction with the RNP-proteins. Although it is clear that as isolated 30S RNPs have additional surface binding sites for RNA, we have not yet determined whether we can induce RNase resistance in added RNA in "reconstituted" particles.

The binding of added RNA to 30S RNP has shed some light on the relative affinities of the proteins for different RNA species. The in vivo binding of poly(A) into a 15S subcomplex and the bulk of the rest of hnRNA sequence in 30S RNP is reflected in cell-free binding experiments (Fig. 6). The relatively low affinities for homopolymer(A) and particularly for double-stranded nucleic acid (Figs. 7 and 8) are taken as suggestive evidence that these sequences would be partly or wholly exposed on or between the 30S subcomplexes of the hnRNP fiber. Thus they would be available as binding sites for other proteins, possibly including specific processing enzymes. It has been suggested that double-stranded RNA may serve as specific cleavage sites in hnRNA (Darnell 1976; Ryskov et al. 1976) and that some mRNA molecules may arise from regions adjacent to oligo(A) (Edmonds et al. 1976; Kinniburgh and Martin 1976b). Some among the heterogeneous population of proteins reported to be present on large hnRNP complexes (Pederson 1974; Stevenin and Jacob 1974) may attach to such exposed sites between the particulate 30S substructures which, as we have pointed out, are multimers of a simple set of 35,000–40,000-dalton polypeptides. Although studies involving putative processing enzymes using naked RNA molecules are perfectly valid, we consider it likely that the duplication of in vivo specificity may require the RNA to be in RNP form.

Clearly, the most important aspects of such simple model building are that it should present to others the potential significance of nuclear proteins as participants in the posttranscriptional regulation of gene expression in eukaryotes and that the inherent inadequacies should send its authors back into the laboratory.

Acknowledgments

We thank Ms. Ljerka Urbas for excellent technical assistance which greatly facilitated the research described. A. K. and P. B. are recipients of U.S. Public Health Service Training Grant HD-00174. J. P. is a student of the Department of Biophysics, University of Chicago, and is supported by the U.S. Public Health Service Medical Scientist Training Program (TO5-GM-01939). The research was funded by U.S. Public Health Service Grant CA-12550 and the University of Chicago Cancer Research Center Grant CA-19265, Project 510.

REFERENCES

BILLINGS, P. B. and T. E. MARTIN. 1977. Proteins of nuclear ribonucleoprotein subcomplexes. *Methods Cell Biol.* **17**: (in press).

DARNELL, J. E. 1976. mRNA structure and function. *Prog. Nucleic Acid Res. Mol. Biol.* **19**: 493.

DUNN, J. J. and F. W. STUDIER. 1973. T7 early RNAs and *Escherichia coli* ribosomal RNAs are cut from large precursor RNAs in vivo by ribonuclease III. *Proc. Natl. Acad. Sci.* **70**: 3296.

EDMONDS, M., H. NAKAZATO, E. L. KORWEK, and S. VENKATESAN. 1976. Transcribed oligonucleotide sequences in HeLa cell hnRNA and mRNA. *Prog. Nucleic Acid Res. Mol. Biol.* **19**: 99.

FEDOROFF, N., P. K. WELLAUER, and R. WALL. 1977. Intermolecular duplexes in heterogeneous nuclear RNA from HeLa cells. *Cell* **10**: 597.

GROSS, A. J. and I. W. SIZER. 1959. The oxidation of tyramine, tyrosine, and related compounds by peroxidase. *J. Biol. Chem.* **234**: 1611.

HUXLEY, H. E. and G. ZUBAY. 1960. Electron microscope observations of the microsomal particles from *Escherichia coli. J. Mol. Biol.* **2**: 10.

KINNIBURGH, A. J. and T. E. MARTIN. 1976a. Detection of mRNA sequences in nuclear 30S ribonucleoprotein subcomplexes. *Proc. Natl. Acad. Sci.* **73**: 2725.

———. 1976b. Oligo(A) and oligo(A)-adjacent sequences present in nuclear ribonucleoprotein complexes. *Biochem. Biophys. Res. Commun.* **73**: 718.

KINNIBURGH, A. J., P. B. BILLINGS, T. J. QUINLAN, and T. E. MARTIN. 1976. Distribution of hnRNA and mRNA sequences in nuclear ribonucleoprotein complexes. *Prog. Nucleic Acid Res. Mol. Biol.* **19**: 335.

KORNBERG, R. D. 1974. Chromatin structure: A repeating unit of histones and DNA. *Science* **184**: 868.

LAIRD, C. D., L. E. WILKINSON, V. E. FOE, and W. Y. CHOOI. 1976. Analysis of chromatin-associated fiber arrays. *Chromosoma* **58**: 169.

LEHRER, S. S. and G. D. FASMAN. 1967. Ultraviolet irradiation effects in poly-L-tyrosine and model compounds. Identification of bityrosine as a photo-product. *Biochemistry* **6**: 757.

LUKANIDIN, E. M., S. OLSNES, and A. PIHL. 1972a. Antigenic difference between informofers and protein bound to polyribosomal mRNA from rat liver. *Nat. New Biol.* **240**: 90.

LUKANIDIN, E. M., E. S. ZALMANZON, L. KOMAROMI, O. P. SAMARINA, and G. P. GEORGIEV. 1972b. Structure and function of informofers. *Nat. New Biol.* **238**: 193.

MALCOLM, D. B. and J. SOMMERVILLE. 1974. The structure of chromosome-derived ribonucleoprotein in oocytes of *Triturus cristatus carnifex* (Laurenti). *Chromosoma* **48**: 137.

MARCHALONIS, J. J. 1969. An enzymatic method for the trace iodination of immunoglobulins and other proteins. *Biochem. J.* **113**: 299.

MARTIN, T. E. and B. J. McCARTHY. 1972. Synthesis and turnover of RNA in the 30S nuclear ribonucleoprotein complexes of mouse ascites cells. *Biochim. Biophys. Acta* **277**: 354.

MARTIN, T., P. BILLINGS, A. LEVEY, S. OZARSLAN, T. QUINLAN, H. SWIFT, and L. URBAS. 1974. Some properties of RNA:Protein complexes from the nucleus of eukaryotic cells. *Cold Spring Harbor Symp. Quant. Biol.* **38**: 921.

MARTINSON, H. G. and B. J. McCARTHY. 1976. Histone-histone interactions within chromatin. Preliminary characterization of presumptive H2B-H2A and H2B-H4 binding sites. *Biochemistry* **15**: 4126.

MILLER, O. L., JR. and A. H. BAKKEN. 1972. Morphological studies of transcription. *Karolinska Symp. Res. Methods Reprod. Endocrinol.* **5**: 155.

PEDERSON, T. 1974. Proteins associated with heterogeneous nuclear RNA in eukaryotic cells. *J. Mol. Biol.* **83**: 163.

QUINLAN, T. J. 1976. "Structure and metabolism of nuclear polyadenylate-protein complexes from mouse ascites cells." Ph.D. thesis, University of Chicago, Illinois.

QUINLAN, T. J., P. B. BILLINGS, and T. E. MARTIN. 1974. Nuclear ribonucleoprotein complexes containing polyadenylate from mouse ascites cells. *Proc. Natl. Acad. Sci.* **71**: 2632.

QUINLAN, T. J., A. J. KINNIBURGH, and T. E. MARTIN. 1977. Properties of a nuclear polyadenylate-protein complex from mouse ascites cells. *J. Biol. Chem.* **252**: 1156.

ROBERTSON, H. D., E. DICKSON, and J. J. DUNN. 1977. A nucleotide sequence from a ribonuclease III processing site in bacteriophage T7 RNA. *Proc. Natl. Acad. Sci.* **74**: 822.

RYSKOV, A. P., O. V. TOKARSKAYA, G. P. GEORGIEV, C. COUTELLE, and B. THIELE. 1976. Globin mRNA contains a sequence complementary to double-stranded region of nuclear pre-mRNA. *Nucleic Acids Res.* **3**: 1487.

SAMARINA, O. P., A. A. KRICHEVSKAYA, and G. P. GEORGIEV. 1966. Nuclear ribonucleoprotein particles containing messenger ribonucleic acid. *Nature* **210**: 1319.

_____. 1967a. Nuclear ribonucleoproteins containing messenger RNA II. Interaction with free mRNA and ribosomes. *Mol. Biol.* **1**: 565.

SAMARINA, O. P., E. M. LUKANIDIN, J. MOLNAR, and G. P. GEORGIEV. 1968. Structural organization of nuclear complexes containing DNA-like RNA. *J. Mol. Biol.* **33**: 251.

SAMARINA, O. P., J. MOLNAR, A. A. KRICHEVSKAYA, E. M. LUKANIDIN, V. I. BRUSKOV, and G. P. GEORGIEV. 1967b. Nuclear ribonucleoproteins containing messenger RNA III. Dissociation of particles to RNA and protein and their subsequent reconstruction. *Mol. Biol.* **1**: 648.

STEVENIN, J. and M. JACOB. 1974. Effects of sodium chloride and pancreatic ribonuclease on the rat-brain nuclear particles: The fate of the protein moiety. *Eur. J. Biochem.* **47**: 129.

VASILIEV, V. D. 1971. Electron microscopy study of 70S ribosomes of *Escherichia coli. FEBS Lett.* **14**: 203.

WEINTRAUB, H., K. PALTER, and R. VAN LENTE. 1975. Histones H2a, H2b, H3, and H4 form a tetrameric complex in solutions of high salt. *Cell* **6**: 85.

YAMAZAKI, I., H. S. MASON, and L. PIETTE. 1960. Identification, by electron paramagnetic resonance spectroscopy, of free radicals generated from substrates by peroxidase. *J. Biol. Chem.* **235**: 2444.

YIP, C. C. and L. D. HADLEY. 1967. Involvement of free radicals in the iodination of tyrosine and thyroglobulin by myeloperoxidase and a purified beef thyroid peroxidase. *Arch. Biochem. Biophys.* **120**: 533.

Free Informofers and Reconstitution of 30S RNP

V. V. Kulguskin, E. M. Lukanidin, and G. P. Georgiev

Institute of Molecular Biology, Academy of Sciences of the USSR, Moscow B-312, USSR

As early as 1968 it was proposed that the nuclear ribonucleoprotein (RNP) complexes containing pre-mRNA possess the beads-on-a-string structure with pre-mRNA wrapped on the surface of the globular protein particles which we call "informofers" (Samarina et al. 1968). The existence of a repeating structural unit was proved directly (Samarina et al. 1968), whereas only some indirect evidence in favor of surface localization of the RNA in the particle was obtained (Georgiev and Samarina 1971). In this paper we present some data confirming the latter statement (these data were obtained by V. V. Kulguskin and E. M. Lukanidin in our laboratory).

Previously we found that the treatment of nuclear 30S particles containing pre-mRNA with 2 M NaCl led to the dissociation of RNA and protein (Samarina et al. 1967). When the particles were subjected to an iodination procedure, most of the labeled protein was recovered in the 30S peak upon ultracentrifugation in a sucrose gradient containing 2 M NaCl. Thus the informofer survived the removal of RNA as a protein particle (Lukanidin 1971, 1973). However,

in experiments where iodination had been omitted, most of the informofers were dissociated into smaller subunits, "informatin" molecules (~4S). The dissociation was found to be concentration-dependent: at low concentrations of protein, only the 4S peak was recovered (Fig. 1a). If the concentration of protein was high enough, a significant proportion of the protein sedimented as 30S particles (Fig. 1c). Special experiments showed that informofer dissociation is a reversible reaction. The conditions which occur during iodination procedures in some way stabilize informofer against dissociation. Similar stabilization also occurs upon the aging of the particles in solution; therefore, the informofers do not dissociate completely into subunits even when diluted. The protein compositions of the 30S RNP of informofers and of dissociated informofers are virtually identical: most of the material is contained in two polypeptides with close apparent molecular weights (38,000 and 41,000 daltons) (Fig. 2). Thus one can study the reconstitution of 30S RNP particles from RNA and either informofers or dissociated material (informa-

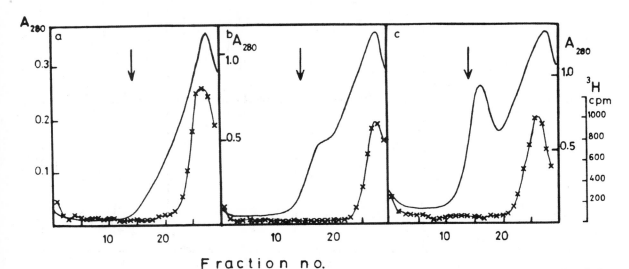

Figure 1. Ultracentrifugation in 5–25% sucrose gradient (in 2 M NaCl) of the material of 30S RNP treated with 2 M NaCl at different concentrations of the protein. The 30S RNPs isolated by sucrose density gradient fractionation were concentrated according to the procedure developed by A. Krichevskaya in our laboratory and processed further as described by Kulguskin (1977). The 30S particles were dissolved in 0.15 M NaCl, 0.01 M EDTA, 0.05 M Tris-HCl, pH 8.5, and then dialyzed against 2 M NaCl; after removal of precipitated RNA, the protein solution was layered on the sucrose gradient prepared in 2 M NaCl, 0.001 M MgCl₂, 0.01 M Na₂HPO₄, pH 6.5. (a) ~2 mg protein/ml; (b) ~7 mg protein/ml; (c) ~16 mg protein/ml; (x——x) [³H]RNA; (——)OD₂₈₀. Vertical arrow indicates the position of a marker (fixed 30S RNP) in a sucrose gradient. Note that 30S protein particles appear upon an increase of the protein concentration in the initial sample.

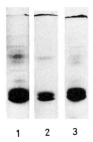

Figure 2. Protein compositions of the 30S RNP particles (gel 1), of informofers (gel 2), and of dissociated informofers (gel 3). Protein was electrophoresed in SDS-polyacrylamide gel according to the method of Laemmli (1970).

tin or its oligomers). The mixture of RNA and protein in 2 M NaCl was dialyzed against 0.1 M NaCl, 0.001 M $MgCl_2$, 0.004 M Na_2HPO_4, pH 7.0–7.2. In both cases, 30S RNP particles with a buoyant density of about 1.40–1.41 g/cm^3 were formed (Fig. 3a,b,e,f). In the electron microscope, particles approximately 200 Å in diameter could be observed.

However, the two types of particles are very different. On one hand, 30S RNPs obtained from RNA and informofers are similar to the original RNPs in that they are very sensitive to RNase action as

well as to repeated treatment with the concentrated salt solution (Fig. 3c,d). On the other hand, 30S RNPs reconstituted from dissociated material are resistant to high salt and to RNase treatment (Fig. 3g,h), being very different from the native RNPs. Finally, the reconstituted particles were repeatedly treated with 2 M NaCl, and the material with a sedimentation coefficient of about 30S was collected and ultracentrifuged in a CsCl density gradient. The particles obtained from RNP reconstituted from informofers and RNA had low density, typical of informofers free from RNA. In the case of particles reconstituted from RNA and informatin, the 30S material possessed a density of 1.4 g/cm^3, typical of undissociated RNP (Fig. 4).

One can conclude that, if only informofers (30S protein aggregates) are used for reconstitution, particles identical to the original ones are formed as a result of RNA wrapping on the informofer surface. When RNA interacts with informatin, the RNP strand is probably formed first and is then folded in such a way that at least a part of the RNA is covered with protein and protected against RNase and high-salt action.

Thus, reconstitution experiments gave additional proof to the idea that RNA in native particles is localized on the informofer surface.

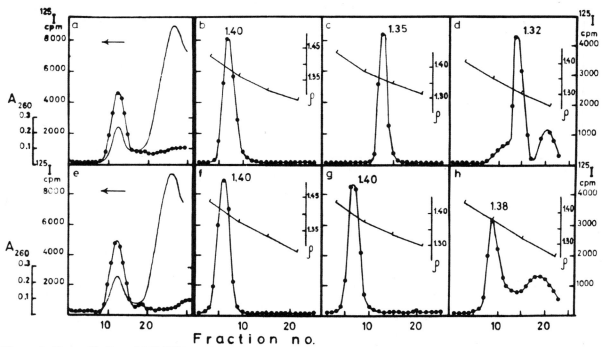

Figure 3. Reconstitution of 30S RNP particles from RNA and informofer *(a–d)* or from RNA and dissociated informofers *(e–h)*. Protein was labeled with ^{125}I after reconstitution according to the method of Kulguskin (1977). *(a,e)* Ultracentrifugation of reconstituted particles in a 5–20% sucrose gradient prepared in 0.1 M NaCl, 0.001 M $MgCl_2$, 0.004 M Na_2HPO_4, pH 7.0–7.2 (SW 25.2 rotor, 18 hr, 18000 rpm, 4°C). *(b,f)* Ultracentrifugation of reconstituted particles in a CsCl density gradient (after fixation with CH_2O). *(c,g)* Same as in *b* and *f*, but before fixation the particles were treated with RNase. In one case *(c)* the RNA in the particles is RNase-sensitive and as a result the buoyant density is lowered. *(d,h)* Same as in *b* and *f*, but the particles were not fixed with CH_2O. Again, in one case *(d)* the RNA was removed from the protein, whereas in another *(h)* it was not. (——) OD_{260}; (●——●) ^{125}I-labeled protein.

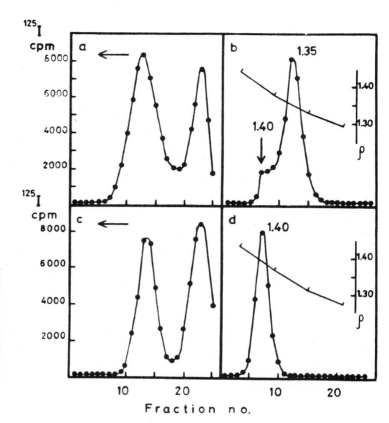

Figure 4. Reisolation of informofers from reconstituted 30S particles. The RNPs used were reconstituted from RNA and either informofers *(a,b)* or dissociated protein *(c,d)*. *(a,c)* The reconstituted RNPs were treated with 2 M NaCl and then centrifuged in a 5–25% sucrose gradient containing 2 M NaCl as in Fig. 1. *(b,d)* The 30S peak isolated in *a* and *c* was ultracentrifuged in a CsCl density gradient. The material with a buoyant density of 1.40 g/cm³ in *b* is the result of reconstitution of 30S RNP from RNA and dissociated informofers.

REFERENCES

GEORGIEV, G. P. and O. P. SAMARINA. 1971. D-RNA containing ribonucleoprotein particles. In *Advances in cell biology* (ed. D. M. Prescott et al.), vol. 2, p. 47. Meredith, New York.

KULGUSKIN, V. V. 1977. Nuclear ribonucleoprotein particles containing premessenger RNA. 12. Studies of the dissociation and reconstruction of 30S particles. *Mol. Biol.* **11**: 620.

LAEMMLI, U. K. 1970. Cleavage of structural proteins during the assembly of the head of bacteriophage T4. *Nature* **227**: 680.

LUKANIDIN, E. M., N. A. AITKHOZHINA, V. V. KULGUSKIN, and G. P. GEORGIEV. 1971. The isolation of informofers free from dRNA. *FEBS Lett.* **19**: 101.

LUKANIDIN, E. M., V. V. KULGUSKIN, N. A. AITKHOZHINA, L. KOMAROMI, A. S. TIKHONENKO, and G. P. GEORGIEV. 1973. Nuclear RNP particles containing premessenger RNA. 10. The isolation of informofers free from dRNA and their properties. *Mol. Biol.* **7**: 360.

SAMARINA, O. P., E. M. LUKANIDIN, J. MOLNAR, and G. P. GEORGIEV. 1968. Structural organization of nucleoprotein complexes containing DNA-like RNA. *J. Mol. Biol.* **33**: 251.

SAMARINA, O. P., J. MOLNAR, E. M. LUKANIDIN, V. I. BRUSKOV, A. A. KRICHEVSKAYA and G. P. GEORGIEV. 1967. Reversible dissociation of nuclear ribonucleoprotein particles containing mRNA into RNA and protein. *J. Mol. Biol.* **27**: 187.

The Cloning of Mouse Globin and Surrounding Gene Sequences in Bacteriophage λ

P. Leder, S. M. Tilghman, D. C. Tiemeier, F. I. Polsky, J. G. Seidman, M. H. Edgell, L. W. Enquist, A. Leder, and B. Norman

Laboratory of Molecular Genetics, National Institute of Child Health and Human Development, National Institutes of Health, Bethesda, Maryland 20014

We have developed a rather straightforward procedure that brings virtually any segment of the mammalian genome within cloning range of a very versatile EK2 bacteriophage λ vector. We illustrate the procedure in terms of cloning a segment of the mouse genome containing a β-like globin gene and its surrounding sequences. Surprisingly, the sequence cloned does not contain a continuous representation of the globin messenger RNA (mRNA) sequence, but instead is interrupted about two-thirds through by an approximately 550-base-long *intervening sequence*. The significance of this structure and its possible generality are discussed together with the notion that single polypeptide chains may be encoded by discrete and separate gene sequences.

The Technical Problem and Its Solution: Purification of Globin Genes

The essential difference between cloning specific segments of a mammalian genome and those of a prokaryotic genome is the 1000-fold greater complexity of the former. When we began this work 3 years ago, we calculated that the available screening techniques would make it possible, though uncomfortable, to screen through several thousand cloned fragments in order to find a sequence of interest. Inasmuch as the mammalian genome is divided into approximately one million fragments by the *Escherichia coli* restriction endonuclease, *Eco*RI, this would require us to purify a given fragment approximately 1000-fold prior to cloning. With this degree of purification in mind, we have adopted two procedures which meet our requirements for high capacity and resolution.

The first dimension is RPC-5, a reverse phase chromatographic system originally developed for the separation of tRNAs by Pearson et al. (1971). Hardies and Wells (1976) and Landy et al. (1976) have shown that this technique can be easily adapted for the purpose of separating restriction fragments of bacteriophage λ DNA. We have applied this medium for the resolution of *Eco*RI fragments of total mouse genomic DNA — derived from the mouse plasmacytoma MOPC-149 — as shown in Figure 1 (Tiemeier et al. 1977). The DNA is eluted over approximately 150 fractions, each having the approximate degree

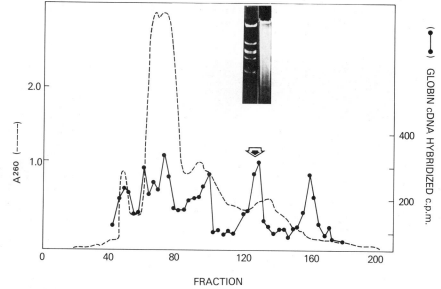

Figure 1. RPC-5 chromatography of *Eco*RI-digested mouse genomic DNA. An *Eco*RI digest of mouse genomic DNA was chromatographed on an RPC-5 column as described in Tilghman et al. (1977). (−−−)A₂₆₀; (●——●) S₁-resistant hybridization of ³²P-labeled globin cDNA. The insert above fractions 124–130 illustrates an ethidium-bromide-stained 1% analytical agarose gel of an *Eco*RI digest of 0.5 μg of wild-type λcI857 DNA (left) (where migration is from top to bottom and the kilobase pair sizes of individual fragments are 21.3, 7.35, 5.79, 5.40, 4.69, and 3.3 kb) and 2 μg of fraction 130 (right).

of complexity shown by the fraction illustrated in the inset in Figure 1. Each fraction is then assayed for the presence of globin gene sequences using hybridization to a mixture of α and β globin ³²P-labeled cDNAs. Several peaks were observed and that migrating at approximately fraction 130 was used for further purification.

As shown in the inset in Figure 1, the fraction containing a globin sequence represents a mixture of *Eco*RI fragments ranging in size from over 20,000 base pairs to less than 1000, which migrate essentially as a smear. Obviously, this size heterogeneity provides a logical basis for a second dimension for purification. To take advantage of it and to achieve the capacity and resolution required, a high-capacity, horizontal slab gel electrophoresis apparatus controlled with an electronic timer was designed (Polsky et al. 1977). With this device, the pooled fractions from the RPC-5 column can be highly enriched according to size, and the globin sequences localized as shown in Figure 2. As can be seen, the single peak fraction from RPC-5 contained two *Eco*RI fragments which hybridized to globin cDNA, one approximately 7000 bases long and the other approximately 15,000 bases long. A rough estimate of the purity of the lower-molecular-weight fraction suggested that the globin-containing sequence was represented approximately once per 2000 fragments (Tilghman et al. 1977). The larger fraction was of even greater purity.

The Vector System and Identification of Positive Clones

Not only did the degree of purification noted above meet our early requirements, but bacteriophage λ proved to be a far more powerful cloning vector than we had originally expected. The vector system used in these studies represents an EK2 derivative of the λgt system originally constructed by Thomas et al. (1974). We have introduced a phenotypically inert fragment of λ DNA (indicated as the B fragment

in Fig. 3), as well as certain mutations designed to meet the guidelines established by the National Institutes of Health Advisory Committee on Recombinant DNA Research. As presently constituted, this λ vector offers both a positive selection for the insertion of a fragment of DNA and a negative selection against the reformation of the parental-type phage (Tiemeier et al. 1977; Leder et al. 1977). The reason for this is that the left and right arms of the phage contain all the genes necessary for lytic phage growth, but lack sufficient DNA for efficient packaging of viral phage particles. This provides a positive selection in that only those phage which have incorporated a foreign fragment of DNA will grow. On the other hand, parental-type recombinants can be selected against by either purifying the left and right arms by RPC-5 chromatography (Tiemeier et al. 1977) or by utilizing endonuclease *Sst*I sites located only in this fragment (cf. Fig. 3). When the vector is digested with both *Eco*RI and *Sst*I, the two cloning arms are freed of the central B fragment which is destroyed by *Sst*I, thus providing a negative selection against parental-type recombinants.

We estimated that one in several thousand of the purified fragments used for cloning would contain globin sequences. Although we originally used several methods for detecting positive clones, by far the most efficient was that recently described by Benton and Davis (1977). This technique allows the simple screening of literally tens of thousands of clones, provided a radioactive hybridization probe is available for their detection. Using this technique, we screened several plates in order to detect several positive clones as illustrated in Figure 4. These were then plaque-purified and preparatively grown for further characterization.

Characterization of the Globin-gene-containing Sequence

To determine whether the sequence cloned was related to an α- or β-globin sequence, positive hybrid

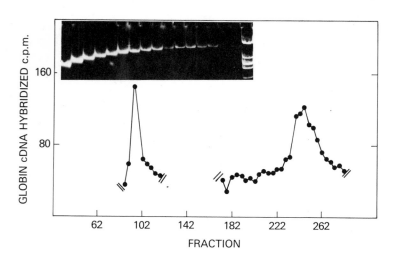

Figure 2. Detection of globin-containing fragments obtained from a preparative discontinuous electroelution gel. Approximately 2 mg of DNA from fractions 124–130 of the RPC-5 chromatography run (Fig. 1) were electrophoresed through a preparative 1% agarose discontinuous electroelution gel (Polsky et al. 1977). (●——●) S_1-resistant hybridization of pooled fractions to ³²P-labeled globin cDNA. The insert contains an ethidium bromide stain of an analytical 1% agarose gel containing 50 μl of individual fractions which correspond to the values at the bottom of the diagram. Size markers are provided by an *Eco*RI digest of λcI857 as in Fig. 1.

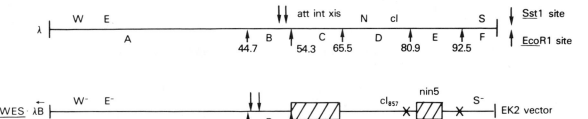

Figure 3. Modified λ phage suitable for cloning DNA from higher organisms. The lines represent the genome of phage λ. The length of the line drawn represents the full length of the genome of wild-type bacteriophage λ. Letters over each line refer to specific λ genes. Letters under each line refer to *Eco*RI restriction fragments of λ, with each arrow indicating an *Eco*RI site. The number under each arrow represents the position of the site as a percentage of the λ genome. Arrows over each line indicate the position of an *Sst*I site (note inversion of the *Eco*RI B fragment as compared to wild type). Scored boxes represent deleted portions of the λ genome. X represents the point at which an *Eco*RI site has been eliminated by mutation. Details of the construction of the two vectors have been described by Enquist et al. (1976) and Tiemeier et al. (1976).

phage were tested for their ability to anneal to plasmids containing the reverse transcript of either α or β mouse and rabbit globin sequences (Rougeon and Mach 1977; Efstratiadis et al. 1977). As shown in Figure 5, positive clones hybridized only to plasmids containing β-globin sequences. The congruity of hybridization (the match between the globin mRNA sequence and the cloned gene sequence) was further tested using resistance to deoxyribonuclease S_1 of a globin [32P]-labeled cDNA annealed to the cloned sequence (Table 1). The sequence contained in the clone was able to protect radioactively labeled globin cDNA to the same extent as the plasmid cloned sequence representing the reverse transcript of the β-globin messenger RNA.

A Surprising Structural Feature of the Cloned Globin Gene Sequence

An initial restriction map of the cloned sequence demonstrated that the approximately 7000-base-

long fragment of mouse DNA contained globin hybridizing sequences approximately 1.5 and 4.5 kilobases (kb) from either end of the fragment (Fig. 6). A restriction-site analysis of the cloned reverse transcript of the β-globin mRNA revealed a single *Bam*HI site in both mouse and rabbit globin sequences (Tilghman et al. 1977; Efstratiadis et al. 1977) corresponding to amino acids 98–100 of the β-globin protein. A single *Bam*HI site was also found to cleave the hybridizing β-globin sequence in this cloned segment of the genome. However, two additional enzymes, *Hin*dIII and *Sst*I, which *do not* cleave the cloned mRNA sequence, *do* split the globin hybridizing sequence, which implies that they occur within the globin coding region. There are several possible explanations for this puzzling observation. First, it is possible that the sequence cloned is different from that of the adult β-globin sequence,

PROBE

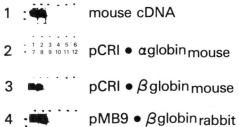

Figure 5. In situ hybridization of λgt*WES* · βG1 and λgt*WES* · βG2 to α- and β-containing chimeric plasmids. Replica filters were prepared from phage suspensions according to Kramer et al. (1976) and hybridized to the [32P]DNA probes as follows: (1) mouse α and β cDNA; (2) pcR1 · mouse α globin; (3) pcR1 · mouse β globin; (4) PMB9 · rabbit β globin. The individual grids, numbered in filter 2, contained: (1) λgt*WES* · βG2; (2) λgt*WES* · βG1; (3–8) λgt*WES* · mouse DNA hybrids which were negative in the original screening; (9) λgt*WES* · mouse rDNA (Tiemeier et al. 1977); (10–12) λgt*WES* · mouse DNA hybrids which contain reiterated sequences.

Figure 4. Identification of hybrid phage containing globin sequences. Plates containing 2000–3000 hybrid phage plaques were blotted and hybridized to [32P]-labeled globin cDNA essentially as described by Benton and Davis (1977). Plate 9 has eight positive phage; plate 12 has none.

Table 1. Hybridization of ^{32}P-labeled Globin cDNA to Hybrid Phage DNA

Hybrid DNA	Percentage hybridization (S$_1$-nuclease-resistant)
λgt*WES* · MβG1.0	45.3
λgt*WES* · MβG2.0	42.5
λgt*WES* · MC20 B4	4.2
λcI857	5.5
pCR1·α globin$_{mouse}$	43.1
pCR1·β globin$_{mouse}$	51.6

Hybridization of 0.1–0.2 μg of phage or plasmid DNA was performed in the presence of 1200 cpm ^{32}P-labeled globin cDNA as described by Tilghman et al. (1977).

for example, an embryonic sequence. Although this remains a possibility, another interpretation seems more likely in view of the electron microscopic data indicated below. Together, these data suggest that the structural globin gene sequence is *divided* by a segment of DNA that *does not* appear in the globin mRNA itself.

That this is indeed the case is indicated by an electron microscopic R-loop analysis (Thomas et al. 1976) of the type illustrated in Figure 7. Here we have taken mouse globin mRNA and annealed it to the cloned fragment. As shown in the figure, two R-loop structures may be visualized virtually adjacent to each other but separated by a looped-out, double-stranded region of DNA for which there is no displaced single-stranded DNA. This structure is seen in three independent isolates (nonsibling) of this 7-kb-long segment of the mouse genome, as well as in the separately cloned 15-kb fragment which also contains a mouse β-globin sequence (D. C. Tiemeier et al., in prep.). The correlation of

detailed electron microscopic analyses of these data together with that of the restriction map above suggests that the globin gene is divided into two coding segments which are separated by an intervening sequence of DNA approximately 550 bases long (Fig. 7) (S. Tilghman et al., in prep.).

Possible Interpretations of the New Finding

It is clear that the existence of the interrupted β-globin gene sequence can be interpreted in several ways. The possibility that this structure represents an artifact of the cloning procedure appears to be ruled out by the fact that three independently isolated, nonsibling clones contain the same R-loop structure. The possibility that this structure is a consequence of chromosomal DNA having been somehow transposed in the plasmacytoma from which the DNA was originally isolated appears less likely in view of our finding the same type of interrupted structure in a second β-globin gene encoded in the 15-kb segment of DNA which we have cloned (D. C. Tiemeier et al., in prep.).

None of the above possibilities allow us to conclude firmly that the sequence we have isolated is the same as that transcribed in globin-producing tissue. On the other hand, the agreed-upon length of a β-globin mRNA precursor, approximately 1600 bases (Curtis and Weissmann 1976; Ross 1976; Kwan et al. 1977; Bastos and Aviv 1977), is certainly long enough to accommodate the globin structural and intervening sequences. Although it is entirely possible that this intervening sequence plays some role in inactivating gene structures not destined to be expressed in a given cell line, this explanation requires more as-

Figure 6. Location of restriction endonuclease sites and globin sequences in λgt*WES* · MβG2. The top line represents the DNA genome of λgtWES · MβG2. The numbers refer to kilobase pairs from the left end of the hybrid. The 7000-base-long insert is illustrated as an open rectangle set off by vertical lines. The symbols in the left and right arms represent restriction enzyme sites: (⊘) *Bam*HI; (↓) *Bgl*II; (↑) *Hind*III; (▽) *Hpa*I; (▼) *Sst*I. Note that there are additional *Hpa*I sites in the arms; only those directly adjacent to the insert are presented here. The insert is presented on an expanded scale in the second through the sixth lines. In this case, the numbers refer to the sizes of fragments, in kilobase pairs, generated by digestion with the restriction enzyme shown at the left. Fragments which hybridize to β-globin nucleic acid probes are represented as open rectangles (Tilghman et al. 1977). Fragments which do

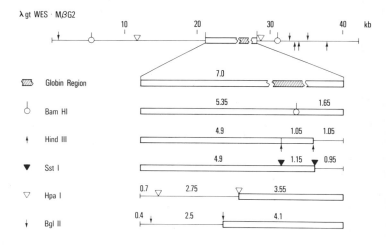

not hybridize are simply represented as horizontal lines. The hatched area within the insert displayed on the first and second lines represents an estimate of the region containing β-globin sequences based on these hybridization data and R-loop mapping in the electron microscope.

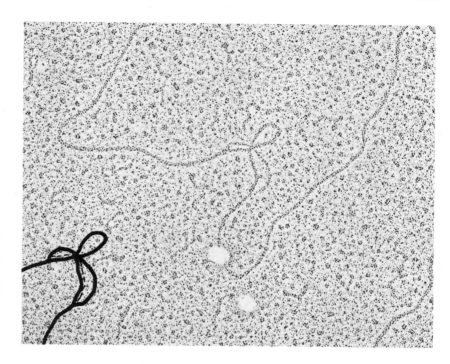

Figure 7. Electron microscopic representation of R loop formed between the cloned mouse globin sequence and mouse globin mRNA. The diagram represents double-stranded DNA (heavy line), displaced single-stranded DNA (solid fine line), and DNA-mRNA hybrid (heavy line and dashed fine line).

sumptions than appear warranted at this point. The greater likelihood seems that this sequence is transcribed into a precursor. If so, it raises the possibility that mRNA production occurs via a bighting-ligating mechanism in which the intervening sequence is clipped out of the RNA after transcription. Such a mechanism immediately raises new possibilities in interpreting physiologic and genetic data. First of all, it raises the possibility that a single polypeptide chain can be coded for by two (or more) distant structural gene sequences. It raises the further possibility that the processing reactions may play a role in the coherent assembly of genetic information. It also raises the possibility that such permutations may greatly amplify the genetic repertoire of a somatic cell. And finally, it offers a new way to think about the special problem of increasing the diversity of immunoglobulin genes and joining immunoglobulin constant- and variable-region gene sequences. Fortunately, the technology seems advanced to a state which should soon permit us to test these interesting possibilities.

Acknowledgments

We are most grateful to Ms. Catherine Kunkle who assisted us and is in large part responsible for the preparation of the manuscript. We are also grateful to Drs. Thomas Maniatis and Bernard Mach for providing plasmid clones used in these studies. We are also most grateful to Dr. Dolph Hatfield who provided us with the original sample of RPC-5 and for his patient advice.

REFERENCES

BASTOS, R. N. and H. AVIV. 1977. Globin RNA precursor molecules: Biosynthesis and processing in erythroid cells. *Cell* **11**: 641.

BENTON, W. D. and R. W. DAVIS. 1977. Screening λgt recombinant clones by hybridization to single plaques *in situ*. *Science* **196**: 180.

CURTIS, P. J. and C. WEISSMANN. 1976. Purification of globin messenger RNA from dimethylsulfoxide-induced Friend cells and detection of a putative globin messenger RNA precursor. *J. Mol. Biol.* **106**: 1061.

EFSTRATIADIS, A., F.C. KAFATOS, and T. MANIATIS. 1977. The primary structure of rabbit β-globin mRNA as determined from cloned DNA. *Cell* **10**: 571.

ENQUIST, L., D. TIEMEIER, P. LEDER, R. WEISBERG, and N. STERNBERG. 1976. Safer derivatives of bacteriophage λgt·λC for use in cloning of recombinant DNA molecules. *Nature* **259**: 596.

HARDIES, S. C. and R. D. WELLS. 1976. Preparative fractionation of DNA restriction fragments by reversed phase column chromatography. *Proc. Natl. Acad. Sci.* **73**: 3117.

KRAMER, R. A., J. R. CAMERON, and R. W. DAVIS. 1976. Isolation of bacteriophage λ containing yeast ribosomal RNA genes: Screening by *in situ* RNA hybridization to plaques. *Cell* **8**: 227.

KWAN, S.-P., T. G. WOOD, and J. B. LINGREL. 1977. Purification of a putative precursor of globin messenger RNA from mouse nucleated erythroid cell. *Proc. Natl. Acad. Sci.* **74**: 178.

LANDY, A., C. FOELLER, R. RESZELBACH, and B. DUDOCK. 1976. Preparative fractionation of DNA restriction fragments by high pressure column chromatography on RPC-5. *Nucleic Acids Res.* **3**: 2575.

LEDER, P., D. TIEMEIER, and L. ENQUIST. 1977. EK2 derivative of bacteriophage lambda useful in the cloning of DNA from higher organisms: The λgt*WES* system. *Science* **196**: 175.

PEARSON, R. L., J. F. WEISS, and A. D. KELMERS. 1971.

Improved separation of transfer RNA's on polychloro-trifluourethylene-supported reversed-phase chromatography columns. *Biochim. Biophys. Acta* **228**: 770.

POLSKY, F., M. H. EDGELL, J. G. SEIDMAN, and P. LEDER. 1977. High capacity gel electrophoresis for purification of fragments of genomic DNA. *Anal. Biochem.* (in press).

Ross, J. 1976. A precursor of globin messenger RNA. *J. Mol. Biol.* **106**: 403.

ROUGEON, F. and B. MACH. 1977. Cloning and amplification of α and β mouse globin gene sequences synthesized *in vitro. Gene* **1**: 229.

THOMAS, M., J. R. CAMERON, and R. W. DAVIS. 1974. Viable molecular hybrids of bacteriophage lambda and eukaryotic DNA. *Proc. Natl. Acad. Sci.* **71**: 4579.

THOMAS, M., R. L. WHITE, and R. W. DAVIS. 1976. Hybridization of RNA to double-stranded DNA: Formation of R-loops. *Proc. Natl. Acad. Sci.* **73**: 2294.

TIEMEIER, D., L. ENQUIST, and P. LEDER. 1976. An improved derivative of a bacteriophage λ EK2 vector useful in the cloning of recombinant DNA molecules: λgt*WES* ·λB. *Nature* **263**: 526.

TIEMEIER, D. C., S. M. TILGHMAN, and P. LEDER. 1977. Purification and cloning of a mouse ribosomal gene fragment in phage λ. *Gene* (in press).

TILGHMAN, S., D. TIEMEIER, F. POLSKY, M. EDGELL, J. SEIDMAN, P. LEDER, L. ENQUIST, B. NORMAN, and A. LEDER. 1977. Cloning specific segments of the mammalian genome: Bacteriophage λ containing mouse globin and surrounding gene sequences. *Proc. Natl. Acad. Sci.* **74**: 4406.

Organization of Immunoglobulin Genes

S. Tonegawa,* C. Brack,* N. Hozumi,* and V. Pirrotta†

*Basel Institute for Immunology, Basel 5, Switzerland; †Department of
Microbiology, Biocenter, University of Basel, Basel, Switzerland.

Both light and heavy chains of antibody (or immunoglobulin) molecules consist of two regions. About 100 residues at the amino-terminal ends of the polypeptide chains determine the specificity of the molecules and compose *variable* (V) regions, whereas the remaining residues determine the *class* or *subclass* of antibodies to which the molecule belongs and compose *constant* (C) regions. Two types of heterogeneities exist among antibody molecules. The first distinguishes about 20 types of molecules with respect to their C regions. In mice, three (κ, λ_I, and λ_{II}) and eight (δ, μ, γ_I, etc.) different C regions are known for the light and heavy chains, respectively. The second type of heterogeneity distinguishes millions of molecules differing with respect to their V regions. This heterogeneity is often called "antibody diversity," and the question of whether the diversity in the structural genes coding for the V regions *(V genes)* arose in evolution or arises in ontogeny has been one of the most debated subjects in modern immunology (Wigzell 1973; Capra and Kehoe 1974).

Recent hybridization studies with purified immunoglobulin mRNA indicated that the number of germ-line V genes is far too small to account for the observed diversity in the V regions. For instance, there are no more than a few (and probably only one) germ-line V genes for the entire repertoire of the V regions which are associated with the λ_I-type light chains (Tonegawa 1976; Honjo et al. 1976). These results strongly suggested the existence of some mechanism by which coding information in the V genes is somatically diversified during differentiation of lymphocytes. In mice, the heterogeneity in the V_κ regions (V regions of κ-type chains) is much greater than in the V_λ regions (V regions of λ-type chains) (McKean et al. 1973; Cohn et al. 1974). The hybridization studies, as well as amino acid sequence analysis of myeloma κ chains, suggested that the mouse V_κ genes as a whole constitute a multigene family of as many as a few hundred unique but related sequences (Cohn et al. 1974; Tonegawa et al. 1977c). As with V_λ genes, the coding content of each of these V_κ genes diversifies somatically in the clones of B (bone-marrow-derived) lymphocytes, such that a large V_κ-region repertoire is generated from no more than a few germ-line genes in an adult mouse (Tonegawa et al. 1974; Rabbitts et al. 1975; Tonegawa 1976).

The number of C genes was also estimated by nucleic acid hybridization. These studies showed that both C_κ and C_λ genes are unique or nearly unique (Faust et al. 1974; Stavnezer et al. 1974; Tonegawa et al. 1974; Honjo et al. 1974). Thus, at least in κ chains, the *total* number of germ-line genes is much greater for V regions than for C regions. This led us to investigate the two-"genes"–one-polypeptide-chain hypothesis for antibody chains, which was put forward a decade ago by Dreyer and Bennett (1965). We analyzed DNA from early embryos and from myelomas by digestion with restriction endonucleases and by hybridization with a purified κ-chain mRNA. These experiments strongly suggested that V_κ and C_κ genes are separate in early embryo DNA and that one of the multiple V_κ genes becomes contiguous to the C_κ gene during differentiation of B lymphocytes (Hozumi and Tonegawa 1976). The rearrangement permits the continuous transcription of a full immunoglobulin gene.

To further study the organization and regulation of this gene family, it is useful to clone chromosomal DNA fragments carrying immunoglobulin genes. By combining biochemical enrichment of such a DNA fragment with screening in situ of phage plaques, we were able to clone a λ phage which carries a mouse V_λ gene DNA insert. In this paper we describe isolation of the phage and characterization of the mouse DNA insert. In addition, further analysis with a restriction enzyme of total mouse DNA of various cellular sources is described.

Enrichment of a V_λ Gene

We previously showed that, in a total *Eco*RI digest of early BALB/c embryo DNA, three fragments of 4.5, 2.7, and 2.0 megadaltons contain λ-chain gene sequences (Tonegawa et al. 1977c). The use of a whole (a probe for V_λ and C_λ sequences) as well as a 3′-end-half (a probe for C_λ sequences) λ-chain mRNA in hybridization permitted us to conclude that the λ-chain gene sequences contained in the 4.5-megadalton fragment are exclusively for a C_λ region, whereas those contained in the 2.7- and 2.0-megadalton fragments are for V_λ regions (Tonegawa et al. 1977c). As a first attempt to clone an immunoglobulin gene, we selected the 2.7-megadalton V_λ-gene fragment as the source of mouse DNA.

We incubated the DNA fragments with excess λ-chain mRNA, purified from HOPC-2020 myeloma, in 70% formamide under the conditions which promote replacement of the coding DNA strand with the RNA, but which do not completely dissociate the two DNA strands (R-loop formation) (White and Hogness 1977). We subjected the nucleic acid mixture to equilibrium centrifugation in CsCl and localized the position of the R-loop structure by hybridization with ^{125}I-labeled HOPC-2020 λ-chain mRNA after removal by alkali of the prehybridized RNA (Fig. 1). When no λ-chain mRNA was added during incubation in formamide (Fig. 1B), the fragment carrying V_λ gene sequences cobanded with the bulk DNA. When λ-chain mRNA was added (Fig. 1A), the major proportion of the V_λ-sequence-carrying DNA banded at a position which was clearly heavier

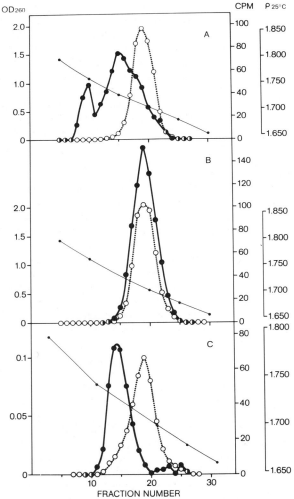

Figure 1. Preparative R-loop formation. *Eco*RI-digested embryonic DNA prefractionated by agarose gel electrophoresis for V_λ gene sequences was incubated in the presence (*A*) and absence (*B*) of HOPC-2020 λ-chain mRNA and centrifuged in CsCl as described elsewhere (Tonegawa et al. 1977a). Fractions 12–16 shown in *A* and another tube which showed profiles nearly identical to those in *A* were pooled and recentrifuged (*C*). (○----○) OD_{260}; (●——●) DNA-RNA hybrid; (——) buoyant density.

than the peak of the bulk DNA. The buoyant density at the hybridization peak in Figure 1A was 0.018 g/cm^3 higher than that in Figure 1B.

In a separate experiment we isolated a λgt*WES* phage which contains as its insert a 6.9-kb mouse DNA fragment (generated by *Eco*RI) carrying ribosomal DNA sequences (S. Tonegawa, in prep.). When the DNA fragment excised from the phage genome was annealed with the purified mouse 18S ribosomal RNA (2.0 kb) and centrifuged to equilibrium in CsCl, the R loop banded at a position 0.023 g/cm^3 denser than the duplex DNA. As a first-order approximation, the density increase of R loops is inversely proportional to the ratio of the lengths of DNA and RNA. The lengths of the V_λ-carrying DNA fragment and the λ-chain mRNA are about 4.8 and 1.2 kb, respectively. We therefore expect that the R loop formed between them is 0.019 g/cm^3 heavier than the DNA duplex. Thus we assume that the major hybridization peak observed in Figure 1A is composed of such an R loop. In Figure 1A, a second hybridization peak is observed in fraction 10. The increment of buoyant density of this peak is approximately twice that of the major hybridization peak, suggesting that this component is a hybrid formed between a single strand of the V_λ DNA fragment and the λ-chain mRNA. This component was not characterized further. The fractions 12–16 shown in Figure 1A were recentrifuged in CsCl. Although the hybridization peak remains at the original position, a large proportion of DNA (i.e., OD_{260} absorbing material) banded at the position where the bulk of the DNA banded in the first centrifugation. The fractions 12–16 shown in Figure 1C were pooled and used for cloning.

Enrichment for DNA fragments carrying the V_λ gene was approximately ten-, nine-, and fourfold by agarose gel electrophoresis and by first and second CsCl density gradient centrifugation, respectively. If we assume that other operations carried out between these steps (e.g., ethanol precipitation and dialysis) did not cause loss of specific DNA fragments, the overall enrichment factor was about 360-fold. The final yield was about 2 μg from 10 mg *Eco*RI-digested embryo DNA.

Cloning of a V_λ Gene

We inserted the partially purified V_λ DNA fragment in the middle of the λgt*WES* phage DNA by using T_4 ligase (Mertz and Davis 1973). The phage vector was developed by P. Leder and his coworkers and was approved as an EK-2 vector by the National Institutes of Health Advisory Committee on Recombinant DNA Research (Enquist et al. 1976). Upon transfection of CaCl$_2$-treated *Escherichia coli* 803 (r_k^- m_k^- Su$_{III}^+$) (Mandel and Higa 1970), we obtained about 6000 plaques from 2 μg of the partially purified V_λ DNA. We screened about 4000 plaques by in situ hybridization with ^{125}I-labeled HOPC-2020 mRNA (Kramer et al. 1976). The 38 plaques which produced gray or black autoradiographic spots of varying strength were reexamined by means of a second in situ hybridization with less RNA probe. Since the λ-chain mRNA probe used was about 90% pure and the impurity is distributed among many different RNA species (Tonegawa 1976), each composing a small fraction of the mRNA preparation, the use of smaller amounts of the RNA probe in hybridization favors detection of the DNA clone complementary to the major mRNA component. Out of the 38 plaques reexamined in this way, one plaque, λgtWES-Ig 13, gave a distinctly stronger autoradiographic spot than the others.

Characterization of the Mouse DNA Insert

Length of the insert. When analyzed by agarose gel electrophoresis, EcoRI digestion of the DNA extracted from the clone λgtWES-Ig 13 yielded, in addition to the left and right arms of the parental λgtWES genome, one fragment of 3.0 megadaltons. The length of the insert was also determined by electron microscopy using two independent methods. λgtWES-Ig 13 DNA was digested with EcoRI, and the three resulting fragments were measured. Taking PM2 DNA (10 kb) as the internal length standard, the left and right arms of the vector were calculated to be 21.2 and 13.9 kb, respectively; the mouse DNA insert was 4.8 kb.

In heteroduplex preparations made between λgtWES-λC and λgtWES-Ig 13 DNA (Davis et al. 1971), these measurements were confirmed. One large heteroduplex region showed up at the expected position: between 53.1 and 65.2 map units from the left end (Fig. 2). The lengths of the two single-

stranded regions were 4.8 kb (mouse DNA) and 5.5 kb (λ EcoRI C fragment).

Hybridization with λ-Chain mRNA and cDNA

In Table 1, hybridization properties of the whole λgtWES-Ig 13 DNA with various RNA probes are shown. The DNA hybridized well with HOPC-2020 whole λ mRNA, whereas the levels of hybridization obtained with the 3'-end-half of the same mRNA were no higher than when the same mRNA was hybridized to a clone carrying mouse ribosomal DNA. The results suggest that λgtWES-Ig 13 DNA contains V_λ sequences and lacks C_λ gene sequences of HOPC-2020 mRNA. The same DNA showed no hybridization with a MOPC-321 κ-chain mRNA. Because these experiments were conducted with DNA baked on a Millipore filter, the efficiency of hybridization was not high. To circumvent this problem, full-length cDNA was synthesized from λ-chain mRNA with the avian myeloblastosis virus reverse transcriptase, and this DNA was hybridized with excess λgtWES-Ig 13 DNA in liquid. The results are shown in Table 2. When assayed by hydroxylapatite, at least 60% of the cDNA, prepared either from MOPC-104E λ-chain mRNA or HOPC-2020 λ-chain mRNA, hybridized with the cloned DNA. When nuclease S_1 was used to remove the tail (and possibly some mismatched bases), the hybridization levels were reduced by about one-half. The mean melting point of the hybrid thus formed was 84°C in 0.12 M NaPO₄, as assayed by the hydroxylapatite method. It is not surprising that the hybridization of cDNA, as assayed by hydroxylapatite, is incomplete (up to 64%), since the λ mRNA used in preparing the cDNA was about 80% pure and the efficiency of hy-

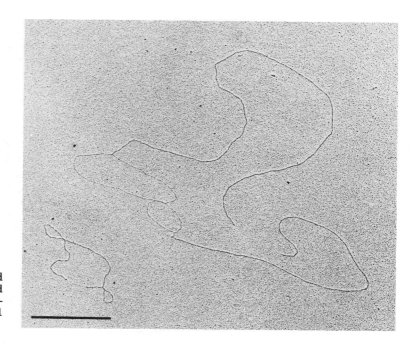

Figure 2. Electron micrograph of a hybrid molecule formed between λgtWES-λC and λgtWES-Ig 13 DNA showing one large heteroduplex region. The bar represents 1 μm.

Table 1. Hybridization of ^{125}I-labeled Light-chain mRNA and Its Fragments with λgtWES-Ig 13 DNA

DNA on filter	μg	[125I]mRNA	Input (cpm)	Hybrid (cpm)	Percentage of input
λgtWES-Ig 13	25	λ chain (whole)	16,000	2,943	15.6
λgtWES-Ig 13	50	λ chain (whole)	16,000	3,397	21.2
λgt-rD	25	λ chain (whole)	16,000	136	0.85
λgt-rD	50	λ chain (whole)	16,000	172	1.08
λgtWES-Ig 13	25	λ chain (3'-end-half)	9,500	136	1.39
λgtWES-Ig 13	50	λ chain (3'-end-half)	9,500	163	1.72
λgt-rD	25	λ chain (3'-end-half)	9,500	125	1.32
λgt-rD	50	λ chain (3'-end-half)	9,500	132	1.39
λgtWES-Ig 13	25	κ chain (whole)	15,000	211	1.35
λgtWES-Ig 13	50	κ chain (whole)	15,000	144	0.93
λgt-rD	25	κ chain (whole)	15,000	158	1.00
λgt-rD	50	κ chain (whole)	15,000	186	1.20

DNA was heat-denatured and fixed on a Millipore filter by the method of Gillespie and Spiegelman (1965). Hybridization was carried out in $2 \times$ SSC buffered with 0.05 M PIPES (pH 7.1) at 69°C for 14 hr. Specific activity of RNA was 3×10^7 cpm/μg for whole and 3'-end-half λ-chain mRNA and 3.2×10^7 cpm/μg for κ mRNA. The hybrid was assayed by RNase treatment (RNase A, 20 μg/ml; RNase T$_1$, 2 units/ml, in $2 \times$ SSC). λgt-rD designates a λWES phage in which the center EcoRI fragment was replaced with a 6.9-kb mouse DNA fragment carrying ribosomal DNA sequences. The ^{125}I-labeled 3'-end-half of the λ-chain mRNA and the whole κ-chain mRNA used in this experiment hybridized well with corresponding cDNA (data not shown).

bridization under these conditions is about 90%. These results are consistent with the hypothesis that the λgtWES-Ig 13 clone carries V_λ gene sequences.

Length and Location of Homology Region

Further evidence for the existence of V_λ gene sequences was obtained by electron microscopic examination of R loops. HOPC-2020 λ-chain mRNA and λgtWES-Ig 13 DNA were incubated under conditions of R-loop formation (Thomas et al. 1976; Tonegawa

Table 2. Hybridization of [32]P λ-Chain cDNA with Ig 13 DNA

[32]P-labeled cDNA prepared on	$C_0 t$	Fraction of ^{32}P counts in hybrid	
		hydroxylapatite assay (%)	nuclease S$_1$ assay (%)
HOPC-2020 mRNA	0	4.4	2.7
HOPC-2020 mRNA	5×10^{-2}	61	27
HOPC-2020 mRNA	6.7×10^{-2}	61	27
MOPC-104E mRNA	2.2×10^{-1}	64	30

Isolation of Ig 13 DNA, synthesis of full transcript cDNA, and annealing were carried out as described in the legend to Fig. 2. For the hydroxylapatite assay, the annealing mixtures were made to 0.12 M NaPO$_4$ by the addition of water, and were loaded on a small hydroxylapatite column preequilibrated with 0.12 M NaPO$_4$, pH 6.9, at 60°C. The column was washed with 9 ml of the same buffer and the hybrid was eluted with 6 ml of 0.4 M NaPO$_4$, pH 6.9. The hybrid fraction was determined by dividing the radioactivity in the 0.4 M NaPO$_4$ fraction by the total radioactivity recovered from the column. For the nuclease S$_1$ assay, the annealing mixture was divided into two equal parts. DNA in one part was directly precipitated with trichloroacetic acid, and DNA in the other part was precipitated after treatment with nuclease S$_1$ (see legend to Fig. 2). The hybrid fraction was determined by subtracting the intrinsic S$_1$-resistant counts (1.2%) from the ratio of the S$_1$-resistant counts and the total TCA precipitable counts. $C_0 t = 0$ sample was prepared by placing the heat-denatured sample directly in a dry-ice ethanol bath. $C_0 t$ refers to the concentration of DNA in moles of nucleotide per liter × incubation time in sec.

et al. 1977a). We observed only a relatively low proportion of hybrid molecules. About 30% of the molecules displayed a small R loop, and some of the R loops had a small RNA tail at one end (Fig. 3). Measurement of 89 such hybrid molecules showed that the R loop is at a unique position: 63 ± 1 map units from the left end (Fig. 4). Since the right EcoRI site on this chimeric molecule is at 65.2 map units, we conclude that the region to which HOPC-2020 λ-chain mRNA hybridizes lies within the mouse DNA.

The lengths of the R loop and the RNA tails were 400 ± 100 nucleotides and 200–600 nucleotides, respectively. We think that the large variations in the length measurements of the RNA tail result from that fact that the mRNA molecules are not fully denatured and do not completely extend under the spreading condition used.

More accurate measurement of the position and length of the homology region was made by electron microscopic examination of R loops formed with the λ-chain mRNA and the purified 4.8-kb EcoRI fragment, as well as hybrid molecules formed with the same DNA fragment and full-length cDNA synthesized on MOPC-104E λ-chain mRNA (in prep.). The results obtained from these experiments confirmed the conclusion drawn above and indicated that the homology region is approximately 400 nucleotides long and that it is localized between 3290 and 3690 (± 200) nucleotide pairs from the left end of the 4.8-kb mouse DNA insert. Figure 4 shows the maps of λgtWES-Ig 13, λgtWES-λC DNA, and the mouse DNA insert as obtained from electron microscopy.

Restriction Enzyme Cleavage Sites

To characterize the mouse DNA insert further, we determined the cleavage sites of several restric-

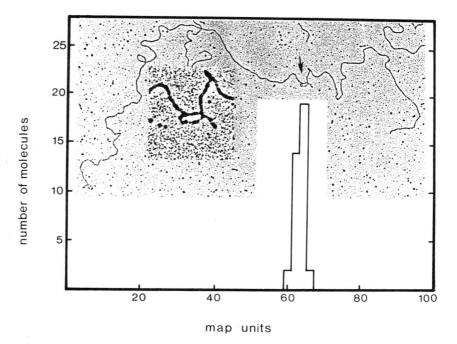

Figure 3. Electron micrograph of an R-loop molecule formed by hybridizing myeloma HOPC-2020 λ-chain mRNA with λgt-*WES*-Ig 13 DNA. The R loop with an RNA tail is shown at higher magnification in the insert. The histogram was made from one representative set of measurements from the R loops made by incubation at 57°C and high salt, and spread from 70% formamide. The R loop is at a unique position 63 ± 1 map units from the left end of the λgt*WES*-Ig 13 molecule.

tion enzymes. Two kinds of experiments were carried out. In one series of experiments, the 4.8-kb *Eco*RI fragment was digested by several restriction enzymes which cleave DNA relatively infrequently. The digests were electrophoresed in 1.2% agarose gel, and the DNA fragments containing sequences homologous to HOPC-2020 λ-chain mRNA were identified by the Southern transfer technique (Southern 1975). In some cases, mere comparison of these results with the homology map obtained by electron microscopy (Fig. 4) allowed us to order the fragments. In other cases, fragment order was deduced from results obtained by single- and multiple-enzyme digestion. Figure 5A summarizes these results.

In the other series of experiments, a more extensive cleavage-site map of the 1.5-kb *Hae*III fragment was obtained by the partial-digestion method of Smith and Birnstiel (1976). The 4.8-kb *Eco*RI fragment was terminally labeled with [γ-³²P]ATP by T₄ polynucleotide kinase. The 1.5-kb *Hae*III fragment was isolated by agarose gel electrophoresis and was partially digested with *Hin*f, *Mbo*II, and *Alu*I. The positions of the cleavage sites determined by this method are summarized in Figure 5B. These results

were confirmed by total digestion of the *Hae*III fragment (data not shown).

Nucleotide Sequences

Ultimate evidence that the mouse DNA insert contains λ-chain gene sequences comes from direct DNA sequencing. We have determined a partial sequence of the *Hin*f C fragment. This fragment is about 290 nucleotides long and includes part of the homology region and the sequences immediately adjacent to it (Fig. 5B). According to the map in Figure 5B, there should be an *Mbo*II restriction site near the center of this fragment. We labeled the 5′ ends of the *Hin*f fragment with polynucleotide kinase, cut it with *Mbo*II, and determined the nucleotide sequences of the two halves by the method of Maxam and Gilbert (1977).

The results of the left and right halves are shown in Figures 6 and 7, respectively. To determine the fit of the observed sequences to that predicted from the amino acid sequence of the λ chains, all possible *Hin*f sites in the predicted sequence were aligned with the DNA sequences. Figure 6 shows that the possible *Hin*f site at amino acid positions 63 and

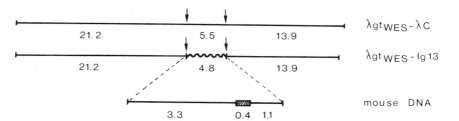

Figure 4. Maps of λgt*WES*-λC and λgt*WES*-Ig 13 DNA. The wavy line indicates the mouse DNA insert. The hatched box represents the region homologous to HOPC-2020 λ-chain mRNA.

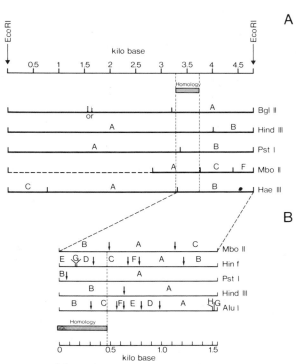

Figure 5. Maps of restriction enzyme cleavage sites on the mouse DNA insert: *(A)* 4.8-kb *Eco*RI fragment; *(B)* 1.5-kb *Hae*III fragment. Fragments are labeled in alphabetical order from large to small. The cross-hatched boxes indicate the region homologous to HOPC-2020 λ-chain mRNA. The exact position of the *Bgl*II cleavage site at the left was not determined. The *Mbo*II map is incomplete. See text for further explanation.

64 of V_λ regions gives an alignment which results in an excellent fit, with only a few exceptions in the sequences so far determined. In Figure 6 the DNA sequences thus aligned are shown, together with the predicted sequences obtained from the MOPC-104E V_λ region and the MOPC-315 $V_{\lambda II}$ region. At positions 85 and 87, the DNA sequences correspond to the predicted sequences of $V_{\lambda II}$ instead of $V_{\lambda I}$. At two more positions, 94 and 95, the sequence correspondence is with $V_{\lambda I}$. Does this mean that embryonic DNA contains a single V_λ gene which is a hybrid of the $V_{\lambda I}$ and $V_{\lambda II}$ types? We believe that a more likely explanation is that there are separate germ-line genes for $V_{\lambda I}$ and $V_{\lambda II}$ regions, and that the gene contained in the cloned mouse DNA is for the $V_{\lambda II}$ region. This is because positions 94 and 95 are in the *hypervariable region,* whereas positions 85 and 87 are *framework residues* (Cohn et al. 1974). The two germ-line V genes code for common amino acids at positions 94 and 95, and the amino acids observed in MOPC-315 λ_{II} at these positions are the result of somatic changes.

Is there a C gene immediately adjacent to the V gene? Since λ_I-type mRNA was used in the hybridization experiments, it is possible that the presence of a $C_{\lambda II}$ gene might have been overlooked because of the relatively large difference (29 out of 102 amino acids) (Dugan et al. 1973; Cohn et al. 1974) of amino acid sequences between the two C regions. The DNA sequence of the right half of the *Hin*f C fragment (Fig. 5B) shows no noticeable similarity to the sequences predicted from either the C_I or C_{II} region. Figure 7 shows the observed DNA sequence, its complementary sequence, and the corresponding amino

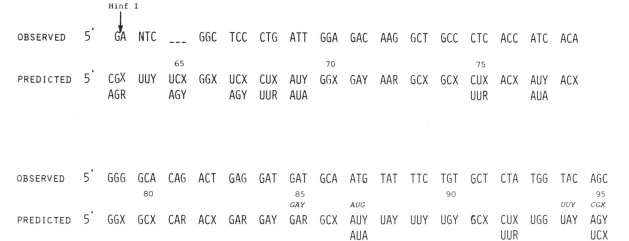

Figure 6. Partial nucleotide sequences of the left half of the *Hin*f C fragment. Experimentally determined nucleotide sequences are shown in comparison with the sequences predicted from amino acids within a $V_{\lambda I}$ region (MOPC-104E λ chain). The numbers designate amino acid positions. The amino acids, and consequently predicted nucleotides of a $V_{\lambda II}$ region (MOPC-315 λ chain), are identical with those of MOPC-104E λ chain, except for positions 85, 87, 94, and 95. Nucleotides predicted from the MOPC-315 λ chain at these positions are shown in italics. Positions 94 and 95 are within a hypervariable region (Cohn et al. 1974). X,N indicates any one of four bases; R indicates purine; Y indicates pyrimidine. Three nucleotides immediately adjacent to the *Hin*f site were not determined.

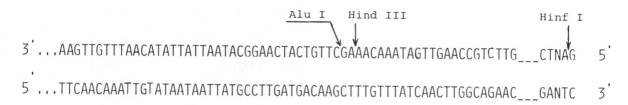

Figure 7. Partial nucleotide sequences of the right half of the *Hinf* C fragment. Experimentally determined nucleotide sequence is shown at the top. In the second row, the complementary sequence is shown. The latter sequence was transformed into amino acid sequences by reading it in all three frames. These amino acid sequences are shown under the complementary nucleotide sequence. Positions of *Alu*I and *Hinf* cleavage sites are indicated by arrows. (✻) Indicates a termination triplet.

acid sequences obtained by reading the complementary sequence in each of the three frames. Amino acid sequences read by two possible frames run into double termination codons. The third frames yielded amino acid sequences which bear no similarity to either the C$_I$ or C$_{II}$ region. Alternatively, we have tried to align the observed and predicted nucleotide sequences using as reference points the *Hinf* and the *Alu*I recognition sequences. No alignment was possible which gave less than 40% mismatching, even using the most favorable codon assignments. Whether or not some residual sequence homology exists between this region and C$_\lambda$ genes remains to be determined. We conclude that a C$_\lambda$ gene is not contiguous to the V$_\lambda$ gene.

In collaboration with A. Maxam, W. Gilbert, R. Tizard, and O. Bernard, we recently determined more extensively the nucleotide sequences in the homology region and the regions immediately adjacent to it (in prep.). The result confirms the sequences shown in Figures 6 and 7.

Rearrangement of Immunoglobulin Genes

The DNA cloning technique is a powerful tool for analyzing the organization and function of eukaryotic genes. There is no doubt that the analysis of various immunoglobulin gene clones from different cellular origins will reveal many microscopic details of the arrangement and rearrangement of DNA sequences. For a better understanding of the various problems associated with this gene family, it would be useful to combine these studies with analysis of total cellular DNA at a more macroscopic level.

When limit *Bam*HI digests of total DNA from BALB/c embryos or MOPC-321 myeloma (a κ-chain producer) were fractionated by agarose gel electrophoresis and the DNA in each fraction was hybridized with a whole κ-chain mRNA (a probe for V$_\kappa$ and C$_\kappa$ gene sequences), profiles of hybridization

were drastically different with the two DNAs (Hozumi and Tonegawa 1976). We interpreted the profile difference as a result of a rearrangement of immunoglobulin genes which takes place during differentiation of lymphocytes. The rearrangement is thought to bring one of the multiple V genes in contiguity to a C gene, thereby permitting the continuous transcription of a full immunoglobulin gene. An alternative explanation of the results, namely, that accumulation of multiple mutations or base modifications leading to either loss or gain of *Bam*HI sites generated the observed pattern difference, was considered to be unlikely (Hozumi and Tonegawa 1976). Another possible, also unlikely, interpretation was to ascribe the pattern difference to massive scrambling of DNA sequences which might accompany generation or propagation of the myeloma cells. To exclude these trivial interpretations, we extended the analysis to DNAs of other sources.

In Figure 8B a *Bam*HI digest of kidney DNA from adult BALB/c mice was analyzed with a whole κ-chain mRNA purified from MOPC-321 myeloma, as well as with its 3′-end-half. Two DNA components of 6.0 and 3.9 megadaltons hybridized with the whole RNA molecule, whereas only the 6.0-megadalton component hybridized with the 3′-end-half. Overall, hybridization patterns were indistinguishable from those obtained with embryonic DNA (Fig. 8A) (Hozumi and Tonegawa 1976). Essentially identical patterns were obtained with DNAs of other adult tissues such as liver and brain. The hybridization patterns of the whole κ-chain mRNA with DNA from J558 myeloma were also indistinguishable from those obtained with embryonic DNA (Fig. 8A). In this myeloma, a λ chain is produced and no κ-chain gene is expressed. DNA from two other λ-chain-producing myelomas (MOPC-104E and HOPC-2020) gave essentially the same result (not shown). Conversely, when DNA from a κ-chain-producing myeloma (MOPC-321) was analyzed with a purified λ-chain mRNA as hybridization probe (HOPC-2020), the hybridization pattern was indistinguishable from that

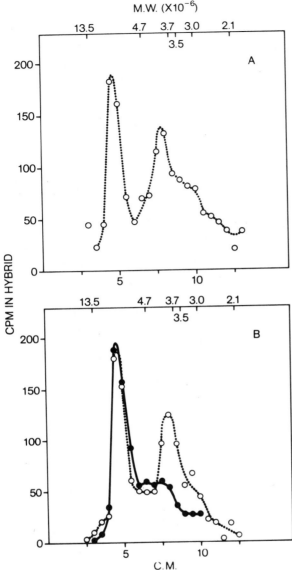

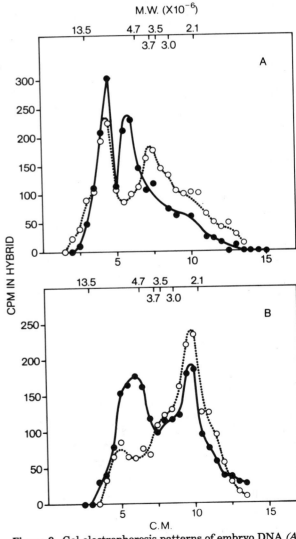

V regions are somatically generated from a few, probably single, germ line gene(s) (Tonegawa et al. 1974; Rabbitts et al. 1975; Tonegawa 1976; Honjo et al. 1976; Tonegawa et al. 1977c). Such a V-region group is best approximated to the subgroup as defined by Cohn et al. (1974). We previously presented direct demonstration of separate germ-line V genes for two V regions of different subgroups (Tonegawa et al. 1977c). The results are reproduced in Figure 9A for the present context. Embryonic DNA digested with *Bam*HI enzyme was analyzed with two mRNAs coding for two κ chains of different subgroups, MOPC-321 and MOPC-21. These two κ chains show little homology in their V regions, whereas they have identical sequences in the C region (Milstein and Svasti 1971; McKean et al. 1973). As expected, both RNAs hybridized with the 6.0-megadalton com-

Figure 8. Gel electrophoresis patterns of λ J558 DNA *(A)* and kidney DNA *(B)* digested with *Bam*HI. Whole MOPC-321 RNA (○-----○) (1250 cpm, 7 × 10⁷ cpm/μg) or its 3′-end-half fragment (●——●) (600 cpm, 7 × 10⁷ cpm/μg) was hybridized to fractionated DNA. Digestion of DNA, preparation of RNA probes, electrophoresis, and hybridization procedures were as described elsewhere (Tonegawa 1976; Hozumi and Tonegawa 1976). The number at the top of each panel in this figure and in Figs. 9 and 10 indicates the molecular weight (in megadaltons) of *Eco*RI-digested phage λ DNA used as migration markers.

obtained with embryo DNA by the same mRNA (Tonegawa et al. 1977b). These results render the trivial interpretations described above unlikely.

An Inactive V Gene Remains Unjoined with the Corresponding C Gene in Plasma Cells

Our earlier hybridization studies, as well as those of others, indicated that a group of closely related

Figure 9. Gel electrophoresis patterns of embryo DNA *(A)* and MOPC-321 DNA *(B)* digested with *Bam*HI. Whole MOPC-321 mRNA (○-----○) (1250 cpm, 7 × 10⁷ cpm/μg) or MOPC-21 mRNA (●——●) (1220 cpm, 8 × 10⁷ cpm/μg) was annealed with extracted DNA.

ponent which carries the C_κ gene (see above). In addition, each of the two RNAs hybridized with a second, but mutually different, DNA component. These DNA components of 5.0 and 3.9 megadaltons should carry MOPC-21 and MOPC-321 V-gene sequences, respectively.

Is a V_κ gene for a given subgroup joined with a C_κ gene in the myeloma which synthesizes a κ chain carrying a V region of another subgroup? That this is probably not the case is shown in Figure 9B, where MOPC-321 DNA was analyzed with the homologous (MOPC-321) and heterologous (MOPC-21) κ mRNA. As we reported previously, the homologous κ mRNA hybridized with a major DNA component of 2.4 megadaltons (Hozumi and Tonegawa 1976). This component carries both $V_{\kappa\ \text{MOPC-321}}$ and C_κ gene sequences. While the 6.0-megadalton, embryonic C_κ gene component disappears, the 5.0-megadalton $V_{\kappa\ \text{MOPC-21}}$ gene fragment remains at the embryonic position (this band was atypically broad in this particular experiment). These results, as well as those described in the last section (i.e., analysis of κ-gene sequences in DNA from adult tissues and from λ-chain-producing myelomas), suggest that there is a strict correlation between the V–C-gene joining and activation of the joined immunoglobulin genes.

Does the V–C-gene Joining Take Place in Both of the Homologous Chromosomes?

The apparent disappearance of the embryonic V- and C-gene components (Hozumi and Tonegawa 1976) in MOPC-321 DNA suggested that the joining event leads to homozygosis. As we pointed out previously (Tonegawa et al. 1977c), homozygosis could result from the loss of one homolog followed by reduplication of the other, or it could result from somatic recombination between the centromere and the immunoglobulin locus. The alternative view—that joining takes place in both chromosomes—presents a problem since there is more than one V gene (see above). There is no intrinsic reason why the same V gene should be joined on both homologs. The above results, however, could be explained by the known abnormality of the karyotype of mouse myelomas. Although MOPC-321 is subtetraploid (S. Tonegawa, unpubl.), we have no information as to the number of chromosomes on which the κ genes lie. Thus it is possible that this myeloma has lost the homolog(s) on which the $V_{\kappa\ \text{MOPC-321}}$ and C_κ genes are located separately.

To investigate this possibility, we analyzed DNA from another myeloma, TEPC-124 (Fig. 10). MOPC-321 and TEPC-124 κ chains are different only by three amino acids in the V regions (McKean et al. 1973) and belong to a single subgroup. The two V regions, therefore, presumably share the same germline V gene. Since nucleotide sequences in the two mRNAs are extensively homologous (Tonegawa et al. 1974; S. Tonegawa, unpubl.), MOPC-321 κ mRNA and its 3'-end-half were used as the hybridization

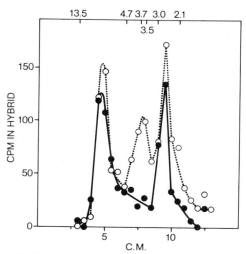

Figure 10. Gel electrophoresis patterns of TEPC-124 DNA digested with *Bam*HI. Whole MOPC-321 mRNA (O-----O) or its 3'-end-half fragment (●——●) was annealed with extracted DNA.

probes instead of the homologous mRNA. Three major DNA components hybridized with the whole κ mRNA, two of which hybridized also with the 3'-end-half fragments (Fig. 10). The size and hybridization properties of these DNA components suggest that the overall pattern is a composite of the two patterns obtained when embryonic and MOPC-321 DNA were analyzed by the same RNA probes. Thus the principal difference in the two hybridization patterns, one of embryonic DNA and the other of TEPC-124 tumor, is the addition of the 2.4-megadalton component in the latter. This component hybridized with both whole and 3'-end half RNA probes. Each of the two components that hybridized with the 3'-end-half should contain a complete C gene sequence, for our recent study with a κ-chain cDNA clone indicates that there is no *Bam*HI cleavage site in the C gene (G. Matthyssens and S. Tonegawa, unpubl.). If the two V regions indeed share the same germ-line V gene, the straightforward interpretation of the above results is that the V gene is joined with the C gene only in one of the two homologs in plasma cells.

For immunoglobulin loci, only one allele is expressed in any given lymphocyte (allelic exclusion) (Pernis et al. 1965). This is not the case for any other autosomal gene studied until now. Considering the strict correlation between joining of a pair of V and C genes and the expression of the joined V gene, the above results with TEPC-124 DNA may be relevant to the mechanism of allelic exclusion. On the other hand, the apparent homozygosis observed in MOPC-321 DNA can also conveniently explain allelic exclusion, if this is a naturally occurring event. In this case, heterozygosis observed in TEPC-124 DNA could result from accidental acquisition of a homolog from a nonlymphatic cell during generation or propagation of the tumor. We feel that the matter will be clarified only by analysis of DNA from natural lymphocyte clones.

CONCLUSIONS

The nucleotide-sequence determination of a cloned, embryonic V_λ gene directly demonstrated that V genes are separate from a corresponding C gene in embryonic cells. Analysis by restriction enzymes of total cellular DNA from various sources strongly suggested that the two separate immunoglobulin genes become continuous during differentiation of B lymphocytes. There seems to be a strict correlation between the joining event and activation of the joined genes. Cloning of more immunoglobulin genes from embryo and plasma cells will not only provide direct demonstration of such a gene-joining event but also help in the elucidation of a possible relationship of the event to gene activation mechanisms.

The gene-cloning experiments were carried out in a P-3 laboratory in accordance with the National Institutes of Health guidelines issued in June 1976. The phage vector used was approved, in January 1976, as an EK-2 vector by the National Institutes of Health Advisory Committee on Recombinant DNA Research. After completion of the work, in April 1977, we were informed that this approval was withdrawn. In compliance with the new rule, we are now growing the λgtWES-Ig 13 clone on *E. coli* DP50. Retention of the original *amber* mutations in this phage clone was kindly confirmed by Dr. B. Hohn at the Biocenter, Basel.

Acknowledgments

We thank Drs. M. Potter, M. Cohn, M. Weigert, J. W. Beard, T. Bickle, P. Leder, M. Sugiura, and M. Takanami for providing strains and enzymes. We thank Dr. B. Hohn for analyzing the genetic markers carried in λ phages. We also thank Dr. G. Matthyssens for a gift of λ-chain cDNA. The technical assistance of Mr. G. Dastoornikoo, Ms. R. Schuller, Ms. P. Riegert, and Mr. A. Traunecker is highly appreciated. The excellent secretarial help of Ms. K. Perret-Thurston is also appreciated.

REFERENCES

CAPRA, J. D. and J. M. KEHOE. 1974. Antibody diversity: Is it all coded for by the germ line genes? *Scand. J. Immunol.* **3**: 1.

COHN, M., B. BLOMBERG, W. GECKLER, W. RASCHKE, R. RIBLET, and M. WEIGERT. 1974. First order considerations in analyzing the generator of diversity. In *The immune system, genes, receptors, signals* (ed. E. E. Sercarz et al.), p. 89. Academic Press, New York.

DAVIS, R. W., M. SIMON, and N. DAVIDSON. 1971. Electron microscope heteroduplex methods for mapping regions of base sequence homology in nucleic acids. *Methods Enzymol.* **21D**: 413.

DREYER, W. J. and J. C. BENNETT. 1965. The molecular basis of antibody formation: A paradox. *Proc. Natl. Acad. Sci.* **54**: 864.

DUGAN, E. S., R. A. BRADSHAW, E. S. SIMMS, and H. N. EISEN. 1973. Amino acid sequence of the light chain of a mouse myeloma protein (MOPC-315). *Biochemistry* **12**: 5400.

ENQUIST, L., D. TIEMEIER, P. LEDER, R. WEISBERG, and N. STERNBERG. 1976. Safer derivatives of bacteriophage λgt·λc for use in cloning of recombinant DNA molecules. *Nature* **259**: 596.

FAUST, C. H., H. DIGGELMANN, and B. MACH. 1974. Estimation of the number of genes coding for the constant part of the mouse immunoglobulin kappa light chain. *Proc. Natl. Acad. Sci.* **71**: 2491.

GILLESPIE, D. and S. SPIEGELMAN. 1965. A quantitation assay for DNA-RNA hybrids with DNA immobilized on a membrane. *J. Mol. Biol.* **12**: 829.

HONJO, T., S. PACKMAN, D. SWAN, and P. LEDER. 1976. Quantitation of constant and variable region genes for mouse immunoglobulin λ chains. *Biochemistry* **15**: 2780.

HONJO, T., S. PACKMAN, D. SWAN, M. NAU, and P. LEDER. 1974. Organization of immunoglobulin genes: Reiteration frequency of the mouse κ chain constant region gene. *Proc. Natl. Acad. Sci.* **71**: 3659.

HOZUMI, N. and S. TONEGAWA. 1976. Evidence for somatic rearrangement of immunoglobulin genes coding for variable and constant regions. *Proc. Natl. Acad. Sci.* **73**: 3628.

KRAMER, R. A., J. R. CAMERON, and R. W. DAVIS. 1976. Isolation of bacteriophage λ containing yeast ribosomal RNA genes: Screening by in situ RNA hybridization to plaques. *Cell* **8**: 227.

MANDEL, M. and A. HIGA. 1970. Calcium-dependent bacteriophage DNA infection. *J. Mol. Biol.* **53**: 159.

MAXAM, A. M. and W. GILBERT. 1977. A new method for sequencing DNA. *Proc. Natl. Acad. Sci.* **74**: 560.

MCKEAN, D., M. POTTER, and L. HOOD. 1973. Mouse immunoglobulin chains. Pattern of sequence variation among kappa chains with limited sequence differences. *Biochemistry* **12**: 760.

MERTZ, J. E. and R. W. DAVIS. 1972. Cleavage of DNA by R_I restriction endonuclease generates cohesive ends. *Proc. Natl. Acad. Sci.* **69**: 3370

MILSTEIN, C. and J. SVASTI. 1971. Expansion and contraction in the evolution of immunoglobulin gene pools. *Prog. Immunol.* **1**: 33.

PERNIS, B., G. CHIAPPINO, A. S. KELUS, and P. G. H. GELL. 1965. Cellular localization of immunoglobulins with different allotypic specificities in rabbit lymphoid tissues. *J. Exp. Med.* **122**: 853.

RABBITTS, T. H., J. M. JARVIS, and C. MILSTEIN. 1975. Demonstration that a mouse immunoglobulin light chain messenger RNA hybridizes exclusively with unique DNA. *Cell* **6**: 5.

SMITH, H. O. and M. L. BIRNSTIEL. 1976. A simple method for DNA restriction site mapping. *Nucleic Acids Res.* **3**: 2387.

SOUTHERN, E. M. 1975. Detection of specific sequences among DNA fragments separated by gel-electrophoresis. *J. Mol. Biol.* **98**: 503.

STAVNEZER, J., R. C. C. HUANG, E. STAVNEZER, and J. M. BISHOP. 1974. Isolation of messenger RNA for an immunoglobulin kappa chain and enumeration of the genes for the constant region of kappa chain in the mouse. *J. Mol. Biol.* **88**: 43.

THOMAS, M., J. R. CAMERON, and R. W. DAVIS. 1974. Viable molecular hybrids of bacteriophage lambda in eukaryotic DNA. *Proc. Natl. Acad. Sci.* **71**: 4579.

TONEGAWA, S. 1976. Reiteration frequency of immunoglobulin light chain genes: Further evidence for somatic generation of antibody diversity. *Proc. Natl. Acad. Sci.* **73**: 203.

TONEGAWA, S., C. BRACK, N. HOZUMI, and R. SCHULLER. 1977a. Cloning of an immunoglobulin variable region gene from mouse embryo. *Proc. Natl. Acad. Sci.* **74:** 3518.

TONEGAWA, S., N. HOZUMI, C. BRACK, and R. SCHULLER. 1977b. Arrangement and rearrangement of immunoglobulin genes. In Regulation of the immune system: Genes and the cells in which they function. *ICN-UCLA Symp.* (in press).

TONEGAWA, S., N. HOZUMI, G. MATTHYSSENS, and R. SCHULLER. 1977c. Somatic changes in the content and context of immunoglobulin genes. *Cold Spring Harbor Symp. Quant. Biol.* **41:** 877.

TONEGAWA, S., C. STEINBERG, S. DUBE, and A. BERNARDINI. 1974. Evidence for somatic generation of antibody diversity. *Proc. Natl. Acad. Sci.* **71:** 4027.

WHITE, R. L. and D. S. HOGNESS. 1977. R-loop mapping of the 18S and 28S sequences in the long and short repeating units of *Drosophila melanogaster* rDNA. *Cell* **10:** 177.

WIGZELL, H. 1973. Antibody diversity: Is it all coded for by the germ line genes? *Scand. J. Immunol.* **2:** 199.

Studies on the Structure of Genes Expressed during Development

G. K. SIM,* A. EFSTRATIADIS,† C. W. JONES,† F. C. KAFATOS,† M. KOEHLER,† H. M. KRONENBERG,‡ T. MANIATIS,* J. C. REGIER,† B. F. ROBERTS,‡ AND N. ROSENTHAL†

*Division of Biology, California Institute of Technology, Pasadena, California 91125; †Cellular and Developmental Biology, Biological Laboratories, Harvard University, Cambridge, Massachusetts 02138; ‡Biology Department, Massachusetts Institute of Technology, Cambridge, Massachusetts 02139

During cellular differentiation, the temporal and quantitative expression of specific genes is precisely controlled. By isolating the genes which code for proteins involved in a particular developmental pathway, the mechanism of differential gene activity could be studied at the molecular level. For example, the chromosomal arrangement of coordinately and/or sequentially expressed genes and of their putative regulatory sequences, as well as the timing and rates of synthesis and turnover of specific mRNA species, could be determined. However, the complexity of the eukaryotic genome is a major obstacle to the isolation of specific structural genes.

One means of circumventing this problem is to identify the protein and mRNA products of developmentally regulated genes and then use the mRNA as a probe to identify and isolate specific regions of chromosomal DNA carrying the genes. The major difficulty of this approach is the identification and isolation of mRNAs corresponding to specific proteins. Although many elegant procedures have been developed for specific mRNA isolation, they are most applicable to a few terminally differentiated cell types where a major fraction of the message is of one or a few species. The problem, of course, is more difficult in the case of a complex developmental pathway where many different kinds of mRNAs are found, some of them in extremely small amounts. One way of isolating individual mRNA sequences in such a system is to convert the entire population of mRNA into double-stranded DNA, and then clone and amplify this DNA using recombinant DNA techniques.

A number of laboratories, including ours, independently developed procedures for the synthesis of double-stranded DNA copies of mRNA (Rougeon et al. 1975; Efstratiadis et al. 1976; Rabbitts 1976; Higuchi et al. 1976). The procedure most commonly used is to synthesize a single-stranded DNA copy (cDNA) with reverse transcriptase, copy this DNA with *Escherichia coli* DNA polymerase I to generate a second DNA strand covalently linked and complementary to the first, and finally break the covalent bond joining the two strands using the single-strand-specific nuclease S_1. Remarkably, by using the appropriate conditions, it is possible to obtain nearly full-length double-stranded DNA copies of mRNA (Efstratiadis et al. 1976). By attaching these molecules to bacterial plasmids using standard in vitro DNA-joining techniques and cloning them in *E. coli*, large amounts of the hybrid DNA molecules can be obtained (Maniatis et al. 1976). By the very nature of the procedure, even if the initial mRNA is a complex mixture, each bacterial clone will carry a homogeneous population of hybrid plasmid DNA molecules representing only one species of mRNA. As discussed below, the cloned DNA is an exact mRNA copy and can be employed in the direct determination of the primary structure of the mRNA or as a probe to identify and isolate the gene and its flanking sequences in chromosomal DNA.

Rabbit β-Globin mRNA

Double-stranded cDNA synthesis and cloning procedures were established using rabbit globin mRNA. One clone, PβG1, which carries a copy of rabbit β-globin mRNA, was characterized in detail using restriction endonucleases (Maniatis et al. 1976). The entire eukaryotic DNA sequence in this clone was determined using the rapid DNA sequencing method of Maxam and Gilbert (1977). The sequence analysis was undertaken to determine whether or not the cloned cDNA was an exact copy of the mRNA. Fidelity could be tested quite stringently in the case of globin, since as much as 97% of the β-globin sequence in PβG1 could be compared with independent sequence data derived from β-globin protein and mRNA (see Efstratiadis et al. 1977 for detailed arguments). We found that the cloned β-globin sequence, which contains all but the first 13 nucleotides at the 5' end of the message, is an exact copy of the mRNA sequence. Thus the utility of this approach to mRNA sequencing was established.

One of the important implications of the demonstration of fidelity is that the primary structure of eukaryotic mRNAs, even those which are available in extremely small amounts, can now be determined. This should provide insights into the functional importance of specific mRNA regions, as well as into the evolution of eukaryotic structural genes.

For instance, the rabbit β-globin mRNA sequence

reveals a highly selective use of synonymous codons (Efstratiadis et al. 1977). Of the six possible leucine codons only two are used: CUC twice and CUG sixteen times. Such a nonrandom use of synonymous codons clearly indicates that, in addition to the well-documented evolutionary pressures on β-globin protein structure (Dayhoff 1972), natural selection also acts at the level of mRNA. Many nucleotide substitutions have not been established, even though the changes would not result in an amino acid replacement (i.e., silent substitutions). This may be related to the observation that codon usage in globin can be correlated with the prevalence of the corresponding isoacceptor tRNAs in rabbit reticulocytes, but not in rabbit liver (D. Hatfield et al., in prep.).

Presumably, the translational efficiency of specific codons within the reticulocyte cytoplasm has provided the selective pressure for their preferential usage in β-globin mRNA. Consistent with this presumption is the observation that human β-globin mRNA shows a nearly identical pattern of codon preference to that found in rabbit (Table 1) (Marrota et al. 1977). The codon usage pattern for other eukaryotic and prokaryotic mRNAs is different but also nonrandom (see Efstratiadis et al. 1977 for discussion). All of these observations are consistent with the hypothesis of functional adaptation of tRNA (Garel et al. 1971; Suzuki and Brown 1972; Garel 1974; Smith 1975). According to this hypothesis, the cytoplasmic tRNA population in highly differentiated cells is adjusted to the amino acid composition of the proteins being synthesized. An alternative hypothesis consistent with the data is that some codons are under selective pressure because their corresponding isoacceptor tRNAs modulate the rate of translation (Itano 1965). In either case, the mechanism by which the levels of specific isoacceptor tRNAs are controlled in reticulocytes and other highly differentiated cells is not known.

A comparison of the rabbit and human β-globin mRNA sequences has provided further evidence that natural selection acts to minimize changes in particular mRNA regions. As shown in Figure 1, the distribution of substitutions within both the coding region and the 3' untranslated region is clearly nonrandom. The distal 65 nucleotides of the 3' untranslated region, nearest to the poly(A), show only ten substitutions; by contrast, the proximal part of the 3' untranslated region, nearest the terminator codon, is very divergent and unequal in length in the two species (30 nucleotides in the rabbit, 70 in the human). Within the coding region, two long segments show particularly low divergence: one segment is 62 nucleotides long and completely invariant

Table 1. Codon Usage in β-Globin mRNA

1\2		U			C			A			G		2\3
U	Phe	UUU$^3_5$	8	Ser	UCU$^3_0$	3	Tyr	UAU$^1_2$	3	(Cys)	UGU$^1_2$	3	U
		UUC$^5_3$	8		UCC$^3_2$	5		UAC$^2_1$	3		UGC	0	C
	Leu	UUA	0		UCA	0	(Term)	UAA$^0_1$	1	(Term)	UGA$^1_0$	1	A
		UUG	0		UCG	0		UAG	0	(Trp)	UGG$^2_2$	4	G
C	Leu	CUU	0	Pro	CCU$^3_5$	8	His	CAU$^4_2$	6	Arg	CGU	0	U
		CUC$^2_3$	5		CCC	0		CAC$^5_7$	12		CGC	0	C
		CUA	0		CCA$^1_2$	3	Gln	CAA	0		CGA	0	A
		CUG$^{16}_{14}$	30		CCG	0		CAG$^4_3$	7		CGG	0	G
A	(Ileu)	AUU$^1_0$	1	Thr	ACU$^2_3$	5	Asn	AAU$^1_4$	5	Ser	AGU$^4_2$	6	U
		AUC	0		ACC$^2_3$	5		AAC$^4_5$	9		AGC	0	C
		AUA	0		ACA$^0_1$	1	Lys	AAA$^3_3$	6	Arg	AGA	0	A
	(Met)	AUG$^1_1$	2		ACG	0		AAG$^9_8$	17		AGG$^3_3$	6	G
G	Val	GUU$^4_3$	7	Ala	GCU$^7_4$	11	Asp	GAU$^1_5$	6	Gly	GGU$^4_4$	8	U
		GUC$^2_2$	4		GCC$^6_7$	13		GAC$^3_2$	5		GGC$^6_8$	14	C
		GUA	0		GCA$^1_1$	2	Glu	GAA$^4_2$	6		GGA	0	A
		GUG$^{12}_{13}$	25		GCG$^1_0$	1		GAG$^6_6$	12		GGG$^1_1$	2	G

The frequency of use of each codon is shown for rabbit (upper left), human (lower left), and the sum of the two (right side). Codons with ambiguous nucleotides in the human sequence are excluded. Underlined amino acids are those for which codon usage in both species combined is significantly nonrandom according to a x^2 test ($p < 0.05$). Codon usage cannot be evaluated statistically for the cases shown in parentheses and is statistically random for the rest.

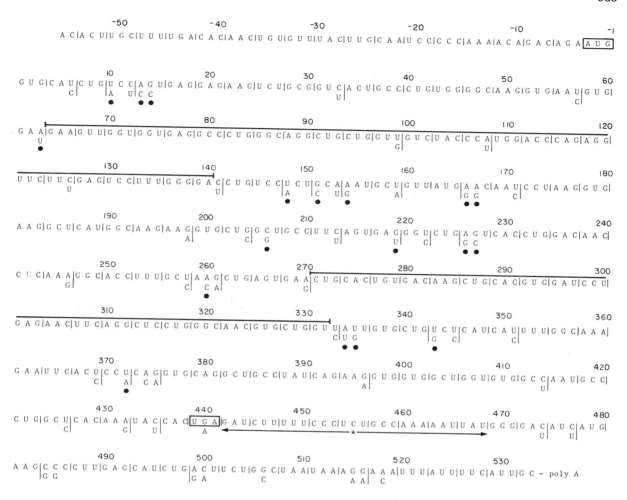

* GCUCGCUUUCUUGCUGUCCAAUUUCUAUUAAAGGUUCCUUUGUUCCCUAAGUCCAACUACUAAACUG

Figure 1. Comparison of rabbit and human β-globin mRNA sequences. The rabbit β-globin mRNA sequence *(top line)* and the nucleotide substitutions (except for the 5′ untranslated region) in the human β-globin mRNA sequence *(bottom line)* are shown. Nucleotide substitutions leading to amino acid replacements are indicated by dots. The initiation and termination codons are in boxes. Triplets in the coding region are separated by vertical lines. Two almost invariant nucleotide stretches corresponding to the functionally most important amino acid residues in β-globin chain are overlined. The region immediately following the termination codon is very divergent, and the longer human sequence (*) is shown separately. Y symbolizes a pyrimidine and X any nucleotide.

(nucleotides 271–332) (see Efstratiadis et al. 1977 for nomenclature), whereas the other is 77 nucleotides long and contains only three substitutions (nucleotides 64–140). It is particularly intriguing that these two segments code for the majority of the 35 functionally most important amino acid residues of β-globin (Goodman et al. 1975). It is possible that the structure of the mRNA has evolved in such a way that selection operating at the mRNA level imposes the greatest constraints on nucleotide regions roughly corresponding to the most important amino acid domains. This might occur, for example, if these domains correspond to nucleotide regions which must remain extensively base-paired for proper mRNA function.

The availability of cloned β-globin genes also pro-

vides the opportunity for determining which regions of the mRNA are required for protein synthesis. As a first step in this direction, we have tested the translatability of RNA transcripts from cloned β-globin DNA. The β-globin DNA insertion was excised from PβG1 with S_1 nuclease using the procedure of Hofstetter et al. (1976), and the globin DNA was separated from plasmid DNA by sucrose gradient centrifugation. When this DNA, which is missing at least 13 nucleotides corresponding to the 5′ end of the message, is added to an in-vitro-linked transcription-translation system from wheat germ (Roberts et al. 1975; Rozenblatt et al. 1976), the incorporation of [35S]methionine into TCA-precipitable material is stimulated more than 20-fold over background (Table 2). This stimulation is not substantially lower

Table 2. [³⁵S]Methionine Incorporation in a Linked Cell-free Transcription-Translation System

Template	Wheat germ RNA polymerase	Incorporation into TCA-insoluble material (cpm/μl)
No DNA	—	515
Globin DNA[a]	—	530
Globin DNA	+	11,880
Globin BglII A DNA[b]	+	5,250
Globin mRNA[c]		30,000
SV40 DNA I	+	36,690

The "linked" system consists of a brief transcription reaction in a small volume, followed by the addition of wheat germ extract optimized for translation (Roberts et al. 1975; Rozenblatt et al. 1976). DNA templates (0.5–1 μg) were transcribed by 10–15 μg of wheat germ RNA polymerase II in a 10-μl reaction containing 20 mM HEPES (pH 7.9), 10 mM Mg acetate, 1 mM MnCl₂, 5 mM DTT, and 0.5 mM of each of the nucleotide triphosphates. After 15 min at 37°C, 40 μl of a translation mixture was added and incubation continued for 3 hr at 22°C. The final 50-μl mixture consisted of 30 mM HEPES (pH 7.0), 3 mM Mg acetate, 90 mM K acetate, 200 μM spermidine, 100 μM GTP, 1 mM ATP, 8 mM phosphocreatine, 8 μg/ml creatine phosphokinase, 1 mM DTT, 2 μM (600 Ci/mmole) [³⁵S]methionine (Amersham), 25 μM of the remaining 19 unlabeled amino acids, and 10 μl of preincubated wheat germ extract digested with micrococcal nuclease (Pelham and Jackson 1976). Aliquots of the reaction mixtures were precipitated with TCA and processed as described (Roberts and Paterson 1973).

[a] Globin DNA refers to the insertion of hybrid plasmid PβG1 excised with S₁ nuclease in the presence of 45% formamide (Hofstetter et al. 1976). The digestion products from 400 μg of PβG1 DNA were extracted with phenol and ethanol-precipitated. They were then dissolved in 200 μl of 20 mM EDTA containing 300 μg/ml ethidium bromide and layered on a 12-ml 5–20% sucrose gradient in 0.1 M NaCl, 10 mM Tris-HCl (pH 7.5), and 1 mM EDTA. The gradient was centrifuged for 19 hr at 250,000g, and the resolved band of eukaryotic DNA was visualized directly and collected by puncturing the side of the gradient tube with a hypodermic needle. Ethidium bromide was removed by three extractions with an equal volume of isoamyl alcohol, and the DNA was ethanol-precipitated.

[b] Globin DNA was excised from PβG1 and further purified as described above. It was then digested with BglII, which cuts the sequence once, exactly after the termination codon (Maniatis et al. 1976). The larger of the two resulting fragments (BglII A DNA : 480 bp) was recovered following electrophoresis under native conditions in a preparative 6% polyacrylamide slab gel (1:30 bis-acrylamide). Its size was estimated from the known length of plasmid PMB9 HaeIII fragments, run in parallel.

[c] The data from the cell-free translation of rabbit globin mRNA (1 μg) included in this table are from different but comparable experiments, and are shown for purposes of comparison.

Figure 2. Autofluorogram of globin polypeptides synthesized in a linked cell-free transcription-translation system. [³⁵S]methionine-labeled polypeptides were electrophoresed in a 12% SDS-polyacrylamide slab gel in phosphate buffer (Weber and Osborn 1969) at 75 V for 10 hr. The gel was exposed at −80°C for 24 hr after impregnation with PPO (Laskey and Mills 1975). The samples were 10-μl aliquots of the cell-free transcription-translation reactions described in Table 2. *(Slot 1)* Globin DNA minus RNA polymerase; *(slot 2)* globin DNA plus RNA polymerase; *(slot 3)* globin mRNA; *(slot 4)* globin BglII A DNA plus RNA polymerase; *(slot 5)* complete reaction mix minus DNA.

than that observed with an equivalent amount of SV40 DNA or of a mixture of α- and β-globin mRNAs. As shown in Figure 2, the majority of the incorporated label comigrates with β-globin on an SDS-polyacrylamide gel which resolves α- and β-globin chains. Moreover, CM-cellulose chromatography, peptide mapping, and direct N-terminal protein sequencing of the protein have identified the product as rabbit β-globin (data not shown).

We also examined the question of whether or not the cloned globin DNA would direct the in vitro synthesis of β-globin after removal of additional defined segments of the noncoding regions. This was possible in the case of the 3′ untranslated region because the enzyme BglII introduces a unique cut in β-globin DNA, immediately following the termination codon (Efstratiadis et al. 1977). S₁-excised globin DNA was cleaved with BglII, and the large frag-

ment containing the coding region (BglII A) was purified by polyacrylamide gel electrophoresis and used as a template in the linked system. Although the stimulation of incorporation of [³⁵S]methionine into acid-precipitable material was approximately half of that observed with an equivalent amount of intact excised β-globin DNA, authentic β-globin was synthesized (Table 1; Fig. 2). We conclude that at least the first 13 nucleotides of the mRNA, the poly(A) tail (Williamson et al. 1974; Bard et al. 1974; Soreq et al. 1974; Sippel et al. 1974) and the entire 3′ noncoding region, is not essential for in vitro translation. This conclusion argues against the possibility that base pairing between the 5′ and 3′ ends of the mRNA plays a critical role in the initiation of protein synthesis (see, e.g., Baralle 1977; and Proudfoot 1977).

The Silk Moth Chorion

The methods discussed above provide the opportunity for studying complicated sets of developmentally regulated genes at the molecular level. We have begun to apply these methods to the study of such a set: the genes which are responsible for formation of the eggshell (chorion) in silk moths (Paul et al. 1972; Paul and Kafatos 1975). The chorion system has been reviewed elsewhere in detail (Kafatos et al. 1977). Here, we shall briefly summarize the information relevant to our studies of chorion genes.

The eggshell is a proteinaceous structure synthesized and deposited around each oocyte, at the end of oogenesis, by an enveloping epithelium. This epithelium and the oocyte it contains are called a follicle. Within the ovary, the follicles are found strung together in eight ovarioles, each of which is a developmental series of progressively more mature follicles. Thus the various stages of choriogenesis can be obtained conveniently by dissection of a single animal. The amount of tissue available is sufficient for biochemical studies.

The chorion has a complicated ultrastructure which is associated with a similarly complicated biochemical composition. At least 80 chorion proteins are resolved by a combination of isoelectric focusing and SDS-polyacrylamide gel electrophoresis (Fig. 3) (O'Farrell 1975). As in the case of other secretory proteins, chorion proteins are processed posttranslationally (G. Thireos and M. Nadel, pers. comm.). However, it is clear that a large number of different chorion genes are responsible for the complexity of the chorion protein pattern. This conclusion is supported by the comparable complexities of protein patterns derived from mature chorions and from experiments involving cell-free translation, in vivo pulse-labeling, or in vivo pulse-chase. However, the strongest evidence comes from protein-sequencing studies: partial sequences from 17 chorion proteins are available to date, and all but one are distinguished by a unique primary structure (J. C. Regier et al., in prep.; G. Rodakis et al., unpubl.).

The chorion system is favorable for studies of gene expression during cell differentiation, primarily because specific chorion genes are expressed in a sequential manner during specific stages of choriogenesis. This is best illustrated by the "protein synthetic profiles" shown in Figure 4. Individual follicular epithelia of successive developmental stages were la-

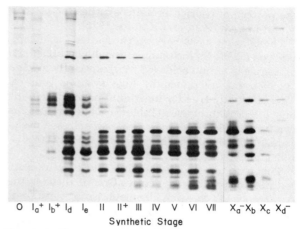

O I_a^+ I_b^+ I_d I_e II II^+ III IV V VI VII X_a^- X_b X_c X_d^-

Synthetic Stage

Figure 4. Changing pattern of protein synthesis during follicular development. A series of follicles from a single ovariole of *A. polyphemus* were labeled with [³H]leucine for 30 min. The follicular epithelial cells plus chorion were recovered, solubilized, and analyzed by electrophoresis on an SDS-urea-polyacrylamide slab gel. The labeled proteins were detected by autofluorography. Maturation is from left to right. The synthetic stages are identified in Paul and Kafatos (1975).

beled with [³H]leucine in organ culture, and the total proteins of each epithelium were analyzed by electrophoresis in an SDS-polyacrylamide slab gel. The newly synthesized proteins were detected by autofluorography. Examination of the synthetic profiles leads to two important conclusions. First, during choriogenesis the follicular cells synthesize almost exclusively chorion proteins. Second, the pattern of chorion protein synthesis changes during development; each chorion protein is produced during a characteristic period. Follicles maintained in organ culture "mature" normally, progressively changing

Figure 3. Two-dimensional fractionation of *Antheraea polyphemus* chorion proteins. The proteins were labeled in vitro with [¹⁴C]iodoacetamide (Efstratiadis and Kafatos 1976) and fractionated by isoelectric focusing (first dimension) followed by SDS-urea-polyacrylamide slab gel electrophoresis (second dimension) (O'Farrell 1975). The isoelectric focusing ampholine range is pH 4–6. Arrows indicate direction of electrophoresis. The labeled proteins were detected by autofluorography. Because of the iodoacetamide labeling procedure, the cysteine-rich A proteins are overrepresented.

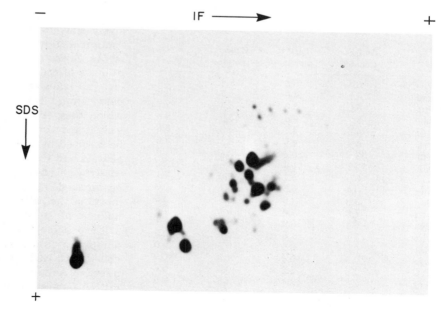

−   IF ⟶   +

SDS ↓

the pattern of chorion proteins synthesized as they would in vivo (Paul and Kafatos 1975).

The chorion has a very unusual amino acid composition (Table 3), including a very high percentage of nonpolar amino acids (68.8%), especially glycine (32.6%), and an unusually high cysteine content. This is equally true for individual purified proteins and for protein classes (Table 3). Chorion proteins can be classified into three major size classes, designated A (7000–10,000 daltons), B (11,000–14,000 daltons), and C (15,000–20,000 daltons).

The proteins are also distinguished by unusual features of primary structure. One of these is the occurrence of repeating oligopeptides. For example, in the typical A protein shown in Figure 5 (protein A4-c1), approximately half of the residues are found in repeating structures. These include a total of 6 tetrapeptide and 11 dipeptide repeats. Although the repeats often vary among proteins, the Gly-Leu-Gly and Cys-Gly peptides are particularly common. The Cys-Gly repeats tend to be located near the ends of the chorion proteins, possibly to facilitate cross-linking.

An important characteristic of chorion proteins is their extensive homology. Thus far, partial sequences have been determined for several members of both the A and B classes (J. C. Regier et al., in prep.; G. Rodakis et al., unpubl.). Numerous homologies exist within each class, whereas the two classes are not detectably related to each other within the region sequenced thus far. Within the A class at least, several subclasses can be defined by the existence of more extensive homology. It seems clear that the chorion genes of each class have evolved by repeated duplications followed by diversification, and thus can be considered members of a multigene

family. Hood et al. (1975) have defined "informational multigene families" more precisely as groups of functionally and evolutionarily related genes which are physically linked and evolve rapidly in species-specific subclasses. The chorion genes appear to meet this definition, since genetic evidence suggests that they are linked (M. R. Goldsmith, pers. comm.).

Production of Nucleic Acid Probes for Individual Chorion Structural Genes

We have seen that chorion structural genes are members of multigene families, which are strictly regulated during development, at least at the level of protein synthesis. The question of whether or not this control is imposed at the level of transcription is unanswered. The incorporation of ^{32}P into electrophoretically defined cytoplasmic chorion mRNAs has been measured in pulse-labeling experiments (Gelinas and Kafatos 1973, 1977). Synthesis of these mRNAs is detectable throughout choriogenesis, but not at earlier developmental stages. Moreover, cell-free translation assays, which reveal accumulated chorion mRNAs throughout choriogenesis, fail to indicate any accumulation of translatable chorion mRNAs during the developmental stages preceding in vivo chorion formation (Gelinas and Kafatos 1977). This evidence is consistent with the hypothesis that choriogenesis is primarily controlled at the transcriptional level, but does not constitute a critical test of that hypothesis. One approach to this question is to prepare specific nucleic acid probes, corresponding to individual mRNAs coding for a single chorion protein, and use these probes to follow the appearance of specific mRNA sequences during development.

Chorion mRNAs are easy to purify as a group because of the high specialization of the cells that produce them. During choriogenesis, the polysomes of the follicular cells, which synthesize almost exclusively the low-molecular-weight chorion proteins, contain a group of low-molecular-weight poly(A)+ RNAs. These are chorion mRNAs, as shown by a number of criteria (Kafatos et al. 1977) including cell-free translation (G. Thireos and M. R. Nadel, pers. comm.). However, since the multiple proteins which they encode are very similar in molecular weight and composition, the chorion mRNAs cannot be resolved by physical means into individual species. At best they can be separated into three broad zones by gel electrophoresis. These zones can in turn be resolved into a total of only 11 bands after removal of the poly(A) tails of variable length (Vournakis et al. 1975).

It is possible, however, to prepare individual chorion gene probes using the cloning procedures established with rabbit globin mRNA (G. K. Sim et al., in prep.). As starting material, we used total chorion mRNA purified from pooled follicles of all develop-

Table 3. Amino Acid Compositions of *A. polyphemus* Chorion Proteins (Residues/100 Residues)

	Unfrac-tionated	Class B and C proteins	Class A proteins	Purified Class A protein (A4-c1)
cm-Cys	6.4	5.6	8.4	9.4
Asp	3.7	4.0	2.5	3.6
Thr	3.0	3.0	3.4	2.7
Ser	3.7	3.8	3.0	3.6
Glu	4.5	4.7	4.2	3.6
Pro	4.4	4.5	4.1	3.6
Gly	32.6	32.2	32.4	31.5
Ala	12.1	11.4	13.8	12.9
Val	6.5	6.1	7.2	7.1
Met	0.4	0.5	0.1	0.0
Ile	3.8	3.6	3.8	3.5
Leu	7.6	7.9	6.4	7.4
Tyr	6.4	6.9	6.2	6.1
Phe	1.4	1.6	1.0	0.9
His	0.0	0.0	0.0	0.0
Lys	0.5	0.4	1.0	0.9
Trp	1.0	1.2	0.5	0.8
Arg	2.3	2.4	1.8	2.5

```
                        5
val   cys   arg   gly   gly   leu   gly   leu   lys   gly   leu   ala   ala   pro   ala   cys   gly   cys   gly
                                                  10                              15

20                      25                        30                              35
gly   leu   gly   tyr   glu   gly   leu   gly   tyr   gly   ala   leu   gly   tyr   asp   gly   leu   gly   tyr

      40                45                        50                              55
gly   ala   gly   trp   ala   gly   pro   ala   cys   gly   ?   tyr   gly   gly   glu   gly   ile   gly   asn

            60                65                        70                              75
val   ala   val   ala   gly   glu   leu   pro   val   ala   gly   thr   thr   ala   val   ala   gly   gln
```

```
                              ~95                         ~100                         ~105
[ca. 17 residues]       cys   thr   gly   cys   gly   cys   gly   cys   gly   ser   ser   (tyr, leu)
```

Figure 5. Partial sequence of *A. polyphemus* chorion protein A4-c1. Dashed and dotted lines indicate two types of tetrapeptide repeats; solid lines indicate the dipeptide repeats Val-Ala, Gly-Leu, Cys-Gly (the latter two are also part of the tetrapeptide repeats).

mental stages. Double-stranded chorion cDNA was synthesized, joined to the bacterial plasmid PML21, and the hybrid DNA used to transform *E. coli* K-12. Several hundred bacterial clones were generated, and those bearing a plasmid with inserted chorion DNA were identified by colony hybridization (Grunstein and Hogness 1975) using [^{32}P]cRNA copied from chorion cDNA as probe. Hybrid plasmid DNA was isolated from many different clones, digested with restriction enzymes, and analyzed by agarose gel electrophoresis. Clones yielding different patterns of restriction fragments were selected for further study.

Detailed Characterization of Two Synthetic Chorion Genes

Two hybrid plasmid DNAs isolated from different clones were subjected to detailed restriction enzyme mapping using procedures similar to those used in constructing the map of the globin DNA insertion in plasmid PβG1 (Maniatis et al. 1976). These clones, designated C-10 and C-401, contain insertions of 500 and 580 base pairs (bp), respectively. Of this, the chorion DNAs constitute approximately 370 and 520 bp, respectively, the remainder being the poly(dA)·(dT) linkers which join the chorion DNA to the plasmid. The detailed maps of the DNA insertions shown in Figure 6 clearly establish the nonidentity of the chorion sequences in these clones.

Direct DNA sequencing was performed to confirm that the sequences of the insertions correspond to those expected for chorion mRNA. The restriction maps were used as guides for generating a variety of overlapping restriction fragments, each uniquely labeled with ^{32}P at a single 5′ end. The end-labeled fragments were then sequenced by the method of Maxam and Gilbert (1977).

To choose among the six possible reading frames of the coding sequence, the orientation of the message strand in clone C-401 was first determined, relative to the restriction map, according to published procedures (Maniatis et al. 1976). The plasmid DNA was cleaved by the restriction enzyme *Bal* near the left end of the chorion insertion, as shown on the map of Figure 6, and one or the other strand was removed by partial digestion with either λ-exonuclease or exonuclease III. Hybridization of the two resulting DNA samples with total chorion cDNA revealed that the message strand of clone C-401 is oriented 5′ to 3′ from left to right in Figure 6.

The partial sequence of the message corresponding to this strand is shown in Figure 7. An AUG triplet is found near the 5′ end of the message sequence (beginning at nucleotide 41, according to the map shown in Fig. 6). This may represent either the initiator codon or the codon for an internal methionine; the other two reading frames are excluded, since each yields a UGA terminator (underlined in Fig. 7). When read in phase from the AUG, the partial nucleotide sequence of C-401 corresponds to an amino acid composition typical of chorion proteins: high in cysteine and glycine and in all nonpolar amino acids combined (cf. Table 3). Moreover, this sequence meets an additional criterion for being chorion mRNA: a substantial portion of the nucleotides is found in repeating units. The nucleotide repeats correspond to dipeptides, tripeptides, and hexapeptides. Particularly notable are two tandem repetitions of an 18-nucleotide segment, with only one substitution. Part of this same segment, GGUCUUGGU, is reiterated twice more elsewhere in the sequence and corresponds to Gly-Leu-Gly, a repeating peptide very common in chorion (see Fig. 5). Other oligonucleotide repeats correspond to the tripeptide Cys-Gly-Gly and the related dipeptide Cys-Gly, which is typical of many chorion proteins (Fig. 5). As in protein A4-c1 (Fig. 5) and other chorion proteins, most of the Cys-Gly (or Cys-Gly-Gly) repeats deduced from C-401 DNA are found near the two ends of the chain, in tandem arrays. In summary, it seems clear that C-401 codes for a chorion protein, although not one that has yet been sequenced.

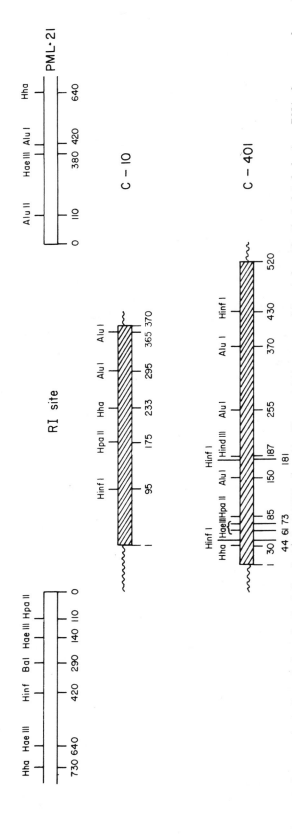

Figure 6. A comparison of two independently isolated chorion cDNA clones are shown. The chorion sequences are indicated by hatched boxes, and the PML21 DNA by open boxes. The wavy lines indicate the poly(dA-dT) sequences which link the chorion sequences to plasmid DNA. Restriction sites in the insertions are identified by a number indicating the distance from the first nucleotide of chorion sequence to the left. Except for the sequenced regions, the fragment lengths are estimated by mobility on polyacrylamide gels (Maniatis et al. 1975b). Restriction sites in PML21 DNA are identified by numbers indicating distances from the *Eco*RI site to the right or to the left. For clone C-401, the polarity of the strand synonymous to the corresponding mRNA is 5′ to 3′ from left to right.

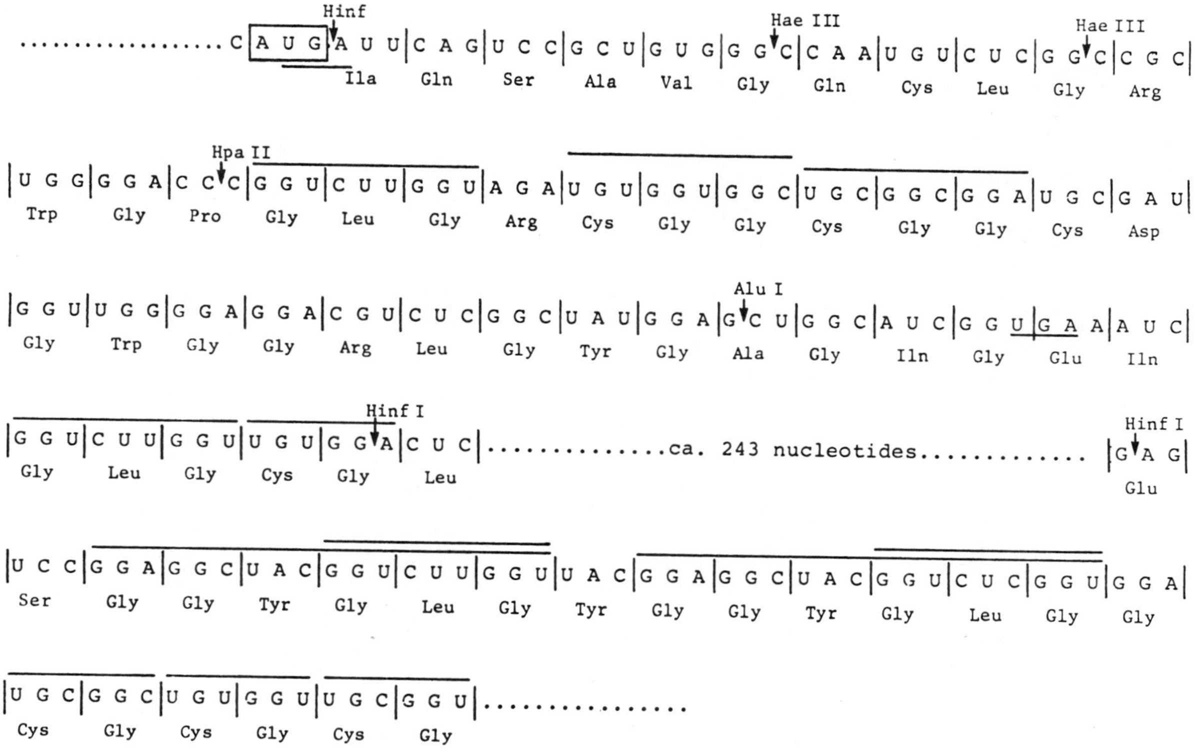

```
                              Hinf                                  Hae III                              Hae III
·················· C | A U G | A U U | C A G | U C C | G C U | G U G | G G C | C A A | U G U | C U C | G G C | C G C |
                          Ila     Gln     Ser     Ala     Val     Gly     Gln     Cys     Leu     Gly     Arg


                   Hpa II
| U G G | G G A | C C G | G U | C U U | G G U | A G A | U G U | G G U | G G C | U G C | G G C | G G A | U G C | G A U |
   Trp     Gly     Pro     Gly     Leu     Gly     Arg     Cys     Gly     Gly     Cys     Gly     Gly     Cys     Asp


                                                              Alu I
| G G U | U G G | G G A | G G A | C G U | C U C | G G C | U A U | G G A | G C U | G G C | A U C | G G U | G A A | A U C |
   Gly     Trp     Gly     Gly     Arg     Leu     Gly     Tyr     Gly     Ala     Gly     Iln     Gly     Glu     Iln


                         Hinf I                                                                           Hinf I
| G G U | C U U | G G U | U G U | G G A | C U C | ·············· ca. 243 nucleotides ············ | G A G |
   Gly     Leu     Gly     Cys     Gly     Leu                                                             Glu


| U C C | G G A | G G C | U A C | G G U | C U U | G G U | U A C | G G A | G G C | U A C | G G U | C U C | G G U | G G A |
   Ser     Gly     Gly     Tyr     Gly     Leu     Gly     Tyr     Gly     Gly     Tyr     Gly     Leu     Gly     Gly


| U G C | G G C | U G U | G G U | U G C | G G U | ················
   Cys     Gly     Cys     Gly     Cys     Gly
```

Partial Composition

11% cysteine

44% glycine

68% non-polar

46% repeating peptides

Figure 7. Nucleotide sequence of the chorion DNA insertion in clone C-401. The sequence was determined using the method of Maxam and Gilbert (1977). One of the three reading frames of chorion mRNA is shown. The other two frames are excluded by the presence of nonsense codons (underlined). An in-phase AUG codon (possibly the initiator) is boxed. Nucleotide stretches corresponding to peptide repeats are overlined. The positions of restriction endonuclease cleavage sites are shown.

Isolation and Identification of Developmental-stage-specific mRNA Sequences

As mentioned above, the chorion consists of as many as 80 distinct proteins. Therefore, the first step in isolating and identifying developmental-stage-specific genes was to compile a catalog of plasmid DNAs corresponding to individual chorion mRNA sequences from the total population. This was accomplished using a variety of techniques including restriction endonuclease cleavage, heteroduplex analysis, and cross-hybridization experiments using in-vitro-labeled chorion cDNA plasmids. At the present time, 19 different chorion cDNA plasmids have been unambiguously identified (G. K. Sim et al., in prep.). An interesting feature related to the multigene properties of the chorion gene family was revealed in cross-hybridization experiments. In these experiments one particular chorion cDNA plasmid was labeled by nick translation, and the chorion DNA insertion excised and purified on a polyacrylamide gel. The excised fragment was then used as a hybridization probe against plasmids which had previously been identified as "different" on the basis of restriction endonuclease cleavage patterns. An example of these data is shown in Figure 8. Seventeen different hybrid cDNA plasmids which hybridize chorion mRNA were digested with the restriction enzyme Hha, fractionated by electrophoresis on a 1.4% agarose gel, and transferred to nitrocellulose paper according to the method of Southern (1975). The filter was then hybridized to the in-vitro-labeled chorion DNA fragment from C-10. As shown in Figure 8B, this fragment not only hybridizes to the plasmid from which it was derived, it also hybridizes to ten other plasmids. Some of the homology is only partial since the hybridization to many of the fragments is lost when the filter is washed in 30% formamide, 2 × SSC, 65°C. The autoradiogram of the filter after this wash is shown in Figure 8C.

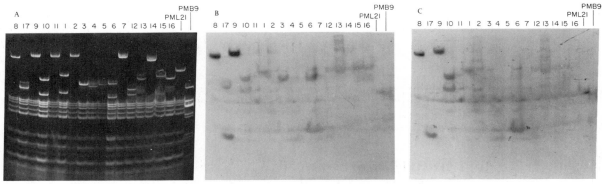

Figure 8. Cross-hybridization between "unique" chorion cDNA plasmids. Independently isolated hybrid plasmid DNA molecules shown to hybridize chorion cDNA and found to differ in their susceptibility to a large number of restriction endonucleases were digested with the enzyme *Hha*, fractionated by electrophoresis on a 1.4% agarose gel, and transferred directly to nitrocellulose paper (Southern 1975). The filter was prepared for DNA-DNA hybridization by the method of Denhardt (1966) and hybridized with in-vitro-labeled chorion DNA excised from the plasmid designated "10" using the procedure of Hofstetter et al. (1976). The plasmid DNA was labeled in vitro by nick translation using the conditions of Maniatis et al. (1975a), phenol-extracted, ethanol-precipitated, resuspended, and subjected to S_1 nuclease digestion in the presence of formamide as described by Hofstetter et al. (1976). The chorion insertion was separated from plasmid DNA by electrophoresis on a 3.5% polyacrylamide gel, and the DNA was recovered from the gel as previously described (Maniatis et al. 1976). *(A)* Ethidium-bromide-stained agarose gel. *(B)* Autoradiogram of the nitrocellulose filter. After hybridization for 16 hr at 65°C in 4 × SSC, the filter was washed extensively with 2 × SSC at 65°C (Botchan et al. 1976). PML21 designates the gel slot containing the plasmid vector DNA. The numbers at the top of each gel slot indicate the number assigned to the various plasmids. *(C)* The filter shown in *B* after further washing with 2 × SSC, 30% formamide, at 65°C.

As can be seen, the bands in the self-hybrid are relatively more prominent than in Figure 8B, and many other bands disappear. The background on the filter is due to the presence of the plasmid PML21 sequences contaminating the excised chorion DNA. The fact that hybrids formed between the probe and some plasmids, which were shown in other experiments to be different, are stable even under these conditions demonstrates the presence of highly homologous sequences in the different chorion genes. This is consistent with amino acid sequence data and the general properties of multigene families.

Having compiled a catalog of 19 different chorion cDNA plasmids, we attempted to assign each of them to the developmental stages at which they are expressed. Stage-specific probes corresponding to individual chorion genes can be used to address the question of linkage and, more importantly, the relationship of linkage, coordinate control, and evolutionary relatedness. Moreover, it should be possible to unambiguously determine whether or not selective gene expression is mediated at the level of transcription.

To classify the cDNA clones according to the devel-

opmental stage from which each arises, we used the following procedure. A series of animals was dissected and one of the eight ovarioles from each animal was pulse-labeled with [³H]leucine. The follicles from the rest of the ovarioles were pooled according to corresponding morphological developmental stage and frozen separately for each animal. The pulse-labeled chorion proteins were electrophoresed in polyacrylamide slab gels, and synthetic profiles were identified and used as a guide to pool follicles (50–150) of the same synthetic stage from different animals. Cytoplasmic RNA was extracted and passed over oligo(dT)-cellulose. The bound material was then eluted and fragmented by heating at alkaline pH, end-labeled with [γ-³²P]ATP and polynucleotide kinase, and used as a stage-specific hybridization probe. Hybridization was carried out with the 19 different plasmid DNAs which were linearized by restriction cleavage, fractionated by electrophoresis on a 1.4% agarose gel, and transferred to nitrocellulose paper according to Southern (1975).

Although the results of this experiment are still preliminary, the data of Figure 9 clearly show that individual chorion cDNA plasmids can be assigned

Figure 9. Hybridization of in-vitro-labeled stage-specific mRNA to individual chorion cDNA plasmids. PML21 and 19 unique hybrid plasmids were linearized by digestion with restriction endonucleases that cleave PML21 DNA once but do not cut within the inserted chorion sequence. The linear DNAs were then electrophoresed on a 1% agarose gel and transferred directly to nitrocellulose filters (Southern 1975). Poly(A)-containing cytoplasmic mRNA was isolated from follicles at different stages of choriogenesis (Efstratiadis and Kafatos 1976), fragmented by heating for 40 sec (pH 9.1 at 82°C), and then labeled at the 5' ends using [γ-³²P]ATP and T₄ polynucleotide kinase. Each stage-specific probe was hybridized to a filter containing DNA from all 19 plasmids. Hybridization was carried out in 4 × SSC, 65°C, for 8 hr. The filters were washed extensively in 2 × SSC 0.2% SDS at 65°C and autoradiographed. *(Slots 1–19)* Hybrid plasmid DNAs; *(slot 20)* PML21 DNA.

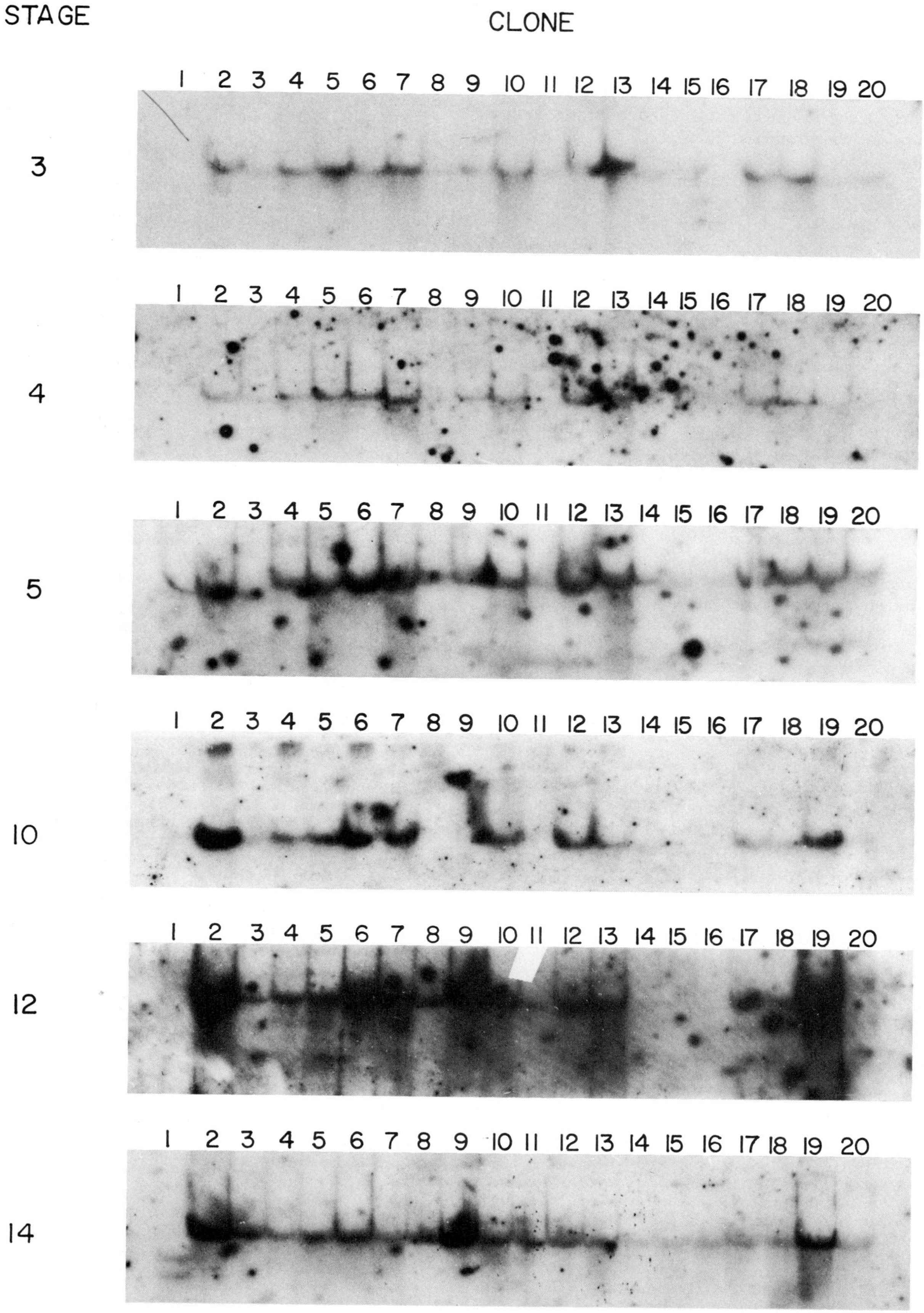

STAGE

CLONE

Figure 9. *(See facing page for legend)*

943

to developmental stages (early, middle, or late). For example, the chorion mRNA sequence represented in clone 2 is present in relatively small amounts early in choriogenesis, but is found in large amounts late in choriogenesis. Similarly, the mRNA sequences in clones 17 and 18 appear early and diminish at late times, whereas clone 19 is absent early and is the most heavily labeled plasmid when probes from the late developmental stages are used. Melting experiments similar to that described in Figure 8 are being carried out to discriminate between perfect and partial sequence homology.

DISCUSSION

An approach has been developed for the study of the structure and organization of developmentally regulated genes. Nearly full-length double-stranded DNA copies of mRNA can be accurately synthesized in vitro and amplified using recombinant DNA techniques. We have shown that reliable nucleotide sequence information can be derived from the cloned DNA and that individual members of a complex family of mRNA sequences involved in differentiation can be purified to homogeneity and amplified. The assignment of individual cDNA plasmids to the developmental stage in which the corresponding mRNA sequence is found provides the opportunity to study the organization and regulation of individual developmental stage-specific genes in chromosomal DNA.

Acknowledgments

Some of the work described here was performed while two of us (G. K. S. and T. M.) were at the Cold Spring Harbor Laboratory. We are grateful to J. D. Watson for support and encouragement. B. R. and H. K. are grateful to A. Rich for support. We thank B. Forget and S. Weissman for communicating unpublished data. This work was supported primarily by a grant from the National Science Foundation to F. C. K. and T. M.

REFERENCES

BARALLE, F. E. 1977. Complete nucleotide sequence of the 5′ noncoding region of rabbit β-globin mRNA. *Cell* 10: 549.

BARD, E., D. EFRON, A. MARCUS, and R. P. PERRY. 1974. Translational capacity of deadenylated messenger RNA. *Cell* 1: 101.

BOTCHAN, M., W. TOPP, and J. SAMBROOK. 1976. The arrangement of simian virus 40 sequences in the DNA of transformed cells. *Cell* 9: 269.

DAYHOFF, M. O. 1972. *Atlas of protein sequence and structure.* vol. 5. National Biomedical Research Foundation, Silver Spring, Maryland.

DENHARDT, D. 1966. A membrane-filter technique for the detection of complementary DNA. *Biochem. Biophys. Res. Commun.* 23: 641.

EFSTRATIADIS, A. and F. C. KAFATOS. 1976. The chorion of insects: Techniques and perspectives. *Meth. Mol. Biol.* 8: 1.

EFSTRATIADIS, A., F. C. KAFATOS, and T. MANIATIS. 1977. The primary structure of rabbit β-globin mRNA as determined from cloned DNA. *Cell* 10: 571.

EFSTRATIADIS, A., F. C. KAFATOS, A. M. MAXAM, and T. MANIATIS. 1976. Enzymatic *in vitro* synthesis of globin genes. *Cell* 7: 279.

GAREL, J. P. 1974. Functional adaptation of tRNA population. *J. Theor. Biol.* 43: 211.

GAREL, J. P., P. MANDEL, G. CHAVANCY, and J. DAELLE. 1971. Functional adaptation of tRNAs to protein biosynthesis in a highly differentiated system. *FEBS Lett.* 12: 249.

GELINAS, R. E. and F. C. KAFATOS. 1973. Purification of a family of specific mRNAs from moth follicular cells. *Proc. Natl. Acad. Sci.* 70: 3764.

———. 1977. The control of chorion protein synthesis in silkmoths: mRNA production parallels protein synthesis. *Dev. Biol.* 55: 179.

GOODMAN, M., G. W. MOORE, and G. MATSUDA. 1975. Darwinian evolution in the geneology of haemoglobin. *Nature* 253: 603.

GRUNSTEIN, M. and D. HOGNESS. 1975. Colony hybridization: A method for the isolation of cloned DNAs that contain a specific gene. *Proc. Natl. Acad. Sci.* 72: 3961.

HIGUCHI, R., G. V. PADDOCK, R. WALL, and W. SALSER. 1976. A general method for cloning eukaryotic structural gene sequences. *Proc. Natl. Acad. Sci.* 73: 3146.

HOFSTETTER, H., A. SCHAMBÖCK, J. VAN DEN BERG, and C. WEISSMANN. 1976. Specific excision of the inserted DNA segment from hybrid plasmids constructed by the poly(dA)·poly(dT) method. *Biochim. Biophys. Acta* 454: 578.

HOOD, L., J. H. CAMPBELL, and S. C. R. ELGIN. 1975. The organization, expression and evolution of antibody genes and other multigene families. *Annu. Rev. Genet.* 9: 305.

ITANO, H. A. 1965. The synthesis and structure of normal and abnormal haemoglobins. In *Abnormal haemoglobins in Africa* (ed. J. H. P. Jonxis), p. 3. Blackwell, Oxford.

KAFATOS, F. C., J. C. REGIER, G. D. MAZUR, M. R. NADEL, H. BLAU, W. H. PETRI, A. R. WYMAN, R. E. GELINAS, P. B. MOORE, M. PAUL, A. EFSTRATIADIS, J. VOURNAKIS, M. R. GOLDSMITH, J. HUNSLEY, B. BAKER, J. NARDI, and M. KOEHLER. 1977. The eggshell of insects: Differentiation-specific proteins and the control of their synthesis and accumulation during development. *Results Probl. Cell Differ.* 8: (in press).

LASKEY, R. A. and A. D. MILLS. 1975. Quantitative film detection of ³H and ¹⁴C in polyadenylamide gels by fluorography. *Eur. J. Biochem.* 56: 335.

MANIATIS, T., A. JEFFREY, and D. G. KLEID. 1975a. Nucleotide sequence of the rightward operator of phage λ. *Proc. Natl. Acad. Sci.* 72: 1184.

MANIATIS, T., A. JEFFREY, and H. VAN DE SANDE. 1975b. Chain length determination of small double and single stranded DNA molecules by polyacrylamide gel electrophoresis. *Biochemistry* 14: 3787.

MANIATIS, T., G. K. SIM, A. EFSTRATIADIS, and F. C. KAFATOS. 1976. Amplification and characterization of a β-globin gene synthesized *in vitro*. *Cell* 8: 163.

MARROTA, C. A., J. T. WILSON, B. G. FORGET, and S. M. WEISSMAN. 1977. Human β-globin mRNA. *J. Biol. Chem.* (in press).

MAXAM, A. M. and W. GILBERT. 1977. A new method for sequencing DNA. *Proc. Natl. Acad. Sci.* 74: 560.

O'FARRELL, D. H. 1975. High resolution two-dimensional electrophoresis of proteins. *J. Biol. Chem.* 250: 4607.

PAUL, M. and F. C. KAFATOS. 1975. Specific protein synthesis in cellular differentiation. II. The program of protein synthetic changes during chorion formation by silkmoth follicles and its implementation organ culture. *Dev. Biol.* 42: 141.

PAUL, M., M. R. GOLDSMITH, J. R. HUNSLEY, and F. C. KA-FATOS. 1972. Specific protein synthesis in cellular differentiation. I. *J. Cell Biol.* **55**: 653.

PELHAM, H. R. B. and R. J. JACKSON. 1976. An efficient mRNA-dependent translation system from reticulocyte lysates. *Eur. J. Biochem.* **67**: 247.

PROUDFOOT, N. J. 1977. Complete 3′ noncoding region sequences of rabbit and human β-globin messenger RNAs. *Cell* **10**: 559.

RABBITTS, T. H. 1976. Bacterial cloning of plasmids carrying copies of rabbit globin messenger RNA. *Nature* **260**: 221.

ROBERTS, B. E. and B. M. PATERSON. 1973. Efficient translation of tobacco mosaic virus RNA and rabbit globin 9S RNA in a cell-free system from commercial wheat germ. *Proc. Natl. Acad. Sci.* **70**: 2330.

ROBERTS, B. E., M. GORECKI, R. C. MULLIGAN, K. J. DANNA, S. ROZENBLATT, and A. RICH. 1975. Simian virus 40 DNA directs synthesis of authentic viral polypeptides in a linked transcription-translation cell-free system. *Proc. Natl. Acad. Sci.* **72**: 1922.

ROUGEON, F., D. KOURILSKY, and B. MACH. 1975. Insertion of a rabbit β-globin gene sequence into an *E. coli* plasmid. *Nucleic Acids Res.* **2**: 2365.

ROZENBLATT, S., R. C. MULLIGAN, M. GORECKI, B. E. ROBERTS, and A. RICH. 1976. Direct biochemical mapping of eukaryotic viral DNA by means of a linked transcription-translation cell-free system. *Proc. Natl. Acad. Sci.* **73**: 2747.

SIPPEL, A. E., J. G. STAVRIANOPOULOS, G. SCHUTZ, and P. FEIGELSON. 1974. Translational properties of rabbit globin mRNA after specific removal of poly(A) with ribonuclease H. *Proc. Natl. Acad. Sci.* **71**: 4635.

SMITH, D. W. E. 1975. Reticulocyte transfer RNA and hemoglobin synthesis. *Science* **190**: 529.

SOREQ, H., U. NUDEL, R. SALOMON, M. REVEL, and U. Z. LITTAUER. 1974. *In vitro* translation of polyadenylic acid-free rabbit globin messenger RNA. *J. Mol. Biol.* **88**: 233.

SOUTHERN, E. 1975. Detection of specific sequences among DNA fragments separated by gel electrophoresis. *J. Mol. Biol.* **98**: 503.

SUZUKI, Y. and D. D. BROWN. 1972. Isolation and identification of the mRNA for silk fibroin from *Bombyx mori*. *J. Mol. Biol.* **63**: 409.

VOURNAKIS, J. N., A. EFSTRATIADIS, and F. C. KAFATOS. 1975. Electrophoretic patterns of deadenylated chorion and globin mRNAs. *Proc. Natl. Acad. Sci.* **72**: 2959.

WEBER, K. and M. OSBORN. 1969. The reliability of molecular weight determinations by dodecyl sulfate-polyacrylamide gel electrophoresis. *J. Biol. Chem.* **244**: 4406.

WILLIAMSON, R., J. CROSSLEY, and S. HUMPHRIES. 1974. Translation of mouse globin messenger ribonucleic acid from which the poly(adenylic acid) sequence has been removed. *Biochemistry* **13**: 703.

Isolation and Characterization of the Silk Fibroin Gene with Its Flanking Sequences

Y. Suzuki and Y. Ohshima

Department of Embryology, Carnegie Institution of Washington, Baltimore, Maryland 21210

The silk fibroin gene of the silkworm *Bombyx mori* reveals several unique features for transcriptional studies of tissue-specific genes (for review, see Suzuki 1976, 1977): (1) Although the fibroin gene exists as one copy per haploid complement (Suzuki et al. 1972; Gage and Manning 1976), its concentration in the *B. mori* genome is quite high (0.004%; Suzuki et al. 1972) because the genome size is small (0.52 pg; Gage 1974; Rasch 1974) and the gene is very large (about 11×10^6 daltons; Lizardi and Brown 1975; Lizardi et al. 1975). (2) More than 80% of the fibroin mRNA sequence is composed of very simple repetitious nucleotides of high GC content (Suzuki and Brown 1972; Suzuki and Suzuki 1974), and the gene can be identified unequivocally by RNase T_1 fingerprint analysis of the hybridized RNA (Suzuki et al. 1972). (3) The gene expression seems to be temporally regulated (Suzuki and Suzuki 1974), and (4) at a certain stage of the posterior silk gland differentiation, nascent mRNA comprises about 6% of the total RNA pulse-labeled in 10 minutes (Suzuki and Giza 1976). (5) A pair of posterior silk glands supplies about 200 μg each of fibroin mRNA and bulk DNA (Suzuki and Giza 1976). These features of the fibroin gene have been useful in cloning the gene with its flanking sequences and will also be useful as we try to understand fibroin gene control by reconstituting its function in vitro with the purified gene and necessary components.

We have obtained 20 fibroin gene clones of genomic origin and have constructed a restriction enzyme map of the gene and its flanking sequences (Ohshima and Suzuki 1977). Some of the clones carry both the mRNA coding sequence and adjacent flanking sequences which are presumed to contain regulatory information for fibroin gene transcription. This is the first example of a tissue-specific gene isolated from the differentiated tissue which transcribes the gene actively. We have also mapped the probable primary initiation site of fibroin gene transcription (Y. Suzuki et al., unpubl.) at or near the site where the 5′ end of mature mRNA was mapped (Y. Tsujimoto and Y. Suzuki, unpubl.). Some of these experiments are summarized below.

Experimental Design for the Cloning of Fibroin Gene

We wanted to clone the complete fibroin gene with its flanking sequences from the posterior silk glands

which transcribe the gene specifically. Since the mRNA coding part was about 11×10^6 daltons by itself (Lizardi et al. 1975), we wished to work with DNA in the range of $20–25 \times 10^6$ daltons. To insert DNA of this size into a plasmid, the poly(dA)-poly(dT) joining method (Lobban and Kaiser 1973) was thought suitable. An alternative method we employed was the ligation of an *Eco*RI fragment with a plasmid (Dugaiczyk et al. 1975). The cloning vehicle we used is pMB9 (Rodriguez et al. 1976), which is a ColE1 derivative and carries a tetracycline-resistance marker. Fibroin gene was enriched by repeated centrifugations of mildly sheared DNA in actinomycin D-CsCl gradients, or by an actinomycin D-CsCl gradient centrifugation of *Eco*RI-digested DNA followed by a size fractionation of the fragments in a sucrose gradient. The enriched fractions were integrated into pMB9, the transformants selected by tetracycline resistance, and fibroin gene clones selected by hybridization of the colonies (Grunstein and Hogness 1975) with ^{125}I-labeled fibroin mRNA.

Safety Precautions

This type of cloning is classified as P-2 + EK1 by the National Institutes of Health guidelines. Microbiological experiments for the current project were done in the laboratory in our department having P-3 facilities.

Preparation of High-molecular-weight DNA, Partial Purification of Fibroin Gene, and Cloning of Fibroin Gene Plasmids

High-molecular-weight DNA was obtained by first isolating posterior silk gland nuclei and then extracting DNA from the nuclei (Suzuki and Giza 1976). When extracted, the DNA had molecular weights of 60×10^6 and 30×10^6 daltons for double and single strand, respectively, indicating that single-strand nicks were not detectable. The high-molecular-weight DNA was sheared mildly in the presence of high salt (Hershey et al. 1962) down to $20–25 \times 10^6$ daltons (double-stranded) and subjected to rounds of actinomycin D-CsCl density gradient centrifugation (Fig. 1). Fibroin gene showed a bimodal distribution in the gradient (Fig. 1a); the intact gene belongs to the size class of the bulk DNA, and the

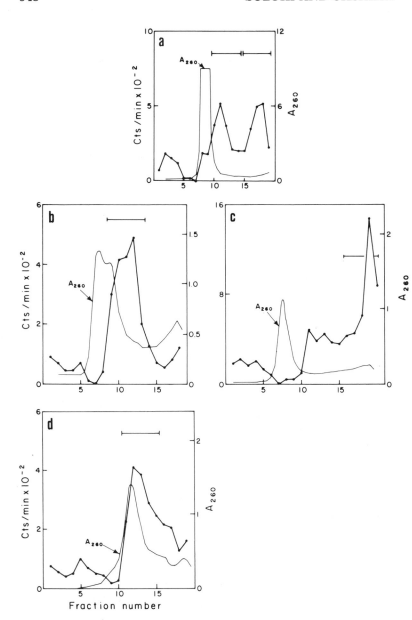

Figure 1. Partial purification of fibroin gene by actinomycin D-CsCl gradient centrifugation. (a) The first cycle of actinomycin D-CsCl centrifugation using 15 mg of *B. mori* DNA from the posterior silk glands in 12 × 20 ml gradients. (b,c) The gene-enriched fractions shown by the brackets in *a* were subjected to the second cycle (fractions 10–14 of *a* for gradient *b* and fractions 15–19 of *a* for gradient *c*). Fractions 16–19 of *c* were used for the cloning. (d) Fractions 9–13 of *b* were subjected to the third cycle. Fractions 11–15 of *d* were used for cloning. (——) A_{260}; (–●–) cpm (hybridization of aliquot fractions with ^{125}I-labeled mRNA).

other class consists of gene fragments smaller than 12×10^6 daltons (not shown). The bimodal distribution was not due to the mild shearing described above, because even unsheared DNA displayed the same proportion of fragmented genes as the mildly sheared DNA preparation (Fig. 2). It is also not due to the fragmentation of the gene in the living cell, because other DNA preparations obtained by a different method ($30–40 \times 10^6$ daltons for double strand with some nicks) showed a single size class of fibroin gene similar to bulk DNA size (see below). The fraction of the fragmented genes varied from preparation to preparation, ranging from 30% to 60% of the total fibroin gene. This fibroin-gene fragmentation could be explained by the observation that active genes in specific tissues are more susceptible to nuclease attack than inactive genes (Wein-

traub and Groudine 1976; Garel and Axel, this volume). Separate experiments showed that the intact size class of fibroin gene was kept unchanged in its single-strand molecular weight (assayed by an alkaline sucrose gradient centrifugation and hybridization of the gene with ^{125}I-labeled mRNA) through at least one cycle of actinomycin D-CsCl centrifugation and DNA recovery from the gradient. Each size class of the gene shown by the brackets in Figure 1a was pooled separately and subjected to a second cycle of actinomycin D-CsCl centrifugation. The fragmented gene fraction (Fig. 1c) was enriched about 1000-fold, giving a gene concentration of about 4%. The intact gene fraction (Fig. 1b) was further subjected to a third cycle of actinomycin D-CsCl centrifugation (Fig. 1d) and enriched about 15-fold overall (gene concentration: 0.06%).

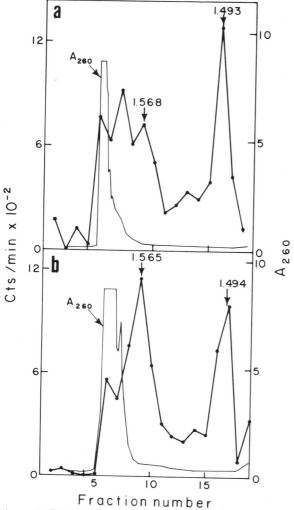

Figure 2. Distribution of fibroin gene in the actinomycin D-CsCl gradients before and after a mild shearing. *(a)* Posterior silk gland DNA (0.76 mg) extracted by the nuclei-isolation method, having a molecular weight of $\geq 60 \times 10^6$ and 30×10^6 daltons for double and single strands, respectively, in a 20-ml gradient. *(b)* After mild shearing in the presence of 0.5 M CsCl, the DNA (1.4 mg) had a molecular weight of 26×10^6 and 14×10^6 daltons for double and single strands, respectively. (——) A_{260}; (-●-) cpm (aliquots hybridized with ^{125}I-labeled mRNA; for *a*, twofold aliquots were used).

Poly(dT) tails were added to the 1000-fold-enriched fraction (Fig. 1c), and the DNA was annealed with poly(dA)-tailed pMB9 DNA. We obtained about 230 tetracycline-resistant transformants, which included 12 fibroin gene clones, at least six of which differed from each other judging from the insertion size and restriction pattern. The largest one, named pFb10, carried about 6 kb of the mRNA coding sequence and 1 kb of a flanking sequence adjacent to the 3' end of the gene as described below (see Figs. 8 and 9).

Next we carried out transformation using the high-molecular-weight fibroin gene fraction (Fig. 1d).

Out of 16,000 tetracycline-resistant transformants, we obtained seven fibroin gene clones, at least six of which were different from each other. The longest insertion was 12×10^6 daltons. This plasmid, named pFb19, carries about 6 kb of the mRNA coding sequence and 12 kb of a flanking sequence adjacent to the 5' end of the gene (see Figs. 8 and 9).

Recently, a restriction map of fibroin gene in bulk DNA of *B. mori* has been constructed by Manning and Gage (1977, and pers. comm.) and by us. We also now know a more detailed restriction map of the gene through the study of the cloned fibroin gene plasmids (see Fig. 9). The restriction enzyme *Eco*RI generates a DNA fragment of about 14×10^6 daltons which includes the entire fibroin gene of 16 kb (messenger coding sequence), 4–5 kb of a flanking sequence outside the 5' end of the gene, and possibly a trace (less than 0.5 kb) of a flanking sequence outside the 3' end. Using the SDS-proteinase K-phenol extraction method, we extracted high-molecular-weight DNA (30–40×10^6 daltons for double strand with some nicks) from the posterior silk glands without fragmentation of the fibroin gene. The DNA was digested with *Eco*RI, and the fragment containing the gene was purified 60-fold (gene concentration: 0.24%) by a banding in actinomycin D-CsCl and sucrose gradient sedimentation (not shown). Using the ligation method, we obtained one fibroin gene clone (named pFb29) out of 20,000 transformants. This plasmid has a 21-kb insertion covering the entire mRNA coding sequence and flanking sequences (Figs. 8 and 9).

Definite Identification of the Fibroin Gene in Cloned Plasmids

Although the ^{125}I-labeled mRNA used for hybridization has been judged as greater than 95% pure, some of the plasmids could correspond to DNA sequences coding for the contaminant RNAs. Therefore, we wanted to identify the gene in the key plasmids unequivocally. First, conditions for hybridizing ^{125}I-labeled mRNA at different mRNA concentrations to a gene plasmid, pFb10, were worked out (Fig. 3). Under certain subsaturation conditions, about 40% of the input mRNA hybridized with the pFb10 DNA (Fig. 3), indicating that pFb10 carries DNA coding for the major component (fibroin mRNA) of the assay probe. Second, the RNA that had hybridized at saturation and had been exposed to RNase T₁ was recovered and digested with RNase T₁, and the digest was fractionated on a DEAE-Sephadex column. The oligonucleotide patterns obtained from pFb10 and pFb19 were indistinguishable from that of the input mRNA (Fig. 4). This is the strongest evidence that these plasmids carry fibroin gene sequences. Third, a thermal denaturation profile of pFb10 was obtained (Fig. 5). About 50% of the hyperchromicity showed a very high one-step melting profile, indicating that about 6 kb of the

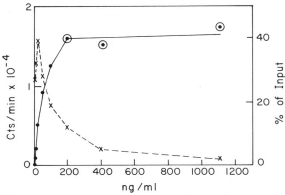

Figure 3. Saturation hybridization of ^{125}I-labeled mRNA to pFb10. About 300 ng of pFb10, which includes 75 ng of mRNA coding sequence, was used for each point. Up to 200 ng mRNA/ml undiluted ^{125}I-labeled mRNA was used. For the points at 200, 400, and 1100 ng/ml (—⊙—), the labeled mRNA was diluted with cold mRNA, and taking the undiluted 200-ng/ml point as a standard, corrections were made and plotted. (—●—) Cpm; (--x--) percent of input mRNA taken up by the gene.

12-kb plasmid represents a high-GC sequence as expected from the high-GC nature of the fibroin gene. This type of analysis reveals the approximate proportion of the mRNA coding sequence in a plasmid. These experiments eliminate any ambiguity about the nature of the cloned gene.

Screening and Restriction Mapping of the Fibroin Gene Plasmids Carrying Both mRNA Coding Sequence and Flanking Sequence(s)

We need a rapid screening method to sort out the fibroin gene plasmids which carry both the mRNA coding sequence and the flanking sequence(s) from the 20 plasmids to be tested. The screening was possible because of the following circumstances.

Manning and Gage (1977) and we have prepared a restriction map of fibroin gene using total DNA from *B. mori*. It was found that the endonuclease *Hind*III restricts at (or near) both ends of the gene, and *Eco*RI restricts close to one end of the gene and at a site away from the other end in a flanking-sequence region (see Fig. 9). *Bam*HI restricts similar to *Eco*RI, but the cleavage sites are the other way around. The vector pMB9 has one restriction site each for *Eco*RI, *Hind*III, and *Bam*HI (Rodriguez et al. 1976), but in our case the *Eco*RI site has been destroyed by the poly(dA)-tail addition (except in the case of pFb29). Using this information, we screened the fibroin gene plasmids first by testing whether they have additional restriction sites for these enzymes and then by checking whether there are any restriction fragments that do not hybridize with ^{125}I-labeled mRNA. This screening method identifies plasmids that carry both mRNA coding and flanking sequences.

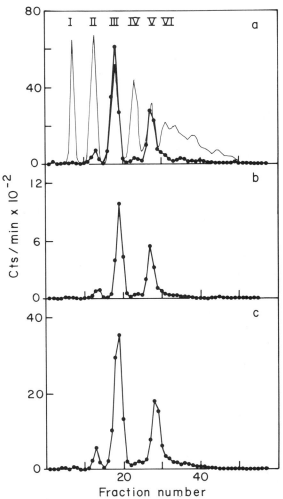

Figure 4. The oligonucleotide profiles of RNase-T₁ digests of ^{125}I-labeled fibroin mRNA, and RNAs hybridized with pFb10 and pFb19. (*a*) ^{125}I-labeled fibroin mRNA was digested by RNase T₁, and the resulting oligonucleotides were fractionated by a DEAE-Sephadex A-25 column. (*b*) ^{125}I-labeled mRNA was hybridized with pFb10, and the hybridized RNA was liberated by heating and digested with RNase T₁. (*c*) Same as in *b*, but pFb19 was used for the experiment. (——) A₂₆₀; (—●—) cpm.

Of 20 plasmids tested, most of them turned out to have intragenic fragments 0.6–11 kb in size. However, as described below, pFb10, pFb19, and pFb29 revealed multiple restriction sites for several endonucleases (see Fig. 8). As shown in Figure 6 (lane 2), pFb10 has three *Hind*III sites giving three fragments, A (11.6 kb), B (0.75 kb), and C (0.3 kb). Only fragment A hybridizes with ^{125}I-labeled mRNA (not shown). ^{32}P-labeled pMB9 DNA prepared by the nick-translation method (Kelly et al. 1970; Maniatis et al. 1975) hybridizes strongly with A and faintly with B (not shown). Therefore, we conclude that all of C and part of B are flanking sequences. The ^{32}P-labeled pMB9 DNA hybridization experiments also assign the location of the flanking sequence next to the poly(dA)-poly(dT) joining region near to the

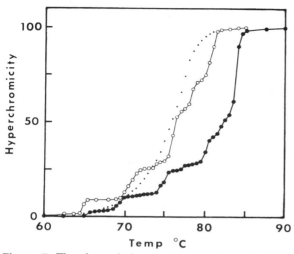

Figure 5. The thermal denaturation profiles of pFb10, pMB9, and *E. coli* DNA. About 5 μg of pFb10, pMB9, and *E. coli* DNA was subjected to thermal denaturation in 0.1 × SSC. pFb10 and pMB9 were made linear by *Hind*III digestion before the denaturation. (–●–) pFb10 DNA; (–○–) pMB9 DNA; (····) *E. coli* DNA

pMB9 *Hind*III site (Fig. 8). There is no additional *Bam*HI site in pFb10 except the one from pMB9 itself (Fig. 6, lane 6). However, pFb10 has three *Eco*RI sites giving three fragments, A (11.6 kb), B (0.9 kb), and C (0.3 kb) (Fig. 6, lane 4). ^{32}P-labeled pMB9 DNA hybridizes only with A (not shown). The approximate size of the flanking sequence is about 1 kb. From these data we have constructed a restriction map for pFb10 (Fig. 8).

The pFb19 has five *Hind*III sites giving five fragments, A (11.8 kb), B (5.7 kb), C (3.4 kb), D (3.1 kb), and E (0.2 kb) (see Fig. 7a, lane 2; E was run off from the gel shown). ^{32}P-labeled pMB9 DNA hy-

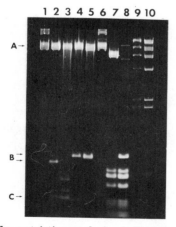

Figure 6. The restriction analysis of pFb10 by an agarose slab gel electrophoresis. Samples shown are (1) intact pFb10, and restriction digests of pFb10 by (2) *Hind*III, (3) *Hind*III and *Eco*RI, (4) *Eco*RI, (5) *Eco*RI and *Bam*HI, (6) *Bam*HI, and (7) *Hae*III, respectively, together with size markers of (8) pMB9 *Hae*III digest and (9, 10) λ DNA *Hind*III digest.

bridized strongly with the A and weakly with the C fragment (Fig. 7b). There was an additional faint hybridization band between the A and C fragments. However, it did not match with B, the appearance was not reproducible with a different batch of DNA, and the band was not detectable at lower concentrations of ^{32}P-labeled pMB9 DNA. It could be due to either partial digestion of the plasmid, microheterogeneity of pFb19, or some contaminant. Thus, we mapped the C fragment next to A in the orientation shown in Figure 8. ^{125}I-labeled mRNA at 10–20 ng/ml generally hybridizes strongly only with the A fragment in the range of 0.3–1.0 μg of the plasmid DNA. When the mRNA concentration was increased to 40–50 ng/ml, additional faint hybridization was recognized at three bands (Fig. 7c); none of them corresponds to any major restriction fragments, and the one denoted B in Figure 7c was close to the restriction fragment B but not exactly on it. Furthermore, these faint bands were not detectable in a different batch of pFb19. If the hybridization band B corresponded to the *Hind*III B fragment, the hybridized RNA might represent sequences of a presumed precursor to the mature mRNA since pFb19 was identified as the one carrying the flanking sequence adjacent to the 5' end of the gene (see next section). Therefore, the hybridized RNA (treated with RNase T$_1$) was recovered and digested with RNase T$_1$, and the resulting oligonucleotides were fractionated on a DEAE-Sephadex column. The pattern was indistinguishable from the one shown in Figure 4c. Therefore, we interpret the hybridization as not representing a unique sequence outside of the 5' end of fibroin gene, but rather an internal gene hybridization with a smaller *Hind*III A fragment derived possibly from the minor heterogeneity of the pFb19 plasmid. pFb19 has at least one *Eco*RI site (Fig. 7a, lane 8), and the *Hind*III B fragment (5.7 kb) was apparently split into two fragments by *Eco*RI digestion (Fig. 7a, lane 6): one about 4.1 kb and the other about 1.2 kb (see legend to Fig. 8 for the size discrepancy). This 1.2-kb fragment, the *Hind*III D fragment (3.1 kb; Fig. 7a, lane 2), and the *Hind*III E fragment were also detected in a *Hind*III-*Eco*RI digest of pFb29 (not shown). The *Hind*III E fragment of pFb19 (the same as the *Hind*III D fragment of pFb29) was mapped next to A (opposite to C) by a hybridization experiment of the nick-translated ^{32}P-labeled E fragment to a *Bam*HI digest of pFb29. From the comparison of the restriction maps of pFb19 and pFb29 and also from the distance between the *Eco*RI site and *Hind*III sites, the *Hind*III D fragment was located next to the *Hind*III E fragment (Fig. 8). Thus, we constructed the restriction map for pFb19 (Fig. 8). The *Hind*III fragments E, D, B, and most of C represent the 5'-end flanking sequence covering about 12 kb. Similarly, a restriction map for pFb29 has been constructed (Fig. 8).

In Figure 9 we summarize the fibroin gene restric-

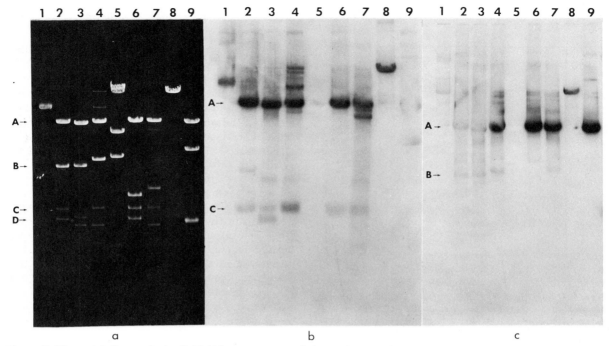

Figure 7. The restriction analysis of pFb19 by an agarose gel electrophoresis. Samples are (1) intact pFb19, and restriction digests of pFb19 by (2) *Hin*dIII, (3) *Hin*dIII and *Bam*HI, (4) *Bam*HI, (6) *Eco*RI and *Hin*dIII, (7) *Eco*RI and *Bam*HI, (8) *Eco*RI, and (9) *Sal*I, respectively, together with size markers of (5) λ DNA *Hin*dIII digest. *(a)* Electrophoresis pattern; *(b)* hybridization of the DNA with ³²P-labeled pMB9 DNA; *(c)* hybridization of DNA derived from a similar but separate experiment with ¹²⁵I-labeled mRNA.

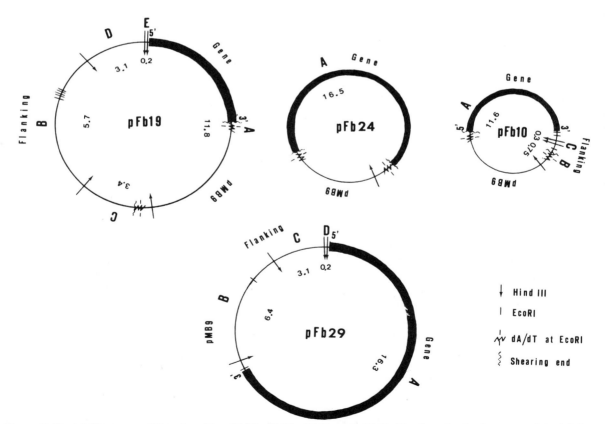

Figure 8. Restriction maps of the plasmids, pFb19, pFb24, pFb10, and pFb29. Numbers in the figure stand for kilobases between *Hin*dIII sites. For pFb19, multiple *Eco*RI sites in the *Hin*dIII B fragment are shown tentatively because *Eco*RI digestion of the B fragment (5.7 kb) gave two fragments (4.1 and 1.2kb) and other fragments larger than 0.2 kb were not observed.

952

tion map constructed mainly from the cloned plasmids and partly from the analysis of total DNA. In the figure we show the restriction maps of only the fibroin DNA part inserted into the vector.

Determination of Orientation of the Cloned Fibroin Genes

*Hin*dIII digestion of pFb10 and pFb19 gives a large fragment composed of approximately half mRNA coding sequence and half pMB9 (see Fig. 8, *Hin*dIII fragment A). Therefore, upon digestion of these fragments with either exonuclease III (3′ exo.) or λ exonuclease (5′ exo.), the coding strand will be digested away or will be conserved and become available for hybridization with ¹²⁵I-labeled mRNA without denaturation of the digested DNA. The coding strand of pFb10 became hybridizable with mRNA by digestion of the DNA with exonuclease III (Table 1, experiment 1), indicating that the 5′ end of the coding strand is extruding out in the *Hin*dIII A fragment. On the other hand, the coding strand of pFb19 became hybridizable upon digestion of the DNA with λ exonuclease, indicating that the 3′ end of the coding strand is extruding out in the *Hin*dIII A fragment (Table 1, experiment 2). Therefore, we con-

clude that pFb19 carries the coding sequence corresponding to the 5′ end of the mRNA and extra adjacent sequences outside of the gene (see Figs. 8 and 9). We also conclude that pFb10 carries sequence coding for the 3′ end of mRNA and an extra sequence adjacent to it (Figs. 8 and 9).

The Site of In Vivo Transcription Initiation on the Fibroin Gene

Before we proceed to the sequence analysis of the presumed regulatory region and/or faithful transcription studies of the gene in vitro, we must pinpoint the site of primary transcription on the gene map.

Hybridization of ¹²⁵I-labeled mRNA (20 ng/ml of mRNA to 1 μg plasmid DNA) to the plasmids pFb19 and pFb29 occurred only to the largest *Hin*dIII fragments (A). From the restriction mapping and the hybridization experiments described above, we concluded that the 5′ terminus of mature mRNA is very close (downstream less than 1 kb) to the *Hin*dIII site of the A fragment (Fig. 9). Our ¹²⁵I-labeled mRNA preparation should include any precursor molecules to the mRNA because in the mRNA purification procedure we always pool all RNA fractions

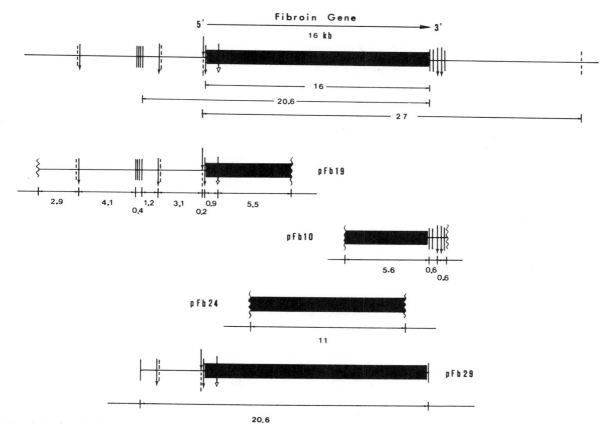

Figure 9. Restriction map of fibroin gene compiled from plasmids and *B. mori* total DNA together with restriction maps of fibroin genes in the plasmids. (↓) *Hin*dIII site: (|) *Eco*RI site; (¦) *Bam*HI site; (ʃ) sheared DNA ends with poly(dA)-poly(dT); (↓) *Hae*III site (Y. Tsujimoto and Y. Suzuki, unpubl.).

Table 1. Determination of Transcription Orientation of the Fibroin Gene in pFb10 and pFb19

Percentage DNA digested	Denaturation of DNA before hybridization	^{125}I-labeled mRNA cpm hybridized	Percentage of hybridization
Experiment 1: pFb10 *Hin*dIII fragments digested by exonuclease III			
0	—	120	0.07
15	—	26,500	15
28	—	45,900	26
0	+	178,000	100
Experiment 2: pFb19 *Hin*dIII fragments digested by λ exonuclease			
0	—	0	0
32	—	13,100	36
50	—	37,300	103
0	+	36,300	100

greater than 40S. Using this mRNA preparation, other *Hin*dIII fragments (B, C, D, and E) did not show any hybridization even after a prolonged incubation and at an mRNA concentration of 50 ng/ml. However, this experiment might not detect a primary transcript if it is present in exceedingly low amounts in the mRNA preparation.

A more direct method for locating the site of transcription initiation involves the use of a complex of capping enzymes isolated from vaccinia virion (Moss 1977). Moss and coworkers have shown that these enzymes will cap any RNA molecule bearing a 5′ di- or triphosphate terminus to form the structure m^7 GpppX^{m}pYpZp. . . . The enzymes will not cap monophosphate or hydroxyl termini. Reeder et al. (this volume) have successfully applied this in vitro capping to 40S rRNA precursor molecules to map the primary transcription site on the rDNA map. A fibroin mRNA preparation, which was obtained from the void volume region of a Bio-Gel A-50m column and should include putative precursors, was capped and labeled with S-adenosyl-[^{3}H]-methionine by these enzymes in vitro (Y. Suzuki and R. Reeder, unpubl.). The labeled mRNA was hybridized to *Hin*dIII fragments from three kinds of fibroin gene plasmids: (1) pFb19, having 12 kb of the 5′-end flanking sequence and 6 kb of mRNA coding sequence; (2) pFb25, having only an intragenic portion; and (3) pFb10 having 1 kb of the 3′-end flanking sequence and 6 kb of mRNA coding sequence. The hybrids were exposed to RNase T$_1$ to eliminate the flanking portion of the mRNA outside of the true hybrid. As shown in Figure 10, the ^{3}H-labeled mRNA hybridized only with the *Hin*dIII A fragment of pFb19, but not with any other fragments of pFb19 or other plasmids. This specificity of hybridization confirmed that the methylation occurred at the 5′ end of the mRNA and/or precursors. It also indicates that the primary site of transcription is located within the *Hin*dIII A fragment, and mRNA precursors, if any, should not be any longer

at the 5′ end than mature mRNA within a limit of the experimental error of size estimation for the 16-kb mRNA. The error should be less than 1 kb, which is about 6% of the mature mRNA length. It should be emphasized that in pFb29 the *Hin*dIII-*Eco*RI A fragment which includes the entire mRNA coding sequence was measured as 16 kb in the electron microscope (see Fig. 9).

More recently, using end-labeling techniques, we have prepared more detailed restriction map near the *Hin*dIII site (downstream 2-kb region) where the 5′ end of mature mRNA has been mapped (Y. Tsujimoto and Y. Suzuki, unpubl.). Furthermore, the site of hybridization of the capped ^{3}H-labeled mRNA has been detected in the 0.9-kb fragment which covers from the 5′-end *Hin*dIII site to the nearest *Hae*III site (downstream; Fig. 9) (Y. Suzuki et al., unpubl.).

Yang et al. (1976) reported that the capped structure of fibroin mRNA was m^7GpppA^{m}pU^{m}pCp-(Xp)Gp. We were concerned that the vaccinia enzymes might be exchanging radioactive label into mature, preexistent mRNA cap rather than labeling initiation termini. If this were occurring, digestion of the capped RNA with RNase T$_2$ would yield labeled fragments of the type m^7GpppX^{m}pY^{m}pZp since fibroin mRNA has trimethylated (cap II) termini. In contrast, the vaccinia enzymes only produce dimethylated (cap I) termini (Moss 1977). Therefore, if we were only capping polyphosphate initiation termini, RNase-T$_2$ digestion would only yield labeled fragments of the structure m^7GpppX^{m}pYp. An aliquot of the in-vitro-capped fibroin mRNA was digested with RNase T$_2$, the digest was fractionated on a DEAE-Sephadex column, and more than 90% of the radioactivity was found in a peak of −5.5 charge and about 7% in a peak of −6 charge. From this result we conclude that in the in vitro reaction we are capping polyphosphate termini and not exchanging label into preexistent caps.

RNase-T$_1$ digestion of both the in-vitro-capped fibroin mRNA and the mRNA hybridized to the

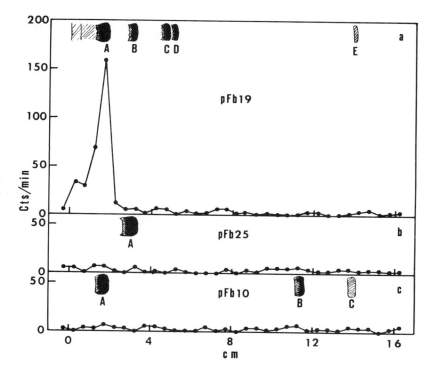

Figure 10. Hybridization of *Hind*III fragments of pFb19, pFb25, and pFb10 with ³H-labeled mRNA capped in vitro by the vaccinia enzymes. Each 10 µg of pFb19, pFb25, and pFb10 was digested with *Hind*III, electrophoresed, transferred to a Millipore filter, and hybridized with ³H-labeled mRNA which was capped in vitro by the vaccinia enzymes (the generous gift of B. Moss). The electrophoresis pattern was shown schematically above the radioactivity profile in each panel.(–●–) cpm (the filter was cut into 0.5-cm pieces and counted in a toluene fluor).

*Hind*III A fragment of pFb19 yields the same charge (−8) as that obtained from digestion of mature mRNA. RNase-A digestion of in-vitro-capped fibroin mRNA yields the structure having −5.0 charge, indicating that the structure is m⁷GpppXᵐpPyp. This structure does not contradict that of mature mRNA (m⁷GpppAᵐpUᵐpCp . . .). These results, coupled with the mapping experiment for the primary initiation site and 5′ terminus of the mature mRNA, suggest that the 5′ ends of the primary transcript and of the mature mRNA are identical for this gene, or both termini are very close to each other on the gene map.

DISCUSSION

The hybrid plasmid molecules containing tandemly repeated *Drosophila* satellite DNA have been found unstable in both *recA* and *recBC* hosts and in both pSC101 and pCR1 vectors (Brutlag et al. 1977). Similar but much less serious recombination events have been observed with the fibroin gene clones in pMB9 vector and in *recA* host. When different batches of pFb19 (or pFb29) were compared by their *Hind*III restriction patterns, only the size of the A fragment which contains the repetitious mRNA coding sequence varied from batch to batch, leaving fragments B, C, D, and E (the 5′-end flanking sequences) unchanged between the batches (Y. Suzuki and Y. Ohshima, unpubl.). For the current analysis, the batch that gave the largest size for the *Hind*III A fragment was used. In spite of this scrambling in the repetitious-sequence part of pFb19,

there appears to be a stability in the flanking sequences on which we wish to focus our attention. A better situation exists in other plasmids such as pFb10 and pFb24, which have been found more stable than pFb19 or pFb29. One of the reasons for the difference in stability between pFb19 (or pFb29) and others could be the difference in the insertion sizes. However, in a practical sense, these plasmids are quite useful for transcriptional studies as well as sequence analysis in the flanking sequences.

The sequence analysis of the fibroin coding part of the gene is not so attractive for us because the coding part is very long (more than 14 kb), and fibroin has a very simple repetitious sequence (85% of which is composed of only three amino acids: glycine, alanine, and serine [see Lucas and Rudall 1968; Suzuki 1977]). Furthermore, we already knew most of the fibroin coding sequence in essence from the partial sequence analysis of fibroin mRNA, and we found that the principal codons for glycine are GGU and GGA, leaving the frequency of GGC and GGG negligible, and that the major alanine and serine codons are GCU and UCA, respectively (Suzuki and Brown 1972). These biased preferences in codon assignments within the fibroin gene, as well as those in the β-globin gene (Sim et al., this volume), imply that the third position of these codons is not established by random or "neutral" mutation (Kimura 1974) but must be influenced by natural selection.

In vitro capping of RNAs having di- or triphosphate at their 5′ termini by the vaccinia enzymes (Moss 1977) offers a powerful tool for identifying the nascent 5′ end of RNAs. This enzyme complex appears absolutely specific for polyphosphate termini (see Reeder et al., this volume). It is not yet resolved whether or not processed, monophosphate termini are rephosphorylated to polyphosphates at a significant rate in living cells. Some data of Schibler and Perry (1976) suggest that this reaction may occur.

Lizardi (1976a,b) described a putative precursor which was about 1×10^6 daltons (1.5 kb) larger than mature mRNA. However, its identification remains tentative. The molecules might have extra sequences at their 5' and/or 3' ends. At the 5' end we did not detect as much as 1.5-kb difference between the primary initiation site and the site of 5' terminus of the mature mRNA. We know that the primary initiation site is in a 0.9-kb region from the HindIII site near the 5' end. Currently, a more detailed restriction mapping and sequence analysis around the primary transcription site are under way in our laboratory.

The ability to map the primary transcription site on the gene and the availability of fibroin gene plasmids covering different regions of the gene have given us a good assay system for in vitro transcription studies. We can now ask whether disrupted nuclei (or chromatin) retain the ability to initiate faithful transcription of fibroin genes by endogenous RNA polymerase. We have already reported that the disrupted nuclei showed extensive fibroin mRNA synthesis, that the fibroin gene was transcribed by RNA polymerase II, and that the system retained some ability to initiate transcription (Suzuki and Giza 1976). If the faithful initiation of fibroin gene transcription in the system were proven, we would try to reconstruct chromatin complexes between cloned gene segments and protein components from the posterior silk gland. By adding exogenous RNA polymerase II from the posterior gland to the reconstituted chromatin, we hope to reconstruct faithful transcription in vitro. Through this type of analysis, we hope to find regulatory factor(s) for fibroin gene transcription.

Note Added in Proof

Recently we have found that the capping reaction we have been using is not the polyphosphate terminus recognition, but the methylation of a capped and unmethylated structure (Y. Suzuki et al., unpubl.). Therefore, the initiation site described here is only a possible candidate, and the problem is still open.

Acknowledgments

We thank P. E. Giza for his excellent technical assistance, E. Suzuki for her occasional assistance, P. Botchan for his help in preparing nick-translated DNAs, L. P. Gage and R. Manning for their help in preparing our first batch of ^{125}I-labeled mRNA, and L. P. Gage for his open personal communication with us. Critical reading of the manuscript by D. D. Brown, R. H. Reeder, and Y. Tsujimoto is also appreciated. This project has been aided in part by National Institutes of Health Grant 1-R01-GM-23469-01 to Y. S.

REFERENCES

BRUTLAG, D., K. FRY, T. NELSON, and P. HUNG. 1977. Synthesis of hybrid bacterial plasmids containing highly repeated satellite DNA. Cell 10: 509.

DUGAICZYK, A., H. W. BOYER, and H. M. GOODMAN. 1975. Ligation of EcoRI endonuclease-generated DNA fragments into linear and circular structures. J. Mol. Biol. 96: 171.

GAGE, L. P. 1974. The Bombyx mori genome: Analysis by DNA reassociation kinetics. Chromosoma 45: 27.

GAGE, L. P. and R. F. MANNING. 1976. Determination of the multiplicity of the silk fibroin gene and detection of fibroin gene-related DNA in the genome of Bombyx mori. J. Mol. Biol. 101: 327.

GRUNSTEIN, M. and D. S. HOGNESS. 1975. Colony hybridization: A method for the isolation of cloned DNAs that contain a specific gene. Proc. Natl. Acad. Sci. 72: 3961.

HERSHEY, A. D., E. BURGI, and L. INGRAHAM. 1962. Sedimentation coefficient and fragility under hydrodynamic shear as measures of molecular weight of the DNA of phage T5. Biophys. J. 2: 423.

KELLY, R. B., N. R. COZZARELLI, M. P. DEUTCHER, I. R. LEHMAN, and A. KORNBERG. 1970. Enzymatic synthesis of deoxyribonucleic acid. J. Biol. Chem. 245: 39.

KIMURA, M. 1974. Gene pool of higher organisms as a product of evolution. Cold Spring Harbor Symp. Quant. Biol. 38: 515.

LIZARDI, P. M. 1976A. The size of pulse-labeled fibroin messenger RNA. Cell 7: 239.

————. 1976b. Biogenesis of silk fibroin mRNA: An example of very rapid processing. Prog. Nucleic Acid Res. Mol. Biol. 19: 301.

LIZARDI, P. M. and D. D. BROWN. 1975. The length of the fibroin gene in the Bombyx mori genome. Cell 4: 207.

LIZARDI, P. M., R. L. WILLIAMSON, and D. D. BROWN. 1975. The size of fibroin messenger RNA and its polyadenylic acid content. Cell 4: 199.

LOBBAN, P. E. and A. D. KAISER. 1973. Enzymatic end-to-end joining of DNA molecules. J. Mol. Biol. 78: 453.

LUCAS, F. AND K. M. RUDALL. 1968. Extracellular fibrous proteins: The silks. In Comprehensive biochemistry (ed. M. Florkin and E. H. Stotz), vol. 26, part B, p. 475. Elsevier, Amsterdam.

MANIATIS, T., A. JEFFREY, and D. G. KLEID. 1975. Nucleotide sequence of the rightward operator of phage λ. Proc. Natl. Acad. Sci. 72: 1184.

MANNING, R. F. and L. P. GAGE. 1977. Restriction mapping of silk fibroin gene DNA. Fed. Proc. 36: 878 (Abstr.).

MOSS, B. 1977. Utilization of the guanylyltransferase and methyltransferases of vaccinia virus to modify and identify the 5'-terminals of heterologous RNA species. Biochem. Biophys. Res. Commun. 74: 374.

OHSHIMA, Y. and Y. SUZUKI. 1977. Cloning of the silk fibroin gene and its flanking sequences. Proc. Natl. Acad. Sci. 74: (in press).

RASCH, E. M. 1974. The DNA content of sperm and hemocyte nuclei of the silkworm, Bombyx mori L. Chromosoma 45: 1.

RODRIGUEZ, R. L., F. BOLIVAR, H. M. GOODMAN, H. W. BOYER, and M. BETLACH. 1976. Construction and characterization of cloning vehicles. In Molecular mechanisms in the control of gene expression (ed. D. P. Nierlich et al.), vol. 5, p. 471. Academic Press, New York.

SCHIBLER, U. and R. P. PERRY. 1976. Characterization of the 5′ termini of hnRNA in mouse L cells: Implications for processing and cap formation. *Cell* **9**: 121.

SUZUKI, Y. 1976. Fibroin messenger RNA and its gene. *Adv. Biophys.* **8**: 83.

————. 1977. Differentiation of the silk gland: A model system for the study of differential gene action. *Results Probl. Cell Differ.* **8**: 1.

SUZUKI, Y. and D. D. BROWN. 1972. Isolation and identification of the messenger RNA for silk fibroin from *Bombyx mori. J. Mol. Biol.* **61**: 409.

SUZUKI, Y. and P. E. GIZA. 1976. Accentuated expression of silk fibroin genes in vivo and in vitro. *J. Mol. Biol.* **107**: 183.

SUZUKI, Y. and E. SUZUKI. 1974. Quantitative measurements of fibroin messenger RNA synthesis in the posterior silk gland of normal and mutant *Bombyx mori. J. Mol. Biol.* **88**: 393.

SUZUKI, Y., L. P. GAGE, and D. D. BROWN. 1972. The genes for silk fibroin in *Bombyx mori. J. Mol. Biol.* **70**: 637.

WEINTRAUB, H. and M. GROUDINE. 1976. Chromosomal subunits in active genes have an altered conformation. *Science* **193**: 848.

YANG, N.-S., R. F. MANNING, and L. P. GAGE. 1976. The blocked and methylated 5′ terminal sequence of a specific cellular messenger: The mRNA for silk fibroin of *Bombyx mori. Cell* **7**: 339.

Studies on the DNA Fragments of Mammals and *Drosophila* Containing Structural Genes and Adjacent Sequences

Y. V. ILYIN,* N. A. TCHURIKOV,* E. V. ANANIEV,† A. P. RYSKOV,* G. N. YENIKOLOPOV,*
S. A. LIMBORSKA,‡ N. E. MALEEVA,‡ V. A. GVOZDEV,† AND G. P. GEORGIEV*

*Institute of Molecular Biology, Academy of Sciences of the USSR; †Institute of Atomic Energy, and ‡Institute of
Medical Genetics, Academy of Medical Sciences of the USSR, Moscow

Several hypothetical models have been proposed to explain the structure and function of the genome in eukaryotic cells (Scherrer and Marcaud 1968; Georgiev 1969; Britten and Davidson 1969; Crick 1971; Georgiev et al. 1972, 1974). However, there has been no experimental approach for the direct verification of these models. Now it becomes possible due to the techniques of cloning and amplification of fragments of eukaryotic DNA in bacteria (Morrow et al. 1974; Thomas et al. 1974; Wensink et al. 1974). Work in this direction has been started in our laboratory along two lines. First, among λgt-Dm clones containing random fragments of *Drosophila melanogaster* DNA, those containing structural genes were selected and the properties of these inserted regions were studied (Ilyin et al. 1976; Georgiev et al. 1977).

The second line of experiments was undertaken to develop a general method for the isolation of long DNA fragments containing structural genes at desired locations in the fragment (Yenikolopov et al. 1976; Georgiev et al. 1977). In this paper the further development of work along these two lines is presented. The successful application of the above-mentioned method for the purification of DNA fragments containing mouse or human globin genes is described. In addition, the study of clones containing *D. melanogaster* DNA fragments efficiently transcribed in cell culture led us to some conclusions about the properties and the role of intercalary heterochromatin in *Drosophila*.

Method for Isolation of Native DNA Fragments Containing Structural Genes and Its Application for Purification of Globin Genes Containing Adjacent Sequences

This method (Yenikolopov et al. 1976; Georgiev et al. 1977) is based on producing short single-stranded regions in long fragments of native DNA, followed by hybridization of these fragments with mRNA and recovery of hybrids on poly(U)-Sepharose or SH-Sepharose (if mRNA is mercurated ((Fig. 1). If high-molecular-weight DNA is first fragmented and then treated with exonuclease III attacking 3' ends, then the fragments selected contain structural genes at the end of the coding strand. If a λ exonuclease attacking 5' ends of DNA is used, the fragments containing genes at the beginning of the coding strand are selected. Finally, it is possible to obtain fragments which contain structural genes mainly in the middle of the chain. For this, very long DNA strands are nicked with DNase I; then the nicks are converted to gaps with the aid of exonuclease III. Subsequently, the DNA is cut by restriction enzymes to shorter fragments, and hybridization with mRNA is performed. Exonuclease III creates not only internal gaps but also single-stranded regions at the ends of DNA. However, if the size of nonfragmented DNA is much larger than that after fragmentation and the number of nicks is optimal, then most of the single-stranded regions are located inside the DNA fragments.

The first two variants are very convenient for analytical purposes. On the other hand, the third technique is more practical for the insertion of fragments into the vector molecules (phages or plasmids) because the fragments obtained contain sticky ends.

Previously these techniques were used for the isolation of fragments containing the population of all mouse genes expressed in Ehrlich carcinoma or liver cells. This was achieved by hybridization of DNA fragments with the total mRNA of these cells. Now the techniques are applied to the purification of fragments containing particular genes, namely, the globin gene of mouse and humans.

In the case of mouse DNA, the first technique was used. About 1 mg mouse Ehrlich carcinoma DNA labeled with [³H]thymidine was treated by restriction endonuclease *Eco*RI and then by exonuclease III. About 10% of the DNA was digested. We hoped that the mouse globin gene (at least β globin) would be contained in the *Eco*RI cut, as is the rabbit globin gene (Maniatis et al. 1976), and therefore that most of the globin gene would become single-stranded after restriction. The material obtained was divided into two parts and hybridized with rabbit globin mRNA mercurated according to Dale et al. (1975) (homologous mouse globin mRNA was unavailable). Then 20 µg of mRNA and 800 µg of DNA were incubated in 10 ml of 0.3 M NaCl, 0.01 M Tris-HCl, pH

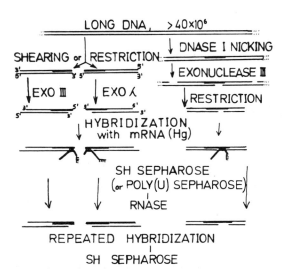

Figure 1. Scheme of the method for isolation of long fragments of DNA containing structural genes at the beginning, in the middle, or at the end of the coding strand.

7.5, at 65°C for 1 hour. R_0t was 0.024 mole · liter^{-1} · second and D_0t was ~0.1 mole · liter^{-1} · second. The mixture was fractionated on a 0.5–1.0-ml column of SH-Sepharose (Pharmacia). Unbound material was exhaustively washed out with 0.05 M NaCl, 0.05 M Tris-HCl, pH 7.5, and the hybrids were eluted with the same solution containing 0.2 M mercaptoethanol. Both fractions were collected, precipitated by ethanol with tRNA carrier, and used for further analysis.

The human unlabeled DNA ($\geq 40 \times 10^6$ daltons on average) was prepared from placenta[1] and then the internal gaps were created by successive DNase-I and exonuclease-III action (Yenikolopov et al. 1976). The number of gaps was about 1 per 4000–5000 base pairs (bp) and about 10% of the DNA was digested. Then the DNA was digested with restriction endonuclease EcoRI and hybridized with human globin mRNA prepared from human reticulocytes as previously described (Limborska et al. 1976).

The conditions of hybridization and chromatography were the same as described above. In some experiments the hybridization with nonmercurated mRNA was performed and the hybrids were isolated by chromatography on poly(U)-Sepharose columns (Ryskov et al. 1976). The samples in 0.1 M NaCl, 0.01 M EDTA, 0.2% SDS, 0.01 M Tris-HCl, pH 7.5, were passed through a 0.3–0.5-ml column of poly(U)-Sepharose (Pharmacia). The column was washed with the solution and the bound hybrids were eluted with 0.2% SDS.

The yield of bound material, which reflects the nonspecific binding, was about 0.5% on poly(U)-Sepharose and about 0.01% on SH-Sepharose. Thus,

the upper limit of purification cannot be higher than 200 and 10^4 times, respectively. In some cases a second cycle of purification was used. For this purpose, RNA was digested by RNase treatment at low ionic strength, and the DNA was deproteinized, dialyzed, and again hybridized with mercurated RNA under the same conditions, but with the DNA/RNA ratio decreased 100–1000-fold.

For the detection of the true degree of purification, the DNA samples before and after chromatography were titrated with iodinated globin mRNA according to the method of Anderson and Schimke (1976). A fixed amount (77 µg) of alkaline-denatured DNA (either total human or mouse; or the same amount of Escherichia coli DNA; or E. coli DNA plus 0.01–0.05 µg of fractionated DNA) was heated for 20 minutes at 100°C in 0.3 M NaOH, neutralized, precipitated by ethanol, and dissolved in 40 µl of 5 × SSC containing various amounts of iodinated ^{125}I-labeled globin mRNA from human or rabbit. The samples were annealed for 20 hours at 68°C. The D_0t value was about 2000 moles · liter^{-1} · second. The hybrids were recovered as acid-insoluble material after RNase treatment. The results, expressed as the difference between the RNase-resistant acid-insoluble material obtained in the presence of the studied DNA and E. coli DNA alone, were plotted against the total input RNA.

It was found in experiments with total DNA that the saturation level is proportional to the quantity of mouse or human DNA present in the reaction. The total amount of RNA bound depended on the RNA used. It was higher in homologous reactions and lower in heterologous. In general, it correlates well with the known content of globin genes in the eukaryotic genome. Figures 2 through 4 demonstrate the results obtained with fractionated DNA samples, and one can see that a very significant purification has been achieved. In the case of the poly-(U)-Sepharose technique applied to hybrids formed by human DNA containing gaps, a 70-fold enrichment with respect to globin sequences takes place, whereas after SH-Sepharose recovery of hybrids, the degree of enrichment is much higher, approximately 2200. Considering the randomness of nicks in DNA, the size of the globin gene (~600 bp), the size of the gap (~500 bp), and the final size of the fragment (4000–5000 bp), one could expect that every fourth or fifth globin gene will be separated by hybridization. Since the maximal enrichments are 200- and 10,000-fold for the two techniques, respectively, one could expect with gapped DNA 40–50-fold and 2000–2500-fold enrichments, respectively. These figures are in good correlation with the experimental results. Even higher enrichment with globin gene, 6800-fold (compare this to a theoretical maximum of ~10,000!), was obtained with EcoRI-restricted mouse DNA bearing single-stranded regions at the 5' ends. This probably depends on the existence of EcoRI sites near the globin gene(s).

---

[1] To obtain labeled material, a small amount of labeled [³H]DNA obtained from HeLa cells was added.

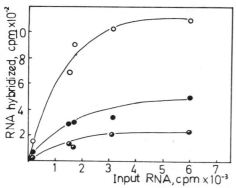

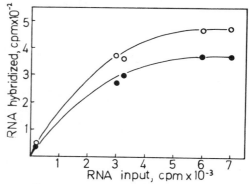

Figure 2. Saturation hybridization curves of total human DNA and DNA fractionated on poly(U)-Sepharose with human globin mRNA. Three different series of samples for hybridization were prepared. The samples of the first series contained 77 μg of total human DNA; samples of the second contained 77 μg of *E. coli* DNA, and samples of the third contained 77 μg of *E. coli* DNA plus 0.2 or 0.5 μg of human DNA fractionated on poly(U)-Sepharose. The DNA was heated for 20 min at 100°C in 0.3 м NaOH, neutralized, precipitated by ethanol, dried, and dissolved in 40 μl 5 × SSC containing various amounts of ^{125}I-labeled globin mRNA isolated from human reticulocytes and purified electrophoretically. Each hybridization series contained two or three parallel probes. The samples were annealed for 20 hr at 68°C. The hybrids were recovered after RNase treatment. The results were expressed as the differences between the RNase-resistant acid-insoluble radioactivity obtained in the presence of the human DNA (or *E. coli* DNA + human DNA) and that obtained in the presence of *E. coli* DNA alone. They were plotted against the total input RNA. (○———○) Hybridization of human globin ^{125}I-labeled mRNA with total human DNA (77μg); (●———●) the same, but with 0.5 μg of human DNA enriched with globin gene by poly(U)-Sepharose technique; (◑———◑) the same, but with 0.2 μg of the enriched DNA.

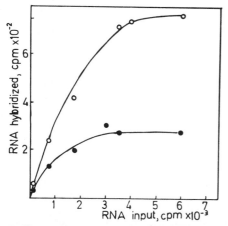

Figure 3. Saturation hybridization curves of total human DNA and DNA fractionated on SH-Sepharose with human globin mRNA. The experiments were performed as described for Fig. 2, but the amount of fractionated DNA added was 0.01 μg per hybridization probe. (○———○) Hybridization of human globin ^{125}I-labeled mRNA with total human DNA (77 μg); (●———●) the same, but with 0.01 μg of human DNA enriched with globin gene by SH-Sepharose technique.

Figure 4. Saturation hybridization curves of total mouse DNA and mouse DNA fractionated on SH-Sepharose with rabbit globin mRNA. The experiments were performed as described for Fig. 3. (○———○) Hybridization of rabbit globin ^{125}I-labeled mRNA with total mouse DNA (77 μg); (●———●) the same, but with 0.01 μg of mouse DNA enriched with globin gene by the SH-Sepharose technique.

The general conclusion is that our technique for the isolation of large DNA fragments containing structural genes applied to the isolation of individual genes gives, even after one step, a very significant purification. The SH-Sepharose technique is much more efficient than the poly(U)-Sepharose technique for hybrid recovery as it gives much lower background. Another advantage is that the background of the former is nonspecific, and in the second cycle of purification it remains of the same order. With the poly(U)-Sepharose technique, however, the background in the second cycle of purification significantly increases. Preliminary results with the two-step purification of globin gene by hybridization of native DNA fragments containing gaps with mercurated mRNA show that about a 10^6-fold purification could be achieved. At this stage of purification, the material obtained can be used for insertion into prokaryotic vectors.

Properties of *D. melanogaster* Fragments Containing Genes Efficiently Transcribed into Cytoplasmic Poly(A)$^+$ RNA in Cell Culture

The second direction of our studies was the investigation of recombinant λgt phage clones (Thomas et al. 1974) containing fragments of *D. melanogaster* DNA. By hybridization with poly(A)$^+$ RNA isolated from the cytoplasm or polysomes of *Drosophila* culture cells, clones containing sequences complementary to this RNA, probably corresponding to the structural genes of *Drosophila*, have been isolated. It was found that in all clones where DNA fragments hybridized with mRNA, the fragments also contained sequences with repetitive DNA of *D. melanogaster* (Ilyin et al. 1976; Georgiev et al. 1977).

Three clones designated as Dm 118, Dm 225, and Dm 234 were selected for further analysis as their DNA bound the highest amounts of mRNA. Dm 225 DNA fragment was also transferred from λgt-Dm

225 to the plasmid pMB9 (a kind gift of Dr. H. W. Boyer) (M. C. Betlach and H. W. Boyer, pers. comm.) that allows us to prepare the higher amounts of Dm 225 DNA.

For the analysis we prepared a high amount of DNA from each of the three clones, immobilized it on nitrocellulose filters, and hybridized it with (1) in-vivo-labeled mRNA, i.e., poly(A)+ cytoplasmic RNA (McKenzie et al. 1975) isolated from culture cells 67J 25D incubated with [³H]uridine (100 µCi/ml) for 15 hours (5 hr before the end of incubation, new [³H]uridine [100 µCi/ml] was added); (2) mRNA labeled in vitro by iodination with ¹²⁵I (Prensky et al. 1973); (3) total DNA labeled with ³²P in vitro by nick translation according to Maniatis et al. (1975). The hybridization was repeated with new filters until zero binding was observed. Due to the high amount of clonal DNA used in the experiment, more than 90% of hybridizable material was usually bound to the first filter. In parallel experiments the size of Dm fragments was detected by gel electrophoresis after restriction of the phage DNA by EcoRI. The results are summarized in Table 1. The sizes of the amplified Dm fragments are rather small (from 1 to 3 kb).

Dm 234 contains two EcoRI fragments about 3.0 and 1.1 kb in size. However, both the repetitive sequence and the sequence hybridizing with mRNA are located in the same fragment, B, and therefore only fragment B will be considered below. Fragment A, which contains the unique sequence, nonhybridizing with mRNA, is located in region 38D (according to Bridge's map). Dm 118 also consists of two EcoRI fragments. Fragment A contains only the repetitive sequence, whereas fragment B contains both the repetitive sequence and the structural gene.

Each of the three clones binds a very significant proportion of the total poly(A)+ RNA of cytoplasm. With purified polysomal poly(A)+ RNA (McKenzie et al. 1975), the same binding was obtained. On the other hand, no hybridization with cytoplasmic poly(A)⁻ RNA could be observed. Therefore it seems unlikely that the DNA of clones hybridize with some nonmessenger RNA, contaminating mRNA preparations. Differences were observed in the binding of mRNAs labeled in vivo and in vitro (the latter is presumably uniformly labeled). An especially high proportion of mRNA was bound by Dm 118 DNA: 1.4% of the ³H-labeled and 0.4% of ¹²⁵I-labeled mRNA. On the other hand, Dm 225 bound a comparable amount of ³H-labeled mRNA (0.95%) and only 0.05% of ¹²⁵I-labeled mRNA. Possibly mRNA transcribed from Dm 118 is more stable than that transcribed from Dm 225. Dm 234 DNA bound a much

Table 1. The Properties of Three Dm Fragments Amplified in λgt Vector

DNA	Length of EcoRI fragments (kb)	Hybridization with ³H-labeled mRNA (1.9 × 10⁵ cpm)		Hybridization with ¹²⁵I-labeled mRNA (1.4 × 10⁷ cpm)		Hybridization with nick-translated [³²P]DNA (375,000)		
		RNA in hybrid on three filters used successively and total cpm	hybridization (%)	RNA in hybrid on two filters used successively and total cpm	hybridization (%)	DNA in hybrid on two filters used successively and total cpm	hybridization (%)	copies per genome
Dm 225	2.9	1420 320 70 — 1810	0.95	7000 200 — 7200	0.05	2600 70 — 2670	0.72	348
Dm 234	1.1	300 120 30 — 450	0.24	3900 100 — 4000	0.03	1160 70 — 1230	0.32	407
Dm 118	3.0	—	—	—	—	346 50 — 396	0.11	35
	1.2	2090 414 128 — 2632	1.4	52,000 3000 — 55,000	0.4			

λgt-Dm DNA was restricted by EcoRI endonuclease and electrophoresed in 1% or 2% agarose gel for determination of molecular weight. As markers, EcoRI restricts of λcI857 DNA or HaeIII restricts of SV40 DNA were used. The λgt-Dm DNA was denatured and immobilized on HA nitrocellulose filters, and the sectors of filters containing 0.5–1.0 µg DNA were incubated either with mRNA labeled in vivo with [³H]uridine (sp. act. ~5 × 10⁵ cpm/µg) in 50 µl of 2 × SSC, 0.1% SDS, or with nick-translated [³²P]DNA from Drosophila culture cells (8 × 10⁶ cpm/µg) in 50 µl 4 × SCC, 0.1% SDS. The sectors of filters with 2.0–3.0 µg DNA were incubated with iodinated ¹²⁵I-labeled mRNA (5–10) × 10⁶ cpm/µg in 50 µl 4 × SCC, 0.1% SDS. The annealing time in all cases was about 20 hr at 65°C. Next the filters were removed, washed, and counted as previously described (Ilyin et al. 1976). Then new filters with DNA were put into hybridization tubes and the hybridization was repeated once or twice. As a blank, empty filters or filters with 1 µg pMB9 plasmid DNA were used. Nick translation of DNA was performed by incubation of 0.5–1 µg DNA in the presence of 180 pmoles of α-³²P-labeled dATP and dTTP and nonlabeled dGTP and dCTP in 100 µl of 50 mM Tris-HCl, pH 7.8, 5 mM MgCl₂, 10 mM β-mercaptoethanol containing 50 µg/ml BSA and DNA polymerase I (2 units) for 1 hr at 15°C. Then the DNA was deproteinized by phenol and chloroform treatment and purified by gel chromatography from the low-molecular-weight fragments.

Table 2. Cross-hybridization of mRNA with DNA of Different Clones

DNA	Dm 225 mRNA (^{125}I)		Dm 118 mRNA (^{125}I)	
	cpm in hybrid (of 7000 cpm input)	hybridization (%)	cpm in hybrid (of 12,500 cpm input)	hybridization (%)
λgt-Dm 225	3607	51.5	99	0.8
λgt-Dm 234	74	1	102	0.8
λgt-Dm 118	342	4.9	7450	59.6
λgt-λC	2	—	4	—

Hybridization was performed in 50 μl 5 × SSC, 0.1% SDS containing filters with either 0.5 μg λgt-Dm 225 DNA or 0.4 μg λgt-Dm 234 DNA or 1.1 μg λgt-Dm 118 DNA for 14 hr at 68°C.

lower fraction of ^{3}H-labeled RNA. The results of the hybridization experiments suggested that all three clones contain different genes, binding different mRNAs.

To confirm this, cross-hybridization experiments were developed. The fraction of mRNA bound to the filter with Dm 225 or with Dm 118 DNA was eluted from the filter by heating in 0.1% SDS containing 100 μg tRNA/ml for 10 minutes at 100°C, treated by pancreatic DNase, deproteinized, and rehybridized with DNA of either one of the three clones. The data show (Table 2) that the homologous hybridization is very high (up to 50–60%) and the cross-hybridization is very low, confirming the differences between the three Dm fragments and mRNA transcribed from them.

From the data on hybridization of the DNA fragments with total highly labeled DNA (Table 1), one can see that Dm 225 and Dm 234 DNAs form hybrids with an unexpectably high fraction of total DNA — 0.7% and 0.3%, respectively. This corresponds to about 400 repeats of the sequences per genome. Dm 118 repetitiveness is much lower.

The important question is whether the structural gene and the repeated sequence are the same or different parts of a fragment. To obtain an answer, we analyzed the competition of Dm mRNA taken in excess with either labeled mRNA or labeled Dm

cRNA (transcribed from *D. melanogaster* DNA with the aid of *E. coli* RNA polymerase) for the binding sites on Dm 225 and Dm 234 DNAs. Table 3 shows that in the both cases cold mRNA strongly inhibited the binding of labeled mRNA. mRNA also strongly competed with cRNA for the binding sites of Dm 234 DNA, but not of Dm 225 DNA. Thus, in the case of Dm 234, the structural gene and the repetitive sequence are the same entity. In Dm 225, this possibility could not be completely excluded, as nonrandom transcription of this particular gene with *E. coli* RNA polymerase may take place. Therefore the experiment was repeated, but intsead of cRNA, nick-translated DNA was used for hybridization with Dm 225 DNA. A fourfold decrease of hybridization in the presence of a competitor mRNA was found.

We can conclude that at least some part of the repetitive sequence in Dm 225 coincides with mRNA coding sequences. In other words, the structural genes present in both Dm 225 and Dm 234 belong to a class of repetitive genes. For the Dm 118 gene, the answer has not been yet obtained. It should be pointed out that the repetitive sequence of Dm 118 is less abundant in the genome than that of Dm 225 and Dm 234.

Next we performed in situ hybridization experiments using either clonal DNA labeled with ^{125}I-

Table 3. The Results of Competitive Hybridization Experiments

DNA on filter	Hybridization of ^{3}H-labeled mRNA			Hybridization of ^{3}H-labeled cRNA or ^{32}P-labeled DNA		
	without competitor (cpm)	with cold mRNA (cpm)	competition (%)	without competitor (cpm)	with cold mRNA (cpm)	competition (%)
Dm 234	118	16	86	1131	270	76
Dm 225	483	82	83	320	95	70

The hybridization mixture contained in 40 μl of 4 × SSC, 0.1% SDS (1) 0.9 μg λgt-Dm 234 or λgt-Dm 225 DNA, (2) ^{3}H-labeled mRNA (10^5 cpm) or ^{3}H-labeled cRNA (5×10^5 cpm) or ^{32}P-labeled nick-translated DNA (0.7×10^5 cpm). As a competitor, 10 μg of nonlabeled mRNA was added. The samples were annealed for 20 hr at 65°C. In experiments with Dm 234 DNA, ^{3}H-labeled cRNA was used; in experiments with DM 225 DNA, ^{32}P-labeled nick-translated DNA was used.

or ³H-labeled cRNA transcribed from DNA. The latter technique is more convenient for precise localization. There were two main observations. First, the localization of Dm 225 and Dm 234 sequences is not fixed and can vary significantly. Second, these sequences are almost always localized in intercalary heterochromatin regions.

In the first experiments, giant polytene chromosomes of hybrid strain gtwᵃ/gt 13ᶻ were used and about 40 hybridization sites could be observed (Fig. 5). Sometimes the chromosomes of this *D. melanogaster* strain contain unpaired regions (Fig. 6). Very often completely different regions of two unpaired

homologs bind labeled material. In other words, the location of the Dm 225 sequence seems to be different in different strains.

For further analysis, the in situ hybridization of Dm 225 cRNA with the polytene chromosomes of the parent stocks gtwᵃ and gt 13ᶻ was studied. The results indicated that the locations of Dm 225 sequences in these two stocks are very different from each other. Only 6 hybridization sites are common for two strains, whereas 33 sites are different. The total number of sites in homozygous parent strains (23 and 22, respectively) is much lower than in hybrid animals (see Table 4). Almost all hybridization

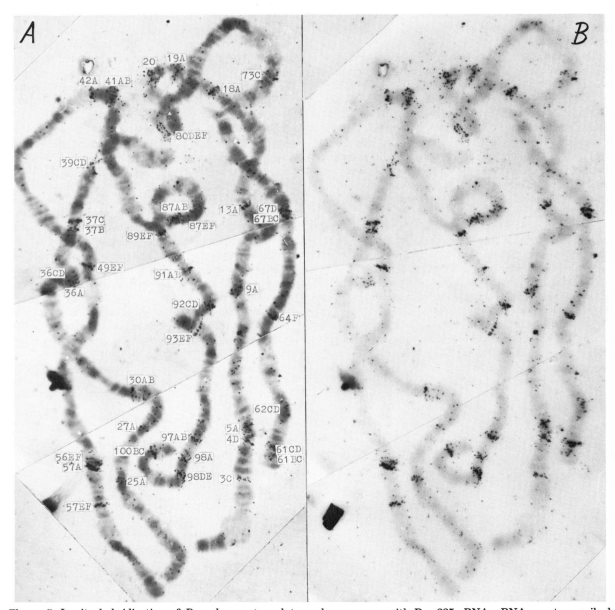

Figure 5. In situ hybridization of *D. melanogaster* polytene chromosomes with Dm 225 cRNA. cRNA was transcribed with the aid of *E. coli* RNA polymerase from λgt-Dm 225 DNA. For hybridization, the chromosomes of hybrid ♀ gtwᵃ/ ♂ gt 13ᶻ were used. *(A)* Stained preparation; *(B)* the same preparation before staining.

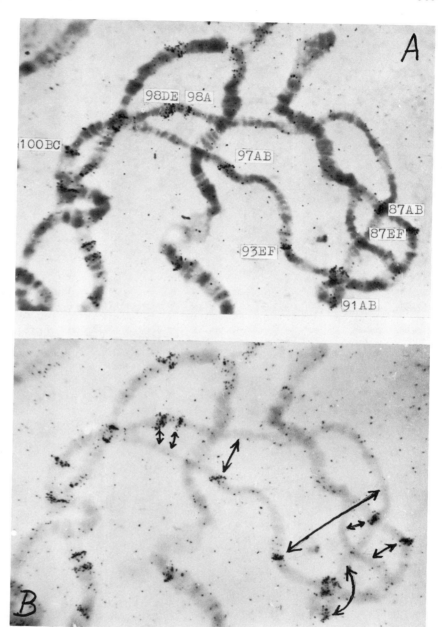

Figure 6. The differences in the localization of Dm 225 hybridization sites in the different strains of *D. melanogaster* as seen in the unpaired regions of chromosomes of a gtw^a/gt 13^z hybrid.

The nonconjugated zones are localized between regions 98 and 92 and between regions 88 and 86. It should be pointed out that distribution of binding sites between homologs corresponds to its distribution in the parent lines (see Table 4). For hybridization, λgt-Dm 225 ^{3}H-labeled cRNA was used. *(A)* Stained preparation; *(B)* the same preparation before staining.

sites present in any of the parent strains are also present in the hybrids.

However, although the variations between different flies of the same hybrid line are not as significant as between different stocks, they still exist. Only 30 of 35–40 hybridization sites are shared by chromosomes of all four gtw^a/gt 13^z animals presented in Table 4. Another example of individual variations between the flies of the same stock was obtained in the course of study of chromosomes from three different animals of the Oregon RC strain. Again the differences in localization of hybridization sites are considerable (Fig. 4), although they are less pronounced than between Oregon RC and gtw^a/gt 13^z.

The distribution of the Dm 234 sequence is also variable. In general, it coincides (but not completely) with the distribution of Dm 225 sequences. Note that the repetitive sequences of Dm 225 and Dm 234 DNAs hybridize with polysomal poly(A)$^+$ RNA, and therefore they very probably represent structural genes, which are repetitive and dispersed throughout the genome. Thus, the location of dispersed structural genes can vary significantly between individual animals and more prominently between different strains.

In addition to variation in the location of the hybridization sites, they also differ in intensity of labeling. The repetitiveness of Dm 225 and Dm 234 se-

Table 4. Results of Experiments on In Situ Hybridization of DNA or cRNA of Dm 225 and Dm 234 Clones with Giant Chromosomes of Different Lines of *D. melanogaster*

Chromosome region	Dm 225					Dm 234			Dm 225		
	gtw^{a}	$gt\,13^{z}$	$gtw^{a}/gt\,13^{z}$	$gtw^{a}/gt\,13^{z}$	$gtw^{a}/gt\,13^{z}$	$gtw^{a}/gt\,13^{z}$	$gtw^{a}/gt\,13^{z}$	$gtw^{a}/gt\,13^{z}$	Oregon	Oregon	Oregon
X											
3C	—	o	o	o	o	o	o	o	—	—	—
4D	—	o	o	o	o	o	o	o	—	—	—
5A	—	o	oo	o	o	o	o	o	o	o	o
7AB	—	—	—	—	—	o	—	—	—	—	—
9A	o	—	o	o	o	o	o	o	—	—	—
13A	0	—	o	o	o	o	o	o	—	—	—
18A	o	—	o	o	o	o	o	o	—	—	—
19A	o	—	o	o	o	o	o	o	o	o	o
20AB	o	—	o	o	o	o	o	o	—	—	o
2L											
21F	—	—	—	—	—	—	—	—	o	—	—
22AB	—	—	—	—	—	o	—	—	o	o	o
25A	o	o	o	o	o	o	o	o	o	o	o
25DE	—	—	—	—	—	—	—	—	o	—	—
26C	—	—	—	—	—	—	—	—	—	—	o
27A	—	o	o	o	o	o	—	—	—	—	—
30AB	—	o	o	o	o	o	o	o	o	o	o
30BC	—	o	—	o	o	o	—	—	—	—	o
32A	—	—	—	—	—	—	—	—	o	—	o
33CD	—	—	—	—	—	o	o	o	o	—	o
34CD	—	—	—	o	—	o	o	o	o	—	—
34F	—	—	—	—	—	—	—	—	o	o	o
36AB	o	—	o	o	o	o	o	o	—	—	—
36CD	o	—	o	o	o	o	o	o	—	—	—
37B	—	—	o	—	—	—	—	—	—	—	—
37C	—	—	o	—	—	—	—	—	—	—	—
38D	—	—	—	—	—	o	o	—	—	—	—
39CD	o	o	o	o	o	o	—	—	—	o	o
2R											
41AB	o	—	o	o	o	o	o	o	o	o	o
42A	o	—	o	o	o	o	o	o	o	o	o
42DE	—	o	—	o	o	—	o	o	—	—	—
47BC	—	—	—	—	—	—	—	—	—	—	o
49EF	o	—	o	o	o	o	o	o	o	—	—
51CDE	—	—	—	—	—	—	—	—	—	o	o
56EF	o	o	o	o	o	o	o	o	o	o	o
57A	o	—	o	o	o	o	o	o	—	o	o
57EF	o	o	o	o	o	o	o	o	o	o	o
59ABC	—	—	—	—	—	—	o	—	o	—	—
60CD	—	—	—	—	—	—	—	—	—	o	—
3L											
61BC	--	o	o	o	o	o	—	—	o	o	o
61CD	—	o	—	o	—	—	o	—	—	—	—
62CD	o	—	o	o	o	o	—	—	—	—	—
63A	—	—	—	—	—	—	—	—	—	—	—
64F	—	o	o	—	o	o	o	o	—	—	—
67BC	o	o	o	o	o	o	o	o	—	—	o
67D	o	—	o	o	o	o	o	o	o	o	o
67EF	—	—	—	—	o	—	—	—	—	—	—
68F	—	—	—	—	—	—	—	—	o	—	o
69DE	—	—	—	—	—	—	o	o	—	—	—
71AB	—	—	—	—	—	—	—	—	o	o	o
73C	—	o	—	—	—	—	—	o	—	—	—
75A	—	—	—	—	—	—	—	—	o	o	o
76F	—	—	—	—	—	—	—	—	—	—	o
79CD	—	—	—	—	—	—	—	—	o	o	o
80DEF	o	o	o	o	o	o	o	o	o	—	o
3R											
83DE	—	o	—	o	—	o	o	—	o	o	o
84A	—	—	—	—	—	—	—	—	—	o	—
84DE	—	—	—	—	—	—	—	—	—	o	—
85F	—	o	—	o	o	o	—	—	—	—	—
87AB	—	o	o	o	o	o	o	o	—	—	—
87EF	—	o	o	o	o	o	o	o	—	—	—
89EF	o	—	o	—	o	o	o	o	o	o	o
91AB	o	—	o	—	o	o	o	o	—	—	—
92AB	—	—	o	—	o	o	o	o	—	—	—
92CD	o	—	—	—	o	—	o	o	o	—	—
93EF	—	o	o	o	o	o	o	o	o	o	o
97AB	—	o	o	o	o	o	o	o	o	o	o
98A	—	o	o	o	o	o	o	—	o	o	o
98DE	—	o	o	o	o	o	o	o	o	o	o
100BC	o	o	o	o	o	o	o	o	o	—	o
No. hybridization sites	23	22	40	37	38	42	39	37	28	28	33

quences is higher than the number of hybridization sites; however, this fact may be explained by the presence of a number of identical sequences in the same site and by their unequal distribution among different sites. In the various nuclei of the same animal, the hybridization patterns are identical.

The second important feature of the hybridization patterns of Dm 225 and Dm 234 is the coincidence of the hybridization sites with the localization of the intercalary heterochromatin in the chromosomes. These are the specific sites of the chromosomes which were delineated by cytogenetics on polytene chromosomes on the basis of their ability for nonhomologous ectopic pairing, high frequency of breakage, existence of weak spots or constrictions,

and late replication (Prokofyeva-Belgovskaya 1941; Hannah 1951; Kaufmann and Iddles 1963; Arcos-Teran 1972; Zhimulev and Kulichkov 1977; Kulichkov and Zhimulev 1976). It was suggested that intercalary heterochromatin contains polygenes which have similar small quantitative effects (Hannah 1951).

Many tandem repetitive genes such as genes for rRNA (28S, 18S, and 5S), histone genes, and heat-shock genes (Pardue et al. 1972; Grigliatti et al. 1974; McKenzie et al. 1975) are present in the regions where intercalary heterochromatin is detected (see Fig. 7).

Finally, when genes are translocated to certain intercalary heterochromatin sites, variegated posi-

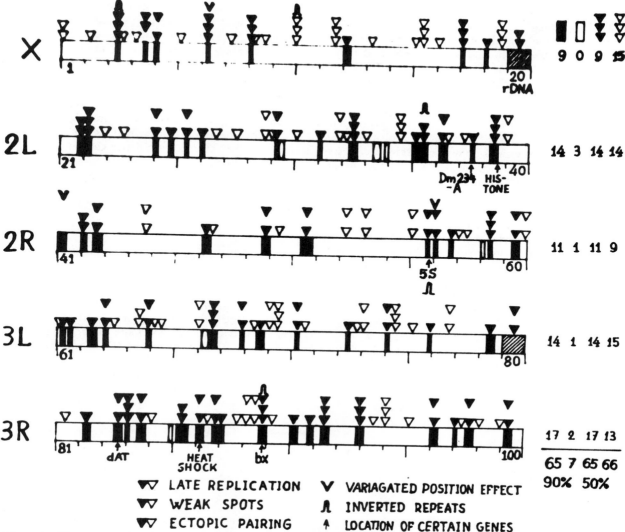

Figure 7. The relationship between intercalary heterochromatin and hybridization sites of Dm 225 and Dm 234. (2L) The chromosomal location of the hybridization site which coincides with intercalary heterochromatin. (2R) The chromosomal location of the hybridization site not coinciding with intercalary heterochromatin. (3L) Intercalary heterochromatin site which hybridizes with Dm 225 or Dm 234 DNA. (3R) Intercalary heterochromatin site which does not hybridize with Dm 225 and Dm 234. The lower triangle indicates sites of ectopic pairing; the middle triangle indicates weak spots; the upper triangle indicates late replication. (V) Sites where variegated position effect was observed; (∩) localization of long inverted repeats detected cytologically or biochemically; (↑) localization of certain genes on chromosome.

tion effects were observed (Hannah 1951; Spofford 1976; Lefevre 1976). By comparing the distribution of hybridization sites of Dm 225 and Dm 234 sequences with the known location of intercalary heterochromatin, a remarkable correlation can be seen. Sixty-five of the 72 (> 90%) hybridization sites are localized in the regions where intercalary heterochromatin is present. About 50% of the latter regions hybridize with these sequences. These sequences also coincide with the regions of intercalary heterochromatin capable of inducing the variegated position effect. The latter possibly depends on the underreplication of adjacent genes (Ananiev and Gvozdev 1974).

The regions containing rDNA, 5S RNA genes, histone genes, the main heat-shock genes, and long poly[d(AT)] sequences (Barr and Ellison 1972; Leibovitsch et al. 1974) are able to hybridize with Dm 225 and 235 sequences. Finally, the hybridization was also detected in the regions containing cytologically detectable inverted repeats (including locus *bx*) (Kaufmann and Iddles 1963).

With Dm 118 DNA, the in situ hybridization was performed less extensively and the only observation was that there are about 40 hybridization sites and many of them do not coincide with the sites for Dm 225 and Dm 234 DNAs.

Two conclusions may be drawn from the analysis

of in situ hybridization data with Dm 225 and Dm 234 DNA fragments. First, these sequences are almost exclusively localized in the regions of intercalary heterochromatin, and second, their localization is extremely unstable. Thus the genes presented in these two DNA fragments seem to be "jumping genes," and intercalary heterochromatin can be considered as the sites at which the jumping genes are inserted. It should be pointed out that these genes are intensively transcribed, at least in cultured cells.

Another possibility — that the differences in hybridization patterns are dependent on the underreplication of DNA during polyteinization — also cannot be excluded.

Experiments to discriminate between these two explanations are in progress now. In favor of the jumping-gene model of intercalary heterochromatin are the following observations concerning the nucleoli. It was found that sometimes nucleoluslike structures of different sizes are located not in region 20 but in some other sections of polytene chromosomes where intercalary heterochromatin occurs (3C, 11A, 12DE). To elucidate the nature of these formations, we used in situ hybridization with ^{125}I-labeled ribosomal RNA. It was purified in the following way. High-molecular-weight cytoplasmic poly-(A)$^-$ RNA was hybridized with filters containing the DNA of one of our clones (λgt-Dm 34) which was found to contain rDNA genes (unpubl. data).

The bound RNA was eluted from filters and in situ hybridization was performed. Normally the label is recovered in the nucleolus only. However, where the nucleoluslike structures are also present, their DNA efficiently binds purified rRNA (Fig. 8). Thus the nucleoluslike structures actually contain genes for rRNA, which have been translocated to abnormal positions localized in intercalary heterochromatin regions. It seems very probable that the intercalary heterochromatin regions are organized in such a way that certain genes can be easily inserted into them and can be easily lost as well. These moving genes mainly represent extensively transcribed repetitive genes whose products are accumulated in high amounts in the cytoplasm of the *Drosophila* cell.

CONCLUSION AND SUMMARY

The previously described method of isolation of long fragments of native DNA bearing structural genes at the beginning, in the middle, or at the end of the coding strand was applied to purification of fragments containing the globin gene. A one-cycle purification procedure gives up to several-thousand-fold enrichment of the globin gene. A repetition of the procedure allows one to obtain almost pure material.

The clones containing *D. melanogaster* genes

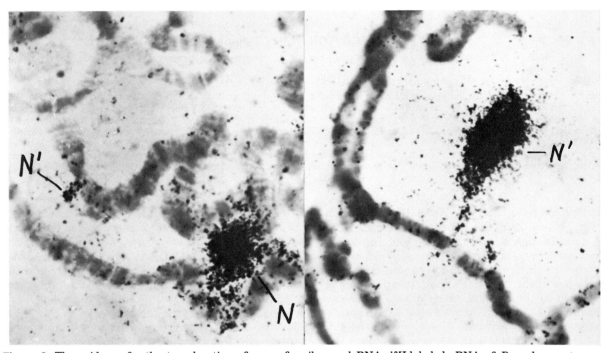

Figure 8. The evidence for the translocation of genes for ribosomal RNA. ^{125}I-labeled rRNA of *D. melanogaster* was hybridized with λgt-Dm 34 DNA which contains ribosomal genes. The hybridized RNA was eluted by heating in 0.2% SDS, treated by DNase, deproteinized, and hybridized to chromosomes containing nucleoluslike structures in unusual positions. The labeled RNA was bound to the DNA of nucleolus in normal position (N) and to the DNA of unusually located nucleoluslike structures (N') in the 3C region *(left)* and in the 2R chromosome *(right)*.

which produce very abundant mRNA in cell culture were characterized. It was found that the structural genes present in two of them are repetitive and represented by approximately 400 copies per genome. In situ hybridization experiments show that Dm 225 and Dm 234 are present in several tens of sites in polytene chromosomes, the great majority of which correspond to the sites of intercalary heterochromatin. The localization of Dm 225 and Dm 234 sites is not fixed. The conclusion has been drawn that the intercalary heterochromatin represents the chromosomal loci where the extensively expressed repetitive genes are localized and that these genes may readily move from one site of intercalary heterochromatin to another.

Acknowledgment

We are greatly indebted to Dr. R. Davis for providing us with λgt phage.

REFERENCES

ANANIEV, E. V. and V. A. GVOZDEV. 1974. Changed pattern of transcription and replication in polytene chromosomes of *D. melanogaster* resulting from eu-heterochromatin rearrangement. *Chromosoma* **45**: 173.

ANDERSON, J. N. and R. T. SCHIMKE. 1976. Partial purification of the ovalbumin gene. *Cell* **7**: 331.

ARCOS-TERAN, L. 1972. DNS-replication und die Natur der spat replizierenden Orte in X-chromosom von *Drosophila melanogaster*. *Chromosoma* **37**: 233.

BARR, H. Y. and J. R. ELLISON. 1972. Ectopic pairing of chromosome region containing chemically similar DNA. *Chromosoma* **39**: 53.

BRITTEN, R. J. and E. H. DAVIDSON. 1969. Gene regulation for higher cells: A theory. *Science* **165**: 349.

CRICK, F. 1971. General model for the chromosome of higher organisms. *Nature* **234**: 25.

DALE, R. M. K., E. MARTIN, D. C. LIVINGSTON, and D. C. B. WARD. 1975. Direct covalent mercuriation of nucleotides and polynucleotides. *Biochemistry* **14**: 2447.

GEORGIEV, G. P. 1969. On the structural organization of operon and the regulation of RNA synthesis in animal cells. *J. Theor. Biol.* **25**: 473.

GEORGIEV, G. P., A. J. VARSHAVSKY, A. P. RYSKOV, and R. B. CHURCH. 1974. On the structural organization of the transcriptional unit in animal chromosomes. *Cold Spring Harbor Symp. Quant. Biol.* **38**: 869.

GEORGIEV, G. P., A. P. RYSKOV, C. COUTELLE, V. L. MANTIEVA, and E. R. AVAKYAN. 1972. On the structure of transcriptional unit in mammalian cells. *Biochim. Biophys. Acta* **259**: 259.

GEORGIEV, G. P., Y. V. ILYIN, A. P. RYSKOV, N. A. TCHURIKOV, G. N. YENIKOLOPOV, V. A. GVOZDEV, and E. V. ANANIEV. 1977. Isolation of eukaryotic DNA fragments containing structural genes and the adjacent sequences. *Science* **195**: 394.

GRIGLIATTI, T. A., B. N. WHITE, G. M. TENER, T. C. KAUFMAN, J. J. HOLDEN, and D. T. SUZUKI. 1974. Studies on the transfer RNA genes of Drosophila. *Cold Spring Harbor Symp. Quant. Biol.* **38**: 461.

HANNAH, A. 1951. Localization and function of heterochromatin in *Drosophila melanogaster*. *Adv. Genet.* **4**: 87.

ILYIN, Y. V., N. A. TCHURIKOV, and G. P. GEORGIEV. 1976. Selection and some properties of recombinant clones of lambda bacteriophage containing genes of *Drosophila melanogaster*. *Nucleic Acids Res.* **3**: 2115.

KAUFMANN, B. P. and M. K. IDDLES. 1963. Ectopic pairing in salivary gland chromosome of *Drosophila melanogaster*. *Port. Acta Biol. Ser. A* **7**: 225.

KULICHKOV, V. A. and I. F. ZHIMULEV. 1976. Analysis of spatial organization of *Drosophila melanogaster* genome based on ectopic pairing of polytene chromosomes. *Genetika* **12**: 81.

LEFEVRE, G. 1976. A photographic representation and interpretation of the polytene chromosomes of *Drosophila melanogaster* salivary glands. In *Genetics and biology of* Drosophila (ed. M. Ashburner and E. Newitsky), vol. 1a, p, 31. Academic Press, New York.

LEIBOVITSCH, B. A., V. A. GVOZDEV, I. F. ZHIMULEV, and E. S. BELYAEVA. 1974. Disproportional incorporation of H³-UTP and H³-ATP in some regions of *Drosophila melanogaster* chromosome in *E. coli* RNA-polymerase reaction mixture. *Drosophila Inform. Serv.* **51**: 79.

LIMBORSKA, S. A., L. Y. FROLOVA, E. E. MAKAROVSKAYA, and S. N. HIL'KO. 1976. Isolation of human globin mRNA and synthesis of complementary DNA with the aid of reverse transcription. *Proc. Acad. Sci. USSR* **230**: 234.

MANIATIS, T., A. JEFFREY, and D. G. KLEID. 1975. Nucleotide sequence of the rightward operator of phage λ. *Proc. Natl. Acad. Sci.* **72**: 1184.

MANIATIS, T., G.-K. SIM, A. EFSTRATIADIS, and F. C. KAFATOS. 1976. Amplification and characterization of a β-globin gene synthesized in vitro. *Cell* **8**: 163.

MCKENZIE, S. L., S. HENIKOFF, and M. MESELSON. 1975. Localization of RNA from heat-induced polysomes at puff sites of *Drosophila melanogaster*. *Proc. Natl. Acad. Sci.* **72**: 1117.

MORROW, J. F., S. N. COHEN, A. C. Y. CHANG, H. W. BOYER, H. M. GOODMAN, and R. B. HELLING. 1974. Replication and transcription of eukaryotic DNA in *Escherichia coli*. *Proc. Natl. Acad. Sci.* **71**: 1743.

PARDUE, M. L., E. WEINBERG, L. H. KEDES, and M. L. BIRNSTIEL. 1972. Localization of sequences coding for histone messenger RNA in chromosome of *Drosophila melanogaster*. *J. Cell Biol.* **55**: 199a.

PRENSKY, W., D. M. STEFFENSEN, and W. L. HUGHES. 1973. The use of iodinated RNA for gene localization. *Proc. Natl. Acad. Sci.* **70**: 1860.

PROKOFYEVA-BELGOVSKAYA, A. A. 1941. Cytological properties of inert region and their bearing on the mechanisms of mosaicism and chromosome rearrangement. *Drosophila Inform. Serv.* **15**: 35.

RYSKOV, A. P., O. V. TOKARSKAYA, and G. P. GEORGIEV. 1976. Direct demonstration of a complementarity between mRNA and double stranded sequences of pre-mRNA. *Mol. Biol. Rep.* **2**: 353.

SCHERRER, K. and L. MARCAUD. 1968. Messenger RNA in avian erythroblasts at the transcriptional and translation level and the problem of regulation in animal cells. *J. Cell. Physiol.* **72** (Suppl.): 181.

SPOFFORD, J. B. 1976. Position-effect variagation in Drosophila. In *Genetics and biology of* Drosophila (ed. M. Ashburner and E. Novitsky), vol. 1c, p. 955. Academic Press, New York.

THOMAS, M., J. R. CAMERON, and R. W. DAVIS. 1974. Viable molecular hybrids of bacteriophage lambda and eukaryotic DNA. *Proc. Natl. Acad. Sci.* **71**: 4579.

WENSINK, P. C., D. J. FINNEGAN, J. E. DONELSON, and D. S. HOGNESS. 1974. A system for mapping DNA sequences in the chromosomes of *Drosophila melanogaster*. *Cell* **3**: 315.

YENIKOLOPOV, G. N., T. NITTA, A. P. RYSKOV, and G. P. GEORGIEV. 1976. Isolation of native DNA fragments containing structural genes at the beginning, in the middle or at the end of the coding strand. *Nucleic Acids Res.* **3**: 2645.

ZHIMULEV, I. F. and V. A. KULICHKOV. 1977. Break sites in polytene chromosomes of *Drosophila melanogaster*, their location and replication features. *Genetika* **13**: 85.

Characterization and Kinetics of Synthesis of 15S β-Globin RNA, a Putative Precursor of β-Globin mRNA

P. J. CURTIS, N. MANTEI, AND C. WEISSMANN

Institut für Molekularbiologie I, Universität Zürich, 8093 Zürich, Switzerland

Several authors (for review, see Darnell et al. 1973) have proposed that mRNA of eukaryotic cells is derived by processing from heterogeneous nuclear RNA (hnRNA), a class of RNA with a high rate of turnover (Soeiro et al. 1968; Brandhorst and McConkey 1974), sedimenting between 20S and 100S (Scherrer et al. 1963; Georgiev et al. 1963; Warner et al. 1966). A number of reports have dealt with the detection of sequences specific for a particular mRNA in high-molecular-weight hnRNA. Polysome-derived globin mRNAs, with a chain length of 600–700 nucleotides, sediment at about 9S (Chantrenne et al. 1967). It has been reported that globin-specific sequences could be detected in duck erythroblast nuclear RNA and mouse fetal liver RNA sedimenting above 30S (Imaizumi et al. 1973; Williamson et al. 1973). In addition, Macnaughton et al. (1974) identified globin-specific sequences in duck erythroblast nuclear RNA sedimenting at 14S. On the other hand, Spohr et al. (1976), using denaturing gels, were unable to detect significant amounts of oversized globin-specific RNA.

All these early studies were performed on unlabeled RNA using more-or-less-pure, labeled globin cDNA probes, or by assaying for the capacity to direct globin synthesis in a protein-synthesizing system.

Clearly, if the steady-state level of a precursor is low compared to that of the mature molecule, it is advantageous to examine RNA from cells labeled for periods that are short compared to the half-life of the precursor, since there is less contamination of high-molecular-weight RNA fractions with labeled mature mRNA. Furthermore, it is desirable to measure the rate of synthesis and degradation of the precursor and to characterize it, in particular by showing that it contains messenger-specific nucleotide sequences and a discrete set of additional sequences. To make this possible, we have used two different techniques for analysis and purification of labeled globin-specific RNA. The first is based on the so-called poly(I)-Sephadex method described by Coffin et al. (1974), and the second makes use of cloned mouse globin cDNA hybrid plasmid attached to nitrocellulose filters. Applying these methods to RNA from dimethylsulfoxide-induced Friend cells (Friend et al. 1971), we have detected and purified a labeled 15S β-globin-specific RNA, measured the kinetics of its synthesis and its steady-state level, and characterized it by fingerprinting and by the analysis of its 5′ terminus. Moreover, we have detected an α-globin-specific RNA of slightly higher molecular weight than the corresponding mature mRNA. A 15S globin-specific RNA has also been detected and studied in mouse fetal liver cells by Ross (1976) and in mouse anemic spleen cells by Kwan et al. (1977).

The Poly(I)-Sephadex Method

In this method, outlined in Figure 1, globin cDNA (Kacian et al. 1971; Ross et al. 1972; Verma et al. 1972) is elongated with dCMP residues, using terminal deoxynucleotidyl terminal transferase. The resulting poly(dC) globin cDNA is hybridized to labeled RNA in the presence of poly(U) and oligo(C). The oligo(C) is required to block G-rich regions on the RNA which could otherwise interact with the poly(dC) tails, and the poly(U) blocks the poly(A) sequences which would otherwise bind all mRNA to the poly(I)-Sephadex. The annealed mixture is passed through a poly(I)-Sephadex column which retains the poly(dC) cDNA and any RNA hybridized to it. For analytical purposes, the column is rinsed with RNase A to remove nonhybridized RNA segments and to reduce the background, and the bound RNA is then eluted with formamide and quantified. To determine the hybridization efficiency, the radioactive sample is mixed with a defined amount of authentic globin mRNA labeled with a different isotope prior to carrying out the assay. The analysis of [³H]RNA from Friend cells from dimethylsulfoxide-induced cells revealed that in dimethylsulfoxide-induced cells about 0.2% of the [³H]RNA was globin-specific and that no globin-specific RNA could be detected in uninduced cells (less than about 0.005%). We have obtained values as high as 0.5% after freshly cloning cells and as low as 0.03% after two to three months of continuous culture (Curtis and Weissmann 1976).

For preparative purposes, the RNase-A treatment of the column-bound material was omitted, and reduction of the background was achieved by an elution schedule involving low-ionic-strength buffer. As shown in Table 1, one cycle of hybridization yielded a preparation of 25% purity (about 80-fold purifica-

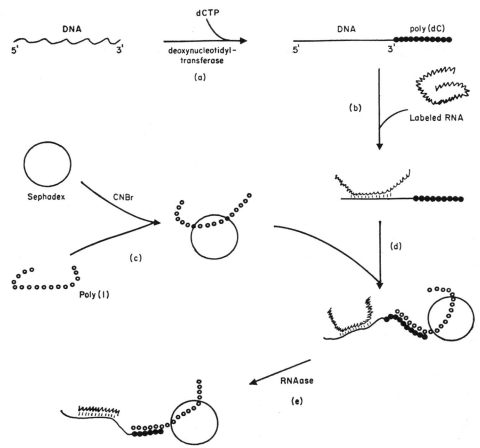

Figure 1. Scheme for the detection of specific RNA by hybridization to complementary poly(dC)-elongated DNA. *(a)* DNA is elongated with terminal deoxynucleotidyl transferase and dCTP to give poly(dC) DNA. *(b)* Labeled RNA complementary to the DNA forms a hybrid. *(c)* Poly(I) is coupled to Sephadex to give poly(I)-Sephadex. *(d)* The RNA–poly(dC)-DNA hybrid is bound to the poly(I)-Sephadex. *(e)* RNase treatment trims off nonhybridized segments of the labeled RNA. The labeled RNA is subsequently removed from the poly(I)-Sephadex by elution with formamide. (Reprinted, with permission, from Coffin et al. 1974.)

Table 1. Radiochemical Purification of Globin-specific RNA from Induced Friend-cell RNA

RNA[a]	Total ^{32}P recovered (cpm $\times 10^{-6}$)	Recovery (%)	radioactivity in hybrid (% of input) ^{32}P	^{125}I	globin-specific RNA (% of total)	Recovery of globin-specific RNA (%)
Induced-cell RNA	98	100	0.20 (0.037)	54.3 (1.0)	0.31	100
RNA eluted from first poly(I)-Sephadex column	0.78	0.80	13.8 (0.85)	52.5 (1.1)	25.2	66
RNA eluted from second poly(I)-Sephadex column	0.035	0.036	42.0 (5.5)	50.6 (1.6)	74.5	8.6

Data from Curtis and Weissmann (1976).

[a] Cells were induced by 1.5% DMSO for 3 days. Then 10^7 cells in 10 ml of phosphate-free medium were labeled with 2 mCi [^{32}P]phosphate for 16 hr. The RNA (sp. act. 1.06×10^6 cpm/μg) was extracted and purified as described in Curtis and Weissmann (1976).

[b] Analysis for globin-specific RNA was carried out as described in Curtis and Weissmann (1976). The values are the average of duplicate determinations. Background values (given in parentheses, not subtracted) were obtained by hybridizing in the presence of 0.1 μg of globin mRNA. The input of ^{32}P-labeled RNA and ^{125}I-labeled globin mRNA (sp. act. 10^6 cpm/μg) was adjusted to give about 500 cpm in the hybrid fraction.

tion); for further purification, a second cycle of hybridization was carried out.

The Filter Hybridization Method

The mouse globin cDNA hybrid plasmids required for filter hybridization were prepared essentially by the approach of Maniatis et al. (1976). Poly(dA)-elongated globin cDNA was annealed with poly(dT)-elongated pCR1 DNA and transfected into streptomycin-dependent *Escherichia coli* N543 (Weissmann and Boll 1976). Colonies containing globin DNA were identified by in situ hybridization to ¹²⁵I-labeled mouse α- and β-globin mRNA, respectively, purified by hybridization to rabbit β-globin plasmid PβG (Maniatis et al. 1976) and rabbit α-globin plasmid ZRαG (P. Curtis and J. van den Berg, unpubl.).

The globin cDNA inserts of ZMαG and ZMβG excised by the nuclease S₁ method of Hofstetter et al. (1976) are about 375 base pairs (bp) (α) and 530

bp (β) long. Both are in the largest fragments resulting from cleavage with *Hha* or *Hap*II. The results of restriction-site mapping and DNA sequencing to characterize and definitively identify the globin cDNA inserts of plasmids ZMαG and ZMβG are shown in Figure 2.

Quantitation of labeled α- and β-globin-specific RNA using filter-bound plasmid DNA was carried out by the technique of Gillespie and Spiegelman (1965), and preparative purification by the method of Weinberg et al. (1972).

Purification of Radioactive Mature Globin mRNA from Friend Cells

The potentiality of the poly(I)-Sephadex method was explored by using it to purify RNA from dimethylsulfoxide-induced Friend cells labeled for 16 hours with [³²P]phosphate. The purified [³²P]RNA was 74% pure globin RNA as judged by hybridiza-

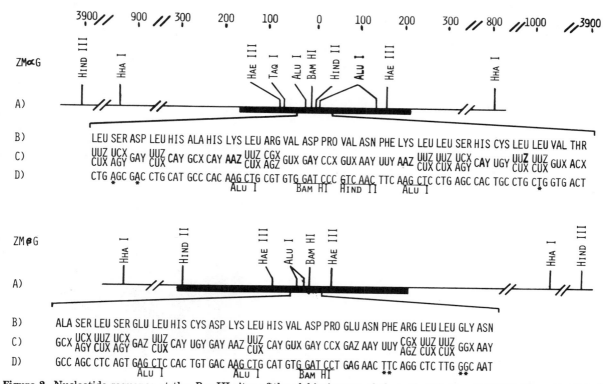

Figure 2. Nucleotide sequence at the *Bam*HI sites of the globin inserts of plasmids ZMαG and ZMβG. Plasmids ZMαG and ZMβG were cleaved with restriction endonuclease *Bam*HI, the ends were labeled with [³²P]phosphate, and the DNAs further cleaved with restriction endonuclease *Hha*I. The two labeled fragments from each plasmid were separated by agarose gel electrophoresis, extracted from the gel, and used for DNA sequencing by the method of Maxam and Gilbert (1977). Restriction-site mapping was by the method of Smith and Birnstiel (1976). The *Alu*I and *Hind*II sites closest to the *Bam*HI site were deduced from the nucleotide sequence. The scale at the top of the figure is numbered in base pairs, with the *Bam*HI site as origin. Globin gene inserts are indicated by heavy bars within the hybrid plasmid map excerpt (*A*). The fragments arising by *Bam*HI cleavage of the S₁ excision product were identified by digestion with *Taq*I in the case of ZMαG and *Alu*I in the case of ZMβG. The orientation of the *Bam*HI-*Hha*I fragments with respect to the (single) *Hind*III site in pCR1 (Armstrong et al. 1977) was determined by digesting the isolated *Bam*HI-*Hind*III fragments with *Hha*I. Neither insert was cleaved by *Eco*RI, *Bgl*II, *Hha*I, *Hap*II, or *Hind*III. The expanded region in the figure shows part of the globin amino acid sequence (*B*) (Dayhoff 1972) with the inferred messenger RNA nucleotide sequence (*C*), and the experimentally determined DNA sequence (*D*) below. X represents any nucleotide; Y represents a pyrimidine; Z represents a purine. Asterisks indicate uncertain assignments.

tion to cDNA (Curtis and Weissmann 1976). The [^{32}P]RNA was further characterized by T$_1$ finger-printing. Figure 3d shows the RNase-T$_1$-resistant oligonucleotides obtained from the ^{32}P-labeled, purified RNA, resolved by electrophoresis in the first dimension and homochromatography in the second dimension. The ^{32}P-labeled oligonucleotides were further analyzed by RNase-A digestion. Since a mixture of α- and β-globin poly(dC) cDNA had been used for purification, the resulting preparation was expected to consist of α- and β-globin RNA in about similar proportions. As shown in Table 2, the partial structures of five oligonucleotides was compatible with the sequence of α mouse globin, and that of four

oligonucleotides with the sequence of β mouse globin.

Hybridization of [^{32}P]RNA from long-term-labeled Friend cells with filter-bound cloned α- and β-globin-pCR1 hybrid DNA yielded two [^{32}P]RNA preparations. The two RNAs give quite different fingerprints (Fig. 3a,b); superposition of the two yields the pattern of large T$_1$ oligonucleotides found with the RNA prepared by the poly(I)-Sephadex method with one exception (Fig. 3d, spot marked X). We conclude that the filter hybridization allows separation of α- and β-globin mRNA. Moreover, with one exception, the T$_1$ oligonucleotides assigned to β-globin mRNA on the basis of their partial sequence were found in the RNA purified with β-globin plasmid, and all as-

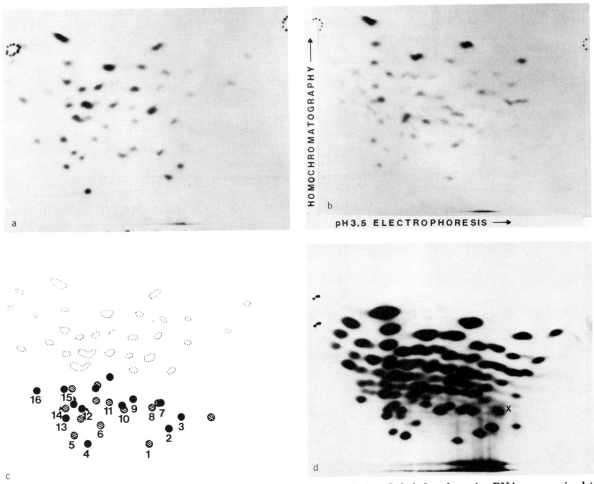

Figure 3. Characterization of globin-specific [^{32}P]RNAs by T$_1$ fingerprinting. Labeled and carrier RNAs were mixed to give about 20 μg and digested with 3 units of RNase T$_1$ in 3 μl of 20 mM Tris-HCl, pH 7.5, 1 mM EDTA at 37°C for 30 min. The products were fractionated by cellulose acetate electrophoresis at pH 3.5, followed by homochromatography on PEI plates (Volckaert et al. 1976). *(a)* α-Globin ^{32}P-labeled mRNA (2500 cpm) purified by hybridization to α-globin ^{32}P-labeled mRNA (2500 cpm) purified by hybridization to α-globin hybrid DNA fixed to a Millipore filter. *(b)* β-globin ^{32}P-labeled mRNA (2500 cpm) purified by hybridization to β-globin hybrid DNA fixed to a Millipore filter. *(c)* Composite tracing of *a* and *b*: (●) characteristic α-specific oligonucleotides; (⊘) β-specific oligonucleotides. The numbering is as in Curtis and Weissmann (1976). *(d)* 10S globin-specific [^{32}P]RNA (40,000 cpm) purified from induced cells labeled for 16 hr with [^{32}P]phosphate by two cycles of hybridization to poly(dC) (α + β) globin cDNA and poly(I)-Sephadex chromatography (cf. Table 1). The spot marked X may represent a partial digestion product or a product not due to globin mRNA (see text). (Reprinted, with permission, from Curtis and Weissmann 1976 and Curtis et al. 1977.)

Table 2. Correlation of T₁ Products with Globin α and β Polypeptides

Product	Relative molar yield[a]	Partial structure[b]	Possible sequence of corresponding T₁ oligonucleotide deduced from amino acid sequence[c]	Possible location of sequence in globin
1	0.52	(A-A-A-C,2A-C,8C,6U)G		
2	0.71	(4A-C,8C,9U)G		
3	0.89	(A-C,10C,8U)G		
4	1.04	(2A-C,7C,2U)A-A-G	U-A-C-U-U-Py-U-C-Py-U-C-Py-U-U-Py-G	β 41–46
5	0.52	(A-A-C,2A-C,5C,U)A-G	U-U-Py-C-Py-C-A-C-C-A-C-C-A-A-G	α 36–40
6	0.52	(A-A-A-U,A-C,C)A-A-A-A-G[d]	A-C-U-A-A-C-C-A-Py-Py-U-A-G	β 75–79
7	0.94	(A-A-A-U-,A-U,A-C,3C,3U)G		
8	1.12	(A-U,A-C,6C,4U)G		
9	0.94	(A-A-C,3C,4U)A-A-G	U-C-A-A-C-U-U-Py-A-A-G	α 96–99
10	1.84	(A-U,A-C,5C,2U)G	A-U-U-U-Py-A-C-Py-C-C-Py-G	α 116–120
11	1.00	(A-A-U,A-C,3C,2U)A-A-G		β 121–125
12	1.04	(A-A-C,A-U,3C)A-A-G	C-A-A-C-A-U-C-A-A-G	α 8–11
13	1.17	(2A-C,5C,U)G	Py-C-A-Py-C-A-Py-C-C-Py-G	α 112–115
14	1.39	(2A-C,4C,U)G	Py-C-A-Py-C-A-Py-Py-C-Py-G	β 117–120
15	1.10	(A-C,3C,U)A-A-G	A-C-Py-U-C-Py-A-A-G	α 88–90
16	1.07	(A-C,4C)A-A-G	C-C-C-A-C-A-A-G	β 142–145

Friend cells were induced by the addition of 1.5% DMSO. Then 4.5×10^7 cells in 15 ml of phosphate-free SF12–10% HS were labeled with 20 mCi of [³²P]phosphate for 16 hr. The extracted RNA (380 µg, specific ³²P-radioactivity 7×10^6 cpm/µg) was purified as described in Curtis and Weissmann (1976), precipitated with ethanol, and dissolved in 100 µl Tris-EDTA buffer. The sample was heated for 45 sec at 100°C in the presence of 1 µg of poly(A) and centrifuged through a sucrose gradient for 2.5 hr. The 10S region (4.6×10^5 cpm) was pooled, 20 µg of Qβ RNA was added as carrier, and the RNA was precipitated with ethanol. Digestion with RNase T₁ and separation of the digestion products were carried out as described in the legend to Fig. 3. The oligonucleotides designated by numbers in Fig. 3c were eluted and digested with RNase A, and the resulting products were characterized as described in Curtis and Weissmann (1976). Data from Curtis and Weissmann (1976).

[a] The average radioactivity per phosphate was determined from all oligonucleotides listed. The relative molar yield of an oligonucleotide was calculated as the radioactivity in the oligonucleotide, divided by the average radioactivity per phosphate, times the number of phosphates in the oligonucleotide.

[b] Deduced from RNase-A digestion products. The quantitation of the products was accurate to ±20%.

[c] Dayhoff (1972). Nucleotide sequences were matched to protein sequences by trial and error.

[d] Presumed sequence on basis of electrophoretic mobility only.

signments of Table 2 were confirmed for the α-globin mRNA.

Detection and Characterization of 15S Globin-specific RNA

Cells induced by dimethylsulfoxide were labeled for 16 hours by [³H]uridine and for 20 minutes with [³²P]phosphate. The RNA was extracted, heat-denatured for 45 seconds at 100°C, and sedimented through a sucrose density gradient. Each fraction was assayed for globin-specific RNA by the poly(I)-Sephadex method. Figure 4 shows that ³H-labeled globin-specific RNA sedimented at 10S, the position of mature mRNA, whereas ³²P-labeled globin-specific RNA sedimented as two peaks, a smaller one at 15S and a larger one at 10S. The possibility that the 15S RNA is an aggregate of 10S RNA is very unlikely for two reasons: (1) The heat treatment to which the sample was subjected before sucrose gradient centrifugation is more stringent than that recommended by McKnight and Schimke (1974) and suffices to denature double-labeled Qβ RNA (unpubl. results from our laboratory; see also Flavell et al. 1974). (2) the ³H-labeled endogenous 10S RNA, which serves as an ideal internal reference, shows no evidence of aggregation.

The RNA sedimenting to the bottom of the tube (fractions 1 and 2) was reexamined for its content of globin sequences. Within the limits of our technique, none could be detected (< 0.005%).

To isolate 15S globin-specific RNA, [³²P]RNA from Friend cells labeled for 20 minutes was denatured and centrifuged through a sucrose gradient. The fractions corresponding to the 15S peak were pooled and the globin-specific RNA was purified by two cycles of purification by the poly(I)-Sephadex method. Comparison of the fingerprint of the purified [³²P]RNA (Fig. 5a) with the patterns of purified α- and β-globin mRNAs (Fig. 3) showed that all typical β-globin-specific oligonucleotides could be accounted for in the pattern of the 15S RNA (as judged by relative mobilities; there was not enough material for further characterization); however, no characteristic α-globin-specific oligonucleotides could be detected. Since, as shown above, the poly(dC) ($\alpha + \beta$) cDNA preparation used in these experiments allowed the purification of almost an equimolar mixture of α- and β-globin mRNA from long-term ³²P-labeled Friend cells, we concluded that there was little if any α-globin-specific RNA in the 15S RNA fraction. This conclusion was confirmed by the experiment described below (Fig. 8), where fractions from a similar product were assayed separately with fil-

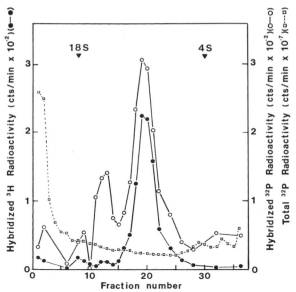

Figure 4. Sucrose gradient centrifugation of labeled Friend-cell RNA; location of globin-specific sequences by hybridization. Induced cells were labeled with [^{32}P]phosphate for 20 min. The extracted RNA (575 μg, specific ^{3}H-radioactivity 1.2×10^5 cpm/μg; specific ^{32}P-radioactivity 2.8×10^5 cpm/μg) was denatured at 100°C for 45 sec in 100 μl Tris-EDTA buffer and centrifuged through a 5–23% sucrose gradient in 50 mM Tris-HCl, pH 7.5, 5 mM EDTA, 0.1% SDS for 16 hr at 15°C and 32,000 rpm in an SW 41 Spinco rotor. From each fraction (350 μl), 15 μl was hybridized with 20 ng poly(dC) globin cDNA and assayed by the poly(I)-Sephadex method. (□) Total [^{32}P]RNA; (○) hybridized [^{32}P]RNA; (●) hybridized [^{3}H]-RNA. (Reprinted, with permission, from Curtis and Weissmann 1976.)

ter-bound α- and β-globin hybrid DNA. The 15S RNA hybridized only to the β-globin filters.

Purified 15S β-globin [^{32}P]RNA yielded about 20 T_1 oligonucleotides not found in the case of α- or β-globin mRNA; some eight of these showed the high mobility in the first dimension characteristic for high UMP content. Some or all of the extra fragments could be derived from the 15S globin-specific RNA; however, it is striking that many spots are more intense than the ones attributed to β-globin mRNA. The unequal intensities could be due to (1) the occurrence of oligonucleotides in more than one molar yield (repetitive sequences), (2) different lengths of the oligonucleotides, (3) the coincidence on the chromatogram of two or more oligonucleotides, (4) unequal labeling along the chain, or (5) unequal specific activities of the four nucleotides. On the other hand, the extra oligonucleotides could be derived from one or few RNA species copurifying with the 15S β-globin-specific RNA and present in higher yield or at higher specific activity than the latter. We consider an unspecific contamination unlikely for the following reasons: (1) The fingerprint of 15S [^{32}P]RNA purified by hybridization to β-globin hybrid DNA fixed to Millipore filters or to poly(dC)

rabbit globin cDNA were indistinguishable from those of Figure 5a. (2) No comparable spots were found when globin-specific RNA was purified from long-term-labeled Friend cells by the same procedures or from 15S short-term-labeled RNA using α-globin hybrid DNA fixed to Millipore filters. Since the purification techniques, in contrast to the analytical procedures, do not involve RNase digestion, specific copurification could come about if β-globin RNA had some sequences in common with other RNA species; these could then hybridize to the cDNA used in the purification procedures. Alternatively, RNA could attach to 15S globin-specific RNA because of partial complementarity to the latter; it could, in addition, form concatenates if it contained extended hairpin loops (Ryskov et al. 1976; see also Fedoroff et al. 1977).

Heat-denatured, purified 15S β-globin [^{32}P]RNA run on a 6% polyacrylamide gel showed a major band (Fig. 6e) and some heterogeneous material of greater mobility, as well as a small amount of label at the position of 10S β-globin RNA derived from short-term-labeled cells (Fig. 6d). The RNA in the major band of the 15S RNA gave a similar fingerprint as shown in Figure 5a; however, some of the spots in the extreme right-hand lower quadrant appeared to be missing. It seems that most large T_1 oligonucleotides are associated with RNA forming a single band on polyacrylamide gel electrophoresis. It is striking that 10S α- and β-globin RNA from short-term-labeled cells (Fig. 6c,d) have a lower mobility than their counterparts from long-term-labeled cells (Fig. 6a,b). This may be due to further processing, either of the poly(A) or of the heteropolymeric sequences.

If the 15S β-globin-specific RNA were a primary transcript, it might yield a 5' terminal pppNp or ppNp sequence on hydrolysis. Purified 15S β-globin [^{32}P]RNA, 10S β-globin [^{32}P]mRNA, and unpurified 8S–12S [^{32}P]RNA from short-term-labeled cells were totally digested with RNase T_2 and the products resolved by two-dimensional electrophoresis. As shown in Figure 7, unpurified 8S–12S [^{32}P]RNA yielded the four nucleoside 3' monophosphates and a large number of faint spots; these probably comprise a variety of nucleoside di-, tri-, and tetraphosphates, as well as of capped and otherwise modified nucleotides, as judged by the results of Adams and Cory (1975). The 10S β-globin [^{32}P]RNA gave rise to two spots besides the four standard nucleotides — a stronger one (K_1) (about 0.4% of the radioactivity) and a weaker one (K_2) (about 0.2%). The 15S β-globin-specific [^{32}P]RNA showed, in addition to the four nucleotides, only one spot (K_1') (about 0.4% of the radioactivity) with a mobility corresponding to nucleotide K_1. After phosphatase treatment, both K_1 and K_1' had the same reduced mobility on paper electrophoresis at pH 3.5. Cheng and Kazazian (1977) and Heckle et al. (1977) have determined that the cap structures of both α- and β-mouse globin mRNA are m^{7}Gpppm^{6}A^{m}pC and

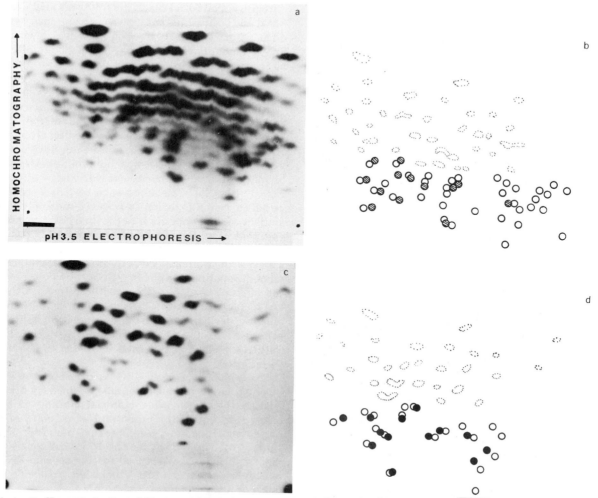

Figure 5. Characterization of short-term-labeled globin-specific [^{32}P]RNA by T$_1$ fingerprinting. Labeled RNA was digested and fractionated as described in the legend to Fig. 3. *(a)* 15S globin-specific [^{32}P]RNA (10^4 cpm) purified from induced cells labeled for 20 min with [^{32}P]phosphate by two cycles of hybridization to poly(dC) ($\alpha + \beta$) globin cDNA and poly(I)-Sephadex chromatography. *(b)* Composite tracing of the T$_1$ fingerprints of purified 15S globin-specific [^{32}P]RNA from *a* (○) and of purified 10S β-globin [^{32}P]RNA from Fig. 3b (∅). *(c)* α-globin [^{32}P]RNA (5600 cpm) from the 10S–12S RNA fraction of induced cells labeled with [^{32}P]phosphate for 20 min, purified by hybridization to α-globin hybrid DNA fixed to Millipore filters. *(d)* Composite tracing of the T$_1$ fingerprints of 10S–12S α-globin [^{32}P]RNA from *c* (○) and 10S α-globin ^{32}P-labeled mRNA from Fig. 3a (●). (Reprinted, with permission, from Curtis et al. 1977.)

m^{7}Gpppm^{6}A^{m}C^{m}pN. Although we have not carried out further analyses, it seems very likely that the cap associated with our preparation of 15S β-globin-specific RNA was the same as one of those of mature β-globin mRNA.

Kinetics of Synthesis of 15S β-Globin-specific RNA: Evidence for a Precursor Role

Convincing evidence for a precursor-product relationship between two RNA species is difficult to obtain in animal cells. A popular approach is to label cells for a short period with a radioactive precursor (pulse), then prevent further incorporation of radioactivity as completely as possible (chase), and then show that labeled product accumulates while the putative precursor diminishes. Ideally, less labeled precursor should be incorporated into total RNA during the chase than appears in the product during this period. This condition is very difficult to attain if the product is only a minor fraction of the RNA being synthesized, since addition of unlabeled nucleoside is usually insufficient to completely prevent incorporation of labeled triphosphates present in the intracellular pool or arising by turnover of short-lived RNA species. Inhibition of RNA synthesis by actinomycin D is quite effective; however, the drug may cause other changes in RNA metabolism (Endo et al. 1971; Scholtissek 1972).

We have therefore limited ourselves to showing

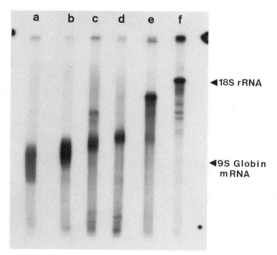

◄18S rRNA

◄9S Globin mRNA

Figure 6. Gel electrophoresis of globin-specific [³²P]RNA. Labeled globin-specific RNA was purified from induced cells labeled for 30 min or 16 hr with [³²P]phosphate by hybridization to α- or β-globin hybrid DNA fixed to Millipore filters. The purity of the 15S β-globin [³²P]RNA was estimated by rehybridization to β-globin hybrid DNA fixed to a Millipore filter followed by RNase-A and T₁ digestion; 20% of the [³²P]RNA was protected. Since the 15S β-globin RNA is about 2.5 times globin mRNA, the purity is at least 50%. Labeled RNA was heated in 7 M urea, 9 mM Tris-borate, pH 8.2, 2.5 mM EDTA at 70°C for 2 min before electrophoresis in 6% polyacrylamide gel, basically as described by Maniatis et al. (1975), but with the gel overlayered with a 3% spacer gel containing 7 M urea, 90 mM Tris-borate, pH 8.2, 2 mM EDTA at 300 V for 16 hr. (a) α-Globin [³²P]RNA (16,800 cpm) from cells labeled for 16 hr, (b) β-globin [³²P]RNA (15,300 cpm) as in a; (c) 10S α-globin [³²P]RNA (17,800 cpm) from cells labeled for 20 min from 10S RNA; (d) 10S β-globin [³²P]RNA (16,000 cpm) as in c; (e) 15S β-globin [³²P]RNA (20,000 cpm) as in c; (f) 18S ribosomal [³²P]RNA (50,000 cpm). The lower arrow indicates the position of unlabeled rabbit 10S globin mRNA run in a separate lane.

that the rate of synthesis of 15S β-globin-specific RNA is about equal to or greater than that of β-globin mRNA, a necessary condition for a precursor-product relationship. Moreover, we have obtained approximate values for the half-life of the 15S RNA and its steady-state level in the cell.

To measure the rate of synthesis of 10S and 15S β-globin-specific RNA, it was necessary to determine both the incorporation of radioactivity into these species and the specific radioactivity of the intracellular nucleoside triphosphates. Dimethylsulfoxide-induced Friend cells were grown for 96 hours on medium containing a low level of [³²P]phosphate of known specific activity and labeled for 5, 13, 25, and 60 minutes with [³H]uridine. RNA was extracted by lysis of the cells with sodium dodecyl sulfate followed by pronase digestion, phenol extraction, and ethanol precipitation. From the supernatant fluid, the ribonucleoside triphosphates were isolated and separated, and the specific radioactivity of each was determined from the ratio of ³H/³²P cpm (Table 3).

After digestion with DNase, the RNA was heat-denatured and sedimented through a sucrose density gradient. A part of each fraction was mixed with a known amount of ³²P-labeled α- and β-globin mRNA as internal standard and hybridized to two filters, one with α globin and the other with β-globin hybrid plasmid fixed to it. The RNase-resistant, filter-bound ³H and ³²P radioactivities were determined, the relative hybridization efficiencies calculated from the ³²P values, and the ³H values corrected accordingly. After 5 minutes of labeling, more than half of the β-globin-specific [³H]RNA was found in a sharp peak at 15S, while most of the remainder was in the 10S region (Fig. 8a). The occurrence of two β-specific peaks in the 10S region is possibly an analytical artifact, since a repetition of the analysis gave only one peak (Fig. 8b). The scatter in the 10S region of the 5-minute sample (which stemmed from about 4×10^7 cells) is probably due to the presence of large amounts of unlabeled globin mRNA, which exceeded the amount of globin DNA on the filter and thus led to large correction factors. After a 13-minute labeling period, the β-globin-specific 10S [³H]RNA peak was about 2.7 times larger than the 15S [³H]RNA peak, and this ratio increased to 15 after 60 minutes.

The α-globin-specific [³H]RNA, after 5 minutes of labeling, gave only a single, albeit relatively broad peak at 11S, with some material trailing to the heavy side, but with no peak at 15S. After 13 minutes of labeling, the peak became narrower and shifted to about 9.5S. Thus the 15S RNA consisted mainly, if not exclusively, of β-globin-specific RNA, in confirmation of the conclusion reached by fingerprinting.

The accumulation of 10S and 15S β-globin [³H]-RNA was estimated by dividing the radioactivity in each peak by the specific ³H-radioactivity of the nucleoside triphosphates averaged over the preceding time interval and then multiplying by 4. This is justified if the U:C and (U + C):(A + G) ratios are close to 1 and if the rate of change of specific radioactivity is small. Since the latter is not true for the 5-minute sample, the value calculated for this time point may not be very accurate.

Figure 9 shows that the amount of 10S β-globin [³H]RNA increased steadily over the entire observation period of 1 hour; the amount of 15S β-globin [³H]RNA remained constant after about 5 minutes. The 10S RNA thus behaved as though it were a steadily accumulating species, either stable or with a half-life that was long compared to the observation period; the 15S RNA, however, showed the labeling kinetics of an unstable species with a comparatively short half-life. If the labeling procedure does not affect the rates of synthesis and turnover, and if there is no significant cell growth, then the time required for replacing half the molecules in the 15S β-globin RNA pool by newly synthesized, labeled 15S

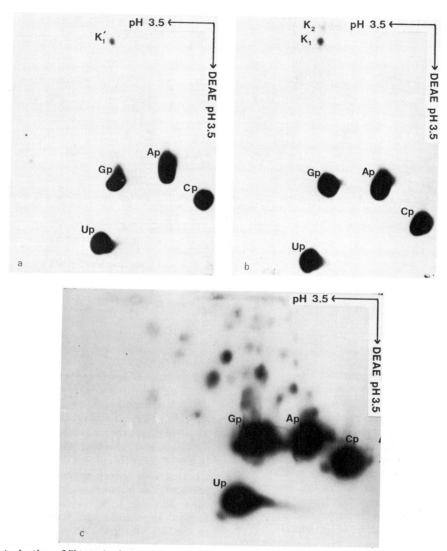

Figure 7. Characterization of 5'-terminal structures of globin-specific [^{32}P]RNA. [^{32}P]RNA containing 20 μg carrier RNA was digested with RNase A and RNase T$_2$ in 3 μl 15 mM sodium acetate (pH 4.5), 1 mM EDTA at 37°C for 60 min. The products were separated by cellulose acetate electrophoresis, followed by electrophoresis on DEAE paper, both at pH 3.5. *(a)* 15S β-globin-specific [^{32}P]RNA (44,000 cpm) purified from induced cells labeled for 30 min with [^{32}P]phosphate by hybridization to β-globin hybrid DNA fixed to Millipore filters. *(b)* 10S β-globin-specific [^{32}P]RNA (65,000 cpm) purified as above from induced cells labeled for 16 hr with [^{32}P]phosphate. *(c)* 10S–12S [^{32}P]RNA (10^6 cpm) from induced cells labeled for 16 hr with [^{32}P]phosphate.

β-globin RNA is equal to the half-life of the RNA in this pool. The data of Figure 9 suggest that this half-life is 2–3 minutes or less. However, this value should be considered as an estimate only, since the 5-minute determination is the least accurate.

The steady-state level of the 15S β-globin-specific RNA is 24 fmoles of total nucleotide in RNA per 10^6 cells, which is equivalent to 24 molecules/cell (since only hybridized sequences are measured, our calculations are based on a chain length of 600; see Table 3). Assuming a constant rate of synthesis, $dc_s/dt = k_s$, and a first-order rate of processing, $dc_p/dt = -k_p \cdot c$, the concentration of precursor is given by $dc/dt = k_s - k_p \cdot c$, from which $c = k_s/k_p(1 - e^{-k_p t})$ and the steady-state concentration $c_{t=\infty} = k_s/k_p$. If the half-life of the 15S RNA is 2.5 minutes, then $k_p = \ln 2/2.5 = 0.28$ min^{-1} and $k_s = c_{t=\infty} \times k_p = 24 \times 0.28 = 6.6$ molecules/min/cell. The rate of incorporation of nucleotide into 10S β-globin mRNA is about 27 fmoles nucleotide/10^6 cells/5 min, or 5.4 molecules/min/cell (Table 3). We consider the two rates to be compatible; however, the imprecision inherent in the determination of the half-life of the 15S β-globin-specific RNA has to be borne in mind.

Table 3. Kinetics of Synthesis of 10S and 15S β-Globin RNA—Specific Radioactivity of Nucleoside Triphosphates

Time of labeling (min)	No. of cells labeled $\times 10^{-7}$	Total ^{3}H-radio-activity (cpm $\times 10^{-7}$)	Total ^{3}H-radio-activity in β-globin-specific peak[a] (corrected cpm)		Total nucleo-tide incor-porated[b] (mole/cell $\times 10^{21}$)		β-Globin molecules synthesized per cell[c]		Ratio of ^{3}H/^{32}P-radioactivities in nucleoside triphosphates $\times 10^{-4}$		Specific radioactivity[d] (dpm/nmole $\times 10^{-6}$)	
			15S	10S	15S	10S			CTP	UTP	CTP	UTP
5	4.0	1.9	1054	942	24.0	21.2	24.0	21.2	1.1	9.9	3.6	31.0
13	1.8	2.4	978	3146	22.8	72.8	22.8	72.8	2.7	10.7	8.6	33.9
25	0.7	1.9	308	2924	15.6	135.2	15.6	135.2	3.4	12.2	10.8	38.8
60	0.6	2.8	438	10,984	22.0	548.0	22.0	548.0	5.1	13.7	16.2	43.5

The data are from Fig. 8.

[a] The radioactivities under the 15S and 10S β-globin RNA were multiplied by 2 (half of each fraction was assayed). The measured cpm were corrected for hybridization efficiency as explained in the legend to Fig. 8.

[b] Nucleotide incorporated (mole per cell) = ^{3}H-cpm $\times \dfrac{1}{\text{no. of cells}} \times \dfrac{100}{\text{counting efficiency (\%)}} \times \dfrac{4}{\text{average specific radioactivity of UTP} + \text{average specific radioactivity of CTP}}$

(specific activity averaged between time point and earlier point).

[c] Molecules per cell = $\dfrac{\text{nucleotides incorporated (mole/cell)}}{600} \times 6 \times 10^{23}$ (assumed chain length of β-globin RNA = 600).

[d] Specific radioactivity = $\dfrac{\text{ratio of } ^3\text{H}/^{32}\text{P}}{3 \times \text{specific radioactivity of } [^{32}\text{P}]\text{phosphate}}$. Specific radioactivity of phosphate was 105 dpm per nmole.

ATP and GTP contained negligible ^{3}H-radioactivity.

Is There a Precursor for Mouse α-Globin mRNA?

As pointed out in the previous section, α-globin-specific RNA labeled for 60 minutes or longer sedimented predominantly at about 9.5S (Fig. 8), whereas after only 5 minutes of labeling, a broad peak around 11S was observed, suggesting the possibility of an α-globin-specific mRNA precursor.

[^{32}P]RNA from induced Friend cells labeled for 20 minutes was fractionated by sucrose gradient centrifugation. The RNA from the heavy side of the 10S peak was hybridized to Millipore-filter-bound α-globin hybrid DNA and fingerprinted as before. All oligonucleotides found in the α-globin mRNA fingerprint were present, with one possible exception (Fig. 5c). In addition, about six oligonucleotides not found in the α-globin mRNA fingerprint appeared, albeit in apparently less than molar yield. These results, along with sedimentation data, suggest that the > 10S fraction contained an α-globin-specific RNA about 30% longer than the mature globin mRNA.

DISCUSSION

Recent studies have demonstrated the existence of a 15S globin-specific RNA, i.e., of a species two to three times the length of mature globin mRNA, in a variety of erythroid cells such as mouse fetal liver (Ross 1976), anemic spleen cells (Kwan et al. 1977), and erythroleukemic (Friend) cells (Curtis and Weissmann 1976; Bastos and Aviv 1977). We have shown by T$_1$ oligonucleotide fingerprinting that the 15S globin RNA contained all characteristic β-globin-specific oligonucleotides, as well as a defined set of additional oligonucleotides, but lacked α-globin-specific sequences. The latter finding was confirmed by hybridization to cloned α and β mouse globin hybrid DNA.

The rate of synthesis of 15S β-globin RNA and 10S β-globin mRNA were estimated to be about 6.6 and 5.4 molecules/min/cell, respectively. A precursor-product relationship is thus possible as far as kinetics of synthesis are concerned.

We estimate the half-life of the 15S β-globin RNA to be 2.5 minutes or less. However, there are uncertainties with regard to the specific activities of the intracellular nucleotides incorporated at short times after labeling; thus a twofold higher value of the half-life is possible. Ross (1976) determined the half-life of 15S RNA to be 45 minutes or less by labeling with [^{3}H]uridine and inhibiting further synthesis with actinomycin D; pulse-labeling followed by a chase with unlabeled uridine did not lead to a complete transfer of label from 15S to 10S within 75 minutes. Bastos and Aviv (1977) have estimated a half-life of about 5 minutes for globin-specific 15S RNA in Friend cells. For comparison, it may be noted that recent estimates for the half-life of heterogeneous nuclear RNA range from 7 minutes for sea urchin embryo (Brandhorst and Humphreys 1972) to 23 minutes for L cells (Brandhorst and McConkey 1974) and 27 minutes for rabbit erythroid cells (Hunt 1976). A half-life of about 3 minutes for hnRNA of HeLa cells had previously been reported by Soeiro et al. (1968).

It is interesting to note that, using the values determined by us for the rate of synthesis of 10S β-globin mRNA (5.4 molecules/min/cell) and by Bastos et al. (1977) for the half-life of globin mRNA (17 hr), the equation given earlier predicts that 24 and 72 hours after beginning full expression of globin genes, Friend cells should contain 5000 and 7500 molecules, respectively, of β-globin mRNA. This is

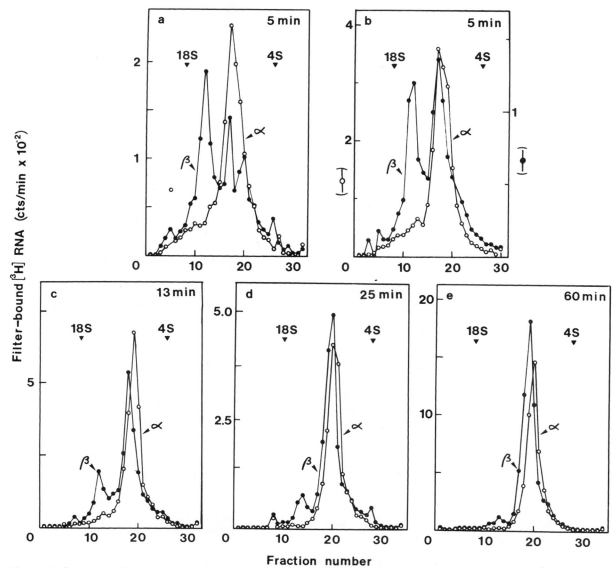

Figure 8. Sucrose gradient centrifugation of [³H]RNA from Friend cells labeled for various times with [³H]uridine; location of globin-specific sequences by hybridization. Friend cells were grown in medium (Curtis and Weissmann 1976) containing [³²P]phosphate (210 cpm/nmole). After 1 day, dimethylsulfoxide was added to 1.5%, and, after 4 days, the cells (7 × 10⁷ in 3.5 ml) were labeled with 30 mCi of [³H]uridine (25–30 Ci/mmole) at 37°C. Samples were mixed with frozen medium at different times and pelleted. Each pellet was resuspended with unlabeled induced cells to give a total of 4 × 10⁷ cells. The nucleic acids were extracted and precipitated with ethanol (Curtis and Weissmann 1976). Nucleoside triphosphates were precipitated from the ethanol supernatant with 16 mM barium acetate in 4 mM NaHCO₃, recovered by treatment of the barium salts with Dowex 50 (H⁺ form), and separated by two-dimensional chromatography (Randerath and Randerath 1967). The appropriate areas of the chromatogram were eluted with 1 M NH₃, the ³²P- and ³H-radioactivities determined in a Triton–toluene-based scintillator cocktail (Patterson and Greene 1965), and the specific ³H-radioactivities were calculated (see Table 3).

After DNase digestion, the recovery of [³H]RNA was 20–30 × 10⁶ cpm in each sample; that of [³²P]RNA varied from 150,000 in the 5-min sample to 18,000 in the 60-min sample. After heating in 20 mM Tris-HCl, pH 7.5, 1 mM EDTA at 100°C for 45 sec, the RNA was centrifuged for 16 hr at 15°C and 32,000 rpm in a SW 41 rotor. One half of each gradient fraction (350 μl) was mixed with purified α- and β-globin ³²P-labeled mRNA (340 cpm each) and hybridized at 66°C in 0.5 M NaCl, 10 mM Tris-HCl, pH 7.5, for 24 hr to two 4-mm filters, each containing about 3.6 μg of α- and β-globin hybrid DNA, respectively. After treatment with RNase A and T₁, the radioactivities were determined.

The hybridization efficiencies were calculated from the filter-bound ³²P-radioactivities; it was assumed that the α and β [³²P]RNAs were 65% and 80% pure, respectively, the maximum values of RNase resistance found after liquid hybridization in globin cDNA excess. The ³H-values were corrected for the hybridization efficiencies. The correction factors were about 2 for the α globin and 2–2.5 for the β globin, except in the 10S region where they reached 4–5. (*b*) A repeat of the hybridization analysis on the same sample as in *a*, but using different α- and β-globin [³²P]RNA standards and a different set of filters. (○) α-Globin-specific [³H]RNA; (●) β-globin-specific [³H]RNA. The arrows indicate the positions of the endogenous 4S and 18S [³²P]RNAs. (Reprinted, with permission, from Curtis et al. 1977.)

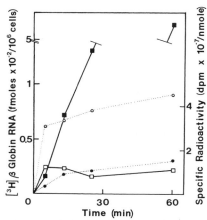

Figure 9. Accumulation of 15S and 10S β-globin [³H]RNA in dimethylsulfoxide-induced Friend cells. The radioactivity in 15S and 10S β-globin [³H]RNA was calculated from the data in Fig. 8, taking into account aliquot size and assuming that the recovery of [³H]RNA was 80% for each sample. The total radioactivity in each peak was calculated as shown in Table 3. (○----○) Specific ³H-radioactivity of UTP; (●---●) specific ³H-radioactivity of CTP; (□——□) 15S β-globin [³H]RNA; (■——■) 10S β-globin [³H]RNA. (Reprinted, with permission, from Curtis et al. 1977.)

in reasonable agreement with the values of 6600 (Ross et al. 1974) and about 10,000 (Harrison et al. 1974) molecules per cell of α- and β-globin mRNA determined experimentally, in view of the fact that erythroid cells contain about equal numbers of α and β chains (Morrison et al. 1974; Cheng and Kazazian 1976). Although it has not been strictly proven, the evidence adduced by several independent investigators strongly suggests that the 15S β-globin mRNA is a precursor of 10S β-globin mRNA.

We have preliminary evidence for the existence of an α-globin-specific RNA about 30% longer than the mature mRNA. It is clear that α- and β-globin mRNAs do not arise in a closely similar fashion: Either the α-globin mRNA does not have a precursor with a length similar to that of its β counterpart, or if it does, the precursor is processed far more rapidly.

The question arises as to whether the 15S β-globin RNA is a primary transcript or whether it is a processed product deprived from an even longer precursor. We have not succeeded in demonstrating the presence of nucleoside tetra-, tri-, or diphosphates in preparations of ³²P-labeled 15S β-globin RNA; however, a high yield of a "cap" indistinguishable from that of mature β-globin mRNA was found. This does not exclude the possibility that 15S β-globin RNA is a primary transcript, since capping might occur rapidly after initiation of transcription, as is the case in reovirus mRNA (Furuichi and Shatkin 1976). Bastos and Aviv (1977) have described a 27S globin-specific RNA and have suggested that this represents a precursor of the 15S molecule. Despite several attempts, we have not been able to find a

peak of globin-specific RNA sedimenting substantially faster than 15S–18S. Since we have also examined RNA from the Friend cell clone used by Bastos and Aviv (kindly made available by them), it seems that some differences in the techniques used are responsible for the discrepancy in the findings.

The work by Bastos and Aviv (1977), as well as preliminary results from our laboratory (unpubl. experiments), suggests that half (or more) of the 15S β-globin mRNA is polyadenylated. Thus the 15S β-globin RNA seems to have the same 5' and 3' termini as mature β-globin mRNA. The evidence suggesting that the formation of several adenovirus-specific mRNAs involves a "splicing" step in which interior sequences of a precursor molecule are excised (see Sambrook 1977) leads us to consider a similar mechanism for the maturation of the 15S β-globin RNA. It has been found that the mouse β-globin chromosomal gene contains an inserted sequence of about 500 nucleotides which is not represented in the mouse β-globin mRNA (S. Tilghman and P. Leder, pers. comm.). A similar conclusion has been reached by A. Jeffreys and R. Flavell (pers. comm.) for rabbit β-globin DNA. It will be of interest to determine whether the 15S β-globin RNA is a copy of the gene containing the insert and whether maturation of this precursor is due to excision of this internal sequence.

SUMMARY

We have applied the techniques of poly(I)-Sephadex chromatography (Coffin et al. 1974) and hybridization to filters with attached α- and β-globin hybrid plasmid DNA to detect and characterize globin-specific RNAs in mouse erythroleukemic (Friend) cells. A 15S species specific for β-globin mRNA was purified and characterized. Fingerprinting revealed oligonucleotides present in β-globin mRNA but not in α globin mRNA, in addition to about 20 oligonucleotides not found in either α- or β-globin mRNA. From kinetic studies, the rate of synthesis of 15S β-globin RNA was calculated as 6.6 molecules/min/cell, a value comparable to that for 10S β-globin RNA, which was 5.4 molecules/min/cell. Thus a precursor-product relationship is possible. The half-life of the 15S species was estimated to be 2.5 minutes or less. We also detected and purified an α-globin-specific 11S species; fingerprinting analysis of 11S α-globin RNA purified from short-term-labeled cells showed six T_1 oligonucleotides in addition to those found in mature α-globin mRNA.

Acknowledgments

We thank Drs. Bastos and Aviv for providing us with the clone of Friend cells used in their experiments. We also thank Dr. Williamson for generous supplies of mouse globin mRNA. The work was supported by the Schweizerische National funds (No. 3.475.75) and the Kanton of Zürich.

REFERENCES

ADAMS, J. M. and C. CORY. 1975. Modified nucleosides and bizarre 5'-termini in mouse myeloma mRNA. *Nature* **255**: 28.

ARMSTRONG, K. A., M. V. HERSHFIELD, and D. R. HELINSKI. 1977. Gene cloning and containment properties of plasmid Col E1 and its derivatives. *Science* **196**: 172.

BASTOS, R. N. and H. AVIV. 1977. Globin RNA precursor molecules: Biosynthesis and processing in erythroid cells. *Cell* **11**: 641.

BASTOS, R. N., Z. VOLLOCH, and H. AVIV. 1977. Messenger RNA population analysis during erythroid differentiation: A kinetical approach. *J. Mol. Biol.* **110**: 191.

BRANDHORST, B. P. and T. HUMPHREYS. 1972. Stabilities of nuclear and messenger RNA molecules in sea urchin embryos. *J. Cell Biol.* **53**: 474.

BRANDHORST, B. P. and E. H. McCONKEY. 1974. Stability of nuclear RNA in mammalian cells. *J. Mol. Biol.* **85**: 451.

CHANTRENNE, H., A. BURNY, and G. MARBAIX. 1967. The search for the messenger RNA of hemoglobin. *Prog. Nucleic Acid Res. Mol. Biol.* **7**: 173.

CHENG, T.-C. and H. H. KAZAZIAN, JR. 1976. Unequal accumulation of α- and β-globin mRNA in erythropoietic mouse spleen. *Proc. Natl. Acad. Sci.* **73**: 1811.

———. 1977. The 5'-terminal structures of murine α- and β-globin messenger RNA. *J. Biol. Chem.* **252**: 1758.

COFFIN, J. M., J. T. PARSONS, L. RYMO, R. K. HAROZ, and C. WEISSMANN. 1974. A new approach to the isolation of RNA-DNA hybrids and its application to the quantitative determination of labeled tumor virus RNA. *J. Mol. Biol.* **86**: 373.

CURTIS, P. J. and C. WEISSMANN. 1976. Purification of globin messenger RNA from dimethylsulfoxide-induced Friend cells and detection of a putative globin messenger RNA precursor. *J. Mol. Biol.* **106**: 1061.

CURTIS, P. J., N. MANTEI, J. VAN DEN BERG, and C. WEISSMANN. 1977. Presence of a putative 15S precursor to β-globin mRNA but not to α-globin mRNA in Friend cells. *Proc. Natl. Acad. Sci.* **74**: 3184.

DARNELL, J. E., W. R. JELINEK, and G. R. MOLLOY. 1973. Biogenesis of mRNA: Genetic regulation in mammalian cells. *Science* **181**: 1215.

DAYHOFF, M. O. 1972. *Atlas of protein sequence and structure*, vol. V. National Biomedical Research Foundation, Washington, D.C.

ENDO, Y., H. TOMINAGA, and Y. NATORI. 1971. Effect of actinomycin D on turnover rate of messenger ribonucleic acid in rat liver. *Biochim. Biophys. Acta* **240**: 215.

FEDEROFF, N., P. K. WELLAUER, and R. WALL. 1977. Intermolecular duplexes in heterogeneous nuclear RNA from HeLa cells. *Cell* **10**: 597.

FLAVELL, R. A., D. L. SABO, E. F. BANDLE, and C. WEISSMANN. 1974. Site-directed mutagenesis: Generation of an extracistronic mutation in bacteriophage Qβ RNA. *J. Mol. Biol.* **89**: 255.

FRIEND, C., W. SCHER, J. G. HOLLAND, and T. SATO. 1971. Hemoglobin synthesis in murine virus-induced leukemic cells in vitro: Stimulation of erythroid differentiation by dimethylsulfoxide. *Proc. Natl. Acad. Sci.* **68**: 378.

FURUICHI, Y. and A. J. SHATKIN. 1976. Differential synthesis of blocked and unblocked 5'-termini in reovirus mRNA: Effect of pyrophosphate and pyrophosphatase. *Proc. Natl. Acad. Sci.* **73**: 3448.

GEORGIEV, G. P., O. P. SAMARINA, M. I. LERMAN, M. N. SMIRNOV, and A. N. SEVERTZOV. 1963. Biosynthesis of messenger and ribosomal ribonucleic acids in the nu-

cleolochromosomal apparatus of animal cells. *Nature* **200**: 1291.

GILLESPIE, D. and S. SPIEGELMAN. 1965. A quantitative assay for DNA-RNA hybrids with DNA immobilized on a membrane. *J. Mol. Biol.* **12**: 829.

HARRISON, P. R., R. S. GILMOUR, N. A. AFFARA, D. CONKIE, and J. PAUL. 1974. Globin messenger RNA synthesis and processing during haemoglobin induction in Friend cells. II. Evidence for post-transcriptional control in clone 707. *Cell Differ.* **3**: 23.

HECKLE, W. L., JR., R. G. FENTON, T. G. WOOD, C. G. MERKEL, and J. B. LINGREL. 1977. Methylated nucleosides in globin mRNA from mouse nucleated erythroid cells. *J. Biol. Chem.* **252**: 1764.

HOFSTETTER, H., A. SCHAMBÖCK, J. VAN DEN BERG, and C. WEISSMANN. 1976. Specific excision of the inserted DNA segment from hybrid plasmids constructed by the poly-(dA)-poly(dT) method. *Biochim. Biophys. Acta* **454**: 587.

HUNT, J. A. 1976. Ribonucleic acid synthesis in rabbit erythroid cells. *Biochem. J.* **160**: 727.

IMAIZUMI, T., H. DIGGELMANN, and K. SCHERRER. 1973. Demonstration of globin messenger sequences in giant nuclear precursors of messenger RNA of avian erythroblasts. *Proc. Natl. Acad. Sci.* **70**: 1122.

KACIAN, D. L., K. F. WATSON, A. BURNY, and S. SPIEGELMAN. 1971. Purification of the DNA polymerase of avian myeloblastosis virus. *Biochim. Biophys. Acta* **246**: 365.

KWAN, S.-P., T. G. WOOD, and J. B. LINGREL. 1977. Purification of a putative precursor of globin messenger RNA from mouse nucleated erythroid cells. *Proc. Natl. Acad. Sci.* **74**: 178.

MACNAUGHTON, M., K. B. FREEMAN, and J. O. BISHOP. 1974. A precursor to hemoglobin mRNA in nuclei of immature duck red blood cells. *Cell* **1**: 117.

MANIATIS, T., A. JEFFREY and H. VAN deSANDE. 1975. Chain length determination of small double- and single-stranded DNA molecules by polyacrylamide gel electrophoresis. *Biochemistry* **14**: 3787.

MANIATIS, T., G. K. SIM, A. EFSTRATIADIS, and F. C. KAFATOS. 1976. Amplification and characterization of a β-globin gene synthesized in vitro. *Cell* **8**: 163.

MAXAM, A. M. and W. GILBERT. 1977. A new method for sequencing DNA. *Proc. Natl. Acad. Sci.* **74**: 560.

McKNIGHT, G. S. and R. T. SCHIMKE. 1974. Ovalbumin messenger RNA: Evidence that the initial product of transcription is the same size as polysomal ovalbumin messenger. *Proc. Natl. Acad. Sci.* **71**: 4327.

MORRISON, M. R., S. A. BRINKLEY, J. GORSKI, and J. B. LINGREL. 1974. The separation and identification of α- and β-globin messenger ribonucleic acids. *J. Biol. Chem.* **249**: 5290.

PATTERSON, M. S. and R. C. GREENE. 1965. Measurement of low energy beta-emitters in aqueous solution by liquid scintillation counting of emulsions. *Anal. Chem.* **37**: 854.

RANDERATH, K. and E. RANDERATH. 1967. Thin-layer separation methods for nucleic acid derivatives. *Methods Enzymol.* **12A**: 323.

ROSS, J. 1976. A precursor of globin messenger RNA. *J. Mol. Biol.* **106**: 403.

ROSS, J., H. AVIV, E. SCOLNICK, and P. LEDER. 1972. In vitro synthesis of DNA complementary to purified rabbit globin mRNA. *Proc. Natl. Acad. Sci.* **69**: 264.

ROSS, J., J. GIELEN, S. PACKMAN, Y. IKAWA, and P. LEDER. 1974. Globin gene expression in cultured erythroleukemic cells. *J. Mol. Biol.* **87**: 697.

RYSKOV, A. P., O. V. TOKARSKAYA, G. P. GEORGIEV, C. COUTELLE, and B. THIELE. 1976. Globin mRNA contains a sequence complementary to double-stranded region of nuclear pre-mRNA. *Nucleic Acids Res.* **3**: 1487.

SAMBROOK, J. 1977. Adenovirus amazes at Cold Spring Harbor. *Nature* **268**: 101.

SCHERRER, K., H. LATHAM, and J. E. DARNELL. 1963. Demonstration of an unstable RNA and of a precursor to ribosomal RNA in HeLa cells. *Proc. Natl. Acad. Sci.* **49**: 240.

SCHOLTISSEK, C. 1972. Unphysiological breakdown of fast-labeled RNA by actinomycin D in primary chick fibroblasts. *Eur. J. Biochem.* **28**: 70.

SMITH, H. O. and M. L. BIRNSTIEL. 1976. A simple method for DNA restriction site mapping. *Nucleic Acids Res.* **3**: 2387.

SOEIRO, R., M. H. VAUGHAN, J. R. WARNER, and J. E. DARNELL, JR. 1968. The turnover of nuclear DNA-like RNA in HeLa cells. *J. Cell Biol.* **39**: 112.

SPOHR, G., G. DETTORI, and V. MANZARI. 1976. Globin mRNA sequences in polyadenylated and nonpolyadenylated nuclear precursor-messenger RNA from avian erythroblasts. *Cell* **8**: 505.

VERMA, I. M., F. G. TEMPLE, H. FAN, and D. BALTIMORE. 1972. In vitro synthesis of DNA complementary to rabbit reticulocyte 10S RNA. *Nat. New Biol.* **235**: 163.

VOLCKAERT, G., W. MIN JOU, and W. FIERS. 1976. Analysis of ^{32}P-labeled bacteriophage MS2 RNA by a minifinger-printing procedure. *Anal. Biochem.* **72**: 433.

WARNER, J. R., R. SOEIRO, H. C. BIRNBOIM, M. GIRARD, and J. E. DARNELL. 1966. Rapidly labeled HeLa cell nuclear RNA. I. Identification by zone sedimentation of a heterogeneous fraction separate from ribosomal precursor RNA. *J. Mol. Biol.* **19**: 349.

WEINBERG, R. A., S. O. WARNAAR, and E. WINOCOUR. 1972. Isolation and characterization of simian virus 40 ribonucleic acid. *J. Virol.* **10**: 193.

WEISSMANN, C. and W. BOLL. 1976. Reduction of possible hazards in the preparation of recombinant plasmid DNA. *Nature* **261**: 428.

WILLIAMSON, R., C. E. DREWIENKIEWICZ, and J. PAUL. 1973. Globin messenger sequences in high molecular weight RNA from embryonic mouse liver. *Nat. New Biol.* **241**: 66.

Globin mRNA Sequences: Analysis of Base Pairing and Evolutionary Implications

W. SALSER

Molecular Biology Institute and Department of Biology, University of California, Los Angeles, California 90024

Over the past several years we attempted to develop in my laboratory general techniques for the sequence analysis of mammalian mRNAs (Salser et al. 1972, 1973, 1974; Paddock et al. 1974; Poon et al. 1974). Existing RNA sequencing techniques were poorly suited for this task, and our first approach was to use reverse transcriptase to make cDNA copies of the mRNAs so that DNA sequencing techniques could be employed (Salser et al. 1973). By analysis of RNA transcripts of the cDNA and the use of deoxysubstitution techniques, we were able to determine the sequences of substantial portions of the rabbit globin cDNAs and establish that reverse transcriptase gave high fidelity when used in this manner (Salser et al. 1973; Poon et al. 1974). These conclusions were independently confirmed by Marrotta et al. (1974), working with human globin cDNAs.

At the same time, it became apparent that RNA polymerase was not transcribing the entire cDNA sequence, and so certain portions of the rabbit globin mRNAs were not easily accessible to such sequencing techniques (Salser et al. 1976a; Paddock et al. 1977). We solved this problem and achieved other advantages by cloning the cDNA sequences in bacterial plasmids (Higuchi et al. 1976a,b; Paddock et al. 1976; Salser et al. 1976a; Browne et al. 1977).

Earlier we had made the unexpected observation that cDNAs made with reverse transcriptase are unique in having small self-complementary foldback regions or "hooks" at their 3' termini. These hooks provided an excellent primer which would permit the efficient synthesis of a duplex gene copy with DNA polymerase I of *Escherichia coli* (Salser 1974). These observations have since been confirmed in a number of laboratories (Efstratiadis et al. 1976; Rougeon and Mach 1976; O'Malley et al. 1976). The priming activity of this hook has been used in many laboratories (Higuchi et al. 1976a,b; Salser et al. 1976a; Rougeon and Mach 1976; Maniatis et al. 1976; O'Malley et al. 1976) to develop efficient techniques for cloning cDNA sequences in bacterial cells. For the sequencer, this is important because it supplies duplex gene copies in high quantity and purity so that restriction enzymes and the new ultra-rapid "ladder" sequencing techniques can be employed. This permitted the rapid elucidation of the entire sequence of the rabbit beta-globin mRNA (Lockard

and RajBhandary 1976; Browne et al. 1977; Baralle 1977; Proudfoot 1977; Efstratiadis et al. 1977; Paddock et al. 1977) and has permitted us to determine 368 nucleotides of the rabbit alpha-globin mRNA sequence (H. Heindell et al., in prep.). With earlier sequencing data (Proudfoot and Brownlee 1976; Proudfoot 1976; Liu et al. 1977; and Baralle 1977), this brings the known alpha-globin mRNA sequence to greater than 90% completion (Fig. 1).

The nucleotide sequences which we have determined for the rabbit alpha-globin gene fit perfectly with the known amino acid sequence of one of the known alleles of the corresponding protein. At its 5' terminus, the sequence we obtained from the rabbit alpha-globin plasmid pHb72 has an 18-nucleotide overlap and agrees perfectly with the 43-nucleotide sequence determined from the 5' end of the mRNA by Baralle (1977).

Base Composition and Dinucleotide Sequences in the Alpha-Globin mRNA

In the more than 90% of the alpha-globin structural gene that has been sequenced thus far, the GC content is 62.2%. This is considerably greater than the 54.2% GC content of the rabbit beta-globin structural gene (shown in Figure 3) and the 44.2% GC content of total rabbit DNA.

The 98 nucleotides of 5' and 3' untranslated alpha sequences thus far determined are also high in GC content (58%), unlike the situation in the rabbit beta-globin mRNA where the untranslated regions are AT-rich (GC = 40.2%; Fig. 3).

Another interesting feature of the rabbit alpha-globin mRNA sequence (Fig. 1) is that the dinucleotide sequence CpG is present 22 times, whereas in the beta-globin mRNA sequence (Fig. 3) only three CpG sequences are found. It has been recognized for some time that the nuclear DNA of vertebrates differs from mitochondrial DNA and most bacterial and lower eukaryotic DNAs in having many fewer CpG dinucleotide sequences than expected in random sequences of the same base composition (much of the available data on nearest neighbor frequencies has been collated by Setlow 1976). Russell et al. (1976) have observed that this bias is not present in vertebrate sequences for tRNAs, 5S RNA, or ribosomal RNA. Moreover, although the level of CpG

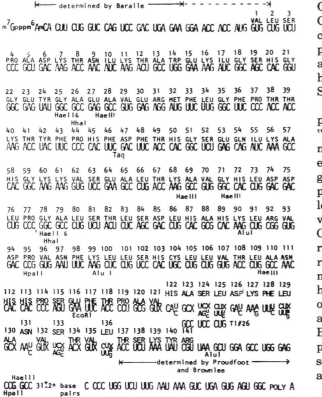

Figure 1. Nucleotide sequence of rabbit alpha-globin mRNA. The sequence shown is that determined by Liu et al. (1977) and H. Heindell et al. (in prep.), except for the 25 nucleotides at the 5′ terminus (determined by Baralle 1977b) and the sequence following amino acid residue 136 (determined by Proudfoot and Brownlee 1976). The gap of "31 ± 2 base pairs" in the 3′ untranslated sequence was reported as known by Proudfoot (1976); however, the sequence was not given in that paper.

Obviously, vertebrate mRNAs with high levels of CpG can be translated with entirely satisfactory efficiencies. A similar conclusion is suggested by the presence of high levels of CpG sequences in some animal viruses (e.g., adenovirus and herpes simplex have, respectively, 6.2 and 10.9% CpG dinucleotides; Setlow [1976]).

As an alternative hypothesis, Russell et al. (1976) proposed that the low of CpG sequences could be "the result of unidentified constraints (vertebrate nucleosome packaging?) which affect all of the DNA except satellites, ribosomal cistrons, and tRNA genes." This hypothesis, like the preceding one, appears to depend upon natural selection to keep the level of CpG sequences low by eliminating those individuals which have acquired increased numbers of CpG sequences by mutations. On the average, roughly 6% of all mutations would be expected to result in a new CpG sequence. According to the model of Russell et al. (1976), most of these would have to be eliminated on the basis of their deleterious effects in order to maintain the CpG frequency at the low level observed in bulk vertebrate DNA. Because of considerations of genetic load discussed previously (Salser and Isaacson 1976) and in a later section of this paper, we believe that this would create an intolerable genetic load.

We therefore present an alternative hypothesis in which it is imagined that CpG sequences are eliminated not because of supposed deleterious effects, but rather because the sequence CpG is a hot spot for mutations. The advantage of this model is that it explains how the low frequency of CpG sequences can be achieved at a much lower genetic cost, consistent with current estimates of the genetic load. For instance, many of the CpG sequences will occur in the large fraction of single-copy DNA which is genetically drifting (for review; see Salser and Isaacson 1976). Mutation of CpG to another sequence in such a region will have no deleterious effect and will not contribute to the genetic load.

It is not difficult to imagine why the sequence CpG might be a mutational hot spot, since it is known that this sequence is highly methylated (Sinsheimer 1955; Doskocil and Sorm 1962), and enzyme activities which can methylate such sequences in vitro have been isolated from mammalian cells (Roy and Weissbach 1975). Only a modest increase in the mutation rate, say five- to ten-fold above the average, is required to account for the observed rarity of the sequence CpG. This could happen if methylation caused the affected base to pair incorrectly during replication, as in base analog mutagenesis, or caused it to be more susceptible to damage.

It was the large number of CpG sequences in the alpha-globin structural gene which prompted us to reexamine and discard the model of Russell et al. In the context of the model presented here, these CpG sequences would be expected to be preserved if they had some role in mRNA function, so that

sequences in two satellite DNAs tested is lower than average for a random sequence, it is considerably higher than the level in bulk vertebrate DNA. Because they found little or no shortage of CpG in several sequences thought not to code for proteins, Russell et al. (1976) have advanced the hypothesis that the avoidance of CpG sequences is "a consequence of natural selective forces which have conferred this particular design on polypeptide-coding DNA directly or indirectly through the machinery of the translation apparatus. As only a small proportion of nuclear DNA is thought to be actively involved in protein coding . . . it must be further argued on this assumption that the bulk of nuclear DNA derives from and maintains the gross sequence characteristics of polypeptide-specifying DNA."

It seems to us that the logic of this hypothesis is compromised by the discovery that a major vertebrate mRNA, such as that coding for rabbit alpha-globin, carries a large number of CpG sequences.

mutations changing CpG to another dinucleotide would be eliminated by natural selection. For instance, in the alpha-globin sequence CpG may be important in base pairing such as that shown for the beta-globin mRNA in Figure 3 or in obeying the codon selection rules discussed below. For example, 13 of the CpG sequences could simply result from the very strong preference for C in the third position of the codons for His, Gly, Ala, Phe, Asp, and Pro (see Table 1).

Codon Assignments in Reticulocyte mRNAs

As observed in several other mRNA sequences, the rabbit alpha-globin mRNA shows nonrandom use of synonymous codons. The data are summarized in Table 1 and compared with corresponding figures for rabbit beta-globin mRNA, mouse immunoglobulin mRNAs, and rat insulin mRNA.

In several respects the alpha- and beta-globin mRNAs show codon preferences which are similar to each other and different from other mammalian mRNAs. This is especially well illustrated by the preferences for the CUG Leu codon (utilized 5.2-fold more often than expected by chance) and for the GUG Val codon (utilized threefold more often than expected by chance), and in the avoidance of the UCR Ser codon (6.3 expected by chance; none observed).

It has been proposed that selective use of certain codons might be a mechanism for control of gene expression (for review see Lodish 1976). The absence of certain codons in the alpha chain and their use in the beta chain suggest, for instance, how it could be possible to regulate the early synthesis of the alpha chain and the later turn-on of beta-chain synthesis by controlling the availability of isoaccepting tRNAs. However, the data of Lodish suggest that such controls cannot be absolute since in vitro protein-synthesizing systems from one eukaryotic tissue can translate mRNAs from diverse tissues.

It has been observed that the abundance of various tRNAs in rabbit reticulocytes correlates with the prevalence of the corresponding amino acids in globin (Smith and McNamara 1971; Smith 1975), and subsequent data obtained by D. Hatfield et al. (in prep.) indicate a positive correlation between the isoacceptor patterns and codon usage. In contradistinction to the idea that limited selections of tRNAs control which mRNAs can be translated, Garel (1974) proposed that differentiated cells may selectively adjust their rates of production of specific tRNAs (including isoacceptors) so that they acquire a tRNA complement appropriate for the synthesis of specific proteins. Although this model explains how tRNA synthesis would correlate with codon usage, it unfortunately does not explain why codon usage should be so strongly biased.

There are also significant differences in the codon utilizations in the alpha- and beta-globin genes. As mentioned in the preceding section, some of the most striking of these differences are due to a strong preference for use of C in the third position in the alpha-globin gene sequence. For instance, there are ten

Table 1. Codon Usage In Alpha- and Beta-Globin, Immunoglobulin, and Insulin mRNAs

1\2		U	α	β	im	ins		C	α	β	im	ins		A	α	β	im	ins		G	α	β	im	ins	2/3
U	Phe	UUU	0	3	1	1	Ser	UCU	3	3	1	0	Tyr	UAU	2	1	2	0	Cys	UGU	0	1	1	2	U
	Phe	UUC	7	5	7	2		UCC	3	3	6	2		UAC	1	2	5	3		UGC	1	0	1	4	C
	Leu	UUA	0	0	0	0		UCA	0	0	4	0	Term	UAA	1	0	0	0	Term	UGA	0	1	0	1	A
	Leu	UUG	1	0	2	1		UCG	0	0	0	0		UAG	0	0	0	0	Trp	UGG	1	2	1	1	G
C	Leu	CUU	0	0	0	2	Pro	CCU	1	3	1	2	His	CAU	0	4	1	0	Arg	CGU	1	0	0	4	U
	Leu	CUC	2	2	1	3		CCC	5	0	4	2		CAC	10	5	2	2		CGC	0	0	0	0	C
	Leu	CUA	0	0	1	0		CCA	0	1	5	1	Gln	CAA	0	0	2	3		CGA	0	0	1	0	A
	Leu	CUG	12	16	0	8		CCG	1	0	0	2		CAG	1	4	2	5		CGG	1	0	2	1	G
A	Ile	AUU	0	1	2	1	Thr	ACU	2	2	5	0	Asn	AAU	1	4	3	0	Ser	AGU	1	4	1	0	U
	Ile	AUC	3	0	2	1		ACC	9	2	10	2		AAC	1	4	8	3		AGC	2	0	5	1	C
	Ile	AUA	0	0	1	0		ACA	0	0	3	1	Lys	AAA	2	3	6	1	Arg	AGA	0	0	0	0	A
	Met	AUG	1	1	1	0		ACG	0	0	2	0		AAG	9	9	10	2		AGG	1	3	1	0	G
G	Val	GUU	0	4	0	1	Ala	GCU	1	7	2	3	Asp	GAU	0	1	3	2	Gly	GGU	1	4	1	3	U
	Val	GUC	0	2	3	2		GCC	9	6	3	3		GAC	6	3	2	1		GGC	8	6	2	2	C
	Val	GUA	0	0	2	0		GCA	0	1	2	1	Glu	GAA	4	4	1	2		GGA	0	0	1	1	A
	Val	GUG	7	12	2	5		GCG	2	1	0	0		GAG	3	6	1	7		GGG	0	1	0	2	G

The frequency of use of each codon is indicated for four mammalian mRNAs. The standard code-table format is used, with further subdivision into four columns, one for each mRNA. The column headings α, β, im, and ins denote, respectively, rabbit alpha mRNA, rabbit beta mRNA, mouse immunoglobulin mRNAs, and rat insulin mRNA. Initiator codons are not included in the tally. The alpha-globin mRNA sequence data are those presented in Fig. 1; the beta-globin sequence data are those shown in Fig. 3; the immunoglobulin data represent the sum of the sequences presented for the immunoglobulin MOPC = 21 light-chain mRNA by C. Milstein et al. (pers. comm.) and the MOPC-21 H-chain mRNA by Cowan et al. (1976); and the insulin mRNA data are taken from the sequence determined by Ullrich et al. (1977). Underlines are used to indicate some amino acids with marked codon preferences in both the rabbit alpha- and beta-globin mRNAs. Other amino acids with marked codon preferences in only one of the two mRNAs are indicated by dotted underlines.

CAC and no CAU His codons, seven UUC and no UUU Phe codons, biases which are not seen in the beta-globin gene (Table 1).

Attempts to Relate mRNA Structure and Function

Many lines of evidence suggest that eukaryotic mRNAs should contain a variety of "signals" involved in functions such as ribosome binding, mRNA processing, transport from nucleus to cytoplasm, and so on. One test of the functional importance of a particular nucleotide sequence is to ask how well the sequence is conserved during various spans of evolution.

As shown earlier by Proudfoot and Brownlee (1976), the rabbit alpha-globin mRNA contains one conserved sequence near its 3' end: the hexanucleotide AAUAAA, which has been found near the 3' poly(A) in all polyadenylated eukaryotic mRNAs analyzed to date. As noted earlier by Lockard and RajBhandary (1976), the hexanucleotide sequence ACACUU, which occurs at the 5' terminus, is common to both the alpha- and beta-globin sequences. It will be interesting to note whether this sequence is repeated in a wide variety of mRNAs, as might be expected of a signal for mRNA processing, or is peculiar to mRNAs of erythroid cells, suggesting a more restricted function. Such a large number of mutations have occurred during the evolutionary divergence of the alpha- and beta-globin mRNAs that their 5' untranslated sequences bear little resemblance other than this conserved hexanucleotide. This is consistent with the high mutation rate computed later in this paper and suggests that the conservation of the 5'-terminal hexanucleotide is significant.

More problematic is the question of specific ribosome binding sites designating the proper AUG codon as a site for the initiation of protein synthesis. Shine and Dalgarno (1975) have proposed that the initiation of translation in prokaryotes involves base pairing between a short sequence near the 3' end of 16S rRNA and a complementary sequence located to the 5' side of the mRNA initiator codon. This hypothesis has received experimental support in *E. coli* (Steitz and Jakes 1975). The existence of an invariant 3'-terminal sequence, 3' AUUACUAG . . . 5', in all eukaryotic 18S rRNAs thus far examined suggests the possibility of a similar mechanism in higher organisms (Hunt 1970; Shine and Dalgarno 1974; Sprague et al. 1975; Eladari and Galibert 1975, 1976).

This hypothesis has been supported by the discovery that the 18S rRNA sequence shows substantial complementarity with sequences preceding AUG intitiation sequences in Brome mosaic virus (BMV), Rous sarcoma virus (RSV), and a mRNA sequence from SV40 (see Table 2a). The correlations seen with globin mRNAs have been less impressive, with Efstratiadis et al. (1977) arguing in favor of interactions

that involve pairing of only three consecutive nucleotides in the beta-globin mRNA. It is difficult to assess whether such interactions are significant since complementarities of three or more base pairs will occur frequently by chance. (The 8-nucleotide 5'-terminal rRNA sequence should show an exact fit of 3 or more nucleotides on the average about once every 10.7 nucleotides along the mRNA sequence and more often if GU pairs are allowed.) To examine the alternative possibilities, we have pictured the most energetically favorable interactions for the entire 5' untranslated regions of both the rabbit alpha- and beta-globin mRNAs in Table 2b. Legon (1976) has proposed that the interaction shown as L in Table 2b is the significant complementarity for the beta-chain mRNA; however, Efstratiadis et al. (1977) discarded this in favor of region J, because they considered it improbable that the 18S rRNA would interact with the initiation codon. They give no reason for discarding the alternative interaction K, which has greater stability and lies at about the same distance from the initiator codon, or alternative interaction I.

In the alpha mRNA sequence there are three possible interactions, denoted F, G, and H in Table 2, which have roughly similar stabilities. Wishing to determine how often interactions of similar stability might appear by chance in an arbitrary sequence, we started with the 59 nucleotides which *follow* rather than precede the AUG codon and reversed the polarity to eliminate any possible real interactions. In this reversed sequence, six interactions were noted, with ΔG calculated as −4.8, −4.0, −3.4, −3.0, −2.7, and −2.4 kcal (not shown) as compared to −6.7, −4.2, −3.9, and −3.0 for the real 5' beta sequence of the same size as shown.

This exercise emphasizes that none of the individual interactions seen is extraordinary on the basis of interaction energy alone. Therefore, if there is a single "real" interaction in each mRNA, then its biological significance must be identified using some other criterion such as a feature common to all of the different mRNAs. From the sequences we have determined for the rabbit alpha-globin mRNA, we note a complementarity (interaction H in Table 2b) which overlaps the AUG initiation codon. The existence of this interaction supports the hypothesis of Legon (1976). We find this attractive despite the caveat of Efstratiadis et al. (1977), since the 18S rRNA sequence interacts strongly with a sequence overlapping the AUG codon in every mRNA thus far examined. This appears to be the most universal mRNA feature observed so far for the interaction with 18S rRNA. This interaction is weaker in the alpha-globin mRNA sequence than in the beta-globin mRNA, a reasonable result since it is known that the alpha-globin mRNA is translated less efficiently in vivo. The overlap of the strong binding sites with AUG initiator codons suggests two-step mechanisms in which the 18S rRNA sequence first binds the mRNA

Table 2. Possible Interactions of 5' Terminal Sequences of Eukaryotic and Globin mRNAs with 18S rRNA

mRNA	Interactions	Reference

(a) Eukaryotic mRNAs

BMV

(A)
HOAU UAC UAG. . .
− 6.9 kcal

mGpppGUAUUAAUUAA(AUG)UCG. . .
HOAUUAC   UAG. . .
− 4.8 kcal
(B)

Dasgupta et al. (1975)

SV40-VP1

. . .GCUUAUGAAGA(AUG)GCC
HOAUUAC UAG. . .
(C)
− 3.9 kcal

Van De Voorde et al. (1976)

RSV

. . .CCUAACGAUUGCGAACACCUGA(AUG)AAG
HOAUUACUAG. . . − 5.3 kcal   AU UAC UAG. . . − 7.2 kcal
(D)   (E)

Haseltine et al. (1977)

(b) Globin mRNAs

Alpha-globin

(F)
mGppp[ACACUU]CUGGUCCCAGUCCGACUGAGAAGGAACCACC(AUG)GUG. . .
HOAU UACUAG. . .   HOAUUAC UAG. . .
− 5.8 kcal   − 5.5 kcal
(G)   (H)

Beta-globin

(I)
mGpppp[ACACUU]CUUUUGACACAACUGUGUUUACUGCAAUCCCCAAAACAGACAGA(AUG)GUG. . .
HOAUUACUAG. . .   HOAU UAC UAG. . .
− 4.2 kcal   − 6.7 kcal
(J)   (L)
− 3.0 kcal
HOAUUACUAG. . .
HOAUUACUAG. . .
− 3.9 kcal
(K)

mRNA sequences are written in 5'→3' orientation. The AUUACUAG rRNA sequence is shown in 3'→5' orientation. Initiation codons are set off by parentheses. Brackets indicate a homology between the 5' termini of the two globin mRNAs. Interactions are indicated (A) through (L). The ΔG values for the interactions were computed using the stabilities given in the legend to Fig. 2, but neglecting end effects. All potential base pairs are shown by underlines, but the energy is computed for the most stable configuration.

and then a translocation or other conformational shift makes the initiator codon available for binding to fMet/tRNA, a possibility consistent with the structure shown in Figure 3.

Other Conserved Features of mRNA Structure

The strategy of searching for features conserved during evolution is most useful for recognizing signals which have a "housekeeping" function—by which we mean signals needed by all or most mRNAs—and which are recognized as a sequence per se rather than as a feature of secondary structure or tertiary folding. It would be more difficult to recognize a signal peculiar to a small class of mRNAs since there are fewer comparisons available. It would be especially difficult to recognize signals peculiar to the globin mRNAs, for instance, if the corresponding signals on the alpha- and beta-globin mRNAs bound to the same site with different partial homologies. Such a case would be analogous. to the situation with the putative rRNA binding sites: it would be extremely difficult to recognize partial homologies if the complementary rRNA sequences were not known to us and if there were few examples to compare. Also, we must bear in mind that the availability of a particular sequence for recognition will be strongly influenced by the secondary structure and tertiary folding of the mRNA chain. If the signal recognized is not a sequence per se, but a feature of the base-paired structure, the problem becomes even more complex. At present there are only tentative steps toward satisfactory methods for predicting the base-pairing arrangements for an mRNA in the cell. Since this configuration may be strongly influenced by the known binding of specific proteins (see, e.g., Morel et al. 1971; Spohr et al. 1972; and Irwin et al. 1975), as well as many other cytoplasmic influences, this is likely to remain a difficult problem.

In the face of such difficulties we believe that one of the most powerful approaches is to compare sequences for a particular mRNA as it is represented in several different species. A careful study of which

sequences are conserved, and especially which base-pairing interactions are preserved for long periods of evolution, may help us to determine which of the pairing interactions are biologically significant. Moreover, as we will show below, a mutational analysis of this sort may be helpful in identifying those features of mRNA secondary structure which are of biological importance.

Our previous comparisons of the beta-globin mRNA sequences which we had determined for the rabbit with human beta-globin mRNA sequences (kindly supplied by S. Weissman, B. Forget, and C. Marrotta) indicate that the natural mutation rate is sufficiently high that such comparisons will prove useful in enabling us to recognize conserved regions (Salser et al. 1976b; Salser and Isaacson 1976). For example, comparison of the 146-nucleotide rabbit beta-globin sequence published by Browne et al. (1977) with the corresponding human beta-globin sequences (Marotta et al. 1976; B. Forget and S. Weissman, pers. comm.) revealed that although the first 56 nucleotides were identical, there was an average of one base substitution every 6–7 nucleotides in the remainder of this region. Subsequent publication of the complete sequence of the rabbit beta-globin mRNA by Efstratiadis et al. (1977) shows that this conserved region totals 62 nucleotides (corresponding to the amino acid sequence from 91 to 111) and that there are other regions of 35 nucleotides (amino acid residues 22–33) and 26 nucleotides (amino acid residues 58–66) whose lack of base substitutions is significant (Fig. 2).

In the past, observation of a region in which the amino acid sequence is strongly conserved has been taken to mean that this amino acid sequence is critical to the function of the protein. If conservation of a crucial amino acid sequence were the only factor involved, then one would expect to find close to the average number of *silent* mutations in the region. For example, see the region from amino acid residues 126–146 in Fig. 2. This region contains no amino acid substitutions but seven silent nucleotide substitutions. The region from amino acid residues 30–50 provides a similar example, suggesting that conservation of the amino acid sequence is impor-

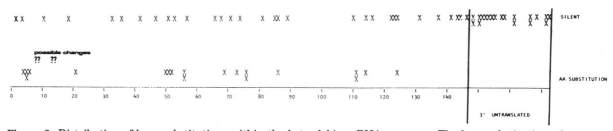

Figure 2. Distribution of base substitutions within the beta-globin mRNA sequence. The base substitutions shown are those which have occurred during the evolutionary divergence of rabbits and primates. Changes resulting in amino acid substitutions are indicated separately from silent base substitutions. Because of uncertainties in the human beta-globin sequence, there are five additional positions where silent base substitutions may occur. These are indicated by question marks. The rabbit beta-mRNA sequence is that shown in Fig. 3. The human beta-mRNA sequence was from B. Forget and S. Weissman (pers. comm.) and has been published, in part, by Marotta et al. (1976).

tant in this region and that conservation of the mRNA sequence is not.

The conspicuous lack of silent base substitutions in the region corresponding to amino acid residues 91–111 argues that here it is the importance of the nucleotide sequence itself which is responsible for its conservation. It is, of course, novel to propose that an amino acid sequence of substantial size might be conserved because it was coded by a critical mRNA sequence rather than because the amino acid sequence was critical to the protein per se. Whether the amino acid sequence in this region may also be important, in and of itself, is impossible to say from the substitution data alone.

Evaluation of mRNA Secondary Structure

In an attempt to ask whether conservation of particular globin mRNA sequences noted above may be due to the need to conserve a particular mRNA secondary structure, we have used a modified version of the computer program of Pipas and McMahon (1975) to search for the most stable base pairing. This program uses the base-pair stacking energies of Borer et al. (1974) and an updated version of the estimations by Tinoco et. al. (1973) of the destabilizing effects of various single-stranded loops to compute the base-paired structure which would be most stable at 25°C in neutral buffer of moderate or high ionic strength (some further modification of the energy calculation is indicated in the legend to Fig. 3). As pointed out earlier, even if this goal is perfectly realized, the interpretation of its biological significance is fraught with peril since interactions with specific binding proteins and other intracellular macromolecules might stabilize any of a large number of alternative structures. Moreover, as explained in the legend to Figure 3, the computer program as it now exists makes simplifying assumptions which may make it difficult to find even that structure which is most stable free in solution.

Despite these reservations, analysis of the proposed rabbit beta-globin mRNA secondary structure shown in Figure 3 suggests that such an approach may be useful when it is combined with an analysis of base substitution data.

Comparison with Previously Published Secondary Structures Based on Partial Sequences

Proudfoot (1976) calculated possible base-pairing configurations for the 75-nucleotide sequence adjacent to the 3' poly(A) sequence of rabbit beta-globin mRNA; Efstratiadis et al. (1977) similarly presented a possible secondary structure for the 80 nucleotides at the 5' terminus of the same mRNA. In neither case does the base-pairing scheme resemble the base pairing we show for the corresponding sequences in Figure 3. Presumably the reason for this difference is that the rest of the structure in Figure 3 provides more stable base pairing for these se-

quences than was found within the sequences themselves in the earlier attempts cited. To assess the differences in stabilities, we have attempted to compute the contribution made by each of these sequences to the total free energy of the structure. In making this apportionment, we ascribed half of the binding energy of each region of base pairing to each of the two strands. The destabilizing effects of loops were similarly apportioned. The energies of the structures proposed by Proudfoot (1976) and by Efstratiadis et al. (1977) were recomputed using the same energy estimates used in computing the structure in Figure 3 (see legend to Fig. 3 for a summary of these rules). According to these calculations, the energy of the most stable configuration given by Proudfoot (1976) for the 75-nucleotide 3' untranslated sequence is about −10 kcal. The same sequence contributes about −28 kcal to our structure (Fig. 3) (about −0.37 kcal/nucleotide). The most stable configuration shown by Efstratiadis et al. (1977) for the sequence at the 5' terminus has a free energy of about −7 kcal, whereas in our structure the 5' 82 nucleotides contribute about −11 kcal (−0.13 kcal/nucleotide). The energy of the total structure in Figure 3 is about −174 kcal. This is an average of about −0.30 kcal/nucleotide in the entire molecule.

This result underscores the importance of considering the whole mRNA sequence when possible secondary structures are computed. The striking increase in stability of pairing of the 3' terminal 75 nucleotides (from 0.13 kcal/nucleotide in the structure proposed by Proudfoot (1976) to 0.37 kcal/nucleotide in that portion of the structure shown in Fig. 3) comes mainly from strong base pairing between the 3' untranslated region and the 5' end of the mRNA sequence, making rabbit beta-globin mRNA assume a functionally circular configuration. It will be interesting to see whether other mRNA sequences have the same complementarity between 5' and 3' untranslated regions.

The structure shown in Figure 3 was the most stable structure found in a search which made no attempt to favor structures with accessible ribosome attachment sites. Therefore it is encouraging that the two most stable complementarities with the 18S rRNA sequence, shown in Table 2 as interactions I and L, are found as single-strand hairpin loops where they are readily available for pairing. Furthermore, it should be noted that the folding of the structure could move interaction I very close to interaction L which includes the initiator codon AUG, facilitating the two-step initiation mechanism mentioned earlier.

Are the Base-pairing Arrangements Shown in Figure 3 Functionally Important?

It is essential to keep in mind that proposed base-pairing arrangements such as shown in Figure 3 are hypothetical. The possibility that the structures obtained are similar to the conformation of the mRNA in the cell depends upon (1) the accuracy and thoroughness of the physical chemistry measurements carried out with model compounds to establish energies of various structures free in solution; (2) the extent to which the computer program can be successfully used to find the configuration of lowest energy according to these rules; and (3) the degree to which conditions inside the cell, including the presence of specific binding proteins, may stabilize alternative configurations. At present there are severe limitations at all three levels of this analysis.

The problem then is to assess which parts of the structure in Figure 3 may actually exist in the complex milieu of the cell. Especially, we would like to know which of these structures might be biologically important, for instance, as binding sites for specific proteins or to direct the proper cleavage of the mRNA precursor by processing enzymes. To aid in this evaluation, we have indicated in Figure 3 each position where the human sequence is known to be different from the rabbit sequence. The nucleotide present in the human sequence is shown in bold-

face type; a boldface line indicates those regions of base pairing which are either conserved in the corresponding human sequence or so little changed that they remain highly stable. It may be seen that the base pairing in arms 2 and 3 of the molecule is highly conserved, the base substitutions occurring predominantly in regions represented as single-stranded. By contrast, in arm 4, containing the termination signal for protein synthesis, the base pairing is poorly conserved. There may be an alternative structure for this region which *is* conserved, or it may be that the secondary structure of arm 4 of the mRNA is not as important for the function of the molecule as that in arms 2 and 3. Less can be said about the ribosome attachment region of the structure (arm 1) because the base-substitution data are incomplete: The 5′ untranslated sequence of human beta-globin mRNA is unknown, and there are four uncertainties in the human sequence between nucleotides 27 and 42 of the structural gene.

It should be emphasized that only the rabbit beta-globin sequence was used to arrive at the structure shown in Figure 3. There may be other base-pairing arrangements of lower stability which show greater

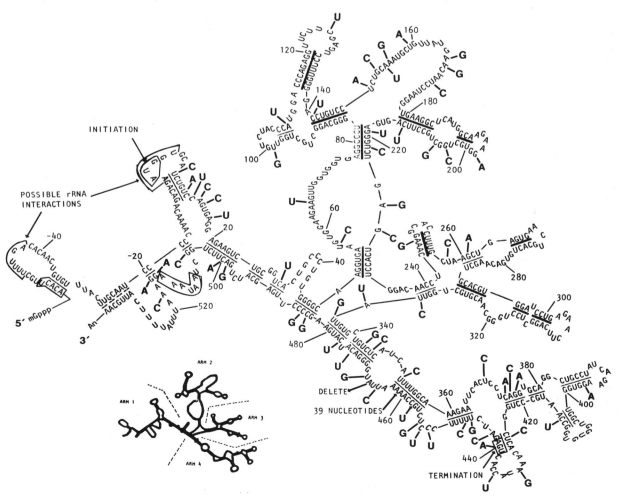

Figure 3. *(See facing page for legend.)*

Figure 3. The computer program of Pipas and McMahon (1975) was used in an attempt to find the most stable base-pairing arrangement for the rabbit beta-globin mRNA sequence (Lockhard and Rojbhandary 1976; Browne et al. 1977; Baralle 1977; Proudfoot 1977; Efstratiadis et al. 1977; Paddock et al. 1977). The sequence shown is the allele studied by Efstratiadis et al. (1977). Letters in boldface type represent the base substitutions by which the human beta-globin mRNA sequence differs from that found in the rabbit (Marotta et al. 1976; B. Forget and S. Weissman, pers. comm.). Because the human beta-globin mRNA sequence is incomplete, there are no data on base substitutions in the 5' untranslated sequence and additional silent mutations in the structural gene are possible at nucleotides 27, 30, 39, 42, and 384. Conserved "signal" sequences (see text) are boxed, and two possible interactions with 18S rRNA (−4.2 kcal and −3.9 kcal; see Table 2b) are bracketed. The label "39 nucleotides" at nucleotides 461–462 indicates a position where the human sequence has an additional 39 nucleotides not present in the rabbit mRNA.

I am indebted to Mr. Ian Cummings for pioneering the use of the Pipas and McMahon computer program at UCLA and for computing the first structure (about −145 kcal) for the beta-globin mRNA. Subsequently, with the advice and technical assistance of Mr. Cummings, Gary Studnicka, Georgia Rahn, and David Finck, I refined the structure to its present, more stable configuration (about −177 kcal). This was accomplished by adopting stratagems to circumvent difficulties which the program has in dealing with large sequences, its inability to consider combinations of overlapping regions with mutually exclusive base pairing, and biases due to the decision by Pipas and McMahon to assign zero energy to every multibranched internal loop. We are presently attempting to devise a new computer program to circumvent such difficulties and make the process less laborious and more reliable (G. Studnicka et al., in prep.).

Since the energies listed by Pipas and McMahon are incomplete, we list in condensed form the energies used in our calculations as taken from Jacobson and Stockmeyer (1950), Gralla and Crothers (1973), Tinoco et al. (1973), Borer et al. (1974), Pipas and McMahon (1975), and J. McMahon (pers. comm.).

Base-Pairing Energies

5' Nucleotide	3' nucleotide			
	A	C	G	U
A	−1.2	−2.1	−2.1	−1.8
C	−2.1	−4.8	−3.0	−2.1
G	−2.1	−4.3	−4.8	−2.1
U	−1.8	−2.1	−2.1	−1.2

The energies for stacking of GC and AU base pairs are given. For example, the paired sequences -C-C- and -G-G- are both shown as 4.8 kcal, which is appropriate since both indicate the energy of two successive GC pairs with the G's on the same strand. In addition, the energies for stacking of internal GU pairs was computed as follows: GU next to GC = −1.3 kcal; GU next to AU or GU = −0.3 kcal. For these calculations, terminal GU pairs are assumed to be unstable.

Destabilizing Effects of Unpaired Loops

Type of loop[a]	Size of loop[b]																
	1	2	3	4	5	6	7	8	9	10	12	14	16	18	20	25	30
B	2.80	3.90	4.45	5.00	5.15	5.30	5.45	5.60	5.69	5.78	5.93	6.05	6.16	6.26	6.35	6.54	6.70
H(G)	x	x	8.40	5.90	4.10	4.30	4.50	4.60	4.80	4.89	5.03	5.16	5.27	5.37	5.46	5.65	5.88
H(A)	x	x	8.00	7.50	6.90	6.40	6.60	6.80	6.90	7.00	7.13	7.25	7.36	7.46	7.55	7.74	7.90
I(GG)	x	0.10	0.90	1.60	2.10	2.50	2.62	2.72	2.82	2.90	3.05	3.18	3.29	3.38	3.47	3.66	3.85
I(GA)	x	0.95	1.75	2.45	2.95	3.35	3.47	3.57	3.67	3.75	3.90	4.03	4.14	4.23	4.32	4.51	4.70
I(AA)	x	1.80	2.60	3.30	3.80	4.20	4.32	4.42	4.51	4.60	4.75	4.88	4.99	5.08	5.17	5.36	5.55

[a]B denotes a bulge-loop; H(G) and H(A) denote hairpins closed by GC or AU pairs, respectively; I(GG), I(GA), and I(AA) denote internal loops closed by two GC pairs, by one GC and one AU pair, or by two AU pairs, respectively. In the case of bulge-loops, the value given in the table must be corrected by subtracting the stacking energy across the bulge as computed by Tinoco et al. (1973). For example, a one-nucleotide bulge closed by a GC pair on one side and an AU pair on the other side has an energy of +2.80 − 2.10 = +0.70, where the +2.80 is due to the bulge and the −2.10 is due to the GC-AU pairing on either side of the bulge. Note that Pipas and McMahon (1975) assigned zero energy to internal loops connecting three or more stems. Although it is true that there are no experimental data indicating the destabilizing effect of such structures, we feel that assuming zero energy may be an especially bad choice. Therefore we have computed the energy of such loops using the values of I(GG), I(GA), and I(AA) for internal loops connecting two stems. I(GA) is used if both GC and AU pairs are represented in the loop closure, and I(GG) and I(AA) if all stems are closed by GC or AU pairing, respectively.

[b]Values beyond 10 are extrapolations according to Jacobson and Stockmeyer (1950).

The same computer program was used to evaluate possible tertiary folding which might result from base pairing between single-stranded regions in the structure shown above. A search was made for all possible complementarities between the single-stranded regions in the structure shown. Some weakly paired regions were included in the search. The energies and sequences involved in the eight strongest interactions are as follows: (1) nucleotides 36–39, 190–193, −11.20 kcal; (2) 71–75, 434–438 −11.10 kcal; (3) 159–165, 186–193, −9.60 kcal; (4) −2–+2, 434–437, −9.0 kcal; (5) 16–19, 452–455, −9.0 kcal; (6) 58–61, 435–438, −9.0 kcal; (7) 72–75, 434–437, −9.0 kcal; (8) 171–174, 514–517, −9.0 kcal. It is interesting that interactions, 2, 4, 6, and 7 all involve the same sequence near the termination codon and are thus mutually exclusive. Such features could give the molecule a metastable tertiary structure with several alternative folded configurations. An obvious drawback of this approach is that it does not evaluate whether a particular structure is sterically possible and does not consider the nonstandard bonding interactions important in the tertiary folding of phenylalanyl tRNA as revealed by X-ray crystallographic studies (Kim et al. 1974; Robertus et al. 1974).

evolutionary conservation of base-paired regions, but no attempt was made to search for these.

Finally, we would like to know the tertiary configurations of the mRNA molecule. The existing computer search programs make no attempt to evaluate tertiary folding. The presence of long single-stranded regions in the structure shown in Figure 3 suggests that the folding of such a structure could be dominated by base pairing between such regions. As explained in the legend to the figure, the computer program was used to search for such interactions; 13 interactions with energies greater than −8.0 kcal were found. It should be recognized that the molecule may have a metastable tertiary structure with several alternative configurations. For instance, the nucleotide sequence 434–438 has four mutually exclusive pairings possible, with the single-stranded regions starting at nucleotides −2, 58, 71, and 72.

Measurements of Mutation Rates during Mammalian Evolution

Mutation rates for mammalian genomes have often been computed from rates of amino acid substitution (Kimura 1968; King and Jukes 1969). Earlier we pointed out how important sources of errors in such calculations can be avoided by considering instead the rates of those nucleotide substitutions which are silent (Salser et al. 1976b; Salser and Isaacson 1976). Using the silent-mutation data from the 116 nucleotides then known for both the rabbit and human sequences (11 silent base substitutions), we computed that there had been an average of about 0.6 base substitutions per nucleotide during the roughly 200 million years of evolution separating rabbits and humans. This mutation rate is about tenfold higher than that estimated from the rate of amino acid substitutions in globin (a protein whose mutation rate is near the average for all proteins used in the mutation-rate calculations of Kimura [1968]). The elimination of this source of bias in the earlier calculations has a dramatic effect on the conclusions.

In their consideration of these results, Salser and Isaacson (1976) concluded that the mutation rate derived from the frequency of silent base substitutions in globin RNA is orders of magnitude greater than the rate at which deleterious mutations can be eliminated from single-copy DNA. It was suggested that there may be different explanations for this dilemma as it pertains to single-copy DNA, clustered repetitious DNA, and interspersed repeat sequences (Salser and Isaacson 1976). In the case of the single-copy DNA, the favored explanation was that only a small fraction of the genome, roughly 0.6%, which is about enough to code for 30,000 globin-size proteins, was devoted to functional single-copy genes. According to this model, mutations in the remainder of the single-copy DNA would not

be deleterious since those regions have no sequence-dependent function and consequently can drift freely. By contrast, it was proposed that the arrays of clustered repetitious sequences (including such diverse examples as ribosomal RNA genes and the highly repetitive satellite DNAs) are kept accurate by a novel version of the old concept of truncation selection. Finally, special mechanisms were discussed which might permit the accuracy of the interspersed repetitive sequences to be maintained by a combination of events of the truncation-selection and gene-conversion types.

Perhaps the most striking prediction of the silent base substitution data is that most of the single-copy DNA has randomly drifting sequences. Fortunately, it was possible to check and confirm the validity of this prediction by comparing observations already available in the literature. Since most of the mutations which cause amino acid substitutions in globin appear to be sufficiently deleterious that they are eliminated, our model predicted that the nucleotide sequences of a functionless nucleic acid sequence should drift about three or more times faster than those for the globin gene. Gummerson and Williamson (1974) and Leder et al. (1974) have measured the ΔT_m for interspecies heteroduplexes between globin mRNAs and globin cDNAs. The ΔT_m is about 7°C for mouse × human and mouse × rabbit (Gummerson and Williamson 1974) and about 9°C for rabbit × mouse (Leder et al. 1974). Since the lagomorph, rodent, and primate lines all are thought to have diverged about the same time, each of these figures corresponds to about 200×10^6 years of evolution (data of MacKenna, cited in Salser and Isaacson 1976). As predicted, the ΔT_m values for the divergence of total single-copy DNA are much higher than those for the globin sequences. This includes a ΔT_m of 20°C per 200×10^6 years for artiodactyls (Laird et al. 1969), 28°C per 200×10^6 years for primates (Kohne et al. 1970, Hoyer et al. 1972), and figures almost an order of magnitude higher for rodents (Laird et al. 1969). Thus, in each case the evolutionary divergence of the total single-copy DNA is severalfold more rapid than that for globin genes, as predicted if the great majority of single-copy DNA sequences are randomly drifting. Note that this argument is based solely on the comparison of different ΔT_m values and is independent of the choice of a conversion factor relating ΔT_m to percent mispairing. This is essential since the value given for this conversion factor has varied rather widely in the literature (e.g., the same laboratory has used values differing by 2.4-fold at different times in interpreting the same data: cf. Fig. 4 of Laird et al. [1969] and Fig. 8 of McCarthy and Farquhar [1972]). If the base-substitution rates which may be computed from the beta-globin mRNA sequence data in Figure 3 are typical of the remainder of the globin mRNA sequences, then comparison with the ΔT_m values of Gummerson and Williamson (1974) suggests that the proper conversion factor is about 1.9% mismatch per 1°C change in the T_m. If so, then the calculations of McCarthy and Farquhar (1972) underestimate the rate of nucleotide sequence divergence by about three fold.

Because of the importance of these questions and the uncertainties of mutation rates estimated from ΔT_m data, it is useful to reexamine the mutation-rate calculations as based upon silent-mutation data. The larger data base (5 times greater than that used in the earlier calculations) permits greater accuracy. It has also been possible to partially correct for the fact that many silent base substitutions appear to be excluded as if they have a deleterious effect on mRNA function.

The Use of Silent Mutations to Calculate the Mutation Rate

Mutations which have no deleterious or beneficial effect are defined as neutral. Neutral mutations will increase or decrease in the population by random drift and will ultimately be eliminated or become the dominant sequence in the population. We will also consider as effectively neutral those mutations whose selective advantage or disadvantage is so slight that the probability that they will take over the population or be eliminated is mainly determined by chance.

In general, over evolutionary time spans the rate of neutral base substitutions in the population must equal the rate of occurrence of neutral mutations in an average individual genome (for a simple mathematical demonstration of this fact, see p. 104 Smith 1972). Consequently, if we were informed that all of the mutations in a particular gene sequence were neutral, we could determine the mutation rate for the genome by measuring the rate of nucleotide substitutions in this gene. Our goal, therefore, is to measure the rate of base substitutions in positions where mutations are most likely to be neutral.

One way we approach this goal is to use silent mutations. The reasons for this choice can be illustrated by considering a hypothetical protein composed of the amino acids valine, alanine, threonine, proline, and glycine. (These amino acids are chosen for the example because each has four possible codons so that, on the average, one-third of the mutations in the sequence will be silent and two-thirds will cause amino acid substitutions.) Suppose that we sequence mRNAs for this protein obtained from two species separated by 10^7 years of evolution, and observe that base substitutions have occurred in 10% of the third nucleotide (silent) positions and in 1% of the other positions. The simplest interpretation of these results is that the mutation rate must be *at least* 10^{-8} per nucleotide per year, the value computed assuming that all silent mutations are neutral, and mutations causing amino acid substitutions are much more likely to be deleterious than

are silent mutations. Two alternative hypotheses must be considered, and we believe that they can be eliminated: the first alternative explains the results in terms of hot-spot mutations, and the second alternative postulates that most base substitutions are fixed in the population because they have a positive selectional advantage.

According to the first alternative hypothesis, silent mutations occur much more frequently than those which cause amino acid substitutions because they are mutational hot spots. Although it is true that some nucleotide sequences appear to be inherently more mutable than others (hot spots), we find it impossible to imagine how hot spots could be preferentially concentrated in the third base positions: For this to occur, mutagenesis occurring at the DNA level would have to be correlated with the reading frame in which the corresponding mRNA is translated. We will therefore assume that, on the average over a large sequence, hot spots are equally probable in silent and nonsilent positions.

The second alternative hypothesis takes note of the obvious fact that any mutation with a strong selectional advantage will show a rapid increase in the population. Consequently, unlike neutral mutations, the rate of fixation of advantageous mutations in the population can be much greater than the average rate at which they occur in typical single genome. According to this hypothesis, it is proposed that most of the base substitutions noted in the example have been fixed in the population because they have beneficial effects. In this case the observed substitution rate could be produced by a mutation rate orders of magnitude less than 10^{-8} per nucleotide per generation. We exclude this hypothesis because analysis of the base-substitution data in Figure 3 shows that base substitutions are concentrated in the silent rather than amino acid changing positions of the rabbit beta-globin gene. Moreover, earlier data cited above show, by comparison of T_m depressions, that base substitutions are fixed in globin genes severalfold less frequently than in the average single-copy DNA.

If the base substitutions which accumulate over the course of evolution are selected for their positive effects (rather than tolerated because they are effectively neutral), then we should have expected instead that the rate of base substitutions would be greater in globin genes than in total single-copy DNA. Thus, these observations are consistent with the idea that most of the observed base substitutions are neutral and are the inverse of what would be expected if the observed base substitutions were chosen for their positive selectional value. More recently, Rosbash and Gummerson (1975) have repeated similar T_m depression studies for the total polysome-bound mRNA of the rat and the mouse versus the total single-copy DNA of rat and mouse. Their results suggest that globin is a typical gene: messenger sequences in general are more highly conserved than total single-copy DNA, as expected if most mutations fixed were of neutral rather than positive selectional advantage.

Finally, it should be pointed out that the observed rates of base substitution are so high as to argue against the "selectionist" hypothesis on purely quantitative grounds. Haldane (1957) introduced the notion of the "cost of natural selection." He estimated that there could not be more than about one substitution (of an allele with a selectional advantage over its predecessor) per 300 generations. This estimate of the cost of positive selection is controversial and has been challenged by several authors (Brues 1964; Smith 1968; Sved 1968). Sved estimates that as many as one substitution per generation might be possible. But our data for globin mRNA, when coupled with comparisons of the depression of T_m due to mismatch in interspecies hybridizations of either globin mRNA or total single-copy DNA, suggest that there are about 30 times as many base substitutions per mammalian genome per year as the maximum suggested by Sved. Since none of Haldane's critics appear to be able to account for positive evolution occurring at this rate, 10,000-fold faster than the limit proposed by Haldane, we can again conclude that the great majority of such substitutions must represent fixation of neutral mutations. As first pointed out by Kimura (1968), there is nothing surprising about such high rates of base substitution so long as genetic drift due to neutral mutations is involved: there is no "genetic cost" of such a process.

Computation of Mutation Rates from Actual Data

The computation of minimum estimates of the mutation rate from silent-mutation data is complicated by the fact that different numbers of silent mutations are possible in different codons. For this reason, we would expect to find silent base substitutions three times more frequently in proline codons (CCN) than in phenylalanine codons (UUY), so long as all transversions and transitions were equally probable and all silent mutations were neutral. In our calculations we have corrected for this by computing the base-substitution rates separately for residues where one or three silent mutations are possible. (Our calculations ignore the very small amount of data from special cases, such as isoleucine codons, where two silent mutations are possible in the third position, and from the first and second positions of serine codons. All other silent-mutation data can be treated as in the case of proline, where all four bases are permitted, or phenylalanine, where two bases related by transition mutations are permitted.) Mutation rates are computed from each of these substitution rates and an average mutation rate is computed and weighted appropriately for the amount of data in each class. These calculations are shown in formal mathematical terms below, first without and then with appropriate corrections for

multiple events at a single site. For instance, we could arbitrarily set $r = 1$ and explain the difference between the 44 and 67% transversion figures discussed above by postulating that transversions are 2.5-fold more likely to be eliminated in the course of natural selection because they are more likely to cause amino acid substitutions and because they may tend to cause more severe disruptions in functionally important base-pairing arrangements. Assuming $r = 2.5$, the average mutation rate computed from the hypothetical mRNA sequence described above is 1.01 mutations per base per 10^8 years per average genome.

Mathematical Formalization and Correction for Multiple Hits

These calculations are shown in more formal mathematical terms below, first without and then with appropriate corrections for multiple events at a single site. The basic equation for computation of the mutation rate M is

$$M = (n_3\alpha_3 + n_1\alpha_1)/(n_3 + n_1)2T, \qquad (1)$$

where: M = average number of mutations per year per nucleotide in a typical genome of the population.

$\alpha_1 = \left(\dfrac{r+2}{r}\right)(n_1/N_1)$ = fraction of those positions where only transition mutations are silent which have actually sustained mutations one or more times. That fraction which has sustained *silent* mutations is n_1/N_1. Multiplication by $(r+2)/r$ gives the fraction which has sustained *any* mutation.

n_1 = number of silent base substitutions which have occurred at base residues where only transition mutations are silent.

N_1 = number of nucleotides where only transition mutations are silent.

r = (average rate of transition mutations)/ (average rate of transversion mutations).

$\alpha_3 = n_3/N_3$ = fraction of positions with three possible silent base substitutions which have actually been mutated one or more times (not corrected for multiple events).

n_3 = number of silent base substitutions which have occurred at base residues where all three of the possible base substitutions are silent.

N_3 = number of nucleotides where three silent base substitutions are possible.

T = years since divergence of the two species which are being compared (hence $2T$ = number of years of evolution separating the two genomes).

The terms α_3 and α_1 consider the fraction of positions with one *or more* mutations. For short evolutionary times, such that there are few cases of two or more successive mutations at the same site, α_3 and α_1 will be proportional to the mutation rate and Equation 1 will give an accurate value for M. The results of this computation are shown in the first line of Table 3.

This treatment fails to correct for multiple events, leading to large errors in the computation of M as α_3 approaches 0.75 or as α_1 approaches $0.5(r+2)/r$ (their theoretical maximal values). Therefore, we define two new terms, α'_3 and α'_1, which avoid such errors by correcting for multiple events. The term α'_3 is a measure of the total number of mutations which have occurred during the evolutionary divergence at a typical position where three silent mutations are possible; α'_1 is a measure of the total number of mutations which have occurred during the evolutionary divergence at an average position where only transition mutations are silent. Because the probability of any particular nucleotide receiving a certain number of mutations is Poisson-distributed, it follows that the probability of a nucleotide receiving zero mutations is equal to e^{-x} (where x is the average number of mutations per nucleotide), and further correcting for those nucleotides which appear to be unchanged as a result of multiple hits cancelling out, we compute the following:

$$\alpha'_3 = -\tfrac{3}{4} \ln(1 - 4n_3/3N_3), \qquad (2)$$

$$\alpha'_1 = -\left(\frac{r+2}{r}\right) \ln(1 - 2n_1/N_1). \qquad (3)$$

The value of M corrected for multiple hits is then:

$$M = \frac{n_3\alpha'_3 + n_1\alpha'_1}{(n_3 + n_1)2T}. \qquad (4)$$

Calculations with Actual Beta-Globin mRNA Sequence Data

From the known amino acid sequences and the genetic code, we may compute N_1 (rabbit) = 85 and N_1 (human) = 80. The value of N_1 to be used in computing M is the average of these two: $N_1 = 82.5$. Similarly, N_3 (rabbit) = 71 and N_3 (human) = 77 so, $N_3 = 74$. (These values have been computed omitting the five uncertain codons from the human sequence, as well as their counterparts from the rabbit sequence; see legend to Fig. 2). From the base substitution data in Figure 3, we compute that $n_1 = 15$ and $n_3 = 16$. (Note that when there are two base substitutions in a codon, one may be silent while the other causes an amino acid substitution. If one of the two amino acids has four possible codons and the other has two, we allocate the silent mutation half to n_1 and half to n_3.) T for the lagomorph-primate divergence is estimated at 10^8 years (for discussion, see Salser and Isaacson 1976). The results of the calculation are shown in the first two lines of Table 3.

Estimates of a Correction Factor for Codon-Utilization Biases

Our observation that there are a disproportionate number of silent base substitutions, as compared with those affecting amino acids, shows that use of silent mutations to estimate mutation rates eliminates an important source of bias and gives a useful minimum estimate of the true mutation rate. But if, for instance, two-thirds of the silent mutations are themselves deleterious, this estimate will be threefold too low. Analysis of the base-substitution data for the betaglobin messenger RNAs suggests that such effects do substantially lower the estimates of mutation rates.

Suppose, for instance, that in the hypothetical gene we are considering it was found that proline

Table 3. Use of Silent-Mutation Data to Compute the Mutation Rate for Mammalian Genomes

	n_1	N_1	n_3	N_3	α_1 or α'_1	α_3 or α'_3	M (for $r=1$)	M (for $r=2.5$)
Uncorrected calculation[a]	15	82.5	16	74	$\left(\dfrac{r+2}{r}\right)$ (0.182)	0.216	1.87×10^{-9}	1.35×10^{-9}
Corrected for multiple hits[b]	15	82.5	16	74	$\left(\dfrac{r+2}{2r}\right)$ (0.45)	0.255	2.29×10^{-9}	1.63×10^{-9}
Corrected for multiple hits and codon utilization —correction A[c]	18.5	103.5	12.5	53	$\left(\dfrac{r+2}{2r}\right)$ (0.44)	0.285	2.5×10^{-9}	1.7×10^{-9}
Corrected for multiple hits and codon utilization —correction B[d]	21	134	3	5.5	$\left(\dfrac{r+2}{2r}\right)$ (0.40)	0.975	3.1×10^{-9}	2.1×10^{-9}

[a] The uncorrected calculation uses Equation 1 of the text.

[b] The calculation corrected for multiple hits uses Equations 2–4 of the text.

[c] The correction for the codon utilization biases is explained in the text and uses definitions 1–4. These new values for n_3, N_3, n_1, and N_1 are substituted into Equations 2–4. Correction A makes very conservative assumptions about the restrictions on codon utilization in the globin mRNAs (see text for details).

[d] Correction is as above; however, for correction B more drastic assumptions are made about the extent of restrictions on codon utilizations (see text for details).

(CCN) was always specified by CCG, that valine (GUN) was always specified by GUU or GUC, and that alanine (GCN) showed utilization of all four codons. If the number of occurrences was sufficiently large to be statistically significant, we could conclude on the basis of codon preferences alone that all three of the silent mutations possible at proline codons are so strongly deleterious that they are always eliminated and that at least two-thirds of the silent mutations at valine codons are deleterious. In computing the mutation rate, we can correct for these codon preferences by discarding the proline codons from the sample and by treating the valine codon as if only the transition mutations were silent. This means changing the definitions of n_1, n_3, N_1, and N_3 used in carrying out the calculations shown in the third line of Table 3. The new definitions, with the changes italicized, are as follows:

1. n_3 is the number of silent base substitutions which have occurred at base residues where all three of the possible base sutstitutions are silent *and permitted according to the codon utilization data.*
2. N_3 is the number of nucleotides where three different silent base substitutions are possible *and permitted according to the codon utilization data.*
3. n_1 is the number of silent base substitutions which have occurred at base residues where only transition mutations are silent *or the codon-utilization data suggest silent mutations other than transitions are eliminated.*
4. N_1 is the number of nucleotides where only transition mutations are silent *or the codon utilization data suggest that silent mutations other than transitions are eliminated.*

Application of Correction for Carbon Utilization to Beta-Globin mRNA Sequence Data

Table 3 (fourth line) shows the results obtained when the data are corrected as described above, using the codon preference information shown in Table 1. Since there are many elements of choice in such corrections, we have carried out two extreme versions of the correction to see whether the results obtained are strongly dependent upon the choices made. Line 3 of Table 3 shows "correction A," in which we have treated the serine (UCN), threonine (ACN), and glycine (GGN) codons as if they were limited to transitions (one silent mutation permitted per codon, rather than three). These three codons provide clear-cut examples of such codon limitations. In the portions of the rabbit alpha- and beta-globin mRNAs now sequenced, there are no UCR

codons (0 out of 12 UCN serines), no ACR codons (0 out of 10 threonines), and only one GGR codon (1 out of 16 glycines). Line 4 of Table 3 shows "correction B," the more drastic version of the correction, in which we have made the additional assumptions that the extreme codon biases in the rabbit alpha- and beta-globin mRNA data for leucine, glutamine, lysine, and aspartic acid can best be interpreted as indicating strong selection against all but one codon in each case. Therefore these codons are withdrawn from the calculation. Moreover, valine was added to the list of amino acids for which it was assumed that only one silent mutation was neutral. The estimates of the average mutation rates obtained with these two extreme versions of the correction range from 2.5×10^{-9} to 3.1×10^{-9} mutation per nucleotide per year if r is assumed to be 1, and from 1.7×10^{-9} to 2.1×10^{-9} mutation per nucleotide per year if r is assumed to be 2.5. Thus we can conclude that the estimate of the mutation rate should be increased by about 20% to account for the restricted codon utilization in globin mRNAs and that the result obtained will vary only ±10% even if we make extreme choices about the details of applying this correction.

Estimates of the Fraction of the Silent Mutations Which Are Deleterious because They Disrupt the Base Pairing or Otherwise Interfere with mRNA Function

We have already commented on the absence of base substitutions of any sort in certain regions of the beta-globin structural gene. A detailed calculation argued that the absence of silent mutations in these regions could not be explained by preferential use of amino acids such as arginine which have very restricted codon utilization. The number of silent mutations which would be expected in a particular region if silent mutations were randomly distributed may be computed from Equation 5 below which corrects for codon utilization biases.

Since exclusion of mutations from these regions appears to be due to some principle other than the codon preference discussed earlier, then it is appropriate to attempt further corrections of the mutation-rate calculation for those nucleotides which seem to have the greatest restrictions on silent mutations to observe how this affects the mutation-rate calculation, and to consider how far the result obtained may differ from the true mutation rate.

The mutation-rate calculations of Table 3 (correction for codon preferences) were repeated eliminating the conserved regions of 35, 26, and 62 nucleo-

$$\begin{pmatrix} \text{number of silent} \\ \text{mutations expected} \\ \text{in a particular} \\ \text{region} \end{pmatrix} = \sum_{n=1}^{20} \begin{pmatrix} \text{number of silent mutations} \\ \text{in codons for amino acid } n \\ \text{throughout the structural} \\ \text{gene} \end{pmatrix} \begin{pmatrix} \text{fraction of the codons} \\ \text{for amino acid } n \text{ which} \\ \text{are found in the region} \\ \text{considered} \end{pmatrix} \quad (5)$$

tides starting, respectively, at amino acid residues 22, 58, and 91. Since these sequences contain no silent nucleotide substitutions, this correction changes N_1 and N_3 but has no effect on n_1 or n_3. When Equation 4 was used with no correction for codon preferences, the effect of withdrawing these regions was to increase the computed mutation rates 66% (to 3.9×10^{-9} for $r = 1$ and to 2.7×10^{-9} for $r = 2.5$). Lesser effects were noted when this correction was applied in combination with the corrections for codon preferences.

What of other nucleotide sequences where silent mutations would disrupt mRNA function by destroying critical base-pairing or recognition sequences? Such conserved nubleotides might not be easy to recognize. If the base pairing in arms 2 and 3 of the structure in Figure 3 is important, then many more base-paired nucleotides will be conserved. There are many other possibilities of this sort. The extent of their effects upon the mutation-rate calculations cannot be estimated at this time, but it is clear that in every case such effects will tend to make the calculated mutation rate underestimate the real value.[1]

Constraints on the Maintenance of Single-Copy DNA Sequences: The Neutralist-Selectionist Controversy

Kimura (1968) and King and Jukes (1969) have argued that most of the nucleotide substitutions occurring during the evolution of eukaryotic genomes are the results of the fixation of selectively neutral mutations through random genetic drift. This so-called neutralist position has been termed "non-Darwinian" evolution to distinguish it from the more conventional view that most evolutionary changes in DNA are preserved because of a positive selectional advantage conferred upon the population. Some of their arguments depend upon assumptions which are seriously in question, as pointed out by many later authors. For instance, Kimura and King and Jukes agreed with Haldane's (1957) estimate of the cost of positive evolution (termed the "substitution load") which concluded that only about one new advantageous allele could be incorporated per 300 generations. However, Li (1963), Brues (1964), Smith (1968), and Sved (1968) have argued that, if

advantageous alleles were available, gene substitutions increasing the fitness of the population might occur at much higher rates. Sved made what may be the strongest statement in this regard: "rates of one completed substitution per generation or more are by no means excluded."

The arguments of Kimura (1968) and King and Jukes (1969) led them to speculate that silent mutations are usually neutral and they proposed that a large portion of amino acid substitutions are neutral as well, saying, "there is very little restriction on the type of amino acid that can be accommodated at most variable sites." A number of authors as early as Richmond (1970) and as recently as Efstratiadis et al. (1977) have adduced evidence to the contrary as proof that the concept of non-Darwinian evolution is wrong.

We believe that the ability to measure the number of silent mutations through nucleotide sequence analysis permits us to resolve this controversy in a convincing way, but with a different conclusion from that of Richmond (1970) and Efstratiadis et al. (1977). We agree with Li (1963), Brues, (1964), Smith (1968), Sved (1968), and others that Haldane's (1957) calculations are questionable and do not provide convincing support for the neutralist hypothesis. On the existence of neutral mutations within the beta-globin structural gene, we support the Darwinist or so-called neo-Darwinist critiques of the neutralist's arguments: mutations which cause amino acid substitutions appear to have a fixation probability about fivefold less than that for silent mutations within the same structural gene. This argues strongly that most amino acid substitutions in this gene are sufficiently deleterious to be eliminated in the course of evolution. Furthermore, many silent mutations in the globin mRNAs appear sufficiently deleterious to be eliminated: There are strong biases shown in the use of the degeneracy of the code; long sequences are conserved with no silent mutations; and finally, the 3' untranslated sequence is preferentially conserved as estimated by comparison of the values of α'_3 computed from Equation 2 for the structural gene and the 3' untranslated region.

Since our data and that which we cite support the arguments used by the neo-Darwinists against the neutralist hypothesis, we may acknowledge that the neutralists supported their major conclusion on partially faulty logic. It is therefore ironic that the silent-mutation data which reveal some of the faults in this logic show also that the major conclusion of the neutralists is correct: a large fraction of the changes in the genome *are* neutral. The resolution of this paradox lies in recognizing that the mutation rate is so high that only a small fraction of the genome can be kept as accurate as the globin genes (roughly 0.6% of the genome, if truncation selection is not important in maintaining the accuracy of single-copy DNA sequences; Salser and Isaacson 1976).

[1] *Note added in proof:* Tilghman et al. (*Proc. Natl. Acad. Sci.* 75:725 [1978]) have shown recently that the processing of mouse beta-globin mRNA involves a joining reaction after excision of a 550-nucleotide sequence, which follows amino acid codon 104. This is in the center of the region (codons 99–111) that we identify as most highly conserved (Fig. 2). The cleavage observed occurs immediately prior to the mispaired nucleotide 312 in the base-pairing structure shown in Figure 3. It seems possible that the strong evolutionary conservation we have noted for the base pairing in the region serves to specify the correct processing of the mRNA precursor.

The rest of the single-copy DNA must be genetically drifting, although the repetitive DNA sequences may be efficiently kept accurate by special stratagems as discussed earlier (Salser and Isaacson 1976).

Although we will not attempt to duplicate the arguments of Salser and Isaacson here, it will be useful to stress that these calculations are totally independent of Haldane's and others' estimates of the substitution load or cost, if any, of positive evolution. Our calculations focused instead on the mutation load, the cost of eliminating deleterious mutations. The important step was to recognize that silent-mutation data permitted the mutation load to be estimated with less questionable assumptions and, when the mutation load involved in maintaining the accuracy of the entire genome was found to be intolerably large, the neutralist hypothesis was supported independently of any substitution load considerations.

This issue has been controversial one. The problem has interested scientists with a broad range of expertise and it may be useful to set down some specific points which we feel any convincing defense of the selectionist hypothesis must consider to account for the observed data:

1. The selectionist would have to convince us that silent mutations in the globin gene are roughly five times more likely than amino acid substitutions to have a positive selectional advantage over the prevalent genotype. It seems more logical to suppose the reverse — that mutations which change the protein gene product are more likely to be advantageous.

2. The selectionist would also have to explain why mutations in the large amount of single-copy DNA which does not code for mRNA are several times more likely to have a positive selectional advantage over the wild-type genome than mutations within mRNA sequences. Both our theoretical calculations and T_m depression experiments suggest that the total single-copy DNA evolves much more rapidly than the globin gene or total mRNA sequences.

3. The selectionist would have to maintain that advantageous mutations are being produced and fixed in the population at what seems an astoundingly high rate: in each species about nine mutations per year of evolution must be advantageous enough to take over.

It impresses many as bizarre that only (roughly) 1% of the mamalian DNA may be devoted to conventional single-copy genes like globin. In fact, 0.6–1.0% of the DNA is enough to code for the entire complement of 30,000–50,000 different gene products estimated by quite different arguments to be active in mammalian systems; so 1% is not embarrassingly low. Our conclusions are, however, incompatible with models such as that proposed by Crick (1971) in which it is supposed that the bulk of the single-copy DNA has a direct control function. Salser and Isaacson (1976) have proposed ways in which large amounts of interspersed repetitious DNA sequences might be kept accurate for function as control signals. These proposals depend upon the repetitious nature of the sequences, and unless truncation selection plays a much larger role than heretofore imagined, we see no way in which more than a few percent of the mammalian single-copy DNA could be maintained with the accuracy expected of a "control" sequence.

How can one explain the presence of a large amount of single-copy DNA whose drifting sequence suggests an absence of any conventional function? Production of such useless DNA would be the consequence of unequal recombination events. In the case of globin genes, such events have been postulated to explain (1) the evolution of the beta-gene cluster including production of the fusion hemoglobins Lepore, anti-Lepore, and Kenya; (2) the fact that some natural populations contain individuals with different numbers of alpha or beta genes; (3) the production of the deletion thalassemias; and (4) the presence of a 39-nucleotide duplication in the 3′ untranslated portion of the human beta-globin mRNA. It seems likely that such unequal-crossover events occur frequently throughout eukaryotic genomes. If the region of the duplication-deletion contained only functionless DNA, the results would be very different from those expected if the region contained an essential gene. In events which involved an essential gene, the deletion would be eliminated from the population because of its *strong selectional disadvantage,* whereas the duplication might frequently be neutral or only *slightly disadvantageous.* The net effect of duplication-deletion events involving essential genes would be an increase in the size of the genome through duplication of functional genes and surrounding sequences. In most cases one of the duplicate functional gene copies would eventually lose its function: mutations which inactivate it would not be strongly selected against since a good gene copy remains. Consequently, a strong drive to duplicate functional genes is, in the long term, a strong drive to create more genetically drifting sequences. Duplication-deletion events which only involve drifting sequences would have a different and compensating effect. The deletion would have a *slight advantage* (through the elimination of useless DNA), and the duplication would be at a *slight disadvantage.* Such events should result in a gradual elimination of useless DNA. In the steady state, the rate of production and elimination of useless DNA must be equal. Since the selection pressure for the creation of drifting DNA sequences is expected to be stronger than the selection for its elimination, it is not surprising that the amount of such single-copy DNA sequences builds up to a level much higher than that of functional genes before equilibrium between production and elimination of the drifting sequences is attained.

According to this view, the randomly drifting sequences in the genome should contain all stages of the deterioration of duplicate gene copies that were once functional. Since the average lifetime of a particular sequence as "useless" DNA may be long and since the point mutation rate is high, most of these drifting sequences may bear little resemblance to their active counterparts.

We do not wish to exclude the possibility that substantial amounts of single-copy DNA arise through the progressive accumulation of mutations by highly repetitive or intermediate repetitive DNA sequences. But we have earlier discussed mechanisms by which such sequences could be kept accurate (Salser and Isaacson 1976) and have shown that the accuracy of satellite sequences can be maintained over very long evolutionary periods (Salser et al. 1976b; Fry and Salser 1977). Consequently, it seems to us that the extent of this contribution to the single-copy DNA may be small.

Acknowledgments

This investigation was supported by Grants GM-18586 and HL-21831 from the National Institutes of Health and by Grant VC-240 from the American Cancer Society. I am indebted to Gary Paddock, Howard Heindell, Alvin Liu, Bernard Forget, and Sherman Weissman for permitting me to refer to their sequence data in advance of publication.

I acknowledge helpful discussions with Judith Isaacson, Gary Studnicka, Sherman Weissman, James McMahon, and Fred Hatch, and assistance with the computer analysis by Georgia Rahn, Ian Cummings, and David Finck.

REFERENCES

BARALLE, F. E. 1977a. Complete nucleotide sequence of the 5′ noncoding region of rabbit beta globin mRNA. *Cell* **10**: 549.

———. 1977b. Structure-function relationship of 5′ noncoding sequence of rabbit alpha- and beta-globin mRNA. *Nature* **267**: 279.

BORER, P. N., B. DENGLER, and I. TINOCO. JR. 1974. Stability of ribonucleic acid and double-stranded helices. *J. Mol. Biol.* **86**: 843.

BROWNE, J. K., G. V. PADDOCK, A. LIU, P. CLARKE, H. C. HEINDELL, and W. SALSER. 1977. Nucleotide sequences from the rabbit beta globin gene inserted into *Escherichia coli* plasmids. *Science* **195**: 389.

BRUES, A. M. 1964. The cost of evolving vs. the cost of not evolving. *Evolution* **18**: 379.

COWAN, N. J., D. S. SECHER, and C. MILSTEIN. 1976. Purification and sequence analysis of the mRNA coding for an immunoglobulin heavy chain. *Eur. J. Biochem.* **61**: 355.

CRICK, F. 1971. General model for the chromosomes of higher organisms. *Nature* **234**: 25.

DASGUPTA, R., D. S. SHIH, C. SARIS, and P. KAESBERG. 1975. Nucleotide sequence of a viral RNA fragment that binds to eukaryotic ribosomes. *Nature* **256**: 624.

DOSKOCIL, J. and F. SORM. 1962. Distribution of 5-methylcytosine in pyrimidine sequences of deoxyribonucleic acids. *Biochim. Biophys. Acta* **55**: 953.

EFSTRATIADIS, A., F. C. KAFATOS, and T. MANIATIS. 1977. The primary structure of rabbit beta-globin mRNA as determined from cloned DNA. *Cell* **10**: 571.

EFSTRATIADIS, A., F. C. KAFATOS, A. M. MAXAM, and T. MANIATIS. 1976. Enzymatic *in vitro* synthesis of globin genes. *Cell* **7**: 279.

ELADARI, M. and E. GALIBERT. 1975. Sequence determination of 5′-terminal and 3′-terminal T$_1$ oligonucleotides of 18-S ribosomal RNA of a mouse cell line (L5178Y). *Eur. J. Biochem.* **55**: 247.

———. 1976. Sequence determination of the 3′ terminal T$_1$ oligonucleotide of 18S ribosomal RNA. *Nucleic Acid Res.* **3**: 10.

FRY, K. and W. SALSER. 1977. Nucleotide sequences of HS alpha-satellite DNA from kangaroo rat *Dipodomys ordii* and characterization of similar sequences in other rodents. *Cell* **12**: 1069.

GAREL, J. 1974. Functional adaptation of rRNA population. *J. Theor. Biol.* **43**: 211.

GRALLA, J. and D. M. CROTHERS. 1973. Free energy of imperfect nucleic acid helices. II. Small hairpin loops. *J. Mol. Biol.* **73**: 497.

GUMMERSON, K. S. and R. WILLIAMSON. 1974. Sequence divergence of mammalian globin messenger RNA. *Nature* **247**: 265.

HALDANE, J. B. S. 1957. The cost of natural selection. *J. Genet.* **55**: 511.

HASELTINE, W. A., A. M. MAXAM, and W. GILBERT. 1977. Rous sarcoma virus genome is terminally redundant with the 5′ sequence. *Proc. Natl. Acad. Sci.* **74**: 989.

HIGUCHI, R., G. V. PADDOCK, R. WALL, and W. SALSER. 1976a. Insertion of rabbit globin sequences into *E. coli* plasmids pSC101 and pMB9. *Fed. Proc.* **35**: 1369.

———. 1976b. A general method for cloning eukaryotic structural gene sequences. *Proc. Natl. Acad. Sci.* **73**: 3146.

HOYER, B. H., N. W. VAN DE VELDE, M. GOODMAN, and R. B. ROBERTS. 1972. Examination of primate relationships by DNA homology. *Carnegie Inst. Wash. Year Book* **71**: 260.

HUNT, J. A. 1970. Terminal-sequence studies of high molecular-weight ribonucleic acid. *Biochem. J.* **120**: 353.

IRWIN, D., A. KUMAR, and R. MALT. 1975. Messenger ribonucleoprotein complexes isolated with oligo (dT)-cellulose chromatography from kidney polysomes. *Cell* **4**: 157.

JACOBSON, H. and W. STOCKMEYER. 1950. Intramolecular reaction in polycondensations. I. The theory of linear systems. *J. Chem. Phys.* **18**: 1600.

KIM, S. H., F. L. SUDDATH, G. J. QUIGLEY, A. McPHERSON, J. L. SUSSMAN, A. H. J. WONG, N. C. SEEMAN, and A. RICH. 1974. Three-dimensional tertiary structure of yeast phenylalanine transfer RNA. *Science* **185**: 435.

KIMURA, M. 1968. Evolutionary rate at the molecular level. *Nature* **217**: 624.

KING, J. L. and T. H. JUKES. 1969. Non-Darwinian evolution. *Science* **164**: 788.

KOHNE, D. E., J. A. CHISCON, and B. H. HOYER. 1970. Nucleotide sequence change in nonrepeated DNA during evolution. *Carnegie Inst. Wash. Year Book* **69**: 488.

LAIRD, C. D., B. L. McCONAUGHY, and B. J. McCARTHY. 1969. Rate of fixation of nucleotide substitutions in evolution. *Nature* **224**: 149.

LEDER, P., J. ROSS, J. GIELEN, S. PACKMAN, Y. IKAWA, H. AVIV, and D. SWAN. 1974. Regulated expression of mammalian genes: Globin and immunoglobulin as model systems. *Cold Spring Harbor Symp. Quant. Biol.* **38**: 753.

LEGON, S. 1976. Characterization of the ribosome-protected regions of ^{125}I labelled rabbit globin messenger RNA. *J. Mol. Biol.* **106**: 37.

LI, C. C. 1963. The way the load ration works. *Am. J. Hum. Genet.* **15**: 316.

LOCKARD, R. E. and U. L. RAJBHANDARY. 1976. Nucleotide

sequences at the 5' termini of rabbit alpha- and beta-globin in RNA. *Cell* **9**: 747.

LODISH, H. F. 1976. Translation control of protein synthesis. *Annu. Rev. Biochem.* **45**: 39.

MANIATIS, T., G. K. SIM, A. EFSTRATIADIS, and F. C. KAFATOS. 1976. Amplification and characterization of a beta globin gene synthesized *in vitro*. *Cell* **8**: 163.

MAROTTA, C. A., B. G. FORGET, M. COHEN-SOLAL, and S. M. WEISSMAN. 1976. Nucleotide sequence analysis of coding and non-coding regions of human β-globin mRNA. *Prog. Nucleic Acid Res. Mol. Biol.* **19**: 165.

MAROTTA, C. A., B. G. FORGET, S. M. WEISSMAN, I. M. VERMA, R. P. MCCAFFREY, and D. BALTIMORE. 1974. Nucleotide sequences of human globin messenger RNA. *Proc. Natl. Acad. Sci.* **71**: 2300.

MCCARTHY, B. J. and M. N. FARQUHAR. 1972. The rate of change of DNA in evolution. *Brookhaven Symp. Biol.* **23**: 44.

MOREL, C., B. KAYIBANDA, and K. SCHERRER. 1971. Proteins associated with globin messenger RNA in avian erythroblasts: Isolation and comparison with the proteins bound to nuclear messenger-like RNA. *FEBS Lett.* **18**: 84.

O'MALLEY, B. W., S. L. C. WOO, J. J. MONAHAN, L. MCREYNOLDS, S. HARRIS, M. TSAI, S. TSAI, and A. MEANS. 1976. The synthesis, isolation, amplification and transcription of the ovalbumin gene. In *Molecular mechanisms in the control of gene expression. ICN-UCLA Symposium on Molecular and Cellular Biology* vol. 5, p. 309. Academic Press, New York.

PADDOCK, G. V., H. HEINDELL, and W. SALSER. 1974. Deoxy-substitution in RNA by RNA polymerase *in vitro*: A new approach to nucleotide sequence determinations. *Proc. Natl. Acad. Sci.* **71**: 5017.

PADDOCK, G. V., R. HIGUCHI, R. WALL, and W. SALSER. 1976. Insertion of rabbit globin sequences into *E. coli* plasmids. In *Proceedings of the 1976 ICN-UCLA Winter Conference on Molecular and Cellular Biology* (D. P. Nierlich ed., et al.). Academic Press, N.Y.

PADDOCK, G. V., R. POON, H. C. HEINDELL, J. ISAACSON, and W. SALSER. 1977. Rabbit globin mRNA: Analysis of T$_1$ digestion fragments. *J. Biol. Chem.* **252**: 3446.

PIPAS, J. M. and J. E. MCMAHON. 1975. Method for predicting RNA secondary structure. *Proc. Natl. Acad. Sci.* **72**: 2017.

POON, R., G. V. PADDOCK, H. HEINDELL, P. WHITCOME, W. SALSER, D. KACIAN, A. BANK, R. GAMBINO, and F. RAMIREZ. 1974. Nucleotide sequence analysis of RNA synthesized from rabbit globin complimentary DNA. *Proc. Natl. Acad. Sci.* **71**: 3502.

PROUDFOOT, N. J. 1976. Sequence analysis of the 3' non-coding regions of rabbit alpha- and beta-globin messenger RNAs. *J. Mol. Biol.* **107**: 491.

———. 1977. Complete 3' noncoding region sequences of rabbit and human beta-globin messenger RNAs. *Cell* **10**: 559.

PROUDFOOT, N. J. and G. G. BROWNLEE. 1976. 3' Non-coding region sequences in eukaryotic messenger RNA. *Nature* **263**: 211.

RICHMOND, R. C. 1970. Non-Darwinian evolution: A critique. *Nature* **225**: 1025.

ROBERTUS, J. D., J. LADNER, J. FINCH, D. RHODES, R. BROWN, B. CLARK, and A. KLUG. 1974. Structure of yeast phenylalanine tRNA at 3 Å resolution. *Nature* **250**: 546.

ROSBASH, M., M. S. CAMPO, and K. S. GUMMERSON. 1975. Conservation of cytoplasmic poly(A)-containing RNA in mouse and rat. *Nature* **258**: 682.

ROUGEON, F. and B. MACH. 1976. Stepwise biosynthesis *in vitro* of globin genes from globin mRNA by DNA polymerase of avian myeloblastosis virus. *Proc. Natl. Acad. Sci.* **73**: 3418.

ROY, P. H. and A. WEISSBACH. 1975. DNA methylase from HeLa cell nuclei. *Nucleic Acid Res.* **2**: 1669.

RUSSELL, G. J., P. M. B. WALKER, R. A. ELTON, and J. H. SUBAK-SHARPE. 1976. Doublet frequency analysis of fractionated vertebrate nuclear DNA. *J. Mol. Biol.* **108**: 1.

SALSER, W. 1974. DNA sequencing techniques. *Annu. Rev. Biochem.* **43**: 923. (See especially pp. 948–949.)

SALSER, W. and J. S. ISAACSON. 1976. Mutation rates in globin genes: The genetic load and Haldane's dilemma. *Prog. Nucleic Acid Res. Mol. Biol.* **19**: 205.

SALSER, W., K. FRY, C. BRUNK, and R. POON. 1972. Nucleotide sequencing of DNA: Preliminary characterization of the products of specific cleavage at guanosine, cytosine or adenine residues. *Proc. Natl. Acad. Sci.* **69**: 238.

SALSER, W., R. POON, R. WHITCOME, and K. FRY. 1973. New techniques for determining nucleotide sequences from eucaryotic cells. In *Virus research. Second ICN-UCLA Symposium on Molecular Biology* (ed. C. F. Fox et al.), p. 545. Academic Press New York.

SALSER, W., J. BROWNE, P. CLARKE, H. HEINDELL, R. HIGUCHI, G. PADDOCK, J. ROBERTS, G. STUDNICKA, and P. ZAKAR. 1976a. Determination of globin mRNA sequences and their insertion into bacterial plasmids. *Prog. Nucleic Acids Res. Mol. Biol.* **19**: 177.

SALSER, W., S. BOWEN, D. BROWNE, F. ELADLI, N. FEDOROFF, K. FRY, H. HEINDELL, G. PADDOCK, R. POON, B. WALLACE, and R. WHITCOME. 1976b. Investigation of the organization of mammalian chromosomes at the DNA sequence level. *Fed. Proc.* **35**: 23.

SETLOW, P. 1976. Nearest neighbor frequencies in deoxyribonucleic acids. In *Handbook of biochemistry and molecular biology: Nucleic acids* (ed. G. D. Fasman), vol. 2, p. 312. CRC Press, Cleveland, Ohio.

SHINE, J. and L. DALGARNO. 1974. Identical 3'-terminal octanucleotide sequence in 18S ribosomal ribonucleic acid from different eukaryotes. *Biochem. J.* **141**: 609.

———. 1975. Determinant of cistron specificity in bacterial ribosomes. *Nature* **254**: 34.

SINSHEIMER, R. L. 1955. The action of pancreatic deoxyribonuclease. II. Isomeric dinucleotides. *J. Biol. Chem.* **215**: 579.

SMITH, D. W. E. 1975. Reticulocyte transfer RNA and hemoglobin synthesis. *Science* **190**: 529.

SMITH, D. W. E. and A. L. MCNAMARA. 1971. Specialization of rabbit reticulocyte transfer RNA content for hemoglobin synthesis. *Science* **171**: 577.

SMITH, J. M. 1968. "Haldane's dilemma" and the rate of evolution. *Nature* **219**: 1114.

———. 1972. *On evolution*. University Press, Edinburgh.

SPOHR, G., B. KAYIBANDA, and K. SCHERRER. 1972. Polyribosome-bound and free-cytoplasmic hemoglobin messenger RNA in differentiating avian erythroblasts. *Eur. J. Biochem.* **31**: 194.

SPRAGUE, K. U., R. KRAMER, and M. JACKSON. 1975. The terminal sequences of *Bombyx mori* 18S ribosomal RNA. *Nucleic Acids Res.* **2**: 2111.

STEITZ, J. A. and K. JAKES. 1975. How ribosomes select mutation regions in mRNA: Base pair formation between the 3' terminus of 16S rRNA and the mRNA during initiation of protein synthesis in *Escherichia coli*. *Proc. Natl. Acad. Sci.* **72**: 4734.

SVED, J. A. 1968. Possible rates of gene substitution in evolution. *Am. Nat.* **102**: 283.

TINOCO, I., P. N. BORER, B. DENGLER, M. D. LEVINE, O. C. UHLENBECK, D. M. CROTHERS, and J. GRALLA. 1973. Improved estimation of secondary structure in ribonucleic acids. *Nat. New Biol.* **246**: 40.

ULLRICH, A., J. SHINE, J. CHIRGWIN, R. PICTET, E. TISCHER, W. J. RUTTER, and H. M. GOODMAN. 1977. Rat insulin genes: Construction of plasmids containing the coding sequences. *Science* **196**: 1313.

VAN DE VOORDE, A., R. CONTRERAS, R. ROGIERS, and W. FIERS. 1976. The initiation region of the SV40 V P$_1$ gene. *Cell* **9**: 117.

Physical Mapping of Repetitive DNA Sequences Neighboring the Rabbit β-Globin Gene

R. A. FLAVELL, A. J. JEFFREYS,* AND G. C. GROSVELD

Section for Medical Enzymology and Molecular Biology, Laboratory of Biochemistry, University of Amsterdam, Amsterdam, The Netherlands

In the last few years considerable information has accumulated on the organization of DNA in various classes of eukaryotic repetitive genes such as ribosomal RNA cistrons (Wellauer et al. 1974), 5S RNA cistrons (Carroll and Brown 1976a,b), transfer RNA cistrons (Clarkson and Kurer 1976), and the histone gene cluster (Kedes 1976; Schaffner et al. 1976; Gross et al. 1976). However, we still know almost nothing about the DNA sequences which surround defined eukaryotic single-copy structural genes. It has been established that the genomes of many animals consist of a large number of repetitive DNA sequences interspersed with single-copy DNA. The role of repetitive DNA in the expression of structural genes is as yet unclear. In DNA of echinoderms, structural-gene sequences are linked to interspersed repetitive elements (Davidson et al. 1975), and Bishop and Freeman (1974) have presented evidence for a repetitive DNA element in the vicinity of the hemoglobin genes in the duck genome.

The availability of recombinant DNA methods and physical DNA mapping techniques makes it possible to probe DNA sequences proximal to structural genes with more precision than was hitherto possible. As a first step in this approach, we have constructed a physical map of restriction endonuclease cleavage sites around the rabbit β-globin gene (Jeffreys and Flavell 1977). This map was determined by cleaving rabbit DNA with various restriction endonucleases and resolving the resulting DNA fragments by electrophoresis in agarose gels. DNA fragments were then transferred by blotting onto a nitrocellulose filter using the methods devised by Southern (1975). Fragments on the filter containing the β-globin gene were hybridized with a radioactive hybridization probe for this gene (plasmid pβG1) which contains a β-globin cDNA insert (Maniatis et al. 1976) and detected by autoradiography.

In this paper we show how these methods have enabled us to localize a segment of intermediate-repetitive DNA in a region 3' of the rabbit β-globin gene. In addition, we show that the 3' extragenic region also contains oligo(dA · dT) cluster(s).

EXPERIMENTAL PROCEDURES

Procedures for detecting rabbit β-globin fragments by filter hybridization, including the preparation of DNA, nick-translation of pβG1 DNA, restriction endonuclease digestions, gel electrophoresis, transfer to nitrocellulose filters, and filter hybridizations, are fully described in Jeffreys and Flavell (1977).

Binding of duplex DNA to poly(U)-Sephadex was performed as described in Flavell and Van den Berg (1975) and Flavell et al. (1977). Poly(U) plus poly(A) Sephadex triple-layer columns (Flavell et al. 1977) were equilibrated in 2 M LiCl, 0.5% SDS, 10 mM Tris-HCl (pH 7.5) and loaded with rabbit DNA which had been digested by restriction endonucleases (5–10 µg DNA per ml packed poly(U)-Sephadex). DNA not bound to the column was eluted with a 2 M LiCl-buffer wash, and bound DNA was recovered by elution with 0.1 M LiCl buffer. The DNA from each fraction was precipitated with ethanol at −70°C for 1 hour, centrifuged at 10,000 rpm in a Sorvall HB-4 rotor for 10 minutes at −20°C, and washed with 70% ethanol to remove excess salt. The DNA was then dissolved in 10 mM Tris-HCl (pH 7.5), fractionated by gel electrophoresis, and assayed for fragments containing β-globin sequences as described in Jeffreys and Flavell (1977).

Reassociation analysis of rabbit DNA cleaved with various restriction endonucleases was carried out at 70°C in 0.6 M sodium phosphate (pH 6.8). The samples were heat-denatured (5 min at 100°C) and annealed either to a C_0t of 6×10^{-3} (a 10-sec incubation in 6.7 ml) or to a C_0t of 60 (a 56-min incubation in 200 µl). Samples were then diluted to 34 ml and the phosphate concentration adjusted to 0.12 M. Next, 4 ml of a 50% v/v hydroxylapatite suspension in 0.12 M phosphate was added and the sample mixed well for 2 minutes at room temperature to facilitate DNA binding. The hydroxylapatite (HAP) was poured into a column which was washed twice at 60°C with 2.5 column volumes of 0.12 M phosphate buffer to elute single-stranded DNA. The duplex DNA was eluted by two washes with 2.5 column volumes of 0.4 M phosphate buffer at 60°C followed by the same elution protocol at 100°C. Fractions were dialyzed against 10 mM Tris-HCl (pH 7.5), 1 mM EDTA for 2 days, passed over Sephadex G-50 to remove residual phosphate, and precipitated with

* Present address: Department of Genetics, University of Leicester, Adrian Building, University Road, Leicester LE1 7RH, United Kingdom.

ethanol. The single-strand and double-strand fractions were then alkali-denatured and analyzed for β-globin DNA fragments as usual (Jeffreys and Flavell 1977). The $C_0 t$ values cited here have been corrected for the salt concentration according to the tables in Britten et al. (1974).

RESULTS

Detection of Fragments Containing a β-Globin Gene in Restriction Enzyme Digests of Rabbit DNA

To detect DNA fragments containing sequences which code for the β-globin chain, we have used as a hybridization probe pβG1 DNA (Maniatis et al. 1976) labeled by nick-translation with ³²P-labeled deoxyribonucleoside triphosphates to a specific activity of 10^7–7.5×10^7 cpm/μg. To detect globin sequences, two criteria must be fulfilled. First, the labeled probe must be capable of detecting picogram amounts of filter-bound β-globin-specific DNA in total digests of rabbit DNA. Second, the hybridization has to be highly selective, since only about 1 in 10^6 DNA fragments in a restriction enzyme digest of mammalian DNA contains β-globin DNA. The limits of detection of complementary sequences by the labeled probe were determined by hybridizing ³²P-labeled pβG1 DNA with decreasing amounts of unlabeled pβG1 DNA linearized by cleavage with restriction endonuclease HindIII. The HindIII linears were electrophoresed in an agarose gel, and the DNA was denatured in the gel and transferred by blotting onto a nitrocellulose filter using the techniques of Southern (1975). The DNA filters were pretreated and hybridized with the probe as described in Experimental Procedures. After washing the filter extensively under hybridization conditions to remove unbound DNA from the nitrocellulose strip, we found that we could readily detect as little as 10 pg of pβG1 DNA × HindIII as a labeled band after overnight autoradiography (see Fig. 1). The specificity of the probe for globin sequences is shown by the positive hybridization with globin cDNA. In contrast, no hybridization was found between labeled pβG1 DNA and microgram amounts of marker DNAs (phages λ and φ29 DNA).

When we hybridized filters containing rabbit DNA digested with endonucleases EcoRI or PstI at 65°C in 3 × SSC with ³²P-labeled pβG1 DNA, a number of discrete labeled components could be detected in addition to a broad smear of labeling which followed the overall distribution of the DNA in the gel. Washing the filters in reduced salt concentrations at 65°C allowed us to remove this aspecific background (1 × SSC instead of 3 × SSC), and further reduction of the salt concentration revealed two classes of globin-related fragments. The more stable class was still present after washing in 0.03 × SSC and represents presumably DNA fragments which contain the β-globin gene. The second class consisted of hybrid-

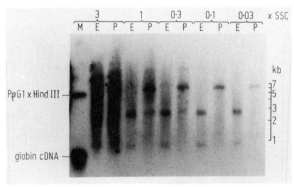

Figure 1. Detection of DNA fragments containing a β-globin gene in digests of rabbit DNA with endonucleases EcoRI or PstI. Rabbit liver DNA (30 μg) cleaved with EcoRI (E) or with PstI (P) was denatured with alkali and electrophoresed in a 1.2% agarose slab gel. pβG1 DNA (10 pg) linearized by cleavage with endonuclease HindIII and 100 pg of globin cDNA copied from total adult globin mRNA using reverse transcriptase was similarly treated with alkali and run as hybridization markers (M). Denatured DNA was subsequently transferred by blotting to a nitrocellulose filter. This filter was cut into strips and hybridized in 3 × SSC at 65°C to ³²P-labeled pβG1 DNA. After hybridization, unbound labeled probe was washed from the strips in 3 × SSC at 65°C. Duplicate strips were further washed at 65°C in lower concentrations of SSC as indicated. Full details of hybridization and washing are given in Jeffreys and Flavell (1977). Labeled components on each strip were subsequently detected by autoradiography. The single-strand molecular-weight scale in kilobases was determined from coelectrophoresed denatured marker DNAs as described in Jeffreys and Flavell (1977). (Reprinted, with permission, from Jeffreys and Flavell 1977.)

ized components which appeared to melt out on reducing the salt concentration from 0.3 × SSC to 0.1 × SSC at 65°C (Fig. 1). These bands therefore contain sequences related to, but not identical with, those in the β-globin gene represented in pβG1 DNA. If labeled pMB9 DNA, the plasmid vector in pβG1 DNA, is used as a filter-hybridization probe, none of these rabbit DNA fragments are detectable (data not shown). This indicates that all fragments detected by pβG1 DNA contain a globin gene rather than sequences which by chance are homologous with the plasmid vector.

A Map of Restriction Enzyme Cleavage Sites Around the Rabbit β-Globin Gene

To detect only the fragments containing the rabbit β-globin gene, all filters were washed in 0.1 × SSC to remove the additional globin-related bands. We have constructed a physical map of restriction enzyme cleavage sites around the rabbit β-globin gene for the restriction endonucleases EcoRI, PstI, and KpnI. Table 1 gives the molecular weights of the components observed with the three enzymes and the molecular weights of all double-digestion products. These data can be accommodated in a map of cleavage sites around a single β-globin gene. A

Table 1. Summary of β-Globin Fragments in Rabbit DNA Cleaved with Various Restriction Endonucleases

EcoRI	PstI	KpnI	PstI + EcoRI	KpnI + EcoRI	KpnI + PstI
	6.3				
		5.1			
					3.6
2.6				2.6	
			1.3		
0.9			0.9	0.9	

All single-strand lengths were estimated from a standard mixture of phage λ, phage φ29, and phage φ29 × EcoRI marker DNAs which was denatured with alkali and run in parallel with rabbit DNA digests (sizes in kilobases).

single band is observed with restriction endonucleases PstI and KpnI, consistent with the observation that the rabbit β-globin gene is not cleaved by these enzymes: The two bands observed after digestion with EcoRI correspond to the 5' and 3' segments of the β-globin gene generated by cleavage in the gene at the EcoRI site corresponding to amino acid positions 121 and 122 of the β-globin polypeptide chain (Efstratiadis et al. 1977). The orientation of the map was established by hybridizing nitrocellulose strips containing Southern replicas of EcoRI-cleaved rabbit DNA with probes for the 5' and 3' segments of the β-globin gene. (The probes were obtained by cleaving pβG1 DNA with EcoRI and AvaI, which provides segments of 2.3 kb containing the 5' segment of the β-globin cDNA insert and 1.8 kb containing the 3' segment. These fragments were separated on agarose gels and labeled in vitro by nick-translation.) Since the 5' probe hybridized exclusively to the 2.6-kb EcoRI fragment and the 3' probe only to the 0.9-kb fragment, the orientation of the map shown in Figure 6 is established (Jeffreys and Flavell 1977).

Repetitive DNA Sequences Flanking the Rabbit β-Globin Gene

To determine whether repetitive DNA sequences are localized in the regions around the β-globin gene contained within our physical map, we cleaved rabbit DNA with restriction endonuclease EcoRI, PstI, or KpnI. The DNA samples were denatured and annealed to C_0t values of 6.0×10^{-3} and 60, respectively. Inspection of the C_0t curve of rabbit DNA (Fig. 2; carried out with DNA of a fragment size of 0.4 kb) shows that under these conditions foldback DNA and the major repetitive DNA of rabbit, respectively, are in duplex form. The duplex DNA was separated from single-stranded DNA by fractionation over hydroxylapatite under standard conditions (see Experimental Procedures). If a fragment containing globin sequences also contains a segment of foldback DNA or repetitive DNA, it will appear in the duplex DNA fraction from the HAP column if it was annealed to a suitable C_0t value. To assay for the fragments containing β-globin sequences,

samples from the HAP column were recovered, denatured with alkali, electrophoresed on an agarose gel, transferred to a nitrocellulose filter, and hybridized with labeled pβG1 DNA as usual. Figure 3 shows the autoradiograms of the EcoRI and KpnI samples from the HAP columns after incubation to these C_0t values. The globin-containing sequences from both digests (and also from PstI digests; unpubl.) are found in the single-stranded fraction at a C_0t of 6.0×10^{-3}, showing that no foldback sequences are found in the regions surrounding the β-globin gene which are contained by our map. However, at a C_0t of 60, the KpnI fragment containing the β-globin gene is found in the duplex DNA fraction, whereas the two EcoRI fragments of 2.6 and 0.9 kb are found in the single-stranded fraction. This localizes a repetitive DNA element on a segment of the KpnI fragment outside the region containing the two EcoRI fragments. The 3.6-kb β-globin fragment generated by cleavage of rabbit DNA with KpnI plus PstI is also found in the duplex fraction

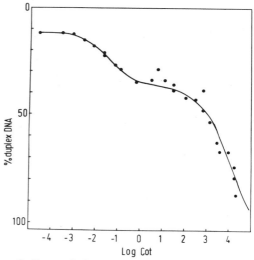

Figure 2. Reassociation curve of 0.4-kb randomly sheared rabbit DNA. Rabbit [32P]DNA was heat-denatured and annealed with excess unlabeled total rabbit DNA. The fraction of DNA reassociated at each C_0t value was determined by HAP chromatography at 60°C. For details, see Experimental Procedures and Flavell et al. (1977).

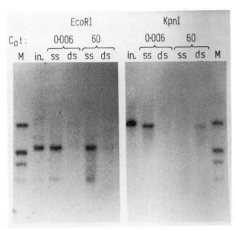

Figure 3. Repetitive DNA on DNA fragments containing β-globin sequences. Rabbit DNA was cleaved with restriction endonuclease *Kpn*I or *Eco*RI, heat-denatured, and annealed to the $C_0 t$ indicated. The samples were then chromatographed on HAP to give single-stranded and (partially) double-stranded DNA fractions. DNA was recovered from these fractions, denatured with alkali, subjected to gel electrophoresis, and assayed for β-globin genes as described in Experimental Procedures and in Jeffreys and Flavell (1977). Each slot contains the equivalent of 30 μg of digested rabbit DNA before fractionation.

after annealing to a $C_0 t$ of 60 (data not shown). These experiments therefore localize a segment of repetitive DNA in the region extending from 0.7 to 2.1 kb on the 3' side of the β-globin gene of rabbit.

Oligo(dA· dT) Clusters Flanking the β-Globin Gene

A specific repetitive sequence that is found in the DNA of all higher organisms is a tract of nonalternating oligo(dA· dT) of length 12–25 base pairs. These sequences can easily be characterized by hybridizing labeled poly(U) (Shenkin and Burdon 1974; Bishop et al. 1974) or poly(A) (Mol and Borst 1976) to the clusters. There are probably about 10^5–10^6 of these clusters per mammalian genome, and at least some of them are transcribed into the oligo(A) (Nakazoto et al. 1974; see also Dubroff and Nemer 1975) and oligo(U) (Molloy et al. 1972) segments that are present in most hnRNAs. The pure $(dA· dT)_{25}$ tracts present in *Dictyostelium* DNA are transcribed into an oligo$(A)_{25}$ segment found at the 3' terminus of most, if not all, mRNA precursors (Jacobson et al. 1974). Although it is not yet clear whether this occurs in higher eukaryotes, some evidence has been presented to suggest that a transcribed oligo$(A)_{25}$ serves as a primer for poly(A) addition in nuclear pre-mRNAs (Edmonds et al. 1976). To investigate whether oligo(dA· dT) clusters are involved in the transcription and processing of defined mRNAs, we devised a method to isolate duplex DNA containing these (dA· dT) clusters (Flavell and Van den Berg 1975). This isolation is achieved by hybridizing the DNA in duplex form to Sephadex-bound poly(U) un-

der conditions where a (dA· dT)rU triple helix is formed. Under suitable conditions, only DNA containing (dA· dT) clusters is retained by the column, and this bound DNA can be further fractionated on the basis of the length of the (dA· dT) cluster it contains (Flavell et al. 1977). Analysis of the effect of DNA size on the percentage of binding of total rabbit DNA to poly(U)-Sephadex suggests that the clusters are spread throughout at least 80% of the rabbit genome with an average of one (dA· dT) cluster every 3 kb of DNA (Flavell et al. 1977).

To determine the linkage of (dA· dT) clusters to the rabbit β-globin gene, we digested rabbit DNA with restriction endonuclease *Pst*I, *Kpn*I, or *Eco*RI. The digested DNAs were then applied to poly(U)-Sephadex columns, and DNA fragments containing (dA· dT) clusters were separated from DNA fragments which lack such clusters. The DNA was then recovered, electrophoresed in agarose gels, transferred to a nitrocellulose strip, and hybridized with ^{32}P-labeled pβG1 DNA to detect the β-globin DNA fragments. The 3' *Eco*RI fragment (Fig. 4), and to a lesser extent the *Kpn*I and *Kpn*I-*Pst*I fragments (Fig. 5), showed a weak but significant binding to poly(U)-Sephadex. In contrast, the 6.3-kb *Pst*I fragment binds strongly to poly(U)-Sephadex, whereas the 2.6-kb 5' *Eco*RI fragment shows no detectable binding (Fig. 4). This localizes a (dA· dT) cluster on the 3' *Eco*RI fragment and further suggests that a second cluster, promoting tight binding to poly(U)-Sephadex, is located in the DNA segment between the 3' *Kpn*I site and the 3' *Pst*I site, i.e., between 2.1 and 4.8 kb from the 3' terminus of the mRNA coding sequences. There seems to be no (dA· dT) cluster in the 5' extragenic region up to the *Eco*RI cleavage site (2.3-kb 5' of the start of the β-globin gene). The localization of the repetitive DNA and oligo(dA· dT) clusters on the physical map is shown in Figure 6.

Analysis of Human Globin Genes

As a system to study the structure and expression of globin genes, the human non-alpha-globin gene cluster offers several advantages over the rabbit genes. The changes in hemoglobin types during embryonic development are well described, and genetic evidence is available showing linkage of the four non-alpha-globin genes in the order GγAγδβ (see Wetherall and Clegg 1972). Moreover, certain thalassemias (inborn errors in which the synthesis of globin polypeptide chains is either reduced or absent) have the characteristics of control element mutants. As a first step in the analysis of these mutations, we have applied our β-globin-gene mapping technology to human DNA. Human placenta DNA was cleaved with restriction endonucleases *Eco*RI, *Pst*I, *Bgl*II, and *Bam*HI, electrophoresed on agarose gels, transferred to nitrocellulose filters, and hybridized with the heterologous ^{32}P-labeled pβG1 probe

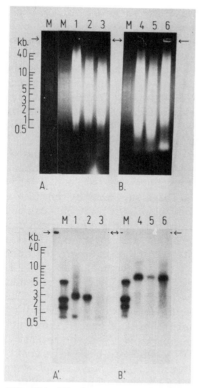

A.   B.

A'.   B'.

Figure 4. Oligo(dA·dT) clusters on DNA fragments containing β-globin sequences. Rabbit DNA was cleaved with restriction endonuclease EcoRI or PstI and fractionated on poly(U)-Sephadex to obtain a bound fraction, which consists of DNA containing (dA·dT) clusters, and an unbound fraction, which contains DNA with no such clusters. *(A,B)* UV photographs of the 1.2% agarose slab gel (stained with ethidium bromide) in which the denatured column fractions were electrophoresed. The DNA from the gel was then transferred to a nitrocellulose filter by blotting and hybridized to ³²P-labeled pβG1 DNA to detect the β-globin gene fragments as usual. *(A',B')* Autoradiogram of the filter. (1) 25 μg rabbit DNA × EcoRI input; (2) rabbit DNA × EcoRI, fraction not bound to poly(U)-Sephadex; (3) rabbit DNA × EcoRI, fraction bound to poly(U)-Sephadex; (4,5,6) corresponding fractions from PstI-cleaved rabbit DNA. *(M)* contains marker phage T7 DNA, phage ϕ29 DNA, and ϕ29 DNA cleaved with EcoRI, plus hybridization markers consisting of 10 pg pβG1 DNA cleaved with HindIII and 30 pg pβG1 DNA cleaved with EcoRI, HindIII, and AvaI.

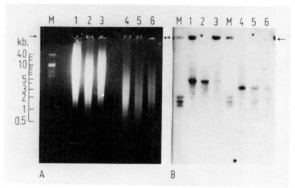

A   B

Figure 5. Oligo(dA·dT) clusters on DNA fragments containing β-globin sequences. Rabbit DNA was cleaved with restriction endonuclease KpnI or KpnI + PstI, fractionated on poly(U)-Sephadex, and analyzed for β-globin gene fragments as described in Experimental Procedures and Fig. 4. *(A)* UV photograph of the gel; *(B)* autoradiogram of the nitrocellulose strip showing the β-globin gene fragments. (1) 30 μg rabbit DNA cleaved with KpnI input; (2) rabbit DNA × KpnI, fraction not bound to poly(U)-Sephadex; (3) rabbit DNA × KpnI, fraction bound to poly(U)-Sephadex; (4,5,6) corresponding fractions from rabbit DNA cleaved with KpnI + PstI. *(M)* contains the same phage markers as in Fig. 4 plus 30 pg pβG1 DNA × EcoRI × AvaI as hybridization markers.

unequivocally the genes present on each band by comparison of these banding patterns with those of DNA from patients with altered globin genes caused either by point mutation or gene fusion.

DISCUSSION

Our results show that rabbit DNA cleaved with various restriction endonucleases contains a limited number of different discrete components which can form well-matched hybrids with the β-globin cDNA plasmid pβG1. The number of components seen correlates with the absence or presence of a cleavage site within the β-globin gene. Thus two contiguous

under the same conditions used for the rabbit DNA. As shown in Figure 7, a discrete number of β-globin-related bands are again seen. In particular, PstI and BglII give two hybridizable components. Since neither enzyme cleaves the human β-globin gene (N. Proudfoot, pers. comm.), these bands are most likely to correspond to fragments containing intact β-related globin genes such as the human β and δ genes. EcoRI and BamHI give at least three bands, consistent with the presence of cleavage sites for both these enzymes in the β-globin gene (Marotta et al. 1976). We are at present establishing the maps of cleavage sites around these globin genes and hope to identify

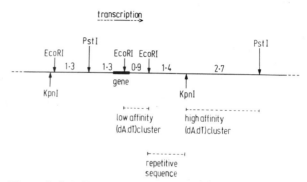

Figure 6. Localization of repetitive DNA and oligo(dA·dT) clusters on the physical map of restriction enzyme cleavage sites flanking the rabbit β-globin gene. The distances between restriction enzyme cleavage sites are given in kilobases.

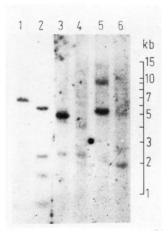

Figure 7. Human DNA fragments which can form hybrids with pβG1 DNA. Human placenta DNA was cleaved with various restriction enzymes, denatured with alkali, subjected to gel electrophoresis, transferred to nitrocellulose filters, and hybridized with pβG1 DNA as described by Jeffreys and Flavell (1977). The filters were washed as described for filters containing rabbit DNA (Jeffreys and Flavell 1977) except that the final wash was in 0.5 × SSC at 65°C instead of 0.1 × SSC. (1) Rabbit DNA × PstI; (2) 10 pg pβG1 DNA × HindIII + 25 pg pβG1 DNA × EcoRI × HindIII × AvaI hybridization markers; (3) human DNA × PstI; (4) human DNA × EcoRI; (5) human DNA × BglII; (6) human DNA × BamHI. Human or rabbit DNA (30 μg) was applied per slot.

components are generated by EcoRI which cleaves the β-globin gene once, whereas endonucleases PstI and KpnI, neither of which cleaves the gene, give a single band. This strongly suggests that these rabbit DNA fragments contain the β-globin structural gene homologous to the β-globin cDNA insert in pβG1 DNA. However, it remains possible that there are β-globin DNA fragments not detected in this analysis. For example, large fragments might not be transferred efficiently out of agarose gels into nitrocellulose filters, and very small fragments might not be detected efficiently on such filters (Southern 1975). We find, in fact, that the small (0.9 kb) β-globin DNA component generated by digestion with endonuclease EcoRI is frequently difficult to detect and can become lost in the general background labeling on filters. This faint labeling may also be a consequence of the relatively short stretch of β-globin DNA in this 3′ gene fragment since the EcoRI site is asymmetrically placed in the β-globin gene 170 base pairs from the 3′ end of the mRNA coding sequence.

Our estimates of DNA fragment sites must be regarded as approximate since they were determined from the electrophoretic mobility of single-stranded DNA in nondenaturing agarose gels (see Experimental Procedures). Aberrant mobilities are possible as a consequence of unusual secondary structures, although all marker DNAs examined have given a completely linear plot of log (fragment size) versus mobility over the range 0.8–10 kb. We have also electrophoresed native rabbit DNA cleaved with PstI, KpnI, EcoRI, or PstI plus EcoRI, followed by denaturation, transfer to a nitrocellulose filter, and assaying for the fragments containing the β-globin gene as described. We find the same pattern of β-globin-DNA-containing fragments as we observed with the single-stranded DNA gels, although labeled bands were frequently distorted as a result of overloading of the agarose gel.

Much evidence has been accumulated over the years to suggest that there is one β-globin structural gene per haploid genome (see, e.g., Bishop and Freeman 1974). This is supported by the observation that our major β-globin DNA fragments can be fitted into a physical map around a single β-globin structural gene homologous to the insert in pβG1. It is of course still possible that there is more than one β-globin gene per haploid genome, each gene being surrounded by identical patterns of endonucleases EcoRI, PstI, and KpnI cleavage sites. This possibility seems unlikely, however, when the extensive gene quantitation data (e.g., Bishop and Freeman 1974) are considered. Also, the intensity of rabbit β-globin-DNA bands we observe is about the same as the intensity of the band found when a single copy of a 1.5-kb piece of pMB9 is added to rabbit DNA digests as an internal control in our hybridizations. However, we clearly find other components capable of forming less stable hybrids with pβG1 DNA which we consider likely to be β-related globin genes in the rabbit, reminiscent of the δ and γ globins in the human. We are at present characterizing these genes in more detail, especially with respect to their physical linkage on the chromosome.

The physical map we have constructed spans a total of about 8 kb around the β-globin gene, and we have already been able to localize a segment of repetitive DNA and a (dA·dT) cluster(s) within this map. The significance of these observations, however, depends to a great extent on the positioning of the primary transcript of the rabbit β-globin gene on this physical map. P. J. Curtis and C. Weissmann (1976, and pers. comm.) have provided evidence for a 15S precursor (ca. 1.8 kb) to mouse β-globin mRNA which subsequently appears to be processed to give mature mRNA. A similar putative precursor has been reported in duck (MacNaughton et al. 1974). If such a precursor were produced in rabbit, then this would have to be contained within the boundaries of our physical map. However, R. N. Bastos and H. Aviv (pers. comm.) have evidence for a 27S (about 5 kb) globin mRNA precursor in mouse which has not yet been assigned to the α or β genes.

Despite these uncertainties, however, certain conclusions can be drawn at this stage. First, the entire 8-kb region in our map contains no detectable foldback DNA. Since processing sites for RNA molecules must exist in this region (e.g., 15S–9S mRNA), it seems unlikely that this processing must occur via a substantial hairpin structure, as has been suggested (see, e.g., Georgiev et al. 1974), which would be coded for by foldback DNA. Second, the DNA region extending more than 2 kb on the 5' side of the β-globin gene appears to consist of DNA with a reiteration frequency of less than 100 and is quite conceivably single-copy DNA. At present we can find no evidence for repetitive DNA in this region which might be expected to fulfill a regulatory function of the type proposed by Britten and Davidson (1969) or Georgiev et al. (1974). We are performing a detailed analysis of this region in a search for DNA of a low repetition frequency. Third, there is a repetitive DNA segment distal to the gene on the 3' side. Since this repeated sequence seems to be completely annealed at a C_0t of 60, we anticipate that it must have a repetition frequency on the order of 10^3–10^4 per haploid genome. We cannot at this stage determine whether this region is involved with the transcriptional unit of the β-globin gene.

Finally, we have localized (dA·dT) clusters in the extragenic region on the 3' side of the gene. A cluster promoting weak binding (up to about 50% under our standard conditions) is found in the region extending from the internal EcoRI site in the β gene to a region about 700 base pairs into the 3' flanking region. A number of arguments suggest that binding to poly(U)-Sephadex requires the presence of (dA·dT) clusters of about 12 nucleotides in length (Flavell and Van den Berg 1975; Flavell et al. 1977). The partial binding observed suggests that the cluster present on this fragment is probably of this minimal length, i.e., 10–12 base pairs. Such a cluster is not present in the β-globin gene sequence determined by Efstratiadis et al. (1977), although an imperfect cluster containing nine (dA·dT) base pairs interrupted by three internal base pairs is present in the region immediately next to the poly(A) tail. Although we have no evidence to exclude the possibility that this site is responsible for this weak binding, we consider it more likely that a longer (dA·dT) cluster would be required. The binding of the KpnI and KpnI plus PstI double-digest fragments is significantly lower than that of the 3' EcoRI fragment. It is likely that the binding of all of these fragments is mediated by the (dA·dT) cluster located on the 3' EcoRI fragment; we attribute the reduced binding efficiency of the KpnI and KpnI plus PstI DNA fragments to steric hindrance caused by the increased length of these DNA fragments. The 6.3-kb PstI β-globin fragment binds tightly to poly(U)-Sephadex, which suggests that a further binding site is present in the region starting at the 3' KpnI site, 2.1 kb from the 3' terminus of the gene, and terminated by the 3' PstI site, 4.8 kb from the 3' terminus. We are at present determining the affinities of the PstI and 3' EcoRI fragments for poly(U)-Sephadex to provide further evidence for two different (dA·dT) clusters in the extragenic region on the 3' side of the gene.

Figure 6 shows the tentative localization of these structures on the physical map. All the unusual sequences appear to be located 3' of the rabbit β-globin gene; until now, the region extending 2.3 kb 5' of the gene has not been shown to contain any unusual sequences. It is as yet unclear what the significance of these findings is, with respect to the mode of expression of the β-globin gene. Two approaches should clarify this issue. First, other globin genes would be expected to be linked to the same types of DNA sequences if these flanking DNA sequences were essential for the functioning of these genes. Since we can detect the human β-related genes using our methodology (Fig. 7), we are currently testing whether the same repetitive sequences are linked to these genes. Second, the availability of β-globin DNA fragments cloned from genomic DNA, together with a suitable assay system (either in vivo or in vitro) for gene function of the isolated fragments, will make it possible to ask whether such sequences are essential for transcription and processing to produce active β-globin mRNA. In vitro site-directed mutagenesis of the sequences of interest, either by deletion (Mertz et al. 1975) or by single base changes (Flavell et al. 1974), should ultimately relate these enigmatic sequences to gene function in higher organisms.

Acknowledgments

We thank Ernie De Boer and Jan M. Kooter for excellent technical assistance, and Profs. P. Borst and C. Weissmann and Drs. R. I. Kamen, O. L. Destree, and H. F. Tabak for helpful discussions. We are grateful for the gifts of restriction endonucleases from Mrs. F. Fase-Fowler (PstI, KpnI), Dr. L. A. Grivell (HindIII), Prof. C. Weissmann, Zurich (EcoRI), Dr. E. Humphries, ICRF, London (BglII), and Dr. M. Van Montagu, Ghent (AvaI). We also thank Mr. P. De Greve and his staff (Centraal Laboratorium van de Bloedtransfusiedienst, Amsterdam) for generous gifts of rabbit liver, and Dr. P. E. Treffers (Wilhelmina Gasthuis, Amsterdam) for gifts of human placenta. A. J. J. is a Postdoctoral Fellow of the European Molecular Biology Organization. This work was supported in part by a grant to P. Borst from the Netherlands Foundation for Chemical Research (SON) and with financial aid from The Netherlands Organization for the Advancement of Pure Research (ZWO).

REFERENCES

BISHOP, J. O. and K. B. FREEMAN. 1974. DNA sequences neighboring the duck haemoglobin genes. *Cold Spring Harbor Symp. Quant. Biol.* **38**: 707.

BISHOP, J. O., M. ROSBASH, and D. EVANS. 1974. Polynucleotide sequences in eukaryotic DNA and RNA that form ribonuclease resistent complexes with polyuridylic acid. *J. Mol. Biol.* **85**: 75.

BRITTEN, R. J. and E. H. DAVIDSON. 1969. Gene regulation for higher cells: A theory. *Science* **165**: 349.

BRITTEN, R. J., D. E. GRAHAM, and B. R. NEUFELD. 1974. Analysis of repeating DNA sequences by reassociation. *Methods Enzymol.* **29E**: 363.

CARROLL, D. and D. D. BROWN. 1976a. Repeating units of *Xenopus laevis* oocyte-type 5S DNA are heterogeneous in length. *Cell* **7**: 467.

————. 1976b. Adjacent repeating units of *Xenopus laevis* 5S DNA can be heterogeneous in length. *Cell* **7**: 477.

CLARKSON, S. G. and V. KURER. 1976. Isolation and some properties of DNA coding for tRNA^{met} from *Xenopus laevis.* *Cell* **8**: 183.

CURTIS, P. J. and C. WEISSMANN. 1976. Purification of globin messenger RNA from dimethylsulphoxide induced Friend cells: Detection of a putative globin messenger RNA precursor. *J. Mol. Biol.* **106**: 1016.

DAVIDSON, E. H., B. R. HOUGH, W. H. KLEIN, and R. J. BRITTEN. 1975. Structural genes adjacent to interspersed repetitive DNA sequences. *Cell* **4**: 217.

DUBROFF, L. M. and M. NEMER. 1975. Molecular classes of heterogeneous nuclear RNA in sea urchin embryos. *J. Mol. Biol.* **95**: 455.

EDMONDS, M., H. NAKAZOTO, E. L. KORWEK, and S. VENKATESAN. 1976. Transcribed oligonucletide sequences in HeLa cell hnRNA and mRNA. *Prog. Nucleic Acid Res. Mol. Biol.* **19**: 99.

EFSTRATIADIS, A., F. KAFATOS, and T. MANIATIS. 1977. The primary structure of rabbit β globin mRNA as determined from cloned DNA. *Cell* **10**: 571.

FLAVELL, R. A. and F. M. VAN DEN BERG. 1975. The isolation of duplex DNA containing (dA·dT) clusters by affinity chromatography on poly(U) Sephadex. *FEBS Lett.* **58**: 90.

FLAVELL, R. A., F. M. VAN DEN BERG, and G. C. GROSVELD. 1977. Isolation and characterization of the oligo (dA·dT) clusters and their flanking DNA segments in the rabbit genome. *J. Mol. Biol.* **115**: 715.

FLAVELL, R. A., D. L. SABO, E. F. BANDLE, and C. WEISSMANN. 1974. Site directed mutagenesis: Generation of an extracistronic mutation in bacteriophage Qβ RNA. *J. Mol. Biol.* **89**: 255.

GEORGIEV, G. P., A. J. VARSHAVSKY, A. P. RYSKOV, and R. B. CHURCH. 1974. On the structural organization of the transcriptional unit in animal chromosomes. *Cold Spring Harbor Symp. Quant. Biol.* **38**: 869.

GROSS, K., W. SCHAFFNER, J. TELFORD, and M. BIRNSTIEL. 1976. Molecular analysis of the histone gene cluster of *Psammechinus miliaris.* III. Polarity and asymmetry of the histone coding sequences. *Cell* **8**: 479.

JACOBSON, A., R. A. FIRTEL, and H. F. LODISH. 1974. Transcription of polydeoxythymidylate sequences in the genome of the cellular slime mould *Dictyostelium discoideum. Proc. Natl. Acad. Sci.* **71**: 1607.

JEFFREYS, A. J. and R. A. FLAVELL. 1977. A physical map of the DNA regions flanking the rabbit β-globin gene. *Cell* **12**: 429.

KEDES, L. H. 1976. Histone messengers and histone genes. *Cell* **8**: 321.

MACNAUGHTON, M., K. B. FREEMAN, and J. O. BISHOP. 1974. A precursor to hemoglobin mRNA in nuclei of immature duck red blood cells. *Cell* **1**: 117.

MANIATIS, T., G. K. SIM, A. EFSTRATIADIS, and F. C. KAFATOS. 1976. Amplification and characterization of a β-globin gene synthesized *in vitro. Cell* **8**: 163.

MAROTTA, C. A., B. G. FORGET, M. COHEN-SOLAL, and S. M. WEISSMAN. 1976. Nucleotide sequence analysis of coding and noncoding regions of human β-globin mRNA. *Prog. Nucleic Acid Res. Mol. Biol.* **19**: 165.

MERTZ, J. E., J. CARBON, M. HERZBERG, R. W. DAVIS, and P. BERG. 1975. Isolation and characterization of individual clones of simian virus 40 mutants containing deletions, duplications, and insertions in their DNA. *Cold Spring Harbor Symp. Quant. Biol.* **39**: 69.

MOL, J. N. M. and P. BORST. 1976. The binding of poly(rA) and poly(rU) to denatured DNA. II. Studies with natural DNAs. *Nucleic Acids Res.* **3**: 1029.

MOLLOY, G., W. L. THOMAS, and J. E. DARNELL. 1972. Occurrence of uridylate-rich oligonucleotide regions in heterogeneous nuclear RNA of HeLa cells. *Proc. Natl. Acad. Sci.* **69**: 3684.

NAKAZOTO, H., M. EDMONDS, and D. W. KOPP. 1974. Differential metabolism of large and small poly(A) sequences in the heterogeneous nuclear RNA of HeLa cells. *Proc. Natl. Acad. Sci.* **71**: 200.

SCHAFFNER, W., K. GROSS, J. TELFORD, and M. BIRNSTIEL. 1976. Molecular analysis of the histone-gene cluster of *Psammechinus miliaris.* II. The arrangement of the five histone-coding and spacer sequences. *Cell* **8**: 471.

SHENKIN, A. and R. H. BURDON. 1974. Deoxyadenylate-rich and deoxyguanylate-rich regions in mammalian DNA. *J. Mol. Biol.* **85**: 19.

SOUTHERN, E. M. 1975. Detection of specific sequences among DNA fragments separated by gel electrophoresis. *J. Mol. Biol.* **98**: 503.

WELLAUER, P. K., R. H. REEDER, D. CARROLL, D. D. BROWN, A. DEUTSCH, T. HIGASHINAKAGAWA, and I. B. DAWID. 1974. Amplified ribosomal DNA from *Xenopus laevis* has heterogeneous spacer length. *Proc. Natl. Acad. Sci.* **71**: 2823.

WETHERALL, D. J. and J. B. CLEGG. 1972. *The thalassaemia syndromes.* 2nd Ed. Blackwell, Oxford.

Gene Organization in *Drosophila*

A. Chovnick, M. McCarron, A. Hilliker, J. O'Donnell, W. Gelbart,* and S. Clark

Genetics and Cell Biology Section, Biological Sciences Group, The University of Connecticut, Storrs, Connecticut 06268

In recent years considerable attention has been directed to questions concerning the structure and function of elementary genetic units in higher organisms. Most of this effort is focused upon studies of whole genome DNA, RNA, and chromatin. Eventually, resolution of these issues will require analysis at the level of specific genetic units, with DNA genetically marked in order to permit identification of structural and regulatory components and to relate these components to their specific functions. Such an approach remains for the future. For some years, major emphasis in this laboratory has been directed toward the development of an experimental system for just such investigation in *Drosophila melanogaster*. This paper summarizes our recent progress in this work.

The Genetic System

The rosy locus in *D. melanogaster* (*ry*:3–52.0) is a genetic unit controlling xanthine dehydrogenase (XDH) activity, located on the right arm of chromosome 3 (Fig. 1) within polytene chromosome bands 87D8-12 (Lefevre 1971). The locus was originally defined by a set of brownish eye-color mutants deficient in drosopterin pigment. Such mutants were shown subsequently to exhibit no detectable XDH activity (Glassman and Mitchell 1959). Figure 2 summarizes reactions used in this laboratory to assay *Drosophila* XDH (Forrest et al. 1956; Glassman and Mitchell 1959). Of particular interest is the fact that zygotes possessing little or no XDH activity are unable to complete development and die before eclosion on standard *Drosophila* culture medium supplemented with purine (Glassman 1965). This fact is the basis for a nutritional selective procedure which has made large-scale genetic fine-structure mapping a routine laboratory exercise (Chovnick et al. 1971; Chovnick 1973).

It is now clear that *Drosophila* XDH is a homodimer (Gelbart et al. 1974) with a subunit molecular weight of 150,000 daltons (Andres 1976; Edwards et al. 1977). Two observations demonstrate that the structural information for XDH is encoded by the rosy locus: (1) Variation in dosage of *ry*+ alleles, from 0–3 doses, appears to be the limiting factor in deter-

mining the level of XDH activity per fly in otherwise wild-type flies (Grell 1962; Glassman et al. 1962). (2) The genetic basis for all of the variation in XDH electrophoretic mobility seen in wild-type strains maps to sites within the rosy locus map of XDH⁻, rosy eye-color mutants (Gelbart et al. 1976).

Cytogenetic Analysis of the Rosy Microregion

The rosy region of chromosome 3 has been the subject of extensive cytogenetic analysis in our laboratory. The precise segment which has been the major focus of this analysis extends from 87D2-4 to 87E12-F1, a chromosomal segment of 23 to 24 polytene chromosome bands (Fig. 1). A detailed description of this study will be reported elsewhere.

The raw material of this study is a group of 142 nonrosy, lethal, apparent site mutants, induced largely with EMS, but including some radiation-induced mutants. These were originally selected as being lethal with one or another of several chromosomes possessing deletions of part or all of this chromosomal segment. Then from a series of experiments involving (1) inter se complementation, (2) tests for lethality with chromosomes carrying smaller overlapping deletions of this region, and (3) some recombination tests, we have been able to identify and order a total of 21 lethal complementation groups within the 87D2-4 to 87E12-F1 segment of 3R. Thus, with the addition of the rosy locus, 22 complementation groups have been identified in this segment of 23 to 24 polytene bands, a result entirely consistent with the one-gene, one-chromomere hypothesis (Bridges 1935; Judd et al. 1972; Hochman 1974). Moreover, if we assume saturation of the map of this region and the validity of the one-gene, one-chromomere hypothesis, then we are able to further define the rosy locus to 87D10–12 (Fig. 1).

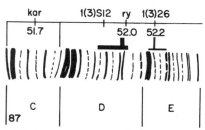

Figure 1. Chromosome map of the rosy region of chromosome 3 in *D. melanogaster* (Bridges 1941).

* Present Address: Cellular and Developmental Biology Group, Harvard University, 16 Divinity Avenue, Cambridge, Massachusetts 02138.

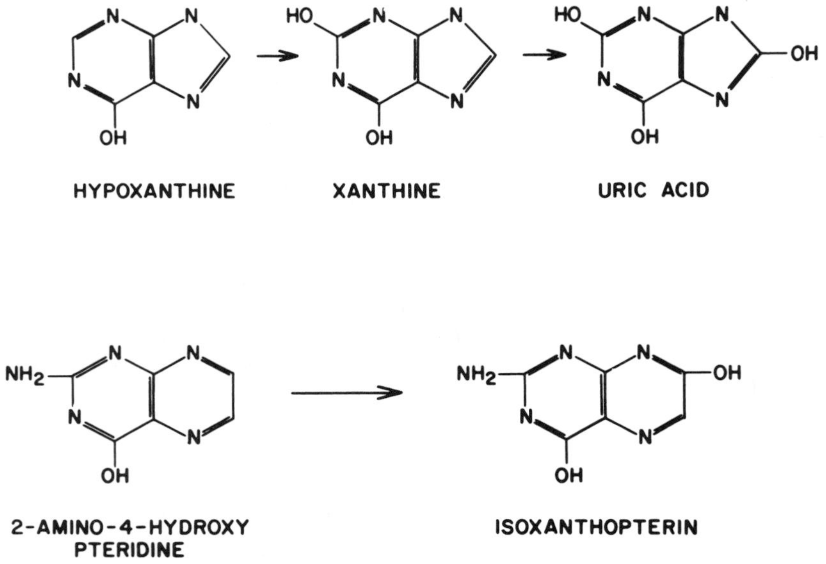

Figure 2. Reactions of *Drosophila* xanthine dehydrogenase.

An additional point of interest concerning the organization of the rosy locus derives from the following observations: (1) The analysis of a large number of EMS and radiation-induced rosy mutants has revealed that none are recessive lethals under standard nutritional conditions, with the exception of bona fide deletions which involve adjacent genetic units to varying extents. (2) All of the lethal site mutants in the rosy region belong to complementation groups that are functionally and spatially distinct from the rosy locus. Taken together, these observations support the contention previously drawn (Schalet et al. 1964; Chovnick 1966) that rosy is a discrete genetic unit, and not part of a functional complex involving any of the adjacent genes.

Rosy Locus Variation

Electrophoretic-mobility variants of XDH are readily isolated from laboratory stocks and natural populations of *D. melanogaster*. From these sources, we have established a number of wild-type isoalleles of the rosy locus. These are maintained as stable lines which possess XDH molecules with distinctive electrophoretic mobilities and thermal properties. Moreover, the XDH enzyme activity level associated with each of these wild-type alleles is also a distinctive, stable phenotypic character. Table 1 summarizes our present array of ry^+ isoalleles. The XDH produced by ry^{+0} serves as a mobility standard and is designated XDH$^{1.00}$. Under standard conditions of electrophoresis (McCarron et al. 1974), all variant XDHs that are slower are designated by relative mobility values less than 1.00, and faster XDHs are designated by superscripts greater than 1.00. The ry^{+4} allele is associated with sharply higher XDH activity than all others and is classified in Table 1 as high (H); ry^{+10} exhibits much lower activity than

all others and is classified as low (L). The activity levels of the remaining alleles are representative of intermediate levels which we presently classify as normal (N).

Enzymatically inactive rosy mutants are readily selected by virtue of their visibly mutant rosy eye-color phenotype. Over a period of many years, X-ray, γ-ray, and EMS mutagenesis experiments have provided us with a large number of such mutants. We have adopted the convention of labeling each mutant with a superscript which identifies the ry^+ isoallele from which the mutant was derived. Thus the XDH$^-$ rosy eye-color mutant, ry^{402}, is the second variant isolated from ry^{+4}, and ry^{1201} is the first mutant derivative of the ry^{+12} allele. Although all of these mutants are XDH$^-$ and exhibit a rosy mutant eye color when homozygous, examination of mutant heterozygote combinations reveals a pattern of allele complementation. The XDH$^-$ mutants fall into eight complementation classes which form a circular map (Fig. 3). Complementing heterozygotes exhibit normal or near-normal eye-color phenotypes, but the restored XDH activity levels are quite low ($<$1–16% of wild-type). Such complementing mutant genotypes also die on purine-supplemented media (Gelbart et al. 1976).

Table 1. Wild-type Isoalleles of Rosy

ry^+ alleles	Mobility	XDH activity[a]
+12, +13	0.90	N
+14	0.94	N
+10	0.97	L
+0, +6, +7, +8	1.00	N
+1, +11, +16	1.02	N
+4	1.02	H
+2	1.03	N
+3, +5	1.05	N

[a] N = normal, H = high, L = low.

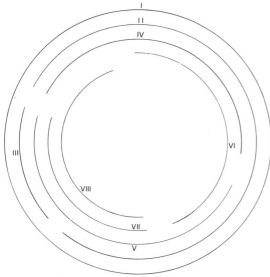

Figure 3. The complementation map of the rosy locus. Mutants are divided into complementation groups I through VIII. Complementation groups whose lines do not overlap are complementary. The mutants in each group are as follows: (I) 56 mutants: *1, 3a, 4, 5, 6, 7, 9, 17, 18, 19, 20, 21, 23, 24, 26, 40, 41, 45, 48, 56, 57, 58, 59, 61, 62, 63, 64, 102, 103, 106, 110, 201, 203, 204, 205, 206, 208, 209, 210, 301, 404, 405, 502, 506, 601, 603, 604, 605, 607, 608, 1001, 1002, L.12, L.14, L.18;* (II) 1 mutant: *602;* (III) 2 mutants: *609, L.19;* (IV) 1 mutant: *606;* (V) 1 mutant: *60;* (VI) 1 mutant: *406;* (VII) 3 mutants: *42, 207, 501;* (VIII) 1 mutant: *1003.*

From the above discussion, one might infer the existence of another class of rosy locus mutants which, when homozygous, would exhibit normal or near-normal eye color, possess low levels of XDH activity, and die on purine-supplemented media. Just such a class of mutants has recently been added to the array of rosy locus variants. These are labeled with superscripts designed to identify the ry^+ isoalleles from which they were derived, and the superscripts are prefixed with *ps* (purine sensitive). Thus ry^{ps214}, the fourteenth-identified variant of ry^{+2}, is a purine-sensitive mutant associated with a low level of XDH activity and a wild-type eye color (Gelbart et al. 1976).

Intragenic Fine Structure

Through the use of purine-supplemented standard *Drosophila* medium, we have been able to carry out fine-structure recombination mapping on a scale approaching that of prokaryote studies. Figure 4A presents a summary map of separable, XDH⁻ mutant sites within the locus resulting from such experiments. Since none of these rosy mutant alleles (Fig. 4A) have been associated with a detectable altered XDH product, this map provides no information about the organization of the locus. Initially we believed that our map of null mutants represented an array of sites spread throughout the locus

including both structural and control regions. Our strategy, then, was to identify variants which clearly produced alterations in XDH structure and to locate the sites of such variation on our standard map of null enzyme mutants of the rosy locus. In this way, we hoped to place genetic limits on that portion of the rosy locus map which coded for XDH. Our first experiments were designed to locate the sites responsible for differences in electrophoretic mobility.

In the absence of a selective procedure, it is impractical to map these genetic sites directly. Instead, we induced null enzyme mutants upon each of the ry^+ isoalleles and then utilized the purine selective system to recover wild-type recombinants in mutant heteroallele mapping experiments. Wild-type recombinants recovered from experiments involving different mutants of the same ry^+ isoallele invariably exhibit the parental electrophoretic class of XDH (McCarron et al. 1974). Recombination experiments involving mutants induced on different ry^+ alleles permit us to follow the electrophoretic character as an unselected marker. Electrophoretic classification of the wild-type recombinant survivors of such experiments leads to the identification of sites responsible for the electrophoretic difference (McCarron et al. 1974). Following this logic, experiments were carried out to identify the genetic sites responsible for the electrophoretic-mobility differences between ry^{+0} (our standard wild-type allele) and the isoalleles ry^{+1}, ry^{+2}, ry^{+3}, ry^{+4}, and ry^{+5} (Gelbart et al. 1974). These sites (Fig. 4C) are designated by numerical superscripts preceded by the letter *e*. Thus ry^{e217} is the seventeenth-identified variant of the ry^{+2} allele. It is an electrophoretic site identified from recombination experiments involving ry^{200}-series mutants tested against the ry^{+0} allele. Since ry^{e217} is the site responsible for the increased mobility of the ry^{+2} enzyme, we describe ry^{+2} as carrying *e217F* while ry^{+0} is marked by the *e217S* alternative.

One feature of these early experiments was the broad distribution of these electrophoretic sites to opposite ends of the map. Clearly, the XDH structural information is not confined to a small portion of the map.

Further definition of the boundaries of the XDH structural element was accomplished by fine-structure mapping experiments that utilized two additional classes of structural variants within the rosy locus (Gelbart et al. 1976). The first class involved the null enzyme, rosy eye-color mutants that exhibit interallelic complementation (see above). The second class of structural mutants is derived from the group of purine-sensitive rosy mutants also discussed earlier. These mutants are associated with low levels of XDH activity, exhibit normal or near-normal eye color, and die on purine-supplemented medium. Within this group we were able to identify a subclass of purine sensitives that we believe to be alterations in the XDH structural element since they possess

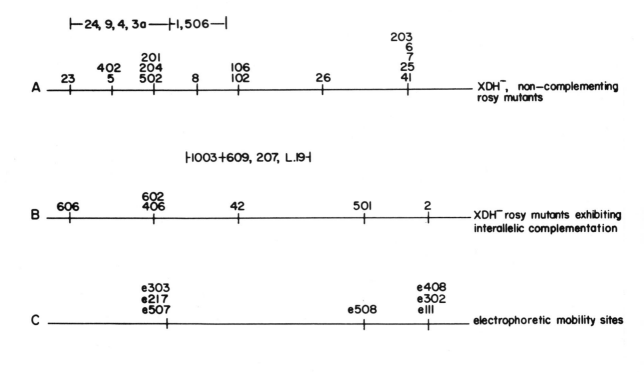

Figure 4. Genetic fine-structure maps of rosy locus sites. Map locations of unambiguous structural-element variants *(B, C, D)* are positioned relative to map of XDH⁻ noncomplementing mutants *(A).*

normal or near-normal levels of a protein that reacts with XDH-specific antibodies (CRM).

Since both the allele-complementing mutants and purine sensitives die on purine-supplemented medium, they are mapped directly against each other and against mutant sites on our standard map. Figure 4 summarizes the present state of our map of the rosy locus relative to the boundaries of the XDH structural element. At the right end of the map, several electrophoretic sites and the complementing mutant, *ry²*, identify the right border of the structural element with no known XDH structural variants beyond them. The complementing rosy eye-color mutant, *ry⁶⁰⁶*, and the purine-sensitive variant, *ry^{ps 214}*, are the leftmost unambiguous structural variants, with *ry⁶⁰⁶* definitely to the left of *ry^{ps 214}*. On the basis of comparative recombination data and the failure of large-scale tests with *ry⁶⁰⁶* to produce recombinants, we conclude that the leftmost member of the standard map, *ry²³*, must also mark the left border of the structural element.

The maps of Figure 4 position 41 sites to the right of our present left boundary of the XDH structural element. Moreover, an additional ten sites not indicated, in fact, map in the structural element. We

hope eventually to relate the genetic boundaries described in Figure 4 to the amino and carboxy termini of the XDH polypeptide chain. However, the extensive genetic data upon which we base the present left boundary suggest that, at least for this end of the structural element, we are close to a terminus.

Distribution of Single-site Inactivations

From an examination of the maps of Figure 4, we now see that the structural element boundaries include the entire array of null-XDH mutants. We have no XDH⁻ mutant sites, either radiation- or EMS-induced, which fall outside this region and which might be candidates for control-element alterations. This fact stands in striking contrast to our expectation at the beginning of this study. We assumed that the distribution of mutants would directly reflect the relative sizes of the control and structural elements. Consider the following: (1) Extensive analyses of EMS-induced mutants in *D. melanogaster* reveal that most, if not all, such mutants are single-site lesions (Lim and Snyder 1974; Hilliker 1976). (2) In contrast to conventional belief, the vast majority of wholly intragenic X-ray-induced mutations are single-site lesions (Gutz 1961, 1963; Hartman et al. 1971; Malling and DeSerres 1973). (3) With the exception of the spontaneous mutants ry^1 and ry^2, all of the rosy eye-color mutants included in Figure 4A and B were recovered from X-ray, γ-ray, and EMS mutagenesis experiments. Moreover, they were included in the fine-structure analysis because there was no evidence to suggest that they were associated with deletions or other rearrangements.

Thus, by virtue of the mutagens used and the selection of wholly intragenic rosy eye-color mutants (which exhibit complete loss of XDH function), the mapping study may be viewed more correctly as an investigation of the distribution (structural vs control) of single-site inactivations of gene function.

Variation in Intensity of XDH Activity

In addition to electrophoretic-mobility differences, we have already noted that the various ry^+ isoalleles are associated with variation in level of XDH activity, which also behaves as a stable phenotypic character. Consider the ry^{+4} and ry^{+10} lines which exhibit much greater and much less activity, respectively, than all of our other wild-type lines (Table 1). These differences are observed by following either the purine or the pteridine reaction (Fig. 2) and are readily classified in cuvette assays (spectrophotometry or fluorometry) or upon gel electrophoresis. Figure 5 illustrates typical fluorometric assays of XDH activity in matched, partially purified extracts of several ry^+ isoallelic stocks including both ry^{+4} and ry^{+10}. Measurements of XDH activity/mg protein, activity/fly, and activity/mg wet weight invariably yield similar relationships. Activity levels associated with ry^{+2}, ry^{+6}, and ry^{+11} exhibit a range of variation (Fig. 5) which we classify as normal (N) in Table 1 and which never overlaps the ry^{+4} and ry^{+10} extremes. Indeed, ry^{+10} homozygotes exhibit such low activity that they may be distinguished from other wild-type strains by their purine sensitivity. A detailed analysis of the basis for the ry^{+4} phenotype is presented elsewhere (Chovnick et al. 1976), and a report on the ry^{+10} character is now in preparation. Together these studies identify a *cis*-acting control element located adjacent to the left (centromere proximal) side of the XDH structural element. The following sections outline the experimental basis for this conclusion.

Further Characterization of the XDH-activity-level Variants

That variation in level of XDH activity (Table 1; Fig. 5) is a property of the rosy locus derives from conventional genetic analysis involving standard mapping procedures (Chovnick et al. 1976). However, the task of distinguishing whether the genetic bases for the level-of-activity differences reside in

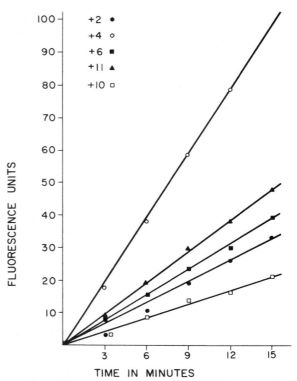

Figure 5. Fluorometric assay of XDH activities of matched extracts of the indicated homozygous wild-type isoallelic stocks.

the XDH structural element, or possibly in a control element, is not as simply resolved. Based upon fine-structure recombination analysis, some of our ry^+ isoalleles differ by as many as five or six structural element sites. Independent studies (Singh et al. 1976) of XDH structural gene polymorphism in natural populations additionally support the view that any two isolated lines will very likely possess structural element differences. Obviously, the genetic polymorphism exhibited by the XDH structural element might also extend into an adjacent control element. Certainly, level of enzyme activity is a superficial phenotypic character that might reflect either structural or control element variation. It thus behooves us to further define this character and to carry out high-resolution fine-structure mapping experiments. Final proof of the existence of a control element demands that putative control-element variants map to the locus of the genetic unit known to carry the structural information and additionally *map adjacent to, but outside the boundaries of, the structural element.*

We have considered the possibility that there are structural differences between the XDH molecules produced by ry^{+4} and those produced by other ry^+ isoalleles that are responsible for the sharply increased activity of ry^{+4} preparations. However, we have been unable to associate this activity difference in any systematic manner with electrophoretic, thermolability, or kinetic differences (Chovnick et al. 1976; Edwards et al. 1977). Similar comparisons involving XDH molecules produced by ry^{+10} and those produced by other ry^+ isoalleles have also failed.

On the other hand, immunological experiments clearly support the notion that the activity differences reflect differences in number of XDH molecules. One approach has used standard antiserum titration experiments. A series of dilutions of an antibody preparation against *Drosophila* XDH are tested for ability to remove XDH activity from matched extracts of the various ry^+ isoallele stocks. Still another approach has used the method of quantitative "rocket electrophoresis" (Laurell 1966; Weeke 1973) to compare these extracts. Both kinds of experiments indicate that ry^{+4} preparations contain more XDH molecules than do matched preparations from normal activity strains, and these, in turn, have more molecules of XDH than does ry^{+10}.

Developmental Profile of XDH

The time course of appearance of enzyme during development has been determined for a number of our wild-type isoallelic lines, as well as for certain interesting variants. Synchronous populations from the various lines are raised to the appropriate developmental stages, at which time extracts are examined for XDH activity, total protein, and response to XDH antibody (CRM). Thus we are able to examine and compare XDH activity per individual, per milligram of protein, or in terms of CRM levels throughout development.

Figure 6 illustrates a normal developmental profile of XDH in the ry^{+0} line. Maximum XDH activity is seen on day 10, and the data are presented as the percent of maximum activity seen at other times during development. No gross differences have yet been detected among the various strains. Although ry^{+4} exhibits relatively increased amounts of enzyme throughout development and ry^{+10} is associated with reduced enzyme, their developmental profiles are otherwise normal.

Genetic Fine-structure Experiments

Let us now consider the genetic bases for the level of enzyme-activity differences described above. The first experiments involve recombination tests of XDH$^-$ rosy eye-color mutants of the ry^{+4} allele against established site mutants within the XDH structural element. Five series of tests were carried out which sampled a total of 4.7×10^6 progeny and yielded 35 crossovers. Figure 7 summarizes the results of these experiments which are divided into two classes in terms of the distribution of rosy mutant sites in the heterozygote. Heterozygote A of Figure 7 illustrates a test of ry^x/ry^{40y}, where ry^x is located to the right of ry^{40y}. Heterozygote B of Figure 7 illustrates a test involving ry^z/ry^{40q}, where ry^z is

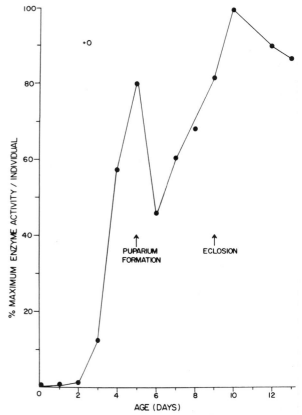

Figure 6. Developmental profile of XDH in ry^{+0} homozygotes.

located to the left of ry^{40q}. In all of these experiments, there is an additional heterozygous site located near the right end of the XDH structural element. This site, designated e^S or e^F in the figure, is responsible for the electrophoretic-mobility difference between the parental wild-type isoalleles (XDH$^{1.00}$ and XDH$^{1.02}$). Consider now the ry^+ recombinant products that are associated with exchange for the flanking outside markers a and b. For heterozygote A each exchange product will recombine the left portion of the ry^x-bearing strand with the right section of the ry^{40y} chromosome. As noted (Fig. 7), all of these exchanges are $a\ ry^+\ +$, and, upon electrophoresis, exhibit an XDH$^{1.02}$ with normal (N) levels of activity. Complementary results are seen in experiments involving heterozygote B, where all crossovers are $+\ ry^+\ b$ and exhibit an XDH$^{1.00}$ with high (H) levels of activity. Fourteen of the 35 crossovers were recovered from tests of heterozygotes of type A, and the remaining 21 were recovered from type-B heterozygotes. It should be noted that in all of these experiments *no intermediates or nonparental classes* with respect to XDH activity level were recovered. Clearly, these experiments identify a site responsible for the variation in intensity of XDH activity. It is different and separable from the electrophoretic site. We designate this

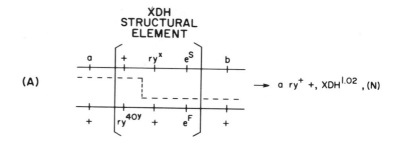

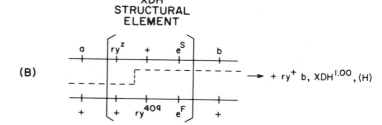

Figure 7. Genetic fine-structure analysis of crossovers within the XDH structural element. Initial experimental localization of the *i409* site.

site as *i409*, with *ry^{+4}* carrying *i409H* and other wild-type isoalleles as *i409N*. Moreover, from this group of 35 crossovers representing exchange points across the XDH structural element, we note that the genetic basis for the difference in intensity of XDH activity *(i409)* falls to the left of all crossover points, whereas the electrophoretic site is located to the right of all exchange points.

Another point should be noted about these observations. Two of the crosses, involving type-B heterozygotes (Fig. 7), tested the leftmost known *ry^{400}*-series mutant against *ry^z* mutants which represent markers for the left border of the XDH structural element. These recombination tests span a region which we estimate to represent the leftmost one-tenth of the XDH structural element map. Yet all of these crossovers (six out of the total study) place the site of *i409* to the left of their exchange points. From this we infer that *i409* is located either to the left of the structural element or just inside the left border, and very close to the cluster of rosy mutant sites at that end of the map (Fig. 4).

One final point should be noted about the experiments described in Figure 7. Progeny survive if they carry a chromosome resulting from an event leading to the production of a *ry^+* recombinant. Among the recombinants are numerous conversions of the XDH^- rosy mutants. These events provide additional information about the location of *i409*. In these ex-

periments, conversions of XDH^- → XDH^+ sites are selected events, whereas *i409* represents an unselected heterozygous marker. We may now ask about the frequency with which conversion of a given mutant allele is accompanied by coconversion of the unselected *i409* site. The significance of the phenomenon of coconversion was recognized in fungal studies as demonstrating that conversion events involve variable-sized segments of DNA. Fogel et al. (1971) have studied frequency of coconversion as a function of distance between the sites and have shown a linear relationship inversely proportional to distance. Although the rosy locus data are not as extensive, our observations are entirely consistent with the fungal data. This point is illustrated in Table 2. From a large number of experiments, observations on the conversion of various rosy mutants are tabulated. Only those experiments are included where coconversion of an unselected electrophoretic site, *e217*, could be scored. A comparison of the map positions of the mutant sites with that of *e217* (Fig. 4) reveals that coconversion frequency is inversely proportional to distance.

Now, consider conversion observations on rosy mutants located at the left end of the XDH structural element map (Table 3) in which the recombination experiment provided opportunity to observe coconversion for the *i409* site. If the *i409* site were located just inside the structural element left border,

Table 2. Coconversions of *e217* in Association with Conversions of the Indicated Mutants

Mutant allele	*ps*214	5	201	204	406	8	106	203	41
Conversions observed	14	6	3	8	3	10	5	7	17
Coconversions of *e217*	3	3	3	7	2	5	0	0	1

Table 3. Coconversions of *i409* in Association with Conversions of the Indicated Mutants

Mutant allele	606	23	*ps*214	402	5	406
Conversions observed	6	8	1	6	4	8
Coconversions of *i409*	1	1	0	0	0	0

then frequent coconversions of *i409* would be expected with conversion of these neighboring sites. In fact, a low frequency of coconversion (2/14) is seen only with the two rosy mutants which mark the present left border of the structural element (*ry*[23], *ry*[606]), and there were no coconversions seen among 19 conversions of mutant alleles just inside that boundary (*ry*[ps214], *ry*[402], *ry*[5], *ry*[406]). Taken together, interpretation of both crossover and coconversion data would place *i409* well to the left of the XDH structural element (Fig. 8).

Confirmation of the Position of *i409*

Further experiments were designed to locate *i409* relative to *l(3)S12*, a marker to the left of rosy (Fig. 8), and the XDH structural element. Note that, despite its proximity to the rosy locus (0.0054 map units), *l(3)S12* has no effect upon rosy locus function. Rather, we believe it to be a site mutant in the next genetic unit to the left of rosy. These experiments selected for survival crossovers between *l(3)S12* and an XDH⁻ rosy mutant located near the left border of the structural element. If *i409* lies to the left of the structural element, then the exchange events involved heterozygotes of the composition $\frac{l(3)S12\ i409N\ +}{+\ i409H\ ry}$, which were mated in a selective system cross designed to kill all *l(3)S12*- and/or *ry*-bearing meiotic products of the heterozygote. Since recombination to generate meiotic products that are *l(3)S12*⁺, *ry*⁺ may take place on either side of the *i409* site, survivors should fall into two classes with respect to level of XDH activity. On the other hand, if *i409* were located within the XDH structural element (i.e., to the right of the rosy mutant marker), the test heterozygote would be $\frac{l(3)S12\ +\ i409N}{+\ ry\ i409H}$, and selection for *l(3)S12*⁺, *ry*⁺ recombinants would be expected to yield only *i409N*-bearing products.

In fact, a total of 16 crossovers were recovered in 1.18 × 10⁶ sampled progeny, and these fell into

the two classes *i409N* (6) and *i409H* (10). These results therefore provide a very strong confirmation of the localization of *i409* inferred from the recombination data presented in the previous section (Fig. 8).

The *ry*[+10] Isoallele

A series of fine-structure recombination experiments were carried out that parallel those described in the previous sections. In these experiments, mutants of the *ry*[+0] allele, with known map positions, were tested against *ry*[+10]. It should be noted that in all of these experiments *no intermediate or nonparental classes* with respect to XDH activity level were recovered among the surviving recombinants. These recombinants identify a site responsible for the low XDH activity associated with the *ry*[+10] allele. We designate this site as *i1005*, with *ry*[+10] carrying *i1005L* (low) and our normal wild-type allele (*ry*[+0]) carrying *i1005N* (normal). Moreover, these crosses position *i1005* to the right of *l(3)S12* and to the left of *ry*[23], a mutant marker of the left end of the XDH structural element. Clearly, these experiments locate *i1005* to the same region as *i409*.

What is the relationship between *i1005* and *i409*? Might they be synonyms designating the same site, or might they mark separable sites? Heterozygotes of the composition $\frac{+\ i1005L\ +}{kar\ i409H\ ry^{406}}$ were mated to an appropriate tester strain, and their progeny reared on medium containing a level of purine sufficient to kill *i1005L*- or *ry*[406]-bearing zygotes. (There were additional markers in the cross to facilitate diagnosis of all survivors.) In a total of 1.25 × 10⁶ progeny sampled, there were 16 surviving *ry*⁺ recombinants. Three were conversions of *ry*[406] and exhibited the *high* level of XDH activity associated with *i409H*. One was a conversion of *i1005L* and possessed a *normal* level of XDH activity. Finally, there were 12 *ry*⁺ progeny associated with exchange for the flanking markers. Nine of these possessed *high* activity, and the remaining three exhibited *normal* XDH activity. The results of this cross indicate that *i1005* and *i409* mark separable sites. However, this experiment is unable to determine the relative positions of the two sites for the following reason. Although it is clear that a *ry*⁺ chromosome carrying *i1005N i409N* will exhibit a normal level of XDH activity, the double variant, *i1005L i409H*, might also appear normal. However, an unambigu-

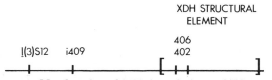

Figure 8. Map location of *i409* site relative to *l(3)S12* and the XDH structural element.

ous resolution of this question awaits the completion of a recombination test involving half-tetrad analysis.

Relationship between *i409*, *i1005*, and the Structural Element

In previous reports we have considered and eliminated the possibility that *i409* might mark a tandem duplication of the XDH structural element (Chovnick et al. 1976, 1977). Although the association of *i1005L* with a reduction in the number of XDH molecules per fly does not seemingly lend itself to the tandem duplication model, precedence for such consideration exists in the case of the Bar duplication in *D. melanogaster* and its associated position effect (Sturtevant 1925; Muller et al. 1936; Bridges 1936; Muller 1936). On such a notion, the ry^{+10} isoallele would be considered to possess two XDH structural elements in tandem. Moreover, because of the resulting change in position of each member of the duplex relative to some adjacent genetic element(s), a disturbed function of both XDH structural elements results. Such a model is precluded on two counts:

1. The ry^{+10} allele is associated with a single XDH electrophoretic band of the mobility class $XDH^{0.97}$ (Table 1). XDH is a homodimer, and the presence of two electrophoretically distinct structural elements will produce individuals possessing three XDH moieties. The tandem duplication model then requires that the ry^{+10} allele possess two XDH structural elements whose peptide products are indistinguishable and of the mobility class $XDH^{0.97}$. Thus *i1005L* should be associated with an $XDH^{0.97}$. On this point, the tandem duplication model fails. In experiments which recombine *i1005L* with an electrophoretically distinct structural element, there is no evidence of the production of an $XDH^{0.97}$ moiety.

2. Tandem duplications are characterized by instability in homozygotes due to increased incidence of unequal exchange events. The ry^{+10} stock has been quite stable. Moreover, fine-structure recombination experiments involving tests of *i1005L* against XDH^- mutants have been characterized by regular exchange events and the complete absence of unequal crossing over.

Having eliminated the possibility that either *i409* or *i1005* might be associated with a tandem duplication of the XDH structural element, it is appropriate at this point to consider that these separable sites mark one or more genetic elements that regulate XDH. Arguments in support of this point are:

1. We have succeeded in mapping more than 50 sites within a region clearly marked by unambiguous XDH structural variants (Fig. 4). The *i409* and *i1005* sites are the only sites which have mapped outside these boundaries.

2. Variation at *i409* and *i1005* leads to alteration in number of molecules of XDH per fly.

3. We now possess stocks carrying *i409N* and *i409H* with structural elements producing $XDH^{0.97}$, $XDH^{1.00}$, $XDH^{1.02}$, $XDH^{1.03}$, and $XDH^{1.05}$. Examination of these lines has failed to associate *i409* with any XDH structural characteristic.

4. On a more limited scale, we have similarly produced stocks carrying *i1005N* and *i1005L* with structural elements producing $XDH^{0.97}$ and $XDH^{1.00}$. Similarly, we are unable to associate *i1005* with any structural characteristic.

5. In previous reports (Chovnick et al. 1976, 1977), evidence was presented that demonstrated the *cis*-acting nature of *i409*. Thus, in the heterozygote $\dfrac{i409H ry^{+4}}{i409N ry^{+12}}$, there is an increased number of $XDH^{1.02}$ molecules (ry^{+4} product) and not an increase in the $XDH^{0.90}$ molecules (ry^{+12} product). Parallel experiments involving *i1005* similarly reveal the *cis*-acting nature of *i1005*.

On the basis of these observations, we are drawn to the possibility that *i409* and *i1005* mark the 5′ control element of the rosy locus. Although it would be premature to specify the nature of the control function associated with each of these sites, it would be reasonable to expect that variants of such a control element might exhibit alterations in DNA sequences which serve as binding sites for regulatory signal(s), sites for RNA polymerase binding and initiation of transcription, transcript processing sites, and sites for ribosome binding and initiation of translation.

Structural- and Control-element Size Estimates

Figure 9 summarizes our present map-length estimates for the rosy locus structural and control elements. The left end of the structural element is based upon extensive mapping experiments (Gelbart et al. 1976) and we anticipate no significant change. However, a similar test of the right end is yet to be accomplished. Moreover, the size of the control element is quite tentative. At present, we suggest that the distance from *i409* to the structural element (0.0034 map units) serve as a minimal estimate, and

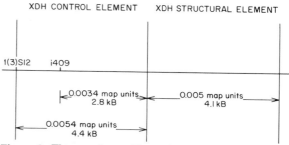

Figure 9. The rosy locus. Size estimates of the structural and control elements.

the distance from *l(3)S12* to the XDH structural element (0.0054 map units) be taken as a maximum estimate of the control element size (Chovnick et al. 1976).

Translation of these estimates into DNA base lengths proceeds from the XDH peptide molecular weight of 150,000 daltons. Assuming an average amino acid molecular weight (adjusted for peptide linkage) to be 110, then the length of DNA in the XDH structural element responsible for such a peptide is approximately 4.1 kb (150,000 × 3/110). Then, from the recombination map length of the structural element (0.005 map units), we may relate map length to physical length (0.01 map unit = 8.2 kb), which should be directly applicable to the adjacent control element. Thus the size of the control element is estimated to be 2.8–4.4 kb, and the total length of the rosy locus from 6.9 to 8.5 kb.

Another estimate of the length of DNA in the rosy locus may be derived directly from its genetic length (0.0084–0.010 map units). With 1.6×10^5 kb as the total genome DNA represented as single-copy and middle repetitive sequences (Rasch et al. 1971; Manning et al. 1975), and 275 map units as the total map length, then 0.01 map unit represents 5.8 kb. By this method, the rosy locus DNA length is estimated at 4.87–5.8 kb. Note that this method ignores regional differences in recombination frequency and that the rosy locus is subject to reduced recombination due to its centromere proximal position (Gelbart et al. 1976). In this context, we view the former method to yield a better estimate.

Figure 10 is a photograph of polytene chromosome segment 87D provided by Dr. George Lefevre. Although the rosy locus has not yet been assigned to a single polytene band, it has been restricted to a region possessing bands, whose haploid DNA content may be estimated by reference to Rudkin (1965). All of the bands that might be associated with rosy are much finer than the average (30 kb) but are certainly greater than the finest bands to be seen (5 kb).

Models of gene organization proposed as solutions of the chromomere paradox in higher organisms postulate control elements an order of magnitude larger than their structural components (Beermann 1972). Although the present data for the rosy locus provide little support for such a model, we note that the size estimate for the rosy locus control element is huge in comparison to comparable prokaryote elements.

Acknowledgments

The authors gratefully acknowledge the technical assistance of F. Johnston, H. Levine, J. Dutson, and L. Yedvobnick. The photograph of the rosy region of chromosome 3 was kindly provided by Dr. George Lefevre of the California State University, Northridge.

This investigation was supported by Research Grant GM-09886 from the Public Health Service and by Research Grant BMS 74–19628 from the National Science Foundation. Some of this work was conducted during the tenure (by S. C.) of a postdoctoral fellowship, F32 GM-05260, from the U.S. Public Health Service.

REFERENCES

ANDRES, R. Y. 1976. Aldehyde oxidase and xanthine dehydrogenase from wild-type *Drosophila melanogaster* and immunologically cross-reacting material from *ma-l* mutants. *Eur. J. Biochem.* **62**: 591.

BEERMAN, W. 1972. Chromomeres and genes. *Cell Differ.* **4**: 1.

BRIDGES, C. B. 1935. Salivary chromosome maps. With a key to the banding of the chromosomes of *Drosophila melanogaster*. *J. Hered.* **26**: 60.

———. 1936. The Bar "gene" a duplication. *Science* **83**: 210.

BRIDGES, P. N. 1941. A revision of the salivary gland 3R-chromosome map of *Drosophila melanogaster*. *J. Hered.* **32**: 299.

CHOVNICK, A. 1966. Genetic organization in higher organisms. *Proc. Roy. Soc. Lond. B* **164**: 198.

———. 1973. Gene conversion and transfer of genetic information within the inverted region of inversion heterozygotes. *Genetics* **75**: 123.

CHOVNICK, A., G. H. BALLANTYNE, and D. G. HOLM. 1971. Studies on gene conversion and its relationship to linked exchange in *Drosophila melanogaster*. *Genetics* **69**: 179.

CHOVNICK, A., W. GELBART, and M. McCARRON. 1977. Organization of the rosy locus in *Drosophila melanogaster*. *Cell* **11**: 1.

CHOVNICK, A., W. GELBART, M. McCARRON, B. OSMOND, E. P. M. CANDIDO, and D. L. BAILLIE. 1976. Organization of the rosy locus in *Drosophila melanogaster*: Evidence for a control element adjacent to the xanthine dehydrogenase structural element. *Genetics* **84**: 233.

EDWARDS, T. C. R., E. P. M. CANDIDO, and A. CHOVNICK. 1977. Xanthine dehydrogenase from *Drosophila melanogaster*. A comparison of the kinetic parameters of the pure enzyme from two wild-type isoalleles differing at a putative regulatory site. *Mol. Gen. Genet.* **154**: 1.

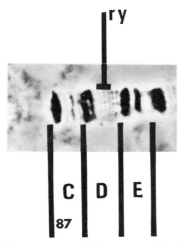

Figure 10. The rosy region of chromosome 3R. (Photograph kindly provided by Dr. G. Lefevre.)

FOGEL, S., D. D. HURST, and R. K. MORTIMER. 1971. Gene conversion in unselected tetrads from multipoint crosses. In *Stadler Genetics Symposia*, (ed. G. Kimber and G. P. Redei), vol. 1–2, p. 89. Agricultural Experiment Station, University of Missouri, Columbia.

FORREST, H. S., E. GLASSMAN, and H. K. MITCHELL. 1956. Conversion of 2-amino-4-hydroxypteridine to isoxanthopterin in *D. melanogaster*. *Science* 124: 725.

GELBART, W., M. McCARRON, and A. CHOVNICK. 1976. Extension of the limits of the XDH structural element in *Drosophila melanogaster*. *Genetics* 84: 211.

GELBART, W. M., M. McCARRON, J. PANDEY, and A. CHOVNICK. 1974. Genetics limits of the xanthine dehydrogenase structural element within the rosy locus in *Drosophila melanogaster*. *Genetics* 78: 869.

GLASSMAN, E. 1965. Genetic regulation of xanthine dehydrogenase in *Drosophila melanogaster*. *Fed. Proc.* 24: 1243.

GLASSMAN, E. and H. K. MITCHELL. 1959. Mutants of *Drosophila melanogaster* deficient in xanthine dehydrogenase. *Genetics* 44: 153.

GLASSMAN, E., J. D. KARAM, and E. C. KELLER. 1962. Differential response to gene dosage experiments involving the two loci which control xanthine dehydrogenase of *Drosophila melanogaster*. *Z. Vererb.* 93: 399.

GRELL, E. H. 1962. The dose effect of $ma-l^+$ and ry^+ on xanthine dehydrogenase activity in *Drosophila melanogaster*. *Z. Vererb.* 93: 371.

GUTZ, H. 1961. Distribution of X-ray and nitrous acid induced mutations in the genetic fine structure of the ad_7 locus of *Scizosaccharomyces pombe*. *Nature* 191: 1125.

———. 1963. Studies on the genetic fine structure of the ad_7 and ad_6 loci of *Schizosaccharomyces pombe*. *Proc. XI. Int. Congr. Genet., The Hague.* 1: 7.

HARTMAN, P. E., Z. HARTMAN, R. C. STAHL, and B. N. AMES. 1971. Classification and mapping of spontaneous and induced mutations in the histidine operon of *Salmonella*. *Adv. Genet.* 16: 1.

HILLIKER, A. J. 1976. Genetic analysis of the centromeric heterochromatin of chromosome 2 of *Drosophila melanogaster*: Deficiency mapping of EMS-induced lethal complementation groups. *Genetics* 83: 765.

HOCHMAN, B. 1974. Analysis of a whole chromosome in *Drosophila*. *Cold Spring Harbor Symp. Quant. Biol.* 38: 581.

JUDD, B. H., M. W. SHEN, and T. C. KAUFMAN. 1972. The anatomy and function of a segment of the X chromosome of *Drosophila melanogaster*. *Genetics* 71: 139.

LAURELL, C.-B. 1966. Quantitative estimation of proteins by electrophoresis in agarose gel containing antibodies. *Anal. Biochem.* 15: 45.

LEFEVRE, G., JR. 1971. Cytological information regarding mutants listed in Lindsley and Grell 1968. *Drosophila Inform. Serv.* 46: 40.

LIM, J. K. and L. A. SNYDER. 1974. Cytogenetic and complementation analyses of recessive lethal mutations induced in the X chromosome of *Drosophila* by three alkylating agents. *Genet. Res.* 24: 1.

MALLING, H. V. and F. J. DeSERRES. 1973. Genetic alterations at the molecular level in X-ray induced *ad-3B* mutants of *Neurospora crassa*. *Radiat. Res.* 53: 77.

MANNING, J. E., C. W. SCHMID, and N. DAVIDSON. 1975. Interspersion of repetitive and nonrepetitive DNA sequences in the *Drosophila melanogaster* genome. *Cell* 4: 141.

McCARRON, M., W. GELBART, and A. CHOVNICK. 1974. Intracistronic mapping of electrophoretic sites in *Drosophila melanogaster*: Fidelity of information transfer by gene conversion. *Genetics* 76: 289.

MULLER, H. J. 1936. Bar duplication. *Science* 83: 528.

MULLER, H. J., A. A. PROKOFYEVA-BELGOVSKAYA, and K. V. KOSSIKOV. 1936. Unequal crossing-over in the Bar mutant as a result of duplication of a minute chromosome section. *C. R. (Dokl.) Acad. Sci. U.R.S.S., N.S.* 1(10): 87.

RASCH, E. M., H. J. BARR, and R. W. RASCH. 1971. The DNA content of sperm of *Drosophila melanogaster*. *Chromosoma* 33: 1.

RUDKIN, G. T. 1965. The relative mutabilities of DNA in regions of the X chromosome of *Drosophila melanogaster*. *Genetics* 52: 665.

SCHALET, A., R. P. KERNAGHAN, and A. CHOVNICK. 1964. Structural and phenotypic definition of the rosy cistron in *Drosophila melanogaster*. *Genetics* 50: 1261.

SINGH, R. S., R. C. LEWONTIN, and A. A. FELTON. 1976. "Alleles" of xanthine dehydrogenase in *Drosophila pseudoobscura*. *Genetics* 84: 609.

STURTEVANT, A. H. 1925. The effects of unequal crossing-over at the Bar locus in *Drosophila*. *Genetics* 10: 117.

WEEKE, B. 1973. Rocket immunoelectrophoresis. In "A manual of quantitative immunoelectrophoresis: Methods and applications" (ed. N. H. Axelson et al.). *Scand. J. Immunol.* 2(Suppl. 1): 37.

The Two-dimensional Fractionation of *Drosophila* DNA

S. S. POTTER AND C. A. THOMAS, JR.

Department of Biological Chemistry, Harvard Medical School, Boston, Massachusetts 02115

A variety of experiments could be done if it were possible to resolve discrete restriction segments from the entire *Drosophila* genome. Restriction endonucleases that find cleavage sites every 1500–5000 base pairs (bp) would be expected to break the *Drosophila* genome of 1.5×10^8 bp into 10^5–3×10^4 DNA segments—a discouragingly large number. Nonetheless, there was some reason for optimism since the two-dimensional procedure, described below, has already proved successful in resolving restriction segments of bacterial DNA which are only 30 times less numerous (Potter et al. 1977). The experiments described in this paper show our progress to date.

Resolution in Two Dimensions

Drosophila melanogaster cells in culture were labeled with ^{32}P for several generations, the DNA isolated and terminally digested with *Bam*HI (Wilson and Young 1975), and the resulting segments electroeluted through a long (30 cm) square (2×2 cm) plastic pipe containing 1% agarose. About 100 5-ml fractions were collected, mixed with carrier DNA, precipitated with isopropanol, redissolved in buffer, and digested with *Eco*RI. These samples were electrophoresed in adjacent channels on long (45 cm) horizontal agarose slabs. After drying and autoradiography, a two-dimensional display of bands appears. Figure 1 shows such a pattern and the legend gives more details of the method.

These patterns display a number of prominent bands — the landmarks — as well as a variety of bands of lighter density. In all, one can see 2000–5000 bands, significantly fewer than the expected 10^4–10^5. These patterns are reproducible provided both the primary and the secondary digestions are complete. To verify this, unlabeled λ DNA was added (usually to each fraction) and the characteristic λ-R segments observed after staining with ethidium bromide.

We believe it is likely that this procedure displays a portion of the total population of *Bam*-RI segments in such a way as to resolve segments of unique length which are only represented once per genome. First, it is apparent that bands are resolved, and the conclusion that the faintest of these represent segments of unique sequence is the simplest since more than half of the *Drosophila* DNA reassociates as single-copy sequences (Manning et al. 1975). Second, the

resolution of unique segments is to be expected provided one makes some reasonable assumptions regarding (1) the distribution and density of restriction sites, and (2) the ability of the electroelution apparatus and the agarose slabs to distinguish segments of similar length. If we suppose that the primary restriction endonuclease finds cleavage sites with a probability p and that these sites are located more or less at random — as appears to be the case (Hamer and Thomas 1975) — then the number of different restriction segments of length t per resolvable fraction, P_t, is given by:

$$P_t = \Lambda p e^{-pt}(1 - e^{-\alpha pt}), \qquad (1)$$

where Λ is the number of base pairs per genome (considered to consist entirely of unique sequences) and the electroeluter can separate segments of length t from those of $t + \alpha t$. The maximum value of Equation 1 occurs when $pt = 1$. Approximating $1 - e^{-\alpha pt} = \alpha pt$, we have

$$P_t(\text{max}) = \Lambda p(0.36)\alpha. \qquad (2)$$

If we assume that $\Lambda = 1.5 \times 10^8$ bp and $p^{-1} = 5000$ (a reasonable estimate for *Bam*HI), then the number of primary segments in the densest fraction will depend on the value of α, which is a measure of the resolving power of the electroelution device. In this range we estimate $\alpha = 0.02$, giving a maximum of 200 different segments in this fraction. Of these, 24% or 48 will be expected to remain unbroken because they contain no RI sites. The remaining 152 segments will be expected to be broken into a total of 438 secondary segments, but 120 of these should be smaller than 700 bp and not be recorded at the bottom of the autoradiographs. Therefore, about 318 segments should be displayed in the densest fractions. Would these be expected to be resolved on the agar slabs?

The number of secondary segments of length u per resolvable fraction, S_u, is given by:

$$S_u = P_t[1 + (t - u)s]e^{-su}(1 - e^{-\beta su}) \qquad (3a)$$

$$N_t^o = P_t e^{-st}, \qquad (3b)$$

where s is the density of secondary restriction sites and βu is the length difference per fraction. N_t^o is the number of primary segments that are not ex-

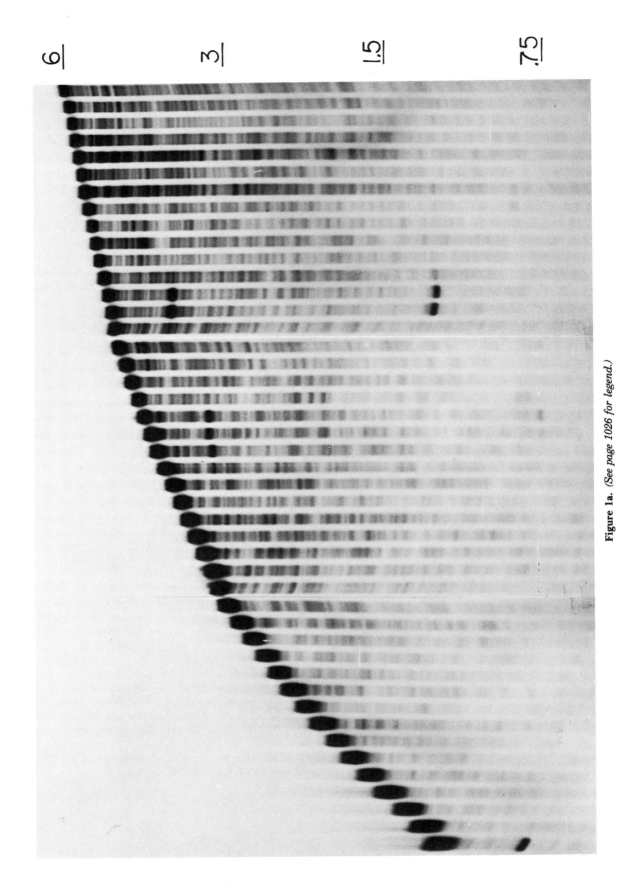

6

3

1.5

.75

Figure 1a. *(See page 1026 for legend.)*

25 9 3 1.5 .8

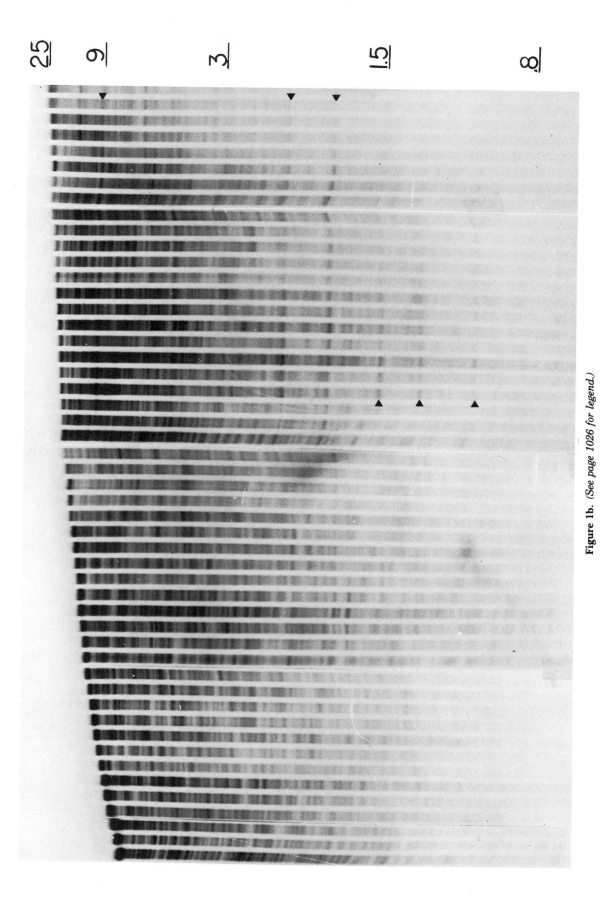

Figure 1b. *(See page 1026 for legend.)*

pected to contain a secondary cleavage site. The maximum value of Equation 3a is when the value of su is near unity.

$$S_u(\text{max}) = P_t(0.36)\beta st. \qquad (4)$$

By measuring the width of the bands in the slab gels, we were able to estimate that the value of β is less than 0.01. Combining this with Equation 2 and substituting the values mentioned, we have

$$S_u(\text{max}) \leq 1.02.$$

This means that although some unique DNA segments will be resolved, there will also be many multiplets or overlapping regions.

Finally, we can demonstrate that the intensity of the lightest bands is approximately that expected for unique segments. This is done by adding ^{32}P-labeled λ DNA to the ^{32}P-labeled *Drosophila* DNA prior to the two-dimensional analysis. To assure unambiguous identification of the λ segments, ^{32}P-labeled λ DNA was added in amounts of radiolabel corresponding to five and ten times the expected radiolabel in unique segments of *Drosophila*. The results for two such λ bands are shown in Figure 2. Here, many neighboring *Drosophila* bands are substantially less dense than the λ bands in the 5X case, indicating that many of these segments are present at less than five copies per genome. It is also clear that some *Drosophila* bands do approach 5X intensity. These multiplets are discussed below.

Taken together, the above evidence suggests that we are resolving segments of unique sequence; however, not all of the genome is displayed with any given pair of restriction endonucleases.

Multiplicity

Perhaps the most striking feature in Figure 1 is the intense band found in each channel at the upper

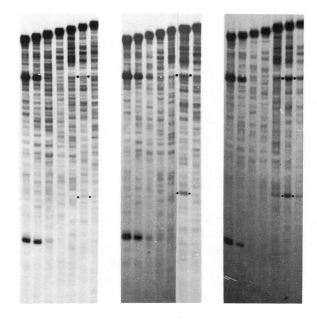

O X   5 X   I0 X

Figure 2. Segmental multiplicity. A mixture of ^{32}P-labeled λ DNA and *Drosophila* DNA was treated as described for Fig. 1. The λ DNA was added in amounts of radioactivity corresponding to 0X, 5X, and 10X λ copies per *Drosophila* genome (i. e., $1X = 4.6 \times 10^4/1.5 \times 10^8$ multiplied by the total cpm in *Drosophila* DNA). The 5.9-kb λ *Bam* segment is broken into 3.3- and 2.6-kb segments which are marked by black dots. From the relative intensity of the *Drosophila* bands to these known bands, one can estimate the multiplicity of the *Drosophila* segments. For purposes of display, both Figs. 1 and 2 have been "dodged" to reduce the apparent density of the larger segments.

limit of the pattern. The resulting line, which curves along the top of the pattern, consists of *Bam*HI DNA segments lacking recognition sites for *Eco*RI. In the early fractions the *Bam*HI segments are small and hence few of them contain *Eco*RI sites; however, in

Figure 1. The *Bam*-RI banding pattern of *Drosophila* DNA (cell line S). A *D. melanogaster* embryo cell line originally established by Echalier and Ohanessian (1970) was obtained from John Sedat (and renamed S) and grown to 75% confluency in D-20 without serum; then 250 μCi/ml ^{32}PO$_4$ was added and growth continued for 3 days. The purified DNA (Hamer and Thomas 1975) was digested completely with *Bam*HI prepared according to Wilson and Young (1975) and 10^7 dpm (< 20 μg) loaded in the sample well of the electroelution device together with a small amount of blue dye. The electroelution pipe (30 × 2 × 2 cm) contained 1% agarose in an electrophoresis buffer (25 mM Tris, pH 7.2, 12 mM Na acetate, 2.5 mM EDTA) and operated at 100 V total corresponding 7 mA/cm² at 4°C. A dialysis membrane was mounted at a distance of 1 mm from the end of the agar pipe, and this chamber was continuously pumped at a rate of 7.5 ml/hr. About 100 5-ml fractions were collected over 3 days, each mixed with 10 μg salmon sperm DNA and occasionally with 2 μg of λ DNA, then precipitated with 8.5 ml isopropanol at −20°C overnight. The DNA was collected by spinning in a Sorvall HS-4 rotor at 7000 rpm for 40 min, resuspended in RI buffer (100 mM Tris, pH 7.4, 50 mM NaCl, 10 mM MgCl₂), and digested with a 10X excess of RI prepared according to Greene et al. (1974). The samples were adjusted to 20% sucrose and loaded into adjacent wells of a large (45 × 16 × 0.3 cm) horizontal slab gel of 1.2% agarose (McDonell et al. 1977), and electrophoresis was continued at 115 V until the blue dye moved 40 cm. The gels were stained with ethidium bromide (1 μg/ml, 1 hr) and observed under UV to verify the complete digestion of λ DNA in each sample. The gels were dried under vacuum onto 3MM filter paper and autoradiographed with Kodak noscreen film (or in Fig. 5 with X-Omat R film with DuPont Lightning Plus fluorescent screens) and developed by standard methods. The approximate size calibration scales (shown in kb) were prepared by comparing numerous landmark bands with λ *Hae*III and λ-RI segments in other patterns, then drawing calibration curves and picking off the values shown. Note that in *a* the size of the primary segments can be determined from the position of the topmost (uncut) band. Six lines (described in text) are pointed out by triangles.

later fractions the longer *Bam*HI segments have a greater probability of being cleaved by *Eco*RI and therefore the top line becomes less intense.

Within the pattern certain bands are significantly denser than their neighbors, a fact that indicates the superposition of different segments of identical or nearly identical length. A certain fraction of these denser bands are accidental overlaps: at an average of *one* segment per resolvable fraction, 36% of the fractions would have no segment and 26% would be expected to have two or more. However, accidental overlap would not account for the many unusually dense bands which probably arise from the middle repetitive fraction that has been studied by Manning et al. (1975) and Finnegan et al. (this volume). Multiply repeated sequences of sufficient nonrepeating length, such as those coding for rRNA, are expected to generate intense bands.

Horizontal Lines

An interesting feature of the patterns is the appearance of horizontal "lines" composed of bands of nearly identical mobility in many contiguous channels. These lines must arise from secondary segments of defined length that arise from primary segments of variable length. In Figure 1, six of such lines are marked. These lines vary in size from 700 bp to about 15,000 bp, and they range over 8 to 23 channels. Microdensitometer scan analysis of the autoradiographs indicates that the DNA forming a line is present in about 20–200 copies per genome. This is based on the assumption that the faint bands near lines represent single-copy DNA.

Manning et al. (1975), using primarily electron microscopic techniques, have shown that the *Drosophila* genome contains long blocks (5–6-kb average length) of interspersed middle repetitive DNA. Sequences such as these would cause lines in the two-dimensional patterns by the mechanism illustrated in Figure 3. Supporting this view, Finnegan et al. (this volume) have shown, however, that some of these blocks consist of a number of elements, each

of which is repeated in the genome but with a different set of elements at each site.

There is other evidence which portends that these lines are not the result of "normal" middle repetitive DNA. We have observed that most of the lines of Figure 2 are not present in two-dimensional patterns of DNA from another *D. melanogaster* cell line. We have also observed that the lines are almost totally absent in two-dimensional patterns of nick-translated DNA from *D. melanogaster* embryos. This might suggest that the lines represent a peculiarity in the structure of the genome acquired during passage in tissue culture. One might speculate that the lines represent interspersed integrated viral genomes, since our *Drosophila* cell lines are known to contain viruslike particles (unpubl. results). Perhaps the lines represent interspersed duplicated genes which facilitate growth in tissue culture. The answer is simply not yet known.

Adult vs Embryo

A fascinating story of flexibility in the primary structure of chromosomes has been emerging during recent years. In prokaryotes, for example, it is now well established that the DNA elements called "insertion sequences" are capable of altering gene expression (turning genes on and off) by physically integrating into operons. The best-characterized examples are found in *Escherichia coli* (for review, see Starlinger and Saedler 1976), but in *Salmonella* it has also been shown that two genes coding for flagellar antigens are regulated by the orientation of a stretch of DNA (Zieg et al. 1977).

There is also abundant evidence that insertion-type sequences might exist in eukaryotes. The controlling elements of maize studied by B. McClintock are probably "mobile" DNA elements which alter gene expression (for reviews, see Fincham and Sastry 1974; Nevers and Saedler 1977), and there is some suggestion that similar regulatory mechanisms might be found in yeast (Hicks et al. 1977) and *Drosophila* (Rasmuson et al. 1974). Hozumi and

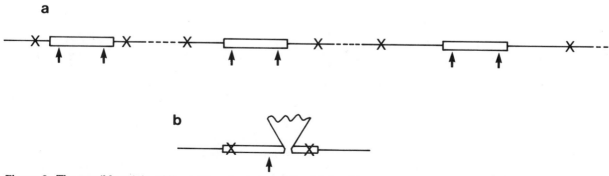

Figure 3. The possible origin of lines. Secondary segments of defined length that are produced from primary segments of variable length could arise from the insertion of a definite sequence (rectangles) into various chromosomal contexts *(a)*, or one or more defined sequences which themselves contains an insertion of variable length *(b)*. (X) *Bam*HI sites; (↑) RI sites. Many variations of these models are possible.

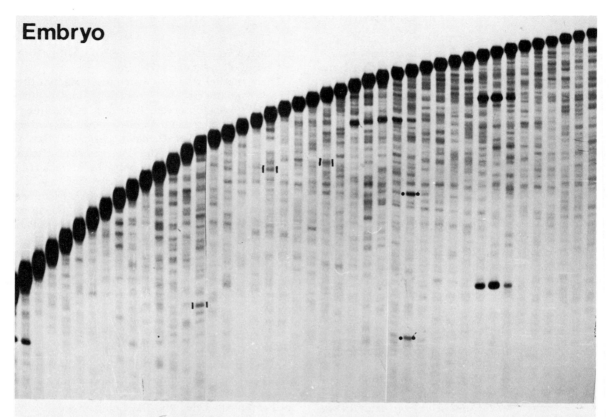

Embryo

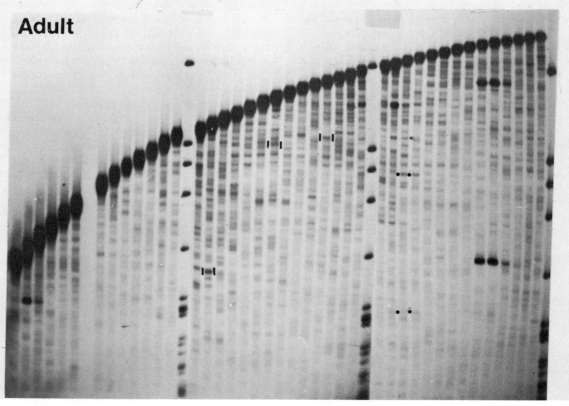

Adult

Figure 4. Embryo and adult *D. melanogaster* (Oregon R) two-dimensional *Bam*-RI restriction banding patterns. DNA was isolated from 6–18-hr embryo by the procedure of Peacock et al. (1974). DNA was isolated from adults in the following manner. Flies (5 g, stored at −70°C) were ground with a mortar and pestle, added to 35 ml of buffer (0.1 M NaCl, 0.2 M sucrose, 0.01 M EDTA, 0.03 M Tris, pH 8), and Dounce homogenized (10 strokes). Chitinous material was then pelleted by centrifugation for 1 min at 1000 rpm in a Sorvoll SS-34 rotor. The supernatant was centrifuged at 7000 rpm for 10 min and the nuclear pellet resuspended in 20 ml of lysis solution (0.15 M NaCl, 0.015 M Na acetate, 0.1 M EDTA, pH 7.5).

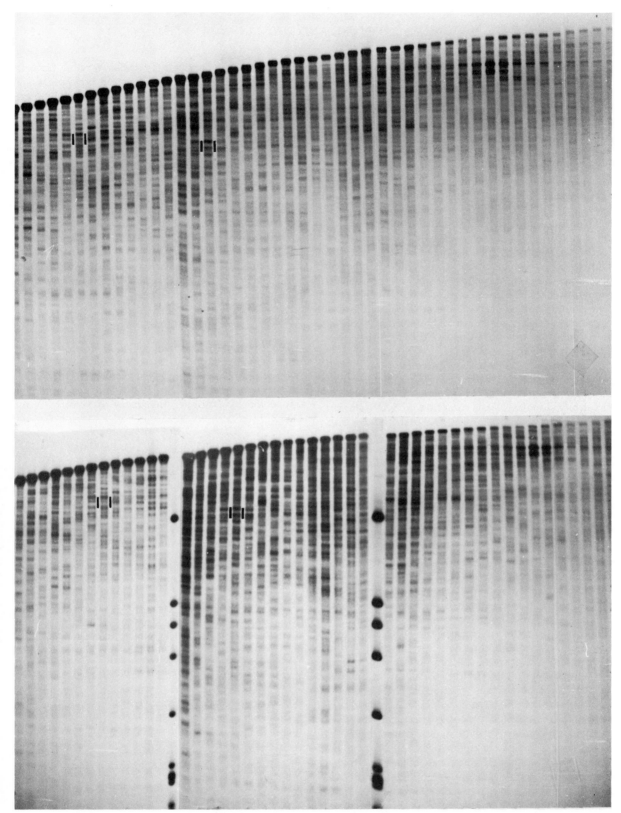

Figure 4 *(continued).* Proteinase K (3 mg) and 1 ml of 25% sodium lauryl sulfate were added and the mixture incubated at 55°C for 1 hr. The solution was then extracted three times with chloroform:isoamyl alcohol (24:1) and the DNA was spooled out. After complete digestion with *Bam*HI, the DNA was labeled at the ends by the following "fill-in" reaction. DNA (100 μg) in 1 ml of buffer (0.02 M Tris, pH 7.5, 0.01 M MgCl$_2$, 0.01 M β-mercaptoethanol, 0.08 M NaCl, 15 μM dATP) with 100 μCi of [α-^{32}P]dGTP (200 Ci/mmole) and 15 units of *E. coli* polymerase I (Boehringer Mannheim) was incubated at 5°C for 30 min. The DNA was then chloroform-extracted as before and precipitated with 2.5 volumes of ethanol before loading onto the electroelution device. Some of these corresponding bands are marked by vertical lines. The bands marked by dots are more intense in the embryo pattern.

Tonegawa (1976) have reported that there is a rearrangement of mammalian immunoglobulin genes during development.

If changes do occur in the genome during differentiation, then differences might be found between the two-dimensional restriction patterns of *Drosophila* embryo and adult DNA. The patterns form very detailed fingerprints and are potentially capable of detecting such DNA rearrangements or the modification (methylation) of a single base in a restriction site.

The adult and embryo *Drosophila* DNAs were purified and the *Bam*HI terminals labeled in vitro by "filling-in" with Pol I as described in the legend to Figure 4. It is important to note that this labeling procedure labeled only the termini of the *Bam*HI segments. This has the advantage of simplifying the two-dimensional patterns since the *Eco*RI-generated segments which lack a labeled terminus will not be detected; however, it has the disadvantage that it monitors fewer bands.

A difficulty in comparing the embryo and adult patterns develops because two *Bam*HI-generated DNA segments that are in the same fraction in one two-dimensional analysis may be found in adjacent fractions in a second analysis, as a result of the fraction collector shifting at different points in the electroelutions. This complicates the comparison since bands in a single channel in one analysis may be in adjacent channels in a separate analysis.

With this problem in mind, we carefully examined every region of the autoradiographs in Figure 4 looking for convincing differences. None were found. There are some places where bands appear more intense in one pattern, but there are no places where a band is unambiguously present in one pattern and absent in the other.

It is perhaps appropriate to mention some of the weaknesses of this approach. First, it is clearly possible that some DNA rearrangements could occur that would not alter the spacings of the restriction sites and therefore would not be detected. Second, if rearrangements were organ-specific, for example, then any single type of DNA alteration might involve such a small percentage of the total DNA that it would not be detected. Nonetheless, the patterns are very detailed and it is probably significant that we can find no conclusive differences. It should be recalled that a variety of differences can be seen between our various cell lines and between embryo DNA and any single cell-line DNA. The *Drosophila* genome certainly does not undergo wholesale DNA rearrangement during development. Any subtle alterations which do occur will require still more sensitive techniques for detection.

Future Opportunities

There are a variety of future experiments. We are currently exploring different pairs of restriction

endonucleases, particularly those having CpG in their recognition sites which would be expected to be influenced by methylation at various developmental stages (Grippo et al. 1968; Holliday and Pugh 1975; Bird and Southern 1977; Bird 1977). These comparisons would be facilitated if a suitable double-label procedure could be developed. A number of bands of high multiplicity have already been identified which need further analysis in terms of their chromosomal location and function. Finally, there is the hope of recognizing segments covered by deficiencies of various lengths which allow the limited growth of embryos. This would eventually allow the assignment of genes to segments. In sum, it may be technically possible to apply the same kind of analysis to the *Drosophila* genome that is now commonplace with viral genomes.

Acknowledgments

We thank Carol Stoner for her excellent assistance during all phases of this work. S. S. P. was the recipient of National Institutes of Health Postdoctoral Fellowship GM-05986. The research was supported by National Institutes of Health Grant GM-21740.

REFERENCES

BIRD, A. P. 1977. Use of restriction enzymes to study eukaryotic DNA methylation. II. The symmetry of methylated sites supports semi-conservative copying of the methylation pattern. *J. Mol. Biol.* (in press).

BIRD, A. P. and E. M. SOUTHERN. 1977. Use of restriction enzymes to study eukaryotic methylation. I. The methylation pattern in ribosomal DNA from *Xenopus laevis*. *J. Mol. Biol.* (in press).

ECHALIER, G. and A. OHANESSIAN. 1970. In vitro culture of *Drosophila melanogaster* embryonic cells. *In Vitro* **6**: 162.

FINCHAM, J. R. S. and G. R. K. SASTRY. 1974. Controlling elements in maize. *Annu. Rev. Genet.* **8**: 15.

GREENE, P. J., M. C. BETLACH, H. M. GOODMAN, and H. W. BOYER. 1974. The *Eco RI* restriction endonuclease. *Meth. Mol. Biol.* **7**: 88.

GRIPPO, P., M. IACCARINO, E. PARISI, and E. SCARANO. 1968. Methylation of DNA in developing sea urchin embryos. *J. Mol. Biol.* **36**: 195.

HAMER, D. H. and C. A. THOMAS, JR. 1975. The cleavage of *Drosophila melanogaster* DNA by restriction endonucleases. *Chromosoma* **49**: 243.

HICKS, J. B., J. N. STRATHERN, and I. HERSKOWITZ. 1977. The cassette model of mating type interconversion. In *DNA insertion elements, plasmids, and episomes* (ed. A. Bukhari et al.), p. 457. Cold Spring Harbor Laboratory, Cold Spring Harbor, New York.

HOLLIDAY, R. and J. E. PUGH. 1975. DNA modification mechanisms and gene activity during development. *Science* **187**: 226.

HOZUMI, N. and S. TONEGAWA. 1976. Evidence for somatic rearrangement of immunoglobulin genes coding for variable and constant regions. *Proc. Natl. Acad. Sci.* **73**: 3628.

MANNING, J. E., C. W. SCHMID, and N. DAVIDSON. 1975. Interspersion of repetitive and non-repetitive DNA sequences in the *Drosophila melanogaster* genome. *Cell* **4**: 141.

McDonell, M. W., M. N. Simon, and F. W. Studier. 1977. Analysis of restriction fragments of T7 DNA and determination of molecular weights by electrophoresis in neutral and alkaline gels. *J. Mol. Biol.* **110**: 119.

Nevers, P. and H. Saedler. 1977. Transposable genetic elements as agents of gene instability and chromosomal rearrangements. *Nature* **268**: 109.

Peacock, W. J., D. Brutlag, E. Goldring, R. Appels, C. W. Hinton, and D. L. Lindsley. 1974. The organization of highly repeated DNA sequences in *Drosophila melanogaster* chromosomes. *Cold Spring Harbor Symp. Quant. Biol.* **38**: 405.

Potter, S. S., K. Bott, and J. E. Newbold. 1977. Two-dimensional restriction analysis of the *Bacillus subtilis*

genome: Gene purification and ribosomal ribonucleic acid gene organization. *J. Bacteriol.* **129**: 492.

Rasmuson, B., M. M. Green, and B. M. Karlsson. 1974. Genetic instability in *Drosophila melanogaster*. Evidence for inserting mutations. *Mol. Gen. Genet.* **133**: 237.

Starlinger, P. and H. Saedler. 1976. IS-elements in microorganisms. *Curr. Top. Microbiol. Immunol.* **75**: 111.

Wilson, G. A. and F. E. Young. 1975. Isolation of a sequence-specific endonuclease (*Bam* 1) from *Bacillus amyloliquefaciens* H. *J. Mol. Biol.* **97**: 123.

Zieg, J., M. Silverman, M. Hilmen, and M. Simon. 1977. Recombinant switch for gene expression. *Science* **196**: 170.

Sequence Homology within Families of *Drosophila melanogaster* Middle Repetitive DNA

P. C. WENSINK

Department of Biochemistry and Rosenstiel Basic Medical Sciences Research Center, Brandeis University, Waltham, Massachusetts 02154

A large fraction of the DNA in higher organisms is repeated 10–100,000 times per nucleus. Although the function of this middle repetitive DNA is unknown, there have been proposals that it allows complex patterns of gene expression (reviewed in Davidson and Britten 1973), chromosome rearrangements (Lee 1975), or specific chromosomal associations such as ectopic pairing (Finnegan et al.; Rubin; both this volume). All of these proposals require that homologous repeats have an identical function so that, for example, the repeats of one sequence may all be recognized and bound by the same repressor protein. If two repeats have such an identical function, then there must be a limit to the difference between their base sequences. Thus the evidence that middle repetitive DNA of many organisms has a broad range of sequence homology (Britten and Davidson 1971; Davidson and Britten 1973) is significant for these proposals because it suggests either that there are many types of functions performed by the middle repeats and each type sets a different limit on sequence divergence or that functional limits apply only to a minor fraction of the middle repetitive DNA. To investigate this issue and to gain a more detailed picture of *Drosophila melanogaster's* middle repetitive DNA, I have examined the distribution of homology within the middle repetitive DNA of this insect.

Heterogeneity in the nucleotide sequence of related repeats has been detected by studies of the reassociation kinetics and thermostability of DNA. The reassociated middle repetitive DNAs from species ranging from sea urchin to human were found to have a broad range of thermostability, suggesting that there is a parallel broad range of base pairing between the repeats (Britten and Davidson 1971; Davidson and Britten 1973). In addition, if the base-pairing criterion of reassociation is lowered to allow formation of more poorly paired duplexes, then a larger fraction of total DNA reassociates at the rate of middle repetitive DNA. This larger fraction has a lower average thermostability (Britten and Kohne 1968; Kohne 1970) and so indicates that the additional repeats are less homologous.

These and other experiments demonstrating imprecise repetition have not investigated the way this heterogeneity is distributed among the repeats.

Since this distribution is important in estimating the likelihood of the functions proposed for middle repetitive DNA, the experiments in this paper have been given a novel design which tests two possible distributions. These distributions are most easily described in terms of the operational definition of a family of repeated sequences as a set of sequences, all of which can reassociate with each other. In the first model of distribution, each family of related sequences has a wide spectrum of homology. This model predicts that the detected size and composition of a family will vary with the reassociation base-pairing criterion. The second model pictures each family with narrow limits to its heterogeneity so that, for example, each member of a family is 18–20% mismatched with all other members. This second model predicts that the detected size and composition of a family should not vary over a broad range of base-pairing criteria. If the second model is accurate, then the heterogeneity observed in all middle repeats is due to differences between families rather than to differences within families. Homogeneity within families accords with the proposals that families have functional significance, for it allows the possibility that all repeats in a family perform the same function. By the same argument, if heterogeneity is within families, then a functional link between family members is less likely.

About 12% of the *D. melanogaster* genome is middle repetitive when measured at the standard reassociation criterion of $T_m - 25°C$ (Laird and McCarthy 1969; Schachat and Hogness 1974). (T_m is the temperature at which 50% of the DNA is melted.) These repeats have a narrow range of repetition frequency that centers around 150. As with most other higher organisms, some of this DNA is arranged in tandem repeats and contains the genes for rRNA, 5S RNA, and the histones. In *D. melanogaster* an unusually large fraction (about 30%) of the middle repetitive DNA is known to be arranged in this way and also to contain these genes. Most of the remaining middle repeats have no known function and are not in long tandem arrays (Manning et al. 1975; Crain et al. 1976), but instead are probably scattered around the chromosomes (Wensink et al. 1974; Crain et al. 1976; Finnegan et al., this volume). In *D. melanogaster* these scattered middle repeats are about ten times longer than the majority of repetitive DNA in most other eukaryotes (Manning et al. 1975; Crain et al. 1976).

This paper describes the homology between members of one repeated family which has a scattered arrangement of repeats. It also describes the homology within all families. These descriptions lead to the conclusion that the single family and all, or almost all, other families of *D. melanogaster* middle repeats have very narrow limits to their heterogeneity. Unexpectedly, they also indicate that all, or almost all, families are well base-paired. Therefore, this higher organism does not have the broad middle-repeat heterogeneity found in other species, and so most of its repeated families may be linked by function.

MATERIALS AND METHODS

Isolation and labeling of DNA. *D. melanogaster* (Oregon R) DNA was purified from frozen 0–18-hour embryos (a generous gift of S. Elgin) as described by Schachat and Hogness (1974). The origin and method of purification of plasmid pDm1 have been described by Wensink et al. (1974). DNA was radiolabeled by the nick-translation activity of *Escherichia coli* DNA polymerase I (purchased from Boehringer Manheim) (Rigby et al. 1977) using the procedure of Maniatis et al. (1975).

Reassociation reactions and thermal denaturation. The reactions were performed as described by Wensink et al. (1974), except that the NaCl concentration was increased to 1.5 M to accelerate the rate of reassociation. Reassociation to duplex form was assayed by the single-strand-specific S_1 endonuclease (purified by the method of Vogt [1973] through the ammonium sulfate precipitation step and generously provided by M. Rosbash) at pH 4.5 in 200 mM NaCl, 30 mM NaAc, 1 mM ZnSO$_4$, 1.3 mM Tris, 0.13 mM EDTA, 5% (w/w) glycerol, and 40 μg/ml each of native and denatured salmon sperm DNA. For the thermal denaturation studies, DNA from these reactions was bound to hydroxylapatite (HAP) by diluting tenfold into 12 mM sodium phosphate buffer (PB), pH 7.0, and passing through water-jacketed columns containing HAP (Bio-Gel HTP DNA grade, from Bio-Rad). Control experiments demonstrated that the diluted components had no effect on the quantity of binding or the elution pattern of double- or single-stranded DNA. The melt curves were obtained by eluting from HAP with 0.12 M PB, pH 7.0, at increasing temperatures. The eluted samples were precipitated onto filters with trichloroacetic acid and counted by liquid scintillation.

Isolation of middle repetitive fractions. These were isolated from embryonic DNA with the aid and advice of M. Izquierdo. The DNA was sonicated and passed through Sephadex SP50-Chelex (Britten et

al. 1973). It was then denatured with alkali and incubated to a C_0t of 0.05 moles of nucleotides × liter^{-1} × sec at 60°C in 0.12 M PB, pH 7.0. Single and double strands were eluted from HAP at 60°C with 0.12 M and 0.4 M PB, respectively. The unreassociated DNA was dialyzed to remove phosphate, ethanol precipitated, and then resuspended in 10 mM Tris, 1 mM EDTA, pH 8.0. This DNA was reassociated at either 50°C or 60°C to a C_0t of 3 moles × liters^{-1} × sec in 1.5 M NaCl, 10 mM Tris, 1 mM EDTA, pH 8.0. If the sample was to be digested with the S_1 enzyme, buffer was added to achieve the reaction conditions described above. Following digestion, the sample was diluted tenfold into 0.12 M PB and the double-stranded DNA isolated on HAP as before. Nick translation of each reassociated middle repetitive fraction gave rise to about 25% labeled foldback structures. These reassociated with zero-order kinetics and were removed by HAP. It is presumed that this foldback DNA is due to the strand-hopping activity of polymerase I (Schildkraut et al. 1964), which is likely to be more frequent in poorly base-paired regions. Since the low- and high-criterion middle repetitive fractions (see below) had the same amount of labeled foldback DNA, I conclude that there was no selective loss of the most poorly base-paired regions.

Determination of the size of reassociated DNA. When reassociated samples were taken for thermostability studies, an additional sample was taken to determine the single-stranded size of the DNA. Samples were centrifuged in 5–20% (w/w) alkaline sucrose gradients in 0.5% SDS, 0.1 N NaOH. The equations of Studier (1965) were used to calculate the weight-average size of the labeled DNA relative to a 650-nucleotide-long marker (a gift from M. Rosbash and F. Gibson). The sizes ranged from 400 to 600 nucleotides.

RESULTS

Sequence Homology within a Single Family of Middle Repeats

The bacterial plasmid pDm1 (Wensink et al. 1974) contains the sequence Dm1 which is a 3-kb insert of *D. melanogaster* DNA. Reassociation experiments demonstrate that at the standard reassociation condition of $T_m - 25°C$,[1] the entire Dm1 sequence appears to be 90-fold repeated in *D. melanogaster*'s embryonic DNA. This family of repeated sequences is scattered around the chromosomes, since Dm1 hybridizes to at least 15 chromomeric regions and a diffuse area of the chromocenter of salivary polytene chromosomes (Wensink et al. 1974).

The experiments described in this section investi-

---

[1] The reassociation criterion will frequently be expressed relative to T_m to allow comparison of results obtained in different buffer systems.

gate this family of scattered repeats by determining the distribution of sequence homology between its members and Dm1. If the Dm1 family has members that range from being exact copies to very poor copies of Dm1, then as the base-pairing criterion of reassociation is lowered to allow the formation of more poorly base-paired structures, there will be an increase in the number of duplexes Dm1 can form and also an increase in the base-pairing mismatch of these duplexes. These changes will be reflected in the rate of duplex formation and in the thermostability of the duplexes. However, if all members of the Dm1 family differ from each other by the same fixed percentage of their sequence, then the repetition frequency and the duplex thermostability should be insensitive to a broad range of this experimental variation.

This experimental strategy was accomplished by radiolabeling pDm1 and allowing it to reassociate with a large excess of unlabeled total *D. melanogaster* DNA. The excess of unlabeled DNA was such that Dm1 formed 0.999 of its duplexes with the unlabeled DNA rather than with the labeled Dm1. Therefore the rate at which Dm1 formed duplexes and the thermostability of these duplexes are both measures of the Dm1 family. These reassociation reactions (Fig. 1) were performed at three criteria: 15°C, 25°C, and 35°C below the T_m of Dm1. In accord with previous experiments (Wensink et al. 1974), about 26% of pDm1, and therefore all of the 3-kb *D. melanogaster* insert, is repeated about 100 times in embryonic DNA. The rate of reassociation at $T_m -$ 15°C and $T_m - 35$°C is two- to fourfold slower than at $T_m - 25$°C (Fig. 1). This would be expected if the rate change were entirely due to the effect of temperature on the reassociation rate (Studier 1969). Therefore these reassociation experiments provide evidence that lowering the criterion does not greatly increase the number of embryonic sequences that can reassociate with Dm1.

A more critical test of the family heterogeneity would be given by an examination of the thermostability of these duplexes. To accomplish this, it was necessary to separate the radiolabeled duplexes from the unreassociated bacterial component of pDm1 (Fig. 1) by HAP chromatography. Samples from each reassociation reaction were allowed to bind to HAP at the lowest reassociation criterion (50°C, 0.12 M PB or $T_m - 35$°C) in order to allow the least stable heteroduplexes to bind. In each case about 40% of the label bound to HAP. Since this is 14% more label than is included in the Dm1 part of pDm1 (Fig. 1 and Wensink et al. 1974), it indicates artifactual binding of single-stranded DNA under these low-criterion conditions. The thermostability of this HAP-bound DNA was determined (Fig. 2). Each melt curve was biphasic, with the first phase reaching a plateau at about 35–40%.

Since the thermostability of the Dm1 duplexes was sought, it was necessary to determine which part

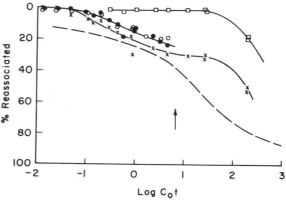

Figure 1. pDm1 reassociation with *D. melanogaster* DNA. A small amount of ³H-labeled pDm1 was added to unlabeled *D. melanogaster* embryonic DNA (2 mg/ml). The denatured DNAs were allowed to reassociate at 60°C (●), 70°C (x), and 80°C (○). The broken line describes a parallel reassociation with ³H-labeled total *D. melanogaster* DNA as tracer. (□) ³H-labeled pDm1 tracer reassociation in the absence of embryonic DNA. With exception of this latter curve, C_0 refers to the unlabeled DNA.

of these curves was due to dissociation of bound single-stranded DNA. For this reason single-stranded pDm1 DNA was passed through HAP at various temperatures, and the fraction binding and its temperature of dissociation were determined. As the binding temperature was decreased, an increasing fraction of single-stranded DNA bound to the HAP (Fig. 3). This is likely to be due to increased foldback secondary structure (Schmid et al. 1975). This result allows a prediction that 35% of the total labeled DNA that binds the HAP at $T_m - 35$°C is due to the unreassociated bacterial component of pDm1. This is just the amount of labeled DNA in the first phase of the melt curve, and so this phase may contain only unreassociated DNA. This interpretation is strengthened by the observation that bound sin-

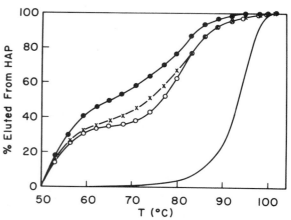

Figure 2. Melt of duplexes with Dm1. ³H-labeled pDm1 DNA reassociated with embryonic DNA at 60°C (●), 70°C (x), or 80°C (○) was eluted from HAP as described in Materials and Methods. The solid line indicates the melt of ³²P-labeled, sonicated, and native pMB9 DNA.

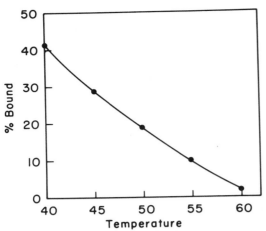

Figure 3. Binding of single strands to HAP. Denatured ^{3}H-labeled pDm1 DNA was passed through HAP at the indicated temperatures in 0.012 M PB. After washing with 0.12 M PB at those temperatures, the temperature was raised to 60°C and the single and double strands were eluted under standard conditions. The percentage of single strands bound is equal to the amount eluted at 60°C, 0.12 M PB divided by total eluted at all temperatures with all solutions. The amount eluting as double strands at 60°C was less than 4% of the total.

gle-stranded DNA disassociates from the HAP column in parallel with this first phase (unpubl. obs.). Finally, if any of the Dm1-duplex reassociation samples are passed through HAP at $T_m - 25°C$, only about 25% of the labeled DNA is retained by the column and the first phase of the melt curve is eliminated (unpubl. obs.). I conclude that the first phase of the melt curves in Figure 2 is the elution of bound single strands that are not Dm1 DNA.

The melt of duplexes between Dm1 and its complements in embryonic DNA is shown in Figure 4,

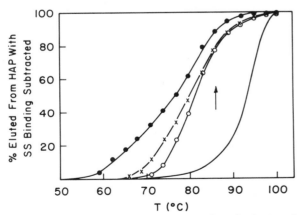

Figure 4. Melt of duplexes with Dm1 after single-strand binding is subtracted. This replots the data of Fig. 2 after subtraction of the calculated single-strand binding. The arrow marks the T_m of native Dm1 DNA. This was calculated from the calibrations of Szybalski and Szybalski (1971) and Marmur and Doty (1962) using the density (1.697 g/cc) of Dm1 determined by Wensink et al. (1974). The solid line indicates pMB9 (see Fig. 2).

where the background single-stranded binding has been subtracted. As the criterion of reassociation was lowered by 20°C, the T_m of Dm1 duplexes was reduced by only 4°C, despite the fact that duplexes with a 26°C lower T_m could have been formed. I conclude that Dm1 can find only a narrow range of mismatched repeats in embryonic DNA. Although an exact correlation cannot be made between the T_m and the percent of mismatch in base pairing, estimates based on model experiments (for review, see Hutton and Wetmur 1973; Thomas and Dancis 1973) suggest 1% mismatch per degree or, in this case, the narrow range of 4%. The data of Figure 4 also demonstrate that at the highest criterion most of the duplexes are still 5° less stable than the native Dm1. This is in contrast with pDm1 which, when reassociated with itself at any of these criteria, forms DNA duplexes with a T_m about 1°C below native pDm1 (unpubl. obs.). From this difference in T_m depression, I conclude that almost all of the 90 Dm1 complements in embryonic DNA are inexact copies and differ in at least 4% of their base pairs. Thus the thermostability studies of Dm1 duplexes demonstrate that this family of middle repeats has a narrow range of heterogeneity which varies from 4% to 8% mismatch. Clearly, these figures may be slightly altered as the correlation between mismatch and T_m depression is refined.

Sequence Homology within Families That Contain Well-base-paired Repeats

Next I wished to determine whether the Dm1 family of *D. melanogaster* DNA is typical of well-base-paired middle repeats. If this is the case, all such repeats have only well-base-paired complements. To isolate these repeats, the rapidly reassociating component ($C_0 t < 0.05$, $T_m - 25°C$) was removed from total embryonic DNA by HAP chromatography. The remaining DNA was incubated to allow middle repetitive sequences to reassociate at optimal rate ($T_m - 25°C$) to a $C_0 t$ of 3. Single-stranded, nonrepetitive DNA and very poorly paired duplexes were then removed by S_1 nuclease digestion and HAP chromotography. This fraction (termed "well-base-paired middle repeats") was labeled with radioactivity by nick translation and then used as a tracer in experiments identical to those described above for Dm1. The reassociation of this tracer DNA with embryonic DNA (Fig. 5) demonstrates that the tracer is not contaminated with more than a few percent of either highly repetitive or unique DNA. As with the Dm1 probe, these curves do not reveal an increased family size at the lowest reassociation criterion. Since almost all of the DNA was reassociated in this sample, no artifactual single-strand binding was detected.

The thermostability of duplexes formed between the well-paired middle repeats and other embryonic DNA (Fig. 6) is also nearly identical to the Dm1

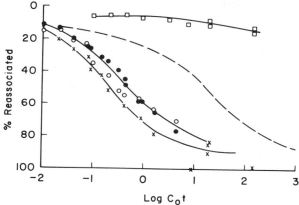

Figure 5. Reassociation of well-base-paired middle repeats with total *D. melanogaster* DNA. A small amount of ³H-labeled well-base-paired middle repetitive DNA was reassociated at 60°C (●), 70°C (x), and 80°C (○) with embryonic DNA (2 mg/ml). The broken line describes a parallel reassociation with ³H-labeled total embryonic DNA as tracer. (□) ³H-labeled middle repetitive DNA tracer reassociated in the absence of embryonic DNA. With exception of this latter curve, C_0 refers to the unlabeled DNA.

experiments described above. The product formed at each criterion has a T_m and half-width (difference between temperatures at which 25% and 75% of the DNA is melted) indistinguishable from the Dm1 duplexes. I conclude that if a family contains some members that are very homologous, then all, or almost all, of the other members are also very homologous. The limits of heterogeneity in these families is ±2% mismatch. The exact T_m of these families is unknown, but if it is the same as that of total *D. melanogaster* DNA (85°C in 0.12 M PB, estimated from Laird and McCarthy 1969), the range of mismatch in these families is 3–7%. This allows the further observation that, like the Dm1 family, all

of the well-base-paired families do not have identical members.

Sequence Homology within Families Selected to Include Poorly Base-paired Repeats

The results detailed above are consistent with a model in which the expected broad middle-repetitive-sequence heterogeneity is between families rather than within families. To determine whether this is the case or whether there is no broad heterogeneity, middle repetitive DNA was isolated under conditions allowing very poorly base-paired repeats to be included. The middle repetitive fraction was isolated as before, with two exceptions. Criterion of reassociation was $T_m - 35°C$ to allow the formation of more poorly paired structures, and the middle repetitive fraction was not enzymatically digested. This DNA was radioactively labeled and used as tracer in experiments identical to those described above. The reassociation reactions were identical to those of Figure 4, and the melt curves (Fig. 7) were only slightly different from those in which well-base-paired repeats were the radioactive tracer (Fig. 6). The T_m's of the two highest-criterion duplexes are 0.5°C lower, and the T_m of the lowest-criterion product is the same as in the well-base-paired tracer experiments. This variation is probably within experimental error and is certainly a minor effect. Thus, in *D. melanogaster* embryonic DNA there are few if any families with mismatch between 10% and 30%.

DISCUSSION

This work was begun with the expectation that *D. melanogaster* middle repetitive DNA would have the broad spectrum of sequence homology that has

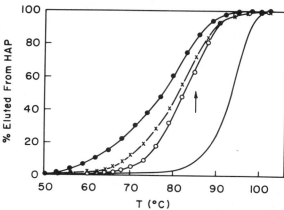

Figure 6. Melt of duplexes with well-base-paired middle repeats. ³H-labeled well-base-paired middle repeats reassociated with embryonic DNA at 60°C (●), 70°C (x), and 80°C (○) (see Fig. 5) were eluted from HAP as described in Materials and Methods. The arrow indicates the T_m of total *D. melanogaster* DNA (see text). The solid line indicates pMB9 DNA (see Fig. 2).

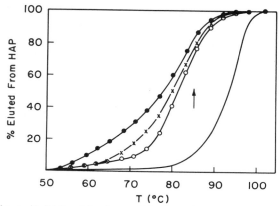

Figure 7. Melt of duplexes with low-criterion middle repeats. ³H-labeled low-criterion middle repeats were reassociated with embryonic DNA at 60°C (●), 70°C (x), and 80°C (○) to a C_0t of 5. The DNA was bound to HAP and eluted with 0.12 M PB at the temperatures shown. The arrow indicates the T_m of total *D. melanogaster* DNA. The solid line indicates pMB9 DNA (see Fig. 2).

been observed in other higher organisms (Britten and Davidson 1971; Davidson and Britten 1973). The results demonstrate, however, that in *D. melanogaster* all or almost all middle repetitive families contain very homologous repeats. Thus, although individual families of related sequences contain few members that are identical, almost all members are limited to a difference of between 3% and 7% in nucleotide sequence. This range of difference is the average for all families as well as that explicitly found in one family. Since the thermal denaturation half-width for total middle repetitive DNA is only twice that found with a pure plasmid DNA (Fig. 7), I conclude that no more than a few of *D. melanogaster*'s 40 middle repetitive families (Schmid et al. 1975) are outside of this range. Also, Figures 4 and 6 clearly demonstrate that this limit applies to repeated families that have different arrangements, namely, the tandem arrangement that accounts for 30% of the total middle repetitive DNA (Glover and Hogness 1977; Wellauer and Dawid 1977; Pelligrini et al. 1977; Hershey et al. 1977; Lifton et al., this volume) and the scattered arrangement found in the Dm1 family. This suggests that the mechanisms proposed to explain family resemblance among tandem repeats, namely, rolling-circle replication (Gilbert and Dressler 1969) of one repeat or unequal crossover between the tandem repeats (Smith 1974), are unlikely to account for all family resemblances.

The close family resemblance could arise in at least two other ways. The first (Ingram 1963; Britten and Kohne 1968; Ohno 1970) pictures an unusual past event which made copies of one sequence. These copies are free to accumulate mutations and so to diverge with increasing time. If this hypothesis is used to explain the divergence of all *D. melanogaster* middle repetitive DNA, then all families must have arisen at about the same and fairly recent time in evolution. The second explanation for close family resemblance is that there are functional limits to divergence. It is easy to imagine broad categories of these limitations. For example, all repeats in a family may code for the same protein and so remain functional only if a limited number of base changes are made. It is also possible that these repeats code for a structural RNA whose function allows limited sequence divergence. Nontranscribed repeats could also be limited if they function as binding sites for a protein or nucleic acid which binds specifically despite limited sequence divergence. The results reported in this paper are consistent with the functional-limit explanation for family resemblance. They suggest that if this explanation is correct, then very similar limits are put on all the families and these limits are narrow.

There is no evidence that either of these explanations accounts for all of the close family resemblance observed in *D. melanogaster* middle repetitive DNA. It is unlikely that all of the tandem and scattered families arose at the same time in evolution. It seems more plausible that functional limits apply, and if this is so, then the very narrow range of sequence divergence suggests that only one type of functional limit applies to all families. In this light, it is suggestive that some of this repetitive DNA is transcribed (rDNA, 5S DNA, and histone DNA).

It is unlikely that the middle repetitive DNA examined by reassociation and thermostability studies could be limited by its ability to bind activator or repressor proteins. These binding sequences are probably less than 50 nucleotides in length (Ptashne et al. 1976) and so are at the lower limit of size that will quantitatively bind to hydroxylapatite (Wilson and Thomas 1973). For this reason they would probably not be detected in the thermostability curves. In addition, if such short sequences are surrounded by nonrepetitive sequences and if they are repeated only 10–20 times, then they will not be detected in the reassociation curves and will not be isolated by the procedure used. This follows from the fact that the length of the reassociated strands is ten times greater than 50 nucleotides (Materials and Methods), so that the hypothetical 50-nucleotide repeat will cause only a twofold increase in the reassociation rate of these strands.

There is a clear difference between these results with *D. melanogaster* middle repetitive DNA and the results with many other species. I find no evidence for a broad heterogeneity in sequence mismatch in *D. melanogaster;* yet, as mentioned previously, this has been found in human, calf, sea urchin, and salmon DNA. Some of the observed breadth in heterogeneity could be attributed to the artifactual binding of single strands to HAP that was discussed in this paper. However, most of these earlier experiments used conditions which should not give this artifact. Thus the difference between the heterogeneity found in *D. melanogaster* middle repetitive DNA and that in other eukaryotes appears real. The *D. melanogaster* middle repeats are not alone, however, since Lee et al. (this volume) report that several middle repetitive DNA families of sea urchin are also well matched. Since this observation differs from the results found with total sea urchin middle repetitive DNA (Britten and Davidson 1971), it is possible that, as discussed above, these few sea urchin DNA families have a distribution of homology similar to *D. melanogaster*'s, whereas other sea urchin families have different distributions.

Acknowledgments

I thank Elizabeth Levine for excellent technical assistance, Marta Izquierdo for help during preparation of middle repetitive fractions, Michael Rosbash for S_1 nuclease, and Sally Elgin for gifts of *D. melanogaster* embryos. I also thank Joel Loewenberg, Marta Izquierdo, Robert Schleif, John Woolford, and Michael Rosbash for their useful criticism of the manuscript. This research was supported by National Institutes of Health research grant GM-21626 and by National Institutes of Health Research Career Development Award GM-00212.

REFERENCES

BRITTEN, R. J. and E. H. DAVIDSON. 1971. Repetitive and non-repetitive DNA sequences and a speculation on the origins of evolutionary novelty. *Q. Rev. Biol.* **46**: 111.

BRITTEN, R. J. and D. E. KOHNE. 1968. Repeated sequences in DNA. *Science* **161**: 529.

BRITTEN, R. J., D. E. GRAHAM, and B. R. NEUFELD. 1973. Analysis of repeated DNA sequences by reassociation. *Methods Enzymol.* **29**: 363.

CRAIN, W. R., F. C. EDEN, W. R. PEARSON, E. H. DAVIDSON, and R. J. BRITTEN. 1976. Absence of short period interspersion of repetitive and non-repetitive sequences in the DNA of *Drosophila melanogaster*. *Chromosoma* **56**: 309.

DAVIDSON, E. H. and R. J. BRITTEN. 1973. Organization, transcription and regulation in the animal genome. *Q. Rev. Biol.* **48**: 565.

GILBERT, W. and D. DRESSLER. 1969. DNA replication: The rolling circle model. *Cold Spring Harbor Symp. Quant. Biol.* **33**: 473.

GLOVER, D. M. and D. S. HOGNESS. 1977. A novel arrangement of the 18S and 28S sequences in a repeating unit of *Drosophila melanogaster* rDNA. *Cell* **10**: 167.

HERSHEY, N. D., S. E. CONRAD, A. SODJA, P. H. YEN, M. COHEN, JR., N. DAVIDSON, C. ILGEN, and J. CARBON. 1977. The sequence arrangement of *Drosophila melanogaster* 5S DNA cloned in recombinant plasmids. *Cell* **11**: 585.

HUTTON, J. R. and J. G. WETMUR. 1973. Effect of chemical modification on the rate of renaturation of deoxyribonucleic acid. Deaminated and glyoxalated deoxyribonucleic acid. *Biochemistry* **12**: 558.

INGRAM, V. M. 1963. *Hemoglobins in genetics and evolution.* Columbia University Press, New York.

KOHNE, D. 1970. Evolution of higher organism DNA. *Q. Rev. Biophys.* **3**: 327.

LAIRD, C. D. and B. J. MCCARTHY. 1969. Molecular characterization of the *Drosophila* genome. *Genetics* **63**: 865.

LEE, C. S. 1975. A possible role of repetitious DNA in recombinatory joining during chromosome rearrangement in *Drosophila melanogaster*. *Genetics* **79**: 467.

MANIATIS, T., A. JEFFREY, and D. G. KLEID. 1975. Nucleotide sequence of the rightward operator of phage λ. *Proc. Natl. Acad. Sci.* **72**: 1184.

MANNING, J. E., C. W. SCHMID, and N. DAVIDSON. 1975. Interspersion of repetitive and nonrepetitive DNA sequences in the *Drosophila melanogaster* genome. *Cell* **4**: 141.

MARMUR, J. and P. DOTY. 1962. Determination of the base composition of deoxyribonucleic acid from its thermal denaturation temperature. *J. Mol. Biol.* **5**: 109.

OHNO, S. 1970. *Evolution by gene duplication.* Springer Verlag, New York.

PELLEGRINI, M., J. MANNING, and N. DAVIDSON. 1977. Sequence arrangement of the rDNA of *Drosophila melanogaster*. *Cell* **10**: 213.

PTASHNE, M., K. BACKMAN, M. Z. HUMAYUN, A. JEFFREY, R. MAURER, B. MEYER, and R. T. SAUER. 1976. Autoregulation and function of a repressor in bacteriophage lambda. *Science* **194**: 156.

RIGBY, P. W. J., M. DIECKMANN, C. RHODES, and P. BERG. 1977. Labeling of deoxyribonucleic acid to high specific activity *in vitro* by nick translation with DNA polymerase I. *J. Mol. Biol.* **113**: 237.

SCHACHAT, F. S. and D. S. HOGNESS. 1974. Repetitive sequences in isolated Thomas circles from *Drosophila melanogaster*. *Cold Spring Harbor Symp. Quant. Biol.* **38**: 371.

SCHILDKRAUT, C. L., C. C. RICHARDSON, and A. KORNBERG. 1964. Enzymic synthesis of deoxyribonucleic acid. XVII. Some unusual physical properties of the product primed by native templates. *J. Mol. Biol.* **9**: 24.

SCHMID, C. W., J. E. MANNING, and N. DAVIDSON. 1975. Inverted repeat sequences in the *Drosophila* genome. *Cell* **5**: 159.

SMITH, G. P. 1974. Unequal crossover and the evolution of multigene families. *Cold Spring Harbor Symp. Quant. Biol.* **38**: 507.

STUDIER, F. W. 1965. Sedimentation studies of the size and shape of DNA. *J. Mol. Biol.* **11**: 33.

————. 1969. Effects of the conformation of single-stranded DNA on renaturation and aggregation. *J. Mol. Biol.* **41**: 199.

SZYBALSKI, W. and E. H. SZYBALSKI. 1971. Equilibrium density gradient centrifugation. In *Procedures in nucleic acid research* (ed. G. L. Cantoni and D. R. Davies), vol. 2, p. 311. Harper and Row, New York.

THOMAS, C. A., JR. and B. M. DANCIS. 1973. Ring stability. *J. Mol. Biol.* **77**: 43.

VOGT, V. M. 1973. Purification and further properties of single-strand specific nuclease from *Aspergillus oryzae*. *Eur. J. Biochem.* **33**: 192.

WELLAUER, P. K. and I. B. DAWID. 1977. The structural organization of ribosomal DNA in *Drosophila melanogaster*. *Cell* **10**: 193.

WENSINK, P. C., D. J. FINNEGAN, J. E. DONELSON, and D. S. HOGNESS. 1974. A system for mapping DNA sequences in the chromosomes of *Drosophila melanogaster*. *Cell* **3**: 315.

WILSON, D. A. and C. A. THOMAS, JR. 1973. Hydroxyapatite chromatography of short double-helical DNA. *Biochim. Biophys. Acta* **331**: 333.

Isolation of a Telomeric DNA Sequence from
Drosophila melanogaster

G. M. RUBIN

Department of Basic Sciences, Sidney Farber Cancer Institute, Harvard Medical School, Boston, Massachusetts 02115

There is substantial genetic and cytological evidence that the ends of a eukaryotic chromosome, the telomeres, are specialized structures which can mediate associations of chromosomes. On one hand, the telomeres appear to prevent the ends of chromosomes from fusing with one another (McClintock 1938, 1939, 1940, 1942). On the other hand, they promote transient, end-to-end associations of nonhomologous chromosomes during interphase and meiotic prophase in plant, insect, and mammalian nuclei (see White 1948; DuPraw 1970). This paper reports the finding of a DNA sequence, contained in the cloned segment Dm356, which is common to the termini of the five major chromosome arms of *Drosophila melanogaster*. Such a sequence may be a component of a common telomere structure and could explain such properties of telomeres as their capacity to associate.

That telomeres may be involved in preventing the fusion of normal chromosome ends was first indicated by McClintock's analysis of the behavior of broken chromosomes in maize (McClintock 1938, 1939, 1940, 1942). These chromosomes, which lack telomeres, were often observed to fuse. The dicentric chromosome generated by this fusion was itself pulled apart and broken at the next anaphase, initiating a continuing breakage-fusion cycle. Similar observations have been made on the behavior of broken chromosomes in *Drosophila,* where there is extensive evidence that the ends of these chromosomes can fuse only with the ends of other broken chromosomes, and never with normal ends (Muller and Herskowitz 1954). These data led Muller (1932, 1942) to propose that telomeres are specific chromosomal "organelles" necessary for chromosome stability. Although terminal deletions of the X chromosome in *Drosophila* have been reported (Demerec and Hoover 1936; Sutton 1940), the results of Muller and Herskowitz (1954), and more recently, of Roberts (1975) strongly suggest that these are not genuine terminal deletions, but rather interstitial deletions which preserve the original chromosome end.

That the transient end-to-end association of chromosomes may be due to sequence homology of telomeres was suggested by White (1961). Such associations have been extensively studied in the interphase polytene nuclei of *Drosophila* salivary glands (Hinton 1945; Warters and Griffen 1950; Poluéktova 1975). All pairwise associations of telomeres occur, but with differing frequencies. The variation in frequency is strain-dependent and appears to be a property of the chromosome ends themselves. These findings suggest that, although all chromosome ends may be homologous, some are more homologous than others.

MATERIALS AND METHODS

D. melanogaster (Oregon R) DNA was isolated from nuclei of 0–15-hr embryos according to the method of Schachat and Hogness (1974) with minor modifications. The restriction endonucleases *Eco*RI, *Bam*HI and *Kpn*I were obtained from New England BioLabs. Hybrid DNA molecules were constructed (D. J. Finnegan et al., in prep.) by the method of Wensink et al. (1974). Agarose gel electrophoresis and electron microscopy of DNA were performed as described by Finnegan et al. (this volume). *Drosophila* embryo nuclei were isolated and digested with micrococcal nuclease following the protocol of A. Prunell and R. D. Kornberg (in prep.).

Hybridization of radioactive RNA to DNA immobilized on nitrocellulose was carried out in 4 × SSCP (1 × SSCP = 0.12 M NaCl, 0.015 M sodium citrate, 0.02 M sodium phosphate, pH 7) containing 0.5% SDS and 500 μg/ml yeast tRNA, for 16 hours at 65°C, followed by washing for 6–8 hours in 2 × SSCP, 0.5% SDS at 65°C. The thermal stability of DNA-RNA hybrids was determined by the method of Church and McCarthy (1968). Embryonic DNA was digested with *Eco*RI, fractionated by agarose gel electrophoresis, and transferred to nitrocellulose strips (Southern 1975). After hybridization with [^{32}P]cRNA made from the 3.0-kb repeating unit of Dm356 (see below), the strips were incubated in 1 × SSCP at 65°C for 30 minutes, and duplicate strips were then removed. The temperature of the buffer was raised, the incubation continued for 20 minutes, and duplicate strips again removed. The extent of hybridization to strips removed at 65°C, 70°C, 75°C, 80°C, and 85°C was estimated from microdensitometer tracings of autoradiograms, and approximate T_m's were determined.

RESULTS AND DISCUSSION

Dm356 Contains a Tandem Repeat

The cloned segment of *D. melanogaster* DNA, Dm356, was originally selected as part of another study because of its homology to an abundant cellular RNA. This RNA coding sequence has been shown to be a member of the *copia* multigene family (see Finnegan et al., this volume). A physical map of the plasmid containing Dm356 is shown in Figure 1. The region of homology to RNA is confined to the extreme left-hand end. The remainder of the DNA consists of four tandem repeats of a 3.0-kb unit, as judged by the repetition of cleavage sites for the restriction enzymes *Eco*RI, *Bam*HI, and *Kpn*I.

Sequences Homologous to the Dm356 Repeat Unit Are Found at the Chromosome Termini and in Fibers Connecting Them

A radioactive probe made from cDm 356, hybridized in situ to the polytene chromosomes, labels the tips of the five major chromosome arms, and also the thin fibers (see Hinton 1945) often seen connecting them (see Fig. 2). These same chromosome regions are also labeled when a probe consisting only of the purified 3-kb repeat unit is used. These results suggest that all the *D. melanogaster* telomeres share at least one property — a DNA sequence homologous to the 3-kb repeat unit found on Dm356. Conversely, this evidence for a common property suggests that the telomere is indeed a distinctive structure. How close does the 3-kb repeat unit actually lie to the chromosome end? The labeling of fibers extending from the ends suggests that the 3-kb sequence lies extremely close and may be truly terminal.

Dm356 Homologous Sequences Are Found in Repeat Units of Several Different Lengths

The pairwise associations of telomeres vary in both strength and frequency of occurrence (Hinton

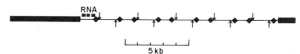

Figure 1. Physical map of cDm356. The thin horizontal line represents *Drosophila* DNA. The thick line represents DNA of the plasmid vector ColE1. The circular plasmid is represented as a linear molecule opened at a *Sma*I restriction enzyme cleavage site. The symbols represent the cleavage sites for the restriction enzymes *Eco*RI (↓), *Bam*HI (↑), and *Kpn*I (♦). A scale in kilobases (1000 base pairs = 1 kb) is shown. The approximate location of the sequences complementary to an abundant cellular RNA is indicated above the map.

1945), suggesting that the sequences found at the ends of individual chromosome arms, although homologous, may not be identical. Such heterogeneity might be revealed as a difference in the spacing of restriction enzyme sites within those sequences in the genome which are homologous to the 3-kb repeat of Dm356. To investigate this possibility, embryonic DNA was digested with a restriction endonuclease, fractionated by electrophoresis in an agarose gel, transferred to a nitrocellulose strip, and hybridized with a cRNA probe made from the 3-kb Dm356 repeat unit.

Several fragments were found which hybridized to the 3-kb Dm356 repeat unit, but which would not be derived from digestions of repeat units of the Dm356 type (see Fig. 3). The presence of these additional fragments shows that there is indeed heterogeneity among those sequences in the genome having homology to the 3-kb repeat. The extent of hybridization to the additional fragments shows that they are repeated in the genome. Our working hypothesis is that different repeat units may be found at the ends of different chromosome arms.

The degree of sequence homology between the Dm356 repeat and the other repeat classes can be estimated by determining the thermal stability of the DNA-RNA hybrids on the nitrocellulose strips described above. For example, among repeat classes revealed by *Eco*RI digestion, those of 3.9, 3.0, 2.15, and 1.9 kb are closely homologous to the Dm356 repeat since the T_m's of these hybrids lie between 80°C and 85°C (see Materials and Methods). However, the 2.4-kb repeat unit is only partially homologous since this hybrid melts between 70°C and 75°C.

The presence of a homologous sequence at all the chromosome ends is an essential feature of several models for the replication of chromosome ends. One model, proposed by Watson (1972), involves the formation of circular or concatemeric replicative intermediates. Cavalier-Smith (1974) and Bateman (1975) have proposed models which suggest that chromosome ends consist of palindromic base sequences. Although we do not yet have any data concerning the structure of the end of the DNA molecule itself, preliminary electron microscopic characterization of the 3.0-kb repeat of Dm356 does not reveal any palindromic sequences.

Dm356 Sequences Are Found in Nucleosomes

Some of the properties of telomeres, such as their tendency to associate with the nuclear membrane (see White 1948), may be due not only to their DNA sequences but also to specific proteins and chromatin structures associated with them. Chromosome termini often appear as distinct structures in both light (Warters and Griffen 1950) and electron micrographs (Berendes and Meyer 1968). Is the DNA in

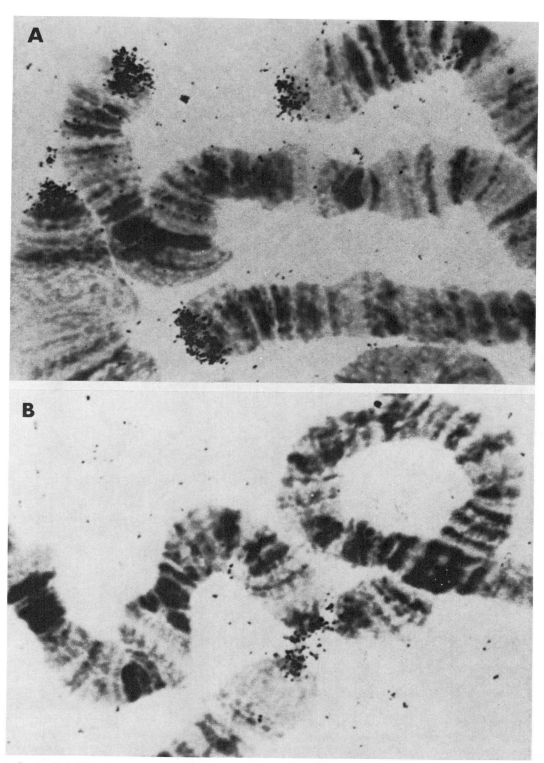

Figure 2. In situ hybridization of cDm356 sequences to *D. melanogaster* polytene chromosomes. cDm356 DNA was labeled with tritium by nick-translation (Rigby et al. 1977) and the hybridization was carried out as described previously (Wensink et al. 1974). The distal regions of four chromosome arms are shown in *A* and several examples of end-to-end associations of chromosome termini are shown in *B, C,* and *D*.

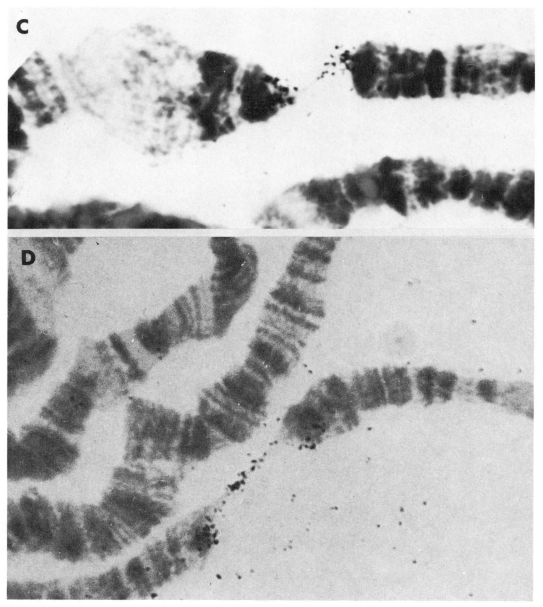

Figure 2 *(continued)*

these structures complexed with histones in the form of nucleosomes? To answer this question, nuclei were isolated from *Drosophila* embryos, digested with micrococcal nuclease, and deproteinized. The DNA fragments were then separated according to size and assayed for the presence of Dm356 repeat-unit sequences (see Fig. 4). These sequences are clearly present in the monomer fragment and its multimers, indicating that they are contained within nucleosomes. This experiment would not detect, however, a small fraction of repeats which were organized in a different, more nuclease-sensitive structure.

CONCLUSION

The telomere, as a cytologically and genetically defined entity, has long been known to be essential for the stability of chromosomes. Like the centromere, the telomere is a chromosomal region whose major functions are mechanical and structural; both aid in preserving the orderly arrangement of nuclear contents and in ensuring the integrity of chromosomes and their normal segregation. We wish to make use of the molecular probe for telomere regions described in this paper to isolate and characterize cloned DNA segments which originate from

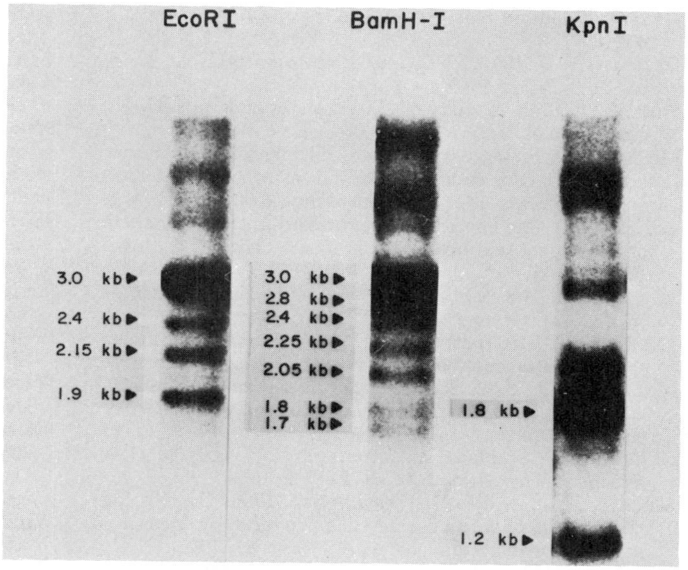

Figure 3. The pattern of restriction enzyme sites in regions of the *Drosophila* genome homologous to the 3-kb repeat of Dm356. *D. melanogaster* embryonic DNA was digested with either *Eco*RI, *Bam*HI, or *Kpn*I. The products were separated by electrophoresis in a 1.5% agarose gel and then transferred to nitrocellulose strips by the procedure of Southern (1975). Fragments containing sequences homologous to the 3.0-kb Dm356 repeat unit were revealed by hybridization to [^{32}P]cRNA made from this repeat unit. Autoradiograms of the strips are shown. The lengths of several of the more prominent fragments are indicated.

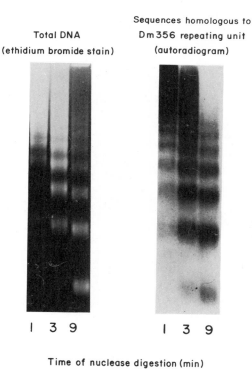

Total DNA
(ethidium bromide stain)

Sequences homologous to
Dm356 repeating unit
(autoradiogram)

Time of nuclease digestion (min)

Figure 4. Dm356 sequences are found in nucleosomes. Nuclei isolated from *Drosophila* embryos were digested with micrococcal nuclease for 1, 3, or 9 min. After deproteinization, the resultant DNA fragments were separated by electrophoresis in a 1.5% agarose gel *(left panel)*. The position of sequences homologous to the Dm356 repeat unit were determined *(right panel)* as described in the legend to Fig. 3.

the telomere of each chromosome arm. Such analysis of the structure of telomeres should aid in defining their functions at the molecular level.

Acknowledgments

The isolation and preliminary characterization of cDm356 was carried out during the tenure of a Helen Hay Whitney Fellowship in the laboratory of Dr. David S. Hogness, Department of Biochemistry, Stanford University. I am particularly grateful to Mariana Wolfner, David S. Hogness, David Finnegan, and Pamela Dunsmuir for many useful discussions and to Sarah Elgin for providing the *Drosophila* embryos used in some of these experiments. This work was supported by National Institute of Health Grant GM-23614.

REFERENCES

BATEMAN, A. J. 1975. Simplification of palindromic telomere theory. *Nature* **253:** 379.

BERENDES, H. D. and G. F. MEYER. 1968. A specific chromosome element, the telomere of *Drosophila* polytene chromosomes. *Chromosoma* **25:** 184.

CAVALIER-SMITH, T. 1974. Palindromic base sequences and replication of eukaryotic chromosome ends. *Nature* **250:** 467.

CHURCH, R. B. and B. J. MCCARTHY. 1968. Related base sequences in the DNA of simple and complex organisms. II. The interpretation of DNA/RNA hybridization studies with mammalian nucleic acids. *Biochem. Genet.* **2:** 55.

DEMEREC, M. and M. E. HOOVER. 1936. Three related X chromosome deficiencies in *Drosophila* differing in extent of deleted material and in viability. *J. Hered.* **27**: 206.

DuPRAW, E. J. 1970. *DNA and chromosomes.* Holt, Rinehart and Winston, New York.

HINTON, T. 1945. A study of chromosome ends in salivary gland nuclei of *Drosophila. Biol. Bull.* **88**: 144.

McCLINTOCK, B. 1938. The fusion of broken ends of sister half-chromatids following chromatid breakage at meiotic anaphases. *Mo. Agric. Exp. Stn. Res. Bull.* **290**: 1.

_____. 1939. The behavior in successive nuclear divisions of a chromosome broken at meiosis. *Proc. Natl. Acad. Sci.* **25**: 405.

_____. 1940. The stability of broken ends of chromosomes in *Zea mays. Genetics* **26**: 234.

_____. 1942. The fusion of broken ends of chromosomes following nuclear fusion. *Proc. Natl. Acad. Sci.* **28**: 458.

MULLER, H. J. 1932. Further studies on the nature and causes of gene mutations. *Proc. VIth Int. Congr. Genet.* (Ithaca) **1**: 213.

_____. 1942. Induced mutations in *Drosophila. Cold Spring Harbor Symp. Quant. Biol.* **9**: 151.

MULLER, H. J. and I. H. HERSKOWITZ. 1954. Concerning the healing of chromosome ends produced by breakage in *Drosophila melanogaster. Am. Nat.* **88**: 177.

POLUÉKTOVA, E. V. 1975. Associations of telomere regions of the salivary gland chromosomes in *Drosophila* species of the *virilis* group. *Genetika* **11**: 54.

RIGBY, P. W. J., M. DIECKMANN, C. RHODES, and P. BERG.

1977. Labelling deoxyribonucleic acid to high specific activity *in vitro* by nick translation with DNA polymerase I. *J. Mol. Biol.* **113**: 237.

ROBERTS, P. A. 1975. In support of the telomere concept. *Genetics* **80**: 135.

SCHACHAT, F. H. and D. S. HOGNESS. 1974. Repetitive sequences in isolated Thomas circles from *Drosophila melanogaster. Cold Spring Harbor Symp. Quant. Biol.* **38**: 371.

SOUTHERN, E. 1975. Detection of specific sequences among DNA fragments separated by gel electrophoresis. *J. Mol. Biol.* **98**: 503.

SUTTON, E. 1940. Terminal deficiencies in the X chromosome of *Drosophila melanogaster. Genetics* **25**: 628.

WARTERS, M. and A. B. GRIFFEN. 1950. The telomeres of *Drosophila. J. Hered.* **41**: 182.

WATSON, J. D. 1972. Origin of concatemeric T7 DNA. *Nat. New Biol.* **239**: 197.

WENSINK, P. C., D. J. FINNEGAN, J. E. DONELSON, and D. S. HOGNESS. 1974. A system for mapping DNA sequences in the chromosomes of *Drosophila melanogaster. Cell* **3**: 315.

WHITE, M. J. D. 1948. *Animal cytology and evolution.* Cambridge University Press, Cambridge, England.

_____. 1961. The role of chromosomal translocations in *Urodele* evolution and speciation in the light of work on grasshoppers. *Am. Nat.* **95**: 315.

The Organization of the Histone Genes in *Drosophila melanogaster:* Functional and Evolutionary Implications

R. P. LIFTON, M. L. GOLDBERG, R. W. KARP, AND D. S. HOGNESS

Department of Biochemistry, Stanford University School of Medicine, Stanford, California 94305

The expression of the genes for each of the five histone proteins is tightly regulated with respect to the cell cycle, the developmental stage, and the expression of the other histone genes. In several species studied, the synthesis of all five proteins occurs exclusively in concert with the DNA synthetic phase of the cell cycle (Robbins and Borun 1967; Kedes et al. 1969; Perry and Kelley 1973); the four histones of the nucleosome core are found in equimolar amounts, whereas H1 protein is found in less than molar quantity (for review, see Kornberg 1974). Variant histone proteins associated with specific developmental stages have been observed (Cohen et al. 1975; Arceci et al. 1976; Newrock and Cohen, this volume), indicating that there must be genes for these variant proteins which are differentially expressed according to precise developmental programs. *Drosophila melanogaster* is an appealing organism in which to study these genes and their expression for several reasons: (1) Its hereditary mechanics are understood in detail; (2) its polytene chromosomes allow high-resolution cytological mapping of the genome; and (3) it is easily accessible to experimental manipulation by virtue of the availability of tissue culture cells, a short life cycle that consists of well-defined developmental stages, and a small genome that facilitates the isolation of specific genes in cloned DNA segments (Dm segments). In this paper we review our knowledge of the organization of the *D. melanogaster* histone genes and discuss some of the functional and evolutionary implications of this organization.

Characterization of the *D. melanogaster* Histone Gene Repeat Unit

The amino acid sequences of the histone proteins are highly conserved across species barriers; consequently, histone mRNAs, readily isolated from sea urchins, can hybridize to DNA from a wide variety of organisms (Kedes and Birnstiel 1971; Farquhar and McCarthy 1973). The histone genes of *D. melanogaster* were therefore first isolated by screening a collection of ColE1 hybrid plasmids containing randomly generated segments of *D. melanogaster* chromosomal DNA (cDm plasmids) for those that can hybridize with sea urchin histone mRNA (Karp and Hogness 1976, and in prep.). The plasmid cDm500 was obtained from this screen. Figure 1A shows a restriction map of the Dm500 segment carried by this hybrid and reveals that it consists of a 4.8-kb sequence repeated in tandem 1.8 times. The reassociation kinetics of this repeat unit in the presence of a vast excess of total *D. melanogaster* DNA indicates that its sequences are repeated approximately 100 times per haploid genome. Virtually all copies of this DNA sequence are located in region 39DE of *D. melanogaster* salivary gland polytene chromosomes, as shown by in situ hybridization using ^{3}H-labeled cDm500 DNA (Karp and Hogness 1976, and in prep.).

The organization of genes within the repeating unit of Dm500 has been elucidated. Five poly(A)$^-$ RNAs present in *D. melanogaster* embryos and tissue-culture cells are complementary to this DNA, and each exhibits a length in the range expected for histone mRNAs. These RNAs were individually purified by annealing them to denatured cDm500 DNA covalently linked to cellulose powder (Noyes and Stark 1975) and subsequent gel electrophoresis. The purified RNAs were then hybridized to sets of restriction fragments from cDm500 to map the sequences in the DNA that are homologous to each of the five classes of RNA (R. Lifton et al., in prep.). Figure 1B shows that all five of the genes coding for these RNAs are confined to a single repeat unit. These genes were then identified by determining the nucleotide sequence of a portion of the DNA within each homology region and showing that such a sequence could be conceptually translated to yield the amino acid sequence of part of one of the histone proteins (M. Goldberg et al., in prep.).

The orientation of each gene, that is, the 5'-to-3' direction of the coding sequence and hence the direction of transcription, was determined by annealing the individual RNA species to separated DNA strands of a hybrid λ phage containing the 4.8-kb unit (R. Lifton et al., in prep.). Figure 1B shows that the direction of transcription of each successive gene alternates as one proceeds to the right across the map from the H3 gene to the H1 gene. Three genes (H3, H2A, H1) are therefore transcribed from one DNA strand, and two (H4, H2B) from the other strand. These directions are also deducible from the nucleotide sequence of the DNA in each gene (M. Goldberg et al., in prep.), and they agree with this direct experimental determination.

Figure 1. Maps of histone gene repeating units. *(A)* Restriction map of cDm500. The open rectangles and horizontal line represent, respectively, the ColE1 and Dm500 segments joined in cDm500 by poly(dA)-poly(dT) connectors according to the method of Wensink et al. (1974). The vertical lines represent restriction enzyme sites in the Dm500 segment. Starting with the leftmost site and progressing to the right, the sites are, respectively, *BglII, BamI, HindIII, HpaI*, and *SstI*. Each of these sites is repeated in the Dm500 segment with a spacing of 4.8 kb. *(B)* Arrangement of the histone genes in *D. melanogaster* repeat units. The two classes of units are represented, one class differing from the other by an insertion in the spacer between H1 and H3. The scale in kilobases is indicated by the double arrow and applies to both *B* and *C*. *(C)* Arrangement of the histone genes in *S. purpuratus* (Cohn et al. 1976).

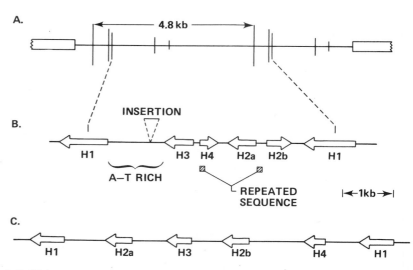

Is the histone gene organization in cDm500 representative of all 100 copies of this sequence in the *D. melanogaster* genome, or is there heterogeneity in the topographies of either the genes or their intervening spacers? Using labeled cDm500 DNA as a probe for homologous sequences in restriction enzyme digests of total *D. melanogaster* DNA blotted onto nitrocellulose by the procedure of Southern (1975), we have found two types of repeating unit. One corresponds to the 4.8-kb unit contained in cDm500, and the other has the same gene organization but contains an additional block of 270 base pairs in the spacer separating H1 from H3 (Fig. 1B). This latter type of repeat unit is three times as frequent as the former in the *D. melanogaster* genome (Karp and Hogness 1976, and in prep.). These two types of units constitute the vast majority of the *D. melanogaster* histone gene sequences.

Models for Histone Gene Transcription

The map of the *D. melanogaster* histone genes enables us to exclude certain models for the mode of expression of these genes. Since the five genes are not all transcribed from the same DNA strand, we can rule out the possibility that coordinate expression of these genes is effected by a single polycistronic transcript that is cleaved to yield the five mature RNAs. Assuming that the RNA polymerase cannot switch from one strand to the other during transcription of these genes, the five RNAs must derive from at least two transcripts. Such transcripts could arise in a variety of ways. For instance, promoters lying outside the cluster of histone gene repeat units could initiate wholesale transcription of both DNA strands such that the individual transcripts cover many repeat units. At the other extreme, two polycistronic transcripts could extend across just enough of each strand to include the

three or two RNAs derived from that strand. Since genes of opposite orientations are interdigitated throughout the repeat unit, any two-transcript model requires the transcription of both sense and antisense strands within the units.

A model in which each of the five genes is individually transcribed to yield RNAs that closely approximate the mature message seems particularly suited to the organization of the *D. melanogaster* histone genes. In such a scheme, transcription would originate in both directions from the spacers lying between gene pairs H3-H4 and H2A-H2B, whose members have their 5′ ends in close proximity. By contrast, the 5′ end and presumptive promoter of H1 mRNA lies in relative isolation from the other genes, separated by approximately 1200 base pairs from the 3′ end of H3. Since histones H2A, H2B, H3, and H4 are coordinately synthesized and utilized in equimolar amounts, one might expect the two proposed sites for the bidirectional promoters to share common DNA sequences. This may in fact be the case. When single strands of a linear DNA consisting of one repeat unit are observed in the electron microscope, hairpin snap-back structures characteristic of a nontandem inverted repeat sequence are seen. The regions of homology are roughly 100 base pairs in length and are at, or very close to, the two proposed locations of bidirectional promoters (Fig. 1B) (R. Lifton et al., in prep.). In this model, H1 would have its own distinct promoter to account for the nonstoichiometric amounts of both H1 protein (Kornberg 1974) and mRNA (R. Lifton et al., in prep.) found in the cell.

Preliminary evidence from our laboratory is consistent with this model. Hybridization of electrophoretically separated total RNA from cultured cells labeled by a 12-minute pulse of [³H]uridine to cDm500 DNA revealed no histone sequences in transcripts perceptibly larger than the mature histone mRNAs. In addition, we have looked for histone-specific transcripts in total cell RNA by the sensitive RNA-blotting technique developed by Alwine et al. (1977), using ^{32}P-labeled cDm500 DNA as the probe. Again, the sequences homologous to cDm500 are confined to molecules the size of the histone mRNAs themselves (R. Lifton, unpubl.). Definitive proof that the primary transcripts are very similar to the mature histone mRNAs will require the demonstration of a nucleoside 5' triphosphate on such species.

Comparison of *D. melanogaster* and Sea Urchin Histone Genes

The organization of the histone genes in several species of sea urchin has been elucidated (Fig. 1C). In each, all five genes are contained in a unit that is tandemly repeated, and all the genes are transcribed in the same direction. The gene order is identical in all species examined (for review, see Kedes 1976). Bearing in mind that the branches of the phylogenetic tree which lead, respectively, to the sea urchins and the fruit flies diverged roughly 600 million years ago (Dickerson 1971), it is of interest to ask what structural and organizational features have been conserved since this evolutionary divergence, since we may suppose that conserved features will be those most crucial for the maintenance of proper gene expression.

From a cursory comparison of the sea urchin and *D. melanogaster* histone genes, one sees that the unit of organization is the same; each of the five genes is represented once per unit. The organization within this unit, however, is quite different. The *D. melanogaster* histone genes are transcribed from both DNA strands, and the gene order is completely rearranged with respect to that in sea urchins.

Although the histone protein sequences are rigidly conserved in evolutionary time, we find it curious that the arrangement of the histone genes within the repeat unit is capable of undergoing extensive changes. In considering these changes, it is of interest to determine the shortest paths of interconversion between the arrangements in sea urchins and *D. melanogaster*. At least five breaks are required to interconvert the two gene orders. Among the various five-break paths for gene-order conversion, only two will also interconvert the transcriptional orientations. These are indicated in Figure 2. Each involves a single intermediate (or common precursor) that can be converted to the sea urchin or *D. melanogaster* arrangement by a two- or three-break path.

The finding that both organisms carry repeat units containing all five genes arranged in such a way that they can be interconverted by relatively simple paths leads us to suggest that the five histone genes were linked in those species whose descendants subsequently diverged to give rise to the Protostomia and Deuterostomia. Assuming that evolution is economical, we might then expect the ancestral gene arrangement to be one of the four shown in Figure 2; i.e., we might expect to find one of these four arrangements in the primitive bilaterally symmetrical organisms whose prototypes evolved into the Protostomia and Deuterostomia. In any case, it would be important to a determination of the time and chromosome mechanics involved in the evolution of the two known arrangements of histone genes to investigate such arrangements in species that lie below *D. melanogaster* and sea urchins on these two branches of the phylogenetic tree.

The rearrangement of the histone genes has been accompanied by the divergence of the spacers which separate these coding sequences. In the sea urchin

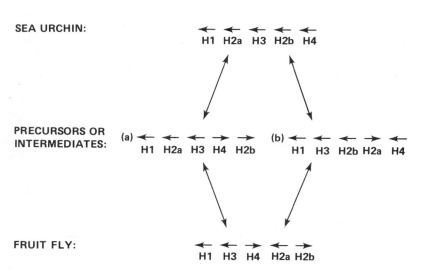

Figure 2. Two possible evolutionary paths for interconverting sea urchin and fruit fly histone gene organizations. The histone gene organization of sea urchin and fruit fly can be interconverted by rearrangements involving five chromosomal breakages involving either of two intermediate topographies. The sea urchin histone gene arrangement can be converted to the arrangement in structure *a* by inverting the H2B-H4 segment with respect to the other genes (two breaks). This structure can then be converted to the fruit fly organization by cutting out H2A and inserting it between H4 and H2B (three breaks). The same end result can be accomplished by cutting out H2A from the sea urchin genes, inverting it, and inserting it between H2B and H4 (three breaks). This structure *(b)* is then converted by an inversion of segment H2B, H2A, and H4 to the fruit fly topography. The path arrows are drawn in both directions since any of the four arrangements could represent the ancestral histone gene topography.

Strongylocentrotus purpuratus, the five spacers vary in length from 600 to 1200 base pairs (Fig. 1C). All of these spacers are A-T-rich in *Psammechinus miliaris,* as indicated by their preferential denaturation in formaldehyde (Portman et al. 1976). In *D. melanogaster,* four spacers are quite short (in the range of 0–250 bp) and the fifth is long (1200 bp). When the *D. melanogaster* repeat unit is spread for electron microscopy in 80% formamide, a single region of denaturation is observed; this region comprises the long spacer that lies between H3 and H1 (M. Goldberg et al., in prep.). These observations therefore yield a rule that applies to the repeat units in sea urchins and in *D. melanogaster:* Whenever adjacent genes are transcribed in the same direction, they are separated by a relatively long A-T-rich spacer. At present we can do little more than note this correlation, since any attempt to evaluate its significance will depend upon the location of the promoters and terminators for the transcription of the respective genes, and these loci have not been determined.

Since the five histone genes in *D. melanogaster* do not function as a single transcriptional unit, one might reasonably ask why these genes are closely linked. One might argue that this linkage simply reflects evolutionary history if, as has been suggested, these genes evolved from a common ancestral gene by repeated tandem duplication followed by divergence (Temussi 1975). However, a strong selection pressure must exist if this linkage has been maintained for roughly 1.2 billion years since these genes arose (Dickerson 1971) or at least, as we have postulated, the last 600 million years. Two functional advantages of this linkage come easily to mind. First, since these genes are reiterated, their linkage in a single unit provides a simple means of maintaining an equal dosage of each gene. Changes in the number of copies of repeat unit arising by unequal crossing-over will equally affect all five genes, thereby avoiding imbalances in dosage that could arise if, for example, each gene were individually reiterated in tandem. Linkage may also facilitate the coordinate expression of these genes. One can imagine that local events, such as the regional unfolding of the chromatin, could render all five previously inaccessible genes simultaneously available for transcription (see Finnegan et al., this volume, for a general consideration of this kind of control). The linkage of the histone genes could thus help to ensure both the simultaneous synthesis of histone proteins and their synthesis in proper molar ratios.

The Organization of Histone Gene Repeat Units in 39DE

The 100 copies of the *D. melanogaster* histone genes appear to span most of 12 chromomeres in the 39DE region in the left arm of chromosome 2 (Pardue 1975; R. Karp and D. Hogness, in prep.). Partial digests of total *D. melanogaster* DNA indicate that these repeats generally occur in tandem; arrays comprising up to five tandem repeats have been observed (R. Karp and D. Hogness, in prep.). How are the two types of repeat units (i.e., the 4.8-kb and 5.0-kb units) arranged with respect to one another in this region? One can imagine several distinct possibilities for this arrangement. The two types could represent a genetic polymorphism having no functional significance. In this case one might expect recombination to lead to the random interspersion of the two types of repeat units. Alternatively, if each type has a different physiological role, one might expect to find these units arranged in some specific manner. To study this problem we have made use of the fact that the 4.8-kb unit contains no *Eco*RI cleavage site, whereas the 5.0-kb unit contains such a site within the insertion shown in Figure 1B. In probing the total *Eco*RI-cut *D. melanogaster* genome for histone gene sequences, we never find the 4.8-kb unit lying immediately between two 5.0-kb units (R. Karp and D. Hogness, in prep.). We conclude that the two types are not randomly interspersed, since roughly 15 of the 25 4.8-kb units would lie between 5.0-kb units if random interspersion occurred, and these would be easily detectable. These two types of units are not, however, completely segregated. We have cloned other histone gene units. Two plasmids each contain three full repeat units in tandem, and each contains both the 4.8-kb and the 5.0-kb units (R. Lifton, unpubl.). These combined results indicate that the two units are interspersed, but that this interspersion is decidedly nonrandom. These findings support the notion that each of the two types of units has physiological significance to the fly, though speculation regarding the different functions each type might serve is thus far unrestrained by data.

As previously mentioned, there are 12 bands and interbands in the 39DE region. Do the histone gene repeat units traverse all of these chromomeres in a single tandem array consisting of all 100 repeats, or are these arrays interrupted by sequences unrelated to the histone repeat unit or by inversions of some repeat units? As indicated above, tandem arrays of at least five repeat units exist. However, two lines of evidence suggest that interruptions do in fact occur. First, recall that the 4.8-kb units are not found between two 5.0-kb units. If uninterrupted tandem arrays are maintained throughout 39DE, then the 4.8-kb units should be found clustered in groups of two, three, or more units in tandem, bounded on either side by 5.0-kb units. We know that this is not the case, since the Southern blots of *Eco*RI-cut total *D. melanogaster* DNA do not reveal distinct fragments containing histone DNA sequences with lengths that are integral multiples of the unit length. Instead, DNA fragments containing homology to 4.8-kb units are found over a wide length range (R. Karp and D. Hogness, in prep.). The simplest explanation of this observation is that the tandem arrays are interrupted so that the clusters of 4.8-kb units are bounded on at least one side by sequences of variable length or composition that are not homologous to those in the histone repeat unit.

In support of this notion, we have observed that when total *D. melanogaster* DNA is cleaved with *Hin*dIII, which cuts both types of units only once, and then blotted to nitrocellulose filters, we detect many fragments other than those of 4.8 kb and 5.0 kb that contain sequences homologous to the histone repeat unit. These fragments are all present at much lower frequencies per genome relative to the 4.8-kb and 5.0-kb fragments, and most cannot be explained by the gain or loss of a *Hin*dIII site in variant repeat units (R. Karp, unpubl.). Since there is no evidence that regions outside of 39DE have homology to histone DNA, we favor the interpretation that at least some of these extra fragments represent sites where the tandem array is interrupted. Indeed, we have recently cloned two Dm segments that contain histone repeat sequences adjacent to blocks of sequences that are not homologous to histone DNA (R. Karp, unpubl.).

What could be the function of these other sequences? If the formation of band-interband boundaries is determined by specific DNA sequences, such sequences must be present somewhere along the histone gene arrays in order to account for the multiple chromomeres in the 39DE region. It is tempting, therefore, to speculate that these intervening sequences could be involved in such a function. Alternatively, the information required for band-interband boundaries must lie in the histone repeat units themselves, for example, in the 270-bp insertion.

There are a variety of indications that individual chromomeres represent single functional units (Beermann 1972). Does this concept hold in the case of the chromomeres that contain the histone genes? Developmental variants of the histone proteins exist in sea urchins, and there is evidence for multiple forms of the *D. melanogaster* H1 protein (see Newrock and Cohen, this volume). Perhaps each chromomere in the 39DE region contains a developmentally specific set of histone genes that is coordinately activated by the local unfolding of the chromatin of that band. In such a case, we might expect to find all members of each type of variant repeat clustered in a single tandem array bounded on both sides by blocks of nonhistone repeat unit sequences. This macroorganization could then allow the independent regulation of expression of the repeat units containing different histone variants. By an approach combining cloning and cytogenetic techniques, we hope to be able to test these hypotheses.

Acknowledgments

This work was supported by research grants from the National Science Foundation and the National Institutes of Health. R. P. L. is a trainee in the Medical Scientists' Training Program of the U. S. Public Health Service. M. L. G. and R. W. K. were predoctoral fellows of the National Science Foundation and are currently supported by a training grant from the National Institutes of Health.

REFERENCES

ALWINE, J. C., D. J. KEMP, and G. R. STARK. 1977. A method for the detection of specific RNAs in agarose gels by transfer to diazobenzyloxymethyl paper and hybridization with DNA probes. *Proc. Natl. Acad. Sci.* **74**: 5350.

ARCECI, R. J., D. R. SENGER, and P. R. GROSS. 1976. The programmed switch in lysine-rich histone synthesis at gastrulation. *Cell* **9**: 171.

BEERMANN, W. 1972. Chromosomes and genes. *Results Probl. Cell Differ.* **4**: 1.

COHEN, L. H., K. N. NEWROCK, and A. ZWEIDLER. 1975. Stage-specific switches in histone synthesis during embryogenesis of the sea urchin. *Science* **190**: 994.

COHN, R., J. C. LOWRY, and L. H. KEDES. 1976. Histone genes of the sea urchin *(S. purpuratus)* cloned in *E. coli*: Order, polarity and strandedness of the five histone-coding and spacer regions. *Cell* **9**: 147.

DICKERSON, R. E. 1971. The structure of cytochrome C and the rates of molecular evolution. *J. Mol. Evol.* **1**: 26.

FARQUHAR, M. N. and B. J. McCARTHY. 1973. Evolutionary stability of the histone genes of sea urchins. *Biochemistry* **12**: 4113.

KARP, R. and D. S. HOGNESS. 1976. Isolation and mapping of histone genes contained in cloned segments of *Drosophila* DNA. *Fed. Proc.* **35**: 1623.

KEDES, L. H. 1976. Histone messengers and histone genes. *Cell* **8**: 321.

KEDES, L. H. and M. L. BIRNSTIEL. 1971. Reiteration and clustering of DNA sequences complementary to histone messenger RNA. *Nat. New Biol.* **230**: 165.

KEDES, L. H., P. R. GROSS, G. COGNETTI, and A. L. HUNTER. 1969. Synthesis of nuclear and chromosomal proteins on light polyribosomes during cleavage in sea urchin embryo. *J. Mol. Biol.* **45**: 337.

KORNBERG, R. D. 1974. Chromatin structure: A repeating unit of histones and DNA. *Science* **184**: 868.

NOYES, B. E. and G. R. STARK. 1975. Nucleic acid hybridization using DNA covalently coupled to cellulose. *Cell* **5**: 301.

PARDUE, M. L. 1975. Repeated DNA sequences in the chromosomes of higher organisms. *Genetics* **79**: 159.

PERRY, R. P. and D. E. KELLEY. 1973. Messenger RNA turnover in mouse L cells. *J. Mol. Biol.* **79**: 681.

PORTMAN, R., W. SCHAFFNER, and M. BIRNSTIEL. 1976. Partial denaturation of cloned histone DNA from the sea urchin *Psammechinus miliaris*. *Nature* **264**: 31.

ROBBINS, E. and T. W. BORUN. 1967. The cytoplasmic synthesis of histones in HeLa cells and its temporal relationship to DNA replication. *Proc. Natl. Acad. Sci.* **57**: 409.

SOUTHERN, E. M. 1975. Detection of specific sequences among DNA fragments separated by gel electrophoresis. *J. Mol. Biol.* **98**: 503.

TEMUSSI, P. A. 1975. Automatic comparison of the sequence of calf thymus histones. *J. Theor. Biol.* **50**: 25.

WENSINK, P. C., D. J. FINNEGAN, J. E. DONELSON, and D. S. HOGNESS. 1974. A system for mapping DNA sequences in the chromosomes of *Drosophila melanogaster*. *Cell* **3**: 315.

Repeated Gene Families in *Drosophila melanogaster*

D. J. Finnegan,* G. M. Rubin,† M. W. Young,‡ and D. S. Hogness

Department of Biochemistry, Stanford University School of Medicine, Stanford, California 94305

This paper is about repeated gene families in *Drosophila melanogaster*. A repeated gene family consists of many identical or nearly identical genes that cohabit a single haploid genome. In some cases the genes are contained within tandemly repeated DNA units. Indeed, the successive observations that the 18S and 28S rRNA genes in *Xenopus laevis* (Brown and Weber 1968; Birnstiel et al. 1968; Miller and Beatty 1969), the 5S RNA genes of this frog (Brown et al. 1971), and the histone genes of sea urchins (reviewed by Kedes 1976) are all tandemly repeated created an illusion that repeated gene families generally assume such tandem topographies. It is true that these three repeated genes have been found in tandem arrays in other eukaryotes — notably in *D. melanogaster,* which is the only species in which each has been isolated and studied in molecular detail. (rRNA genes: Glover and Hogness 1977; White and Hogness 1977; Wellauer and Dawid 1977; Pellegrini et al. 1977. 5S RNA genes: Procunier and Tartof 1976; Artavanis-Tsakonas et al. 1977. Histone genes: Lifton et al., this volume.) It is not true, however, that tandem repetition constitutes a general rule applicable to all repeated gene families.

We emphasize this lack of generality because we have recently discovered two repeated gene families whose members are widely scattered over the *D. melanogaster* genome. The genes that characterize these families are called *412* and *copia,* and each codes for an abundant mRNA present in a variety of cell types. The first part of this paper consists of a review of the salient properties of these families. Our purpose is to define this new dispersed class so that it may be contrasted to the tandem class in the second part of the paper, where we consider the significance of these two topographies for repeated gene families.

TWO DISPERSED FAMILIES OF REPEATED GENES

The Isolation of Genes That Code for Abundant mRNAs and the Finding of *copia* and *412* Genes among Them

Genes belonging to the *copia* and *412* families can be isolated by a general procedure designed to obtain cloned segments of *D. melanogaster* DNA (Dm segments) that contain genes coding for the abundant mRNAs that are present in any given population of *D. melanogaster* cells (Young and Hogness 1977). The procedure depends upon a random or nearly random set of cloned Dm segments that can be rapidly screened for those segments that will hybridize with any of the abundant mRNAs in heterogeneous preparations obtained from the cells of interest. The set that we use consists of hybrid plasmids constructed from randomly sheared fragments of embryonic nuclear DNA by the dA:dT connector method of Wensink et al. (1974), using the ColE1 plasmid as the vector (D. J. Finnegan et al., in prep.). These hybrid or cDm plasmids are cloned in *Escherichia coli* K-12, where they can be screened in situ by the colony hybridization method of Grunstein and Hogness (1975). The hybridization probe used for the primary screen referred to here consists of ^{32}P-labeled cytoplasmic polyadenylated RNA prepared from Eschalier's Kc$_0$ line of *D. melanogaster* cells. Two sorts of RNAs in this probe are at sufficiently high concentration to yield a positive hybridization response: the abundant mRNAs and the contaminating 18S and 28S rRNAs; hence, two sorts of cDm plasmids are selected by this primary screen: those containing the desired genes and those consisting of rDNA. These can be distinguished by a second colony hybridization with a [^{32}P]cRNA probe obtained by in vitro transcription of a cloned rDNA repeating unit.

When this screening procedure was applied to approximately 3000 independently cloned cDm plasmids, about 3% (103/3050) hybridized with the abundant mRNAs (Young and Hogness 1977). Eight of these exhibited an exceptionally strong response to the colony hybridization, and each was subsequently shown to carry a *copia* gene (M. W. Young and D. S. Hogness, in prep.). Evidently, *copia* codes for a particularly abundant mRNA — a conclusion that was confirmed by quantitative filter hybridization. Thus the *copia* sequences account for at least 3% of the cytoplasmic polyadenylated RNA probe, whereas the cDm DNAs in the remaining positively responding clones hybridized between 0.6% and somewhat less than 0.1% of this probe (mean = 0.3%), as judged from a representative sample of seven clones from this second group (Young and Hogness 1977).

The *412* gene sequences account for approximately 0.6% of the RNA in the probe. This was

Present addresses *Department of Molecular Biology, University of Edinburgh, Edinburgh, Scotland; †Department of Basic Sciences, Sidney Farber Cancer Institute, Harvard Medical School, Boston, Massachusetts; ‡The Rockefeller University, New York, New York 10021.

known before the above screen was initiated from studies on the cDm412 plasmid, which contains the first of the *412* genes to be identified and the one that has been used to define the family (Rubin et al. 1976, and in prep.). A [³²P]cRNA probe containing sequences derived from the *412* gene in cDm412 was used to identify six additional *412*-containing plasmids among the clones provided by the screen.

The number of *412*- and *copia*-containing plasmids obtained in the above screen is large compared to that expected for genes that are not repeated within the *D. melanogaster* genome. The average number of cloned Dm segments that are expected to contain 50% or more of a given nonrepetitive gene in a random set of 3000 segments is 0.2, and the probability of finding two or more segments in the set that contain 50% or more of the same gene is 0.02. (The mean length of the Dm segments in the screen and that used for this calculation is 10 kb; the effective genome size was taken as 132,000 kb, since 20% of the 165,000 kb in the total haploid genome consists of highly repetitive satellite sequences that are not cloned by our procedures [D. J. Finnegan et al., in prep.].) The observed numbers of Dm segments containing *412* and *copia* genes are therefore 30- and 40-fold greater than this single-copy expectation and are indicative of the repetition of these genes within the *D. melanogaster* genome.

Properties of the *412* Gene Family

Mapping the **412** *gene in cDm412.* Figure 1 shows a restriction map of the Dm412 segment in cDm412. To identify sequences in Dm412 that are homologous to the *412* mRNA, restriction fragments generated by consecutive digestion of cDm412 DNA with the *Hind*III and *Eco*RI endonucleases were separated by agarose gel electrophoresis, transferred to a nitrocellulose filter by the method of Southern (1975), and then hybridized to ³²P-labeled cytoplasmic polyadenylated RNA obtained from Kc₀ cells. Only the

contiguous fragments labeled A through F in Figure 1 hybridized with the RNA (Rubin et al. 1976, and in prep.).

Quantitative filter hybridization of this RNA to individual restriction fragments isolated by cloning confirmed these results and allowed us to map the approximate endpoints of the mRNA homology region in the A and F fragments (G. M. Rubin et al., in prep.). Approximately 1.9 kb of A and 1.0 kb of F are homologous to the mRNA, and when these lengths are added to the sum of the B, C, D, and E fragments (4.4 kb), a value of 7.3 kb is obtained for the total length of the mRNA region shown on the map in Figure 1.

Are all of the sequences in this region contained in a single mRNA molecule? Although we cannot yet answer this question with precision, we have shown that most of the cytoplasmic polyadenylated RNAs that hybridize with the DNA in this region exhibit electrophoretic mobilities in formamide acrylamide gels that are characteristic of molecules with lengths greater than 5–6 kb (Rubin et al. 1976, and in prep.). We have also shown that short polyadenylated fragments of these long molecules, obtained by alkaline degradation and subsequent oligo(dT)-cellulose chromatography, hybridize preferentially to the A restriction fragment. The simplest interpretation of these results is that the nucleotide sequence in the majority of the hybridizing RNA molecules matches that for the complete 7-kb mRNA region, with the 3′ and 5′ termini of the RNA corresponding to the endpoints in the A and F fragments, respectively. However, we do not wish to imply that these results eliminate the possibility that sequences in short stretches of DNA within the region may be absent in these RNA molecules.

This mapping of the mRNA region constitutes one definition of the *412* gene in Dm412. We call it an mRNA region and identify *412* as a structural gene because we suppose that the RNA molecules used to map the region are translated. We think this is a reasonable supposition since this RNA accounts for approximately 0.6 and 1.0% of the long-term (8–14 hr) ³²P-labeled polyadenylated RNA in cytoplasmic and polysomal preparations, respectively, and less than one-tenth of it is found in the cytoplasmic RNA fraction that fails to bind to oligo(dT)-cellulose (Rubin et al. 1976). Nevertheless, until translation is demonstrated, this identification must be considered tentative.

Mapping the **412** *genes in polytene chromosomes.* In situ hybridization of ³H-labeled cDm412 DNA to the polytene chromosomes of larval salivary glands results in the labeling of about 70 sites on the chromosome arms and of the chromocenter (G. M. Rubin et al., in prep.). To determine which of these sites contain *412* genes, we first carried out an in situ hybridization with [³H]cRNA to fragment E, which had been purified by cloning. This resulted in the

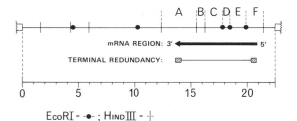

EcoRI = -•- ; HᵢₙₒⅢ = -|-

Figure 1. Map of cDm412 sites cleaved by restriction endonucleases *Eco*RI and *Hind*III. The thin horizontal line represents the Dm412 segment, and the open blocks at the ends of the map represent the ColE1 DNA. The positions of the terminal redundancy described in the text are indicated by the two hatched blocks. An arrow indicates the position of sequences homologous to mRNA and the direction of their transcription. The scale at the bottom of the figure is in kilobases.

labeling of about 30 sites on the chromosome arms and no detectable labeling of the chromocenter. A similar result was obtained after in situ hybridization with [³H]cRNA to the HindIII restriction fragment from cDm412 that contains the C, D, E, and F regions (see Fig. 1); this is shown in Figure 2A.

If the entire 412 gene, as defined in Dm412, is present at each of the sites labeled by the E fragment, we should expect that [³H]cRNA to the A fragment will label each of them. We have tested this expectation for the 11 sites on the X chromosome which are labeled by fragment E and find that each is also labeled by fragment A (G. M. Rubin et al., in prep.). These results suggest that each of the 30 or so sites labeled by fragment E contains a complete 412 gene, and indicate that the Dm412 segment derives from one of them. The number of these sites is consistent with the frequency at which other, independently cloned Dm segments containing 412 sequences were detected in the screen for cDm plas-mids carrying abundant mRNA genes, which was described earlier.

The analysis of restriction fragments from 412 genes at different chromosomal sites. To obtain more precise information about the similarity of the 412 genes at the different chromosomal sites, the restriction fragments formed from them by cleavage with EcoRI and HindIII were examined by two different methods (Rubin et al. 1976, and in prep.). The first consists of an analysis of six independently cloned cDm plasmids that appear to contain the entire 412 gene, some of which were obtained from the screen referred to above and some by screening additional clones from our collection. If these plasmids contain copies of the 412 gene as it exists in cDm412, then we expect each to generate four fragments identical to the internal B, C, D, and E fragments that are formed by cleavage at the EcoRI and HindIII sites within the gene (Fig. 1). Gel electrophoresis of EcoRI-

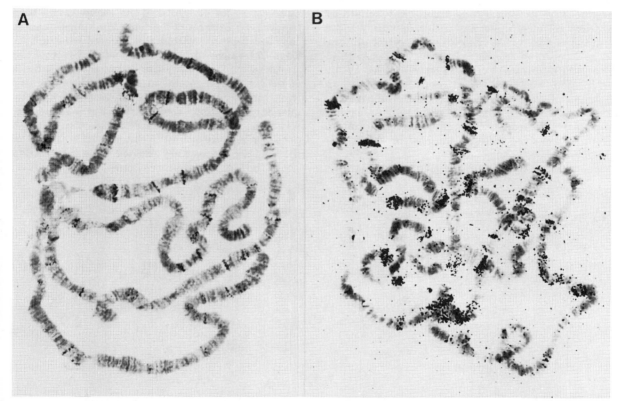

Figure 2. *(A)* Localization of the 412 genes in polytene chromosomes. This autoradiograph was obtained by hybridizing polytene chromosomes, in situ, with [³H]cRNA transcribed from ckDm626DH. The Dm segment in this plasmid consists of the HindIII restriction fragment that encompasses the C, D, E, and F regions of cDm412 (see Fig. 1). (Reprinted, with permission, from Rubin et al. 1976.) *(B)* Mapping the *copia* sites in polytene chromosomes by in situ hybridization. [³H]cRNA was transcribed from pkDm1215DH in which the Dm segment consists of the 4.2-kb *Hha* fragment from within the mRNA complementary region of cDm1142 (Fig. 5). It is curious that in this photomicrograph the end of the left arm of the second chromosome (2L) is not paired and the pattern of labeling on the two homologs clearly differs at three positions *(upper right)*. Another chromosome complement, which was obtained from the same larva, was also unpaired along an even larger region of 2L (not shown), but in this case a difference at only two of these three positions was found. When the homologs are fully paired, all three sites for hybridization are always seen. We are not convinced that these observations reflect sequence differences between these homologs (see Ilyan et al., this volume).

HindIII digests of the plasmids and subsequent hybridization of the separated fragments with ^{32}P-labeled cytoplasmic polyadenylated RNA from Kc₀ cells by the method of Southern (1975) confirmed this expectation; each plasmid yields four fragments that comigrate with the B, C, D, and E fragments and hybridize the RNA. By contrast, we expect the lengths of the terminal fragments that contain the endpoints of the *412* gene to be different for Dm segments derived from different chromosomal sites if those sites have different sequences flanking the gene. None of the plasmids generate fragments that both comigrate with the A or F terminal fragments from cDm412 and hybridize the RNA; rather, they reveal the expected new hybridizing fragments, indicating that the Dm segments they carry derive from different sites with different flanking sequences.

In the second method, we examine the internal and terminal restriction fragments generated from all of the *412* genes contained in the total *D. melanogaster* DNA isolated from the nuclei of embryos. This DNA was digested with *Eco*RI, *Hin*dIII, or both of these restriction endonucleases, and the resulting fragments were fractionated according to length by agarose gel electrophoresis. After denaturing the fragments and transferring them to nitrocellulose, those containing sequences present in the A, B, C, E, and F fragments of cDm412 were assayed by hybridization with [32P]DNA from the appropriate fragment. Figure 3 shows that the autoradiographic patterns generated by this hybridization divide into two classes according to whether the hybridization probe was derived from an internal or a terminal fragment.

The internal B, C, and E fragments yield patterns in which most of the hybridization is confined to fragments of the same length as that used for the probe—more than 95% in the case of the B and C probes and about two-thirds with the E probe. This result, like that obtained with the cloned genes, indicates that most sites contain *412* genes that are very much alike. Although the observation of minor length classes which hybridize with the E probe suggests the existence of some *412* variants that exhibit different sequences in this region near the 5′ end of the gene, we have not yet shown that these minor fragments are in fact linked to the other internal fragments within the genome.

In contrast to the simple pattern obtained with the internal fragments, the terminal fragments generate complex hybridization patterns in which many different length classes are labeled, and the one corresponding to the Dm412 fragment accounts for only a small fraction of the hybridization. This is the expected result if the sequences that flank the *412* genes divide into many different classes, as the preceding analysis of the cloned genes indicates. However, this part of the experiment suffers from the fact that about one-third of the sequences in both the A and F fragments consist of flanking sequences that are repeated within the genome—and not always in linkage with the *412* gene (G. M. Rubin et al., in prep.; D. J. Finnegan et al., in prep.). Some of the labeled fragments in the patterns generated by A and F will therefore result from flanking sequences in the probe, some from *412* gene sequences, and some from both. Thus, although the A and F patterns are consistent with multiple classes of flanking sequences, they provide little information about the number of these classes.

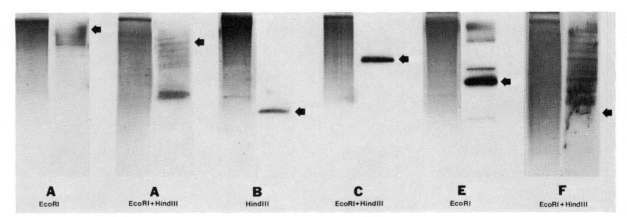

Figure 3. Comparison of the restriction fragments in cDm412 with restriction fragments from total *D. melanogaster* DNA which contain *412* sequences. For each set of results, the left-hand panel shows the ethidium-bromide staining pattern of *D. melanogaster* (Oregon R) embryonic nuclear DNA which has been digested with either *Hin*dIII, or *Eco*RI, or *Hin*dIII and *Eco*RI. (*A,F*) The restriction fragments were separated on a 1.4% agarose gel; (*B,C,E*) the separation was on 1.0% gels. The right-hand panel for each set shows the autoradiograph obtained when these restriction fragments are subsequently hybridized with ^{32}P-labeled restriction fragments derived from cDm412, either A, B, C, E, or F (see Fig. 1), according to the method of Southern (1975). An arrow is used to show the position at which the A, B, C, E, or F restriction fragment from cDm412 migrates, except in the case of the left-hand pair of panels (*A, Eco*RI), where the arrow represents the position of the *Eco*RI fragment from cDM412 that contains the A region (see Fig. 1).

The 412 gene is terminally redundant. During the course of this work we observed that the A and F restriction fragments both can hybridize a [³²P]DNA probe made from one of them, indicating that these terminal fragments carry a common sequence. In attempting to map this sequence repetition, we first asked whether it is represented by an inverted or a direct repeat. Single strands from linear molecules carrying an inverted repeat contain complementary sequences which rapidly interact ("snap back") to form characteristic hairpin structures that are easily recognized in the electron microscope; however, molecules with direct repeats yield single strands that cannot snap back in this manner because the two sequences are identical rather than complementary. Linear cDm412 molecules were therefore denatured and the single strands allowed to anneal at low DNA concentrations prior to examination in the electron microscope. No snap-back structures attributable to interaction of sequences in the A and F regions were observed under conditions where the poly(dA) and poly(dT) sequences of the connectors joining the Dm412 and ColE1 segments do interact to form stable hairpins.

That A and F contain a direct repeat was shown by the simple device of constructing a DNA molecule in which one of these two regions is inverted relative to the other so as to convert any direct repeat into an inverted repeat (D. J. Finnegan et al., in prep.). Figure 4 shows the nature of the rearrangement by which this conversion was effected. Single strands from linear DNA molecules that contain the rearranged segment snap back to yield hairpin structures from which we have deduced the length and position of the direct repeat in the Dm412 segment (Fig. 1). The length of the repeat is 0.49 kb. In A, its left end is located 1.84 kb from the A/B boundary,

and in F, its right end is located 0.96 kb from the E/F boundary.

A comparison of these values with the lengths of the mRNA homology region in A (1.9 kb) and F (1.0 kb) suggests that the ends of this region coincide with the repeat, in which case we should expect other genes in the *412* family to contain this terminal redundancy. We have examined two other cloned *412* genes for the presence of the repeat by hybridizing [³²P]DNA from fragment F to the restriction fragments generated by cleavage of the respective cDm plasmids with *Eco*RI and *Hind*III. In both cases, two fragments hybridize with this probe, and these are the ones we had previously inferred to contain the ends of the mRNA homology region, i.e., the terminal fragments. Although more extensive and precise mapping studies are required to determine the nature and uniformity of the relationship between the ends of the mRNA region and the repeat, the present evidence strongly suggests that the genes in the *412* family are characterized by a terminal redundancy of approximately 500 base pairs, representing 7% of the gene.

Properties of the *copia* Gene Family

The general characteristics of the *copia* and *412* families are remarkably similar. Most of the *copia* characteristics were initially determined for the cloned gene in cDm1142, to which other members of the family were then compared (M. W. Young and D. S. Hogness, in prep.). A total of ten different members have been cloned and used in this comparison. Two of these (pPW220 and pPW221) derive from Pieter Wensink's collection of Dm segments cloned with the pMB9 plasmid vector (Wensink, this volume) and were independently recognized to contain sequences homologous to an RNA from a dispersed gene family (S. Henikoff et al., pers. comm.). The remaining eight derive from our screen for abundant mRNA genes described earlier in this paper, and cDm1142 belongs to this group.

Figure 5 shows a restriction map of the Dm1142

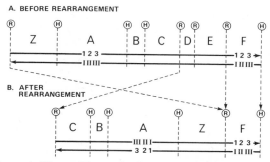

Figure 4. The position and orientation of terminal redundancy in cDm412. That cDm412 contains a direct repeat sequence in the A and F segments was demonstrated as follows: *(A)* by excising the *Eco*RI fragment which carries regions Z, A, B, and C; *(B)* by replacing this fragment in reverse order. In this process, regions D and E, both *Eco*RI fragments, were deleted. Single-stranded DNA is self-complementary for a part of regions A and F only after this rearrangement. The encircled R and H represent *Eco*RI and *Hind*III sites, respectively.

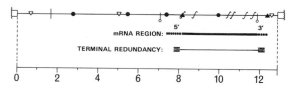

Figure 5. Restriction map of cDm1142. The thin horizontal line represents the Dm1142 segment, and the open blocks at the ends of the map represent ColE1 DNA. The positions of *Hha*I and *Hinf*I restriction endonuclease sites are given only for that part of Dm1142 which is homologous to mRNA. The positions of the terminal redundancy in cDm1142 are indicated by the two hatched blocks. The scale at the bottom of the figure is in kilobases.

segment. The sequences homologous to the *copia* mRNA were mapped by hybridization of ^{32}P-labeled cytoplasmic polyadenylated RNA from Kc$_0$ cells to restriction fragments of cDm1142, using the blotting method of Southern (1975) in much the same manner as described above for the analysis of cDm412 (M. W. Young and D. S. Hogness, in prep.). The *copia* mRNA region is between 4 and 5 kb long (Fig. 5), and most of the RNAs that hybridize to this region are at least this long, as judged by their electrophoretic mobilities.

We have examined the restriction fragments derived from the ten different *copia* genes that have been cloned and from all members of the family contained in the total nuclear DNA from embryos, again using the techniques described above for the analysis of the *412* family. The restriction fragments derived from within the gene are, with a few exceptions, conserved in length, whereas the lengths of fragments containing the endpoints of the gene are highly varied (M. W. Young and D. S. Hogness, in prep.). Thus the *copia* family, like *412*, appears to consist of genes of very similar structure located at multiple sites that are characterized by different flanking sequences.

That these sites are scattered through the genome is indicated by the in situ hybridization of polytene chromosomes with [^{3}H]cRNA to the 4.2-kb *Hha*I fragment from Dm1142. Figure 2B shows that this probe labels the chromocenter and approximately 35 sites in the chromosome arms (M. W. Young and D. S. Hogness, in prep.).

There is a strong overlap between this pattern of in situ labeling and that reported by Spradling et al. (1975) for loci labeled by the total polysomal polyadenylated RNA prepared from cell cultures. For example, all but one of the seven regions on the X chromosome labeled by their general mRNA probe are also labeled by our *copia* cRNA probe. By contrast, only one of these seven regions exhibits a possible overlap with the eleven *412* sites in the X chromosome, and since this region, 3C, also contains a *copia* gene, it is likely that this overlap is spurious. In limiting their scoring to "prominent labeled regions," Spradling et al. (1975) evidently missed identifying most of the different kinds of genes coding for the abundant mRNAs (i.e., genes like *412* that represent ≤0.6% of the RNA probe) and scored only those genes, such as *copia,* that code for particularly abundant mRNAs. We mention their results because they indicate that the *copia* genes account for most, and perhaps all, of the polytene DNA in larval salivary glands that is homologous to these particularly abundant mRNAs. This conclusion can also be drawn from the results of the colony hybridization screen for cDm clones containing abundant mRNA genes, except that here it applies to the embryonic nuclear DNA from which these clones derive. In this case, we made the analogous observation that each of the eight independent

clones that gave an intense response to hybridization with the cytoplasmic polyadenylated RNA probe carried a *copia* gene.

The observations of Levy W. and McCarthy (1975) are also of interest in regard to this particularly abundant mRNA class, since they suggest that most of the RNAs in this class derive from *copia*—a conclusion that is not provided by the above results, since one can imagine that several single-copy genes each contribute an mRNA to this class, yet together comprise only a minority of the homologous DNA. They observed that the kinetics of hybridization of cytoplasmic polyadenylated RNA from cell cultures to the homologous cDNA could best be fit by dividing this RNA into three abundancy classes that exhibit complexities of approximately 5 kb, 2×10^2 kb, and 8×10^3 kb and represent, respectively, about 5%, 60%, and 35% of the RNA. Although the approximate nature of these values must be stressed, it is interesting that the *copia* mRNA can nicely account for the entire first class. The second class would then consist of all the other abundant mRNAs in these cells and would require on the order of 100 different kinds of genes, each contributing an average of approximately 0.5% of the RNA, and of which *412* would be but one. The third class consists of thousands of different rare mRNAs that do not concern us here, because their concentration would be too low to be effective in either colony or in situ hybridizations.

We have saved till last what is perhaps the most striking correlation between the *copia* and *412* families, namely, that a direct repeat also exists at or near the termini of *copia* genes, suggesting that terminal redundancy may be a general characteristic of dispersed families. We first mapped this direct repeat in cDm351, another of the eight plasmids from our collection that contains a *copia* gene (D. J. Finnegan et al., in prep.). Mapping was accomplished by the same trick of converting a direct repeat to an inverted repeat and then examining the resulting hairpin structure in the electron microscope. In this case, we found that the repeat is 0.30 kb long and that its center-to-center distance is 4.7 kb, i.e., within the range of lengths determined for the *copia* mRNA region. Although the repeat is close to the ends of this region, we cannot be precise about their relationship in Dm351, since these ends have been mapped to only a low level of resolution in this segment (i.e., ±0.45 kb for one end and ±1.4 kb for the other).

We have, however, also mapped a repeat in Dm1142 by cross-hybridization of sequences from various of its restriction fragments (M. W. Young and D. S. Hogness, in prep.). This has placed the repeat sequences at or very near the two *Hha*I sites in this segment, indicating that each end of the mRNA region is within a few hundred base pairs of a repeat sequence (Fig. 5). Thus it appears that the *copia* genes are also characterized by a terminal redundancy which, though shorter than that for *412*, also represents about 7% of the gene. It should be emphasized that cross-hybridization tests have failed to reveal any sequence homology between the *copia* and *412* terminal repeats or, for that matter, between any other parts of these two dispersed families.

WHY TANDEM? WHY DISPERSED?

The finding of repeated gene families whose members are scattered over the *D. melanogaster* genome raises the fundamental question of why some repeated genes are dispersed in this manner, while others are clustered in tandem arrays. This question has been largely ignored in the past for the obvious reason that all well-defined families belonged to the tandem class. Hypotheses about this class that are directly dependent upon the tandem topography have tended to be concerned with mechanisms for generating and maintaining a high degree of sequence homogeneity among the family members. These include Callan's (1967) master-slave hypothesis, its derivatives (Whitehouse 1967; Thomas 1970, 1974), and the more tenable unequal-crossover models (Edelman and Gally 1970; Tartof 1974; Brown and Sugimoto 1973, 1974; Smith 1974). Although these hypotheses are not stated in exclusive terms, i.e., they do not exclude the possibility that sequence homogeneity could be obtained by mechanisms not dependent upon tandem arrays, the apparent universality of the tandem arrangement induced the notion that high levels of sequence homogeneity might depend upon this topography.

The properties of the *412* and *copia* genes that we have described are inconsistent with that notion. Although there may be variants in both the *412* and *copia* families, just as there are in tandem families (see, e.g., White and Hogness 1977; Lifton et al., this volume), the sequence conservation observed for the vast majority of the members is high—high enough at least to eliminate sequence maintenance as a likely function for determining topographic class. The problem of how dispersed genes do maintain a common sequence is, however, of general interest. Because we think it may be related to another common characteristic of the *412* and *copia* families—their terminal redundancies—we will consider it before taking up the question of the topographic determinants.

Speculation I: How the Terminal Redundancies of Dispersed Families Can Yield Transposable Elements That Maintain Sequence Homogeneity

An intriguing possible role of the direct repeats which comprise the terminal redundancy is that they provide sites for recombinational events that allow the *412* and *copia* genes to move around the genome. Figure 6 shows how a single crossover event could lead to the insertion of a gene, consisting of

Figure 6. A model for gene transposition.

a circular DNA containing one copy of the repeat, at an "empty" chromosomal site containing another copy, to yield a "filled" site that exhibits the terminal redundancy which we observe in the *412* and *copia* genes. Deletion of a gene from a filled site could occur by the reverse reaction, yielding the circular gene and an empty site.

One can account for sequence homogeneity if it is imagined that germ-line cells contain only one filled site. Conversion of such a germ-line pattern to a somatic pattern, where many sites are filled, might then occur by the excision of the gene in the germ-line site and, after its amplification by replication, insertion of the replicas at empty sites. The special replication required for amplification can be avoided if one supposes that the single germ-line site is filled with a large number of tandemly repeated genes and that these can be excised in the manner described for one-gene sites. Sequence homogeneity among the germ-line genes would then be preserved by whatever mechanisms operate in tandem families, perhaps by unequal-crossover fixation. Such a mechanism for the insertion and excision of these genes would clearly allow different patterns of filled and unfilled sites in different cell types. All empty sites might be available, or "open," for this insertion; alternatively, only specific subsets controlled by the determined state of the somatic cell might be open.

A difficulty with this simplest form of the excision-integration model is that there is no reason why an open site should not be subject to successive integration events resulting in the creation of a tandem array of genes at the site. Since we failed to observe such tandem arrays of *412* or *copia* genes in any of the cloned Dm segments that were examined, we would add the restriction that the first integration event "closes" the site to further integration. For example, closure might be effected by a change in chromatin structure following the first integration. Alternatively, closure might be obtained by a site-specific recombination mechanism analogous to that employed to restrict the integration of bacteriophage λ into the *E. coli* chromosome (Landy and Ross 1977). In this case, the DNA segments that interact to effect the integration (represented by the hatched and open triangles in Fig. 6), although exhibiting considerable sequence homology, would not be identical. The terminal regions of the integrated genes would then not be identical to each other or to either of the segments required for integration, although they again would exhibit strong sequence homologies.

The speculative nature of this model for sequence maintenance by gene migration is emphasized by the fact that there is as yet only indirect evidence for the existence of transposable elements in eukaryotic genomes (McClintock 1957; Hozumi and Tonegawa 1976; Tonegawa et al., this volume). However, this speculation has the advantage that much of it is readily testable, given the availability of the isolated *412* and *copia* genes. The increase from one filled site in germ-line DNA to many filled sites in somatic DNA can be tested by the analysis of terminal restriction fragments obtained from the genes in the two kinds of DNAs. Such an analysis would employ hybridization procedures like those shown in Figure 3 A and F, using probes that are specific for terminal gene sequences other than those in the direct repeat. The reciprocal change in unfilled sites can be tested by similar hybridization experiments with probes specific for sequences in the direct repeats. As a last example, we note that the covalently closed, circular DNAs which have been isolated from *D. melanogaster* nuclei exhibit a size range that includes the *412* and *copia* gene lengths (Stanfield and Helinski 1976). It would clearly be of interest to test for the presence of these genes in the circular DNAs isolated from embryos at different stages of development, again by the hybridization procedures indicated in Figure 3.

Speculation II: Two Reasons for Gene Repetition and Their Assignment to the Two Repetition Topographies

Why are genes repeated? The usual thought is that genes are repeated to allow the synthesis of more transcripts of a given kind per minute per cell. We shall call this mode "gene-dosage repetition." Genes that are repeated according to this mode are expected to be under common control.

A second possibility is that genes are repeated so that their transcription can be regulated by a large number of different control elements. Here we suppose that the different members of a repeated gene family are independently controlled by interaction between linked regulator sequences and control elements specific to each member. We shall call this mode "regulatory repetition." The advantages of independent multiple controls on a single gene activity in complex developing systems have been amply discussed in the context of nonrepeated genes (Britten and Davidson 1969; Davidson and Britten 1973). The same principles apply here; only the manner of applying the controls differs. Britten and Davidson envisaged multiple control of a nonrepeated gene via a series of different regulatory sequences all linked to that gene. There are certain obvious limitations inherent in this method of multiple control that are relieved by increasing the number of copies of the gene. In addition, by repeating the gene, new

methods for coordinating the expression of that gene with many different sets of genes become available.

We consider here the possibility that the tandem and dispersed topographies for repeated genes correspond, respectively, to the gene-dosage and regulatory modes of repetition. There are good reasons for thinking that the tandemly repeated 18S and 28S rRNA genes, 5S RNA genes, and histone genes represent examples of gene-dosage repetition. The best evidence is for the 18S and 28S rRNA genes which can be visualized in transcriptionally active and inactive states by electron microscopy (Miller and Beatty 1969). It is a common observation in many species that many of the repeated units are simultaneously active within a given tandem array (for an extensive list of these species, see Foe et al. 1976). More specifically, a general correlation has been observed between the fraction of active units in these arrays and the overall rate of rRNA synthesis in developing newt oocytes (Scheer et al. 1976), and in embryos of the milkweed bug, *Oncopeltus fasciatus* (Foe, this volume), and of *D. melanogaster* (McKnight and Miller 1976).

Although the evidence is less direct, the organization of the 5S RNA genes (Brown and Sugimoto 1973, 1974) and the histone genes (Kedes 1976) also appears to reflect gene-dosage repetition. The arrangement of the 5S RNA genes in *X. laevis* is of special interest. This family can be divided into at least two subfamilies which appear to be subject to different controls (Brown and Sugimoto 1973, 1974). The vast majority of the 24,000 gene copies belong to the oocyte subfamily that codes for the 5S RNAs synthesized in young oocytes, whereas the minority or somatic subfamily codes for the 5S RNAs found in cultured somatic cells. The *X. laevis* 5S family appears to conform to both the gene-dosage and regulatory modes of repetition. It is of considerable interest to find that the family topography is both tandem and dispersed in this instance. Pardue et al. (1973; also see Pardue 1974) mapped these 5S RNA genes at or near the telomeres of the long arms of most metaphase chromosomes. The in situ hybridization procedures used for this mapping required the presence of large numbers of the genes at each chromosomal site, and since isolated *Xenopus* 5S DNAs consist of tandemly repeated gene-spacer units (Brown and Sugimoto 1973, 1974), it may be presumed that the genes clustered at the different chromosomal sites are in tandem arrays. This 5S RNA gene family therefore appears to be divided into several clusters of tandemly repeated genes that are dispersed in the genome. If our hypothesis is correct, each of these clusters should include either oocyte or somatic 5S RNA genes.

How might the gene-dosage mode be facilitated by tandem arrays? Unequal-crossover events between arrays in sister strands can act to adjust (i.e., increase or decrease) the number of copies according to demands for different overall transcription rates

(Tartof 1974). Not only can the number of identical genes be adjusted in this way, but if the repeating unit contains the promoter(s) for the gene(s) — as is certainly the case for the 18S–28S rRNA and 5S RNA genes (see above references) and is probably the case for the five histone genes that are contained within a unit, at least in *D. melanogaster* (Lifton et al., this volume) — then the transcription of these identical genes is expected to be subject to the same control mechanism. This is an important requirement for gene-dosage repetition.

We suggest another way by which tandem arrays may facilitate the gene-dosage function. We suppose that the regions in the chromatin fiber that can and cannot be transcribed exhibit different structures which we loosely refer to, respectively, as "noncompacted" and "compacted," without specifying the exact difference in chromatin structure that differentiates the two states. In *D. melanogaster,* the compacted state may correspond to the independent supercoiled loops of Benyajati and Worcel (1976) or to the chromatin in the bands of polytene chromosomes. In any case, we further suppose that conversion from the compacted state to the noncompacted state is required before the promoters and adjacent sequences can be made available to the appropriate RNA polymerases and control elements. The conversion process is therefore viewed as the focus of primary events which must occur before secondary controls that modulate the activity level of promoters within the region can operate. If conversion is initiated by a controlled triggering event and proceeds to completion by cooperative processes (see Alberts et al. 1977 for a discussion of this general concept), then this primary control could operate most efficiently and accurately on genes exhibiting dosage repetition if the copies are clustered within a region, i.e., tandemly arrayed.

Although we have not obtained a direct measure of the number of *412* and *copia* genes at each of the chromosomal regions identified by in situ hybridization, the present evidence indicates a lack of tandem repetition and suggests that most regions contain only one gene: (1) The genes are not tandemly repeated in any of the cloned Dm segments that were examined. (2) The sequences surrounding the genes in the different Dm segments are different. (3) The frequency at which we obtained such segments in the screen of 3000 independent clones correlates with the number of *412* and *copia* regions labeled by in situ hybridization. (4) The level of this hybridization is, with the exception of the chromocentral labeling by the *copia* probe, not markedly different among the different sites within a family, and is about the same as that observed for a nonrepeated gene (Young and Hogness 1977).

There is also little reason to suppose that gene dosage forms the basis for the repetition of these genes. Repetition is certainly not a prerequisite for the synthesis of abundant mRNAs. We have cloned

a gene called *sam* that produces an abundant mRNA present in *D. melanogaster* cell cultures at about half the concentration of the *412* mRNA; yet *sam* is confined to one chromosomal site where it does not appear to be repeated (Young and Hogness 1977). Similarly, the kinetic studies of Levy W. and McCarthy (1975) suggest that about half of the abundant mRNAs in these cultures are coded by nonrepeated genes. There is also a lack of correspondence between the approximately equal number of chromosomal regions containing *copia* and *412* genes and the high ratio of *copia* to *412* mRNA concentrations observed in several different cell types. These considerations do not, of course, eliminate the possibility that gene dosage plays some role in the repetition of these genes, but they do indicate that the *412* and *copia* mRNAs may result from the activities of only a small fraction of the total genes in each family — perhaps only one.

One consequence of regulatory repetition is that the transcripts in a given cell type should derive from a specific fraction of the total family complement; in different cell types different subsets of the family should be active. We have recently obtained preliminary evidence that only a fraction of the *copia* sites in larval salivary gland polytene chromosomes contain *copia* transcripts (M. W. Young, unpubl.). In addition, Bonner and Pardue (1977) have recently observed that some of the sites which are labeled by in situ hybridization with [3H]-labeled nuclear RNA isolated from salivary glands are not labeled by direct [3H]uridine incorporation. A number of these sites correspond to those labeled by the mRNA from cultured cells (Spradling et al. 1975), which, it will be recalled, exhibit a strong overlap with the *copia* sites. Both of these observations suggest that a specific subset of the *copia* family may be transcribed in larval salivary glands. This important point requires more detailed confirmation and extension, both to other cell types and to the *412* family.

Another consequence of regulatory repetition is that sequences flanking the different genes in a family will be different. In both the *412* and *copia* families, the present evidence indicates that the sequences which differ from site to site are not very distant from, and may be adjacent to, the ends of the mRNA region. We are attempting to map these boundaries between sequences common to the family members and those specific to the site at higher resolution by electron microscopic examination of the heteroduplexes formed between Dm segments derived from different sites and by direct DNA sequencing.

The next critical question concerns the nature of the flanking sequences and, in particular, whether they contain site-specific promoters. The experimental answer to this question is related to the problem of determining which sites are actively transcribed in any given cell type, since the mapping of a promoter in a particular Dm segment (for methods, see Reeder et al., this volume) clearly requires a source of transcripts for the site from which the segment is derived. Although we have as yet no indication of the position of these promoters, the possibility that they lie outside of boundaries determined by the direct repeat is intriguing since this would encourage the hypothesis that the genes are inserted at sites adjacent to specific promoters.

The object of regulatory repetition is to provide multiple, independent controls on a given gene activity. The activity of each member of such a repeated gene family can then be coordinated with that of a different set of dissimilar genes, and each gene set would be coordinately regulated by a common control element. The model for regulatory repetition that we have presented does not depend upon any particular spatial relationship among the genes in a set; they can be scattered throughout the genome or clustered at a single site. In concluding this paper, we will briefly consider a special case of the clustered arrangement.

We shall call a set that is clustered at one site a "gene pack." In the special case considered here, the gene pack is contained within a region of chromatin that can be converted from a compacted, inactive state to a noncompacted, active state. This condition is directly analogous to the tandem arrays that we previously imagined to be contained in such regions, with this difference: the genes in the pack are of different kinds, whereas in tandem arrays they are of one kind. As in the tandem arrays, each gene in the pack has its own promoter, but these are assumed to be different from one another so as to allow different frequencies of transcription for the different genes in the pack. Similarly, the different members of a repeated gene family that belong to different packs can be assumed to have promoters that exhibit different efficiencies of transcription initiation. Coordinate regulation is achieved by the conversion process and is controlled at the level of the triggering event that initiates the conversion.

If all genes in a pack belong to dispersed repeat families, then a pack will represent a block of repetitive sequence elements and different packs will represent blocks containing different combinations of such elements. We have shown that the *D. melanogaster* genome contains blocks of repeated elements that exhibit these properties (D. J. Finnegan et al., in prep.). Indeed, the Dm412 segment is derived from such a block since it consists of several different repetitive elements that are dispersed over the genome, one of which is the *412* gene. Do the other repetitive elements in Dm412 represent other genes that together with *412* form a gene pack? We do not know the answer to this question, since we have tested for the ability of these other repeated sequences to hybridize with mRNAs from only one cell source, i.e., Dm412 might represent a gene pack that is not active in the cell cultures we used, but

is active in some other cell type. We have similarly not determined whether the non-*412* regions in Dm segments derived from other *412* sites in the *D. melanogaster* genome consist of dispersed repeated elements. It should, however, be emphasized that there is no a priori reason why all genes in a pack must be repetitive.

Acknowledgments

This work was supported by research grants from the National Science Foundation and the National Institutes of Health. D. J. F. was a Fellow of the Jane Coffin Childs Memorial Fund and a Senior Fellow of the American Cancer Society, California Division. G. M. R. was a Fellow of the Helen Hay Whitney Foundation, and M. W. Y. is a National Institutes of Health Research Fellow.

REFERENCES

ALBERTS, B., A. WORCEL, and H. WEINTRAUB. 1977. On the biological implications of chromatin structure. In *Organization and expression of the eukaryotic genome. Proceedings of the International Symposium, Teheran, 1976* (ed. E. M. Bradbury and K. Jaraherian), p. 165. Academic Press, New York.

ARTAVANIS-TSAKONAS, S., P. SCHEDL, C. TSCHUDI, V. PIRROTTA, R. STEWARD, and W. J. GEHRING. 1977. The 5S genes of *Drosophila melanogaster. Cell* 12: 1057.

BENYAJATI, C. and A. WORCEL. 1976. Isolation, characterization, and structure of the folded interphase genome of *Drosophila melanogaster. Cell* 9: 393.

BIRNSTIEL, M., J. SPEIRS, I. PURDOM, K. JONES, and U. E. LOENING. 1968. Properties and composition of the isolated ribosomal DNA satellite of *Xenopus laevis. Nature* 219: 454.

BONNER, J. J. and M. L. PARDUE. 1977. Polytene chromosome puffing and *in situ* hybridization measure different aspects of RNA metabolism. *Cell* (in press).

BRITTEN, R. J. and E. H. DAVIDSON. 1969. Gene regulation for higher cells: A theory. *Science* 165: 349.

BROWN, D. D. and K. SUGIMOTO. 1973. The 5S DNAs of *Xenopus laevis* and *Xenopus mulleri:* The evolution of a gene family. *J. Mol. Biol.* 78: 397.

———. 1974. The structure and evolution of ribosomal and 5S DNAs in *Xenopus laevis* and *Xenopus mulleri. Cold Spring Harbor Symp. Quant. Biol.* 38: 501.

BROWN, D. D. and C. S. WEBER. 1968. Gene linkage by RNA-DNA hybridization. II. Arrangement of the redundant gene sequences for 28S and 18S ribosomal RNA. *J. Mol. Biol.* 34: 681.

BROWN, D. D., P. C. WENSINK, and E. JORDAN. 1971. Purification and some characteristics of the 5S DNA from *Xenopus laevis. Proc. Natl. Acad. Sci.* 68: 3175.

CALLAN, H. G. 1967. The organization of genetic units in chromosomes. *J. Cell Sci.* 2: 1.

DAVIDSON, E. H. and R. J. BRITTEN. 1973. Organization, transcription, and regulation in the animal genome. *Q. Rev. Biol.* 48: 565.

EDELMAN, G. M. and J. A. GALLY. 1970. Arrangement and evolution of eukaryotic genes. In *The neurosciences: Second study program* (ed. F. O. Schmidt), p. 962. Rockefeller University Press, New York.

FOE, V. E., L. E. WILKINSON, and C. D. LAIRD. 1976. Comparative organization of active transcription units in *Oncopeltus fasciatus. Cell* 9: 131.

GLOVER, D. M. and D. S. HOGNESS. 1977. A novel arrangement of the 18S and 28S sequences in a repeating unit of *Drosophila melanogaster* rDNA. *Cell* **10**: 167.

GRUNSTEIN, M. and D. S. HOGNESS. 1975. Colony hybridization: A method for the isolation of cloned DNAs that contain a specific gene. *Proc. Natl. Acad. Sci.* **72**: 3961.

HOZUMI, N. and S. TONEGAWA. 1976. Evidence for somatic rearrangement of immunoglobulin genes coding for variable and constant regions. *Proc. Natl. Acad. Sci.* **73**: 3628.

KEDES, L. H. 1976. Histone messengers and histone genes. *Cell* **8**: 321.

LANDY, A. and W. ROSS. 1977. Viral integration and excision: Structure of the lambda *att* sites. *Science* **197**: 1147.

LEVY W., B. and B. MCCARTHY. 1975. Messenger RNA complexity in *Drosophila melanogaster*. *Biochemistry* **14**: 2440.

MCCLINTOCK, B. 1957. Controlling elements and the gene. *Cold Spring Harbor Symp. Quant. Biol.* **21**: 197.

MCKNIGHT, S. L. and O. L. MILLER, JR. 1976. Ultrastructural patterns of RNA synthesis during early embryogenesis of *Drosophila melanogaster*. *Cell* **8**: 305.

MILLER, O. L. and B. R. BEATTY. 1969. Visualization of nucleolar genes. *Science* **164**: 955.

PARDUE, M. L. 1974. Localization of repeated DNA sequences in *Xenopus* chromosomes. *Cold Spring Harbor Symp. Quant. Biol.* **38**: 475.

PARDUE, M. L., D. D. BROWN, and M. L. BIRNSTIEL. 1973. Localization of the genes for 5S ribosomal RNA in *Xenopus laevis*. *Chromosoma* **42**: 191.

PELLEGRINI, M., J. MANNING, and N. DAVIDSON. 1977. Sequence arrangement of the rDNA of *Drosophila melanogaster*. *Cell* **10**: 213.

PROCUNIER, J. D. and K. D. TARTOF. 1976. Restriction map of 5S RNA genes of *Drosophila melanogaster*. *Nature* **263**: 255.

RUBIN, G. M., D. J. FINNEGAN, and D. S. HOGNESS. 1976. The chromosomal arrangement of coding sequences in a family of repeated genes. *Prog. Nucleic Acid Res. Mol. Biol.* **19**: 221.

SCHEER, U., M. F. TRENDELENBURG, and W. W. FRANKE. 1976. Regulation of transcription of genes of ribosomal RNA during amphibian oogenesis: A biochemical and morphological study. *J. Cell Biol.* **69**: 465.

SMITH, G. P. 1974. Unequal crossover and the evolution of multigene families. *Cold Spring Harbor Symp. Quant. Biol.* **38**: 507.

SOUTHERN, E. M. 1975. Detection of specific sequences among DNA fragments separated by gel electrophoresis. *J. Mol. Biol.* **98**: 503.

SPRADLING, A., S. PENMAN, and M. L. PARDUE. 1975. Analysis of *Drosophila* mRNA by *in situ* hybridization: Sequences transcribed in normal and heat shocked cultured cells. *Cell* **4**: 395.

STANFIELD, S. and D. R. HELINSKI. 1976. Small circular DNA in *Drosophila melanogaster*. *Cell* **9**: 333.

TARTOF, K. D. 1974. Unequal mitotic sister chromatid exchange and disproportionate replication as mechanisms regulating ribosomal RNA gene redundancy. *Cold Spring Harbor Symp. Quant. Biol.* **38**: 491.

THOMAS, C. A., JR. 1970. The theory of the master gene. In *The neurosciences: Second study program* (ed. F. O. Schmidt), p. 973. Rockefeller University Press, New York.

———. 1974. The rolling helix: A model for the eukaryotic gene? *Cold Spring Harbor Symp. Quant. Biol.* **38**: 347.

WELLAUER, P. K. and I. B. DAWID. 1977. The structural organization of ribosomal DNA in *Drosophila melanogaster*. *Cell* **10**: 193.

WENSINK, P. C., D. J. FINNEGAN, J. E. DONELSON, and D. S. HOGNESS. 1974. A system for mapping DNA sequences in the chromosomes of *Drosophila melanogaster*. *Cell* **3**: 315.

WHITE, R. L. and D. S. HOGNESS. 1977. R loop mapping of the 18S and 28S sequences in the long and short repeating units of *Drosophila melanogaster* rDNA. *Cell* **10**: 177.

WHITEHOUSE, H. L. K. 1967. A cycloid model for the chromosome. *J. Cell Sci.* **2**: 9.

YOUNG, M. W. and D. S. HOGNESS. 1977. A new approach for identifying and mapping structural genes in *Drosophila melanogaster*. In *Eucaryotic genetics system*. *ICN-UCLA Symposia on Molecular and Cellular Biology* (ed. G. Wilcox et al.), vol. 8, p. 315. Academic Press, New York. (In press.)

QUESTIONS/COMMENTS

Question by:
S. C. R. ELGIN
Harvard University

Response by:
D. FINNEGAN
University of Edinburgh

Is the 500-base-pair sequence found in fragments A and F found similarly associated with the mRNA sequence at other locations in the genome?

Unfortunately, we cannot answer this question definitively. Fragment F does hybridize to the fragments of cDm454 and cDm468, which we have inferred to contain sequences complementary to the 3' and 5' ends of mRNA$_{412}$. We do not know, however, that the sequence binding to these fragments is the same as that carried by fragment A of cDm412. The same is true of three other copies of the mRNA$_{412}$ sequence and it can only be suggested that it is the case for the remainder.

Short-period Repetitive-sequence Interspersion in Cloned Fragments of Sea Urchin DNA

A. Shiu Lee, R. J. Britten, and E. H. Davidson

Division of Biology, California Institute of Technology, Pasadena, California 91125

Interspersion of short repetitive DNA sequences with longer single-copy regions appears to be a dominant form of genomic organization in the animal kingdom. In earlier publications we termed this form of sequence organization "short-period interspersion." Most current knowledge of sequence organization derives from studies on the DNA of whole genomes, in which a variety of physical methods have been applied to partially renatured DNA fragments. The approaches used have included direct visualization of the renatured structures in the electron microscope and measurements of their ability to bind to hydroxylapatite, their optical hyperchromicity, and their sensitivity to single-strand-specific nucleases.

On the basis of such studies, the following general features are ascribed to the short-period interspersion pattern: (1) A large fraction of the repeated DNA sequence, usually over 50%, is located in sequence elements whose lengths appear rather narrowly distributed in the range of 200–500 nucleotides. (2) There are typically well over 10^5 such short sequence elements, representing at least several thousand diverse (i.e., nonhomologous) repetitive-sequence families per haploid genome. (3) The short repetitive sequences alternate with single-copy DNA regions ranging in length from hundreds to several thousands of nucleotides per sequence, so that typically over 70 or 80% of randomly chosen DNA fragments 3000–4000 nucleotides long include at least one repetitive-sequence element. (4) Longer repetitive sequences also exist and, at least in some organisms, some of the short repeats appear to be interspersed in an environment of other repetitive sequences while some are interspersed with single-copy DNA. These statements are based mainly upon data presented in the papers of Davidson et al. (1973, 1974, 1975), Graham et al. (1974), Angerer et al. (1975), Chamberlin et al. (1975), Goldberg et al. (1975), and Eden et al. (1977). Short-period interspersion is characteristic of mammalian genomes (Britten and Smith 1970; Bonner et al. 1974; Pearson 1977), including human (Schmid and Deininger 1975; Deininger and Schmid 1976) as well as amphibian, echinoderm, molluscan, coelenterate, nemertean, chelicerate, and a number of insect genomes (Graham et al. 1974; Angerer et al. 1975; Goldberg et al. 1975; Efstratiadis et al. 1976; Crain et al.

1976a). However, it is clear that this pattern of sequence arrangement is not universal since the genomes of some (though not all) insects display a different form of sequence arrangement which we have termed "long-period interspersion." In these DNAs, both the repetitive-sequence and single-copy-sequence elements are many thousands of nucleotides in length (Manning et al. 1975; Crain et al. 1976a,b; Wells et al. 1976).

The discovery of short-period interspersion as a general pattern of animal DNA sequence organization raises a new set of basic issues. The most interesting of these are directly related to the underlying question of the functional significance of moderately repetitive DNA. One such problem is whether there are local regularities in the sequence interspersion pattern. It is important to determine, for example, whether interspersed repetitive sequences within a confined region of the genome tend to belong to the same repetitive-sequence family and whether the spacing of repetitive sequences, i.e., the length of the intervening single-copy DNA regions, tends to be uniform in each local area (see, e.g., Bonner and Wu 1973; Laird et al. 1974).

The local character of sequence interspersion patterns can be determined in cloned DNA fragments. In this paper we describe the first such analysis to be carried out on DNA derived from an organism whose genome displays the typical short-period interspersion pattern of sequence organization. The clones used for this investigation were CSO859, which contains a 5500-nucleotide sea urchin DNA insert, and pSC34, which contains a 7200-nucleotide sea urchin DNA insert. The sea urchin DNA inserts are terminated by natural *Eco*RI sites and are inserted at the *Eco*RI sites of plasmid vectors RSF2124 and pSC101 in CSO859 and pSC34, respectively. The construction and some specific features of these clones were reported earlier (Lee et al. 1976, 1977). At least three-quarters of the genome of the sea urchin *(Strongylocentrotus purpuratus)* used in our work consists of single-copy DNA interspersed with short repetitive sequences, according to Graham et al. (1974). Here we show that the two inserts used for the present study are typical of the overall pattern of sequence organization. Thus each contains three short interspersed repeats spaced by single-copy DNA. We also describe the physical location and partial characterization of the interspersed repetitive sequences on the cloned inserts.

Restriction Map of the Sea Urchin DNA Insert in Clone CSO859

An obvious approach to locating the repetitive-sequence elements in the cloned DNA is to subdivide the insert into smaller restriction enzyme fragments, order the fragments, and determine which bear repetitive-sequence elements by reaction with sea urchin DNA. Data determining a map of restriction enzyme sites for the CSO859 insert are shown

in Figure 1. Figure 1A shows the terminal digestion products of the 5500-nucleotide insert resolved in agarose gels after treatment with various restriction enzymes. The sizes of the individual fragments generated are listed in the legend. The single sea urchin DNA insert released by EcoRI digestion of the recombinant plasmid is shown in Figure 1A (i). This insert contains a single HindIII site, situated 3500 nucleotide pairs from one end (Fig. 1A, ii). HincII cleaves the sea urchin fragment three times, generating four subfragments (Fig. 1A, iii). HaeIII produces ten fragments ranging in size from 2900 to 70 nucleotide pairs. HaeIII fragments A, B, C, D,

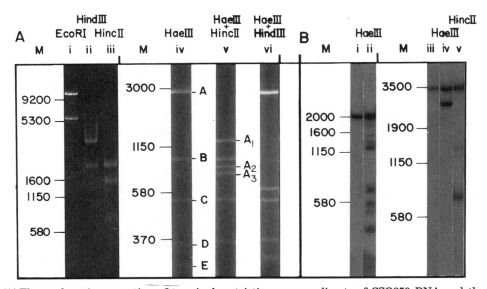

Figure 1. *(A)* Electrophoretic separation of terminal restriction enzyme digests of CSO859 DNA and the purified sea urchin DNA insert. The molecular-weight markers (M) used were restriction fragments of φX174 replicative form, λ DNA, and RSF2124. These lengths are indicated in nucleotide pairs. (i–iii) Analysis on 1.5% agarose slab gels. (i) CSO859 DNA was digested with EcoRI, yielding the linear RSF2124 molecule and the 5500-nucleotide sea urchin DNA insert. (ii) The 5500-nucleotide EcoRI fragment of CSO859 was digested with HindIII, yielding two fragments of 3500 and 2100 nucleotides. (iii) The 5500-nucleotide EcoRI fragment of CSO859 was digested with HincII, yielding four fragments of sizes 2250, 1600, 800, and 750 nucleotides. (iv–vi) Analysis on 3% agarose cylindrical gels of terminal digests of the 5500-nucleotide EcoRI fragment from CSO859. (iv) HaeIII cleaves the 5500-nucleotide fragment into fragments A (2900 nucleotides), B (850 nucleotides), C (550 nucleotides), D (350 nucleotides), E (250 nucleotides), F (150 nucleotides), G (130 nucleotides), H (120 nucleotides), I (100 nucleotides), and J (70 nucleotides). HaeIII fragments F–J are not present on the gel shown here but were detected in 5% polyacrylamide gels, though it is difficult to define them accurately. However, in the partial digest analysis shown in B, the existence of these small fragments is clearly demonstrated. (v) The 5500-nucleotide EcoRI fragment was digested with HaeIII and HincII. HaeIII fragment A is cleaved by HincII into three fragments of sizes 1300 nucleotides (A_1), 800 nucleotides (A_2), and 750 nucleotides (A_3). HaeIII fragment E is also cleaved by HincII. The other HaeIII fragments are uncleaved since there are only three HincII sites in the 5500-nucleotide fragment (see iii). (vi) The 5500-nucleotide EcoRI fragment was digested with HaeIII and HindIII. The single HindIII site is located within the HaeIII B fragment, generating two new fragments of 600 and 270 nucleotides.

(B) Autoradiographs of the "ladder" experiments for the localization of the restriction sites in the 5500-nucleotide EcoRI fragment of CSO859. The digestion products were electrophoresed on 1.5% or 1.75% agarose slab gel (14 × 12 × 0.12 cm), along with molecular-weight markers. Electrophoresis was carried out at 30 V until the bromphenol-blue marker was at 13 cm. The gel was dried by the method of Smith and Birnstiel (1976) and then placed in direct contact with Kodak XR-5 film for exposure. (i) The 2100-nucleotide fragment generated from HindIII digestion of the 5500-nucleotide EcoRI fragment was isolated and rerun on a 1.5% agarose slab gel. (ii) The same fragment as in gel i was partially digested with HaeIII. Some of the 2100-nucleotide fragment remains undigested and partials of the following sizes were produced: 1800 nucleotides, 1550 nucleotides, 1400 nucleotides, 1300 nucleotides, 700 nucleotides, 580 nucleotides, and 510 nucleotides. The smallest fragment produced was about 380 nucleotides long. This is identified as HaeIII fragment D (see A, iv). (iii) The 3500-nucleotide fragment generated from HindIII digestion of the 5500-nucleotide EcoRI insert was isolated and rerun on a 1.75% agarose slab gel. (iv) The same fragment as in gel iii was partially digested with HaeIII. Some of the 3500-nucleotide fragment remained undigested. The other fragment produced was 2900 nucleotides long and is identified as HaeIII fragment A (see A, iv). (v) The same fragment as in gel iii was partially digested with HincII. In addition to the 3500-nucleotide fragment, a partial of 1500 nucleotides was detected. The terminally labeled fragment is 800 nucleotides and is identified as HincII A_2.

and E can be conveniently separated on a 3% agarose gel (Fig. 1A, iv). *Hae*III fragments F–J can be resolved on 5% polyacrylamide gels. Due to their small size and the difficulty of obtaining them quantitatively, these were mapped but were not analyzed further. When digested with *Hinc*II, the large *Hae*III A fragment yields three subfragments, A_1, A_2, and A_3, as shown in Figure 1A (v). The *Hae*III and *Hinc*II fragments were ordered by a modification of the procedure described by Smith and Birnstiel (1976). The 5500-nucleotide *Eco*RI fragment was labeled at the 5′ termini with [^{32}P]phosphoryl groups using polynucleotide kinase and [γ-^{32}P]ATP (Chaconas et al. 1975). The labeled fragment was then cleaved with *Hind*III to yield the 3500- and 2100-nucleotide fragments (Fig. 1A, ii). Each of these, now labeled only at one end, was redigested with *Hae*III or *Hinc*II to various extents to produce a series of partial digests. The two overlapping sets of partials, each sharing a common labeled terminus, were fractionated by agarose gel electrophoresis and analyzed by autoradiography (Fig. 1B). The relative mobilities of the labeled fragments locate the respective restriction sites in order of distance from the labeled terminus. The *Hae*III sites on the 2100-nucleotide *Hind*III fragment were ordered from partial digests such as that shown in Figure 1B (ii), and the *Hae*III and *Hinc*II sites on the 3500-nucleotide *Hind*III fragment were ordered as shown in Figure 1B (iv) and (v).

A map of the *Hae*III, *Hind*III, and *Hinc*II sites on the sea urchin insert of CSO859 is presented in Figure 2. There is no *Bam*HI site located within this fragment. A statistically unlikely feature of the restriction map is the distribution of the *Hae*III sites. There are eight *Hae*III sites 5′(GG ↓ CC)3′ clustered at the left of the fragment as shown, whereas no *Hae*III site occurs over a 2900-nucleotide distance on the right-hand end of the molecule.

Restriction Map of the 7200-Nucleotide Sea Urchin DNA Insert in Clone pSC34

When pSC34 DNA is digested with *Eco*RI, two sea urchin DNA fragments are released (Fig. 3A, i). These are 7200 and 1900 nucleotides in length. It is not known whether these two *Eco*RI fragments are contiguous to each another in the sea urchin

genome or whether they were ligated during the construction of the recombinant DNA molecules. The following studies concern only the 7200-nucleotide insert. *Hpa*II cleaves the 7200-nucleotide *Eco*RI fragment two times to yield large fragments of 5500 and 1400 nucleotides, plus one small fragment of about 300 nucleotides (Fig. 3A, ii). The result of digesting the 7200-nucleotide *Eco*RI fragment with *Hha*I is shown in Figure 3A (iv). Seven distinct bands are obtained. These range from 2000 nucleotides (fragment A) to 390 nucleotides (fragment G). *Hha*I band E actually consists of three fragments, all about 520 nucleotides in length. Two other small *Hha*I fragments, H and I, can be detected when the terminal *Hha*I digest of the 7200-nucleotide *Eco*RI fragment is analyzed on 5% polyacrylamide gels. Due to their small size these were not further studied. The double-digest experiment of Figure 3A (v) shows that the *Hpa*II sites all lie within the *Hha*I A fragment. It follows that the *Hha*I A fragment is located on one terminus of the molecule. We found also that the enzyme *Pst*I cleaves the 7200-nucleotide *Eco*RI fragment at three sites. Since one of the *Pst*I sites is very close to the *Eco*RI site, only three large fragments are produced. One of the *Pst*I sites is close to the middle of the 7200-nucleotide fragment and lies within the *Hha*I B fragment, as is evident in the *Hha*I + *Pst*I double-digestion experiment (Fig. 3A, vi). One of the other *Pst*I sites is located within the *Hha*I A fragment and another lies within one of the *Hha*I E fragments (data not shown).

The *Hha*I sites were mapped by the same partial analysis method used for CSO859. In this case, *Hpa*II was used to generate two fragments, each bearing one labeled 5′ terminus. Autoradiographs of the partial digestion products are shown in Figure 3B, and their lengths are listed in the legend. Combining the double-digestion and partial-analysis data summarized in Figure 3A, the physical order of the *Hha*I sites can be deduced. The resulting map is shown in Figure 4. *Hha*I fragment pairs C, D and F, G are very similar in size, and it is difficult to assign their positions unequivocally. The assignment shown is derived from our best estimates of the partial-digest fragment sizes based on averages of three different experiments. There are no *Xma*I, *Kpn*I, *Bam*HI, and *Hind*III sites on this fragment of sea urchin DNA.

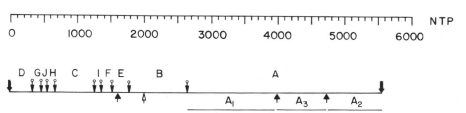

Figure 2. Restriction map of the 5500-nucleotide *Eco*RI insert of CSO859. The sites of *Hae*III (⦶), *Eco*RI (↓), *Hinc*II (↟), and *Hind*III (⦶) are indicated. *Hinc*II cleaves the *Hae*III A fragment into the three subfragments A_1, A_2, and A_3.

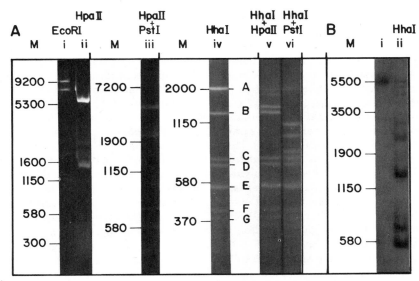

Figure 3. (A) Agarose gel electrophoresis of terminal restriction enzyme digests of pSC34 DNA. (i) pSC34 was digested with *Eco*RI and the products electrophoresed on a 1.5% agarose slab gel. Three fragments were produced, with sizes of 9200 nucleotides (the unit length of a pSC101 linear molecule), 7200 nucleotides, and 1900 nucleotides. The latter two fragments are sea urchin DNA inserts. (ii) The 7200-nucleotide sea urchin *Eco*RI fragment was digested with *Hpa*II. Two large fragments, 5500 nucleotide pairs and 1400 nucleotide pairs, were produced, plus a small fragment of about 300 nucleotides. The two faint bands are contaminants from pSC101 DNA. (iii) The 7200-nucleotide *Eco*RI fragment was digested with *Pst*I and *Hpa*II and the products were electrophoresed on a 1.75% agarose gel. Four large fragments were produced, with sizes of 3500 nucleotides, 2200 nucleotides, 820 nucleotides, and 580 nucleotides. The 300-nucleotide fragment also generated was not shown in this gel. (iv) *Hha*I cleaves the 7200-nucleotide *Eco*RI fragment into fragments A (2000 nucleotides), B (1300 nucleotides), C (700 nucleotides), D (640 nucleotides), E (520 nucleotides), F (430 nucleotides), G (390 nucleotides), H (130 nucleotides), and I (60 nucleotides). Band E actually contains three fragments of about the same size. Fragments H and I are detected only in 5% polyacrylamide gels, and their existence was confirmed in the partial analysis experiment shown in *B*. (v) The 7200-nucleotide *Eco*RI fragment was digested with *Hha*I and *Hpa*II. The two *Hpa*II sites are located within the *Hha*I A fragment, generating new fragments of 1400 nucleotides and 300 nucleotides (two fragments). The first two faint bands on top of the gel are partial cleavage products of *Hha*I fragment A. (vi) The 7200-nucleotide *Eco*RI fragment was digested with *Hha*I and *Pst*I. The sites for *Pst*I are located within *Hha*I fragments A, B, and E. *Hha*I fragment A was cleaved into two fragments, 1100 nucleotides and 820 nucleotides long. *Hha*I fragment B was also cleaved into two fragments, 1000 nucleotides and 300 nucleotides long. One of the E fragments is also cleaved by this enzyme near one end.

(B) Autoradiographs of the "ladder" experiment for the determination of the *Hha*I restriction sites in the 7200-nucleotide *Eco*RI insert of pSC34. Conditions for electrophoresis and autoradiography are the same as in Fig. 1B. (i) The 5500-nucleotide fragment generated from *Hpa*II digestion of the original 7200-nucleotide *Eco*RI fragment was isolated and rerun on a 1.75% agarose slab gel. (ii) The same fragment as in gel i was partially cleaved with *Hha*I. Some of the 5500-nucleotide fragment remained undigested, and the sizes of the partials generated are as follows: 5120 nucleotides, 4620 nucleotides, 4150 nucleotides, 2970 nucleotides, 2400 nucleotides, 1800 nucleotides, 1740 nucleotides, 1360 nucleotides, and 690 nucleotides. The smallest fragment produced was 545 nucleotides long and is identified as one of the *Hha*I E fragments.

Location of Repetitive Sequences in the pSC34 and CSO859 Inserts

Lee et al. (1977) showed earlier that the repetitive sequences included on the cloned inserts are all around 300–500 nucleotides in length. This was dem-

onstrated by labeling the cloned DNA in vivo and reacting it with excess sea urchin DNA to a C_0t of 20. The partially renatured fragments were digested with S_1 nuclease under conditions known to spare most or all repetitive sea urchin DNA duplexes (see Davidson et al. 1973; Chamberlin et al. 1975; Britten

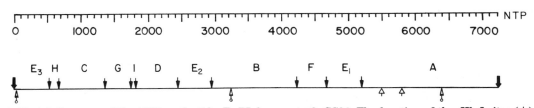

Figure 4. Restriction map of the 7200-nucleotide *Eco*RI fragment of pSC34. The location of the *Hha*I sites (↓), *Eco*RI sites (↓), *Pst*I sites (⚲) and *Hpa*II sites (⚲) are shown.

et al. 1976). The range of lengths of the repetitive-sequence elements was then obtained by gel electrophoresis of the labeled enzyme-resistant duplexes formed from each insert. The total amount of repetitive sequence was also estimated as the fraction of the insert which became S_1-nuclease-resistant after the C_0t 20 reaction with sea urchin DNA. The pSC34 insert was found to include 900–1000 nucleotides of repetitive sequence and the CSO859 insert about 850 nucleotides of repetitive sequence. Thus, on the basis of the S_1 nuclease experiments, each clone appeared to contain two or possibly three repetitive-sequence elements separated by other (probably single-copy) sequences remaining single-stranded after low-C_0t incubation with sea urchin DNA.

The location of the short repetitive-sequence ele-

ments on the restriction map of the cloned inserts was established as follows. The HhaI fragments of the pSC34 insert and the HaeIII fragments of the CSO859 insert were labeled by the kinase method and individually isolated from electrophoretic gels. These fragments were then reacted to a C_0t of 20 and to a C_0t of 11,000 or more using a sufficient excess of sea urchin DNA to prevent any significant self-reaction at a driver DNA C_0t of 20. The amount of renaturation was measured by hydroxylapatite binding, and the results are summarized in Table 1. Here it can be seen that the high-C_0t reactions were in the range of 60–95%. These reactions occur mainly between the sea urchin DNA and the cloned fragments, although a minor self-reaction component is also included. However, this is of no signifi-

Table 1. Identification of Repetitive Sequences in Restriction Fragments of CSO859 and pSC34

	CSO859[a]				pSC34[b]			
	reaction with sea urchin driver (%)		contains repetitive sequence		reaction with sea urchin driver (%)		contains repetitive sequence
tracer	C_0t 20	C_0t 12,000–15,000		tracer	C_0t 20	C_0t 11,000–15,000	
HaeIII ^{32}P-labeled fragments				HhaI ^{32}P-labeled fragments			
A	51	93	+	A	2	60	—
B	10	N.D.	—	B	1	65	—
C	73	90	+	C	56	80	+
D	66	92	+	D	49	82	+
E	5	N.D.	—	E (3X)	39	82	++—
				F	62	74	+
				G	63	78	+
				E(terminus)	5	95	—
HincII redigestion of HaeIII ^{3}H-labeled fragments							
A$_1$	67	85	+				
A$_2$	5	93	—	2200-Nucleotide			
A$_3$	1	95	—	HpaII + PstI ^{3}H-labeled fragment	76	94	+

[a] About 10 μg of the 5500-nucleotide EcoRI insert of clone CSO859 was terminally digested with HaeIII. The digestion mixture was extracted twice with a mixture of 80% phenol and 20% chloroform-isoamyl alcohol (24:1), equilibrated with 10 mM Tris (pH 7.4), 1 mM EDTA, and then extracted once with ether. RSF2124 DNA was added to the aqueous phase as carrier to a final concentration of 10 μg/ml. The aqueous phase was then adjusted to 0.3 M sodium acetate (pH 6.8), passed through a 0.4-ml Chelex column, and isopropanol-precipitated at −20°C. The precipitate was resuspended in 0.3 ml of 10 mM Tris-HCl (pH 8) and reacted with 1 μl of bacterial alkaline phosphatase (180 units/ml; Worthington, BAPF grade) at 37°C for 30 min. At the end of the reaction, 0.1 vol. of 3 M sodium acetate (pH 6.8) was added. The solution was deproteinized as above and precipitated with isopropanol. The 5′ termini were labeled in the denatured condition with polynucleotide kinase according to Maxam and Gilbert (1977) with [γ-^{32}P]ATP (1200 Ci/mmole; ICN). The kinase reaction mixtures were deproteinized as above and precipitated with 2 vol. of 100% ethanol at −20°C, in a final concentration of 0.3 M sodium acetate. The dried precipitate was resuspended in 50 μl of 0.1 M NaCl, 50 mM Tris (pH 7.2) and incubated at 60°C for 17 hr in a sealed glass capillary to allow the denatured fragments to reanneal. The solution was then loaded on a 3% agarose gel and electrophoresed at 60 V for 3.5 hr, when the bromphenol-blue marker was at the end of the 9.5-cm cylindrical gel. Gels were stained with ethidium bromide (0.5 μg/ml) and autoradiographed to locate the HaeIII fragments. These were eluted from the gel by dissolving in 5 M NaClO$_4$ at 37°C for 1 hr. The DNA was then bound to hydroxylapatite in 0.12 M phosphate buffer at 23°C and eluted with 0.5 M phosphate buffer. The labeled fragments (sp. act. about 2 × 10[6] cpm/μg) were reassociated to C_0t 20 with excess 450-nucleotide sheared sea urchin DNA (1 × 10[6] mass ratio) in 0.12 M phosphate buffer at 60°C. The fraction of the [^{32}P]DNA reassociated was determined by hydroxylapatite chromatography in 0.12 M phosphate buffer 60°C. Self-reactions of the labeled fragments were monitored in otherwise identical reaction mixtures lacking the sea urchin driver DNA. C_0t 20 values shown in the table have been corrected for the small amount of self-reaction which could have occurred; this ranged from 2–10%. The high C_0t reactions shown were carried out to determine the reactivity of the labeled fragments, and therefore the possible self-reaction has not been subtracted from the observed values (see text). The HincII redigestion products of the HaeIII fragment were prepared from in-vivo-labeled ^{3}H-labeled supercoils of CSO859 (sp. act. 2 × 10[5] cpm/μg). These were eluted from agarose gels and reassociated with excess sea urchin DNA (4 × 10[5] mass ratio) as above. N.D. denotes values not determined.

[b] The 7200-nucleotide EcoRI insert of clone pSC34 was digested to completion with HhaI. The fragments produced were kinased and reassociated with excess total sea urchin DNA as described above for CSO859. The 2200-nucleotide fragment in the last experiment was generated by digestion of in-vivo-labeled [^{3}H]DNA. The EcoRI insert was isolated and treated with HpaII and PstI. The 2200-nucleotide fragment displayed a specific activity of 4 × 10[5] cpm/μg. The reactivity of these fragments was determined by reacting with excess total sea urchin DNA to a C_0t of 11,000–15,000.

cance for the present purposes since we already know that the inserts consist of sea urchin DNA, and the high-C_0t reactions are required only to show that the labeled fragments are capable of renaturing. For many of the fragments tested, large reactions were obtained by C_0t 20. Some fragments, however, reacted only at high C_0t, and no more than a few percent of these molecules were present in the duplex fraction at C_0t 20. These fragments must consist of sequences represented once or a very few times in the sea urchin genome. We refer to them as single-copy fragments, though we cannot exclude the possibilities that some of them are actually a few-fold repetitive or that they contain unrecognizably short lengths of repetitive sequence (e.g., < 50 nucleotides) as well as single-copy sequence. However, *Hha*I fragment A reacts with sea urchin DNA with a $C_0t_{1/2}$ of about 1000 $M^{-1} \cdot sec^{-1}$ and thus displays the expected single-copy kinetics (data not shown).

In the CSO859 insert, *Hae*III fragments A, C, and D clearly contain repetitive sequences, whereas *Hae*III fragments B and E appear to be single copy. As discussed below, it is important to determine how much of *Hae*III fragments C and D are included in repetitive sequence. C and D fragments ^{3}H-labeled in vivo were reacted separately with sea urchin DNA and the fraction of these fragments rendered resistant to S_1 nuclease digestion at driver DNA C_0t 5 and C_0t 20, respectively, were measured. This experiment showed that about 55% or 300 nucleotides of the 500-nucleotide fragment C and 80% or 280 nucleotides of fragment D are repetitive. *Hae*III fragment A is a relatively large fragment (2900 nucleotides) and we wished to further localize the repetitive sequence within it. When digested with *Hinc*II, the *Hae*III A fragment yields three new subfragments (see Fig. 2). Table 1 shows that, of these, only the *Hinc*II A$_1$ fragment contains a repetitive sequence; *Hinc*II A$_2$ and A$_3$ behave as single-copy-sequence elements.

The two largest *Hha*I fragments in the pSC34 insert, A and B, appear to consist of single-copy sequences, whereas all the other *Hha*I fragments (C–G) contain repetitive sequences (Table 1). As noted above, the *Hha*I E band actually includes three fragments of approximately the same size. The intermediate level of reactivity of band-E DNA with the excess sea urchin DNA at C_0t 20 suggests that one of these three fragments may be a single-copy sequence (Table 1). From the restriction map data, we also know that one of these fragments (E$_3$) is located at the left-hand terminus in the orientation shown, and that the other two E fragments are internal. Fragment E$_3$ was selectively labeled by kinase treatment of the whole 7200-nucleotide *Eco*RI insert and subsequent isolation of the *Hha*I E band. When reacted with sea urchin DNA, the labeled member of the E triplet, E$_3$, contained no detectable repetitive sequence (Table 1, pSC34). It follows that both

termini of the 7200-nucleotide *Eco*RI insert consist of single-copy sequences. We show below that the internal members of the *Hha*I E triplet contain repetitive sequences.

Given the order of the restriction fragments shown in Figures 2 and 4 and the fact that each repetitive-sequence element is only 300–500 nucleotides in length, the information in Table 1 approximately locates the repetitive-sequence elements. It is clear that some of these sequences occur in widely separated positions in the interior of each cloned sea urchin DNA insert. A detailed interspersion map of these inserts is presented later.

Interspersed Repetitive Sequences in the pSC34 Insert Viewed in the Electron Microscope

An independent approach to locating the repetitive sequences was to visualize them as sites of heteroduplex formation with sea urchin DNA. The procedure used was described earlier by Lee et al. (1977) and can only be briefly summarized here. The 7200-nucleotide sea urchin DNA insert of pSC34 was isolated from an *Eco*RI digest, and poly(A) tracts a few hundred nucleotides long were added to the 3' termini of the fragments by the terminal transferase method of Roychoudhury et al. (1976). These fragments were then reacted to C_0t 20 with excess sea urchin DNA of average fragment length 1400 nucleotides, and the partially renatured molecules were bound to an oligo(dT)-cellulose column. This step removes most of the excess sea urchin DNA. The bound fraction was recycled twice over oligo(dT) columns to further reduce the driver DNA background, and then concentrated and prepared for electron microscopy. Molecules of approximately the length of the pSC34 insert were observed in the spreads, along with much shorter DNA driver complexes. Both the long and short molecules displayed the typical interspersed branched structures described by Chamberlin et al. (1975), who demonstrated that when interspersed repetitive sequences on fragments longer than 1000 nucleotides react with each other, structures resembling an "H" are frequently formed (or multiples thereof for multifragment complexes). The bar of each H is usually 300–400 nucleotides long and is the renatured interspersed repetitive sequence. The four arms of the H are thus the flanking single-copy regions, which of course are generally nonhomologous.

A micrograph showing the complex structures formed by the partial renaturation of the pSC34 insert with sea urchin DNA driver was published earlier (Lee et al. 1977). The contours and lengths of 39 such molecules were recorded, and these data are arranged as line drawings in Figure 5A. A vertical mark indicates that a branch (or branches) was observed at that position on the pSC34 insert. All molecules encountered on the grids that were in the size range of 5000–8000 nucleotides are included

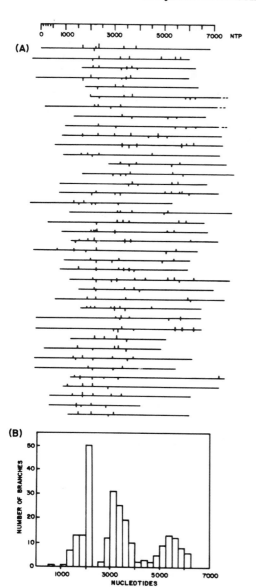

Figure 5. Summary of electron microscope observations on pSC34–sea urchin DNA renaturation products. *(A)* pSC34 molecules renatured with total sea urchin DNA to $C_0t\,20$ were spread for electron microscopy. The contours and lengths of these molecules were measured and graphed as straight lines. Each branched structure protruding from the long linear molecules is represented by a short vertical bar on the line drawings. The branched structures are driver sea urchin DNA tails (see Chamberlin et al. 1975). The molecules were aligned arbitrarily as described in the text. *(B)* Histogram showing the occurrence of branched structures along the 7200-nucleotide insert of pSC34 assuming the alignment shown in *A*.

length heterogeneity in Figure 5A. These strand scissions derived mainly from the commercial terminal transferase preparation we used, which proved to be contaminated to an undesirable extent with endonuclease. In a few of the cases included in Figure 5A, the left-hand branch complex had apparently been interrupted by an internal strand scission, so that it was no longer explicitly recognizable. However, these molecules could be aligned by the position of the right-hand branch complexes.

No quantitative conclusions can be drawn as to sequence organization from the positions of the driver DNA branches plotted in Figure 5A, since the resolution of the method is not high, and uncertainty is introduced by the endonucleolytic cleavages which severely decreased the yield of unbroken molecules. Furthermore, many of the branches seen appear to derive from driver DNA complexes, and the pSC34 insert cannot be followed easily. Nonetheless, Figure 5 shows that most of the structures scored are consistent with a particular arrangement of repetitive sequences. The main feature is that one end of the insert, oriented to the left as shown, contains at least one and probably two interspersed repetitive-sequence elements. One- to two-thousand nucleotides to the right, another such element appears. On the right-hand end, a relatively long single-copy tail extends to the terminus of the insert.

The main features of the interspersion arrangement indicated in Figure 5B agree well with the restriction enzyme map shown in Figure 4 and the data of Table 1. Thus the left-hand repetitive region of Figure 5B can be aligned with the left-hand region of Figure 4. Here are located *Hha*I fragments C and G, which according to Table 1 contain repetitive sequences. The second repetitive sequence suggested in Figure 5B would consist of portions of *Hha*I fragments D and E_2, and the right-hand repetitive sequence, of parts of *Hha*I fragments F and E_1. The latter repetitive fragments are separated from repetitive fragments D and E_2 by *Hha*I fragment B, according to Figure 4. Table 1 shows that this fragment, which is 1300 nucleotides in length, contains only single-copy sequences. Similarly, the right-hand end of the pSC34 insert lacks repetitive sequences according to Figure 5B, and this region is mainly accounted for by the long single-copy *Hha*I fragment A.

To confirm by an independent test the existence of the right-hand repetitive sequence in the pSC34 insert, we isolated the 2200-nucleotide fragment from a double digest made with *Pst*I and *Hpa*II (see Fig. 3A, iii, and Fig. 4). This fragment contains portions of the single-copy *Hha*I fragments A and B as well as the entire length of *Hha*I fragments F and E_1. Therefore it should include the right-hand repetitive sequence indicated in both the electron microscope study and the data of Table 1. After recovery from an agarose gel, this fragment was renatured with excess sea urchin DNA. The result is

except for about ten which appeared as uninterpretable tangles. The molecules in Figure 5A are aligned arbitrarily by the position of the innermost of the left-hand cluster of driver DNA branches. This method of alignment was chosen because many of the molecules observed were terminated by strand scissions at one end or the other, as shown by the

shown in the last line of Table 1 (pSC34), where it can be seen that the 2200-nucleotide PstI-HpaII fragment does indeed include a repetitive-sequence element.

Frequency of Occurrence in the Sea Urchin Genome of the Cloned Repetitive Sequences

Those HhaI fragments of pSC34 and HaeIII fragments of CSO859 which contain repetitive-sequence

elements were labeled by the kinase procedure and reacted with excess sea urchin DNA. The kinetics of the renaturation reactions were measured. Thus the approximate number of times each of the repetitive sequences represented on the cloned inserts occurs in the sea urchin genome could be calculated. In Figure 6A the kinetics of the reactions of CSO859 HaeIII A, C, and D fragments with excess sea urchin DNA are illustrated, and in Figure 6B similar reactions with the pSC34 HhaI fragments C, D, E, F,

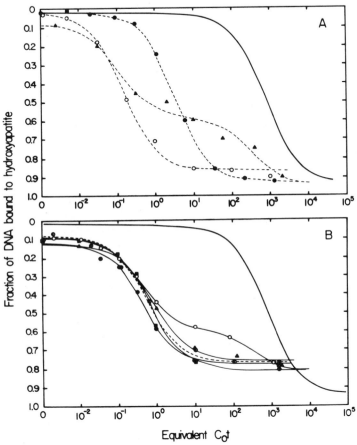

Figure 6. Reassociation kinetics of kinase-labeled restriction fragments of cloned sea urchin DNA with excess sea urchin DNA. (A) Clone CSO859. HaeIII ^{32}P-labeled fragments A, C, and D of the 5500-nucleotide EcoRI insert were prepared as described in Table 1. These DNA fragments were reassociated with excess 400-nucleotide sea urchin DNA in 0.12 M phosphate buffer at 55°C at a mass ratio of 5×10^5. Single-copy [^{3}H]DNA 300 nucleotides long was included as a kinetic standard. The amount of [^{32}P]DNA and single-copy [^{3}H]DNA in the duplex fraction was determined by hydroxylapatite chromatography in 0.12 M phosphate buffer at 55°C. The solid curve is the least-squares solution for the [^{32}P]DNA, assuming second-order kinetics. (▲‑ ‑ ‑▲) HaeIII A; (○———○) HaeIII C; (●———●) HaeIII D: all driven by total sea urchin DNA. The rate constants determined for these reactions are listed in Table 2. The solid curve represents the reaction of the single-copy sequences in the driver DNA. The biphasic HaeIII A reaction displays a fast-reacting component as well as a nearly single-copy component. This is probably due to the occurrence of single-strand nicks in the HaeIII A fragment, which is 2900 nucleotides long, during the denaturation and reassociation processes. Such nicks generate end-labeled single-copy fragments which are separated from the repetitive-sequence element and therefore reassociate as a single-copy sequence. As demonstrated above, at least 1550 nucleotides of the HaeIII A fragment comprise a single-copy sequence (Table 1, CSO859).

(B) Clone pSC34. Kinase-labeled HhaI fragments of the 7200-nucleotide EcoRI insert of pSC34 were prepared as described in Table 1. These were reassociated with total sea urchin DNA and assayed by hydroxylapatite chromatography as described in A. The curves with symbols are the least-squares solutions for [^{32}P]DNA. (●———●) HhaI C; (■———■) HhaI D; (○———○) HhaI E; (▲———▲) HhaI F; (◆———◆) HhaI G: all driven by total sea urchin DNA. The rate constants for each of these reactions are listed in Table 2. The solid curve represents the reaction of single-copy sequences in the driver DNA. The reaction of the triplet HhaI band-E DNA with excess sea urchin DNA indicates that two of the E fragments contain repetitive sequences and one fragment is a single-copy sequence or contains a sequence repeated only a very few times (see text).

and G are shown. The repetition frequency of each fragment was calculated with reference to the reaction of ^{3}H-labeled single-copy tracers included in the same renaturing mixtures. The single-copy tracers thus served as internal kinetic standards. The data were fit by least-squares methods on the assumption of second-order kinetics. Results are listed in Table 2. All of the pSC34 repetitive sequences appear to be of approximately equal repetition frequency. This statement is secure within a factor of about 2, a reasonable error considering possible length corrections and other minor sources of variations not taken into account in this analysis. On the other hand, the repetition frequencies present on the CSO859 inserts are dissimilar. Though very close to one another on the restriction map (Fig. 2), the HaeIII C and D fragments contain repetitive sequences whose copy numbers in the sea urchin genome differ by more than a factor of 25.

Are the Repetitive Sequences of Each Fragment Homologous?

We now ask whether the same repetitive-sequence family is represented in the various interspersed repetitive elements on each cloned insert. If this is the case, the various restriction fragments bearing repetitive sequences should cross-react. The appropriate restriction fragments were isolated from electrophoretic gels and labeled by the kinase procedure. They were then renatured with unlabeled DNA of the same or different restriction fragments. The unlabeled DNA was present in large excess to prevent

tracer self-reaction, and the extent of reaction with the homologous fragment served as a control for the reactivity of the kinased tracer molecules. These experiments are summarized in Table 3. The homologous reactions were about 80% terminated in the case of CSO859. On incubation to the same driver fragment C_0t, 0–4% of nonhomologous repetitive tracers reacted. Hence the three repetitive sequences of the CSO859 insert share no detectable sequence homology. The same appears to be true for the interspersed repetitive sequences on the pSC34 insert. In this case, the homologous reactions were carried only to 47–67% completion. Except for the fragment pairs C and D, E and F, and F and G, no cross-reactions were observed. About a third of the homologous amount of reaction was measured when fragment pair C and D and fragment pair F and G were reacted. However, these fragments cannot be separated clearly in our gel system, and since they are of approximately equal reactivity when tested with the homologous driver, it seems apparent that the small cross-reactions observed are due to cross-contamination rather than sequence homology. The same is true of bands E and F. Furthermore, in this case the reciprocal reaction of E tracer with F driver gave only a 10% signal (Table 3). Even though the E band contains three different fragments, 10% is only about a third of the amount of reaction that would be expected if one of the E fragments were homologous with the sequence carried on the F fragment. We believe that the three repetitive-sequence elements on each cloned insert belong to three distinct nonhomologous repetitive-sequence families.

Table 2. Reassociation Kinetics of Restriction Enzyme Fragments of pSC34 and CSO859 with Excess Sea Urchin DNA

	Length (nucleotides)	Restriction fragment rate constant ($M^{-1} \cdot sec^{-1}$)	Calculated frequency of occurrence per haploid genome[a]
CSO859			
*Hae*III fragment			
A	2900	9.5[b]	7900
C	550	6.4	5300
D	350	0.3	250
pSC34			
*Hha*I fragment			
C	700	2.3	1900
D	640	1.5	1300
E_1, E_2	520	2.5	2100
E_3	520	3.4×10^{-3c}	1–3[c]
F	430	1.2	1000
G	390	1.8	1500

[a] The reiteration frequencies are calculated by dividing the observed rate constants for the restriction fragments by 1.2×10^{-3} $M^{-1} \cdot sec^{-1}$. The latter value is the rate constant for the reaction of single-copy sequence in sea urchin DNA of the length used at 55°C in 0.12 M phosphate buffer. No attempt has been made to further correct the observed rates for the tracer length, since these corrections would be small and the actual fragment lengths during the reaction are not known.

[b] Rate constant fit to the rapidly renaturing portion of the A fragment case (Fig. 6A).

[c] See text for experimental evidence supporting the assignment of E_3 as the terminal, single-copy member of the E-band triplet.

Table 3. Percentage of Cross-hybridization of Restriction Fragments Which
Contain Repetitive Sequence Elements

	CSO859[a]				pSC34[b]				
	tracer					tracer			
driver	A*	C*	D*	driver	C*	D*	E*	F*	G*
A	81	0	0	C	65	23	0	0	0
C	4	80	0	D		50	0	0	0
				E			47	19	0
				F			10	67	21

[a] Clone CSO859. Kinase-labeled *Hae*III fragments A, C, and D (denoted by asterisk) were prepared as described in the footnote to Table 2. These were renatured with excess unlabeled *Hae*III A and C fragments purified from agarose gels using an estimated mass ratio of 100. Reactions were carried out in 0.12 M phosphate buffer at 60°C, to about 10 times the estimated $C_0t_{1/2}$ of the DNA driver fragments. The fraction of [^{32}P]DNA renatured was determined by hydroxylapatite chromatography in 0.12 M phosphate buffer at 60°C. Data shown have been corrected by subtraction of the tracer self-reaction. This was measured as the percent of [^{32}P]DNA bound to hydroxylapatite after a similar incubation in the absence of unlabeled DNA and was always ≤10%.

[b] Clone pSC34. Kinase-labeled *Hha*I fragments C*, D*, E*, F*, and G* were renatured with unlabeled *Hha*I fragments C, D, E, and F as described for CSO859. The estimated mass ratio was about 50 and the renaturation reactions were carried out to about 10 times the estimated $C_0t_{1/2}$ of the DNA driver fragment. The numbers shown represent the percent of the labeled fragment bound to hydroxylapatite after correction for the self-reaction of the labeled fragment.

Repetitive-sequence Interspersion in the CSO859 and pSC34 Inserts

The data presented in this paper are summarized in Figure 7. Each of the cloned inserts contains three interspersed repetitive-sequence elements separated by single-copy or nearly single-copy DNA. The single-copy nature of the DNA lying between adjacent repetitive sequences was shown explicitly for the longer fragments in the experiments of Table 1. The shorter, interspersed single-copy regions were inferred from several sources of information as detailed earlier: These are the restriction maps shown in Figures 2 and 4, the electron microscope study summarized in Figure 5, and the S$_1$ nuclease experiments of Lee et al. (1977), showing that the individual repetitive sequence elements are only 300–500 nucleotides in length and that the total amount of repetitive sequence is 900–1000 nucleotides in the pSC34 insert and about 850 nucleotides in the CSO859 insert (see legend to Fig. 7). The data included in this paper provide a specific demonstration of short-period repetitive-sequence interspersion in the cloned DNA inserts (see Fig. 7).

It is clear that the sequence organization in these inserts is typical of that of sea urchin DNA as a whole. Thus, in accord with the dominant pattern in this and many other animal genomes, short repetitive-sequence elements are interspersed with single-copy regions ranging from several hundred to at least 2000 nucleotides. Several particular aspects of the sequence organization of these cloned samples of the sea urchin genome deserve comment. The repetitive-sequence elements cannot be located more precisely than the boundaries of the restriction fragments allow, as indicated by the dashed lines in Figure 7. Nonetheless, even taking into account the possible range in the single-copy sequence

lengths, it is clear that very different lengths of single-copy DNA may adjoin one another. No local regularity in single-copy spacer lengths exists, at least not in the two interspersed regions of the genome examined here. Furthermore, in the CSO859 insert it is evident that sequences belonging to families of widely diverse repetition frequencies may be found in close proximity (Fig. 7). Here the results with the two cloned inserts differ, since the repetition frequencies of the pSC34 repeats are all fairly similar. Further experiments will be necessary before we can decide whether it is likely that adjacent repetitive sequences belong to families of similar repetition frequencies.

An important finding is that the interspersed repetitive sequences of each local region are nonhomologous. Suppose the genome consisted of a series of domains, each bearing tens or hundreds of homologous interspersed repetitive sequences. If such domains were dominant, fragments of the length studied here would not be expected to contain repetitive sequences which all belong to distinct families of repeats. Therefore, if the results obtained here prove to be general, simple models of this kind would be ruled out. It is, of course, always possible that the minute sample of the genome examined here (about 0.002%) is peculiar. However, the same conclusion was drawn by Chamberlin et al. (1975) in their electron microscope study of partially renatured unfractionated *Xenopus* DNA. These authors argued that nearby interspersed repetitive sequences are generally not homologous. Thus, in areas where the participating strands were of sufficient length, the structures expected if there is regional localization of repetitive sequence families were not observed. The predicted structures are renatured duplexes alternating with regularly occurring single-stranded "bubbles." Instead, multiple,

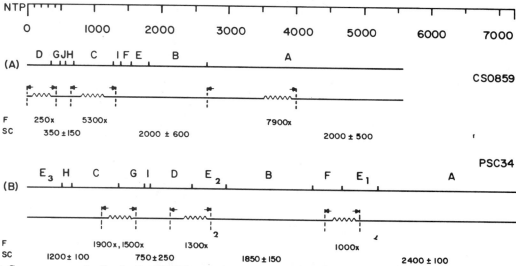

Figure 7. Sequence organization in the CSO859 and pSC34 inserts. *(A)* CSO859 insert. The linear order of *Hae*III fragments A–J and *Hinc*II fragments A_1–A_3 are reproduced from Fig. 2. The repetitive sequences are positioned on the map according to the data in Table 1 and the text, assuming that each repetitive-sequence element is around 300 nucleotides in length. Thus the total amount of repetitive sequence is about 900 nucleotides, as required by the results of Lee et al. (1977). Since they were not investigated per se, we do not know whether the small *Hae*III fragments F–J are single-copy or repetitive. However, as described in the text, both fragments C and D contain about 300 nucleotides of repetitive sequence. Therefore there must exist two repetitive sequences on the left-hand side of the molecules, with the distance between them being at least 200 nucleotides and possibly 500 nucleotides. The vertical bars indicate the limits which are set by assuming that lengths of repetitive sequence ≥50 nucleotides will cause the restriction fragment to bind to hydroxylapatite after renaturation with total sea urchin DNA. The lengths of the single-copy sequence elements (SC) and their repetition frequencies (F) (from Table 2) are also indicated. *(B)* pSC34 insert. The linear order of *Hha*I fragments A–I of the 7200-nucleotide fragment is reproduced from Fig. 4. *Hha*I fragments C, D, E_1, E_2, F, and G have been shown to contain repetitive sequences (Table 1). The interspersion diagram shown derives from these data and the earlier results of Lee et al. (1977). The single-copy space shown as separating the two left-hand repetitive sequences is necessitated by the following argument. The repetitive sequences are only 300–500 nucleotides long whereas the total distance from the beginning of fragment C to the right-hand end of fragment E_2 is over 2300 nucleotides (see Fig. 4). Since fragments E_2 and C both contain repetitive sequences, a significant amount of single-copy DNA must separate the repetitive sequences carried on *Hha*I fragments C and E_2. The resulting lengths of the interspersed single-copy sequences (SC) and the calculated repetition frequencies (F) of the sequences carried on the various restriction fragments (from Table 2) are also indicated.

branched H structures formed along the renaturing strands. This is expected if each nearby repeat belongs to a different repetitive-sequence family, as shown here for the pSC34 and CSO859 inserts. With respect to the latter finding, a further possibility should be briefly considered, i.e., that the regions of the sea urchin genome from which the cloned inserts derived originally contained some locally homologous repetitive sequence elements, but that these were lost during plasmid growth or transformation, by means of some recombinational event. In the case of pSC34 the host strain of bacteria (HB101) is RecA⁻, which renders this form of alteration unlikely. The CSO859 plasmid has been carried in a RecA⁺ strain (C600). However, this plasmid and others of the same origin also studied in our laboratory have been absolutely stable over a 2-year period. Thus CSO859 always yields the same single 5500-nucleotide sea urchin band on *Eco*RI digestion. The most important point, however, is that even if homologous repetitive sequences had been lost during cloning, the nonhomologous repeats now present on the inserts were also present in the original in-

sert. Thus the general conclusion that diverse repetitive sequences exist in local regions of the DNA remains unaffected.

Sequence organization of the pSC34 fragment is interesting from another point of view as well. According to information which is still preliminary, this insert includes a single-copy sequence represented to a significant extent in the RNA of mature sea urchin oocytes. This sequence is probably a structural gene for one of the maternal mRNAs stored in the oocyte. The transcribed sequence lies within a few hundred or a thousand nucleotides of one or more interspersed repetitive sequences. Clone pSC34 will clearly be useful in exploring the functional meaning of short-period repetitive-sequence interspersion.

SUMMARY

In this paper we described a study of repetitive-sequence interspersion in two recombinant DNA molecules containing inserts of sea urchin DNA. The cloned inserts are terminated by natural *Eco*RI sites and are 5500 and 7200 nucleotides long. Restriction maps of the cloned inserts were prepared, and the individual subfragments tested for the presence of repetitive sequences by reaction with excess sea urchin DNA. Each cloned insert contains three repetitive-sequence elements 300–500 nucleotides in length, interspersed with longer single-copy DNA regions. The repetitive sequences in the local regions of the genome represented in each clone are nonhomologous. Thus they belong to distinct repetitive-sequence families, and several frequencies of occurrence in the sea urchin genome are represented.

Acknowledgments

We thank Mr. Tom Estes for technical assistance and Dr. William Klein and Mr. Cary Lai for measuring the renaturation kinetics. This research was supported by National Institutes of Health Grants HD-05753 and GM-20927 and by National Science Foundation Grant BMS-75–07359. A. S. L. was supported by a fellowship from the American Cancer Society, California Division (J-289) and by a National Institutes of Health Fellowship. R. J. B. is also a staff member of the Carnegie Institution of Washington.

REFERENCES

ANGERER, R. C., E. H. DAVIDSON, and R. J. BRITTEN. 1975. DNA sequence organization in the mollusc *Aplysia californica*. *Cell* **6**: 29.

BONNER, J. and J.-R. WU. 1973. A proposal for the structure of the *Drosophila* genome. *Proc. Natl. Acad. Sci.* **70**: 535.

BONNER, J., W. T. GARRARD, J. GOTTESFELD, D. S. HOLMES, J. S. SEVALL, and M. WILKES. 1974. Functional organization of the mammalian genome. *Cold Spring Harbor Symp. Quant. Biol.* **38**: 303.

BRITTEN, R. J. and J. SMITH. 1970. A bovine genome. *Carnegie Inst. Wash. Year Book* **68**: 378.

BRITTEN, R. J., D. E. GRAHAM, F. C. EDEN, D. M. PAINCHAUD, and E. H. DAVIDSON. 1976. Evolutionary divergence and length of repetitive sequences in sea urchin DNA. *J. Mol. Evol.* **9**: 1.

CHACONAS, G., J. H. VAN DESANDE and R. B. CHURCH. 1975. End group labeling of RNA and double stranded DNA by phosphate exchange catalyzed by bacteriophage T₄ induced polynucleotide kinase. *Biochem. Biophys. Res. Commun.* **66**: 962.

CHAMBERLIN, M. E., R. J. BRITTEN, and E. H. DAVIDSON. 1975. Sequence organization in *Xenopus* DNA studied by the electron microscope. *J. Mol. Biol.* **96**: 317.

CRAIN, W. R., E. H. DAVIDSON, and R. J. BRITTEN. 1976a. Contrasting patterns of DNA sequence arrangement in *Apis mellifera* (honeybee) and *Musca domestica* (housefly). *Chromosoma* **59**: 1.

CRAIN, W. R., F. C. EDEN, W. R. PEARSON, E. H. DAVIDSON, and R. J. BRITTEN. 1976b. Absence of short period interspersion of repetitive and non-repetitive sequences in the DNA of *Drosphila melanogaster*. *Chromosoma* **56**: 309.

DAVIDSON, E. H., G. A. GALAU, R. C. ANGERER, and R. J. BRITTEN. 1975. Comparative aspects of DNA sequence organization in metazoa. *Chromosoma* **51**: 253.

DAVIDSON, E. H., B. R. HOUGH, C. S. AMENSON, and R. J. BRITTEN. 1973. General interspersion of repetitive with non-repetitive sequence elements in the DNA of *Xenopus*. *J. Mol. Biol.* **77**: 1.

DAVIDSON, E. H., D. E. GRAHAM, B. R. NEUFELD, M. E. CHAMBERLIN, C. S. AMENSON, B. R. HOUGH, and R. J. BRITTEN. 1974. Arrangement and characterization of repetitive sequence elements in animal DNA's. *Cold Spring Harbor Symp. Quant. Biol.* **38**: 295.

DEININGER, P. L. and C. W. SCHMID. 1976. An electron microscope study of DNA sequence organization of the human genome. *J. Mol. Biol.* **106**: 773.

EDEN, F. C., D. E. GRAHAM, E. H. DAVIDSON, and R. J. BRITTEN. 1977. Exploration of long and short repetitive sequence relationships in the sea urchin genome. *Nucleic Acids Res.* **4**: 1553.

EFSTRATIADIS, A., W. R. CRAIN, R. J. BRITTEN, E. H. DAVIDSON, and F. C. KAFATOS. 1976. DNA sequence organization in the lepidopteran *Antheraea pernyi*. *Proc. Natl. Acad. Sci.* **73**: 2289.

GOLDBERG, R. B., W. R. CRAIN, J. V. RUDERMAN, G. P. MOORE, T. R. BARNETT, R. C. HIGGINS, R. A. GELFAND, G. A. GALAU, R. J. BRITTEN, and E. H. DAVIDSON. 1975. DNA sequence organization in the genomes of five marine invertebrates. *Chromosoma* **51**: 225.

GRAHAM, D. E., B. R. NEUFELD, E. H. DAVIDSON, and R. J. BRITTEN. 1974. Interspersion of repetitive and non-repetitive DNA sequences in the sea urchin genome. *Cell* **1**: 127.

LAIRD, C. D., W. Y. CHOOI, E. H. COHEN, E. DICKSON, N. HUTCHINSON, and S. H. TURNER. 1974. Organization and transcription of DNA in chromosomes and mitochondria of *Drosophila*. *Cold Spring Harbor Symp. Quant. Biol.* **38**: 311.

LEE, A. S., R. J. BRITTEN, and E. H. DAVIDSON. 1977. Interspersion of short repetitive sequences studied in cloned sea urchin DNA fragments. *Science* **196**: 189.

LEE, A. S., F. COSTANTINI, R. J. BRITTEN, and E. H. DAVIDSON. 1976. Initial studies of sea urchin DNA sequence organization by molecular cloning. In *Molecular mechanisms in the control of gene expression* (ed. D. P. Nierlich et al.), vol. 5, p. 553. Academic Press, New York.

MANNING, J. E., C. W. SCHMID, and N. DAVIDSON. 1975. Interspersion of repetitive and non-repetitive DNA sequences in the *Drosophila melanogaster* genome. *Cell* **4**: 141.

MAXAM, A. M. and W. GILBERT. 1977. A new method for sequencing DNA. *Proc. Natl. Acad. Sci.* **74**: 560.

PEARSON, W. R. 1977. "Studies on the arrangement of repeated sequences in DNA." Ph.D. thesis, California Institute of Technology, Pasadena, California.

ROYCHOUDHURY, R., E. JAY, and R. WU. 1976. Terminal labeling and addition of homopolymer tracts to duplex DNA fragments by terminal deoxynucleotidyl transferase. *Nucleic Acids Res.* **3**: 101.

SCHMID, C. W. and P. L. DEININGER. 1975. Sequence organization of the human genome. *Cell* **6**: 345.

SMITH, H. O. and M. L. BIRNSTIEL. 1976. A simple method for DNA restriction site mapping. *Nucleic Acids Res.* **3**: 2387.

WELLS, R., H. D. ROYER, and C. P. HOLLENBERG. 1976. Non-*Xenopus*-like DNA sequence organization in the *Chironomus tentans* genome. *Mol. Gen. Genet.* **147**: 45.

Transcription of *Xenopus* tDNA$_I^{met}$ and Sea Urchin Histone DNA Injected into the *Xenopus* Oocyte Nucleus

A. KRESSMANN, S. G. CLARKSON, J. L. TELFORD, AND M. L. BIRNSTIEL

Institut für Molekularbiologie II, der Universität Zürich, 8057 Zürich, Switzerland

Anyone interested in understanding the *functional* organization of DNA sequences in isolated eukaryotic genes is faced with the problems that in most cases the transcriptional units are poorly defined and that the individual gene units contain much spacer DNA of unknown function. Although it is now possible to sequence any cloned eukaryotic gene by quite rapid methods (Sanger and Coulson 1975; Maxam and Gilbert 1977), the problem remains of how to distinguish between those sequences which are relevant to gene regulation and those which are not.

We are using two kinds of approaches in an attempt to make such a distinction. The first approach is to ask whether we can identify eukaryotic DNA sequences of potential importance by their locations and by the homologies they may possess with comparably located sequences. The 6-kb histone DNA repeat unit of the sea urchin *Psammechinus miliaris,* over one-third of which by now has been sequenced (Birnstiel et al. 1977a; W. Schaffner et al., in prep.), contains genes for each of the five histone proteins together with intervening spacer DNA (Portmann et al. 1976; Schaffner et al. 1976). We would therefore expect such a DNA molecule to contain some internal structural homologies, for example, sequences coding for the ribosomal attachment sites of the various histone mRNAs, as well as signals for gene regulation at both the transcriptional and translational levels. Indeed, such comparative sequence data have already revealed that sequences of similar topology relative to the structural genes often show striking similarities (W. Schaffner et al., in prep.), and this is taken as an indication of their importance for the regulation of histone gene expression. We are now extending the sequence studies to those clones of histone DNA which are considered most likely to include histone *variants* which show developmental and tissue-specific expression in the sea urchin. By such sequence comparisons, we hope to discover both the common features of histone gene structure and those particular features that are associated with tissue-specific expression.

With similar aims in mind, we have begun to determine the sequence of cloned fragments of *Xenopus laevis* DNA coding for some tRNA species.

The typical tDNA$_I^{met}$ repeat unit comprises 3180 base pairs and appears to contain two genes coding for tRNA^{met} and at least one gene coding for a second 4S RNA species (Clarkson and Kurer 1976; S. G. Clarkson et al., in prep.). Any homologies in the sequences flanking these various coding regions would be of obvious interest.

This first comparative sequence approach enables us to identify DNA sequences of potential interest, and we are now seeking to understand their role at a more fundamental level. The basic plan is to study their expression in a living cell nucleus into which cloned DNA fragments, in an unmodified form or after suitable sequence alteration, have been transported by mechanical injection (Birnstiel et al. 1977). Here we present a much simplified method for injecting cloned DNA into the germinal vesicle (GV) of the *Xenopus* oocyte. We report on the results obtained for the transcription of both homologous cloned tDNA$_I^{met}$ and for heterologous sea urchin histone DNA injected into the *Xenopus* GV.

EXPERIMENTAL PROCEDURES: SIMPLIFIED INJECTION TECHNIQUE

DNA and labeled GTP. DNA (1–5 µg) and 50–100 µCi of [³H]GTP (10 Ci/mmole) or [α-³²P]GTP (100–200 Ci/mmole) are resuspended in 10 µl of 15 mM Tris-HCl, pH 7.5, 88 mM NaCl.

Oocytes. Individual stage-V–VI oocytes are isolated with a platinum loop and transferred to a 5-cm diameter plastic petri dish containing 6–8 ml of modified Barth's medium. On the bottom of the petri dish is fixed (with chloroform) a plastic grid with a mesh size of 0.8 mm. The oocytes, generally 50–150, are turned manually so that the (brown) animal pole at each oocyte is facing upwards. The petri dish is then centrifuged in an MSE Coolspin 6 × 1 liter swing-out rotor at 2000 rpm for 10 minutes at 18°C. This fixes the oocytes to the grid and moves the nuclei to the top of the oocytes. A dark-brown pigment ring is now visible within each animal pole, and the oocytes are ready for injection.

Injection. The DNA and labeled GTP solution is taken up in a glass micropipette having a needle diameter of 20–30 μm. This is connected via Teflon tubing to a micromanipulator, a microsyringe, and a foot-controlled motor. The petri dish, mounted under a binocular microscope, is moved so that an oocyte is close to the needle, and the micromanipulator is then adjusted so that the needle pierces the surface with its point in the center of the dark-brown pigment ring. Then 20 nl is injected per oocyte, equivalent to approximately 0.1 μCi of labeled GTP and 2–10 ng of DNA. When all oocytes have been injected, they are detached from the grid by gently blowing medium over them with a pasteur pipette. They are then collected and incubated in approximately 2 ml of fresh modified Barth's medium at 18°C for the desired time.

RESULTS

Injection of DNA into the Oocyte Nucleus (Germinal Vesicle) of *X. laevis*

Nuclear injection was pioneered by Gurdon (1976) who, together with his colleagues, was able to show that plasmid, viral, and cloned *Drosophila* histone DNAs are extensively transcribed when injected into the *Xenopus* oocyte GV (Mertz and Gurdon 1977). Transcription of SV40 DNA is considered to be faithful on the basis of the appearance of two specific SV40 proteins on two-dimensional gel analysis (De Robertis and Mertz 1977). Conclusive evidence for the production of typical cellular products comes from the elegant work of Brown and Gurdon (1977), who inserted *Xenopus* erythrocyte 5S DNA into the oocyte GV and showed that its transcription is to a very large extent faithful.

The major problem encountered in the nucleus injection experiments is that of locating the nucleus inside the opaque oocyte. As demonstrated by Gurdon, the relatively large GV can be found and injected by piercing the animal pole of an oocyte with a glass needle while aiming it at the center of the oocyte and squeezing both sides of the oocyte with forceps (Gurdon 1976). Clearly, this procedure requires great manual skill. An alternative, much simpler technique was developed in our laboratory about a year ago.

When *Xenopus* oocytes are centrifuged gently, the oocyte nuclei float toward the surface of these cells. In so doing they displace some pigment granules, so that after centrifugation the position of the cell nucleus is clearly demarcated by a neat dark ring made up of the pigment granules. Centrifuged oocytes retain this typical target-ring pattern for about 1–2 hours. The nucleus can thus be injected with great ease and accuracy by inserting the injection needle and introducing the material at the very center of the target ring just below the surface of the oocyte. To facilitate and to speed up the injection, individual oocytes, a hundred or so at a time, are placed on a nylon grid prior to centrifugation. During centrifugation the oocytes become fixed to the grid, and this ensures uniform orientation of the target rings which now all face the injector. Further details are given in Experimental Procedures.

Transcription of Cloned *X. laevis* tDNA$_T^{met}$ in the Oocyte GV

Rate of tDNA$_T^{met}$ transcription. tDNA$_T^{met}$ cloned in a λ vector can be reclaimed from the recombinant DNA by *Hin*dIII digestion and separation of the DNA fragments by preparative gel electrophoresis (S. G. Clarkson et al., in prep.). Oocytes injected with 2 ng of linear tDNA$_T^{met}$ fragments and 0.1 μCi [α-^{32}P]GTP (sp. act. 100 Ci mM) synthesize labeled 4S RNA at a rate of 2–5×10^3 dpm/oocyte during a 3-hour incubation. The 4S RNA fraction accounts for between 30% and 70% of all newly synthesized RNA. Discounting the dilution of the [α-^{32}P]GTP by cellular pools, it may be calculated that this represents the synthesis of some 2×10^8 4S RNA molecules/hour/oocyte. The 2 ng of tDNA$_T^{met}$ is equivalent to around 6×10^8 molecules, so the utilization of injected DNA at first sight appears to be rather poor. However, this quantity of injected DNA is greater by some 5×10^5 than the normal complement of tDNA$_T^{met}$ of each tetraploid oocyte, and it may well be that the GV is able to support the same rate of 4S RNA transcription with considerably lower amounts of injected tDNA$_T^{met}$.

Nature of the tDNA$_T^{met}$ transcripts. The typical 3.18-kb tDNA$_T^{met}$ repeat unit contains genes for both tRNA$_T^{met}$ and another 4S RNA species (Clarkson and Kurer 1976; S. G. Clarkson et al., in prep.). When the cloned tDNA$_T^{met}$ is hybridized with long-term 4S [^{32}P]RNA from *X. laevis* tissue-culture cells, the RNA can be recovered by melting of the hybrid. This RNA can be fractionated by two-dimensional electrophoresis into two spots (Fig. 1c,f). Fingerprint analysis of the RNase T$_1$ oligonucleotides has revealed that the faster-migrating component is tRNA$_T^{met}$, whereas the more slowly moving RNA is apparently homogeneous in sequence, but of unknown function (V. Pirrotta and S. G. Clarkson, unpubl.). Thus, the sequence complexity of the 4S RNA capable of hybridizing to tDNA$_T^{met}$ is very small and that of the total kidney cell 4S RNA is obviously rather large (see Fig. 1a).

The 4S RNA synthesized by oocytes injected with cloned tDNA$_T^{met}$ was recovered by sucrose gradient centrifugation and then subjected to two-dimensional electrophoresis as described above. The 4S RNA separated into at least four electrophoretic components (Fig. 1b,e). T$_1$ fingerprints of this oocyte component have revealed that it is indeed tRNA$_T^{met}$ (Kressmann et al. 1978). No identification of the other components has been made as yet, although

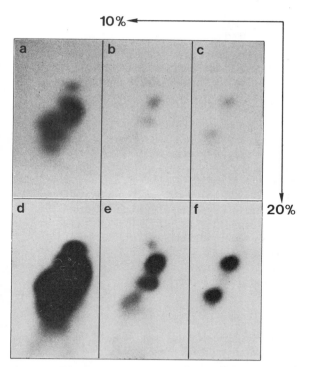

10% ←

20%

Figure 1. Two-dimensional polyacrylamide gel electrophoresis of the total 4S RNA from *Xenopus* tissue-culture cells *(a,d)*, the RNA recovered after hybridization of the same RNA to cloned tDNA$_1^{met}$ *(c,f)*, and the 4S RNA transcribed by *Xenopus* oocytes after nuclear injection of cloned tDNA$_1^{met}$ *(b,e)*. Panels *a,b,* and *c* are short-exposure, and *d,e,* and *f,* long-exposure autoradiographs of the same second-dimension (20%) gel. Conditions of electrophoresis were as described by Ikemura and Dahlberg (1973).

they appear to comprise unique RNA species of low sequence complexity. The conclusion, therefore, is that the tDNA$_1^{met}$ mediates the synthesis of a limited number of 4S RNA species, one of which is tRNA$_1^{met}$.

Transcription of Sea Urchin Histone DNA in the *Xenopus* Oocyte GV

Rate of histone DNA transcription. The 6-kb histone DNA repeat units contain genes for each of the five histone proteins (Schaffner et al. 1976; Portmann et al. 1976), but the number and locations of "promoters" are unknown. To obviate the disadvantage that the promoter might fortuitously lie near the end of the linear molecule, both linear and circular repeat units were injected into *Xenopus* oocyte GV (see Table 1). The repeats were recovered from λh22 phage recombinant DNA (Clarkson et al. 1976), and monomer circles were isolated by ligation and ethidium-bromide–CsCl gradient centrifugation.

To determine how much of the newly synthesized RNA derived from the inserted sea urchin histone DNA, the RNA was hybridized to sea urchin histone DNA under stringent conditions. In test systems, sea urchin histone mRNA was able to back-hybridize to its template with an efficiency of 60%, the hybridization being apparently unaffected by the presence of *Xenopus* histone mRNA (E. Triplett and J. L. Telford, unpubl.). As indicated in Table 1, after injection of histone DNA, up to 23% of the newly synthesized total cellular RNA hybridized to histone DNA. Circular DNA was a much better template than linear

Table 1. Hybridization of the RNA Transcribed after Injection of *P. miliaris* Histone DNA and/or SV40 DNA into the Nucleus of *X. laevis* Oocytes

Sample	Source of DNA	Injected DNA (ng/oocyte)	Incubation time (hr)	Acid-precipitable cpm hybridizable (%)
1	SV40 (form I) plus	5	6	4.5
	linear h22	5		3.9
2	SV40 (form I)	5	5	6.6
3	linear h22	5	5	3.7
4	—	—	4	0.08
5	circular h22	10	3	23
6	linear h22	10	3	2.5
7	linear h22 injected into the cytoplasm	10	3	0.1

The h22 refers to the cloned 6-kb histone DNA fragments recovered from the recombinant phage λh22 (Clarkson et al. 1976). DNA and labeled GTP were injected simultaneously at time zero. Samples 1–4 were injected with 0.1 μCi [³H]GTP/oocyte, and samples 5–7 with 0.2 μCi [α-³²P]GTP/oocyte. RNA recovered from each sample by proteinase-K–SDS treatment and phenol-chloroform extraction was incubated with nitrocellulose filters, each containing 5 μg of either SV40 or λh22 DNA, in 1 × SSC at 70°C for 4 hr. Hybrids were treated with 10 μg/ml of RNase A in 2 × SSC at 23°C for 20 min.

histone DNA. No hybridization was detectable when oocytes were injected with GTP alone. From the hybridization data, it may be calculated that the rate of synthesis of histone-DNA-specific RNA corresponds to approximately 10^8 histone mRNA equivalents/hour/oocyte.

Nature of the histone DNA transcripts. Histone mRNA of the cleaving sea urchin sediments as a well-defined peak as 9S RNA. A comparable 9S peak is not found by sucrose gradient fractionation of the newly synthesized RNA of *Xenopus* oocytes injected with either linear or circular histone DNA. In fact, the distribution of the labeled RNA is the same as that of control oocytes, except for a small increase in the labeled 4S–5S RNA region. Hybridization of RNA from each fraction of the sucrose gradient reveals that about 40% of the sea-urchin-specific RNA sediments in a polydisperse manner, with molecules as large as 30S being found; another 40% sediments in the 5S RNA region, and the remaining 20% is found in the 5S–14S range (Kressmann et al., in prep.).

Under suitable conditions of electrophoresis (Gross et al. 1976a), labeled histone mRNA from cleaving sea urchin may be fractionated into five components representing the five histone mRNAs H4, H2A, H2B, H3, H1 in order of decreasing electrophoretic mobility (see Fig. 2b). When electrophoresed under identical conditions, the 5S–14S RNA from the sucrose gradient showed a *Xenopus*-specific component moving just ahead of the H4 mRNA of the sea urchin. Its synthesis was detectable in nuclei injected with GTP alone and enhanced in those injected with histone DNA or SV40 DNA.

Sea-urchin-specific RNA bands, on the other hand, appeared with electrophoretic mobilities identical to those of H2A and H2B histone mRNAs. These bands must be derived from GC-rich sequences of histone DNA containing the structural genes, on the basis of the high melting temperature of the RNA-DNA hybrids they form (Kressmann et al., in prep.). However, they represent only a few percent of the sea-urchin-specific 5S–14S RNA, the remainder being distributed along the whole length of the gel. Distinct sea-urchin-specific RNA bands may also occur in the 3% stacking gel; but since this region is heavily contaminated by *Xenopus*-specific RNA molecules, the situation here is not as yet clear. The main conclusion from these experiments is that, although some histone mRNA-like molecules appear after injection of the sea urchin DNA into the *Xenopus* oocyte GV, the bulk of the newly synthesized RNA is less well defined in electrophoretic and sedimentation properties.

DISCUSSION

Centrifugation of the oocytes of *X. laevis* affords the possibility of inserting probes into the nucleus with great ease and high score averages. Such a technique is not only potentially interesting for transcriptional studies, but may also be of value for chromatin reconstitution experiments and the study of phenomena governing the transport of materials across the nuclear membrane.

In our experiments we have used relatively large volumes of injection medium (20 nl), the equivalent of about 50% of the nuclear volume. As evident from cytological sectioning, not all nuclei survive this treatment unscathed. We have confirmed the observation (Mertz and Gurdon 1977) that insertion of the DNA inside the nucleus is essential for the transcriptional activity of that DNA. We have also found that 30–40% of the injected oocytes produce RNA molecules from added template at a high rate, whereas the remainder do not synthesize any at all (Kressmann et al. 1978). We therefore conclude that they were not injected properly or that the nucleus was damaged during the process.

Centrifuged oocytes are apparently little impaired by the centrifugation itself. They synthesize RNA in a linear fashion for at least 24 hours, although at only about 60% the rate of control oocytes. In our hands, sea urchin histone DNA and SV40 DNA appear to be transcribed at a rate similar to that of SV40 DNA, and of other DNAs presumed to be transcribed by RNA polymerase II, injected by the Gurdon technique (Mertz and Gurdon 1977; De Robertis and Mertz 1977). The efficiency of transcription of injected cloned tDNA$_1^{met}$ is also similar to that measured for injected erythrocyte 5S DNA (Brown and Gurdon 1977). Moreover, the 4S RNA synthesized shows low sequence complexity (Fig. 1). Hence, at face value, our technique of injecting

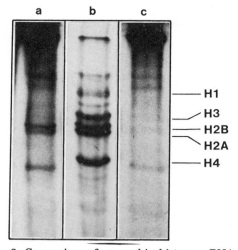

Figure 2. Comparison of sea urchin histone mRNA with the RNA synthesized by *Xenopus* oocytes injected with cloned sea urchin histone DNA. *(a)* Total RNA synthesized by oocytes injected with cloned circular histone DNA; *(c)* total oocyte RNA after injection with GTP only; *(b)* purified *P. miliaris* histone mRNA. Electrophoresis was through a 6% polyacrylamide slab gel at 6 V/cm for 5 hr at 35°C, as described by Gross et al. (1976a).

oocyte nuclei yields results very comparable to those obtained using the Gurdon technique (Gurdon 1976).

Endogenous synthesis remains unaffected by injection of foreign DNA, as indicated, for example, by the normal transcription and processing of ribosomal precursor RNA to 28S and 18S RNA in injected oocytes. However, the injection of heterologous histone DNA, in contrast to the 5S DNA and tDNA of *Xenopus,* yields RNA with rather poorly defined properties. Hybridization of total RNA to strand-separated histone DNA shows that this transcription is largely symmetrical and cannot all represent bona fide mRNA (Birnstiel et al. 1977b). Since all structural genes lie on the same DNA strand in this cloned fragment (Gross et al. 1976b), these results suggest that transcription is more random with this heterologous DNA. Furthermore, only a small fraction (1–2%) of all the sea-urchin-specific RNA is mRNA-like in its electrophoretic mobility. Whether these specific RNA molecules or the high-molecular-weight polydisperse RNA can be translated into proteins is not as yet known. However, it is interesting to relate our calculation that circular histone DNA (or SV40 DNA) is transcribed at a rate of approximately 10^8 mRNA equivalents/hour/injected oocyte to the recent estimate that this stage oocyte contains approximately 10^{11} mRNA (poly[A]$^+$) molecules (Cabada et al. 1977). It follows that a 100-hour incubation of oocytes injected with SV40 or histone DNA would give rise to the mass equivalent of 10% of the poly(A)$^+$ mRNA present in those cells. If only 1% of those molecules transcribed during such an incubation period are functional, the translated products should be detectable in sensitive protein assays. Hence, the detection of gene expression at the protein level need not be a good indicator of fidelity of transcription.

It might be argued that histone DNA, in contrast to tDNA, requires more time to reconstitute an active chromatin from preexisting protein molecules and that our failure to obtain specific RNA molecules, except for a small minority, might derive from the fact that we injected the DNA and the label simultaneously. However, when histone DNA was injected into the oocyte nucleus and left there for 24 hours prior to the application of labeled GTP, the quality of sea-urchin-specific RNA did not improve. We consider that the problems encountered with transcription of injected heterologous histone DNA in the oocyte probably stem from the failure of regulatory proteins to interact with the signals of the heterologous DNA or, more simply, that such proteins are present in too small amounts as a result of gene dosage effects.

We are encouraged by the transcription of the injected homologous tDNA, which appears to be both extensive and faithful. Since the injection technique is so simple and large amounts of radioactive RNA may be obtained, more sophisticated injection experiments can now be contemplated involving fragments with sequence alterations and deletions. In this way it should be possible to learn more about the transcription of this DNA in the living cell, and about the nature of its regulatory sequences.

Note Added in Proof

When the size of the oocyte GTP pool is taken into account, the absolute rate of transcription of injected DNAs may be calculated to be about 150 times greater than that quoted in the text (cf. Kressmann 1978). Hence, the number of molecules synthesized from injected templates is even more appreciable than was at first realized.

Acknowledgments

We thank V. Kurer, A. Binkert, and H. Dätwyler for expert assistance. This work was supported by the Swiss National Fund for Scientific Research, grant No. 3.602.075.

REFERENCES

BIRNSTIEL, M. L., W. SCHAFFNER, and H. O. SMITH. 1977a. DNA sequences coding for the H2B histone of *Psammechinus miliaris. Nature* **266**: 603.

BIRNSTIEL, M. L., A KRESSMANN, W. SCHAFFNER, R. PORTMANN, and M. BUSSLINGER. 1977b. Aspects of the regulation of histone genes. *Phil. Trans. R. Soc. Lond. B* (in press).

BROWN, D. D. and J. B. GURDON. 1977. High-fidelity transcription of 5S DNA injected into *Xenopus* oocytes. *Proc. Natl. Acad. Sci.* **74**: 2064.

CABADA, M. O., C. DARNBROUGH, P. J. FORD, and P. C. TURNER. 1977. Differential accumulation of two size classes of poly(A) associated with messenger RNA during oogenesis in *Xenopus laevis. Dev. Biol.* **57**: 427.

CLARKSON, S. G. and V. KURER. 1976. Isolation and some properties of DNA coding for tRNA^{met} from *Xenopus laevis. Cell* **8**: 183.

CLARKSON, S. G., H. O. SMITH, W. SCHAFFNER, K. W. GROSS, and M. L. BIRNSTIEL. 1976. Integration of eukaryotic genes for 5S RNA and histone proteins into a phage lambda receptor. *Nucleic Acids Res.* **3**: 2617.

DE ROBERTIS, E. M. and J. E. MERTZ. 1977. Coupled transcription-translation in DNA-injected *Xenopus* oocytes. *Cell* (in press).

GROSS, K. W., E. PROBST, W. SCHAFFNER, and M. BIRNSTIEL. 1976a. Molecular analysis of the histone gene cluster of *Psammechinus miliaris.* I. Fractionation and identification of five individual histone mRNAs. *Cell* **8**: 455.

GROSS, K., W. SCHAFFNER, J. TELFORD, and M. BIRNSTIEL. 1976b. Molecular analysis of the histone gene cluster of *Psammechinus miliaris.* III. Polarity and asymmetry of the histone-coding sequences. *Cell* **8**: 479.

GURDON, J. B. 1976. Injected nuclei in frog oocytes: Fate, enlargement, and chromatin dispersal. *J. Embryol. Exp. Morph.* **36**: 523.

IKEMURA, T. and J. E. DAHLBERG. 1973. Small ribonucleic acids of *Escherichia coli.* I. Characterization by polyacrylamide gel electrophoresis and fingerprint analysis. *J. Biol. Chem.* **248**: 5024.

KRESSMANN, A., S. G. CLARKSON, V. PIRROTTA, and M. L. BIRNSTIEL. 1978. Transcription of cloned tRNA gene fragments and subfragments injected into the oocyte nucleus of *Xenopus laevis. Proc. Natl. Acad. Sci.* (in press).

MAXAM, A. M. and W. GILBERT. 1977. A new method for sequencing DNA. *Proc. Natl. Acad. Sci.* **74**: 560.

MERTZ, J. E. and J. B. GURDON. 1977. Purified DNAs are transcribed after microinjection into *Xenopus* oocytes. *Proc. Natl. Acad. Sci.* **74**: 1502.

PORTMANN, R., W. SCHAFFNER, and M. BIRNSTIEL. 1976. Partial denaturation mapping of cloned histone DNA from the sea urchin *Psammechinus miliaris*. *Nature* **264**: 31.

SANGER, F. and A. R. COULSON. 1975. A rapid method for determining sequences in DNA by primed synthesis with DNA polymerase. *J. Mol. Biol.* **94**: 441.

SCHAFFNER, W., K. GROSS, J. TELFORD, and M. BIRNSTIEL. 1976. Molecular analysis of the histone gene cluster of *Psammechinus miliaris:* II. The arrangement of the five histone-coding and spacer sequences. *Cell* **8**: 471.

The Histone H4 Gene of *Strongylocentrotus purpuratus:* DNA and mRNA Sequences at the 5′ End

M. GRUNSTEIN AND J. E. GRUNSTEIN

Molecular Biology Institute and the Department of Biology, University of California, Los Angeles, California 90024

The histone genes and histone messenger RNAs of the sea urchin are excellent candidates for sequence analysis from the standpoint of understanding both their genetic control during embryogenesis and the evolution of their particular genetic organization. The function of the histone genes is modulated at distinct developmental stages. Just after fertilization of the egg, the histone genes are selectively activated in a dramatic fashion, with over 60% of new polysomal RNA being histone message (Kedes and Gross 1969; Grunstein and Schedl 1976). As the embryo develops, histone (H1, H2A, H2B) protein "variants" are synthesized. These variant proteins show distinct primary structure differences and occur as a result of the activation of variant histone genes (Ruderman and Gross 1974; Cohen et al. 1975; Newrock et al., this volume). Sequence analysis of the histone genes may uncover the nucleotide signals underlying this regulation.

Study of these genes may also help us to understand the general principles determining the evolution of a gene family. Histone genes are repetitive in a wide variety of sea urchin species and have been shown to be clustered on the chromosome (Kedes and Birnstiel 1971; Weinberg et al. 1972; Grunstein and Schedl 1976). In a particular species the multiple copies of each gene are similar to one another in sequence. Between species, however, the entire gene family evolves in tandem (Weinberg et al. 1972; Grunstein et al. 1976). A comparison may reveal the evolutionary constraints on the various gene and spacer sequences. In all species examined, the main sea urchin histone gene repeat is a DNA fragment approximately 6500 bases in length. Each such molecule contains the genes coding for histones H1, H4, H2B, H3, and H2A, in that order, separated by spacer sequences of unknown function (Kedes 1976; Schaffner et al. 1976). In the sea urchin *Strongylocentrotus purpuratus,* two of the genes, H3 and H2B, have been identified by DNA sequence analysis (Sures et al. 1976). The identities of the rest are inferred from histone mRNA-DNA hybridization studies (Wu et al. 1976; Cohn et al. 1976).

The histone messenger RNAs have also been characterized. They have been separated from one another (Grunstein et al. 1974; Gross et al. 1976), and all five main histone mRNA species have been identified by their capacity to direct histone protein synthesis in cell-free systems (Gross et al. 1973; Levy et al. 1975; Gross et al. 1976). Of the mRNAs, the histone H4 messenger has been identified by sequence analysis in two different sea urchin species (Grunstein and Schedl 1976; Grunstein et al. 1976). The general conclusions of this latter study are that (1) the histone H4 messenger RNA can be isolated in high purity; (2) the mRNA is approximately 400 nucleotides long, of which 306 nucleotides code for the 102 amino acids of histone H4; and (3) the coding sequence of the mRNA diverges extensively in evolution in the third, degenerate position of the codon.

In this paper we report the DNA sequence of most of the histone H4 gene of the sea urchin *S. purpuratus.* We have sequenced a region of DNA, approximately 500 nucleotides in length, overlapping the 5′ end of the RNA molecule. We conclude that most of the untranslated H4 mRNA sequence must be at the 5′ end of the RNA. Within this region adjacent to the initiator (AUG) codon, there is a repeating (AUC) sequence A U C A A U C A U C A U C A U G which shows complementarity to the 3′ end of 18S ribosomal RNA and may be the ribosome attachment site. We also discuss the evolution of the 5′ end of the H4 messenger RNA, nonrandom codon utilization for H4 amino acids, and the processing and evolution of the histone H4 protein.

MATERIALS AND METHODS

Bacteria and plasmids. pMB8-Sp2 is a colicin-resistant plasmid hybrid which was amplified in the *Escherichia coli* K-12 strain HB101 (*hsm⁻, hrs⁻, recA⁻, gal⁻, pro⁻, str ʳ*) (Boyer and Roulland-Dussoix 1969) using chloramphenicol at 150 μg/ml. The procedures used are essentially as described by Glover et al. (1975).

Enzymes. Most of the restriction enzymes used were purchased from, and the enzymic conditions used were as described by, New England BioLabs (Beverly, Mass.). The endonuclease *Sal*I was purified from *Streptomyces albus* according to the procedure of J. Upcroft (unpubl.). The conditions used for enzymic digestion with *Sal*I were 6 mM Tris-HCl, pH 8.0, 6 mM $MgCl_2$, 100 mM NaCl. T_4 polynucleotide kinase was purchased from P. L. Biochemicals (Milwaukee, Wis.).

DNA labeling and sequence analysis. All the procedures used were essentially as described by Maxam and Gilbert (1977).

mRNA isolation and sequence analysis. All the procedures used were as described by Grunstein and Schedl (1976).

RESULTS

Sequence Analysis of the Histone H4 Gene

The sequencing technique we have used is that reported by Maxam and Gilbert (1977). A brief description of this procedure follows: A restriction-endonuclease-generated fragment is labeled at the 5′ ends with polynucleotide kinase and [γ-³²P]ATP. Cleavage with a second restriction enzyme liberates two molecules terminally labeled at one end only. These uniquely labeled DNAs are then separated on polyacrylamide gels. The eluted DNAs are subjected to chemical modification, resulting in specific base removal and strand scission. Four different reactions are run simultaneously, representing G, A, T, and C scissions, respectively. In each reaction the molecule is cleaved at a single specific base whose position is random. This generates a series of fragments, each labeled at the same 5′ end, but terminating at a different scission point. The products of these four reactions are then separated in adjacent wells on polyacrylamide gels which are capable of resolving molecules differing in length by one base. The gel is analyzed by autoradiography, and the sequence is read directly.

In *S. purpuratus* the main histone gene repeating unit contains five histone genes as shown in Figure 1. The RI fragment Sp2 was isolated using the plasmid pSC101 (approximately 9000 bases or 9 kb in length) and subculture cloning in *Escherichia coli* (Kedes et al. 1975a). The Sp2 fragment was then transferred to the small colicinogenic plasmid pMB8 (2.5 kb) by RI cleavage and ligation. This new hybrid, pMB8-Sp2, was kindly donated to us by H. Heineker and H. Boyer (University of California, San Francisco).

The restriction endonuclease *Sal*I cleaves Sp2 DNA in the vicinity of the H4 coding region (Cohn et al. 1976). No other *Sal*I site is present in the rest of the Sp2 fragment, nor in pMB8. Accordingly, pMB8-Sp2 DNA was digested with *Sal*I and labeled with [γ-³²P]ATP and polynucleotide kinase as described by Maxam and Gilbert (1977). The labeled DNA was deproteinized, precipitated with ethanol, and digested with the endonuclease *Hae*III to liberate the two labeled *Sal*I ends. These were then separated on an 8% polyacrylamide slab gel and analyzed by autoradiography (Fig. 2). Conversely, pMB8-Sp2 DNA was first cleaved with *Hae*III, labeled, and then digested with *Sal*I (Fig. 2). This yields the S-1, S-2, and H-2 fragments (Fig. 1). Similarly, the Hh-2 and Hp-2 fragments are obtained by *Hha*I and *Hpa*II cleavage, respectively. In each case, the labeled DNA is then digested with *Sal*I and separated on an 8% polyacrylamide slab gel

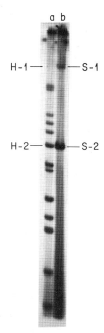

Figure 2. Autoradiograph of an 8% polyacrylamide gel illustrating the restriction fragments to be sequenced. *(a)* pMB8-Sp2 DNA was cleaved with restriction endonuclease *Hae*III. The resulting DNA molecules were labeled with polynucleotide kinase and [γ-³²P]ATP and then digested with *Sal*I enzyme. The two fragments, H-1 and H-2, are labeled at the *Hae*III-generated end only. *(b)* pMB8-Sp2 DNA labeled at the *Sal*I-generated ends. The DNA was cleaved with *Sal*I, labeled at the 5′ ends, and then digested with *Hae*III. This results in the S-1 and S-2 fragments.

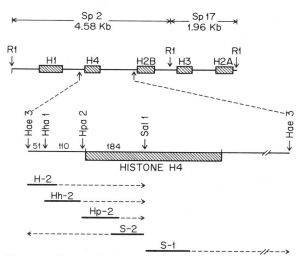

Figure 1. Schematic diagram of a histone gene repeat unit of *S. purpuratus*. The region analyzed by DNA sequencing is represented by the heavy lines beneath the expanded histone H4 gene segment. The arrows from these lines represent unlabeled (3′) ends of the single-stranded DNA molecules, each of which is labeled at the 5′ end.

(Fig. 3). The appropriate labeled DNAs were then eluted for sequencing. An example of such a sequence analysis is shown in Figure 4. The total sequence we have determined represents the overlapping H-2, Hh-2, Hp-2, S-2, and S-1 fragments (Fig. 1) and is shown superimposed on the histone H4 amino acid sequence of the calf (Fig. 5). The DNA sequence is colinear with the protein sequence and includes DNA which is not transcribed into the mature mRNA, DNA which is transcribed but not translated, and, finally, most of the H4 protein coding region.

Sequence Analysis of Histone H4 mRNA

We have previously reported a catalog of sequences resulting from ribonuclease T_1 digestion of histone H4 messenger RNA (Grunstein et al. 1976). Since it is difficult to label sea urchin mRNA in vivo to high specific activities with ^{32}P, the catalog contained many oligonucleotides which were not completely sequenced and therefore unassigned as to their positions with respect to coding or noncoding functions. Now that we have sequenced the DNA, however, many of these oligonucleotides can be assigned with greater confidence on the basis of their lengths, partial sequencing data, and correspondence to oligonucleotides predicted from the DNA sequence.

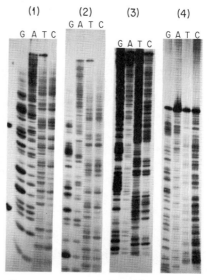

Figure 4. Sequence of the H-2 restriction fragment. The autoradiograph illustrates the 20% polyacrylamide denaturing gels used to sequence the H-2 fragment. The four different lanes represent the products of G (strong guanine, weak adenine cleavage), A (strong adenine, weak guanine cleavage), T (cleavage at thymine and cytosine), and C (cleavage at cytosine) (Maxam and Gilbert 1977). The four gels were electrophoresed for varying periods of time at 1000 V, ranging from 12 hr (1) to 36 hr (4) as described by Maxam and Gilbert (1977). The sequence obtained is shown in Fig. 5 and represents the first 80 nucleotides from the HaeIII end.

Figure 6 shows an autoradiograph of 9S (histone) messenger RNAs separated on a 3%/6% polyacrylamide slab gel. The H4 mRNA species C-2 was eluted and digested with T_1 ribonuclease. The oligonucleotides generated were separated in the first dimension on cellulose acetate (pH 3.5) and in the second, on polyethylene imine thin-layer chromatograms by homochromatography (Brownlee 1972). Since this procedure involves successive displacement of labeled oligonucleotides by unlabeled oligonucleotides, it provides a measure of the approximate length of the nucleotide sequence in question. An example of such a "fingerprint" is shown in Figure 7. The revised catalog of T_1 oligonucleotide sequences is summarized in Table 1. It includes many oligonucleotides previously reported (Grunstein et al. 1976), some of which have been sequenced since then by standard RNA sequence analysis, and some determined by comparison of partial sequences to the oligonucleotides predicted from the DNA sequence. Most of the T_1 oligonucleotides predicted from the DNA sequence can be accounted for by the messenger RNA oligonucleotides (Fig. 5).

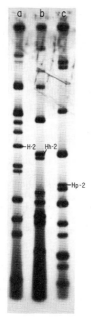

Figure 3. Autoradiograph of an 8% polyacrylamide gel illustrating restriction fragments to be sequenced. (a) pMB8-Sp2 DNA was digested with HaeIII, labeled with polynucleotide kinase and [γ-^{32}P]ATP, and then digested with SalI, to yield the H-2 fragment. (b) pMB8-Sp2 DNA was digested with HhaI, labeled, then digested with SalI to yield the Hh-2 fragment. (c) pMB8-Sp2 DNA was digested with HpaII, labeled, then digested with SalI, to yield the Hp-2 fragment.

DISCUSSION

Identity and Map Position of the H4 Gene

Figure 5 summarizes the sequence data obtained from our analysis of the Sp2 region which hybridizes

```
Hae 3                                                    Hha 1
.....ACATAAATGCATGTACTAATGCTAGCGAATACTCGCCACAAGGGGGCGCACTCAGAA
.....TGTATTTACGTACATGATTACGATCGCTTATGAGCGGTGTTCCCCGCGTGAGTCTT
```

```
TGGGGAGTCTCCGCCACTCCAGTCCCGCATTACCGTAACGCATGCCGCAATCTCGTTCACCC
ACCCCTCAGAGGCGGTGAGGTCAGGGCGTAATGGCATTGCGTACGGCGTTAGAGCAAGTGGG
```

```
                              Hpa 2
AAGTCCGCAATGTGTGTTAACAACTACTCGGTGCAACCGGTTGAGGCATCATTCGCTT
TTCAGGCGTTACACACAATTGTTGATGAGCCACGTTGGCCAACTCCGTAGTAAGCGAA
                                                       (36)
                                                  .....GCUU
```

```
AGCGTAATATCCAGTCTACAGGATCACACAGAACTCGCTCTCAACTATCAATCATCATCATG
TCGCATTATAGGTCAGATGTCCTAGTGTGTCTTGAGCGAGAGTTGATAGTTAGTAGTAGTAC
     (39)      (32)     (25)    (22)                 (44)
AGCGUAAUAUCCAGUCUACAGGAUCACACAGAACUCGCUCUCAACUAUCAAUCAUCAUCAUG
```

Figure 5. Sequence of the histone H4 gene complementary to and surrounding the 5' end of the H4 messenger RNA. The sequence was obtained from analysis of five overlapping fragments as shown in Fig. 1. Beneath the DNA sequence is shown the known H4 messenger RNA sequence and the amino acid sequence of the calf histone H4 protein. The circled numbers denote oligonucleotides, longer than 4 based, produced by RNase T₁ digestion (Table 1).

```
TCA GGT AGA GGA AAA GGA GGA AAG GGA CTC GGA AAG GGT GGT GCC AAA
AGT CCA TCT CCT TTT CCT CCT TTC CCT GAG CCT TTC CCA CCA CGG TTT
(17)    (11)            (9)         (20)        (9)         (10)
UCA GGU AGA GGA AAA GGA GGA AAG GGA CUC GGA AAG GGU GGU GCC AAA
ser-gly-arg-gly-lys-gly-gly-lys-gly-leu-gly-lys-gly-gly-ala-lys
 1                             10
```

```
CGT CAT CGC AAG GTT CTA CGA GAT AAC ATC CAA GGC ATC ACC AAG CCT
GCA GTA GCG TTC CAA GAT GCT CTA TTG TAG GTT CCG TAG TGG TTC GGA
(33)  (7)   (40)          (30)            (25a)            (16)
CGU CAU CGC AAG GUU CUA CGA GAU AAC AUC CAA GGC AUC ACC AAG CCU
arg-his-arg-lys-val-leu-arg-asp-asn-ile-gln-gly-ile-thr-lys-pro
             20                          30
```

Confidence in the DNA sequence: Each of the DNA fragments has been sequenced at least three times. Nevertheless, there are certain portions of the sequence which must be termed "less certain" than others. The initial 50–75 bases extending from a labeled end are normally quite secure in their identification, since the bands on the gel representing these bases (Fig. 4, gels 1 and 2) are easily discernible. When the region to be discriminated is 75–150 bases from the labeled end, the gels must be electrophoresed for longer periods, resulting in more compressed and less discernible bands (Fig. 4, gels 3 and 4). In this case there are bound to be errors in the sequence. Corroborating data, such as the opposite DNA strand sequence, an overlapping sequence, or the colinear protein sequence, are necessary. In this way one may extend the sequence determined 100 bases, or as much as 150 bases, from the labeled end. Consequently, we are most confident of the regions which are close to the labeled 5' end, or overlap with other sequences, or are coding regions. For example, the S-1 sequence extends 150 bases from the *Sal*I end towards the 3' end of the messenger RNA. Much of this sequence (65–150 bases from the *Sal*I end) was determined using the protein sequence as a guide. The sequence was then confirmed independently by sequencing from the *Mbo*II site which is in the middle of the S-1 sequence (above amino acid 52) (data not shown). Arbitrarily, we would estimate that most of the sequence presented contains less than one error in 50 bases since it has been determined through overlaps and colinearity. The only region which we do not have equal confidence in is the terminal 25–35 bases of the Hh-2 sequence immediately preceding the *Hpa*II restriction site (Fig. 5). This short region may contain a higher proportion of errors.

```
              Sal 1
GCA ATC CGT CGA CTN GCT AGA AGG GGA GGT GTC AAG AGG ATC TCT GGT
CGT TAG GCA GCT GAN CGA TCT TCC CCT CCA CAG TTC TCC TAG AGA CCA
(24)              (17)          (21)          (40)
GCA AUC CGU CGA CUN GCU AGA AGG GGA GGU GUC AAG AGG AUC UCU GGU
ala-ile-arg-arg-leu-ala-arg-arg-gly-gly-val-lys-arg-ile-ser-gly
                        40
```

```
CTC ATC TAC GAA GAG ACA CGC GGT GTA CAG AAG GTC TTC CTG GAG AAT
GAG TAG ATG CTT CTC TGT GCG CCA CAT GTC TTG CAG AAG GAC CTC TTA
(42)              (8)                    (46)              (19)
CUC AUC UAC GAA GAG ACA CGC GGU GUA CAG AAG GUC UUC CUG GAG AAU
leu-ile-tyr-glu-glu-thr-arg-gly-val-leu-lys-val-phe-leu-glu-asn
        50                          60
```

```
GTC ATC CGT GAT GCA GTC ACC TAC TGT GAG CAC GCT AAG CGT AAG ACC
CAG TAG GCA CTA CGT CAG TGG ATG ACA CTC GTG CGA TTC GCA TTC TGG
(31)              (38)          (5)   (21)
GUC AUC CGU GAU GCA GUC ACC UAC UGU GAG CAC GCU AAG CGU AAG ACC
val-ile-arg-asp-ala-val-thr-tyr-cys-glu-his-ala-lys-arg-lys-thr
             70                          80
```

```
GTC ACA GCC ATG GAC
CAG TGT CGG TAC CTG
(23)  (20)
GUC ACA GCC AUG GAC
val-thr-ala-met-asp-val-val-tyr-ala-leu-lys-arg-gln-gly-arg-thr
                              90
```

```
leu-tyr-gly-phe-gly-gly
         100           COOH
```

1086

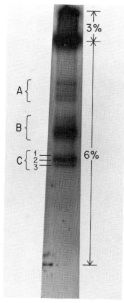

Figure 6. Separation of *S. purpuratus* histone messenger RNAs on a 3%/6% polyacrylamide gel.

to H4 messenger RNA (Kedes et al. 1975b). The sequence derived from the S-1 and S-2 fragments is almost completely colinear with the calf histone H4 protein sequence as determined by De Lange et al. (1969). We have therefore confirmed, by direct sequencing, the map position of the H4 gene on Sp2 DNA (Wu et al. 1976; Cohn et al. 1976). The only divergence we find between the calf protein and the *S. purpuratus* nucleotide sequence is at amino acid position 73. However, this is expected. In the sea urchin *(Parechinus angulosus)*, cysteine has replaced threonine at amino acid 73 of the H4 histone protein (Strickland et al. 1974). This is also true for the *S. purpuratus* H4 protein, as inferred from our H4 mRNA sequence data (Grunstein et al. 1976). Colinearity between DNA and protein is therefore complete.

Evolution of the H4 Protein

Pea and cow histones H4 are almost identical. The only amino acid sequence differences are at positions 60 and 77. At these positions, there have been changes in the pea from valine to isoleucine and from lysine to arginine, respectively. Note that the sea urchin H4 DNA sequence is colinear with the animal protein rather than the plant at these positions.

The cysteine change at position 73 represents a third variable amino acid in histone H4. However, we must keep in mind that we are making a direct comparison between the *embryonic* sea urchin and the *adult* cow histone H4. As Newrock et al. (this volume) show, primary sequence variants of histones are synthesized at different developmental stages.

H1, H2A, and H2B proteins certainly show such developmental differences. It is not known whether more discrete variants of H3 and H4 exist as well. If this is the case, it is possible that embryonic calf H4 also contains cysteine and, conversely, that the histone H4 gene we are sequencing may code for the embryonic H4 protein and an adult (variant) histone H4 gene lacking cysteine exists elsewhere in the sea urchin chromosome.

The H4 Messenger RNA

In Figure 5 we have summarized the RNA sequencing data and its relationship to the DNA sequence. The oligonucleotides produced as a result of RNase T_1 digestion are shown lined and numbered beneath the DNA sequence. On the whole, the relationship is very good, although we do not yet have the sequence of all the T_1 oligonucleotides in the mRNA.

The T_1 oligonucleotide sequences are especially important as they allow us to map the boundaries of the DNA molecule within which the mature H4 mRNA is transcribed. The H4 protein contains 102 amino acids. The mRNA contains approximately 400 nucleotides. Therefore, some 94 nucleotides are untranslated. Certain oligonucleotides (e.g., 44, 22, 25, 32, 39, 36) are not colinear with the coding portion of the DNA molecule. They can, however, be ordered within the 5' noncoding region of the messenger RNA (Fig. 8). As we examine the DNA sequence "upstream" of these oligonucleotides, we find that none of the longer sequences predicted are present in the mRNA T_1 catalog. We believe that the mRNA start may be within the sequence G C A T C A T T C G or displaced in either direction by no more than 6–8 nucleotides. For example, the oligonucleotide (G) C A U C A U U C G may be present superimposed on some other oligonucleotide on the fingerprint (Fig. 7). However, this would not be possible for the next oligonucleotide, (G) U U G, whose position above (G) U G is vacant on the autoradiograph (Fig. 7). Also, in the other direction, it is possible that, although the oligonucleotide 36, (G) C U U A G, is colinear with this position on the DNA sequence, it may in fact be transcribed from some other region. However, this is unlikely to be the case for the next, rather long oligonucleotide, (G) U A A U A U C C A G. We can conclude that most of the untranslated H4 mRNA is at the 5' end of the message.

The GC content of H4 messenger RNA is higher than that of total sea urchin DNA (40% GC). The mRNA coding region we have determined here has a GC content of 53%; the mRNA at the 5' noncoding end, 42% GC; and the DNA not transcribed into mature mRNA, 54% GC. This is consistent with initial observations (Grunstein et al. 1974; Birnstiel et al. 1974) that the sea urchin histone genes are of high GC content.

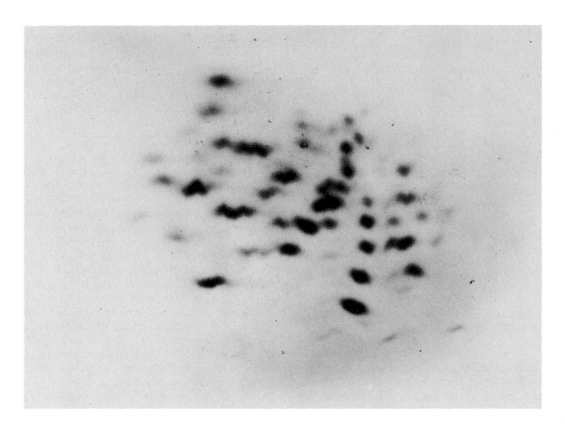

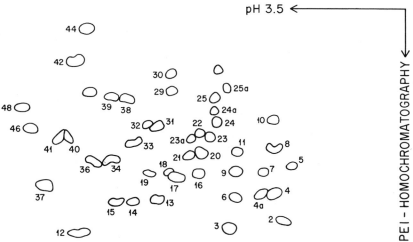

Figure 7. Two-dimensional fingerprint of *S. purpuratus* histone H4 messenger RNA. The first dimension represents high-voltage electrophoresis on cellulose acetate paper at pH 3.5; the second dimension represents homochromatography on a polyethyleneimine thin-layer sheet.

Codon Utilization

With 83% of the coding region sequenced, we can examine the randomness of codon utilization. The data are summarized in Table 2. Valine is most frequently coded for by GUC, lysine by AAG, isoleucine by AUC, glycine by either GGU or GGA, and threo-

nine by either ACC or ACA. In contrast, the arginine codons show very little selectivity. The other amino acids do not occur often enough for any comment to be warranted concerning their respective codon selection. This strong preference for certain synonymous codons has been observed in both prokaryotes and eukaryotes and may be correlated with the pres-

Table 1. Sequences of Oligonucleotides Generated by Ribonuclease T₁ Digestion of *S. purpuratus* Histone H4 Messenger RNA

Oligonucleotide		Molar yield
SP-T₁-		
1	(G)G	24.5
2	(G)CG	3.8
3	(G)AG	7.4
4	(G)CAG + (G)ACG	3.9
5	(G)CACG	0.5
6	(G)AAG	2.8
7	(G)CAAG	0.7
8	(G)CCAAG + (G)ACACG	1.1
9	(G)AAAG	2.5
10	(G)CCAAACG	0.7
11	(G)AAAAG	0.7
12	(G)UG	10.2
13	(G)UCG + (G)CUG	3.2
14	(G)UAG	0.7
15	(G)AUG	1.3
16	(G)CCUG	1.2
17	(G)CUAG + (G)UACG + (G)UCAG	3.5
18	(G)ACUG	1.2
19	(G)AAUG	1.3
20	(G)CCAUG + (G)ACUCG	2.4
20a	(G)CCUAG	0.7
21	(G)CUAAG + (G)UCAAG	1.9
22	(G)AACUCG	0.8
23	(G)UCACAG	0.9
24	(G)CAAUCCG	1.0
25	(G)AUCACACAG	0.8
25a	(G)CAUCACCAAG	0.3
28	(G)AAUAACUAAACCG	0.5
30	(G)AUAACAUCCAAG	0.8
31	(G)UCAUCCG	1.2
32	(G)UCUACAG	1.1
33	(G)UCAUCG	1.0
34	(G)CUUCG	1.3
36	(G)UACUG + (G)ACUUG + (G)UCUAG + CUUAG	3.8
37	(G)UAUG	1.2
38	(G)UCACCUACUG	0.9
39	(G)UAAUAUCCAG	0.8
40	(G)AUCUCUG + (G)UUCUACG	2.2
41	(G)UACAUUG + ?	1.4
42	(G)UCUCAUCUACG	1.4
44	(G)CUCUCAACUAUCAAUCAUCAUCAUG	0.4
46	(G)UCUUCCUG	1.3

```
Hpa II
  ↓
5'...CCGGTTGAGGCATCATTCGCTTAGCGTAATATCCAGTCTACAGGATCACACAGAACTCGCTCTCAACTATCAATCATCATCATGTCAGGTAGA
3'...GGCCAACTCCGTAGTAAGCGAATCGCATTATAGGTCAGATGTCCTAGTGTGTCTTGAGCGAGAGTTGATAGTTAGTAGTAGTACAGTCCATCT
```

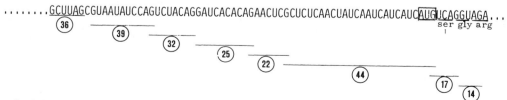

Figure 8. The 5' end of the histone H4 messenger RNA. The top two rows represent the DNA sequence from the *Hpa*II restriction site. The circled numbers refer to RNase T₁ oligonucleotides of the histone H4 mRNA (Table 1) that are complementary to the DNA sequence.

Table 2. Codon Utilization in Histone H4 Messenger RNA

U	phe	UUU UUC 1	ser	UCU 1 UCC	tyr	UAU UAC 2	cys	UGU 1 UGC	U C
	leu	UUA UUG		UCA 1 UCG	term	UAA UAG	term trp	UGA UGG	A G
C	leu	CUU CUC 2 CUA 1 CUG 2	pro	CCU 1 CCC CCA CCG	his gln	CAU 1 CAC 1 CAA 1 CAG	arg	CGU 3 CGC 3 CGA 2 CGG	U C A G
A	ile met	AUU AUC 6 AUA AUG 1	thr	ACU ACC 3 ACA 2 ACG	asn lys	AAU 1 AAC 1 AAA 2 AAG 8	ser arg	AGU AGC AGA 2 AGG 2	U C A G
G	val	GUU 1 GUC 5 GUA 1 GUG	ala	GCU 2 GCC 2 GCA 2 GCG	asp glu	GAU 2 GAC 1 GAA 1 GAG 3	gly	GGU 6 GGC 1 GGA 6 GGG	U C A G

ence of parallel isoacceptor tRNA patterns (Efstradiatis et al. 1977). The functional significance of specific codon utilization is, however, still not understood.

The 5′ Noncoding Region

Of approximately 96 untranslated bases in the H4 messenger RNA, approximately two-thirds are at the 5′ end. The most likely function of at least a portion of this sequence is in ribosome binding and the initiation of translation. Prokaryotic and eukaryotic mRNAs have been complexed to ribosomes and treated with pancreatic ribonuclease. The region of the RNA protected from digestion, i.e., the initiation complex, has certain features in common. The initiation complex contains five or six triplets 5′ to the initiator codon (usually AUG) and some three triplets into the coding region, 3′ to the initiator codon (for review, see Steitz 1977).

In some cases, this sequence can be placed in a configuration suggesting secondary structure formed by pairing of complementary nucleotides within the strand. However, this is rare and in most cases is not a characteristic of a ribosome initiation complex (Steitz 1977). The one common feature that these complexes share is a sequence, 5′ to the initiator codon, which shows complementarity to the invariant 3′ end of the smaller (16S) ribosomal RNA (Shine and Dalgarno 1975; Steitz and Jakes 1975). In eukaryotes, the 3′ end of 18S rRNA is also invariant (Hunt 1970; Dubin and Shine 1976). What complementarity exists between it and the 5′ end of the histone H4 message? Directly next to the first (serine) amino acid codon is an AUG codon, usually the initiator codon in prokaryotes and eukaryotes. No other AUG is present in the 5′ noncoding sequence we have determined. The position of this codon next to the serine codon makes it extremely likely that this AUG is in fact the initiator triplet for the H4

messenger RNA. A repeating (AUC) sequence, A U C A A U C A U C A U C A U G, is located 5′ to this triplet. This sequence shows considerable complementarity to the invariant 3′ end of 18S ribosomal RNA with six of the eight bases in the 18S 3′ end complementary (Figure 9). Therefore we believe it is very likely that this region forms part of the initiation complex actually in contact with the ribosome.

Evolution of the 5′ End of H4 Messenger RNA

The histone H4 protein has been strongly conserved in 2 billion years of evolution. We have shown that, in contrast, the H4 messenger RNA has diverged considerably in but 60 million years of evolution between the two sea urchin species, *Strongylocentrotus purpuratus* and *Lytechinus pictus* (Grunstein et al. 1976). This work examined only the coding region of the message and we concluded that as many as one in two codons have diverged, but only in the degenerate third position. It would be interesting to know the extent of selective pressure on the 5′ noncoding region of these mRNAs.

The DNA sequence of the *S. purpuratus* H4 gene dictates the T_1 oligonucleotides which must exist in the 5′ noncoding mRNA. We have seen that these are in fact present in the *S. purpuratus* H4 messenger RNA. We can then ask whether these expected oligonucleotides are present in the *L. pictus* H4 messenger molecule described previously (Grunstein and Schedl 1976). The result is that not a single predicted *S. purpuratus* oligonucleotide is present

```
H4 mRNA      5′.....AACUAUCAAUCAUCAUCAUGUCAGGUAGA
                          |||  |||
                          AUUACUAG........
18S rRNA        HO
```

Figure 9. Complementarity between the 3′ end of 18S ribosomal RNA and the 5′ end of histone H4 messenger RNA adjacent to the initiator codon.

in the *L. pictus* message. The most obvious homology visible is to the *L. pictus* T_1 oligonucleotide (G) U A A C A U C C A G (Grunstein and Schedl 1976). For example:

(G) U A A $\boxed{\text{U}}$ A U C C A G   *S. purpuratus*
(G) U A A $\boxed{\text{C}}$ A U C C A G   *L. pictus.*

These sequence homologies are long enough so that it is highly likely they occur in the same relative position of the 5′ end of the RNA. They differ in a change from U to C in the oligonucleotide. This type of comparison gives us a minimum estimate for divergence in the 5′ end of the RNA molecule. If we assume that at least one base has changed in each oligonucleotide to result in the lack of homology of T_1 sequences between the two species, this results in a *minimum* divergence of 6/66 or 9%. The rate of divergence in the coding region is approximately 11.5% (Grunstein et al. 1976). We must conclude that the 5′ noncoding sequence has diverged almost as much or perhaps even more than the coding region of these two species.

The Precursor to Histone H4 Protein

The initiator codon, AUG, is directly next to the first codon, UCA, the triplet for serine. We conclude that the precursor to the protein contains but one added methionine residue at the amino terminus. Similarly, hemoglobin also lacks a lengthy precursor sequence (Efstradiatis et al. 1977). In contrast, secretory proteins often are first synthesized containing a lengthy, often strongly hydrophobic, leader sequence. Blobel and Dobberstein (1975a,b) have suggested that the leader or signal sequence may be necessary for the transfer of a protein across a membrane (signal hypothesis). However, histones are synthesized in the cytoplasm and must then be transferred across the nuclear membrane. It follows that either a hydrophobic leader sequence is unnecessary for transfer of a protein across a membrane or the nuclear membrane (possibly by virtue of its nuclear pores) is a very different barrier as compared to the microsomal membranes.

Multiple Histone H4 Genes and the Multiple H4 Messenger RNAs

In the data presented above, the sequenced DNA was shown to be the histone H4 gene, complementary to the partially sequenced histone H4 messenger RNA. However, this must be viewed in the context of the work of Ruderman and Gross (1974) and most recently that of Newrock et al. (this volume), describing the variants of histone proteins. They observe that primary structure variants of H1, H2A, and H2B are synthesized at different stages in early development. One form (α) is synthesized through the preblastula stages. The other forms (β, γ, δ) are first synthesized later, specifically during the morula, blastula, and mesenchyme blastula stages and then on through to the prism stage. They have not yet observed variants of H3 or H4 proteins. This may be due to the highly conserved nature of these proteins.

The 9S blastula mRNAs are divided on the basis of size into three groups, A, B, and C. If we examine the C region of the 9S mRNA banding pattern, it is evident that at least three C mRNAs are present, two of which are partially superimposed. We have previously shown, in a different sea urchin species (*L. pictus*), that the C mRNAs are very similar, although not identical, in the fingerprint pattern produced by T_1 RNase digestion and two-dimensional high-voltage electrophoresis (Grunstein et al. 1974). This was confirmed by digesting each of the T_1 oligonucleotides with a second enzyme, pancreatic RNase, and examining the secondary digestion products (M. Grunstein and P. Schedl, unpubl.). Therefore, it is very likely that even though these RNAs differ in several oligonucleotides, all the C mRNAs are histone H4 mRNAs. This may mean that despite the absence, to date, of H4 variant proteins, H4 variant *mRNAs* may exist. This could be explained if we assume that "chromosome remodeling" occurs when a different class or cluster of histone genes, containing all five different genes (H1, H4, H2B, H3, H2A), is activated. In this case one might expect the H4 mRNAs produced by these variant genes to differ, possibly in some third-base degenerate positions or in untranslated nonessential sequences. The H4 mRNA, and therefore the H4 gene we have sequenced, is probably that which activated earliest in development, since the mRNA is present in largest quantity (of the C mRNAs) even as early as the 16-cell stage (M. Grunstein, unpubl.). If, however, the other C mRNAs represent the products of variant histone gene tandem repeats or clusters, one would predict that (1) the (9S) mRNA banding pattern should change during development to coincide with the appearance of α, β, σ, and δ proteins and (2) isolation of random histone gene clones by recombinant DNA procedures should uncover DNA fragments containing all five histone genes but whose H4 genes are complementary to either the C1, C2, or C3 histone H4 messenger RNAs. Both these predictions are currently being tested in our laboratory.

Acknowledgments

We thank Drs. L. Kedes, E. Weinberg, and I. Sures for the free flow of unpublished information between their laboratories and our own. We also are grateful to Alan Maxam for making available to us details of the DNA sequencing procedure prior to publication. This work was supported by the National Institutes of Health (GM-23674).

REFERENCES

BIRNSTIEL, M., J. TELFORD, E. WEINBERG, and D. STAFFORD. 1974. Isolation and some properties of the genes coding for histone proteins. *Proc. Natl. Acad. Sci.* **71**: 2900.

BLOBEL, G. and B. DOBBERSTEIN. 1975a. Transfer of proteins across membranes. I. Presence of proteolytically processed and unprocessed nascent immunoglobulin light chains on membrane-bound ribosomes of murine myeloma. *J. Cell Biol.* **67**: 835.

———. 1975b. Transfer of proteins across membranes. II. Reconstitution of functional rough microsomes from heterologous components. *J. Cell Biol.* **67**: 852.

BOYER, H. W. and D. ROULLAND-DUSSOIX. 1969. A complementation analysis of the restriction and modification of DNA in *Escherichia coli. J. Mol. Biol.* **41**: 459.

BROWNLEE, G. G. 1972. In *Determination of sequences in RNA* (ed. T. S. Work and E. Work), p. 130. North-Holland/American Elsevier, New York.

COHEN, L. H., K. M. NEWROCK, and A. ZWEIDLER. 1975. Stage-specific switches in histone synthesis during embryogenesis of the sea urchin. *Science* **190**: 994.

COHN, R. H., J. C. LOWRY, and L. H. KEDES. 1976. Histone genes of the sea urchin *(S. purpuratus)* cloned in *E. coli:* Order, polarity and strandedness of the five histone-coding and spacer regions. *Cell* **9**: 147.

DE LANGE, R. J., D. FAMBROUGH, E. L. SMITH, and J. BONNER. 1969. Calf and pea histone IV. *J. Biol. Chem.* **244**: 319.

DUBIN, D. T. and J. SHINE. 1976. The 3′ terminal sequence of mitochondrial 13S ribosomal RNA. *Nucleic Acids Res.* **3**: 1225.

EFSTRADIATIS, A., F. C. KAFATOS, and T. MANIATIS. 1977. The primary structure of rabbit β-globin mRNA as determined from cloned DNA. *Cell* **10**: 571.

GLOVER, D. M., R. L. WHITE, D. J. FINNEGAN, and D. S. HOGNESS. 1975. Characterization of six cloned DNAs from *Drosophila melanogaster,* including one that contains the genes for rRNA. *Cell* **5**: 149.

GROSS, K., E. PROBST, W. SCHAFFNER, and M. BIRNSTIEL. 1976. Molecular analysis of the histone gene cluster of *Psammechinus miliaris:* Fractionation and identification of five individual histone mRNAs. *Cell* **8**: 455.

GROSS, K., J. RUDERMAN, M. JACOBS-LORENA, C. BAGLIONI, and P. R. GROSS. 1973. Cell-free synthesis of histones directed by messenger RNA from sea urchin embryos. *Nat. New Biol.* **241**: 272.

GRUNSTEIN, M. and P. SCHEDL. 1976. Isolation and sequence analysis of sea urchin *(Lytechinus pictus)* histone H4 messenger RNA. *J. Mol. Biol.* **104**: 323.

GRUNSTEIN, M., P. SCHEDL, and L. H. KEDES. 1976. Sequence analysis and evolution of sea urchin *(Lytechinus pictus* and *Strongylocentrotus purpuratus)* histone H4 messenger RNAs. *J. Mol. Biol.* **104**: 351.

GRUNSTEIN, M., S. LEVY, P. SCHEDL, and L. H. KEDES. 1974. Messenger RNAs for individual histone proteins: Fingerprint analysis and *in vitro* translation. *Cold Spring Harbor Symp. Quant. Biol.* **38**: 717.

HUNT, J. A. 1970. Terminal sequence studies of high molecular weight ribonucleic acid: The 3′ termini of rabbit reticulocyte ribosomal RNA. *Biochem. J.* **120**: 353.

KEDES, L. H. 1976. Histone messengers and histone genes. *Cell* **8**: 321.

KEDES, L. H. and M. L. BIRNSTIEL. 1971. Reiteration and clustering of DNA sequences complementary to histone messenger RNA. *Nat. New Biol.* **230**: 165.

KEDES, L. H. and P. R. GROSS. 1969. Identification in cleaving embryos of three RNA species serving as templates for the synthesis of nuclear proteins. *Nature* **223**: 1335.

KEDES, L. H., A. C. Y. CHANG, D. HOUSEMAN, and S. N. COHEN. 1975a. Isolation of histone genes from unfractionated sea urchin DNA by subculture cloning in *E. coli. Nature* **255**: 533.

KEDES, L. H., R. H. COHN, J. C. LOWRY, A. C. Y. CHANG, and S. N. COHEN. 1975b. The organization of sea urchin histone genes. *Cell* **6**: 359.

LEVY, S., P. WOOD, M. GRUNSTEIN, and L. H. KEDES. 1975. Individual histone messenger RNAs: Identification by template activity. *Cell* **4**: 239.

MAXAM, A. M. and W. GILBERT. 1977. A new method for sequencing DNA. *Proc. Natl. Acad. Sci.* **74**: 560.

RUDERMAN, J. V. and P. R. GROSS. 1974. Histones and histone synthesis in sea urchin development. *Dev. Biol.* **36**: 286.

SCHAFFNER, W., K. GROSS, J. TELFORD, and M. BIRNSTIEL. 1976. Molecular analysis of the histone gene cluster of *Psammechinus miliaris.* II. The arrangement of the five histone-coding and spacer sequences. *Cell* **8**: 471.

SHINE, J. and L. DALGARNO. 1975. Determinant of cistron specificity in bacterial ribosomes. *Nature* **254**: 34.

STEITZ, J. A. 1977. Genetic signals and nucleotide sequences in messenger RNAs. In *Biological regulation and control* (ed. R. Goldberger). Plenum Press, New York. (In press.)

STEITZ, J. A. and K. JAKES. 1975. How ribosomes select initiator regions in mRNA: Base pair formation between the 3′ terminus of 16S rRNA and the mRNA during initiation of protein synthesis in *E. coli. Proc. Natl. Acad. Sci.* **72**: 4734.

STRICKLAND, M., W. N. STRICKLAND, W. F. BRANDT, and C. VON HOLT. 1974. Sequence of the cysteine-containing portion of histone F29, from the sea urchin *Parachinus angulosus. FEBS Lett.* **40**: 346.

SURES, I., A. MAXAM, R. COHN, and L. H. KEDES. 1976. Identification and location of the histone H2A and H3 genes by sequence analysis of sea urchin *(S. purpuratus)* DNA cloned in *E. coli. Cell* **9**: 495.

WEINBERG, E. S., M. L. BIRNSTIEL, I. F. PURDOM, and R. WILLIAMSON. 1972. Genes coding for polysomal 9S RNA of the sea urchin: Conservation and divergence. *Nature* **240**: 225.

WU, M., D. S. HOLMES, N. DAVIDSON, R. H. COHN, and L. H. KEDES. 1976. The relative positions of sea urchin histone genes on the chimeric plasmids pSp2 and pSp17 as studied by electron microscopy. *Cell* **9**: 163.

Histone Gene Heterogeneity in the Sea Urchin
Strongylocentrotus Purpuratus

E. S. WEINBERG,* G. C. OVERTON,* M. B. HENDRICKS,* K. M. NEWROCK,† AND L. H. COHEN†

*Department of Biology, Johns Hopkins University, Baltimore, Maryland 21218; †The Institute for Cancer Research, Fox Chase, Philadelphia, Pennsylvania 19111

Histone genes in sea urchins are reiterated several hundred to a thousand times (Kedes and Birnstiel 1971; Weinberg et al. 1972; Grunstein et al. 1974; Grunstein and Schedl 1976) and are arranged in a unit which contains the genes for each of the five histones interspersed with noncoding spacer sequences (Weinberg et al. 1975; Kedes et al. 1975b; Birnstiel et al. 1975; Schaffner et al. 1976; Cohn et al. 1976; Wu et al. 1976; Portmann et al. 1976; Holmes et al. 1977). This repeating unit was found to be 6–7 kb in length by hybridization to filter-transferred (Southern 1975) DNA enriched in histone sequences which had been digested with restriction enzymes (Weinberg et al. 1975; Birnstiel et al. 1975; Cohn et al. 1976; Schaffner et al. 1976), or by characterization of plasmids or λ phage containing histone DNA fragments (Kedes et al. 1975b; Cohn et al. 1976; Wu et al. 1976; Portmann et al. 1976; Holmes et al. 1977). The repeating histone gene unit of *Strongylocentrotus purpuratus* is diagrammed in Figure 1. Multiple copies of the histone genes appear to be quite similar in sequence by the criteria of the high T_m and sharp melting transition of RNA-DNA hybrids formed with histone mRNA (Weinberg et al. 1972). Fingerprint analysis of the H4 mRNA transcribed from these genes (Grunstein et al. 1974; Grunstein and Schedl 1976; Grunstein and Grunstein, this volume) also indicates that the RNA population is fairly homogeneous, but the presence of spots in less than molar yield suggests some microheterogeneity.

Despite indications of similarity of the multiple gene copies, we already know that there are more than five distinct histone species in the sea urchin. Sperm histones appear to be quite distinct in mobility on acetic-acid–urea gels from embryonic histones (Easton and Chalkley 1972). Translation experiments in which mRNAs from morula and gastrula are used as a template in cell-free systems show that there are at least two H1 species synthesized on different mRNAs at specific times during development (Ruderman et al. 1974; Arceci et al. 1976). When histones from various stages of sea urchin embryos are analyzed on Triton-acid-urea gels, the H2A and H2B classes can be resolved into several components, and the synthesis of these species follows a stage-specific pattern (Cohen et al. 1975; New-

rock et al., this volume). We will show here that these H2A and H2B variants are different proteins synthesized on distinct sets of mRNAs. The multiplicity of H2A, H2B, and H1 proteins demands that there be sequence heterogeneity of the multiple histone gene copies. We have surveyed the extent of heterogeneity of the repeating gene structure to find candidates for sequences coding for the variant histones, as well as to explore the evolution of these genes.

EXPERIMENTAL PROCEDURES

Preparation of RNA. Polysomal pellets were prepared from various embryonic stages of *S. purpuratus* (Pacific Biomarine) as previously described (Weinberg and Overton 1977). If RNA was to be labeled, 2–5 ml of [³H]uridine (45 Ci/mole; Amersham-Searle) was added to 100–200-ml cultures of embryos at 4 hours after fertilization and the polysomal pellets were prepared 4 hours later. The pellets were dissolved in 0.01 M Tris, pH 7.4, 0.001 M EDTA, 0.1 M NaCl, 0.5% sodium dodecyl sulfate (SDS) and extracted with phenol-chloroform (1:1) and chloroform. The RNA was precipitated with 0.1 volume of sodium acetate and 2 volumes of absolute ethanol at −20°C. The pellets were resuspended in the Tris-SDS buffer and centrifuged on 15–30% sucrose gradients. The 8S–14S fractions were pooled and precipitated with sodium acetate and ethanol. The RNA, used for in vitro translation, was washed twice with absolute ethanol, air-dried, and resuspended in distilled water.

Highly labeled RNA complementary to histone DNA was prepared in vitro using the plasmids pSp2 (the kind gift of L. Kedes) and pSR1 (isolated in our laboratory by D. Wall) as templates in an *Escherichia coli* RNA polymerase reaction. pSp2 contains the large *Eco*RI fragment of the histone repeat unit linked to the vector pSC101 (Kedes et al. 1975a,b); pSR1 contains the small *Eco*RI fragment linked to pMB9 (see Fig. 1).

Cell-free translation of RNA. Wheat germ, a gift of Dr. D. Ish-Horowicz, was the source of an S-30 fraction prepared by the procedure of Marcu and Dudock (1974). Incubation reactions for the cell-free

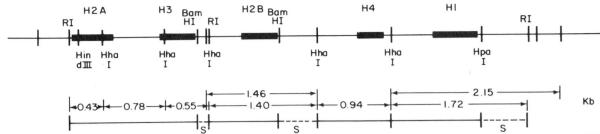

Figure 1. Structure of the histone gene repeat unit of *S. purpuratus*. Restriction enzyme sites (Weinberg et al. 1975; Kedes et al. 1975a,b; Cohn et al. 1976; Weinberg and Overton 1977) have been mapped with respect to the regions coding for the histones (Cohn et al. 1976; Wu et al. 1976; Holmes et al. 1977). The solid rectangles correspond to coding sequences and the light lines are putative spacer sequences. Values indicated below the map are the sizes of *Hha*I and *Hha*I + *Eco*RI digestion products (our values, which correspond to those of Cohn et al. [1976]). Fragments containing only spacer sequences are denoted with an S. This map refers only to the predominant fragments seen on digestion of enriched histone DNA or on digestion of the recombinant plasmids pSp2 and pSp17 (Kedes et al. 1975a,b).

protein synthesis contained 15 µl of S-30 and 5 µCi of [³H]lysine (80 Ci/mmole; Amersham-Searle) and were brought to a final concentration of 20 mM HEPES (pH 7.6), 2 mM dithiothreitol, 1 mM ATP, 40 µM GTP, 8 mM creatine phosphate, 40 µg/ml creatine phosphokinase (Sigma), 60 mM KCl, 2.5 mM magnesium acetate, 30 µM of each amino acid with the exception of lysine, and 55 µM spermine. The reaction volumes were 50 µl. About 1 µg of 8S–14S polysomal RNA was added to each incubation mixture, except for the endogenous control. Reactions were carried out for 90 minutes at 26°C. Additions of 1 µg of the RNA routinely resulted in eight- to tenfold stimulation of [³H]lysine into trichloroacetic-acid-precipitable material.

Preparation of in-vivo-labeled histones. Histones synthesized by the mesenchyme blastula were prepared from embryos of *S. purpuratus* grown as described previously (Cohen et al. 1975). Embryos derived from 5 ml of eggs, suspended in 3.5 liters of seawater, were labeled with 1 mCi of [³H]leucine (56 Ci/mM; Amersham-Searle) for a period from 20 to 24 hours after fertilization. At this time the embryos were washed and cultured to prism stage in the presence of unlabeled leucine (2.5 µM). Fractionation of the H1 and non-H1 histones was done by addition of 10% perchloric acid to a 4 mg/ml solution of the histones (Johns 1964; Oliver et al. 1972).

Triton-acid-urea gel electrophoresis. Electrophoresis in Triton-acid-urea gels (0.9 M acetic acid, 8 M urea, and 6 mM Triton X-100) was performed as previously described (Alfageme et al. 1975; Cohen et al. 1975) on 15 × 0.1-cm slab gels. The in vitro products were prepared for electrophoresis by removing insoluble components by centrifugation and solubilizing the histones in a solution containing 8 M urea (K. Newrock et al., in prep.). Gels were stained with Coomassie brilliant blue R and then fluorographed (Bonner and Laskey 1974; Laskey and Mills 1975).

Preparation of DNA. The DNA was prepared from sea urchin sperm by procedures previously described

(Weinberg and Overton 1977). A preparation of DNA from 5–10 sea urchins was enriched in histone DNA by actinomycin-CsCl and Hg-Cs₂SO₄ equilibrium density gradient centrifugation (Weinberg and Overton 1977). The sample used in these experiments was about 30-fold purified.

Restriction, electrophoresis, and hybridization of DNA. Restriction enzymes (purchased from New England BioLabs and Bethesda Research Laboratory) were used to digest 5 µg of enriched histone DNA or 10 µg of total DNA prepared from individual sea urchins. Reaction conditions were essentially those recommended by New England BioLabs. For the sequential *Hpa*I and *Eco*RI digestion, the *Hpa*I reaction was carried out for 3 hours, the solution was made 50 mM in NaCl, *Eco*RI was added, and the reaction was allowed to proceed for another 3 hours. Reactions were stopped by addition of 0.1 vol. of 0.5 M EDTA, and the DNA was precipitated with 2 vol. of ethanol and then resuspended in 20 µl of 10 mM Tris-HCl, pH 7.4, 1 mM Na-EDTA, 10% glycerol, and 0.01% bromphenol blue. Electrophoresis was carried out in 1.2% agarose gels, and gels were stained in 1 µg/ml ethidium bromide for 30 minutes (Sharp et al. 1973). Transfer of DNA from the gels to Millipore nitrocellulose filters was as described by Southern (1975). Hybridization of the filters with 9S [³H]RNA or [³H]cRNA from pSp2 or pSR1 was carried out in 4 × SSC, 44% formamide, and 1 mg/ml yeast RNA (Sigma frac. XI) for 48–72 hours at 37°C. Filters were washed extensively in 6 × SSC and treated for 1 hour with 10 µg/ml RNase A in 2 × SSC, dried, and fluorographed (Laskey and Mills 1975).

RESULTS

Multiple Species of H2A, H2B, and H1

It was previously shown that the early sea urchin blastula synthesizes the α subtypes of H1, H2A, and H2B, whereas the gastrula synthesizes different subtypes, termed β, γ, and δ, of these histone classes

(Cohen et al. 1975; Newrock et al., this volume). Figure 2 a and b shows the histones synthesized at the early mesenchyme blastula stage (20–24 hr), a time at which all the subtypes, both early and late, are made. The H1 (Fig. 2b) and non-H1 (Fig. 2a) fractions are resolved with perchloric acid precipitation (Johns 1964; Oliver et al. 1972). The H3 and H4 appear as multiple bands in this gel, presumably because of acetylated forms (Fig. 2a).

When mRNAs isolated from the polysomes of 17-hour blastula, 26-hour late mesenchyme blastula, and 36-hour gastrula are translated in a wheat germ system, the products seen in Figure 2c through e are resolved. The patterns of synthesis are similar to those reported for the in vivo synthesis of histones at these stages (Cohen et al. 1975; Newrock et al., this volume). The 17-hour mRNA supports the synthesis of $H2A_\alpha$ and a minor protein that runs close to, but slightly faster than, $H2A_\delta$. This minor product, which we call Y_1, appears to be different from the $H2A_\delta$ synthesized in vitro in response to mesen-

chyme blastula or gastrula RNA in that its mobility is altered after oxidation with H_2O_2 and it can be labeled with [^{35}S]methionine (K. Newrock et al., in prep). Y_1 has not been detected, however, in the histones synthesized in vivo in the blastula. The RNAs from 26-hour mesenchyme blastula and 36-hour gastrula direct the synthesis of the β, γ, and δ, but not the α, subtypes of H2A (Fig. 2d,e). The specific differences in the in vitro synthesis of the H2B subtypes are seen as well. The blastula mRNA is a template for $H2B_\alpha$, whereas the later mRNAs support the synthesis of $H2B_\gamma$ and $H2B_\delta$.

Resolution of histones on Triton-acid-urea gels shows that there is a shift in the H1 proteins made on the different mRNA preparations, as has been previously reported from studies using SDS and acetic-acid-urea gels (Ruderman et al. 1974; Arceci et al. 1976). A small amount of gastrula-type H1 is synthesized, in addition to the $H1_\alpha$, when blastula mRNA is used as the template. The mRNAs from mesenchyme blastula and gastrula no longer direct the synthesis of $H1_\alpha$, but make the $H1_{\beta,\gamma}$. The late H1 proteins sometimes can be resolved into two species by electrophoresis on SDS gels as well as on Triton-acid-urea gels. The H3 and H4 proteins made in vitro electrophorese in Triton-acid-urea gels differently than the in vivo products, as expected if the known acetylations of these proteins do not occur in the wheat germ system. In addition, the H4, made in vitro, runs faster than the naturally occurring H4, as expected if there is no N-terminal acetylation in the cell-free system. Differences are not seen in the mobility of the cell-free H4 products made from the three mRNA preparations. The proteins in the H3 region do, however, show some variation in the three sets of in vitro products. A faint band with a faster mobility than the major H3 protein can be seen in the 26-hour RNA products, whereas this minor species is absent in the 36-hour RNA products. If these differences are due to H3 heterogeneity, they would be difficult to identify in the in vivo products since the acetylated forms would obscure the slight differences in mobility. Subtypes of H3 have, however, been identified in mammalian histone preparations (Marzluff et al. 1972; Franklin and Zweidler 1977), and a two-dimensional Triton-acid-urea gel can resolve additional H3 spots in preparations of histones synthesized in vivo from sea urchin embryos (F. Zweidler and L. Cohen, unpubl.).

The shifts in H2A and H2B subtypes can also be identified in SDS polyacrylamide gels (K. Newrock et al., in prep.). The $H2A_\alpha$ and $H2B_\alpha$ run together and cannot be distinguished. The gastrula H2A and H2B products run as two bands, each with a different mobility from the blastula products. As with the Triton-acid-urea gels, the SDS gels show that the stage-specific transitions in the H2A, H2B, and H1 histones made in vivo are also seen in the in vitro products. We have also translated egg RNA in vitro and electrophoresed the products in both gel sys-

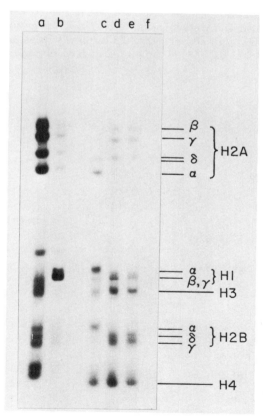

Figure 2. Triton-acid-urea gel of histones synthesized in vivo and of products made in vitro in response to 9S polysomal RNA. (a) [^{3}H]leucine-labeled histones made in sea urchin mesenchyme blastula, perchloric-acid-insoluble fraction (non-H1); (b) same as in a, but perchloric-acid-soluble fraction (H1); (c) in vitro [^{3}H]lysine-labeled products made with RNA from 17-hr embryos (blastula); (d) same as in c, but with RNA from 26-hr embryos (late mesenchyme blastula); (e) same as in c, but with RNA from 36-hr embryos (gastrula); (f) endogenous wheat germ control.

tems. The α forms of H2A, H2B, and H1 are synthesized, whereas the β, γ, and δ forms are not made (E. Weinberg et al., in prep.). The mRNAs for H2A$_\alpha$ and H2B$_\alpha$ are therefore stored in the egg, a result already obtained for H1$_\alpha$ (Arceci et al. 1977).

The multiple species of H2A and H2B histones resolved by the Triton X 100-urea gels appear to be different proteins. Not only are all the forms of H2A, H2B, and H1 made in vitro under conditions in which modifications would not be expected, but also the stage-specific shifts seen in vivo are reproduced in the in vitro products. Posttranslational modifications of H2A, H3, and H1 are an unlikely explanation for the observed variation since the same translational system makes different histone subtypes in response to different mRNA preparations, and modifications of H3 and H4 known to occur in vivo do not appear to occur in the cell-free products. Recent work on mRNA from HeLa cells also demonstrates that the subtypes resolved on Triton-acid-urea gels can be synthesized in cell-free systems (Borun et al. 1977).

Heterogeneity of Histone Gene Repeat Length

A careful analysis of the DNA that hybridizes to labeled histone mRNA shows that, although the major histone gene complex is 6–7 kb in length, there is extensive length heterogeneity. When enriched histone DNA prepared from 5–10 individual male sea urchins is digested with a variety of restriction enzymes, electrophoresed on agarose gels, and hybridized by the Southern transfer procedure (Southern 1975), DNA fragments of many sizes hybridize to histone mRNA (Fig. 3). Since HindIII has been found to cut the histone repeat only once, the major band at 6.3 kb is expected (Weinberg et al. 1975; Kedes et al. 1975a,b; Cohn et al. 1976). Yet, in addition to the major hybridization at 6.3 kb, DNA fragments of 2–15 kb show homology with the histone mRNA (Fig. 3f). Enzymes which cut the major repeat twice, EcoRI and BamHI, also give rise to fragments of heterogeneous length (Fig. 3g,h). EcoRI digests the DNA to form a group of bands at 4.05–4.52 kb and a 1.85-kb fragment as the major products. A substantial amount of a fragment at 6.3 kb is produced as well as minor amounts of fragments of diverse size. Digestion with BamHI produces a group of bands at 4.78–5.20 kb and a band at 1.28 kb. As with EcoRI, a 6.3-kb fragment and a broad distribution of other fragments are formed. If EcoRI-treated DNA is codigested with HpaI, the group of bands originally at 4.05–4.52 kb now run at 3.42–3.85 kb, the 6.3-kb band runs at 5.85 kb, and the

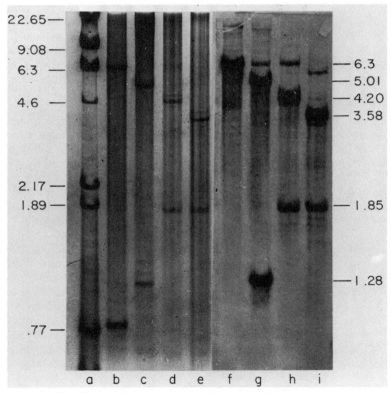

Figure 3. Restriction enzyme digestion products of enriched histone DNA run on a 1.2% agarose gel. *(a–e)* Ethidium-bromide-stained gels; *(f–i)* gels after transfer to Millipore filters, hybridization with 9S [³H]RNA, and fluorography. *(a)* HindIII-digested λ and HaeIII-digested pMB9 DNA run as markers; *(b,f)* HindIII-digested histone DNA; *(c,g)* BamHI-digested histone DNA; *(d,h)* EcoRI-digested histone DNA; *(e,i)* EcoRI + HpaI-digested histone DNA.

1.85-kb fragment is unchanged. The 0.62-kb fragment produced on codigestion with EcoRI and HpaI (Fig. 1) contains only spacer sequences and does not hybridize with the mRNA. In all these digestions, a family of bands differing from one another by 400–500 nucleotides is produced. This is most clearly seen in hybridization of the products of EcoRI and EcoRI + HpaI digestion. The fragments which run slightly faster and slower than the main large digestion product represent a reasonably substantial amount of the histone DNA satellite since they can be seen in the stained gel (Fig. 3b–e).

The region that is common to the 5.01-kb BamHI, 4.20-kb EcoRI, and 3.58-kb EcoRI + HpaI fragments is the area between the H2B gene and the HpaI site adjacent to the H1 gene. Although a more precise mapping of the regions of heterogeneity is not complete, one of the length variants seen in the 5.01-, 4.20-, and 3.58-kb fragments has been found to have an insert of 200-300 bases in the spacer sequence between the H2B and H4 genes. Digestion of the histone DNA satellite with HhaI produces the hybridizable bands expected from studies on a recombinant plasmid (Cohn et al. 1976) of 2.15, 1.46, 0.94, 0.55, and 0.78 kb (see Fig. 1). There are additional bands, however, which hybridize with histone mRNA, a major one of which is at 1.73 kb. On subsequent digestion with EcoRI, the 1.46- and 1.73-kb fragments are converted to 1.40 and 1.67 kb, respectively, a loss in each case of about 60 nucleotides. Hybridization with purified H2B mRNA shows that the 1.40- and 1.67-kb fragments both contain H2B genes. Since the region common to the 1.40-kb HhaI fragment and the 5.01-, 4.20- and 3.58-kb fragments described above is the region between the H2B gene and the HhaI site between the H2B and H4 genes, the extra 200-300 nucleotides are probably within this spacer region. A more extensive description of these experiments will be published elsewhere (G. C. Overton and E. S. Weinberg, in prep.).

Variation of the size of restriction enzyme digestion products might be due to contraction or expansion of the distance between two restriction sites due to gain or loss of blocks of DNA, as suggested above. However, fragment-length heterogeneity might also be due to nucleotide substitution, resulting in the addition or deletion of restriction sites. Candidates for this type of heterogeneity are the 6.3-kb fragments seen after either BamHI or EcoRI digestion of the histone DNA (Fig. 3g,h). If one of the two BamHI sites or EcoRI sites were absent, a full 6.3-kb repeat unit would be produced on treatment with either enzyme. Another form of heterogeneity would result if some histone genes were organized into a completely different kind of unit. The restriction enzyme digestion patterns of such a unit would show little if any homology to the patterns already described. These three examples of origin of heterogeneity are not mutually exclusive; intermediate cases might exist in which part of the repeat

unit is the standard organization (Fig. 1) and part is quite different.

The gene heterogeneity is also observed in DNA from individual sea urchins. Figures 4 and 5 illustrate EcoRI digestion patterns of DNA from five different individual sea urchins. In Figure 4 the DNA is hybridized to [³H]cRNA made from a plasmid containing a large EcoRI histone fragment, in Figure 5 the hybridization probe is [³H]cRNA made from a small EcoRI histone gene fragment. Most individuals show heterogeneity in either their large or their small EcoRI fragments. Comparison of slots 1 through 5 of Figure 4 shows that the large EcoRI fragment does vary in size from individual to individual. Each urchin has from one to four distinct fragment sizes, but there may be small amounts of other fragments visible as a smear of hybridization near the major bands. The broad bands in individuals 2 and 3 (Fig. 4) are probably poorly resolved doublets. The size range of the bands that are well resolved is from 3.95 to 4.88 kb. Similar results are obtained for the smaller type of EcoRI fragment (Fig. 5). Individuals 1 and 4 in Figure 5 clearly show heterogeneity; the broad band of individual 5 is a doublet; and the rather diffuse bands in individuals 2 and 3 are also probably due to heterogeneous size. The range of length of these fragments is 1.12–2.26 kb. There is, however, a 6.9-kb fragment which is present in each individual tested. This cleavage product shows complementarity to only the smaller of the

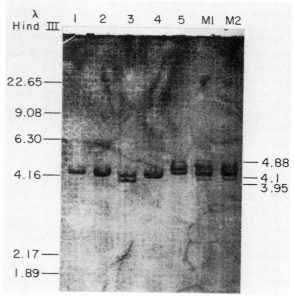

Figure 4. DNA from individual sea urchins digested with EcoRI, electrophoresed on a 1.2% agarose gel, transferred to a Millipore filter, and hybridized with [³H]cRNA made from pSp2 DNA (contains the large EcoRI fragment) as the template. Numbers 1–5 refer to the five different individuals; M2 is a mixture of DNA from these individuals; M1 is a mixture of DNA from five different individuals. Sizes were determined by running HindIII-treated λ DNA in an adjacent slot.

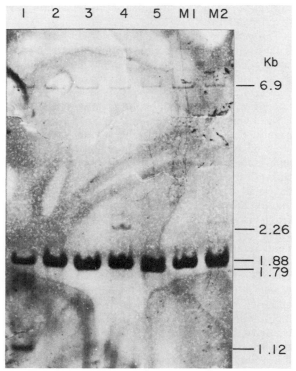

Figure 5. DNA from individual sea urchins digested with *Eco*RI and hybridized with [³H]cRNA made from pSR1 DNA (contains the small *Eco*RI fragment). Numbering and procedures are the same as in Fig. 4.

cloned *Eco*RI fragments. There is no one size of large *Eco*RI fragment that is required to occur in each individual. For example, the band at about 4.1 kb in individual 1 clearly is larger than the two bands in individual 3 (Fig. 4). Similarly, the bands in individual 3 do not coincide with any of the bands in individual 5. An abstract summarizing our findings on length heterogeneity has been previously published (Overton and Weinberg 1976).

DISCUSSION

We have shown that there must be heterogeneity of genes coding for the H2A and H2B, as well as the H1, histones. We have also demonstrated that the multiple copies of the histone gene unit are not all of one size, but can vary considerably in length. Although some of the variation in length may be due to nucleotide substitution resulting in the appearance and disappearance of restriction enzyme sites, at least some of the length difference is due to expansion or contraction of the amount of DNA between defined restriction enzyme sites.

The identification of any one of the variant length fragments as a sequence coding for any particular histone subtype must await either experiments using hybridization probes which can distinguish between genes coding for the subtypes of any one his-

tone class, or sequencing information of histone coding regions on the DNA fragments. Since variants of H2A, H2B, and H3 from mammalian tissues, distinguishable by Triton-acid-urea gels, have been found to differ from one another by only one or two amino acids (Franklin and Zweider 1977), it is likely that the sea urchin subtypes of any one histone class are similar to one another in sequence. Depending on how long ago the genes coding for the sea urchin histone subtypes were established, the degree of third-base substitution may or may not allow differential hybridization with the mRNAs of related variants.

The genes coding for the different subtypes might be organized in a variety of ways. Adjacent repeat units in a cluster might contain different subtype genes, yet still be composed of sequences coding for all five histone classes. In this model, the genes might be arranged in batteries which could be coordinately controlled so that particular histone subtypes can be synthesized together at the proper embryonic stage. One such unit transcribed at cleavage and early blastula might contain genes for H3, H4, H2A$_\alpha$, H2B$_\alpha$, and H1$_\alpha$, whereas a unit expressed at gastrula might contain genes for H3, H4, H2A$_\gamma$, H2B$_\gamma$, and H1$_\gamma$. Differences between the repeat units could be as minor as the few base changes necessary to code for the different subtypes. Since the histone subtypes are apparently expressed in each individual urchin, a requirement for the different coding sequences is that, whatever their organization, they be present in the DNA from each individual. Since the major repeat units we have identified vary in length from individual to individual, most of the length heterogeneity we see probably has no relationship to the organization of different subtype genes. Instead, it is most likely due to the variation of the size of spacer regions in the unit or to sequence changes at restriction enzyme recognition sites. The distribution of units must evolve rapidly so that the repeats within a single individual are of a limited number of lengths but are distinct in size from the units of another individual. The sets of units of different lengths in the different individuals might be thought of, in this model, as alleles of similarly organized gene clusters in which large blocks of repeat units evolve together. If all histone subtypes are coded in similar repeat units, it follows that, despite the different sequences necessary to code for the various subtypes, the units all change lengths in parallel, whatever the evolutionary mechanism.

On the other hand, genes coding for the subtypes might be organized in very different kinds of units with their own characteristic lengths and restriction enzyme digestion patterns, perhaps completely unlinked to each other. If this model is true and if the probes we are using cross-hybridize with the genes coding for related subtypes, we should be able to identify additional fragment lengths present in each individual. Further speculations on the organization of the different subtype genes must await a more complete characterization of the fragments we have identified.

Previous reports on the histone genes of *S. purpuratus* have not discussed repeat-length heterogeneity. Those studies based on analysis of cloned fragments (Kedes 1975a,b; Cohn et al. 1976. Wu et al. 1976; Holmes et al. 1977) presented the repeat unit as typified by two cloned *Eco*RI fragments. The hybridization patterns seen on filter-transferred digestion products of native enriched histone DNA (Weinberg et al. 1975; Cohn et al. 1976) were interpreted with respect to the major fragments. In one case (Weinberg et al. 1975), the filters were cut into strips for scintillation counting — a method far less sensitive than fluorography in the discrimination of bands. In retrospect, the asymmetry of the hybridization peaks and the presence of minor fluctuations are consistent with our findings presented here. In the other case (Cohn et al. 1976), fluorographs were either too faint or indistinct to have detected heterogeneity. In similar experiments with *P. miliaris* (Schaffner et al. 1976), it can be seen, in retrospect, that there are additional minor bands in the tracings of some of the fluorographs. The histone genes of *Echinus esculentus* (K. Gross and E. Southern, pers. comm.) and *Lytechinus pictus* (G. C. Overton and E. Weinberg, unpubl.) also are heterogeneous in length. In some species of sea urchin, at least, the histone gene repeat unit is not as homogeneous as previously believed.

The histone genes are highly regulated in the sea urchin embryo. During cleavage and blastula stages of embryogenesis, only the α forms of H2A, H2B, and H1 are synthesized. During the mesenchyme blastula stage the synthesis of these subtypes ceases and a new set of H2A, H2B, and H1 sequences is made. For these three histone classes, there is a definite developmental program of appearance of the protein product and the presence of mRNA on polysomes. Thus the histone gene system is not only interesting from the standpoint of structure, but also is characterized by a high degree of developmental regulation. Until the discovery of the developmental program of sea urchin histone subtype synthesis (Cohen et al. 1975), work on mammalian cells, as well as sea urchin embryos, indicated that regulation of histone synthesis was mainly a matter of expression during periods of DNA synthesis (for review, see Borun et al. 1975; Kedes 1976; Stahl and Gallwitz 1977). Now, however, it is clear that in the sea urchin embryo, each subtype is synthesized for a discrete period during development. It will be of interest to learn whether there is a program of transcriptional activity for each of the subtype genes.

Acknowledgments

This work was supported by grants GM-22155 and CA-12544 from the National Institutes of Health. G. C. O. and M. B. H. received support from National Institutes of Health Training Grants and K. M. N. from a National Institutes of Health postdoctoral fellowship.

REFERENCES

ALFAGEME, C. R., A. ZWEIDLER, A. MAHOWALD, and L. H. COHEN. 1974. Histones of *Drosophila* embryos. Electrophoretic isolation and structural studies. *J. Biol. Chem.* **249**: 3729.

ARCECI, R. J., D. R. SENGER, and P. R. GROSS. 1977. The programmed switch in lysine-rich histone synthesis at gastrulation. *Cell* **9**: 171.

BIRNSTIEL, M., K. GROSS, W. SCHAFFNER, and J. TELFORD. 1975. Biochemical dissection of the histone gene cluster of sea urchin. *FEBS Symp.* **31**: 735.

BONNER, W. M. and R. A. LASKEY. 1974. A film detection method for tritium labeled proteins and nucleic acids in polyacrylamide gels. *Eur. J. Biochem.* **46**: 83.

BORUN, T. W., K. AJIRO, A. ZWEIDLER, T. W. DOLBY, and R. E. STEPHENS. 1977. Studies of human histone messenger RNA. II. The resolution of fractions containing individual human histone messenger RNA species. *J. Biol. Chem.* **252**: 173.

BORUN, T. W., F. GABRIELLI, K. AJIRO, A. ZWEIDLER, and C. BAGLIONI. 1975. Further evidence of transcriptional and translational control of histone messenger RNA during the HeLa S3 cycle. *Cell* **4**: 59.

COHEN, L. H., K. M. NEWROCK, and A. ZWEIDLER. 1975. Stage-specific switches in histone synthesis during embryogenesis of the sea urchin. *Science* **190**: 994.

COHN, R. H., J. C. LOWRY, and L. H. KEDES. 1976. Histone genes of the sea urchin (*S. purpuratus*) cloned in *E. coli*: Order, polarity, and strandedness of the five histone-coding and spacer regions. *Cell* **9**: 147.

EASTON, D. and R. CHALKLEY. 1972. High resolution electrophoretic analyses of the histones from embryos and sperm of *Arbacia punctulata*. *Exp. Cell Res.* **72**: 502.

FRANKLIN, S. G. and A. ZWEIDLER. 1977. Non-allelic variants of histones 2a, 2b and 3 in mammals. *Nature* **266**: 273.

GRUNSTEIN, M. and P. SCHEDL. 1976. Isolation and sequence analysis of sea urchin (*Lytechinus pictus*) histone H4 messenger RNA. *J. Mol. Biol.* **104**: 323.

GRUNSTEIN, M., S. LEVY, P. SCHEDL, and L. KEDES. 1974. Messenger RNAs for individual histone proteins: Fingerprint analyses and *in vitro* translation. *Cold Spring Harbor Symp. Quant. Biol.* **38**: 717.

HOLMES, D. S., R. H. COHN, L. H. KEDES, and N. DAVIDSON. 1977. Positions of sea urchin (*Strongylocentrotus purpuratus*) histone genes relative to restriction endonuclease sites on the chimeric plasmids pSp2 and pSp17. *Biochemistry* **16**: 1504.

JOHNS, E. W. 1964. Studies on histones. 7. Preparative methods for histone fractions from calf thymus. *Biochem. J.* **92**: 55.

KEDES, L. H. 1976. Histone messengers and histone genes. *Cell* **8**: 321.

KEDES, L. H. and M. B. BIRNSTIEL. 1971. Reiteration and clustering of DNA sequences complementary to histone messenger RNA. *Nat. New Biol.* **230**: 165.

KEDES, L. H., A. C. Y. CHANG, D. HOUSEMAN, and S. N. COHEN. 1975a. Isolation of histone genes from unfractionated sea urchin DNA by subculture cloning in *E. coli*. *Nature* **255**: 533.

KEDES, L. H., R. H. COHN, S. C. LOWRY, A. C. Y. CHANG, and S. N. COHEN. 1975b. The organization of sea urchin genes. *Cell* **6**: 359.

LASKEY, R. A. and A. D. MILLS. 1975. Quantitative film detection of ³H and ¹⁴C in polyacrylamide gels by fluorography. *Eur. J. Biochem.* **69**: 3665.

MARCU, K. and B. DUDOCK. 1974. Characterization of a highly efficient protein synthesizing system derived from commercial wheat germ. *Nucleic Acids Res.* **1**: 1385.

MARZLUFF, W. F., JR., L. A. SANDERS, D. M. MILLER, and K. S. McCARTY. 1972. Two chemically and metabolically distinct forms of calf thymus H3. *J. Biol. Chem.* **224**: 2026.

OLIVER, D., K. R. SOMMER, S. PANYIM, S. SPIKER, and R. CHALKLEY. 1972. A modified procedure for fractionating histones. *Biochem. J.* **129**: 349.

OVERTON, G. C. and E. S. WEINBERG. 1976. Histone gene length heterogeneity in the sea urchin. *Fed. Proc.* **36**: 662.

PORTMANN, R., W. SCHAFFNER, and M. BIRNSTIEL. 1976. Partial denaturation mapping of cloned histone DNA from the sea urchin, *Psammechinus miliaris. Nature* **264**: 31.

RUDERMAN, J. V., C. BAGLIONI, and P. R. GROSS. 1974. Histone mRNA and histone synthesis during embryogenesis. *Nature* **247**: 36.

SCHAFFNER, W., K. GROSS, J. TELFORD, and M. BIRNSTIEL. 1976. Molecular analysis of the histone gene cluster of *Psammechinus miliaris.* II. The arrangement of the five histone-coding and spacer sequences. *Cell* **8**: 471.

SHARP, P. A., B. SNYDER, and J. SAMBROOK. 1973. Detection of two restriction endonuclease activities in *Haemophilus parainfluenzae* using analytical agarose-ethidium bromide electrophoresis. *Biochemistry* **12**: 3055.

SOUTHERN, E. M. 1975. Detection of specific sequences among DNA fragments separated by gel electrophoresis. *J. Mol. Biol.* **98**: 503.

STAHL, H. and D. GALLWITZ. 1977. Fate of histone messenger RNA in synchronized HeLa cells in the absence of initiation of protein synthesis. *Eur. J. Biochem.* **72**: 385.

WEINBERG, E. S. and G. C. OVERTON. 1977. Enrichment and purification of sea urchin histone genes. *Methods Cell Biol.* **19**: 273.

WEINBERG, E. S., M. L. BIRNSTIEL, I. F. PURDOM, and R. WILLIAMSON. 1972. Genes coding for polysomal 9S RNA of sea urchins: Conservation and divergence. *Nature* **240**: 5378.

WEINBERG, E. S., G. C. OVERTON, R. H. SHUTT, and R. H. REEDER. 1975. Histone gene arrangement in the sea urchin, *Strongylocentrotus purpuratus. Proc. Natl. Acad. Sci.* **72**: 4815.

WU, M., D. S. HOLMES, N. DAVIDSON, R. H. COHN, and L. H. KEDES. 1976. The relative positions of sea urchin histone genes on the chimeric plasmids pSp2 and pSp17 as studied by electron microscopy. *Cell* **9**: 163.

The Localization of the Genes Coding for Histone H4 in Human Chromosomes

L. C. Yu, P. Szabo, T. W. Borun,* and W. Prensky

*Memorial Sloan Kettering Cancer Center, New York, New York 10021; *Wistar Institute of Anatomy and Biology, Philadelphia, Pennsylvania 19104*

The development of in situ hybridization (Gall and Pardue 1969; John et al. 1969) made possible the chromosomal localization of many repetitive DNA sequences (Steffensen and Wimber 1972; Pardue and Gall 1972; Hennig 1973). In the case of species which have polytene chromosomes, in situ hybridization is a powerful cytological mapping tool because even unique DNA sequences are found in up to 1000 copies per chromosomal site. In species lacking polytene chromosomes, a similar sequence will be represented by two copies in a metaphase chromosome, making its detection by present methods difficult at best. The use of ^{125}I for labeling RNA was thought to extend the usefulness of the technique to the mapping of sites containing relatively few copies of a particular gene sequence (Wimber and Steffensen 1973). In particular 5S and 18S + 28S rRNA genes in various diploid chromosomes became amenable to mapping by RNA:DNA hybridization techniques (Steffensen et al. 1974; Wimber et al. 1974; Henderson et al. 1974).

In man, the histone genes are present in fewer copies than ribosomal genes (Wilson and Melli 1977). These authors estimate that the human genome contains a maximum of 40 copies of histone DNA. In this study we are utilizing ^{125}I-labeled mRNA coding for histone H4 to find the location of the histone genes in human chromosomes. The H4 mRNA is about 400 nucleotides long, and therefore the size of the target in each cell's chromosome complement is about 32,000 nucleotides long. For the H4 genes to be mapped successfully, the experimental system must be able to detect homologous DNA present no more frequently than 1 part in 10^5, the approximate fraction of human DNA homologous to H4 histone message. This paper presents the approach we utilized to identify chromosome 7 as the major site of histone H4 genes in man.

METHODS

Preparation and characterization of H4 mRNA. The preparation and translation analysis of the HeLa cell mRNA used in these studies has been described (Stephens et al. 1977; Borun et al. 1977). Briefly, HeLa H4 mRNA was purified from polyribosomes by preparative sucrose gradient centrifugation. The RNA sedimenting as 4S–18S was separated into polyadenylated [poly(A)$^+$] and nonpolyadenylated [poly(A)$^-$] fractions by oligo(dT)-cellulose chromatography. The poly(A)$^-$ fraction that contained the bulk of the templates for histone polypeptides was further fractionated into discreet RNA species by preparative polyacrylamide gel electrophoresis. The template activity of each RNA peak was assayed using the Krebs II ascites cell-free translation system (Borun et al. 1977). The homogeneity of the RNA in each peak was analyzed by ^{125}I-labeling and fingerprint analysis as described by Robertson et al. (1973). For the hybridization studies, we selected two independent isolates of 8.6S RNA which coded exclusively for histone H4 by translation assay. Figure 1 shows the oligonucleotide pattern of one of these RNAs. The sample was labeled with ^{125}I, digested with T$_1$ ribonuclease, and separated by electrophoresis on cellulose acetate strips at pH 3.5 in the first dimension, and homochromatography on DEAE-cel-

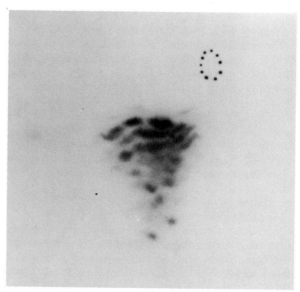

Figure 1. Ribonuclease T$_1$ fingerprint of ^{125}I-labeled HeLa cell poly(A)$^-$ mRNA coding for histone H4 protein. Details of the procedure for preparing this fingerprint are given elewhere (Robertson et al. 1973). The origin is at the lower right; the first dimension was from right to left and the second from the bottom to the top of the figure.

1101

lulose thin-layer plates in the second dimension (Barrell 1971). The pattern seen here is more complex than that of iodinated HeLa cell 5S RNA and less complex than that of iodinated duck globin mRNA (Robertson et al. 1973). The complexity of the iodinated human H4 RNA is roughly equivalent to the complexity of ^{32}P-labeled sea urchin H4 mRNA (Grunstein et al. 1976). Both human H4 mRNA samples had similar fingerprint patterns.

Since the amino acid sequence of histone H4 is highly conserved in evolution, it is not unreasonable to expect that HeLa mRNA might hybridize to the DNA of even distantly related species. Thus, sea urchin histone RNA was used by Pardue et al. (1972) to find the bands coding for histone in *Drosophila* salivary gland chromosomes. To show that our sample of mRNA contained conserved RNA sequences, in situ hybridization experiments were carried out with leopard frog *(Rana pipiens)* chromosomes prepared from a haploid cell line (Freed 1962).

In situ hybridization. The H4 mRNA was labeled with ^{125}I using Commerford's reaction as described by Prensky (1976). The two samples used for the in situ hybridization experiments had final specific activities of 1.5×10^8 and 4×10^8 dpm/µg respectively. After iodination, the RNA was purified by two successive passages over CF-11 cellulose (Franklin 1966), hydroxylapatite chromatography, extensive dialysis against distilled water, and ethanol precipitation. The pH 7.0 hydrization solution contained 50% formamide; $2 \times$ SSC ($1 \times$ SSC = 0.15 M NaCl, 0.015 M Na citrate), 10^{-4} M KI, 1 mg/ml *E. coli* tRNA. The concentration of [^{125}I]RNA was varied from 0.01 to 1 µg/ml. In some experiments the hybridization solution also contained a 100-fold excess by weight of HeLa cell ribosomal RNA.

Using the procedure described by Szabo et al. (1977b), the [^{125}I]RNA was hybridized to standard human metaphase spreads prepared from phytohemagglutinin stimulated peripheral lymphocytes. Some slides were denatured in 0.2 N HCl for 20 minutes at 25°C; denaturation was stopped by immersion in $2 \times$ SSC. Other slides were denatured by treatment with $2 \times$ SSC in 70% formamide for 2 minutes at 70°C, followed by immersion in 70% ethanol. Hybridization solution (10–20 µl) was added to slides and sealed under a coverslip. The slides were then incubated in moist chambers at 43°C. The set of slides hybridized with the first RNA sample was incubated for 6 hours at concentrations ranging from 0.08 to 0.8 µg/ml. The product of the RNA concentration and time $(C_r t)$, which defines the extent of the hybridization reaction, was thus varied from 5.1×10^{-3} to 5.1×10^{-2} moles · second/liter. Similarly, a range of concentrations (0.01 to 1 µg/ml) was used for the second experiment, where the incubation time was 6 hours, yielding $C_r t$ values from 6.5×10^{-4} to 6.5×10^{-2} moles · second/liter. The frog cells were incubated at 0.15 µg/ml for 10 hours yielding a $C_r t$ of 1.8×10^{-2} moles · second/liter.

After hybridization, the slides were processed through autoradiography as described by Gall and Pardue (1971). Exposure time was 30–60 days, and the cells were stained with Giemsa for photographic analysis.

Analysis of silver grain distribution. It was evident after a preliminary analysis of the initial slides that the site of the H4 genes in man was most probably on one of the C-group (6–12) chromosomes. Since these were difficult to classify after hybridization and autoradiography, we used procedures designed to minimize a priori bias in subsequent chromosome classification.

Well-spread metaphase cells were photographed and their locations recorded. Karyotype analysis was done sequentially by two naive cytotechnicians, each cell was then reexamined on the microscope, and the grain count over each chromosome was verified and recorded.

The relative grain count was obtained by dividing the fraction of all the silver grains assigned to a chromosome by the fraction of DNA for that chromosome as measured by Mayall et al. (1976). If all grains were due to random noise the values thus obtained would cluster around 1.0. This ratio was plotted (Fig. 3) as a function of the absolute frequency of labeling events for each chromosome. These coordinates define a labeling index for each human chromosome.

RESULTS AND DISCUSSION

On the basis of its migration rate through acrylamide gels (Borun et al. 1977) and the complexity of its RNase T_1 fingerprint (Fig. 1), human H4 mRNA is about 400 nucleotides long. The rate constants of RNA:DNA hybridization have been measured in cytological systems for both 5S RNA (120 nucleotides) and 18S + 28S RNA (±6000 nucleotides) (Szabo et al. 1977b). Using these rate constants and correcting for the relative complexity of H4 mRNA, we estimated that the $C_r t$ value needed to saturate half the available H4 sequences ($C_r t_{1/2}$) is about 5×10^{-3} moles · second/liter.

Figure 2 shows an autoradiograph of a metaphase cell of *R. pipiens* following hybridization with ^{125}I-labeled human H4 mRNA. One of the shortest chromosomes, either chromosome 12 or 13, is labeled in every cell with at least four grains. Neither 18S + 28S nor 5S ribosomal RNA hybridize to these chromosomes (P. Szabo et al. unpubl.). The extent of hybridization was limited to $3.6 \times C_r t_{1/2}$ to preclude extensive hybridization by all but the dominant RNA species in the sample (Birnstiel et al. 1972; Szabo et al. 1977a,b). We therefore conclude that the RNA species represented by the oligonucleotide pattern in Figure 1 is also the species responsible for the positive hybridization signal in Figure 2. The RNA seen in the fingerprint therefore represents human H4 mRNA.

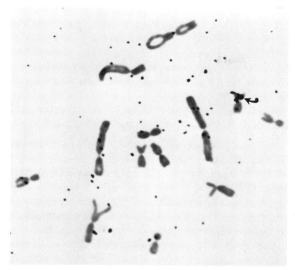

Figure 2. Hybridization of ¹²⁵I-labeled HeLa cell H4 mRNA to the metaphase chromosomes from a haploid cell line of *R. pipiens*.

The human slides we analyzed in detail were hybridized to $C_r t$ values which should have resulted in 50–100% saturation of the available histone H4 DNA. The data in Figure 3 A, B, and C are based on slides which were hybridized to 5X, 1.4X, and 13X the $C_r t_{1/2}$ value, respectively. A consistent feature of all these analyses is the higher than average labeling index associated with chromosome 7. The data in Figure 3B indicates that the labeling of chromosome 7 represents specifically the hybridization of H4 mRNA. The extent of this reaction was low enough to insure that only H4 mRNA would hybridize to a significant fraction (60–70%) of the available homologous DNA sites on the chromosomes. In Figure 3, A and C are based on the analysis of two different preparations of mRNA, each hybridized to saturate the available H4 DNA sites. Both samples showed a high frequency of chromosome 7 labeling.

The H4 mRNA probe contains sequences which hybridize to chromosomes 21 and/or 22 at a slower rate than the major sequence which hybridizes to chromosome 7. Since a 100-fold excess of rRNA was used in the hybridization reaction analyzed in Figure 3C, the RNA responsible for the observed signal is not rRNA. Other features of the data, like the inconsistent labeling of chromosomes 1, 2, and 14, cannot be readily explained. Furthermore, there is no way to infer from this data whether only the largest single block or all of the H4 genes are located on chromosome 7.

Our conclusion that the histone H4 genes are located on chromosome 7 in man can be most readily tested by utilizing segregants from mouse-human somatic cell hybrids which retain only human chromosome 7 in a mouse background. Such cell lines have been produced (Khoury and Croce 1975). RNA:DNA annealing studies utilizing immobilized

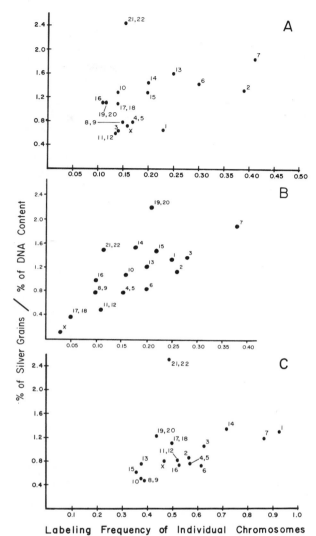

Figure 3. The labeling index plot of human chromosomes hybridized to ¹²⁵I-labeled HeLa cell H4 mRNA. The chromosomes were hybridized to a $C_r t$ of (A) 28×10^{-3}, (B) 6.6×10^{-3}, and (C) 66×10^{-3} moles · second/liter. The specific activity of the iodinated RNA was 1.5×10^8 dpm/μg in A and 4×10^8 dpm/μg in B and C.

DNA extracted from these and suitable control cells support our conclusion. Analysis of hybridized metaphase chromosomes from these hybrids is complicated by the infrequent presence of other human chromosomes. This makes a statistical comparison between the labeling index of chromosome 7 and other human chromosomes difficult. However, the human chromosomes are readily identifiable in these cells and their cytological analysis is useful for determining the approximate location of the H4 genes. Preliminary data indicate that the silver grains are concentrated over the middle of the long arm. In addition to our own studies, independent in situ hybridization experiments utilizing ³H-labeled cRNA prepared from sea urchin histone DNA plasmids indicate that there is specific hybridization between this probe and one C-group human chromosome (D. M. Steffensen, pers. comm.).

The efficiency of hybridization to chromosome 7 was very low. The two samples of mRNA had about 1.0% and 2.4% of their bases, respectively, labeled with ^{125}I. Assuming a target of 32,000 nucleotides at a single site, one might expect the deposition of about 320 and 770 ^{125}I atoms per chromosome. Allowing for an autoradiographic efficiency of about 20% (Ada et al. 1966; Prensky et al. 1973; Prensky 1976), a 60-day exposure should result in 32 and 77 grains per chromosome at saturation. The actual number of grains over chromosome 7 reflected a hybridization efficiency of 1.5–2%. Wimber and Steffensen (1973) have estimated that the efficiency of the RNA:DNA annealing reaction to *Drosophila* polytene chromosomes is about 5%. The H4 mRNA probes were also hybridized to human DNA filter disks, and we observed a fourfold lower saturation value than expected on the basis of the published gene copy number. The low amount of hybridization observed in both types of experiments could be due to the poor quality of the RNA resulting from a vigorous iodination reaction, to an overestimate of the iteration frequency of the histone genes, or to both.

There were other technical problems besides low hybridization efficiency which complicated this analysis. It was difficult to distinguish the different C-group chromosomes from each other. In addition, chromosomes other than chromosome 7 were often labeled. The combination of high background and low grain count over chromosome 7 meant that there was no ready means of identifying the location of the H4 genes except by the methods outlined above.

The problem of background noise over human chromosomes could be due to a number of factors, including the following:

(a) Iodinated RNA yields an inherently high background. Hybridization with RNA iodinated in vitro probably results in higher background than hybridization with RNA synthesized enzymatically, using ^3H- or ^{125}I-labeled precursors. This cannot be the sole reason for the background observed over human chromosomes, since the use of the identical H4 mRNA probes resulted in less background over chromosomes from other species.

(b) All histone H4 genes are not clustered on one chromosome. The possibility that minor H4 gene sites are scattered in the genome and therefore mistaken as background noise cannot yet be dismissed.

(c) The probe contains RNA contaminants resulting in high background. While the probes must contain some contaminants which are inactive in translation assays, the amount and number of contaminating RNA species cannot be very high. Such contaminants should contribute a high background only over chromosomes hybridized to high C_rt values

(Szabo et al. 1977a). Some of the background noise, especially that seen in Figure 3C, is probably due to this source.

(d) Short stretches of histone RNA from either the coding or noncoding regions of the molecule are partially homologous to sequences elsewhere in the genome.

(e) Human chromosomal proteins interact with human mRNA, increasing the background noise.

The last two factors cannot be evaluated without the use of RNA probes prepared from other species or from plasmids containing mammalian histone DNA. We are convinced that mRNA isolated from a given organism produces higher background noise when hybridized to homologous chromosomes. A much lower background was observed over frog (Fig. 2) and over mouse chromosomes (data not shown) than over human chromosomes hybridized with the same ^{125}I-labeled mRNA probes. D. M. Steffensen (pers. comm.) has also observed high backgrounds over human chromosomes hybridized with human β-globin mRNA. Noise level problems were also encountered by us while analyzing the hybridization of feline leukemia virus (FeLV) RNA to different mammalian chromosomes; the background was considerably higher over feline chromosomes. A more precise understanding of the causes of background noise will be necessary if in situ hybridization is to become a general tool for localizing mRNA gene sequences.

Acknowledgments

The authors wish to thank Dr. D. M. Steffensen for permission to refer to unpublished data from his laboratory, Dr. H. D. Robertson for preparing the fingerprint shown in Figure 1, and Dr. M. Siniscalco for his interest and advice. We also thank Drs. K. Ajiro, T. W. Dolby, R. E. Stephens and A. Zweidler for aid in preparing and identifying some of the RNA preparations examined in the course of this study. Thanks are due to Ms. Maria Velivasakis and Ms. Anna Yu for expert technical assistance with the cytological aspects of this project. The work of T. W. B. was supported by U.S. Public Health Service grants CA11463, CA17865, and CA15135 from the National Cancer Institute, and AG00368 from the National Institute of Aging, and Research Career Development Award CA00088. The work by the other authors was supported in part by grants CA16599, CA17085, and CA08748, and by contract N01-CM-53820 from the National Cancer Institute, and by a National Institutes of Health postdoctoral fellowship (GM05645) to L. C. Y.

REFERENCES

Ada, G., J. Humphrey, B. Askonas, H. McDevitt, and G. Nossal. 1966. Correlation of grain counts with radioactivity (^{125}I and tritium) in autoradiography. *Exp. Cell Res.* 41: 557.

BARRELL, B. G. 1971. Fractionation and sequence analysis of radioactive nucleotides. In *Procedures in nucleic acid research* (ed. G. L. Cantoni and D. R. Davies), vol. 2, p. 751. Harper and Row, New York.

BIRNSTIEL, M. L., B. H. SELLS, and I. F. PURDOM. 1972. Kinetic complexity of RNA molecules. *J. Mol. Biol.* **63**: 21.

BORUN, T. W., K. AJIRO, A. ZWEIDLER, T. W. DOLBY, and R. E. STEPHENS. 1977. Studies of human histone messenger RNA. II. The resolution of fractions containing individual human histone messenger RNA species. *J. Biol. Chem.* **252**: 173.

FRANKLIN, R. M. 1966. Purification and properties of the replicative intermediate of the RNA bacteriophage R17. *Proc. Natl. Acad. Sci.* **55**: 1504.

FREED, J. J. 1962. Continuous cultivation of cells derived from haploid *Rana pipiens* embryos. *Exp. Cell Res.* **26**: 327.

GALL, J. and M. L. PARDUE. 1969. The formation and detection of RNA-DNA hybrid molecules in cytological preparations. *Proc. Natl. Acad. Sci.* **63**: 378.

————. 1971. Nucleic acid hybridization in cytological preparations. *Methods Enzymol.* **21**: 470.

GRUNSTEIN, M., P. SCHEDL, and L. KEDES. 1976. Sequence analysis and evolution of sea urchin (*Lytechinus pictus* and *Strongylocentrotus purpuratus*) histone H4 messenger RNAs. *J. Mol. Biol.* **104**: 351.

HENDERSON, A. S., E. M. EICHER, M. T. YU, and K. C. ATWOOD. 1974. The chromosomal location of ribosomal DNA in the mouse. *Chromosoma* **49**: 155.

HENNIG, W. 1973. Molecular hybridization of DNA and RNA *in situ*. *Int. Rev. Cytol.* **36**: 1.

JOHN, H., M. BIRNSTIEL, and K. JONES. 1969. RNA-DNA hybrids at the cytological level. *Nature* **223**: 582.

KHOURY, G. and C. M. CROCE. 1975. Quantitation of the viral DNA present in somatic cell hybrids between mouse and SV40-transformed human cells. *Cell* **6**: 535.

MAYALL, B. H., A. V. CARRANO, D. H. MOORE, II, L. K. ASHWORTH, D. E. BENNETT, E. BOGART, J. L. LITTLEPAGE, J. L. MINKLER, D. L. PILUSO, and M. L. MENDELSOHN. 1976. Cytophotometric analysis of human chromosomes. In *Automation of cytogenetics* (ed. M. L. Mendelsohn), Conf. 751158, p. 135. U.S. Energy Research and Development Administration, Washington, D.C.

PARDUE, M. L. and J. G. GALL. 1972. Molecular cytogenetics. In *Molecular genetics and developmental biology* (ed. M. Sussman), p. 65. Prentice-Hall, Englewood Cliffs, New Jersey.

PARDUE, M. L., E. WEINBERG, L. KEDES, and M. BIRNSTIEL. 1972. Localization of sequences coding for histone messenger RNA in the chromosomes of *Drosophila melanogaster*. *J. Cell Biol.* **55**: 199a.

PRENSKY, W. 1976. The radioiodination of RNA and DNA to high specific activities. *Methods Cell Biol.* **13**: 121.

PRENSKY, W., D. M. STEFFENSEN, and W. L. HUGHES. 1973. The use of iodinated RNA for gene localization. *Proc. Natl. Acad. Sci.* **70**: 1860.

ROBERTSON, H. D., E. DICKSON, P. MODEL, and W. PRENSKY. 1973. Application of fingerprinting techniques to iodinated nucleic acids. *Proc. Natl. Acad. Sci.* **70**: 3260.

STEFFENSEN, D. M. and D. E. WIMBER. 1972. Hybridization of nucleic acids to chromosomes. *Results Probl. Cell Differ.* **3**: 47.

STEFFENSEN, D. M., P. DUFFEY, and W. PRENSKY. 1974. Localisation of 5S ribosomal RNA genes on human chromosome I. *Nature* **252**: 741.

STEPHENS, R. E., C-J. PAN, K. AJIRO, T. W. DOLBY, and T. W. BORUN. 1977. Studies of human histone messenger RNA. 1. Methods for the isolation and partial characterization of RNA fractions containing human histone message from HeLa S3 polyribosomes. *J. Biol. Chem.* **252**: 166.

SZABO, P., L. YU, and W. PRENSKY. 1977a. Kinetic aspects of *in situ* hybridization in relation to the problem of gene localization. In *Human cytogenetics* (ed. R. S. Sparkes et al.), vol. 7, p. 283. Academic Press, New York.

SZABO, P., R. ELDER, D. M. STEFFENSEN, and O. C. UHLENBECK. 1977b. Quantitative *in situ* hybridization of ribosomal RNAs to polytene chromosomes of *Drosophila melanogaster*. *J. Mol. Biol.* **115**: 539.

WILSON, M. C. and M. MELLI. 1977. Determination of the number of histone genes in human DNA. *J. Mol. Biol.* **110**: 511.

WIMBER, D. E. and D. M. STEFFENSEN. 1973. Localization of gene function. *Annu. Rev. Genet.* **7**: 205.

WIMBER, D. E., P. A. DUFFEY, D. M. STEFFENSEN, and W. PRENSKY. 1974. Localization of the 5S RNA genes in *Zea mays* by RNA-DNA hybridization *in situ*. *Chromosoma* **47**: 353.

Regulation of Histone Gene Expression in HeLa S$_3$ Cells

G. S. STEIN, J. L. STEIN,* W. D. PARK, S. DETKE, A. C. LICHTLER, E. A. SHEPHARD, R. L. JANSING, AND I. R. PHILLIPS

*Department of Biochemistry and Molecular Biology, and * Department of Immunology and Medical Microbiology, University of Florida, Gainesville, Florida 32610*

Throughout the cell cycle of continuously dividing cells, as well as after the stimulation of nondividing cells to proliferate, a complex and interdependent series of biochemical events occurs requiring modifications in expression of information encoded in the genome. Hence the cell cycle provides an effective biological system for studying the regulation of gene readout. For the past several years our laboratory has been focusing on the cell-cycle-stage-specific regulation of the genes that code for histones. In the present paper, several lines of evidence are presented which suggest that (a) regulation of histone gene expression resides at least in part at the transcriptional level, and (b) a subset of the nonhistone chromosomal proteins associated with the genome during the S phase of the cell cycle is involved in the regulation of histone gene transcription.

Histone mRNAs

In HeLa cells there exist five classes of histones coded by at least five mRNA species. Histone mRNAs lack poly (A) at their 3'-OH termini (Adesnik and Darnell 1972) and contain capped 5' termini of the types m^{7}GpppX^{m}pYp and m^{7}GpppX^{m}pY^{m}pZp (J. Stein et al. 1977; Moss et al. 1977).

Recently, we have observed that two distinct mRNA species in S-phase HeLa cells code for histone H4 (Lichtler et al. 1977). When ^{32}P-labeled 4S–18S RNA from S-phase HeLa cells was fractionated electrophoretically on a 6% polyacrylamide gel according to the method of Grunstein and Schedl (1976), the profile shown in Figure 1A was obtained. The individual bands were excised and the RNAs were eluted electrophoretically. The RNAs were then translated in a wheat germ protein-synthesizing system containing [^{3}H]lysine, and the translation products were electrophoresed with unlabeled marker histones on acetic acid-urea polyacrylamide gels (Fig. 1B). No preliminary purification to separate the histones from other translation products was carried out prior to electrophoresis. The difference in electrophoretic mobility between the two H4 histone mRNAs does not appear to be due to the presence of AMP residues at the 3'-OH termini of one of the RNA species, since both mRNAs were recovered in the unbound fraction during oligo(dT)-cellulose chromatography. When the two H4 histone mRNAs were eluted from a 6% acrylamide-0.2% sodium dodecyl sulfate (SDS) gel and rerun under denaturing conditions in parallel wells of an 8% acrylamide-95% formamide gel, both RNA species retained their distinct electrophoretic mobilities (Fig. 1C). The latter result indicates that these H4 histone mRNA species are of different molecular weights; therefore, separation in the aqueous gel system was not simply because of differences in secondary structure or because of aggregation with smaller RNA species.

There are several possible explanations for the apparent differences in molecular weight between the two RNA species that code for histone H4. One is that the smaller molecular weight RNA represents a cleavage product of the higher molecular weight species; another possibility is that since histone genes are reiterated in human cells, the different RNAs could represent transcripts from different copies of the gene. Studies are presently underway to characterize further the two H4 mRNA species and the proteins for which they code.

Evidence for Transcriptional Control of Histone Gene Expression

It has been established in many systems that histone synthesis and the deposition of these proteins on DNA is restricted to the S phase of the cell cycle (Spalding et al. 1966; Robbins and Borun 1967; Stein and Borun 1972), both in continuously dividing populations of cells and after stimulation of nondividing cells to proliferate. It has also been observed that inhibition of DNA replication results in a rapid and complete shutdown of histone synthesis (Spalding et al. 1966; Robbins and Borun 1967; Borun et al. 1967; Gallwitz and Mueller 1969; Stein and Borun 1972; Stein and Thrall 1973). These findings suggest that expression of histone genes is confined to the S phase of the cell cycle, and the coupling of histone and DNA synthesis is consistent with a functional relationship between these two events. We have been examining the regulation of histone gene expression and the level at which control is mediated. The presence of histone mRNA sequences on HeLa S$_3$ cell polyribosomes, in the postpolyribosomal cytoplasmic fraction, and in the nucleus during the G$_1$, S, and G$_2$ phases of the cell cycle has been examined.

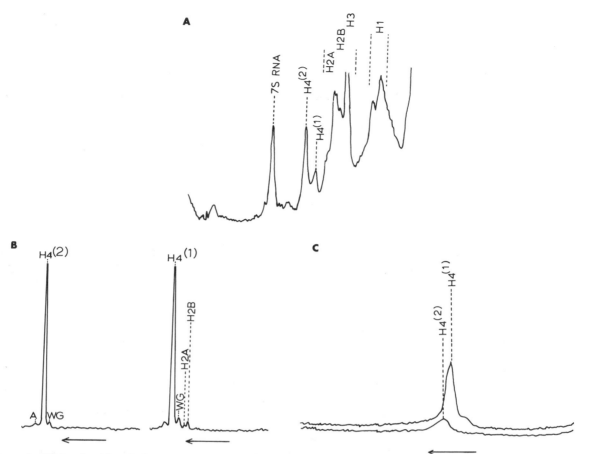

Figure 1. *(A)* Acrylamide gel electrophoretic fractionation of 4S–18S polysomal RNA from S-phase HeLa S₃ cells. Unlabeled RNA (75 µg) was combined with 7 × 10⁵ cpm of ³²P-labeled, 4S–18S RNA from S-phase HeLa cells, loaded on 0.3 × 0.4-cm wells of a 6% acrylamide gel and electrophoresed as described by Grunstein and Schedl (1976). The gel was analyzed autoradiographically and a densitometric tracing of one of the wells is shown. Details of the procedures have been reported elsewhere (Lichtler et al. 1977). *(B)* Acetic acid-urea acrylamide gel electrophoretic analysis of in vitro translation products of RNA extracted from bands H4(1) and H4(2) shown in *(A)*. A 15-µl sample of [³H]lysine-labeled wheat germ translation products was electrophoresed in the presence of marker histones, and fluorography was performed as described by Bonner and Laskey (1974) and Laskey and Mills (1975). *(C)* Electrophoretic analysis under denaturing conditions of RNAs coding for histone H4. ³²P-labeled RNAs extracted from an acrylamide gel similar to that shown in *(A)* were electrophoresed in adjacent wells on an 8% acrylamide-98% formamide gel as described by Maniatis et al. (1975). Densitometric scans were superimposed to facilitate comparison.

In addition, in vitro transcription of histone mRNA sequences from nuclei and chromatin isolated from HeLa cells at various times during the cell cycle has been assayed.

Since these studies require a high-resolution probe for identification of histone mRNA sequences, we have synthesized a ³H-labeled single-stranded DNA complementary to histone mRNAs. Histone mRNAs were isolated from polyribosomes of S-phase HeLa cells and chromatographed on oligo(dT)-cellulose to remove poly(A)-containing material. Poly(A) was then added to the 3′-OH ends of the histone mRNAs with an ATP:polynucleotidylexotransferase isolated from maize seedlings (Mans and Huff 1975), and the polyadenylated mRNAs were transcribed with RNA-dependent DNA polymerase isolated from avian myeloblastosis virus, using dT₁₀ as a primer in the presence of [³H]dCTP and [³H]dGTP. Tran-

scription was carried out in the presence of actinomycin D to insure that the DNA copy was single-stranded. Isolation, purification, and characterization of histone mRNAs, as well as synthesis and properties of the histone cDNA probe, have been reported (Thrall et al. 1974; G. S. Stein et al. 1975; J. L. Stein et al. 1975; Thrall et al. 1977). Identification and quantitation of histone mRNA sequences synthesized in vivo or transcribed in vitro from nuclei or chromatin were based on the kinetics of hybridization to ³H-labeled histone cDNA. Hybridization was carried out in RNA excess in the presence of 50% formamide and 0.5 M NaCl, and hybrid formation was assayed by resistance to single-strand-specific S₁ nuclease.

Histone mRNA sequences in cellular fractions. Our initial attempts to assess the level(s) at which regulation of histone gene expression resides involved determination of the representation of histone mRNA sequences on polysomes during the cell cycle of synchronized HeLa S_3 cells (J. L. Stein et al. 1975). Two methods were employed to achieve cell synchrony. S- and G_2-phase cells were obtained by two cycles of 2mM thymidine block. Three hours after release from the second thymidine block, 98% of the cells are in S phase (Fig. 2). At 7.5 hours after release from thymidine block, when G_2 cells are harvested, approximately 20% of the cells are still undergoing DNA replication, as assayed by thymidine labeling followed by autoradiography. This high background of S-phase cells in the G_2 population complicates interpretation of G_2 nucleic acid hybridization studies. Unfortunately, better methods are not available for obtaining a pure population of G_2-phase HeLa cells. Double thymidine synchronization is even less suitable for obtaining G_1 cells because, as Figure 2 clearly indicates, when cells synchronized by this procedure reach G_1 (11 hr after release), 25% of the cells are undergoing DNA replication (are in S phase). Therefore, we routinely obtain G_1 cells by mitotic selective detachment — a procedure which yields 97% G_1 cells 2 hours after harvest of mitotic cells, without detectable levels of S-phase cells.

Formation of hybrids between S-phase polyribosomal RNA and histone cDNA indicates the presence of histone-specific sequences on polyribosomes of S-phase cells (Fig. 3). In contrast, the absence of G_1 polyribosomal RNA hybridization demonstrates that histone mRNA sequences are not components of G_1 polyribosomes. Comparison of the kinetics of the hybridization reaction between S-phase polyribosomal RNA and histone cDNA ($C_{r0}t_{1/2} = 1.8$) with the kinetics of the histone mRNA-cDNA hybridization reaction ($C_{r0}t_{1/2} = 1.7 \times 10^{-2}$) indicates that histone mRNA sequences account for 0.9% of the RNA from S-phase, non-membrane-bound polyribosomes (J. L. Stein et al. 1975). This value is consistent with the situation in vivo where approximately 10–15%

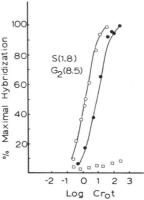

Figure 3. Kinetics of annealing of histone cDNA to RNA isolated from non-membrane-bound polyribosomes of G_1-, S-, and G_2-phase HeLa S_3 cells. ^{3}H-labeled cDNA (27,000 dpm/ng) and unlabeled RNA were hybridized at 52°C in sealed glass capillary tubes containing, in a volume of 15 μl, 50% formamide, 0.5 M NaCl, 25 mM HEPES (pH 7.0), 1 mM EDTA, 0.04 ng cDNA, and 3.75 or 7.5 μg polyribosomal RNA from phase G_1 (□), S (○), or G_2 (●) HeLa S_3 cells. Samples were removed at various times and incubated for 20 min in 2.0 ml of 30 mM Na acetate, 0.3 M NaCl, 1 mM ZnSO$_4$, 5% glycerol (pH 4.6), containing S$_1$ nuclease at a concentration sufficient to degrade at least 96% of the single-stranded nucleic acids present. The amount of labeled DNA resistant to digestion was determined by trichloroacetic acid precipitation. Polyribosomal RNA was isolated as reported previously (J. L. Stein et al. 1975). $C_{r0}t$ = moles ribonucleotides · sec/liter.

of the protein synthesis in S-phase HeLa cells is histone synthesis (Stein and Borun 1972). Furthermore, the absence of hybrid formation between G_1 polyribosomal RNA and histone cDNA establishes the absence of ribosomal RNA (5S, 18S, and 28S) and tRNA complementary sequences in the histone cDNA probe.

Determination of the presence or absence of histone mRNA sequences on G_2 polyribosomes is complex. The kinetics of the hybridization reaction between G_2 polyribosomal RNA and the histone cDNA ($C_{r0}t_{1/2} = 8.5$) suggests that the amount of histone mRNA sequences present on the polyribosomes of G_2-phase cells is 21% of that present on S-phase polyribosomes. However, as discussed previously, 20% of the G_2-phase cell population consists of S-phase cells. It is therefore reasonable to conclude that the histone mRNA sequences present in the G_2 polyribosomal RNA are due to the S-phase cells in the G_2 population. This implies that histone mRNA sequences are not associated with polyribosomes during the G_2 phase of the cell cycle.

These results demonstrate that in HeLa cells histone mRNA sequences become associated with polyribosomes during the transition from the G_1 to the S phase of the cell cycle. Such findings are in agreement with in vitro translation studies from several laboratories, which indicate that RNA isolated from polyribosomes of S-phase HeLa cells supports the synthesis of histones, whereas the RNA from polyribosomes of G_1 cells or of S-phase cells treated with

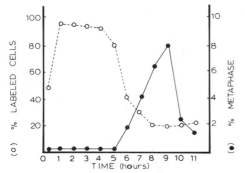

Figure 2. Percentage of HeLa S_3 cells synthesizing DNA and mitotic index at various times after release of HeLa S_3 cells from two cycles of 2 mM thymidine block. Cells were pulse-labeled for 15 min with 5 μCi of [^{3}H]thymidine/ml and the percentage of cells synthesizing DNA was determined autoradiographically (○). The mitotic index (●) was also determined from the autoradiographic preparations.

inhibitors of DNA synthesis does not (Borun et al. 1975). The hybridization studies eliminate the possibility that histone mRNAs are components of the polyribosomes during periods of the cell cycle other than S phase, but at such times are in some way rendered nontranslatable. These findings suggest that histone gene expression in HeLa cells is not regulated at the translational level and transcriptional control is implied. This interpretation is supported by other data from our laboratory suggesting that histone mRNA sequences are present in the nuclei of S-phase cells, but not in the nuclear RNA of G_1-phase cells (Fig. 4), and that histone mRNA sequences are not sequestered in the post-polysomal cytoplasmic fraction of G_1-phase cells.

It should be emphasized that the type of regulation of histone gene expression observed during the cell cycle of HeLa cells may not be universal. For example, there is evidence that during early stages of embryonic development control of histone synthesis may be mediated, at least in part, posttranscriptionally (Farquhar and McCarthy 1973; Skoultchi and Gross 1973; Gross et al. 1973; Gabrielli and Baglioni 1975). In such circumstances, histone mRNA sequences appear to be components of a stored maternal mRNA population which become templates for protein synthesis after fertilization.

In vitro transcription of nuclei. Another line of evidence suggesting that regulation of histone gene expression resides at least in part at the transcriptional level is provided by in vitro transcription of isolated nuclei (S. Detke et al., in prep.). Nuclei were isolated by a modification of the method of Sarma et al. (1976) and transcribed using the endogenous RNA polymerase. In this system, incorporation of [^{3}H]UMP into RNA is linear for 45 minutes and is dependent upon the addition of exogenous ribonucleoside triphosphates. In a 45-minute incubation, nuclei from S-phase HeLa cells synthesize 0.17 pg of RNA per nucleus. The isolated nuclei retain activity representative of all three classes of DNA-dependent RNA polymerase. If transcription is inhibited by incubating the nuclei with increasing amounts of α-amanitin, a three-component inhibition curve is obtained. Based on known sensitivities of the solubilized polymerases from HeLa cells to α-amanitin (Hossenlopp et al. 1975; Benecke and Siefart 1975; Weil and Blatti 1976), it is apparent that the nuclei possess all three classes of RNA polymerase. Under our conditions, the class I polymerase comprises 35% of the total RNA synthesizing activity; polymerase II, 58%; and polymerase III, 7%.

Results of hybridization of ^{3}H-labeled histone cDNA with nuclear RNAs indicate that histone mRNA sequences are being actively synthesized in isolated S-phase nuclei. The RNA from transcriptionally active nuclei incubated in the presence of all four ribonucleoside triphosphates hybridizes with histone cDNA with a $C_{rot_{1/2}}$ of 2.7, whereas RNA from nuclei which are not actively transcribing due to the absence of the four ribonucleoside triphosphates hybridizes with a $C_{rot_{1/2}}$ of 10 (Fig. 4). The increase in the representation of histone mRNA sequences which we have observed ranges from two- to fourfold. The histone mRNA sequences comprise 0.38–0.63% of the total RNA of active nuclei and 0.13–0.17% of the endogenous pool of nuclear RNA. Although nuclei isolated by this method have been reported to possess the capability for initiation of transcription (Sarma et al. 1976), it is not known whether the observed stimulation is due to the de novo initiation of histone mRNA synthesis or whether we are merely detecting the completion of preinitiated histone mRNAs. In contrast, G_1 nuclei did not synthesize detectable amounts of histone mRNA sequences (Fig. 4), although the general transcriptional activity of G_1- and S-phase nuclei was similar. Neither RNA of the endogenous pool of G_1 nuclei nor the RNA of active G_1 nuclei was found to hybridize with the histone cDNA probe, even at a C_{rot} of 320. It thus appears that the histone genes are transcribed only during the S phase portion of the cell cycle.

The polymerase responsible for the synthesis of histone mRNA sequences in S-phase nuclei can be determined by incubating the nuclei in the presence of varying concentrations of α-amanitin. The complete inhibition of the synthesis of histone mRNA sequences at as little as 1 μg α-amanitin/ml (Fig. 4) indicates that the class II RNA polymerase is responsible for the transcription of histone genes. It

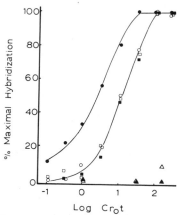

Figure 4. Kinetics of hybridization with ^{3}H-labeled histone cDNA of RNAs transcribed from isolated nuclei of G_1 and S-phase HeLa S_3 cells. The transcription reactions contained 0.4 mM GTP, CTP, UTP, and ATP, 5 mM Mg acetate, 70 mM KCl, 25 mM HEPES (pH 7.6), 0.04% 2-mercaptoethanol, 12.5% glycerol, and 1–5 × 10^7 nuclei/ml. Incubation was at 25°C for 45 min. RNA transcribed from G_1 (▲) and S-phase (●) nuclei. RNA isolated from G_1 (△) and S-phase (○) nuclei that were incubated in the absence of ribonucleoside triphosphates. RNA from S-phase nuclei transcribed in the presence of α-amanitin at concentrations of 1 μg/ml (□) and 100 μg/ml (■).

is the only class of polymerase which is inhibited completely at this concentration, whereas the other two polymerase classes are inhibited only slightly or not at all. Since the $C_0t_{1/2}$ of the hybridization reaction between histone cDNA and RNA isolated from nuclei transcribed in the presence of either 1 μg or 100 μg α-amanitin/ml is equal to 10, a value identical to that obtained with endogenous S-phase nuclear RNA, no new histone mRNA sequences are synthesized in the presence of α-amanitin at a concentration of 1 μg/ml or more. Thus the synthesis of histone mRNA sequences appears to be performed by the class II RNA polymerase.

Chromatin transcription. Chromatin from G_1- and S-phase cells was transcribed with *E. coli* RNA polymerase in a cell-free system, the RNA molecules were isolated, and their ability to form hybrids with histone cDNA was determined (G. S. Stein et al. 1975). The kinetics of the hybridization of histone cDNA and RNA transcripts from G_1- and S-phase chromatin are shown in Figure 5. Although transcripts from S-phase chromatin hybridize with histone cDNA with a $C_0t_{1/2}$ value of 2×10^{-1} compared with a value of 1.7×10^{-2} for the histone mRNA-histone cDNA hybridization reaction, there is no evidence of hybrid formation between histone cDNA and G_1-phase transcripts, even at a C_0t value of 100, indicating at least a 1000-fold increase in availability of histone genes for transcription. Since the overall template activity in vitro for RNA synthesis of G_1- and S-phase chromatin is similar, it is unlikely that the failure to detect histone mRNA sequences in G_1 chromatin transcripts results from a dilution effect. The maximal hybrid formation (65%) between histone cDNA and S-phase transcripts is the same as that observed between histone cDNA and histone mRNA. Fidelity of the hybrids formed between his-

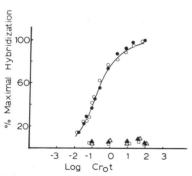

Figure 5. Hybridization of histone cDNA to in vitro transcripts from native and reconstituted HeLa S_3 cell chromatin. ^{3}H-labeled cDNA and 0.15 or 1.5 μg of RNA were hybridized as described in the legend to Fig. 3. RNA transcripts from native S-phase chromatin ($\bullet$), native G_1-phase chromatin ($\blacktriangle$), chromatin reconstituted with S-phase nonhistone chromosomal proteins ($\bigcirc$), and chromatin reconstituted with G_1-phase nonhistone chromosomal proteins ($\triangle$). Histone ^{3}H-cDNA was also annealed to RNA isolated from native S-phase chromatin in the presence of *E. coli* RNA as carrier.

tone cDNA and transcripts from S-phase chromatin is suggested by the fact that the T_m (melting temperature) of these hybrids is identical to the T_m of histone mRNA-cDNA hybrids (65°C in 50% formamide, 0.5 M NaCl, 25 mM HEPES [pH 7.0], 1 mM EDTA). It should be noted that the T_m value obtained under these conditions is consistent with an RNA-DNA hybrid having a GC content of 54%, which is the nucleotide composition of histone mRNA reported by Adesnik and Darnell (1972) and Thrall et al. (1974).

RNAs synthesized in intact cells may remain associated with chromatin during isolation and in part account for hybrid formation between RNA transcripts formed in vitro and cDNA for specific genes. Undoubtedly, the extent to which this phenomenon occurs varies with the tissue or cell and the method of chromatin preparation. To determine if such endogenous RNAs account for histone-specific sequences, which are detected in transcripts from S-phase chromatin, the following controls were carried out. S-phase chromatin was placed in the transcription mixture without RNA polymerase, and an amount of *E. coli* RNA equivalent to the amount of RNA transcribed from S-phase chromatin was added. When RNA was extracted and annealed with histone cDNA, no significant hybridization was observed (Fig. 5). Furthermore, RNA isolated from S-phase chromatin in the absence of carrier RNA showed no hybrid formation with histone cDNA. These results indicate that endogenous histone-specific sequences associated with S-phase chromatin are not contributing significantly to the hybridization observed with S-phase transcripts. It is therefore reasonable to conclude that histone sequences present in S-phase transcripts can be accounted for by synthesis in vitro.

When G_1 chromatin was transcribed in the presence of an amount of histone mRNA equivalent to that transcribed from S-phase chromatin, and the mixture of G_1 transcripts and added histone mRNA was subsequently isolated, hybridization with histone cDNA occurred at the expected $C_0t_{1/2}$ value (2×10^{-1}) (Park et al. 1976). This result suggests that the absence of histone mRNA sequences among RNA transcripts from G_1 chromatin is not attributable to a specific nuclease associated with chromatin during the G_1 phase of the cell cycle. The possibility that histone mRNA sequences were present in G_1 transcripts but were not detected because they were in a double-stranded form due to symmetric transcription is unlikely, since heating the hybridization mixture at 100°C for 10 minutes before incubation had no effect on the hybridization of histone cDNA with the transcripts (Park et al. 1976).

The results from these studies indicate that histone sequences are available for transcription from chromatin during S phase but not during G_1. Such findings are consistent with the restriction of histone synthesis to the S phase of the cell cycle and the presence of histone mRNAs on polyribosomes, in the postpolysomal cytoplasmic fraction and in the nucleus only during S phase. Taken together with results from the in vitro nuclear transcription studies, this evidence suggests that in continuously dividing HeLa S₃ cells expression of histone genes is regulated, at least in part, at the transcriptional level, and that readout of these genetic sequences occurs only during the period of DNA replication. It is also reasonable to conclude that chromatin is a valid and effective model for studying the regulation of cell-cycle-stage-specific transcription of histone genes.

Coupling of Histone Gene Expression and DNA Replication

Inhibition of DNA synthesis is associated with a rapid and complete inhibition of histone synthesis (Spalding et al. 1966; Robbins and Borun 1967; Borun et al. 1967; Gallwitz and Mueller 1969; Stein and Borun 1972; Stein and Thrall 1973). Although a definitive explanation for the coupling of histone synthesis and DNA synthesis cannot be provided at this time, it is reasonable to speculate that histones are required to complex with newly replicated DNA. Neither nucleoplasmic nor cytoplasmic pools of histones are present, and histones are needed for repression of DNA sequences which are not to be immediately transcribed and for imposition of the appropriate structure to the genome, i.e., packaging of the newly replicated DNA.

To examine the level at which the coupling of histone gene expression and DNA replication resides, we have pursued the following approach (G. S. Stein et al. 1977). S-phase HeLa S₃ cells were treated for 30 minutes with cytosine arabinoside (40 µg/ml) or hydroxyurea (10 mM) — conditions which result in greater than 98% inhibition of semiconservative DNA synthesis. Both inhibitors also effectively block histone synthesis. We then assayed the influence of these inhibitors on the levels of histone mRNA sequences present in the various intracellular RNA fractions by hybridization to histone cDNA. Consistent with in vitro translation data from several laboratories (Butler and Mueller 1973; Breindl

and Gallwitz 1974; Borun et al. 1975), cytosine arabinoside and hydroxyurea bring about a drastic reduction (>99%) in the representation of histone mRNA sequences on polyribosomes (Table 1). In contrast, neither inhibitor reduces in vitro transcription of histone mRNA sequences from chromatin (Table 1), and no reduction is observed in the level of histone mRNA sequences in nuclei of cells treated with hydroxyurea or cytosine arabinoside.

These results suggest that coupling of histone gene expression and DNA replication is not mediated at the transcriptional level, and posttranscriptional or translational control is strongly implied. This interpretation is further supported by a tenfold increase in the representation of histone mRNA sequences in the postpolysomal cytoplasmic fraction (Table 1). Accumulation of histone mRNA sequences in the cytoplasm after inhibition of DNA synthesis may be the result of release of histone mRNAs from polysomes or may reflect processing of histone mRNAs from the nucleus. In previous studies in which histone mRNAs were assayed by in vitro translation, an elevated level of histone mRNA sequences in the cytoplasm was not observed (Stahl and Gallwitz 1977). However, in vitro translation does not eliminate the possibility that histone mRNAs are present in nontranslatable states.

Nonhistone Chromosomal Proteins in the Regulation of Histone Gene Expression

A role for nonhistone chromosomal proteins in the regulation of histone gene expression during the cell cycle has been suggested by several lines of evidence. Variations observed in the composition and metabolism of the nonhistone chromosomal proteins during G₁, S, G₂, and mitosis, and their correlation with changes in transcription are consistent with a regulatory function for these proteins (reviewed by Stein and Baserga 1972; Baserga 1974; Stein et al. 1974b; Elgin and Weintraub 1975). Further evidence that nonhistone chromosomal proteins may be responsible for specific transcription at various stages of the cell cycle comes from a series of chromatin reconstitution studies, which indicate that non-

Table 1. Effect of Hydroxyurea and Cytosine Arabinoside on Representation of Histone mRNA Sequences in Chromatin Transcripts and in Various Subcellular Fractions of S-phase HeLa Cells

	Percentage of untreated S-phase control	
	hydroxyurea	cytosine arabinoside
Chromatin transcripts	100	100
Nuclear RNA	100	100
Polysomal RNA	<0.5	0.5
Postpolysomal cytoplasmic RNA	1100	1100

S-phase HeLa cells were treated with hydroxyurea (10 mM) or cytosine arabinoside (40 µg/ml) for 30 min. RNAs were then isolated from nuclei, polysomes, and the postpolysomal cytoplasmic fraction. Chromatin was prepared and transcribed with *E. coli* RNA polymerase, and transcripts were isolated. The kinetics of hybridization of the RNAs with ³H-labeled histone cDNA were measured.

histone chromosomal proteins determine the quantitative differences in availability of DNA as template for RNA synthesis during the cell cycle of continuously dividing cells (Stein and Farber 1972), as well as after stimulation of nondividing cells to proliferate (Stein et al. 1974a). To examine directly the involvement of nonhistone chromosomal proteins in the control of the cell-cycle-stage-specific transcription of a defined set of genetic sequences, the histone genes, we initially pursued the following approach.

Chromatin isolated from G_1- and S-phase cells was dissociated in 3 M NaCl, 5 M urea, and each chromatin preparation was fractionated into DNA, histones, and nonhistone chromosomal proteins. Chromatin preparations were then reconstituted by the gradient dialysis method of Bekhor et al. (1969) using DNA and histones pooled from G_1- and S-phase cells, and either G_1- or S-phase nonhistone chromosomal proteins. Figure 5 indicates that RNA transcripts from chromatin reconstituted with S-phase nonhistone chromosomal proteins hybridize with histone cDNA ($C_{r0}t_{1/2} = 2 \times 10^{-1}$), whereas those from chromatin reconstituted with G_1 nonhistone chromosomal proteins do not exhibit a significant degree of hybrid formation (G. S. Stein et al. 1975). It should be emphasized that the kinetics and extent of hybridization with the cDNA are the same for transcripts of native S-phase chromatin and chromatin reconstituted with S-phase nonhistone chromosomal proteins. Furthermore, the amount of RNA transcribed and the recovery during isolation of transcripts from native and reconstituted chromatin preparations were essentially identical. These results suggest a functional role for nonhistone chromosomal proteins in regulating the availability of histone sequences for in vitro transcription during the cell cycle. Such a regulatory role for the nonhistone chromosomal proteins is in agreement with results from other laboratories which have indicated that these proteins are responsible for the tissue-specific transcription of globin genes (Paul et al. 1974; Barrett et al. 1974; Chiu et al. 1975) and the hormone-induced transcription of ovalbumin genes (Tsai et al. 1976). However, the present results indicate that nonhistone chromosomal proteins are involved in the regulation of genes that are transiently expressed.

We then addressed the question of whether the difference in the transcription of histone genes in vitro from G_1- and S-phase chromatin is due to a component(s) of the S-phase nonhistone chromosomal proteins which renders histone genes transcribable or to a specific inhibitor of histone gene transcription present among the G_1 nonhistone chromosomal proteins (Park et al. 1976). When G_1 chromatin is dissociated and then reconstituted in the presence of increasing amounts of S-phase nonhistone chromosomal proteins, hybrid formation between transcripts from these chromatins and histone cDNA is seen at progressively lower $C_{r0}t$ values

(Fig. 6), indicating a dose-dependent increase in the transcription of histone genes from the G_1 chromatin. By comparing the kinetics of the hybridization of histone cDNA with transcripts from S-phase chromatin ($C_{r0}t_{1/2} = 2 \times 10^{-1}$), and the kinetics of the hybridization of histone cDNA with transcripts from G_1 chromatin reconstituted with a 1:1 ratio of S-phase nonhistone chromosomal proteins to DNA ($C_{r0}t_{1/2} = 3 \times 10^{-1}$), it can be seen that the histone genes are transcribed from the reconstituted chromatin to approximately the same extent as from native S-phase chromatin. This level of histone gene transcription is the maximum that can be achieved, even if the added S-phase nonhistone chromosomal protein to DNA ratio is increased above 1:1. The T_m of the hybrids and the maximal level of hybridization (65%) are in all cases identical to those of the hybrids formed between histone mRNA and histone cDNA. When G_1 chromatin was dissociated and then reconstituted in the presence of S-phase histones, even at a 1:1 ratio of S-phase histone to DNA, no stimulation of transcription of histone genes was observed (Fig. 6). It should be noted that there were no significant differences among the various chromatin preparations in the yield or recovery of RNA, even though the presence of S-phase nonhistone

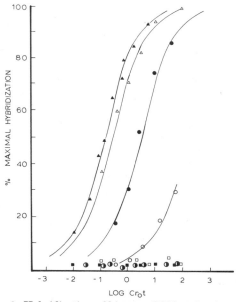

Figure 6. Hybridization of histone cDNA to in vitro transcripts from G_1-phase HeLa chromatin reconstituted in the presence of various amounts of S-phase HeLa nonhistone chromosomal proteins. cDNA was annealed to RNA transcripts from G_1 chromatin reconstituted in the presence of 0.01 mg (○), 0.10 mg (●), or 1.00 mg (△) of S-phase nonhistone chromosomal protein or 1.0 mg of S-phase histones (□)/mg of G_1 DNA as chromatin. cDNA was also annealed to RNA transcripts from G_1 chromatin reconstituted in the presence of 1.0 mg of G_1 total chromosomal protein/mg of G_1 DNA as chromatin (■) and to RNA transcripts from native chromatin of G_1 (◑) and S (▲) phase cells.

chromosomal proteins during reconstitution could cause a greater than 1000-fold stimulation in the amount of histone sequences transcribed from G_1 chromatin. Therefore, the observed increase in representation of histone mRNA sequences cannot be attributed to nonspecific alteration of template activity. Stimulation of histone gene transcription was not observed when G_1 chromatin was dissociated and then reconstituted in the presence of additional G_1 chromosomal proteins, even at a 1:1 ratio of additional G_1 protein to DNA (Fig. 6). This result suggests that specific chromosomal proteins are required to elicit histone gene readout.

To determine whether G_1 chromatin contains an inhibitor of histone gene transcription that is degraded or inactivated as the cells progress from the G_1 to the S phase of the cell cycle, chromatin from S-phase cells was dissociated and reconstituted in the presence of total chromosomal proteins from G_1 cells. Even at a 1:1 ratio of total G_1 chromosomal proteins to DNA, histone gene transcription from S-phase chromatin was not significantly inhibited (Park et al. 1976). This result would suggest that any specific inhibitor of histone gene expression is lost during isolation, dissociation, fractionation or reconstitution of chromatin, or that any inhibition of histone gene transcription by G_1 chromosomal proteins can be overridden by S-phase nonhistone chromosomal proteins. Results from chromatin reconstitution experiments described above are consistent with a direct role for nonhistone chromosomal proteins in dictating availability of histone genes for transcription from chromatin of continuously dividing HeLa S_3 cells. Similar experiments indicate that transcription of histone mRNA sequences from chromatin of human diploid fibroblasts after stimulation of these cells to proliferate is also mediated by a component of the S-phase nonhistone chromosomal proteins (Jansing et al. 1977). Other results suggest that phosphorylation of nonhistone chromosomal proteins may be an important component of the mechanism by which histone gene readout is regulated (Kleinsmith et al. 1976; Thomson et al. 1976).

Fractionation of Nonhistone Chromosomal Proteins

To purify the molecules responsible for the regulation of specific genes, it is necessary to determine not only whether a given fraction has activity but also now much activity is present. We have been using the techniques of chromatin reconstitution and in vitro transcription to assay and quantitate the activity of nonhistone chromosomal protein fractions for their involvement in the control of histone gene transcription from chromatin. G_1 chromatin, which does not serve as a template for histone gene transcription, is dissociated in 5 M urea, 3 M NaCl,

10 mM Tris (pH 8.3) and then reconstituted in the presence of added S-phase nonhistone chromosomal protein fractions. The reconstituted chromatin preparations are transcribed in a cell-free system with *E. coli* RNA polymerase, and the transcripts are assayed for their ability to hybridize with histone cDNA. Since, as discussed above, this system responds to added nonhistone chromosomal proteins with a dose-dependent, but saturable, increase in the transcription of histone genes, we have a viable method for monitoring nonhistone chromosomal protein fractionation.

We have recently been able to achieve a substantial purification of the S-phase nonhistone chromosomal protein(s) which exhibits the ability to render histone mRNA sequences transcribable from chromatin. This fractionation of S-phase HeLa nonhistone chromosomal proteins was accomplished by ion-exchange chromatography on QAE-Sephadex, followed by SP-Sephadex ion-exchange chromatography and then gel-filtration chromatography on Sephadex G-100 (W. D. Park et al., in prep.).

Ion-exchange chromatography of S-phase HeLa chromosomal proteins on QAE-Sephadex A-25 was carried out as follows. Chromosomal proteins from which nucleic acids had been removed by ultracentrifugation were dialyzed against 5 M urea, 10 mM Tris (pH 8.3) and were loaded on a column of QAE-Sephadex A-25 equilibrated previously with the same buffer. The proteins were then eluted with two column volumes each of 5 M urea, 10 mM Tris (pH 8.3), containing 0, 0.1, 0.25, 0.5, and 3 M NaCl. The histones and approximately 10% of the nonhistone chromosomal proteins were not bound and were eluted in the void volume (Fig. 7), whereas a complex but electrophoretically distinct class of nonhistone chromosomal proteins was eluted by each salt concentration (Park et al. 1977). Total recovery of proteins from the column was approximately 85%. To determine the ability of each of the QAE fractions

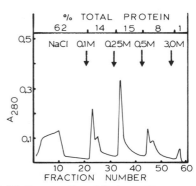

Figure 7. Elution profile of S-phase HeLa chromosomal proteins from QAE-Sephadex. Proteins were loaded in 5 M urea, 10 mM Tris (pH 8.3) and were eluted with this buffer containing 0.10 M, 0.25 M, and 3.0 M NaCl. The percentage of protein eluted in each peak is shown in the upper panel.

to render histone genes available for transcription, 3 mg of G_1 chromatin (containing ~1 mg of DNA) were dissociated and then reconstituted in the presence of 100 μg of each of the QAE fractions. Transcripts from G_1 chromatin reconstituted in the presence of the unbound fraction or of the material eluted with 0.1, 0.25, or 3.0 M NaCl did not show significant hybrid formation with histone ^{3}H-labeled cDNA (Fig. 8A). In contrast, even though the total amount of RNA transcribed was similar, transcripts of G_1 chromatin reconstituted in the presence of the 0.5 M QAE fraction hybridized efficiently with histone cDNA ($C_0t_{1/2} = 4 \times 10^{-1}$).

As discussed above, when G_1 chromatin is reconstituted in the presence of various amounts of added S-phase chromosomal proteins, there is a dose-dependent but saturable activation of histone gene transcription. Specifically, transcripts from G_1 chromatin reconstituted in the presence of 1000 μg of S-phase total chromosomal protein per mg of G_1 DNA (as chromatin) contain approximately 10 times more histone mRNA sequences than transcripts from the same amount of G_1 chromatin reconstituted in the presence of 100 μg of these proteins. Since the 0.5 M QAE fraction contains only approximately 8% of the total chromosomal protein, one would anticipate that 100 μg of the 0.5 M QAE fraction should activate histone gene transcription from G_1 chromatin to the same degree as 1000 μg of the total S-phase HeLa chromosomal protein. As seen in Figure 8A, there are no significant differences in the kinetics of hybrid formation with histone cDNA between transcripts from G_1 chromatin reconstituted in the presence of 100 μg of the 0.5 M QAE fraction and 1000 μg of the total HeLa chromosomal proteins per mg of G_1 DNA (as chromatin), indicating that *at least* a tenfold purification of the S-phase nonhistone chromosomal protein(s) involved in transcription of histone genes has been achieved.

Additional purification of the S-phase nonhistone

chromosomal protein(s) involved with transcription of histone genes was obtained by chromatography on SP-Sephadex C-25. The proteins eluted from QAE-Sephadex by 0.5 M NaCl were titrated to pH 5.2 with 1 M sodium acetate (pH 4.5), dialyzed against 5 M urea, 0.1 M NaCl, 0.2 M sodium acetate (pH 5.2), and loaded on a column equilibrated previously with the same buffer. The proteins were eluted with two column volumes each of 5 M urea, 0.2 M sodium acetate (pH 5.2) containing 0.2 M and 0.4 M NaCl and then with two column volumes of 5 M urea, 3 M NaCl, 100 mM Tris (pH 8.3). Total recovery of protein from the column was approximately 50%. To assay the ability of the SP fractions to render histone genes available for transcription, 1 mg of DNA as chromatin was dissociated and then reconstituted in the presence of 10 μg of each of the fractions. Only the transcripts from the chromatin reconstituted in the presence of the 0.4-M SP fraction showed a significant level of hybridization with histone cDNA ($C_0t_{1/2} = 2 \times 10^{-1}$) (Fig. 8B). The 0.4-M SP fraction contains 10% of the protein loaded on the column. Since 1 μg of the SP fraction is as effective as 10 μg of this same fraction in dictating availability of histone genes for transcription from G_1 chromatin (Fig. 8B), it is evident that the extent to which histone genes can be rendered transcribable by this fraction is saturable. This result would not be expected if endogenous histone mRNA sequences were present in the SP fraction and were responsible for the observed hybridization with the histone cDNA.

To ascertain the molecular weight of the component of the 0.4-M SP fraction that affects transcription of histone genes, we chromatographed the fraction on a 1.5 × 27-cm column of Sephadex G-100. The proteins were titrated to pH 8.3 with 1 M Tris and solid NaCl was added to a final concentration of 3 M. The proteins were eluted from the column with 5 M urea, 3 M NaCl, 10 mM Tris (pH 8.3) and

Figure 8. Hybridization of histone cDNA to in vitro transcripts from G_1 HeLa chromatin reconstituted in the presence of S-phase HeLa cell chromosomal protein fractions. (*A*) Transcripts from 1 mg of G_1 DNA as chromatin reconstituted in the presence of 100 μg of S-phase chromosomal proteins eluted from QAE-Sephadex A-25 by 5 M urea, 10 mM Tris (pH 8.3) containing 0 M (□), 0.1 M (■), 0.25 M (○), 0.5 M (●), and 3.0 M (△) NaCl or in the presence of 1000 μg of S-phase total chromosomal proteins (▲). (*B*) Transcripts from 1 mg of G_1 DNA as chromatin reconstituted in the presence of 10 μg of S-phase chromosomal proteins eluted from SP-Sephadex by 5 M urea, 0.2 M sodium acetate (pH 5.2) containing 0.1 M (○), 0.2 M (●), and 0.4 M (□) NaCl and by 5 M urea, 10 mM Tris (pH 8.3), 3 M NaCl (■) or in the presence of 1 μg (△) or 0.1 μg (▲) of the 0.4-M fraction.

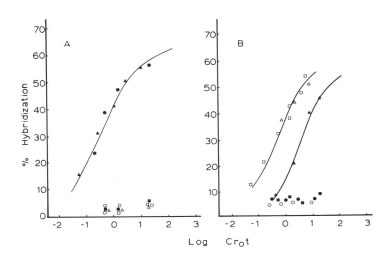

fractions were assayed for their ability to render histone genes transcribable. The activity was contained in the fractions with elution constants between 0.16 and 0.27, corresponding to an apparent molecular weight of 40,000–60,000 daltons. The Sephadex G-100 column was calibrated using bovine serum albumin, ovalbumin, and whale skeletal muscle myoglobin in 5 M urea, 3 M NaCl, 10 mM Tris (pH 8.3).

It is important to determine whether molecules that are involved in the regulation of histone gene transcription from chromatin are protein or nucleic acid in nature. Therefore, the 0.4-M SP fraction was centrifuged in 5 M urea, 10 mM Tris (pH 8.3) containing 0.41 mg CsCl/ml for 80 hours at 37,500 rpm in a Beckman SW 50.1 rotor, conditions which separate greater than 99.5% of the nucleic acids from chromosomal proteins. Each component of the gradient was then assayed for ability to influence availability of histone genes for transcription in G_1 chromatin. We found that the activity resided solely with the material contained in the protein region of the gradient — material with a density between 1.25 and 1.28 g/cc (W. D. Park et al., in prep.). While this result is consistent with the component responsible for transcription of histone mRNA sequences from chromatin being a protein, we cannot dismiss the possibility that small pieces of nucleic acid, covalently or otherwise tightly bound to nonhistone chromosomal proteins, are involved with regulation of histone gene readout.

To examine further the possibility that nucleic acids associated with chromosomal proteins are responsible for determining availability of histone genes for transcription, the activity of the 0.4-M NaCl fraction from the SP-Sephadex column was assayed for sensitivity to micrococcal nuclease. Under the conditions we employed for nuclease di-

gestion, the enzyme effectively degrades RNA and both single- and double-stranded DNA. Since the enzyme activity is Ca^{++}-dependent, it is readily inactivated by EGTA. A 3.4-μg sample of the SP fraction was incubated for 30 minutes at 37°C with 0.06 μg of micrococcal nuclease (in a parallel reaction containing the SP fraction, 3 μg of ^{3}H-labeled λ DNA was rendered 99% TCA-soluble). After the incubation, EGTA was added to a final concentration of 5 mM, and 0.3 μg of the nuclease-treated fraction was then reconstituted with 1 mg of G_1 DNA as chromatin in the presence of 2 mM EGTA. As shown in Figure 9A, there is no significant difference in the kinetics of hybridization with histone cDNA of transcripts from G_1 chromatin reconstituted in the presence of 0.3 μg of nuclease-treated 0.4-M SP fraction and transcripts from G_1 chromatin reconstituted in the presence of 0.3 μg of untreated 0.4-M SP fraction. To examine the possibility that increased histone gene transcription from G_1 chromatin was due to the action of EGTA or of the nuclease itself, G_1 chromatin was dissociated and then reconstituted in the presence of EGTA or EGTA-inactivated micrococcal nuclease. Transcripts from these reconstituted preparations do not hybridize to a significant extent with histone cDNA (Fig. 9A). These results suggest that the component of the 0.4-M SP fraction which has the ability to render histone genes transcribable is not a nucleic acid. However, we cannot eliminate the possibility that the 0.4-M SP fraction contains a small amount of nucleic acid complexed with protein or in a configuration such that it is not susceptible to digestion by micrococcal nuclease.

To approach directly the question of whether the component of the SP-Sephadex fraction that renders histone genes transcribable in G_1 chromatin is a protein, we examined the sensitivity of the 0.4-M

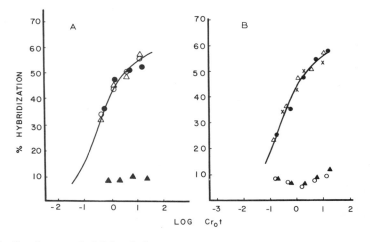

Figure 9. (A) Hybridization of histone cDNA to in vitro transcripts from HeLa cell G_1 chromatin reconstituted in the presence of micrococcal nuclease-treated S-phase HeLa cell nonhistone chromosomal proteins. Transcripts from 1 mg of G_1 DNA as chromatin reconstituted with 0.3 μg of nuclease-treated (O) or untreated (●) 0.4-M SP-Sephadex fraction. G_1 chromatin was also reconstituted in the presence of EGTA and EGTA-inactivated micrococcal nuclease in the absence of additional chromosomal protein (▲) or with 0.3 μg of the 0.4-M SP-Sephadex fraction (△). (B) Kinetics of annealing of histone cDNA to in vitro transcripts from HeLa cell G_1 chromatin reconstituted in the presence of chymotrypsin-treated S-phase HeLa cell nonhistone chromosomal proteins. Transcripts from 1 mg of G_1 DNA as chromatin reconstituted with: 0.3 μg of chymotrypsin-treated (O) or untreated (x) 0.4-M SP-Sephadex fraction; PMSF-containing buffer which had been incubated with immobilized chymotrypsin either in the absence of additional chromosomal protein (▲) or in the presence of 0.3 μg of untreated 0.4-M SP-Sephadex fraction (△); or 0.3 μ of the SP-Sephadex nonhistone chromosomal protein fraction that had been incubated with PMSF-inactivated chymotrypsin (●).

SP fraction to chymotrypsin (Fig. 9B). A 3.4-μg sample of the fraction was incubated for 60 minutes at 22°C with 30 μg of chymotrypsin covalently bound to agarose beads (Miles Laboratories) in 2 M urea, 0.1 M Tris (pH 8.3) — conditions which result in the digestion of at least 50 μg of chromosomal proteins. After incubation, the immobilized enzyme was removed by filtration, and phenylmethylsulfonyl fluoride (PMSF) was added to the 0.4-M SP fraction to a final concentration of 1 mM to inactivate any remaining enzyme. PMSF was not added during reconstitution and transcription. RNA transcripts from G_1 chromatin reconstituted in the presence of 0.3 μg of untreated 0.4-M SP fraction hybridized with tritiated histone cDNA with a $C_rot_{1/2}$ of 3×10^{-1}, whereas transcripts from G_1 chromatin reconstituted with the same amount of chymotrypsin-treated fraction did not hybridize with histone cDNA to a significant level. The ability of the 0.4-M SP fraction to influence transcription of histone genes in G_1 chromatin was not affected when G_1 chromatin was reconstituted in the presence of 0.4-M SP fraction which had been incubated with PMSF-inactivated chymotrypsin. Treatment of the fraction with buffer which had been incubated in the presence of immobilized chymotrypsin also did not reduce the effectiveness of the fraction in rendering histone sequences transcribable. Reconstitution of G_1 chromatin in the presence of buffer which had been incubated with immobilized chymotrypsin and then treated with PMSF did not stimulate transcription of histone mRNA sequences. Taken together, these results suggest that the component of the 0.4-M SP fraction that renders histone genes transcribable from G_1 chromatin is a protein.

Activation of Histone Gene Transcription in Chromatin from Human Diploid Cells or Mouse Liver by a Nonhistone Chromosomal Protein Fraction from HeLa S_3 Cells

It is not known whether the mechanism by which histone gene transcription is regulated is the same in different tissues and species. It is of particular interest to determine whether a transformed, continuously dividing cell such as HeLa contains components that can render histone genes transcribable from other cells which have greater degrees of growth control. To examine these questions, chromatin preparations from both contact-inhibited WI38 human diploid fibroblasts and adult mouse liver (both nonproliferating) were dissociated and then reconstituted in the presence of added chromosomal proteins from S-phase HeLa cells; the reconstituted chromatins were then transcribed in vitro and the transcripts assayed for histone mRNA sequences by hybridization with histone cDNA. These studies show that S-phase HeLa cell nonhistone chromosomal proteins can render histone genes of chromatin from contact-inhibited WI38 human diploid fibro-

blasts (Fig. 10A) or from nondividing mouse liver (Fig. 10B) available for transcription (Park et al. 1977). Specifically, these studies show that when the S-phase HeLa chromosomal proteins are fractionated on QAE-Sephadex in the presence of 5 M urea, only the fraction eluted by 0.5 M NaCl can activate histone gene transcription from chromatin of G_1 HeLa cells, contact-inhibited WI38 fibroblasts, or mouse liver — indicating that activation of histone genes in heterologous chromatins is not elicited by S-phase nonhistone chromosomal proteins in general. Several lines of evidence also suggest that activation of histone gene transcription in mouse liver chromatin by S-phase HeLa nonhistone chromosomal proteins is not a random phenomenon. Addition of the HeLa proteins to mouse liver chromatin does not significantly modify chromatin template activity. More specifically, the HeLa chromosomal proteins do not render mouse globin sequences transcribable (assayed by hybridization of chromatin transcripts with mouse globin ^{3}H-labeled cDNA).

It is well established that histone proteins are similar in different mammalian species and in different cell types of the same species. Our data would seem to suggest that the mechanism by which the transcription of histone genes in chromatin is regulated by the nonhistone chromosomal proteins in HeLa cells, WI38 cells, and mouse liver may be the same or similar. This can be accounted for by postulating

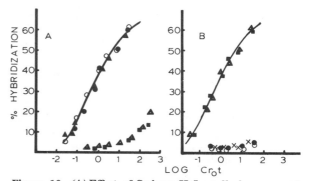

Figure 10. *(A)* Effect of S-phase HeLa cell chromosomal proteins on transcription of histone mRNA sequences from chromatin of WI38 human diploid fibroblasts. ^{3}H-labeled histone cDNA was annealed to transcripts from 1 mg of DNA as chromatin from contact-inhibited WI38 cells reconstituted with no additional chromosomal proteins ($\triangle$), 100 μg of the 0.5 M QAE-Sephadex fraction of S-phase HeLa chromosomal proteins ($\bigcirc$), or 1000 μg of total chromosomal proteins from S-phase WI38 cells ($\bullet$). ^{3}H-labeled cDNA was also annealed to transcripts from native chromatin of contact-inhibited WI38 cells ($\blacksquare$) or S-phase WI38 cells ($\blacktriangle$). *(B)* Effect of S-phase HeLa cell chromosomal proteins on transcription of histone mRNA sequences from mouse liver chromatin. ^{3}H-labeled cDNA was annealed to transcripts from 1 mg of DNA as chromatin from adult mouse liver ($\bigcirc$) and from 1 mg of adult mouse liver chromatin dissociated and reconstituted in the presence of no additional protein ($\bullet$), 1000 μg of S-phase HeLa cell total chromosomal proteins ($\triangle$), 100 μg of the S-phase HeLa cell 0.5-M QAE-Sephadex fraction ($\blacksquare$), or 100 μg of the S-phase HeLa cell 0.25-M QAE-Sephadex fraction (x).

that the DNA sequences with which certain nonhistone chromosomal proteins interact, perhaps regulatory sequences, are conserved between mouse and humans. Alternatively, the DNA sequences involved with activation of histone gene transcription may differ between mouse and humans, but both types of sequences may be recognized by the HeLa nonhistone chromosomal proteins. However, our results do illustrate that a transformed, continuously dividing cervical carcinoma cell such as HeLa contains components necessary to make the histone genes of contact-inhibited tissue culture cells or nondividing cells from an intact organism available for transcription from chromatin by *E. coli* RNA polymerase.

Acknowledgments

These studies were supported by grants from the National Science Foundation (BMS 75–18583) and the National Institutes of Health (GM 20535).

REFERENCES

ADESNIK, M. and J. DARNELL. 1972. Biogenesis and characterization of histone messenger RNA in HeLa cells. *J. Mol. Biol.* **67**: 397.

BARRETT, T., D. MARYANKA, P. HAMLYN, and H. GOULD. 1974. Nonhistone proteins control gene expression in reconstituted chromatin. *Proc. Natl. Acad. Sci.* **71**: 5057.

BASERGA, R. 1974. Nonhistone chromosomal proteins in normal and abnormal growth. *Life Sci.* **15**: 1057.

BEKHOR, I., G. KUNG, and J. BONNER. 1969. Sequence-specific interaction of DNA and chromosomal proteins. *J. Mol. Biol.* **39**: 351.

BENECKE, B. J. and K. H. SIEFART. 1975. DNA-directed RNA polymerase from HeLa cells. Isolation, characterization and cell cycle distribution of three enzymes. *Biochim. Biophys. Acta* **414**: 44.

BONNER, W. M. and R. A. LASKEY. 1974. A film detection method for tritium-labelled proteins and nucleic acids in polyacrylamide gels. *Eur. J. Biochem.* **46**: 83.

BORUN, T. W., M. D. SCHARFF, and E. ROBBINS. 1967. Rapidly labeled, polyribosome-associated RNA having the properties of histone messenger. *Proc. Natl. Acad. Sci.* **58**: 1977.

BORUN, T. W., F. GABRIELLI, K. AJIRO, A. ZWEIDLER, and C. BAGLIONI. 1975. Further evidence for transcriptional and translational control of histone messenger RNA during the HeLa S₃ cycle. *Cell* **4**: 59.

BREINDL, M. and D. GALLWITZ. 1974. On the translational control of histone synthesis. *Eur. J. Biochem.* **45**: 91.

BUTLER, W. B. and G. C. MUELLER. 1973. Control of histone synthesis in HeLa cells. *Biochim. Biophys. Acta* **294**: 481.

CHIU, J.-F., Y.-H. TSAI, K. SAKUMA, and L. S. HNILICA. 1975. Regulation of *in vitro* mRNA transcription by a fraction of chromosomal proteins. *J. Biol. Chem.* **250**: 9431.

ELGIN, S. C. R. and H. WEINTRAUB. 1975. Chromosomal proteins and chromatin structure. *Annu. Rev. Biochem.* **44**: 725.

FARQUHAR, M. and B. MCCARTHY. 1973. Histone mRNA in eggs and embryos of *Strongylocentrotus purpuratus*. *Biochem. Biophys. Res. Commun.* **53**: 515.

GABRIELLI, F. and C. BAGLIONI. 1975. Maternal messenger

RNA and histone synthesis in embryos of the surf clam *Spisula solidissima*. *Dev. Biol.* **43**: 254.

GALLWITZ, D. and G. C. MUELLER. 1969. Histone synthesis *in vitro* on HeLa cell microsomes. *J. Biol. Chem.* **244**: 5947.

GROSS, K. W., M. JACOBS-LORENA, C. BAGLIONI, and P. GROSS. 1973. Cell-free translation of maternal messenger RNA from sea urchin eggs. *Proc. Natl. Acad. Sci.* **70**: 2614.

GRUNSTEIN, M. and P. SCHEDL. 1976. Isolation and sequence analysis of sea urchin (*Lytechinus pictus*) histone H4 messenger RNA. *J. Mol. Biol.* **104**: 323.

HOSSENLOPP, P., D. WELLS, and P. CHAMBON. 1975. Animal DNA-dependent RNA polymerases. *Eur. J. Biochem.* **58**: 237.

JANSING, R. L., J. L. STEIN, and G. S. STEIN. 1977. Activation of histone gene transcription by nonhistone chromosomal proteins in WI-38 human diploid fibroblasts. *Proc. Natl. Acad. Sci.* **74**: 173.

KLEINSMITH, L. J., J. L. STEIN, and G. S. STEIN. 1976. Dephosphorylation of nonhistone proteins specifically alters the pattern of gene transcription in reconstituted chromatin. *Proc. Natl. Acad. Sci.* **73**: 1174.

LASKEY, R. A. and A. D. MILLS. 1975. Quantitative film detection of ³H and ¹⁴C in polyacrylamide gels by fluorography. *Eur. J. Biochem.* **56**: 335.

LICHTLER, A. C., G. S. STEIN, and J. L. STEIN. 1977. Isolation and characterization of two mRNAs from HeLa S₃ cells coding for histone H4. *Biochem. Biophys. Res. Commun.* **77**: 845.

MANIATIS, T., A. JEFFREY, and H. VAN DESANDE. 1975. Chain length determination of small double- and single-stranded DNA molecules by polyacrylamide gel electrophoresis. *Biochemistry* **14**: 3787.

MANS, R. and N. HUFF. 1975. Utilization of ribonucleic acid and deoxyoligomer primers for polyadenylic acid synthesis by adenosine triphosphate: Polynucleotidylexotransferase from maize. *J. Biol. Chem.* **250**: 3672.

MOSS, B., A. GERSHOWITZ, L. WEBER, and C. BAGLIONI. 1977. Histone mRNAs contain blocked and methylated 5′ terminal sequences but lack methylated nucleosides at internal positions. *Cell* **10**: 113.

PARK, W. D., J. L. STEIN, and G. S. STEIN. 1976. Activation of *in vitro* histone gene transcription from HeLa S₃ chromatin by S phase nonhistone chromosomal proteins. *Biochemistry* **15**: 3296.

PARK, W. D., R. L. JANSING, J. L. STEIN, and G. S. STEIN. 1977. Activation of histone gene transcription in quiescent WI-38 cells or mouse liver by a nonhistone chromosomal protein fraction from HeLa S₃ cells. *Biochemistry* **16**: 3713.

PAUL, J., R. S. GILMOUR, N. AFFARA, G. BIRNIE, P. HARRISON, A. HELL, S. HUMPHRIES, J. WINDASS, and B. YOUNG. 1974. The globin gene: Structure and expression. *Cold Spring Harbor Symp. Quant. Biol.* **38**: 885.

ROBBINS, E. and T. W. BORUN. 1967. The cytoplasmic synthesis of histones in HeLa cells and its temporal relationship to DNA replication. *Proc. Natl. Acad. Sci.* **57**: 409.

SARMA, M. H., E. R. FEMAN, and C. BAGLIONI. 1976. RNA synthesis in isolated HeLa cell nuclei. *Biochim. Biophys. Acta* **418**: 29.

SKOULTCHI, A. and P. R. GROSS. 1973. Maternal histone messenger RNA: Detection by molecular hybridization. *Proc. Natl. Acad. Sci.* **70**: 2840.

SPALDING, J., K. KAJIWARA, and G. MUELLER. 1966. The metabolism of basic proteins in HeLa cell nuclei. *Proc. Natl. Acad. Sci.* **56**: 1535.

STAHL, H. and D. GALLWITZ. 1977. Fate of histone messenger RNA in synchronized HeLa cells in the absence of initiation of protein synthesis. *Eur. J. Biochem.* **72**: 385

STEIN, G. S. and R. BASERGA. 1972. Nuclear proteins and the cell cycle. *Adv. Cancer Res.* **15**: 287.

STEIN, G. S. and T. W. BORUN. 1972. The synthesis of acidic chromosomal proteins during the cell cycle of HeLa S₃ cells. I. The accelerated accumulation of acidic residual nuclear proteins before the initiation of DNA replication. *J. Cell Biol.* **52**: 292.

STEIN, G. S. and J. FARBER. 1972. Role of nonhistone chromosomal proteins in the restriction of mitotic chromatin template activity. *Proc. Natl. Acad. Sci.* **69**: 2918.

STEIN, G. S. and C. L. THRALL. 1973. Uncoupling of non-histone chromosomal protein synthesis and DNA replication in human diploid WI-38 fibroblasts. *FEBS Lett.* **34**: 35.

STEIN, G. S., S. C. CHAUDHURI, and R. BASERGA. 1974a. Gene activation in WI-38 fibroblasts stimulated to proliferate. Role of nonhistone chromosomal proteins. *J. Biol. Chem.* **249**: 3918.

STEIN, G. S., T. C. SPELSBERG, and L. J. KLEINSMITH. 1974b. Nonhistone chromosomal proteins and gene regulation. *Science* **183**:817.

STEIN, G. S., W. PARK, C. THRALL, R. MANS, and J. L. STEIN. 1975. Regulation of cell cycle stage-specific transcription of histone genes from chromatin by nonhistone chromosomal proteins. *Nature* **257**: 764.

STEIN, G., J. STEIN, E. SHEPHARD, W. PARK, and I. PHILLIPS. 1977. Evidence that the coupling of histone gene expression and DNA synthesis in HeLa S₃ cells is not mediated at the transcriptional level. *Biochem. Biophys. Res. Commun.* **77**: 245.

STEIN, J. L., G. S. STEIN, and P. M. McGUIRE. 1977. Histone messenger RNA from HeLa cells: Evidence for modified 5'-termini. *Biochemistry* **16**: 2207.

STEIN, J. L., C. L. THRALL, W. D. PARK, R. J. MANS, and G. S. STEIN. 1975. Hybridization analysis of histone messenger RNA: Association with polyribosomes during the cell cycle. *Science* **189**: 557.

THOMSON, J. A., J. L. STEIN, L. J. KLEINSMITH, and G. S. STEIN. 1976. Activation of histone gene transcription by nonhistone chromosomal phosphoproteins. *Science* **194**: 428.

THRALL, C. L., A. C. LICHTLER, J. L. STEIN, and G. S. STEIN. 1977. *In vitro* synthesis of single-stranded DNA complementary to histone messenger RNAs. *Methods Cell Biology* **19**: (in press).

THRALL, C. L., W. D. PARK, H. W. RASHBA, J. L. STEIN, R. J. MANS, and G. S. STEIN. 1974. *In vitro* synthesis of DNA complementary to polyadenylated histone messenger RNA. *Biochem. Biophys. Res. Commun.* **61**: 1443.

TSAI, S. Y., M. J. TSAI, S. E. HARRIS, and B. W. O'MALLEY. 1976. Effects of estrogen on gene expression by nonhistone proteins. *J. Biol. Chem.* **251**: 6475.

WEIL, P. A. and S. P. BLATTI. 1976. HeLa cell deoxyribonucleic acid dependent RNA polymerases: Function and properties of the class III enzymes. *Biochemistry* **15**: 1500.

QUESTIONS/COMMENTS

Question by:
M. M. SMITH
The Johns Hopkins University

You have reported a three- to fourfold increase in the concentration of histone mRNA sequences in nuclear RNA following incubation of G₂-phase nuclei in a synthesis reaction containing nucleotides. From our experience with nuclear transcription in vitro, it is of interest to us that your nuclei can synthesize sufficient RNA in vitro to result in an increased concentration of histone mRNA sequences above endogenous in vivo levels in the nuclei. Could you tell us specifically the following: (1) What is the concentration of histone mRNA sequences in unincubated nuclear RNA? (2) What is the concentration of histone mRNA sequences in incubated nuclear RNA? (3) How much RNA do your nuclei contain per nucleus? (4) How much RNA is synthesized in vitro per nucleus, and thus (5) what is the concentration of histone mRNA sequences in the in vitro transcribed RNA required to account for the increased levels of histone mRNA seen following incubation.

Response by:
G. STEIN and J. STEIN
University of Florida

It should be pointed out that these experiments were carried out with S-phase nuclei.

(1) The concentration of histone mRNA sequences in unincubated nuclei of S-phase HeLa cells is 0.13% of the total RNA (6.5×10^{-3} pg/nucleus).

(2) The concentration of histone mRNA sequences in incubated nuclei of S-phase HeLa cells is 0.4% of the total RNA (2.0×10^{-2} pg/nucleus).

(3) The amount of RNA in our S-phase nuclear preparation is approximately 5 pg/nucleus.

(4) The S-phase nuclei synthesize approximately 0.17 pg of RNA per nucleus during the 45-minute incubation period.

(5) Based on the above figures, ~ 8% of the newly transcribed RNA can be accounted for by histone mRNA sequences. This result is consistent with our results from in vitro transcription of histone mRNA sequences from S-phase chromatin.

Question by:
   M. ZASLOFF
   National Institutes of Health

Response by:
   G. STEIN and J. STEIN
   University of Florida

Comment by:
   G. PIECZENIK
   Rutgers University

Have you measured the histone RNA content of a transcript from chromatin reconstituted with DNA, such as of prokaryotic origin, lacking histone gene sequences?

No.

There appears to be an interesting paradox arising in the work presented in this session. The degree of individual variation observed for histone nucleotide sequences is much greater than the variation observed between species. It suggests a convergence of genotype yet a divergence of phenotype.

Fine Structure and Evolution of DNA in Heterochromatin

W. J. Peacock, A. R. Lohe, W. L. Gerlach, P. Dunsmuir,* E. S. Dennis, and R. Appels

Division of Plant Industry, Commonwealth Scientific and Industrial Research Organization, Canberra, A.C.T., Australia

Heterochromatin was recognized in the eukaryote chromosome some 50 years ago (Heitz 1929). This original definition contrasted heterochromatin to euchromatin in terms of heteropycnosis, heterochromatin remaining in a deeply staining, condensed state during the entire mitotic cycle. Subsequent research has shown other differences in that this ubiquitous chromosome component generally replicates later than the euchromatic regions of the genome, and in general shows no evidence of transcriptional activity. The existence of obvious heterochromatic polymorphisms in many species, and the observations in *Drosophila* that gross deficiencies of heterochromatin are tolerated with little or no phenotypic effects, have led to the erroneous conclusion that it is a genetically inert region of the chromosome.

In recent years it has become clear that specific genetic properties are attributable to heterochromatic regions of chromosomes and that the different segments of heterochromatin in a genome may have different properties. The most telling data stem from genetic analyses in *Drosophila melanogaster* and maize. In *Drosophila,* Cooper (1964) has shown that some segments of heterochromatin of the X and Y chromosomes are important for meiotic conjunction in spermatocytes, whereas other heterochromatin segments in these same chromosomes play no such role. The heterochromatic Y chromosome is also known to control several functions essential to male fertility (Brosseau 1960). Deletion of much of the X-chromosome heterochromatin has pronounced effects on pairing and disjunction of the X and Y chromosomes and also results in severe disturbances of spermatid differentiation (Peacock et al. 1975). Autosomal heterochromatin has also been shown to harbor elements controlling vital genetic functions (Hilliker 1976). In a series of elegant cytogenetic analyses, Rhoades and Dempsey (see Rhoades 1977) have shown that different heterochromatic regions of the maize genome are responsible for a number of specific genetic effects, including levels of recombination, chromosome segregation, and chromatin elimination.

A molecular dimension to heterochromatin was provided when in situ hybridizations showed that highly repeated DNA sequences were predominantly located in this form of chromatin (Pardue and Gall 1970; Jones and Robertson 1970). Many subsequent studies in a variety of organisms have confirmed this correlation but have tended to foster a concept of uniformity of heterochromatin in any given organism. In this paper we present data, primarily from *Drosophila,* to show that the heterochromatin of each chromosome has a unique, segmental identity, and that DNA sequences in heterochromatin have, as do DNA sequences in euchromatin, defined patterns of conservation and change during evolution. We show that the properties discovered in *Drosophila* apply to other eukaryotes, including plants and mammals.

Highly Repeated DNA Sequences in
Drosophila melanogaster

DNA isolated by buoyant density procedures from nuclei obtained from 8-hour embryos of *D. melanogaster* yields highly repeated sequence species accounting for approximately 20–25% of the total genome (Peacock et al. 1974, 1977; Brutlag et al. 1977). This amount of highly repeated sequence DNA is 2–3 times greater than earlier estimates based either on reassociation kinetic analysis or on yields of satellite or highly repeated DNA fractions (Rae 1970; Botchan et al. 1971; Wu et al. 1972). The difference in yield is explicable in terms of three major factors: (1) Highly repeated DNA sequences are grossly underrepresented in polytene nuclei, which form a significant component of the nuclear population in larvae and adult flies. (2) Simple sequence DNA can be preferentially lost in extraction procedures using organic solvents. (3) The thermal stabilities of different sequence species can differ so widely that any particular regime used in reannealing experiments may not enable all sequences to be recovered.

Six distinct density satellites (in addition to mainband DNA), of densities 1.672, 1.688, 1.697, and 1.705 g/cc, are detected in analytical CsCl analysis of fractions taken from preparative actinomycin D-CsCl gradients (Brutlag et al. 1977; Peacock et al. 1977). Each satellite is separable from the others by centrifugation in gradients in which another antibiotic, netropsin sulfate, having a DNA binding specificity different to actinomycin D, is included (Brutlag et

* Present address: Biological Laboratories, Harvard University, Cambridge, Massachusetts 02138.

al. 1977). Four of these satellites, the 1.672 g/cc, 1.686 g/cc, 1.688 g/cc, and 1.705 g/cc density species, have been analyzed extensively. Sequence analysis has revealed the primary repeating sequence of nucleotides in the 1.672 g/cc, 1.686 g/cc, and 1.705 g/cc satellites (Brutlag and Peacock 1975; Endow et al. 1975). Two sequences were detected in the 1.672 g/cc satellite after rigorous purification in antibiotic CsCl and alkaline CsCl gradients, a pentamer $\left(\begin{smallmatrix} 5'AATAT \\ TTATA\ 5' \end{smallmatrix}\right)$ and a septamer $\left(\begin{smallmatrix} 5'\ AATATAT \\ TTATATA\ 5' \end{smallmatrix}\right)$, and these "isomers" have been shown to be in separate, long, tandem arrays (Brutlag and Peacock 1975). The pentamer tracts are homogeneous, with GC substitutions absent over stretches of several hundred nucleotides, but interspersed GC pairs are present in low frequency in the septamer arrays.

Alternate sequence forms were also detected in the 1.686 g/cc satellite (Brutlag and Peacock 1975), but since 90% of the cRNA to a purified single-strand template is digested with T_1 RNase to a decamer sequence, this DNA approximates to a poly-$\left(\begin{smallmatrix} 5'\ AATAACATAG \\ TTATTGTATC\ 5' \end{smallmatrix}\right)$ molecule.

The 1.705 g/cc satellite has a pentamer repeat $\left(\begin{smallmatrix} 5'\ AAGAG \\ TTCTC\ 5' \end{smallmatrix}\right)$ as its major component (Endow et al. 1975), and this sequence may extend uninterrupted for tracts of average lengths of 750 base pairs (Birnboim and Sederoff 1975). Brutlag and Peacock (1975) detected a decamer sequence $\left(\begin{smallmatrix} 5'\ AAGAAGAGAG \\ TT\ CTTCTCTC\ 5' \end{smallmatrix}\right)$ in the satellite, and a septamer repeating unit $\left(\begin{smallmatrix} 5'\ AAGAGAG \\ TTCTCTC\ 5' \end{smallmatrix}\right)$ has been described by Brutlag et al. (this volume) as a minor component in *Mbo*II digests of the 1.705 satellite.

The occurrence of the pentamer and septamer repeats in both the 1.672 g/cc and 1.705 g/cc satellites stresses the relatedness of the sequences of the three satellites; a general formula of $(AAN)_m (AN)_n$ is applicable to the 1.672, 1.686, and 1.705 g/cc satellite sequences. A close sequence relationship between satellites was first detected in *D. virilis* by Gall and Atherton (1974), who suggested that the relation could reflect common origin and/or sequence selection imposed by functional requirements within the nucleus. However, the fourth satellite examined in *D. melanogaster*, the 1.688 g/cc species, does not fit to the formula presented above. This satellite was known to have a greater complexity than the three discussed above. Brutlag et al. (1977) determined a complexity of 234 base pairs from reassociation kinetics, and a repeating unit of 365 base pairs has been detected by restriction enzyme analysis (Manteuil et al. 1975). Sequence analysis has now shown that a predominant repeat sequence in the satellite is $\left(\begin{smallmatrix} 5'\ TTTCC \\ AAAGG\ 5' \end{smallmatrix}\right)$, which is not simply related to other satellite sequences (Brutlag et al., this volume). Certainly, in other species where several satellites have been analyzed, there is no simple relationship evident between their sequences (Southern 1972; Salser et al. 1976; Altenburger et al. 1977).

The remaining two satellites have not been analyzed in detail. The 1.690 g/cc satellite contrasts with the other highly repeated sequences in that it binds more actinomycin D than main-band sequences and is presumably rich in $\left(\begin{smallmatrix} GC \\ CG \end{smallmatrix}\right)$ dinucleotides. It has a buoyant density increase of 0.002 g/cc after reassociation in 0.2 M Na$^+$; further details of its sequence or physical characteristics have not been examined. The 1.697 g/cc species is not homogeneous. A highly repeated component does not show any change in T_m after reassociation in 0.2 M Na$^+$, but a significant proportion of the satellite has a heterogeneous density profile after reassociation. Furthermore, upon shearing, the amount of DNA reassociating to a density of 1.697 g/cc is drastically reduced, emphasizing complexity within this satellite species (Peacock et al. 1977). The component with a heterogeneous reassociation profile presumably contains the ribosomal genes and spacer repeating units since the ribosomal genes can be shown by hybridization analyses to be present in this satellite. The satellite demonstrates a buoyant density shift when reannealed in the presence of an excess of ribosomal RNA (Fig. 1). The extent of the density shift is not very great, and we interpret this to be due to aggregate formation involving a simple sequence, which results in a low RNA:DNA ratio. A proportion of the ribosomal genes (26S assayed here) are included in the aggregate. The satellites 1.686 and 1.688 show no density shift although, as discussed later, they are present in the nucleoli of polytene nuclei. We are currently isolating the highly repeated DNA sequences from this satellite.

The amount of each of the isolated density satellites has been determined from yields in preparative gradients, and the genome equivalents of the 1.672, 1.686, 1.688, and 1.705 satellites was determined by both saturation hybridization and competition hybridization procedures (Brutlag et al. 1977). These estimations are shown in the last column of Table 1; from the known amounts of the two classes of rDNA repeats in the genome (Glover and Hogness 1977), it can be calculated that approximately 40% of the 1.697 g/cc density species is attributable to the ribosomal genes and spacer repeating units.

The yields of the 1.672, 1.686, and 1.705 satellites in antibiotic-CsCl gradients show only minor perturbations when compared between DNA populations of different mean molecular weight (Goldring et al. 1975). Using this data, Brutlag et al. (1977) have calculated that the average size of regions of the 1.705 g/cc sequence, largely poly$\left(\begin{smallmatrix} AAGAG \\ TTCTC \end{smallmatrix}\right)$, is 750,000 base pairs and that approximately ten regions occur in the genome. Regions of comparable, or even greater, length apply to the 1.672 g/cc and 1.686 g/cc satellites. The 1.688 g/cc and 1.690 g/cc have not been analyzed in this way due to complexi-

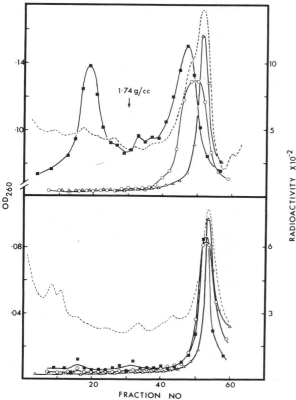

Figure 1. Density shift of 1.697 sequences after the formation of rRNA-DNA hybrids. The methods for isolating DNA from 6–12-hr *D. melanogaster* embryos and its fractionation in antibiotic-CsCl gradients have been described in detail by Brutlag et al. (1977). The DNA fraction from an actinomycin D-CsCl gradient containing most of the ribosomal genes (see Fig. 2) was denatured and renatured in the presence of an excess of ribosomal RNA (rRNA contained both 18S and 28S rRNA, ratio of RNA:DNA was 10:1) in 0.2 M Na$^+$ to a $C_0 t$ of 1.0. Following RNase treatment, the renatured material was phenol-extracted, the phenol removed by ether, dialysed overnight against 10 mM Tris-HCl, pH 8.4, 1 mM EDTA and centrifuged to equilibrium in CsCl of initial density 1.74 g/cc. Fractions from the gradients were then assayed for 28S rDNA, 1.697 (in the presence of excess rRNA), and 1.688 satellite sequences by the nitrocellulose filter technique of Birnstiel et al. (1972). (- - -) OD$_{260}$, (○) 1.697, (△) 1.688, (■) 28S rDNA. *(Top)* DNA renatured in presence of rRNA; *(bottom)* DNA renatured without added rRNA.

ties imposed by their densities in antibiotic gradients, but their yield characteristics indicate that they, too, must have a long, tandem array organization of their component repeats.

Another feature of the highly repeated DNA sequence species that has emerged from gradient analyses and that is pertinent to their chromosomal organization is evidence of covalent attachment of different classes of highly repeated DNA (Goldring et al. 1975; Brutlag et al. 1977). In DNA of mean molecular length approximately 4 kb, molecules can be detected which contain two of the satellite species (e.g., 1.705 g/cc and 1.672 g/cc). One type of experiment used to demonstrate these covalent attachments involves hybridization analysis across

Table 1. Percentage Distribution of Satellite DNA Species on a Chromosomal Basis

DNA species	X	Y	2	3	4	?[a]	Recovered yield (%)[b]
1.672[c]	5	68	1	4	22		4.2
1.686	7	50	24	18	1		3.3
1.688	48	44	4	3	1		4.6
1.705	5	52	33	9	1		4.7
1.697	46	38	10	6	—		4.5
1.690						100	3.0

[a] No in situ data available for this satellite sequence.

[b] Approximations of genome content of the satellites obtained from preparative gradients, saturation hybridization, and/or competitive hybridization (see text).

[c] There is evidence to indicate that the pentamer and septamer isomers present in this satellite do not have identical chromosomal distributions as discussed in detail by Peacock et al. (1977). The figures shown here refer to the pentamer isomer. The content of 1.672 in the genome refers to both the pentamer and septamer (in the calculations for Table 3, the genome content of the pentamer is assumed to be half of this on the basis of data presented by Brutlag and Peacock [1975]).

buoyant density gradients and indicates that complementary sequences to a highly purified probe sequence (one of the satellites) can be detected at uncharacteristic density position (Fig. 2). These junction molecules have also been detected in other ways. For example, isolation of highly repeated DNA

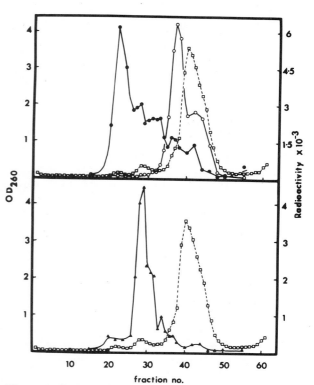

Figure 2. Covalent linkage of satellites. *D. melanogaster* DNA was fractionated in an actinomycin D-CsCl gradient and fractions were assayed for sequences complementary to 1.705 and 1.672 satellites and ribosomal RNA (18S and 28S). The hybridization technique used has been described by Birnstiel et al. (1972). *(Top)* (□) OD$_{260}$, (●) 1.705 sequences, (○) ribosomal gene sequences. *(Bottom)* (▲) 1.672 sequences, (□) OD$_{260}$.

from unfractionated DNA by the use of the hydroxylapatite technique (Britten et al. 1974) under conditions which should prevent the recovery of the 1.672 sequences as a renatured species (renaturation conducted above the melting point (T_m) of this species) still results in the recovery of 1.672 sequences which band at a density of 1.705 g/cc in CsCl gradients. The single-stranded length of DNA used in this experiment was 1.5 kb, so that in a limited number of sites in the genome, the 1.705 and 1.672 repeat sequences must occur on the same 1.5-kb fragment of DNA. The detailed sequence structure of the satellite junction molecules cannot be determined from the available data, but it will be possible to do this by using cloned fragments. The principal conclusion from these data is that the long arrays of tandem repeats of the various satellites are likely to be in close proximity in the genome.

Chromosomal Locations of the Satellites in
D. melanogaster

Early in situ hybridization experiments in *D. melanogaster* showed highly repeated sequence DNA to be localized in the chromocenter of polytene chromosomes in salivary gland cells (Rae 1970; Botchan et al. 1971; Gall et al. 1971), and Rae (1970) also found one banded region of one chromosome arm that showed a positive autoradiogram. Detailed chromosomal analyses of satellite locations have become possible with refinements of the in situ technique and the development of a suitable method for preparing mitotic metaphases from larval neuroblasts. These cells permit the analysis to be made in specific, cytologically defined stocks and provide a means of placing cytogenetic coordinates to the in situ map. The technique and results are reported in detail for the 1.705 and 1.672 satellites by W. J. Peacock et al. (in prep.), and preliminary locations for five satellites have been summarized elsewhere (Peacock et al. 1977). The satellites of the *D. melanogaster* genome are particularly suited to an in situ analysis because appropriate hybridization conditions can be chosen to obviate cross-contamination. Although there is some variability in the in situ technique, even within areas on one slide, we have found that it provides a quantitative and precise hybridization procedure. Using freshly prepared slides, we have determined relative chromosomal contents of a particular sequence (within the range of linear response of the emulsion) by in situ hybridization and have produced thermal dissocation profiles identical to those obtained from filter hybridization experiments using the same conditions (Peacock et al. 1977). Conditions for accurate hybridization were determined for each satellite by a series of filter experiments before proceeding with the in situ analysis.

Examples of hybridization of the 1.705 g/cc and

1.672 g/cc cRNA probes to neuroblast metaphases from a male larva are shown in Figure 3a, together with diagrams of their chromosomal locations (Fig. 3b). Such diagrams are prepared following examination of a large number of cells from several experiments and often incorporate information obtained from polytene nuclei hybridization. Quantitative data on amounts of the satellite at different sites is determined from shorter exposures than that shown in Figure 3, before the photographic emulsion is saturated. In some experiments, quantitation of the smaller sites is obtained with longer exposures than used for the major sites, and then compared to the major sites assuming that the emulsion has a linear response with time (Peacock et al., in prep.).

A summary of the quantitative distribution of five of the satellites is given in Table 1. The proportional values are calculated from grain counts over the sites on each chromosome of mitotic metaphase preparations from male neuroblasts and are calculated on a single chromosome basis. In all cases the locations of the highly repeated DNA sequences complementary to the probes were restricted to the pericentric heterochromatin of the three autosomes, the proximal heterochromatin of the X, and the entire heterochromatic Y chromosome. A composite distribution map is given in Table 2.

In addition to the heterochromatic sites, analysis of polytene chromosomes reveals one minor euchromatic site for the 1.688 g/cc sequence at the tip of the polytene X chromosome (Fig. 4a). Three satellites (1.672, 1.686, and 1.705) also give a positive autoradiogram on band 21D 1,2 on the left arm of chromosome 2 (Fig. 4b). The 1.705 g/cc probe hybridized to the 21D 1,2 region only under conditions of lower fidelity of base pairing (Peacock et al. 1977). The only other euchromatic location is a band in region 81 on the right arm of chromosome 3 which hybridized the 1.672 sequence. The map (Table 2) does not contain any rigorous indication of the juxtapositions of sequences in given regions. Stocks containing chromosomal rearrangements will be used to provide this information. The locations of the sequences associated with the nucleolus organizer region on both the X and Y chromosomes has been checked by hybridization to the nucleoli of polytene cells of larvae with genotypes constructed such that the only nucleolus organizer regions present were either from the X chromosome or the Y chromosome alone (e.g., normal females for X organizers, sc^4sc^8/Y males for Y chromosome organizer). A series of X chromosome inversions has been used to determine the left terminus of the block of 1.705 g/cc satellite, which is located near the centromere of the X chromosome — we find it to be proximal to the sc^4, sc^{L8}, sc^8, and sc^{S1} breakpoints but distal to the sc^{V2} breakpoint (Lindsley and Grell 1968).

A detailed comparison of the maps of different wild-type strains of *D. melanogaster* has not been made but preliminary experiments have not re-

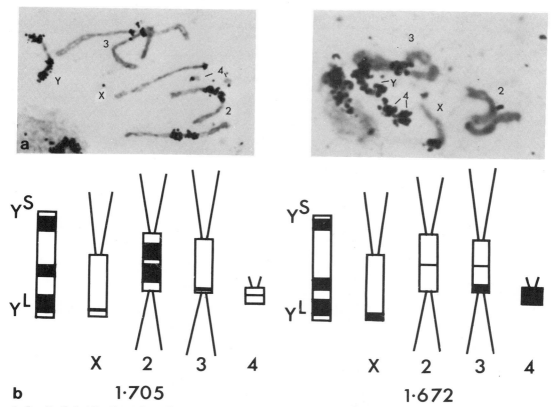

Figure 3. In situ hybridization of satellite sequences to mitotic chromosomes. *(a)* Autoradiographic localization of 1.705 and 1.672 to neuroblast metaphase chromosomes. *(b)* Diagrammatic representation of the distribution of sequences within the heterochromatin of the chromosomes. The localization of the sequence is indicated by the filled areas; the euchromatin is represented as a line only.

Salivary glands and neural tissue from *D. melanogaster* larvae (3rd instar) were used to obtain squashes of polytene and mitotic chromosomes, respectively. Tissues were squashed in 45% acetic acid after a preliminary treatment in a lactic acid:acetic acid (1:1) solution for 5 min. To accumulate metaphase chromosomes, the neural tissue was incubated for 15 min in a 0.05% colchicine solution, followed by 15 min in 1% sodium citrate prior to treatment in fixative. Chromosome squashes were dehydrated in absolute alcohol and air dried before use.

Chromosomal DNA was denatured by incubation in 0.2 N HCl at 37°C for 30 min, then dehydrated through an alcohol series and air dried. A 5-μl sample of ³H-labeled cRNA, synthesized from purified DNA templates (using *E. coli* RNA polymerase [holo] prepared as described by Burgess and Jendrisak 1975) in 3 × SSC, 50% formamide (10⁴–10⁵ cpm/μl) was placed on each slide, covered with an 18-mm² cover slip and sealed with rubber gum. The preparations were heated to 70°C for 3 min to ensure that all chromosomal sites were available for hybridization and then incubated for 3 hr at the appropriate hybridization temperature (determined by nitrocellulose filter hybridization experiments). Cover slips were removed and slides were washed twice in 3 × SSC, 50% formamide (10 min each wash, at the hybridization temperature), twice in 2 × SSC (10 min each) before RNase treatment (10 μg/ml Sigma Pancreatic RNase, 1.25 units/ml Calbiochem T₁ RNase in 2 × SSC, 30 min at 37°C). Following further washes in 2 × SSC (6 washes, 10 min each), the slides were ethanol dehydrated, air dried and coated with Ilford K2 emulsion (50% dilution).

vealed any differences in Oregon R when compared with the Canberra wild type.

Using the chromosomal distribution values and the genome content of each particular satellite (Table 1), it is possible to calculate the content of the highly repeated sequences in the heterochromatin of each chromosome. The 8-hour embryos used in the DNA extractions are diploid, with equal numbers of males and females, and therefore each autosome will be represented four times and the X chromosome three times for every representation of the Y chromosome. Therefore, the calculations of the chromosomal values have been transformed according to the formula, $3 \cdot X + 1 \cdot Y + 4 \cdot Chr2 + 4$ $\cdot Chr3 + 4 \cdot Chr4 = DNA$ pool. The haploid genome was assumed to contain 170,000 kb DNA with the X chromosome containing 39,000 kb and chromosomes 2, 3, and 4 containing 50,000 kb, 65,000 kb and 6,000 kb, respectively. These figures are based on the cytophotomeric measurements of Rasch et al. (1971) and Rudkin (1964). In the absence of direct determinations, it has been assumed that the Y chromosome contains an amount of DNA equivalent to the X chromosome (39,000 kb) even though their relative cytological lengths are quite variable in the different stages of mitosis. The transformed satellite DNA values are given in Table 3.

The major points shown by the data in Tables

Table 2. Composite Map of Satellite Distribution in the Mitotic Chromosomes of *D. melanogaster*

X Chromosome		Y Chromosome		Chromosome 2		Chromosome 3		Chromosome 4
Centromere region	1.672 1.686 1.705	short arm of Y	1.672 1.705	euchromatin-heterochromatin junction region	1.697 1.705 1.672	euchromatin-heterochromatin junction region	1.697	1.705[a] 1.672[a] 1.688[a]
Nucleolus organizer region	1.688 1.686 1.697	nucleolus organizer region	1.697 1.686 1.688	centromere region	1.688 1.686 1.672 1.705	centromere region		1.688 1.686 1.672 1.686
Heterochromatin-euchromatin junction region	1.686	centromere region	1.697 1.705 1.697 1.672 1.686	heterochromatin-euchromatin junction region	1.697	heterochromatin-euchromatin junction region		1.705
		long arm of Y	1.705 1.688 1.672					

[a] No region specified.

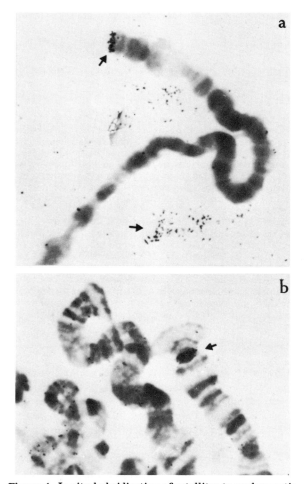

Figure 4. In situ hybridization of satellites to euchromatic regions. Polytene chromosomes are required to show the euchromatic location of satellite sequences. *(a)* The 1.688 sequences are shown by autoradiography to be localized at the tip of the X chromosome and the nucleolus (arrows indicate sites). *(b)* Sequences related to 1.705 are shown to be located at 21D 1,2 (arrow indicates site).

1–3 are as follows: (1) Each satellite is present in only a few locations in the genome, this finding being concordant with the molecular data cited in the preceding section. (2) The different satellites are all located, almost entirely, in the heterochromatic regions of the genome; the heterochromatic regions are, therefore, domains rich in highly repeated sequences, whereas the euchromatin is almost devoid of these sequences. (3) No satellite is restricted to any one chromosome, but each chromosome has a quantitatively and qualitatively unique segmental makeup of its heterochromatin. (4) In the X chromosome, for which there is the best estimate of heterochromatic DNA content, the calculated amounts of satellite blocks, together with the ribosomal cistrons, argue that few other DNA sequences are located in the heterochromatin. The other chromosomes also have a predominant highly repeated sequence DNA component in their heterochromatin, but no definite conclusions can be made as to the likely extent of other sequences becaue direct measurements of DNA content of the heterochromatin are not available.

Highly Repeated DNA Sequences in *D. simulans*

To compare the distribution and types of satellites found in *D. melanogaster* with a closely related species, a study of highly repeated DNA sequences in *D. simulans* has been carried out. The comparison extends the generality of the observations made in the preceding section and allows conclusions about the evolution of this type of DNA to be made.

A combination of antibiotic-CsCl centrifugation and melting point analysis of fractions from gradients has defined eight major satellites in *D. simulans* of neutral CsCl densities 1.672, 1.686, 1.694, 1.695, 1.696, 1.697, and 1.707 (contains two sequences) g/cc (Lohe 1977). Chromosomal locations

Table 3. Distribution of Satellite DNA in *D. melanogaster* (kb)

DNA species	X	Y	2	3	4	?
1.672[a]	380	5140	75	300	1650	
1.686	640	4620	2220	1660	91	
1.688	6830	6260	570	430	560	
1.697	4000[c]	3400[c]	1300	765	—	
1.705	660	6820	4320	1180	130	
1.690						
rDNA	2800	2200				(5100)
Total	15,310	28,440	8490	4340	2430	(5100)
Est. het.[b]	13,000–16,000	39,000	8000–15,000	13,000–16,000	3000–4500	

[a] Data refer to the pentamer isomer of this satellite (see legend to Table 1).

[b] The value for the X chromosome is based on the observations of Rudkin (1964) in his comparisons of the sc^4sc^8 and sc^8 chromosomes. The lower limit does not take cognizance of the heterochromatin remaining in the sc^4sc^8 chromosome. We feel the upper value is more probably correct. The value for the Y chromosome is an assumption based on the cytology of the X and Y (see text). The values of the autosomes are also based on cytological measurements of the proportions of heterochromatin and euchromatin in mitotic metaphase and prometaphase chromosomes.

[c] The 1.697 values have been corrected to allow for the observation that 66% of the ribosomal genes are isolated with the satellite (Fig. 2).

for four of these (1.672, 1.695, 1.696, and 1.707 g/cc species) have been determined and are predominantly in heterochromatin. The 1.695 and 1.696 satellites are also found in the nucleoli of polytene nuclei from salivary glands. As in *D. melanogaster,* each satellite is found on more than one chromosome, and the heterochromatin of each of the *D. simulans* chromosomes has a characteristic array and amount of satellite DNA sequences.

Even though sequence information is not available for most of the *D. simulans* satellites, their buoyant densities in neutral CsCl and in antibiotic-CsCl gradients (Peacock et al. 1977) show that these two sibling species have markedly different highly repeated DNA complements, a situation reported for other taxonomic groups (see review by Walker 1971). There is some overlap possible between the complements since both species have satellites with buoyant densities of 1.672 g/cc and 1.686 g/cc. However, the 1.686 g/cc satellites do not cross hybridize and have different alkaline CsCl properties. On the other hand, the 1.672 g/cc satellites do have at least some identical sequences. The pentamer repeat $\left(\begin{smallmatrix} 5' & AATAT \\ & TTATA & 5' \end{smallmatrix}\right)$ is present in long arrays in both species, as evidenced by identical thermal stabilities of heterologous and homologous hybrids.

Hybridization analyses with other probes have shown that the two species, despite their different profiles of major satellites, do have many highly repeated sequences in common. Thus, satellites such as the 1.695 and 1.696 g/cc species which are present in *D. simulans* and apparently absent in buoyant density profiles of *D. melanogaster* were detected in the latter by both nitrocellulose filter hybridization and in situ hybridization experiments. Similarly, major *D. melanogaster* satellites can be detected in small amounts, but with apparently identical sequences as judged by thermal stability of hybrids, in *D. simulans.* Although minor components, the sequences still occur in tandem arrays: for example, hybridization of a 1.695 g/cc probe from the *D. simulans* genome (approximately 3400 kb of this sequence per genome) to an actinomycin gradient of *D. melanogaster* DNA detects a peak of this sequence (approx. 340 kb) at the density position expected on the basis of its location in a *D. simulans* gradient.

The interspecific comparison has included in situ hybridization analyses. An example in which cytologically differentiated cells of the two species were mixed and hybridized with a probe is shown in Figure 5. The grain counts show that *D. melanogaster* has 4.3 times more than the amount of 1.705 sequences occurring in *D. simulans,* a factor in excellent agreement with the ratio of 3.9:1 determined by filter competition hybridization. This suggests that the in situ procedure is detecting most, if not all, of the sites of the 1.705 sequence in the *D. simulans* genome and confirms the conclusion made above that the sequences are in segmental arrays. The chromosome data show three blocks of 1.705 sequences on the *D. simulans* Y chromosome in positions similar to the three sites on the *D. melanogaster* Y chromosome. However, each site is relatively reduced in *D. simulans,* the extent of the interspecific difference being different at each site. A dramatic difference is seen in chromosome 2, which contains 52% of the 1.705 sequence in *D. melanogaster,* but where only long exposures give any evidence of a site in *D. simulans.* Thus, although there is sequence and probably chromosome site conservation, large scale modulation of sequence frequency occurs and to different extents at given sites. These same characteristics apply to the other highly repeated sequences that have been examined (1.672, 1.686, 1.688 and 1.697 from *D. melanogaster;* 1.695

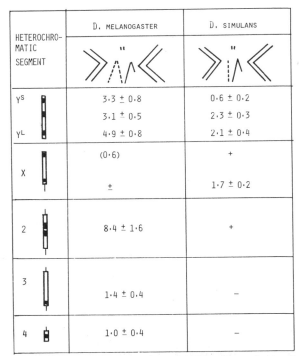

Figure 5. In situ hybridization of 1.705 g/cc probe to *D. melanogaster* and *D. simulans*. Larval neuroblast cells of *D. melanogaster* females (attached X/Y) and *D. simulans* males (X/Y) were intermixed, squashed, and hybridized with cRNA prepared from a *D. melanogaster* 1.705 g/cc satellite template. Grain counts were made at low grain density while there was still linear emulsion response on *D. melanogaster* chromosome 2. The X chromosome value is calculated using the data of Table 1.

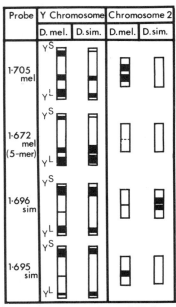

Figure 6. Interspecific comparison of Y chromosome and chromosome 2 locations of four satellites. Diagram of heterochromatin locations of two *D. melanogaster* and two *D. simulans* highly repeated sequence probes. The site bars are proportional to intraspecific, but not interspecific, sequence contents.

and 1.696 from *D. simulans*). These results will be reported in detail elsewhere, but the examples given in Figure 6 show that each satellite has a similar site pattern in the two species, despite there being large interspecific differences in sequence frequency. In a particular experiment, competition hybridization measurements indicated that the *D. melanogaster:D. simulans* content for 1.672 (pentamers only), 1.695, and 1.696 are 3.8%:1.7%, 2.0%:0.2%, and 1.6%:less than 0.1%, respectively. The diagram reflects the relative site sizes within species but is not calibrated to show the large interspecific differences.

The comparison of *D. simulans* and *D. melanogaster* satellites has emphasized that these sibling species show different arrays of *major* satellite sequences, but that the differences are of a quantitative nature and the species have a common representation of highly repeated sequences.

Distribution of Satellite DNA in the Chromosomes of Cereal Plants

In asking whether the *Drosophila* "rules" for heterochromatin apply in other eukaryotes, we have investigated the highly repeated sequence DNA of cereal plants. The few available data for plants are conflicting in that although a correspondence of highly repeated sequences with heterochromatin has been shown for *Scilla* (Timmis et al. 1975a), it is claimed to be lacking in the cereal, rye (Timmis et al. 1975b). We have examined rye (Imperial rye, *Secale cereale*, $2n = 14$) and two other cereals, wheat and barley.

Two different, highly repeated sequence probes have been isolated from Imperial rye. In situ hybridization analysis has shown that one of the probes, a rapidly renaturing (C_0t 0.02) fraction isolated by hydroxylapatite chromatography and containing a mixture of highly repeated sequences, is predominantly located in the large blocks of heterochromatin that occur near the telomeres of the rye chromosomes. Grain counts from RNA-excess hybridizations show that 80% of the highly repeated sequences of the rye genome are located in these distal heterochromatic segments, a finding which contradicts the earlier report of Timmis et al. (1975b). The remaining grains appear to be scattered over the chromosomes, but probably their distribution reflects a pattern of segments of much smaller dimensions than those in the telomeric regions. For example, the other probe we have studied, long pyrimidine tracts isolated according to Birnboim et al. (1975), has major chromosomal locations at interstitial sites on three of the rye chromosomes (R. Appels and W. J. Peacock, in prep.) but represents only a minor component of the total highly repeated DNA.

Unique chromosome identities are given by the

segmental patterns obtained by combining these two probes. We expect that a more elaborate, segmental differentiation, equivalent to that in *Drosophila* heterochromatin, will emerge when the DNA species constituting the C_0t 0.02 probe are mapped individually.

A satellite DNA has been isolated in Ag^+-Cs_2SO_4 buoyant density gradients from hexaploid wheat, *Triticum aestivum* var. Chinese Spring ($2n = 42$, genome constitution AABBDD). This particular variety has been chosen because a number of cytogenetic tools are available for it; these include defined aneuploid stocks and lines containing individual rye chromosomes. Renaturation studies show that the satellite is a highly repeated sequence with a complexity for the repeat of 5–10 base pairs. This has been verified by digests of ^{32}P-labeled cRNAs made from separated strands of the satellite. The localization of this satellite by in situ hybridization (Fig. 7a) shows it to have major sites on nine chromosomes, with smaller amounts of the sequence on other chromosomes. Telocentric tester stocks, in which each metacentric chromosome in turn is replaced by two telocentric chromosomes produced by centric fission, have been used to identify the distinctively labeled chromosomes. All seven B-genome chromosomes and two A-genome chromosomes (Fig. 7b) have unique segmental placements of the probe sequence.

A satellite DNA has also been isolated from Clipper barley in Ag^+-Cs_2SO_4 density gradients. It is located by in situ autoradiography at discrete sites on all chromosomes, the sites being predominantly pericentromeric. Each chromosome is differentiated from the others by the size and distribution of the sequence blocks. This satellite has a closely related, if not identical, sequence to the wheat satellite DNA: (a) both native and renatured satellites have the same thermal stabilities and buoyant densities; (b) they show the same renaturation kinetics in 0.03 M Na^+; (c) the complementary strands can be separated and have the same buoyant densities in alkaline CsCl gradients; and (d) a cRNA made from the barley satellite as template hybridizes to wheat DNA and this heterologous hybrid has the same thermal stability as the homologous cRNA-barley DNA hybrid.

As in the *Drosophila* interspecies comparison, the quantities of this conserved, repeating sequence differs markedly in the two plant species. In wheat the satellite accounts for 0.3% of the genome, whereas in barley this same sequence represents 4% of the genome.

Thus, our studies of highly repeated DNA sequences in cereal plants reinforce the conclusions reached from the studies of satellite DNAs in *Drosophila* species, in that (a) particular satellite DNAs are located at major chromosomal sites on more than one chromosome of the complement and individual chromosomes appear to have distinctive segmental patterns of highly repeated sequences; and (b) a satellite DNA sequence appears to have been stringently conserved during evolution, the two species compared being from different taxonomic subtribes.

Segmental Arrays in Satellite DNA of a Marsupial

The individuality of the heterochromatin of a particular chromosome shown by the distribution of satellite DNA sequences in *Drosophila* and cereal plants is an aspect of the structure of heterochromatin which may appear not to apply in species such as *Mus musculus*. In this case only one major satellite DNA species can be identified by buoyant density analysis; it is located on all chromosomes except the Y (Pardue and Gall 1970; Jones and Robertson 1970) and accounts for the bulk of the heterochromatin. This satellite is heterogeneous when examined by restriction enzymes (Hörz et al. 1974; Southern 1975; Hörz and Zachau 1977), raising the possibility that certain subpopulations of DNA within this satellite may show chromosome specificity. An analogous situation has been found in the marsupial species *Macropus rufogriseus* (red-necked wallaby). This species has two satellites which are located in the centromeric heterochromatin of each autosome (Dunsmuir 1976). Most chromosomes can be distinguished in terms of their content of these highly repeated sequences but the quantitative differences are small. A detailed analysis of the major satellite (1.708 g/cc), accounting for 20% of the genome, is discussed here with the view of strengthening the possibility that chromosome-specific, segmental arrays of sequence repeats occur generally in eukaryotes.

Restriction enzyme digestion of the 1.708 satellite gives various sorts of patterns (Fig. 8). The enzymes can be divided into two classes; first are those which do not differentiate any subpopulations in the satellite DNA (*Bam*H and *Pst*).

Cleavage of purified 1.708 g/cc satellite by *Bam*H results in a series of DNA segments in a monomer, dimer, trimer, etc., series based on a monomer of 2500 base pairs. Simultaneous hybridization to monomer and dimer DNA with differentially labeled cRNAs prepared against monomer and dimer templates shows that the two segment lengths are derived from the same sequence population. Furthermore, relative frequencies of different fragments are consistent with the assumption that the larger segments result from random inactivation of the *Bam*H sites by mutation or modification. Other restriction enzymes also cleave the satellite DNA with segments of 2500 base pairs as the unit length. *Pst* generates almost all monomers with only a small population of dimers and higher multimers. Double digestion experiments with *Bam*H and *Pst*, or the subsequent digestion of *Bam*H monomers with *Pst*, show that there is only one *Bam*H and one *Pst* site per monomer and that these sites are 450 base pairs apart.

Partial digests with either of these two enzymes show up to 11-mers, and subsequent complete digests

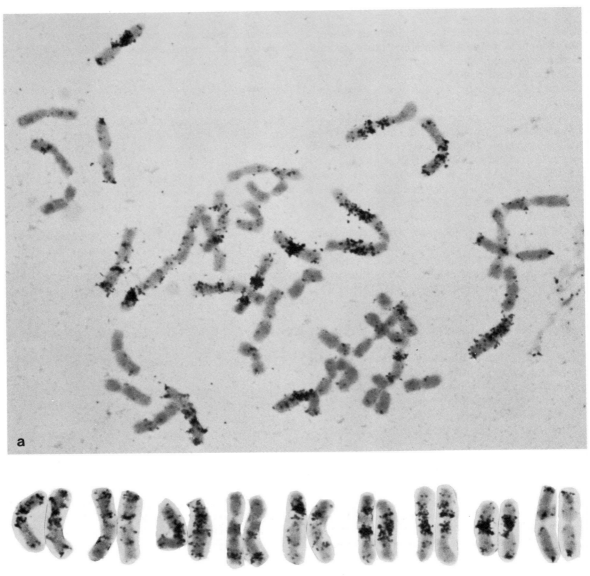

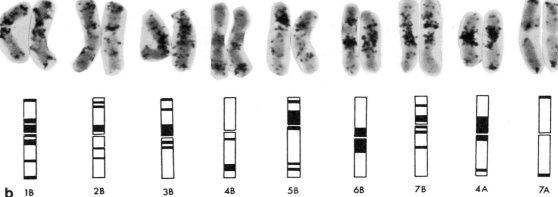

Figure 7. Location of wheat satellite in wheat chromosomes. *(a)* A complete metaphase spread of *Triticum aestivum* var. Chinese Spring $(2n = 42)$ showing the location of the satellite recovered from Ag^+-Cs_2SO_4 gradients by autoradiography. *(b)* A karyotype of the chromosomes in *a* with identifications being determined using ditelocentric tester stocks.

Cereal DNA was isolated according to R. Appels and W. J. Peacock (in prep.) and fractionated in Ag^+-Cs_2SO_4 buoyant density gradients as originally described by Corneo et al. (1968). For analytical ultracentrifugation, 250 μl native wheat DNA (120 μg/ml in TE) was mixed with 150 μl 25 mM borate buffer (50 mM Na_2SO_4, 25 mM $Na_2B_4O_7$, pH 9.1). While vortexing, $AgNo_3$ (either 0.5×10^{-2} M or 0.5×10^{-3} M) was then added dropwise until the desired R_F was achieved (R_F = mole ratio of silver ions to DNA phosphate). Samples were then made to a final volume of 650 μl with H_2O and Cs_2SO_4 (Merck) added to densities of 1.48–1.53 g/cm^3. For preparative isolation of satellite DNA, similar conditions to those showing separation of satellite from main band DNA in analytical ultracentrifugation were used. Fractions containing satellite DNA were located by absorbance at 260 nm, pooled, and dialyzed against 0.01 M Tris-HCl, pH 8.4, 0.001 M EDTA. It was found that DNA isolated in this way was contaminated with material which showed a strong absorbance at 250 nm. This was overcome by ethanol precipitation of the DNA from a 5 M perchlorate solution.

Root tip metaphases from cereals were accumulated by treating 3-day-old seedlings grown at 27°C with 0.25% colchicine for 3 hr prior to fixation in 1:3 acetic:ethanol. Root tip squashes were prepared in 45% acetic acid. In situ hybridization was carried out as detailed in the legend to Fig. 3.

of multimers from the partial digests demonstrate that monomer and higher order lengths are interspersed with each other.

The second class of enzymes differentiate the 1.708 DNA into different classes (*Xma, HindIII, EcoRI, HaeII, HhaI*) (see Fig. 8). These enzymes do not

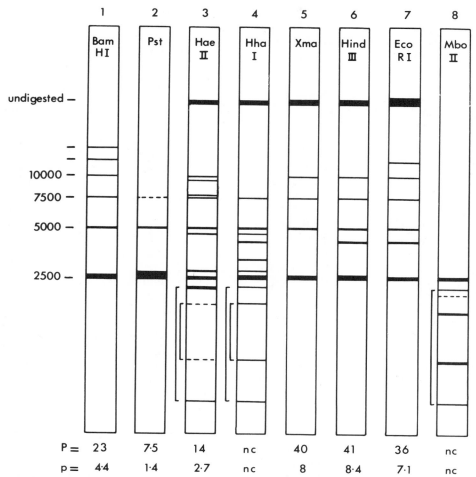

Figure 8. Summary of restriction endonuclease patterns from a marsupial satellite. Patterns obtained by digestion of *M. rufogriseus* 1.708 g/cc satellite DNA digested with eight restriction endonucleases. The patterns are qualitative with regard to the amounts present in each band although these were accurately determined for the calculation of P and p values. The latter were calculated only from the digested portion of DNA as described by Southern (1975); nc indicates no meaningful value could be calculated because of the complexity of the pattern. Brackets indicate bands which together add to form a monomer length.

DNA was isolated and fractionated as described by Dunsmuir (1976) except that only one preparative actinomycin D-CsCl gradient was required when fractions were collected from the top of the gradient and the speed of centrifugation was 39,000 rpm instead of 44,000 rpm.

EcoRI, PstI, HindIII, HaeII, and *HhaI* were purchased from BioLabs. *Bam*HI and *Hpa*II were gifts from Dr. Bentley Atchison. *Sma* was purchased from Boehringer Mannheim. *Xma* was a gift from Dr. R. Herrmann and *Mbo*II a gift from Dr. D. L. Brutlag.

The buffers used for the digestions were as follows (all included 6 mM β-mercaptoethanol except that used for *EcoRI*): *Bam*HI, 6 mM Tris-HCl, pH 7.4, 6 mM MgCl₂; *EcoRI*, 100 mM Tris-HCl, pH 7.5, 150 mM NaCl, 5 mM MgCl₂, 100 μg/ml gelatin; *HhaI*, 6 mM Tris-HCl, pH 7.4, 50 mM NaCl, 6 mM MgCl₂, 100 μg/ml BSA; *HaeII*; 6 mM Tris-HCl, pH 7.4, 10 mM MgCl₂, 6 mM KCl; *Xma* and *Sma,* 30 mM Tris-HCl, pH 7.5, 20 mM NaCl, 5 mM MgCl₂; *Pst,* 6 mM Tris-HCl, pH 7.4, 50 mM NaCl, 6 mM MgCl₂, 100 μg/ml gelatin; *HindIII,* 6 mM Tris-HCl, pH 7.4, 50 mM NaCl, 5 mM MgCl₂, 100 μg/ml BSA; *Mbo*II, 10 mM Tris-HCl, pH 7.5, 7 mM MgCl₂, 60 mM KCl.

Digested DNA was analyzed in agarose gels which were prepared by boiling Sigma (analytical) or Sea Plaque (preparative) agarose in running buffer (40 mM Tris, 20 mM Na acetate, 2 mM EDTA, pH 8.4). On cooling to 60°C, ethidium bromide (0.5 μg/ml) was added. Concentrations of agarose varied from 0.8–1.5%. Electrophoresis was for 3–4 hr at a constant current of 4 mA/gel. Running buffer contained 0.5 μg/ml ethidium bromide. Sometimes ethidium bromide was omitted and the gel stained in 1 μg/ml ethidium bromide. The gels were illuminated with a UV light (Mineralight UVS-11) and photographed on Polaroid 105 sheet film with a Kodak yellow filter (#15). Negatives were scanned on a Joyce Loebl microdensitometer.

Note: The band moving faster than 5000 bp in channels 6 and 7 (estimated size 4100 bp) was shown not to hybridize with a [³H]cRNA complementary to purified *Bam*HI monomers. This result suggests that the band is due to another highly repeated DNA sequence copurified with the 1.708 satellite.

cleave the DNA uniformly. In all cases, a multimeric series is produced but some DNA remains undigested; neither increased digestion times nor additional enzyme has any effect on the resistant population. Restriction of the satellite with HaeII or HhaI, in addition to showing the 2500-bp series, displays segments that result from an additional site in some of the monomer fragments.

An example of the clear demonstration of segmental differentiation within the satellite DNA population is given in patterns 5 and 6 of Figure 8 from digests with Xma and HindIII, respectively. Both of these enzymes detect sites, regularly spaced, in only some of the satellite molecules, Xma leaving approximately 40% of the DNA undigested and HindIII leaving approximately 60% undigested. Double digests of the total DNA population and sequential digests, with the second enzyme challenging the restricted and spared populations of the first digest, have shown that the respective restricted populations are nonoverlapping. Thus, the HindIII monomers do not contain any Xma sites and vice versa, indicating that the populations containing Xma and HindIII sites have a specific arrangement with respect to each other.

Partial digests with both HindIII and Xma yield molecules consisting of decamers which, with further digestion, demonstrate that monomers, dimers, and trimers are interspersed. Since decamers are equivalent to the lengths of DNA molecules used in the experiments, it is clear that lengths of DNA of at least 25 kb are differentiated by HindIII and Xma. A tandem repeat pattern of the different monomer classes is emphasized by the disclosure, by HindIII digestion of BamH monomers, that there are four different monomers with repeat to the position of a HindIII site along the 2500-bp segment. Since HindIII digestion of the total satellite population yields the 2500-bp segment series, it follows that each HindIII monomer class occurs in long, tandem arrays, separate from the other monomer types. Thus, Xma and HindIII define six sub-populations within the satellite, and further segmental differentiations must exist; for example, Xma sites are found in the 1950-bp band of HaeII but not in the monomer (2500 bp) band.

For each of the subpopulations of the 1.708 g/cc satellite, it must be assumed that a process, as yet undefined, generated a localized tandem array from a unit segment variant. It is clear from the frequencies of monomer types that at least some occur on more than one chromosome. We might expect, also, that these different subpopulations will show independent frequency modulation through evolution, just as the sequence variants do in Drosophila.

Differentiation within the 2500-bp monomer. A composite map of the monomers can be constructed from double enzyme digests and from second enzyme digests of the BamH monomer (Fig. 9). It shows that the enzyme cleavage sites that are close to each other do not necessarily have the same pattern of distribution within the satellite DNA population; e.g., the all-pervasive Pst site is within 100 bp of the HaeII site, which has limited representation. The restriction patterns have not shown any indication of an internal repeat substructure to the monomer although, since reannealed monomer forms large aggregates, the presence of shorter repeating segments seems likely. This satellite may have evolved in the complex manner inferred by Brutlag et al. (this volume) for the Drosophila 1.688 g/cc DNA, rather than by the stepwise duplication scheme proposed by Southern (1975) for the mouse satellite.

Reference to the values calculated for the probabilities of site inactivation and the monomer map (Figs. 8 and 9) does indicate that regions of the monomer may be more prone to divergence than others. In the MboII digest (column 8, Fig. 8) the monomer is the largest segment in the digest, the series of fragments smaller than the monomer reflecting additional MboII sites within the same monomer. The absence of segments larger than the monomer indicates that at least one MboII site has a low probability of inactivation; this site may be in a region of the monomer that must be conserved for some basic recognition function. It may be a general rule that long repeat sequences will show relatively conserved and relatively variable sequence regions.

DISCUSSION

Segmental Organization of Heterochromatin

Our results show that heterochromatin has a sequence fine structure with respect to the disposition of the various highly repeated DNA species occurring in the genome. Each chromosome is characterized by its segmental pattern of the satellite DNAs, and no satellite is restricted to any one chromosome. These facts apply to D. melanogaster and to its sibling species, D. simulans, which has a different

Figure 9. Restriction endonuclease map of monomer repeating unit. The monomer repeating unit of the 1.708 g/cc *Macropus rufogriseus* satellite defined by *Bam*HI contains several other restriction endonuclease sites. The map summarizes the relative positioning of a number of these sites; *not* all sites occur in the same DNA segment.

complement of major satellite DNAs. In effect, heterochromatin is differentiated along a chromosome.

Highly repeated DNA sequences certainly constitute the bulk of heterochromatin, and the identification of "junction" molecules shows that different sequence segments can be in close proximity. However, other sequence classes, single-copy or moderately repeated DNA, may be interspersed with long segments of highly repeated sequences — our data do not exclude this possibility. In fact, the demonstration of functional genetic loci, in low density, in the proximal heterochromatin of both arms of chromosome 2 of *D. melanogaster* (Hilliker 1976) and the autoradiographic demonstration of transcribed sequences in the chromocenter of salivary gland nuclei (Spradling et al. 1975) suggest that sequences other than the satellites do occur in heterochromatin.

The in situ studies of highly repeated sequences in the cereal plants, eukaryotes far removed from *Drosophila* in a phylogenetic sense, indicate that the properties of multichromosome distribution with unique chromosome patterning are likely to be generally true of heterochromatin. Even where a single buoyant density satellite accounts for a major amount of a genome's heterochromatin, restriction enzyme analyses in a number of species have shown that the satellite should actually be considered as a family of related, highly repeated DNA species. Thus the heterogeneity demonstrated in the 1.708 g/cc satellite of *Macropus* contrasts with the apparent homogeneity as judged by its buoyant density and initial melting profiles, but is consistent with its characteristics observed after reannealing (Dunsmuir 1976). Since the sequence heterogeneity is not random but exists in several defined populations, it is possible that these segments are distributed in a nonrandom way amongst chromosomes, as are the satellite DNA species in *Drosophila* and the cereals. The genetic stocks of wheat which contain one particular rye chromosome (all seven stocks exist) should permit the restriction enzyme-type analysis to be extended to a study of chromosome pattern for any complex satellite of the rye genome.

Evolution of Heterochromatin Sequence Structure

The analyses of the complements of highly repeated DNA sequences of *D. melanogaster* and *D. simulans* show that the length of tandem arrays of a repeat, rather than its nucleotide sequence, is the evolutionary variable in heterochromatin structure. The cereal plants also show frequency modulation for a conserved, highly repeated sequence. In *Macropus,* the restriction analysis gives no obvious clues as to the sequence evolution of the 2500-bp unit, but it emphasizes that localized segmental amplification of sequence variants was an important feature in the development of the present-day satellite.

Thus any model for the evolution of satellite DNA must encompass a mechanism for amplifying a DNA segment into a longer (or reducing it to a shorter) tandem array of these segments. Unequal sister chromatid exchange (Smith 1974, 1976) and saltatory replication (Southern 1970; Walker 1971) represent general classes of such mechanisms, but we are unaware of any present data which give conclusive evaluations of these or alternative processes. In the *Drosophila* interspecific comparison, the data suggest that, whereas the extent of amplification is a site-specific process, sites often show a coordinate direction of change: for example, the data show the Y chromosome and the autosomes to have parallel differences, though the differences on the Y tend to be less than those in the autosomes. Further data on other species are needed to determine if the Y chromosome does in fact have a relative conservatism in its sequence content.

Sturtevant and Novitski (1941) pointed out the evolutionary integrity of chromosome "elements," showing that in a wide range of *Drosophila* species, the bands (and gene markers) of any one chromosome arm tended to remain together as a unit, even though their arrangement within the unit might be substantially altered by inversions. Our observations suggest that, similarly, heterochromatin has an evolutionary integrity but with an important difference; whereas in euchromatin the evolutionary variable is the primary nucleotide sequence, the variable in heterochromatin is the amount (number of copies) of a primary sequence unit (Fig. 10). Our data also show a conservation in chromosomal sites of particular sequences in the heterochromatin, however we would only be able to detect major inversions; in the euchromatin the polytene chromosomes provide high resolution of rearrangement detection.

The probable sequence identity, in the two *Drosophila* species, of at least seven highly repeated DNAs show that satellite sequences are not as free to accumulate mutations as was previously thought. Comparisons of heterologous and homologous single copy sequences from these two species show easily identifiable sequence divergence (Laird and McCarthy 1968), while the satellites display no measurable sequence changes in similar analyses. Our data con-

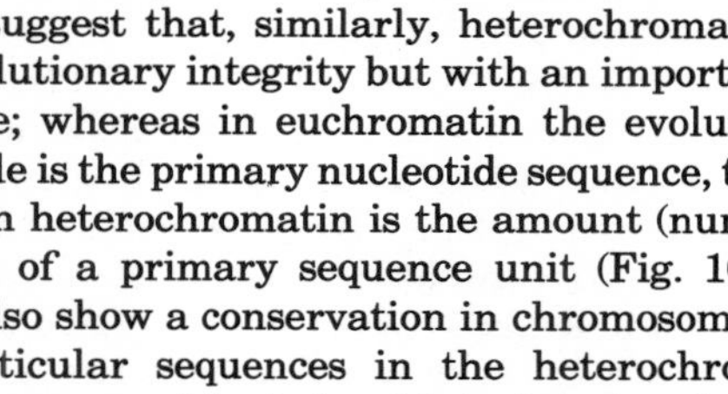

EUCHROMATIN		HETEROCHROMATIN	
SPECIES 1	SPECIES 2	SPECIES 1	SPECIES 2
A	A	A	A
B ↕	C	B <	B
C ↕	B	C >	C

Figure 10. Evolution of euchromatin and heterochromatin. A, B, and C represent unit sequences — bands (genes) in euchromatin, and segments of tandem arrays of a highly repeated DNA sequence in heterochromatin. The arrow signifies inversions in the euchromatin and arrowheads represent sequence amplification in the heterochromatin.

form to the "library" hypothesis of Salser et al. (1976), which postulates that satellite sequences are preserved over long evolutionary times rather than arising de novo during speciation. Thus, apart from a mechanism for segmental amplification, we need a process for maintaining sequence homogeneity. This latter problem is compounded in that it applies to situations where a particular sequence is a minor component, as well as where it is of major representation.

And finally, any evolutionary scheme for heterochromatin has to provide the means for the establishment of segments of identical sequences on several different chromosomes.

Functions of Heterochromatin

It was stressed in the Introduction that heterochromatin is known to have a range of genetic effects, with different segments of heterochromatin having different properties. Our data have shown that satellite DNAs are arranged so as to differentiate blocks of heterochromatin; however, there are no data which unequivocally put highly repeated DNA sequences into a biological context. It is possible that some or all of the functions of heterochromatin are attributable to the *arrangement* of the highly repeated sequences with *other* DNA sequences that may be interspersed or associated with them, rather than being due to the highly repeated sequences per se. A putative example is that the same satellites are located near the ribosomal RNA cistrons in the nucleolus organizer regions of both the X and Y chromosomes of *D. melanogaster*.

Nevertheless, the observations of chromosome-specific distribution of sequences, of disjunct multisite occurrence of sequence species, and of sequence conservation argue strongly for functional roles of highly repeated sequence DNA in the molecular structure and evolution of heterochromatin.

Acknowledgments

We wish to thank Professor C. Driscoll and Dr. K. Shepherd for supplying stocks of cereal plants and for many helpful discussions. Janice Norman and Georgina Koci were indispensable, and the assistance of Kay Faulkner in preparing *Drosophila* material for in situ hybridization is gratefully acknowledged. We thank Yvonne Hort for her help with the marsupial satellite analysis, and we wish to especially thank Professor D. M. Steffensen for his stimulating presence in the laboratory while part of the work was being undertaken. A. R. L. is a research student, R. S. B. S., and W. L. G. a Rothmans fellow, R. S. B. S., at Australian National University, Canberra.

REFERENCES

ALTENBURGER, W., W. HORZ, and H. G. ZACHAU. 1977. Comparative analysis of three guinea pig satellite DNAs by restriction nucleases. *Eur. J. Biochem.* **73**: 393.

BIRNBOIM, H. C. and R. SEDEROFF. 1975. Polypyrimidine segments in *Drosophila melanogaster* DNA. I. Detection of a cryptic satellite containing polypyrimidine/polypurine DNA. *Cell* **5**: 173.

BIRNBOIM, H. C., N. A. STRAUS, and R. R. SEDEROFF. 1975. Characterization of polypyrimidines in *Drosophila* and L-cell DNA. *Biochemistry* **14**: 1643.

BIRNSTIEL, M. L., B. H. SELLS, and I. F. PURDOM. 1972. Kinetic complexity of RNA molecules. *J. Mol. Biol.* **63**: 21.

BOTCHAN, M., R. KRAM, C. W. SCHMID, and J. E. HEARST. 1971. Isolation and chromosomal localization of highly repeated DNA sequences in *Drosophila melanogaster*. *Proc. Natl. Acad. Sci.* **68**: 1125.

BRITTEN, R. J., D. E. GRAHAM, and B. R. NEUFELD. 1974. Analysis of repeating DNA sequences by reassociation. *Methods Enzymol.* **29**: 363.

BROSSEAU, G. E., JR. 1960. Genetic analysis of the male fertility factors on the Y chromosome of *Drosophila melanogaster*. *Genetics* **45**: 257.

BRUTLAG, D. L. and W. J. PEACOCK. 1975. Sequences of highly repeated DNA in *Drosophila melanogaster*. In *The eukaryote chromosome* (ed. W. J. Peacock and R. D. Brock), p. 35. Australian National University Press, Canberra.

BRUTLAG, D. L., R. APPELS, E. DENNIS, and W. J. PEACOCK. 1977. A highly repeated DNA in *Drosphila melanogaster*. *J. Mol. Biol.* **112**: 31.

BURGESS, R. R. and J. J. JENDRISAK. 1975. A procedure for the rapid, large-scale purification of *Escherichia coli* DNA-dependent RNA polymerase involving polymin-P precipitation and DNA-cellulose chromatography. *Biochemistry* **14**: 4634.

COOPER, K. W. 1964. Meiotic conjunctive elements not involving chiasmata. *Proc. Natl. Acad. Sci.* **52**: 1248.

CORNEO, G., E. GINELLI, S. SOAVE, and G. BERNARDI. 1968. Isolation and characterization of mouse and guinea pig satellite deoxyribonucleic acids. *Biochemistry* **7**: 4373.

DUNSMUIR, P. 1976. Satellite DNA in the kangaroo *Macropus rufogriseus*. *Chromosoma* **56**: 111.

ENDOW, S. A., M. L. POLAN, and J. G. GALL. 1975. Satellite DNA sequences of *Drosophila melanogaster*. *J. Mol. Biol.* **96**: 665.

GALL, J. G. and D. D. ATHERTON. 1974. Satellite DNA sequences in *Drosophila virilis*. *J. Mol. Biol.* **85**: 633.

GALL, J. G., E. H. COHEN, and M. L. POLAN. 1971. Repetitive DNA sequences in *Drosophila*. *Chromosoma* **33**: 319.

GLOVER, D. M. and D. S. HOGNESS. 1977. A novel arrangement of the 18S and 28S sequences in the long and short repeating units of *Drosophila melanogaster* rDNA. *Cell* **10**: 177.

GOLDRING, E. S., D. L. BRUTLAG, and W. J. PEACOCK. 1975. Arrangement of the highly repeated DNA of *Drosophila melanogaster*. In *The eukaryote chromosome* (ed. W. J. Peacock and R. D. Brock), p. 47. Australian National University Press, Canberra.

HEITZ, E. 1929. Heterochromatin, chromocentren, chromomeren. *Der. Deats. Bot. Gay.* **47**: 274.

HILLIKER, A. J. 1976. Genetic analysis of the centromeric heterochromatin of chromosome 2 of *Drosophila melanogaster*: Deficiency mapping of EMS-induced lethal complementation groups. *Genetics* **83**: 765.

HÖRZ, W. and H. G. ZACHAU. 1977. Characterization of distinct segments in mouse satellite DNA by restriction nucleases. *Eur. J. Biochem.* **73**: 383.

HÖRZ, W., I. HESS, and H. G. ZACHAU. 1974. Highly regular

arrangement of a restriction nuclease sensitive site in rodent satellite DNAs. *Eur. J. Biochem.* **45**: 501.

JONES, K. W. and F. W. ROBERTSON. 1970. Localization of reiterated nucleotide sequences in *Drosophila* and mouse by in situ hybridization of complementary RNA. *Chromosoma* **31**: 331.

LAIRD, C. D. and B. J. McCARTHY. 1968. Magnitude of interspecific nucleotide sequence variability in *Drosophila*. *Genetics* **60**: 303.

LINDSLEY, D. L. AND E. H. GRELL. 1968. *Genetic variations of* Drosophila melanogaster. Carnegie Institution of Washington Publication No. 627.

LOHE, A. R. 1977. "Highly repeated sequence DNA in *Drosophila simulans:* Chromosomal organization and evolutionary stability." Ph.D. thesis, Australian National University, Canberra.

MANTEUIL, S., D. H. HAMER, and C. A. THOMAS. 1975. Regular arrangement of restriction sites in *Drosophila* DNA. *Cell* **5**: 413.

PARDUE, M. L. and J. G. GALL. 1970. Chromosomal localization of mouse satellite DNA. *Science* **168**: 1356.

PEACOCK, W. J., G. L. G. MIKLOS, and D. J. GOODCHILD. 1975. Sex chromosome meiotic drive in *Drosophila melanogaster* chromosome. *Genetics* **79**: 613.

PEACOCK, W. J., R. APPELS, P. DUNSMUIR, A. R. LOHE, and W. L. GERLACH. 1977. Highly repeated DNA sequences: Chromosomal localization and evolutionary conservatism. In *International cell biology 1976–1977* (ed. B. K. Brinkley and K. R. Porter), p. 494. Rockefeller University Press, New York.

PEACOCK, W. J., D. L. BRUTLAG, E. GOLDRING, R. APPELS, C. W. HINTON, and D. L. LINDSLEY. 1974. The organization of highly repeated DNA sequences in *Drosophila melanogaster* chromosomes. *Cold Spring Harbor Symp. Quant. Biol.* **38**: 405.

RAE, P. M. M. 1970. Chromosomal distribution of rapidly reannealing DNA in *Drosophila melanogaster*. *Proc. Natl. Acad. Sci.* **67**: 1018.

RASCH, E. M., H. J. BARR, and R. W. RASCH. 1971. The DNA content of sperm of *Drosophila melanogaster*. *Chromosoma* **33**: 1.

RHOADES, M. M. 1977. Genetic effects of heterochromatin in maize. In *1975 International Maize Symposium: Genetics and breeding* (ed. David B. Walden). Wiley Interscience, New York. (In press.)

RUDKIN, G. T. 1964. The structure and function of heterochromatin. In *Genetics today,* p. 359. Pergamon Press, Oxford.

SALSER, W., S. BOWEN, D. BROWNE, F. E. ADLER, N. FEDEROFF, K. FRY, H. HEINDELL, G. PADDOCK, G. POON, B. WALLACE, and P. WHITCOME. 1976. Investigation of the organization of mammalian chromosomes at the DNA sequence level. *Fed. Proc.* **35**: 23.

SMITH, G. P. 1974. Unequal crossover and the evolution of multigene families. *Cold Spring Harbor Symp. Quant. Biol.* **38**: 507.

———. 1976. Evolution of repeated DNA sequence by unequal crossover. *Science* **191**: 528.

SOUTHERN, E. M. 1970. Base sequence and evolution of guinea pig satellite DNA. *Nature* **227**: 794.

———. 1972. Repetitive DNA in mammals. *Symp. Med. Hoechst* **6**: 19.

———. 1975. Long range periodicities in mouse satellite DNA. *J. Mol. Biol.* **94**: 51.

SPRADLING, A., S. PENMAN, and M. L. PARDUE. 1975. Analysis of *Drosophila* mRNA by in situ hybridization sequences transcribed in normal and heat shocked cultured cells. *Cell* **4**: 395.

STURTEVANT, A. H. and E. NOVITSKI. 1941. The homologies of the chromosome elements in the genus *Drosophila*. *Genetics* **26**: 517.

TIMMIS, J. N., B. DEUMLING, and J. INGLE. 1975a. Localization of satellite DNA sequences in nuclei and chromosomes of two plants. *Nature* **257**: 152.

TIMMIS, J. N., J. INGLE, J. SINCLAIR, and R. NEIL JONES. 1975b. The genomic quality of rye B chromosomes. *J. Exp. Bot.* **26**: 367.

WALKER, P. M. B. 1971. "Repetitive" DNA in higher organisms. In *Progress in biophysics and molecular biology* (ed. J. A. V. Butler and D. Noble), vol. 23, p. 145. Pergamon Press, Elmsford, New York.

WU, J. R., J. HURN, and J. BONNER. 1972. Size and distribution of the repetitive segments of the *Drosophila* genome. *J. Mol. Biol.* **64**: 211.

DNA Sequence Organization in *Drosophila* Heterochromatin

D. Brutlag, M. Carlson, K. Fry, and T. S. Hsieh

Department of Biochemistry, Stanford University, Stanford, California 94305

Drosophila melanogaster is an ideal organism for the study of the structure and function of heterochromatin because it has only four chromosomes and 25% of its genome is heterochromatic. All of the heterochromatin is constitutive and much of it is located in the sex chromosomes, which have been well characterized genetically and cytogenetically (Cooper 1959; Ashburner and Novitski 1976). The study of *Drosophila* heterochromatin is also attractive at the molecular level since the bulk of the DNA of these regions can be isolated as several discrete satellite DNAs in CsCl equilibrium gradients (Peacock et al. 1974; Endow et al. 1975; Brutlag et al. 1977a). These unusual DNAs consist of short nucleotide sequences (5–378 base pairs [bp]) repeated in tandem arrays over 1,000,000 bp long (Goldring et al. 1975; Brutlag et al. 1977a).

The biological roles of satellite DNA and heterochromatin have long been controversial. The molecular properties of satellite DNA make it unlikely that it is involved in transcription or the normal regulation of genetic expression. The misconception that satellite DNA has no essential function is based on the lack of genes in heterochromatin and on the viability of individuals with large heterochromatic deletions. However, individual *Drosophila* males carrying deletions of X heterochromatin have abnormal meiosis, defective spermatogenesis, and a markedly reduced fertility (Gershenson 1933; Sandler and Braver 1953; Peacock et al. 1975). These germ-line aberrations are directly correlated with the failure of deficient X chromosomes to pair properly during meiosis. Recent evidence shows that females carrying heterochromatic deletions also have a defective meiotic mechanism resulting in a reduced level of recombination (Yamamoto and Miklos 1977). These results indicate a strong selective pressure against chromosomes carrying heterochromatic deletions. They also argue strongly for a role of heterochromatin in the germ line rather than in somatic cells. Heterochromatin, therefore, may be dispensable for the survival of a cell or an individual, but it is essential for the survival of a chromosome in the germ line. A germ-line function is also consistent with the large variations of heterochromatin and satellite DNA between closely related species (Hennig and Walker 1970; Sutton and McCallum 1972; Gall et al. 1974). Indeed, the necessity for proper meiotic pairing of heterochromatin for successful gametogenesis or recombination suggests that variations in satellite DNA could lead to speciation.

To begin an analysis of these meiotic functions at the molecular level, we have studied the organization of the satellite DNA sequences that compose the bulk of the heterochromatin of *D. melanogaster*. We describe here the properties of two simple-sequence satellites (1.705 and 1.672 g/cm³) and of a complex satellite (1.688 g/cm³) which has been located primarily in the regions of the X and Y chromosomes essential for proper meiotic pairing (Gershenson 1940; Peacock et al., this volume).

Simple-sequence Satellite DNAs of *D. melanogaster*

Figure 1 shows a summary of the highly repeated DNA sequences present in the heterochromatin of *D. melanogaster*. Each of these satellite DNA species constitutes 3–5% of the *Drosophila* genome; altogether, 19% of the haploid genome is highly repeated (Brutlag et al. 1977a). The three simple-sequence satellites (1.672, 1.686, and 1.705 species) have been isolated in a very-high-molecular-weight form, which suggested that chromosomal regions containing a single satellite sequence may be larger than 1000 kb (Goldring et al. 1975). The physical homogeneity of each of the purified simple-sequence satellites has been demonstrated by neutral CsCl gradients, thermal denaturation, renaturation kinetics, and alkaline CsCl gradients which separate the complementary strands (Brutlag et al. 1977a).

The repeating units of each satellite have been determined by sequence analysis of either pyrimidine tracts or RNA complementary to the satellite DNA (cRNA). The sequence poly$\left(\begin{smallmatrix} AAGAG \\ TTCTC \end{smallmatrix}\right)$ has been reported for the 1.705 satellite by Sederoff and Lowenstein (1975) from pyrimidine tract analysis and by Endow et al. (1975) from cRNA. Birnboim and Sederoff (1975) showed that pyrimidine tracts isolated from 1.705 DNA averaged 750 nucleotides in size. Transversion base substitutions must therefore be very rare in this DNA. The major repeating units of the 1.686 satellite were determined primarily from cRNA, and the arrangement of these three sequences with respect to each other is not known

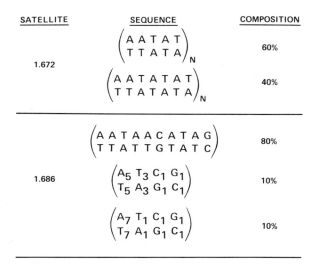

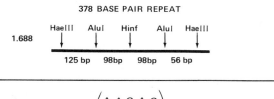

SATELLITE	SEQUENCE	COMPOSITION

Figure 1. Sequences of highly repeated heterochromatic DNAs of *D. melanogaster*. The original sources for these sequences are: 1.672, Peacock et al. (1974); 1.672 and 1.686, Brutlag and Peacock (1975); 1.686 and 1.705, Endow et al. (1975); 1.705, Sederoff and Lowenstein (1975); 1.688, Carlson and Brutlag (1977); 1.705 and 1.688, this work.

(Fig. 1) (Brutlag and Peacock 1975; Endow et al. 1975).

Many results suggested a predominantly alternating arrangement of A and T in the 1.672 DNA (Fansler et al. 1970; Peacock et al. 1974; Endow et al. 1975). However, both physical and template properties of the separated complementary strands of the 1.672 DNA indicated that this satellite actually contained two distinct DNA species with the closely related sequences $poly\left(\begin{smallmatrix} AATAT \\ TTATA \end{smallmatrix}\right)$ and $poly\left(\begin{smallmatrix} AATATAT \\ TTATATA \end{smallmatrix}\right)$ (Brutlag and Peacock 1975). The efficient replication of the 5-bp repeat, $poly\left(\begin{smallmatrix} AATAT \\ TTATA \end{smallmatrix}\right)$, by DNA polymerase in the complete absence of dGTP and dCTP indicates that very few GC substitutions have occurred within this sequence. Thus the simple-sequence satellites, 1.672 and 1.705, both seem to be very homogeneous in their repeating units.

The physical properties of the two 1.672 DNAs containing the 5-bp or 7-bp repeat are so similar

as to make them inseparable by classical physical procedures. Moreover, the finding of two such closely related sequences within a single homogeneous satellite suggested to us that the sequence variation observed in many satellites may result from mixtures of sequences, sequences so similar that they might even cross-hybridize (Sutton and McCallum 1972; Blumenfeld 1974). We therefore developed procedures for cloning individual molecules of satellite DNA in recombinant plasmids to resolve individual sequences and to study sequence variations within a single molecule of satellite DNA.

Cloning Reveals Two DNA Species in the 1.705 Satellite

In our attempts to resolve satellite DNA into individual cloned molecules, we encountered instability of regions containing tandem repeats even in recombination-deficient bacteria (Brutlag et al. 1977b). Regions of satellite DNA in hybrid plasmids are often heterogeneous in length (Fig. 2). Subcloning bacteria containing the hybrid plasmids or transforming other bacteria with a single hybrid plasmid DNA molecule eliminates the heterogeneity, but it is often regenerated upon bacterial growth. These results show that DNA repeated in tandem may be subject to rearrangements or recombinational instability within *recA* bacteria. Nevertheless, the cloned satellite DNA in these plasmids still hybridizes with satellite DNA from *Drosophila*, and we show below that it still contains the same repeating sequences.

To facilitate our analysis of the 1.705 sequences in each of these hybrid plasmids, we have utilized the restriction endonuclease *Mbo*II from *Moraxella bovis* (R. Roberts, unpubl.). Endow (1977) discovered that this enzyme cleaves 1.705 DNA into a series of short oligonucleotides 5 bp, 10 bp, and 15 bp in size. At first we felt that this indicated three repeating sequences in 1.705, one 5 bp, one 10 bp, and one 15 bp in length, each containing a single *Mbo*II recognition site. The mechanism of *Mbo*II cleavage, however, indicated that all these fragments could arise from a single repeating sequence. *Mbo*II recognizes the sequence $\left(\begin{smallmatrix} GAAGA \\ CTTCT \end{smallmatrix}\right)$ but cleaves the DNA eight nucleotides to the right of the terminal AT base pair of the recognition sequence (Sanger et al. 1977) (see legend to Fig. 3). Cleavage of $poly\left(\begin{smallmatrix} AAGAG \\ TTCTC \end{smallmatrix}\right)$ by this enzyme would result in products that were all integral multiples of 5 bp (Fig. 3). Products 20 bp or longer would contain both a recognition site and a cleavage site and would be cleaved further. Products 15 bp or smaller should be resistant to digestion. Cleavage of 1.705 satellite DNA with *Mbo*II produces the fragments shown in Figure 4. Most of the fragments are integral multiples of 5 bp and form a pattern which extends from

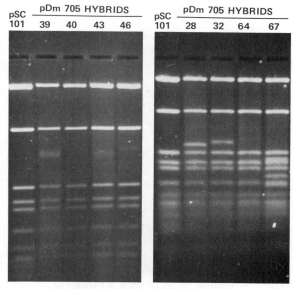

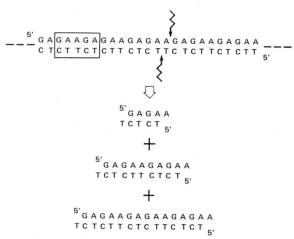

Figure 2. Heterogeneity of satellite DNA regions in hybrid plasmids. Hybrid plasmids containing 1.705 satellite DNA inserted into the vector pSC101 were constructed and the circular plasmid DNA was isolated as previously described (Brutlag et al. 1977b). Each plasmid DNA was cleaved with the restriction endonuclease *Hae*III and the DNA fragments were separated by agarose gel electrophoresis. The gel at the left of each panel shows DNA fragments of the vector pSC101, the largest of which are 2700 bp, 1250 bp, 670 bp, and 570 bp long. The vector fragment containing the *Eco*RI site into which the satellite DNA was inserted is very short (180 bp) and does not appear on these gels. Hybrid plasmids usually show a new DNA fragment in addition to the vector fragments. These new fragments are often heterogeneous in length. The hybrid DNA molecules had replicated between 30 and 40 generations in the bacterial host prior to their isolation. Each of these hybrid plasmids contained 1.705 satellite DNA as determined by the Grunstein and Hogness (1975) colony hybridization procedure, and all except pDm705.67 contained the 5-bp repeating sequence poly$\left(\begin{smallmatrix}AAGAG\\TTCTC\end{smallmatrix}\right)$ as determined by digestion with *Mbo*II as in Fig. 4. The hybrid pDm705.67 contained no detectable satellite DNA by *Mbo*II digestion and also had the weakest response in the colony hybridization test.

Figure 3. Fragments produced from the 1.705 satellite upon *Mbo*II digestion. The recognition site for the *Mbo*II restriction nuclease is shown in the box and the cleavage site is shown by the broken arrows. Note that the cleavage site is 10 bp or exactly one turn of the DNA duplex from the center of the recognition site. The enzyme leaves 3' hydroxyl and 5' phosphoryl termini. Cleavage of poly$\left(\begin{smallmatrix}AAGAG\\TTCTC\end{smallmatrix}\right)$ by this mechanism releases oligonucleotides which are integral multiples of 5 bp and which begin with $\begin{smallmatrix}5'\ GAGAA\\TCTCTT\end{smallmatrix}$ ... and end with ... $\begin{smallmatrix}GAGAA\ 3'\\CTCT\end{smallmatrix}$. Cleavage near the 3' end of such a fragment results in the release of either a 5-bp or 10-bp or 15-bp fragment, each of which is too short to contain both a recognition and a cleavage site. The shortest oligonucleotide that can be produced by cleavage at the 5' end is a 15-bp fragment. Oligonucleotides 20 bp or longer resulting from cleavage at internal sites can always be cleaved further to yield 5-bp, 10-bp, and 15-bp fragments.

The *Mbo*II fragments longer than 15 bp have several possible origins. Control experiments in which SV40 DNA was included or more enzyme was added indicated that digestion was complete. However, due to the proximity of restriction sites in 1.705 DNA, we cannot eliminate the possibility that the binding of one molecule of *Mbo*II blocks recognition or cleavage by another molecule at a nearby site. A third alternative is that longer oligonucleotides result from the presence of sequence alterations in the 1.705 DNA which eliminate some *Mbo*II recognition sites.

Digestion of 15 of our hybrid plasmids with *Mbo*II confirms that they contain the sequence poly$\left(\begin{smallmatrix}AAGAG\\TTCTC\end{smallmatrix}\right)$. However, one of the plasmids (crDm705.5; see Fig. 4) contained a different repeating sequence. *Mbo*II digestion of this plasmid produced no short DNA fragments which were integral multiples of 5 bp. Instead, oligonucleotides 7, 14, 21, and 28 bp in size were present. Fragments of these sizes can also be seen in digests of purified satellite DNA. The recovery of this cloned fragment shows that the 1.705 satellite, like the 1.672 satellite,

the digestion mechanism proposed in Figure 3 by sequence analysis of several of these fragments. Figure 5 shows the sequence of both strands of the 10-bp oligonucleotide. These sequences were found to be complementary except for the terminal 3' nucleotide. Each oligonucleotide therefore has a single protruding 3' nucleotide, either an A or a T. Similar termini were found on the 5-bp and 15-bp products. These findings indicate that *Mbo*II makes a staggered cut as shown in Figure 3. No matter how these three major sequences are arranged next to one another, only the five-nucleotide repeating sequence, poly$\left(\begin{smallmatrix}AAGAG\\TTCTC\end{smallmatrix}\right)$, is obtained. This is strong evidence confirming the sequence determined by other methods.

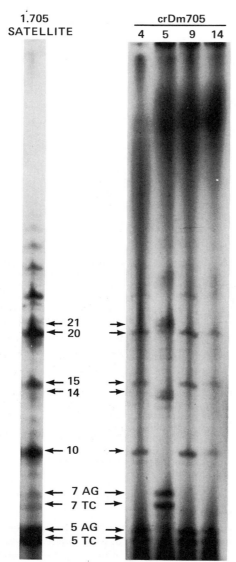

Figure 4. Fragments produced by *Mbo*II endonuclease digestion of 1.705 satellite DNA and recombinant plasmid DNA containing 1.705 sequences. Satellite and plasmid DNAs were purified and digested as previously described (Brutlag et al. 1977b). After *Mbo*II digestion, samples were heated to 100°C for 2 min, cooled to 37°C, and bacterial alkaline phosphatase added. After 30 min the reaction was diluted with appropriate salts and [γ-^{32}P]ATP to give final concentrations of 70 mM Tris-HCl (pH 8.95), 20 mM NaCl, 6.7 mM K$_2$HPO$_4$, 12.3 mM MgCl$_2$, 16.6 mM 2-mercaptoethanol, 0.33 mM DTT, and 1.6 μM ATP. Polynucleotide kinase (Richardson 1965) was added to 33 units/ml and the reaction incubated at 37°C for 30 min. The *Mbo*II fragments were resolved by electrophoresis in a 20% polyacrylamide-urea gel at 15 V/cm (Maxam and Gilbert 1977) and detected by autoradiography. Indicated lengths of fragments were determined by external standards (not shown) and by direct sequence analysis.

consists of two discrete DNAs with different but closely related repeating sequences.

We have determined the sequence of the 7-bp and 14-bp fragments both from the cloned segment and from the total satellite DNA. Figure 6 shows a com-

parison of the sequences of the 7-bp pyrimidine strand of plasmid and of satellite origin. Unlike the fragments in the 5-bp series, these oligonucleotides contain an unpaired terminal G or C nucleotide. Thus the 7-bp and 14-bp fragments can only be arranged adjacent to each other, forming the repeating sequence poly$\left(\begin{smallmatrix}\text{AAGA GAG}\\\text{T TCTCTC}\end{smallmatrix}\right)$ (Fig. 7). All the hybrid plasmids containing either 1.705 sequence have satellite regions about 1 kb average size. This implies that poly$\left(\begin{smallmatrix}\text{AAGAG}\\\text{TTCTC}\end{smallmatrix}\right)$ and poly$\left(\begin{smallmatrix}\text{AAGAGAG}\\\text{TTCTCTC}\end{smallmatrix}\right)$ must be segregated in chromosomal regions of at least this length. Using these hybrid plasmids as hybridization probes for in situ hybridization, we should be able to test whether each sequence has a distinct chromosomal location.

In addition to the major sequence obtained for the 7-bp fragment from satellite, one or more minor sequences appear to be present in this oligonucleotide (Fig. 6). By comparison, the 7-bp fragment from the cloned satellite shows no minor sequences despite prolonged overexposure of the autoradiogram. The minor sequences in the 7-bp oligonucleotide could arise either from sequence alteration of the 7-bp repeat or from insertion or deletion mutations in the 5-bp repeat, giving rise to unusual restriction-site spacing. Regardless of the origin of the variations in satellite DNA, it appears that we have cloned an extremely homogeneous region of the 7-bp repeated sequence.

The similarities between the two sequences in 1.672 DNA and 1.705 DNA (Fig. 1) are reminiscent of the relationship between repeating sequences observed in *Drosophila virilis* by Gall and Atherton (1974). Their analysis of three separable satellite DNAs showed each to consist of a repeating heptanucleotide which was related to the others by one or two base substitutions (TA → CG transitions). The 1.672 and 1.705 sequences described here are related by two or three base substitutions (TA → GC transversions). These two satellites of *D. melanogaster* share several other characteristics. The 5-bp repeat of each satellite contains very few base substitutions as described above. During purification these two species are the most widely resolved in all the gradients utilized, and yet 1.672 DNA is the major contaminant of the 1.705 satellite (Goldring et al. 1975). DNA molecules with 1.705 and 1.672 sequences covalently linked indicated that these species are adjacent in the genome (Brutlag et al. 1977a). This conclusion is supported by the cytological arrangement of the satellite DNAs (Peacock et al., this volume). Using clonally purified sequences of 1.705, we can now test whether the 5- and 7-bp repeats of each satellite are adjacent to each other.

The Complex 1.688 Satellite

The 1.688 g/cm^3 satellite is much more complex than the three simple-sequence satellites of *D. mela-*

AG STRAND

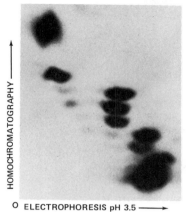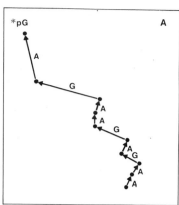

Figure 5. Sequence analysis of the 10-bp fragment from *Mbo*II digestion of 1.705 satellite. These chromatograms are two-dimensional fractionations of snake venom phosphodiesterase partial digests of separated strands of the 10-bp fragment shown in Fig. 4. The mobility changes in the first and second dimensions are characteristic of the nucleotide lost from the 3' end as shown in the line drawings (Galibert et al. 1974; Jay et al. 1974). The sequences determined are arranged to show complementarity. After elution from the polyacrylamide gel, the strands were separated by homochromatography in one dimension. The strand composed of pyrimidines migrates faster than the one of purines. Transfer to the second dimension was carried out as described previously (Southern 1974).

TC STRAND

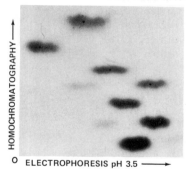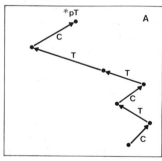

5' GAGAAGAGAA
   TCTCTTCTCT 5'

7 mer FROM 1.705 SATELLITE DNA

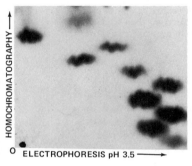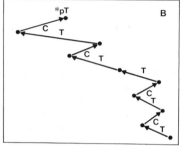

Figure 6. Sequence comparison between the 7-bp fragments of 1.705 satellite and cloned 1.705 DNA. These two-dimensional fractionations resolve snake venom phosphodiesterase partial digests of the pyrimidine strands of the 7-bp fragments from either satellite DNA *(A)* or from cloned 1.705 DNA in crDm705.5 *(B)*. Methods are as described in Fig. 5. The two sequences determined are identical; however, note the minor sequences present in satellite and absent in cloned DNA. The sequence shown is at the bottom of the figure along with the complementary sequence determined independently.

7 mer FROM CLONED 1.705

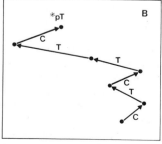

5' AGAAGAG
   CTCTTCT 5'

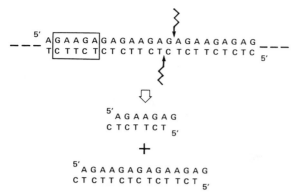

Figure 7. Fragments produced from poly$\left(\begin{smallmatrix}AAGAGAG\\TTCTCTC\end{smallmatrix}\right)$ upon *Mbo*II digestion. When *Mbo*II cleaves poly $\left(\begin{smallmatrix}AAGAGAG\\TTCTCTC\end{smallmatrix}\right)$ by the same mechanism described in Fig. 3, all the products are integral multiples of 7 bp that begin with $^{5'}\begin{smallmatrix}AGAAG\\CTCTTC\end{smallmatrix}\cdots$ and end with $\cdots\begin{smallmatrix}AAGAG\\TTCT\end{smallmatrix}{}^{3'}$. Fragments 21 bp or longer contain both a recognition site and a cleavage site and are subject to further degradation, whereas fragments 7 bp and 14 bp in length are resistant.

Figure 8. 1.688 satellite DNA cleaved by either *Hae*III or *Hin*f. 1.688 satellite DNA was purified by several CsCl gradients and a cesium formate gradient and then cleaved with either *Hae*III or *Hin*f restriction endonucleases as described by Carlson and Brutlag (1977). The DNA fragments were resolved by electrophoresis on a 1% agarose gel. In each case, the DNA in slot 2 was treated with twice as much enzyme for twice the time as the DNA in slot 1. (Reprinted, with permission, from Carlson and Brutlag 1977.)

nogaster, as demonstrated both by renaturation studies and by restriction-site mapping (Carlson and Brutlag 1977). Several workers have shown that digestion of 1.688 DNA with either *Hae*III or *Hin*f endonuclease generates a series of DNA fragments which are integral multiples of a 365-bp monomer (Fig. 8) (Manteuil et al. 1975; Shen et al. 1976; Carlson and Brutlag 1977). It was proposed that the arrangement of restriction sites in this satellite DNA evolved by random alteration of sites in an originally homogeneous array of 365-bp tandem repeats. Random alterations which varied the sequence of 25% of the *Hae*III sites or 10% of the *Hin*f sites would quantitatively produce the pattern of fragments shown in Figure 8.

We have shown that the restriction sites are not altered in a random fashion by cloning segments of 1.688 satellite DNA. One hybrid plasmid (pDm688.23; see Carlson and Brutlag 1977) contained a 5.8-kb region of 1.688 satellite with no altered *Hae*III or *Hin*f sites in 16 tandem 365-bp repeats. On the other hand, we also showed that long regions of 1.688 DNA contained no *Hae*III sites or no *Hin*f sites. In the *Hae*III and the *Hin*f digests of Figure 8, we detected oligomers as long as 15 monomer units by hybridization with cloned monomer DNA. Moreover, the long DNA fragments near the top of the gel are resistant to these enzymes and also hybridize with cloned monomer DNA. These data suggest that there are long regions of 1.688 DNA missing every *Hae*III site or every *Hin*f site and other regions which contain every site intact.

By cloning segments of 1.688 DNA we have demonstrated that the altered *Hae*III sites are variant sequences rather than restriction sites with modified

bases. Any bases specifically modified in *Drosophila* would not be similarly modified after propagation in bacteria. We have recovered several randomly cloned segments of 1.688 DNA which yield oligomer-length fragments upon *Hae*III digestion. One cloned region yielded only monomers upon *Hin*f digestion but gave two large DNA fragments, in addition to a few monomer fragments, after digestion with *Hae*III (pDm688.66 in Carlson and Brutlag 1977). Thus some of the repeats in this plasmid clearly had variant sequences at the expected positions for *Hae*III sites. More direct proof that the missing *Hae*III sites are due to sequence alterations comes from cloning the dimer fragments produced by *Hae*III digestion of 1.688 satellite DNA. Seventeen independent plasmids contained *Hae*III dimers which were still resistant to *Hae*III after propagation in *Escherichia coli*. We also cloned the long *Hae*III-resistant DNA at the top of the gel in Figure 8. Six independent hybrid plasmids contained long *Hae*III-resistant 1.688 regions. These data show that sequence alterations are primarily responsible for the pattern of restriction fragments shown in Figure 8. Furthermore, they confirm that there are long regions of 1.688 satellite DNA which contain no *Hae*III sites.

The Repeating Sequence of 1.688 Satellite DNA

We have examined the sequence of the 1.688 repeat unit and variations within it by direct nucleotide sequence analysis of 15 tandem monomers in the plasmid pDm688.23. In addition to the *Hinf* and *Hae*III sequences, two *Alu*I sites facilitate sequence determination (Fig. 1). Using the technique of Maxam and Gilbert (1977), we have determined a preliminary sequence around both the *Hae*III and *Hinf* sites and in one direction from each *Alu*I site. Some of the data for the sequence around the *Hinf* site are shown in Figure 9 and the preliminary sequence is presented in Figure 10.

The 15 tandem repeats present in pDm688.23 do not all have the same sequence. The procedure of Maxam and Gilbert (1977) usually results in a single possible nucleotide at any given distance from the restriction site. The sequence in Figure 10 is ambiguous, with at least two nucleotides possible at many positions. A variant sequence must be rather common among the 15 monomer units of pDm688.23 to be detected by this sequencing method. Also, the procedure of Maxam and Gilbert (1977) can only detect sequence variants that are related by transversion, i.e., only purine-pyrimidine mixtures can be detected. These technical limitiations mean that there may be more variation among these 15 repeats than is shown in Figure 10. There also appear to be rather long regions which are free of sequence variation, in particular, around the *Hae*III and *Hinf* sites. This may explain the conservation of each of the 16 *Hae*III and *Hinf* sites in this cloned region of 1.688 DNA. The fact that we can obtain relatively unique sequences over long distances suggests that insertion and deletion mutations are rare. The primary mechanism of sequence variation appears to be base substitution. The presence of sequence variations among tandem repeats argues against hypothetical rectification mechanisms in the maintenance of 1.688 satellite (Thomas 1974).

This partial sequence shows that the basic repeating unit of the 1.688 satellite is 378 ± 5 bp and is not composed of a number of shorter repeats or inverted subunits. There is no relationship between the sequence of 1.688 satellite and those of the three simple-sequence satellites of *D. melanogaster*. Instead of the alternating AT or AG sequences found in 1.672 and 1.705 satellites, the 1.688 satellite contains runs of (dA) · (dT) or (dG) · (dC) up to 9 bp long.

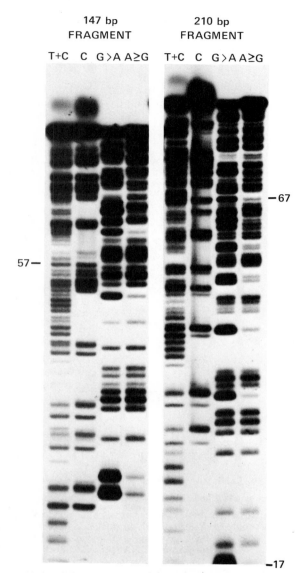

Figure 9. Sequence analysis of cloned 1.688 satellite near the *Hinf* site. DNA of the plasmid pDm688.23 was cleaved with *Hinf* endonuclease and the 15 copies of the 1.688 monomer were purified by polyacrylamide gel electrophoresis. The fragments were then labeled at their 5′ ends by a modification of the method of Maxam and Gilbert (1977) and further cleaved with *Hae*III endonuclease. The 147-bp and 210-bp fragments, each labeled only at their 5′ *Hinf* ends, were again separated by polyacrylamide electrophoresis. Each fragment was then partially cleaved as described by Maxam and Gilbert (1977) specifically at the T and C residues, C residues, preferentially at G residues, or preferentially at A residues. All of the partial digests were resolved by several polyacrylamide gels over ranges extending from 2 bp to over 100 bp. Sequence ambiguity is seen most clearly in the 147-bp fragment beginning at nucelotide 17 at the bottom and reading up: T, T, T, C, C or G, G, etc. Because of the sequence ambiguity in these 15 monomer units, we did not analyze any sequences more than 70 nucleotides from any restriction site. The lengths of the 147-bp and 210-bp fragments, originally determined by polyacrylamide gel electrophoresis (Carlson and Brutlag 1977), are more precisely 154 ± 5 bp and 223 ± 5 bp, based on the partial sequence in Fig. 10.

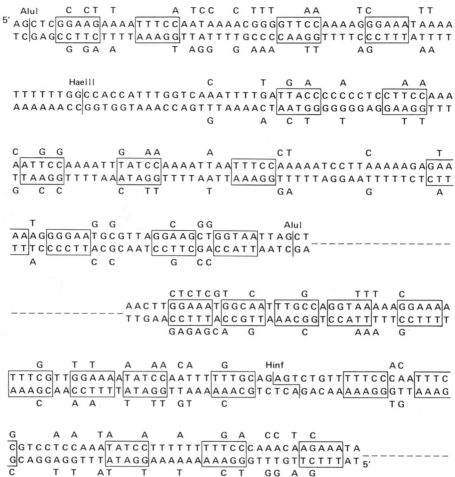

Figure 10. Partial nucleotide sequence of the cloned 1.688 monomer repeating unit. This figure shows the sequence obtained as described in Fig. 9. The boxes locate the occurrences of $\left(\begin{smallmatrix}\text{TTTCC}\\\text{AAAGG}\end{smallmatrix}\right)$ or single-base changes of that sequence. When a particular nucleotide was ambiguous, one base-pair was placed above and below the central sequence. No attempt could be made to quantitate which nucleotide predominated due to the strong sequence effects on the rates of partial cleavage. The sequence about the *Hinf* site was determined in Fig. 9. The sequence about the *Hae*III site was determined similarly, except the plasmid DNA (pDm688.23) was cleaved first with *Hae*III and the monomer was purified, labeled at the 5' end, and then cleaved with *Hinf*. The sequence from the *Alu*I sites was determined by cleaving pDm688.23 DNA with *Alu*I, purifying the resulting 196-bp and 181-bp fragments on polyacrylamide gels, and then labeling their 5' ends. The smaller fragment was cleaved with *Hae*III, resulting in 56-bp and 125-bp fragments whose sequences provided overlap with the sequence about the *Hae*III site. The 196-bp fragment is cleaved by *Hinf* into two fragments of similar size which we have not yet resolved. Therefore, there are about 29 bp missing in the center of the sequence and about 39 bp missing at the end. The first 60 nucleotides have had both strands analyzed (one strand from the *Hae*III site and the other from an *Alu*I site). The rest of the sequence, although determined by multiple gel analysis, must remain tentative. This is especially true of nucleotides within 7 bp of either *Alu*I site.

Despite the lack of an internal repeating structure, there is some evidence for short-range periodicity. The sequences $\left(\begin{smallmatrix}\text{TTTCC}\\\text{AAAGG}\end{smallmatrix}\right)$ and $\left(\begin{smallmatrix}\text{AAATTT}\\\text{TTTAAA}\end{smallmatrix}\right)$ and single-base changes of them are very frequent components of this complex sequence. The sequence $\left(\begin{smallmatrix}\text{AAATTT}\\\text{TTTAAA}\end{smallmatrix}\right)$ is present eight times and single-base variations of it occur an additional eight times. There are nine exact repeats and 18 single-base changes of $\left(\begin{smallmatrix}\text{TTTCC}\\\text{AAAGG}\end{smallmatrix}\right)$, which compose 43% of all the

nucleotides in the repeat (Fig. 10). The precise frequencies with which these sequences are present in the 15 tandem repeats of pDm688.23 are difficult to determine because of the sequence variations observed.

The presence of short repeats within a more complex satellite DNA has often been taken as evidence for an ancestral simple-sequence satellite from which the more complex repeats have derived (Sutton and McCallum 1972; Cooke 1975; Biro et al. 1975; Southern 1975). However, the evolution of the 378-bp repeat of 1.688 from poly$\left(\begin{smallmatrix}\text{TTTCC}\\\text{AAAGG}\end{smallmatrix}\right)$ does not seem very likely. Since the sequence 5' (TTTCC) 3' occurs on both strands of 1.688 DNA, inversions would be necessary in addition to single-base changes in any hypothetical evolutionary scheme.

Moreover, the distribution of $\left(\begin{smallmatrix} TTTCC \\ AAAGG \end{smallmatrix}\right)$ sequences indicates that either multiple overlapping inversions or insertions and deletions occurred as well. All such inversions, insertions, and deletions must have occurred prior to any amplification event that established the regular arrangement of restriction sites. Subsequent to the formation of this regular pattern, only single-base changes could be permitted since inversions, insertions, and deletions would alter the regular pattern. Although it does not rule out such an unusual series of events, this partial sequence suggests that the complex 1.688 satellite did not evolve from a simple-sequence satellite such as the 1.672 or 1.705 species. The presence of 5' (TTTCC) 3' on both strands also argues against the 1.688 satellite being an intermediate in the evolution of a simple-sequence satellite by unequal sister-chromatid exchange as proposed by Smith (1976).

An alternative proposal for the origin of short repeats within the 378-bp monomer is that sequences such as $\left(\begin{smallmatrix} TTTCC \\ AAAGG \end{smallmatrix}\right)$ or $\left(\begin{smallmatrix} AAATTT \\ TTTAAA \end{smallmatrix}\right)$ may have a functional significance and be subject to selection. This complex satellite has been located adjacent and proximal to the nucleolus organizers on both the X and Y chromosomes (Peacock et al., this volume). These are precisely regions that have been implicated as essential for proper meiotic pairing, chromosome segregation, and spermatogenesis (Gershenson 1940; Sandler and Braver 1953; Cooper 1964; Peacock et al. 1975). Were this satellite to be involved in recognition between the predominantly nonhomologous X and Y chromosomes, then the shorter repeats might be essential for sequence-specific DNA–protein interactions. Sequence-specific proteins could readily utilize sequences on both strands. Moreover, such interactions might be modulated by sequence variation, allowing different regions of satellite DNA to react differently, just as they do with the sequence-specific restriction endonucleases. We are currently using our knowledge of DNA sequence organization in heterochromatin to isolate chromatin fractions containing exclusively a single satellite sequence. By developing as stringent criteria for the homogeneity of chromatin fractions as we described here for DNA fractionation, we hope to demonstrate the presence or absence of sequence-specific proteins in heterochromatin.

Acknowledgments

This work was supported by grants from the National Foundation March of Dimes (Basil O'Connor Starter Grant) and from the National Institute of General Medical Science. M. C. is a National Institutes of Health predoctoral fellow; K. F. and T. S. H. are National Institutes of Health postdoctoral fellows.

REFERENCES

ASHBURNER, M. and E. NOVITSKI, eds. 1976. *The genetics and biology of* Drosophila, vol. I. Academic Press, New York.

BIRNBOIM, H. C. and R. SEDEROFF. 1975. Polypyrimidine segments in *Drosophila melanogaster* DNA: I. Detection of a cryptic satellite containing polypyrimidine/polypurine DNA. *Cell* 5: 173.

BIRO, P. A., A. CARR-BROWN, E. SOUTHERN, and P. M. B. WALKER. 1975. Partial sequence analysis of mouse satellite DNA: Evidence for short range periodicities. *J. Mol. Biol.* 94: 71.

BLUMENFELD, M. 1974. The evolution of satellite DNA in *Drosophila virilis*. *Cold Spring Harbor Symp. Quant. Biol.* 38: 423.

BRUTLAG, D. L. and W. J. PEACOCK. 1975. Sequences of highly repeated DNA in *Drosophila melanogaster*. In *The eukaryotic chromosome* (ed. W. J. Peacock and R. D. Brock), p. 35. Australian National University Press, Canberra.

BRUTLAG, D., R. APPELS, E. S. DENNIS, and W. J. PEACOCK. 1977a. Highly repeated DNA in *Drosophila melanogaster*. *J. Mol. Biol.* 112: 31.

BRUTLAG, D., K. FRY, T. NELSON, and P. HUNG. 1977b. Synthesis of hybrid bacterial plasmids containing highly repeated satellite DNA. *Cell* 10: 509.

CARLSON, M. and D. BRUTLAG. 1977. Cloning and characterization of a complex satellite DNA from *Drosophila melanogaster*. *Cell* 11: 371.

COOKE, H. J. 1975. Evolution of the long range structure of satellite DNAs in the genus *Apodemus*. *J. Mol. Biol.* 94: 87.

COOPER, K. W. 1959. Cytogenic analysis of major heterochromatic elements in *Drosophila melanogaster*, and the theory of heterochromatin. *Chromosoma* 10: 535.

————. 1964. Meiotic conjunctive elements not involving chiasmata. *Proc. Natl. Acad. Sci.* 52: 1248.

ENDOW, S. A. 1977. Analysis of *Drosophila melanogaster* satellite IV with restriction endonuclease *Mbo*II. *J. Mol. Biol.* 114: 441.

ENDOW, S. A., M. L. POLAN, and J. G. GALL. 1975. Satellite DNA sequences of *Drosophila melanogaster*. *J. Mol. Biol.* 96: 665.

FANSLER, B. S., E. C. TRAVAGLINI, L. A. LOEB, and J. SCHULTZ. 1970. Structure of *Drosophila melanogaster* dAT replicated in an *in vitro* system. *Biochem. Biophys. Res. Commun.* 40: 1266.

GALIBERT, F., J. SEDAT, and E. ZIFF. 1974. Direct determination of DNA nucleotide sequences: Structure of a fragment of bacteriophage ϕX174. *J. Mol. Biol.* 87: 377.

GALL, J. G. and D. D. ATHERTON. 1974. Satellite DNA sequences in *Drosophila virilis*. *J. Mol. Biol.* 84: 633.

GALL, J. G., E. H. COHEN, and D. D. ATHERTON. 1974. The satellite DNAs of *Drosophila virilis*. *Cold Spring Harbor Symp. Quant. Biol.* 38: 417.

GERSHENSON, S. 1933. Studies on the genetically inert region of the X-chromosome of *Drosophila*. *J. Genet.* 28: 297.

————. 1940. The nature of the so-called genetically inert parts of chromosomes. *Videnskap. Akad. Nauk. U.R.R.S.*, p. 3.

GOLDRING, E. S., D. BRUTLAG, and W. J. PEACOCK. 1975. Arrangement of highly repeated DNA of *Drosophila melanogaster*. In *The eukaryote chromosome* (ed. W. J. Peacock and R. D. Brock), p. 47. Australian National University Press, Canberra.

GRUNSTEIN, M. and D. S. HOGNESS. 1975. Colony hybridization: A method for the isolation of cloned DNAs that contain a specific gene. *Proc. Natl. Acad. Sci.* 72: 3961.

HENNIG, W. and P. M. B. WALKER. 1970. Variations in
the DNA from two rodent families (*Cricetidae* and
Muridae). *Nature* **225**: 915.

JAY, E., R. BAMBARA, R. PADMANABHAN, and R. WU. 1974.
DNA sequence analysis; a general, simple and rapid
method for sequencing large oligodeoxyribonucleotide
fragments by mapping. *Nucleic Acids Res.* **1**: 331.

MANTEUIL, S., D. H. HAMER, and C. A. THOMAS. 1975. Regular arrangement of restriction sites in *Drosophila* DNA.
Cell **5**: 413.

MAXAM, A. M. and W. GILBERT. 1977. A new method for
sequencing DNA. *Proc. Natl. Acad. Sci.* **74**: 560.

PEACOCK, W. J., G. L. G. MIKLOS, and D. J. GOODCHILD.
1975. Sex chromosome meiotic drive in *Drosophila
melanogaster*. *Genetics* **79**: 613.

PEACOCK, W. J., D. BRUTLAG, E. GOLDRING, R. APPELS, C.
W. HINTON, and D. L. LINDSLEY. 1974. The organization
of highly repeated DNA sequences in *Drosophila melanogaster* chromosomes. *Cold Spring Harbor Symp.
Quant. Biol.* **38**: 405.

RICHARDSON, C. C. 1965. Phosphorylation of nucleic acid
by an enzyme from T4 infected *E. coli. Proc. Natl. Acad.
Sci.* **54**: 158.

SANDLER, L. and G. BRAVER. 1953. The meiotic loss of unpaired chromosomes in *Drosophila*. *Genetics* **39**: 365.

SANGER, F., G. M. AIR, B. G. BARRELL, N. L. BROWN,

A. R. COULSON, J. C. FIDDES, C. A. HUTCHINSON III,
P. M. SLOCOMBE, and M. SMITH. 1977. Nucleotide sequence of bacteriophage ϕX174 DNA. *Nature* **265**: 687.

SEDEROFF, R. and L. LOWENSTEIN. 1975. Polypyrimidine
segments in *Drosophila melanogaster* DNA: II. Chromosome location and nucleotide sequence. *Cell* **5**: 183.

SHEN, C. K. J., G. WEISEHAHN, and J. E. HEARST. 1976.
Cleavage patterns of *Drosophila melanogaster* satellite
DNA by restriction enzymes. *Nucleic Acids Res.* **3**: 931.

SMITH, G. P. 1976. Evolution of repeated DNA sequences
by unequal crossover. *Science* **191**: 528.

SOUTHERN, E. M. 1974. An improved method for transfer-
ing nucleotides from electrophoresis strips to thin layers of ion exchange cellulose. *Anal. Biochem.* **62**: 317.

———. 1975. Long range periodicities in mouse satellite
DNA. *J. Mol. Biol.* **94**: 51.

SUTTON, W. D. and M. McCALLUM. 1972. Related satellite
DNA's in the genus *Mus. J. Mol. Biol.* **71**: 633.

THOMAS, C. A., JR. 1974. The rolling helix; a model for
the eukaryotic gene? *Cold Spring Harbor Symp. Quant.
Biol.* **38**: 342.

YAMAMOTO, M. and G. L. G. MIKLOS. 1977. Genetic dissection of heterochromatin in *Drosophila:* The role of basal
X heterochromatin in meiotic sex chromosome behavior. *Chromosoma* **60**: 283.

Mammalian Repetitive DNA and the Subunit Structure of Chromatin

P. R. MUSICH, F. L. BROWN, AND J. J. MAIO

Department of Cell Biology, Albert Einstein College of Medicine, Bronx, New York 10461

The nucleosome is now considered to be the basic organizational mode of eukaryotic chromatin, and intranuclear variations in nucleosome structure appear directly related to the controlled expression of the cell's genetic information (this volume). However, there is little information bearing upon such questions as whether there are specific interactions between the nucleosomal proteins and the DNA sequences with which they are associated. If such specific interactions occur, what is the nature of the proteins that confer specificity, and are such proteins responsible for the transcriptional activation or repression of a given DNA sequence? In addition, the question of whether the structure of the nucleosome has any direct influence on DNA sequence evolution has not as yet been approached. To resolve these questions, we have characterized the repressed heterochromatin portion of the nuclear chromatin and the associated repetitive DNA sequences by an enzymological dissection of the nuclei and of the purified repetitive DNAs of several mammalian species. In this work, we used both nonspecific nucleases and site-specific restriction endonucleases as our primary tools.

The results reported here indicate that (1) there is a direct correspondence between the length of DNA contained within a heterochromatic nucleosome and the periodicity of restriction endonuclease cleavage sites in the highly repetitive DNA sequences. (2) There appears to be a specific register between the nucleosomal proteins and the DNA sequence in the positioning of the highly repetitive DNA onto the nucleosome. (3) Heterochromatin containing only the highly repetitive component α DNA of the African green monkey has a protein composition distinctly different from the bulk of the chromatin: the H1 histones are deficient and are apparently replaced by a class of low molecular weight nonhistone proteins. (4) Nucleosome structure appears to have been an essential element in determining the periodicities of the highly repetitive DNA sequences during the evolution of the mammalian species studied.

MATERIALS AND METHODS

Primary tissue culture cells were obtained from male African green monkey, calf, and sheep kidney cells/ml by Flow Laboratories. All experiments with cells and supplied as cell suspensions at 10^6 viable the primary cell lines were performed on the second to fifth passages in vitro. CV1 cells are an established heteroploid cell line derived from African green monkey kidney (AGMK) and have been maintained in continuous culture for about 12 years. All cells were grown in monolayer cultures in Eagle's minimum essential medium supplemented with nonessential amino acids, 5% calf serum and 5% fetal calf serum. Mouse and rat DNAs were purified from nuclei isolated from liver; human DNA was obtained from isolated placental nuclei. DNA was purified from all nuclear preparations and the various satellite DNAs were isolated by preparative ultracentrifugation in Ag^+-Cs_2SO_4, Hg^{2+}-Cs_2SO_4, and CsCl as described in several publications (Kurnit and Maio 1973; Kurnit et al. 1973; D. M. Kurnit et al., in prep.)

Chromatin digestions in intact nuclei with micrococcal nuclease were carried out as follows: the cells were washed twice in Earle's balanced salts solution and once in 5 mM sodium phosphate buffer (pH 7.0), plus 0.025 mM $CaCl_2$ in 0.34 M sucrose. The cells were resuspended in this buffer and the nuclei released by homogenization with a Dounce tissue homogenizer. Following centrifugation through 2.2 M sucrose in the sodium phosphate-$CaCl_2$ buffer just described, the nuclei were washed once in 5 mM sodium phosphate buffer (pH 7.0), plus 0.25 mM $CaCl_2$. Digestion with micrococcal nuclease (Worthington) was in this buffer at 37°C.

Chromatin digestions in intact nuclei with restriction enzymes were carried out as follows: nuclei were isolated as just described. Following the centrifugation in 2.2 M sucrose, the nuclei were resuspended in 20 volumes of the buffer appropriate to the restriction enzyme under study and then centrifuged at $750g$ for 5 minutes. For enzyme digestion, the nuclear pellets were resuspended in the appropriate buffer to 2.5–5 × 10⁷ nuclei/ml as determined by phase contrast hemocytometer counts. DNA was purified after the enzyme digestions with either micrococcal nuclease or restriction enzymes by adding Tris-HCl, pH 8.4, to 100 mм, EDTA to 50 mм, and SDS to 1%. The contents of the tubes were mixed with a Vortex mixer, NaCl added to 1 м, and the tubes mixed again. The samples were deproteinized twice (without precipitation of the DNA) with phenol equilibrated against the Tris-EDTA-NaCl-SDS buffer just described, and once with chloroform-isoamyl alcohol (24:1).

DNA samples were precipitated overnight at −20°C in one volume of 0.3 м ammonium acetate and four volumes of absolute ethanol. The precipitates were collected by centrifugation in siliconized polyallomer tubes at 175,000g for 45 minutes at 4°C and dissolved in 20 μl of one-third strength electrophoresis buffer. Full strength electrophoresis buffer contains 89 mм Tris, 82 mм boric acid, 2 mм EDTA and has a pH of 8.3. Sucrose was added to 7.5% to the samples for layering in the sample wells and 0.025% bromophenol blue or Orange G was used as tracking dye. Electrophoresis in agarose slab gels (1.3 mm thick) was with full strength electrophoresis buffer at 50 V for 3.5 hours at room temperature. Acrylamide slab gels contained 5.68% (w/v) acrylamide, 0.32% (w/v) N,N'-methylenebisacrylamide, 0.4% (w/v) ammonium persulfate and 0.26% (v/v) 3-dimethylaminopropionitrile. After the gels had polymerized at room temperature for at least 1 hour, the lucite comb was removed and the gel was then pre-electrophoresed for 2 hours at 200 V to remove quenching substances. The buffer was then discarded and fresh buffer added to both buffer chambers. Electrophoresis of the samples was at 75 V for about 4 hours.

The fluorescent DNA bands were photographed through a Kodak Wratten 23A filter with Polaroid Type 55 P/N film on a Model C5 Mineralight ultraviolet table (Ultra-Violet Products, Inc., San Gabriel, Calif.). Molecular weights were calculated from the microdensitometer tracings of the film negatives. The relative amounts of DNA in individual bands were calculated from areas under the curves of microdensitometer tracings of negatives obtained from photographs in the linear range. In experiments with [³H]thymidine-labeled DNA (F. L. Brown and J. J. Maio, unpubl.), ethidium bromide fluorescence is proportional to DNA concentration in the band if the amount of DNA is low (about 0.2 μg or less per band) or if the band is relatively broad and diffuse. At higher DNA concentrations, fluorescence measurements consistently underestimated the amount of DNA present.

The sizes of the restriction segments and micrococal nuclease DNA fragments of mammalian DNA and nuclei were determined by comparing their mobilities in both 1.8% agarose and 5% polyacrylamide gels with the mobilities of marker DNA segments of known size. The restriction segments of φX174 RF DNA produced by HindII and HaeIII were employed as marker DNA segments. The sizes of some of these φX174 restriction segments have been accurately determined by direct DNA sequencing (Sanger et al. 1977).

RESULTS AND DISCUSSION

The Component α Sequence

Figure 1 shows that purified component α DNA (Maio 1971) is attacked by at least four different restriction enzyme activities: EcoRI, EcoRI*, HindIII, and HaeIII. An estimated 5–10% of the total DNA is reduced to multimeric segments based upon a repeat unit of 176 ± 4 nucleotide base pairs (nbp) by EcoRI or HaeIII (Fig. 1A,D). In contrast,

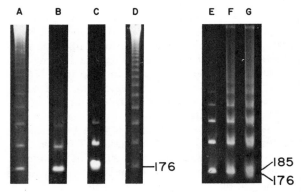

Figure 1. Repeat periodicity of African green monkey component α DNA and chromatin subunit DNA. Panels A–D are photographs in ultraviolet light of 1.4% agarose gels after staining with ethidium bromide. The direction of electrophoresis was from top to bottom in all gels. Numbers refer to lengths of segments in nbp ± 4. (A) The 1.2 μg of component α was treated for 4 hr with 120 units of EcoRI in 100 μl of EcoRI buffer (Polisky et al. 1975); 120 units of EcoRI were again added and incubation continued for 4 hr. (B) Conditions were the same as in A except that the digestions were carried out in EcoRI* buffer (Polisky et al. 1975). (C) The 1.2 μg of component α DNA in 100 μl of HindIII buffer was treated with 12 units of HindIII for 4 hr; 12 units were again added for a total incubation time of 8 hr. (D) The 1.2 μg of component α DNA in 100 μl of HaeIII buffer was treated with 5 units of HaeIII; after 4 hr another 5 units were added for a total incubation time of 8 hr.

Panels E–G are photographs of 1.8% agarose gels. (E) A partial digest of component α after limited treatment with EcoRI* (120 units, 2 hr) to reveal the higher multimers. (F) A mixture of this EcoRI* digest and DNA purified from a micrococcal nuclease partial digest (15 units/ml, 5 min at 37°C) of total CV1 nuclear chromatin. Note that the component α restriction segments are distinctly visible in the chromatin bands and remain roughly in phase at least as high as the hexamer. However, the component α segments (176 nbp) are slightly shorter than the average length of the chromatin fragments (185 nbp) and migrate to the front of each of the chromatin multimers. (G) The same micrococcal nuclease partial digest of CV1 nuclei after electrophoresis in the absence of added component α/EcoRI* restriction segments.

EcoRI* and HindIII effectively reduce the sequence to monomers of the basic repeat unit of 176 ± 4 nbp (Fig. 1B,C). Secondary EcoRI* sites within the repeat units also produce minor, intermediate series of bands in partial component α/EcoRI* digests (Fig. 2A). Oligomeric segments (predominantly dimers and some trimers) produced in the limit digests are perhaps due to random base substitutions in the EcoRI* and HindIII restriction sites. From appropriate plots of the amount of DNA in the various bands (Slack 1974), the sequence divergence is 2.6–3.0% per base pair for both the HindIII sites (6 nbp) and the EcoRI* sites (4 nbp). Interestingly, similar sequence divergence has been calculated for restriction sites in satellites of Mus (Southern 1975), Apodemus (Cooke 1975), and Drosophila (Shen et al. 1976), organisms with widely different life cycles, evolutionary histories, and repetitive DNAs.

The finding that component α is composed of tandemly repeating units of 176 nbp was intriguing because this DNA length is within the range of chromatin subunit periodicities of the entire genome. The ranges described in higher eukaryotes span 155 nbp (Aspergillus) to 240 nbp (sea urchin sperm) (Lohr and Van Holde 1975; Compton et al. 1976). Accordingly, we prepared a partial EcoRI* digest of component α to obtain an arithmetic series of multimers (Fig. 1E) and also a partial micrococcal nuclease digest of intact monkey nuclei (CV1 cells) to obtain an arithmetic series of DNA bands representative of the total chromatin (Fig. 1G). We next added equal amounts of the two DNAs in the same sample well before electrophoresis (Fig. 1F). The component α restriction segments and the total chromatin fragments of DNA released by micrococcal nuclease showed similar mobilities at least as high as the hexamer. We found that, although the component α restriction segments remained within the chromatin DNA bands, they migrated at the front of these bands of heterogeneous fragments produced by micrococcal nuclease (Fig. 1F). The repeat unit (176 nbp) of component α (and of constitutive heterochromatin, see below) is therefore about 10 nbp shorter than the average periodicity of the bulk chromatin, which we estimated at about 185 nbp. Thus, heterogeneity in nucleosome spacing exists within the nuclei of homogeneous cell populations. A specific class of nucleosomes contributing to this heterogeneity in CV1 cells is in constitutive heterochromatin which contains highly repetitive DNA sequences and distinctive nuclear proteins (see below). It is likely that such constitutively heterochromatic subunits form a large, homogeneous class within an intranuclear collection of heterogeneous classes.

Mouse satellite DNA. We wondered whether this close correspondence of the repeat structure of component α DNA with chromatin subunit structure was coincidental, or whether it was a characteristic shared with other mammalian highly repetitive

DNAs. The work of Hörz et al. (1976) indicates that such a relationship might exist between mouse satellite and mouse chromatin. Southern (1975) has shown that EcoRII restriction endonuclease cleaves mouse satellite DNA into segments of a basic periodicity of 244 nbp. In addition to the major series of restriction segments produced by EcoRII, mouse satellite yields a minor series of intermediate bands seemingly based upon the ½-mer of the major series and another fainter series based upon the ¼-mer (Southern 1975). Any attempt to understand the repeat structure and origins of mouse satellite and their possible relation to chromatin subunit structure must take into account not only the dominant repeat organization indicated by the major series of restriction segments but also the two minor series of intermediate bands.

We treated mouse satellite with EcoRII and component α with EcoRI* and compared the restriction segments of the two DNAs (Fig. 2A, B). Our revised estimate of the length of the mouse satellite monomer, based upon calibrations with the sequenced restriction segments of φX174, suggested a correction of about 6% of Southern's (1975) estimate, that is, to 235 ± 6 nbp for the mouse satellite monomer. As shown in Figure 2 (A, B), one major correspondence between the two DNAs occurs at the component α tetramer and the mouse satellite trimer, another at the component α octamer and the mouse satellite hexamer, and yet another at the component α dodecamer and the mouse satellite nonomer. Correspondences with the component α segments were also observed in the minor series of mouse satellite bands: for example, the size of the 1½-mer of mouse satellite corresponds with the component α dimer (Fig. 2A, B). In 6% polyacrylamide gels, EcoRII digests of mouse satellite reveal a faint band, 176 ± 4 nbp in length, which corresponds to the monomer of the component α sequence (Fig. 2C). Both the major and minor series of mouse satellite restriction segments are related to the simple organization of component α DNA, but the minimum size at which major segments of both DNAs may be superimposed was about 700 nbp — that is, at the tetramer of constitutive heterochromatin subunit DNA.

Sheep satellite II. These correspondences in the molecular weights of mouse satellite and component α restriction segments could be a trivial consequence of the simple numerical relationships between the monomeric segments of the two DNAs (length of component α monomer/length of mouse satellite monomer = 0.75). Accordingly, we examined a third satellite, sheep satellite II restricted with HaeIII. In the limit digests, both agarose (Fig. 2D) and acrylamide (Fig. 2E) gel electrophoresis resolved three major segments. The longest was the same length as the mouse satellite monomer, i.e., 235 nbp; the next longest segment was the same length of the component α monomer, or 176 nbp. The third and

Figure 2. Recurrent molecular weights of restriction segments of highly repetitive mammalian DNAs: monkey, mouse, sheep, and human. Panels *A, B, D,* and *E* are photographs of DNA segments in 1.4% agarose gels; panels *C* and *F* are of 6% polyacrylamide gels. *(A)* Component α DNA (1.2 μg) was treated with 120 units of *Eco*RI* for 4 hr at 37°C. *(B)* Mouse satellite (1.2 μg) was treated with 4 units of *Eco*RII for 4 hr; another 4 units were added and incubation at 37°C continued for 4 hr. Arrows and arrowheads mark the coincidences of the major and minor bands, respectively, of mouse satellite and component α/*Eco*RI* restriction segments. *(C)* An *Eco*-RII digest of mouse satellite after electrophoresis in a 6% polyacrylamide gel to resolve

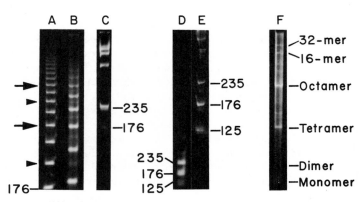

the shorter restriction segments. Note that the monomer (235 ± 6 nbp) resolves into two closely spaced bands and that there is a minor band (176 ± 4 nbp) below the monomer segment. *(D,E)* Sheep satellite II DNA (D. M. Kurnit et al., in prep.) was treated with 5 units of *Hae*III for 4 hr; 5 units were again added and incubation continued for 4 hr. The restriction segments were electrophoresed in a 1.4% agarose gel *(D)* and a 6% polyacrylamide gel *(E)*. From microdensitometry, the three major segments of lengths 235, 176, and 125 nbp occur in the molar ratios of 1:2:1, respectively. This suggests an overall length of 712 ± 16 nbp for the entire repeat unit (a nucleosome tetramer). *(F)* An *Eco*RI digest of a human satellite DNA (ρ = 1.701 g/ml in CsCl) illustrating a geometric series of segment lengths. The monomeric segments of 176 ± 4 nbp upon which the dominant geometric series is based are only faintly visible.

shortest segment was 125 nbp in length and did not correspond to either the mouse satellite or the component α monomer. From microdensitometer tracings of the limit digests in polyacrylamide gels (Fig. 2E), the three segments occur in the mole ratios of 1:2:1, respectively, so that the sum of their molecular weights is equivalent to a component α tetramer, that is, about 4.6×10^5 daltons or 700 nbp. This tetrameric organization makes sheep satellite II resemble mouse satellite despite the dissimilarities in the organization of the individual segments.

Human satellite DNA. We isolated a human satellite from human placental DNA through preparative ultracentrifugation in Ag^+-Cs_2SO_4 and Hg^{2+}-Cs_2SO_4 density gradients (F. L. Brown and J. J. Maio, in prep.). In CsCl at pH 8.4 it has a density of 1.701 g/ml; in alkaline CsCl at pH 12.5 it separates into two strands of densities 1.766 and 1.759 g/ml. Possibly, it is equivalent to human satellite IV previously described by Corneo et al. (1972), but our density measurements in alkaline CsCl differ from those reported by Corneo et al. for satellite IV. Restriction of this satellite with *Eco*RI produced a multimeric series of bands in agarose gels unlike that of any highly repetitive DNA so far analyzed. Rather than the arithmetic series of multimeric bands commonly observed, this sequence produced a series that was predominantly geometric (Fig. 2F). The basic monomer length of about 176 nbp was the lowest molecular weight segment: the multimeric series increased by progressive doublings of this basic unit at least as high as the 32-mer and possibly beyond (Fig. 2F).

Rat repetitive DNA fractions. Rat total DNA restricted with *Eco*RI yields a prominent band of 9.2×10^5 daltons in agarose gels (Fig. 3A,B). This band migrated with the octamer (1400 nbp) of com-

ponent α DNA when the restricted rat DNA was coelectrophoresed with a partial *Eco*RI* digest of component α. In this respect, the repetitive sequence in the rat resembles calf satellite I (see below). However, unlike calf satellite I, *Eco*RI* restriction did not reduce the rat octameric sequence to a subset of shorter segments. *Hin*dIII restriction of rat total DNA also revealed a set of multimeric segments presumably derived from yet another repetitive sequence in the rat genome (Fig. 3C,D). The multimers appeared based upon even-numbered (dimeric) multiples of a unit repeat of 188 nbp, which was faintly visible in the gels (Fig. 3C). The monomer of this series is therefore longer by about 12 nbp than the basic and recurrent periodicity of 176 nbp that underlies the repeat structure of the other highly repetitive DNAs we examined. If this sequence is indeed a satellite DNA, then it is possible that there are variant forms of constitutive heterochromatin with different chromatin structure. The possibility that variant forms of constitutive heterochromatin exist is suggested by the occurrence of both Q-band and G-band positive and negative constitutive heterochromatin in the same mammal (Lau and Arrighi 1976) and of constitutive heterochromatin that does not appear to contain highly repetitive sequences (Arrighi et al. 1974). It was reported (Botchan et al. 1976) that *Hin*dIII restriction of rat DNA releases a segment that hybridizes with 5S RNA. The observations reported here raise the possibility that chromatin subunit structure also figures in the organization of the 5S DNA sequences which, in *Xenopus*, contain long stretches of satellitelike sequences (Fedoroff and Brown, this volume).

Calf satellite I. This DNA, the major satellite sequence in the calf genome (Kurnit et al. 1973), is reduced by *Eco*RI to a repeat segment 1400 nbp in

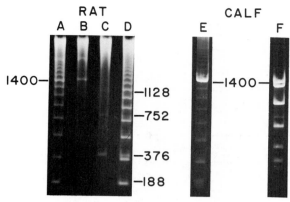

Figure 3. Recurrent molecular weights of restriction segments of rat and calf repetitive DNAs. All photographs are of 1.4% agarose gels. In panels A–D, individual satellite fractions were not isolated; restriction enzyme digests were of unfractionated rat total nuclear DNA. *(A)* A partial digest of component α with *Eco*RI* was mixed with rat total DNA restricted with *Eco*RI before electrophoresis. Restriction segments of both DNAs are superimposed at the component α octamer (about 1400 nbp); another superimposition occurs at the component α 16-mer. *(B)* A similar *Eco*RI digest of rat total DNA without added component α restriction segments. These bands in rat total DNA were not altered by treatment with *Eco*RI* (not shown). *(C)* Rat total DNA after restriction with *Hin*dIII. The repetitive sequence revealed by *Hin*dIII digestion appears based upon a monomer of 188 ± 4 nbp, which is only faintly visible in the gel. Dimer (376 nbp), tetramer (752 nbp), and hexamer (1128 nbp) segments based upon this periodicity are clearly visible. *(D)* A similar *Hin*dIII digest of rat total DNA mixed with component α/*Eco*RI* restriction segments. Numbers to the right of the gel refer to the lengths in nbp of the rat/*Hin*dIII restriction segments. Note that the rat segments are not in phase with the component α segments. *(E)* Calf satellite I digested with *Eco*RI. The digest was mixed with a partial *Eco*RI* digest of component α DNA (faint multimeric bands) before electrophoresis. Note that the calf satellite I segment migrates with the octamer of the component α series. *(F)* Calf satellite I treated with *Eco*RI*. Component α DNA was not added to this sample.

length (Botchan 1974). We treated component α in partial digests with *Eco*RI* and calf satellite I in limit digests with *Eco*RI. We then electrophoresed the two DNAs in the same sample well in 1.4% agarose gels (Fig. 3E). The octamer of the component α/*Eco*RI* partial digest and the unit segment of the calf satellite I/*Eco*RI digest migrated with the same mobility and indicated a molecular weight, based upon the component α octamer, of 9.2×10^5 daltons (1400 nbp). The higher order organization of calf satellite I therefore appears based upon octameric lengths of constitutive heterochromatin subunit DNA. In contrast with the uniform segment produced by *Eco*RI restriction of calf satellite I, the *Eco*RI* activity (Polisky et al. 1975) produced a complex series of bands, indicating at least six *Eco*RI* sites within the octameric repeat unit (Fig. 3F). In partial digests the length of the longest segment is 1400 nbp and that of the shortest segment is 176

nbp. In Figure 3F, note an especially prominent segment 352 nbp in length and another that is only 60 nbp shorter than the octameric segment of 1400 nbp. It will be shown that the *Eco*RI* sites giving rise to these bands and the other *Eco*RI* sites within the octamer are protected from restriction attack in intact nuclei.

It is not known whether the sequence structure of all highly repetitive mammalian DNAs is organized in an apparent relation to constitutive heterochromatin subunit structure. In other work (F. L. Brown and J. J. Maio, unpubl.), sheep and goat satellites I are quite similar to one another, but their higher order organization cannot be superimposed upon simple multiples of 176 nbp. In this respect, they resemble the uncharacterized rat fraction shown in Figure 3C. On the other hand, satellites II of both animals do show such an organization. The recurrent molecular weights in the repeat organization of the highly repetitive DNAs of such diverse mammalian species described above are not coincidental: rather, they reflect a common mechanism at play in determining the evolutionary development and long range organization of the sequences. Whatever the nature of this mechanism, it involves some basic unit or structure in the mammalian constitutive heterochromatin that we studied. Component α DNA probably expresses this unit or structure in one of its simplest forms. Furthermore, with the exception of component α DNA which shows an arithmetic series based on the monomeric unit, the highly repetitive DNAs examined repeatedly showed a distinct accent on even-numbered or geometric multiples of this fundamental unit in their long range organization. Figure 4 schematizes the higher-order organization of the various highly repetitive DNAs suggested by our data.

Highly Repetitive DNAs, Chromatin Subunit Structure, and Nucleosome Phasing

The higher-order organization of the mammalian highly repetitive DNA sequences is modulated or constrained by some common mechanism involved in the evolution of the sequences. The experiment shown in Figure 1E–G suggested that such a mechanism involves the subunit structure of chromatin and that a direct relation might exist between the DNA repeat unit and the nucleosomal proteins. To test this hypothesis, it was first necessary to construct a restriction map of the component α sequence (Fig. 5A). This mapping (Brown et al. 1977) led to the findings that (1) nearly 100% of the component α repeat sequences have either an *Eco*RI* or a *Hin*dIII site and about 70% of the sequences have both sites. (2) The *Eco*RI* and *Hin*dIII sites are 36 nbp apart at the near spacing and 140 nbp apart at the far spacing. (3) The *Hae*III sites are located near the middle of the segments released by *Eco*RI*. The map data were useful in two tests to determine

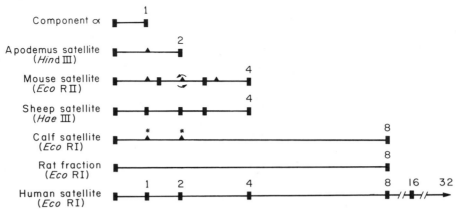

Figure 4. Tentative arrangement of major and minor restriction sites in various highly repetitive mammalian DNAs. The repeat unit of component α DNA (176 ± 4 nbp, top line) is taken as the reference standard for other DNA repeat units and for constitutive heterochromatin subunits. Major restriction sites in the sequences are indicated by rectangles; sites that occur infrequently and which give rise to minor, intermediate series of bands in the gel analysis are indicated by triangles. The suggested dimeric organization of the *Apodemus* satellite is based upon the restriction analysis of satellite DNAs of this genus presented by Cooke (1975). The length of the sequences were conserved among different families although the sequence itself changed. We suggest that a dinucleosome length of the repeat sequence dominated its recent expansion. In Cooke's (1975) study, minor series of segments based upon half this length were also generated in *Hind*III digests of the *A. microps* satellite. The minor sites suggest that remnants of earlier events involving mononucleosome lengths in the hierarchical development of the sequence are still in evidence.

The scheme for the organization of major and minor restriction sites in mouse satellite DNA is based upon Southern's (1975) study and our results with *Eco*RII. As proposed here, the entire repeat unit, which in the modern sequence comprises three major *Eco*RII sites, recapitulates as a nucleosome tetramer (Fig. 2B). Thus, four nucleosome lengths of the sequence participated in its recent expansion. Southern (1975) also described two minor, intermediate series of bands in which the 1½-mers and 1¼-mers could be distinguished. The series based upon the 1½-mer was twice as intense as the series based upon the 1¼-mer. The scheme proposed here accounts for all three series of major and minor bands in a simple and symmetrical manner apposed to the basic repeat unit of component α DNA (Fig. 2C). It also accounts for the relative molar proportions of the series based upon 1½-mers and 1¼-mers because there are two ways of generating 1½-mers but only one way to produce the 1¼-mer within the tetranucleosome DNA repeat unit. Note that the symmetry in the spacing of the major and minor *Eco*RII restriction sites might also derive from alternating inversions (arrows in the figure) of dinucleosome lengths of the sequence in a recent stage of its expansion. Perhaps the genera *Mus* and *Apodemus* have had a recent, common ancestor in which an AT-rich satellite DNA showed a basic repeat unit based upon dinucleosome lengths of the repetitive sequence.

The scheme for the arrangement of *Hae*III sites in sheep satellite II is derived from our gel analysis (Fig. 2D,E) and from microdensitometer tracings of the gel photographs. There are several possible arrangements of the segments within the repeat unit. The arrangement shown here is arbitrary but represents the three segments comprising a tetramer in the correct molar ratios of 1:2:1.

*Eco*RI restriction analysis of an uncharacterized rat fraction (Fig. 3A,B) and calf satellite I (Fig. 3C,E) suggests a superimposed higher order organization based upon nucleosome octamers. We again presumed that this organization represents a recent development in the evolution of the sequences. Minor bands of DNA corresponding in length to component α monomers and dimers were also detected in *Eco*RI* digests of calf satellite I (Fig. 3F and Maio et al. 1977). The *Eco*RI* sites giving rise to these segments are indicated by asterisks in the Figure.

The spacings of *Eco*RI sites giving rise to a geometric series of bands in human satellite (Fig. 2F) are indicated schematically in the bottom diagram. It is likely that the series extends beyond the 32-mer (see Fig. 2F).

whether there exists a fixed and specific alignment of chromatin subunit proteins with DNA repeat sequences — that is, whether nucleosome phasing occurs in component α chromatin.

In the first test, we treated intact monkey nuclei with each of the four restriction enzymes, *Eco*RI, *Eco*RI*, *Hind*III and *Hae*III, known to attack the component α sequence (Fig. 1A–D). We then purified the DNA from each of the digests and analyzed the products by gel electrophoresis to see which of the enzymes attacked the sequence in intact nuclear chromatin and which enzymes did not. If there is a phase relation between the restriction sites (and hence, the DNA repeat units) and chromatin subunit proteins, then some of the restriction sites will be cleaved by the corresponding enzymes and other sites will be protected. If the chromatin subunit pro-

teins are randomly oriented with respect to the DNA repeat units, then all restriction sites will be randomly exposed to attack. When this experiment was performed, we found that only the *Eco*RI-RI* sites were readily accessible to restriction attack in intact monkey nuclei: the *Hind*III and *Hae*III sites were relatively or completely protected (Musich et al. 1977b). The results of this experiment suggested the scheme shown in Figure 5B, in which the *Eco*RI-RI* sites in component α chromatin occur at nucleosomal interstices — that is, the same sites between nucleosomes that are readily accessible to attack during the initial stages of micrococcal nuclease digestion of the total chromatin. The *Hind*III and *Hae*III sites were protected from attack and are located within the relatively protected regions of nucleosomal cores.

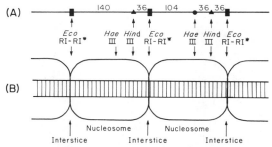

Figure 5. Restriction map of component α DNA and the proposed relation between restriction sites and nucleosome subunit structure. *(A)* The restriction map of component α is based upon sequential digestions of component α repeat segments generated by *Eco*RI, *Eco*RI*, *Hind*III or *Hae*III (Fig. 1) with a second, heterologous enzyme (Brown et al. 1977). The map is in good agreement with data on the component α sequence presented by Gruss and Sauer (1975) and Fittler (1977). Note especially the spatial arrangement of the two major restriction sites in the sequence, *Hind*III and *Eco*RI*, which are separated at the further distance by 140 nbp. Component α DNA also contains secondary *Eco*RI* sites (Brown et al. 1977), which produce several minor, intermediate series of bands upon gel analysis (Fig. 2A). To simplify the scheme, these secondary *Eco*RI* sites are not included in the map.

(B) Proposed arrangement of restriction sites and nucleosome structure in component α chromatin. For simplicity, DNA (hatched strip) and nucleosomes are shown diagrammatically as linear structures rather than globular ones. Each vertical bar represents about 10 nbp. Interstices denote internucleosomal regions that are readily accessible to attack during the initial stages of micrococcal nuclease digestion of chromatin. Because component α DNA in intact nuclei is attacked by *Eco*RI and *Eco*RI* (Musich et al. 1977b), these sites are positioned at or near the nucleosome interstices. *Hind*III and *Hae*III sites are relatively or completely protected from restriction attack in intact monkey nuclei (Musich et al. 1977b). These sites are positioned in relatively inaccessible regions of nucleosome structure, that is, on or near the nucleosome cores in accordance with their map distances from the *Eco*RI-RI* sites. This tentative arrangement of restriction sites and nucleosome structure was based upon restriction analysis of the sequence in intact nuclei and independently confirmed by experiments described in Figs. 6 and 7. Note that this scheme implies a phase relation between the DNA repeat units and chromatin subunits.

There are many objections to this type of experiment in which restriction enzymes are used to digest differentially specific sequences in intact nuclei: some restriction enzymes may be unable to pass the permeability barrier of the nuclear membrane; others may be differentially inactivated by adsorption to nuclear proteins, etc. It was therefore necessary to take an approach from another angle. We treated isolated monkey nuclei first with micrococcal nuclease to cleave the regions between nucleosomes and to release chromatin monomers and cores. We fractionated and purified the subunits, deproteinized the DNA and then treated the DNA segments with the various restriction enzymes that attack component α DNA. If the tentative scheme shown in Figure 5B for the orientation of the repeat sequences and chromatin subunit proteins is correct, then subsequent *Eco*RI-RI* digestion of the released chroma-

tin subunit DNA will not release any major secondary segments of component α. This is because the *Eco*RI-RI* sites are too near the sites that are cleaved by micrococcal nuclease. But, from the restriction map (Fig. 5A) and the phasing scheme in Figure 5B, treatment with *Hind*III should release a conspicuous segment of 140 nbp. On the other hand, if the restriction sites shown in the map (Fig. 5A) are randomly oriented with respect to chromatin subunit proteins, then subsequent digestion of the micrococcal nuclease segments should release a random assortment of secondary segments upon treatment with any of the restriction endonucleases.

CV1 nuclei were treated with micrococcal nuclease and the chromatin monomer and core subunits were separated on isokinetic sucrose gradients (Fig. 6). The DNA was purified from region I (chromatin monomers and some dimers) and region II (nucleosome cores) and treated in separate reaction mixtures with *Eco*RI, *Eco*RI* and *Hind*III (Fig. 7). Only *Hind*III released a prominent secondary segment from the chromatin monomer region, and it was 140 nbp in length. Neither *Eco*RI nor *Eco*RI* released any prominent secondary segments, and heterogeneous DNA fragments were not detected after treatment with any of the restriction enzymes (Fig. 7A–D). A similar pattern was obtained when the DNA segments of the nucleosome cores were treated with the restriction enzyme activities (Fig. 7F–H). Again, only *Hind*III released a prominent secondary segment from the core DNA. Heterogeneous DNA fragments resulting from a possible random orientation of the restriction sites with chromatin subunit proteins were not produced. The results are consistent with the scheme shown in Figure 5B and indicate a phase relation between the chromatin subunit proteins and the component α repeat sequence.

Lipchitz and Axel (1976) reported that about 60% of the *Eco*RI restriction sites in calf satellite I chromatin are accessible to attack in intact calf nuclei. Although those authors interpreted their results as indicative of random phasing, they pointed out that more complex models of nucleosomal organization that are nonrandom with respect to DNA sequence are also in accord with their data. Figure 8A–C confirms Lipchitz and Axel's result showing that in intact nuclei, calf satellite I chromatin has accessible *Eco*RI sites. Calf satellite I DNA has at least six *Eco*RI* sites included within the segment of 1400 nbp released by *Eco*RI (Fig. 3F). One of these sites is only 60 nbp distant from the regular *Eco*RI site, but this is a critical distance because in intact nuclei, this, and the other *Eco*RI* sites within the repeat segment, are protected from restriction attack (Fig. 8F). This result strongly suggests that some *Eco*RI* sites in calf satellite I chromatin are specifically blocked by chromatin proteins, whereas other *Eco*RI-RI* sites are exposed. Since we showed earlier in this article that other mammalian highly repetitive DNAs share elements of their repeat structure

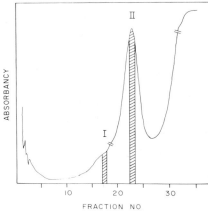

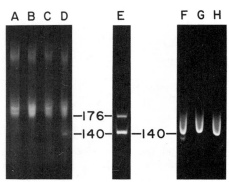

Figure 6. Fractionation of CV1 nucleosomes released by micrococcal nuclease digestion. Nuclei were isolated from CV1 cells as described in Materials and Methods and washed once in 20 volumes of 25 mM Tris-HCl, pH 8.5, 2 mM MgCl$_2$ and 0.25 mM CaCl$_2$. They were resuspended in this buffer to 5×10^7 nuclei/ml and treated with 30 units/ml of micrococcal nuclease for 5.5 min at 37°C. The reaction was stopped by chilling in ice and centrifugation at 5000g for 7 min at 4°C. EDTA was added to the supernatant to 5 mM and aliquots of the suspension were layered onto isokinetic sucrose gradients buffered with 2 mM Tris-HCl, pH 7.0, 1 mM EDTA. The gradients had the following parameters: $c_t = 5\%$ sucrose, $c_r = 28.8\%$ sucrose, $v_m = 33$ ml. Centrifugation was at 25,000 rpm for 20 hr at 4°C in a Beckman SW27 rotor. The direction of sedimentation is from right to left in the Figure. The gradient fractions were collected from the bottom and simultaneously scanned at 254 nm with a Pharmacia UV monitor, model 110 (Piscataway, N.J.). The break in the curve between peaks I and II denotes a fourfold increase in the optical density scale. This fractionation procedure results in the isolation of mono- and dinucleosomes (Peak I) and nucleosome cores (Peak II) deficient in H1 histone. DNAs from the peak fractions I and II shown were purified by treatment with 50 μg/ml of protease K (E.M. Biochemicals) in 0.1% sodium dodecyl sulfate at 37°C for 4 hr. NaCl was then added to 0.5 M and EDTA to 25 mM and the mixtures deproteinized twice with chloroform-isoamyl alcohol. (It was essential to remove the protease K for subsequent treatment of the DNA with restriction enzymes.) The DNAs were dialyzed against two changes of 0.01 M Tris-HCl, pH 8.4, plus 1 mM EDTA, precipitated with ethanol, and resuspended in the appropriate restriction enzyme buffers before treatment with the enzymes (Fig. 7).

Figure 7. Treatment with restriction endonucleases of DNA extracted from nucleosomes and from core segments released by micrococcal nuclease digestion of CV1 nuclei. DNAs from peak fractions I and II of the isokinetic sucrose gradients were prepared for restriction endonuclease digestion as described in Fig. 6. Panels A–D show the control and restricted segments isolated from region I (predominantly nucleosome monomers) after electrophoresis in 6% polyacrylamide slab gels. Panels F–H show control and restricted segments isolated from region II (nucleosome cores). The gels contained ethidium bromide and were also poststained. For molecular weight comparisons, a double digest of purified component α with *Hin*dIII and *Eco*RI* (Brown et al. 1977) is shown in panel E. All incubations with *Eco*RI and *Eco*RI* were with 120 units of *Eco*RI for 6.5 hr at 37°C. Incubations with *Hin*dIII were with 6 units of enzyme for 6.5 hr at 37°C. All electrophoreses were in 6% polyacrylamide slab gels in the presence of ethidium bromide. Numbers refer to lengths of segments in nbp.

DNA from region I: *(A)* unrestricted control; *(B)* restricted with *Eco*RI; *(C)* restricted with *Eco*RI*; *(D)* restricted with *Hin*dIII. Note the uniform band embedded within the more heterogeneous nucleosome monomer DNA in panels A–C. This band disappears in the *Hin*dIII digest (panel D) with the concomitant appearance of the band of DNA 140 nbp in length. We suggest that this homogeneous component within the heterogeneous chromatic band in panels A–C represents segments of component α cut sharply and uniformly by the previous treatment with micrococcal nuclease. *(E)* For molecular weight comparisons of secondary segments, purified component α DNA was digested with both *Hin*dIII and *Eco*RI*. In accordance with the restriction map (Fig. 5), a conspicuous secondary segment 140 nbp in length is released.

DNA from region II: *(F)* restricted with *Hin*dIII; *(G)* restricted with *Eco*RI*; *(H)* unrestricted control. These gels were slightly overloaded and also poststained, accounting for the edge effects in the core segments. Note the absence of DNA segments of heterogeneous lengths in all of the panels. Such segments would be expected if the chromatin subunits were randomly oriented with respect to the various restriction sites.

in common with component α and calf satellite I, it seems likely that similar phase relations or specific blockage of restriction sites will emerge in future studies of other highly repetitive mammalian DNA sequences.

Nucleoproteins Associated with a Constitutively Repressed Heterochromatic Mammalian DNA Sequence

The transcriptional inertness of the highly repetitive DNAs, their unusual cytological properties, apparent phase relations with chromatin subunit structure, and the basic conservatism of their nucleotide sequence organization all suggested some

special nucleoprotein structure associated with these DNAs. We have tried to gain an insight into the nucleoprotein composition of constitutive heterochromatin by one of the gentlest methods available for isolating nucleosomal arrays of a specific DNA sequence — restriction enzyme excision from intact nuclei (Musich et al. 1977a). Harsh treatments such as sonication, shearing, or salt extraction were avoided because such treatments may disrupt the native structure of chromatin and induce protein migrations or exchanges.

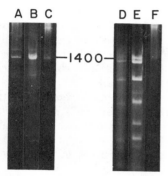

Figure 8. Restriction of purified calf total nuclear DNA, calf satellite I, and calf nuclei with *Eco*RI and *Eco*RI*. Purified calf total nuclear DNA and calf satellite I were prepared from calf kidney cells as described previously (Kurnit et al. 1973). The DNAs were treated with 120 units of *Eco*RI or *Eco*RI* for 4 hr; 120 units of *Eco*RI or *Eco*RI* were again added and incubation continued for 4 hr. Nuclei from primary cultures of calf kidney cells were prepared for restriction in *Eco*RI or *Eco*RI* buffers as described in Materials and Methods. The nuclei were incubated in 250 μl of reaction mixtures at a concentration of 2.5×10^7 nuclei/ml in the presence of 600 units/ml of *Eco*RI or *Eco*RI* for 6.5 hr at 37°C. DNAs from the nuclear digests were purified by phenol deproteinization and analyzed in 1.4% agarose gels.

(A) Calf total nuclear DNA after restriction with *Eco*RI. *(B)* Calf satellite I after restriction with *Eco*RI. *(C)* Isolated calf nuclei treated with *Eco*RI. *(D)* Total calf nuclear DNA after restriction with *Eco*RI*. *(E)* Calf satellite I after restriction with *Eco*RI*. *(F)* Isolated calf nuclei treated with *Eco*RI*. Note that a segment 1400 nbp in length was released by *Eco*RI* treatment of intact nuclei and that none of the other *Eco*RI* sites (panel *E*) included within the *Eco*RI segment (panel *B*) were attacked. This suggests a specific blockage of *Eco*RI* sites by chromatin proteins in intact nuclei.

We treated isolated monkey nuclei with high concentrations of *Eco*RI, an activity which attacks component α sequences in intact nuclei (Fig. 5B). We then lysed the nuclei in low-ionic-strength EDTA buffer and fractionated the chromatin on isokinetic sucrose gradients (Fig. 9). Region I in Figure 9 contained mostly bulk chromatin and some very high molecular weight nucleosomal arrays of component α chromatin. Region II contained only component α oligonucleosomes, ranging from the monomers near the top of the gradient to about the 15-mer near the middle of the gradient. The purity of these fractions was demonstrated by two independent tests: (1) re-restricting the DNA with *Hind*III, which reduces component α, but not bulk DNA, predominantly to monomeric segments, and (2) analytical CsCl buoyant density centrifugation of the heated and annealed DNA (Maio 1971).

Several changes in the protein gel patterns occurred in the transition region of the sucrose gradients from predominantly bulk DNA chromatin (region I) to pure component α nucleosomal arrays in region II (Fig. 9). The most conspicuous change involved the H1 histones, whereas the relative proportions of the core histones, H2A, H2B, H3, and H4,

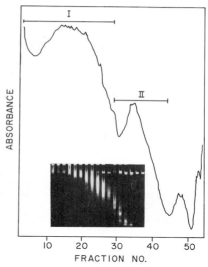

Figure 9. Sedimentation pattern of nucleosomal arrays excised from CV1 nuclei by *Eco*RI restriction endonuclease. Isolated CV1 nuclei (5×10^7/ml) in *Eco*RI restriction buffer were treated with *Eco*RI endonuclease (1000 units/ml) at 37°C for 2.5 hr. The digested nuclei were sedimented and the nucleosomes released by resuspension in 0.5 mM EDTA (pH 7.0). Aliquots of the suspension were layered onto isokinetic sucrose gradients and centrifuged at 22,000 rpm for 18 hr at 4°C in a Beckman SW27 rotor. The gradients were buffered with 2 mM Tris-HCl, pH 7.0, 0.1 mM EDTA and had the following parameters: $c_t = 5\%$ sucrose, $c_r = 28.8\%$ sucrose, $v_m = 33$ ml. The direction of sedimentation is from right to left in the figure. The gradients were collected from the bottom and scanned at 254 nm with a model 110 Pharmacia UV monitor. Region I of the gradient contains main band chromatin, whereas region II consists of component α nucleosomal arrays.

The insert (not in scale with the gradient profile) shows an analysis of the DNA extracted from nucleosomal fractions spanning the entire gradient. To reduce the number of samples, the fractions were pooled in consecutive groups of three before deproteinization and electrophoresis in 1.4% agarose gels. The high molecular weight DNA, to the left in the insert, represents mostly bulk DNA and some component α sequences (region I). The distinct bands to the right in the insert (region II) are component α restriction segments, which range from the monomer to about the 15-mer. Nucleosomal proteins associated with fractions from region II of the gradient are shown in Fig. 10.

remained unchanged throughout the gradient. The H1 histones were depleted and possibly absent in the oligonucleosomes containing only component α DNA (Fig. 10). Coincident with the reduction in the H1 histones and in the same molecular weight range (24,000 to 43,000 daltons), there appeared five minor nonhistone proteins which were not detected in chromatin from fractions containing bulk DNA (Fig. 10). The minor, low molecular weight nonhistone proteins represented 12.4% of the protein in component α nucleosomes and thus replaced the H1 histones (13.7% of the protein) nearly quantitatively. The resistance to salt (0.6 to 2.0 M) and sulfuric acid (0.4 N) extraction indicated that the low molecular weight nonhistone proteins are truly nonhistone proteins, rather than modified forms of H1 or

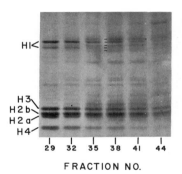

FRACTION NO.

Figure 10. Gel electrophoresis pattern of the proteins associated with the component α nucleosomes released by *Eco*RI digestion of CV1 nuclei. The figure shows the change in the gel pattern between bulk chromatin (fraction 29, region I of the sucrose gradient of Fig. 9) and pure component α nucleosomal arrays (fractions 32–44 of region II). To reduce the number of samples, fractions were pooled in consecutive groups of three. The fraction numbers shown refer to the middle fractions of each pool, that is, fraction #29 is the pool of fractions 28, 29, and 30 in Fig. 9. The proteins from the various fractions were precipitated with 20% TCA, washed with acetone, and then analyzed by SDS-polyacrylamide gel electrophoresis (Weintraub et al. 1975). The photograph encompasses only the molecular-weight region of 12,000–50,000 daltons. Five low-molecular-weight nonhistone proteins in the molecular-weight range 24,000–43,000 daltons are indicated between the gel patterns of fractions 35 and 38.

of the core histones, and that they are tenaciously bound to the component α chromatin.

These results corroborate earlier experiments in which it was shown that high-salt extraction of isolated mouse nucleoli (Schildkraut and Maio 1968) or metaphase chromosomes (Maio and Schildkraut 1969) resulted in the solubilization of bulk DNA sequences and histones. The insoluble pellet fractions became progressively enriched in satellite DNA with repeated extractions with 2 M NaCl. These results are interesting in view of the report by Gorovsky et al. (this volume) that the transcriptionally inactive micronuclear chromatin of *Tetrahymena* also lacks H1 histones, although such proteins are found in the transcriptionally active macronucleus. The results presented here also pose the question of whether such nonhistone proteins account for the phasing relation we observed between component α nucleosomes and the DNA repeat units. Is it possible that these nonhistone proteins are capable of recognizing specific DNA sequences and thereby orient the deposition of the core histones which are incapable of such stereospecific recognition?

The Development of Highly Repetitive DNAs

Several models have been proposed for the origin, evolution, and higher-order organization of highly repetitive DNAs (for a review, see Southern 1974). We suggest that at least two stages have characterized the development of many, but not all, highly repetitive DNAs:

In stage 1, sequences which may be as short as a d(A-T) dinucleotide or a heptanucleotide (Sueoka and Cheng 1962; Gall and Atherton 1974) formed the basic repeating unit which was reiterated many times within a relatively short stretch of DNA (< 200 nbp). The mechanism through which such sequences arise remains extremely obscure: slippage replication (Wells et al. 1967), mutation from simple homopolymerlike sequences, and unequal crossovers (Smith 1974, 1976) have been proposed by various authors, but direct proof of these hypotheses is lacking.

Stage 2, the principal subject of this paper, is characterized by the long-range or higher-order development of the sequence, which may involve hundreds (component α and mouse satellite) or thousands of nucleotide base pairs as in calf satellites I and IV (Botchan 1974; F. L. Brown and J. J. Maio, unpubl.). It is this latter stage that may account for (1) the large-scale effects in the expansion (or contraction) of highly repetitive DNAs that have been observed in closely related mammalian species, (2) the rapid dissemination of such sequences throughout the genome during evolution, and (3) the sequence organization that is commonly revealed by restriction enzyme analysis.

It is not known whether the mechanisms of stages 1 and 2 are separate, or whether each stage represents a continuation of the same basic mechanism. But because satellite DNAs vary greatly in complexity, it is possible that one mechanism accounts for the short-range expansion and homogeneity of a highly repetitive sequence (stage 1), and still another for the higher-order organization (stage 2). We presented evidence that the basic repeat unit of 176 nbp of component α is not internally highly repetitious and that the sequence may be unique throughout its length (Brown et al. 1977). The corrected sequence complexity of about 200 nbp reported for several mammalian highly repetitive DNAs is in accord with this hypothesis. Conceivably, component α by-passed the short-range development (stage 1) involving oligonucleotide repeats characteristic of the crab and *D. virilis* satellites and expanded directly as a 176 nbp repeat unit throughout the genome. What is the nature of this mechanism that determines the higher-order organization and massive development of the sequences?

We have stressed throughout this paper that the higher-order organization of mammalian highly repetitive DNAs is mediated in some way by chromatin subunit structure, and that certain preferred repeat arrays — seemingly based upon even-numbered or geometric multiples of a nucleosome length of the sequences — are generated. Rolling circle formation (for which there is little or no direct evidence in mammalian cells) and slippage replication seem inadequate to account for these distinctive patterns, which frequently involve sequences of millions of daltons in molecular weight. Among many possible models, we briefly discuss two that seem particularly attractive because they are in reasonable accord with much of our data.

Unequal crossover model. Smith (1974, 1976) proposed that a segment of DNA whose sequence is not maintained by selection can be deleted or tandemly duplicated through random, unequal crossovers between two daughter molecules produced by the replication of a parental DNA molecule. Such events must occur in the germ-line cells in order to be transmitted to progeny. By computer simulation, Smith (1974) demonstrated that such a mechanism could give rise to repetitive DNA sequences with the characteristics of satellite DNAs. In Smith's model, the pattern of repeats and the lengths of the periodicities depend upon the exact nature of the unknown crossover mechanism.

Smith's basic proposals are plausible explanations for many of the phenomena observed with mammalian highly repetitive DNAs; however, our data suggest that the repeat periodicities of the highly repetitive sequences are biased by the subunit structure

of constitutive heterochromatin and perhaps entirely determined by that structure. Thus, the element of randomness in the mechanism proposed by Smith (1974, 1976) is greatly modulated by the influence of chromatin subunits in determining the permissible sites of crossovers. In Figure 11, we have revised Smith's model in accordance with our observations. Two essential features of this revision are (1) the distribution of restriction sites in the repeat sequence is a reflection of the repeat periodicity produced by multiple cycles of unequal crossovers occurring at defined regions of the chromatin subunits. These sites of crossover are taken to be at nucleosomal interstices. (2) Entire blocks of the repetitive sequence equivalent to integral multiples of nucleosome lengths participate in the crossover mechanism. From the schemes in Figure 11, it is apparent that the length of the arrays may quickly become quite large: thus, the appearance and fixation of a repeat sequence such as component α (comprising nearly 10^{12} daltons of DNA in each AGMK cell) might then occur after a number of crossover events smaller by orders of magnitude than the mil-

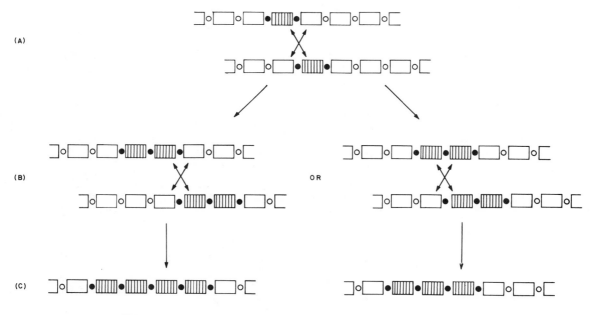

TETRAMER   TRIMER

Figure 11. Tentative scheme for the expansion of nucleosome lengths of repetitive sequences by unequal crossovers at nucleosome interstices. Entire blocks of the sequence equivalent to integral multiples of nucleosome lengths of DNA (hatched blocks) participate in the unequal crossover mechanism. The critical crossover points at nucleosome interstices are indicated by closed circles. It is suggested that these regions contain accessible, short, repeat sequences or inverted repeats that are peculiarly adapted for recombination events. The crossovers between daughter molecules of the same parental molecule shown in *(A)* involve mononucleosome lengths of a sequence that is flanked at one or both ends by the short repeat. This gives rise to dinucleosome lengths of the repeat sequence *(B)*. If end-to-end crossovers involving whole nucleosome lengths predominate *(B, left)*, the sequence may expand geometrically. Divergence in one or more preexisting restriction sites, or the appearance of a restriction site during these early events, may later reflect such an expansion (e.g., the human satellite). Alternatively, if blocks of mononucleosomes undergo crossovers more or less at random or by intercalation of single mononucleosome lengths *(B, right)*, then an arithmetic expansion of repeat units based upon mononucleosome lengths may be generated (e.g., component α DNA). According to this scheme, changes such as base substitutions introduced locally during the later evolution of the sequence will be subordinate to the massive and rapid changes in the amounts of the repetitive sequence caused by crossovers between blocks of nucleosomes and large polynucleosomal arrays.

lions of events required by computer-simulated, random models and on a reduced evolutionary time-scale.

Replicative loop hypothesis. Keyl (1965) noted a geometric expansion of DNA content in 30 selected homologous chromomeres in different subspecies of *Chironomus thummi.* This process was accompanied by a prolongation of DNA synthesis in the affected chromomeres and, frequently, by the formation of cytologically detectable inversions. Keyl suggested that the DNA of the chromomere formed a replication loop, but the daughter strands of the replicated DNA did not separate. Instead, they joined end-to-end and doubled the DNA content of the affected chromomere such that recurrences of this process during the evolution of different subspecies of *Chironomus* produced a geometric series of DNA content in a given chromomere. This hypothesis could account for the higher-order organization of the repetitive DNAs seemingly based upon recurrent, geometric multiples of nucleosome-length sequences. However, to be cytologically detectable, the variations in chromomere DNA content described by Keyl presumably involved DNA segments in the molecular weight range of millions of daltons, whereas the periodicities we describe involve sequences one or two orders of magnitude shorter. Conceivably, both phenomena are due to the same basic mechanism: restriction enzyme analysis reveals this mechanism operating at short range in the hierarchical development of the highly repetitive DNAs.

We have modified Keyl's hypothesis according to the scheme shown in Figure 12. In this scheme, the looped replicative sequence may or may not be internally repetitious, but the loop comprises a nucleosome length of DNA or a multiple thereof. As in the previous unequal crossover model (Fig. 11), simple, short repeats or short palindromes uniquely adapted for crossover (Sobell 1972; Wagner and Radman 1975) occur at nucleosome interstices. These sequences are brought into proximity when they occur at the base of the twisted loops and may participate in intrastrand crossover events as shown in the upper diagram of Figure 12. There are several possible modes for such crossover events, two of which are shown in Figure 12. One of these (Fig. 12, left) may generate tandem repeats; another may generate inverted, palindromic repeats (Fig. 12, right). (Other possible modes, not shown here, may result in the excision of the loop as a circular, extra-chromosomal molecule, or the reduplication of the sequence in one of the daughter strands and its loss in the other daughter strand upon completion of replication of the entire chromosome.)

The replicative loop hypothesis is an attractive alternative to Smith's unequal crossover model for several reasons: (1) It is based upon somewhat more convincing cytological evidence. (2) It could account

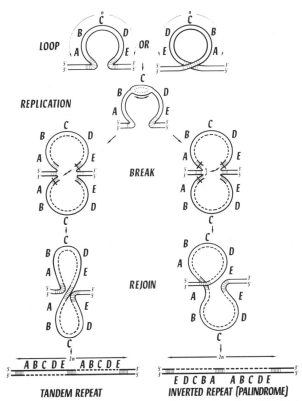

Figure 12. The expansion of nucleosome lengths of highly repetitive sequences by replicative loop formation. In the top diagram the arbitrary sequence, ABCDE, to be expanded forms an open *(left),* or in chromatin, a supercoiled loop *(right)* as illustrated. (It is convenient to schematize the subsequent events with an open loop as shown.) The unit length of DNA in the loop ($n=1$) is the length of DNA in a single nucleosome; larger loops correspond to integral multiples of this nucleosome length. The hatched areas at the base of the loop indicate a short repeat sequence, or inverted repeat, at which intrastrand crossovers might readily occur (cf. Fig. 11). The sequence occurs at some fixed and accessible site in nucleosome structure, presumably at nucleosome interstices (also indicated by the hatched areas at the base of the loops). The DNA replicates but the daughter strands do not separate. Instead, intrastrand crossovers occur which are facilitated by the short repeat sequences located at the base of the loops and between nucleosomes. Two possible crossovers resulting in an expansion of the sequence are shown: one crossover results in a tandem repeat *(left),* double crossovers produce an inverted, palindromic repeat *(right).* In either case, the sequence is doubled. If this process recurs, then a quadrupling, octupling, etc., of the sequence results and a geometric series forms. Note that the events schematized here imply a disproportionate replication of DNA in the loops.

for the observations we report here and the geometric increments in the doses of rRNA cistrons reported in different species of amphibia (Buongiorno-Nardelli et al. 1972). (3) It is consistent with the generation of long, inverted palindromic sequences that characterize *Tetrahymena* macronuclear rDNA (Karrer and Gall 1976). (Note: this model might predict the existence of similar, short repetitive sequences at both termini of such long palindromic sequences.) (4) It is consistent with recent observations that chromatin of eukaryotic cells is arranged in vivo in negatively supercoiled configurations (twisted loops in Fig. 12), in which the long supercoils of up to 85 kb in length are themselves maintained by toroidal supercoils of nucleosome-length stretches of DNA (Benyajati and Worcel 1976).

CONCLUDING REMARKS

The hypothesis that the unit segments and higher-order organization of highly repetitive sequences develop from chromatin subunit structure has implications for the development, organization, and nuclear distribution of repetitive DNAs other than satellites. Repetitive DNAs or short satellitelike sequences are increasingly found as flanking or spacer sequences in nontranslated portions of a variety of structural genes as well as in repetitive cistrons coding for various types of cellular RNAs. Possibly, the development and mechanics of such repetitive sequences are similar to those of satellites and involve integral arrays of repeat units based on chromatin subunit structure. If such repetitive sequences flank transcribed, structural sequences, their presence may influence the number, distribution, and perhaps the intranuclear rearrangements of the transcribed sequences they enclose or with which they are associated. A suggestive example is presented by the genes coding for the 5S RNA of *Xenopus laevis*. Like satellite DNAs, the 20,000 copies of these genes are located on many chromosomes and, indeed, a major portion of the repeat unit is an AT-rich, satellitelike sequence (Fedoroff and Brown, this volume). Interestingly, the repeat unit, which is about 700 nbp long, approximates a tetranucleosome length of DNA (cf. the organizational units proposed for mouse satellite and sheep satellite II, Fig. 4).

Recombinant DNA technology is proving extremely valuable in understanding the complex nucleotide structure of eukaryotic genes. However, little is known of the nucleoprotein configurations associated with the presumptive control sequences or with the transcribed and translated portions of the genes. Although several previous reports indicated random phasing between the bulk of the DNA sequences and nucleosomal proteins, our results indicate that phasing occurs with at least some highly repetitive sequences and that the nucleoprotein composition of chromatin containing such sequences differs from that of the bulk chromatin. Conceivably, the short repetitive sequences associated with nontranscribed or nontranslated regions of several eukaryotic genes are in fact control sequences and do not differ substantially from highly repetitive DNAs except in their transcriptional flexibility and the lengths of the tandem repeats. However, because such repetitive sequences constitute a small part of the complete cistron, nucleosome phasing and altered nucleoprotein composition (indicative of regulatory elements?) will be difficult to detect. Accordingly, the study of highly repetitive DNAs and their associated nucleoproteins may provide the first glimpse of the mechanism for transcriptional activation or inactivation of eukaryotic structural genes.

Acknowledgments

These studies were supported by a grant (CA 16790) from the National Institutes of Health and in part by an NIH Institutional grant (GM 19100).

REFERENCES

ARRIGHI, F. E., T. C. HSU, S. PATHAK, and H. SAWADA. 1974. The sex chromosomes of the Chinese hamster: Constitutive heterochromatin deficient in repetitive DNA sequences. *Cytogenet. Cell Genet.* 13: 268.

BENYAJATI, C. and A. WORCEL. 1976. Isolation, characterization, and structure of the folded interphase genome of *Drosophila melanogaster*. *Cell* 9: 393.

BOTCHAN, M. R. 1974. Bovine satellite I DNA consists of repetitive units 1,400 base pairs in length. *Nature* 251: 288.

BOTCHAN, M., W. TOPP, and J. SAMBROOK. 1976. The arrangement of simian virus 40 sequences in the DNA of transformed cells. *Cell* 9: 269.

BROWN, F. L., P. R. MUSICH, and J. J. MAIO. 1977. Subunit structure of chromatin and the organization of eukaryotic highly repetitive DNA. I. Restriction analysis of African green monkey component α DNA. *J. Mol. Biol.* (in press).

BUONGIORNO-NARDELLI, M., F. AMALDI, and P. A. LAVA-SANCHEZ. 1972. Amplification as a rectification mechanism for the redundant rRNA genes. *Nat. New Biol.* 238: 134.

COMPTON, J. L., M. BELLARD, and P. CHAMBON. 1976. Biochemical evidence of variability in the DNA repeat length in the chromatin of higher eukaryotes. *Proc. Natl. Acad. Sci.* 73: 4382.

COOKE, H. J. 1975. Evolution of the long range structure of satellite DNAs in the genus *Apodemus*. *J. Mol. Biol.* 94: 87.

CORNEO, G., L. ZARDI, and E. POLLI. 1972. Elution of human satellite DNAs on a methylated albumin kieselguhr chromatographic column: Isolation of satellite DNA IV. *Biochim. Biophys. Acta* 269: 201.

FITTLER, F. 1977. Analysis of the α-satellite DNA from African green monkey cells by restriction nucleases. *Eur. J. Biochem.* 74: 343.

GALL, J. G. and D. D. ATHERTON. 1974. Satellite DNA sequences in *Drosophila virilis*. *J. Mol. Biol.* 85: 633.

GRUSS, P. and G. SAUER. 1975. Repetitive primate DNA containing the recognition sequences for two restriction endonucleases which generate cohesive ends. *FEBS Lett.* 60: 85.

HÖRZ, W., T. IGO-KEMENES, W. PFEIFFER, and H. G. ZACHAU. 1976. Specific cleavage of chromatin by restriction nucleases. *Nucleic Acids Res.* 3: 3213.

KARRER, K. M. and J. G. GALL. 1976. The macronuclear ribosomal DNA of *Tetrahymena pyriformis* is a palindrome. *J. Mol. Biol.* 104: 421.

KEYL, H. G. 1965. A demonstrable local and geometric increase in the chromosomal DNA of *Chironomus*. *Experientia* 21: 191.

KURNIT, D. M. and J. J. MAIO. 1973. Subnuclear redistribution of DNA species in confluent and growing mammalian cells. *Chromosoma* 42: 23.

KURNIT, D. M., B. R. SHAFIT, and J. J. MAIO. 1973. Multiple satellite deoxyribonucleic acids in the calf and their relation to the sex chromosomes. *J. Mol. Biol.* **81**: 273.

LAU, Y. F. and F. E. ARRIGHI. 1976. Studies of the squirrel monkey, *Saimiri sciureus*, genome. I. Cytological characterizations of chromosomal heterozygosity. *Cytogenet. Cell Genet.* **17**: 51.

LIPCHITZ, L. and R. AXEL. 1976. Restriction endonuclease cleavage of satellite DNA in intact bovine nuclei. *Cell* **9**: 355.

LOHR, D. and K. E. VAN HOLDE. 1975. Yeast chromatin subunit structure. *Science* **188**: 165.

MAIO, J. J. 1971. DNA strand reassociation and polyribonucleotide binding in the African green monkey, *Cercopithecus aethiops*. *J. Mol. Biol.* **56**: 579.

MAIO, J. J. and C. L. SCHILDKRAUT. 1969. Isolated mammalian metaphase chromosomes. II. Fractionated chromosomes of mouse and Chinese hamster cells. *J. Mol. Biol.* **40**: 203.

MAIO, J. J., F. L. BROWN, and P. R. MUSICH. 1977. Subunit structure of chromatin and the organization of eukaryotic highly repetitive DNA: Recurrent periodicities and models for the evolutionary origins of repetitive DNA. *J. Mol. Biol.* **117**: 637.

MUSICH, P. R., F. L. BROWN, and J. J. MAIO. 1977a. Subunit structure of chromatin and the organization of eukaryotic highly repetitive DNA. IV. Nucleosomal proteins associated with a highly repetitive mammalian DNA. *Proc. Natl. Acad. Sci.* **74**: 3297.

MUSICH, P. R., J. J. MAIO, and F. L. BROWN. 1977b. Subunit structure of chromatin and the organization of eukaryotic highly repetitive DNA. III. Indications of a phase relationship between restriction sites and chromatin subunits in African green monkey and calf nuclei. *J. Mol. Biol.* **117**: 657.

POLISKY, B., P. GREENE, D. E. GARFIN, B. J. MCCARTHY, H. M. GOODMAN, and H. W. BOYER. 1975. Specificity of substrate recognition by the *Eco* RI restriction endonuclease. *Proc. Natl. Acad. Sci.* **72**: 3310.

SANGER, F., G. M. AIR, B. G. BARRELL, N. L. BROWN, A. R. COULSON, J. C. FIDDES, C. A. HUTCHISON III, P. M.

SLOCOMBE, and M. SMITH. 1977. Nucleotide sequence of bacteriophage ϕX 174 DNA. *Nature* **265**: 687.

SCHILDKRAUT, C. L. and J. J. MAIO. 1968. Studies on the intranuclear distribution and properties of mouse satellite DNA. *Biochim. Biophys. Acta* **161**: 76.

SHEN, J. C., G. WIESEHAHN, and J. E. HEARST. 1976. Cleavage patterns of *Drosophila melanogaster* satellite DNA by restriction enzymes. *Nucleic Acids Res.* **3**: 931.

SLACK, J. M. W. 1974. The interpretation of oligonucleotide maps. A theoretical study of nucleic acid digests with special reference to repeat diverged sequences. *Biopolymers* **13**: 2241.

SMITH, G. P. 1974. Unequal crossover and the evolution of multigene families. *Cold Spring Harbor Symp. Quant. Biol.* **38**: 507.

———. 1976. Evolution of repeated DNA sequences by unequal crossover. *Science* **191**: 528.

SOBELL, H. M. 1972. Molecular mechanism for genetic recombination. *Proc. Natl. Acad. Sci.* **69**: 2483.

SOUTHERN, E. 1974. Eukaryotic DNA. In *Biochemistry of nucleic acids* (ed. K. Burton), vol. 6, p. 101. Butterworth, London.

———. 1975. Long range periodicities in mouse satellite DNA. *J. Mol. Biol.* **94**: 51.

SUEOKA, N. and T. CHENG. 1962. Natural occurrence of a deoxyribonucleic acid resembling the deoxyadenylate-deoxythymidylate polymer. *Proc. Natl. Acad. Sci.* **48**: 1851.

WAGNER, R. E. and M. RADMAN. 1975. A mechanism for initiation of genetic recombination. *Proc. Natl. Acad. Sci.* **72**: 3619.

WEINTRAUB, H., K. PALTER, and F. VAN LENTE. 1975. Histones H2a, H2b, H3, and H4 form a tetrameric complex in solutions of high salt. *Cell* **6**: 85.

WELLS, R. D., H. BÜCHI, H. KÖSSEL, E. OHTSUKA, and H. G. KHORANA. 1967. Studies on polynucleotides. LXX. Synthetic deoxyribonucleotides as templates for the DNA polymerase of *Escherichia coli:* DNA-like polymers containing repeating tetranucleotide sequences. *J. Mol. Biol.* **27**: 265.

Random Phasing of Polypyrimidine/Polypurine Segments and Nucleosome Monomers in Chromatin from Mouse L Cells

H. C. Birnboim,* R. M. Holford,* and V. L. Seligy†

*Biology and Health Physics Division, Atomic Energy of Canada Limited, Chalk River Nuclear Laboratories, Chalk River, Ontario K0J 1J0, Canada; †National Research Council of Canada, Cell Biochemistry Group, Ottawa, Ontario K1A 0R6, Canada

DNA from many higher organisms contains appreciable amounts of long runs of pyrimidine nucleotides (and complementary purine runs) greater than 25 residues in length (Birnboim et al. 1973; Birnboim and Straus 1975; Case and Baker 1975; Szala et al. 1976; Straus and Birnboim 1976). In the case of mouse cells, these polypyrimidine/polypurine segments range in length up to about 250 nucleotide pairs, contain both thymine and cytosine residues in the pyrimidine strand, and account for about 0.5% of the total nucleotides. They are not localized to a satellite component; instead they are widely distributed throughout the majority of the DNA at intervals averaging 7 kilobases (kb) (Birnboim 1976, 1978). Polypyrimidines are a component of middle repeated DNA (Straus and Birnboim 1974) and preliminary evidence suggests that they may be transcribed (Straus and Birnboim 1976). The function of polypyrimidine/polypurine segments in DNA is unknown, but the fact that they are found in DNA from a variety of unrelated eukaryotes suggests that they have been conserved because they are involved in some important cell function.

We have used micrococcal nuclease-digested chromatin (1) to determine the sensitivity of polypyrimidines relative to bulk DNA and (2) to see if polypyrimidines are arranged in any special phase relationship to nucleosomes. The bulk of single-copy DNA seems to be associated with nucleosomes in a random way with respect to base sequence (Prunell and Kornberg, this volume), but this does not exclude the possibility that a minor DNA component (polypyrimidines) may be arranged in a specific fashion. The experiments we have carried out strongly indicate that polypyrimidines are associated with nucleosomes, but that they are not arranged in any specific phase relationship.

THEORETICAL CONSIDERATIONS

Three possible models for the arrangement of polypyrimidines with respect to nucleosomes within DNA were selected for examination (Fig. 1). We assume that when a nucleosome is associated with DNA, it protects a piece, w nucleotides in length, from digestion by micrococcal nuclease, and we have considered the probable fate of polypyrimidine segments of length l predicted by these models. In case (a), the segments are randomly arranged with respect to nucleosomes; some parts will be protected from digestion and some parts will not. The longest pyrimidine fragments remaining after digestion will be of length w; these arise where, by chance, a nucleosome was completely contained within a pyrimidine segment. Partial overlapping is more likely and in this case the original segment will be broken to one or more fragments of variable length.

Let $q(l)$ be the quantity of material (total number of nucleotides) contained in pyrimidine segments of length l before digestion, and let $p(l,x)$ be the quantity contained in fragments of length x recovered after digestion of the DNA and derived from these segments. The number of segments is $q(l)/l$ and the number of fragments of length x is $p(l,x)/x$. If the nucleosomes are assumed to be regularly spaced with only negligible gaps between them, then, because breaks must occur between nucleotides, there are only w significantly different arrangements of the nucleosomes and a polypyrimidine segment. By counting the fragments produced in every case, and assuming all arrangements are equally likely, we find that from a single segment of length l the average number of fragments of length $x < w$ is $2/w$, while for fragments of length w it is $(l - w + 1)/w$, provided $l > w$. If $l < w$ the average for length $x < l$ is still $2/w$, while the probability of a segment escaping digestion and remaining at length l is $(w - l + 1)/w$. Therefore, the quantity $p(l,x)$ is defined by the following equations:

$$p(l,x) = x \cdot (2/w) \cdot q(l)/l, \text{ for } x < w \text{ and } l;$$
$$p(l,w) = (l - w + 1) \cdot q(l)/l, \text{ if } l > w;$$
$$p(l,l) = (w - l + 1) \cdot q(l)/w, \text{ if } l \leqslant w;$$
$$p(l,x) = 0 \text{ for } x > w \text{ or } l.$$

Losses due to complete digestion of "linker" DNA between the nucleosomes have been ignored, but it can be shown that, if the linker has a length g, the effect is merely to reduce recovery of all size classes by a constant factor $w/(w + g)$.

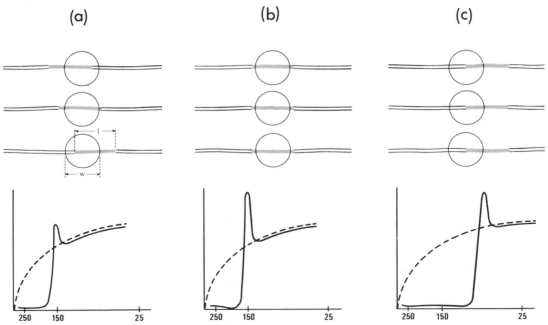

Figure 1. Schematic representation of possible phase relationships between polypyrimidine/polypurine segments in DNA and nucleosomes. The circles are nucleosomes which protect a length of DNA, w, from digestion by micrococcal nuclease. The shaded areas are polypyrimidine/polypurine segments in DNA of length l. The lower part of each figure is a polyacrylamide gel electropherogram of polypyrimidines; nucleotide length is shown. The dashed line indicates the heterogeneous size distribution of polypyrimidines from high molecular weight DNA. The solid line indicates the expected redistribution of sizes of polypyrimidines isolated from nucleosome monomer DNA. In a, polypyrimidine/polypurine segments are considered to be randomly phased with respect to nucleosomes. In b, the two elements are in phase, a nucleosome always being centered on a polypyrimidine/polypurine segment. In c, the two elements are in phase but a nucleosome is always centered on the end of a polypyrimidine/polypurine segment.

The polypyrimidine segments are found experimentally to be heterogeneous in length, but provided their size distribution, described by the function $q(l)$, is known, the result of micrococcal nuclease digestion can be predicted by summing $p(l,x)$ over all values of l. The "theoretical" results in Figure 3a were prepared from the "control" data in this way, using a programmable calculator.

In the case of Figure 1b, a regular arrangement of the pyrimidine segments and nucleosomes is considered, such that a nucleosome is always contained within a segment. All segments of length l are reduced to size w and a large spike at w would be expected in the distribution pattern of polypyrimidines after digestion. In this case, $p(l,w)$ for $l > w$ is $w/l \cdot q(l)$. A schematic representation of the distribution pattern expected is shown in the lower part of Figure 1b. Another simple regular arrangement is shown in Figure 1c, but since the predicted distribution pattern after digestion bears little resemblance to the observed pattern (Fig. 3c), it is unlikely to be valid.

EXPERIMENTAL PROCEDURES

Mouse L cells (obtained from R. P. Perry) were grown in spinner culture in minimal essential medium (Eagle) supplemented with 4% fetal calf serum

and 4% calf serum. For the preparation of labeled chromatin and DNA, 60 ml of culture (2×10^5 cells/ml) was incubated for 18 hours in the presence of [$methyl$-^{3}H]thymidine (1 μCi/ml; 19 Ci/mmole). Further steps were carried out at 0°C. Cells were collected by low-speed centrifugation and washed once with cold Earle's salt solution and once with 5 ml of a solution containing 0.1 M NaCl, 0.01 M Tris-HCl, pH 7.5. The cells were suspended in 2.0 ml of 0.1 M NaCl, 0.01 M Tris-HCl, pH 7.5, 0.5% Triton X-100 and held at 0°C for 5 minutes. The suspension was centrifuged (3000 rpm for 2 minutes) and the pellet was suspended in 2.0 ml of NaCl/Tris/Triton X-100. After 5 minutes, nuclei which were liberated by the detergent treatment were collected by low-speed centrifugation and suspended in 0.1 M NaCl, 0.01 M Tris-HCl, pH 8.0, at an optical density (as measured after lysis of an aliquot in sodium dodecyl sulfate) of 10 at 260 nm. Cyclohexanediamine tetraacetate (CDTA), a chelating agent, was added to a concentration of 3 mM to allow nuclei to swell slightly for 10 minutes at 0°C. CaCl₂ was added to 4 mM and micrococcal nuclease (Worthington NFCP) to 30 units/ml. Nuclei were incubated at 37°C for 5 minutes; digestion was terminated by adding CDTA to 5 mM and chilling the sample on ice. Nucleosome monomers were isolated by sucrose gradient centrifugation. One-half ml of the digested nuclear suspension was layered onto each of three sucrose gradients (5–20% w/v linear gradients containing 0.1 M NaCl, 0.01 M Tris-HCl, pH 7.5, 0.1 mM CDTA) and centrifuged in Beckman SW40 rotor at 31,000 rpm for 16 hours at 5°C. The optical density and distribution of radioactivity in the gradients were monitored and fractions corresponding to the nucleosome monomer were pooled. DNA was purified by extraction with phenol/chloroform, then collected by ethanol precipitation.

The size of double-stranded monomer DNA was determined by agarose gel electrophoresis, with reference to DNA recovered from chromatin which had been partially digested with micrococcal nuclease. A horizontal electrophoresis apparatus was used. The agarose slab contained 1.5% agarose in 0.016 M Tris base, 0.015 M NaH$_2$PO$_4$, 0.0005 M EDTA. Following electrophoresis, bands were detected by staining with ethidium bromide (1 μg/ml) and visualized under a short-wavelength ultraviolet lamp.

Polypyrimidines were isolated from high molecular weight DNA and from nucleosome monomer DNA and separated according to chain length in a 10% polyacrylamide gel containing 7 M urea (Birnboim et al. 1975; Birnboim 1978). Sizes were determined with reference to a series of markers [(pT)$_{10}$]n (van deSande and Kalisch 1976) kindly provided by J. H. van deSande. When nucleosome monomer DNA was examined on a denaturing gel, it was first denatured by heating at 100°C for 2 minutes.

Randomly degraded DNA of approximately nucleosome monomer size was prepared by extensively sonicating a sample, degrading it lightly with spleen deoxyribonuclease II, and collecting those fragments which sedimented on an alkaline sucrose gradient at the position of monomer DNA.

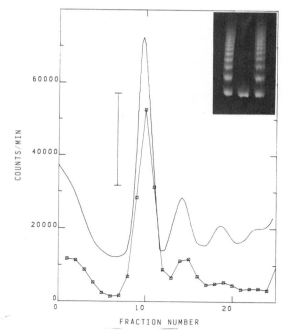

Figure 2. Sucrose gradient centrifugation analysis of nucleosomes from mouse L cells. Nuclei were isolated, digested with micrococcal nuclease, and layered directly onto 5–20% (w/v) sucrose gradients, as described in the text. Following centrifugation, gradients were collected and the distribution of radioactivity in [³H]thymidine-labeled DNA (□) and the absorbance profile (——) (at 260 nm) were determined. The top of the gradient is to the left. The vertical bar indicates 0.2 absorbance units. *(Inset)* agarose gel electropherogram of DNA purified from the major (monomer) sucrose gradient peak. The sample is shown along with marker DNA prepared from chromatin partially digested with micrococcal nuclease. DNA samples are double-stranded and visualized under UV light after staining with ethidium bromide.

RESULTS

Nucleosome monomers to be used as a source of DNA for isolating polypyrimidines were recovered in high yield from mouse L-cell nuclei using the digestion conditions described in Experimental Procedures (Fig. 2). The distribution of radioactivity applied to the sucrose gradient in this experiment was 42.8% in the monomer peak, 13.3% in small DNA fragments including acid-soluble material, and 19.4% in the pellet at the bottom of the centrifuge tube. Nuclei that were treated with micrococcal nuclease without pretreatment with a chelating agent (CDTA) showed a lower yield of monomers and an increased yield of short fragments (data not shown). The size of DNA isolated from the nucleosome monomer peak was examined in an agarose gel; its mobility was very similar to the fastest moving peak of DNA prepared from partially digested chromatin (inset of Fig. 2).

Polypyrimidines were isolated from high molecular weight DNA as a control to compare their size distribution with polypyrimidines isolated from monomer DNA. They were analyzed by electrophoresis in a 10% polyacrylamide gel in 7 M urea (Fig. 3). Control polypyrimidines range from approximately 25 nucleotides to 250 nucleotides in length. (About 99% of the pyrimidine nucleotides in DNA are degraded to shorter tracts and most of these are removed by ethanol precipitation prior to electrophoresis.) The size of polypyrimidines recovered from monomer DNA is also shown (Fig. 3c), as well as a theoretically calculated distribution in Figure 3a and polypyrimidines recovered from randomly broken DNA (Fig. 3b). The theoretical calculation is based upon the random model of Figure 1a; the figure shows the expected redistribution of radioactivity if polypyrimidines in high molecular weight DNA were randomly protected within nucleosomes (where $w = 150$ nucleotides) against micrococcal nuclease digestion.

In Figure 3b, DNA was randomly broken by sonication and deoxyribonuclease II action to a size approximating monomer DNA. Its size after denatur-

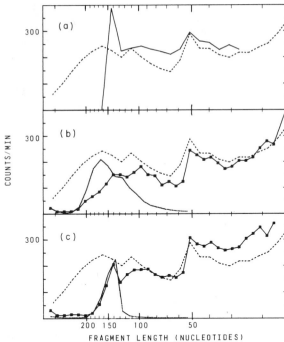

Figure 3. Polyacrylamide gel electropherograms of polypyrimidines and of denatured DNA. (----) Distribution of radioactivity in polypyrimidines prepared from high molecular weight (control) DNA; it is redrawn for comparison in the three figures. *(a)* (——) Theoretical redistribution of the radioactivity in control polypyrimidines, calculated according to the random model of Fig. 1a; the "protected" size of DNA in monomers, w, is taken to be 150 nucleotides. *(b)* (—■—) The distribution of radioactivity in polypyrimidines, prepared from randomly degraded DNA, of the size shown (·····). *(c)* (—■—) The distribution of radioactivity in polypyrimidines, prepared from nucleosome monomer DNA, whose (single strand) size is shown (·····). Counts/min in polypyrimidines have been normalized by a factor equal to radioactivity in total control DNA, divided by radioactivity in total monomer DNA.

ation, as shown in the figure, is rather heterogeneous. Polypyrimidines were prepared from this randomly degraded DNA and their size distribution is also shown.

In Figure 3c, the size of denatured monomer DNA and of polypyrimidines derived from it are shown. Monomer DNA has a much narrower size distribution than randomly broken DNA (Fig. 3b). As expected, polypyrimidines from monomer DNA have a much sharper size maximum than polypyrimidines from randomly broken DNA. The maximum size of polypyrimidines closely follows the maximum size of monomer DNA. There is a small peak corresponding in position to the peak of monomer DNA and thereafter the polypyrimidine pattern is similar to the "control" pattern. In these three respects the observed pattern of polypyrimidines from monomer DNA is very similar to the predicted pattern of Figure 3a.

DISCUSSION

Polypyrimidine/polypurine segments are a special subset of middle repeated DNA. They comprise about 0.5% of total mouse DNA and are quite regularly distributed in main band DNA at intervals of about 7 kb (Birnboim 1978). The pyrimidine strand can readily be isolated following depurination of DNA with formic acid (Burton 1967). Their size is heterogeneous, ranging up to about 250 nucleotides. Because their maximum length is close to the length of DNA recovered from nucleosome monomers, they are a favorable set of sequences for determining if they are in some special phase relationship to nucleosomes. As outlined in Theoretical Considerations, where l (polypyrimidine length) and w (monomer DNA length) are close to the same, distinct differences can be expected depending upon whether nucleosomes are randomly arranged along polypyrimidine/polypurine segments in DNA or whether they are arranged in some regular fashion (Fig. 1). Several conclusions can be drawn from the experimental results (Fig. 3c). First, most polypyrimidines are likely to be in nucleosomes since their maximum size corresponds precisely to the size of monomer DNA. If they were not in nucleosomes, they could be either more sensitive to micrococcal nuclease, in which case their recovery would be low, or more resistant, in which case longer polypyrimidines would not be reduced to monomer DNA size. Second, they are not significantly phased with respect to nucleosomes, else a large peak of polypyrimidines corresponding to the "protected" fragment would appear (as outlined schematically in Fig. 1b). The observed profile is close to the theoretically predicted one and the profile obtained using randomly degraded DNA, if allowance is made in the latter case for greater heterogeneity of randomly degraded DNA. Finally, it appears that polypyrimidines in chromatin are liberated as nucleosome monomers by micrococcal nuclease to about the same extent as bulk DNA; i.e., monomer DNA is neither enriched nor depleted in polypyrimidines relative to total DNA. In this experiment, nucleosomes were extensively digested to give a high proportion of monomers. It is possible that a difference in the rate of release of polypyrimidines into monomers might be detected at early stages of digestion.

Acknowledgments

The skillful assistance of J. J. Jevcak and A. G. Knight is gratefully acknowledged.

REFERENCES

Birnboim, H. C. 1976. Arrangement of polypyrimidine sequences in the mouse genome as studied by poly (AG) binding. *Fed. Proc.* **35:** 1676.

———. 1978. Spacing of polypyrimidine regions in mouse DNA as determined by poly (AG) binding. *J. Mol. Biol.* (in press.)

BIRNBOIM, H. C. and N. A. STRAUS. 1975. DNA from eukaryotic cells contain unusually long pyrimidine sequences. *Can. J. Biochem.* **53**: 640.

BIRNBOIM, H. C., R. E. J. MITCHEL, and N. A. STRAUS. 1973. Analysis of long pyrimidine polynucleotides in HeLa cell nuclear DNA: Absence of polydeoxythymidylate. *Proc. Natl. Acad. Sci.* **70**: 2189.

BIRNBOIM, H. C., N. A. STRAUS, and R. R. SEDEROFF. 1975. Characterization of polypyrimidines in *Drosophila* and L-cell DNA. *Biochemistry* **14**: 1643.

BURTON, K. 1967. Preparation of apurinic acid and of oligodeoxyribonucleotides with formic acid and diphenylamine. *Methods Enzymol.* **12A**: 222.

CASE, S. T. and R. F. BAKER. 1975. Detection of long eukaryote-specific pyrimidine runs in repetitive DNA sequences and their relation to single-stranded regions in DNA isolated from sea urchin embryos. *J. Mol. Biol.* **98**: 69.

STRAUS, N. A. and H. C. BIRNBOIM. 1974. Long pyrimidine tracts of L-cell DNA: Localization to repeated DNA. *Proc. Natl. Acad. Sci.* **71**: 2992.

———. 1976. Polypyrimidine sequences found in eukaryotic DNA have been conserved during evolution. *Biochim. Biophys. Acta* **454**: 419.

SZALA, S., B. BIENIEK, J. MICHALSKA, and M. CHORAZY. 1976. Interspersion and transcription of repeated sequences of rat DNA. *Biochim. Biophys. Acta* **432**: 129.

VAN DESANDE, J. H. and B. W. KALISCH. 1976. Polymerization of oligodeoxythymidylates and oligoriboadenylates catalyzed by T4 polynucleotide ligase and their use as analytical markers in polyacrylamide-gel electrophoresis. *Anal. Biochem.* **75**: 509.

Ribosomal Genes and Their Proteins from *Xenopus*

R. H. Reeder, H. L. Wahn, P. Botchan, R. Hipskind, and B. Sollner-Webb

Department of Embryology, Carnegie Institution of Washington, Baltimore, Maryland 21210

For the past several years our laboratory has been studying the structure and function of the genes coding for 18S and 28S rRNA from the frog *Xenopus*. In this article we will summarize two recent developments from this work. One development is the use of a novel procedure for identifying the 5′ termini of primary transcripts from rDNA. This is a procedure that should have wide application for the study of other genes in addition to rDNA. A second area we will discuss is the isolation of highly purified rDNA chromatin and the initial characterization of the proteins that are associated with it. Other studies concerning the arrangement of repetitive DNA sequences in the nontranscribed spacer have been published recently (Botchan et al. 1977) and will not be discussed in this article.

RNA CHAIN INITIATION SITES ON RIBOSOMAL DNA

The size of the primary transcription product is still unknown for most eukaryotic genes that have been studied. Only in the case of the 5S RNA of ribosomes has it been possible to demonstrate a triphosphate terminus on the 5′ end of the RNA (Hatlen et al. 1969; Brownlee et al. 1972; Wegnez et al. 1972) and, therefore, to pinpoint the site of chain initiation. In all other cases, the approach has been to isolate the largest RNA that can be detected after brief labeling and assume that it is the primary transcript. Experience with prokaryotes (Pace 1973) has shown, however, that processing may occur so rapidly for some genes that under normal circumstances the primary transcript never appears intact and the earliest discernable products are already processed intermediates.

We have been using a novel method for locating the 5′ ends of primary transcripts that is independent of size or kinetics of labeling. The method involves use of a complex of capping enzymes from vaccinia virus which catalyze the reaction

pppG + *S*-adenosyl methionine + (p)ppXpYpZp . . .
→ ^{7m}GpppX^{m}pYpZp . . . (Ensinger et al. 1975). This enzyme complex will cap 5′ termini bearing a di- or triphosphate terminus but will not cap monophosphate or hydroxyl termini. On the reasonable assumption that 5′ polyphosphate termini are the result of primary RNA chain initiation events, we have used this reaction to radioactively cap nucleolar RNA from *Xenopus* and to locate sites of transcription initiation on the rDNA.

Capping of Nucleolar RNA

Nucleoli from immature oocytes of *X. laevis* were isolated as described in the second half of this article and their RNA was fractionated on an SDS-sucrose gradient (Fig. 1). Nucleoli isolated by this procedure contain ribosomal DNA and no detectable bulk chromosomal DNA. Their major RNA species is the 40S rRNA precursor with a second component at 30S, the size of one of the more stable processing intermediates. Little, if any, mature 28S and 18S is observed. The high UV absorbance at the top of the gradient is mostly due to metrizamide carried over from the preceding density gradient step used to purify the nucleoli. Pooled fractions from the gradient were extracted with phenol, precipitated with ethanol, and tested for their ability to serve as substrate for the vaccinia capping enzymes. Using ^{3}H-labeled *S*-adenosyl methionine as the radioactive donor, capping acceptor activity was found in a broad peak

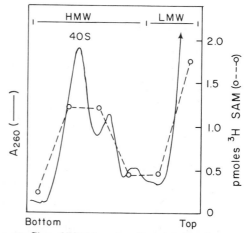

Figure 1. Size of RNA bearing di- or triphosphate termini from nucleoli of *X. laevis* oocytes. Nucleoli were isolated from immature oocytes of *X. laevis* (Higashinakagawa et al. 1977), dissolved in SDS, and their RNA fractionated on a sucrose gradient. Pooled fractions from the gradient were then tested for their relative ability to accept 5′ terminal caps using ^{3}H-labeled *S*-adenosyl methionine as the label donor. Reaction conditions are given in the legend to Fig. 2. (——) A$_{260}$; (o----o) pmoles *S*-adenosyl methionine incorporated.

from 30S and 40S and also in the low-molecular-weight fractions.

Evidence that the label is in cap structures. Figure 2 shows the time course of capping of RNA taken from the heavy side of the 40S peak. The incorporation of *methyl*-³H-label is completely dependent upon GTP as is expected from the known mechanism of these enzymes (Ensinger et al. 1975). The GTP dependence is good evidence that the methylation is in cap structures and not at internal sites along the RNA chain. The reaction shown in Figure 2 reached a plateau at 0.44 mole of *methyl*-³H-groups incorporated per mole of 40S rRNA. Assuming that each cap contains two methyl groups, this means that 22% of the 40S termini were able to accept caps. Results shown in Figure 3 demonstrate that this assumption is correct. The level of incorporation is high enough to make it likely that the 40S itself is being capped rather than some minor contaminant. The less than stoichiometric amount of capping is not unexpected in view of the apparent lability of polyphosphate termini in vivo (Slack and Loening 1974; Kominami and Muramatsu 1977).

An independent method for checking that the label is really in cap structures is to digest the labeled RNA with RNase P₁. This nuclease cleaves 5′ phosphoryl termini and is not blocked by methylation of the 2′ hydroxyl on the adjacent ribose. However, it cannot cleave the 5′ → 5′ phosphodiester linkage within the cap itself and, thus, from capped structures should generate a labeled fragment with −2.5 charges. If, on the other hand, the methylation were occurring at internal sites along the RNA

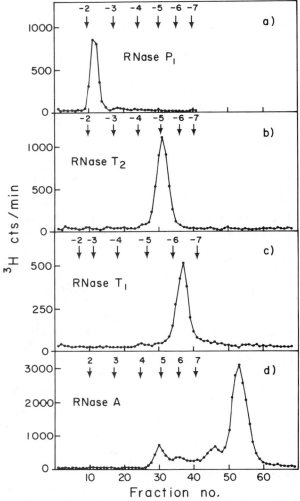

Figure 3. Digestion of capped high-molecular-weight (HMW) nucleolar RNA with various RNases. The HMW region of the gradient shown in Fig. 1 was pooled and capped using ³H-labeled *S*-adenosyl methionine as the label. The capped RNA was then digested with various RNases and the resulting oligonucleotides were fractionated on columns of DEAE-Sephadex (0.6 × 25 cm, gradient was from 0.1 M to 0.5 M NaCl in 7 M urea, 0.01 M Tris-HCl (pH 8). (*a*) RNase P₁; (*b*) RNase T₂; (*c*) RNase T₁; (*d*) RNase A. The elution position of unlabeled marker oligonucleotides of known charge is indicated for each column profile.

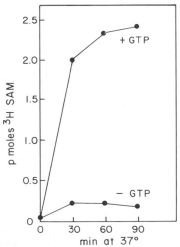

Figure 2. GTP dependence of ³H-labeled *S*-adenosyl methionine incorporation into caps. Each reaction in 100 μl contained 8.5 μg 40S rRNA, 50 mM Tris (pH 7.6), 1 mM β-mercaptoethanol, 2 mM MgCl₂, 65 pmoles ³H-labeled *S*-adenosyl methionine (7700 cpm/pmole), 5 μl of enzyme and 2 mM GTP as indicated. Aliquots were taken at intervals to determine incorporation of *methyl*-³H into acid-precipitable material.

chain, P₁ should generate fragments with −2 or fewer charges. Figure 3a shows that P₁ RNase liberates a single fragment of about −2.5 charges.

We have also checked to see if the capped 40S is methylated on the 2′ hydroxyl of the terminal ribose, since Moss and coworkers found this position methylated with the substrates they were using (Ensinger et al. 1975). If the 2′ hydroxyl is methylated, digestion with RNase T₂, which leaves 3′ phosphoryl termini and is blocked by methylation of the 2′ hydroxyl on the adjacent ribose, should yield a single fragment with a charge of −5.5. Figure 3b shows this result is obtained.

We conclude from these results that all of the *methyl*-³H-label is incorporated into structures of the dimethylated cap 1 variety (Shatkin 1976) and is not incorporated into internal sites along the RNA chains.

Specificity for polyphosphate termini. Only molecules which are known to terminate in di- or triphosphate termini have been observed to accept caps. Martin and Moss (1975) showed that ppA(pA)n will cap, whereas pA(pA)n will not. We have found that processed molecules such as *E. coli* transfer RNA and *Xenopus* 18S and 28S rRNA will not accept caps, but *Xenopus* 5S rRNA will cap (experiments not shown).

In an alternate approach to the specificity question, total nucleolar RNA was capped using ³H-labeled *S*-adenosyl methionine and [β,γ-³²P]GTP as radioactive donors in the same reaction. With the enzymes from vaccinia, Ensinger et al. (1975) have shown that only the α-phosphate from GTP is incorporated into the cap, and incorporation of the G residue is obligatory before the methyl groups can be added. Incorporation of only the α-phosphate agrees with the enzyme's specificity for polyphosphate termini. Enzymes that do cap monophosphate termini have been shown to incorporate both the α and β phosphates from GTP (Abraham et al. 1975). Data in Table 1 show that for every mole of termini labeled with methyl groups, at most only 0.01 mole of termini could have been labeled with the β and γ phosphates. We feel confident, therefore, that the capping enzymes we are using are highly specific for di- or triphosphate termini.

Partial Sequencing and Hybridization of Capped RNA

High-molecular-weight nucleolar RNA (30S to 40S) was pooled, capped using a *methyl*-³H-donor, digested with RNase A, and the oligonucleotides were separated according to charge (Fig. 3d). As we will show below (Fig. 5), the two oligonucleotides that have greater than −7 charges are also recovered from capped RNA which was hybridized to a DNA fragment containing the 5′ end of the 40S sequence. The oligonucleotides with less than −7 charges apparently do not result from rDNA transcription and have not been characterized further.

Table 1. Capping of Nucleolar RNA

Radioactive group	pMole incorporated
methyl-³H	0.63
³²P (β or γ)	0.003

Reaction in 100 μl contained 23 μg unfractionated nucleolar RNA, 50 mM Tris-HCl (pH 7.6), 1 mM β-mercaptoethanol, 2 mM MgCl₂, 65 pmoles ³H-labeled *S*-adenosyl methionine (7700 cpm/pmole), 2.6 nmoles [β,γ-³²P]GTP (3.68 × 10⁵ cpm/pmole), and 5 μl of enzyme. Incubation was for 90 min at 37°C.

The largest RNase A fragment was recovered from a DEAE column and redigested with RNase T₁. This yielded a single fragment with −6.5 charges (data not shown). RNase T₁ digestion of total capped 30S–40S RNA also yielded a single fragment of −6.5 charges (Fig. 3c). This number of charges is consistent with the structure being ⁷ᵐGpppXᵐpApGp. The second nucleotide was deduced to be A, since RNase T₂ cleaves at this point (Fig. 3b) but RNase T₁ and A do not.

Further sequencing of the 5′ end of the 40S RNA should proceed rapidly. Since capping can be used to introduce a ³²P-label on the 5′ end, partial cleavage with base-specific RNases should allow sequencing by procedures analogous to those described by Maxam and Gilbert (1977) for DNA sequencing.

To determine which of the RNase-A oligonucleotides were associated with rDNA transcription, capped nucleolar RNA was hybridized to cloned DNA fragments that span the entire repeating unit of rDNA. DNA from the chimeric plasmid pXlr11 was digested with *Eco*RI endonuclease and combined with an equimolar amount of pXlr13 that had been doubly digested with *Bam*HI and *Eco*RI. These plasmids and their restriction enzyme sites have been previously described (Botchan et al. 1977). Figure 4 shows the location on the rDNA of each of the fragments used in this study. The combined fragments were electrophoresed on 1% agarose and transferred onto a strip of nitrocellulose filter paper (Southern 1975). One such filter strip was then hybridized with capped 30S to 40S nucleolar RNA; another strip was hybridized with capped low molecular weight RNA. The filters were then treated with RNase A to remove nonhybridized RNA, dried, cut into pieces, and counted. The result for the 30S to 40S RNA is shown in Figure 5a. A closely similar

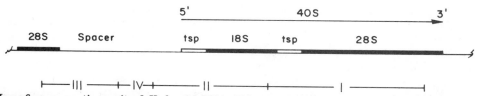

Figure 4. Map of one repeating unit of *X. laevis* rDNA. This map shows the location of fragments I, II, III, and IV which were used for the hybridization experiment shown in Fig. 5. Fragments were derived from clones Xlr11 and Xlr13 using *Bam*HI and *Eco*RI (see Botchan et al. 1977 for more detailed restriction maps).

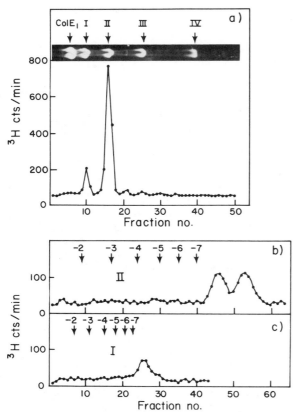

Figure 5. Hybridization of capped nucleolar RNA to rDNA restriction fragments. ³H-labeled capped nucleolar RNA (HMW fraction from Fig. 3) was hybridized to restriction fragments immobilized on a nitrocellulose filter (50,000 cpm ³H-labeled capped RNA in 1 ml of 0.6 M NaCl, 0.2 M Tris [pH 8], 50% formamide at 40°C overnight). The map location of the fragments is shown in Fig. 4. After trimming hybrid with RNase A, the filter was cut in pieces and counted. Hybrid RNA was then removed from the filter pieces, digested with RNase A, and the labeled oligonucleotides were fractionated on DEAE-Sephadex as described in Fig. 3. *(a)* Photograph of the agarose gel used to separate the rDNA fragments; underneath is represented the amount of RNase-resistant radioactivity hybridizing to each fragment. *(b)* DEAE chromatography of RNase-A oligonucleotides recovered from hybrid to fragment II. *(c)* Oligonucleotides recovered from hybrid to fragment I.

result was obtained with the low-molecular-weight RNA and is not shown.

About 85% of the hybridization occurred with fragment II, which spans the 5′ end of the 40S precursor. There was no hybridization to the fragment spanning the 3′ end (III), to the center of the nontranscribed spacer (IV), nor to the ColE1 vehicle. The RNA bound to fragment II was recovered, digested with RNase A, and the oligonucleotides run on a DEAE column (Fig. 5b). Two peaks were observed, corresponding in charge to the two largest peaks obtained from RNase-A digestion of unhybridized 30S to 40S rRNA (Fig. 3D). These results strongly support the conclusion that the majority of capped

structures are actually on the 5′ ends of 40S molecules. This leads to the further conclusion that the large majority of initiation events on rDNA occur close to or at the 5′ end of the 40S. The recovery of two oligonucleotides also suggests that the 5′ end of the 40S may be heterogeneous.

A somewhat surprising result, also shown in Figure 5a, is that 15% of the hybridization is associated with fragment I, a fragment from the middle of the transcribed gene. RNase-A digestion of RNA from this hybrid yields a single peak of greater than −7 charges (Fig. 5c). The significance of this apparent initiation event is unknown at present.

Some Implications of Capping Results

Rungger and Crippa (1977) have recently summarized experiments from several laboratories which suggest that transcription may occur in the "nontranscribed" spacer. For instance, Slack and Loening (1974) have searched for triphosphate termini on 40S rRNA from *Xenopus* cultured cells and report they can only detect 5′ monophosphate termini with guanine as the terminal base. This monophosphate terminus could result from cleavage of a larger precursor or it could represent the true transcription initiation site with two of its phosphates removed. There is evidence that 40S processing in oocytes is slower than in cultured cells (Gall 1966). This fact, coupled with the much greater sensitivity of the capping method, provides a very plausible explanation as to why we have been able to detect polyphosphate termini where they could not. Another line of evidence concerning possible RNA chain initiation sites on rDNA comes from the observation by Scheer et al. (1977) that structures resembling RNA polymerase and small transcription complexes are occasionally seen in electron micrographs of the "nontranscribed" spacer. Rungger and Crippa (1977) have made a related observation that treatment of oocytes with high concentrations of 5-fluorouridine results in the appearance of large amounts of transcription in the spacer region. Both results could be due to polymerase initiation in the spacer region. In view of our own data, however, it seems more likely that their results are due to a failure of the proper termination at the 3′ end of the 40S, coupled with a variable degree of processing of the aberrant transcript. We obviously cannot rule out the possibility that initiation does occur in the spacer with these 5′ termini being extrasensitive to degradation and thus not detected. But, we can say that at least 85% of the detectable initiation events occur close to the 5′ end of the 40S sequence, and we are therefore justified in concentrating on this region to find the promoter and other possible control sequences.

The data of Scheer and coworkers (1977) and of Rungger and Crippa (1977) demonstrate that RNA polymerase molecules can occasionally traverse the "nontranscribed spacer" of *Xenopus* rDNA. Therefore, the spacer presumably has the structure of transcriptionally active chromatin; we propose that it is only the presence of a termination site upstream that normally results in the absence of transcription.

ISOLATION OF RIBOSOMAL DNA CHROMATIN

Approximately two-thirds of the mass of a eukaryotic chromosome is composed of proteins. Although it is certain that some of these proteins regulate gene expression, how many proteins are involved in this regulation and how they act is still largely unknown. In an effort to answer these questions, we have developed a procedure for isolating pure ribosomal gene chromatin from cells that are actively synthesizing rRNA. This isolation is possible due to the fact that at the pachytene stage of oogenesis, oocytes of *Xenopus* specifically amplify their rDNA several thousandfold (Reeder 1974 for review). This results in a nucleus containing over twice as much rDNA (30 pg) as it does bulk chromosomal DNA (12 pg, since the chromosomes are 4C at this stage). This amplified rDNA eventually becomes packaged in about 1500 extrachromosomal nucleoli, which lie just under the nuclear membrane and by mid-diplotene (stage III; Dumont 1972) are highly active in rRNA synthesis.

Procedure for Isolating Amplified Nucleoli

The detailed method for isolating amplified nucleoli from *Xenopus* oocytes has been published (Higashinakagawa et al. 1977). In outline, the method involves digesting ovaries from adult frogs with collagenase to liberate a single-cell suspension of oocytes. The oocytes are then fractionated by several settling steps in a separatory funnel to enrich for immature oocytes. Finally, the immature oocytes are layered over 1.7 M sucrose and centrifuged at low speed. Unpigmented oocytes just beginning vitellogenesis (stages II and III) pellet under these conditions, while more mature, pigmented oocytes and follicle cells remain floating. The bright yellow pellet of immature oocytes is then homogenized and used for nucleolar isolation.

The homogenate of immature oocytes is centrifuged at high speed in a preformed buoyant density gradient of Metrizamide (Birnie et al. 1973). The time of centrifugation is long enough for large particles (nucleoli, nuclei, mitochondria) to reach their equilibrium density position while ribosomes and soluble proteins do not. Figure 6a shows a typical light-scattering profile from a first-round Metrizamide gradient (Metrizamide absorbs strongly in the ultraviolet). Examination of each fraction by phase-contrast light microscopy shows that the most dense peak of light scattering contains free nucleoli, while nuclei (from follicle cells), whole follicle cells and

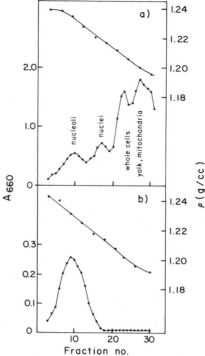

Figure 6. Equilibrium banding of nucleoli in Metrizamide density gradients. *(a)* Immature oocytes from *Xenopus laevis* (stages II and III) were homogenized and the homogenate centrifuged for 1 hr (26,000 rpm, SW27 rotor) in a preformed Metrizamide buoyant density gradient. The gradient was fractionated and light scattering at A_{660} was monitored in each fraction. Organelles were identified by phase-contrast light microscopy. *(b)* Nucleoli from the gradient in *(a)* were rebanded in a second-round preformed Metrizamide gradient and assayed as before.

other contaminants band at lighter densities. Oocyte nuclei are large and fragile and break during the initial homogenization. To completely remove bulk chromosomal DNA contaminants, the nucleolar peak is pooled and rerun on a second-round Metrizamide gradient (Fig. 6b). Figure 7 shows an analytical CsCl density gradient of DNA from such nucleoli, demonstrating that at this point they contain no DNA other than rDNA.

RNA Polymerase and Nicking-Closing Enzyme in Nucleoli

Nucleoli isolated by this method are in an apparently native configuration and have the characteristics one would expect of transcriptionally active genes. By both light and electron microscopy, they retain the same morphology seen in the intact cell. The great majority of their RNA content is 40S precursor rRNA or its processed products (see Fig. 1). Isolated nucleoli also contain endogenous RNA polymerase activity as shown in Figure 8. The activity is resistant to both high and low levels of α-amanitin and therefore has the characteristics of the

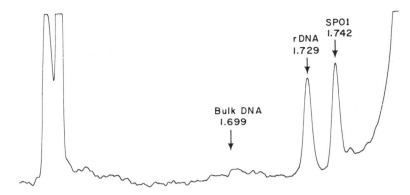

Figure 7. Analytical CsCl gradient of DNA from isolated nucleoli. Nucleoli from a second-round Metrizamide gradient (Fig. 6b) were pelleted through 2 M sucrose, dissolved in Sarkosyl, and the DNA centrifuged to equilibrium in the Model-E analytical ultracentrifuge. Phage SPO1 DNA was used as a buoyant density marker.

type-I polymerase previously shown to be responsible for rRNA synthesis (Reeder and Roeder 1972; Weinman and Roeder 1974). Figure 8 also shows that the endogenous activity is stimulated by addition of heparin. Other experiments have shown that 0.5% Sarkosyl also stimulates activity, whereas exogenously added rDNA does not stimulate. From these results, we conclude that the observed synthesis represents primarily chain elongation by preinitiated polymerase and that little, if any, reinitiation is occurring in vitro. The relatively low amount of synthesis (about one-half gene equivalent of RNA per gene copy after 3 hr) suggests that chain elongation may also be impaired.

On repeated occasions we have observed that intact nucleoli are heterogeneous with respect to endogenous RNA polymerase activity. Figure 9a shows an experiment in which fractions from a second-round Metrizamide gradient were analyzed for light scattering (to locate nucleoli) and for endogenous

polymerase activity. Fractions on the dense side of the nucleolar peak were more active than fractions on the light side. We suspect that this heterogeneity comes from the fact that the immature oocytes we isolate are also heterogeneous; some of them are stage II and are just beginning rRNA synthesis, while others are stage III and in rapid synthesis. (As we will show below [Fig. 10], dissociation of the RNP cortex surrounding the nucleoli causes the remaining DNA-containing cores to band as a single component.)

Nucleoli contain an enzyme which behaves like the nicking-closing enzymes described in other eukaryotes (Baase and Wang 1974; Keller 1975; Vos-

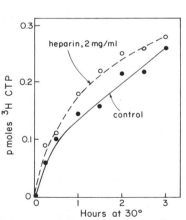

Figure 8. Effect of heparin on endogenous RNA synthesis by isolated nucleoli. Nucleoli from a second-round Metrizamide gradient were incubated in reactions of 40 μl containing 28 ng rDNA as nucleoli; 25% glycerol, 0.15 M NaCl, 5 mM MgCl₂, 0.01 M Tris-HCl (pH 8), 0.1 M EDTA, 5 mM β-mercaptoethanol, 0.6 mM each of ATP, UTP, and GTP, 1.25 μCi of [³H]CTP (24.5 Ci/mmole) and 2 mg/ml heparin as indicated. A separate reaction was used for each time point and all points were done in duplicate. Incubation was at 30°C and reactions were terminated by acid precipitation.

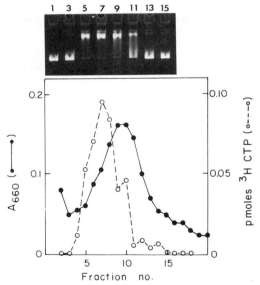

Figure 9. Assay of endogenous RNA polymerase and nicking-closing activity across a Metrizamide gradient. Fractions from the nucleolar region of a second-round Metrizamide gradient were assayed for A₆₆₀ (●——●) and endogenous RNA polymerase activity (○----○). Alternate fractions were also assayed for nicking-closing enzyme by incubating an aliquot with supercoiled ColE1 DNA as described elsewhere (Higashinakagawa et al. 1977). After incubation, the ColE1 DNA was electrophoresed on a 1% agarose gel as shown in the inset. Active fractions relax the supercoiled DNA and cause it to migrate more slowly upon electrophoresis.

berg et al. 1975). The activity operates in the presence of EDTA and can relax super-helical turns in exogenously added DNA. We have assayed it by incubating supercoiled ColE1 DNA to nucleoli and then electrophoresing the ColE1 DNA on an agarose gel. Figure 9b shows the results when such an assay is applied to fractions across a nucleolar peak. The relative nicking-closing activity tracks closely with the distribution of endogenous RNA polymerase activity and does not follow the bulk distribution of nucleoli. These nucleoli underwent their last DNA replication weeks to months before they were isolated and they never undergo any further replication. Therefore, it is unlikely that the presence of nicking-closing enzymes is due to replication. More likely, we propose that the enzyme plays an essential role in normal transcription of these genes. This role could possibly be to release strain due to polymerase initiation, to ease the passage of polymerase through nucleosomes, or to simply provide swivels so that the polymerase with its attached mass of nascent RNP need not spiral around and around the DNA during transcription.

Electrophoresis of Nucleolar Proteins

We have begun cataloging the proteins associated with transcriptionally active rDNA. Initial experiments, in which intact nucleoli were dissolved in SDS and all their proteins displayed on a gel, yielded a bewildering number of polypeptides (Higashinakagawa et al. 1977). Some of these proteins comigrated with proteins from cytoplasmic ribosomes, suggesting that some of the complexity was due to proteins in the RNP shell surrounding the nucleolar core. In an attempt to reduce this complexity, nucleoli were dissociated by suspending them in 10 mM Tris (pH 8), 1 mM EDTA, and the DNA-containing cores were rebanded in Metrizamide. As shown in Figure 10, this treatment abolishes the density heterogeneity of nucleoli and causes the rDNA and endogenous polymerase activity to band together. Protein from each fraction of the gradient shown (Fig. 10) was run on a 10% acrylamide-SDS gel. A photograph of the stained gel is shown above the gradient profile. Rebanding of the dissociated nucleolar cores caused the loss of some protein which

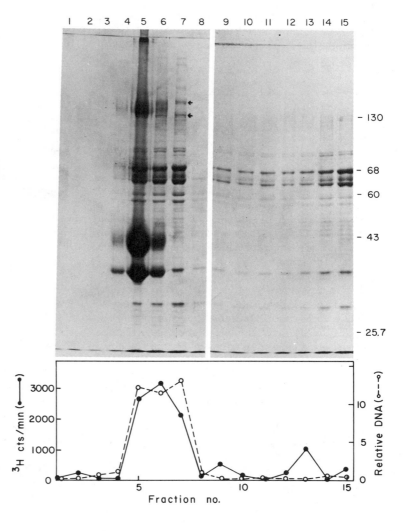

Figure 10. Metrizamide buoyant gradient of nucleolar cores. Intact oocyte nucleoli (isolated by a previous Metrizamide gradient) were dissociated by resuspension in 5 mM Tris (pH 8), 5 mM EDTA and spun to equilibrium in a second Metrizamide gradient. Aliquots from each fraction were assayed for endogenous RNA polymerase activity (●——●) and relative rDNA content (○ - - ○), and their proteins were electrophoresed on a 10% acrylamide gel in SDS. The bottom (dense) end of the gradient is on the left. Molecular weights are in daltons; arrows indicate putative RNA polymerase-I subunits.

is at the top of the gradient (fractions 9–15). However, most of these same proteins are still represented among the proteins banding with the DNA and RNA polymerase activity in fractions 5–7. The arrows in fraction 7 point to two polypeptides that are enriched in the active fractions and have the correct molecular weight to be the two large subunits of RNA polymerase I. Fraction 5 (Fig. 10) contains three very abundant, fuzzy bands plus material stuck at the top of the gel. We assume these proteins come from pigment granules, which were observed to band sharply in fraction 5, and we have not studied them further.

It is clear that more work is necessary to sort out which, if any, of these "nonhistone proteins" are interacting directly with genes and which are involved in RNA packaging, processing, or other as yet unknown functions.

Active fractions from a parallel gradient were pooled, extracted with 0.4 N H_2SO_4, and electrophoresed on a 15% acrylamide-SDS gel to look for low-molecular-weight basic proteins. Figure 11 shows a photograph of such a gel. Four bands are visible which migrate close to the four nucleosomal histones extracted from *Xenopus* cultured cells. We have not yet identified any protein that behaves as an H1-type histone. It will obviously be of interest to examine the degree of secondary modification of these histones in the near future.

APPENDIX

Contraction Ratio of the Nontranscribed Spacer of *Xenopus* rDNA Chromatin

R. H. REEDER, S. L. McKNIGHT,[*]
AND O. MILLER, JR.[*]

Figure 12a shows ribosomal genes from amplified nucleoli of *Xenopus* that were spread for electron microscopy from a hypotonic medium (Miller and Beatty 1969). Figure 12b shows a high-magnification micrograph of a nontranscribed spacer region which separates two transcribed genes. Several lines of evidence argue that nucleosomal histones are associated with this spacer region. (1) In the preceding section, it was shown that transcriptionally active nucleolar cores contain proteins that comigrate with the four nucleosomal histones; (2) the spacer regions between active ribosomal genes in *Drosophila* chromatin react specifically with anti-histone antibodies (S. L. McKnight and O. L. Miller, Jr., unpubl.); and (3) under certain spreading conditions the *Xenopus* spacer has a beaded appearance consistent with the presence of histones (Fig. 12b and Woodcock et al. 1976). A limited number of particles in the spacer may be RNA polymerase molecules that have failed to terminate at the 3' end of the 40S sequence

* Address: Department of Biology, University of Virginia, Charlottesville, Virginia 22903.

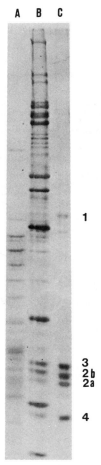

Figure 11. Acid-extractable proteins from nucleolar cores. Fractions containing rDNA from the gradient in Fig. 10 were pooled, extracted with 0.4 N H_2SO_4, and run on a 15% acrylamide gel in SDS. *(A)* Acid-extractable proteins from cytoplasmic ribosomes; *(B)* acid-extractable proteins from nucleolar cores; *(C)* histones from cultured cells of *X. laevis*.

(Franke et al. 1976). However, most of the beads are smaller, in the size range of nucleosomes on inactive chromatin. It may be that beads on the spacer are more easily destroyed during spreading than are beads on inactive chromatin, since other authors have reported not finding them (Franke et al. 1976; Scheer et al. 1977).

If the spacer is associated with the four nucleosome histones, do these histones contract the DNA to the same degree as seen in inactive chromatin? We have tried to answer this question in the following way. Several oocytes were removed from the ovary of a single female frog; amplified genes were prepared for microscopy and their apparent spacer lengths were measured. A histogram of the observed spacer lengths is shown in Figure 13b. DNA was then extracted from the rest of the ovary and amplified rDNA was isolated from it by neutral CsCl gradient centrifugation (Dawid et al. 1970). The amplified rDNA was digested with *Eco*RI, electrophoresed

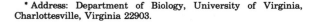

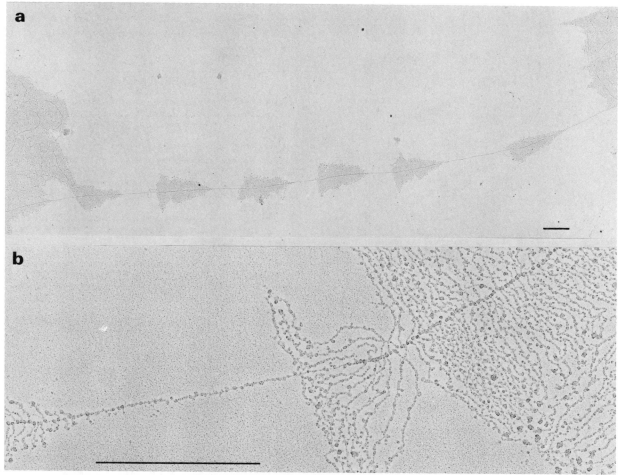

Figure 12. Electron micrographs of amplified ribosomal genes from *Xenopus laevis*. *(a)* Low-magnification micrograph showing several tandemly linked transcription units separated by nontranscribed spacer regions. *(b)* Higher-magnification view of the nontranscribed spacer between two transcribed regions. Bar = 1 micron.

on an agarose gel, and the sizes of the various spacer classes in that individual frog were measured (a densitometer trace of the stained gel is shown in Fig. 13a). This procedure was necessary since spacer lengths are heterogeneous in *Xenopus* rDNA and their lengths can vary significantly from frog to frog (Wellauer et al. 1976). This particular frog had two main spacer size classes as shown by gel electrophoresis (Figure 13a). Assuming that these two classes correspond to the two size classes seen by electron microscopy, we have calculated the apparent spacer contraction ratio as described in the legend to Figure 13. Contraction ratios of 1.20 and 1.24 were obtained for the shorter and longer classes, respectively. These figures are probably more accurate than previous estimates in which spacer variation among individual frogs was not determined (Scheer et al. 1977). However, our figures agree with their conclusion that the contraction ratio of the rDNA spacer chromatin is relatively low.

To interpret the observed contraction ratio of the rDNA spacer properly, we need to know the contrac-

tion ratio of inactive chromatin spread under identical low ionic strength conditions. It has previously been pointed out that the ionic strength of the spreading medium has a significant effect on the apparent contraction ratio of chromatin (Griffith 1975). Accordingly, inactive *Xenopus* chromatin was spread under conditions identical to those used in Figure 13. The contraction ratio was calculated by counting the number of beads per micron of contour length and assuming that each bead contains 185 base pairs of DNA (R. Reeder, unpubl.). Inactive *Xenopus* chromatin was observed to have 31 ± 4 beads per micron, which yields a contraction ratio of about 2.0. This value is considerably lower than the value assumed by previous authors (Scheer et al. 1977) but is still higher than that of the rDNA spacer.

What is the significance of a decreased contraction ratio in the rDNA spacer chromatin? The spacer chromatin of metabolically active nucleoli may exist in an active or transcribable configuration characterized by a reduced contraction ratio and/or an

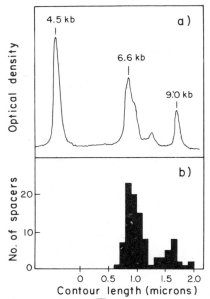

Figure 13. Contraction ratio of the nontranscribed spacer. *(a)* Amplified rDNA from a single frog was digested with *Eco*RI and electrophoresed on a 1% agarose gel. The gel was stained with ethidium bromide, photographed, and the negative was traced with a densitometer. DNA sizes (in kb) were determined by coelectrophoresis of λ *Hin*dIII fragments of known size. The amplified rDNA of this frog yielded the usual 4.5-kb *Eco*RI fragment that comes from the middle of the gene plus two major classes of *Eco*RI spacer-containing fragments. Allowing for the known amount of gene sequence on each fragment (Botchan et al. 1977), the 6.6-kb and 9.0-kb fragments contain nontranscribed spacers of 3.6 kb and 6.0 kb, respectively.

(b) Amplified ribosomal genes were spread for microscopy from several oocytes of the same frog used in *a.* Contour lengths of 98 nontranscribed spacers between active genes were measured and are shown as a histogram. Two major size classes were observed with lengths of about 0.85 and 1.65 microns. Assuming one micron of B-form DNA contains 2.94 kb, and assuming further that the two size classes seen in the histogram correspond to the two size classes seen on the agarose gel, we calculate contraction ratios of 1.20 and 1.24 for the shorter and longer spacers, respectively.

increased lability to spreading under low ionic strength conditions. As was pointed out in the main body of this article, several laboratories have observed occasional transcription complexes in the "nontranscribed" spacer (Scheer et al. 1977; Rungger and Crippa 1977). We suggest that the spacer is normally nontranscribed only because of efficient polymerase termination at the 3' end of the 40S sequence, and we propose that there is nothing intrinsic in the structure of spacer chromatin that prevents transcription. If this view is correct, it becomes of interest to analyze in detail the composition and structure of spacer chromatin.

Acknowledgments

We thank Dr. Bernard Moss for generous gifts of vaccinia capping enzymes which made this work possible. We also thank Eileen Hogan for expert technical assistance.

This work was partially supported by a grant from the National Institutes of Health to R.H.R.

REFERENCES

ABRAHAM, G., D. P. RHODES, and A. K. BANNERJEE. 1975. The 5' terminal structure of the methylated mRNA synthesized *in vitro* by vesicular stomatitis virus. *Cell* **5:** 51.

BAASE, W. A. and J. C. WANG. 1974. An omega protein from *Drosophila melanogaster. Biochemistry* **13:** 4299.

BIRNIE, G. D., D. RICKWOOD, and A. HELL. 1973. Buoyant densities and hydration of nucleic acids, proteins and nucleoproteins in metrizamide. *Biochim. Biophys. Acta* **331:** 283.

BOTCHAN, P., R. H. REEDER, and I. B. DAWID. 1977. Restriction analysis of the non-transcribed spacers of *Xenopus laevis* ribosomal DNA. *Cell* **11:** 599.

BROWNLEE, G. G., E. CARTWRIGHT, T. MCSHANE, and R. WILLIAMSON. 1972. The nucleotide sequence of somatic 5S RNA from *Xenopus laevis. FEBS Lett.* **25:** 8.

DAWID, I. B., D. D. BROWN, and R. H. REEDER. 1970. Composition and structure of chromosomal and amplified ribosomal DNA's of *Xenopus laevis. J. Mol. Biol.* **51:** 341.

DUMONT, J. N. 1972. Oogenesis in *Xenopus laevis* (Daudin): Stages of oocyte development in laboratory maintained animals. *J. Morphol.* **136:** 153.

ENSINGER, M. J., S. A. MARTIN, E. PAOLETTI, and B. MOSS. 1975. Modification of the 5' terminus of mRNA by soluble guanylyl and methyl transferases from vaccinia virus. *Proc. Natl. Acad. Sci.* **72:** 2525.

FRANKE, W. W., U. SCHEER, M. F. TRENDELENBURG, H. SPRING, and H. ZENTGRAF. 1976. Absence of nucleosomes in transcriptionally active chromatin. *Cytobiologie* **13:** 401.

GALL, J. G. 1966. Nuclear RNA of the salamander oocyte. In *The nucleolus, its structure and function* (ed. W. S. Vincent and O. L. Miller, Jr.), p. 475. National Cancer Institute, Bethesda, Maryland.

GRIFFITH, J. D. 1975. Chromatin structure: Deduced from a mini-chromosome. *Science* **187:** 1202.

HATLEN, L. E., R. AMALDI, and G. ATTARDI. 1969. Oligonucleotide pattern after pancreatic ribonuclease digestion and the 3' and 5' termini of 5S ribonucleic acid from HeLa cells. *Biochemistry* **8:** 4989.

HIGASHINAKAGAWA, T., H. L. WAHN, and R. H. REEDER. 1977. Isolation of ribosomal gene chromatin. *Dev. Biol.* **55:** 375.

KELLER, W. 1975. Characterization of purified DNA relaxing enzyme from human tissue culture cells. *Proc. Natl. Acad. Sci.* **72:** 2550.

KOMINAMI, R. and M. MURAMATSU. 1977. Heterogeneity of 5' termini of nucleolar 45S, 32S and 28S RNA in mouse hepatoma. *Nucleic Acids Res.* **4:** 229.

MARTIN, S. A. and B. MOSS. 1975. Modification of RNA by mRNA guanylyl transferase and mRNA (guanine-7-)methyltransferase from vaccinia virions. *J. Biol. Chem.* **250:** 9330.

MAXAM, A. M. and W. GILBERT. 1977. A new method for sequencing DNA. *Proc. Natl. Acad. Sci.* **74:** 560.

MILLER, O. L., JR. and B. R. BEATTY. 1969. Visualization of nucleolar genes. *Science* **164:** 955.

PACE, N. R. 1973. Structure and synthesis of the ribosomal RNA of prokaryotes. *Bacteriol. Rev.* **37:** 562.

REEDER, R. H. 1974. Ribosomes from eukaryotes: Genetics. In *Ribosomes* (ed. M. Nomura et al.), p. 489. Cold Spring Harbor Laboratory, Cold Spring Harbor, New York.

REEDER, R. H. and R. G. ROEDER. 1972. Ribosomal RNA synthesis in isolated nuclei. *J. Mol. Biol.* **67:** 433.

RUNGGER, D. and M. CRIPPA. 1977. The primary ribosomal

DNA transcript in eukaryotes. *Prog. Biophys. Mol. Biol.* **31**: 247.

SCHEER, U., M. F. TRENDELENBURG, G. KROHNE, and W. W. FRANKE. 1977. Lengths and patterns of transcriptional units in the amplified nucleoli of oocytes of *Xenopus laevis. Chromosoma* **60**: 147.

SHATKIN, A. J. 1976. Capping of eukaryotic mRNAs. *Cell* **9**: 645.

SLACK, J. M. W. and U. E. LOENING. 1974. 5' Ends of ribosomal and ribosomal precursor RNAS from *Xenopus laevis. Eur. J. Biochem.* **43**: 59.

SOUTHERN, E. M. 1975. Detection of specific sequences among DNA fragments separated by gel electrophoresis. *J. Mol. Biol.* **98**: 503.

VOSBERG, H., L. GROSSMAN, and J. VINOGRAD. 1975. Isolation and partial characterization of the relaxation protein from nuclei of cultured mouse and human cells. *Eur. J. Biochem.* **55**: 79.

WEGNEZ, M., R. MONIER, and H. DENIS. 1972. Sequence heterogeneity in the 5S RNA in *Xenopus laevis. FEBS Lett.* **25**: 13.

WEINMAN, R. and R. G. ROEDER. 1974. Role of DNA-dependent RNA polymerase III in the transcription of the tRNA and 5S RNA genes. *Proc. Natl. Acad. Sci.* **71**: 1790.

WELLAUER, P. K., R. H. REEDER, I. B. DAWID, and D. D. BROWN. 1976. The arrangement of length heterogeneity in repeating units of amplified and chromosomal ribosomal DNA from *Xenopus laevis. J. Mol. Biol.* **105**: 487.

WOODCOCK, C. L. F., L. L. FRADO, C. L. HATCH, and L. RICCIARDIELLO. 1976. Fine structure of active ribosomal genes. *Chromosoma* **58**: 33.

A Study of Early Events in Ribosomal Gene Amplification

A. P. BIRD

MRC Mammalian Genome Unit, King's Buildings, Edinburgh, Scotland

During most of oogenesis in *Xenopus laevis,* the ribosomal RNA genes are amplified well above their normal chromosomal level (reviewed by Gall 1969; Macgregor 1972; Tobler 1975). The first burst of extra rDNA synthesis occurs in both sexes when only 9–16 primordial germ cells are present (Kalt and Gall 1974), and the resulting 10- to 40-fold increase in ribosomal RNA gene number is maintained throughout gonial cell proliferation until meiosis begins (Pardue and Gall 1972). At meiosis the sexes differ. In the male, all detectable amplified rDNA is lost, but in the female there is a transient reduction in rDNA followed by a dramatic second wave of amplification to about 2500 times the chromosomal rDNA level (Gall 1968; Macgregor 1968; Evans and Birnstiel 1968; Brown and Dawid 1968). It is this second wave of amplification that has been most thoroughly studied and is now known to involve a rolling-circle replication mechanism (Gilbert and Dressler 1969; Hourcade et al. 1974; Rochaix et al. 1974) giving rise to extrachromosomal rings of rDNA (Miller 1964). Little is known in molecular terms about the earlier gonial wave of amplification and, more importantly, the mechanism leading to the first extrachromosomal rDNA copies remains a complete mystery. This paper attempts to shed light on these early events in amplification by means of restriction-enzyme analysis of the rDNA complement in individual oocyte nuclei and by electron microscopy of gonial and somatic rDNAs.

EXPERIMENTAL PROCEDURES

DNA from individual oocyte nuclei. Nuclei (germinal vesicles) were dissected from pigmented oocytes in 0.1 M KCl solution and transferred to 50 μl glycine buffer (0.1 M glycine, 0.2 M NaCl, 1 mM EDTA, pH 9.6). After lysis with 2 μl 20% sodium dodecyl sulfate (SDS), carrier DNA (0.4 μg *Micrococcus luteus* DNA) was added. After several hours, lysates were phenol-extracted and ethanol-precipitated at −20°C. DNA was recovered in an Eppendorf microcentrifuge at 8000g for 10 minutes. Nuclear DNAs and 0.03 μg of total DNA from the same female were digested overnight with restriction endonuclease *Eco*RI (Green et al. 1974) (Boehringer) in 50 mM Tris (pH 7.6), 20 mM MgCl$_2$. Digests were run on a 20-cm-square 1% agarose gel, blotted onto Millipore filters,

and hybridized (Southern 1975) with ^{32}P-labeled nick-translated (according to Maniatis et al. 1975) plasmid pCD4 DNA (Morrow et al. 1974). This plasmid contains one *Xenopus laevis* rDNA repeat unit integrated into pSC101. Hybridization conditions were 0.2 μg ^{32}P]DNA in 10 ml Denhardt's solution (Denhardt 1966) containing 0.45 M NaCl, 0.045 M Na citrate, and 1% SDS at 65°C for 3 days. Washing of the filter was in 2 × SSC, at 65°C, for 3 hours, followed by 0.1 × SSC, 65°C, for 30 minutes. The dried filter was exposed to flashed X-ray film backed by a fast tungstate intensification screen at −70°C for 1 day (Laskey and Mills 1977).

Testis rDNA. Two pairs of testes were washed, homogenized in glycine buffer (see above), lysed with SDS, and the DNA extracted. Ribosomal DNA was purified using actinomycin D-CsCl gradients as described previously (Bird and Southern 1977).

Sperm purification. Adult testes were teased apart with jewelers forceps in 20% Steinberg's solution. The cell suspension was then loaded onto a 0.25– 2.0 M sucrose gradient in a 30-ml Corex tube and centrifuged at 1500 rpm for 15 minutes. Two turbid bands were resolved, containing tailed and untailed sperm cells. These were pooled and their DNA extracted as above.

CsCl gradients. DNA samples were mixed with solid CsCl to a density of 1.715 g/ml and spun at 30 krpm for 3 days (18°C) in a Spinco A1–50 rotor. Fractions were denatured, bound to Millipore filters, and hybridized with ^{3}H]rRNA (Birnstiel et al. 1966).

Electron microscopy. rDNA was spread according to the method of Davis et al. (1971). PM2 DNA (molecular size 9.76 kb) was used as a size marker.

RESULTS AND DISCUSSION

Selectivity of Amplification in Individual Oocytes

Digestion of *Xenopus* rDNA with restriction endonucleases has clearly shown that the tandemly arranged repeat units are not homogeneous in length (Morrow et al. 1974; Wellauer et al. 1974, 1976). Instead, the population contains a spread of repeat sizes due to heterogeneity in the nontranscribed spacer, and any frog displays a selection of size classes from within the spread (Buongiorno-Nardelli et al. 1977). Using the length of the repeat unit as a marker, we can ask whether an oocyte amplifies some or all of the repeats present in the female chromosomes. Fortunately, the 30 pg of amplified rDNA in each oocyte (Macgregor 1968) is sufficient to be detected by blotting the restricted fragments from an agarose gel (Southern 1975) and hybridizing with ^{32}P-labeled nick-translated rDNA. (Chromosomal rDNA from a single oocyte, on the other hand, falls well below the limits of detectability.) Figure 1 compares the *Eco*RI digestion patterns of rDNA from 21 individual oocyte nuclei and from blood DNA of a single female. The lowest band, common to all digests which were complete, is a homogeneous rDNA fragment from within the transcribed region, while the upper bands each contain the rest of the repeat unit, including the variable spacer. Comparing the range of spacers in the chromosomes of this female with the amplified spacers, it is clear that only a limited subset of repeats is amplified by each oocyte. Furthermore, oocytes differ in their choice of subset.

What do these results tell us about the early events in gene amplification? First, the possibility that a single rDNA repeat unit is "excised" in each germ cell and is the source for all amplified copies can be ruled out. Most oocytes display one or more minor spacer classes in addition to a predominant class, suggesting that either there is more than one "excision" event, or that the "excised" segment contains a number of repeats. Second, it is clear that each oocyte chooses independently which repeat

units it will amplify, since out of 21 oocytes examined, 18 different spacer combinations were seen.

In this female, the spacers amplified most often are those most frequently represented in the chromosomes. However, studies of rDNA from whole ovaries indicate that spacers which are a minor component of the chromosomal rDNA may sometimes predominate in total amplified rDNA from the same frog (Reeder et al. 1976). This discrepancy between amplified and chromosomal spacer patterns cannot be explained by preferential selection of repeats from one of the two chromosomal ribosomal gene clusters, since it is also seen in animals heterozygous for a deletion of the ribosomal genes. Rather, it appears that certain repeats in a cluster are preferentially amplified—perhaps by virtue of their position within the tandem gene array. The distinctive spacer pattern in each oocyte could then reflect the original choice of a few chromosomal repeats from the preferred region. Alternatively, the oocyte pattern may arise by chance variations in replication frequency among an initially large variety of preferred repeat classes.

Electron Microscopy of Premeiotic Amplified rDNA

Cytological hybridization has detected amplified ribosomal RNA genes in nuclei of premeiotic germ cells (Gall 1969; Pardue and Gall 1972; Kalt and Gall 1974), so it is of interest to isolate this DNA and determine its molecular state. Amplified rDNA from oocytes is not methylated and consequently bands in cesium chloride gradients at a significantly

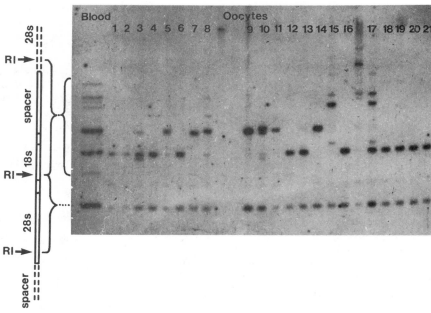

Figure 1. *Eco*RI restriction analysis of amplified rDNA in individual oocyte nuclei, as compared with blood-cell rDNA. All oocytes and blood DNA are from the same female frog. The figure is a 1-day autoradiograph of the blotted gel after hybridization to [32P]rDNA. On the left is a schematic drawing of an rDNA repeat unit, showing fragments generated by *Eco*RI (Wellauer et al. 1974) and their positions on the gel. Slots containing no DNA or showing incomplete digestion have been omitted from the numbering.

higher density than the heavily methylated somatic cell rDNA (Dawid et al. 1970). It has been reported that DNA from mature *Xenopus* testes also displays a small band at the unmethylated rDNA density, and this has been tentatively attributed to spermatogonial amplification (Kalt and Gall 1974). When rDNA is purified from testis, it is indeed found to form a split band in CsCl gradients, and both peaks hybridize to rRNA (Fig. 2a,b). In the electron microscope, 36% of high-density molecules (fraction I, Fig. 2a) are circular compared to 3% circles in the low-density peak (fraction II). Low-density circles and many high-density linears are attributable to overlap of the two peaks. As in the amplifying oocyte (Hourcade et al. 1974), circle length varies in steps equal in size to an average rDNA repeat unit, the smallest circles being of unit length. Of the cell types in the testis, it is most likely that premeiotic spermatogonia, which are known to contain amplified rDNA, are the source of high-density, circle-containing rDNA. This is supported by rRNA hybridization across CsCl gradients since neither sperm DNA (Fig. 2d) nor somatic DNAs (e.g., blood, liver; Fig. 2c) display high-density rDNA fractions (see also Dawid et al. 1970).

Premeiotic amplified rDNA was also detected in unsexed gonads dissected from stage 53–55 tadpoles (Nieuwkoop and Faber 1967). Light microscopy of

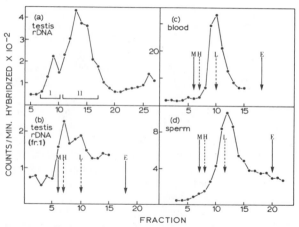

Figure 2. Buoyant density of rDNA from various cell types assayed by hybridization across CsCl gradients. *(a)* Preparative gradient of testis rDNA showing the split peak. High-density fraction I and low-density fraction II were examined in the electron microscope. *(b)* Rerun of some of fraction I (from above gradient) in the presence of density markers, showing increased proportion of high-density rDNA. *(c)* Blood DNA containing no detectable high-density rDNA peak. *(d)* Sperm DNA showing only low-density rDNA. Solid arrows M and E indicate the buoyant densities of *Micrococcus luteus* DNA ($\rho = 1.731$ g/ml) and *E. coli* DNA ($\rho = 1.710$ g/ml) included in each gradient. Dotted arrows denote computed densities of low-density somatic rDNA (L) and high-density amplified rDNA (H) ($\rho = 1.724$ g/ml and 1.729 g/ml, respectively).

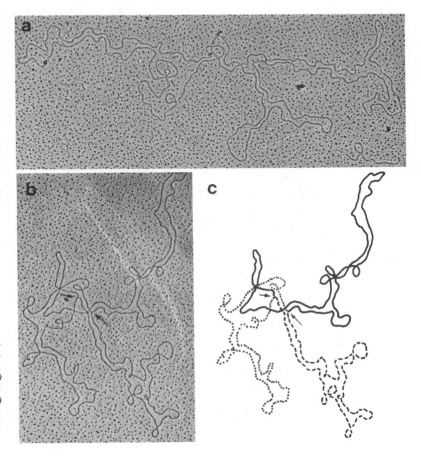

Figure 3. Electron micrographs of gonial rDNA. *(a)* Unequal-sized interlocked circles from premeiotic gonad rDNA (sizes 27.8 kb and 21.1 kb). *(b)* A Cairns' form from testis high-density rDNA. Arrows point out the forks. *(c)* Diagram of the Cairns' form shown in *b*. The two dotted loops are identical in length.

squashed preparations confirmed that, at these early stages, spermatogonia and oogonia are proliferating, and meiosis has not yet begun (Kalt and Gall 1974). By pooling one hundred gonads, sufficient purified rDNA was obtained for a single electron microscope spread. Once again, circles were frequent, and their lengths showed them to be oligomers of the rDNA repeat unit. Thus there is a correlation between high-density, circular rDNA and the presence of amplified premeiotic germ cells in both adult testes and premeiotic larval gonads. Together the results suggest that the amplified rDNA present in oogonia and spermatogonia is in the form of extrachromosomal rings of unmethylated DNA.

A feature common to testis high-density rDNA and premeiotic gonad rDNA is the high frequency of circular molecules — several times higher than in rDNA from amplifying oocytes (Hourcade et al. 1974; Rochaix et al. 1974). Since, by contrast, small oligomeric rings are scarce in diplotene oocytes of *Xenopus* (Scheer et al. 1977), it is likely that oocytes

in the process of amplification represent an intermediate state, and that, following prolonged rolling circle replication in early meiotic prophase, amplified rDNA becomes organized in rings much larger than those common in gonial cells.

Unlike oocytes, gonia maintain their amplified rDNA while dividing mitotically. This means that either the rDNA must duplicate and segregate to the daughter cells or the rDNA level is continuously replenished by reamplification of chromosomal genes. Unfortunately, only a few possible gonial replication intermediates were seen in the electron microscope, amounting to two Cairns' forms (Fig. 3b,c) (Cairns 1964), two-tailed circles with tails shorter than the circle, and two-tailed circles with much longer tails. Tentatively, it appears that Cairns' forms are more frequent during gonial amplification than during oocyte amplification (Hourcade et al. 1974; Rochaix et al. 1974) and may be involved in gonial amplified rDNA synthesis. Several interlocked circles in equal- and unequal-sized pairs were

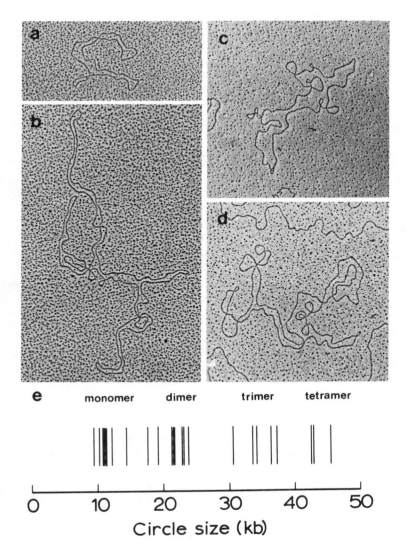

Figure 4. Circular rDNA molecules from *Xenopus* blood cells. *(a)* A repeat-unit monomer-size circle (10.9 kb); *(b)* a trimer-size circle (36.4 kb); *(c)* a dimer-size circle (21.4 kb); *(d)* another trimer-size circle (34.2 kb). *(e)* Distribution of blood rDNA circle sizes. Each vertical line represents one measured circle.

found in rDNA from both sources (Fig. 3a). The prevalence of unequal interlocked circles favors a recombinational origin for these structures since the other possible origin (failure of rings to separate after a round of Cairns-type replication) would lead to only equal pairs.

Somatic rDNA Circles

Unexpectedly, rDNA from *Xenopus* tissue-culture cells and from blood cells has been found to contain a low percentage of circles (Rochaix and Bird 1975) (Fig. 4). Circle lengths suggest that they are not contaminants, but are rDNA oligomers (Fig. 4e). Furthermore, their stability in high concentrations of formamide argues that the circles are not formed by annealing of single-stranded ends generated during DNA extraction. The significance of these rare extrachromosomal rDNA molecules is not clear; they may, for example, represent chance recombination events within the linear rDNA cluster. Their presence, however, suggests that the difference in the state of rDNA between a somatic cell and an amplified germ cell may only be one of degree.

Acknowledgments

I am grateful to E. Southern for a critical reading of the manuscript. This work has been supported by the Medical Research Council (London) and previously by Grant N.38630.72SR to Professor M. L. Birnstiel (Institute für Molekularbiologie II, Zürich).

REFERENCES

BIRD, A. P. and E. M. SOUTHERN. 1977. Use of restriction enzymes to study eukaryotic DNA methylation. I. The methylation pattern in ribosomal DNA from *Xenopus laevis. J. Mol. Biol.* **118**: 27.

BIRNSTIEL, M. L., H. WALLACE, J. L. SIRLIN, and M. FISCHBERG. 1966. Localization of the ribosomal DNA complements in the nucleolar organizer region of *Xenopus laevis. Natl. Cancer Inst. Monogr.* **23**: 431.

BROWN, D. D. and I. DAWID. 1968. Specific gene amplification in oocytes. *Science* **160**: 272.

BUONGIORNO-NARDELLI, M., F. AMALDI, E. BECCARI, and N. JUNAKOVIC. 1977. Size of ribosomal DNA repeating units in *Xenopus laevis:* Limited individual heterogeneity and extensive population polymorphism. *J. Mol. Biol.* **110**: 105.

CAIRNS, J. 1964. The chromosome of *Escherichia coli. Cold Spring Harbor Symp. Quant. Biol.* **28**: 43.

DAVIS, R., M. SIMON, and N. DAVIDSON. 1971. Electron microscope heteroduplex methods for mapping regions of base sequence homology in nucleic acids. *Methods Enzymol.* **21**: 413.

DAWID, I. B., D. D. BROWN, and R. H. REEDER. 1970. Composition and structure of chromosomal and amplified ribosomal DNAs of *Xenopus laevis. J. Mol. Biol.* **51**: 341.

DENHARDT, D. 1966. A membrane filter technique for the detection of complementary DNA. *Biochem. Biophys. Res. Commun.* **23**: 641.

EVANS, D. and M. L. BIRNSTIEL. 1968. Localization of amplified ribosomal DNA in the oocyte of *Xenopus laevis. Biochim. Biophys. Acta* **166**: 274.

GALL, J. G. 1968. Differential synthesis of genes for ribosomal RNA during amphibian oogenesis. *Proc. Natl. Acad. Sci.* **60**: 553.

———. 1969. The genes for ribosomal RNA during oogenesis. *Genetics* (Suppl.) **61**: 1.

GILBERT, W. and D. DRESSLER. 1969. DNA Replication: The rolling circle model. *Cold Spring Harbor Symp. Quant. Biol.* **33**: 473.

GREEN, P. J., M. C. BETLACH, H. W. BOYER, and H. M. GOODMAN. 1974. The Eco RI restriction endonuclease. *Meth. Mol. Biol.* **7**: 87.

HOURCADE, D., D. DRESSLER, and J. WOLFSON. 1974. The nucleolus and the rolling circle. *Cold Spring Harbor Symp. Quant. Biol.* **38**: 537.

KALT, M. R. and J. G. GALL. 1974. Observations on early germ cell development and premeiotic ribosomal DNA amplification in *Xenopus laevis. J. Cell Biol.* **62**: 460.

LASKEY, R. A. and A. D. MILLS. 1977. Enhanced autoradiographic detection of ³²P and ¹²⁵I using intensifying screens and hypersensitized film. *FEBS Lett.* (in press).

MACGREGOR, H. C. 1968. Nucleolar DNA in oocytes of *Xenopus laevis. J. Cell Sci.* **3**: 437.

———. 1972. The nucleolus and its genes in amphibian oogenesis. *Biol. Revs.* **47**: 177.

MANIATIS, T., A. JEFFREY, and D. G. KLEID. 1975. Nucleotide sequence of the rightward operator of phage lambda. *Proc. Natl. Acad. Sci.* **72**: 1184.

MILLER, O. L. 1964. Extrachromosomal nucleolar DNA in amphibian oocytes. *J. Cell Biol.* **23**: 60A.

MORROW, J. F., S. N. COHEN, A. C. Y. CHANG, H. W. BOYER, H. M. GOODMAN, and R. B. HELLING. 1974. Replication and transcription of eukaryotic DNA in *E. coli. Proc. Natl. Acad. Sci.* **71**: 1743.

NIEUWKOOP, P. D. and J. FABER. 1967. *Normal table of* Xenopus laevis *(Daudin)*, 2nd Ed. North-Holland, Amsterdam.

PARDUE, M. L. and J. G. GALL. 1972. Molecular cytogenics. In *Molecular genetics and developmental biology* (ed. M. Sussman), p. 65. Prentice Hall, Englewood Cliffs, New Jersey.

REEDER, R. H., D. D. BROWN, P. K. WELLAUER, and I. B. DAWID. 1976. Patterns of ribosomal DNA spacer lengths are inherited. *J. Mol. Biol.* **105**: 507.

ROCHAIX, J-D. and A. P. BIRD. 1975. Circular ribosomal DNA and ribosomal DNA replication in somatic amphibian cells. *Chromosoma* **52**: 317.

ROCHAIX, J-D., A. P. BIRD, and A. BAKKEN. 1974. Ribosomal RNA gene amplification by rolling circles. *J. Mol. Biol.* **87**: 473.

SCHEER, U., M. F. TRENDELENBURG, G. KROHNE, and W. W. FRANKE. 1977. Lengths and patterns of transcription units in the amplified nucleoli of oocytes in *Xenopus laevis. Chromosoma* **60**: 147.

SOUTHERN, E. M. 1975. Detection of specific sequences among DNA fragments separated by gel electrophoresis. *J. Mol. Biol.* **98**: 503.

TOBLER, H. 1975. Occurrence and developmental significance of gene amplification. In *Biochemistry of animal development*, vol. III. Academic Press, New York.

WELLAUER, P. K., I. B. DAWID, D. D. BROWN, and R. H. REEDER. 1976. The molecular basis of length heterogeneity in ribosomal DNA from *Xenopus laevis. J. Mol. Biol.* **105**: 461.

WELLAUER, P. K., R. H. REEDER, D. CARROLL, D. D. BROWN, A. DEUTCH, T. HIGASHINAKAGAWA, and I. B. DAWID. 1974. Amplified ribosomal DNA from *Xenopus laevis* has heterogeneous spacer lengths. *Proc. Natl. Acad. Sci.* **71**: 2823.

Ribosomal DNA and Related Sequences in *Drosophila melanogaster*

I. B. DAWID* AND P. K. WELLAUER†

* Department of Embryology, Carnegie Institution of Washington, Baltimore, Maryland 21210; † Swiss Institute for
Experimental Cancer Research, Epalinges s/Lausanne, Switzerland

This paper summarizes our present knowledge about rDNA in *Drosophila melanogaster* and reports that sequences related to a particular region of rDNA occur in other parts of the *Drosophila* genome. Ribosomal genes occur in multiple copies that form one or more clusters of tandemly repeated sequences in the nuclear genomes of all animals studied. In other organisms where this has been studied, the sequence organization of rDNA is relatively simple in that the only heterogeneity refers to length heterogeneity in the nontranscribed spacer regions (Wellauer et al. 1976). In *D. melanogaster,* a similar sequence arrangement has been found which is complicated by the fact that many of the repeating units contain a DNA region within the 28S gene sequence which interrupts this sequence and which has been called the "ribosomal insertion" (Glover and Hogness 1977; White and Hogness 1977; Wellauer and Dawid 1977; Pellegrini et al. 1977). In rDNA from the strains of *D. melanogaster* that have been studied, insertion-containing repeating units account for one-half to two-thirds of all rDNA repeats. This is true for DNA from at least three stages of development, embryos, pupae, and adults. Insertion sequences occur in different size classes, with a major size class of 5 kb.

Because of the complexity of the rDNA population with respect to different sizes of insertions and additional size heterogeneity in the nontranscribed spacer, we proceeded to analyze cloned homogeneous rDNA fragments. We will compare the properties of our cloned rDNA fragments with those of total rDNA isolated directly from the animal. This is an important aspect of studying repetitive genes by cloning methodology. When a single-copy gene is cloned, the cloned material should be identical to the only gene present, unless polymorphism or rearrangement during cloning complicates the situation. This is not necessarily true in the case of repetitive DNA. In particular, most cloning techniques favor smaller fragments over larger fragments, and it is quite possible to obtain an atypical sample.

Cloned rDNA Fragments from *D. melanogaster*

Highly purified or partially purified rDNA was obtained from the stock of *D. melanogaster* Oregon R that we used earlier (Wellauer and Dawid 1977) and from samples of DNA which contained only X chromosomal or Y chromosomal rDNA. The rDNA was digested with *Eco*RI and the fragments were inserted into the plasmid pMB9 (Rodriguez et al. 1976) by techniques similar to those described earlier (Morrow et al. 1974; Wellauer et al. 1976).

We have studied cloned rDNA fragments with the sizes and properties summarized in Table 1. As shown previously (White and Hogness 1977; Wellauer and Dawid 1977), all repeating units of rDNA have an *Eco*RI site in the 18S gene. Most of the clones therefore contained a complete repeating unit. A number of clones were obtained which did not have any insertion in the 28S gene and which varied in length only slightly due to differences in the sizes of their nontranscribed spacers. Among the clones with insertions, we found three size classes. One class corresponds to the abundant type of insertion of 5 kb. This insertion contains sites for the enzymes *Sma*I and *Bam*HI in an arrangement that will be described below. The other two size classes contained 1.0- and 0.5-kb insertions. These insertions had no *Sma*I site but had two and one *Bam*HI sites, respectively. Furthermore, we obtained four clones which carried less than a full repeat of rDNA. These clones derived from rDNA repeats that contain a second *Eco*RI site in addition to the standard site in the 18S gene. As reported earlier (Wellauer and Dawid 1977), a fraction of rDNA repeats have such an extra *Eco*RI site. We will return to the properties of these "half-repeats" later.

The *Sma*I and *Bam*HI sites in the rDNA clones were mapped by the analysis of digests of the ribosomal fragments themselves or the total plasmid DNAs by gel electrophoresis. We have shown previously (Tartof and Dawid 1976; Wellauer and Dawid 1977) that *Sma*I produces three distinct bands and one region of heterogeneous material from total *Drosophila* rDNA. We have also shown which of these bands hybridize with 18S–28S RNA. This information, together with the measurement of the sizes of *Sma*I fragments, is sufficient to construct a map. The construction of these maps is aided by the map of the vector pMB9 which has been presented by Rodriguez et al. (1976) and to which we have added the map position for the single *Sma*I

Table 1. Summary of Properties of Cloned rDNA Fragments
from *D. melanogaster*

Class	No. of clones	Size range	Insertion size (kb)	Restriction sites in insertion
I	3	15.5–17	5	*Sma*I (2); *Bam*HI (2)
II	2	11–12	1.0	*Bam*HI (2)
III	5	11–12	0.5	*Bam*HI (1)
IV	9	10–11	none	
V	4	4.3–7.4		*Eco*RI site in insertion giving half-repeats; no *Sma* or *Bam* site

The class designation allows reference to Fig. 2. In addition to the
clones listed, we have studied one additional clone that appears to contain
a large deletion; this clone will be described elsewhere (P. K. Wellauer
and I. B. Dawid, in prep.).

site. Figure 1 a and b shows two maps for recombinant plasmids in which the rDNA is inserted in either of the two possible polarities. The two examples
include one rDNA fragment without an insertion
and one fragment with a 1.0-kb insertion. The latter
insertion has two *Bam*HI sites which were mapped

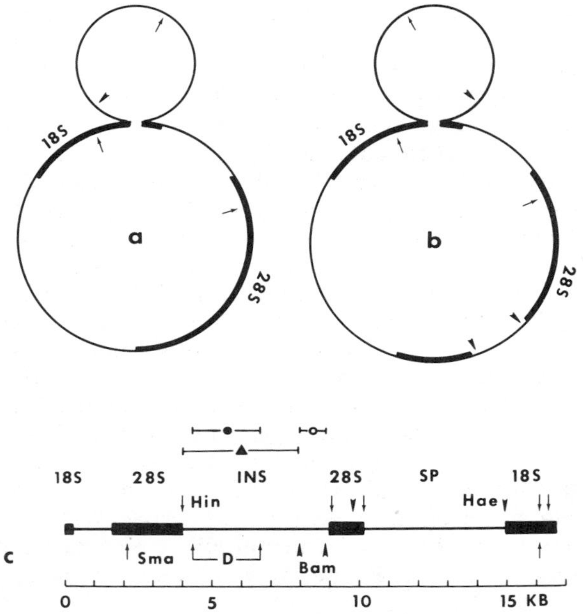

Figure 1. Maps of cloned *Drosophila* rDNA fragments.
(a,b) Recombinant plasmids, with the smaller circle representing the pMB9 vector. The junctions between pMB9
and rDNA are *Eco*RI sites. Gene regions for rRNA are
indicated by heavy lines. Sites for *Sma*I are indicated by
arrows and *Bam*HI sites are indicated by arrowheads. *(a)*
pDmrc51, which is a continuous repeat (no insertion); *(b)*
pDmre55 with a 1.0-kb insertion. The two rDNA fragments
in *a* and *b* are inserted in opposite polarities. *(c)* The map
of the rDNA fragment Dmra56 which contains a 5-kb insertion. *Sma*I (arrows) and *Bam*HI (arrowheads) sites are
shown below; *Hin*dIII (arrows) and *Hae*III (arrowheads)
sites are shown above. There are additional *Hae*III sites
which have not been mapped, but there are no *Hae*III sites
between the two shown; therefore, a large spacer fragment
is obtained with this nuclease. The *Sma*I D fragment is
indicated. Lines above the insertion refer to fragments that
were used in the hybridization experiment of Fig. 2.

similarly and quite simply because in rDNA only
insertions (no other regions) are cut by this enzyme.
The presence of a single *Bam* site in pMB9 facilitated the analysis (Fig. 1a,b).

To ascertain directly the presence of an insertion
and to measure its size, the clones were also subjected to electron microscopic analysis. This was
done by RNA-DNA hybridization and by heteroduplex analysis of various combinations of these clones
under the conditions described earlier (Wellauer and
Dawid 1977). Some of the results of this analysis
are incorporated into Table 1.

A more detailed map for an rDNA clone with a
5-kb insertion is presented in Figure 1c. This clone,
Dmra56, has been used extensively as a reference
clone for the comparison of different materials. The
*Bam*HI and *Sma*I sites were established as described
above. The *Hin*dIII sites are at analogous positions
to those mapped in a different rDNA clone by Glover
and Hogness (1977) and have been confirmed in our
work by the analysis of double digests of *Hin*dIII
and *Sma*I or *Bam*HI. The two *Hae*III sites shown are
not the only sites for this enzyme in the rDNA; many
sites occur elsewhere in the rDNA, but no additional
sites occur in the spacer between the two sites
shown. Therefore, *Hae*III is a useful enzyme to obtain a large fragment of the entire nontranscribed
spacer. A double digest with *Hae*III and *Hin*dIII
gives almost exactly the full spacer region, including
the nontranscribed and external transcribed spacer
of *Drosophila* rDNA. The two other rDNA clones
with 5-kb insertions that we studied have the same
arrangement of *Hin*dIII, *Sma*I, and *Bam*HI sites as
Dmra56.

Nucleotide composition of regions in rDNA. We studied this question using fragments of Dmra56. There
is considerable base-composition heterogeneity in
the rDNA repeating unit. Earlier work (Tartof and
Perry 1970; Klukas and Dawid 1976) has shown that
the mature rRNA has a GC content of 39%. The
same composition is indicated by the density and
melting temperature of the gene-containing *Hin*dIII
fragment which includes the region for 0–4 kb on
the map on Figure 1c. The insertion as represented

in the *Hind*III fragment from 4 to 9 kb in Figure 1c has a density and T_m suggesting a content of 55% GC. The nontranscribed spacer as represented by the *Hind*III-*Hae*III double-digest fragment from 10.3 to 15 kb of the map has a density and T_m suggesting a composition of 32% and 37% GC, respectively. The fact that density and T_m give significantly different GC contents for this region suggests that it may contain a repeated-sequence characteristic of some satellite DNAs.

Melting curves of several other entire cloned rDNA repeats were recorded. The curves reflect the regions of different composition described above. The high-GC segment of the melting curve is missing from repeats which do not contain an insertion. The repeats with short insertions have such a high-GC component, but it is proportionately smaller.

Homology between different insertions. We wondered whether insertions of different sizes contained homologous sequences. Insertions of closely similar sizes have similar restriction maps, and heteroduplex analysis and some cross-hybridization studies showed that the insertions within each class of Table 1 are homologous and probably identical. We wondered whether small insertions were homologous to a certain region of the larger insertions. To study this question, we adsorbed rDNA repeats with different insertions to filters and hybridized them with labeled fragments derived from the large repeat of

Dmra56 (Fig. 1c). As an internal standard, we used differently labeled rRNA. Figure 2 shows the result of this experiment. The figure gives the hybridization ratio between [³²P]DNA containing different regions of the insertion and [¹²⁵I]rRNA. The rDNA clones are classified into groups as in Table 1. Class I contains 5-kb insertions; these clones are homologous to all parts of the insertion of Dmra56 and are considered entirely homologous to it. Class II includes clones with a 1-kb insertion; these clones are homologous to the *Bam*HI fragment and to the *Sma*-*Bam* double-digest fragment of Dmra56, but not to the *Sma*I D fragment. Therefore, we conclude that these insertions correspond to the rightmost 1000 bp of Dmra56. There is no homology between the *Bam*HI and the *Sma*I D fragment of Dmra56. Class-III clones have a 500-bp insertion and a single *Bam*HI site; they are homologous to the *Bam* fragment of Dmra56 but not to the *Hin*d-*Bam* double-digest fragment. This finding suggests (1) that the insertion of these clones is homologous to the rightmost 500 bp of Dmra56 and (2) that the sequences immediately to the left of the *Bam*HI site at 8 kb (see Fig. 1c) of Dmra56 is not homologous to sequences between the two *Bam*HI sites. These results suggest that the 5-kb insertion is not internally repetitious, at least at the level detected by the *Bam*HI, *Hind*III, and *Sma*I enzymes. Since the left half and the rightmost 1000 bp are not homologous to each other, there cannot be a repetitious

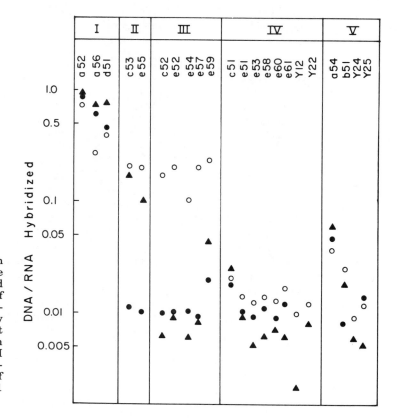

Figure 2. Hybridization of rDNA clones with various parts of the insertion of Dmra56. The various cloned rDNA fragments were loaded on filters and hybridized with mixtures of [¹²⁵I]rRNA and [³²P]DNA under conditions described by Dawid (1977). DNA was labeled by nick-translation as described by Maniatis et al. (1975). The ratio of hybridization between [³²P]DNA and [¹²⁵I]RNA is given. (●), *Sma*I D fragment; (○) *Bam*HI fragment; (▲) *Hind*III-*Bam*HI double-digest fragment. The origin of the three fragments is indicated in Fig. 1 above the map of Dmra56.

element throughout the entire insertion. However, it remains possible that the 1-kb insertion is a duplication of the 0.5-kb insertion. It is also possible that repetitious elements occur within the *Sma*I D fragment which are not recognized by any of the restriction enzymes that have been used so far.

Intermediate-size insertions containing an Eco*RI *site are a separate class. The last class in Table 1 and Figure 2 contains four clones which are composed of about half of an rDNA repeating unit. These are due to a secondary *Eco*RI site within the repeat. In three of the clones it is clear that the extra *Eco*RI site is in the insertion, whereas in one clone (DmrY24) it has not been decided whether the site is in an insertion or in the nontranscribed spacer. Three of these clones contain the left half of the repeat and the fourth (Dmrb51) contains the right half (as shown in Fig. 1c). This assignment is based on the analysis by radioactivity and electron microscopy of hybrids with 18S and 28S rRNA and by hybridization with the *Hin*dIII-*Hae*III spacer fragment of Dmra56. Since Dmrb51 hybridizes with the latter fragment, it follows that the spacers in repeats with and without a second *Eco*RI site are homologous. In contrast, there is little hybridization between the four half-repeat clones and any of the insertion fragments of Dmra56 (Fig. 2). There may be some homology with Dmra54, but the other three clones are entirely negative. Although the structure of the half-repeat clones needs further study, it appears that *Eco*RI-restrictable insertions are not homologous to the major class of insertions that lack an *Eco*RI site. Thus two sequence classes of insertions occur in *Drosophila* rDNA, each with multiple size classes.

The rDNA clones are representative of the rDNA population. We feel that the clones we studied are representative of the majority of the rDNA repeats in *Drosophila* because of the results of analysis by electrophoresis and microscopy of total rDNA purified from the animal. These results are illustrated in part in Figure 3. After digestion of rDNA with *Eco*RI, we see the major two bands at 11 and 16–17 kb which correspond to continuous repeats and repeats with the major 5-kb insertion, respectively. Material between these two bands corresponds to repeats with intermediate-size insertions, and material below the 11-kb band is representative of repeats with a second *Eco*RI site. The distribution of these different kinds of repeats has been analyzed previously (Wellauer and Dawid 1977), and it is clear that our cloned sample is fairly representative of the proportion of repeats with long, short, and *Eco*RI-restrictable insertions (Table 1). The absence of repeats with intermediate-size insertions in the cloned sample is due, at least in part, to the fact that many of these intermediate repeats have an extra *Eco*RI site. Repeats with 5-kb insertions are underrepresented, which may be due to technical reasons.

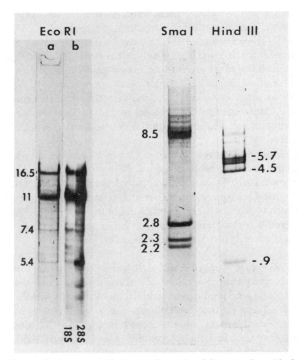

Figure 3. Agarose gel electrophoresis of digests of purified rDNA from *D. melanogaster*. The restriction enzymes used are indicated in the figure. Gels stained with ethidium bromide are shown for *Eco*RI *(a)* and for the other two endonucleases. *Eco*RI *(b)* is the corresponding autoradiograph after transfer of the fragments to a filter and hybridization with labeled rRNA. Sizes of fragments are shown in kilobases. The gels were run separately so that mobilities are not comparable.

The pattern obtained from total rDNA with *Sma*I (see Tartof and Dawid 1976; Wellauer and Dawid 1977) makes the following points. The sharp, strong band at 2.8 kb is derived from a transcribed region of the genome. It spans from the *Sma*I site in the 18S gene to the site in the 28S gene (see Fig. 1c). Most, if not all, rDNA repeats generate this precise 2.8-kb band; it is clearly visible even in stained gels of unfractionated *Drosophila* DNA digested with *Sma*I (Wellauer and Dawid 1977). The two distinct bands at 2.2 and 2.3 kb are derived from repeats with long insertions and are due to the two *Sma*I sites within the insertion (see Fig. 1c). These two distinct bands are also visible in unfractionated *Drosophila* DNA after *Sma*I digestion and show that a substantial number of rDNA repeats have a precisely identical structure with respect to *Sma*I sites in the 28S gene and in the insertion.

The *Hin*dIII digest of total rDNA is highly compatible with the distribution of *Hin*d sites in repeats with 5-kb insertions and without insertions, as shown in Figure 1c and as mapped originally by Glover and Hogness (1977). The distinct 0.9-kb band which is generated from the right end of the 28S gene is clearly visible in total rDNA and shows that the structure of the spacer region just beyond the

28S gene is invariant in most, if not all, rDNA repeats.

Digestion of total rDNA with *Bam*HI (not shown) displays a distinct 0.9-kb band as well. This band is generated from the right end of long insertions and from short insertions, as shown in Figure 1 b and c. This fact demonstrates that insertions which have two *Bam* sites have these two sites at a constant distance from each other.

Another point may be made about the homogeneity within the transcribed region of rDNA repeats. Although we cannot exclude the possibility that some repeats are mutated in this region, most repeats have a constant structure. This conclusion is based on the following observations. The 0.5-kb fragment between the *Sma*I site in the 18S region and the *Eco*RI site in that gene (see Fig. 1c) is constant in all clones that we studied. This distance is also constant in the majority of rDNA repeats since uncloned rDNA after double digestion with *Eco*RI and *Sma*I yields a distinct sharp band of 500 bp which is derived from that region. The distance between the *Eco*RI site and the *Sma*I site in the 28S gene is also constant, as seen in the same kind of experiment.

We have, however, obtained one cloned rDNA repeat with an atypical structure. This repeat appears to be the result of a deletion of part of the gene and spacer regions. We will describe the structure of this repeat elsewhere (P. K. Wellauer and I. B. Dawid, in prep.).

Because of these gel electrophoretic analyses, we feel that the collection of cloned repeats that we studied are largely representative of the *D. melanogaster* rDNA locus.

Non-rDNA Insertion Sequences

We wondered whether insertion sequences were restricted to the ribosomal locus or whether members of the same sequence family did occur in other regions of the *Drosophila* genome. The existence of such non-rDNA insertion sequences was reported by Dawid and Botchan (1977). We present here additional evidence and further details on non-rDNA insertion DNA.[1]

We detected insertion sequences by hybridization with labeled restriction fragments derived from the cloned rDNA repeat Dmra56 (Fig. 1c). From this insertion we isolated the *Sma*I D fragment, which represents the "left" half, and the 0.9-kb *Bam*HI fragment from the "right" half. In some cases we also used other parts of the insertion as hybridization probes.

We found (Dawid and Botchan 1977) that rDNA could be separated from non-rDNA insertion DNA in several types of density gradients. Figure 4a shows an experiment in which total embryo DNA[2] was banded in Cs_2SO_4 in the presence of mercury. The rDNA is visualized by hybridization with rRNA and bands slightly to the dense side of the main band. Hybridization of each fraction with the *Sma*I D fragment of Dmra56 revealed that insertion sequences are distributed in the rDNA region and in a broad peak to the low-density side. A clear separation was obtained between insertion sequences in rDNA and homologous sequences not linked to rDNA. Since the DNA used in this gradient is, on average, more than twice as large as one rDNA repeat, all insertion regions derived from rDNA are still linked to the other segments of this DNA. Thus, sequences homologous to the ribosomal insertion occur in other parts of the genome.

Linkage of non-rDNA insertion sequences with DNA of different composition. Separation of rDNA from non-rDNA insertion DNA could also be achieved in CsCl gradients containing actinomycin D. In the gradient of Figure 4b, we added to total adult DNA a small amount of ^{32}P-labeled *Sma*I D fragment as a marker for the density of pure insertion sequences. This fragment separated well from rDNA and main-band DNA, but also separated from most of the DNA components that hybridized with insertion probes. In this gradient, we localized DNA molecules homologous to the insertion both with the *Sma*I D and with the *Bam*HI fragments (see Fig. 1c). Although these two fragments are not homologous to each other, their hybridization tracks along the gradient. Non-rDNA molecules containing sequences homologous to the insertion band at densities intermediate between the main band and the insertion marker. We interpret this result to mean that most if not all non-rDNA insertion DNA molecules are not pure insertion sequences, but rather contain such se-

---

[1] The nomenclature of the sequences poses a problem. Insertions in rDNA are named as such because they interrupt the 28S rRNA coding region. Sequences homologous to ribosomal insertions outside the rDNA locus are not known to be inserted into any genes, but are identified as "insertion sequences" in a purely operational sense without prejudice as to their possible function and origin.

[2] *Drosophila* DNA was prepared from whole embryos or embryonic nuclei. To obtain high-molecular-weight DNA from larvae, pupae, or adults, a crude nuclear pellet was always used. Nuclei were made by homogenization in 0.25 M sucrose, 0.03 M Tris-HCl, pH 7.5, 1 mM EDTA (or 2.5 mM EDTA and 2.5 mM $CaCl_2$). A loose homogenizer was used for embryos and larvae; pupae and adults were ground in a chilled mortar and pestle. The homogenate was filtered through four layers of cheese cloth and centrifuged at 2000 rpm for 5 min. The pellet was resuspended and washed once. DNA was prepared either according to Peacock et al. (1974) or as follows. The pellet was suspended in 10–20 vol. of 0.15 M NaCl, 0.1 M EDTA, 0.05 M borate, titrated to pH 9.6 with NaOH. Sodium lauryl sulfate was added to 2% and the solution mixed with 1 vol. of phenol by gentle shaking for 1–2 days at 4°C. The aqueous phase was collected, the DNA precipitated with ethanol and spooled out if possible. Alternately, the precipitate was collected by centrifugation. The DNA was dissolved in 0.15 M NaCl, 0.03 M Tris-HCl, pH 8, 1 mM EDTA and digested for 30 min at 37°C with 20 μg/ml of pancreatic RNase; sodium lauryl sulfate (0.5%) and Pronase (0.5 mg/ml) were added and incubation continued for 1–2 hr. After another long phenol extraction, the DNA was recovered by precipitation with ethanol. To obtain very large DNA, the precipitation steps were omitted and replaced by prolonged dialysis against large volumes of buffer.

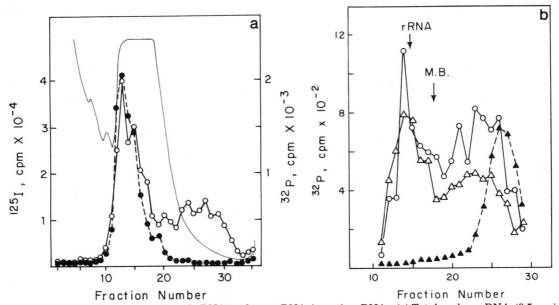

Figure 4. Density gradient separation of rDNA and non-rDNA insertion DNA. *(a)* Total embryo DNA (0.5 mg) of a size of 20–40 kb was banded in an 18-ml gradient of Cs_2SO_4 in the presence of 0.2 mole $HgCl_2$ per mole nucleotide. The thin continuous line is the absorbance as recorded in a flow-through cell at 260 nm. Small parts of each fraction were hybridized with [^{125}I]rRNA (●---●) and ^{32}P-labeled *Sma*I D fragment of Dmra56 (○——○) (see Fig. 1c). *(b)* Total adult DNA (0.2 mg) was mixed with a small amount (10,000 cpm, about 1 ng) of ^{32}P-labeled *Sma*I D fragment of Dmra56, and the DNA was banded in an 18-ml gradient in the presence of an equal weight of actinomycin D. After collecting fractions, the banding position of the added *Sma*I D fragment was located by Cerenkov counting (▲----▲). The position of the absorbance peak is indicated as M. B. (main band), and the position of rRNA hybridization is shown by an arrow. Aliquots of 0.05 of each fraction were hybridized with ^{32}P-labeled *Sma*I D (○——○) or *Bam*HI (△——△) fragments of Dmra56.

quences linked to other DNA segments of a different base composition. Other possible interpretations will be discussed below. Figure 4b further shows that non-rDNA insertion DNA occurs in adult DNA.

Problems of linkage were investigated further in the experiment shown in Figure 5. Embryo DNA mixed with a small amount of ^{32}P-labeled *Sma*I D fragment as marker was banded in CsCl-actinomycin D. Insertion sequences as detected with both probes separated from rDNA and banded mostly between the main band and the position of the *Sma*I D fragment (Fig. 5a). The DNA at intermediate density position (fractions 14–17) was recovered, sonicated to a size between 1 and 2 kb, and banded again in CsCl-actinomycin D after addition of fresh ^{32}P-labeled *Sma*I D fragment. As shown in Figure 5b, the position of the insertion sequences shifted to lower density so that most of these sequences now banded close to, though not coincident with, the marker *Sma*I D fragment (low density in actinomycin D implies high GC content). This experiment shows that insertion-DNA banding at intermediate density in the original gradient is composed of insertion sequences of a density similar to that of pure ribosomal insertion linked to DNA of a lower GC content. The alternate hypothesis would hold that non-rDNA insertion sequences themselves have a much lower GC content than do rDNA insertions,

while preserving some homology. The lower-GC sequences would then band close to the main band in the gradient of Figure 5a. The shift in density after sonication negates such a hypothesis and strongly supports the conclusion that non-rDNA insertion sequences have a high GC content and are linked to lower-GC DNA.

Additional gradient separations with embryo, pupae, and adult DNA were done under the conditions of Figures 4 and 5 and also in gradients of CsCl alone or CsCl and the dye Hoechst 33258 (see also Dawid and Botchan 1977). Non-rDNA insertion sequences occur in the DNA from all three stages of development in about the same proportion. Linkage to other types of DNA sequences was indicated for all three stages as well.

rDNA insertions and non-rDNA insertions are closely related sequences. Non-rDNA insertion DNA was purified from rDNA and partly enriched from main band in gradients such as those shown in Figures 4 and 5. This purified DNA was hybridized with the *Hind*III-*Bam*HI fragment of Dmra56 (4–8 kb in Fig. 1c); the hybrid was then melted. Figure 6 shows that the duplex between this cloned ribosomal insertion fragment and non-rDNA insertion sequences is just as stable as the duplex with its parent clone, Dmra56. A similar experiment with labeled *Bam*HI

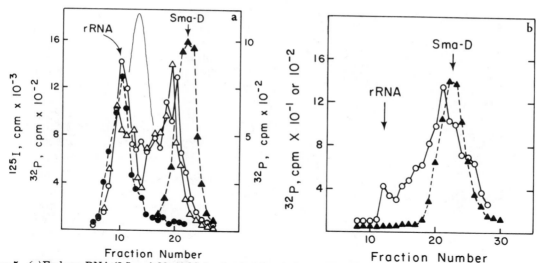

Figure 5. *(a)* Embryo DNA (0.5 mg) 20–40 kb in size was banded in a CsCl-actinomycin D gradient as in Fig. 4b. Labeled *Sma*I D fragment of Dmra56 was added at the outset. Hybridization with ^{32}P-labeled *Sma*I D and *Bam*HI fragments was carried out with small aliquots. *(b)* Fractions 14–17 of *a* were pooled, and the DNA was sonicated for 30 sec at a setting of 4 of the Branson Sonifier. The size of the DNA decreased to between 1 and 2 kb. Fresh ^{32}P-labeled *Sma*I D fragment was added and the DNA banded again in CsCl-actinomycin D. Fractions were hybridized with rRNA (arrow) and with ^{32}P-labeled *Sma*I D fragment. The symbols in this figure are the same as in Fig. 4.

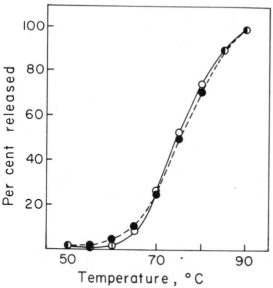

Figure 6. Melting of insertion DNA hybrids. Non-rDNA insertion DNA was prepared from a gradient, such as the one in Fig. 4a, and absorbed to membrane filters. Control filters were prepared with the cloned rDNA fragment Dmra56. Both types of filters were hybridized together with a solution of ^{32}P-labeled *Hind*III-*Bam*HI insertion fragment of Dmra56, under the conditions described previously (Dawid 1977). Filters were suspended in 15 mM NaCl, 5 mM Tris-HCl, pH 8.0, 0.5 mM EDTA and heated for 5 min at increasing temperatures. The solutions were collected at each temperature and counted. Release of label is shown for filters with Dmra56 (●- - - -●) and non-rDNA insertion DNA (○———○).

fragment showed a difference of about 2°C in the melting temperatures of the homologous duplex and the hybrid with non-rDNA insertions.

Restriction analysis of non-rDNA insertion DNA. Non-rDNA insertion DNA was purified by density gradient centrifugation, recovered, and digested with several restriction endonucleases. The fragments were separated on agarose gels and transferred to membrane filters (Southern 1975). The filters were then hybridized with rRNA or with different insertion probes. To test for complete digestion, we added phage λ DNA to each reaction mixture and visualized the fragments by staining; the smear derived from *Drosophila* DNA does not obscure the bands. Figure 7 (top panel) shows the stained gel containing *Hind*III fragments. The characteristic bands produced from λ DNA are present in all lanes. Lanes containing the *Hind*III fragments of total *Drosophila* DNA also show a heavy smear; lanes with purified non-rDNA insertion DNA were loaded with less material, which is barely visible after staining. The autoradiograph of the lanes from the same gel after transfer and hybridization is shown in Figure 7 (bottom panel). Lane a shows the fragments generated from rDNA as visualized with [^{125}I]rRNA. The heterogeneity is caused by the presence of the insertion in some repeats of rDNA and by length variation in the spacer (see Figs. 1c and 3). When total *Drosophila* DNA is digested with *Hind*III and the fragments challenged with insertion probes, we see a number of fragments not present in rDNA

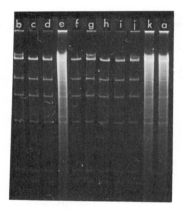

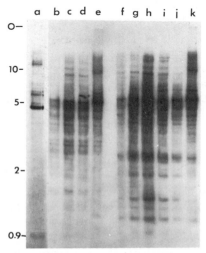

Figure 7. Fragment patterns after digestion of *Drosophila* DNAs with *Hind*III. To test for complete digestion, 0.5 μg phage λ DNA was added to each digest and the gel stained with ethidium bromide *(top panel)*. The characteristic λ-*Hind*III fragments are visible in every lane. Lanes in the stained gel are labeled to correspond to the autoradiograph *(bottom panel)*. The DNA in all lanes was transferred to a membrane filter (Southern 1975) and hybridized (Dawid 1977) with rRNA *(a)*, *Bam*HI fragment of Dmra56 *(b–e)*, or *Sma*I D fragment of Dmra56 *(f–k)*. Total embryo DNA was used in lanes *a*, *e*, and *k*. The other lanes contained non-rDNA insertion DNA from embryos *(c,d,h,i,j)*, pupae *(b,g)*, and adults *(f)*. Fragment sizes are given in kilobases.

(lanes f, k). These fragments arise from non-rDNA insertion DNA. In lanes b–e, the fragments carrying insertion sequences were visualized with ³²P-labeled *Bam*HI fragment; for lanes f–k, ³²P-labeled *Sma*I D fragment was used. Figure 7 leads to several conclusions: (1) Non-rDNA insertion DNA is heterogeneous, yielding many sizes of restriction fragments with different stoichiometry. (2) The fragments as visualized with the nonhomologous *Sma*I D and *Bam* HI insertion fragments are similar, especially among longer fragments. (3) In the area of smaller DNA pieces, the patterns diverge. (4) The restriction patterns derived from embryo, pupae, and adult DNA are similar, if not identical. (5) Different puri-

fication methods yield non-rDNA insertion DNA with identical restriction patterns.

These conclusions were also reached by examining the patterns obtained with the endonucleases *Eco*RI (Fig. 8), *Sma*I (Fig. 9), and *Bam*HI (Dawid and Botchan 1977). With every enzyme, many bands were generated from non-rDNA insertion DNA and the distribution of these bands was entirely different from the pattern derived from rDNA. No differences in restriction patterns were observed between different developmental stages with any of these enzymes.

Quantitation and arrangement of non-rDNA insertion sequences. From gradients such as those shown in Figures 4 and 5, we estimate that about 0.2% of the *Drosophila* genome is represented by non-rDNA insertions. This corresponds to about 400 kb of DNA. The distribution of this DNA in the genome is not known, but some possibilities are suggested by the available observations. Essentially all insertion sequences are linked to other DNAs in molecules of about 20 kb. This implies that the blocks of insertion sequences themselves are smaller than 20 kb and could frequently have a size of about 5 kb as in the ribosomal locus. Since the nonhomologous sequences represented by the *Sma*I D and *Bam*HI fragments of Dmra56 are linked to each other in larger restriction fragments derived from non-rDNA insertions (Figs. 7–9), it appears that at least a substantial fraction of non-rDNA insertions are larger than 3–4 kb. We therefore suggest that a large fraction of non-rDNA insertions occur in blocks about 5 kb in size. It is possible and even likely that smaller insertions occur.

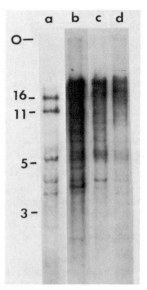

Figure 8. Fragment patterns after digestion of embryo DNA with *Eco*RI. Autoradiographs are shown after hybridization with rRNA *(a)*, *Sma*I D of Dmra56 *(b)*, or *Bam*HI fragment of Dmra56 *(c,d)*. Total embryo DNA *(a,d)* or purified non-rDNA insertion DNA *(b,c)* was used.

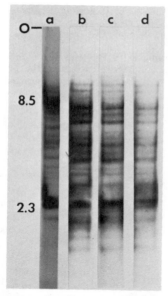

Figure 9. Patterns after digestion of DNA with *Sma*I. All lanes were hybridized with *Sma*I D fragment of Dmra56. Lane *a* contains a fraction enriched for rDNA and depleted of non-rDNA insertion sequences in a mercury-Cs$_2$SO$_4$ gradient. The other lanes contain purified non-rDNA insertion DNA from embryos *(b,c)* or adults *(d)*.

The restriction patterns as visualized by hybridization with insertion probes (Figs. 7–9) suggest further that non-rDNA insertion sequences are flanked by many different kinds of sequences. The different kinds of bands that occur could in part be due to slight sequence variations within the insertions themselves, leading to variable positions of restriction sites. However, such sequence variation is relatively small (Fig. 6), and it is more likely that most of the multiplicity of bands arises from sites in heterogeneous flanking sequences. At the same time, some regions of insertion sequences, together with their flanking sequences, appear to be conserved precisely, since strong bands occur in the restriction patterns which are due to multiple copies within the genome of the same restriction fragment. Other bands occur that are quite weak; the weakest bands in the patterns of Figures 7 through 9 are likely to represent single copies per genome. The overall impression of the arrangement of non-rDNA insertions is thus one of great variability both in the nature of the flanking sequences and in the multiplicities of different sequence blocks. In contrast, the sequence of insertion regions themselves is conserved quite precisely, as shown by melting experiments (Fig. 6).

CONCLUSIONS

Drosophila rDNA constitutes a particularly complex group of sequences. Repeating units vary in the length of their nontranscribed spacers and with

respect to the presence or absence of an insertion in the 28S gene. The insertion occurs always at the same location in the gene. There are two distinct sequence classes of insertions: those that do and those that do not have an *Eco*RI site. Both sequence classes occur in different size classes. The distribution of repeats with insertions differs in the nucleolus organizers of the X and Y chromosomes (Tartof and Dawid 1976; P. K. Wellauer et al. unpubl.).

Sequences homologous to ribosomal insertions of the class lacking an *Eco*RI site occur in regions outside the ribosomal gene cluster. Such sequences appear to be localized in the chromocenter (W. J. Peacock, unpubl., quoted in White and Hogness 1977; P. K. Wellauer and I. B. Dawid, unpubl.). The possible function of non-rDNA insertion sequences is not known. Experiments are in progress to determine whether insertions within or outside the ribosomal cluster are transcribed and whether regions flanking non-rDNA insertions are transcribed. Such information may indicate whether insertion sequences in *Drosophila* are more likely to be regulatory elements or evolutionary accidents.

Acknowledgments

We thank Peter Botchan for his contributions to some of these experiments, Martha Rebbert for excellent assistance, and Ron Reeder for gifts of restriction nucleases.

REFERENCES

DAWID, I. B. 1977. DNA/DNA hybridization on membrane filters: A convenient method using formamide. *Biochim. Biophys. Acta* **477**: 191.

DAWID, I. B. and P. BOTCHAN. 1977. Sequences homologous to ribosomal insertions occur in the *Drosophila* genome outside the nucleolus organizer. *Proc. Natl. Acad. Sci.* **74**: 4233.

GLOVER, D. M. and D. S. HOGNESS. 1977. A novel arrangement of the 18S and 28S sequence in a repeating unit of *Drosophila melanogaster* rDNA. *Cell* **10**: 167.

KLUKAS, C. K. and I. B. DAWID. 1976. Characterization and mapping of mitochondrial ribosomal RNA and mitochondrial DNA in *Drosophila melanogaster*. *Cell* **9**: 615.

MANIATIS, T., A. JEFFREY, and D. KLEID. 1975. Nucleotide sequence of the rightward operator of phage lambda. *Proc. Natl. Acad. Sci.* **72**: 1184.

MORROW, J. F., S. N. COHEN, A. C. Y. CHANG, H. W. BOYER, H. M. GOODMAN, and R. B. HELLING. 1974. Replication and transcription of eukaryotic DNA in *Escherichia coli*. *Proc. Natl. Acad. Sci.* **71**: 1743.

PEACOCK, W. J., D. BRUTLAG, E. GOLDRING, R. APPELS, C. W. HINTON, and D. L. LINDSLEY. 1974. The organization of highly repeated DNA sequences in *Drosophila melanogaster* chromosomes. *Cold Spring Harbor Symp. Quant. Biol.* **38**: 405.

PELLEGRINI, M., J. MANNING, and N. DAVIDSON. 1977. Sequence arrangement of the rDNA of *Drosophila melanogaster*. *Cell* **10**: 213.

RODRIGUEZ, R. L., F. BOLIVAR, H. M. GOODMAN, H. W. BOYER, and M. BETLACH. 1976. Construction and characterization of cloning vehicles. In *ICN-UCLA Symposium*, vol. V, p. 471. Academic Press, New York.

SOUTHERN, E. M. 1975. Detection of specific sequences

among DNA fragments separated by gel electrophoresis. *J. Mol. Biol.* **98**: 503.

TARTOF, K. D. and I. B. DAWID. 1976. Similarities and differences in the structure of X and Y chromosome rRNA genes in *Drosophila. Nature* **263**: 27.

TARTOF, K. D. and R. P. PERRY. 1970. The 5S RNA genes of *Drosophila melanogaster. J. Mol. Biol.* **51**: 171.

WELLAUER, P. K. and I. B. DAWID. 1977. The structural organization of ribosomal DNA in *Drosophila melanogaster. Cell* **10**: 193.

WELLAUER, P. K., I. B. DAWID, D. D. BROWN, and R. H. REEDER. 1976. The molecular basis for length heterogeneity in ribosomal DNA from *Xenopus laevis. J. Mol. Biol.* **105**: 461.

WHITE, R. L. and D. S. HOGNESS. 1977. R loop mapping of the 18S and 28S sequences in the long and short repeating unit of *Drosophila melanogaster* rDNA. *Cell* **10**: 177.

The Nucleotide Sequence of the Repeating Unit in the Oocyte 5S Ribosomal DNA of *Xenopus laevis*

N. V. FEDOROFF AND D. D. BROWN

Department of Embryology, Carnegie Institution of Washington, Baltimore, Maryland 21210

To understand a gene from a developmental perspective, it is necessary to identify DNA sequences which influence the quantity and timing of its expression and, eventually, to elucidate the underlying molecular control mechanisms. In tandemly redundant gene clusters, such as the 5S ribosomal RNA genes, the repeating unit of DNA is longer than the transcription unit. The extra sequences, designated "spacers," probably contain important regulatory signals. The periodicity of the DNA clearly delimits the DNA sequences associated with a single transcription unit, facilitating the task of locating and studying potentially significant extragenic sequences in the immediate vicinity of the gene. Here we summarize the results of nucleotide sequencing studies on the repeating unit of DNA containing the oocyte 5S rRNA genes (5S DNA) in *Xenopus laevis*. We discuss some evolutionary and functional implications of the sequence organization observed in 5S DNA.

Two different sets of 5S rRNA genes are expressed during the life cycle of *Xenopus*. One of them, designated the "oocyte type," is expressed only during oogenesis. A different set of genes is expressed during the rest of the developmental cycle (Brown and Sugimoto 1974). There are roughly 24,000 copies of the oocyte 5S rRNA genes, and they occur in tandem clusters at the telomeres of most or all of the chromosomes (Pardue et al. 1973). Oocyte 5S DNA has proved especially amenable to analysis because the repeating unit is short and the total population of repeating units is rather homogeneous.

Organization of the Repeating Unit

The repeating unit in oocyte 5S DNA averages about 700 nucleotides in length and can be roughly divided into an AT-rich half (region A) and a GC-rich half (region B) as indicated in Figure 1. The restriction endonuclease *Hin*dIII cleaves once per repeating unit at one of the junctures between the AT-rich and GC-rich portions of the repeating unit (Carroll and Brown 1976a). The fragments obtained upon digestion of 5S DNA with *Hin*dIII are heterogeneous in length (Carroll and Brown 1976a,b). This heterogeneity is confined to region A (Fig. 1), which can be further subdivided into regions which are more (A_2) or less ($A_1 + A_3$) variable in length within

the population of repeating units. Region B contains the gene and a related sequence termed the "pseudogene" (Jacq et al. 1977). Region B is constant in length between repeating units. The restriction enzyme *Hae*III cleaves at three sites in the majority of repeating units — twice in the gene and once in the pseudogene (Carroll and Brown 1976a; Jacq et al. 1977).

Primary Structure of the Repeating Unit

The primary sequence of the 5S DNA repeating unit has been deduced from several different kinds of sequencing studies. The 5S RNA species present in oocytes have been sequenced by Wegnez et al. (1972) and by Ford and Southern (1973). These represent the sequences of active genes. Transcripts of purified 5S DNA from frogs were used in initial studies of the AT-rich spacer (Brownlee et al. 1974) and, more recently, to determine the sequence of the pseudogene (Jacq et al. 1977). We have used direct DNA sequencing procedures on restriction enzyme fragments of 5S DNA purified from the frog and individual repeating units cloned in bacterial plasmids (N. V. Fedoroff and D. D. Brown, in prep.). From these data we have constructed a general picture of the population of repeating units in 5S DNA.

Region B

The 5S DNA population is sufficiently homogeneous so that a unique sequence was obtained for most of the repeating unit. The presence of several different sequences in the population of 5S RNA molecules found in oocytes attests to the presence of variants in the gene population (Wegnez et al. 1972; Ford and Southern 1973). However, these are present in sufficiently small amounts not to obscure the dominant sequence. The sequence of the GC-rich portion of the repeating unit (region B) was determined on restriction enzyme fragments derived from total frog DNA and therefore represents the dominant sequence in the population (N. V. Fedoroff, unpubl.). The sequence of the noncoding strand of region B is shown in the lower portion of Figure 1. The dominant gene sequence detected in the frog DNA corresponds to the most common 5S RNA sequence in oocyte ribosomes (Wegnez et al. 1972). Im-

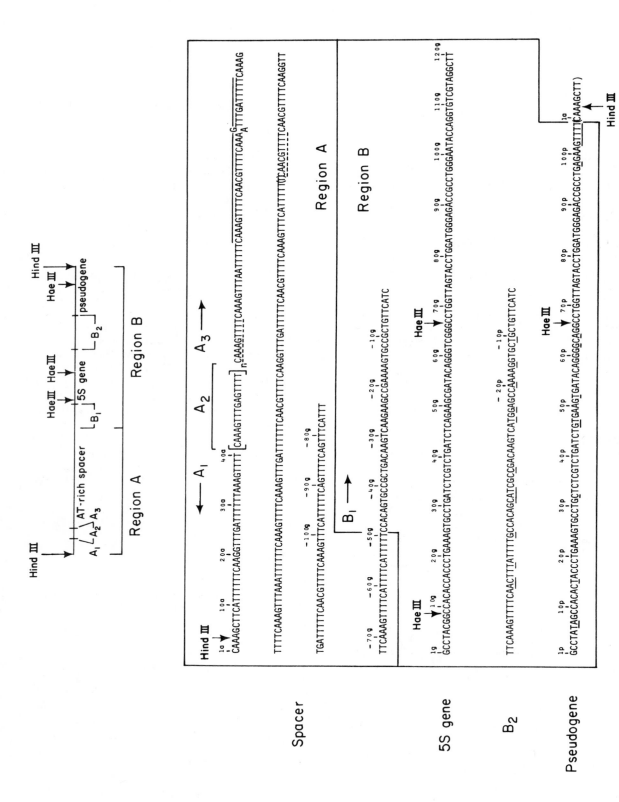

Figure 1. *(See facing page for legend.)*

mediately adjacent to the gene on the 5' side in the sense of transcription is a relatively nonrepetitive and GC-rich sequence 49 nucleotides in length (region B_1). The repetitive AT-rich spacer (region A) commences 50 nucleotides away from the gene on the 5' side. Region B_1 is quite strikingly different from the remainder of the spacer and is much less repetitive than the AT-rich spacer. It does contain several more or less perfectly repeated oligonucleotides, but these are very different in character from the repetitive sequences in the AT-rich spacer (Fig. 2). Also found in this region are several palindromes (Fig. 2); this kind of symmetry does not occur elsewhere in the spacer. Regions of dyad symmetry (self-complementarity) are not found in this sequence. Since transcription starts at the 5' terminus of the gene itself, this region is likely to contain sequences necessary for correct initiation.

At the 3' end, the terminal nucleotide of the gene (residue 120) is the second of a cluster of four T residues. The sequence extending away from the gene on the 3' side is AT-rich and strongly resembles the portion of the AT-rich spacer immediately adjacent to region B_1, extending from nucleotide −73g to −50g on the 5' side of the gene. The remainder of the region between the gene and pseudogene (region B_2) resembles region B_1. The sequences of regions B_1 and B_2 have been aligned vertically in Figure 1. The underlined nucleotides in B_2 are those that differ from the corresponding sequence on the 5' side of the gene. There is also extensive sequence homology between the gene and the pseudogene. These sequences have also been aligned vertically in Figure 1 to emphasize their similarity. Of the first 100 nucleotides in the pseudogene sequence, only ten differ from their counterparts in the gene sequence. These are underlined in the pseudogene sequence. The pseudogene sequence given here, determined by direct sequencing of DNA fragments, largely confirms the sequence data reported by Jacq et al. (1977) obtained by sequencing transcripts of 5S DNA. The single discrepancy is at nucleotide 15

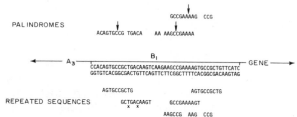

Figure 2. Symmetry elements in the 49-nucleotide sequence adjacent to the gene on the 5' side in the sense of transcription. The noncoding strand of the sequence is shown in the center. Sequences comprising direct repeats are reproduced below it; x indicates an imperfection in the repetition. Palindromic sequences are reproduced on the top lines; arrows indicate the centers of symmetry. Gaps represent nucleotides that must be looped out in either type of symmetry element.

(see legend to Figure 1). The gene and pseudogene sequences diverge beyond nucleotide 101, and we have designated nucleotide 106 as the 3'-terminal nucleotide of the pseudogene. We regard the adjacent oligonucleotide (CAAAGCTT), which contains the single *Hin*dIII cleavage site of the repeating unit, as the beginning of the AT-rich spacer (region A), since it resembles spacer oligonucleotides. In summary, the portion of region B adjacent to the gene on the 3' side, including region B_2 and the pseudogene, strongly resembles the sequence commencing at nucleotide −73g on the 5' side of the gene and extending almost to the end of the gene. This homology suggests that the GC-rich portion of the repeating unit originated as a duplication.

Region A

The sequence for the AT-rich spacer of 5S DNA (region A) was assembled from several kinds of sequencing data (N. V. Fedoroff and D. D. Brown, in prep.). There are few cleavage sites for known restriction enzymes in region A, and we were unable to obtain restriction enzyme fragments small

Figure 1. A composite sequence of the repeating unit in *X. laevis* 5S DNA, representing the dominant sequence types. The noncoding strand is shown. The total sequence is subdivided into regions as shown on the diagram. Sequences are numbered with respect to the first nucleotide of the gene (e.g., 10g is the tenth nucleotide of the gene and −10g is the tenth nucleotide preceding the gene on the 5' side in the sense of transcription), the pseudogene (similarly, 10p or −10p), and the left end of the AT-rich spacer (similarly, 10a). The T residue enclosed in parentheses designates a T cluster of uncertain length, and the position at which two residues appear indicates high levels of heterogeneity at that nucleotide. In the sequence extending from the left end of region A_3 to nucleotide −120g, the A and T cluster lengths were assigned from published RNase T_1 oligonucleotide data derived from spacer transcripts (Brownlee et al. 1974) and are therefore tentative. The underlined oligonucleotides in region A_3 were missing from one of the six cloned 5S DNA fragments analyzed, and the overlined oligonucleotide was missing from another. Region A_2 is represented by a single copy of the most common variant of the redundant 15-nucleotide sequence, enclosed in square brackets. Its variable redundancy is indicated by the subscript *n;* *n* varied from 2 to 12 in the six cloned repeating units analyzed, and its modal value in the total population is about 5. The variants CAAAGTTCGAGTTTT, CAAAGTTCGAGTATT, and CGAAGTTTGAGTTTT have been observed both in cloned fragments and in the total population of fragments. Region B_1 and the adjacent portion of region A_3 are aligned vertically with region B_2 two lines below to emphasize homology. Nucleotides in region B_2 which differ from the corresponding ones at the right end of A_3 and in B_1 are underlined. The gene is similarly aligned with the pseudogene two lines below. Again, nucleotides in the pseudogene which differ from their counterparts in the gene are underlined. The pseudogene sequence is identical to that reported by Jacq et al. (1977) except at nucleotide 15p, where we find a T residue, rather than a C residue.

enough to permit complete sequence determination. However, since the sequence of region A consists almost entirely of A and T clusters, punctuated most commonly by single G and C residues, the overall sequence pattern for the HaeIII-HindIII fragment containing the spacer could be determined if the lengths of A and T clusters were left unspecified (e.g., CA_nGT_n). The RNase T_1 oligonucleotide data obtained from spacer transcripts (Brownlee et al. 1974) were then used to assign precise cluster lengths (N. V. Fedoroff and D. D. Brown, in prep.) in the central portion of the sequence (region A_3 to the left of nucleotide −100g), as well as to confirm the sequences determined for region A_2 where necessary. Accurate direct sequence information was obtained for the terminal 100–150 nucleotides at each end of the spacer.

The sequence shown in Figure 1 for region A is a composite of the information obtained from analyses of the spacers in six single cloned repeating units of 5S DNA and that obtained from spacer fragments of total frog 5S DNA (N. V. Fedoroff and D. D. Brown, in prep.). There is a single dominant sequence for most of region A in the total population of repeating units. A unique sequence could be determined for the total spacer population in the regions designated A_1 and A_3, but not in region A_2, which commences 41 nucleotides from the left end of the spacer. At this point in the spacer sequence, cloned repeating units were found to differ in the number of copies of the oligonucleotide CAAAGTTT-GAGTTTT. The six cloned fragments analyzed contained 2, 4, 5, 6, 6, and 12 copies of this sequence. In Figure 1, region A_2 is represented by one copy of the oligonucleotide in square brackets; its variable redundancy is designated by the subscript n. The distribution of spacer lengths in the total 5S DNA population suggests that n is most commonly 5, but can exceed 16. The sequence shown for the redundant oligonucleotide is the most common one, but not the only one. The several variants found are indicated in the legend to Figure 1 and comprise an estimated 20% of all copies of the redundant oligonucleotide (N. V. Fedoroff and D. D. Brown, in prep.).

Region A_1 is quite homogeneous in the total population of repeating units and almost invariant in the six cloned repeating units analyzed. A single base change from the most common nucleotides T at nucleotide 27 to an A residue was observed in two different cloned repeating units. There are at least two types of heterogeneity — deletions and single base changes — within the population of repeating units in region A_3 of the spacer. Two of the six cloned fragments analyzed differed in length from the rest and from the dominant sequence in the population. One of the cloned fragments lacked the two short oligonucleotides, CAAAGTTTT and CAACGTTTT, which are underscored by dashed lines in Figure 1. The other lacked the longer oligo-

nucleotide overscored in region A_3 of Figure 1. Single base changes were relatively common in this region in cloned fragments. We cannot be entirely certain that all of the variants observed in cloned fragments reflect variants present in 5S DNA. That variants which alter the length of region A_3 exist in 5S DNA is supported by the observation that many of the spacer fragments obtained when frog 5S DNA is digested with HindIII and HaeIII do not fall into the discrete length classes expected for spacers which vary only by exact multiples of the 15-nucleotide sequence in region A_2 (Carroll and Brown 1976a).

Transcription of the Repeating Unit

Upon injection into X. laevis oocyte nuclei, 5S DNA is transcribed correctly and efficiently (Brown and Gurdon 1977). Recently it has become apparent that single cloned repeating units are also transcribed faithfully in oocyte nuclei (D. D. Brown and J. Gurdon, unpubl.). This means that all of the sequence information required for correct initiation and termination of transcription is contained within a single repeating unit. The weight of evidence, both from analyses of 5S RNA 5' termini and from pulse-labeling experiments in injected oocyte nuclei, favors the notion that the 5S RNA gene is the transcription unit. Although longer transcripts containing the 5S RNA sequence have been detected both in 5S RNA isolated from oocytes and in 5S RNA synthesized by oocyte nuclei injected with 5S DNA (D. D. Brown and J. Gurdon, unpubl.), the kinetics of labeling suggest that these longer RNA molecules constitute an independent fraction of transcripts, rather than an obligatory precursor of 5S RNA. High levels of pseudogene transcripts have not been detected in the RNA synthesized in oocyte nuclei injected with 5S DNA (D. D. Brown and J. Gurdon, unpubl.). In addition, there is no evidence that pseudogene transcripts are present in the stable population of oocyte RNA (Jacq et al. 1977). It is possible that the pseudogene is transcribed infrequently or that the transcripts are quite unstable. At present, however, it appears likely that the pseudogene is a transcriptionally inactive portion of the repeating unit. If this is true, then most of the DNA in the 5S repeating unit comprises "nontranscribed" spacer.

DISCUSSION

The most striking aspect of 5S DNA in X. laevis is its hierarchical redundancy. Each repeating unit contains a gene and a similar sequence, the pseudogene. The similarity of these and sequences adjacent to them suggests that the entire GC-rich portion of the repeating unit originated as a duplication. The basic repetitive units in the AT-rich portion of the spacer are variants of the sequences CA_3-GT_{3-4} and $\overset{C}{\underset{A}{G}}AT_n$. Each spacer within the population of repeating units contains many copies of a relatively constant arrangement of the basic units. Most commonly, spacers differ by the redundancy of the oligonucleotide CAAAGTTTGAGTTTT at a single position near the left end of the AT-rich spacer (region A_2). In the several cloned repeating units examined, the number of copies of this oligonucleotide varied from 2 to 12. The modal number of copies per spacer is about five in the total population of repeating units in frog 5S DNA.

The simplest explanation for the maintenance of length heterogeneity is that repetitive regions, primarily region A_2, undergo crossing-over in a variety of alignments with respect to each other. The cluster of variable length (region A_2) probably originated as a short duplication resulting from a crossover between regions of limited homology. Subsequent misalignments could have generated repeating units with progressively more copies of the oligonucleotide. It is curious, however, that the remainder of the repeating unit is rather constant in length, despite other evident internal repetitions capable, at least in principle, of misalignment. The two atypical cloned repeating units lacking oligonucleotides in region A_3 may represent the outcome of crossover events between repeating units misaligned in region A_3 by virtue of its internal redundancy. However, the fraction of repeating units in the 5S DNA population having such deletions is small enough not to obscure the dominant spacer pattern in which regions A_1 and A_3 are constant and only region A_2 varies. This suggests that crossover events in which these regions are misaligned are less frequent than in region A_2, even though the sequence of regions A_1 and A_3 are internally repetitive and closely related to the sequence of region A_2. Crossover events between gene and pseudogene appear to be very infrequent. At least one of the altered restriction enzyme fragments expected to issue from such a crossover event is below the level of detection in total frog DNA (Carroll and Brown 1976a). Hence repeating units generated by unequal crossing-over between gene and pseudogene can comprise no more than a small fraction of repeating units.

If one accepts the postulated explanation for the heterogeneity in length of region A_2, it is necessary to explain the relative constancy of the remainder of the repeating unit, despite internal redundancy. It is possible that crossing-over between gene and pseudogene is specifically suppressed or strongly selected against when it occurs. It is also possible that the enzymes involved in recombination act preferentially on spacer sequences. On the other hand, even without invoking such specificities, the observed heterogeneity is not entirely unexpected, given the organization of the repeating unit. If crossing-over re-

quires some form of trial-and-error matching to align homologous sequences, and if a relatively short region of sequence identity between two duplexes suffices to initiate crossing-over, then a sequence which is repetitious will participate in crossing-over more frequently than a sequence of equal length having a lower internal reiteration frequency. Moreover, the greater the internal redundancy, the greater the fraction of crossover events whose outcome will change the length of the repetitive region. Therefore, the AT-rich portion of the repeating unit may be more heterogeneous in length in the total population in some measure because it is more repetitive than the GC-rich portion.

As with other eukaryotic genes, our understanding of function lags somewhat behind our understanding of structure. Recent success in transcribing exogenous 5S DNA in oocyte nuclei (Brown and Gurdon 1977) gives us hope of more rapid progress in understanding the function of the nontranscribed portions of the 5S DNA repeating unit. It is already clear that a single repeating unit contains the information required for faithful initiation and termination of transcription. There is a striking discontinuity in the nature of the nontranscribed spacer sequence at a distance of 50 nucleotides from the gene on the 5′ side. The initial 49-nucleotide sequence (region B_1) differs both in base composition and in level of internal redundancy from the adjacent AT-rich portion of the spacer. There are three more or less perfectly repeated short sequences in this region, and these are quite different from the repetitive sequences in the AT-rich spacer (Fig. 2). Region B_1 also contains several palindromes, but no sequences which show significant self-complementarity (Fig. 2). From its location and character, we believe region B_1 is likely to contain sequences important for correct initiation of transcription.

The available evidence indicates that the pseudogene, despite its similarity to the gene, is not transcribed. Brownlee (1976) has suggested that the pseudogene is an evolutionary relic which originated as a gene duplication and subsequently diverged. He proposed a current function for the pseudogene in order to explain its homogeneity and conservation by selective pressure (Brownlee 1976; Jacq et al. 1977). We subscribe to the notion that the pseudogene is an evolutionary relic, but suggest that its continued presence need not imply the existence of positive selection. Of the several hypotheses that have been formulated to explain the sequence homogeneity observed within multigene clusters, those postulating a major role for recombination appear most readily to accommodate the actual structure of 5S DNA (Callan 1967; Britten and Kohne 1968; Edelman and Gally 1970; Thomas 1970; Buongiorno-Nardelli et al. 1972; Brown and Sugimoto 1974). Monte Carlo simulations of multigene families allowed to undergo unequal crossing-over show that, in time, certain of the original genes come to dominate the population numerically, while others are lost (Smith 1973). In 5S DNA, the gene appears to be represented by the longer repeating unit described here. From variations in the structure of repeating units within the population, we have argued that "out-of-register" recombination within the internally redundant repeating unit is confined largely to the AT-rich spacer. The uniformity in length of the GC-rich region within the population of repeating units suggests that duplications and deletions in this region are very rare. Hence gene and pseudogene are rarely or never separated by recombination events and appear to have been propagated as a unit. We suggest, then, that the pseudogene may be a fellow traveler of an evolutionarily successful gene, rather than an independently selected entity. It may indeed have originated as a gene duplication and subsequently accumulated mutations which inactivated it, thereby removing it from the library of active 5S genes subject to selective pressure. However, the continued presence and homogeneity of the pseudogene in the contemporary population of repeating units is perhaps incidental to the process which achieves and maintains homogeneity in the gene cluster, rather than an indication of current function.

Acknowledgments

This research was supported in part by a National Institutes of Health fellowship (5-F22 AI-00844) to N. V. F. and a National Institutes of Health grant (GM-22395–02) to D. D. B. We are grateful to Drs. B. Murr, L. Korn, and J. Doering for their critical comments on this manuscript.

REFERENCES

BRITTEN, R. J. and D. E. KOHNE. 1968. Repeated sequences in DNA. *Science* **161**: 529.

BROWN, D. D. and J. GURDON. 1977. High-fidelity transcription of 5S DNA injected into *Xenopus* oocytes. *Proc. Natl. Acad. Sci.* **74**: 2064.

BROWN, D. D. and K. SUGIMOTO. 1974. The structure and evolution of ribosomal and 5S DNAs in *Xenopus laevis* and *Xenopus mulleri*. *Cold Spring Harbor Symp. Quant. Biol.* **38**: 501.

BROWNLEE, G. G. 1976. Sequence analysis of DNA. In *Organization and expression of chromosomes* (ed. V. G. Allfrey et al.), p. 179. Dahlem Konferenzen, Berlin.

BROWNLEE, G. G., E. M. CARTWRIGHT, and D. D. BROWN. 1974. Sequence studies of the 5S DNA of *Xenopus laevis*. *J. Mol. Biol.* **89**: 703.

BUONGIORNO-NARDELLI, M., F. AMALDI, and P. A. LAVA-SANCHEZ. 1972. Amplification as a rectification mechanism for the redundant rRNA genes. *Nat. New Biol.* **238**: 134.

CARROLL, D. and D. D. BROWN. 1976a. Repeating units of *Xenopus laevis* oocyte-type 5S DNA are heterogeneous in length. *Cell* **7**: 467.

———. 1976b. Adjacent repeating units of *Xenopus laevis* 5S DNA can be heterogeneous in length. *Cell* **7**: 477.

CALLAN, H. G. 1967. The organization of genetic units in chromosomes. *J. Cell Sci.* **2**: 1.

EDELMAN, G. M. and J. A. GALLY. 1970. Arrangement and evolution of eukaryotic genes. In *The neurosciences: Second study program* (ed. F. O. Schmitt), p. 962. Rockefeller University Press, New York.

FORD, P. J. and E. M. SOUTHERN. 1973. Different sequences for 5S RNA in the kidney cells and ovaries of *Xenopus laevis*. *Nat. New Biol.* **241**: 7.

JACQ, C., J. R. MILLER, and G. G. BROWNLEE. 1977. A pseudogene structure in 5S DNA of *Xenopus laevis*. *Cell* **12**: 109.

PARDUE, M. L., D. D. BROWN, and M. L. BIRNSTIEL. 1973. Location of the genes for 5S ribosomal RNA in *Xenopus laevis*. *Chromosoma* **42**: 191.

SMITH, G. P. 1973. Unequal crossover and the evolution of multigene families. *Cold Spring Harbor Symp. Quant. Biol.* **38**: 507.

THOMAS, C. A. 1970. The theory of the master gene. In *The neurosciences: Second study program* (ed. F. O. Schmitt), p. 973. Rockefeller University Press, New York.

WEGNEZ, M., R. MONIER, and H. DENIS. 1972. Sequence heterogeneity of 5S RNA in *Xenopus laevis*. *FEBS Lett.* **25**: 13.

Simple Mendelian Inheritance of the Repeating Yeast Ribosomal DNA Genes

T. D. Petes,*‡ L. M. Hereford,† and D. Botstein*

*Department of Biology, Massachusetts Institute of Technology, Cambridge, Massachusetts 02139; †Rosensteil Basic Medical Sciences Research Center and Department of Biology, Brandeis University, Waltham, Massachusetts 02154

The yeast *Saccharomyces cerevisiae* has approximately 100–140 copies per haploid genome of the genes which code for ribosomal RNA (rRNA) (Schweizer et al. 1969). Each ribosomal gene codes for four species of ribosomal RNA: the 25S, 18S, 5.8S, and 5S rRNA (Udem and Warner 1972; Maxam et al. 1977; T. Petes et al., in prep.). In this paper we examine the inheritance of these repeating sequences.

To analyze the ribosomal DNA (rDNA) sequences genetically, we first identified two haploid strains of yeast that had ribosomal DNA genes which differed in base sequence between the two strains. This sequence difference was detected by the different patterns of DNA fragments produced by the digestion of the rDNA from each strain with the site-specific endonuclease *Eco*RI. The two haploid strains with the different types of rDNA genes were mated to form a diploid strain. The diploid, therefore, was heterozygous for the rDNA heterogeneity, containing about 100 rDNA genes from each of the haploid parental strains. This diploid was induced to undergo meiosis. By examining the segregation of the rDNA heterogeneity into the spores, we found that the repeated copies of the yeast rDNA genes were usually inherited as a single Mendelian unit. This result suggests that all of the rDNA genes are located in a single tandem array in which meiotic recombination is suppressed.

EXPERIMENTAL PROCEDURES

Yeast strains. Three strains were used in these experiments: 2262 (provided by C. McLaughlin), A364a and +D4 (both provided by L. H. Hartwell). The genotype of 2262 is α ade1 ura1 gal1 his5 lys11 leu2. The genotype of A364a is a ade1 ura1 gal1 ade2 tyr1 his7 lys2. The diploid +D4 was constructed by mating A364a and 2262.

Genetic analysis. The diploid was induced to sporulate using the conditions described by Brandriss et al. (1975). Tetrad dissection was done by standard procedures (Mortimer and Hawthorne 1969). The genotypes of the spores were checked either by mating to tester strains of known genotype or by plating on minimal plates lacking various amino acids. The marker lys11 was not checked in these experiments.

Isolation and restriction analysis of yeast ribosomal DNA. Individual strains were inoculated into 400 ml of YEPD medium (Mortimer and Hawthorne 1969) and grown at 30°C to a cell density of about 5×10^7 cells/ml. The cells were collected by centrifugation and washed once with 0.01 M Tris, 0.005 M EDTA (pH 8). Then they were resuspended in 5 ml 0.05 M Tris, 0.05 M EDTA (pH 8), and an equal volume of acid-cleaned glass beads (Superbrite, 340-μm diameter) was added. The mixture was agitated for 6 minutes using a Vortex mixer (Vortex-Genie) at maximum setting. Approximately 50% of the cells were broken by this procedure (D. Mills and S. Oliver, pers. comm.).

The cell suspension was then centrifuged at 15,000 rpm for 30 minutes (Sorvall SS-34 rotor). Aliquots of the supernatant (1–4 ml) were centrifuged to density equilibrium in a fluorescent-dye–CsCl density gradient as described previously (Williamson and Fennel 1975), except that the dye Hoechst 33258 was used (D. H. Williamson, pers. comm.). When the gradients were illuminated with long-wavelength ultraviolet radiation, three bands were observed. The bottom band, which contains the yeast rDNA, was removed by side puncture. The dye was removed from the DNA by three extractions with CsCl-saturated isopropyl alcohol. The DNA was then dialyzed against 0.001 M Tris-EDTA (pH 8).

The isolated rDNA was treated with the restriction enzyme *Eco*RI (Miles Research Products). Ribosomal DNA was incubated with the enzyme for 1 hour at 37°C in a buffer containing 0.1 M Tris, 0.05 M NaCl, and 0.01 M $MgCl_2$ (pH 7.5). The DNA fragments produced by the enzyme treatment were analyzed on 1.4% agarose gels containing 0.5 μg/ml of ethidium bromide (Sharp et al. 1973). Gels were photographed under short-wavelength ultraviolet illumination with Polaroid 55 or 57 film.

RESULTS

Identification of Two Forms of Yeast Ribosomal DNA

To study its inheritance, it is necessary to have two alternative forms of a gene (e.g., mutant and

---

‡ Present address: Department of Microbiology, University of Chicago, Chicago, Illinois 60637.

wild-type alleles). Rather than selecting alternative forms of the rDNA genes on the basis of function, we identified two forms of the rDNA that had a change in base sequence which did not affect function in any obvious way. This difference was detected by treating rDNA with the site-specific endonuclease *Eco*RI.

Ribosomal DNA, which accounts for about 5% of the yeast genome (Schweizer et al. 1969), can be isolated in preparative density gradients since it differs in base composition from nonribosomal nuclear DNA (Williamson and Moustacchi 1971; Cramer et al. 1972). When rDNA from the haploid strain A364a was treated with *Eco*RI and analyzed on agarose gels, seven bands were observed (A, B, C, D, E, F, and G in Fig. 1b). When rDNA from the haploid strain 2262 was analyzed, only six bands were observed (X', A, C, D, F, and G in Fig. 1c). The two restriction patterns contained five fragments in common (A, C, D, F, and G). The seven-fragment restriction pattern observed for A364a (which will be referred to as the type-I rDNA pattern) contained two fragments, B and E, not found in the rDNA from 2262 (type-II rDNA). The X' fragment observed in type-II rDNA was not seen in type-I rDNA. As discussed later, the X' fragment contains DNA sequences homologous to both the B and E fragments.

In heterothallic strains of *S. cerevisiae*, cells can be grown as either stable haploids or stable diploids. There are two mating types (m.t.) of haploid strains, *a* and *α*. Fusion between haploids of opposite mating types generates *a/α* diploids (summarized in Mortimer and Hawthorne 1969). Since the haploid strains with the type-I and type-II rDNA genes were of opposite mating types, a diploid (+D4) between them was constructed (L. H. Hartwell, University of Washington). As expected, when rDNA was isolated from this diploid and treated with *Eco*RI, all eight bands (X', A, B, C, D, E, F, and G) present in the two haploid parental strains were observed (Fig. 1a). Figure 1a also shows that the bands X', B, and E are present in fewer copies per cell than are the A, C, D, F, and G fragments. This is the expected result since X' is present only in type-II rDNA genes, B and E are present only in type-I rDNA genes, and A, C, D, F, and G are present in both types. The diploid, therefore, is heterozygous for two forms of the repeated rDNA genes, containing about 100 rDNA genes with the type-I pattern and 100 rDNA genes with the type-II pattern.

To further characterize the type-I and type-II rDNA genes, we made restriction maps of each type of rDNA. For this analysis, randomly sheared fragments of nuclear DNA isolated from the heterozygous diploid +D4 were inserted into the plasmid pMB9 using homopolymers of dA's and dT's as linkers (Lobban and Kaiser 1973; Wensink et al. 1974; T. Petes et al. in prep.). Since the pMB9 plasmid contained genes which coded for tetracycline resistance (H. Boyer, pers. comm.), these recombinant molecules were transformed into CaCl₂-treated *Escherichia coli*, and transformants were selected by tetracycline resistance. Two thousand of the transformants were then screened by colony hybridization (Grunstein and Hogness 1975) using ³²P-labeled yeast rRNA as a probe. Approximately 100 of the 2000 transformants hybridized to the probe and therefore contained insertions of yeast rDNA.

Seventy-five of these plasmids were characterized by analysis with *Eco*RI or other restriction enzymes (T. Petes et al., in prep.). Plasmid DNA from each transformant was isolated, treated with *Eco*RI, and analyzed on agarose gels. Almost all (73/75) recombinant plasmids contained one or more *Eco*RI-generated restriction fragments having the same electrophoretic mobility as X', A, B, C, D, E, F, or G. Most of the recombinant plasmids examined contained only a few of these eight fragments. By determining which fragments were neighbors in such clones, we generated a restriction map for the *Eco*RI sites within the type-I rDNA as shown in Table 1. For example, the first plasmid in Table 1 contains only two fragments, with the sizes of B and G. This suggests that B and G are adjacent fragments within the type-I rDNA gene; we know we are mapping the type-I rDNA gene because the type-II rDNA does not have a fragment with the mobility of B. The second plasmid in Table 1 had the four fragments A, B, E, and F. Since, from the previous plasmid, we knew that G was next to B on one side, the second plasmid indicates that the three fragments A, E,

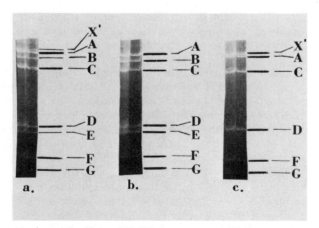

Figure 1. Analysis of *Eco*RI fragments of rDNA from the diploid strain +D4 and its two haploid parental strains A364a and 2262. Yeast rDNA was isolated from each strain by density gradient centrifugation, digested with *Eco*RI, and analyzed on 1.4% agarose gels containing ethidium bromide. Next to each gel photograph, a schematic drawing of the bands is given since the low-molecular-weight bands are not well visualized in the photograph. *(a) Eco*RI digest of rDNA from +D4. Thin lines in the schematic drawing for bands X', B, and E indicate that these fragments are present in smaller molar amounts than the other bands. *(b) Eco*RI digest of rDNA from the haploid parent A364a. *(c) Eco*RI digest of rDNA from the haploid parent 2262. (From T. Petes and D. Botstein, in prep.)

Table 1. Mapping of *Eco*RI Restriction Fragments of Type-I Ribosomal Genes

Clone designation	Fragments produced after *Eco*RI treatment	Derived map[b]
pY1rF2	GB	GB
pY1rF10	ABEF	GB(AEF)
pY1rG12	BCEG	CGBE(AF)
pY1rB3	BEFG	CGBEFA
pY1rF1	ABDEFG	CGBEFAD
pY1rB1	ACDEF	

[a] Excluding fragments containing pMB9 DNA.

[b] Order not determined for fragments within parentheses (T. Petes et al., in prep.).

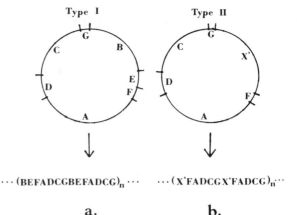

Figure 2. *Eco*RI restriction maps of the two types of rDNA genes present in the diploid +D4. These maps were constructed by analysis of recombinant DNA molecules (made in vitro) which contained insertions of rDNA from +D4. The maps of the rDNA genes are circular because the rDNA genes are arranged in tandem arrays on the chromosome. (*a*) Type-I rDNA gene (seven *Eco*RI sites per repeating unit); (*b*) type-II rDNA gene (six *Eco*RI sites per repeating unit).

and F must be on the other side of B (although the order within these three fragments cannot be determined without additional information). By an extension of these arguments and additional data from other plasmids, an unambiguous map of the type-I rDNA gene was constructed (Table 1).

No recombinant plasmids were observed which contained the X′ fragment along with either B or E. This is the expected result since the observations on rDNA isolated from the haploid yeast strains indicated that X′ was present only in type-II rDNA and that B and E were present only in type-I rDNA genes. There were not enough recombinant plasmids to make an unambiguous map of the type-II rDNA using the same technique employed for the type-I genes. Therefore the map of the type-II genes was constructed by analyzing a recombinant plasmid which had all six *Eco*RI fragments of the type-II rDNA (X′, A, C, D, F, and G) with other restriction enzymes (T. Petes et al., in prep.). The results of this analysis are shown in Figure 2b.

All recombinant plasmids that were analyzed were consistent with one of the two restriction maps shown in Figure 2. This suggests that within the type-I and type-II classifications, the rDNA genes are homogeneous repeating units. In addition, no size heterogeneity was observed for any of the restriction fragments, indicating that the nontranscribed portion of the rDNA gene may also be homogeneous.

Several other features of the maps should be noted. The restriction maps obtained were circular maps. This circularity occurs not because individual rDNA genes are small circular molecules, but because of the tandem arrangement of rDNA genes on the chromosome (Cramer et al. 1972). The restriction maps indicate that X′ occupies in the type-II pattern the position occupied by the B and E fragments in the type-I rDNA. This confirms the conclusion that X′ is an alternative arrangement of the

sequences represented by fragments B and E. This conclusion is also confirmed by the observation that the molecular weight of X′ equals, within experimental error, the sum of the molecular weights of B and E. Finally, it has been observed (T. Petes and L. Hereford, unpubl.) that X′ hybridizes to fragments B and E. The difference between the type-I and type-II rDNA genes, therefore, can be explained by a base change of the *Eco*RI site at the B-E junction. It is interesting that this region is equivalent to the nontranscribed region of the yeast rDNA gene (R. W. Davis, pers. comm.). It should be mentioned that the map for the type-I rDNA is identical to that obtained by Cramer et al. (1977) for the rDNA of a different haploid strain of yeast.

Genetic Analysis of Yeast Ribosomal DNA

The results discussed above showed that the diploid +D is heterozygous for an rDNA heterogeneity; approximately 100 of the rDNA genes have the type-I pattern (seven *Eco*RI sites per repeating unit) and about 100 rDNA genes have the type-II pattern (six *Eco*RI sites per repeat). To analyze how this heterogeneity segregated at meiosis, we induced the diploid to sporulate. The four haploid spores formed from each diploid were separated by standard tetrad dissection procedures. Ribosomal DNA was isolated from cultures grown from each spore, treated with *Eco*RI, and analyzed on agarose gels. In addition, each spore was analyzed for other segregating genetic markers (mostly auxotrophic markers). The results of this analysis are summarized in Table 2.

As expected, all of the heterozygous auxotrophic markers segregated two mutant spores: 2 wild-type

Table 2. Segregation of Markers in Tetrads from the Diploid +D4

	Chromosome III		Chromosome II			IX	XV	rDNA		Chromosome III		Chromosome II			IX	XV	rDNA
Spore	m.t.	*leu2*	*lys2*	*tyr1*	*his7*	*his5*	*ade2*	pattern	Spore	m.t.	*leu2*	*lys2*	*tyr1*	*his7*	*his5*	*ade2*	pattern
1a	a	+	−	−	+	+	−	I ≫ II	8a	a	+	−	−	−	−	−	I
1b	a	+	−	−	−	−	+	I ≫ II	8b	α	−	−	−	−	+	+	II
1c	α	−	+	+	+	+	+	I	8c	α	−	+	+	+	+	−	I
1d	α	−	+	+	−	−	−	I	8d	a	+	+	+	+	−	+	II
2a	α	−	−	+	+	+	−	I	9a	a	+	+	+	−	−	−	I
2b	a	+	+	−	−	+	+	II	9b	α	−	−	−	+	−	−	II
2c	a	+	−	−	+	−	+	I	9c	α	−	−	−	−	+	+	I
2d	α	−	+	+	−	−	−	II	9d	a	+	+	+	+	+	+	II
3a	α	+	−	−	+	+	+	II	10a	α	+	+	+	+	+	+	I ≫ II
3b	a	+	−	−	+	+	−	I	10b	a	+	+	−	−	−	−	I ≫ II
3c	a	−	+	+	−	−	+	I	10c	a	−		+	+	−	+	II ≫ I
3d	α	−	+	+	−	+	−	II	10d	α	−		−	−	+	−	II ≫ I
4a	a	+	+	+	+	+	−	I	11a	α	−	−	−	+	−	+	II
4b	α	−	−	+	+	−	−	I	11b	a	+	−	+	−	+	−	I
4c	a	+	−	+	+	−	+	II	11c	a	+	+	+	+	+	+	I
4d	α	−	+	+	+	+	+	II	11d	α	−	+	−	−	−	−	II
5a	α	−	−	−	−	+	−	II	12a	a	−	+	+	+	+	+	I
5b	α	−	+	+	+	+	−	II	12b	a	+	−	−	−	−	−	II
5c	a	+	−	−	−	−	+	I	12c	α	+	+	+	−	−	+	I
5d	a	+	+	+	+	−	+	I	12d	α	−	−	−	−	+	−	II
6a	α	+	−	−	−	−	−	II	13a	a	+	−	+	+	+	−	I
6b	α	−	+	+	+	+	−	II	13b	a	+	−	−	−	−	+	II
6c	α	+	+	−	−	−	+	I	13c	α	−	+	−	−	−	+	II
6d	α	−	−	+	+	+	+	I	13d	α	−	+	+	+	+	−	I
7a	α	−	−	+	−	−	−	II	14a	α	+	−	+	+	−	+	II
7b	a	+	+	+	+	+	−	I	14b	α	−	+	−	−	−	+	I
7c	α	−	+	−	−	−	+	II	14c	a	+	+	−	−	+	−	I
7d	a	+	−	−	+	+	+	I	14d	a	−	+	+	+	+	−	II

Genotype of +D4 = $\dfrac{a \quad ade2 \quad + \quad gal1 \quad lys2 \quad try1 \quad his7 \quad ura1 \quad ade1 \quad + \quad +}{\alpha \quad + \quad leu2 \quad gal1 \quad + \quad + \quad + \quad ura1 \quad ade1 \quad his5 \quad lys11}$. Methods of analysis are described in Methods. I ≫ II means primarily type-I pattern with a minor amount (less than 25%) of type II. (Data from T. Petes and D. Botstein, in prep.)

spores in at least 13 of 14 tetrads (the exceptions probably resulting from mitotic recombination). The rDNA heterogeneity also segregated 2:2 — 2 spores with type-I rDNA:2 spores with type-II rDNA — in 12 of 14 tetrads. This segregation pattern is shown by the gel in Figure 3. Reconstruction experiments indicate that as little as 10% contamination of one pattern with the other could be detected in these experiments. A more sensitive method, which could detect 3% contamination of one pattern with the other, also failed to show a mixture of type-I and type-II rDNA in the spores of four tetrads examined (T. Petes and D. Botstein, in prep.).

As discussed later, a 2:2 segregation pattern is expected if a trait is controlled by a single gene. The observation that the rDNA genes segregated 2:2 in 12 of 14 tetrads indicates that these 100 repeating units usually behave as a single genetic locus. In two tetrads (tetrad 1 and tetrad 10 of Table 2), the rDNA heterogeneity did not segregate 2:2 into the spores. In tetrad 1, two spores were found which had only type-I rDNA, and two spores were found which contained mostly type-I rDNA but also had

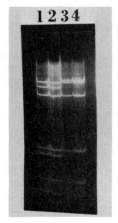

Figure 3. Analysis of *Eco*RI fragments of rDNA isolated from four spores of a yeast tetrad. The spores were cultured individually (tetrad 4 in Table 2). rDNA was then isolated, treated with *Eco*RI, and analyzed on agarose gels. Spores 1 and 2 show the type-I rDNA pattern (seven fragments) and spores 3 and 4 show the type-II rDNA pattern (six fragments) (T. Petes and D. Botstein, in prep.).

type-II rDNA. In tetrad 10, two of the spores had mostly type-I rDNA with a small amount of type-II rDNA, and two spores were of the reciprocal class. As discussed in more detail later, the segregation patterns observed in tetrads 1 and 10 are probably the result of a mitotic rather than a meiotic recombination event which occurred within the rDNA gene cluster.

DISCUSSION

We have identified two types of repeating rDNA genes and have used this heterogeneity to study the inheritance of yeast rDNA. Our genetic analysis can be summarized by the statement that yeast rDNA usually segregates as a single Mendelian unit. When a diploid strain is heterozygous for different forms of the repeated rDNA sequences, the different forms usually segregate 2:2 into the four haploid spores. A more detailed interpretation of these results is given below.

Most of the Yeast rDNA Genes Are on a Single Chromosome

If the ribosomal DNA genes are located in a single gene cluster (and the genes within the cluster are not separated by meiotic recombination), the expected segregation pattern of the rDNA heterogeneity in meiosis is 2:2. If the rDNA is organized into two unlinked clusters of genes, then in half of the tetrads the expected segregation will still be 2 type-I spores:2 type-II spores. In the other half of the tetrads, since the unlinked clusters should segregate independently, all four spores should contain mixtures of both types of rDNA. If there are more than two unlinked clusters of rDNA genes, the probability of a 2:2 segregation pattern becomes even smaller. Therefore, the finding of 2:2 segregation in 12 of 14 tetrads suggests that only one of the approximately 17 yeast chromosomes (Byers and Goetsch 1975) contains a substantial fraction (more than 10%) of the cellular rDNA genes. Our results suggest the possibility that all of the approximately 100 rDNA genes of the cell may be arranged in a single tandem array. Alternatively, several smaller rDNA gene clusters on a single chromosome could be linked by nonribosomal DNA which does not undergo frequent meiotic recombination.

The 2:2 segregation of yeast rDNA genes also makes it unlikely that the rDNA genes are extrachromosomal. We cannot rule out the possibility, however, that extrachromosomal rDNA is synthesized after spore germination and degraded before the next meiosis.

Yeast rDNA Is Not Strongly Centromere-linked

The location of a marker relative to its centromere can be measured by the frequency of second-division segregation (Mortimer and Hawthorne 1966). The frequency of second-division segregation is assayed by measuring the frequency of tetratype asci between the gene of interest and a known centromere-linked marker. In a cross of strains which are heterozygous for two markers, AB × ab, tetrads containing spores of the genotype ab, Ab, aB, and AB are tetratype asci.

The marker leu2 was heterozygous in +D4 and was used as the centromere-linked marker in this analysis. This marker shows 13% second-division segregation (Mortimer and Hawthorne 1966). Table 2 shows that 9 of 14 tetrads (64%) were tetratype asci with respect to the rDNA heterogeneity and leu2. Since leu2 shows 13% second-division segregation, the rDNA genes show 51% (64–13%) second-division segregation. This is close to the value that would be expected (67%) if the rDNA genes were unlinked to the centromere of any chromosome.

Since the previous discussion indicated that the rDNA genes are inherited as a single unit and since this unit is not located near the centromere, the rDNA genes must be located on a single arm of a single chromosome. Thus the rDNA gene cluster does not go through the centromere.

Meiotic Recombination of rDNA Is Less Frequent Than for Nonribosomal Nuclear DNA

The 2:2 segregation of rDNA indicates that the cluster of yeast rDNA genes does not frequently recombine during meiosis. If all the rDNA genes were on a single chromosome, a single meiotic crossover between nonsister chromatids would produce a tetrad in which one of the spores has only type-I rDNA, one has only type-II rDNA, and two spores have mixtures of type-I and type-II rDNA. This pattern of segregation was not observed for any of the 14 tetrads.

The expected frequency of meiotic exchanges within the rDNA cluster can be estimated. The nuclear DNA within each yeast cell has been estimated to recombine about 70 times during each meiosis (Byers and Goetsch 1975). Since the rDNA is about 5% of the total cellular DNA (Schweizer et al. 1969), the rDNA genes should recombine about three to four times in each cell in each meiosis. Since none of the 14 tetrads examined showed the segregation patterns expected for a single meiotic crossover, meiotic recombination for rDNA may be much less frequent (50-fold less) than meiotic recombination for comparably long regions of nonribosomal nuclear DNA. This may be a reflection of differences in chromosome structure in regions coding for rRNA, or it may indicate sequence specificity of the enzymes catalyzing meiotic recombination.

Alternative explanations for the experimental results have not been excluded. For example, it is possible that the type-I and type-II rDNA genes are so different in base sequence that pairing during meiosis cannot occur effectively, thereby suppressing recombination. We think this possibility unlikely since six of seven *Eco*RI sites are conserved between the two types of rDNA genes. Additional restriction mapping with the enzymes *Bgl*II and *Hin*dIII (T. Petes, unpubl.) also indicates conservation of these sites between the different rDNA genes. It is also unlikely that recombination is suppressed by an inversion of one cluster of rDNA genes with respect to another. Crossing-over between such clusters should give very poor spore viability, and we found that segregants from +D4 had relatively good spore viability (14 of 30 tetrads had four viable spores).

More complicated models to explain the results can also be proposed. For example, it is possible that although 100 copies of the rDNA genes are present in vegetatively growing yeast cells, only a few copies of rDNA are present in the "germ line" (the DNA which is present in meiosis). This model would require that most of the cellular rDNA be degraded before meiosis and resynthesized during spore germination. Other models such as "master-slave" correction of ribosomal genes after meiotic crossing-over also cannot be ruled out. However, we favor the relatively simple idea that meiotic recombination of the rDNA genes is suppressed.

Do Yeast rDNA Genes Recombine Mitotically?

In two of the tetrads (1 and 10), the rDNA genes did not segregate 2:2. In tetrad 1, two spores contained only type-I rDNA and two spores contained more type-I than type-II rDNA. Since unequal

amounts of each type of rDNA were recovered in this tetrad, it is unlikely that this segregation pattern was that result of a reciprocal meiotic event or independent segregation of more than one rDNA-containing chromosome. One simple explanation for the observed segregation pattern is shown in Figure 4. A mitotic exchange in a cell before induction of meiosis could yield a diploid which would produce the observed segregation patterns when induced to sporulate. The segregation pattern observed for tetrad 10 could also be explained by a mitotic exchange (Fig. 4, bottom panel). This segregation pattern, however, could also be explained by a four-strand double crossover during meiosis or by the independent segregation of two chromosomes containing unequal amounts of rDNA. Since double meiotic crossovers should be less frequent than single crossovers (which were not observed in these experiments), this explanation seems unlikely. Similarly, two independently segregating chromosomes containing the rDNA genes should have been observed in more than 1 of 14 tetrads. It is likely, therefore, that yeast rDNA genes can recombine mitotically. Further studies are in progress to show that the mixture of the two types of rDNAs seen in the spores of tetrads 1 and 10 is the result of mitotic recombination rather than a different type of mitotic event.

A high frequency of mitotic recombination within the rDNA gene cluster would be consistent with observations made by Kaback and Halvorson (1977). They found that yeast strains which initially had low levels of rDNA, after several hundred genera-

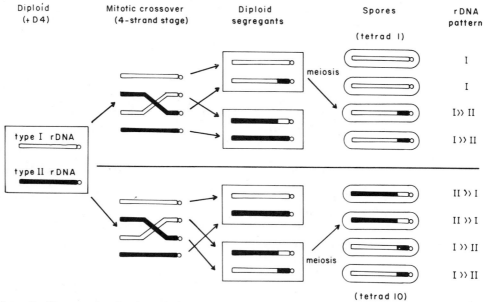

Figure 4. Schematic diagram showing how mitotic recombination can account for the distribution of rDNA in tetrads 1 and 10 of Table 2. For convenience, the rDNA is shown to comprise an entire chromosome, although this is not expected to be the case. I >> II indicates that the rDNA is largely type-I rDNA with a minor amount of type-II rDNA (T. Petes and D. Botstein, in prep.).

tions in culture, could amplify the amount of rDNA to normal levels. One mechanism to explain amplification is that the rDNA genes underwent unequal mitotic crossing-over followed by selection of those cells with nearly diploid amounts of rDNA (Tartof 1974; Kaback and Halvorson 1977).

On Which Chromosome Are the Yeast rDNA Genes Located?

Work from several laboratories (Finkelstein et al. 1972; Oyen 1973; Kaback et al. 1973) suggested that many of the rDNA genes were located on chromosome I of yeast. In these experiments, the amount of rDNA (measured by saturation hybridization to rRNA) in a strain monosomic for chromosome I $(2n - 1)$ was found to be 35% less than in a normal diploid strain. It was therefore concluded that 70% of the yeast rDNA genes were located on chromosome I.

In an experiment designed to test this conclusion using the rDNA heterogeneity, we constructed a diploid which was heterozygous for a genetic marker on chromosome I (ade1) and heterozygous for the two forms of the rDNA. In a preliminary analysis of segregants from this diploid (T. Petes, unpubl.), no linkage between ade1 and the rDNA gene cluster has been detected. Since only eight tetrads have been analyzed, a loose genetic linkage could obviously not be detected.

It was not unexpected that the rDNA genes were unlinked to ade1 since ade1 is located near the centromere of chromosome I (Mortimer and Hawthorne 1969) and our earlier experiments had shown that the rDNA was not detectably centromere-linked. It should be stressed, therefore, that our results do not show that the rDNA is not on chromosome I. If the genes are on chromosome I, however, they are genetically distant from ade1. We are currently attempting to map the rDNA genes using mitotic recombination procedures.

In conclusion, we have found that most of the rDNA genes of yeast are located on a single arm of one of the yeast chromosomes. Meiotic recombination within this cluster of rDNA genes is suppressed, although preliminary evidence suggests that mitotic recombination may occur.

Acknowledgments

We thank Drs. L. Hartwell, C. McLaughlin, and J. Haber for contributing strains used in this study. T. D. P. was supported by a National Institutes of Health postdoctoral fellowship. L. M. H. was supported by a grant from the Medical Foundation and by a National Institutes of Health grant (GM-23549) to Michael Rosbash. The work was also supported by National Institutes of Health grants to D. B. (GM-18973 and GM-21253) and a Research Career Development Award to D. B.

REFERENCES

BRANDRISS, M. C., L. SOLL, and D. BOTSTEIN. 1975. Recessive lethal amber suppressors in yeast. *Genetics* **79**: 551.

BYERS, B. and L. GOETSCH. 1975. Electron microscopic observations on the meiotic karyotype of diploid and tetraploid *Saccharomyces cerevisiae*. *Proc. Natl. Acad. Sci.* **72**: 5056.

CRAMER, J. H., M. M. BHARGAVA, and H. O. HALVORSON. 1972. Isolation and characterization of γ DNA in *Saccharomyces cerevisiae*. *J. Mol. Biol.* **71**: 11.

CRAMER, J. H., F. W. FARRELLY, J. T. BARNITZ, and R. H. ROWND. 1977. Construction and restriction endonuclease mapping of hybrid plasmids containing *Saccharomyces cerevisiae* ribosomal DNA. *Mol. Gen. Genet.* **151**: 229.

FINKELSTEIN, D. B., J. BLAMIRE, and J. MARMUR. 1972. Location of ribosomal RNA cistrons in yeast. *Nat. New Biol.* **240**: 279.

GRUNSTEIN, M. and D. S. HOGNESS. 1975. Colony hybridization: A method for the isolation of cloned DNA's that contain single genes. *Proc. Natl. Acad. Sci.* **72**: 3961.

KABACK, D. B. and H. O. HALVORSON. 1977. Magnification of genes coding for ribosomal RNA in *Saccharomyces cerevisiae*. *Proc. Natl. Acad. Sci.* **74**: 1177.

KABACK, D. B., M. M. BHARGAVA, and H. O. HALVORSON. 1973. Location and arrangement of genes coding for ribosomal RNA in *Saccharomyces cerevisiae*. *J. Mol. Biol.* **79**: 735.

LOBBAN, P. E. and A. D. KAISER. 1973. Enzymatic end-to-end joining of DNA molecules. *J. Mol. Biol.* **78**: 453.

MAXAM, A. M., R. TIZARD, K. G. SKRYABIN, and W. GILBERT. 1977. Promoter region for yeast 5S ribosomal RNA. *Nature* **269**: 643.

MORTIMER, R. K. and D. C. HAWTHORNE. 1966. Genetic mapping in *Saccharomyces*. *Genetics* **53**: 165.

———. 1969. Yeast genetics. In *The yeasts* (ed. A. Rose and J. S. Harrison), vol. 1, p. 386. Academic Press, London.

OYEN, T. 1973. Chromosome I as a possible site for some rRNA cistrons in *Saccharomyces cerevisiae*. *FEBS Lett.* **30**: 53.

SCHWEIZER, E., C. MACKECHNIE, and H. O. HALVORSON. 1969. The redundancy of ribosomal and transfer RNA genes in *Saccharomyces cerevisiae*. *J. Mol. Biol.* **40**: 261.

SHARP, P. A., B. SUGDEN, and J. SAMBROOK. 1973. Detection of two restriction endonuclease activities in *Haemophilus parainfluenzae* using analytical agarose-ethidium bromide electrophoresis. *Biochemistry* **12**: 3055.

TARTOF, K. 1974. Unequal mitotic sister chromatid exchange and disproportionate replication as mechanisms regulating ribosomal DNA gene abundancy. *Cold Spring Harbor Symp. Quant. Biol.* **38**: 491.

UDEM, S. A. and J. WARNER. 1972. Ribosomal RNA synthesis in *Saccharomyces cerevisiae*. *J. Mol. Biol.* **65**: 227.

WENSINK, P. C., D. J. FINNEGAN, J. E. DONELSON, and D. S. HOGNESS. 1974. A system for mapping DNA sequences in the chromosomes of *Drosophila melanogaster*. *Cell* **3**: 315.

WILLIAMSON, D. H. and D. J. FENNEL. 1975. The use of fluorescent DNA binding agents for detecting and separating yeast mitochondrial DNA. *Meth. Cell Biol.* **12**: 335.

WILLIAMSON, D. H. and E. MOUSTACCHI. 1971. The synthesis of mitochondrial DNA during the cell cycle in the yeast *Saccharomyces cerevisiae*. *Biochem. Biophys. Res. Commun.* **42**: 195.

Summary: The Molecular Biology of the Eukaryotic Genome Is Coming of Age*

P. Chambon

Laboratoire de Génétique Moléculaire des Eucaryotes du CNRS, U.44 de l'INSERM, Institut de Chimie Biologique Faculté de Médecine, 67085 Strasbourg-Cédex, France

L'obscur acharnement des hommes pour recréer le monde n'est pas vain parce que rien ne redevient présence au delà de la mort, à l'exception des formes recréés.

ANDRÉ MALRAUX
"La Création Artistique"

"There are times in every science when the outline of future progress seems predictable, straightforward, and perhaps a little boring. Inevitably, this leads many of its practitioners to wonder whether they are in the right field." Doubtless, these first sentences of Jim Watson's foreword to the 1970 Cold Spring Harbor Symposium do not apply to those who are presently working on structure and function of the eukaryotic genome. All of those who have struggled for years with frustrating "dirty" eukaryotic systems, passing through depressing moments where they wished they were working on "clean" and "elegant" prokaryotic models, now have their reward. What they were feeling, mainly from indirect evidence—namely, that the structure and function of higher eukaryote genomes are not just those of a big *Escherichia coli*—is now established. Clearly, neurobiology is not the only new frontier in biology! This year's Symposium marked the beginning of a new phase where we can foresee the day when eukaryotic developmental problems will be elucidated at the molecular level. The contrast is striking when we compare the present situation with the situation 4 years (Swift 1974) or even 1 year ago (Allfrey et al. 1976). This Symposium offered a good example of how knowledge in biology progresses in a stepwise manner thanks to the emergence of new concepts and/or to the discovery of new techniques. There is no doubt that the nucleosome concept (Hewish and Burgoyne 1973; Kornberg, 1974, 1977) has, in providing a framework for future work, acted as a catalyst for studies on chromatin structure and function. But where would the state of eukaryotic gene structure and function be without the remarkable technical breakthroughs of these last years? We have been lucky in having all these advances in rapid DNA and RNA sequencing, in electron microscopy, in nucleic acid and protein separation, in nucleic acid hybridization after blot-ting, in restriction enzymes, and so on, including the *Xenopus* oocyte "test tube." But the decisive technical breakthrough was the discovery of the in vitro recombinant DNA techniques which, by decreasing the complexity of the eukaryotic genome by a factor of 10^6, paved the way to the molecular biology era of the eukaryotic genome. The Symposium also illustrated the usefulness of simpler model systems: the viruses which have diverted some of the cell mechanisms for their own use.

This Symposium on Chromatin covered, in fact, all aspects of structure and function of the eukaryotic nuclear genome corresponding to genetic organization and function both at the DNA level and at the level of DNA-histone (and other protein) interactions, the latter corresponding to "classical" chromatin. I have summarized separately these different aspects of eukaryotic genome structure and function. In view of the enormous output of new results, I have had to ignore some data, even though very interesting. Particularly, I have not given a detailed account of gene organization at the DNA primary sequence level. One of the reasons for this choice is that very little is presently known about the relationship between DNA primary structure and chromatin organization. On the other hand, I have summarized the post-Symposium data demonstrating that the prokaryotic principle of colinearity between DNA and protein sequences is not generally applicable to higher organisms because I consider this a correlate of the discovery of the viral leader mRNA sequences.

Whenever possible, I have integrated data in the literature with data presented at the meeting to make generalizations possible. Other references can be found in the recent reviews of Kornberg (1977), Thomas (1977), and the Dahlem Conference book (Allfrey et al. 1976). No attempt has been made to summarize or mention every paper. References cited without date are to papers included in this volume.

GENE ORGANIZATION AND PRIMARY RNA TRANSCRIPTS

An Amazing Viral Gene Arrangement: Leader or Spliced Segments at the 5′ Termini of Adenovirus and SV40 Messenger RNAs

In view of the difficulty in defining eukaryotic mRNA transcriptional units (Darnell 1977, and re-

---

* Dedicated to the memory of Jacques Monod who encouraged me and many others to work on eukaryotic systems.

lated articles in the same volume), it was thought by several groups of workers that animal DNA viruses, such as adenovirus and simian virus 40 (SV40), could provide a simpler model system for studying the organization of these units in much the same way that studying phage transcription has illuminated bacterial transcription; indeed, they have succeeded beyond our hopes.

The announcement of the discovery of mosaic adenovirus and SV40 mRNAs was the highlight of this meeting. It is clear from the results presented during the meeting and published afterwards (for references, see Berget et al. and Broker et al.) that many late adenovirus mRNAs consist of four separate blocks of sequences. In every case, three small RNA fragments, coded by three different segments of the genome at map positions about 17, 20, and 27 units remote from the DNA coding on the same rightward strand for the main body of the mRNAs, are joined to form a common leader sequence of 150–200 nucleotides at the 5' end of these late mRNAs. Similarly, early adenovirus mRNA transcripts do not represent simpler linear transcripts of their genes, as shown by Westphal and his group (Westphal and Lai; Kitchingman et al. 1977). However, in contrast to the common tripartite leader sequence joined to the main body of the late RNAs transcribed from the rightward strand of DNA, each of the four early adenovirus-2 mRNAs, whether it is transcribed from the rightward or the leftward DNA strand, consists of only two blocks of sequences. Each of these early messengers has a unique starting sequence which is also much longer than the tripartite leader, since it can represent as much as half of the individual early mRNAs.

Late in infection, the situation appears to be analogous in SV40, where the late 16S and 19S RNAs consist of two blocks of sequence (Aloni et al.; Hsu and Ford; Thimmappaya et al.; Hsu and Ford 1977; Lavi and Groner 1977). The "leader" sequence present at the 5' end is common to both 16S and 19S RNAs, but in contrast to that of the late adenovirus mRNA, it appears to be a 150–200-nucleotide RNA segment coded by a single region of the SV40 genome.

These unexpected findings, implying some regulatory processes in eukaryotic gene expression different from those known in prokaryotic cells, raise a number of interesting questions. Four possible ways to synthesize these composite mRNAs were discussed during the meeting. First, the RNA polymerase B (or II) could jump with the attached nascent RNA across looped-out intervening sequences, transcribing the various DNA segments in an ordered fashion to yield directly the different mRNAs. Second, the mosaic RNA molecules could be transcribed from rearranged DNA templates in which the different coding regions have become contiguous. Third, each of the mRNA coding segments may be transcribed independently, and the mature RNA generated by intermolecular ligation of the transcripts. Fourth, both mRNA coding regions and intervening sequences could be transcribed, resulting in a mRNA precursor. The mature mRNA would then be formed by excision of the intervening RNA sequences and ligation (splicing) of the mRNA sequences.

In the case of late adenovirus mRNA, the fourth mechanism seems the most likely (although it is difficult at present to reject definitely the second possibility because the fraction of viral DNA actually engaged in transcription is only a very small fraction of the total viral nuclear DNA). Darnell and his group (for references, see Darnell et al.; Evans et al. 1977) have indeed provided strong evidence that all of the late adenovirus mRNAs sharing the same leader sequence also belong to the same very long rightward transcriptional unit which begins around 16 map units (see also Meissner et al. 1977). By contrast, early in infection there seems to be four independent transcription units (Darnell et al.; Evans et al. 1977; Berk and Sharp 1977), in agreement with the existence of the four separate starting sequences described above. However, there is some indication that in this case also the fourth mechanism discussed above could be involved (Kitchingman et al. 1977). Such an excision-ligation mechanism can account for the large amount of adenovirus-coded RNA which, late in infection, is synthesized in the nucleus but never appears in the cytoplasm (Philipson et al. 1971; Darnell et al. 1973). It is very likely that for SV40 late mRNAs the fourth mechanism, i.e., processing of a larger primary transcript via excision-ligation, is also responsible for the synthesis of the composite 16S and 19S RNA, since Aloni et al. have presented some evidence which appears to exclude the three other alternatives. In any case, it appears that there is no requirement for a chromatin structure for generating mosaic mRNA molecules since, late in infection, the SV40 genome is in the form of a minichromosome (see below), whereas the adenovirus DNA is not associated to histones (Kédinger et al. 1978).

At present, the possible mechanism responsible for the intramolecular ligation event which would result in the deletion of the intervening sequence(s) present in the primary transcript is a matter of speculation. Klessig (1977) has proposed that the intervening RNA sequences could specifically loop out by either base pairing or nucleic-acid–protein interaction. However, such a base-pairing mechanism is not evident from inspection of the SV40 DNA sequences coding for the putative primary transcript of the 16S RNA (Thimmapaya et al.). In any case, and regardless of the details of the mechanism that might be involved in intramolecular RNA ligation, it is very likely that a cellular enzyme must be involved in view of the low coding capacity of the SV40 genome. This brings us to the major problem: to elucidate the biological significance of the synthesis of composite RNAs and find out whether the same phenomenon occurs during cellular mRNA genesis. In SV40 neither the leader sequences nor the intervening deleted sequences code for known functions (Aloni et al.; Celma et al. 1977) and the possible function of the adenovirus leader sequences is a matter of speculation (Broker et al.; Klessig 1977). Could the cellular mRNAs be mosaic molecules? This possibility was discussed during the meeting and was supported by three arguments. First, the previously puzzling results of Perry's and Darnell's groups (Perry et al. 1977; Darnell 1977), showing that the 5'-terminal cap and some of the 3'-terminal poly(A) structures of heterogeneous nuclear RNA (hnRNA) molecules are conserved during processing to mRNA chains which are shorter, could easily be interpreted if one assumes that at least some eukaryotic mRNAs could be processed from hnRNA by an excision-ligation process. This mechanism would also provide a potential explanation for the as yet unexplained old observation that a significant fraction of the rapidly labeled nuclear RNA never appears in the cytoplasm. Second, it was thought that such a process could also explain how the variable and constant parts of the immunoglobulin messenger, which are coded by different portions of the genome (Tonegawa et al.), become associated. Third, DNA intervening sequences are known to occur in regions of the *Drosophila* genome coding for 28S ribosomal RNA (Glover and Hogness 1977; Pelligrini et al. 1977; Wellauer and Dawid 1977; White and Hogness 1977; Dawid and Wellauer). However, since intact *Drosophila* ribosomal cistrons are also found, it is not clear whether the ribosomal cistrons containing the insertion sequence are in fact transcribed.

By the end of the meeting, it was generally believed that the discovery of viral mosaic mRNAs would act as an illuminating stimulus for those groups engaged in the study of eukaryotic gene structure and that it would not be long before we know whether the viruses are using a normal cellular mechanism for their own purposes. Again, the success has been beyond our expectations. Because of their importance, I will briefly summarize some of the very recent evidence (not presented during the meeting) demonstrating that noncolinearity between DNA and product sequences could be a common feature of eukaryotic cells, in marked contrast with prokaryotic cells.

An Amazing Post-meeting Event: Cellular Genes Are Interrupted by Intervening Sequences (Split Genes)

Jeffreys and Flavell (1977) have found a 600-base-pair DNA segment inserted somewhere within the coding sequence for amino acid residues 101–120 of the 146-residue rabbit β-globin chain. A similar insert (intervening sequence), approximately 550 base

pairs in length, interrupts the mouse β-globin gene immediately following the codon corresponding to amino acid 104 (Leder et al.; Tilghman et al. 1978). In retrospect, it is amusing to note again that the way we interpret an experiment is strongly influenced by what we already know. For instance, at the meeting, Leder et al. interpreted their R-loop data as indicating the juxtaposition of two nonallelic β-globin genes. An intervening sequence of 1250 base pairs separating the variable and constant gene sequences was also found within the coding sequence of a cloned mouse plastocytoma λ light chain (Brack and Tonegawa 1977). Taken together with the previous results of Tonegawa's group (Tonegawa et al.), this result indicates that DNA translocation occurs during differentiation of B lymphocytes. However, the mechanism of translocation is such that the two DNA segments coding for the variable and constant regions of the continuous mRNA come closer, but are not contiguous. Rabbits and Forster (1978) have also obtained evidence that the rearrangement of mouse light-chain genes which occurs in myeloma cells may not produce contiguous variable and constant genes in the DNA. A 93-base-pair insertion was detected and sequenced in the 5' region of the protein-coding sequence of a cloned immunoglobulin variable gene of mouse embryo (S. Tonegawa, pers. comm.). Breathnach et al. (1977, and unpubl.) have shown that the ovalbumin gene is interrupted several times (at least twice, and very likely as many as six times) in the sequence coding for ovalbumin mRNA. That intervening sequences could also interrupt viral DNA sequences coding for protein is likely in view of some primary sequence data for SV40 early genes (Thimmappaya and Weissman 1977). Intervening DNA sequences are not confined to mRNA coding genes. In addition to the *Drosophila* rDNA insertions (Dawid and Wellauer), Goodman et al. (1977) and Valenzuela et al. (1978) have found that yeast tyrosine and phenylalanine tRNA genes contain a short intervening sequence (14–20 base pairs) located very close to the nucleotides coding for the tRNA anticodon. It appears, therefore, that split genes characterized by intervening sequences interrupting structural sequences may have some generality in eukaryotic cells, which apparently have transgressed the prokaryotic rule of colinearity between DNA and product sequences. These results pose the semantic problem of defining a structural eukaryotic gene. It should be stressed that all of the present evidence indicates that, at least for globin, ovalbumin, and tyrosine tRNA genes, the intervening sequences are present in "normal" genes which are actually expressed.

These results raise many questions, some identical to those already discussed for mosaic viral RNAs. Undoubtedly, if the intervening sequences were transcribed together with the coding sequences to yield an RNA precursor molecule and then processed by excision-ligation, we would understand the significance of at least some hnRNA molecules and of their peculiar metabolism. The existence of precursor for the messengers of globin (Curtis et al.; Bastos and Aviv 1977; Kwan et al. 1977; Strair et al. 1977) and immunoglobulin (Gilmore-Herbert and Wall 1978) suggests very strongly that the intervening sequences are indeed transcribed and that the mature globin and immunoglobulin mRNAs could derive from these precursors by an excision-ligation mechanism. In fact, the groups of C. Weissmann and P. Leder (pers. comm.) have obtained direct electron microscopic evidence that the putative 15S globin mRNA precursor is a continuous transcript of the globin-coding sequences linked to the transcript of the intervening sequence. This observation represents the first example of a derivation of a specific mature mRNA from a longer hnRNA precursor. I have no doubt that similar derivations will be established very soon for other mRNAs (see Goldberg et al. 1977) and will stimulate the search for an excision-ligation enzyme system. Perhaps the excision specificity is mediated in some way by some of the proteins present in heterogeneous nuclear ribonucleoproteins (hnRNP) (Martin et al.; LeStourgeon et al.; Kulguskin et al.) or by some of the small nuclear RNAs (Benecke and Penman 1977) which could be associated with hnRNP (Deimel et al. 1977; Guimont-Ducamp et al. 1977).

It is remarkable that, until now, all of the intervening sequences found in genes coding for cellular mRNAs interrupt protein-coding sequences. It is not known whether a situation similar to that of the supposedly noncoding tripartite leader sequence common to many late adenovirus mRNAs exists in some cellular mRNAs. I am confident that the physiological significance of the split-gene organization, which is now a matter of speculation, will be the subject of sessions at the next Cold Spring Harbor meeting on the eukaryotic genome. It could be a way specific to eukaryotic cells to generate additional genetic information from otherwise separated DNA regions. Combined with somatic gene reassortment, it could generate somatic gene diversity as in the case of immunoglobulin genes. However, it is probably not generally involved in cell differentiation, since for globin and ovalbumin genes there is no apparent difference between different tissues, whether they do or do not make the protein (Jeffreys and Flavell 1977; Breathnach et al. 1977). It is not known whether long intervening sequences such as those found in the split ovalbumin gene could contain other structural genes. Obviously, the transcribed intervening sequences are good candidates for transcriptional (similar, for instance, to the prokaryotic attenuator [Bertrand et al. 1975]) and posttranscriptional sites of regulation of eukaryotic gene expression. In this respect, the similar positions and sizes of the globin-gene intervening sequences in rabbit and mouse (see Jeffreys and Flavell 1977) and the perfect conservation of intervening DNA seg-

ments in yeast tyrosine (Goodman et al. 1977) and phenylalanine (Valenzuela et al. 1978) tRNA genes suggest that the intervening sequences play some conserved role in the function of these genes.

No doubt we shall soon know whether the intervening sequences are repeated in the genome and whether some of them correspond to prokaryotic insertion-sequence elements (for references, see Bukhari et al. 1977). (This uncertainty is a good reason to use the term "intervening sequence" rather than "insertion sequence.") Finally, we could ask, How widespread is the split structural gene organization in the eukaryotic genome? Although much more work will be required to answer this question, it is already clear that the histone genes of sea urchin (Kressman et al.; Grunstein and Grunstein) and *Drosophila* (Lifton et al.), and possibly the fibroin gene (Suzuki and Ohshima) and the gene coding for the Balbiani ring 2 75S mRNA (Dan/holt et al.; Edström et al.), are not split. Clearly, we are at the very beginning of a long story. Fortunately, the technology has advanced to a state which should soon allow us to answer most of these questions.

Other Aspects of Eukaryotic Gene Organization: The Importance of New Techniques

Another striking new discovery discussed at this meeting was that of gene amplification in methotrexate-resistant cultured mouse cells (Schimke et al.). For the first time, it was shown that a gene coding for a protein can be amplified in animal cells. Thus it appears that, even though such amplification does not occur in highly differentiated cells (e.g., those making large amounts of globins, ovalbumin, myosin, or fibroïn), the machinery required to amplify structural genes is there. One wonders why it was not used during evolution when new proteins made in large quantities emerged. It will be interesting to learn how this amplification is achieved and how it is related to the general process of gene duplication and to the particular structure of the dihydrofolate reductase gene. It remains to be seen whether such an amplification occurs only in cells probably infected by viruses and grown in vitro under selection pressure, or whether it can be taken as further evidence that, under physiological conditions, the eukaryotic genome is not static (Ilyin et al.; for references, see Schimke et al.; Potter and Thomas). Although the present evidence (Potter and Thomas; see also above) indicates that, in higher organisms, differentiation can occur without any apparent changes in the genome organization, the observations of Schimke et al. and those on immunoglobulin genes (Tonegawa et al.) are there to incite us to keep an eye open for possible genome rearrangement during development. Thanks to gene cloning, further studies on the developmental systems that were presented during this meeting (Sim et al.; Suzuki and Ohshima; Lifton et al.; Weinberg et al.) will provide an answer to the above question.

Since the introduction of in vitro recombinant DNA techniques, our approach to the study of genome organization has drastically changed. Questions are now formulated in precise molecular terms (Finnegan et al.; Wensink; Lee et al.). Studies on the significance of repetitive DNA sequences in the genome have taken on a new dimension which hopefully will lead to the understanding of their function and of how the eukaryotic genome has evolved. It will also be interesting to know how representative the *Drosophila* genome is for other eukaryotic organisms. Can the notion of regulatory repetition (Finnegan et al.) as opposed to tandem repetition be generalized?

I will not discuss in any detail, or even attempt to summarize, all of the data which have been accumulated on DNA primary sequence. Again, molecular cloning has been an invaluable tool which has led to new information on genome organization and evolution (see, e.g., Fedoroff and Brown; Salser). It could be argued that we will accumulate a large body of sequence information which we will be unable to interpret. However, I am quite optimistic that, provided we know something about primary RNA transcripts, we will be able to determine the function of the sequences with the aid of the *Xenopus* nucleus (Kressman et al.; Laskey et al.), virus vectors, and good in vitro reconstitution systems (see below). Obviously, most of the time traditional genetics will be missing, but the possibility of in vitro genetics (surrogate genetics [Birnstiel and Chipchase 1977]), by introducing deletions, insertions, etc. into cloned DNA fragments, will certainly be exploited.

STRUCTURE OF BULK (INERT) CHROMATIN

Chromatin may be considered at different structural levels, corresponding to increasing order of organization, and at different levels of function, essentially transcription and replication. The two are closely interrelated, and for simplicity, I will summarize separately what was discussed at the meeting about "bulk," "inert," or "passive" chromatin and "active" chromatin, with respect to both transcription and replication.

There is now a vast body of evidence (for references, see Kornberg 1977; Thomas 1977) that the organization of eukaryotic DNA into nucleosomes, the repeating nucleoprotein units of chromatin, constitutes the first level in a stepwise compaction mechanism which leads to packaging of DNA at high concentration in the interphase nucleus or metaphase chromosome. It is very likely that the nucleosomal periodicity exists in vivo and is not an artifact generated in vitro from an in vivo nonperiodic structure during the purification of nuclei or chromatin or during the nuclease digestion which yields the nucleosomes. In this respect the psoralen photoreac-

tion is a unique tool for probing the structure of chromatin in vivo (Hanson et al. 1976). By covalently cross-linking the DNA in vivo and studying the distribution of cross-links in the purified DNA, Wiesehahn et al. (1977) and Cech and Pardue (1977) have recently demonstrated that the cross-links occur in vivo at nucleosomal length intervals. Furthermore, Cech et al. have shown at this meeting that most, if not all, of the mouse L-cell DNA is in nucleosome structures. In any case, should the nucleosomes be an artifact generated in vitro from an in vivo nonparticulate structure, we would just have to decide that they are quite interesting artifacts reflecting the way DNA and histones interact! Some definitions will be useful (Kornberg 1977; Thomas 1977) before reviewing the present state of knowledge of the nucleosome structure.

Nucleosomes are particles about 110–125 Å in diameter as seen in micrographs of the native compact state of chromatin (Kornberg 1977). They contain a *histone core* (two each of the four histones H2A, H2B, H3, and H4) and the *entire* length of the DNA repeat *(DNA repeat length)* generated at early times of micrococcal nuclease digestion (usually about 200 base pairs), but there are marked variations (see below). A molecule of H1 or H1-like histone is associated with a nucleosome. The packing ratio of DNA in nucleosomes is about 5 to 7.

Nucleosome cores or *core particles* are derived from nucleosomes by further micrococcal digestion. They contain the histone core (but not H1) and a constant amount of DNA (140 base pairs) regardless of the amount of DNA originally contained in the nucleosome.

The remaining DNA which connects one nucleosome core to the next and which is variable in length is termed the *linker*. It should be regarded as a linker and not as a spacer because it is not extended and therefore not seen in native chromatin, where the nucleosomes appear as individual particles in close apposition (Olins; Bram et al.; Oudet et al. 1975; Finch et al. 1975; Franke et al. 1976;). Determination of the mass per unit length of the chromatin fiber by low-angle X-ray scattering (Sperling and Tardieu 1976) also supports the compact model for chromatin structure (Kornberg 1974, 1977) as opposed to an extended one, where the DNA connecting one nucleosome core to the next is regarded as a spacer (Van Holde et al. 1974).

Obviously, knowledge of the internal structure of the nucleosome is of general interest with respect to packaging mechanisms for DNA and is fundamental to an understanding of the mechanisms of eukaryotic gene regulation. In spite of the numerous studies presented at this meeting, we still don't know exactly how a nucleosome is organized. However, we have learned a significant amount of new information and it is clear that crystallographic studies will play a key role in the elucidation of the nucleosome structure.

Histone Content of Nucleosomes: Four Histones in All Nucleosomes

There is no doubt that in almost all eukaryotic species all nucleosomes contain two each of the four histones, the highly conserved H3 and H4 and the more variable H2A and H2B. Strong evidence, of two sorts, supporting this conclusion has been presented. First, immunological studies indicate that probably all in-vivo-assembled nucleosomes contain all four histones (Bustin et al.). Second, in-vitro-reassociation experiments indicate that all four histones in equimolar amounts are required to assemble a nucleosome (Oudet et al.). These results, taken together with those of previous studies (e.g., histone cross-linking in chromatin showing an octamer [Thomas and Kornberg 1975a,b] and careful measurements of individual histone-to-DNA ratios [Olins et al. 1976; Joffe et al. 1977]), indicate that there are two each of the four histones *in every nucleosome*. It should, however, be kept in mind that there could be exceptions. For example, the *Tetrahymena* micronuclei apparently lack H3 but appear to contain a histone, termed HX, which in some respects resembles H2A but is more conserved and readily associates in vitro with H4, suggesting that it could be a hybrid histone molecule replacing both H2A and H3 (Gorovsky et al.).

There is a fifth histone associated with the nucleosome, namely, histone H1 or one of its variants. It is very likely that there is only one H1 molecule per nucleosome (Goodwin et al. 1977a) in spite of various claims in the literature that the real stoichiometry could be 1 to 1 for the five histones.

DNA Content of Nucleosomes: Variability of the DNA Repeat Length and the Phase Problem

The DNA repeat length of a given chromatin, and therefore the amount of DNA contained in a nucleosome, can be determined by digestion of chromatin in nuclei under appropriate conditions and electrophoresis of the DNA in the presence of marker DNA fragments of known sizes (for discussion, see Kornberg 1977; Thomas 1977). The nuclease digestion generates a set of fragments that are multiples of a unit size. Initially, and in agreement with Kornberg's (1974) proposals, a value of about 200 base pairs was found in various tissues such as rat liver (Noll 1974a). However, there are remarkable variations which were not really discussed during this meeting, probably because nobody has a satisfactory explanation for them. As shown in Table 1, the values for the DNA repeat length vary from 154 base pairs in *Aspergillus* to 241 in sea urchin sperm. In higher eukaryotes, there are variations between different animal species, but more surprisingly there are also variations in the same animal within neighboring cells (162 and 200 base pairs for rabbit cerebral cortex neurons and nonastrocytic glial cells,

Table 1. DNA Content of Nucleosomes

Cell type	DNA repeat length (base pairs)
Aspergillus	154[a]
Yeast	165, 163[a]
Rabbit cortical neuron	162[a]
Neurospora	170[a]
Physarum	171,[a] 173[b]
Tetrahymena micronucleus	175[c]
Cells grown in culture	
CHO	177[a]
HeLa	183, 188[a]
hepatoma	188[a]
teratoma	188[a]
P815	188[a]
myoblast	189[a]
CV1, exponentially growing or confluent	189[a]
BHK	190[a]
rat kidney primary culture	191[a]
myotube	193[a]
C6, exponentially growing or confluent	198[a]
Rat bone marrow	192[a]
Rat fetal liver	193[a]
Rat liver	198, 196[a]
Rat kidney	196[a]
Syrian hamster liver	196[a]
Syrian hamster kidney	196[a]
Chick oviduct	196[a]
Tetrahymena macronucleus	202[c]
Rabbit cerebellar neuron	200[a]
Rabbit nonastrocytic glial cells	200[a]
Stylonychia micronucleus	202[a]
Chicken erythrocyte	207, 212[a]
Sea urchin gastrula	218[a]
Stylonychia macronucleus	220[a]
Sea urchin sperm	241[a]

[a] For references, see Kornberg (1977).
[b] Allfrey et al.
[c] Gorovsky et al.

respectively; 200 base pairs for rabbit cerebellar neurons). Martin et al. (1977) very recently reported results suggesting that there are variations in the linker DNA length in calf thymus chromatin.

No general rules can be deduced from examination of these variable repeats. The occurrence of short repeats does not appear to be related to a high rate of cell division (Compton et al. 1976a; Thomas 1977). Although a correlation sometimes seems to exist between a short repeat length and a high level of transcriptional activity, and conversely between a long repeat and transcriptional inactivity (Thomas and Thompson 1977), it is certainly not a rule. Transcriptionally inert micronuclei of *Stylonychia* (Lipps and Morris 1977) and *Tetrahymena* (Gorovsky et al.) have shorter repeat lengths than the active macronuclei. In any case, it is remarkable that the variation resides exclusively in the length of the linker DNA, since the size of the DNA in the nucleosome core obtained by prolonged micrococcal nuclease digestion is always 140 base pairs. It is therefore not surprising that changes in histone H1 have also been invoked (Noll 1976; Morris 1976) to explain the variability of the DNA repeat length, since H1 appears to be bound to at least a part of the linker DNA.

However, there is no clear correlation with phosphorylation of histone H1 (Compton et al. 1976a; Spadafora et al. 1976) or with the replacement of histone H1 by another lysine-rich histone. For instance, Morris (1976) has proposed that the more basic H5 could specify a longer linker. But Wilhelm et al. (1977) have shown that there is no direct relationship between the amino acid sequence of the lysine-rich histone (H1 or H5) and the DNA repeat length of chromatin. In addition, Varshavsky et al. have shown that H1 and H5 protect the same length of the linker region against nuclease digestion.

It was recently suggested that the DNA repeat length could also be heterogeneous in a given cell type. Prunell and Kornberg (1977) have found extensive variation in the length of the linker of nucleosome dimers isolated from rat liver. That at least some of the variations exist in vivo and are not generated during the nuclease treatments is suggested by psoralen cross-linking experiments (Shen and Hearst), which indicate that the linker DNA of *Drosophila melanogaster* satellites I and II is larger than that of the main-band DNA.

The question of whether there is a specific phase relation between nucleosomes and base sequence in DNA (nucleosome phasing), i.e., whether a linear array of nucleosomes occurs at a unique location along the sequence of nucleotides in DNA, is very important for many possible roles of nucleosomes in the control of gene expression. (This definition of nucleosome phasing does not exclude that phasing could occur even though the linker DNA length is variable.) The function of a certain DNA sequence may indeed depend on whether it is present in the nucleosomal linker region, which is presumably more accessible than the nucleosome core to possible regulatory molecules since it is readily accessible to nucleases. A fixed and specific alignment of nucleosomes would then be required in all cells of a given type. Prunell and Kornberg have demonstrated that the location of nucleosomes is random for the single-copy sequences in rat liver cells. The results of Garel and Axel suggest that the assembly of nucleosomes about a unique gene — the chicken ovalbumin gene — is also random with respect to DNA sequence. Together with the results of previous studies concerning the location of nucleosomes on the genome of SV40 (Polisky and McCarthy 1975; Crémisi et al. 1976), these results appear to rule out a unique location of nucleosomes on single-copy DNA sequences. However, the possibility that the nucleosomes could occupy a small number of distinct alternative positions is not ruled out by the above studies, and Ponder and Crawford (1977) have recently suggested that this is indeed the case in minichromosomes from polyoma virus and SV40. Unexpectedly, Musich et al. have clearly shown that nucleosome phasing occurs with the genetically inert, highly repetitive sequences of component α DNA of African green monkey cells. They also suggest that this nucleosome phasing could be related to an H1 histone deficiency in the heterochromatin-containing component α and to its apparent replacement by a class of low-molecular-weight nonhistone proteins. This nucleosome phasing in highly repetitive sequences raises the interesting possibility that the structure of nucleosomes might have an influence on DNA sequence evolution.

Although it is tempting to draw from all these results (see also Birnboim et al.) the conclusion that, in addition to its structural purpose, the nucleosomal packaging of DNA does not serve a functional role in the control of genetic expression, it is by no means excluded that some nucleosomes could be associated with specific nucleotide sequences at key points on chromatin. More work with probes for various segments of specific genes is clearly required before reaching the definite conclusion that this structural aspect of chromatin is truly random and has no regulatory counterpart.

It should also be pointed out that the possibility that random location of nucleosomes is an artifact generated by sliding during the necessary step of micrococcal nuclease digestion is not excluded at the present time. In this respect, the report of Cech et al. showing that the mere exposure of living cells to low temperature may produce some changes in chromatin structure — either changes in accessibility of nucleosome core DNA to psoralen or redistribution of nucleosomes on the DNA — indicates that artifacts generated during the isolation of nuclei are not a remote possibility in in vitro studies of chromatin structure. It is possible that the phasing in the African green monkey satellite chromatin, where H1 is apparently replaced by other proteins (Musich et al.), represents the rule rather than the exception, histone H1 being less efficient in preventing nucleosome sliding during the necessary manipulation. In my mind the problem of nucleosome phasing is still an open question.

Path Followed by the DNA in Nucleosomes

The exact path followed by the DNA in a nucleosome is at present unknown; however, there are several points which are now well established. It was particularly gratifying for those of us who have attended previous meetings, where the physicists were in strong disagreement most of the time, to learn that they are now coming to similar models. Information on the arrangement of DNA in nucleosomes has come from studies on whole chromatin, on isolated nucleosomes, and more recently on isolated nucleosome cores homogeneous in DNA length.

There is no doubt that the DNA is located at the surface of the nucleosome (for reviews, see Kornberg 1977; Thomas 1977). The high accessibility of chromatin to digestion by DNase I (resulting in a series of DNA fragments differing in size by 10 bases) in a denaturing gel provided the first evidence supporting this location (Noll 1974b). Further evidence for the location of the DNA came from neutron diffraction of nucleosomes (Hjelm et al. 1977), or nucleosome core particles (Bradbury et al.; Pardon et al.; Pardon et al. 1975; Suau et al. 1977) using the technique of "contrast variation" in H_2O/D_2O mixtures. The detailed analyses of the DNase-I digestion pattern reported at this meeting, along with the results of Simpson et al. (see also Stein et al. 1977) demonstrating that it is possible to extract the cross-linked histone octamer core from core particles and to reconstitute the nucleosome core from isolated DNA and the cross-linked histone core under nondenaturing conditions, make it also unlikely that there could be any portion of the DNA buried inside the protein core of the nucleosome. The results of Simpson et al. also suggest that the mechanism of packaging DNA by histones does not necessarily involve a histone distribution along the length of DNA to be compacted, followed by a DNA folding which would be driven by histone-histone interactions, as proposed, for instance, by Camerini-Otero and Felsenfeld (1977).

Since X-ray or neutron scattering in solution, which yields spherical averages of intensities, cannot lead to a unique structural model for the nucleosome core, it was very exciting to learn that the MRC Cambridge group (Finch and Klug; Finch et al. 1977) has succeeded in elucidating some of the three-dimensional features of the nucleosome core. Even though the analysis is restricted to details larger than about 20 Å, their very elegant combination of X-ray crystallography and electron microscopy leads to the following picture. The nucleosome core is roughly a disk-shaped particle of diameter 110 Å and thickness 57 Å, somewhat wedge-shaped, and strongly divided into two layers along its short axis. Although at the present resolution one cannot distinguish DNA and protein, the results are consistent with the DNA being wound in a flat superhelix of pitch about 28 Å and average diameter about 90 Å along the outside of a histone core which may consist of two layers (or turns). For unknown reasons, better crystals were obtained from core particles containing cleaved histones. However, cleaved cores do not appear to change in their gross structure from the intact ones. In particular, the thickness of the nucleosome core and its bipartite nature are identical in crystals of intact or cleaved cores. The wedge-shaped appearance suggests that there is less than two complete turns of DNA superhelix. This is consistent with the observed average diameter (about 90 Å) and the known length of the DNA (140 base pairs), since for DNA in the B form these numbers lead to about 1.75 turns of superhelix with about 80 base pairs per superhelical turn.

The Finch and Klug model of the nucleosome core is consistent with the results of previous solution-scattering studies. Although such studies cannot

lead to a unique model, it is striking that the crystal work has led to a model similar to that previously proposed by the Searle group to explain their low-angle X-ray- and neutron-scattering data in solution (Pardon et al.; Richards et al. 1977). The Finch and Klug model is also supported by the neutron-scattering studies of the Portsmouth group (Suau et al. 1977). It therefore appears that the solution and crystal structures of the nucleosome core are very similar. A disk shape for the nucleosome core was also previously proposed by Langmore and Wooley (1975) on the basis of electron microscopic studies, and the electric dichroism studies of Klevan et al. are also in very good agreement with a disk model.

The results of the nuclease digestion studies reported previously and during this meeting support the deductions of the MRC group concerning the path followed by the DNA around the histone core and give further information about the organization of the DNA in the nucleosome core. The DNase-I (Noll; Lutter; Camerini-Otero et al.; Simpson et al.; Noll 1974b; Simpson and Whitlock 1976) and DNase-II (Camerini-Otero et al.) studies indicate that the DNA must be arranged regularly outside the histone core, since each strand of the 140-base-pair fragment is exposed every 10 nucleotides. This may be achieved by a smooth, continuous bending of the DNA (see comment by Crick following Keller; Finch and Klug; Finch et al. 1977; Harrington 1977) or by kinks in the DNA alternating with straight regions of multiples of 10 base pairs (Sobell et al.; Crick and Klug 1975; Sobell et al. 1976). In principle, these two types of models could lead to different predictions concerning the DNase "cutting stagger" on the two strands (Camerini-Otero et al.; Noll; Lutter). The unfortunate general conclusion of very sophisticated studies with DNase I (Noll; Lutter; Sollner-Webb and Felsenfeld 1977) or DNase II (Camerini-Otero et al.) is that the interpretation of the DNase cutting staggers must await more information about DNase structure and function, since the exact location of cutting sites appears to be determined not only by the nucleosome core structure but also by the mechanism by which the nuclease cuts the DNA (Camerini-Otero et al.; Lutter). Clearly, further studies are required to establish whether the DNA is smoothly bent or kinked around the histone core. However, the above studies, and particularly those of Camerini-Otero et al., reveal that nearly half of the nucleotides of the nucleosome core are in some way accessible to exogenous enzymes, which could be important for chromatin function. On the other hand, according to Keller et al., the DNA seems to be rigidly held in place by the histones, since it is unable to undergo the temperature-dependent rotations observed with protein-free DNA in solution.

The careful quantitative analysis of the distribution of DNase-I cuts (Noll; Lutter) not only demonstrates that DNA of the core particle is accessible along its entire length and therefore cannot be significantly buried inside the structure, but also indicates that there are wide variations in the cutting frequencies. Within this variable pattern, there is a tendency for similar types of sites to be about 80 bases apart. This periodicity of 80 is particularly clear for the points 30 and 110, for instance, which are infrequently cut, as previously reported by Simpson and Whitlock (1976; see also Whitlock et al. 1977 for digestion with other nucleases). This cutting pattern is fully consistent with the DNA wound in a flat superhelix with about 80 bases per superhelical turn and a pitch of about 28 Å, as proposed by the MRC group: points on the DNA double helix 80 bases apart are closest together on the superhelix, less than 30 Å apart, and can therefore be similarly protected (Finch and Klug; Lutter; Noll). As discussed by Finch and Klug and by Lutter, this model can also be used to explain the kinetic "pause" at 160 base pairs seen during micrococcal nuclease digestion of nuclei (Noll and Kornberg 1977) by extending the 1.75 turns to 2 full superhelical turns.

The DNase-I cutting pattern gives some indication of the symmetry of the nucleosome core and is consistent with the assumption that the nucleosome core, which has a strong bipartite character, possesses a dyad (Lutter; Finch and Klug). Other results also suggest the presence of a dyad, which was one of the essential features of the nucleosome model proposed by Weintraub et al. (1976). Apart from the finding of a pair of each type of histone in the nucleosome, the existence of a possible axis of twofold rotational symmetry in nucleosomal DNA has been supported by the observation that DNase II cleaves the chromatin DNA at 100-base-pair intervals under certain conditions (Altenburger et al. 1976) and by the electron microscopic evidence that under appropriate conditions a nucleosome can open up into two separate half-nucleosomes (Oudet, Spadafora, and Chambon; Oudet et al. 1977; see below).

The crystal work (Finch and Klug) and the polarity of DNase-I cutting frequencies (Lutter) suggest that the DNA of the nucleosome core is wound in 1.75 turns of a left-handed superhelix with 80 base pairs per turn. There is an apparent discrepancy with the previous estimations of the constraint imposed on the DNA duplex in the nucleosome which gave values equivalent to −1 to −1.25 superhelical turns per nucleosome (Keller et al.; Germond et al. 1975; Keller 1975; Shure and Vinograd 1976). However, the latter number measures the change in linkage number (see comment by Crick following Keller; Crick 1976; Finch et al. 1977), which is the same as the number of superhelical turns, provided the number of base pairs per helix turn of DNA (screw of the double helix) is the same in chromatin and in the reference state (DNA free in solution). It is therefore possible to reconcile the two results by assuming that the helical screw of the DNA duplex changes when free DNA in solution is wrapped around the histone core. Since the DNase-I digestion

studies strongly suggest (but do not prove, as pointed out by F. Crick during the discussion) that the screw of the DNA duplex is very close to 10 base pairs in the nucleosome core, it can be easily calculated (Finch and Klug; Finch et al. 1977) that an increase in the number of base pairs per helix turn to a value of 10.4–10.7 would be required for DNA in solution to reconcile the two sets of results. In fact, a value close to 11 was previously proposed (Bram et al.; Bram 1971) and there is some recent theoretical evidence (see comment by Crick following Keller; Finch et al. 1977) supporting a value of around 10.7 base pairs per helix turn for straight DNA in solution. Although more work is required to fully elucidate these points, it now appears that all data from crystal structure, DNase-I digestion, and supercoiling studies support a model in which the DNA of the nucleosome core is wound in about 1.75 turns of a left-handed superhelix.

The path followed by the linker DNA is at present unknown. The results of Pardon et al. support a model where in dilute solution the chromatin unit thread is envisioned as a repeating structure consisting of core particles about 50 Å in height separated by regions of similar length occupied by intercore DNA associated with proteins (see also Sperling and Tardieu 1976). Such a model agrees with previous electron microscopic data which show that the nucleosomes are in close apposition in native chromatin. At least some of the linker DNA appears to be arranged in a regular way and associated with histone H1 (or an H1-like histone) and also possibly with other histones which could measure out the linker DNA length. Although there is some indication from nuclease digestion studies that 20 base pairs of the linker could be wound regularly to yield, together with the core DNA, two full superhelical turns, it is not known whether the rest of the linker DNA is also supercoiled. Such a linker DNA supercoiling has been proposed in some models (Worcel; Sobell et al.), but there is no evidence in SV40 chromatin that the linker DNA is under constraint whether the viral chromatin is compacted in the presence of H1 or unfolded in its absence (Keller et al.).

Arrangement of Histones and DNA-Histone Interactions in Nucleosomes

The location of histones in the nucleosome may be analyzed both in relation to one another and in relation to the nucleosomal DNA.

Cross-linking experiments (for reviews, see Kornberg 1977; Thomas 1977) have shown that there are two regions of histones in the nucleosome, one containing the histone octamer core, the other containing H1. This observation agrees with nuclease digestion studies which have shown that H1 is released during the formation of the core particle from the nucleosome (Varshavsky et al. 1976; Simpson and Whitlock 1976; Noll and Kornberg 1977), and therefore is probably bound to the linker DNA (Bakayev et al. 1977), which does not exclude that the other histones could also be bound to the linker (see below). In addition, the cross-linking studies have shown that the contacts between histones in the nucleosome are the same as in whole chromatin. Although the octameric structure of the protein core of the nucleosome is well established and the octamer can be isolated from cross-linked chromatin, the question of whether the octameric core exists in solution when the DNA is removed in 2 M NaCl is still a matter of controversy (Pardon et al.; Thomas and Butler). It seems likely that the observed discrepancies are related to instability of the octamer when free in solution (Thomas and Butler; Thomas 1977). It should be stressed that, although the histone core is an octamer in 2 M NaCl, this does not exclude that in combination with DNA, and under appropriate conditions, the nucleosomal octamer histone core might split into two symmetrical heterotropic tetramers, yielding the half-nucleosomes (Oudet, Spadafora, and Chambon; see below) predicted by a nucleosome model which attempts to relate nucleosome structure to its function (Weintraub et al. 1976).

There are very few hard facts concerning the question of interactions between the various histones and DNA in the nucleosome core (for review, see Thomas 1977). Past and present analyses of "subnucleosomal particles" (Varshavsky et al.; Rill and Nelson) have not been very informative. In a more promising approach, Mirzabekov et al. have cross-linked the histones and the partly depurinated DNA of nucleosome cores to study the histone-DNA interactions. Their still preliminary study leads to the conclusion that there is one histone molecule of each type bound to each DNA strand and that two molecules of each histone H2A, H2B, H3, and H4 are arranged symmetrically on the two strands of the DNA duplex in the nucleosome core. H4 and H3 appear to be bound at the 5' and 3' ends of each DNA strand, respectively. This last result is in disagreement with a previous report (Simpson 1976), but Simpson's remark that he was misled by the presence of a contaminating protein kinase may partially explain this discrepancy. As pointed out by Mirzabekov et al., such a preferential interaction of histone molecules with one DNA strand, which supports the Weintraub et al. (1976) model, may not interfere with the transcription and replication processes. In a parallel study Mirzabekov et al. have shown that histones lie mainly outside the DNA grooves, leaving the minor groove well exposed, which may be important for interaction with regulatory proteins involved in genetic expression.

In another promising approach, Simpson et al. (see also Whitlock and Simpson 1977; Lilley and Tatchell

1977) have attempted to define the role of the interaction between the DNA and NH_2-terminal ends of the histones by removing these ends with trypsin. These studies lead to the important conclusion that the trypsin-resistant regions of the different histones not only interact with each other to form the protein core, but also must have important interactions with the nucleosomal DNA, since the removal of the NH_2-terminal ends leaves the DNase-I-susceptible sites at 10-nucleotide intervals. However, the nuclease susceptibility of certain sites along the nucleosome core DNA is increased after trypsin digestion. This should allow one to define the DNA sites which interact with the NH_2-terminal regions of the four histones, provided no secondary structural rearrangement is induced by the removal of these ends. It would certainly be very interesting to complement these studies by attempting to reconstitute nucleosome cores with various combinations of specific histone fragments (Bradbury et al.).

It was previously suggested that the H3 and H4 histones could play a unique role in nucleosome structure (Kornberg 1974; Camerini-Otero et al. 1976; Boseley et al. 1976; Oudet et al. 1977). Therefore, it was especially interesting to learn (Bradbury et al.; Camerini-Otero et al.; Oudet et al.; Simpson et al.) that the association of H3-H4 with DNA results in the formation of a discrete particle smaller than the nucleosome, about 80 Å in diameter, in which 130–140 base pairs of DNA are compacted about fivefold over their extended length under almost the same conformational constraint as in the nucleosome. In addition, X-ray diffraction studies show that one is dealing with a chromatinlike structure (Bradbury et al.). It therefore appears very likely that H3-H4 alone can organize about 140 base pairs of DNA in a manner very similar to that in the core particle. However, it must still be demonstrated unequivocally that each H3-H4 particle contains a tetramer $(H3)_2(H4)_2$ (for discussion, see Oudet et al.; Camerini-Otero et al.; Simpson et al.). If this turns out to be the case, the H3-H4 tetramer could span the diameter of the nucleosome core, measuring out the appropriate length of DNA and protecting it against nucleases as proposed by Finch et al. (1977). Such a structure would also agree with the results of Mirzabekov et al. (see above).

What then is the role of H2A and H2B histones? I would like to propose the following possibility, which is supported by several observations but is by no means demonstrated (Compton et al. 1976a; Oudet et al. 1977; Kornberg 1977). It appears that H2A and H2B histones could play a double role: First, they could interact with the H3-H4 subnucleosomal particle helping to stabilize the DNA fold in the nucleosome core. Second, they could bind, perhaps together with histone H1, to the linker DNA. Studies on SV40 minichromosomes (Bellard et al. 1976) and electron microscopic analysis of chromatin reconstituted with the four histones (Oudet et al. 1975) have indeed shown that both core DNA and at least a part of the linker DNA are contained in nucleosomes devoid of H1. In addition, the nuclease digestion results of Noll and those of Camerini-Otero et al. (1976) and Sollner-Webb et al. (1976) on native chromatin and chromatin stripped of the lysine-rich histones suggest that the DNA of the linker region is also associated with histones and arranged in some regular way (Lohr et al. 1977). In this context, the report at this meeting of the remarkable variability of histone H2A and H2B during sea urchin embryogenesis (Newrock et al.; Weinberg et al.) and the variability of H2A and H2B histones from other sources (Laine et al. 1976; Franklin and Zweidler 1977; for additional references, see Kornberg 1977; Thomas 1977) leads to the interesting possibility that these variabilities could contribute, at least to some extent, to the observed variations in length of DNA associated with a nucleosome, i.e., to nucleosome diversity. H2A and H2B molecules appear to be hybrid molecules with a hydrophobic region highly conserved and a more variable amino-terminal basic region (Hayashi et al. 1977; W.N. Strickland et al. 1977; M. Strickland et al. 1977). Spiker and Isenberg have shown that the histone-histone binding sites have been conserved (see also Spiker and Isenberg 1977). Evidently, the conserved region of H2A and H2B is important for the histone-histone interactions which hold the histone core together, whereas the more variable basic regions which interact with the DNA could define, perhaps in part and together with H1, the linker length. Such a hypothesis could be tested by reconstitution experiments (see below). Nevertheless, it appears from the above results that it is more the nature of the protein-protein interactions than the DNA structure which imposes the invariability of the histones. An interesting question is also raised: Why are the H3 and H4 histones so highly conserved in all parts of the molecule? Is it because they define the length of the DNA of the nucleosome core which is highly conserved? Or is it because their basic region is also engaged in very specific and vital functions, as yet unknown (Spiker and Isenberg)?

Nucleosome Heterogeneity: Histone Variants, Histone Modification, and Nonhistone Proteins

Crystallization of nucleosome cores (Bakayev et al. 1975; Finch et al. 1977) prepared from rat liver and Ehrlich ascites tumor cell nuclei indicates that, at least in these cells, the bulk of the nucleosome core population is homogeneous enough to yield crystals. However, in view of the existence of histone variants and histone modifications, it remains to be demonstrated that different crystal forms cannot be obtained. The observation of Finch and Klug (Finch

et al. 1977) that some histone proteolysis is helpful for crystallizing core particles could reflect an initial heterogeneity of the particles.

Heterogeneity of nucleosomes was suggested during this meeting by the immunological studies of Bustin et al., by the pattern of unfolding of chromatin in the presence of urea (Woodcock and Frado), and by the study of chromatin of transcribing genes (see below). In view of the linker DNA length variability and since a nucleosome is defined as the particle which contains the entire length of the DNA repeat generated at early times of micrococcal nuclease digestion, nucleosome heterogeneity clearly exists. In fact, some level of heterogeneity was expected from a variety of data which include: (1) the finding of histone variants of H1, H3, H2A, and H2B, not only when comparing different organisms or different tissues in the same organism, but also within a given tissue and during embryogenesis (for references, see Newrock et al.; Elgin and Weintraub 1975; Zweidler 1976; Franklin and Zweidler 1977; Thomas 1977). These are important findings because until recently histones (with the exception of H1) were considered to be nonspecific structural elements; (2) the well-known existence of a variety of sequence-specific postsynthetic modifications of histones, including acetylation, phosphorylation, methylation, and poly(ADP ribosyl)ation (for references, see Elgin and Weintraub 1975; Dixon 1976; Thomas 1977; Johnson and Allfrey 1977; Allfrey 1977); and (3) the presence of HMG (high-mobility group) proteins which are found associated with nucleosomes at a level of about 1 molecule per 15 nucleosomes (Goodwin et al. 1977c; for references, see Goodwin et al. 1977b) and localized to a limited number of sites in polytene chromosomes (Alfageme et al. 1976). Of course, the other nonhistone proteins are also potential candidates to contribute to nucleosome heterogeneity, provided it can be demonstrated that they are not artifactually bound to nucleosomes or subnucleosomal particles (Varshavsky et al., Rill and Nelson) during their preparation. In this respect, it would be particularly interesting to learn whether the puzzling A24 hybrid molecule, which represents as much as 0.10 of histone H2A, could contribute to nucleosome diversity (Busch et al.).

The central underlying questions of whether (and how) nucleosome heterogeneity could be related to specific modulations of the basic structure of chromatin (variability of the DNA repeat length, phasing of nucleosomes, transitions of the nucleosome structure, subtle changes in the higher-order coiling of nucleosomes) and how such modulations could possibly be involved in differential gene expression remain subjects for future investigations. In this respect the coordinate regulation of histone-variant synthesis during sea urchin embryogenesis suggests a strong correlation between histone diversity, changes in chromatin structure, and expression of a developmental program (Newrock et al.; Weinberg et al.; see also below).

Higher-order Structure in Chromatin and Chromosomes: Bacterial Chromosome

The nucleosomal filament has to be folded to account for the overall packing of DNA in the interphase nucleus and in the metaphase chromosome. When the nucleosomes are in close apposition, the packing ratio of the DNA is about 5 to 7. This is about sixfold lower than the packing ratio of the DNA in the 200–300-Å fiber of interphase chromatin and about 1000-fold lower than the packing ratio of DNA in metaphase chromosomes. A study of this folding is very important, not only for understanding how the DNA is further packaged, but also to elucidate whether, and how, this folding is related to the expression of a specific transcriptional program in the interphase nucleus of a given cell. In other words, is there some kind of structural differentiation at the nuclear three-dimensional level, functionally linked to cell differentiation? Obviously, this is a much more difficult problem than elucidating the fundamental nucleosome structure because the higher-order structures are likely to be damaged when nuclei are isolated and of course during preparation of chromatin. Although almost all remains to be done, we have learned during this meeting some new, exciting information concerning the way DNA is folded in interphase and metaphase chromosomes.

Previous electron microscopic studies of the structure of chromatin fibers in interphase nuclei and mitotic chromosomes have revealed a "thick fiber," about 200–300 Å in diameter, which very likely represents the native chromatin fiber characteristic of inactive chromatin (for references, see Renz et al.; Olins; Bram; Worcel; Franke et al.; Oudet et al. 1975). Several attempts have been made to explain how this thick fiber is derived from the basic nucleosomal chain. Mainly on the basis of electron microscopic studies, Finch and Klug (1976) have proposed that the nucleosome chain of in vivo assembled chromatin is coiled (about 6 nucleosomes per turn) to give a solenoidal structure 300 Å in diameter (packing ratio about 40), whereas others have proposed that the 200–300-Å fiber is discontinuous and made up of 200-Å "superbead" units (about 8 nucleosomes per superbead; packing ratio about 25) (Rentz et al.; Kiryanov et al. 1976; Hözier et al. 1977; Vengerov and Popenko 1977). Physical studies have also suggested that nucleosomes could be arranged in helical arrays (Carpenter et al. 1976; Carlson and Olins 1976; Pardon et al.). From what I have seen and read, it seems to me that there is no more evidence supporting the in vivo existence of a solenoid structure than that of a superbead structure. In fact, it was clearly shown that it is possible to visualize both types of structures from the same chromatin

preparation, depending on the environmental conditions (in particular, ionic strength and divalent cations) (see Hözier et al. 1977), and Olins has also shown that the two structures are found in the same material (erythrocyte nuclei) under identical conditions. Both structures might be artifacts arising from the same higher-order structure. Alternatively, the two structures could be real (see Franke et al.) and functionally relevant, reflecting the nucleosome heterogeneity discussed above. An apparently irregular folding could be required to allow the transcriptional machinery to reach the DNA when it is necessary and/or for the higher-order coiling of the 200–300-Å fiber. Thus it is interesting that Martin et al. (1977) have observed that nucleosomes with similar linker DNA lengths are clustered (see Cech et al. for another example of clusters in the nucleosome chains). In any case, we are dealing here with a very difficult technical problem and further studies using both electron microscopic and biochemical approaches, possibly with specific gene probes, are required to study whether this first higher-order coiling is as regular as it was initially thought. Although such a regularity is undoubtedly aesthetically very appealing, we have to remember that nature does not work necessarily under an Arts Council grant (Robson 1977; Jacob 1977). Stressing the lack of regularity of this first higher order of coiling, Worcel has presented an astutely revised "minisolenoïd" version of his previous continuous solenoïd model (Worcel and Benyajati 1977), which incorporates most of the recent results and will undoubtedly be stimulating for those working on higher-order structure. However, some discrepancies are already evident between some features of the model and some results of Finch et al. (1977) and Keller et al. That, in interphase nuclei, the thick 200–300-Å fiber is further coiled in some higher-order structures in a rather ordered fashion is clear from the beautiful pictures of Sedat and Manuelidis, but the elucidation of how this is achieved and whether the whole architecture is cell-specific will require technical breakthroughs.

It was already known before this meeting that the interphase chromatin fiber is organized into loops or domains involving cross-linking of the chromatin fiber (for references, see Igó-Kemenes and Zachau). In *D. melanogaster* cells (Benyajati and Worcel 1976), the domains contain on average 85,000 base pairs, and protein and RNA seem to be involved in the cross-linking. The yeast chromosome also appears to be organized in similar domains (Piñon and Salts 1977). The micrococcal nuclease and restriction enzyme digestion studies of Igó-Kemenes and Zachau further support the existence of domains in rat liver nuclei and provide a possible tool to investigate their structural organization and their functional significance.

Strikingly similar domains have been demonstrated in metaphase chromosomes. The very elegant work of Laemmli et al., which opens a new era in chromosome structure studies, clearly shows that structural nonhistone proteins are involved in the higher-order coiling of the thick 200–300-Å fiber. The involvement of nonhistone proteins in metaphase and interphase chromosome architecture was previously suggested by Stubblefield and Wray (1971) and by Comings and Okada (1976). Laemmli et al. have found that, following removal of histones, the DNA of metaphase as well as of interphase chromosomes remains in a highly folded state organized by nonhistone proteins (see also Adolph et al. 1977a; Paulson and Laemmli 1977; Brown et al. 1977). These proteins (about 20–30 species), not covalently attached to the DNA, are arranged in a central core (scaffold) which has been isolated from mitotic chromosomes as a structurally independent entity, retaining the size and the shape of metaphase chromosomes (Adolph et al. 1977b). The DNA is organized in loops, 45,000–90,000 base pairs in length, anchored to the scaffold. The major proteins of the metaphase scaffold are also found in the histone-depleted interphase chromosomes. This raises the fascinating possibility that the same domain structure is conserved throughout the cell cycle with different forms of the scaffold. Thus it is interesting that the structural organization of the metaphase chromosome with DNA loops anchored on a scaffold is analogous to that of meiotic "lampbrush" chromosomes. Whether a domain represents a functional unit and whether the genome is distributed in identical functional domains in different cell types are obvious questions which could possibly be answered using in vitro recombinant DNA techniques. Hopefully, in situ hybridization will become a tool for localizing mRNA gene sequences in nonpolytene chromosomes (Yu et al.).

Obviously, the DNA attached to the scaffold have to be packed to achieve the metaphase compaction ratio. The work of Laemmli et al. demonstrates that histones and scaffold are independent levels of organization in chromosomes. However, if the histones play no major role in maintaining the higher-order structure of the metaphase chromosome, they are clearly involved in packaging the DNA in nucleosomes similar to those found in interphase chromatin (Compton et al. 1976b; for references, see Worcel). Whether some nonhistone proteins are then involved in folding the nucleosomal chain in the 250-Å-thick fiber is unknown (see Wray et al.). It must also be established how, starting from 250-Å fibers attached to the scaffold, the final metaphase chromosome architecture is generated to yield the various structures which can be visualized (Sedat and Manuelidis; Bak and Zeuthen; Daskal et al. 1976). The electron microscopic pictures looked to me sufficiently irregular to suggest that it will be some time before we will be in a position to understand the rules governing these very high orders of DNA coiling. However, the problem will eventually be tackled

at the molecular level as it was already done for the structure of telomeric DNA (Rubin) and of heterochromatin (Peacock et al.; Brutlag et al.) in *Drosophila*.

An interesting parallel can be drawn, at two levels of folding, between the packing of DNA in eukaryotic chromosomes and the packing in prokaryotic nucleoids. In both cases the chromosome appears to be organized in domain loops of about the same size (for references, see Pettijohn 1976). Furthermore, the prokaryotic DNA appears to be packed in a regularly condensed chromatinlike fiber with about the same compaction ratio as in nucleosomes (Griffith 1976). There are no histones in prokaryotes, but it is likely that the folding of DNA in *E. coli* is produced by some histonelike protein(s) (HU protein of Rouvière-Yaniv; BH proteins of Varshavsky et al.). Moreover, there also seems to be one equivalent negative supercoil per 200 base pairs (strictly speaking, a measured change of −1 of the linkage number; [see comment by Crick following Keller]) in the bacterial nucleoid (for references, see Benyajati and Worcel 1976). These observations raise the possibility that the DNA is similarly coiled in about two superhelical turns in the prokaryotic beads and in the nucleosome core. The rather unstable bacterial condensed structure would have been better stabilized by histones as eukaryotes developed, but in both cases the basic coiling of the DNA would correspond to the best configuration for bending DNA in solution (see comment by Crick following Keller; Finch and Klug).

That histone H1, ionic strength, and/or divalent cations are implicated in the maintenance of higher-order chromatin structures was already known before this meeting (Littau et al. 1965; Mirsky et al. 1968; Bradbury et al. 1973; Oudet et al. 1975; Bellard et al. 1976; Noll and Kornberg 1977; Renz et al. 1977). It was therefore not surprising to hear that the presence of H1 in chromatin, or its addition to an H1-depleted chromatin, results in a compaction of the nucleosomal chain. What we now need to know is precisely how H1 is organized with respect to the other histones and to DNA. It should be kept in mind that rearrangement of H1 during manipulations could occur easily, since H1 is the least tightly bound of the histones and can readily exchange with exogenous DNA (for references, see Cole et al.). At present (Cole et al.; Gaubatz et al.; Griffith and Christiansen) it is not clear what the location of H1 is with respect to the nucleosome core and the DNA linker, and whether H1 is bridging between nucleosomal fibers or lying along them. Cross-linking studies (for references, see Thomas 1977) have suggested that chains of H1 molecules may exist, but it is unknown whether the juxtaposition arises from the coiling of the nucleosomal fiber (see also Worcel). Hopefully, methods similar to those used by Mirzabekov et al. for studying the relation between the other histones and DNA will be developed to answer these questions. We would also like to know more about histone H1 phosphorylation, which occurs at very specific sites and which has been implicated in metaphase chromosome condensation (Bradbury et al. 1974; Matthews and Bradbury 1977), and how H1 variants are possibly involved in the modulation of the higher-order structure of chromatin.

Viral Chromatin as a Model System for Cellular Chromatin Structure

The isolation of minichromosomes, corresponding to genomes of well-defined size, from the nuclei of SV40- or polyoma-infected cells and from virions has undoubtedly been very helpful in supporting the nucleosome idea and in revealing the nucleosomal torsional constraint imposed on the DNA (Keller et al.; Griffith 1975; Germond et al. 1975; for references, see Thomas 1977; Kornberg 1977). However, we still don't know whether, from a structural standpoint, the SV40 minichromosome can be considered as the exact counterpart of a cellular chromatin segment. There are uncertainties concerning the exact number of nucleosomes per minichromosome, their DNA content, and their arrangement, and whether H1 is or is not a natural component of the nuclear SV40 minichromosome (Oudet, Spadafora, and Chambon; Griffith and Christiansen; Keller et al.; Varshavsky et al.; Bellard et al. 1976). In this respect, it is interesting to note that when H1 is found associated with the minichromosome, it can be completely released at 0.2–0.3 M NaCl (for references, see Oudet, Spadafora, and Chambon), which corresponds to the salt concentration at which weakly bound surplus H1 present in the pellet of nuclear micrococcal nuclease digests can be washed off (Gaubatz et al.; Gaubatz and Chalkley 1977; see also above for the well-known mobility of histone H1). Whether examination of the SV40 minichromosome is more informative than that of the cellular chromatin in elucidating the higher-order coiling (or folding) of the nucleosomal chain is at least dubious at present. It is clear from the comparison of the reports of Keller et al., Varshavsky et al., Oudet, Spadafora, and Chambon, and those of Griffith (1975) and Christiansen and Griffith (1977) that almost any kind of electron microscopic picture could be obtained from what is supposed to be identical material! Finally, the puzzling micrococcal nuclease resistance of the "native" minichromosome raises the question of whether it is an artifact occurring during the isolation of the minichromosome or whether it does indicate that the folding of the SV40 nucleosomal chain is actually different, as proposed by Varshavsky et al., from that of cellular chromatin. We can hope that all of these discrepancies and inconsistencies will be resolved and that they will turn out to be correlated with differences in the methods used to prepare the viral chromatin. In this way we will probably learn what should be done and what should be avoided when preparing cellular chromatin.

Reconstitution of Nucleosomes and Chromatin

Undoubtedly, the elucidation of the structure and function of chromatin at the molecular and submolecular levels depends at least in part on accurate reconstitution procedures. In the last 2 years, reconstitution of a nucleosomelike structure was achieved by associating DNA and the four histones, as judged by a number of criteria including electron microscopy, nuclease digestion, X-ray diffraction, and histone cross-linking patterns (for references, see Oudet et al.; Woodcock and Frado; Thomas 1977). Nucleosome core particles identical in all physical properties with native particles have been recently obtained by associating purified 140-base-pair DNA fragments and the four histones (Tatchell and Van Holde 1977). A regular repeating DNA structure, as judged by micrococcal nuclease digestion, has been obtained by associating pure DNA with pure histones or by reassociation from 2 M NaCl-dissociated chromatin (Thomas and Butler; Yaneva et al. 1976; Steinmetz et al. 1977). In all cases the DNA repeat length appears to be 140 base pairs, suggesting that the repeating structure consists of closely packed nucleosome cores lacking most of the linker DNA. These results suggest that addition of H1 could be necessary to obtain longer repeats. However, essentially the same results were obtained when H1 was added during the reconstitution. At first sight there is some contradiction between these nuclease digestion results and the previous electron microscopic measurements of Oudet et al. (1975) which have shown that association of adenovirus DNA with the four histones results in nucleosomelike beads containing about 193 base pairs of DNA. This discrepancy may arise from two different approaches, since in the electron microscopic studies the reconstituted beads were widely spaced, whereas in the nuclease digestion studies the repeating structure is generated only when the beads are closely packed. Therefore, the nuclease results do not exclude that, in addition to the 140 base pairs of DNA contained in the core particles, the four histones could organize an additional DNA segment in a looser way. In fact, at early time of nuclease digestion, the monomer of reconstituted chromatin appears to be longer than 140 base pairs (Steinmetz et al. 1977).

In view of these difficulties in reconstituting a repeating structure similar to that present in vivo, the cell-free catalytic nucleosome assembly system described by Laskey et al. is causing considerable interest. This partially purified system isolated from unfertilized *Xenopus* eggs actively assembles minichromosomes from SV40 DNA and either endogenous or exogenous histones under physiological salt conditions. The assembled minichromosome chromatin seems to have a regular DNA repeat length of about 200 base pairs (Laskey et al. 1977).

Further purification of this cell-free system should tell us how nucleosomes are assembled in vivo and solve the controversy of whether or not the assembly of a nucleosome could occur in steps: the arginine-rich histones H3 and H4 being bound first, with the addition of H2A and H2B completing the nucleosome at a later stage (Thomas and Butler; Oudet et al.; Simpson et al.; Ruiz-Carillo and Jorcano; Camerini-Otero and Felsenfeld 1977). Obviously, this physiological system will be extremely useful in answering a number of questions related to nucleosome diversity and raised throughout this summary; for example: What are the respective roles, if any, of H1 and the other histones in determining the length of the linker DNA and therefore the repeat length? What is the importance of histone modifications in chromatin structure and assembly? Are nonhistone proteins implicated in the organization of the nucleosomal chain, and if so, how? In addition, the chromatin assembly system, in conjunction with the *Xenopus* oocyte system (Laskey et al.), looks very promising for functional studies aimed at defining the molecular components involved in gene expression.

The Nucleosome: A Dynamic Structure?

Can the nucleosome structure be viewed as a dynamic structure? This was certainly one of the important issues of this meeting, since it is obviously related to the problem of chromatin structure during transcription and replication.

From electron microscopic and nuclease digestion evidence it seems clear that, at very low ionic strength, the compact nucleosomal structure can be converted to half-nucleosomes and to extended structures (open nucleosomes) (Oudet, Spadafora, and Chambon). The finding that cellular chromatin can be visualized as chains of nucleosomes with diameter about 125 Å, or as flexible chains of half-nucleosomes with diameter about 90 Å, raises the obvious question as to the relationship between nucleosomes, half-nucleosomes, and ν bodies (Olins and Olins 1974). It should be mentioned that Tsanev and Petrov (1976) have found large-size (120 Å) and small-size (80 Å) particles in rat liver chromatin in various preparations, depending on the environmental conditions. There is a good correlation between the structural transitions as visualized by electron microscopy and the effect of different salt concentrations on chromatin structure as detected with fluorescent probes (Zama et al.; Dieterich et al.).

A remarkable property of these structural transitions at low ionic strength is that they can apparently occur without the release of histones from the DNA and that they are reversible, a nucleosome structure being regenerated from a structure which had lost its particulate morphology (Oudet, Spadafora, and Chambon). Such reversible structural transitions support the model of Weintraub et al. (1976) and agree with the results of Mirzabekov et al. and of Hardison et al. (1977) concerning nucleosomal DNA-histone interactions. An important con-

clusion from these structural transitions is that the finding of a DNA repeat after nuclease digestion does not necessarily mean that the DNA was compacted in a nucleosomal structure. In other words, a periodic structure, presumably reflecting the binding of histones to DNA, may be conserved even when the chromatin is in an extended nonparticulate state. Nuclease digestion studies on chromatin extended in the presence of urea (Woodcock and Frado; for references, see Zama et al.; Oudet, Spadafora, and Chambon) and on *Tetrahymena pyriformis* rDNA transcriptional units (Mathis and Gorovsky; see below) have led to a similar proposal.

STRUCTURE OF TRANSCRIBING CHROMATIN

Elucidating a structure could be fascinating by itself, but for molecular biologists, the central question always is: What is the significance of a given structure in terms of its function? Intuitively, and in agreement with the classical concept that histones play a role in the nonspecific repression of gene activity (Stedman and Stedman 1951; Huang and Bonner 1962), the packaging of the DNA template into compact nucleosomes by tight association with histones seems to preclude efficient transcription by the transcriptional machinery. (Somewhat controversial evidence supporting this intuition can be found in the report of Felsenfeld et al.; see also Crémisi et al. [1977].) Since any model of chromatin structure should ultimately also account, at the molecular level, for the selective transcription of only a part of the genome, it is not surprising that many groups have attempted during these last 2 years to establish whether the nucleosomal organization is ubiquitous in the genome. Obviously, one cannot validly study the structure of active chromatin and compare it with that of inactive (inert) chromatin without knowing whether the expression or nonexpression of a given gene in a given tissue or cell is actually directly related to regulation of its rate of transcription and not to posttranscriptional events. Therefore, I will first very briefly review some recent evidence demonstrating that, at least in some systems, gene expression is regulated at the level of RNA chain initiation.

Gene Expression Is Regulated at the Transcriptional Level

The visualization of active ribosomal transcription units (TUs) and their modulation at different stages of development (Foe; Franke et al.; McKnight et al.) provides strong evidence that rDNA transcription is regulated. Other electron microscopic studies suggest that the expression of nonribosomal genes could also be regulated at the level of transcription, since nonribosomal TUs delimited by initiation and termination sites have been visualized (Foe; McKnight et al.). However, these latter results do not exclude that, in a given organism, a given gene (the globin or ovalbumin genes, for instance) could be transcribed in all cell types, but that its RNA transcripts are very rapidly degraded and do not accumulate unless other posttranscriptional events occur. Such a possibility appears unlikely, at least in the case of the chicken ovalbumin and globin genes, since transcribing RNA polymerase molecules are found on these genes only in the cells where they are expressed (Schutz et al.; Bellard et al.; Orkin and Swerdlow 1977). The immunological studies of Jamrich et al. and of Elgin et al., which show that the distribution of RNA polymerase B (or II) is correlated with the transcriptional state (active and inactive) of a given locus in the *Drosophila* genome, also support the conclusion that not all of the genes which can be expressed in an organism are transcribed at all times in a given cell. (Incidently, the study of Jamrich et al. restarts the debate on the band-interband gene localization controversy; see also Daneholt et al.; Skaer [1977].)

Transcribing Genes Are Contained in an Altered Nucleosomal Structure

Although their interpretation could be complicated by the fact that not all rRNA genes are necessarily active at a given time, studies based on micrococcal nuclease digestion of chromatin have indicated that actively transcribed ribosomal genes are present in the bulk nucleosomal structure (Mathis and Gorovsky; Allfrey et al.; Reeves; for references, see Mathis and Gorovsky; Bellard et al.). Similar studies on the bulk of transcribed genes or on specific transcribed genes (globin or ovalbumin) have also shown that the DNA of these genes is found after micrococcal nuclease digestion in particles sedimenting like nucleosomes or in DNA fragments of nucleosomal core length (Garel and Axel; Camerini-Otero et al.; Bellard et al.; for references, see Bellard et al.). That a periodic structure similar to that of bulk chromatin exists in transcribing chromatin is unequivocally established by the results of Mathis and Gorovsky (1976) for ribosomal DNA chromatin, and those of Bellard et al. for ovalbumin gene chromatin. Since histones are responsible for the repeating pattern, it appears very likely that histones are associated with actively transcribed genes. This conclusion is further supported by the psoralen studies of Cech et al., which show the absence of long (>400 base pairs) protein-free (presumably histone-free) regions of DNA in L cells, and by immunological (Elgin et al.; McKnight et al.; see below) and biochemical (Reeder et al.) studies which indicate that histones, with the possible exception of H1 (Elgin et al.; Jamrich et al.), are present on actively transcribed ribosomal and nonribosomal genes.

Although all of the above results suggest that transcribing chromatin is organized in a way very similar to that of bulk chromatin, several lines of evidence, both biochemical and electron microscopic, indicate that actively transcribed genes are not packaged in regular compacted nucleosomes as defined previously. Transcribed genes are much more rapidly degraded to nonhybridizable fragments than are inactive genes by DNase I, which can be considered as a probe of the internal structure of the nucleosome. This was first demonstrated for the globin genes in chicken embryo red blood cells (Weintraub and Groudine 1976) and confirmed during this Symposium for a number of systems, e.g., actively transcribed ribosomal genes (Mathis and Gorovsky), expressed ovalbumin genes (Garel and Axel; Palmiter et al.; Bellard et al.), genes coding for oviduct rare mRNA sequences (Garel and Axel; Garel et al. 1977), genes selectively expressed in *Drosophila* polytene chromosomes (Biessman et al.), genes coding for total mRNA sequences of Friend cells (Paul et al.), and expressed adenovirus genes integrated in the genome of transformed cells (Weintraub et al.; Flint and Weintraub 1977). In the last case, it was clearly shown that the increase in DNase-I sensitivity is limited to the transcription unit and perhaps to a few nucleosomes at both ends. The altered structure of transcribing chromatin is also revealed by digestion with DNase II, which attacks it at a much faster rate than for bulk chromatin (Bonner et al.; Gottesfeld and Butler 1977). This allows one to prepare a fraction enriched in transcriptionally active chromatin subunits which contain the same length of DNA and the same histone complement as bulk nucleosomes, but are much more rapidly digested by DNase I. Similarly, Bellard et al. have shown that the chromatin repeating subunit of an actively transcribed gene (ovalbumin gene in laying-hen oviduct) is released faster by micrococcal nuclease digestion than those of nontranscribed genes. Whether the actively transcribed ribosomal genes exhibit a similar sensitivity to micrococcal nuclease remains to be established (see Allfrey et al. and Reeves for one view, and Mathis and Gorovsky [1976] for a conflicting view).

At present it is not clear whether increased DNase-I sensitivity is conserved in subunits isolated from transcribing chromatin by digestion with micrococcal nuclease or whether it is a unique property of the undigested transcribing chromatin fiber (see Weintraub and Groudine [1976] for one view, and Garel and Axel [1976] for a conflicting view). It is also not clear (Camerini-Otero et al.) whether the increased rate of digestion reflects an increased cutting rate at the normal 10-nucleotide intervals or the appearance of new cutting sites, although for active ribosomal genes no 10-nucleotide (or multiple of 10-nucleotide) DNA fragments have been found at any time of the digestion (Mathis and Gorovsky).

Using the Miller and Beatty (1969) technique, all electron microscopic studies (Foe; McKnight et al.; Franke et al.; Reeder et al.; Woodcock et al. 1976) agree on the absence of nucleosomes on actively transcribed ribosomal genes, since measurements of ribosomal TUs indicate that the DNA must be extended in the B form. However, the DNA is by no means naked, but is associated with basic proteins (Foe). There is some disagreement, possibly related to variations in the spreading conditions (Reeder et al.), on whether or not nucleosomes are present in the spacer regions (Reeder et al.; Foe; Franke et al.). It has been suggested that some of the particles seen in the spacer regions could be RNA polymerase molecules (Franke et al.).

We have seen electron microscopic immunological evidence that histones remain associated with transcriptionally active nonribosomal TUs (McKnight et al.). Whether nucleosomes are present or not in nonribosomal TUs is still a debated question (Franke et al.; McKnight et al.; Foe et al. 1976; Laird et al. 1976), since there are difficulties in determining accurately the actual compaction ratio of undefined nonribosomal TUs. But in any case, the chromatin segments associated with nascent nonribosomal ribonucleoprotein fibers exhibit a reduced bead periodicity when compared with inactive chromatin, and no nucleosomelike beads are visible on the very active transcriptional units of lampbrush chromosome loops (for references, see Franke et al.).

How can we account for all of the above results which at first appear rather contradictory? On one hand, the presence of nucleosomes on transcribing chromatin is suggested by the finding of active genes in the same DNA repeat length as that of bulk chromatin and by the association of histones with these genes. On the other hand, DNase-I digestion studies demonstrate that the nucleosomal structure is altered, and electron microscopy observations indicate that the DNA in transcribing chromatin is not compacted in nucleosomes. It appears to me (and I am sure to others also) that this apparent paradox could be explained on the basis of the dynamic properties of open nucleosomes as follows: In transcribed regions, the nucleosome compact structure is opened (extended); the histones are still present on the DNA and still protect the DNA to a large extent against micrococcal nuclease and DNase-II digestions, but not against DNase-I digestion. Such an interpretation could account for the electron microscopic controversy concerning the presence or absence of nucleosomes on transcribing chromatin, since the in vivo state may be an equilibrium which can be pushed in one way (the extended form) or the other (the compact nucleosomal form), depending on the conditions used for chromatin isolation and visualization.

Several question are raised by the existence of an altered nucleosomal organization in transcribing chromatin. First, although it appears that the association of histones with a gene is not sufficient per se to repress its transcription, it remains to be elucidated how the RNA polymerase could transcribe the DNA with histones still attached to it. Even though the present results and those of Mirzabekov et al. support the model proposed by Weintraub et al. (1976), much more has to be done to establish its validity.

The second problem concerns the mechanism promoting the opening of the nucleosome. At present, although (and also because) the candidates are numerous, we have no information identifying the factors responsible for the induction and maintenance of an altered "active" nucleosomal conformation about very specifically located chromatin regions (for hypothetical models, see Weintraub et al.). Do the structural changes precede or result from the transcriptional process? Several lines of evidence indicate that it is unlikely that the components of the transcription machinery are responsible for the altered nucleosomal structure of transcribing chromatin. The electron microscopic observations of Foe and Franke et al. suggest that, at least in the case of ribosomal TUs, the transition from a compact nucleosomal structure to a nonbeaded, extended 70-Å fiber may precede initiation of transcription. In addition, the DNase-I sensitivity of nonribosomal transcribed genes is not correlated with the rate of transcription (Garel and Axel; Biessman et al.) and, in the case of ovalbumin, appears to be retained even after cessation of transcription (Palmiter et al.). Similarly, the distribution of some nonhistone proteins in Drosophila polytene chromosomes has suggested to Elgin et al. that loci which will be, are, or have been active in transcription in a given developmental period could have a chromatin state different from that of inert, never-transcribed chromatin. However, the generality of a structural pattern which would be necessary but not sufficient to allow transcription is questionable in view of the results of Camerini-Otero et al., which disagree with the previous report of Weintraub and Groudine (1976) establishing that globin genes are specifically digested by DNase I in mature erythroid nuclei in which they are no longer transcribed.

Besides the components of the transcription machinery, many other possible factors could be implicated in the generation and maintenance of the altered nucleosomal structure in transcribing chromatin. Active subunits are highly enriched in nonhistone proteins (Gottesfeld and Butler 1977), and there is circumstantial immunological (Elgin et al.) and biochemical evidence that some specific nonhistone proteins could be involved, including HMG proteins (Levy W. et al.; Vidali et al. 1977). It is obviously tempting to implicate histone modifications, notably acetylation and phosphorylation, since there are many correlations between these modifications and changes in chromatin activity (Allfrey 1977; Johnson and Allfrey 1977). Most of these modifications occur in the NH_2-terminal regions of histones and could therefore be involved in the increased sensitivity to DNase I, since cleaving the NH_2-terminal ends with trypsin increases the rate of DNase-I digestion. In vitro acetylation of chromatin drastically increases its susceptibility to DNase-I digestion (see above) (Bonner et al.). However, D. Mathis (unpubl.) has found that the DNase-I susceptibility of highly acetylated chromatin of butyrate-treated HeLa cells (Riggs et al.) is not dramatically increased, and it is not yet clear whether nuclease digestions preferentially release subunits enriched in acetylated histones (Levy W. et al. 1977; Davie and Candido 1977).

Clearly, to solve these problems of structural transitions in transcribing chromatin, we need an efficent way to purify transcribing chromatin fibers (up to now, all attempts have failed), as well as methods to isolate well-defined fractions of active chromatin (see Parker et al.; Hill and Watt). Undoubtedly, chromatin reconstitution systems from specific cloned DNA and specific nonhistone proteins (see Weideli et al. for a possible approach) will also be very useful.

IN VITRO TRANSCRIPTION OF CHROMATIN

Isolation of chromatin in a state resembling that in vivo and its faithful reconstitution after dissociation of its components are the logical and necessary (particularly because genetic evidence is usually lacking) correlates of any structural study of transcribing chromatin aimed at defining the molecular components involved in selective gene transcription in eukaryotes. Obviously, in addition to structural studies, the "native" state of isolated chromatin and the fidelity of its reconstruction can be assayed by transcription with exogenously added RNA polymerases. It is hoped that such studies will reveal the relative importance of chromatin and RNA polymerase components in the specificity of transcription and lead to their characterization. There are several difficulties in performing such experiments which are related to the state of the template (see Chambon 1975), to the assessment of the selectivity of transcription, and to the nature of the RNA polymerase which should be used.

There are several levels at which the selectivity of transcription can be assessed. One certainly wants to know whether some sequences of a given RNA are present in the collection of in-vitro-synthesized RNAs. However, even at this low level of specificity, to reach a meaningful conclusion, it should be unambiguously demonstrated that the chains of the specific RNA under study are not only elongated, but also initiated in vitro (see below). To assess whether transcription is actually faithful, one has to investigate whether the synthesized RNA molecules are initiated and terminated at the proper sites, and also whether only RNA species found in vivo are transcribed in vitro, which can be tested to some extent by analyzing whether only the sense strand of the DNA is transcribed. Clearly, a prerequisite

for such studies is the characterization of the in vivo primary transcripts. On this front the situation is reasonably good for tRNA and 5S RNA (for references, see Darnell et al.; Parker et al.) and to a lesser extent for rRNA. Until very recently, the situation was less hopeful for mRNAs, in spite of the electron microscopic evidence of well-defined nonribosomal transcriptional units. However, as already mentioned, the recent studies of Darnell et al. suggest very strongly that nuclear hnRNAs are indeed the precursors of mRNAs. There is no doubt that, thanks to molecular cloning, the isolation and characterization of many mRNA precursors will very soon be reported, as outlined at the meeting for the β-globin mRNA (Curtis et al.), possibly with the help of the elegant labeling technique of the 5′ ends of putative mRNA precursors described by Reeder et al.

Up to now, most of the studies aimed at evaluating the possible specificity of in vitro transcription of native or reconstituted chromatin have been performed with a prokaryotic enzyme, mainly E.coli RNA polymerase, because it is much easier to purify than the eukaryotic enzymes. Although such a choice was unavoidable before the characterization of eukaryotic RNA polymerases, it was certainly no longer justified after the discovery of the multiplicity of eukaryotic RNA polymerases. Their distinct subcellular localization, their complex and specific subunit structure, and their specific in vivo functions suggested very strongly that they could play an active role in the specificity of transcription (for reviews, see Chambon 1975; Roeder 1976). In addition, there was no reason to expect that, if promoter and termination sequences were present in eukaryotic DNA, they should be identical to the prokaryotic ones. Indeed, no DNA sequences characteristic of prokaryotic promoter regions have been revealed by the recent published or unpublished sequence studies of eukaryotic DNA (see, e.g., Fedoroff and Brown; Valenzuela et al. 1977). Thus it appears that, at best, E.coli RNA polymerase can only be used as a nonspecific probe, much in the same way as nucleases, to investigate whether transcribing chromatin is in an altered nucleosomal structure more accessible to a transcribing enzyme than the bulk of the chromatin.

For a given mRNA, there is no reason to expect that its in vitro rate of transcription in chromatin by E.coli RNA polymerase will reflect its in vivo rate of transcription since, for genes which are expressed in a given tissue, the altered structure of transcribing chromatin is independent of the rate of transcription (Garel and Axel.). This remark applies, for instance, to the transcription of the reticulocyte globin and oviduct ovalbumin genes on which RNA chain initiation takes place in vivo much more frequently than on other genes expressed in the same cells, as indicated by the high rate of transcription of these two genes by endogenous polymerase in isolated reticulocyte or oviduct nuclei (Fodor and Doty 1977; Schutz et al.; Bellard et al.). From the above considerations there is no more reason to expect that, in reconstructed chromatin systems, the prokaryotic enzyme could be prompted to initiate preferentially in vitro at putative eukaryotic promoter sites by some eukaryotic regulatory components which might possibly act in vivo at the level of transcription (as was found for the prokaryotic factors involved in the negative or positive controls of transcription).

In the light of all these difficulties, it is not surprising that, until very recently, studies of transcription of native or reconstructed chromatin by an exogenous RNA polymerase have been rather disappointing and frustrating (for reviews, see Chambon 1975; Roeder 1976). We therefore all heaved a sigh of relief when we learned from Parker et al. that studying transcription of chromatin was not a deadend. Their studies have clearly demonstrated that all the information required to transcribe selectively the highly reiterated 5S genes of *Xenopus laevis* oocytes is contained in isolated chromatin and RNA polymerase III (or C), which is responsible for their in vivo transcription. The in vitro gene transcripts are initiated de novo, and the natural initiation and termination signals appear to be recognized. No such specificity is observed with the corresponding deproteinized DNA, demonstrating the importance of some chromatin components and suggesting a positive type of control. *E.coli* RNA polymerase and eukaryotic RNA polymerases belonging to the two other classes (I or A, and II or B) do not contain the information for selective transcription of the 5S genes, indicating that recognition of the chromatin regulatory components is achieved only with the class of RNA polymerase which is responsible for 5S RNA synthesis in vivo. However, RNA polymerases of class III from different species act apparently in the same way as the homologous enzyme (Sklar and Roeder 1977), which is not too surprising in view of a very similar structure of all class-III enzymes (Roeder 1976). If it is also true for other eukaryotic RNA polymerase classes, it is good news, since it is not easy to prepare large amounts of RNA polymerase from cells in culture or from salivary glands! Whether the specificity of transcription is mediated through specific interactions between the DNA and/or RNA polymerase and specific chromatin components remains to be elucidated. Such studies will require dissociation-reconstitution experiments which are already under way and seem promising. Of course, the reiterated 5S genes in *X. laevis* oocytes are especially suitable for in vitro chromatin transcription studies, but similar results have been obtained for the selective transcription of pre-tRNA genes by RNA polymerase III added to isolated nuclei (Parker et al.). The recent study of Yamamoto et al. (1977) also suggests that HeLa cell RNA polymerase C (or III) synthesizes specifically 5S RNA from isolated HeLa cell chromatin. All of these results are extremely encouraging, and I am convinced that it will not be long before the

molecular mechanisms involved in the selective transcription of these genes are dissected and characterized.

If the above studies demonstrate unequivocally that the specificity and accuracy of transcription of 5S RNA reside both in some chromatin components and in RNA polymerase class III, we are still far from similar conclusions for transcription of rDNA and even more so for unique genes coding for mRNA. There is no clear-cut evidence that RNA polymerase belonging to class A (or I) transcribes *specifically* and *accurately* deproteinized DNA or chromatin, mainly because the primary transcripts of the ribosomal cistrons are not yet fully characterized. However, it has been suggested that some specificity is exhibited in transcription studies involving the homologous RNA polymerase A (or I) and either deproteinized DNA (Holland et al. 1977; Van Keulen and Retèl 1977) or chromatin (Matsui et al. 1977; Daubert et al. 1977; for other positive or negative reports, see Chambon 1975). Further studies are required to establish unambiguously whether, in the case of the ribosomal genes, the specificity and accuracy reside only in the DNA and the specific RNA polymerase A (or I), or whether chromatin factors are also involved. However, it is already clear that the homologous eukaryotic enzyme, and not *E.coli* polymerase, should be used.

Up to now, all studies aimed at demonstrating a specific in vitro transcription of viral or cellular purified DNA by purified RNA polymerase class B (or II) have failed (for review, see Chambon 1975; Roeder 1976). This suggests that chromatin components could be involved in the specificity of transcription. Indeed, during these last years, there have been many reports of in vitro chromatin transcription implicating nonhistone proteins as important factors for regulating differential gene expression in eukaryotic cells (for references, see Paul et al.; O'Malley et al.; Stein et al.). In some cases, elaborate "prokaryotic-inspired" molecular mechanisms have been proposed (see, e.g., Stein et al.; O'Malley and Schrader 1976). More specifically, it has been repeatedly claimed that, either with native chromatin or with chromatin reconstituted in the presence of appropriate "regulatory" nonhistone protein fractions, the genes which are preferentially transcribed in vitro appear to correspond to the genes which are actively expressed in vivo (ovalbumin, histone, or globin genes in specific tissues or cells). In addition to the fact that the use of a bacterial RNA polymerase in almost all of these studies creates some difficulty in envisioning how specific eukaryotic transcriptional regulatory factors could operate in conjunction with a prokaryotic enzyme, there are many other problems associated with these experiments.

First, since the nature of the primary transcript is unknown, the possible specificity of in vitro transcription is tested only at the lowest level, namely, that of the presence of some sequences of the specific mRNA expressed in the tissue or cells from which chromatin is prepared. Second, the use of labeled cDNA probes for titrating the specific mRNA sequences has raised many difficulties because such an assay does not allow one to demonstrate that the RNA chains are initiated de novo, which is crucial for the interpretation of the results. Even worse, this cDNA assay does not distinguish the specific RNA sequences possibly synthesized in vitro from the specific RNA molecules which could contaminate the chromatin template. Surprisingly, this problem was not fully appreciated until recently. To circumvent this problem, mercurated nucleotides have been employed in the RNA synthesis reaction with the hope that the newly synthesized RNA could be isolated from other nonmercurated RNA by its affinity for a column containing free thiol groups (for references, see Huang et al.). However, this technique still does not allow one to demonstrate whether the in-vitro-synthesized RNA is initiated in vitro, and it was recently demonstrated that it could be responsible for multiple artifacts (due mainly to RNA-dependent RNA synthesis by E.coli RNA polymerase) leading to erroneous conclusions (Schutz et al.; Felsenfeld et al.; O'Malley et al.; Shih et al. 1977; Zasloff and Felsenfeld 1977; Konkel and Ingram 1977; Giesecke et al. 1977). However, with certain precautions, the mercurated nucleoside triphosphates can be used, and it was convincingly demonstrated by Felsenfeld et al. (see also Zasloff and Felsenfeld 1977) that there is no apparent DNA-dependent synthesis of mRNA globin sequences, above random expectation, when reticulocyte chromatin is transcribed in vitro by E.coli RNA polymerase. Schutz et al. (see also Giesecke et al. 1977) have reached a similar conclusion concerning the in vitro transcription of the ovalbumin gene when laying-hen oviduct chromatin was transcribed with the prokaryotic enzyme. For reasons which are not clear at present, O'Malley et al. (see also Towle et al. 1977) found some preferential transcription of the ovalbumin gene when oviduct chromatin was transcribed in vitro by E.coli RNA polymerase. However, it should be pointed out that this preference was only tenfold when compared with globin RNA synthesis in the oviduct system, and not more than fourfold when compared with ovalbumin mRNA sequences transcribed from reticulocyte chromatin. (Similar results were obtained with reconstituted chromatin.) In the oviduct system, the ovalbumin sequences represented only 0.003–0.005% of the synthesized RNA (O'Malley et al.; Towle et al. 1977) whereas Schutz et al. found that they represent as much as 0.1% of the RNA synthesized in isolated nuclei. Moreover, the transcription of the ovalbumin gene from chromatin templates appears to be partially symmetrical, whereas in isolated oviduct nuclei it is highly asymmetrical (Bellard et al.). The interpretation of some results as supporting the view that at lower enzyme-to-DNA ratio, chromatin DNA transcription could be more asymmetrical, and therefore more accurate, is questionable in view of the problems created by the RNA-dependent RNA synthesis reaction (Zasloff and Felsenfeld 1977). It appears, therefore, that there is little selectivity in the in vitro transcription of reticulocyte or oviduct chromatin by E.coli RNA polymerase and that this conclusion holds true for reconstituted chromatins. In a parallel series of studies, Stein et al. have reported a selective transcription of the histone genes in native or reconstituted chromatins prepared from HeLa cells during S phase, but not during G_1. That this apparent selectivity could be related to unexpected artifacts due to chromatin contamination by endogenous histone mRNA is strongly suggested by the recent results of Melli et al. (1977) which indicate that histone mRNA is in fact synthesized in HeLa cells throughout the cell cycle.

At the present time, and in light of all the above studies, in vitro transcription of chromatin, either native or reconstituted, by E.coli RNA polymerase does not appear to me as a valid and effective system for investigating the molecular factors which control the selectivity and accuracy of in vivo transcription of mRNA coding genes. As discussed above and as proposed by Paul et al. and by Biessman et al. (see also Gjerset and McCarthy 1977), it seems that, at best, the heterologous enzyme acts as a nonspecific probe (like nucleases) for the altered nucleosomal structure of transcribing chromatin, probably because it is more accessible to it. There is no doubt that, in the case of unique genes coding for mRNA, the respective roles of chromatin and RNA polymerase components in the molecular mechanisms regulating transcription will be elucidated by using eukaryotic RNA polymerase B (or II), pure cloned genes, affinity-purified nonhistone proteins, and in vitro or "in Xenopus oocyte" reconstitution systems (see above). Several very promising systems which were presented during this Symposium are amenable to such studies (Yamamoto et al.; Feigelson and Kurtz; Sim et al.; Suzuki and Oshima). It is apparent for many reasons that working on Drosophila presents some selective advantages (Bonner et al.; Elgin et al.; Jamrich et al.; Mirault et al.; Biessmann et al.; Compton and Bonner; Chovnick et al.). From the number of groups working on it, it looks already like a flying E.coli!

CHROMATIN REPLICATION

Until recently, replication of chromatin has been "the poor relation" of chromatin studies, most likely for two reasons: First, in prokaryotic organisms transcription was better understood than replication. Second, the idea that elucidating the mechanisms involved in chromatin replication could be very important in understanding the regulation of gene expression during development and differentiation was not widely appreciated until the notions of histone variants, nucleosome heterogeneity, and altered nucleosomal structure in transcribing chromatin became evident. Indeed, one way to explain why the same differentiated state of a cell is often found in its daughter cells and can be stable over many generations (for references, see Alberts et al. 1977; Leffak et al. 1977) is to assume that some information is contained in the structure of the chromosome and can be segregated with the dividing daughter chromosomes (Tsanev and Sendov 1971; Alberts et al. 1977; for references, see Leffak et al. 1977). Such a mechanism, which implies a nonrandom distribution (segregation) of chromosomal proteins during cell replication, raises many questions, some of which were discussed during this meeting. How are new histone octamers assembled into nucleosomes? How are they segregated during several division cycles? Are newly synthesized histones preferentially associated with newly replicating DNA?

No doubt the nucleosomal pattern is rapidly restored after DNA replication (McKnight et al.; Crémisi et al.; Seale). Electron microscopic observations of replicating SV40 and *Drosophila* embryo chromatins indicate that the nucleosomes are not removed or dissociated significantly prior to replication and that they are reformed on both daughter chromatins very rapidly after replication. Weintraub et al. (see also Leffak et al. 1977) have conclusively demonstrated that newly synthesized histones do not mix with old histones in new nucleosomes. Not only do new histone octamers contain new histones, but they also appear to be located next to each other. In addition, new histone octamers appear to segregate conservatively, i.e., they remain stable over at least three to four generations, and adjacent histone octamers seem to segregate together. Although the results of Weintraub et al. indicate that there is no requirement for disassembling the nucleosome histone cores during replication (Alberts et al. 1977), they do not exclude it, and the mechanism by which the conservative segregation of histones is achieved remains to be established. In this respect, it is interesting to note that excision repair in DNA can occur in the absence of any apparent nucleosome redistribution along the DNA (Cleaver 1977).

There has been some disagreement on whether or not newly made histones are associated with newly synthesized DNA (for references, see Crémisi et al.; Leffak et al. 1977). However, the present results of Crémisi et al. clearly show that newly synthesized histones are preferentially associated with SV40 replicating minichromosomes, and Freedlender et al. (1977) have reported that in Chinese hamster ovary cells a significant amount of chromatin proteins segregates nonrandomly. These results agree with those of the previous study of Tsanev and Russev (1974) suggesting that newly made histones are associated with newly synthesized DNA. In addition, it appears (Newrock et al.) that H1 histones could also be nonrandomly distributed during replication in *Drosophila* embryos.

In view of the technical difficulties inherent in such studies, much more remains to be done to confirm all of the above results, to establish their generality, and to elucidate the underlying mechanisms. However, at this time it is not unreasonable to conclude from what we have learned (newly synthesized nucleosomes located preferentially next to each other; preferential association of newly made histones with new DNA) that, at least over a given region of the genome, the parental histones could go to one daughter DNA molecule and new histones could go to the other daughter DNA double helix. Furthermore, it is not excluded that some nonhistone proteins could segregate in a similar way (Freedlender et al. 1977). Such an asymmetrical distribution of histone and of some nonhistone components during replication of chromatin could play an important role not only in generating different chromosomal structures (and therefore different patterns of potentially transcribing chromatin) during development, but also in perpetuating these structural differences in daughter cells during subsequent cell generations (Weintraub et al.; Newrock et al.; Tsanev and Sendov 1971; Alberts et al. 1977; Freedlender et al. 1977). The programmed variability of histones H1, H2A, and H2B during early development (for references, see Newrock et al.) is in keeping with these suggestions.

Acknowledgments

I am grateful to many of my collaborators for their helpful editorial assistance, particularly Drs. D. Mathis and R. Breathnach. I am greatly indebted to Brigitte Chambon for her patience and help in preparing the manuscript. I thank all those who sent me preprints of their work. My laboratory is supported by grants from the CNRS, the INSERM, the DGRST, and the Fondation pour la Recherche Médicale Française.

REFERENCES

ADOLPH, K. W., S. M. CHENG, and U. K. LAEMMLI. 1977a. Role of nonhistone proteins in metaphase chromosome structure. *Cell* **12**: 805.

ADOLPH, K. W., S. M. CHENG, J. R. PAULSON, and U. K. LAEMMLI. 1977b. Isolation of a protein scaffold from mitotic HeLa cell chromosomes. *Proc. Natl. Acad. Sci.* **74**: 4937.

ALBERTS, B., A. WORCEL, and H. WEINTRAUB. 1977. On the biological implications of chromatin structure. In *The organization and expression of the eukaryotic genome* (ed. E. M. Bradbury and K. Javaherian), p. 165. Academic Press, New York.

ALFAGEME, C. R., G. T. RUDKIN, and L. H. COHEN. 1976. Location of chromosomal proteins in polytene chromosomes. *Proc. Natl. Acad. Sci.* **73**: 2038.

ALLFREY, V. G. 1977. Post synthetic modifications of histone structure: A mechanism for the control of chromosome structure by the modulation of histone-DNA interactions. In *Chromatin and chromosome structure* (ed.

H. J. Li and R. A. Eckhardt), p. 167. Academic Press, New York.

ALLFREY, V. G., E. K. F. BAUTZ, B. J. McCARTHY, R. T. SCHIMKE, and A. TISSIÈRES, eds. 1976. *Organization and expression of chromosomes*. Dahlem Konferenzen: Berlin.

ALTENBURGER, W., W. HÖRZ, and H. ZACHAU. 1976. Nuclease cleavage of chromatin at 100-nucleotide pair intervals. *Nature* **264**: 517.

BAKAYEV, V. V., T. G. BAKAYEVA, and A. J. VARSHAVSKY. 1977. Nucleosomes and subnucleosomes: Heterogeneity and composition. *Cell* **11**: 619.

BAKAYEV, V. V., A. A. MELNICKOV, V. D. OSICKA, and A. J. VARSHAVSKY. 1975. Studies on chromatin. II. Isolation and characterization of chromatin subunits. *Nucleic Acids Res.* **2**: 1401.

BASTOS, R. N. and H. AVIV. 1977. Globin RNA precursor molecules: Biosynthesis and processing in erythroid cells. *Cell* **11**: 641.

BELLARD, M., P. OUDET, J. E. GERMOND, and P. CHAMBON. 1976. Subunit structure of simian virus 40 minichromosome. *Eur. J. Biochem.* **70**: 543.

BENECKE, B. J. and S. PENMAN. 1977. A new class of small nuclear RNA molecules synthesized by a type I RNA polymerase in HeLa cells. *Cell* **12**: 939.

BENYAJATI, C. and A. WORCEL. 1976. Isolation, characterization and structure of the folded interphase genome of *Drosophila melanogaster*. *Cell* **9**: 393.

BERK, A. J. and P. A. SHARP. 1977. Ultraviolet mapping of the adenovirus 2 early promoters. *Cell* **12**: 45.

BERTRAND, K., L. KORN, F. LEE, T. PLATT, C. L. SQUIRES, C. SQUIRES, and C. YANOFSKY. 1975. New features of the regulation of the tryptophan operon. *Science* **189**: 22.

BIRNSTIEL, M. and M. CHIPCHASE. 1977. Current work on histone operon. *Trends Biochem. Sci.* **2**: 149.

BOSELEY, P. G., M. BRADBURY, G. S. BUTLER-BROWNE, B. G. CARPENTER, and R. M. STEPHENS. 1976. Physical studies of chromatin. The recombination of histones with DNA. *Eur. J. Biochem.* **62**: 21.

BRACK, C. and S. TONEGAWA. 1977. Variable and constant parts of the immunoglobulin light chain gene of a mouse myeloma cell are 1250 nontranslated bases apart. *Proc. Natl. Acad. Sci.* **74**: 5652.

BRADBURY, E. M., B. G. CARPENTER, and H. W. E. RATTLE. 1973. Magnetic resonance studies of deoxyribonucleoprotein. *Nature* **241**: 123.

BRADBURY, E. M., R. J. INGLIS, and H. R. MATTHEWS. 1974. Control of cell divisions by very lysine rich histone (F1) phosphorylation. *Nature* **247**: 257.

BRAM, S. 1971. The secondary structure of DNA in solution and in nucleohistone. *J. Mol. Biol.* **58**: 277.

BREATHNACH, R., J. L. MANDEL, and P. CHAMBON. 1977. Ovalbumin gene is split in chicken DNA. *Nature* **270**: 314.

BROWN, K., K. W. ADOLPH, and U. K. LAEMMLI. 1978. Isolation and characterization of compact histone-depleted interphase chromosomes. *J. Mol. Biol.* (in press).

BUKHARI, A. I., J. A. SHAPIRO, and S. L. ADHYA. 1977. *DNA insertion elements, plasmids, and episomes*. Cold Spring Harbor Laboratory, Cold Spring Harbor, New York.

CAMERINI-OTERO, R. D. and G. FELSENFELD. 1977. Supercoiling energy and nucleosome formation: The role of the arginine-rich histone kernel. *Nucleic Acids Res.* **4**: 1159.

CAMERINI-OTERO, R. D., B. SOLLNER-WEBB, and G. FELSENFELD. 1976. The organization of histones and DNA in chromatin: Evidence for an arginine-rich histone kernel. *Cell* **8**: 333.

CARLSON, R. D. and D. E. OLINS. 1976. Chromatin model calculations. Arrays of spherical ν bodies. *Nucleic Acids Res.* **3**: 89.

CARPENTER, B. G., J. P. BALDWIN, E. M. BRADBURY, and

K. IBEL. 1976. Organization of subunits in chromatin. *Nucleic Acids Res.* **3**: 1739.

CECH, T. R. and M. L. PARDUE. 1977. Cross-linking of DNA with trimethylpsoralen is a probe for chromatin structure. *Cell* **11**: 631.

CELMA, M. L., R. DHAR, J. PAN, and S. M. WEISSMAN. 1977. Comparison of the nucleotide sequence of the messenger RNA for the major structural protein of SV40 with the DNA sequence encoding the amino acids of the protein. *Nucleic Acids Res.* **4**: 2549.

CHAMBON, P. 1975. Eukaryotic RNA polymerases. *Annu. Rev. Biochem.* **44**: 613.

CHRISTIANSEN, G. and J. GRIFFITH. 1977. Salt and divalent cations affect the flexible nature of the natural beaded chromatin in structure. *Nucleic Acids Res.* **4**: 1837.

CLEAVER, J. E. 1977. Nucleosome structure controls rate of excision repair in DNA of human cells. *Nature* **270**: 451.

COMINGS, D. E. and T. A. OKADA. 1976. Nuclear proteins. III. The fibrillar nature of the nuclear matrix. *Exp. Cell Res.* **103**: 341.

COMPTON, J. L., M. BELLARD, and P. CHAMBON. 1976a. Biochemical evidence of variability in the DNA repeat length in the chromatin of higher eukaryotes. *Proc. Natl. Acad. Sci.* **73**: 4382.

COMPTON, J. L., R. HANCOCK, P. OUDET, and P. CHAMBON. 1976b. Biochemical and electron microscopic evidence that the subunit structure of CHO interphase chromatin is conserved in mitotic chromosomes. *Eur. J. Biochem.* **70**: 555.

CRÉMISI, C., P. F. PIGNATTI, and M. YANIV. 1976. Random location and the absence of movement of the nucleosomes and SV40 nucleoprotein complexes isolated from infected cells. *Biochem. Biophys. Res. Commun.* **73**: 548.

CRÉMISI, C., A. CHESTIER, C. DAUGUET, and M. YANIV. 1977. Transcription of SV40 nucleoprotein complexes *in vitro*. *Biochem. Biophys. Res. Commun.* **78**: 74.

CRICK, F. H. C. 1976. Linking numbers and nucleosomes. *Proc. Natl. Acad. Sci.* **73**: 2639.

CRICK, F. H. C. and A. KLUG. 1975. Kinky helix. *Nature* **255**: 530.

DARNELL, J. E. 1977. mRNA structure and function. *Prog. Nucleic Acid Res. Mol. Biol.* **19**: 493.

DARNELL, J. E., W. R. JELINEK, and G. R. MOLLOY. 1973. Biogenesis of mRNA: Genetic regulation in mammalian cells. *Science* **181**: 1215.

DASKAL, Y., M. L. MACE, W. WRAY, and H. BUSCH. 1976. Use of direct current sputtering for improved visualization of chromosome topology by scanning electron microscopy. *Exp. Cell Res.* **100**: 204.

DAUBERT, S., D. PETERS, and M. E. DAHMUS. 1977. Selective transcription of ribosomal sequences *in vitro* by RNA polymerase I. *Arch. Biochem. Biophys.* **178**: 381.

DAVIE, J. R. and E. P. M. CANDIDO. 1977. Chromatin subunits contain normal levels of major acetylated histone species. *J. Biol. Chem.* **252**: 5962.

DEIMEL, B., C. LOUIS, and C. E. SEKERIS. 1977. The presence of small molecular weight RNAs in nuclear ribonucleoprotein particles carrying HnRNA. *FEBS Lett.* **73**: 80.

DIXON, G. H. 1976. The biological role of basic chromosomal protein modulations. In *Organization and expression of chromosomes* (ed. V. G. Allfrey et al.), p. 197. Dahlem Konferenzen: Berlin.

ELGIN, S. R. C. and H. WEINTRAUB. 1975. Chromosomal proteins and chromatin structure. *Annu. Rev. Biochem.* **44**: 725.

EVANS, R. M., N. FRASER, E. ZIFF, J. WEBER, M. WILSON, and J. E. DARNELL. 1977. The initiation sites for RNA transcription in Ad2 DNA. *Cell* **12**: 733.

FINCH, J. T. and A. KLUG. 1976. Solenoidal model for superstructure in chromatin. *Proc. Natl. Acad. Sci.* **73**: 1897.

FINCH, J. T., M. NOLL, and R. D. KORNBERG. 1975. Electron

microscopy of defined lengths of chromatin. *Proc. Natl. Acad. Sci.* **72**: 3320.

FINCH, J. T., L. C. LUTTER, D. RHODES, R. S. BROWN, B. RUSHTOWN, M. LEVITT, and A. KLUG. 1977. Structure of nucleosome core particles of chromatin. *Nature* **269**: 29.

FLINT, S. J. and H. M. WEINTRAUB. 1977. An altered subunit configuration associated with the actively transcribed DNA of integrated adenovirus genes. *Cell* **12**: 783.

FODOR, E. J. B. and P. DOTY. 1977. Highly specific transcription of globin sequences in isolated reticulocyte nuclei. *Biochem. Biophys. Res. Commun.* **77**: 1478.

FOE, V. E., L. E. WILKINSON, and C. LAIRD. 1976. Comparative organization of active transcription units in *Oncopeltus fasciatus. Cell* **9**: 131.

FRANKE, W. V., U. SCHEER, M. F. TRENDELENBERG, H. SPRING, and H. ZENTGRAF. 1976. Absence of nucleosomes in transcriptionally active chromatin. *Cytobiologie* **13**: 401.

FRANKLIN, S. G. and A. ZWEIDLER. 1977. Non-allelic variants of histones 2a, 2b and 3 in mammals. *Nature* **266**: 273.

FREEDLENDER, E. F., L. TAICHMAN, and O. SMITHIES. 1977. Non-random distribution of chromosomal proteins during cell replication. *Biochemistry* **16**: 1802.

GAREL, A. and R. AXEL. 1976. Selective digestion of transcriptionally active ovalbumin genes from oviduct nuclei. *Proc. Natl. Acad. Sci.* **73**: 3966.

GAREL, A., M. ZOLAM, and R. AXEL. 1977. Genes transcribed at different rates have a similar conformation in chromatin. *Proc. Natl. Acad. Sci.* **74**: 4867.

GAUBATZ, J. W. and R. CHALKLEY. 1977. Distribution of H1 histone in chromatin digested by micrococcal nuclease. *Nucleic Acids Res.* **4**: 3281.

GERMOND, J. E., B. HIRT, M. GROSS-BELLARD, and P. CHAMBON. 1975. Folding of the DNA double helix in chromatin-like structures from SV40. *Proc. Natl. Acad. Sci.* **72**: 1843.

GIESECKE, K., A. E. SIPPEL, M. C. NGUYEN-HUN, B. GRONER, N. E. HYNES, T. WURTZ, and G. SCHUTZ. 1977. A RNA-dependent RNA polymerase activity: Implications for chromatin transcription experiments. *Nucleic Acids Res.* **4**: 3943.

GILMORE-HEBERT, M., and R. WALL. 1978. Immunoglobulin light chain mRNA is processed from large nuclear RNA. *Proc. Natl. Acad. Sci.* **75**: 342.

GJERSET, R. A. and B. J. MCCARTHY. 1977. Limited accessibility of chromatin satellite DNA to RNA polymerase from *Escherichia coli. Proc. Natl. Acad. Sci.* **74**: 4337.

GLOVER, D. M. and D. S. HOGNESS. 1977. A novel arrangement of the 18S and 28S sequences in a repeating unit of *Drosophila melanogaster* rDNA. *Cell* **10**: 167.

GOLDBERG, S., H. SCHWARTZ, and J. E. DARNELL. 1977. Evidence from UV transcription mapping in HeLa cells that heterogeneous nuclear RNA is the messenger RNA precursor. *Proc. Natl. Acad. Sci.* **74**: 4520.

GOODMAN, H. M., M. V. OLSON, and B. D. HALL. 1977. Nucleotide sequence of mutant eukaryotic gene: The yeast tyrosine-inserting ochre suppressor SUP4-0. *Proc. Natl. Acad. Sci.* **74**: 5453.

GOODWIN, G. H., R. H. NICOLAS, and E. W. JOHNS. 1977a. A quantitative analysis of histone H1 in rabbit thymus nuclei. *Biochem. J.* **167**: 485.

GOODWIN, G. H., J. M. WALKER, and E. W. JOHNS. 1977b. The high mobility group (HMG) non-histone chromosomal proteins. In *The cell nucleus* (ed. H. Busch) Academic Press, New York. (In press.)

GOODWIN, G. H., J. L. WOODHEAD, and E. W. JOHNS. 1977c. The presence of high mobility group non-histone chromatin proteins in isolated nucleosomes. *FEBS Lett.* **73**: 85.

GOTTESFELD, J. M. and P. J. G. BUTLER. 1977. Structure of transcriptionally active chromatin subunits. *Nucleic Acids Res.* **4**: 3155.

GRIFFITH, J. D. 1975. Chromatin structure: Deduced from a minichromosome. *Science* **187**: 1202.

———. 1976. Visualization of prokaryotic DNA in a regularly condensed chromatin-like fiber. *Proc. Natl. Acad. Sci.* **73**: 563.

GUIMONT-DUCAMP, C., J. SRI-WIDADA, and P. JEANTEUR. 1977. Occurrence of small molecular weight RNAs in HeLa nuclear ribonucleoprotein particles containing HnRNA (HnRNP). *Biochimie* **59**: 755.

HANSON, C. V., C. K. J. SHEN, and J. E. HEARST. 1976. Cross-linking of DNA in situ as a probe for chromatin structure. *Science* **193**: 62.

HARDISON, R. C., D. P. ZEITHER, J. M. MURPHY, and R. CHALKLEY. 1977. Histone neighbors in nuclei and extended chromatin. *Cell* **12**: 417.

HARRINGTON, R. E. 1977. DNA chain flexibility and the structure of chromatin ν-bodies. *Nucleic Acids Res.* **4**: 3519.

HAYASHI, H., K. IWAI, J. D. JOHNSON, and J. BONNER. 1977. Pea histones H2A and H2B. Variable and conserved regions in the sequences. *J. Biochem.* **82**: 503.

HEWISH, D. R. and L. A. BURGOYNE. 1973. Chromatin substructure. The digestion of chromatin DNA at regularly spaced sites by a nuclear deoxyribonuclease. *Biochem. Biophys. Res. Commun.* **52**: 504.

HJELM, R. P., G. G. KNEALE, P. SUAU, J. P. BALDWIN, E. M. BRADBURY, and K. IBEL. 1977. Small angle neutron scattering studies of chromatin subunits in solution. *Cell* **10**: 139.

HOLLAND, M. J., G. L. HAGER, and W. J. RUTTER, 1977. Transcription of yeast DNA by homologous RNA polymerases I and II: Selective transcription of ribosomal genes by RNA polymerase I. *Biochemistry* **16**: 16.

HÖZIER, J., P. NEHLS, and M. RENTZ. 1978. The chromosome fiber. Evidence for an ordered superstructure of nucleosomes. *Chromosoma* (in press).

HSU, M. T. and J. FORD. 1977. Sequence arrangement of the 5′ ends of simian virus 40 16S and 19S in RNAs. *Proc. Natl. Acad. Sci.* **74**: 4982.

HUANG, R. C. C. and J. BONNER. 1962. Histone, a suppressor of chromosomal RNA synthesis. *Proc. Natl. Acad. Sci.* **48**: 1216.

JACOB, F. 1977. Evolution and tinkering. *Science* **196**: 1161.

JEFFREYS, A. J. and R. A. FLAVELL. 1977. The rabbit β-globin gene contains a large insert of the coding sequence. *Cell* **12**: 1097.

JOFFE, J., M. KEENE, and H. WEINTRAUB. 1977. Histones H2A, H2B, H3 and H4 are present in equimolar amounts in chick erythroblasts. *Biochemistry* **16**: 1236.

JOHNSON, E. M. and V. G. ALLFREY. 1977. Post-synthetic modifications of histone primary structure: Phosphorylation and acetylation as related to chromatin conformation and function. In *Biochemical actions of hormones* (ed. G. Litwack), vol. 5. Academic Press, New York. (In press.)

KÉDINGER, C., O. BRISON, F. PERRIN, and J. WILHELM. 1978. Structural analysis of viral replicative intermediates isolated from adenovirus type 2 infected HeLa cell nuclei. *J. Virol.* (in press).

KELLER, W. 1975. Determination of the number of superhelical turns in simian virus 40 DNA by gel electrophoresis. *Proc. Natl. Acad. Sci.* **72**: 4876.

KIRYANOV, G. I., T. A. MANAMSHJAN, V. YU. POLYAKOV, D. FAIS, and JU. S. CHENTSOV. 1976. Levels of granular organization of chromatin fibers. *FEBS Lett.* **67**: 323.

KITCHINGMAN, G. R., S. P. LAI, and H. WESTPHAL. 1977. Loop structures in hybrids of early mRNA and the separated strands of adenovirus DNA. *Proc. Natl. Acad. Sci.* **74**: 4392.

KLESSIG, D. F. 1977. Two adenovirus mRNAs have a common 5′ terminal leader sequence encoded at least

10 kb upstream from their main coding regions. *Cell* **12**: 9.

KONKEL, D. A. and V. M. INGRAM. 1977. RNA aggregation during sulfhydryl-agarose chromatography of mercurated RNA. *Nucleic Acids Res.* **4**: 1979.

KORNBERG, R. D. 1974. Chromatin structure: A repeating unit of histones and DNA. Chromatin structure is based on a repeating unit of eight histone molecules and about 200 DNA base pairs. *Science* **184**: 868.

———. 1977. Structure of chromatin. *Annu. Rev. Biochem.* **40**: 931.

KWAN, S. P., T. G. WOOD, and J. B. LINGREL. 1977. Purification of a putative precursor of globin messenger RNA from mouse nucleated erythroid cells. *Proc. Natl. Acad. Sci.* **74**: 178.

LAINE, B., P. SAUTIÈRE, and G. BISERTE. 1976. Primary structure and microheterogeneities of rat chloroleukemia histone H2A. *Biochemistry* **15**: 1640.

LAIRD, C. D., L. E. WILKINSON, V. E. FOE, and W. Y. CHOOI. 1976. Analysis of chromatin-associated fiber arrays. *Chromosoma* **58**: 169.

LANGMORE, J. P. and J. C. WOOLEY. 1975. Chromatin architecture: Investigation of a subunit of chromatin by dark field electron microscopy. *Proc. Natl. Acad. Sci.* **72**: 2691.

LASKEY, R. A., A. D. MILLS, and N. R. MORRIS. 1977. Assembly of SV40 chromatin in a cell-free system. *Cell* **10**: 237.

LAVI, S. and Y. GRONER. 1977. 5' Terminal sequences and coding regions of late simian virus 40 mRNAs are derived from noncontiguous segments of the viral genome. *Proc. Natl. Acad. Sci.* **74**: 5323.

LEFFAK, I. M., R. GRAINGER, and H. WEINTRAUB. 1977. Conservative assembly and segregation of nucleosomal histones. *Cell* **12**: 837.

LEVY W., B., R. A. GJERSET, and B. J. McCARTHY. 1977. Acetylation and phosphorylation of *Drosophila* histones. Distribution of acetate and phosphate groups in fractionated chromatin. *Biochim. Biophys. Acta* **475**: 168.

LILLEY, D. M. J. and K. TATCHELL. 1977. Chromatin core particles unfolding induced by tryptic cleavage of histones. *Nucleic Acids Res.* **4**: 2039.

LIPPS, H. J. and N. R. MORRIS. 1977. Chromatin structure in the nuclei of the ciliate *Stylonychia mytilus*. *Biochem. Biophys. Res. Commun.* **74**: 230.

LITTAU, V. C., C. J. BURDICK, V. G. ALLFREY, and A. E. MIRSKY. 1965. The role of histones in the maintenance of chromatin structure. *Proc. Natl. Acad. Sci.* **54**: 1204.

LOHR, D., K. TATCHELL, and K. E. VAN HOLDE. 1977. On the occurrence of nucleosome phasing in chromatin. *Cell* **12**: 829.

MARTIN, D. Z., R. TODD, D. LANG, P. N. PEI, and W. T. GARRARD. 1977. Heterogeneity in nucleosome spacing. *J. Biol. Chem.* **252**: 8269.

MATHIS, D. and M. A. GOROVSKY. 1976. Subunit structure of rDNA-containing chromatin. *Biochemistry* **15**: 750.

MATSUI, S., M. FUKA, and H. BUSCH. 1977. Fidelity of ribosomal RNA synthesis by nucleoli and nucleolar chromatin. *Biochemistry* **16**: 39.

MATTHEWS, H. R. and M. BRADBURY. 1977. The role of H1 histone phosphorylation in the cell cycle: Turbidity studies of H1-DNA interaction. *Exp. Cell Res.* **111**: 343.

MEISSNER, H. C., J. MEYER, J. V. MAIZEL, and H. WESTPHAL. 1977. Visualization and mapping of late nuclear adenovirus RNA. *Cell* **10**: 225.

MELLI, M., G. SPINELLI, and E. ARNOLD. 1977. Synthesis of histone messenger RNA of HeLa cells during the cell cycle. *Cell* **12**: 167.

MILLER, O. L. and B. R. BEATTY. 1969. Extra-chromosomal nucleolar genes in amphibian oocytes. *Genetics* (Suppl.) **61**: 134.

MIRSKY, A. E., C. J. BURDICK, E. H. DAVIDSON, and V. C.

LITTAU. 1968. The role of lysine-rich histone in the maintenance of chromatin structure metaphase chromosomes. *Proc. Natl. Acad. Sci.* **61**: 592.

MORRIS, N. R. 1976. A comparison of the structure of chicken erythrocyte and chicken liver chromatin. *Cell* **9**: 627.

NOLL, M. 1974a. Subunit structure of chromatin. *Nature* **251**: 249.

———. 1974b. Internal structure of the chromatin subunit. *Nucleic Acids Res.* **1**: 1573.

———. 1976. Differences and similarities in chromatin structure of *Neurospora crassa* and higher eucaryotes. *Cell* **8**: 349.

NOLL, M. and R. D. KORNBERG. 1977. Action of micrococcal nuclease on chromatin and the location of histone H1. *J. Mol. Biol.* **109**: 393.

OLINS, A. L. and D. E. OLINS. 1974. Spheroid chromatin units (ν-bodies). *Science* **183**: 330.

OLINS, A. L., R. D. CARLSON, E. B. WRIGHT, and D. E. OLINS. 1976. Chromatin ν bodies: Isolation, subfractionation and physical characterization. *Nucleic Acids Res.* **3**: 3271.

O'MALLEY, B. W. and W. T. SCHRADER. 1976. The receptors of steroid hormones. *Sci. Am.* **234**: 32.

ORKIN, S. H. and P. S. SWERDLOW. 1977. Globin RNA synthesis *in vitro* by isolated erythroleukemic cell nuclei. Direct evidence for increased transcription during erythroid differentiation. *Proc. Natl. Acad. Sci.* **74**: 2475.

OUDET, P., M. GROSS-BELLARD, and P. CHAMBON. 1975. Electron microscopic and biochemical evidence that chromatin structure is a repeating unit. *Cell* **4**: 281.

OUDET, P., J. E. GERMOND, M. BELLARD, C. SPADAFORA, and P. CHAMBON. 1977. Nucleosome structure. *Philos. Trans. R. Soc. Lond. B* (in press).

PARDON, J. F., D. L. WORCESTER, J. C. WOOLEY, K. TATCHELL, K. E. VAN HOLDE, and B. M. RICHARDS. 1975. Low-angle neutron scattering from chromatin subunit particles. *Nucleic Acids Res.* **2**: 2163.

PAULSON, J. R. and U. K. LAEMMLI. 1977. The structure of histone depleted metaphase chromosomes. *Cell* **12**: 817.

PELLEGRINI, M., J. MANNING, and N. DAVIDSON. 1977. Sequence arrangement of the rDNA of *Drosophila melanogaster*. *Cell* **10**: 213.

PERRY, R. P., E. BARD, B. D. HAMES, D. E. KELLEY, and U. SCHIBLER. 1977. The relationship between hnRNA and mRNA. *Prog. Nucleic Acid Res. Mol. Biol.* **19**: 275.

PETTIJOHN, D. E. 1976. Prokaryotic DNA in nucleoid structure. *Crit. Rev. Biochem.* **4**: 175.

PHILIPSON, L., R. WALL, G. GLICKMAN, and J. E. DARNELL. 1971. Addition of polyadenylate sequences to virus-specific RNA during adenovirus replication. *Proc. Natl. Acad. Sci.* **68**: 2806.

PIÑON, R. and Y. SALTS. 1977. Isolation of folded chromosomes from yeast *Saccharomyces cerevisiae*. *Proc. Natl. Acad. Sci.* **74**: 2850.

POLISKY, B. and B. J. McCARTHY. 1975. Location of histones on simian virus 40 DNA. *Proc. Natl. Acad. Sci.* **72**: 2895.

PONDER, B. A. J. and L. V. CRAWFORD. 1977. The arrangement of nucleosomes in nucleoprotein complexes from polyoma virus and SV40. *Cell* **11**: 35.

PRUNELL, A. and R. D. KORNBERG. 1977. Relation of nucleosomes to nucleotide sequences in the rat. *Philos. Trans. R. Soc. Lond. B* (in press).

RABBITS, T. H. and A. FORSTER. 1978. Evidence for noncontiguous variable and constant region genes in both germ line and myeloma DNA. *Cell* **13**: 319.

RENZ, M., P. NEHLS, and J. HOZIER. 1977. Involvement of histone H1 in the organization of the chromosome fiber. *Proc. Natl. Acad. Sci.* **74**: 1879.

RICHARDS, B., J. PARDON, D. LILLEY, R. COTTER, and J. WOOLEY. 1977. The subunit structure of nucleosomes. *Cell Biol. Int. Rep.* **1**: 107.

ROBSON, B. 1977. Biological macromolecules: Outmoding the rigid view. *Nature* **267**: 577.

ROEDER, R. G. 1976. Eukaryotic nuclear RNA polymerases. In *RNA polymerase* (ed. R. Losick and M. Chamberlin), p. 285. Cold Spring Harbor Laboratory, Cold Spring Harbor, New York.

SHIH, T. Y., H. A. YOUNG, W. P. PARKS, and S. SCOLNICK. 1977. In vitro transcription of Moloney leukemia virus genes in infected cell nuclei and chromatin. *Biochemistry* **16**: 1795.

SHURE, M. and J. VINOGRAD. 1976. The number of superhelical turns in native virion SV40 DNA and minicol DNA determined by the band counting method. *Cell* **8**: 215.

SIMPSON, R. T. 1976. Histones H3 and H4 interact with the ends of nucleosome DNA. *Proc. Natl. Acad. Sci.* **73**: 4400.

SIMPSON, R. T. and J. P. WHITLOCK, JR. 1976. Mapping DNAase I-susceptible sites in nucleosomes labeled at the 5' ends. *Cell* **9**: 347.

SKAER, R. J. 1977. Interband transcription in *Drosophila. J. Cell Sci.* **26**: 251.

SKLAR, V. E. F. and R. G. ROEDER. 1977. Transcription of specific genes in isolated nuclei by exogenous RNA polymerases. *Cell* **10**: 405.

SOBELL, H. M., C. TSAI, S. G. GILBERT, S. C. JAIN, and T. D. SAKORE. 1976. Organization of DNA in chromatin. *Proc. Natl. Acad. Sci.* **73**: 3068.

SOLLNER-WEBB, B. and G. FELSENFELD. 1975. A comparison of the digestion of nuclei and chromatin by staphylococcal nuclease. *Biochemistry* **14**: 2915.

――――. 1977. Pancreatic DNAase cleavage sites in nuclei. *Cell* **10**: 537.

SOLLNER-WEBB, B., R. D. CAMERINI-OTERO, and G. FELSENFELD. 1976. Chromatin structure as probed by nucleases and proteases: Evidence for the central role of histones H3 and H1. *Cell* **9**: 179.

SPADAFORA, C., M. BELLARD, J. L. COMPTON, and P. CHAMBON. 1976. The DNA repeat lengths in chromatins from sea urchin sperm and gastrula cells are markedly different. *FEBS Lett.* **69**: 281.

SPERLING, L. and A. TARDIEU. 1976. The mass per unit length of chromatin by low-angle X-ray scattering. *FEBS Lett.* **64**: 89.

SPIKER, S. and I. ISENBERG. 1977. Cross-complexing pattern of plant histones. *Biochemistry* **16**: 1819.

STEDMAN, E. and E. STEDMAN. 1951. The basic proteins of cell nuclei. *Philos. Trans. R. Soc. Lond. B* **235**: 565.

STEIN, A., M. BINA-STEIN, and R. T. SIMPSON. 1977. A cross-linked octamer as a model of the nucleosome core. *Proc. Natl. Acad. Sci.* **74**: 2780.

STEINMETZ, M., R. E. STUCK, and H. G. ZACHAU. 1977. Reconstituted histone-DNA complexes. *Philos. Trans. R. Soc. Lond. B* (in press).

STRAIR, R. K., A. I. SKOULTCHI, and D. A. SHAFRITZ. 1977. A characterization of globin mRNA sequences in the nucleus of duck immature red blood cells. *Cell* **12**: 133.

STRICKLAND, M., W. N. STRICKLAND, W. F. BRANDT, and C. VON HOLT. 1977. The complete amino-acid sequence of histone H2B(1) from sperm of the sea urchin *Parechinus angulosus. Eur. J. Biochem.* **77**: 263.

STRICKLAND, W. N., M. STRICKLAND, W. F. BRANDT, and C. VON HOLT. 1977. The complete amino-acid sequence of histone H2B(2) from sperm of the sea urchin *Parechinus angulosus. Eur. J. Biochem.* **77**: 277.

STUBBLEFIELD, E. and W. WRAY. 1971. Architecture of the Chinese hamster metaphase chromosome. *Chromosoma* **32**: 262.

SUAU, P., G. G. KNEALE, G. W. BRADDOCK, J. P. BALDWIN, and E. M. BRADBURY. 1977. A low resolution model for the chromatin core particle by neutron scattering. *Nucleic Acids Res.* **4**: 3759.

SWIFT, H. 1974. The organization of genetic material in eukaryotes: Progress and prospects. *Cold Spring Harbor Symp. Quant. Biol.* **38**: 963.

TATCHELL, K. and K. E. VAN HOLDE. 1977. Reconstitution of chromatin core particles. *Biochemistry* **16**: 5295.

THIMMAPPAYA, B. and S. M. WEISSMAN. 1977. The early region of SV40 DNA may have more than one gene. *Cell* **11**: 837.

THOMAS, J. O. 1977. Chromatin structure. In *International review of biochemistry* (ed. B. F. C. Clark), Vol. 17. University Park Press, Baltimore, Maryland. (In press.)

THOMAS, J. O. and R. D. KORNBERG. 1975a. An octamer of histones in chromatin and free in solution. *Proc. Natl. Acad. Sci.* **72**: 2626.

――――. 1975b. Cleavable cross-links in the analysis of histone-histone associations. *FEBS Lett.* **58**: 353.

THOMAS, J. O. and R. J. THOMPSON. 1977. Variation in chromatin structure in two cell types from the same tissue: A short repeat length in cerebral cortex neurons. *Cell* **10**: 633.

TILGHMAN, S. M., D. C. TIEMEIER, J. G. SEIDMAN, B. M. PETERLIN, M. SULLIVAN, J. V. MAIZEL, and P. LEDER. 1978. Intervening sequence of DNA identified in the structural portion of a mouse β-globin gene. *Proc. Natl. Acad. Sci.* **75**: 725.

TOWLE, H. C., M. J. TSAI, S. Y. TSAI, and B. W. O'MALLEY. 1977. Effect of oestrogen on gene expression in the chick oviduct. Preferential initiation and asymmetrical transcription of specific chromatin genes. *J. Biol. Chem.* **252**: 2396.

TSANEV, R. and P. PETROV. 1976. The substructure of chromatin and its variations as revealed by electron microscopy. *J. Microsc. Biol. Cell.* **27**: 11.

TSANEV, R. and G. RUSSEV. 1974. Distribution of newly synthesized histones during DNA replication. *Eur. J. Biochem.* **43**: 257.

TSANEV, R. and B. SENDOV. 1971. Possible molecular mechanism for cell differentiation in multicellular organisms. *J. Theor. Biol.* **30**: 337.

VALENZUELA, P., B. G. I. BELL, F. R. MASIARZ, L. J. DEGENNARO, and W. J. RUTTER. 1977. Nucleotide sequence of the yeast 5S ribosomal gene and adjacent putative control regions. *Nature* **267**: 641.

VALENZUELA, P., A. VENEGAS, F. WEINBERG, R. BISHOP, and W. J. RUTTER. 1978. Structure of yeast phenylalanine-tRNA genes. An intervening DNA segment within the region coding for the tRNA. *Proc. Natl. Acad. Sci.* **75**: 190.

VAN HOLDE, K. E., C. G. SAHASRABUDDHA, B. R. SHAW, E. F. J. VAN BRUGGEN, and A. C. ARNBERG. 1974. Electron microscopy of chromatin subunit particles. *Biochem. Biophys. Res. Commun.* **60**: 1365.

VAN KEULEN, H. and J. RETÈL. 1977. Transcription specificity of yeast RNA polymerase A. *Eur. J. Biochem.* **79**: 579.

VARSHAVSKY, A. J., V. V. BAKAYEV, and G. P. GEORGIEV. 1976. Heterogeneity of chromatin subunits in vitro and location of histone H1. *Nucleic Acids Res.* **3**: 477.

VENGEROV, Y. Y. and V. I. POPENKO. 1977. Changes in chromatin structure induced by EDTA treatment and partial removal of histone H1. *Nucleic Acids Res.* **4**: 3017.

VIDALI, G., L. C. BUFFA, and V. G. ALLFREY. 1977. Selective release of chromosomal proteins during limited DNase I digestion of avian erythrocyte chromatin. *Cell* **12**: 409.

WEINTRAUB, H. and M. GROUDINE. 1976. Chromosomal subunits in active genes have an altered conformation. *Science* **193**: 848.

WEINTRAUB, H., A. WORCEL, and B. ALBERTS. 1976. A model for chromatin based upon two symmetrically paired half nucleosomes. *Cell* **9**: 409.

WELLAUER, P. K. and I. B. DAWID. 1977. The structural organization of ribosomal DNA in *Drosophila melanogaster. Cell* **10**: 193.

WHITE, R. L. and D. S. HOGNESS. 1977. R loop mapping of the 18S and 28S sequences in the long and short repeating units of *Drosophila melanogaster* rDNA. *Cell* **10**: 177.

WHITLOCK, J. P., JR. and R. T. SIMPSON. 1978. Localization of the sites along nucleosome DNA which interact with amino terminal histone regions. *J. Biol. Chem.* (in press).

WHITLOCK, J. P., JR., G. W. RUSHIZKY, and R. T. SIMPSON. 1977. DNase-sensitive sites in nucleosomes: Their relative susceptibilities depend on nuclease used. *J. Biol. Chem.* **252**: 3003.

WIESEHAHN, G. J., J. E. HYDE, and J. E. HEARST. 1977. The photoaddition of trimethylpsoralen to *Drosophila melanogaster* nuclei: A probe for chromatin substructure. *Biochemistry* **16**: 925.

WILHELM, M. L., A. MAZEN, and F. X. WILHELM. 1977. Comparison of the DNA repeat length in H1 and H5-containing chromatin. *FEBS Lett.* **79**: 404.

WOODCOCK, C. L. F., L. L. Y. FRADO, C. L. HUTCH, and L. RICCIARDIELLO. 1976. Fine structure of active ribosomal genes. *Chromosoma* **58**: 33.

WORCEL, A. and C. BENYAJATI. 1977. Higher order coiling of DNA in chromatin. *Cell* **12**: 83.

YAMAMOTO, M., D. JONES, and K. SEIFART. 1977. Transcription of ribosomal 5S RNA by RNA polymerase C in isolated chromatin from HeLa cells. *Eur. J. Biochem.* **80**: 243.

YANEVA, M., B. TASKEVA, and G. DESSEV. 1976. Nuclease digestion of reconstituted chromatin. *FEBS Lett.* **70**: 67.

ZASLOFF, M. and G. FELSENFELD. 1977. Analysis of in vitro transcription of duck reticulocyte chromatin using mercury-substituted ribonucleoside triphosphates. *Biochemistry* **16**: 5135.

ZWEIDLER, A. 1976. Complexity and variability of the histone complement. In *Organization and expression of chromosomes* (ed. V. G. Allfrey et al.), p. 187. Dahlem Konferenzen:Berlin.

Name Index

Italics indicate where full reference can be found; **boldface** type designates where author's article is located in this volume.

Subject Index

A

Actinomycin, 90, 150
Adenovirus. *See also* RNA; Transcription
  adeno-SV40 hybrids, 534, 539
  α-amanitin sensitivity of late products, 518
  late RNA transcribed from one promoter, 515
  leader sequence location, 524–525, 543–547
  mRNA has one major 5' terminus, 532–533
  sensitivity to DNase I, 402
  splicing of noncontiguous early and late mRNAs, 523–528, 531–551
  transcription patterns, 523–528, 531–551

B

Bacteriophage λ, use as cloning vector, 915, 921, 959, 961
n-Butyrate, induction of Friend erythroleukemia, 817–819

C

Capping
  in vitro reaction, 954
  nucleolar RNA, 1168–1169
  use in identifying 5' triphosphate termini, 1167
Cap structure, 532
Chromosome. *See also* Cloning; DNA; Nucleosome; SV40; Transcription
  active chromatin detected by nuclease digestion, 69–70, 401–402, 599, 643–645, 704–707, 777–779, 784–788, 796–797, 851
  altered ovalbumin chromatin after estrogen withdrawal, 643–645
  compaction of nontranscribed *Xenopus* ribosomal spacer chromatin, 1174–1176
  domains of supercoiling, 111–116
  freeze-fracture studies, 23
  H1-mediated compaction, 215–225, 229–240, 245–251, 253–262, 265–270
  higher-order structures, 215–225, 229–240, 331–350, 351–360, 361–366, 367–377, 469–471
  histone-depleted chromosomes, 46–48
  isolation of *Drosophila* salivary chromosomes by microdissection, 859–864
  isolation of histone-depleted metaphase chromosomes, 352–353
  metaphase, 345, 352, 361, 380
  mitotic, 369
  order in interphase, 342–345
  polytene, 333, 381, 389, 805–811
  puffs not associated with histone determinants, 381–383
  reconstitution from histone octomer, 122–124
  reconstitution studies, 50–51
  scaffolding observed in histone-depleted chromosomes, 354–357
  stage specificity related to histone content, 401–408, 426–430
  transcriptionally active chromatin as seen by EM, 727–740, 743–755, 757–772
Cloning. *See also* Bacteriophage λ; Repetitive DNA
  characterization of *Drosophila* histone repeat, 1047
  fidelity of reverse transcriptase method, 933–936
  of β-hemoglobin gene, 915–917, 933–935, 959–961
  mRNA using reverse transcriptase, 933
  Oxytricha macronuclear DNA, 484–485
  ribosomal DNA from *Drosophila*, 1185–1186
  silk fibroin gene, 947–955
  structure of repeated *Drosophila* gene copia, 1057–1061
  telomeric DNA, 1040–1044
  use of mercurated RNA, 959, 961
  Vλ mouse immunoglobulin gene, 921
Codon utilization. *See* Cloning; Hemoglobin; Histone
Concanavalin A, 385
Crosslinkings
  DNA, 179, 191–198
  protein, 120, 133

D

DNA. *See also* Chromosome; Cloning; Repetitive DNA
  amplification selectivity in *Xenopus* oocytes, 1179–1183
  breathing, 87, 95
  compaction in chromatin, 351–367, 370
  dichroism of superhelix, 211
  gene enrichment using mercurated RNA, 959–961
  genetic analysis of yeast ribosomal DNA inheritance and localization, 1203–1205
  insertion sequences within *Drosophila* ribosomal DNA, 1187–1189
  intercalation, 87–101
  maintenance of length heterogeneity in *Xenopus* 5S gene, 1199
  methylation, 150
  organization of *Xenopus* 5S gene, 1195
  quantitation and arrangement of nonribosomal insertion sequences in *Drosophila*, 1190–1193
  repetitive satellites, 180, 273–275, 397–400, 1121–1135
  ribosomal insertion sequences located outside the nucleolus organizer locus, 1189–1190
  sequence composition of *Drosophila* ribosomal DNA, 1186–1187
  sequence of *Xenopus* 5S oocyte gene, 1195–1197
  solenoid models and compaction, 319–321, 375
  solvent effects, 87
  specific DNA-binding protein from *Drosophila*, 694–698
  structure as function of hydration, 27–28
  superhelicity, 87, 109, 132, 144, 172, 212, 289, 303, 314
  transition angles by electric dichroism, 208
  two forms of ribosomal DNA in yeast, 1201–1203
Drosophila. *See also* Chromosome; Cloning; DNA; Heat shock; Nucleosome; Repetitive DNA; RNA; Transcription
  isolation of specific ribosomal DNA protein, 693–699
  the rosy locus, 1011
  transcription complexes visualized by EM, 743–755
Drug resistance, methotrexate resistance results from specific gene amplification, 649–656

E

Electric dichrism, 207–214
Electron microscopy. *See also* Chromosome; Nucleosome; Transcription
  chromatin appearance, 218–222, 230–233, 246–247, 288, 303, 326–328
  H1-depleted chromatin, 218–220
  H3-H4 DNA complexes, 289
  interphase chromosomes, 338–345
  native transcriptional complexes, 725–740, 743–755, 757–772